FAO Plant Production and Protection Series   No. 5

REPORT

of the

FAO GLOBAL SURVEY OF PESTICIDE SUSCEPTIBILITY

OF STORED GRAIN PESTS

By B.R. Champ and C.E. Dyte

FOOD AND AGRICULTURE ORGANIZATION OF THE UNITED NATIONS
Rome, 1976

The designations employed and the presentation of material in this publication do not imply the expression of any opinion whatsoever on the part of the Food and Agriculture Organization of the United Nations concerning the legal status of any country, territory, city or area or of its authorities, or concerning the delimitation of its frontiers or boundaries.

ISBN 92-5-100022-0

The copyright in this book is vested in the Food and Agriculture Organization of the United Nations. The book may not be reproduced, in whole or in part, by any method or process, without written permission from the copyright holder. Applications for such permission, with a statement of the purpose and extent of the reproduction desired, should be addressed to the Director, Publications Division, Food and Agriculture Organization of the United Nations, Via delle Terme di Caracalla, 00100 Rome, Italy.

© FAO 1976

FOREWORD

FAO was first formally alerted to the rapidly emerging problems of pesticide resistance by its Conference on Pesticides in Agriculture in 1962. As a result, the Director General of FAO established in 1963 a Working Party of Experts on Resistance of Pests to Pesticides which held annual meetings between 1965 and 1974 "to advise and assist the Director General on all matters related to pest resistance to pesticides". In 1975 the Working Party became the Panel on Pest Resistance to Pesticides and Crop Loss Assessment, with correspondingly wider responsibilities.

At two of its meetings (1965 and 1968) the Working Party carried out world reviews of pesticide resistance in agricultural pests and, on each occasion emergence of important resistances in stored grain pests was noted. At the seventh meeting in 1971 the steadily progressing development of resistance in stored grain pests was viewed with grave concern. Over the previous decade or two, many countries had come to rely almost exclusively on the grain protectants lindane and malathion and the fumigants methyl bromide and phosphine, not only to permit grain to move freely in international trade but also to protect it against serious loss in storage. Already there were reports of important resistance to these pesticides and to a number of other possible alternative materials. However, there was inadequate data available to determine whether control failures due to resistance were likely to be occurring or how widespread resistance problems had become.

The Director General of FAO accepted the recommendation of the seventh meeting of the Working Party that FAO should sponsor a global survey of resistance in stored product pests, the principal objectives being -

" (a)   to map the distribution of resistance in the major pest species,

(b)   to provide information on resistance levels and on cross-resistances present

and

(c)   to make this information available to interested nations."

As part of its programme the Working Party has also sponsored the development of standardised tests for resistance (three of which concern stored product pests) so that results from different parts of the world can be compared directly.

Arrangements were made in 1972 for a consultant to visit a wide selection of member countries of FAO in order to make a world-wide survey of resistance in pests of stored food. Where appropriate, he also demonstrated the use of the relevant standardised tests. The project was established as a joint undertaking between (a) FAO, (b) the Commonwealth Scientific and Industrial Research Organisation of Australia's Division of Entomology, Stored Grain Research Laboratory, which provided staffing including Dr B.R. Champ as a consultant to carry out the survey and (c) the Pest Infestation Control Laboratory, United Kingdom Ministry of Agriculture, Fisheries and Food, which provided testing laboratory facilities under the direction of Mr C.E. Dyte.

The following text is the report of the survey and a serious situation is revealed which clearly demands effective and urgent attention. Relevant sections of the chemical industry were alerted to the situation at a meeting specially convened by the Working Party during its 9th seession in Rome in 1973. Since this time the search for effective new compounds had been greatly intensified but the situation continues to give grave concern.

The Working Party wishes to express its sincere thanks to the large number of people, too numerous to list individually, who, throughout many countries of the world have assisted in providing grain samples and in countless other ways. It believes that the information provided in this report will be of great value in decisions that face us, not only imminently but also in the future.

D.F. Waterhouse
CHAIRMAN, FAO Panel on Pest Resistance
to Pesticides and Crop Loss Assessment
Division of Entomology, CSIRO, Australia.

FAO GLOBAL SURVEY OF PESTICIDE SUSCEPTIBILITY OF STORED GRAIN PESTS

FOREWORD		i, ii
TABLE OF CONTENTS		iii – vii
PREFACE		1

A. CONCLUSIONS ARISING FROM THE REPORT 2
    A.1 Implications of the results of the survey   2
    A.2 Use of resistant strains in evaluation of alternative
         materials   2
    A.3 The need for a rational approach to storage pest
         control programmes   3

B. INTRODUCTION 5

C. CONDUCT OF THE SURVEY 7
    C.1 Species involved   7
    C.2 Pesticides included   8
    C.3 Survey detail   8
        C.3.1 Establishing contacts and arrangements during
               visits   8
        C.3.2 Countries visited   8
        C.3.3 Collection of information from countries visited   10
        C.3.4 Collection and forwarding of samples   11
        C.3.5 Processing of samples   11
        C.3.6 Countries and areas not visited during the survey   12

D. THE MAJOR CEREALS AND THEIR IMPORTANT STORAGE PESTS 13
    D.1 Classification of pest importance   14
    D.2 Collection and interpretation of pest importance data   16
    D.3 Comments   18
        D.3.1 Wheat   18
        D.3.2 Barley   21
        D.3.3 Millet + Sorghum   21
        D.3.4 Rice   26
             D.3.4a Paddy   26
             D.3.4b Milled rice   26
        D.3.5 Maize   29
        D.3.6 Cereals generally and milled products   32
             D.3.6a *Sitophilus* spp.   32
             D.3.6b *Sitotroga cerealella*   37
             D.3.6c *Rhyzopertha dominica*   38
             D.3.6d *Tribolium* spp.   38
             D.3.6e *Oryzaephilus* spp.   39
             D.3.6f Phycitid moths and *Corcyra cephalonica*   39
             D.3.6g *Trogoderma granarium*   41
             D.3.6h Other species   41
             D.3.6i Summary   41

E.	THE PESTICIDES IN COMMON USE IN GRAIN STORAGE		42
	E.1 Malathion		42
	E.2 Lindane		50
	E.3 Other insecticides		51
	E.4 Methyl bromide		53
	E.5 Phosphine		53
	E.6 Other fumigants		54

F.	OCCURRENCE OF RESISTANCE IN STORED CEREAL INSECTS			55
	F.1	Monitoring for resistance		55
		F.1.1	Factors associated with and affecting tolerance to pesticides in monitoring tests	55
			F.1.1a Variability in response	55
			F.1.1b Species specificity	56
			F.1.1c Time for preadult development	57
			F.1.1d Stage of development	58
			F.1.1e Age within the developmental stage	62
			F.1.1f Sex	66
			F.1.1g Morphological and behavioural differences	66
			F.1.1h Fecundity	67
			F.1.1i Temperature	69
			F.1.1j Humidity	70
			F.1.1k Light intensity and time of day	71
			F.1.1l Diet	71
			F.1.1m Stability of resistance	72
		F.1.2	The basis and design of monitoring tests	73
			F.1.2a General considerations	73
			F.1.2b Methods for insecticide monitoring	75
			F.1.2c Methods for fumigant monitoring	77
		F.1.3	Assessment of response	78
			F.1.3a Relationship between knockdown and mortality in tests with insecticides	78
			F.1.3b Exposure and response times	79
		F.1.4	Standardised test methods	81
	F.2	Methods used in the survey		82
		F.2.1	Monitoring for insecticide resistance	82
		F.2.2	Typing of malathion resistance	83
		F.2.3	Monitoring for fumigant resistance	84
	F.3	*Sitophilus oryzae*		86
		F.3.1	Recorded insecticide resistances	86
		F.3.2	Recorded fumigant resistances	87
		F.3.3	Survey data	87
		F.3.4	Comments	88
			F.3.4a Malathion resistance	88
			F.3.4b Lindane resistance	89
			F.3.4c Methyl bromide resistance	89
			F.3.4d Phosphine resistance	90
	F.4	*Sitophilus zeamais*		100
		F.4.1	Recorded insecticide resistances	100
		F.4.2	Recorded fumigant resistances	101
		F.4.3	Survey data	102
		F.4.4	Comments	102
			F.4.4a Malathion resistance	102
			F.4.4b Lindane resistance	103
			F.4.4c Methyl bromide resistance	103
			F.4.4d Phosphine resistance	103

F.5	*Sitophilus granarius*		112
	F.5.1 Recorded insecticide resistances		112
	F.5.2 Recorded fumigant resistances		113
	F.5.3 Survey data		114
	F.5.4 Comments		114
		F.5.4a Malathion resistance	114
		F.5.4b Lindane resistance	115
		F.5.4c Methyl bromide resistance	116
		F.5.4d Phosphine resistance	116
F.6	*Rhyzopertha dominica*		125
	F.6.1 Recorded insecticide resistances		125
	F.6.2 Recorded fumigant resistances		125
	F.6.3 Survey data		126
	F.6.4 Comments		126
		F.6.4a Malathion resistance	126
		F.6.4b Lindane resistance	127
		F.6.4c Methyl bromide resistance	127
		F.6.4d Phosphine resistance	128
F.7	*Tribolium castaneum*		139
	F.7.1 Recorded insecticide resistances		139
	F.7.2 Recorded fumigant resistances		146
	F.7.3 Survey data		147
	F.7.4 Comments		147
		F.7.4a Malathion resistance	147
		F.7.4b Lindane resistance	148
		F.7.4c Methyl bromide resistance	148
		F.7.4d Phosphine resistance	149
F.8	*Tribolium confusum*		163
	F.8.1 Recorded insecticide resistances		163
	F.8.2 Recorded fumigant resistances		163
	F.8.3 Survey data		164
	F.8.4 Comments		164
		F.8.4a Malathion resistance	164
		F.8.4b Lindane resistance	165
		F.8.4c Methyl bromide resistance	166
		F.8.4d Phosphine resistance	167
F.9	*Oryzaephilus surinamensis*		175
	F.9.1 Recorded insecticide resistances		175
	F.9.2 Recorded fumigant resistances		176
	F.9.3 Survey data		176
	F.9.4 Comments		176
		F.9.4a Malathion resistance	176
		F.9.4b Lindane resistance	178
		F.9.4c Methyl bromide and phosphine resistances	178
F.10	*Oryzaephilus mercator*		187
	F.10.1 Recorded insecticide resistances		187
	F.10.2 Recorded fumigant resistances		187
	F.10.3 Survey data		187
	F.10.4 Comments		188
		F.10.4a Malathion resistance	188
		F.10.4b Lindane resistance	188
		F.10.4c Methyl bromide and phosphine resistances	188

F.11	Other Coleoptera		195
	F.11.1	*Latheticus oryzae*	195
	F.11.2	*Tenebroides mauritanicus*	195
	F.11.3	*Trogoderma granarium*	195
	F.11.4	*Dermestes maculatus*	195
	F.11.5	Bruchidae	196
F.12	Lepidoptera		197
	F.12.1	*Ephestia cautella*	197
	F.12.2	*Plodia interpunctella*	198
	F.12.3	*Ephestia elutella*	200
	F.12.4	*Sitotroga cerealella*	200
	F.12.5	Tineidae	200
F.13	Movement of resistant strains in world trade		201
	F.13.1	Movement of insects in international trade	201
	F.13.2	Occurrence of resistant insects in ships and their cargoes	202
F.14	Multiple resistances and cross-resistances		206
	F.14.1	Introduction and definitions	206
	F.14.2	DDT	206
	F.14.3	Lindane and cyclodienes	211
	F.14.4	Pyrethroids	212
	F.14.5	Organophosphorus compounds	218
	F.14.6	Juvenile hormone mimics	221
	F.14.7	Other non-gaseous chemicals	222
	F.14.8	Fumigants	223
		F.14.8a Recorded fumigant cross-resistances	223
		F.14.8b Survey data	225
		F.14.8c Types of fumigant resistance	227
		F.14.8d Relationship with insecticide resistance	227
		F.14.8e Lepidoptera	229
	F.14.9	Cross-resistance to non-chemical measures	229
F.15	The inheritance of resistance in stored grain insects		231
	F.15.1	*Sitophilus* spp.	231
	F.15.2	*Rhyzopertha dominica*	232
	F.15.3	*Tribolium castaneum*	232
	F.15.4	*Tribolium confusum*	233
	F.15.5	Other *Tribolium* species	234
	F.15.6	Other Tenebrionidae	234
	F.15.7	Silvanidae	235
	F.15.8	Bruchidae	235
	F.15.9	*Cryptolestes* spp.	236
	F.15.10	Dermestidae	236
	F.15.11	Other Coleoptera	237
	F.15.12	*Ephestia cautella*	237
	F.15.13	*Ephestia elutella*	238
	F.15.14	*Ephestia kuehniella*	238
	F.15.15	*Plodia interpunctella*	238
	F.15.16	*Sitotroga cerealella*	239
	F.15.17	Comments	239
F.16	Ionising radiations		244

G.	THE GENERAL PROBLEM OF RESISTANCE IN STORED CEREAL INSECTS		247
	G.1 Summary of resistance status of individual species		247
		G.1.1 *Sitophilus oryzae*	247
		G.1.2 *Sitophilus zeamais*	250
		G.1.3 *Sitophilus granarius*	250
		G.1.4 *Rhyzopertha dominica*	251
		G.1.5 *Tribolium castaneum*	251
		G.1.6 *Tribolium confusum*	252
		G.1.7 *Oryzaephilus surinamensis*	253
		G.1.8 *Oryzaephilus mercator*	253
		G.1.9 Other species	253
	G.2 The current status of the pesticides concerned in the survey		254
		G.2.1 Lindane	254
		G.2.2 Malathion	255
		G.2.3 Methyl bromide	256
		G.2.4 Phosphine	259
H.	REFERENCES		260

PREFACE

There has been increasing concern throughout the world on the problems of pesticide resistance in pests of stored grain. Resistant strains of these pests have been shown to be moving actively in world trade and there is a serious threat to effective use of chemicals in maintaining stored foodstuffs in a sound and insect-free condition.

The need for adequate protection of stored grain against insect attack needs no emphasis and losses of 10 per cent or more in developing countries still occur far too frequently. Insecticides have played a major role in reducing losses and malathion, the most widely used material, has become almost indispensable in maintaining the standards that are demanded in international trade and becoming accepted as the general requirement in domestic consumption.

Because of the stringest requirements with respect to chemical residues, there are comparatively few materials that can be used for the control of grain pests. These few materials, however, after repetitive use have led in many instances to development of resistance. Stored grain insects are unique in that most of the major species are cosmopolitan and readily move about in domestic and international trade. Hence, resistant strains are also moving throughout the world thus reaching countries where resistance has not been suspected.

Since a full appreciation of the resistance problem is required for rational planning of insecticide replacement and supplementary control measures, and in accordance with the recommendation of the 7th Session of the FAO Working Party of Experts on Pest Resistance to Pesticides (Rome 1971), FAO undertook a global survey of stored grain pests, in cooperation with CSIRO, Australia, and the Pest Infestation Control Laboratory of the United Kingdom Ministry of Agriculture, Fisheries and Food. Dr Bruce R. Champ, CSIRO, served as FAO Consultant to undertake the actual field survey during 1972-1973; 61 countries were visited to collect strains of major stored grain pests for pesticide resistance monitoring. A total of 799 samples was collected and these, together with samples from another 24 countries were cultured at the Pest Infestation Control Laboratory for testing for resistance to malathion, lindane, methyl bromide and phosphine. Altogether, 1685 strains of the major pests were involved. Mr C.E. Dyte was responsible for testing the level and extent of resistance of processed samples.

The results of this joint undertaking were reviewed during the 10th Session of the FAO Working Party on Resistance to Pests to Pesticides (Rome, October 1974) and it was recommended that FAO should arrange for the early publication of this report. Since the major objective of this survey - to determine the world pattern of the distribution of resistance on stored grain insect pests - has been accomplished beyond expectations, it is very rewarding to present for consideration of member nations, the scientific community and people associated with stored grain management, this report which will provide an important reference and work guide.

William R. Furtick
Chief
Plant Protection Service
Plant Production and Protection Division

Rome, April 1976

A. CONCLUSIONS ARISING FROM THE REPORT

(A.1) *Implications of the results of the survey*

Control failures involving both malathion and lindane have been reported from many countries. Since standard test methods have been lacking, it has seldom, previously, been possible to attribute any of these unequivocally to resistance. The results of the present survey show that resistance to malathion and to lindane, as revealed by discriminating dose tests, is widespread. Control failures involving *Tribolium castaneum* and *Sitophilus oryzae* in the field have been associated with positive diagnosis of resistance in monitoring tests and confirmed by laboratory assays using treated grain.

The emergence of resistance to fumigants under practical conditions is a matter for particular concern. With major world dependence on fumigation both as a routine disinfestation treatment and as a means of combating insecticide-resistant strains, the occurrences reported, although as yet limited in number and often at matginal resistance levels, are of considerable significance and pose a real threat to a continued ability to store grain safely.

The minimum effective doses necessary to control a pest are often considerably less than those applied in practice. This safety margin sometimes ensures that low levels of resistance do not result in immediate control breakdown, although they may well shorten the effective life of the treatment. Monitoring tests will reveal these low levels of resistance before the threat they represent is fully realised. Every effort must be made to promote use of these tests in a general campaign to delay development of the resistance problem, not only for the species concerned in the survey, but for all species of economic significance.

Although the survey was restricted broadly to cereal grains, similar pest problems were apparent in the various legume seeds which constitute a significant part of the diet of many millions of people in developing countries. Lack of basic data prevented their inclusion in the present survey but a similar examination of pest status and susceptibility levels to pesticides is necessary and urgent.

The extent to which resistance can be met by increased dosages is severely limited by the level of residues acceptable internationally in stored foodstuffs. Countermeasures to resistance will thus necessitate the use of alternative insecticides and possibly fumigants.

(A.2) *Use of resistant strains in evaluation of alternative materials*

Insecticide resistance is now so widespread in certain species that it has become essential that resistant strains are used in the evaluation of candidate alternative materials. Only eleven strains of *Tribolium castaneum* susceptible to both malathion and lindane have been found among over 500 tested in the survey, and two of these eleven were from laboratory cultures maintained without exposure to pesticides. Thus, resistance has now become a 'normal' characteristic of this pest.

It is important too, that resistant strains should be used in studies of novel chemical approaches to the control of stored product pests. These may be new insecticides chemically unrelated to those currently in use, or chemicals with less orthodox effects on insects, for example pheromones and hormone mimics. A malathion-resistant strain of *Tribolium castaneum* has been found to be cross-resistant to a juvenile hormone as well as to four juvenile hormone mimetic compounds.

The problem is complicated by the fact that, although few studies of cross-resistance have been undertaken, already more than one type of malathion resistance has been demonstrated in three of the ten pests of stored cereals in which malathion resistance is known. Moreover, except for the malathion-specific type of resistance, comparisons of resistance patterns in different species show few similarities.

(A.3) *The need for a rational approach to storage pest control programmes*

It can be argued logically that there is sufficient knowledge relating to the techniques of storing grain safely and of integrating chemical pesticides into the pest control programmes, to reduce the pest problem in grain storage to one of very minor status. Economic considerations must be taken into account but even within such constraints, it would be possible to achieve a comparable goal if the knowledge was made available and conscientiously interpreted and put into practice.

The past 10-20 years have seen spectacular controls exercised by lindane and malathion often under conditions of poor hygiene and similar misuse. The situation, however, has now changed and it is apparent that the ease with which control could be achieved when these materials were first introduced. fostered attitudes to pest control that allowed inadequate attention to be given to basic principles and so accelerated development of resistance and consequent ineffectiveness of the control measures practised. The lesson to be learned from this is that these principles must be embodied in all control measures if long term efficacy is to be achieved. Thus it is mandatory that controls be based only on:

(i) A high standard of hygiene, that is storage of grain and other infestable commodities under circumstances where applied control measures are expedited, and removal of any spillage, residues or other materials that provide harbourage for pests or otherwise reduce the effectiveness of applied measures.

(ii) Use of suitable storage and transport facilities and if necessary their improvement to an acceptable standard.

(iii) Regular inspection of the commodity and storage for infestation or other causes of deterioration to allow prompt remedial action in the early stages of development of infestations.

(iv) Minimum use of chemicals to reduce incidence of residues, selection for resistance, environmental hazards, and general overhead costs. Whilst this should come primarily from hygiene and good warehousekeeping, aeration, refrigeration and other physical control methods should be used if practicable and economical.

(v) Use of chemical methods that give as complete a kill as possible to delay population build-up and reduce rate of selection for resistance. Wherever possible, fumigants should be used for control of established infestations and contact insecticides to prevent reinfestation.

Additionally where appropriate
(vi) Use of varieties of grain that have resistance to insect attack, harvesting techniques that reduce liability to field infestation and damage to the grain that increases susceptibility to insect attack, and adequate drying of the commodity before or during storage.

It seems unlikely that there will be any spectacular break-through in new methods of pest control and improbable that materials as successful in the short term as malathion will appear again. However, current knowledge of the control of grain pests is more than adequate if properly used to contain and possibly eliminate the problem. Elimination would be costly as it would involve remodelling storage and transport systems for effective utilisation of physical control methods. Nevertheless, it is highly desirable that the introduction of these physical methods with their in-built permanency, be given high priority in long-term control strategies even though economics and circumstances may dictate that it be a gradual process. Containing the problem, however, is an immediately accessible goal if proper attention is given to hygiene and prompt action taken to implement appropriate control measures. These programmes may have to operate within the constraints of management which may see priorities differently, and of reinfestation problems imposed by movement of infestable commodities in trade and cross-infestation from nearby storage or handling operations not under the same control. Nevertheless the programmes will reach their maximum effectiveness only when there is general acceptance of the basic principles involved and thus industry and management education must also be an integral part of the programme should the necessity arise.

A real deficiency exists in communication throughout the world of information on storage pest problems and their remedies. High priority must be given to assessing the situation and correcting the deficiency if the scarce resources available in the storage pest control field are to be used to maximum advantage.

B. INTRODUCTION

FAO has had a long association with efforts to promote increased productivity in agriculture and minimise the post-harvest losses which account for a significant part of total world production of cereals. Losses in storage, although considerably less than losses before harvest, are most tangible, concerning either actual reductions in the quantity of available foodstuffs, a depreciation in their marketable value, or indirectly, the cost of preventative measures that must be taken where there is a reasonable expectation that the direct forms of loss will occur. The last fifteen years have seen a dramatic reduction in overall losses and the importance of maintaining and improving this situation needs no emphasis. Moreover, increasing attention is now being given to establishing national and international buffer stocks of foodstuffs to guard against irregularities in production due to the vagaries of climate, thus creating circumstances in which storage losses can be potentially greater.

Pests are the major cause of these losses and current storage systems and technology dictate that control of these pests is based largely on use of chemical pesticides. Two basic types of storage pest control are recognisable. The first is at the local level and is essentially concerned with preventing physical damage or loss of commodities as for example in subsistence agriculture. As pest populations must be reasonably large or infestations of long-standing for appreciable damage to be done, the level of control required is not necessarily high and light infestations may be tolerated. The second type of control is concerned with contamination by insects of commodities in commerce particularly in international trade. There is usually negligible physical loss of foodstuffs but partly for aesthetic reasons and particularly because of a growing appreciation of the potential damage and economic losses that insects represent, such contamination has become unacceptable in international trade. It is becoming increasingly apparent also that technological advances in agriculture have created surpluses for marketing in both large and small-scale farm production, and standards of insect freedom applicable to international trade are now becoming accepted as a general requirement throughout the industry - both in local and export trade.

There are comparatively few materials that are safe for use in the control of pests of foodstuffs in storage. They comprise essentially a limited range of persistent insecticides and non-persistent fumigants. These few materials have been used widely and intensively and it is probable that the present requirements of freedom from insects would have been unattainable if these materials and particularly malathion had not been available. International trade in many of the world's basic foodstuffs such as cereals, does in fact have almost complete dependence on pesticides to meet the insect tolerance limits, and there must be serious repercussions on the industries concerned if existing levels of control cannot be maintained. The emergence of resistance to pesticides is thus a matter of considerable concern as there is no chemical that can immediately replace malathion in existing programmes, and those that may be cleared for international use in the near future are unlikely to have similar spectra of activity, cheapness, and other attributes. Inevitably resistance will impose a considerable economic burden on the industries concerned - a burden which will increase as resistances spread and intensify and infestation levels tend towards those that existed before malathion was introduced.

Fumigation has long been regarded as a basic method of controlling stored product insects and one which would be of material assistance in delaying the development of resistance to the unrelated residual pesticides. The method itself, by virtue of the low variability in response of individual insects to the commonly-used materials such as methyl bromide, is also usually considered to be less prone to resistance development than the normal methods with residual pesticides. It is disturbing then to note the increasing prevalence of resistance to fumigants and the resultant weakening of one of the most powerful tools available in stored product pest control for delaying or preventing development of resistance.

The major pest species of stored product insects are cosmopolitan and are moved extensively about in domestic and international trade in foodstuffs and other stored products. It has been shown conclusively that resistant strains of these major species are present in shipping involved in international trade, and it is probable that resistances developing in one area will be distributed to other areas before the resistance is detected or even becomes detectable. Because of international ramifications and opportunities for spread of resistances, a full appreciation of the resistance problem can only come from a knowledge of the distribution of the resistances present throughout the world and the various expressions of these resistances in terms of resistance levels and the cross resistances present. Workers throughout the world have given little attention to resistance in stored product insects, and the information available on a world basis is fragmentary. Our knowledge on the distribution of resistance is, in fact, a direct reflection of the distribution of the few individuals who have an interest in the subject rather than of the distribution of the resistances themselves. Nevertheless sufficient is known to give grounds for considerable concern, and there can be little doubt that in terms of continuing availability of foodstuffs, that resistance is as serious a threat to protection of commodities in storage as it is to agricultural production generally. It is thus appropriate that FAO, through its Working Party of Experts on Resistance of Pests to Pesticides which was convened to meet the challenge of pesticide resistance, should accept responsibility for determining the extent and implications of the problems that are emerging in the control of storage pests.

The following text is the report of the survey. Data on the occurrence of resistance is presented together with a summary of the more important background information available on the general subject of pesticide resistance in stored product insects. It seemed appropriate and timely to assemble this information so that the state of knowledge on the subject would be clearly apparent, and in the hope that not only would further research be encouraged but that those whose responsibility lies in the design and implementation of control measures, would gain a sounder appreciation of the bases of resistance, its detection and its implications.

C. CONDUCT OF THE SURVEY

(c.1) *Species involved*

The relative importance of the various storage pests varies from one area to another and is dependent on the commodity involved. Data was not available to make objective assessments of their relative world importance and eight species, *Sitophilus oryzae, S. zeamais, S. granarius, Rhyzopertha dominica, Tribolium castaneum, T. confusum, Oryzaephilus surinamensis* and *O. mercator*, were chosen arbitrarily for resistance monitoring as representing those species considered to be consistently of major economic significance in storage of cereals throughout the world and for which some base-response data for pesticides was available. *O. mercator* was a doubtful but expedient inclusion - the already documented occurrence of resistance, the confusion in the field between it and *O. surinamensis* which undoubtedly warranted inclusion, and the frequent cohabitation of the two species were taken as sufficient justification. The *Cryptolestes* species as a group may have warranted inclusion, but the dearth of species-definitive base-response data and the range of species involved together with the attendant confusion of field records made it doubtful whether meaningful results could be obtained without diverting resources from testing of unquestionably more important species.

Lepidoptera were not included in the resistance survey because appropriate base-response data and methods of monitoring for resistance were unavailable. Opportunity was taken, however, to collect material during the survey for preliminary studies on development of methods for such monitoring.

Resistance becomes of consequence only when the species involved is a significantly important pest in a particular area. Specific information on importance is available for individual species for limited areas but an overall appreciation of the resistance problem can only come if the occurrence of resistance in particular areas can be related to defined importance of the species in the various commodities stored or handled in those areas. To this end a list was compiled of 15 of the more important Coleoptera and Lepidoptera pests of stored cereals and assessments solicited from individual countries of the pest's importance in terms of the frequency with which losses occurred or measures had to be taken to prevent these losses. It was considered that data on real losses would be impossible to obtain or unreliable, and that estimates of potential losses from local surveys and experimentation would be academic and equally could not be relied on to reflect the true situation in the field. Data on importance are included in Section D and tended to confirm the original choice of species for monitoring in the survey. As data for some countries were unavailable at the time of compilation of this report, a sample consisting of 50 countries was used to establish the importance of different pest species for a consideration of the resistance problem. Notable omissions are France, the U.S.S.R. and the U.S.A. It is unlikely, however, that inclusion of these large grain producers would radically change the picture presented with the possible exception of *Sitophilus granarius* which is more prevalent and has a higher pest status in France and the U.S.S.R. than in many of the countries included.

(C.2) *Pesticides included*

When the survey began, provisional methods were available for monitoring only for malathion and lindane resistance in a range of storage Coleoptera. The widespread use and significance of these materials in stored product pest control and the relatively much more restricted use of other residual pesticides prompted the decision to restrict the survey to these materials.

Methods for fumigant testing were not available but the importance of fumigants and the threat to safe grain storage that resistance would impose, dictated that methods had to be developed and fumigants included in the survey. Strains collected during the survey were used to establish base-response data and world usage of materials indicated that for raw cereals, consideration could be confined to methyl bromide and phosphine.

(C.3) *Survey detail*

(C.3.1) *Establishing contacts and arrangements during visits*

Initial contact with countries where relevant FAO projects were operating was made by the Opérations Service of either the Plant Production and Protection Division or the Agricultural Services Division of FAO. With countries where FAO contact was not possible, requests for co-operation were sent directly to the organisation or persons who, from personal knowledge or enquiry, seemed most appropriate for the purposes of the survey. Arrangements for collection of information and samples were made almost invariably through official government agencies, usually local Departments of Agriculture or their equivalent. There was opportunity during most visits to meet the local specialists in stored product entomology and to visit representative storage and other facilities.

(C.3.2) *Countries visited*

Country	Dates	Centres visited	No. of samples collected
NORTH AMERICA			
Canada	7-14.v.72	London (Ont.), Vancouver, Winnipeg	19
Mexico	17-20.v.72	Mexico City, Pantaco, Tlalnepantla	9
U.S.A.	14-17.v.72	Fresno, Riverside, San Francisco	17
	29,30.vi.72	Savannah	43
CENTRAL AMERICA AND CARIBBEAN			
Colombia	24-28,v.72	Barranquilla, Bogota	6
El Salvador	20-23.v.72	La Libertad, San Salvador	8
Guatemala	23,24.v.72	Barcena, Escuintla, Guatemala City	7
Guyana	18-21.vi.72	East Coast Demerara, Georgetown	9
Jamaica	28,29.vi.72	Kingston, Spanish Town	13
Trinidad	21-25.vi.72	Curepe, Port of Spain	17
Venezuela	25-28.vi.72	Caracas, La Encrucijada, La Morita, Maracay, Valencia	8

Country	Dates	Centres visited	No. of samples collected
(continued)			
SOUTH AMERICA			
Argentina	4-10.vi.72	Buenos Aires, La Plata	34
Brazil	10-18.vi.72	Anapolis, Brasilia, Manaus, Porto Alegre, Rio de Janeiro, Santos, Sao Paulo	22
Chile	1-4.vi.72	El Monte, Linderos, Maipo, San Bernado, Santiago, Talagante	11
Peru	28-31.v.72	Callao, Chiclayo, Lambeyeque, Lima	7
EUROPE AND NORTH AFRICA			
Algeria	18-22.ii.73	Algiers, Berrouaghia, Blida, Dar-es-Abeida	8
Egypt	9-12.i.72	Cairo	3
France	4-10.iii.73	Paris	-
Greece	3-6.xii.72	Athens, Lamia, Piraeus	10
Italy	20-22.iv.72	Rome	-
Libya	7-9.i.73	Tripoli	7
Morocco	22-25.ii.73	Casablanca, Rabat, Sale	11
Poland	29.x-1.xi.72	Poznan, Swarzedez, Warsaw, Wronczyn Zieliniec	38
Portugal	25-28.ii.73	Ceiras, Lisbon, Queluz	15
Romania	1-4.xi.72	Bucharest, Budesti, Calarsi	8
Spain	28.ii-4.iii.73	Alcala, Aranda de Deuro, Burgos, Lerma, Madrid	8
Yugoslavia	6-9.xii.72	Belgrade	4
AFRICA - SOUTH OF SAHARA			
Benin	9,10.ii.73	Cotonou	3
Central African Republic	3-9.ii.73	Bangui	13
Ethiopia	12-16.i.73	Addis Ababa, Holetta, Nazret	16
Kenya	21-24.i.73	Nairobi	17
Malawi	24-26.i.73	Blantyre, Bvumbwee, Limbe	8
Mozambique	26-30.i.73	Beira, Lourenco Marques	16
Nigeria	10-15.ii.73	Lagos	11
Senegal	16-18.ii.73	Bambey, Dakar	13
Somalia	16-21.i.73	Algoi, Mogadishu	11
South Africa	30.i.-2.ii.73	Delmas, Eloff, Johannesburg, Pretoria, Radium, Settlers	10
Tanzania	23.i.73	Dar-es-Salaam	8
WEST ASIA			
Afghanistan	2-4.viii.72	Kabul	2
Bahrain	19-21.xi.72	Bahrain	9
Cyprus	8-11.xi.72	Athienou, Famagusta, Nicosia, Pyrga, Zyyi	12
India	4-12.viii.72	Bombay, Borivli, Chamau, Chhizarsi, Delhi, Faridabad, Hapur, Kaili, Karnal, Mathura, Meerut, Naraina	28

Country	Dates	Centre visited	No. of samples collected
WEST ASIA (continued)			
Iran	12-15.xi.72	Abadan, Ahwaz, Karadj, Khorramshar, Teheran	11
Iraq	16-19.xi.72	Baghdad	59
Kuwait	15,16.xi.72	Kuwait	4
Lebanon	11.xi.72	Beirut	2
Pakistan	12.viii.72	Karachi, Lahore	9
Saudi Arabia	21-24.xi.72	Riyadh	4
	27-30.xi.72	Jeddah	9
Syria	30.xi-3.xii.72	Damascus	4
Turkey	4-8.xi.72	Ankara	4
Yemen Arab Republic	24-27.xi.72	Hodeida	10
EAST ASIA			
Burma	12-15.viii./2	Rangoon	8
China	6-13.viii.73	Peking, Shanghai	14
	6-10.ix.72	Taiwan: Keelung, Taipei, Su-Line	11
Indonesia	15-18.viii.72	Djakarta	7
Japan	10-13.ix.72	Chiba City, Tokyo, Yahata City, Yokohama	23
Korea Rep.	13-16.xi.72	Inchon, Seoul, Sosa, Suwon, Youngdungpo,	12
Malaysia	12-16.iv.72	Alor Setar, Kuala Lumpur, Penang	10
Philippines	3-6.ix.72	Los Banos, Manila, Pagsanjan, Pila, Santa Cruz, Tanauan	36
Singapore	8-12.iv.72	Bukit Timah	11
Thailand	16-20.iv.72	Bangkok, Rangsit, Thonburi	22
U.S.S.R.	28.vi-3.vii.73	Leningrad, Moscow, Tashkent	-
Australia	19.viii-3.ix.72	Brisbane, Canberra	-

Number of countries visited - 61   Total samples   799

(C.3.3) *Collection of information from countries visited*

General statements were obtained as to the importance of grain pests locally (Section D) and the chemicals used in their control (Section E). Lists of relevant publications, copies of unpublished data and inventories of the stored product pest fauna of the country were requested and provided in most instances.

(C.3.4) *Collection and forwarding of samples*

Collecting visits were made almost invariably in the company of local personnel. Preference was given to collecting from central storages and animal feed and flour mills, as the probability of finding resistant strains if present in the country would be highest in such premises. Nevertheless, samples of the species concerned were taken from any available habitat and from any commodity harbouring them.

Collections were made with a battery-operated aspirator and transferred to ventilated polycarbonate sample tubes (7.5 cm x 2.5 cm) containing a feeding and breeding medium suitable for all species and the free space filled with crumpled paper. Two or three hundred adult Coleoptera of mixed species were placed in each tube or in the case of Lepidoptera 20-25 adults of a single species. With Coleoptera, a minimum of 80-100 individuals of each of the species concerned in the testing was collected if possible, but the number varied above or below this depending on the infestation, the time available and the numbers of other species to be taken in the sample.

Representative collections of all species were included in each sample but with minor species numbers were kept at levels appropriate for identification only. With Lepidoptera, carbon dioxide anaesthesia was used to facilitate transfer to the sample tubes and to stimulate oviposition. Although the adults died prematurely in the tubes, there was usually sufficient oviposition to enable a satisfactory culture to be established from the sample medium by addition of further medium when the sample was received at the laboratory.

Sample tubes were placed in padded posting bags and forwarded as soon as possible to the laboratory at Slough, usually by airmail postage.

(C.3.5) *Processing of samples*

When received at the laboratory, the samples were sorted as soon as possible and the species required for testing taken into culture. If living material of the species indicated on the label was not evident, the medium was incubated and re-examined in due course. Identifications were checked for all species. Sufficient insects were then bred to enable discriminating tests to be carried out for resistance to malathion, lindane, methyl bromide and phosphine. Parallel tests were carried out with synergised malathion (with triphenyl phosphate) to give a qualitative indication of the type of malathion resistance that might be present.

After the species required for testing were removed from the samples, any dead insects of these species, all living and dead specimens of other species, and the media (after incubation if this was necessary) were passed on to Dr D.G.H. Halstead of the Biology Department of the Pest Infestation Control Laboratory who identified all insect material present and took into culture species of interest for general investigation. This has enabled confirmation and supplementary recording of the occurrence of many species from a wide area of the world.

Results of tests were passed on periodically to cooperators progressively or when tests were complete.

(C.3.6) *Countries and areas not visited during the survey*

Requests were made to appropriate authorities in countries and areas not visited during the survey for information similar to that requested from countries visited and for samples of pests for monitoring for resistance. These requests were channelled through the Pest Infestation Control Laboratory, Slough, for samples and information from the United Kingdom; Dr R. Davis of the U.S. Department of Agriculture Stored Products Insects Research and Development Laboratory at Savannah, Georgia, for the U.S.A.; and Mr S.W. Bailey of C.S.I.R.O., Canberra, for the States of Australia. Some 50 other individual countries were approached. As a result, samples were obtained from a further 24 countries which together with additional material sent from the 61 countries visited yielded a total of 1685 strains of the 8 species involved.

D. THE MAJOR CEREALS AND THEIR IMPORTANT STORAGE PESTS

Occurences of resistance in storage pests must be viewed in their correct perspective and this can only be done by relating them to pest status of the individual species in the countries concerned and in the world generally.

Quantitative data on the importance of the major stored grain pests and the losses they cause are not available from most countries. As the situation can vary markedly from place to place within a country, and from year to year because of differences in weather, changes in varieties grown, management of grain stocks, and pest control procedures and their effectiveness, precise figures are unobtainable. The value of grain in monetary terms is variable and can be many times greater in an importing country than at the point of production particularly if in a developing country. This value, however, is relative to local conditions, and the lower value at the point of production can have much greater social significance, so that in effect, any consideration of grain and real or potential losses on a monetary basis, although appropriate on the local scene, can be misleading and has little relevance to the practical situation on a world basis.

However, some general classification of the importance of pests must be made so as to establish priorities in consideration of the overall problem. Such a classification would permit assessment of the world-wide importance of a species and comparisons with other species, based on the frequency of occurrence as a pest of particular status and the amount of commodity potentially involved. The classification must take into account that losses can assume many different but equally significant forms.

The principal direct effects of insect attack are loss in weight (sometimes masked by changes in moisture content), loss of germinative power (seed and malting grain), and the changes resulting from spontaneous heating due to insect activity which may result in mould attack and sprouting. Loss of nutritive value and palatability may also occur.

Indirect financial loss may result if the presence of insects and damage causes the commodity to be placed in a lower grade or be rejected entirely - this is particularly so when official grading systems are operated in national and international commerce. Infested grain may yield less flour and may contain excessive amounts of insect fragments. In some countries, grain products so contaminated may be condemned by public health authorities.

The presence of insects may lead to grain or grain products being rejected if phytosanitary standards are applied on export or on import into certain countries. Some commercial contracts may give the buyer the right to reject grain if insects are found, particularly in grain for malting. In other instances buyers may impose penalties operable under threat of not negotiating further sales. The cost of preventative and curative measures must also be taken into account.

A reliable measure of the potential of a species in a particular region is loss of weight determined over a period of time by sampling from stocks of grain in current use by farmers or in central storage, the samples being taken and analysed by standardised procedures. Unfortunately, such detailed information is lacking for most countries and the only common factor of which we have some information is the occurrence of the pests causing damage. Thus, in the light of present knowledge, assessment of the importance of the various pests can best be made by the persons in each region who are considered to be most informed on the general subject.

(D.1) *Classification of pest importance*

A small proportion only of the major cereals produced throughout the world enters international trade, for example in 1970, $99_6 \times 10^6$ m.t. were exported from an estimated world production of 1056×10^6 m.t. This is less than 1% and indicates that the major problem in preventing losses of cereals in storage relates to their protection in the country of their origin.

Thus, major emphasis has been placed on the importance of pests in the areas of production. Nevertheless, assessments of importance necessarily are influenced by market pressures both local and international Moreover, because of increasing consumer demand for high quality grain, the operative factor in these assessments is as much the likelihood that damage or loss will occur if preventative measures are not taken, as actual losses occurring. Imports into a country have been regarded as being assimilated with locally-produced grain into the storage handling and distribution system of that country and for practical purposes subjected to similar pest pressures. The relative importance of export trade varies among countries and the importance of pests in areas of production may be influenced considerably by the degree of pest freedom required by the countries to which the grain is exported, as although the amount of grain exported from a particular country may not be great, it may represent a substantial proportion of foreign currency earnings essential to the country's economy. It is as equally valid to accept this reaction to market pressure as a basis for upgrading importance of particular pests as to accept considerations of direct loss or damage. The world trend is to increasing levels of insect freedom in grain and in practice there is little differentiation in importing countries between infestations that would have economic significance and those that would be of no consequence. The literature on movement of insects in world trade is extensive and is considered in Section F.13.1. This occurrence of infestations in imported commodities must be taken into account, but the frequencies of these interceptions must be interpreted cautiously as different pest status may be accorded different species in the importing and exporting countries and there may be differential effectiveness against individual species of any control measures applied. Cross-infestation in transit must also be taken into account.

The following categories of pest importance are intended to apply to particular commodities within a country or region within a country and concern all grain produced or imported whether stored on farms, in trade or in central storages. They are intended as definitions of general importance and only incidentally as a ranking of importance in a particular region - thus a commodity may have pests of minor importance only in that region. The classifications are not intended to refer to particular situations or events but to take into account all circumstances in storage and handling of the commodities between production or import and consumption or export. The problem is a dynamic one and as stated previously, pest abundance in addition to showing seasonal variation may vary from year to

year as well as from place to place within a region. Nevertheless, there
is usually a recognisable general pattern of behaviour of a species in the
region.

Infestations often comprise complexes of pest species of varying
importance but in most instances, particular species are clearly dominant.
Experience of infestations in a particular region, together with a knowledge
of the biology and damage potential of the component species should permit
a reasonably objective assessment of the category of importance of each
species occurring to be made. Probably the greatest source of bias would
be from placing undue emphasis on observations of spectacular outbreaks
of pests or losses which occur towards the end of the storage season and
involve relatively insignificant proportions of total grain stored or handled.
It must be recognised also that the introduction of new high-yielding
varieties that are often more susceptible to insect attack and the changing
patterns of agricultural practices, pest control measures and consumer demands,
can influence the occurrence and potential of the species present in a
particular region so that pests of minor importance may emerge as major
pests or at least assume greater significance, or *vice versa*.

* *Pests of major importance*

<u>Pests that regularly cause significant losses of the commodity
or require control measures to prevent such losses.</u>

Those pests which regularly and frequently cause losses of grain
or grain products that are readily demonstrable as economically
significant in trade or that reduce significantly the amount of
commodity available from storage for consumer use including that
available from subsistence farm storage. These losses may be
either from reduction in weight or germinative power due to the
commodity being consumed or damaged by the insects, or from spoilage
due to spontaneous heating and mould activity resulting from the
activity of the insects.

or   Pests which from previous experience within the region and from
a knowledge of the potential of the species would reasonably be
expected to cause losses regularly and frequently if specific
preventative or remedial measures were not carried out. The
control measures carried out must represent a significant component
of handling and storage procedures and/or charges taking into
account materials used, labour, management costs, depreciation of
equipment and fixed facilities, and extra storage and handling
costs. These costs should include those resulting from detention
of commodities for control measures.

or   Pests where the presence or possibility of infestation, damage
resulting from an infestation, or residues or chemicals used in
preventative or control measures or their effects on the grain
or grain products, result in a decrease or downgrading of quality
of the commodity or subsequently processed products that
significantly affects acceptance or suitability for consumption
by man or animals, and/or value for trade purposes, or incurs
penalties from buyers, or has adverse effects on trade relations.

* *Pests of moderate importance*

> ### Pests that occasionally cause significant losses or require control measures.
>
> Those pests which are only occasionally responsible for significant losses or direct or indirect reduction in value of grain and grain products as outlined above usually but not necessarily at marginal levels, and which do not require specific control measures to be taken regularly. These species may often be present as part of a complex dominated by one or more species of major importance and their control may be achieved incidentally with control of the major species.

* *Pests of minor importance*

> Those pests which are established within a region but do not cause demonstrably significant losses of any nature and rarely or never require specific control measures in that region.

* *Pests which are rare or absent*

> Pests which are recorded rarely, usually on imported commodities, or are not known to occur within the region.

(D.2)   *Collection and interpretation of pest importance data*

Fifteen species of stored grain insects were included in the survey. These were:

> *Sitophilus oryzae, S. zeamais, S. granarius*
> *Rhyzopertha dominica, Trogoderma granarium*
> *Tribolium castaneum, T. confusum*
> *Oryzaephilus surinamensis, O. mercator*
> *Sitotroga cerealella*
> *Ephestia cautella, E. elutella, E. kuehniella*
> *Plodia interpunctella, Corcyra cephalonica*

Oryzaephilus mercator which is regarded primarily as an oil-seed pest was included for comparative purposes because of confusion in the field with *O. surinamensis*. Notwithstanding, *O. mercator* is regarded as the dominant *Oryzaephilus* species in infestations in grain in West Africa. The *Cryptolestes* species warranted inclusion as serious pests in many parts of the world. The difficulties in their definitive identification to species combined with the different climatic requirements of the major species (particularly *C. ferrugineus, C. pusilloides,* and *C. pusillus*) indicated problems in attributing data to the correct species involved - accordingly *Cryptolestes* were not specifically included though data was collected from some areas where species of *Cryptolestes* were of particular significance.

The cereal grains considered were wheat, barley, millet and sorghum, paddy rice, maize and milled products generally irrespective of whether destined for animal or human consumption. Other grains such as

oats and rye, for example, were not considered individually because of
their low levels of world production, minimal involvement in international
trade and often only localised production.

Requests were made to most countries of the world for a listing
of the importance of each of the 15 pest species in each of the cereal
commodities. Within larger countries, states or provinces were considered
individually or as regional groupings. The assessments of pest importance
were obtained wherever possible from local personnel with extensive experience
of local storage practice and its problems. Levels of expertise varied and
a basic interpretation of the classification was therefore provided outlining
in very general terms the categories of importance in which the insects
were to be placed.

As collection of data from all countries was not complete and
as the total amount of data involved and its analysis would have been
excessive for a general consideration of the resistance problem, a sample
of countries was taken whose total production was approximately 50% of
world production estimates for each commodity. The sample of countries
was based on earliest returns of data. The data concerned are given in
Tables D1 to D6 for wheat, barley, millet+sorghum and paddy, rice, maize,
and milled cereal products. Production, export and import estimates for
1970 are given also. Where estimates were not available for 1970, estimates
for earlier years are given. The listing of importance for larger countries
where importance data were available on a regional basis (e.g. China), is
given as a country listing based on regional listings and commodity
distribution between these regions. This was done because production-
export-import data on a regional basis was not available. A detailed
consideration on a regional basis that includes all data available will be
presented in a later report.

The relative importance of the various species is shown for each
commodity in Figures D1 to D5. Comparisons are made on the basis of the
total amount of grain produced in those countries where each species is
considered to be a pest of major, moderate or minor importance. Production
data are supplemented by import estimates of the commodity. Using the
ranking established from production data, the relative importance in
export trade is also shown.

A summary of the importance of the species in the major cereal
grains in terms of total cereal production (wheat, barley, dehusked paddy,
millet+sorghum and maize) is given in Figure D6 and for milled products
in Figure D7. Milled product production estimates were based on total
cereal production (rice included as dehusked paddy) adjusted for exports
and imports of unmilled wheat and meslin, barley, rice, maize, oats, rye
and cereals of not-export standard. Import and export estimates were based
on trade in meal and flour of wheat or of meslin and bran, pollard, sharps
and other by-products (Data from FAO Production and Trade Yearbooks 1970).
This assumed that all harvested cereals other than seed were processed in
some manner before consumption by man or animals. More precise total
estimates are not available.

(D.3)   *Comments*

(D.3.1)   *Wheat*   (Table D1, Fig. D1)

Sitophilus oryzae was the most important species both in terms of production and as a consideration in export trade.

Sitotroga cerealella ranked next as a major pest particularly because of its importance in China and India (which accounted for 41% of the total production in the sample). *S. cerealella*, usually a short term storage pest with infestations originating in the field, was not of major importance in exporting countries where combine harvesting and storage methods have minimised field infestations.

Rhyzopertha dominica, *S. granarius*, *Oryzaephilus surinamensis* and *Tribolium castaneum* were of comparable importance but usually, in different circumstances. *R. dominica* was associated particularly with the warmer, drier wheat areas of Australia, India and Pakistan whereas *S. granarius* was restricted largely to the cooler wheat-producing areas such as in Europe. In export trade, *R. dominica* was more significant, particularly because of its importance in Australia. *O. surinamensis* and *T. castaneum* occur over a wide range of climatic conditions though the latter species is not found extensively in areas with cold winters except in heated premises. The significant production of wheat in these cold areas results in *O. surinamensis* being more important generally than *T. castaneum* although in terms of international trade, *T. castaneum* has greater significance because of the greater frequency with which it appears as a major pest in the warmer exporting countries such as Australia.

Of the phycitid moth pests, *Ephestia elutella* was listed most frequently as a major pest. This species however, only occurs in wheat in cooler areas where pest problems are generally less acute - the major contributions to its overall importance came from the grain-exporting countries of central Europe. *E. cautella*, a tropical and sub-tropical pest of many cereals is found throughout the warmer areas of wheat production and is probably of greater general significance than *E. elutella*, though as conditions for *E. cautella* become optimal, conditions for wheat become marginal. The related galleriid *Corcyra cephalonica* complements and sometimes replaces *E. cautella* in hotter, more humid areas. *Plodia interpunctella* occurs over a wider range of conditions than any other species in this group. Outbreaks often appear to be of a localised nature and to be of greatest significance in cool areas. *E. kuehniella* is essentially a pest of milled products.

Trogoderma granarium assumes major local importance in countries such as Cyprus, Iraq, Pakistan and the Sudan but in many other areas where it occurs such as India and North Africa it is regarded less seriously. Fear of its introduction into countries where it is not currently established has resulted in international trade restrictions lending it an importance not consistent with its demonstrated pest status in areas where it now occurs.

Of the other species, *Sitophilus zeamais* is distributed throughout the warm, more humid regions which usually are not significant wheat producing areas and hence it is not a particularly important pest of wheat. Most problems with this pest arise from cross-infestation in storage. *Tribolium confusum* tends to replace *T. castaneum* in cooler areas in wheat spillage and residues around storages and in mills particularly. *Oryzaephilus mercator* is rarely found associated with wheat.

PRODUCTION, EXPORTS AND IMPORTS** OF WHEAT DURING 1970 IN REPRESENTATIVE COUNTRIES OF THE WORLD AND THE IMPORTANCE OF THE MAJOR STORAGE PESTS IN THESE COUNTRIES.

A = Major importance
B = Moderate importance
C = Minor importance
- = Occur rarely or absent

So - *Sitophilus oryzae*
Sz - *Sitophilus zeamais*
Sg - *Sitophilus granarius*
Rd - *Rhyzopertha dominica*
Tg - *Trogoderma granarium*

Tc - *Tribolium castaneum*
Tco - *Tribolium confusum*
Os - *Oryzaephilus surinamensis*
Om - *Oryzaephilus mercator*
Sc - *Sitotroga cerealella*
Ec - *Ephestia cautella*
Ee - *Ephestia elutella*
Ek - *Ephestia kuehniella*
Pi - *Plodia interpunctella*
Cc - *Corcyra cephalonica*

Country	Production 1000 mt	Exports 1000 mt	Imports 1000 mt	So	Sz	Sg	Rd	Tg	Tc	Tco	Os	Om	Sc	Ec	Ee	Ek	Pi	Cc
Afghanistan	1915	-	20	-	-	B	B	B	-	B	B	-	-	-	-	-	-	-
Algeria	1435	-	421	B	-	B	C	B	B	-	C	-	C	-	-	-	-	-
Argentina	4250	2302	-	A	B	B	B	-	B	B	B	-	B	C	-	C	C	-
Australia	7988	6886	-	A	C	B	A	-	A	C	B	-	B	B	C	C	B	-
Canada	9023	10746	-	-	-	-	-	-	C	-	C	-	-	-	-	-	-	-
Cent. Afr. Rep.	-	-	6	A	-	B	-	A	-	-	-	-	-	-	-	-	-	-
China	31000	-	4980	A	-	-	-	-	C	C	B	-	A	B	-	B	B	-
Colombia	50	-	280	A	A	-	A	-	A	-	-	-	A	C	-	C	-	-
Cyprus	43	-	44	A	-	A	B	A	A	A	A	-	A	A	-	-	B	-
Czechoslovakia	3174	-	1026	A	-	A	B	-	C	B	A	-	B	-	A	C	B	-
Denmark	512	29	5	-	-	B	-	-	-	-	-	-	-	-	-	-	-	-
Dominican Rep.	-	-	76	B	-	-	-	-	B	-	-	-	-	A	-	-	-	B
Egypt	1516	-	851	A	-	C	A	C	A	C	B	-	B	B	B	-	-	-
Ethiopia	840	-	7	A	B	C	A	-	B	C	C	C	C	A	-	C	-	-
Germany F.R.	5662	1450	2209	-	-	A	-	-	-	C	A	-	B	A	C	B	-	-
Netherlands	640	490	1594	-	-	B	-	-	C	-	C	-	-	C	-	-	-	-
India	26093	-	3587	A	C	-	A	C	B	-	B	C	A	B	-	-	-	B
Iraq	1059	-	90	A	-	A	A	A	B	C	A	C	-	C	C	C	C	-
Israel	125	-	418	A	-	-	A	B	A	B	A	-	C	A	-	-	C	-
Italy	9689	639	1164	A	B	A	B	-	-	-	A	-	-	-	-	-	-	-
Japan	474	-	4685	A	A	-	B	-	-	-	C	-	-	-	-	-	-	-
Kenya	205	44	-	A	C	-	-	-	B	-	B	-	C	B	-	C	C	-
Korea Rep.	357	-	1178	-	-	-	-	-	-	-	-	-	-	-	-	-	-	-
Lebanon	50	-	363	A	-	-	-	-	B	B	B	-	B	-	-	-	-	-
Libya	21	-	70*	A	-	-	A	B	C	-	-	-	B	-	-	-	C	-
Mozambique	10	-	74	A	C	-	C	-	A	-	B	-	A	-	-	-	-	-
Nigeria	-	-	259	-	-	-	C	B	C	-	C	-	-	-	-	-	-	-
Norway	12	1	420	-	-	-	-	-	-	-	-	-	-	-	-	-	-	-
Pakistan	7399	-	1684	B	-	C	B	A	A	-	C	-	B	B	-	-	-	B
Paraguay	41	-	72	A	-	A	-	-	A	A	-	-	A	-	-	-	-	-
Poland	4608	-	1104	-	-	A	-	-	-	C	B	-	-	-	-	C	-	-
Senegal	-	-	1124	-	-	-	-	-	-	-	-	-	-	-	-	-	-	-
Singapore	-	22	275	-	A	-	-	-	A	-	-	-	C	-	-	-	..	-
Somalia	-	-	1	-	-	-	-	-	-	-	-	-	-	-	-	-	-	-
Sth Africa	1396	-	121	A	A	A	A	-	B	C	B	-	B	B	-	B	B	-
Spain	4064	123	1	-	-	B	C	C	C	B	C	-	C	-	-	-	C	-
Sudan	115	-	195	-	-	B	A	B	-	-	-	-	-	-	-	-	-	-
Sweden	962	280	40	B	-	A	B	-	C	C	A	-	-	-	-	-	-	-
Switzerland	348	-	497	-	-	B	-	-	-	C	C	-	-	-	-	C	-	-
United K'dom	4236	7	4928	-	-	B	-	-	-	-	A	-	-	-	-	-	-	-

** Data from 1971 FAO Production Yearbook and 1971 FAO Trade Yearbook. *1969 estimate.

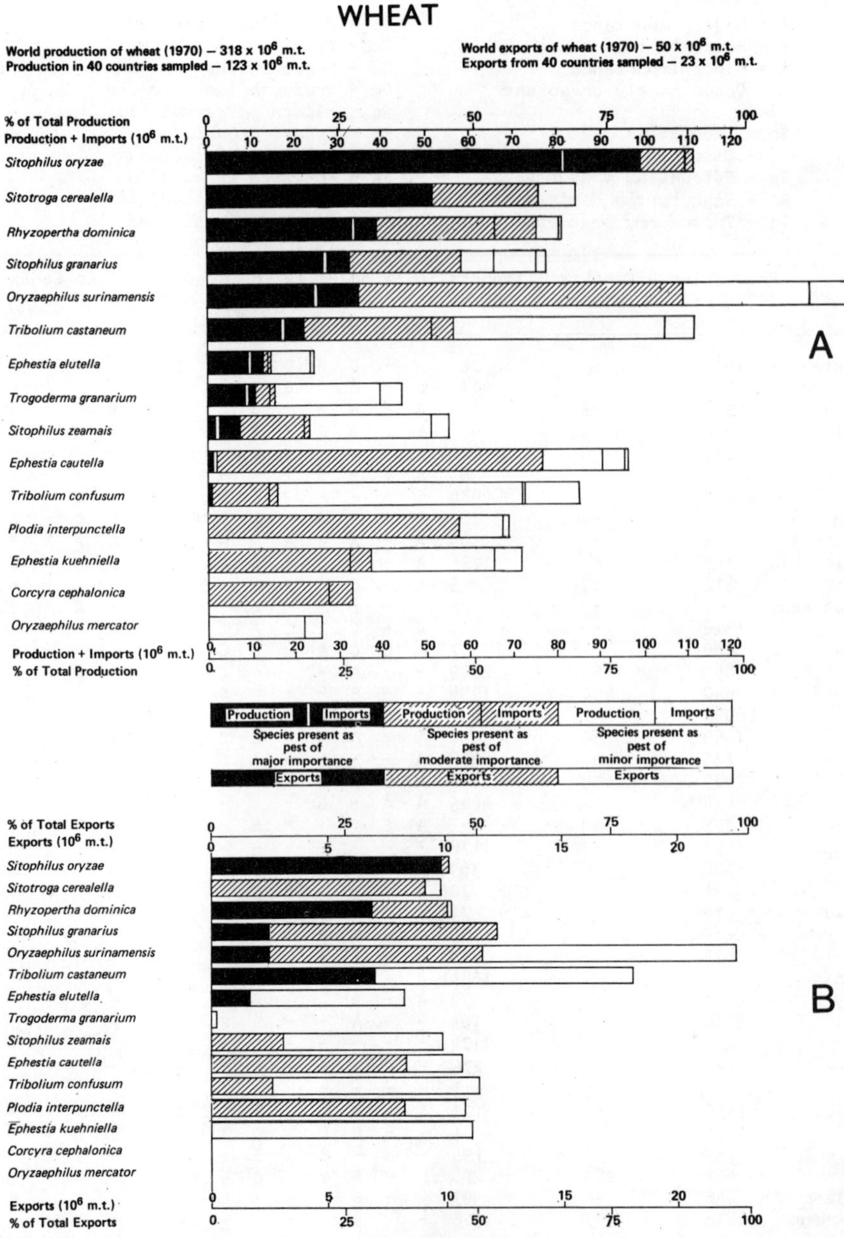

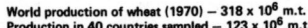

Fig.D1 — The percentage of wheat at risk from damage by insect pests in 40 representative countries of the world.
A — Total production + import estimates.   B — Export estimates.

(D.3.2) *Barley* (Table D2, Fig. D2)

World production of barley is less than half that of wheat and exports approximate 20%. Production tends to be proportionately greater than wheat in cooler countries and this was reflected in a considerable decrease in the importance of most pests including *S. oryzae*. This species however, despite its virtual absence from cool countries, remained the most significant pest of stored barley. Exceptions to this decrease in importance were *S. granarius*, *O. surinamensis* and *E. elutella* which are the species best adapted to survival in cold climates. The overall importance of the two beetles in terms of percentage production was similar for wheat and barley (approximately 20%) while *E. elutella* showed a slight increase for barley. All these species, and *R. dominica* and *T. castaneum*, were significant major pests in exporting countries though they had this status in only 10-15% of total export barley.

T. granarium is also concerned in trade. It is endemic in areas with cold winters (e.g. in north Africa and the Middle East) as well as hot areas and may infest shipments of malting and feed barley. Endemic infestations are known from the artificially heated environment of breweries in cold countries in which it would not normally survive.

S. cerealella as in wheat, assumed significance in terms of local production with the major contribution coming from China. Again this species was not significant in export trade. *S. zeamais*, *O. mercator* and *C. cephalonica* do not occur as pests in normal areas of barley production.

(D.3.3) *Millet+Sorghum* (Table D3, Fig. D3)

The millets and sorghums are essentially cereals of the warmer, drier areas and are mostly consumed locally. Their importance in export/import trade is minimal and has not been considered.

S. oryzae was again the most significant species. It was a major pest in most producing areas (approximately 95%) and assumed lesser importance only in cool areas marginal for production of these grains. In regions where summer day temperatures are very high, mortality can be considerable and this has been suggested as the reason for the absence of *S. oryzae* from areas such as the Southern Sudan. *R. dominica*, the next most important species, is more tolerant of hotter and drier conditions but is considerably less important than *S. oryzae* because of its poorer adaptation to the cooler range of temperatures suitable for millet and sorghum, and its greater susceptibility to disturbance in storage.

S. cerealella is a major pest of sorghums particularly in areas where it is stored unthreshed. Other Lepidoptera infesting the heads in the field also contribute to losses in this type of storage. *T. castaneum* is ubiquitous following attack by the primary species and mechanical damage to grain during harvesting. In the hot areas, *E. cautella* and *T. granarium* often cause serious damage as may *C. cephalonica* and *O. surinamensis*.

Sorghum varieties differ markedly in their susceptibility to insect attack with the harder varieties being significantly more resistant particularly at lower moisture contents. Consumer demand, however, frequently favours the softer varieties.

TABLE D2 : PRODUCTION, EXPORTS AND IMPORTS** OF BARLEY DURING 1970 IN REPRESENTATIVE COUNTRIES OF THE WORLD AND THE IMPORTANCE OF THE MAJOR STORAGE PESTS IN THESE COUNTRIES.

A = Major importance
B = Moderate importance
C = Minor importance
- = Occur rarely or absent

So - *Sitophilus oryzae*
Sz - *Sitophilus zeamais*
Sg - *Sitophilus granarius*
Rd - *Rhyzopertha dominica*
Tg - *Trogoderma granarium*

Tc - *Tribolium castaneum*
Tco- *Tribolium confusum*
Os - *Oryzaephilus surinamensis*
Om - *Oryzaephilus mercator*
Sc - *Sitotroga cerealella*
Ec - *Ephestia cautella*
Ee - *Ephestia elutella*
Ek - *Ephestia kuehniella*
Pi - *Plodia interpunctella*
Cc - *Corcyra cephalonica*

Country	Production 1000 mt	Exports 1000 mt	Imports 1000 mt	So	Sz	Sg	Rd	Tg	Tc	Tco	Os	Om	Sc	Ec	Ee	Ek	Pi	Cc
Afghanistan	360	-	-	-	-	-	B	B	-	-	-	-	-	-	-	-	-	-
Algeria	571	-	-	C	-	B	C	B	C	-	C	-	B	-	-	-	-	-
Argentina	367	92	-	A	-	B	C	-	B	B	B	-	C	C	-	C	C	-
Australia	2472	631	-	A	C	B	A	-	A	C	B	-	B	C	C	C	B	-
Canada	9051	2988	-	-	-	-	-	-	C	-	C	-	-	-	-	-	-	-
China	14700*	-	-	A	-	-	-	-	-	-	C	-	A	-	-	-	C	-
Cyprus	56	-	88	A	-	A	B	A	A	A	A	-	A	A	-	-	B	-
Czechoslovakia	2280	54	139	A	-	A	B	-	-	-	B	-	-	-	A	A	-	-
Denmark	4813	280	91	-	-	B	-	-	-	-	-	-	-	-	-	-	-	-
Egypt	83	-	-	A	-	C	A	C	A	C	B	-	B	B	B	-	-	-
Ethiopia	1525	1*	8*	C	C	B	C	-	B	-	-	-	B	C	-	-	-	-
Germany F.R.	4754	574	2074	-	-	A	-	-	-	C	A	-	B	A	C	B	-	-
India	2716	-	-	A	-	-	A	B	B	-	B	C	B	B	-	-	-	-
Iraq	691	36	-	A	-	A	A	A	-	C	-	-	-	-	C	C	C	-
Israel	14	-	165	A	-	-	A	C	A	B	A	-	C	A	-	-	C	-
Kenya	13	-	-	A	C	-	-	-	-	B	-	C	-	C	B	-	C	C
Korea Rep.	1974	-	11	-	-	-	-	-	-	-	-	-	-	-	-	-	B	-
Lebanon	8	-	121	A	-	-	-	-	B	B	B	-	B	-	-	-	-	-
Libya	53	-	93*	A	-	-	A	B	B	-	-	-	B	C	-	-	C	-
Mozambique	-	-	-	-	-	-	A	-	A	-	A	-	-	B	-	-	-	-
Netherlands	350	127	169	-	-	C	-	-	-	-	-	-	-	-	-	-	-	-
Nigeria	-	-	-	-	-	-	-	-	-	B	-	-	-	-	-	-	-	-
Norway	581	8	237	-	-	-	-	-	-	-	-	-	-	-	-	-	-	-
Pakistan	128	-	-	C	-	-	C	-	-	-	C	-	C	B	-	-	-	B
Poland	2149	145	1094	-	-	A	-	-	-	-	-	-	-	-	-	-	-	-
Sth Africa	33	-	14	A	A	A	A	-	B	B	B	-	B	C	-	C	C	-
Spain	3069	20	6	-	-	A	C	C	A	A	-	C	-	-	-	C	-	-
Sweden	1904	110	19	-	-	B	-	-	C	C	B	-	-	-	-	-	-	-
Switzerland	136	-	456	-	-	-	-	-	-	-	-	-	-	-	-	-	-	-
United K'dom	7529	112	1211	-	-	B	-	-	-	-	A	-	-	-	-	-	-	-
Country	Production 1000 mt	Exports 1000 mt	Imports 1000 mt	So	Sz	Sg	Rd	Tg	Tc	Tco	Os	Om	Sc	Ec	Ee	Ek	Pi	Cc

** Data from 1971 FAO Production Yearbook and 1971 FAO Trade Yearbook
* Data for China 1965, Ethiopia and Libya 1969.

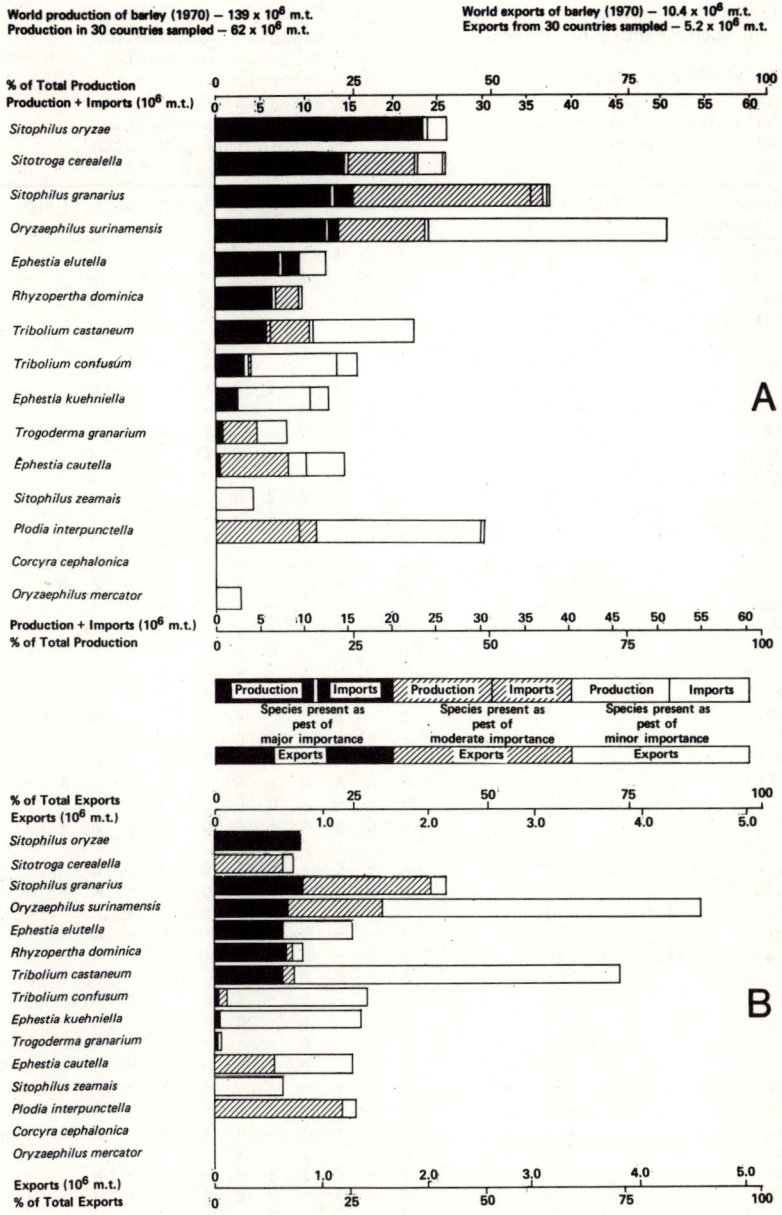

Fig.D2 — The percentage of barley at risk from damage by insect pests in 30 representative countries of the world.
A — Total production + import estimates. B — Export estimates.

TABLE D3: PRODUCTION** OF MILLET + SORGHUM AND PADDY DURING 1970 IN REPRESENTATIVE COUNTRIES OF THE WORLD AND THE IMPORTANCE OF THE MAJOR STORAGE PESTS IN THESE COUNTRIES.

- A = Major importance
- B = Moderate importance
- C = Minor importance
- - = Occur rarely or absent

- So - *Sitophilus oryae*
- Sz - *Sitophilus zeamais*
- Sg - *Sitophilus granarius*
- Rd - *Rhyzopertha dominica*
- Tg - *Trogoderma granarium*
- Tc - *Tribolium castaneum*
- Tco - *Tribolium confusum*
- Os - *Oryzaephilus surinamensis*
- Om - *Oryzaephilus mercator*
- Sc - *Sitotroga cerealella*
- Ec - *Ephestia cautella*
- Ee - *Ephestia elutella*
- Ek - *Ephestia kuehniella*
- Pi - *Plodia interpunctella*
- Cc - *Corcyra cephalonica*

Country	Production 1000 mt	So	Sz	Sg	Rd	Tg	Tc	Tco	Os	Om	Sc	Ec	Ee	Ek	Pi	Cc
MILLET + SORGHUM																
Argentina	4193	A	C	B	B	-	B	B	B	-	C	C	-	C	C	-
Australia	1292	A	C	C	A	-	A	C	B	C	B	B	-	C	B	C
Botswana	9	A	C	-	B	-	A	C	C	C	A	B	-	-	C	-
Cent. Afr. Republic	50	A	-	-	A	-	A	-	-	-	-	-	-	-	-	-
China	17120*	A	-	-	-	-	-	-	C	-	C	C	-	C	B	-
Colombia	165	A	-	-	-	-	-	-	-	-	A	-	-	-	-	-
Egypt	874	A	-	-	A	C	A	C	B	-	C	C	C	-	-	-
El Salvador	147	C	-	-	-	-	-	-	-	-	-	-	-	-	-	-
Ethiopia	2700	A	B	-	B	-	A	C	C	C	B	A	-	-	C	-
Gambia	30	C	C	-	-	-	B	C	-	C	B	C	-	-	C	C
Guatemala	46	B	-	-	-	-	-	-	-	-	-	-	-	-	-	-
India	20262	A	-	-	A	B	B	-	B	-	A	B	-	-	-	B
Iraq	8	-	-	-	-	A	-	-	-	-	-	-	-	-	-	-
Israel	11	A	-	-	B	-	A	B	A	-	C	A	-	-	C	-
Japan	14	C	-	-	-	-	-	-	-	-	-	-	-	-	-	-
Kenya	320*	A	C	-	-	-	B	-	C	-	C	B	-	C	C	C
Korea Rep.	52	-	-	-	-	-	-	-	-	-	-	-	-	-	-	-
Lebanon	1	-	-	-	-	-	B	B	B	-	-	-	-	-	-	-
Libya	1	C	-	-	-	C	-	-	-	-	-	-	-	-	C	C
Mozambique	190	A	-	-	A	-	A	-	-	-	A	-	-	-	-	-
Nigeria	6300	A	C	-	A	B	B	B	-	B	A	B	-	-	B	B
Paraguay	7	B	-	B	-	-	B	B	-	-	B	-	-	-	-	-
Senegal	405	-	-	-	B	C	C	-	-	C	C	-	-	-	C	-
Somalia	50	A	-	-	C	-	A	-	C	-	A	A	-	-	C	A
South Africa	460	A	A	C	-	B	-	B	A	-	A	A	-	B	B	-
Sudan	1989	-	-	-	A	A	B	B	B	-	A	B	-	-	-	-
Zambia	250	A	C	-	A	-	A	C	C	C	A	-	-	-	C	-
PADDY																
Argentina	407	A	-	C	-	-	C	C	B	-	B	C	-	C	C	-
Australia	247	A	C	C	A	-	A	C	A	-	A	-	-	C	A	-
Cent. Afr. Republic	13	A	-	-	A	-	A	-	-	-	-	-	-	-	-	-
China	102000	A	-	-	B	-	-	-	C	-	A	-	-	-	-	-
Dominican Republic	210	B	-	-	B	-	B	-	-	-	A	-	-	-	-	-
Egypt	2605	A	-	C	A	-	A	C	C	-	B	B	B	-	-	-
Gambia	50	-	-	-	B	-	B	C	-	B	-	-	-	-	C	C
Guyana	222	C	-	-	A	-	A	C	-	-	A	-	-	-	-	-
India	63672	-	-	-	A	-	A	-	-	-	A	-	-	-	-	-
Kenya	26	B	C	-	B	-	B	-	B	-	B	B	-	C	C	C
Korea Republic	5476	-	-	-	-	-	-	-	-	-	-	-	-	-	-	-
Malaysia	1429	B	-	-	-	-	-	-	-	-	A	A	-	-	-	-
Mozambique	160	B	-	-	A	-	A	-	B	-	A	C	-	-	-	B
Nigeria	550	C	-	-	B	-	-	-	-	-	-	-	-	-	-	-
Pakistan	20014	B	-	C	B	B	B	-	C	-	B	-	-	-	-	B
Paraguay	45	A	-	A	-	-	B	B	-	-	B	-	-	-	-	-
Senegal	98	B	-	-	-	C	-	-	-	-	-	-	-	-	-	-
Singapore	-	-	A	-	B	-	-	-	-	-	B	-	-	-	-	B
Sudan	6	-	-	-	A	A	B	-	-	-	-	-	-	-	-	-

** Data from 1971 FAO Production Yearbook
* Data from China and Kenya 1965.

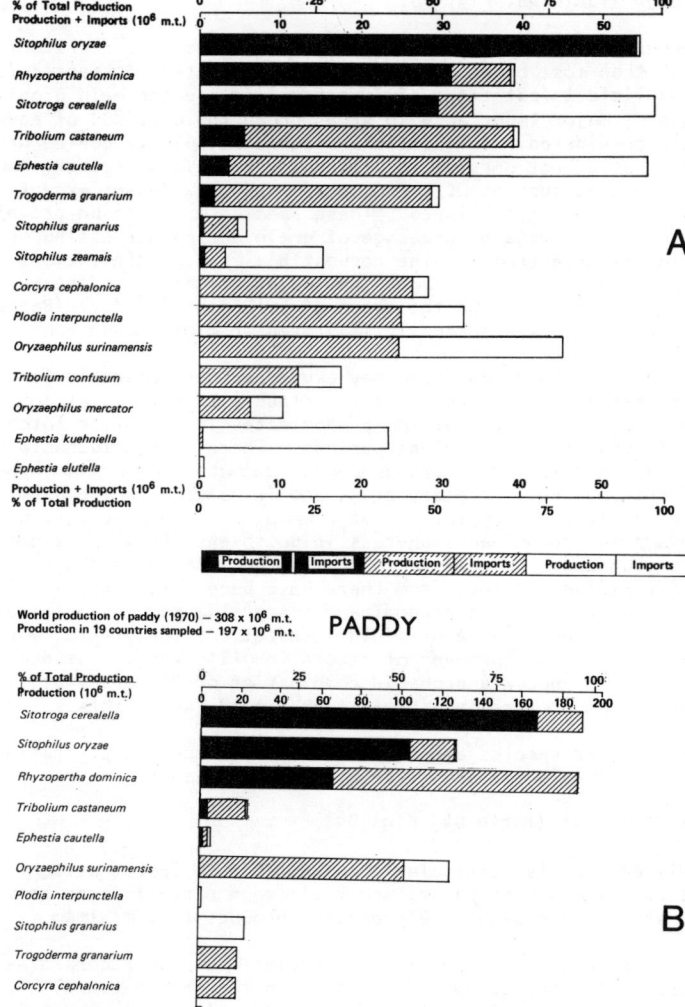

Fig. D3 -- The percentage of millet and sorghum and paddy at risk from damage by insect pests in 27 representative countries of the world.
A — Millet + sorghum. B — Paddy.

(D.3.4)   *Rice*

(D.3.4a)   *Paddy* (Table D3, Fig. D3)

Unhusked rice (paddy) is usually considered more resistant to insect attack than most other cereals. Moisture content is particularly important and field infestations of *S. cerealella* are the main problem. This pest was of major importance in areas producing over 80% of paddy in the sample considered. Infestations occur particularly during drying after harvest but do not persist in storage. The initial attack, however, allows other species such as *O. surinamensis* and to a lesser extent *T. castaneum* to become established. These species may also be prevalent as a result of husk damage or breakage of grain during harvesting or threshing, for example from combine harvesting. Where infestations of *S. cerealella* do persist in storage, they occur primarily in the surface layers and particularly around the edges of bulk stored paddy. Bagged paddy is less liable to attack, especially if stored in close-woven bags.

S. oryzae and *R. dominica* may cause serious damage in parboiled paddy and in varieties less resistant to attack particularly at comparatively high moisture contents. *R. dominica* predominates in the drier batches when these are left undisturbed for long periods. There is considerable confusion as to the relative roles of *S. oryzae* and *S. zeamais*. *S. oryzae* appears to be the dominant species in the dry areas and *S. zeamais* in the warm moist areas. *S. zeamais*, for example, is not common in the Middle East or in India where *S. oryzae* is predominant, whereas in South-east Asia, *S. zeamais* is particularly abundant. *S. zeamais* is not listed from China though most samples of *Sitophilus* obtained from there have been this species. In many areas of South-east Asia, the predominant *Sitophilus* species in rice mills is *S. zeamais*. The occurrence of field infestation, the greater mobility of *S. zeamais*, the rapid turnover of stocks in mills and the associated production of maize appear a probable combination of factors that, together with high local temperatures and humidities, would favour this.

Most other species are of little consequence except that occasional outbreaks of *C. cephalonica* and *T. granarium* may occur.

(D.3.4b)   *Milled rice* (Table D4, Fig. D4)

Milled rice is particularly susceptible to insect attack. It is usual to store the grain as paddy, and mill it or parboil it as required keeping the storage period of the processed product to a minimum.

S. oryzae was the predominant species. It was recorded as a major pest in areas that produced 80% of the rice in the sample taken. *S. zeamais* did not receive a high rating but, as with paddy, its real importance was probably much greater than appreciated because of its being confused with *S. oryzae* particularly in eastern Asia and in Central America and the Caribbean.

R. dominica was recorded as a major pest in areas producing 50% of the sample but because of the limited storage periods usually involved, this species would not be expected to cause damage of the same order as *S. oryzae*. *C. cephalonica* and *T. castaneum* which had frequencies of occurrence similar to *R. dominica*, would be of greater significance. *O. surinamensis* and *E. cautella* must also be taken into account as significant pests though they do not occur as frequently as the species listed earlier.

TABLE D4: PRODUCTION, EXPORTS AND IMPORTS** OF RICE DURING 1970 IN REPRESENTATIVE COUNTRIES OF THE WORLD AND THE IMPORTANCE OF THE MAJOR STORAGE PESTS IN THESE COUNTRIES.

A = Major importance
B = Moderate importance
C = Minor importance
- = Occur rarely or absent

So - *Sitophilus oryzae*
Sz - *Sitophilus zeamais*
Sg - *Sitophilus granarius*
Rd - *Rhyzopertha dominica*
Tg - *Trogoderma granarium*

Tc - *Tribolium castaneum*
Tco- *Tribolium confusum*
Os - *Oryzaephilus surinamensis*
Om - *Oryzaephilus mercator*
Sc - *Sitotroga cerealella*
Ec - *Ephestia cautella*
Ee - *Ephestia elutella*
Ek - *Ephestia kuehniella*
Pi - *Plodia interpunctella*
Cc - *Corcyra cephalonica*

Country	Production 1000 mt	Exports 1000 mt	Imports 1000 mt	So	Sz	Sg	Rd	Tg	Tc	Tco	Os	Om	Sc	Ec	Ee	Ek	Pi	Cc
Australia	161	129	2	A	C	C	A	-	A	B	A	-	C	B	-	B	B	-
Burma	5305	640	-	A	B	-	-	B	A	-	A	-	-	A	-	B	A
Canada	-	-	49	-	-	-	-	-	-	-	C	-	-	-	-	-	-	-
Cent. Afr.Rep.	8	-	1	B	-	-	-	-	A	-	B	-	-	-	-	-	-	-
China	66300	850*	5*	A	-	-	B	-	C	C	B	-	C	B	-	B	B	-
Colombia	489	-	-	B	B	-	-	-	B	-	B	-	-	-	-	-	-	-
Czechoslovakia	-	1	77	A	-	B	-	-	B	B	A	C	-	-	-	A	C	-
Denmark	-	-	7	-	-	-	-	-	-	-	-	-	-	-	-	-	-	-
Dominican Rep.	137	-	13*	-	-	-	-	-	-	-	-	-	-	-	-	-	-	A
Egypt	1693	655	-	C	-	C	B	-	A	C	C	-	B	B	B	-	-	-
El Salvador	29	-	1	B	-	-	-	-	B	-	-	-	-	-	-	-	-	-
Ethiopia	-	-	2*	-	-	-	-	-	C	-	-	-	C	-	-	-	-	-
Gambia	33	-	14	-	-	-	B	C	C	C	-	C	-	-	C	C
Germany F.R.	-	25	170	-	-	-	-	-	-	-	-	-	B	A	-	B	-	-
Guatemala	10	1	1	B	-	-	-	-	-	-	-	-	-	-	-	-	B	-
Guyana	144	61	-	C	-	-	-	-	-	-	-	-	-	-	-	-	-	-
India	41387	27	331	A	-	-	A	C	A	-	B	-	C	B	-	-	-	A
Iraq	133	-	2	A	-	A	A	-	B	-	A	C	-	C	-	-	-	-
Israel	-	-	38	A	-	-	B	-	A	B	A	-	C	B	-	-	C	-
Italy	-	345	4	A	-	B	-	-	-	-	-	-	-	-	-	-	-	-
Japan	10711	630	19	A	A	-	A	-	C	-	C	-	-	-	-	-	-	-
Kenya	17	1	1	B	C	-	C	-	B	-	B	-	C	B	-	C	C	B
Korean Rep.	3559	-	770	-	B	-	-	-	-	-	-	-	-	-	-	-	-	-
Lebanon	-	-	17	A	-	-	-	-	B	B	B	-	-	-	-	-	-	-
Libya	-	-	23*	-	-	-	-	-	A	-	C	-	-	-	-	-	-	-
Malaysia	929	4	272	A	-	-	-	-	B	-	-	-	-	A	-	-	-	A
Mozambique	104	1	-	A	-	-	A	-	A	-	B	-	A	C	-	-	-	B
Netherlands	-	20	52	-	-	-	-	-	-	-	C	-	-	-	-	-	-	-
Nigeria	358	-	2	-	-	-	-	-	-	B	-	-	-	-	-	-	-	-
Norway	-	-	6	-	-	-	-	-	-	-	-	-	-	-	-	-	-	-
Pakistan	13009	129	216	-	-	-	-	B	B	-	C	-	-	B	-	-	-	B
Philippines	3473	1	-	B	B	-	C	-	B	-	B	-	B	B	C	C	B	A
Senegal	64	-	119	C	-	-	C	C	C	-	C	-	-	-	-	-	-	-
Singapore	-	46	281	-	A	-	-	-	C	-	-	-	-	-	-	-	-	B
Somalia	-	-	23	-	-	-	-	-	-	-	-	-	-	-	-	-	-	-
Sudan (S)	4	-	11	-	-	-	A	A	B	-	-	-	-	-	-	-	-	-
Sweden	-	-	13	-	-	-	-	-	C	C	C	-	-	B	-	-	-	-
Switzerland	-	-	28	-	-	-	-	-	-	-	-	-	-	-	-	-	-	-
Country	Production 1000 mt	Exports 1000 mt	Imports 1000 mt	So	Sz	Sg	Rd	Tg	Tc	Tco	Os	Om	Sc	Ec	Ee	Ek	Pi	Cc

** Data from 1971 FAO Production Yearbook and 1971 FAO Trade Yearbook.
 * Data from China 1969, Dominican Rep. 1968, Ethiopia 1969 and Libya 1969.

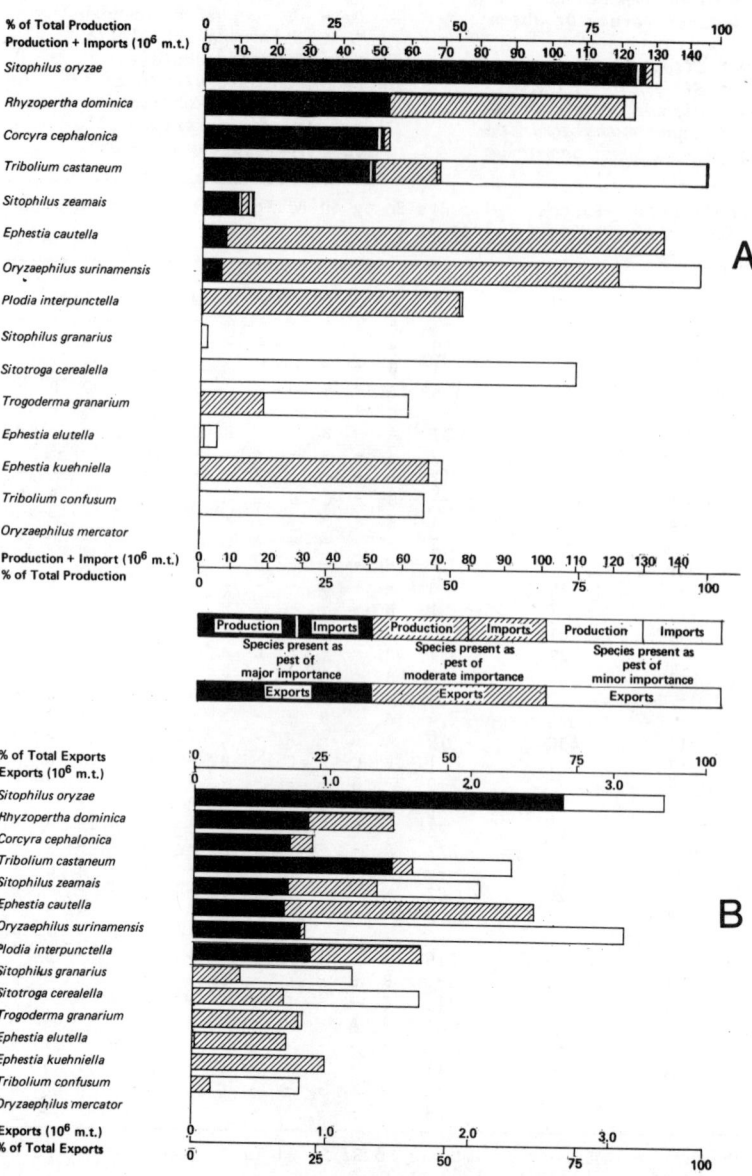

Fig. D4 — The percentage of rice at risk from damage by insect pests in 38 representative countries of the world.
A — Total production + import estimates. B — Export estimates.

Rice bran is a commodity notoriously susceptible to storage pests. It provides a serious source of infestation in many rice mills - *T. castaneum* is the principal species involved and is a common contaminant of rice bran in international trade.

(D.3.5)   *Maize* (Table D5, Fig. D5)

World production of maize (260×10^6 m.t., 1970) is of the same order as rice (considered as dehusked paddy) and not substantially less than wheat (318×10^6 m.t., 1970). The export trade is second only to wheat but because maize has a greater association with subsistence farming and local animal feeding, a lower proportion enters international trade. The two grains together account for approximately 80% of the world export trade in cereals. The sample of 40 countries examined represents considerably less of world production (25%) than other grains considered, primarily because the U.S.A. was not included.

Maize, more than any other cereal, is subject to field infestation. The only other cereals in which field infestation is a problem today are the sorghums and paddy where infestations primarily concern *S. cerealella* attacking unthreshed grain left drying in the field after harvest. Maize, in addition to being attacked by *S. cerealella*, is heavily attacked when standing in the field at an early stage of ripening by *Sitophilus zeamais*. This species may complete one or even two life cycles before harvest. Such field infestations which are often accompanied by a complex of Nitidulidae, Lathridiidae, Mycetophagidae and sometimes in later stages species such as *T. castaneum*, reflect varietal characteristics of the maize plant such as husk length, and their effect may be augmented, sometimes independently, by damage from birds and field crop insects such as *Heliothis*. The problem may be accentuated by crib drying and storing of maize cobs. As with sorghum, varietal differences in hardness of the grain also can have a marked influence on severity of attack.

Sitophilus oryzae was again shown to be the most important pest. *S. zeamais* was considerably less important though its importance probably would have been somewhat greater than is shown if its omission from data from some maize-producing countries where it is known to be a pest were taken into account. The pattern of occurrence of the two species conforms reasonably with what could be expected from climatic considerations - in the warmer often more humid areas, *S. zeamais* is more important than *S. oryzae* (e.g. Australia, Philippines, Malaysia, Burma, Zambia, Nigeria and El Salvador), with *S. zeamais* becoming less important as the humidity decreases (e.g. Egypt, India and Botswana), or the temperature decreases (e.g. Argentina, Czechoslovakia). Both species appear absent in very hot areas such as in the Sudan but in most instances the lower temperature limits for maize appear not to preclude either species being present at least as pests of minor importance. *S. granarius* can also be regarded as a significant pest of stored maize particularly in these cooler areas, and in storage of imported maize in countries in which it cannot be grown.

S. cerealella was of comparable importance to the *Sitophilus* spp. As with sorghum stored on the head, maize stored on the cob is liable to reinfestation in storage increasing the potential of the species. The larger grain size and hence intergranular spaces of maize allow some movement of *S. cerealella* in shelled grain and thus increase the chance of reinfestation in grain bulks. This is not common with the smaller more tightly-packed grains but when it occurs infestations are usually confined to the surface and the periphery of bulks.

TABLE D5: PRODUCTION, EXPORTS AND IMPORTS** OF <u>MAIZE</u> DURING 1970 FROM REPRESENTATIVE COUNTRIES OF THE WORLD AND THE IMPORTANCE OF THE MAJOR STORAGE PESTS IN THESE COUNTRIES.

A = Major importance
B = Moderate importance
C = Minor importance
- = Occur rarely or absent

So - *Sitophilus oryzae*
Sz - *Sitophilus zeamais*
Sg - *Sitophilus granarius*
Rd - *Rhyzopertha dominica*
Tg - *Trogoderma granarium*

Tc - *Tribolium castaneum*
Tco- *Tribolium confusum*
Os - *Oryzaephilus surinamensis*
Om - *Oryzaephilus mercator*
Sc - *Sitotroga cerealella*
Ec - *Ephestia cautella*
Ee - *Ephestia elutella*
Ek - *Ephestia kuehniella*
Pi - *Plodia interpunctella*
Cc - *Corcyra cephalonica*

Country	Production 1000 mt	Exports 1000 mt	Imports 1000 mt	So	Sz	Sg	Rd	Tg	Tc	Tco	Os	Om	Sc	Ec	Ee	Ek	Pi	Cc
Afghanistan	770	-	-	-	-	-	B	B	-	B	-	B	-	-	-	-	B	-
Argentina	9360	5233	-	A	B	B	C	-	B	B	C	-	A	C	-	C	C	-
Australia	192	1	-	B	A	-	B	-	B	C	B	-	B	B	-	C	B	C
Botswana	5	-	-	A	C	-	B	-	A	C	C	C	A	C	-	-	C	-
Burma	69	8	-	B	A	-	-	B	B	-	-	-	-	-	-	-	-	-
Canada	2564	3	473	C	C	C	-	-	C	-	C	-	-	-	-	-	C	-
Cent. Afr. Rep.	48	-	-	-	A	-	-	A	-	B	-	-	-	-	-	-	-	-
China	22720*	2	198*	A	-	-	-	-	C	C	B	-	A	B	-	B	B	-
Colombia	800	5*	21*	A	A	-	A	-	-	-	-	-	A	A	-	-	-	-
Czechoslovakia	513	-	123	A	C	A	B	-	C	C	B	-	A	-	B	-	B	-
Denmark	-	1	258	-	-	-	-	-	-	-	-	-	-	-	-	-	-	-
Dominican Rep.	45	-	30*	A	-	-	B	-	B	-	B	-	B	-	-	-	-	-
Egypt	2393	-	73	A	-	C	A	C	A	C	B	-	B	B	B	-	-	-
El Salvador	363	15	-	-	A	-	-	-	A	-	-	-	-	A	-	-	-	-
Ethiopia	950	-	-	A	A	-	C	-	A	C	B	-	A	A	-	-	-	-
Gambia	1	-	-	-	C	-	C	-	C	-	-	C	C	-	-	-	C	C
Germany F.R.	507	72	2601	-	-	A	-	-	-	C	A	-	B	A	C	B	-	-
India	7413	-	325*	A	B	-	A	B	B	-	B	-	A	A	-	-	-	B
Iraq	6	-	2	A	-	A	A	-	-	-	-	-	-	-	-	-	-	-
Israel	5	-	110	A	A	-	B	-	A	B	A	-	C	A	-	-	C	-
Italy	4754	8	4216	A	A	A	-	-	-	-	-	-	-	-	-	-	-	-
Kenya	1500	5	14	C	A	-	-	-	A	-	C	-	B	B	-	C	C	-
Korea Rep.	68	-	214	-	B	-	-	-	-	-	-	-	-	-	-	B	-	-
Lebanon	1	-	87	A	-	-	-	-	B	B	B	-	-	-	-	-	-	-
Libya	1	-	-	C	-	-	C	C	-	-	-	-	C	-	-	-	C	C
Malaysia	6	-	183	B	A	-	-	-	-	-	-	-	-	-	-	-	-	-
Mozambique	450	12	35	B	A	-	B	-	B	-	-	-	A	B	-	-	B	B
Netherlands	4	548	2467	-	-	-	-	-	-	-	-	-	-	-	-	-	-	-
Nigeria	1220	-	9	B	A	-	B	B	A	B	-	B	C	C	-	-	B	-
Norway	-	-	112	-	-	-	-	-	-	-	-	-	-	-	-	-	-	-
Pakistan	720	-	5	-	-	C	B	-	C	-	C	-	B	-	-	-	-	-
Paraguay	259	23	-	A	-	A	-	-	B	B	-	-	A	-	-	-	-	-
Philippines	2005	-	1	-	A	-	A	-	A	-	-	-	-	C	C	C	B	B
Senegal	39	-	5	-	B	-	C	-	C	-	C	-	-	-	-	-	C	-
Singapore	-	14	10	-	A	-	B	-	C	-	-	B	-	C	-	-	-	B
Somalia	35	-	1	A	-	-	C	-	A	-	C	-	A	A	-	-	C	B
South Africa	6133	1201	200	A	A	A	-	-	B	C	B	-	A	A	-	B	B	-
Sudan (S)	23	-	-	-	-	-	A	A	B	B	-	-	-	B	-	-	-	-
Switzerland	55	-	31	-	-	-	-	-	-	-	-	-	C	-	-	-	-	-
Zambia	550	8*	26*	B	A	-	C	-	A	B	B	B	A	B	-	C	C	C

| Country | Production 1000 mt | Exports 1000 mt | Imports 1000 mt | So | Sz | Sg | Rd | Tg | Tc | Tco | Os | Om | Sc | Ec | Ee | Ek | Pi | Cc |

** Data from 1971 FAO Production Yearbook and 1971 FAO Trade Yearbook.
* Data from China 1965 and 1968, Colombia 1969, Dominican Rep. 1968 and India 1969.

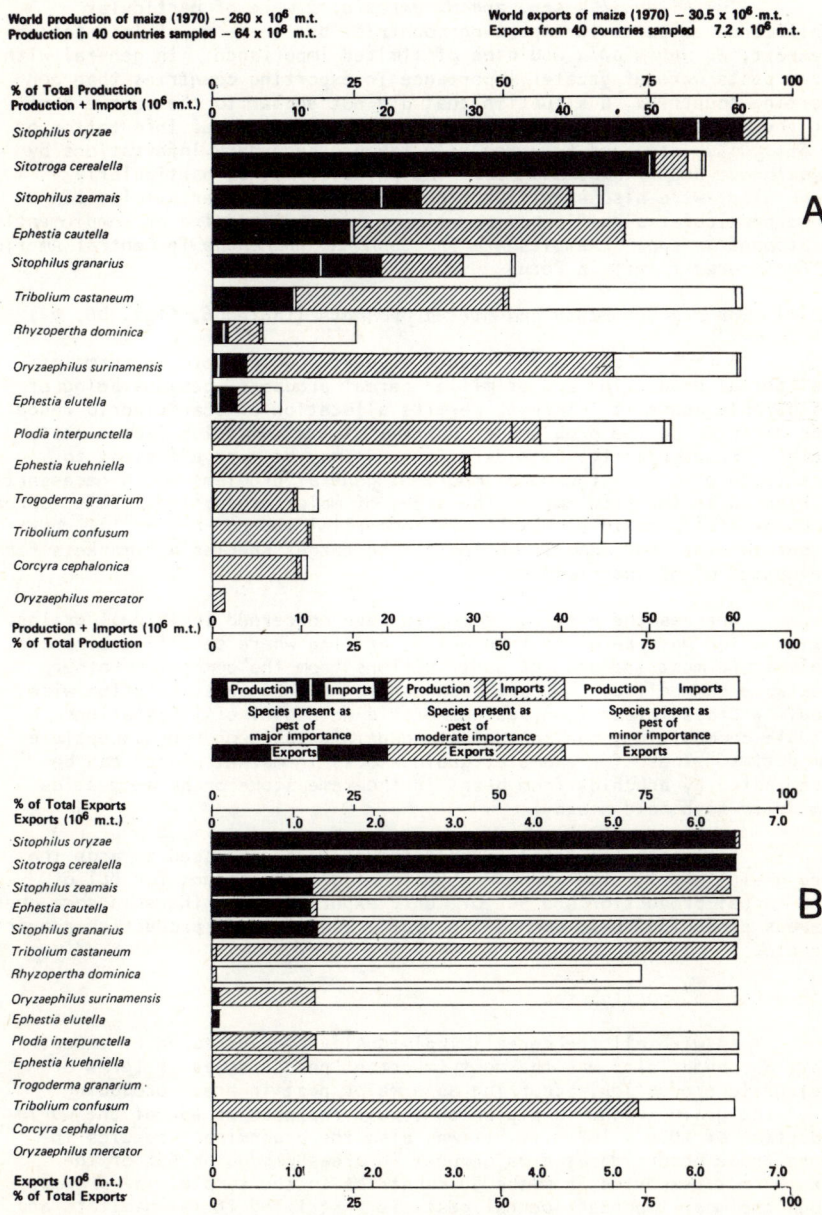

Fig. D5— The percentage of maize at risk from damage by insect pests in 40 representative countries of the world.
A — Total production + import estimates.   B — Export estimates.

The *Sitophilus* spp. and *S. cerealella* were of particular
significance as pests in exporting countries but in terms of infestation
at export, *S. cerealella* would be of limited importance. In general with
maize, pests were of greater importance in exporting countries than non-
exporting countries, a situation that did not appear to the same extent
with the other cereals considered. The high incidence of infestation by
the *Sitophilus* spp. and *S. cerealella* favours secondary infestations by
T. castaneum and *O. surinamensis*. The phycitid moths, particularly
E. cautella, were also significant pests. Locally, other species can
assume particular significance as primary pests displacing or complementing
the *Sitophilus* spp. Examples are *Prostephanus truncatus* in Central America
and *Pagiocerus fiorii* in Peru.

(D.3.6) *Cereals generally and milled products* (Table D6, Figs. D6, D7)

A ranking of the importance of pest species both in terms of
total cereal production and of milled cereal products, besides being of
considerable academic interest, permits allocation of scarce world resources
to be directed to the problems of control of the most significant pest
species. From objective considerations, it may be more efficient to
concentrate effort on the more important general problems which necessarily
are biassed to the problems of the areas of major production. This approach
can be helpful also in planning both commercial and non-commercial research
and development, for example in indicating target species and markets for
development of new pesticides.

Whereas the previous comments have concerned individual grains,
consideration must be given to problems arising where several grains are
involved and must include the contributions from the complex of other
infestable commodities that are grown and/or stored in association with
cereals. Cross infestation besides providing nuclei of infestations in
suitable commodities, can contaminate non-infestable or non-susceptible
commodities. Thus, for example, grain stocks in market places can be
contaminated by bruchids from beans in the same store or by dermestids
from dried fish held nearby.

Estimates for total cereal production are biassed towards the
three grains wheat, rice and maize, which together account for 80% of
world cereal production and 90% of world export trade. There is some overlap
of areas producing these commodities but areas of major production appear
separate.

(D.3.6a) *Sitophilus* spp.

As with all the cereals individually considered here, except
paddy, *S. oryzae* also was the most important pest species in terms of
total grain production, occurring as a major pest in areas producing
2/3 of the grain in the sample which itself represented 45% of the world
production of 1056×10^6 m.t. It was also the predominant species in
export trade occurring as a major pest in areas producing 55% of the
39×10^6 m.t. exported from the 50 countries in the sample. *S. oryzae*,
though the most important cereal pest, is restricted in its habitats and
in terms of its world abundance is recorded comparatively rarely breeding
in commodities other than cereal grains. It was found during the survey
attacking split peas in Trinidad and infestations occasionally are recorded
from grain products of appropriate consistency such as pasta (e.g. spaghetti)

THE MAJOR CEREALS

World production of the major cereals (1970) — 1056 x 10⁶ m.t.
Production in 50 countries sampled — 467 x 10⁶ m.t.
(Wheat, barley, dehusked paddy + rice, maize, millet + sorghum).

World exports of the major cereals (1970) — 99 x 10⁶ m.t.
Exports from 50 countries sampled — 39 x 10⁶ m.t.
(Wheat, barley, rice, maize).

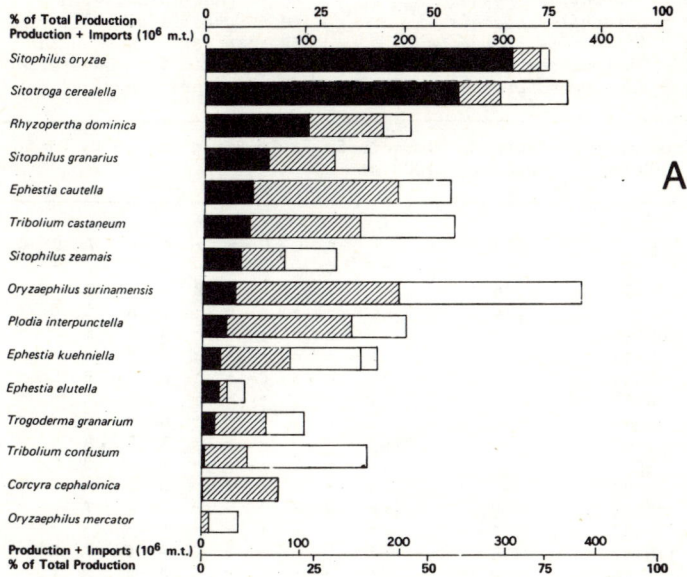

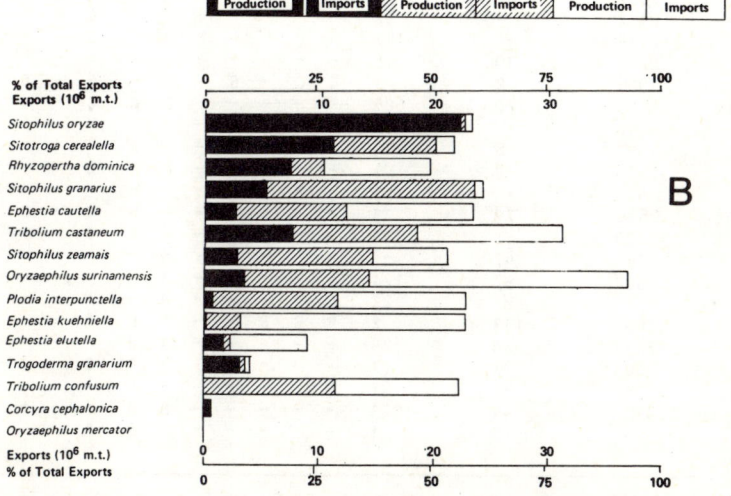

Fig. D6 — The percentage of the major cereals (wheat, barley, dehusked paddy + rice, maize and millet + sorghum) at risk from damage by insect pests based on data from 50 representative countries of the world.
A — Total production + import estimates.   B — Export estimates.

TABLE D6: LOCAL PRODUCTION, EXPORTS* AND IMPORTS OF <u>MILLED CEREAL PRODUCTS</u> DURING 1970 IN REPRESENTATIVE COUNTRIES OF THE WORLD AND THE IMPORTANCE OF THE MAJOR STORAGE PESTS IN THESE COMMODITIES.

A = Major importance
B = Moderate importance
C = Minor importance
- = Occur rarely or absent

So - *Sitophilus oryzae*
Sz - *Sitophilus zeamais*
Sg - *Sitophilus granarius*
Rd - *Rhyzopertha dominica*
Tg - *Trogoderma granarium*

Tc - *Tribolium castaneum*
Tco- *Tribolium confusum*
Os - *Oryzaephilus surinamensis*
Om - *Oryzaephilus mercator*
Sc - *Sitotroga cerealella*
Ec - *Ephestia cautella*
Ee - *Ephestia elutella*
Ek - *Ephestia kuehniella*
Pi - *Plodia interpunctella*
Cc - *Corcyra cephalonica*

Country	Production 1000 mt	Exports 1000 mt	Imports 1000 mt	So	Sz	Sg	Rd	Tg	Tc	Tco	Os	Om	Sc	Ec	Ee	Ek	Pi	Cc
Afghanistan	3720	-	-	-	-	-	-	-	-	A	B	-	-	-	-	-	-	-
Australia	5211	414	-	B	C	C	B	-	A	A	A	C	C	B	C	A	A	C
Botswana	14	-	-	C	-	-	-	-	A	C	C	-	-	-	-	-	C	-
Burma	7682	6	-	-	-	-	-	B	A	-	A	-	-	A	-	-	B	A
Canada	14150	784	4	-	-	-	-	-	C	A	C	C	-	-	-	-	C	-
China	209636	7	-	-	-	-	-	-	A	B	B	-	-	B	-	B	B	-
Colombia	2180	-	1	-	-	-	A	-	A	-	B	-	-	A	-	-	-	A
Czechoslovakia	8633	-	13	B	-	B	-	-	A	A	B	C	-	-	C	A	B	-
Denmark	6269	7	5	-	-	-	-	-	-	-	B	C	-	-	-	A	-	-
Dominican Rep.	346	-	2	B	-	-	-	-	A	-	B	-	-	-	-	-	-	-
Egypt	7740	90	2753	-	-	-	-	-	A	C	A	-	-	-	-	A	B	A
Ethiopia	6025	-	13	B	C	-	-	-	A	C	B	C	-	C	-	B	-	-
Gambia	95	-	3	-	-	-	-	-	B	C	-	C	-	-	-	-	-	-
Germany F.R.	22845	499	246	B	-	-	-	-	-	A	A	-	-	-	B	A	B	-
Guyana	198	2	4	-	-	-	-	-	A	-	-	-	-	-	-	-	-	-
India	118147	128	3	-	-	-	B	C	A	C	A	B	-	B	-	C	B	A
Iraq	2026	7	8	-	-	-	-	-	C	A	B	A	C	-	-	C	C	B	-
Israel	1415	-	40	-	-	-	-	-	A	B	A	-	-	B	-	A	B	-
Italy	21707	268	95	-	-	-	-	-	B	-	B	-	-	-	-	-	-	-
Kenya	2052	64	4	C	C	-	-	-	A	-	B	C	-	B	-	B	C	C
Korea Rep.	10122	-	27	-	-	-	-	-	-	-	B	-	-	-	-	-	-	-
Libya	115	-	147	-	-	-	-	B	A	C	C	-	-	-	-	B	B	-
Malaysia	2218	30	55	-	-	-	-	-	A	-	-	-	-	-	-	-	-	-
Mozambique	908	2	1	C	C	-	B	-	A	-	B	-	B	B	-	-	B	B
Netherlands	4856	170	606	-	-	-	-	-	C	C	C	-	-	-	-	C	-	-
Nigeria	8340	-	6	-	-	-	-	-	-	-	-	-	-	C	-	-	-	-
Norway	1527	8	3	-	-	-	-	-	-	B	-	-	-	-	-	A	-	-
Pakistan	30720	3	1	-	-	-	-	-	-	A	-	C	-	-	-	-	-	B
Paraguay	401	-	1	-	-	-	-	-	-	-	-	-	-	-	-	-	-	-
Philippines	7846	79	13	-	-	-	-	-	A	-	C	-	C	-	-	-	-	-
Poland	18739	-	306	-	-	-	-	-	-	B	B	-	-	-	B	A	C	-
Senegal	780	47	-	-	-	-	-	-	B	-	-	-	-	-	-	C	-	-
Singapore	502	97	121	-	-	-	-	-	A	-	-	-	-	-	-	-	-	-
Somalia	111	-	22	C	-	-	-	-	A	-	-	-	-	A	-	-	-	-
Sth Africa	7400	133	2	B	B	B	-	-	A	B	A	-	-	C	C	A	B	-
Sudan	2337	29	25	-	-	-'	-	-	A	B	-	-	-	-	-	-	-	-
Sweden	4606	27	11	-	-	-	-	-	B	C	B	-	-	C	-	B	-	-
Switzerland	2038	-	41	-	-	-	-	-	-	-	-	-	-	-	-	C	-	-
United K'dom	22510	44	323	-	-	-	-	-	B	B	-	-	-	-	-	B	-	-
Zambia	870	10	1	-	-	-	-	-	A	B	A	-	-	-	-	-	-	-

| Country | Production 1000 mt | Exports 1000 mt | Imports 1000 mt | So | Sz | Sg | Rd | Tg | Tc | Tco | Os | Om | Sc | Ec | Ee | Ek | Pi | Cc |

* Data from <u>1971 FAO Production Yearbook</u> and <u>1971 FAO Trade Yearbook</u>.

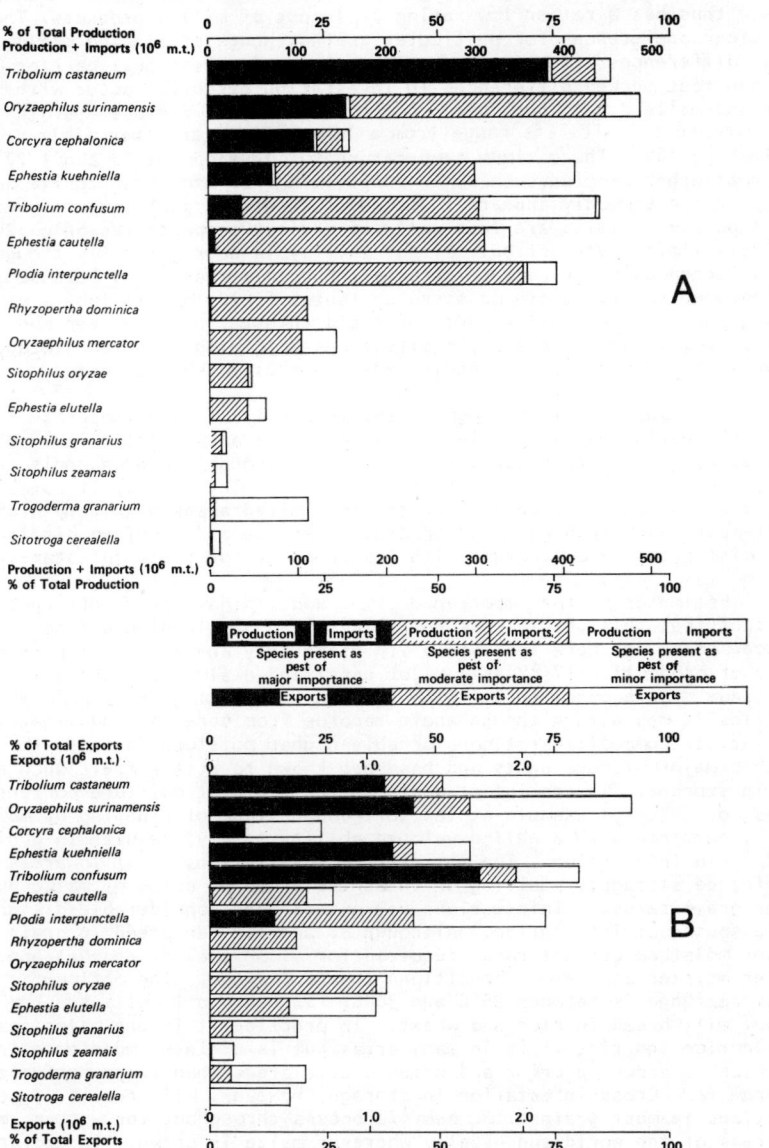

Fig. D7 — The percentage of milled cereal products at risk from damage by insect pests in 40 representative countries of the world.
A — Total production + import estimates.   B — Export estimates.

It is a common contaminant of processed animal foods from cross-infestation in provender mills and storages but will breed only if there are whole grains present or broken grains large enough to support complete larval development. *S. oryzae* thus has a rather low rating as a pest of milled products. There are no clear preferences for particular grains though strain and grain varietal differences affect reproduction. Apart from the husk barrier of paddy, the most marked differences in infestation potential occur with sorghum and maize due to both hardness and biochemical factors. Grain moisture content limits its range from a lower 10% to an upper limit approximating 16%. The optimum temperature for development is about 27°C but whereas other species such as *R. dominica* and *T. castaneum* thrive above 30°C, *S. oryzae* commonly appears to be under stress at such temperatures. Lower temperature limits are less well defined but temperatures below 20°C effectively limit reproduction. Flight ability is poor generally though marked differences occur between strains. For this reason and because it is not adapted to high grain moisture contents, field infestation is of no consequence. Occasional reports of field infestation have been made but apart from those in particular situations adjoining infested storages, the levels of proof necessary require most records to be discounted.

S. oryzae occurs throughout the tropical and warm temperate areas of the world and extends into cool temperate areas particularly along coastlines with warm ocean currents. Movement through trade channels takes it regularly into colder areas but it can survive winters only if established in heated situations. It appears absent in limited areas with very high summer temperatures such as in north central Africa and is often displaced in hot moist areas by *S. zeamais* with its capacity for field infestation.

Estimates of the importance of *S. zeamais* have been influenced by the fact that this species has not always been distinguished from *S. oryzae*. Rankings here indicate a 7th place only in terms of importance in areas of production (7-8% as a major pest) and a similar ranking in export trade. *S. zeamais* seems to be less restricted than *S. oryzae* in the commodities it may attack though again records from cereals predominate. It will attack commodities at considerably higher moisture contents than most other major storage pests and has been known to attack fruit such as apples in storage. It can infest grain in the field at moisture contents in excess of 20%, for example at the soft dough stage of ripening of maize, and this, combined with a well-developed ability to fly, results in a high rate of field infestation. The most severe infestations occur in areas near infested storages, shelling or threshing areas or dumps of maize husks or other grain refuse. Infestations may extend over considerable distances from the source of infestation. Although *S. zeamais* can breed in grain in the lower moisture content range required for *S. oryzae*, *S. zeamais* appears to prefer moister and warmer conditions than *S. oryzae*. The optimum temperature range is between 25°C and 30°C. *S. zeamais* is well adapted to maize but will breed in rice and wheat. In practice it is the major pest of stored rice and rice mills in many areas but is of less importance in wheat which is grown in drier and often cooler areas than those preferred by *S. zeamais*. Cross-infestation in storage, however, will result in serious infestations in most grains. *S. zeamais* occurs throughout the warmer, more humid areas of the world and usually wherever maize is grown. In cooler and drier areas marginal for *S. zeamais* it is usually restricted to maize as a host. Serious infestations occur throughout eastern Asia through the East Indies to north-east Australia. Though occurring through the Indian sub-continent, the drier climate and absence of maize in many areas restrict its abundance. It is virtually absent in much of the Arabian Peninsula area except in imported grain in distribution centres such as Kuwait, Bahrain and Jeddah. Maize is a staple subsistence food throughout Africa and *S. zeamais* is prevalent in most areas other than the hotter drier zones of the Sahara. In the Americas, its distribution extends from the maize growing areas of the Argentine through Brazil where it appears to be considerably more prevalent than *S. oryzae*, through the maize growing areas of Central America, to Mexico particularly and on to the southern states of the U.S.A., appearing also as a less important pest through to the maize growing areas of Ontario, Canada.

S. granarius is principally a pest of wheat, barley and related cool temperate grain crops. It is shown as somewhat more important than *S. zeamais* but will probably be shown to be of less or of at least comparable importance to this species as more is learnt of the true distribution of *S. zeamais*. The inclusion of major grain-producing and exporting countries such as France, however, would have increased the pest status of *S. granarius*. There is little overlap in their distribution as significantly important pests. *S. granarius* is found throughout the cool temperate areas of the world and is of particular importance in Europe, extending through the Mediterranean to North Africa and the Middle East. Although adapted to these cool areas, climate places some restriction on its pest potential when compared with that of a warm climate species such as *S. oryzae*. In warm regions, its occurrence is restricted to the high altitudes - thus for example in Ethiopia it usually is found above 2000 m., in Pakistan it is restricted to the higher areas of the north and into Afghanistan, and in South America apart from the cooler areas of Argentina, it is found mostly in Bolivia and Paraguay. It is virtually absent from South-east Asia and appears only sporadically, again in higher areas, through to southern Australia, where it is of moderate importance as a pest.

S. granarius like some other cool climate species such as *Tribolium confusum* is flightless and dispersal is mainly restricted to passive movement in infested commodities. Being unable to fly probably aids survival in cold countries by restricting dispersal to the storage environment where climatic extremes are less severe. *S. granarius* appears to be the most important and often the only species infesting grain in areas with cold but not severe winters, particularly in continental climates.

As with *S. oryzae*, *S. zeamais* and *S. granarius* are not of particular importance in milled products. They frequently occur as contaminants from cross-infestation and may attack grain products with appropriate physical characteristics such as pasta and pelletted materials though losses are not high.

(D.3.6b) *Sitotroga cerealella*

S. cerealella ranked next in importance to *S. oryzae* as a pest of the major cereals. It was of primary importance in paddy and ranked second or nearly so with all other grains. It is essentially a field pest and most infestations in storage originate in the field. Standing grain may be attacked but the severest outbreaks occur in grain harvested, but not threshed and stored, or in grain left to dry in the field. High moisture contents favour oviposition, and rain during drying or subsequent storage as unthreshed grain in the open will accentuate the problem. Infestations may survive threshing but they do not thrive when the grain is stored, as reinfestation is restricted by the difficulty moths have in moving about grain bulks in search of oviposition sites and by the fact that only the first stage larva is mobile. Thus infestations in bulk grain are largely restricted to the periphery except in maize where the grains are large, and sometimes paddy where more general surface infestations occur. Bagging of grain tends to suppress reinfestation and effective control can be achieved by covering the surface, for example by spreading bags of rice husks over paddy bulks.

S. cerealella is one of the most widely-distributed storage pests. The introduction of combine harvesting and threshing has reduced its importance particularly in the developed countries but it remains a very serious pest in areas of subsistence farming. It has no significance as a pest of milled products.

(D.3.6c)   *Rhyzopertha dominica*

R. dominica while ranking third in importance to *S. oryzae* and *S. cerealella*, was of considerably less significance in practice. It has higher temperature optima than many other major stored grain pest species (30°C+ range) and breeds with difficulty at lower temperatures (e.g. 25°C). It can survive and reproduce in grain at moisture contents of 8-9% which are considerably lower than levels needed for most other stored grain pests. The free-living adults and first and second instar larvae are very susceptible to physical damage during handling of grain. Later stage larvae usually bore into grains to complete development and are protected in some measure though it is not uncommon where considerable frass and debris are present, for all stages to be completed outside the grains. Nevertheless, all these stages and the eggs, which are deposited in cracks in grains, are subject to physical stresses and disturbance with any movement of the grain. *R. dominica* can fly freely and will do so at temperatures above 25°C. It cannot however, cling readily to many surfaces and when flying usually come to rest by colliding with an object and falling to the surface below. This behaviour appears to prevent significant field infestation so delaying build-up of infestations in storage. Thus *R. dominica* reaches its greatest potential as a pest in the warm, drier areas when grain stocks are left undisturbed for long periods. It is of greatest importance in Australia and India, and through the Arabian Peninsula. As with the *Sitophilus* species it is not a pest of any consequence in milled products.

(D.3.6d)   *Tribolium* spp.

T. castaneum is the commonest stored product insect in the world. It is essentially a warm climate species and will survive in colder countries only in heated premises. The closely related *T. confusum* tends to replace *T. castaneum* in colder areas but never reaches similar infestation levels. Though *T. castaneum* ranked sixth only as a pest of major importance in grain which involved 10% only of total grain production, it ranked third in terms of export grain and was considerably more important than any other species as a pest of milled or processed cereal products. It did not achieve the same status however, in exports of milled cereal products much of which tended to originate from the cooler countries. *T. confusum* was the most significant species in this export trade which, however, was small in comparison with total grain production and accounted only for 0.7% of the total estimate for milled cereal products. *T. confusum* did not rate highly as a pest of milled products generally, following after *T. castaneum*, *O. surinamensis*, *C. cephalonica* and *E. kuehniella*. As a pest of grain, it showed a comparatively low frequency of occurrence and was of moderate importance only.

T. castaneum will breed effectively in grain provided broken or cracked grain is present. Grain which appears sound can support high populations and attack by primary pests such as the *Sitophilus* species, provides very satisfactory conditions for multiplication. Milled products provide the ideal breeding medium, and bran and other offals of cereal milling are a serious source of infestation. In rice producing countries

rice bran deserves special mention. *T. castaneum* has been recorded from
a wide range of commodities other than cereals and cereal products. Oil
seeds are particularly susceptible and very high populations develop in
their storage unmilled and as cakes and meals. The adult insects are
very mobile and fly readily at temperatures above 27°C. They have been
recorded infesting maize in the field following primary attack by *S. zeamais*.
T. castaneum moves around the world in trade much more readily than any
other species, principally because of its wide range of foods and its
resilience to physical stress and disturbance. It is very long-lived
and will breed through most of its life.

T. confusum is less a pest of grain than *T. castaneum*. It cannot
fly and appears more in grain refuse and spillage than in grain bulks. It
is a particular pest of flour mills in cooler countries where infestations
become established in mill machinery. As the climate becomes warmer
however, *T. castaneum* tends to replace it in mills.

(D.3.6e) *Oryzaephilus* spp.

O. surinamensis is more widely distributed than *T. castaneum* and
is common in cool temperate areas where *T. castaneum* cannot survive outside
heated premises. It does not occur however, as frequently as, or as
regularly in the large numbers found commonly with that species. Both
species have free-living larvae but whereas *T. castaneum* will pupate
directly in the food medium and may be disturbed and subjected to
reasonable physical stresses, *O. surinamensis* forms a pupal cell for
protection of the comparatively fragile pupa which is easily damaged by
disturbance of the cell. This extreme susceptibility of the pupa is a
limiting factor in the development of infestations which effectively are
favoured by minimal disturbance. Because of this susceptibility and as
adults are long-lived and do not fly readily, quite heavy infestations
containing mainly adults are common. Infestations disperse throughout
grain masses aided by the flat, slender shape of the adults. Larvae
destroy the germ of the cereal kernels and may cause the grain to heat.
O. surinamensis when considered as a pest of major importance in grain
was not of particular significance and was ranked considerably lower
than the grain-attacking species such as the *Sitophilus* spp. and
R. dominica as well as *T. castaneum*, both in terms of production and
of exports involving some 5% and 3% respectively of total estimates.
In terms of milled cereal products however, it ranked second to
T. castaneum at a considerably higher level of involvement as a pest
of major importance (30% for *O. surinamensis* and 74% for *T. castaneum*).

O. mercator is much less common than *O. surinamensis* and is often
regarded primarily as a pest of oilseeds. It has been reported, however,
as the dominant *Oryzaephilus* species in West Africa. It was of little
importance in the grains considered herein, or in milled products, though
it achieved limited local importance principally in tropical Africa and India.

(D.3.6f) Phycitid moths and *Corcyra cephalonica*

The phycitid moths as a group add very significantly to the pest
complex of stored grain and milled products and apart from local importance,
are a major consideration in trade as a contaminant of a wide range of
commodities as well as grain.

Ephesta cautella is a tropical/subtropical species and attacks most grains and milled products as well as a range of other commodities. It is a pest of both bulk and bagged grain and of flour mills. Infestations in bulk grain are confined to the surface layers, a tendency also evident to a lesser degree in bag stacks depending on how tightly the bags are packed in the stacks. Infestations penetrate deepest into maize with its large intergranular space which is a major factor in the greater importance of *E. cautella* in maize than in other cereals. *E. cautella* is replaced in Mediterranean and temperate climates by *E. kuehniella* and to a lesser extent *E. elutella*. In hot and moist areas, *Corcyra cephalonica* tends to become the dominant species.

Plodia interpunctella is the most widely distributed phycitid occurring throughout most grain growing areas except those with severely cold winters. Infestations are more localised and larvae will penetrate much deeper into bulk grain than those of other phycitids. It is particularly common in stored maize but in general is more a pest of milled products than grain.

E. kuehniella is essentially a pest of milled products and milling premises but it is not uncommon in grain particularly in spillage and grain debris, and associated with infestations of the whole-grain attacking species. It reaches its greatest importance in temperate areas and is particularly important in mills and milled products in exporting countries.

E. elutella is a cool climate species of considerably reduced importance when compared with *E. cautella* as a pest of grain and related commodities. It is more restricted in its distribution than the other phycitid species and is of greatest importance in Europe. It does occur in warmer areas but only as a pest of stored tobacco which is immune from attack by *E. cautella*.

Corcyra cephalonica is restricted to tropical and subtropical areas and infestations principally concern rice, to a lesser extent millet and sorghum and a range of other commodities but particularly peanuts. As well as tending to replace *E. cautella* in grain in hotter and moister areas particularly in Australia and India, it also is a significant pest of flour mills in these areas.

The stored grain Lepidoptera are seasonal in occurrence, usually with increasing adult population peaks through the summer months that become more separated as the climate becomes cooler. The populations are often characterised by alternate periods when adults only and then larvae only are present. Diapause during winter can occur in cool countries particularly in *E. elutella* and *P. interpunctella* and is often associated with the very high autumn larval populations which disperse for overwintering. Disease and other biological control agents operate more efficiently than in the Coleoptera and may effectively regulate high populations particularly in autumn. High populations usually appear necessary for these factors to operate limiting their practical importance. In some circumstances populations may be retained at low densities but the levels concerned are usually incompatible with modern requirements for insect control. The phycitids represent a serious threat to processed packaged products as well as to grain and milled products, because adults will deposit eggs in the vicinity of foodstuffs and the very mobile newly-hatched larvae have a remarkable ability to find entry into apparently closed containers.

(D.3.6g) *Trogoderma granarium*

T. granarium appears endemic to an area of Asia west from Burma, Africa and Europe spreading from the Equator to approximately 37° North. Outbreaks have occasionally been reported also from the Caribbean, North America, the East Indies and southern Africa within similar climatic limits though in most instances measures have been taken to eradicate these outbreaks. In addition to these records, infestations are frequently recorded from heated premises in Europe particularly in maltings in breweries.

T. granarium prefers hot, dry conditions but will survive in unheated storages in areas with moderately severe winters. Infestations appear slow to establish but are difficult to control because of larvae sheltering in cracks, entering prolonged diapause and withstanding long periods of starvation, and because of their tolerance to many control agents. It is essentially a pest of unsatisfactory storages, such as those made from mud bricks, and usually is not regarded any more seriously than other primary grain pests in countries where it is endemic. However, the threat of its introduction has been taken very seriously in recent years by countries where the species does not occur. This is despite the fact that effective quarantine is of very recent origin and follows a period of several centuries during which the introduction of insects through trade has been ignored and the insect has not become established. Nevertheless its presence is a consideration in trade today, justifiable on the grounds of the high proportion of unsatisfactory storages in most countries and the consequent difficulties that could be experienced in its control.

(D.3.6h) *Other species*

In areas with very cold winters such as in the prairies of Canada, *Cryptolestes ferrugineus* may have a predominant role as a grain pest to the exclusion of most other species. In northern Europe *Tribolium destructor* becomes a more important pest of milled products, and in cold areas generally, Ptinidae assume significance. Similarly, other species such as *Latheticus oryzae* in India and Burma and various pests of maize may reach serious pest status locally but on a world basis these species mostly do not warrant inclusion. A good case may exist however, for including the widely-distributed species of *Cryptolestes* particularly *C. ferrugineus, C. pusillus* and *C. pusilloides* but as discussed earlier, because of the difficulties in identification, definitive data are not available on local occurrence related to the correct species.

(D.3.6i) *Summary*

In summary, the *Sitophilus* spp. and to a lesser extent *Sitotroga cerealella*, represent the most serious threat to harvested grain on a world basis, and *Tribolium castaneum* and *Oryzaephilus surinamensis* to production and storage of milled products. The Lepidoptera complex supplements this very significantly although the mode of infestation, population growth potential and the spoilage rate of the commodity are different. This situation prevails in most grain producing and consuming areas of the world. Under dry and very hot conditions, the balance of importance shifts towards *Rhyzopertha dominica* and *Trogoderma granarium* in association with *T. castaneum*.

E.   THE PESTICIDES IN COMMON USE IN GRAIN STORAGE

Most facilities that are currently available throughout the world for storing grain are simply containers and do not exclude pests or create an environment unsuitable for them. In most instances the use of pesticides is the only reliable method of preventing the loss or damage that will usually occur if adequate protective measures are not taken. Their use is widespread and the intensity of use of the various materials generally increases in warm climates. The types of pesticide involved, the application methods and the extent of their use depend on many factors:

  the pest complex present or expected to occur if protective treatment
    is not carried out,
  the pesticides available and the regulations governing their use,
  type of cereal, its temperature, moisture content and amount of
    dockage present,
  type of storage container or structure,
  projected storage period,
  resources available for expenditure on pest control,
and ultimate use of the grain, whether as an animal feedstuff, for
    milling for human consumption, or as seed for planting or malting.

The situation is an ever-changing one in response to changing market situations and commodity values, harvest expectations and realisations, pest prevalence, and local availability of pesticides and the resources to purchase them. Data collected during the survey related principally to the period of the survey whereas in considerations of pesticide resistance, the long term use (and misuse) must be taken into account. Moreover because of the active movement of stored products insects in trade, this must be interpreted in terms of the trade patterns which determine the extent and spread of any resistances appearing. Definitive information on actual pesticide usage is difficult to obtain. Although general statements on the materials used and the mode of use can be more or less reliable, dose rates and extent of use are often less accurate. They frequently reflect what are local recommendations for use rather than what is actually used. Total quantities used are often misleading as other usages of particular materials are usually included in these estimates and it is difficult to apportion quantities where different modes of application are used. Definition of target species is also unreliable in many instances but adequate descriptions of these are available from a consideration of the pest status of individual species in particular commodities in the area. Although considerable data relating to pesticide use were obtained during the survey, it seems desirable to discuss this in general terms rather than detail specific uses in particular countries. Specific uses and histories of use if available, are quoted where relevant in considerations of occurrences of resistance in Section F.

(E.1)   *Malathion*

Malathion is without doubt the most widely used material in stored grain pest control. The earliest significant investigations on its application for this use were made by Lindgren *et al.* (1954) and Gore (1958). Subsequently Parkin (1958) gave a provisional assessment of malathion for control of stored product pests, heralding an era of pest suppression in stored cereals in which freedom from insects became readily attainable

under a wide range of conditions. One of the first commercial applications
for malathion was routine treatment of groundnut storage pyramids in
Northern Nigeria (Anon. 1958d). This began in 1957 and subsequently the
material was introduced on an extensive scale to the peanut industries of
U.S.A. (Anon. 1961) and Australia (Champ and Cribb 1965b). Integration of
malathion into the grain industry was rapid also and usage became intensive
in some areas such as Australia where by 1964 all bulk wheat taken into
central storage was admixed at intake with malathion at the rate of 10-12
ppm (Winks and Bailey 1965).

The introduction of malathion into peanut pest control programmes
in Nigeria in 1958 was rapidly followed by the development of resistance
there (by 1961) and in other areas where malathion was used extensively
in peanut storage (*vide* Section F.7.1) Nevertheless malathion still
retains a place in control of pests of stored peanuts. In Australia
where malathion was introduced in 1963 as an admixture treatment for all
peanuts at intake to storage and where resistance appeared in *Tribolium
castaneum* in 1968, the treatment is still of value but only because it is
part of an integrated pest control scheme. There peanuts are harvested
in late autumn and treated before intake to storage as nut-in-shell. The
crop then is shelled as quickly as possible and transferred to refrigerated
storages. By mid- to late summer when beetle and moth activity would be
expected to be at their maximum for the year, the nut-in-shell storages are
empty and cleaned, and all stocks held in refrigerated storage thus breaking
the cycle of reinfestation of the new season's crop. Similarly with cereals,
although resistance is widespread, it does not necessarily invalidate use of
malathion in control measures. Both the occurrence of resistance and the
levels of resistance present may vary considerably among strains even within
localised areas. Thus it is possible to obtain effective suppression of
infestations with malathion at normal dose rates if its use is programmed to
ensure maximum effectiveness of the material applied, for example, removal of
harbourages for residual infestations and adequate application techniques.
The first manifestation of resistance in the field is reduction in the
effective residual life of admixture treatments or deposits on building fabric.
The margin between minimum effective doses and applied doses, however, often
requires that quite significant resistance levels must be present before fresh
deposits or direct application to the insects become ineffective for control.
Nevertheless the ease and reliability of achieving control with malathion has
gone.

The most intensive use of malathion has been in Australia where almost
all grain is handled in bulk and where since 1964 all wheat and considerable
quantities of other grains have been admixed with malathion at intake to
central storage and supplementary treatments applied to the surface of the
bulks and to the fabric of the buildings. Grain is not stored to any great
extent on farms in Australia and although malathion is recommended for use
in protection of grain held on the farm for farm use, its use on farms is
discouraged in grain intended for delivery to central storage. When resistance
to malathion developed, fenitrothion was introduced for treatment of structures
where resistance was present and dichlorvos for disinfestation of grain when
fumigation was not practicable. Bioresmethrin and synergised pyrethrins have
been introduced subsequently as countermeasures to the dichlorvos resistance
now apparent. Most major pest species have been involved. Before malathion
was introduced *Sitophilus oryzae* and to a lesser extent *Rhyzopertha dominica*

were the most damaging species and large populations of *Tribolium castaneum* and *Oryzaephilus surinamensis* were very common. Pest suppression was at a high level after introduction of malathion. *R. dominica*, however, was not completely controlled reflecting the characteristically variable response shown by this species to organophosphorus compounds; this deficiency has been accentuated by the development of resistance, first to malathion and then to dichlorvos following its introduction to combat malathion resistance. *R. dominica* then became the most important species in most areas of Australia. Malathion resistance was prevalent in *S. oryzae* but resistance levels were generally not high and there was a strong correlation between outbreaks of highly-resistant *S. oryzae* and use of unsatisfactory storages and poor hygiene. Resistance first appeared in *T. castaneum* and it now shares prime importance with *R. dominica* in terms of prevalence. *O. surinamensis* was much more effectively suppressed and increases in tolerance have been considerably less common than in other species. There was an apparent increase in pest status of the phycitid Lepidoptera particularly *Ephestia cautella* following the introduction of malathion. The reasons appear three-fold - the suppression of other species increased its relative pest status and infestations were more noticeable. *E. cautella* has a comparatively high level of natural tolerance to malathion, and the parasite-predator complex could not operate as efficiently in the presence of malathion and the absence of other host species. *Plodia interpunctella*, however, almost disappeared in the early period of malathion use probably reflecting the difference in tolerance of malathion between the species. Despite widespread resistance in most species malathion remains of use in stored grain pest control in Australia although blanket use has been reconsidered in favour of strategic use determined by monitoring programmes for resistance and replacement or supplement with dichlorvos, fenitrothion and bioresmethrin where appropriate.

In Argentina, also a major grain-exporting country, malathion has also been used extensively but not on the same scale as in Australia. Maize accounts for a considerable part of production and exports in Argentina and the *Sitophilus* spp. and *Sitotroga cerealella* are the more important pest species. In general grain has been treated only if it remained in storage for several months and use thus has been limited both on farms and in country storage. Malathion was introduced into the grain industry about 1963 and subsequently there appears to have been a reduction in its effectiveness. Argentina has greater dependence on fumigation than Australia and resistance, although present, has been less of a limiting factor to malathion efficacy.

Malathion has a variety of applications in practically every country of the world. Some of these uses are given in Table E1 for countries visited during the survey. It is emphasised that this is a general outline only based on data supplied during the survey and there are undoubtedly many omissions and errors but it is unlikely that these alter the overall picture. The general pattern appeared to be one of most intensive use of malathion in developed exporting countries, with less intensive use where dependence on bag storage increased, per capita national income decreased or cooler climates prevailed. Whereas malathion was the most important material for admixture with grain, increasing use of dichlorvos was evident. There appeared to be a significant move away from malathion for disinfestation of the fabric of buildings and a less marked but noticeable change to use of alternative materials for surface treatment of bags. This presumably reflected the increased incidence of malathion resistance with the corresponding difficulty in controlling residual infestations under conditions where contact with insecticide deposits was marginal for pick-up of a lethal dose. Additionally the replacement of lindane with malathion did not provide

TABLE E1 : THE PRINCIPAL CHEMICALS USED TO CONTROL PESTS OF STORED CEREALS IN SOME COUNTRIES VISITED DURING THE SURVEY

Country	Principal Pests	Principal Insecticide Used			Principal Fumigant Used		
		Malathion	Lindane	Other	Methyl bromide (g/m^3)	Phosphine (g/ton)	Other
THE AMERICAS							
Argentina	So,Sz,Sc	ADS,SBG**	-	Dichlorvos SB*	**30-50	***4	HCN*
Brazil	Sz,So,Ts Sc,Ec,Pi	SB*	SB*	-	*35-40	**2-4	-
Chile	So,Sg,Ts Sc,Ek	AD**SBG*	SM*	DDT FT* Dichlorvos FW*ST* Fenthion SG*	*3	*6-10	HCN**
Colombia	So,Sz,Ts Ec	SBG*	-	Dichlorvos SB* Mal + dichlorvos SB* Pyrethrins ST*	***15-30	-	-
Dominica	So,Ts,Sc Ec,Cc	SB*	-	Diazinon SB*	**22	***10	-
Guatemala	So,Sz,Ts Ec	-	-	Pyrethrins FT,ST*	**75-100	*2	CS_2*
Guyana	Sc,Rd,So Ts	SB**	SB**	Fenthion SB* Pirimiphos-meAS,SB* Pyrethrins SB,ST*	**16	**1.5	-
Jamaica	So,Ts	SB*	SB*	Propoxur SB* Pyrethrins SB,ST*	*32	*5	-
Mexico	Sz,So,Ts Os,Sc	-	-	-	**	***3	-
Paraguay	So,Sg,Ts Sc	SB*	-	-	**16-32	*5-8	-

So - *Sitophilus oryzae*
Sz - *Sitophilus zeamais*
Sg - *Sitophilus granarius*
Rd - *Rhyzopertha dominica*
Tg - *Trogoderma granarium*
Ts - *Tribolium spp.*
Os - *Oryzaephilus surinamensis*
Sc - *Sitotroga cerealella*
Ec - *Ephestia cautella*
Ek - *Ephestia kuehniella*
Cc - *Corcyra cephalonica*
Pi - *Plodia interpunctella*

AD,AS - Admixture as dust or spray
FT - Building fabric treatment
FW - Floor wash
SB - Surface treatment of bags and associated building fabric
SG - Surface treatment of grain and associated building fabric
SBG - SB + SG
SM - Smoke
ST - Space treatment

* Occasional use   ** Moderate use   *** Intensive use

TABLE E1 (contd.)

Country	Principal Pests	Principal Insecticide Used			Principal Fumigant Used		
		Malathion	Lindane	Other	Methyl bromide (g/m3)	Phosphine (g/ton)	Other

Country	Principal Pests	Malathion	Lindane	Other	Methyl bromide (g/m3)	Phosphine (g/ton)	Other
THE AMERICAS (cont'd)							
Peru	So,Ts,Sc Ec	SBG*	-	Aldrin, carbaryl SB* Fenthion SG*	**16-40	*7-12	-
Salvador	Sz,Ts,Sc Ec	SBG**	-	Pirimiphos-me SB*	**24	**6	-
Venezuela	So,Rd,Ts	SB**	-	Pyrethrins ST*	***32	**5	EDB*
EUROPE AND NORTH AFRICA							
Algeria	So,Tg,Rd Ts,Os	SB**	FT*	DDT*	-	*	CCl$_4$*
Czechoslovakia	Sg,Ts,Os Sc	-	ST*	Fenitrothion FT***	**10-40	**4-8	HCN*
Denmark	Sg,Ek	ADS*	-	-	**	-	-
Egypt	So,Rd,Ts Os	SB*	-	Katelsousse*	*24	*2-3	CS$_2$*
England	Os,Ts,Sg	AD,FT**	FT**	Dichlorvos ST* Fenitrothion FT* Pyrethrins ST*	**	*2-6	CCl$_4$* CCl$_4$/ EDC*
France	Sg,So,Ts Os,Sc	ADS*	-	Bioresmethrin AS* Chlorpyriphos-me AS* Dichlorvos AS* Pirimiphos-me AS* Pyrethrins AS*	-	*6	-
Germany FR	Sg,Ts,Os Ek	ADS,ST**	-	Dichlorvos SBG,ST* Pyrethrins SBG,ST*	**	***	HCN*
Greece	Sg,So,Os Ts	FT*	FT*	DDT FT*	-	**6-10	EDC* CCl$_4$*
Libya	So,Rd,Tg Ts,Sc	FT**	FT*	-	*20-60	***1-3	-
Morocco	So,Ts,Os	-	SB*	-	-	*10-12	CS$_2$* CCl$_4$/ EDC*
Netherlands	Sg	FT,ST*	SM*	Dichlorvos ST* Pyrethrins FT,ST*	***15-50	**1-3	HCN*

TABLE E1 (contd.)

Country	Principal Pests	Principal Insecticide Used			Principal Fumigant Used		
		Malathion	Lindane	Other	Methyl bromide (g/m^3)	Phosphine (g/ton)	Other

EUROPE AND NORTH AFRICA (cont'd)

Country	Principal Pests	Malathion	Lindane	Other	Methyl bromide (g/m^3)	Phosphine (g/ton)	Other
Poland	Sg,Ts	FT*	SM*	DDT/BHC SM* Dichlorvos FT* Pyrethrins AD* Trichlorfon*	**20-50	**10-30	EtO/CO$_2$* HCN*
Romania	Sg,So,Ts	FT**,SM*	FT,ST,SM*	DDT FT,SM,ST*	**40	***6-12	CCl$_3$· NO$_2$*
Scotland	Os	AD,FT**	FT,SM*	Fenitrothion FT*	*	*	CCl$_4$* EDC/EDB/ CCl$_4$*
Spain	Sg,Os,Ts	SBG**	-	Pyrethrins SBG*	-	**12	CCl$_4$/ CS$_2$*
Sweden	Sg,Os	FT**	-	Pyrethrins SG*FT**	**20	**12	-
Switzerland	Sg	-	-	-	*	-	-
Yugoslavia	Sg,Os,Ts,Sc	SBG*	-	-	**16-32	***1-3	-

AFRICA - SOUTH OF SAHARA

Country	Principal Pests	Malathion	Lindane	Other	Methyl bromide (g/m^3)	Phosphine (g/ton)	Other
Botswana	So,Ts,Sc	AD*	AS*	DDT AD* Wood & dung ash	-	*15	CS$_2$*
Cent. Afr. Republic	So,Sz,Rd Ts	-	-	Pyrethrins ST*	-	-	-
Ethiopia	So,Sz,Rd Ts,Sc,Ec	AD*	AD**	-	*	*3	-
Gambia	Ts	AD*	AD*	Bromophos FT* Fenitrothion FT*	**	*20	-
Kenya	So,Sz,Ts Ec	AD***	AD***	Dichlorvos ST* Pyrethrins SBG,ST*	***2	**5-10	-
Malawi	Sz,So,Ts	SB*	-	Dichlorvos ST*	*32	*1-2	-
Mozambique	So,Rd,Ts Os,Sc	SB**	AD*	Dichlorvos ST*	*35-40	*7	-
Nigeria	So,Sz,Rd Ts,Sc	AD,SB*	SM*	Dichlorvos FT* Jodfenphos FT*	*32	*2-4	-

TABLE E1 (contd.)

Country	Principal Pests	Principal Insecticide Used			Principal Fumigant Used		
		Malathion	Lindane	Other	Methyl bromide (g/m^3)	Phosphine (g/ton)	Other

AFRICA - SOUTH OF SAHARA (cont'd)

Country	Principal Pests	Malathion	Lindane	Other	Methyl bromide	Phosphine	Other
Senegal	So,Sz,Rd Ts,Sc	SBG*	AD*	BHC AD** (Seed)	*	*5	CCl_4*
Somalia	So,Ts,Sc Ec	SB*	SB*	-	-	*4-6	-
South Africa	So,Sz,Ts Os,Sc,Ec	AS*** SB***	-	Bromophos AS* Pyrethrins AS,ST**	***48	***10	-
Tanzania	So,Sz,Rd Ts	SB*	-	Dichlorvos**	*32	*4-5	-

ASIA AND AUSTRALIA

Country	Principal Pests	Malathion	Lindane	Other	Methyl bromide	Phosphine	Other
Afghanistan	Sg,Rd,Tg Ts,Os	FT*	SB*	-	-	**7-10	-
Australia	So,Rd,Ts Os,Ec	AS,SG*** AD**	ADS** (Seed)	Dichlorvos AS,SG*** Fenitrothion FT*** Pyrethrins AS,ST*	**32	***2-10	CS_2* HCN*
Bahrain	Ts,So,Rd Tg	FT*	SM*	Diazinon SB* Propoxur SB*	*48	-	-
Burma	So,Sz,Rd Ts,Os,Cc	-	AD,FT*	-	*24-32	-	-
China	So,Sz,Ts Os,Sc,Ec	AD*	AD,FT* (Seed)	Dichlorvos FT* Trichlofon FT*	*15-20	*2-3	$CCl_3\text{-}NO_2$*
Cyprus	So,Tg,Sg Ts,Os,Ec	ADS**	-	Bromophos AD* Dichlorvos ST*	**36	*5-10	CS_2*
India	So,Rd,Ts Os,Sc,Ec	AS,FT*	FT*	Dichlorvos FT* Pyrethrins SB*	*22	*2-3	EDB** EDC/ CCl_4*
Indonesia	Sz,So,Ts Rd,Os	-	-	-	*32	*2-3	-
Iran	So,Rd,Tg Ts,Sc	FT*	-	-	*20-30	*10	-
Iraq	Tg,So,Rd Ts	FT*	SM*	-	*16-32	*3-5	-

TABLE E1 (contd.)

Country	Principal Pests	Principal Insecticide Used			Principal Fumigant Used		
		Malathion	Lindane	Other	Methyl bromide (g/m^3)	Phosphine (g/ton)	Other

ASIA AND AUSTRALIA (cont'd)

Country	Principal Pests	Malathion	Lindane	Other	Methyl bromide (g/m^3)	Phosphine (g/ton)	Other
Israel	So,Rd,Ts Os,Ec	SBG**	SBG,SM**	Diazinon SBG* Dichlorvos ST*	**20-60	**2-8	-
Japan	So,Sz,Rd Os,Ts,Ec	SBG*	-	Dichlorvos FT,ST* Pyrethrins FT,ST*	**	**	CCl$_3$-NO$_2$** HCN*
Korea Rep.	So,Ts,Os	-	-	-	*12-18	**2-3	-
Kuwait	So,Ts,Os	-	FT*	Dichlorvos FT* DDT FT*	-	*10	-
Malaysia	So,Sz,Ts Sc,Ec,Cc	SB**	SB**	-	**48	-	-
Pakistan	Tg,So,Rd Sc	SB*	-	Fenitrothion SB*	*16-64	*2-9	EDC/ CCl$_4$**
Philippines	Sz,So,Rd Ts,Cc	AD,SB*	AD,SB*	Carbaryl SB*	*48	*6	-
Singapore	Sz,So,Ts Os	-	SB**	-	**	-	-
Syria	So,Sg,Tg Rd,Ts	SB*	FT*	Carbaryl AD,SBG*	*40-60	-	-
China (Taiwan)	So,Sz,Rd Ts,So	SGB*	-	-	-	*4-6	EDB*
Turkey	So,Sg,Tg Rd,Ts	AD,FT**	-	BHC FT* Bromophos AD,FT**	*35	***5-15	-

comparable residual activity or moth control - these characteristics, singly or together, are possessed by alternative materials such as dichlorvos, fenitrothion and pirimiphos-methyl.

In cool countries, malathion resistance has been less a limiting factor and malathion has remained effective against species such as *O. surinamensis* for example in England. The low temperatures have tended to increase its residual life but are often offset by the high grain moisture contents which may be characteristic of harvests in cool countries. Nevertheless use of malathion in cool countries has not been extensive.

Although malathion is a safe material to use in terms of both operator and consumer hazards, a conservative approach to its use on grain has been evident in some countries. Thus in India, widespread adoption of admixture techniques has been opposed because cereals comprise a high proportion of the diet and local milling techniques do not remove residues to the extent occurring in large-scale milling operations. In Germany a residue tolerance of 2 ppm is advocated compared with a Codex Alimentarius tolerance of 8 ppm presumably reflecting caution and no great need for use of malathion on the local scene. A 2 ppm residue tolerance is incompatible with effective pest control and efficient grain stock management. The well-researched and soundly-based Codex tolerance appears adequate for normal use of the material.

A matter of some concern is the formulation of malathion used in some countries. Whereas premium-grade solvent-free formulations of malathion have generally been used, in many instances 50% and 57% emulsifiable concentrates and similar preparations were the only formulations available. It is desirable that only premium-grade malathion be used for treatment of grain. Moreover as inclusion of solvent makes formulation more difficult and expensive and results in higher freight charges for the increased bulk of the concentrate, can lead to lower stability of the concentrate, may cause damage to equipment particularly rubber belting and complicates the toxicity hazard to operators, it is desirable also that only solvent-free formulations be used. A similar problem existed with other materials such as dichlorvos.

(E.2) *Lindane*

Lindane replaced DDT which found a limited use in stored product pest control in the immediate post-war years. It had many of the desirable characteristics required at the time *viz*. a wide spectrum of insecticidal activity including both beetle and moth pests and reasonable effectiveness against mites, stability under a wide range of conditions, a significant vapour pressure allowing some fumigant effect and in some circumstances repellency, and a comparatively low mammalian toxicity. Early applications included admixture with food and feed grains but these uses have now been largely discontinued except for occasional use in protecting pulses. Seed treatment is still common but the widespread occurrence of resistance has tended to invalidate this use although the incompatibility of most replacement materials (which are usually organophosphorus compounds) with heavy-metal based fungicides, have restricted the choice and delayed change. Pre-emergent effects on plants have been common but have little influence on later growth.

Lindane, because of lack of an equivalent replacement and in part tradition, will continue to find use in many warm countries although resistance is now almost a normal characteristic of many pests. In the cool countries where resistance is less common it remains useful although in some areas for example France, its use has been discontinued because of residue consideration.

The major uses for lindane apart from seed dressings have been in surface treatment of bag stacks and particularly treatment of the fabric of buildings and ships' holds where there is no contact with the commodity in storage.

(E.3) *Other insecticides*

Pyrethrins remain one of the more widely used materials in stored product pest control. The rapid knockdown effect, a wide spectrum of activity against insect pests, a general consumer acceptance of their use associated with foodstuffs, and an established Codex tolerance have been the principal reasons for their use. This use, although widespread, has not been intensive as their cost has been prohibitive for many applications. Resistance has not presented a serious problem but the natural tolerance of some beetle pests including the ubiquitous *Tribolium castaneum*, has been a limitation to their use in situations where resistance to other materials was present in other species. The limitation caused by natural tolerance in some species has been compounded by the recovery that may take place after the initial knockdown. Although synergists such as piperonyl butoxide are necessary for control of beetles but not for moths, they have usually been included in formulations to give as wide a coverage of storage pests as possible.

The *synthetic pyrethroids* have found increasing acceptance over pyrethrins because of their high levels of activity against a wide range of pests and a cost advantage. Resmethrin in the U.S.A. and bioresmethrin in many other countries appear to be those most commonly used. There have been deficiencies as with pyrethrins, for example resistance is known although not extensive and breakdown to malodorous decomposition products has presented a problem where repeated applications have been made to the same surface. In addition, control of *T. castaneum* has been less than desirable. A residue tolerance is available only for bioresmethrin (5 ppm) as a result of a recent recommendation of the Joint FAO/WHO Meeting of Experts on Pesticide Residues. Several other synthetic pyrethroids showing promise against storage pests are under development.

The range of organophosphorus compounds in use was extensive. Of these *dichlorvos* was the next most commonly used material after malathion. It is an extremely effective stored product insecticide and a Codex residue tolerance of 2 ppm has been operative for many years enabling its widespread integration in the grain industry. Dichlorvos has a high vapour pressure compared with conventional insecticides and is used as a fumigant rather than as a contact insecticide. Enclosed spaces such as warehouses, storages and grain bins allow build-up of air concentrations of vapour to levels toxic to most flying and crawling stored product insects. The vapour will not penetrate into grain masses, other commodities, or the fabric of buildings. The principal uses of dichlorvos were in:

- aerosol-dispensing units which could be programmed for automatic daily release, usually at dusk;
- space sprays or fogs;
- slow release formulations in which dichlorvos was dissolved in solid strips (or beads) of polyvinyl chloride plastic which were suspended in the free space of storages;
- surface application of concentrates which were sprinkled on wooden floors of buildings (usually flour mills) or dunnage from which the dichlorvos vapour quickly vaporised;

surface treatment for protection of bagged commodities;
and direct application to grain either as a surface treatment
   for moth control and to a lesser extent to prevent reinfestation,
   or for admixture with grain as a disinfestation treatment, alone
   for control of secondary pests, or sometimes mixed with residual
   protectants for control of infestations containing both primary
   and secondary species.

Dichlorvos has the advantage over the more residual insecticides of being considerably more active against immature stages of pests that develop within individual grains - whereas malathion, for example, will kill only first-instar larvae of *Sitophilus oryzae*, dichlorvos will give a significant kill of all larval stages except late final-instar larvae and the pupal stage. Resistance to dichlorvos has been present at low levels for a considerable time in a range of species but recently higher levels associated with control failure have been detected in *Rhyzopertha dominica*.

Fenitrothion has been used for a considerable time as a building fabric treatment particularly where malathion resistance was present. Its acceptance in the grain industry was delayed by lack of adequate toxicological data for establishing a residue tolerance. This was because no commercial company had exclusive patent rights but data are now available and a tolerance of 10 ppm on wheat has been recommended by the Joint FAO/WHO Meeting of Experts on Pesticide Residues. Fenitrothion is considerably more effective than malathion against *Sitophilus* spp. and Lepidoptera and of comparable effectiveness against *Tribolium* spp. and *R. dominica*. Although used occasionally for admixture treatment, fenitrothion is finding increasing application for surface treatment of bag stacks and of bulk grain where residual protection is also required to control moth infestations, as well as the general hygiene treatments for which it has been used previously.

Pirimiphos-methyl, although developed considerably later than fenitrothion is at a comparable stage of introduction into the grain industry. The Joint FAO/WHO Meeting of Experts on Pesticide Residues has recommended a 10 ppm tolerance for this material also and there is an active and extensive programme of evaluation in a large number of countries throughout the world. These studies include use in admixture with food grain and as a seed dressing as well as for normal hygiene applications. Although pirimiphos-methyl has a significant vapour effect, residues are very stable and little degradation occurs in treated grain held under normal storage conditions. This persistence is compatible with the tolerance of 10 ppm in wheat and corresponding tolerances in milled products (e.g. bran - 20 ppm, wholemeal flour - 5 ppm). Moreover it shows considerable promise as a seed protectant, and it is probable that there will be a significant expansion of present limited use.

Other materials that have been evaluated by the Joint FAO/WHO group for the grain industry are bromophos and chlorpyriphos-methyl for which tolerances recommended are 10 ppm and 4 ppm respectively in wheat. *Bromophos* was used on a limited scale commercially in areas such as South Africa, The Gambia and Turkey. It is less toxic to insects than most other materials in common use. There is less background information on storage pest control applications available for *chlorpyriphos-methyl* and currently commercial use is minimal. Chlorpyriphos-methyl like bromophos has been used in the field for control of non-storage pests for a considerable time and although both materials have been included from time to time in evaluations of prospective grain protectants, it is only in recent years that they have been actively promoted for use in grain.

A range of other pesticides found occasional use. *Fenthion* was used on a limited scale in South America for surface treatment of bags and grain bulks, jodfenphos in Nigeria for control of *Trogoderma granarium* particularly in export storages, and *trichlorfon* in China for disinfestation of empty granaries and equipment. Few carbamates were used. *Carbaryl* was used for control of *Sitotroga cerealella* in Peru and as a bag treatment in the Philippines, and *propoxur* in grain stores in Jamaica.

(E.4) *Methyl bromide*

Methyl bromide usage is very extensive throughout the world and in many countires it is the mainstay of pest control programmes. Its correct use needs a greater level of expertise than most other pest control methods used in stored products and probably for this reason, methyl bromide fumigation in general was the most efficient pest control operation observed throughout the survey. The availability of appropriate fumigation manuals such as the "FAO Manual of Fumigation for Insect Control" (1961, 1969) and the United Kingdom Ministry of Agriculture, Fisheries and Food's Bulletin on "Fumigation with Methyl Bromide under Gas Proof Sheets" (Brown 1954, 1959; Anon 1974c) has also made a significant contribution. Data obtained during the survey are summarised in Table E1 indicating the extent of use and some of the dose rates employed.

The greatest single deficiency in use of methyl bromide was not in the operation itself but in its integration into pest control programmes. In many instances the major reliance was placed on methyl bromide fumigation and hygiene and stock management neglected so that reinfestation occurred immediately the fumigation was completed. Thus frequent fumigation was necessary and many parcels of grain inspected undoubtedly would have had bromide residues in excess of the permissible tolerance set by Codex Alimentarius (one stack of bagged paddy had been fumigated 13 times in 18 months storage). Monitoring of gas concentrations was not practised regularly and an operator hazard was evident in many places in the methods of removing sheets and airing stacks. Attention to safety equipment left much to be desired in most storages inspected.

Most usage concerned fumigation of bag stacks under gas-proof sheets although fumigation of grain in a variety of small containers was not uncommon. Fumigation of large bulks without recirculation was observed only in Australia where conditions precluded use of other materials. There, gas concentrations were monitored carefully, and although basically the method was undesirable, reasonable pest control was obtained. Recirculation fumigation facilities were available in a few countries only and were usually restricted to port installations either associated with export grain shipping programmes (Australia, Guyana) or grain imports (Japan, Taiwan, United Kingdom). Facilities away from ports also were not common and usually were associated with flour mills although a variety of temporary arrangements including forced distribution systems were in use.

(E.5) *Phosphine*

Phosphine is a very efficient fumigant and its use complements that of methyl bromide. Thus methyl bromide is used where short exposure

* (FAO Agricultural Studies No. 56, 1961; No. 79, 1969)

periods only are practical and grain has a moisture content less than 11%. Methyl bromide is also the preferred material for bag stacks and bulk storage at all temperatures and moisture contents when recirculation or forced distribution of fumigant is possible. Phosphine is the preferred material in horizontal bulks of grain that can be probed with tablets and vertical storage where methyl bromide can not be used. The extent of use of phosphine, although comparable with that of methyl bromide, is expanding more rapidly and because of its efficiency and ease of use will continue to do so. Phosphine, however, is less well researched than methyl bromide and it is perhaps unfortunate that it was introduced into commercial use before many of its basic properties as a fumigant were understood. It is now known that high concentrations of gas may induce a protective narcosis, that low concentrations for a protracted period are particularly effective, and that insects may develop tolerance to the material very readily. Thus exposures during fumigation must be of adequate duration and minimum concentrations of gas must be maintained over this period. Early recommendations for use of phosphine indicated use in circumstances where there was not adequate gas-tight enclosure and this combined with a lack of acceptance of minimum recommended exposure periods, still persists in the industry and is probably largely responsible both for the high failure rate in phosphine fumigations and more importantly the emergence of significant phosphine resistances in some of the major grain pests. There is an obvious correlation between misuse of phosphine and the resistances which will discussed later.

The phosphine-liberating formulations used depended on which manufacturer was servicing the particular area of the world. The sources of supply were principally Germany (F.R.) and Germany (D.R.) with local production and consumption in China and India and the latter country exporting some material to eastern Africa. Tablets were most common but pellets were finding increasing use in vertical storage. The sachets were used on a limited scale only for particular applications in small fumigations.

(E.6) *Other fumigants*

Materials other than methyl bromide and phosphine accounted for a small proportion only of world fumigant use. *Hydrogen cyanide* was used in many countries on a limited scale. In Argentina it was used in flour mills. In Chile hydrogen cyanide was the major fumigant as phosphine was regarded as too expensive, fumigation facilities were not available and methyl bromide was obtainable only in 1 lb (454 g) cans. Hydrogen cyanide was used to fumigate ships and grain entering the country to prevent outbreaks of *Sitotroga cerealella* following a serious outbreak in 1963. In Australia hydrogen cyanide was used for disinfestation of export grain in Western Australia and to a limited extent in small scale fumigation in South Australia. It was also used commonly in Europe for example in Czechoslovakia, Germany, Holland and Poland.

Carbon disulphide also was used on a limited scale for small fumigations in Africa and Central America, for example in Egypt, Morocco, Botswana and Guatemala, and in Australia.

Other materials used occasionally for grain were *ethylene dibromide, chloropicrin, methallyl chloride* and *carbon tetrachloride*. Liquid fumigant mixtures were in use on farms and other small scale applications including use in flour mills. These were of varying composition which appeared to be determined principally by local availability from particular manufacturers.

F. OCCURRENCE OF RESISTANCE IN STORED CEREAL INSECTS

(F.1) *Monitoring for resistance*

Failures to control insect pests can occur for many reasons and confirmation of the presence or absence of pesticide resistance may prove valuable if only by eliminating resistance as a contributing factor. Pesticide resistance is manifested in the field as a breakdown of pest control in circumstances where control previously could be maintained satisfactorily by similar use of the pesticide. When highly resistant strains are involved, this breakdown of control is often as spectacular and definitive as was the level of control that was obtained when the material was first introduced. In such circumstances detection is simple, in effect confirming the obvious, and can be achieved by a variety of methods providing simultaneous tests are made with reference susceptible material. This, however, is not always the case and low levels of resistance or low frequencies of resistant phenotypes may escape detection or produce results from which unequivocal diagnoses are not possible. Low resistance factors, for example x2, can be very important, particularly when fumigants are involved and the margin is small between the minimum effective dose and the dose at which damage will occur (e.g. with seed and malting barley). Of comparable importance is the situation with residual pesticides where the first indication of a change in tolerance may be a reduction in the period for which effective protection can be achieved rather than an immediate breakdown of control. All these expressions of resistance involve changes in tolerance to the pesticides and, however slight, are detectable provided appropriate methods are used to compare either response levels at similar doses, or equitoxic doses as with graded dose-response tests. The importance of such methods is partly their role in providing early warning of resistance before control failure is evident, thus providing the opportunity to implement countermeasures to avoid some of the undesirable consequences of resistance such as loss or damage to the infested commodity and the excessive use of pesticides.

(F.1.1) *Factors associated with and affecting tolerance to pesticides in monitoring tests*

Numerous factors influence, to varying degrees, the tolerance of stored grain insects to pesticides and it is useful, therefore, as a background to consideration of monitoring for resistance, to outline some of the effects of the more important factors both intrinsic and environmental on the development and expression of resistance. The following summary is concerned only with our knowledge of stored product insects and does not include the wealth of relevant data that is available for insect pests generally. Such information is available in texts such as Busvine's (1971) "A Critical Review of Techniques for Testing Insecticides."

(F.1.1a) *Variability in response*

Variability in response to external influence is an intrinsic characteristic of all living organisms and provides the basis for selection of the fitter individuals that must exist. The extent of variability in pesticide responses is a characteristic of the insect species and the pesticide but also may vary between different populations of the same species. Populations living in geographic isolation or exposed to different environmental stresses may show heritable differences in response to pesticides but unless the differences are determinable as more or less specific to pesticides, they usually are not regarded as pesticide resistance.

The slope of the regression line relating dosage and response is a relevant measure of the variability in response to pesticides. Dyte and Blackman (1967b, Dyte 1970) have discussed the significance of changes in slopes of these lines in determining resistance. They caution that resistance factors established at LD_{50} levels without reference to slopes may lead to spurious reports of resistance unless there are corresponding increases in tolerance at high mortality levels. Similarly with resistant strains which often have flatter slopes than susceptible strains, resistance factors taken at the LD_{50}'s will give under-estimates of the magnitude of the resistant problem in practical control measures aimed at complete kills. These authors illustrate the former point with data from adults of two lindane susceptible strains described by Champ and Campbell-Brown (1970a) which have corresponding LC_{99}'s but show a x2.5 change in tolerance at the LC_{50}. The strain with the lower LC_{50} had been in laboratory culture for approximately one year whereas the other strain had been in culture for four years and it appeared that selection of this strain, either in the field before collection or in the laboratory, had increased the frequency of the more tolerant (though still susceptible) individuals in the strain. This indicates that caution must be exercised in choosing reference strains that have been in culture for some time because laboratory culturing regimes inevitably produce stresses and selection pressure. Furthermore for similar reasons, these laboratory-cultured strains often show greater homogeneity in response when compared with field strains allowing significant changes in tolerance to be established more readily.

Strong *et al.* (1969) examined the variability in response to malathion of adults of 8 strains of *Tribolium confusum* and 39 strains of *T. castaneum* from California. The differences in susceptibility of the reference strain of *T. castaneum* ranged from 1.0 to 1.1 at the LC_{50} and from 1.0 to 1.4 at the LC_{95}. The field strains, when compared with the susceptible strain, showed variation from 0.8 to 1.3 at the LC_{50} and from 0.8 to 1.5 at the LC_{95} with one value of 2.2. Similar results were found with *T. confusum*. These authors rightly considered that resistance was not indicated by the results of the study. In effect each strain was considered as a population and they were not tested for the occurrence of low frequency resistant phenotypes whose resistance status could have been examined by selective breeding from survivors from treatment.

Fumigants characteristically give steep regression lines as compared with insecticides. This demonstrates a low variability in response of individuals within populations which appears to be reflected in low variability between strains and may allow dose rates in use to have small margins over minimum effective dosages. Selection for resistance may also be reduced. If we compare methyl bromide and phosphine, we find that responses to methyl bromide are generally uniform both within and between populations over series of tests whereas with phosphine, dosage-mortality lines are less steep and there is a considerably greater variation between strains and repetitive tests. In practice, increases in tolerance of methyl bromide do not appear to occur readily in the field or from laboratory selection programmes, whereas the converse appears to hold for phosphine.

(F.1.1b) *Species specificity*

The variability in response between species must also be taken into account. The pest fauna of stored grain is extensive and as discussed in Section D infestations often comprise complexes of pest species of varying importance although in most instances, particular species are clearly dominant. Control programmes must take into account

the range of species present or from experience expected to be present.
Each species has a characteristic pattern of level of responses to the
available insecticides and fumigants and frequently within the same pest
complex, component species are present which show marked differences in
response to particular materials. Thus for example *Sitophilus oryzae* is
considerably less tolerant of malathion, diazinon and fenitrothion than
Rhyzopertha dominica and has less variation in the response levels of
individuals within a normal population (Champ et al. 1969, Fig. 4). These
differences occur at all levels - the phycitid Lepidoptera appear more
tolerant of malathion than Coleoptera and do not respond to the addition
of synergists to pyrethrins, whilst within genera, closely related
species may differ markedly in response both in terms of toxicity and
rate of toxic action (e.g. DDT and the *Tribolium* spp., Dyte and Daly
(1969, 1970)). Nevertheless the patterns of response of individual
species, apart from a few empirical observations of similarities do not
allow any usefully-predictable grouping of species and it is necessary to
test all new materials against all potential target species. A relevant
extension of this is into cross-resistance, where although there are
logical groupings based on the toxicants, it does not extend to obvious
groupings of species. The differences in tolerance that exist between
species do provide different degrees of selection pressure within a pest
complex exposed to a pesticide and these differences in selection pressure
can interact with the differences between species in variability among
individuals, so influencing the expression of resistance, particularly
where complex multifactorial resistances are involved or where modifying
factors are concerned.

(F.1.1c) *Time for preadult development*

The time taken for development to the adult stage is a critical
parameter of culturing programmes defining the maximum periods that can
elapse before progenies are removed to ensure that the slower developing
individuals are included without contamination by individuals of subsequent generations. It is important that individuals representing the
full range of development times be included in progenies as differences
in response have been recorded in comparisons of early maturing individuals
with those appearing later. Thus a DDT-susceptible strain of *Tribolium
castaneum* has been selected by Bhatia and Pradhan (1968) by breeding from
early-maturing individuals, though these authors were unable to demonstrate
any correlation between lindane susceptibility and developmental period
(Bhatia and Pradhan 1971). Maeda (1958) reported an increase from 6 weeks
to 10 or more weeks in the development time during laboratory-selection of
a DDT-resistant strain of *T. confusum*. Similarly with malathion-resistant
strains of *T. castaneum* (Champ and Campbell-Brown 1970a), *Sitophilus oryzae*
and *S. granarius* (unpublished data), the faster-developing individuals
have significantly lower tolerances to malathion than individuals
developing more slowly. A similar though less marked trend was evident
with *Rhyzopertha dominica* but differences were not discernible with
Tribolium confusum and *Oryzaephilus surinamensis*. Shaw and Lloyd (1969),
however, demonstrated a significant increase from 44 days to 52 days in
the time taken for pre-adult development of *Dermestes maculatus* in a
selection programme for lindane resistance which raised the tolerance x26
over 14 generations of selection. By contrast Pasalu (1974) reported more
rapid development of a triphenyl phosphate suppressible malathion-resistant
strain when compared with a susceptible strain of *T. castaneum*. Thus,
though the trend appears to indicate that resistance is generally associated
with an increase in the time spent in pre-adult development, this cannot be
assumed as a general association.

(F.1.1d) *Stage of development*

Adult stored-grain insects are generally the most suitable stage for use in monitoring tests because they are convenient to handle and are relatively tolerant of physical stress. Moreover being the terminal stage of development, they are often long-lived and relatively stable physiologically compared with other stages, and hence more readily available in adequate numbers for toxicological work. A pesticide resistance monitoring test based on adult responses is of considerably more value if all resistances present in immature stages can be detected in adults. Fortunately there is little evidence to suggest that there may be resistances in immature stages that are not expressed in adults, but where resistance is present in adults, the data available does not allow generalisation as to the occurrence of resistance in immature stages. Where differences in tolerance exist between adults and larvae, an intensification of tolerance in the adults, usually the more susceptible stage, to the levels in the immature stages will have practical significance only in the case of residual treatments. With regard to fumigants, Monro *et al.* (1972) considering the results of Upitis *et al.* (1973), stated that "Studies on immature stages of one of our methyl bromide-resistant strains of *S. granarius* have shown that in a programme where all of the selection was done on the adult stage the immature stages did develop some additional tolerance but to a lesser degree than that obtained in the adults". With phosphine however, it has been reported from the same laboratory that "where adults are used for selection some resistance may occur in the pupal stage but other immature stages do not possess the tolerance character." (Anon. 1974b). Other evidence available indicates that resistances that have been selected in the laboratory by culture in insecticide-treated media, are present in adults. In this type of selection the adults themselves may undergo selection on emergence in the medium, though the degree of adult selection would be minimal because adults have a much greater natural tolerance to specified levels of pesticide when these are impregnated in materials such as the flours used in breeding media. Some of the many examples are DDT-resistant strains of *Tribolium castaneum* and *T. confusum* (Maeda 1958, Dyte and Blackman 1967b, Bhatia and Pradhan 1968), lindane-resistant strains of *T. castaneum* (Bhatia and Pradhan 1971) and *Dermestes maculatus* (Shaw and Lloyd 1969), a carbaryl-resistant strain of *Oryzaephilus surinamensis* (Dyte and Wilkin 1963), and a dieldrin-resistant strain of *Tineola bisselliella* (Kühne and Becker 1965).

Differences in tolerance between developmental stages can be marked (Table F1) and consequently of significance in the development of resistance. Field use of pesticides usually involves exposing the immature stages as well as the adults to pesticide, and the differences in response between the different developmental stages is compounded by the varying availability of the pesticide. This contributes to any reduction in the probability of achieving complete kills and thus increases the chances that selection will operate.

The information available for insecticides is not extensive. Potter (1938) showed that larvae of *Ephestia kuehniella* were much more tolerant of pyrethrins than adults. Susceptible *T. castaneum* were shown by Cichy (1969), using water suspensions of synergised pyrethrins, to be most susceptible as 1-5 day old larvae (LC_{50} at 72 hr - 0.006%) with tolerance increasing with larval development (0.05% - 10 day old) to a maximum (0.09%) in the pupal stage but decreasing approximately six times in the adult stage. Similarly Parkin and Forster (1965d) found that the larvae of 11 of 13 species tested were more tolerant of dichlorvos vapour than adults. These species included *Anthrenus flavipes*, *A. verbasci*, *Lasioderma serricorne*, *Oryzaephilus surinamensis*, *Sitophilus granarius*, *Trogoderma granarium*, *Ephestia cautella* and *Plodia interpunctella* but not *Ptinus tectus* and *Stegobium paniceum*. Champ *et al.* (1969) examined the tolerances to malathion, diazinon and fenitrothion alone and in combination with dichlorvos, of immature stages of *Sitophilus oryzae* and *Rhyzopertha dominica in situ* in grains. Under these circumstances they

TABLE F1: THE DIFFERENCES IN TOLERANCE TO FUMIGANTS AT THE LD_{50} BETWEEN THE DEVELOPMENTAL STAGES OF SOME REPRESENTATIVE STORED GRAIN PESTS

Species	Fumigant	Exposure (hr)	Temp.	Stage of Development				Reference
				Eggs	Larva	Pupa	Adult	
Susceptible strains								
S. oryzae	Acrylonitrile	2	27°C	x0.5	x0.5	x1.3	x1	
"	Carbon disulphide	2	27°C	x3.5	x0.9	x1.7	x1	
"	Chloropicrin	2	27°C	x1.0	x0.8	x0.9	x1	
"	Ethylene dibromide	2	27°C	x0.2	x0.5	x1.6	x1	Krohne & Lindgren (1958)
"	Ethylene dichloride	2	27°C	x1.1	x1.1	>x1.5	x1	
"	Hydrogen cyanide	2	27°C	x0.2	x0.5	x0.9	x1	
"	Methyl bromide	2	27°C	x0.7	x0.8	x1.9	x1	
"	Phosphine	16	27°C	x50	x5.0	x17	x1	
S. granarius	Ethylene dibromide	6	24°C	-	x0.8	x1.1	x1	Bang & Telford (1966)
"	Phosphine	16	27°C	x14	x1.0	x10	x1	Lindgren & Vincent (1966b)
T. castaneum	Carbon tetrachloride	6	24°C	-	x1.3	x2.9	x1	
"	Chloropicrin	6	24°C	-	x1.9	x1.9	x1	
"	Ethylene dichloride	6	24°C	-	x0.7	x1.2	x1	Bang & Telford (1966)
"	Hydrogen cyanide	6	24°C	-	x1.9	x3.2	x1	
"	Methyl bromide	6	24°C	-	x1.4	x3.0	x1	
"	Phosphine	6	24°C	-	x3.7	x5.9	x1	
"	Phosphine	20	25°C	x4.7	x0.5-x0.9	x3.2	x1	Winks (1971)
T. confusum	Hydrogen cyanide	6	24°C	-	x1.1	x2.5	x1	Bang & Telford (1966)
"	Phosphine	16	27°C	x18-x2	x0.4	x1.0	x1	Lindgren & Vincent (1966b)
O. surinamensis	Chloropicrin	6	24°C	-	x1.0	-	x1	Bang & Telford (1966)
"	Methyl bromide	6	24°C	-	x0.5	-	x1	
"	Phosphine	6	24°C	-	x0.5	-	x1	
"	** Phosphine	24	27°C	x2.1	x0.9	x2.7	x1	Vincent & Lindgren (1972b)
"	** Ethyl formate	6	27°C	x1.0	x1.5	x2.2	x1	
G. cornutus	Phosphine	24	27°C	-	x1.0	-	x1	Lindgren & Vincent (1966b)
S. paniceum	Phosphine	24	27°C	-	x0.8	-	x1	

* at LD_{99}
** at LD_{95}

TABLE F1
(cont'd)

Species	Fumigant	Exposure (hr)	Temp.	Eggs	Larva	Pupa	Adult	Reference
Susceptible strains (cont'd)								Lindgren & Vincent
L. serricorne	Phosphine	24	27°C	-	x1.8	-	x1	(1966b)
	* Phosphine	16-24	-	x96-x8.5	x0.8-x0.7	x5.0	x1	Cooper & Bengston (1974)
A. megatoma	** Phosphine	24	21°C	x36	x3.9	x2.8	x1	
T. glabrum	** Phosphine	24	21°C	x20	x2.4	x2.2	x1	Vincent & Lindgren (1972a)
T. sternale	** Phosphine	24	21°C	x60	x2.5	x1.4	x1	
T. variable	** Phosphine	24	21°C	x24	x8.8	x2.4	x1	
H. luteolus	** Phosphine	24	27°C	x25	x1.4	x1.2	x1	
U. humeralis	** Phosphine	24	27°C	x7.8	x3.5	x4.5	x1	
C. mutilatus	** Phosphine	24	27°C	x24	x25	x3.6	x1	
C. hemipterus	** Ethyl formate	6	27°C	x1.6	x1.1	x2.4	x1	Vincent & Lindgren (1972b)
"	** Phosphine	24	27°C	x10	x2.7	x11	x1	
E. figuliella	** Ethyl formate	6	27°C	x1.4	x3.7	x4.6	x1	
"	** Phosphine	24	27°C	x6.7	x0.9	x4.1	x1	
P. interpunctella								
	** Ethyl formate	6	27°C	x1.2	x3.6	x4.9	x1	
"	** Phosphine	24	27°C	x56	x1.8	x9.4	x1	
T. mauritanicus	Phosphine	Varying	25°C	>x7-<x0.6	x0.3-x0.5	x0.6	x1	Qureshi et al. (1965)
Susceptible vs resistant strains								
S. granarius	Methyl bromide	Susceptible		x0.9	x1.0-x1.7	x1.6	x1	Upitis et al. (1973)
		Resistant		x1.0	x1.2-x2.7	x2.4	x4.7	

* LD_{99}
** LD_{95}

found that late-stage larvae and pupae were most tolerant and could, under field conditions, be expected to survive commercial treatment with these materials at least to the adult stage when susceptibility increased as did availability of the residual materials.

Data for strains susceptible to fumigants are more extensive (Table F1). There is no set pattern in order of tolerance of the developmental stages - eggs of the same species for example can be the most tolerant and the least tolerant stage when considering different fumigants.

Cotton (1932) demonstrated that the order of tolerance of *Tribolium confusum* and *Ephestia kuehniella* to carbon disulphide was pupa>larva>adult. Lindgren (1935), however, though finding little difference between responses to carbon disulphide of larvae and adults of *Tribolium* determined the complete order for the life-history stages as egg>pupa>adult>larva for this material and for chloropicrin. The responses for carbon disulphide were obtained under conditions of low relative humidity - at a high relative humidity pupae became more tolerant than eggs. With ethylene oxide, Lindgren found the order to be pupa>adult>larva>egg which corresponded with Gough's (1939) results for hydrogen cyanide with *T. confusum*. This pattern of pupa>adult>larva with eggs most tolerant for carbon disulphide and chloropicrin and most susceptible for hydrogen cyanide and ethylene oxide was confirmed for *T. confusum* with these four fumigants by Sun (1947). With ethyl formate, Vincent and Lindgren (1972b) found pupae of two beetle and two moth species to be most tolerant with larvae≥eggs>adults in three of the species, *Oryzaephilus surinamensis*, *Ephestia figuliella* and *Plodia interpunctella*, and eggs more tolerant than larvae in *Carpophilus hemipterus* (Table F1). Sulphuryl fluoride is a fumigant that is effective against all life-history stages of insects except eggs Kenaga (1957) - Outram (1966) has demonstrated a x10 difference in tolerance between eggs and the other stages of *Tenebrio molitor* and Reynolds et al. (1964) still greater differences for various other species. In contrast the eggs of *Tenebroides mauritanicus* (Bond and Monro 1961) and *Tribolium confusum* (Fisk and Shepard 1938, Sylvester and Goodship 1960) are the most susceptible stage to methyl bromide. Krohne and Lindgren (1958) compared the tolerance of the developmental stages of *Sitophilus oryzae* to acrylonitrile, carbon disulphide, chloropicrin, ethylene dibromide, ethylene dichloride, hydrogen cyanide and methyl bromide. The patterns of response are included in Table F1. With this species and a mixture of carbon tetrachloride and methallyl chloride, however, eggs were intermediate in tolerance (Pedersen 1960). Thus Adkisson (1957) showed the order of tolerance to carbon tetrachloride as pupa>larva>egg>adult and to ethylene dichloride as pupa>egg>larva>adult. Similar studies have been made with methyl bromide and *S. granarius* although in this species the prepupae then pupae have been shown to be most tolerant followed in order by the eggs and adults with larvae very susceptible especially the young ones (Howe et al. 1965, Howe and Hole 1966). This corresponds closely with *T. castaneum* where 2-4 day old pupae were most tolerant (Godden et al. 1965, Godden and Howe 1965). Barker (1967) found that adults of *Cryptolestes ferrugineus* and *C. turcicus* were slightly more tolerant than eggs. Further data for methyl bromide are given in Anon. (1974c; Brown 1954, 1959) in a series of estimates of the minimum CT products for 99.9% kill of a range of developmental stages of 11 species of Coleoptera and 4 species of phycitid Lepidoptera at temperatures from 10°C to 30°C, illustrating a general trend of tolerance increasing through the early pre-adult stages to a maximum in the pupal stage and then decreasing to the adult stage.

With phosphine, Özer (1961) found pupae as the most tolerant stage and reported eggs as susceptible. Subsequently, however, the order of tolerance for *S. oryzae* and *S. granarius* was determined as pupa>egg>larva>adult (See Table F1) (Lindgren and Vincent 1966b). This is supported by the extensive tests of Howe and Wagner (1969) and Howe (1970, 1973) with *S. granarius* which established peaks of tolerance to phosphine with the main peak corresponding with the prepupal-pupal stage and a lesser peak with 1-4 day old eggs. Supplementing this Reynolds et al. (1967) showed that tolerance to phosphine increased through the larval stages of *S. granarius* and Mori and Kawamoto (1968) recorded survival of both eggs and pupae when exposed for 4-5 days at comparatively high concentrations of the fumigant. A similar pattern of change in tolerance to phosphine has been reported for the immature stages of *Sitophilus zeamais* (Mori and Kawamoto 1968, Mori et al. 1969). Winks (1971) established the order of tolerance for *T. castaneum* and phosphine as eggs> pupae>adults>20 day larvae>15 day larvae>10 day larvae. This order agrees in general with that reported by Lindgren and Vincent (1966b) for *T. confusum* which was eggs>pupae>adults>larvae. This differs from the order obtained by Bang and Telford (1966) for larvae, pupae and adults of *T. castaneum* whose data suggest that larvae are more tolerant than adults though the larvae dosed were 21-23 days old and a high proportion probably were prepupae. In *Tenebroides mauritanicus*, Qureshi et al.(1965) reported eggs as the stage most tolerant to phosphine followed by adults then pupae, with small differences only between the larval instars. Thus phosphine-tolerant egg stages have been reported from *S. oryzae*, *S. zeamais*, *S. granarius*, *T. confusum* and *T. mauritanicus* (op. cit.), from *T. castaneum* (op. cit., Muthu 1973), *Corcyra cephalonica* and *Callosobruchus chinensis* (Muthu 1973), *Attagenus megatoma* and three species of *Trogoderma* (Vincent and Lindgren 1972a), *Oryzaephilus surinamensis*, four species of *Nitidulidae*, *Ephestia figuliella* and *Plodia interpunctella* (Vincent and Lindgren 1972b) and *Lasioderma serricorne* (Childs et al. 1973) (vide Table F1 also). Only in *O. surinamensis*, *C. chinensis* and sometimes *L. serricorne*, were pupae more tolerant than the eggs. Pupae were generally more tolerant than larvae except in *A. megatoma* and two species of *Trogoderma*, and adult tolerances varied.

The only pattern that emerges is that pupae generally appear more tolerant than larvae. This agrees also with the findings of Busvine (1942) that pupae of *Ephestia kuehniella* were more tolerant than larvae of carbon tetrachloride, toluene, ethylene oxide and hydrogen cyanide. A notable exception to this is *Trogoderma granarium* which has a higher respiratory rate in the pupal stage than in the larval stage (Burges 1960) and is less tolerant in the pupal stage to carbon disulphide and ethylene dichloride (Lindgren et al. 1955), and these materials and ethylene dibromide and an ethylene dichloride-carbon tetrachloride mixture (Punj and Girish 1969). These responses of *T. granarium* are consistent with those for other Dermestidae listed above (Vincent and Lindgren 1972a). In general, the differences in tolerance between the developmental stages are larger with phosphine than methyl bromide.

(F.1.1e) *Age within the developmental stage*

Insects collected in the field for monitoring tests will usually be of unknown age and if a test is to be acceptable, it must make allowance for any variations in response that could result from this cause. The obvious solution would be to make tests only on known-age progeny of field-collected material but this has the inherent serious disadvantage that diagnoses would be delayed. Accepting that the testing of field-collected material is desirable, an allowance for variation in response

with age must be made by adjusting dosages appropriately if required.
This must be done without decreasing appreciably the sensitivity of the
test by increasing dosages to levels that would discriminate against
strains showing low levels of resistance as well as susceptible strains.
Thus a knowledge of the change of response with age is a prerequisite to
final determination of the concentration of pesticide to be used in
discriminating tests.

Monitoring tests with insecticides currently concern adult stages
of stored grain beetles. In a malathion-resistant strain of *Tribolium
castaneum*, the tolerance of adults increased approximately three times to
a relatively stable level of x10 during the first 4-5 days after emergence
(Champ and Campbell-Brown 1970a). Susceptible insects showed minor changes
over the same period with the tolerance increasing slightly with age.
Subsequently (unpublished data) these authors found x4 and x20 increases
respectively in tolerance to malathion of susceptible and resistant
Oryzaephilus surinamensis over the first 14 days of adult life. With
Sitophilus granarius, apart from an increased tolerance of malathion-
resistant insects during their first day after emergence as adults,
tolerance of both susceptible and resistant strains remained remarkably
stable at least from the first to the 100th day from emergence. *S. oryzae*
however, showed a gradual and continual increase in tolerance from
emergence as adults until at least 180 days later. This concerned both
susceptible and malathion-resistant strains, though susceptible strains
nevertheless did not show, over this period, any survival at the
discriminating concentration set for this species. The resistance in
these strains was not suppressed by triphenyl phosphate (see later Section
F.14.5) on cross-resistance. In contrast, a strain of *T. confusum* with
TPP-suppressible malathion resistance showed very high tolerance one day
after emergence decreasing 20 times in the first 10 days of adult life to
a level of x15 which was maintained over a subsequent 180 days. This
reflected a half reduction in tolerance of the susceptible strain over
the first 10 days. A TPP-suppressible malathion-resistant strain of
Rhyzopertha dominica showed a similar though less marked decrease in
tolerance in early adult life as did a corresponding susceptible strain.
Parkin (1944) reported an increase with age in the tolerance of susceptible
T. castaneum and *S. granarius* to dehydrating dusts that reached a peak
three to four weeks after emergence and then decreased. Subsequently
similar patterns of tolerance were found with pyrethrins in these species
and these findings were confirmed in a further series of tests 16 years
later in which results corresponded except for minor differences only in
the first two weeks after emergence - corresponding peaks of tolerance
were found also from exposure to starvation, heat and desiccation (Parkin
and Ball 1962). Cichy (1969, 1971) showed that 1-3 day old adults of a
susceptible strain of *T. castaneum* were half as tolerant to synergised
pyrethrins as 15 day old insects and that this increased tolerance was
lost by the 30th day. Resistant insects showed a similar x4 increase
over the corresponding period to 15 days but subsequent levels were not
monitored. Goos (1961) found a similar increase in tolerance to DDT from
the fifth to the 15th day in adults of *S. granarius* and Mathlein (1952) a
decrease from the 40th to the 85th day. This increase in tolerance in the
first 2-4 weeks of adult life followed by a decline afterwards appears to be
common - *Tenebrio molitor* increased its tolerance x3.5 in the first four
weeks of adult life and by the eighth week tolerance had returned to
normal (Mukerjea 1953) while with *Ahasverus advena* and rotenone the peak
was reached in 10-14 days (Craufurd-Benson 1938). With *Tribolium confusum*
and DDT there was a slight decline from the first week to the eighth week
(Collins and King 1953).

It is apparent that strain differences occur and the available evidence suggests that adult age will influence tolerance of most species. If adults are at least 5-10 days old, however, a reasonably stable response can be expected with most species allowing use of appropriately fixed discriminating doses for resistance monitoring tests. Nevertheless, this should not be assumed and the pattern of change of tolerance with age must be established for each species taking into account, where possible estimates of strain differences also.

The problems of considering age are compounded with larval and pupal stages by the unequal rates of development of individuals which result in differences in the total age at which these individuals enter given stages of development. Because of the marked changes that occur in the periods leading up to and during moulting, age can only be taken definitively as within the particular developmental stage under defined conditions. This often creates problems in toxicity testing of immature stages as it may be difficult to obtain sufficient material entering a particular developmental stage at the same time. The difficulties are intensified with species where immature stages feed within grain. These problems with immature stages are aggravated by the large differences in tolerance with age that may occur - thus in *Tenebrio molitor*, tolerance to DDT and pyrethrins increased x15 from the beginning of pupation to the middle of the period and then decreased to approximately x3 (Mukerjea 1953). Eggs do not have this disadvantage but differences in tolerance with age are often quite marked and vary with the material used. Thus increases in tolerance with age were found with eggs of *Ephestia kuehniella* and DNOC and decreased with this species and mineral oil (Dierick 1942), and eggs of *Sitotroga cerealella* and rotenone and nicotine (Richardson 1943).

The influence of age on susceptibility to fumigants is well-defined. Gough (1939) reported that older eggs of *T. confusum* were more tolerant of hydrogen cyanide than young eggs, while Sun (1947) demonstrated a decline in tolerance with carbon disulphide. Similarly Mostafa et al. (1972) found a decline in tolerance of methyl bromide and carbon disulphide in 1-3 day old eggs of this species and of *Sitophilus oryzae*, *Sitotroga cereallella* and *Ephestia kuehniella*. Sylvester and Goodship (1960), however, found that 3-5 day old eggs of *T. confusum* were most tolerant of methyl bromide and Sylvester et al. (1963) reported 1-3 day old eggs of *Gnathocerus cornutus* as more tolerant to methyl bromide and ethylene dichloride than 0-1 or 3-6 day old eggs. These authors also reported maximum tolerance of *G. cornutus* to sulphuryl fluoride in 0-1 day old eggs with a secondary peak of tolerance in 3-4 day old eggs. Bimodal tolerance frequencies were also reported in eggs of *Tenebroides mauritancius* by Bond and Monro (1961) who found that one day eggs were most susceptible with tolerance increasing to the third day then declining before rising to a maximum in 6-day old eggs. Barker (1967) also found a slight decrease in tolerance of methyl bromide with age in *Cryptolestes ferrugineus* and *C. turcicus*. Sylvester (1964) working with *Tenebrio molitor* and sulphuryl fluoride, found that 4 day old eggs were most tolerant and this was supported by Outram's (1967) studies on the uptake of this fumigant by eggs.

Most studies with fumigants and larvae concern differences between instars rather than larval age as often described. With pupae, Gough (1939) reported for *T. confusum* an initial increase in tolerance to hydrogen cyanide followed by a decrease which continued through the adult stage, while Sun (1947) demonstrated similar findings with carbon disulphide. This slow decline in tolerance of carbon disulphide in adult *T. confusum* was found also in *Sitophilus oryzae* and *S. granarius*. Pupae of *T. castaneum* between 1 and 3 day old were more tolerant to methyl bromide than older pupae over a wide range of concentrations (Reynolds et al. 1968, Bennett 1969, Bennett et al. 1969) supplementing earlier findings of Godden *et al.* (1965, Godden and Howe 1965) who showed a corresponding decrease with age in the tolerance of eggs and a gradual increase in tolerance with larval age. Howe *et al.* (1965) found the pre-adult stages of *S. granarius* responded to methyl bromide in basically the same manner and Loschiavo (1960) reported 2 day-old pupae of *T. confusum* more tolerant of ethylene dibromide than 8 day-old pupae.

With phosphine, the tolerance of eggs of *Ephestia kuehniella* and *T. confusum* decreased (See Table F1) during development of the egg (Bell and Glanville 1970, Lindgren and Vincent 1966b). Similarly Qureshi *et al.* (1965) examined the responses of the egg and adult stages of *Tenebroides mauritanicus* and *S. granarius* respectively. With *T. mauritanicus*, tolerance of the eggs increased slightly from the first to the second day after oviposition but then gradually decreased until hatching on the seventh day. With adult *S. granarius* there was a marked decrease in tolerance from the first day after emergence to the third day after which tolerance decreased gradually over the succeeding 32 days essentially agreeing with the later findings of Howe (1970, 1973). With pupae of *S. granarius*, Reynolds *et al.* (1967) presented data suggesting a decrease in tolerance to phosphine with age, whereas Howe (1970) reported a steady increase through the prepupal and pupal stage. Reynolds *et al. (op. cit.)*, however, found that with all stages of development of *S. granarius* a susceptible phase was reached during the prolonged exposures to phosphine (10 days) that currently appear desirable for maximum insecticidal efficiency (Lindgren *et al.* 1958, Reynolds *et al.* 1967, Brown *et al.* 1969, Howe 1973).

The extension of testing to Lepidoptera will create more complex problems as meaningful tests may more appropriately concern immature stages rather than adults. The cycle of feeding stages followed by pre-moulting and moulting phases with the accompanying marked variations in tolerance, will necessitate considerably greater attention to age within the developmental stage if a reasonable degree of sensitivity is to be retained in the tests. Another aspect of age within the development stage concerns diapause that occurs in a range of stored product pests (Howe 1962a, Burges 1962, Hagstrum and Sharp 1975). This may occur in adult or immature stages and is characterised by a change to a relatively stable physiological state which may affect toxicity. Thus diapausing *Plodia interpunctella* are more tolerant of methyl bromide than non-diapausing larvae although tolerance of hydrogen cyanide is unaffected (Sardesai 1972). Similarly resting larvae of *Trogoderma granarium* can be appreciably more tolerant of methyl bromide than those showing normal activity, particularly at higher temperatures (Anon. 1974c),and diapausing *Ephestia elutella* larvae are considerably more tolerant of methyl bromide and phosphine than feeding larvae (Bell and Glanville 1973a, Bell and Walker 1973). An additional complication particularly in immature stages concerns assessment of responses which may be unacceptably delayed or spasmodic in occurrence. Information is not available on the comparative tolerance levels of normal and diapausing individuals of resistant strains but this would be a prerequisite for interpretation of monitoring tests with species that enter diapause, as control programmes must aim for kill of both types of individuals.

Exposure to pesticide may delay development of individuals enabling them to remain in a more tolerant state longer than would occur normally. Thus Reynolds *et al.* (1967) have reported that survivors from phosphine treatment of 0-1 week old *Sitophilus granarius* were delayed a few days in their development, and Cooper and Bengston (1974) referred to retarded development of phosphine-treated eggs of *Lasioderma serricorne*. In this latter species the LD_{99} decreases from 56 mg hr/l in "young" eggs to 0.3 mg hr/l in 3 day old eggs.

(F.1.1f) *Sex*

Male and female stored grain insects often differ in their tolerance to pesticides but the differences are usually of a low order. Though females are often larger than males, the differences in response are not necessarily a result of size differences and may be affected by such factors as rates of uptake and levels of respiratory metabolism. The sex ratio normally approximates to unity in stored grain insects and sex is usually not an important source of variation in monitoring tests.

Females generally appear more susceptible than males conflicting with the usual situation observed in insects (Busvine 1971), for example: in *Sitophilus granarius*, DDT, BHC, toxaphene and chlordane (Tielecke 1960), phosphine (Qureshi et al. 1965), chloropicrin, methallyl chloride, methyl bromide (Brudnaya et al. 1966); *Tribolium castaneum*, synergised pyrethrins (Cichy 1969), phosphine (Winks 1971); *T. confusum*, DDT (Loschiavo 1955), hydrogen cyanide (Gough 1939), ethylene dibromide (Loschiavo 1960); and *Mylabris obtectus*, ethylene oxide (Horsfall 1934).

Instances where the reverse holds are in the response of susceptible strains of *Sitophilus granarius* to synergised allethrin (Sevintuna and Musgrave 1961), and *Tribolium castaneum* to unsynergised pyrethrins and DDT (Hewlett 1974), lindane (Champ and Campbell-Brown 1969) and demeton-methyl (Cichy 1971), and of malathion-resistant strains of this species to malathion (Champ and Campbell-Brown 1970a). Hashimoto and Fukami (1964) also recorded higher tolerances to methyl-parathion in females of *Ephestia cautella* throughout a selection programme from susceptibility to a x7 increase in tolerance. Sun (1947), however, was scarcely able to detect any difference in tolerance to carbon disulphide between sexes of *T. confusum* and *Zabrotes subfasciatus* and Bennett (1969) could not detect any difference with methyl bromide and *T. castaneum*. Similarly differences in response to DDT were not apparent between sexes of DDT-susceptible and resistant strains of *Sitophilus oryzae* though in test-crossing of this sex-linked resistant strain, males appeared more susceptible in susceptible phenotypes and more tolerant in DDT-resistant phenotypes (Champ 1967). Conversely with methyl bromide-resistant *S. granarius*, females have developed slightly more tolerance than males (Upitis et al. 1973). Such sex-linkage in resistance either as major factors or in resistance-modifying factors, is probably the most significant role that sex can play.

(F.1.1g) *Morphological and behavioural differences*

Morphological and behavioural differences between susceptible and resistant strains have been observed and though these may be heritable associations with resistance, it seems prudent to assume, unless proved otherwise, that they are independent and that any correlations between the factors concerned is fortuitous. There is a wide range of extrinsic and intrinsic factors that can affect morphological characters independently of their effects on pesticide tolerance. Laboratory culturing methods particularly can affect density- and food-dependent characters such as size of progeny and these must be considered, along with the effects of other factors, such as the age of reproducing females on size of progeny (*vide* Howe 1967). Sun (1947) has shown that the tolerance of *S. granarius* to carbon disulphide increases with population density and size considered as a separate factor is well documented as influencing response (*vide* Busvine 1971).

Surtees (1966) recorded differences in locomotory behaviour between susceptible and pyrethrins-resistant *S. granarius*. This strain was heavier than normal (Lloyd 1969a) and otherwise unaccountably showed a x2 increase in tolerance to phosphine (Rajak and Hewiett 1971). In fumigation, however, there is evidence to suggest that tolerance is related to level of respiratory metabolism and weight. These effects are valid expressions of resistance but unless the causal factors are heritable or capable of intensification, care must be taken in their interpretation. Thus Winks (personal communication) has selected a x5 phosphine-resistant strain of *Tribolium castaneum* that was significantly smaller than the parent stock. Upitis et al. (1973, Monro et al. 1961) showed that body weight and the length of the life cycle of a laboratory-selected methyl bromide-resistant strain of *S. granarius* had increased appreciably above that of normal insects while the respiratory rate dropped more than 25%. Conversely Sevintuna and Musgrave (1961) could not detect any significant change in weight in a laboratory-selected allethrin-resistant strain. Tyler and Binns (1973) recorded a malathion-resistant strain of *Oryzaephilus surinamensis* that was less active, smaller in size and darker than the susceptible reference strain. Control mortality associated with laboratory handling was also lower in the resistant strain. Pinniger (1975), in comparing the behaviour patterns of susceptible and resistant *Tribolium castaneum* reported that the resistant strain showed a greater tendency to remain inside a food refuge in an untreated environment, and when malathion or fenitrothion particularly was introduced to the environment, this tendency was considerably reinforced and wandering activity further reduced. The susceptible strain tended to exhibit a high rate of movement and even a small disturbance initiated a period of intense activity. In contrast, the rate of movement of the resistant strain appeared to be slower, the insects spent long periods of time apparently motionless and were affected much less by disturbance. Differences in activity, size and colour are very common among both susceptible and resistant strains of stored product pests. Thus differences in activity between strains will affect rate of pick-up of insecticide and so response, complicating interpretation of results from tests using exposures to residual films of insecticide. Strains with marked differences in activity create particular problems and may have to be examined by alternative methods though in terms of practical control the differences become less important as they reflect basic physiological differences in the strains that influence pick-up, absorption and metabolism irrespective of whether exposure takes place in the laboratory or the field. Behaviour traits such as refuge seeking as described by Pinniger (*op. cit.*), provide a pesticide avoidance mechanism which has to be considered in control situations where such refuges exist.

Dyte et al. (1968a) examined the DDT-tolerance of six species of *Tribolium*, *T. confusum*, *T. castaneum*, *T. anaphe*, *T. madens*, *T. brevicornis* and *T. destructor*, and found in general that DDT-tolerance was correlated with body colour, those species with darker cuticles being the least susceptible. Strains of *T. castaneum* however, that were homozygous for the body colour mutants *mahogany*, *tawny*, *sooty*, *jet* and *black* all of which have darker cuticles than the normal *T. castaneum*, were little different from the normal reference stock in responses to DDT. Erdmann (1970) reported subsequently a correlation between DDT resistance and *sooty*.

(F.1.1h) *Fecundity*

Changes in fitness may be associated with pesticide resistance. In particular, the fecundity of resistant strains may show differences when compared with normal susceptibles strains. Brower (1974b) found

that females of a strain of *Tribolium castaneum* resistant to malathion
(x51) and DDT (x10) (of Speirs *et al.* 1971) produced an average of 24
progeny over a 14 day oviposition period at 27°C, compared with an
average of 123 progeny from a laboratory susceptible strain. A laboratory
selected lindane-resistant strain of the same species was found by Bhatia
and Pradhan (1971) to show a 37% reduction in the number of eggs produced
throughout the oviposition-period although the percentage egg-hatch and
the duration of the oviposition-period were unchanged. These authors also
reported a reduction in fecundity but normal egg hatch rate in a laboratory-
selected DDT-resistant strain (Bhatia and Pradhan 1968) and subsequently
Bansode (1974) working in the same laboratory with a selected
malathion-resistant strain (x49) of *Sitophilus oryzae* reported a 20%
reduction in the number of eggs laid and an 18% reduction in adult emergence
over a one month period. A more significant decrease in fecundity was
reported by Shaw and Lloyd (1969) who found with *Dermestes maculatus* that
a x26 increase in tolerance to lindane developed during 14 generations of
selection, was associated with a reduction in the number of progeny per
female from 33 to 4. With phosphine, Winks (1971) recorded from *T. castaneum*
a mean of 34 progeny per female from a strain that had been selected in six
generations to a resistance level of x6.8 compared with a mean of 46 progeny
per female from the parent stock. These observations do not appear to be
artifacts of the particular techniques of assessment - thus Lloyd and Parkin
(1963) have demonstrated a reduction of fecundity in pyrethrins-resistant
Sitophilus granarius over a range of test temperatures and Dyte and Blackman
(1967b) with DDT-resistant *T. castaneum* and various foodstuffs. Such reductions
in fecundity associated with resistance may give a small measure of assistance
in reducing the impact of resistance but there is *a priori* no reason to assume
that reduced fecundity will be a general phenomenon. Thus Champ and Cribb
(1965a) reported little difference in progeny yields in untreated grain between
susceptible and lindane-resistance *S. oryzae*, and Pasalu (1974) demonstrated
that egg hatch and larval survival of a susceptible strain of *T. castaneum*
were 3.1% and 11.9% lower when compared with a triphenyl phosphate-suppressible malathion-resistant strain in which the resistance had been intensified
in the laboratory during six generations of selection of field-collected
resistant material. Similarly Brudnaya *et al.* (1966) reported an increase
in the number of progeny from strains of *Sitophilus granarius* tolerant to
methyl bromide and chloropicrin. Of particular interest, are the observations
of Winks (1973) who compared a susceptible and a phosphine-resistant strain
of *T. castaneum* at concentrations in the lower range of those encountered in
practical fumigations. He found that there was no clear evidence of
inhibition of reproduction in the susceptible strain, whereas though the
resistant strain showed a x5 tolerance in terms of lethal response, there was
an increasing inhibition of reproductive capacity as mortality increased.
Hence, low level survival in a resistant strain may be associated with a
very low reproductive potential. Thus, increased lethal tolerance may overestimate the practical significance of this type of resistance if their
tolerance in terms of inhibition of reproduction remains unaltered.

Reports of effects of insecticides on fecundity of susceptible
insects are numerous and have been reviewed by Moriarty (1969). There is no
predictable pattern though most reports concern a reduction in fecundity - thus
Kuenen (1958) has shown that *Sitophilus granarius* reared in wheat containing
DDT produced 20% more progeny than controls whereas Loschiavo (1955) found
that *T. confusum* females surviving exposure to surface deposits of DDT
produced significantly fewer eggs than untreated females. Similarly Zettler
and Le Cato (1974) observed reductions in fecundity of *Attagenus megatoma*
in survivors from treatment with dichlorvos or malathion. These effects usually

are dosage-dependant, a characteristic that is well-defined in fumigants also, for example, in *Mylabris obtectus*, ethylene oxide (Horsfall 1934); *S. granarius*, chloropicrin, methallyl chloride and methyl bromide (Brudnaya et al. 1966), methyl bromide (Howe and Hole 1967), phosphine (Reynolds et al. 1967); *T. castaneum*, phosphine (Winks 1971, 1973); and *T. confusum*, ethylene dibromide (Loschiavo 1960). The occurrence of reduced fecundity associated with pesticide use in susceptible strains as well as in resistant strains unexposed to pesticide suggests some association of effects though there is no evidence for this.

(F.1.1i) *Temperature*

Temperature is probably the most significant extrinsic factor affecting responses of insects in monitoring tests. Its effect are threefold.

First the temperature at which insects are reared can influence their response. Thus Cichy (1969) comparing *Tribolium castaneum* cultured for 5-7 generations at 25°C or 30°C found they were four times more tolerant to synergised pyrethrins when bred at the higher temperature. She did not specify the temperature at which the tests were carried out though in a later report of associated experiments, she indicated that all tests were carried out under identical conditions at room temperature. Conversely, Craufurd-Benson (1939) found that *Ahasverus advena* reared at 25°C were more susceptible than when reared at 20°C, and Sun (1947) established similar effects with carbon disulphide and *Tribolium confusum* and *Sitophilus granarius*.

Second, the temperature at which insects are held, before, during and after the test will influence activity of the insects, pick-up of insecticide in tests with residual deposits, toxication and detoxication of the insecticides, the concentration of insecticides at sites of action, the sensitivity of these sites of action, and any eventual recovery from intoxication. Pick-up and spread of insecticides in tests with residual deposits are strongly correlated with activity, particularly where non-volatile carrier oils are used (Pradhan 1949a, Busvine 1962a). There is no consistent pattern of change of response with temperature. Iordanou and Watters (1969) using a method based on 24 hr exposure to filter papers impregnated with insecticides and a 72 hr mortality response, established temperature relationships over the range 10°C to 26.7°C for *T. castaneum*, *T. confusum*, *Oryzaephilus surinamensis*, *O. mercator* and *Cryptolestes ferrugineus*. DDT responses showed a negative correlation with temperature with all species at higher temperatures and a positive correlation at lower temperatures. Lindane gave similar results with *O. mercator*, a converse pattern with *O. surinamensis* and the *Tribolium* spp. and a positive relationship for all temperatures with *C. ferrugineus*. Malathion and bromophos responses also showed positive relationships with temperature over all temperature ranges with all species. Working with *T. castaneum*, others have shown a negative relationship (Potter and Gillham 1946) and a positive relationship (Pradhan 1949a) for DDT, a negative relationship for synergised pyrethrins (Cichy 1969) and carbaryl (Teotia and Pandey 1967), and a positive relationship for malathion (Teotia and Pandey 1967, Champ and Campbell-Brown 1970a). Using treated grain, Champ et al. (1969) demonstrated an approximate x1.5 increase in toxicity to *Sitophilus oryzae* of malathion, diazinon, and fenitrothion in a temperature change from 25°C to 30°C. With fumigants, positive correlations between temperature and toxicity are usually observed over the approximate range 10°C-35°C with a reverse trend becoming evident at lower temperatures with materials other

than methyl bromide (Busvine 1971). These observations concerned
T. confusum with carbon tetrachloride, carbon disulphide, ethylene
dichloride, methyl bromide and chloropicrin, *T. castaneum* with carbon
disulphide and *S. granarius* with methyl bromide and hydrogen cyanide
(Bertand et al. 1919, Fisk and Shepard 1938, Govindan and Cutkomp 1961,
Peters 1938, Peters and Gantner 1935, Shepard et al. 1937, Sun 1947).
Further examples of the correlation between increasing temperature and
toxicity concern *S. oryzae* and carbon disulphide (Hinds and Turner 1910),
ethylene dichloride-carbon tetrachloride mixtures (Muthu and Pingale 1956)
and phosphine (Lindgren et al. 1958, Özer 1961), *S. granarius* and phosphine
(Lindgren et al. 1958, Özer 1961, Mori and Kawamoto 1968), *Trogoderma
granarium* and carbon disulphide, ethylene dibromide, ethylene dichloride and
an ethylene dichloride-carbon tetrachloride mixture (Punj and Girish 1969)
and phosphine (Lindgren et al. 1958), *Lasioderma serricorne* and phosphine
(Childs et al. 1973), and *Callosobruchus chinensis* and phosphine (Mori and
Kawamoto 1968). With phosphine, the increase in toxicity at higher temperatures has been related directly to oxygen consumption in *Callosobruchus
chinensis* over the range 20°C to 40°C so that the product of the LD_{50} (as a
CT product) and the oxygen consumption is a constant (Sato et al. 1973).
The general applicability of this relationship has not been investigated but
it would certainly increase the scope of resistance testing if constants
characteristic for each species could be obtained. Data for methyl bromide
against a range of developmental stages of Coleoptera and Lepidoptera at
temperatures from 10°C to 30°C are given in Anon. (1974c; Brown 1954,
1959) illustrating the 2 to 3 fold increase in toxicity of methyl bromide at
high compared with low temperatures. In summary it is thus mandatory to
maintain a constant temperature during treatment as otherwise the relationships between dose and response become quite complex as illustrated by
Pradhan and Govindan (1954). Effects of post-treatment temperature are
also significant for example with pyrethrins and *T. granarium* (Mookherjee
et al. 1964).

Third, temperature will affect the rate of development of
resistance complementing the shorter life-cycle periods at high temperatures. Cichy (1971) showed that with laboratory selection of adults
of *S. oryzae* at the LD_{50} level using responses for continuous exposure
to impregnated filter papers at room temperature, resistance to DDT,
synergised pyrethrins and demeton methyl developed faster in lines bred at
25°C than at 30°C. She demonstrated a similar influence of temperature on
development of resistance to DDT and demeton methyl in *T. castaneum* but was
unable to demonstrate a similar effect with synergised pyrethrins.

The effect of change of temperature on resistance factors has
not been investigated.

(F.1.1j) *Humidity*

The relative humidity prevailing during toxicity testing does
not appear to be particularly critical. Differences in tolerance have
been recorded, however, in *Tribolium castaneum* in which tolerance to DDT
and DNOC was lower at higher humidities (Pradhan 1949b), whereas in
Ahasverus advena tolerance to rotenone was higher at higher humidities
(Craufurd-Benson 1938).

Small differences have been detected with fumigants also, as in
T. castaneum where adults were more sensitive to methyl bromide at high
humidities (Fisk and Shepard 1938) and where eggs, though their tolerance
to methyl bromide was unaffected by humidity, were more tolerant of carbon
disulphide at intermediate humidities (Sun 1947) and carbon disulphide and
chloropicrin at very low humidities (Lindgren 1935).

These effects on toxicity have concerned the exposure and post-treatment holding periods only. Pre-treatment holding conditions may also have an influence (Sun 1947), and differences in tolerance have been recorded from culturing of test insects at different humidities (Cichy 1969), and humidity has been shown to influence rate of development of resistance (Cichy 1971).

(F.1.1k) *Light intensity and time of day*

The primary effect of light intensity and the time of day is to influence the activity of insects. This is particularly noticeable under field conditions where with many species, spectacular flight activity may take place at dusk. Other peaks of activity may occur also with mating in Lepidoptera well before dusk and in pre-dawn oviposition. Photoperiod must also be taken into consideration with some species, particularly those where diapause occurs. These cycles of activity are a characteristic of laboratory cultures also, although they are less pronounced because of the restriction of culture containers and the lack of associated well-defined temperature changes. As levels of activity have a marked influence on monitoring tests using residual deposits, it is essential that light regimes be standardised and that tests always be conducted at the same time of day. Changes in activity reflect physiological changes in insects and it is prudent to apply this restriction to all forms of toxicity testing.

(F.1.1l) *Diet*

The quantity and quality of food in which insects develop influences their size, general fitness and survival and may affect their tolerance to some pesticides. Their tolerance to pesticides and other adverse agents may be influenced also by whether they have fed recently.

Insects in the field may feed and breed in a wide range of commodities and when material is collected for monitoring tests there is often no clear indication of the commodity in which the insects have developed. The primary monitoring tests use, and are based on the responses of reference strains which have been bred in standardised media thus establishing a point of basic difference between the test strains and the reference strain. Any effects of the difference are minimised however, in confirmatory tests on the progeny of the field material which are obtained by the standardised breeding method.

Data on the influence of different foodstuffs on pesticide responses is limited though sufficient to indicate that diet is a factor to be taken into consideration at least in the interpretation of results. In *Tribolium castaneum* for example, adults have been shown to be most susceptible to synergised pyrethrins when bred on crushed peeled barley with tolerance increasing x2 with ground unpeeled barley and with ground rye, x8 with wheat flour, to x12 with ground rice (Cichy 1969). Devaraj Urs and Mookherjee(1966) found that *T. castaneum, Trogoderma granarium* and *Corcyra cephalonica* were more susceptible to pyrethrins when reared on oilseeds than on cereals. Similarly *T. granarium* has been shown to be more susceptible to malathion, lindane, pyrethrins and carbaryl when reared on maize, Bengal gram or cowpeas than when reared on wheat, green gram, bajra or peanuts (Rattan Lal and Attri 1967) and *T. castaneum* was most susceptible to the same materials when reared on flour of mung and tolerance increased progressively on diets of juar, wheat and maize flours (Rattan Lal and Singh 1966). Cichy (1971) also found that diet played a significant role in the development of resistance in *Sitophilus oryzae* particularly with DDT and demeton-methyl and to a lesser extent with synergised pyrethrins. Strains bred on rye and barley gave highest and lowest levels respectively of DDT resistances (x15 difference), on hulled barley and wheat highest and lowest resistances to demeton-methyl (x37 difference) and on wheat and hulled barley highest and lowest resistances to syngergised pyrethrins (x6 difference).

With fumigants, Sun (1947), in comparison of the response to carbon disulphide of larvae of *T. confusum* reared for three weeks on eight different diets, found that the most tolerant group required 3 times the dose required by the most susceptible group and that these differences resulted from the differences in growth rates on the different diets. With adult *T. confusum* and *Sitophilus granarius* differences in tolerance were smaller as were differences in size. Other reports of increases in tolerance to fumigants associated with suboptimal food intake concern *Corcyra cephalonica* and carbon disulphide (Chatterji 1955); *Trogoderma granarium*, methyl bromide and ethylene oxide (Punj 1970, 1971); *T. castaneum*, carbon disulphide (Punj and Girish 1968) and *Callosobruchus maculatus* and a range of fumigants (Murthy and Srivastava 1971). Similarly Punj (1970) reported that *Trogoderma granarium* reared on pulses was, in general, less susceptible to ethylene dichloride, carbon disulphide, methyl bromide and ethylene dibromide than when reared on cereals. Differences in susceptibility between insects reared on oilseeds compared with other seed are also apparent, for example *T. castaneum* and *C. cephalonica* were more susceptible to methyl bromide when reared on wheat or sorghum than when reared on oilseeds - the opposite held for *T. granarium* (Devaraj Urs and Mookherjee 1967) although Punj (1970) reported to the contrary. Punj (1970), however, did show a two- to five-fold increase in susceptibility to ethylene dibromide in *T. granarium* reared on peanuts compared with progeny reared from a range of other legumes and cereals. With phosphine, larvae of *T. granarium* and *Ephestia cautella* were more susceptible when reared on peanuts than when reared on wheat or Bengal gram, although eggs of both species reared from peanuts were more tolerant (Baskaran and Mookherjee 1971).

The magnitude and diversity of these changes in tolerance outlined above emphasise the importance of considering diet in design of pesticide-resistance monitoring programmes.

(F.1.1m) *Stability of resistance*

Resistances in stored product insects appear comparatively stable. This is supported by numerous reports of resistance levels showing comparatively rapid initial decreases after removal of selection pressure, followed by a slow and protracted loss of resistance. Thus resistant strains have been maintained in laboratory culture for up to 10 years without selection and still show high degrees of resistance.

Speirs and Zettler (1969) selected a malathion resistant strain of *Tribolium castaneum* for two generations at the 60-70% level increasing the resistance three times, and then removed selection pressure for a further 10 generations during which the resistance fell to approximately one half of the original level. A further three generations selection pressure increased the level three times again to a new level approaching half the previous maximum. This level was maintained over a further three generations of selection.

Pierterse and Schulten (1974) have stated that "resistance genes in *T. castaneum* remain easily integrated in the genome of unselected populations" on the basis of the occurrence of γ-BHC and malathion-resistant strains on farms where insecticides had not been used. The abundance of data on the movement of insects through trade channels and the occurrence of resistance in areas where selection could not have taken place, indicate that if this was not the case, resistance would not have spread as rapidly or become the serious problem it is today.

Experiments have been conducted to determine the frequency of resistant phenotypes in successive generations of progeny from parental stock containing different proportions of a pure susceptible strain artificially mixed with comparatively pure lines of a specific- and a non-specific malathion-resistant strain of *T. castaneum*. The resistance in each of these strains was determined by one major factor. After seven generations of breeding from original mixtures containing 75%, 50% and 25% of susceptibles, the malathion-specific strain mixtures contained 82%, 73% and 33% of susceptibles respectively, and the non-specific strain mixtures contained 68%, 56% and 34% respectively (Champ, unpublished data).

With fumigants, resistances appear very stable. Thus resistance to methyl bromide in *Sitophilus granarius* once selected appears to persist indefinitely. Bond and Upitis (1972, Monro et al. 1961) reported for example, a decrease from x8 to x5.6 resistance in 25 generations of culture without selection, and a decrease from x2.3 to x1.7 in another strain cultured for 83 generations over 16 years without selection. Phosphine resistances appear stable also. Thus a phosphine resistant strain of *T. castaneum* selected by Winks (1971) for 6 generations to a resistance level of x6.8 was tested three years later and gave a resistance factor of x5.

The above examples generally concern inbred laboratory lines that may have been selected for susceptibility or resistance indirectly through linked characters such as time of pre-adult-development and have not been subject to continued infusion of susceptible (or resistant) individuals as may occur with field populations. Nevertheless many field populations may remain genetically isolated for considerable periods and irrespective it seems reasonable to assume that once resistances have developed, they will persist unless there is considerable dilution with susceptible insects from outside populations.

(F.1.2) *The basis and design of monitoring tests*

(F.1.2a) *General considerations*

The basis of monitoring for resistance is exposing under defined and standard conditions, samples of pests to single doses of pesticide that would be expected to kill normally susceptible insects. Survival of insects in such discriminating tests is indicative of resistance and the need for more detailed confirmatory tests. The discriminating doses in these primary tests are established from the dosage-mortality relationships of known susceptible strains taking into account normal strain variability. The 99.9% response level is normally used as the basic discriminating dosage level thus giving a probability of 0.1 (i.e. once in 10 tests) of a single insect in a batch of 100 being unaffected due to the chance occurrence of the more tolerant individuals from a normal population. The chances of adventitious failure to respond by a single individual in each of successive repeat tests declines progressively (0.01, 0.001, 0.0001) and the continued appearance of unaffected individuals in tests where few insects survive, can be considered as proof of resistance.

The object in monitoring is to detect resistance efficiently and obtain adequate measures of the level of resistance. All available tests should be compared on a basis of reproducibility, and objective criteria such as the heterogeneity of data obtained, the distribution of tolerances to the insecticide made available to the insect, and the dosage level required for a particular response. The most satisfactory type of test can be defined as that which in regression analysis of dosage-mortality data gives the least heterogeneity (heterogeneity factor, HF) the steepest line slope (b) and the lowest median lethal dose or concentration (LD_{50} or LC_{50}). The heterogeneity factor defines the goodness of fit to linearity and so the appropriateness of the dosage-mortality relationship used. The line slope or regression coefficient is a primary measure of the intrinsic variability within a population and provides in effect a measure of the residual variability of responses in a test that are attributable to dosing, absorption and metabolism of an insecticide. This allows a direct comparison of the dosing techniques available and determines the value of the dosage-mortality relationship for establishing discriminating doses for use in preliminary sorting of samples - steeper slopes increase the efficiency of discriminating doses and lowering the $LD_{99.9}/LD_{50}$ factor. Lower LD_{50}s are often associated with steeper slopes and are desirable because they indicate a more sensitive response to the toxicant and increase the range over which susceptible and resistant strains can be compared before linearity between increasing dose and response is lost.

The procedures to be followed in the development of a resistance test method are:-

(a) Selection of an appropriate life history stage to obtain responses representative of the tolerance status of the species concerned. Adults usually satisfy the requirements most closely (See Section F.1.1d).

(b) Selection as described above of an appropriate type of assay to measure responses.

(c) Establishment of base response data from known susceptible strains.

(d) Use of discriminating doses to screen samples for resistance either on the field-collected material or their progeny, and supported by supplementary discriminating tests on the progeny of the few survivors from the primary test if the results were not considered unequivocal. Susceptible reference strains are included in all tests as a check on procedures.

(e) Complete definition of resistance by comparison of graded dosage-mortality responses of susceptible and resistant strains.

The requirements of a test method in practical terms are:-

(a) The method should be sufficiently sensitive to measure responses over the wide range of concentrations often necessary to determine degrees of resistance.

(b) It should be relatively rapid so that, where necessary remedial measures or other action can be implemented quickly. Pesticide users should not have to wait several months for a statement of whether or not resistance occurs, though precise definition of levels of resistance can wait until later.

(c) It must give consistent results and be capable of being conducted by inexperienced personnel with a minimum of instruction. Apparatus must be simple and readily available.

(d) It must be capable of dealing with all sizes of samples including those where only one or two insects are available. In the early stages of development of resistance, low numbers only of insects may survive treatment and so attract attention for testing. It is important that these insects be tested. Conversely, if a method is chosen that restricts the number of insects that can be tested, the probability of detecting low frequencies of phenotypes is lowered and the effectiveness of the test method is reduced.

(e) Operators must be able to conduct a large number of tests on different samples in a short period of time if circumstances so dictate.

(f) The strains under test must be preserved irrespective of the sample size and whether all available insects are used in the test. This is necessary as further tests may be required to confirm resistance, to compare the resistance with other resistances to the same material, to check cross-tolerance and to determine any other resistances that may be present.

(F.1.2b) *Methods for insecticide monitoring*

Stored grain insects are readily collected from the field and cultured in the laboratory providing a basis for obtaining the reproducible responses that are available under controlled environment conditions. It is desirable that the laboratory tests simulate field use of the chemicals but this should not be done at the expense of sensitivity and reproducibility. An additional complication is that pesticides are often used in the field in very different formulations and ways, and for convenience some common ground must be sought. Essentially, monitoring for esistance is determining changes in physiological responses to pesticides, and thus the ideal test is one in which the only variable factor is the dosage of pesticide. It follows then that the use of grain, for example, as a carrier for insecticide, introduces a further variable which must be standardised or at least taken into account, so complicating the tests. In practice, simulating field conditions is very difficult. There are, however, a number of artificial laboratory techniques for exposing insects to pesticides that can be used as the basis for developing methods that can be readily standardised. The different methods that are available have been summarised by Champ and Campbell-Brown (1970a). Essentially, they are exposure to insecticide-impregnated grain, direct spraying, the topical application of insecticide and exposure to residual films. All these methods have analogy with different methods of field use of insecticides but as discussed this is not essential in resistance monitoring, though an essential part of subsequent investigations to determine the significance of any resistance found.

The relative merits of the various methods are discussed hereunder.

Methods using *insecticide impregnated dust mixed with grain* are not considered satisfactory as (i) there is difficulty in precise reproduction of dusts with respect to flow and absorption characteristics of the carrier; and (ii) grain particle size (Godavaribai *et al.* 1964), absorption and adsorption of insecticide by grain and grain dust, and in the case of organophosphorus compounds, instability associated with enzymatic breakdown (Rowlands 1967) may all influence the availability of insecticide for toxic action. Thus susceptible *S. oryzae* and *T. castaneum* will succumb in 3 days to 2 and 5 ppm of malathion respectively applied to whole wheat but both will survive for extended periods in wheat dust and frass containing 50 ppm.

With *direct spraying of test insects* there are inherent difficulties in the reproducibility of spray application rates and retention of uniform doses by insects. Specialised apparatus also is required.

Topical treatment of individual insects offers the most precise means of measuring susceptibility by using known doses of insecticide. It is very important as a research tool and is the preferred method generally for detecting and measuring resistance (Anon. 1969a) but has practical disadvantages in monitoring tests on stored grain beetles (Champ and Campbell-Brown 1970a). Thus with topical treatment, insects are handled individually restricting the number of insects that can be handled conveniently. Anaesthesia with its attendant physiological effects is usually necessary involving extra equipment and handling and the droplet sizes required may make it difficult to apply sufficient insecticide in tests on resistant strains. With the larger adults and larvae of stored product Lepidoptera, however, topical treatment does have advantages.

Exposure to residual films is a very convenient method for the treatment of large numbers of small insects and, provided that the deposits are reproducible and other variables standardised, results are correspondingly reproducible. The most satisfactory method of producing films is by impregnating filter paper circles with solutions of insecticide in non-volatile oil. Such films have the advantage of being easily prepared and have a high degree of reproducibility. The inclusion of the carrier oil improves the dispersion of the insecticide on the paper circles and the oil will diffuse over the entire surface of insects confined on the papers from tarsal contact only. Exposure to residual films has the disadvantage that dosing is a continuing process during the whole of the exposure period and hence the rate at which dosing occurs is readily influenced by many extrinsic and intrinsic factors. The type of oil used is arbitrary but has a marked influence on the responses obtained. Shell Risella 17, a light spindle oil of negligible aromatic content, is the most widely used carrier oil and meets the general specifications of being relatively non-toxic to most insects, having good solubility characteristics for many insecticides, and being readily available. The pick-up and spread of the carrier oil is temperature dependent but if temperature is controlled, uptake of insecticide is a linear function of concentration and exposure time (Pradhan 1949a, Pennell *et al*. 1964, Champ and Campbell-Brown 1970a). Thus exposure of insects to impregnated paper can provide a precise and convenient method of treatment. Insects are not handled individually, anaesthesia is not necessary and the number of insects and of samples that can be handled in a convenient time is large.

A comparison of the regression coefficients determined for residual film tests and topical dosing tests of *Tribolium castaneum* with a number of different insecticides did not show a pattern that allowed generalization as to which method was superior (Champ and Campbell-Brown 1970a). The impregnated paper test, however, gave steeper regression lines for malathion and lindane which are the insecticides used most commonly in stored product pest control and for which resistance test methods are necessary. Lloyd and Williams (1973) proposed the substitution of a hardened less-absorbent (No. 544) filter paper for the normally-used No 1 filter paper and reported close agreement between the resistance factors obtained by this film test with lindane, DDT, malathion, and synergised pyrethrins and factors obtained in topical application tests - unsynergised pyrethrins however, showed a large difference (x18, x149).

The residual film tests have their limitations under certain circumstances and with some insecticides. Carbamates for example may be sorbed on the cellulose fibre of the filter paper and so unavailable for uptake by test insects - glass fibre paper has been suggested as an alternative substrate to overcome this problem (Georghiou and Gidden 1965). Champ and Campbell-Brown (1970a) were unable to establish a discriminating dose for DDT resistance in *Tribolium castaneum* using the impregnated paper method because of a high LC_{50} and a low regression coefficient, whereas Dyte and Blackman (1967b) using a topical method obtained a usefully low LD_{50} and a steep regression line. Similarly Georghiou and Gidden (1965) obtained an eight-fold increase in toxicity to mosquitoes by using dry deposits of DDT rather than oil films on test papers. This, however, also resulted in a considerable flattening of the dosage mortality line. Though the problem with DDT and *T. castaneum* was later found to be compounded by the first authors' difficulty in finding a truly DDT-susceptible strain, the point remains valid and indicates the need for caution in extrapolating the test to other species and insecticides, and the need to establish by experimentation the test most appropriate for the individual circumstances.

Test methods for detecting insecticide resistance in stored product insects have been proposed from Macdonald College, Quebec, for *Tribolium confusum* and *Cryptolestes* sp. using paper which had been impregnated with insecticide (Kumar and Morrison 1963, 1964); from the Pest Infestation Laboratory, England, as a general test based on insecticide-impregnated dust mixed with grain (Parkin 1965) and as a test for *O. surinamensis* based on Fluon-coated glass rings confining insects on filter paper treated with a solution of insecticide in butyl phthalate (Wilkin 1966); from Australia, for *Sitophilus oryzae*, based similarly on glass rings confining insects on insecticide-impregnated filter paper circles (Champ 1967); and subsequently further similar methods have been described for *Tribolium castaneum* (Champ and Campbell-Brown 1970a), *S. zeamais* (Dyte and Forster 1970b), and *T. castaneum* and *S. oryzae* (Rajak et al. 1973). A simple test is available to detect organophosphorus compound resistance in leaf hoppers by crushing insects onto filter paper impregnated with α- naphthyl acetate and so determining any differences in levels of esterase activity by colour change. This has proved unsuccessful in *T. castaneum* (Coveney and Corban 1970).

(F.1.2c) *Methods for fumigant monitoring*

Choice of a test for detecting fumigant resistance does not present the same problems associated with selection of an appropriate insecticide test. Essentially the method must involve exposure of insects to discrete atmospheres containing known amounts of fumigant. The choice of container follows from a consideration of dosing methods to be used, stability of gas concentration, cost and availability. Consideration of specific detail of fumigation containers and fumigant sources and dispensers is beyond the scope of this report and will be included in a separate research paper on the fumigant test method used in the survey.

As with insecticides the most appropriate stage for testing appears to be the adult. There is very little information available on the relationships between fumigant resistance in adults and that in the developmental stages, particularly the more tolerant eggs and pupae. The general suitability of the adult for toxicity testing, however, provides reasonable justification for its use in primary testing at least until there is evidence to the contrary.

Fumigants have characteristically steep dosage-mortality regression lines and small shifts in tolerance can be significant. Thus it would be desirable to measure gas concentrations associated with all tests because of the inherent problems in handling small quantities of gases and the comparatively high probability of error. There are difficulties in this as a routine procedure, however, and if it were to be a requirement of the test method, the usefulness of the test would be severely limited by restricting the number of laboratories that could conduct tests. The inclusion of reference strains (or species) in all tests provides a check on the problems associated with dosing. Survival of susceptible reference strains indicates low concentrations while complete kill of strains (or species) of slightly greater tolerance indicates abnormally high concentrations. An alternative approach is to use the susceptible strain only and to use three concentrations, one at the discriminating concentration, one at the approximate LD_{90} level and the other an equivalent level above the discriminating concentration.

In general terms, a fumigant test method could be developed on the basis of either exposure time or concentration as the fixed dosage variable. With both approaches the method developed should confine dosage errors to the dependent variable of dosage. Thus if the concentration is fixed, all exposures should be made in the same gas environment or alternatively if exposure time is fixed, all exposures should be of equal duration and within the same time period (e.g. same day). To satisfy these requirements in fixed concentration tests, somewhat specialised or large volume chambers are required providing suitable cage removal facilities so that cages may be removed at the required times with negligible loss of fumigant. Further, fixed-concentration tests almost invariably involve terminating some exposures outside normal working hours when a full series of dosages is required. On these grounds a fixed time test is considered most suitable for routine monitoring of fumigant resistance.

There have not been any test methods proposed in the literature for monitoring for fumigant resistance.

(F.1.3)   *Assessment of response*

(F.1.3a)   *Relationship between knockdown and mortality in tests with insecticides*

The choice of a response in a monitoring test is arbitrary provided that the response is correlated with the change in tolerance resulting from the resistances concerned. With conventional insecticides, the two criteria normally used are mortality and knockdown which is usually defined as inability to stand or walk normally and is usually associated with uncoordinated movements. These criteria, however, are not always applicable, for example in determining resistance to synthetic juvenile hormones (Dyte 1972), or where reproductive capacity is used as a measure of pesticidal effect from sublethal dosages of insecticides or sterilising dosages of ionising radiation or chemosterilants.

Lloyd and Williams (1973), when considering responses of *Sitophilus granarius* to pyrethrins, warn of errors in interpretation of resistance measurements based on knockdown estimates and subsequently (Lloyd *et al.* 1973), mechanical disturbance, changes in light intensity and photoperiod, and morphological and behavioural differences between strains have been cited as factors which may influence results in the usually short-term, knockdown-based tests. These factors operate through changes in activity and the resultant effects on metabolism and the toxication process. This applies particularly when insects are exposed to residual films of insecticide where pick-up is related to activity (*vide* Busvine 1968b). The effects of mechanical disturbance and light should be reduced to acceptable levels by experimental technique and then relevant morphological and behavioural differences that persist with standardised laboratory procedures may be regarded as potential contributing factors to resistance.

Frequently knockdown and mortality appear correlated as for example in *Aedes aegypti* where with DDT, resistance to knockdown and resistance to kill are controlled at the same gene locus (Hitchen and Wood 1974). Whereas knockdown is prerequisite to mortality, it does not follow that death is a necessary sequel to knockdown, or alternatively that should death follow knockdown, the causal mechanism of death is the same as that for knockdown. Conversely, separate genetic factors controlling mechanisms resulting in knockdown and in death (preceded by a separate knockdown) have been reported in *Musca domestica* from exposure to DDT (Milani 1963). Champ and Campbell-Brown (1970a,b) examined some relationships between knockdown and mortality in susceptible and malathion- and lindane-resistant strains of *Tribolium castaneum* and concluded that there was sufficient evidence to use knockdown responses estimated after 5 hr and 24 hr respectively for this species and these materials. These authors did not find any evidence to suggest that knockdown was a phenomenon separate from mortality and accepted knockdown as a measure of acute toxic effects of the insecticides before death or before detoxication could reverse the toxication process.

Thus if the relationship is established between knockdown and mortality for known resistances as was done with *Tribolium castaneum* by Champ and Campbell-Brown (1970b), rapid monitoring tests based on knockdown can be used. In the absence of resistant strains because they have not appeared or are otherwise unavailable, it seems expedient to monitor also for the acute toxic effects indicated by knockdown. It is an attendant disadvantage that the resistances encountered in the field may not correspond with those, if any, on which the test was based. Whereas it is desirable that follow-up tests be carried out using more appropriate criteria and possibly having greater analogy with field conditions, the opportunities for this are usually limited. Selection of a response criterion and time of assessment appropriate for the particular circumstances, the insect and the insecticide, should minimise this problem of encountering new types of resistance - a problem that does not seems to be of significance in practice.

(F.1.3b) *Exposure and response times*

The type and level of response is influenced by the time between the beginning of exposure and assessment of response. This period may involve continuous exposure of insects to the pesticide or a limited time of exposure followed by a recovery period usually of sufficient duration to allow responses to stabilise.

With insecticides, continuous exposure of insects to the deposit is very satisfactory, but some workers have found a fixed exposure time followed by a fixed recovery period to be reliable, for example with *Sitophilus oryzae* (Rajak et al. 1973) and *S. zeamais* (De Lima 1972). However, fixed exposure times with a recovery period involve additional handling of test insects. The usual object of test assays is to determine whether any changes have occurred in the susceptibility of the insects to the acute toxic action of the insecticide. Changes in toxic action may concern both rate of response and the final response level reached for a given concentration. Choice of an exposure time depends on the threshold times for absorption of insecticide and its action and for concentration-response relationships to be expressed. The quick-acting organophosphorus compounds give a clear and reproducible knockdown response after 5 hr whereas DDT requires an extended period for a clear pattern of response to emerge. The time taken for the full effects of poisoning to be manifested varies between species. Dyte and Daly (1969, 1970) found that with topical application 9-day mortality assessments were necessary for end-point estimates of the toxicity of DDT to *Tribolium castaneum*, *T. confusum* and *T. anaphe* and 20-day periods were necessary for *T. destructor* and *T. audax*. Such slow rates of intoxication are particularly associated with the chlorinated hydrocarbons and the less toxic organophosphorus and carbamate compounds. They become more pronounced with increasing dosage when resistance is present. Significantly, Dyte and Forster (1973b) using an impregnated-paper graded-concentration test based on 24 hr knockdown, were unable to demonstrate lindane resistance in one of their strains of *O. mercator* but when a 48 hr mortality response was used a x7.6 factor was obtained. A parallel discriminating concentration test, however, did indicate a suspect resistance (99% knockdown response) after 24 hr.

Exposure times in insecticide tests, which require excessively high or low concentrations should be avoided as availability to the insect is affected and the linear relationship between concentrations and mortality may be lost particularly when measuring responses of resistant strains. Thus lindane can be assayed without difficulty using a 5 hr exposure although a 24 hr exposure results in a more satisfactory LC_{50} (e.g. 0.172% 0.091% respectively) - longer exposure times do not give commensurately more information (Champ and Campbell-Brown 1970a).

The response of an insect to a fumigant is a function of the concentration of the fumigant and the time of exposure. In the simplest case the response is obtained with an exposure time that is inversely proportional to the concentration used. The relationship may be expressed as a concentration-time (CT) product for a specific level of response. Under given conditions particularly of temperature, the nature of the relationship obtained is a characteristic of the insect and the fumigant used. The general form of the relationship may be expressed as $C^n T = k$ which in the simplest situation is $CT = k$. With some fumigants it has been expedient to use the latter relationship, thus with methyl bromide this relationship is used in practice to vary concentration and time to obtain complete kill of insects. The information on the concentration-time relationship for methyl bromide is mainly available for practical fumigation levels and caution should be exercised in working outside this range as there are thresholds for both concentration and time below which the CT product does not apply (Whitney and Walkden 1961). Thus test methods for methyl bromide should be designed within this range.

With phosphine, additional caution is necessary. Short exposure times must be avoided as there is evidence to suggest that with high concentrations at short exposure periods, insects become narcotised which to some extent protects them from the lethal effects of phosphine and produces a characteristic flattening of the dosage-mortality line above 90% mortality (Ozer 1961, Qureshi *et al.* 1965, Lindgren and Vincent 1966b, Barker 1969, Vincent and Lindgren 1972a, Monro *et al.* 1972, Nakakita *et al.* 1974). Concentration appears to be the critical factor and with *T. castaneum*, 0.5 mg/l has been shown as the upper concentration limit of the linear dosage-mortality relationship at 25°C which remained valid from 0.005 to 0.5 mg/l with a range of exposure times from 1 to 100 hours (Winks 1973). Similarly the concentration-time relationship has been shown to hold up to 0.15 mg/l for measurements of median knockdown doses for *Callosobruchus chinensis* (Sato *et at.* 1973) and to 0.32 mg/l for diapausing larvae of *Ephestia elutella* at 20°C (Bell and Glanville 1973b). This restriction on the upper limit of concentration becomes particularly important when examining resistant strains and it is clear from data available that response estimates for resistant strains should be obtained primarily by increasing the exposure period if any meaningful comparison with susceptible strains is to be obtained. Although narcosis can have serious practical implications and could be a basis for insects acquiring resistance, there is insufficient information available on its occurrence to attempt to include the concept and implications of narcosis in the framework of present monitoring tests.

A post-treatment holding period is necessary with fumigants to allow full expression of responses which currently are measured as mortality. A simple, expedient, response criterion that is comparable with knockdown in insecticide tests and correlated with end-point mortality, has not been found for the fumigants. Data available for methyl bromide suggest that end-point mortalities are reached relatively quickly whereas with phosphine they are more prolonged. In tests with *T. castaneum* undertaken to establish the fumigant test methods used in the survey, little change in mortality from methyl bromide was recorded after 4 days whereas with phosphine, the time for a definitive end-point to be reached approximated to a maximum of 14 days from the end of exposure. Variation in response time characteristics of different species could have considerable bearing on the detection of resistance and on the significance of apparent resistance particularly in discriminating tests. Mortality end-points have yet to be determined for many species and in the interim a general post-treatment holding period of 14 days seems desirable.

(F.1.4) *Standardised test methods*

Resistance is a confirmed phenomenon in stored product insects and adequate methods for its detection and monitoring are, as stated previously, a prerequisite to interpreting control situations and maintaining a satisfactory level of control. For maximum utility, the methods must come into general use so that meaningful comparisons can be made between results of different workers.

The FAO Working Party of Experts on Resistance of Pests to Pesticides at their First Meeting (Rome, 1965) outlined the principles governing the detection and measurement of insecticide resistance in arthropods and advocated standardised test methods (Anon. 1967). Busvine (1968a,b) developed their general arguments and the principles were reiterated in an FAO Monograph on Pest Resistance to Pesticides in Agriculture (Anon. 1970b) and in the introductory paper of a series of Recommended Methods for the Detection and Measurement of Resistance of Agricultural Pests to Pesticides (Anon. 1969a) published in the FAO Plant Protection Bulletin. The *Tribolium* spp. were the only stored product species included in the first list of pests drawn up by the Working Party (Rome, September 1967) for which it was recommended that tests be devised (Anon. 1968).

Tentative methods for monitoring for resistance to malathion, lindane, carbaryl and synergised pyrethrins in adults of *Tribolium castaneum* appeared in October, 1970 based on exposure of test insects to insecticide-impregnated filter paper (Anon 1970a). The test was a synthesis of proven techniques that has been in extensive use in many laboratories for at least 25 years. The principles involved in and the arguments for this test were outlined by Champ and Campbell-Brown (1970a) on a basis of applicability to a range of the smaller stored product beetles. Subsequently, as a result of collaborative work between the Pest Infestation Control Laboratory England, and the Stored Grain Research Laboratory of the Commonwealth Scientific and Industrial Research Organization, Australia the test was extended to include monitoring for malathion, lindane, methyl bromide and phosphine resistance in *T. castaneum, T. confusum, Sitophilus oryzae, S. zeamais, S. granarius, Rhyzopertha dominica, Oryzaephilus surinamensis* and *O. mercator* as representing the more important pesticides and general pests of stored grain. Detail of the procedures for these tests have appeared in the FAO series of Recommended Test Methods (Anon. 1974a, 1975).

(F.2)   *Methods used in the survey*

An outline of the general procedures followed in the collection and processing of samples is given in Section C.

(F.2.1)   *Monitoring for insecticide resistance*

The methods used for monitoring for insecticide resistance and confirming its presence followed the Tentative Method for adults of some major beetle pests of stored cereals - FAO Method No. 15 (Anon. 1974a). The insecticides concerned were malathion and lindane and the species *Sitophilus oryzae, S. zeamais, S. granarius, Rhyzopertha dominica, Tribolium castaneum, T. confusum, Oryzaephilus surinamensis* and *O. mercator*. The toxicity testing involved continuous exposure of adult insects at 25°C to filter papers impregnated with solutions of insecticide in Risella 17 oil. The response criterion used was knockdown.

The results reported for the survey were those from tests to discriminate between susceptible and resistant insects. In many instances, tests were repeated to confirm doubtful results and where appropriate selective breeding from survivors unequivocally established resistances. Some response data from susceptible beetles and the exposure periods and discriminating concentrations used are given in Table F2.

TABLE F2:   SOME RESPONSE DATA FROM SUSCEPTIBLE BEETLES AND DISCRIMINATING CONCENTRATIONS FOR DETECTING RESISTANCE

	Exposure Period (hr)	Effective Concentrations			Discriminating Concentrations
		KD50	KD99	KD99.9	
MALATHION					
Sitophilus oryzae	6	0.89	1.38	1.60	1.5
Sitophilus zeamais	6	0.35	0.69	0.85	1.5
Sitophilus granarius	6	0.63	1.25	1.57	1.5
Rhyzopertha dominica	24	0.41	1.42	2.15	2.5
Tribolium castaneum	5	0.15	0.32	0.41	0.5
Tribolium confusum	6	0.24	0.43	0.52	0.5
Oryzaephilus surinamensis	5	0.28	0.62	0.80	1.0
Oryzaephilus mercator	5	0.10	0.66	1.20	1.0
LINDANE					
Sitophilus oryzae	24	0.04	0.10	0.14	0.2
Sitophilus zeamais	24	0.05	0.13	0.18	0.2
Sitophilus granarius	24	0.05	0.18	0.27	0.3
Rhyzopertha dominica	24	0.02	0.06	0.09	0.1
Tribolium castaneum	24	0.29	0.42	0.48	0.5
Tribolium confusum	24	0.11	0.31	0.43	0.5
Oryzaephilus surinamensis	24	0.65	1.49	1.95	2.0
Oryzaephilus mercator	24	0.44	1.80	2.80	2.0

(F.2.2) *Typing of malathion resistance*

There are a number of different mechanisms by which insects can degrade malathion and so develop resistance. Dyte and Rowlands (1968) demonstrated in *Tribolium castaneum* a carboxyesterase type of resistance that was specific to malathion-type compounds and could be suppressed diagnostically by the synergist triphenyl phosphate. Subsequently a second type was demonstrated that involved enhanced detoxication of phosphates (Dyte *et al.* 1970, 1973c) but could not be suppressed by triphenyl phosphate. This type has a wide range of cross-resistances (see Section F14). Thus in *T. castaneum* two general types of malathion resistance can be recognised separable by their responses to triphenyl phosphate. One type is specific to malathion whereas in the second type, various patterns of cross-resistance to other organophosphorus insecticides may occur. The FAO Recommended Test Method for *T. castaneum* includes a diagnostic test with triphenyl phosphate.

The other species of stored product Coleoptera have not been checked biochemically for the presence of the triphenyl phosphate sensitive carboxyesterase system in either susceptible or resistant insects. However, the widespread occurrence of this system in insects has been well established and both malathion-specific and nonspecific resistances could occur in all species. Resistances to malathion suppressible with triphenyl phosphate are known from *T. confusum* and *Rhyzopertha dominica* and in both there appears to be negligible resistance present to a range of other organophosphorus compounds. All species tested however, including *T. confusum* and *R. dominica* have shown other malathion resistances not suppressible with triphenyl phosphate, and which are associated with resistances to other organophosphorus compounds. Thus it seems probable that triphenyl phosphate may prove to be useful in distinguishing different types of malathion resistance in species other than *T. castaneum*. However, it is known that triphenyl phosphate and similar compounds may antagonise malathion in some species and may also be effective synergists for malathion in some susceptible strains. This creates problems in the interpretation of results obtained with triphenyl phosphate. These points are illustrated in Figure F1 in which graded concentration responses to malathion and malathion with triphenyl phosphate added, are presented for strains carrying malathion resistances representative of those known to be present in *Sitophilus oryzae*, *S. granarius*, *R. dominica*, *T. castaneum*, *T. confusum* and *Oryzaephilus surinamensis*. These data from single strains only, show the type of response that can be observed from use of triphenyl phosphate with susceptible strains. The level of the triphenyl phosphate suppressible carboxyesterase activity in susceptible strains would be expected to show normal strain variability, which when compounded with the variability in response to malathion alone from other causes, would reduce the sensitivity of a discriminating diagnostic test based on comparison of equitoxic doses of malathion alone and malathion + triphenyl phosphate, that were established for reference strains.

With these limitations it was considered expedient during the survey to include for all species, discriminating tests for triphenyl phosphate suppressible malathion resistance in the general search for new malathion resistances, and when considered in conjunction with cross-resistance data in Section F14, to allow some appreciation of the extent of the problems with the different types of malathion resistance.

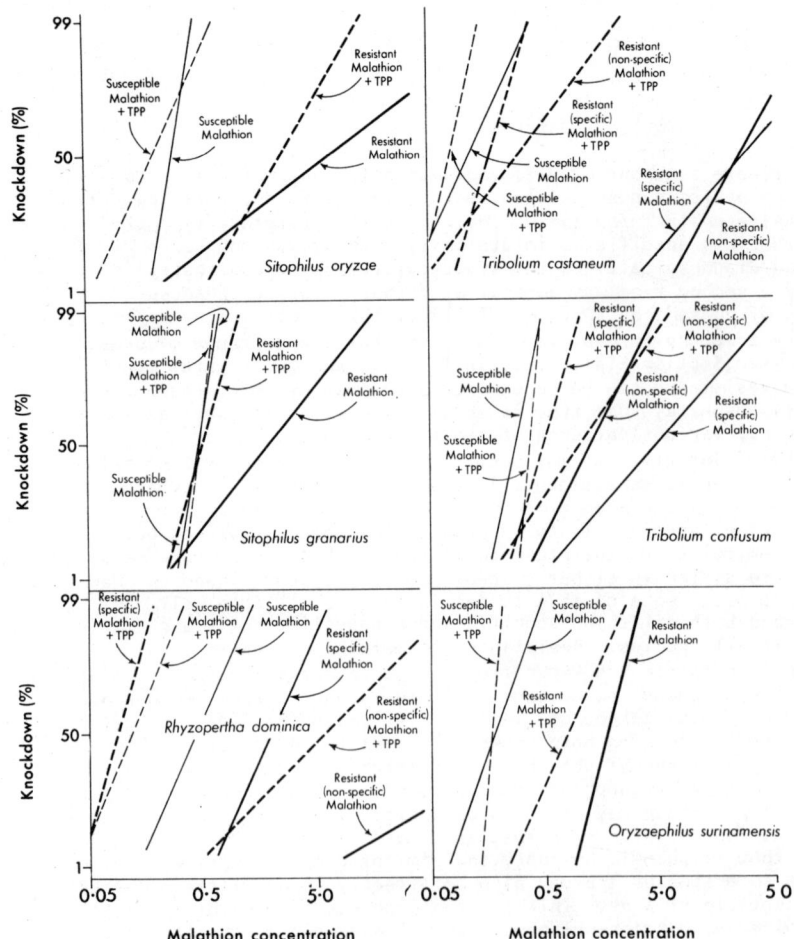

Fig. F1 - The responses of susceptible and malathion-resistant strains of *Sitophilus oryzae, S. granarius, Rhyzopertha dominica, Tribolium castaneum, T. confusum* and *Oryzaephilus surinamensis* to malathion and malathion synergised with triphenyl phosphate(1:10).

(F.2.3) *Monitoring for fumigant resistance*

        Monitoring tests for fumigant resistance were not available when the survey started. A method was synthetised from data and techniques provided by the Stored Grain Research Laboratory, C.S.I.R.O., Australia, the Pest Infestation Control Laboratory, Slough, England, the Canada Department of Agriculture Research Stations at London, Ontario and Winnipeg, The University of California, Riverside and the Stored Products Insects Research and Development Laboratory at Savannah, Georgia. The method essentially involved exposure of insects to fumigant in desiccators (~ 4-6 litres) at 25°C for the required dosing period followed by a 14 day holding period after which mortality was assessed. The fumigants used were methyl bromide and phosphine, and general handling and culture methods were similar to those used for the insecticide tests.

As with the insecticide resistance monitoring, the results reported for the survey are for discriminating tests only. Again in many instances, tests were repeated to confirm doubtful results and this was a routine procedure in most strains in which resistance was detected. Representative strains were examined further by selective breeding from survivors and graded concentration/time mortality tests. Some response data from susceptible beetles and the exposure periods and discriminating concentration-time products used are given in Table F3. Because of the difficulty in separating live *Sitophilus oryzae* and *S. zeamais*, a common discriminating dose was chosen to allow use in primary tests on field-collected material. Thus definitive identifications can be made after the test. The data on normal strain variability in response from which the discriminating concentrations were established, and the rationale of the test will be published in a separate research paper. Detail of the test method has appeared in the series of FAO Recommended Methods in the FAO Plant Protection Bulletin (Anon. 1975).

TABLE F3: SOME NORMAL SUSCEPTIBILITY DATA OBTAINED FOR METHYL BROMIDE AND PHOSPHINE TOGETHER WITH DISCRIMINATING CONCENTRATIONS

	LD_{50} (mg hr/l)	$LD_{99.9}$ (mg hr/l)	Discriminating Dose (mg hr/l)
METHYL BROMIDE			
Exposure period 5 hr			
Sitophilus oryzae	18	24	30
Sitophilus zeamais	16	27	30
Sitophilus granarius	26	38	45
Rhyzopertha dominica	20	37	35
Tribolium castaneum	42	59	60
Tribolium confusum	43	56	65
Oryzaephilus surinamensis	29	43	45
Oryzaephilus mercator	29	43	45
PHOSPHINE			
Exposure period 20 hr			
Sitophilus oryzae	0.22	0.78	0.80
Sitophilus zeamais	0.14	0.26	0.80
Sitophilus granarius	0.26	0.82	1.40
Rhyzopertha dominica	0.16	0.56	0.60
Tribolium castaneum	0.18	0.56	0.80
Tribolium confusum	0.22	0.58	1.00
Oryzaephilus surinamensis	0.24	0.72	1.00
Oryzaephilus mercator	0.22	0.68	1.00

(F.3) *Sitophilus oryzae*

Sitophilus oryzae is the most important stored grain pest in the world and thrives in the tropical and subtropical areas where pest potential is at its maximum level and chemical control measures are almost mandatory for successful grain storage. There has been widespread and intensive use of residual pesticides in these control measures commencing with DDT in the post-war period, followed by lindane and the pyrethrins and subsequently malathion. The extent of use of these materials made development of resistance very probable.

(F.3.1) *Recorded insecticide resistances*

Lindane resistance in *S. oryzae* appears widespread though there are previous records only from Trinidad, Kenya, Tanzania, Bangladesh, Australia, U.S.A., China (Taiwan) and England. DDT resistance does not appear to be common and malathion resistance has been reported from China (Taiwan), England and India only. Fumigant resistance has not been detected previously.

Lindane resistance in *S. oryzae* was first reported by Davey and Amos (1960) from Trinidad where living insects were taken from experimental samples of paddy containing 18 p.p.m. lindane. Tests on this strain at the Pest Infestation Control Laboratory showed 4% survival at 16 p.p.m. whereas a lindane level of < 1 p.p.m. was expected to control susceptible strains. Subsequently Parkin *et al.* (1962a) using a dusted grain assay and a 72 hr mortality obtained an approximate resistance factor of x17 and reported a further lindane resistant strain of approximately x12 resistance from a collection from maize in Kenya. These levels were confirmed later as x13.2 and x9.5 respectively and records of x2 and x2.6 were added from material from sorghum in Tanzania (Parkin 1965, 1967). Using the dusted grain technique again, Parkin and Forster (1966b, Parkin 1969) also reported x6 to x8 lindane resistances from three strains collected from wheat in stores in East Pakistan where lindane was used for grain protection. They found also that these strains had normal susceptibility to malathion as had the Trindidad strain (Parkin and Forster 1967a).

Parallel investigations by Champ and Cribb (1963, 1965a; McDougall 1964, Sloan 1967) showed lindane resistance in *S. oryzae* from sorghum in Queensland where, by local regulation since 1951, sorghum seed had been required to be treated after harvest with 15.6 p.p.m. lindane as a dust. Using an impregnated filter paper technique and a knockdown response they reported resistance factors in field-collected strains up to x240, and related a strain showing a x77 resistance by this method to a resistance factor of x50 using a dusted grain technique and end-point mortality (0.8 p.p.m. for effective control of the susceptible, 39 p.p.m. for resistant). Parkin and Forster (1965b, Parkin 1965) rated the factor for this latter strain as x72 using a 72 hr mortality response. The strain was taken from a seed cleaning and dressing facility and subsequent 70 further strains, directly or indirectly concerned with the sorghum seed trade were shown to be lindane resistant (Champ 1967). Many of these strains were also found to be DDT-resistant - levels in excess of x10 were common but the more resistant strains did not respond to the dosing technique (unpublished data, Champ and Cribb 1965a, Champ 1967, Sloan 1967). DDT was used as a seed protectant in this trade in the immediate post-war period (1946 to 1951) after which it was replaced with the more efficacious lindane except for occasional use as a fabric spray in storage premises. There was no suggestion of DDT resistance at the time of the change (Champ 1967). Parkin and Forster (1967a) were unable to detect any DDT resistance in a lindane-resistant strain from Trinidad. Cichy (1971) using filter papers impregnated with water emulsions of insecticide and 72 hr mortalities, selected strains showing x40 resistance to DDT and also from separate selections, strains showing x6 to pyrethrins+piperonyl butoxide and x434 to demeton-methyl.

Champ and Cribb (1965a, Sloan 1967) demonstrated very high
dieldrin resistance (x600) in a lindane-resistant strain of *S. oryzae*
presumably a cross-resistance as well as a low level resistance to
fenchlorphos (x4) but found that responses to malathion and diazinon
were normal. Neither of the two former materials had been used in the
field in Australia. Parkin and Forster (1967a, 1968b) obtained a
similar result for this strain with malathion and for the lindane
resistant strain from Trinidad, and were unable to detect any change to
bromophos in either strain. Newsom (1967) without supporting data also
reported lindane resistance from the U.S.A. while in Egypt, Toppozada *et al.*
(1969) using treated grain, could not detect resistance to either malathion
or lindane. In China (Taiwan), however, Lin (1973) also using a dusted
grain technique recorded lindane resistance from 37 storehouses of the
various agricultural societies. He also studied the development of lindane
and malathion resistance in treated coarse rice and demonstrated control of
a lindane-resistant strain with malathion and a malathion- and synergised
pyrethrins-resistant strain with phoxim. In England, Dyte *et al.* (1973a)
using an impregnated paper method confirmed both lindane and malathion
resistance in a strain from Southampton and in a survey of the incidence
of resistance in stored product insects in the United Kingdom in 1973,
reported 12 of 15 strains resistant to lindane and 1 of 16 strains
resistant to malathion (Dyte *et al.* 1975), and in Australia Greening *et al.*
(1975) using similar techniques diagnosed after May 1970, malathion
resistance in 19 centres throughout New South Wales involving 52% of samples
collected and two export terminal elevators, country bulk grain storages,
a grain farm, provender mills and stores, and flour mills. The only other
reports of malathion resistance are from India. Rajak *et al.* (1973), in a
survey of malathion resistance in storage insects in India, recorded levels
of x6 from Kanpur and x2.3 from Hapur in Utter Pradesh. They used a filter
paper technique similar to the FAO *Tribolium* method but substituted
linseed oil for Risella oil and used a 6 day end-point mortality. Bansode
(1974) selected a strain showing a x49 resistance to malathion which was
not suppressible with triphenyl phosphate and confirmed this with cross-
resistance studies.

Champ (1968) suggested a test method for detecting insecticide
resistance in *S. oryzae* based on exposure of test insects to impregnated
filter paper as in the FAO methods and listed discriminating concentration
levels for DDT, lindane and malathion. Though malathion resistance was
not recorded, a x3 increase in tolerance to malathion was reported
associated with the presence of DDT and lindane resistances which had to
be taken into account in setting the malathion discriminating concentration.

(F.3.2) *Recorded fumigant resistances*

The use of fumigants for control of *Sitophilus oryzae* has also
been very extensive. Information on fumigant resistance however, is almost
completely lacking except for the report by Lindgren and Vincent (1965,
1966a) of a laboratory-selected strain showing a x2.5 resistance to hydrogen
cyanide and a very slight increase in tolerance to ethylene dibromide and
methyl bromide after 36 generations of selection with hydrogen cyanide.
Kamel and Fam (1962) from Egypt obtained a x1.4 change in tolerance at the
LC_{50} level from four generations of selection with carbon disulphide, which
was accompanied by only a x1.1 change at the $LC_{99.9}$ level suggesting they
had merely eliminated the more susceptible individuals in the population.

(F.3.3) *Survey data*

Results from discriminating tests for malathion, lindane, methyl
bromide and phosphine resistance in *Sitophilus oryzae* are given in Table F4
and Figures F2-F5 together with a summary of the occurrence in Table F5.
The distribution maps (Figs. F2-F5) also show the classification of
importance currently attributed to *S. oryzae* in the different countries.

(F.3.4)   *Comments*

(F.3.4a)   *Malathion resistance*

Malathion resistance was detected in 33 of the 257 strains of *Sitophilus oryzae* tested. It was known previously from four countries, England (presumably an importation), Australia, India and China (Taiwan), all of which are very recent (1974) records. Survey records include India (2 additional states), England, Australia (5 states) and a further seven countries - New Guinea, Central African Republic, Mozambique, Nepal, Pakistan, Colombia and Peru, giving a current total of 11 countries. The four strains collected from China (Taiwan) during the survey were all susceptible.

The previous records from India were from Uttar Pradesh only and now the states of Maharashtra and the Punjab are implicated as well as the neighbouring countries of Nepal and Pakistan, indicating a much more general distribution in the Indian region. Malathion has been used widely in India for a comparatively short time only and the recommended application rates have been considerably less than the levels normally regarded as necessary for adequate control and residual activity - 0.2% (3 l/100m^2) on introduction in the mid-1960's increased because of a diminishing level of control to 0.5% currently, compared with 1.5% elsewhere. Use of these suboptimal dose rates would predispose to resistance development.

The highest frequency of malathion resistance in samples was from Australia though these results are not directly comparable with data from samples from elsewhere for two reasons. Firstly, the Australian strains were tested in Australia using a discriminating concentration of 1.0% malathion and a 7 hr exposure period which would detect smaller increases in tolerance than would the discriminating concentration of 1.5% with a 6 hr period used in the survey. Secondly, most samples were collected from infestations which had developed in grain storages where malathion was used as a routine admixture and surface treatment, whereas in the survey samples were collected at random without regard to previous use of malathion. Nevertheless a widespread problem in Australia is evident. In Australia, all wheat and large quantities of other grains are admixed with malathion at intake to central storage over a wide range of climatic and storage conditions. Approximately 6-7 years elapsed between the time when malathion was introduced on this scale in Australia and when resistance was detected in *S. oryzae* in the Newcastle grain terminal in 1970, although malathion had been in more limited use during the previous 4-5 years. Subsequently Champ (unpublished data) has recorded malathion resistance from all states of Australia. Resistance levels, though of economic significance, were not high (to x8) except in Western Australia where x24 and x68 increases in tolerance were recorded in strains from wooden wheat storage bins associated with a flour mill at Guildford. These resistance levels were determined using procedures similar to those of the FAO Test Methods - complementary assessments after 3 days exposure in treated grain gave a resistance factor at the LC50 of x5.9 (at LC50, 10 days -x8; LC99.9, 3 and 10 days -x22) for one of the more tolerant strains (x24) associated with minimum nominal LC99.9's of 83 p.p.m. and 31 p.p.m. malathion based on 3 and 10 day mortalities respectively and a concentration of >20 p.p.m. required to prevent reproduction. There has been a strong correlation between the outbreaks of malathion-resistant *S. oryzae* in Australia and use of unsatisfactory storages and poor hygiene.

The occurrence of malathion resistance in the Central African Republic where insecticides are used rarely, probably has resulted from importation of resistant strains. The resistance picture in the Republic represents a particular type of situation where considerable amounts of grain are imported sometimes in the form of aid from the developed countries. The risk that infestation is present is higher in such grain and this is compounded by the favourable conditions for cross-infestation and for development of infestations during the long period of transport and storage associated with unloading from ships at Point Noire and river transport from Brazzaville to Bangui. Thus one strain with malathion resistance was collected from wheat ex France that was being unloaded at a flour mill in Bangui and had been discharged at Point Noire six months previously. It will be shown as the various pest species are discussed, that a wider range of resistances have appeared in the Central African Republic than in many other countries.

Malathion has been used occasionally since 1964 in Colombia and Peru for surface treatment of bag stacks, grain bulks and associated building fabric. In Mozambique, it has had moderate use for similar purposes but the situations in these countries seem little different from that prevailing in many countries throughout the world and apart from the limitation of the small number of samples examined, there is not any apparent reason why malathion resistance has appeared in these countries particularly.

Malathion resistance is now a significant though still limited problem of increasing importance in *S. oryzae*. When compared with other species such as *Tribolium castaneum*, it does not appear, however, to be particularly prone to development of resistance to malathion. Nevertheless the resistances exist and though there are indications that in some strains the resistance may be suppressed with triphenyl phosphate, the presence of such strains requires confirmation and the resistance can be regarded generally as of a non-specific type involving cross-resistance to a range of other organophosphorus compounds.

(F.3.4b) *Lindane resistance*

Lindane resistance in *Sitophilus oryzae* is widespread occurring in 53 of the 58 countries sampled and 75% of the 235 strains tested compared with records from seven countries previously. Lindane has been used extensively throughout the world for the past 25 years and it seems probable that lindane resistance appeared as an economic problem at least 15 years ago. The movement of insects in international trade since then would be expected to have distributed the resistance extensively as shown in the survey. Susceptible strains are becoming increasingly difficult to locate and now occur mostly in cooler areas such as Poland and southern Australia.

The addition of lindane to food grains is not recommended or practiced to the extent today that it was some years ago having been replaced in many areas with the less toxic malathion. This has resulted from the increasing incidence of resistance and toxicological considerations. The long residual life of lindane has contributed considerably to its value as a seed protectant but the general occurrence of lindane resistance now has vitiated this use and it seems unlikely that there will be another comparable material.

(F.3.4c) *Methyl bromide resistance*

Methyl bromide resistance has not been conclusively demonstrated in this species in the laboratory or in samples from the field. The single diagnosis of suspected resistance was from a sample from Mombasa and this occurrence is not outside the probability of a normal chance occurrence. Nevertheless with the lack of strain variability that characterises methyl bromide responses, the strain requires further examination.

(F.3.4d) *Phosphine resistance*

The variability in response between strains of *Sitophilus oryzae* appears greater with phosphine than with methyl bromide. The discriminating dosage for resistance to phosphine has thus been set correspondingly higher and while sensitivity has been reduced there seems little doubt that the resistances diagnosed in the survey are real changes in tolerance. These have concerned eight strains from six countries and show clearest expression in India where approximate resistance factors of x2.5 x2 and x1.5 respectively, have been recorded from Karnal in the southern Punjab, and Bombay and nearby Borivli in Maharashtra state. The resistances appear to have developed in association with extensive use of phosphine. The strain from Karnal was collected from paddy residues in a rice mill and the strains from the Bombay area were collected from wheat, rice and maize in godowns fumigated 3 or 4 times a year at the rate of 2g phosphine/ton. The godowns at Borivli were associated with silos which were fumigated once a year, supplemented with surface fumigation, at the rate of 1g phosphine/ton. The total annual usage on the site approximated 0.5 metric tons phosphine equivalent. Similarly the strain from Guyana showing phosphine resistance was collected from residues in an empty bin of wheat in a flour mill where phosphine was used in the empty bins often for 24 hr total exposures only. The strain from Kenya which was the only strain also showing any change in tolerance to methyl bromide, again was collected from a flour mill where in this instance phosphine and methyl bromide were used fairly regularly. Of the remaining occurrences of phosphine resistance, the strain from Portugal was found in a bag store with a history of phosphine usage, whereas that from the Yemen was from imported grain of indeterminate origin in a bag storage where, as was general practice, chemical controls were not used. Resistance in this latter strain and in the strain from Indonesia cannot be regarded as confirmed occurrences at this stage.

Though limited in number, the known occurrences of phosphine resistance indicate an emerging problem with this material. Dosages used in practice should be considerably in excess of the minimum effective dosages needed for *S. oryzae* and these together with the long exposure periods that are often used, reduce the potential for resistance. Factors in excess of the maximum of x2.5 currently recorded, will be necessary for a general breakdown of control but even at levels such as x2.5, a reduction in control could result under the conditions of poor distribution of gas that would be associated with the sub-standard fumigations with this material that regrettably appear to be common.

TABLE F4: RESPONSES OF REPRESENTATIVE WORLD STRAINS OF *SITOPHILUS ORYZAE* TO MALATHION, MALATHION + TPP, LINDANE, METHYL BROMIDE AND PHOSPHINE IN DISCRIMINATING TESTS FOR RESISTANCE

Locality	Mal	M+T	Lin	MeB	PH_3	Locality	Mal	M+T	Lin	MeB	PH_3
NORTH AMERICA						**SOUTH AMERICA** (Cont'd)					
MEXICO						ARGENTINA (Cont'd)					
Mexico City	100	-	52	-	-	Ent. Rios - Villaguay	100	-	6	100	100
" "	-	-	-	100	100	Sta Fe - Galvez	100	-	28	-	-
						San Francisco	100	-	43	-	-
U.S.A.						**BRAZIL**					
Cal. - Riverside	100	-	100	100	100	Goias - Anapolis	100	-	64	-	-
Kans.	100	-	100	-	-	"	-	-	-	100	100
Tex. - Beaumont	100	-	42	-	-	R.G.do Sul-Livramento	100	-	1	-	-
Kelly	100	-	100	-	-	Porto Alegre	100	-	11	-	-
						S.P. - Sao Paulo	100	-	100	100	100
CENTRAL AMERICA AND CARIBBEAN											
						PERU					
COLOMBIA						Callao	98	100	26	-	-
Barranquilla	100	-	100	-	-	"	89	92	9	100	100
"	83	97	99	100	100	"	100	-	66	100	100
						"	100	-	21	100	100
EL SALVADOR						Chiclayo	83	97	7	-	-
San Martin	100	-	8	100	100	"	-	-	-	100	100
FRENCH GUIANA						Lambayeque	83	91	30	100	100
Cayenne	100	-	60	100	100	**URUGUAY**					
GUATEMALA						Rivera	100	-	0	100	100
Guatemala City	100	-	-	-	-						
" "	100	-	98	100	100	**EUROPE AND NORTH AFRICA**					
" "	100	-	-	100	100						
" "	100	-	-	-	-	**ALGERIA**					
GUYANA						Algiers	100	-	3	100	100
Georgetown	-	-	-	100	91	"	100	-	41	100	100
						Blida	100	-	1	-	-
JAMAICA						"	100	-	59	100	100
Kingston	-	-	-	100	100	Dar-es-Abeida	100	-	30	-	-
NICARAGUA						" "	99	99	2	100	100
Samulali	100	-	98	-	-						
"	100	-	0	100	100	**BULGARIA**					
						Sofia	100	-	88	-	-
TRINIDAD						**EGYPT**					
Port of Spain	100	-	82	-	-	Behira	100	-	0	100	100
" "	100	-	65	100	100	"	-	-	-	100	100
" "	100	-	46	100	-	Cairo	100	-	72	-	-
" "	100	-	79	100	100						
" "	100	-	83	-	-	**ENGLAND**					
						Camb. - Cambridge	100	-	80	-	-
SOUTH AMERICA						Devon. - Barnstaple	100	-	37	100	99
ARGENTINA						" - Bideford	100	-	-	-	-
B.A. - Buenos Aires	100	-	5	-	-	Essex - Braintree	100	-	100	-	-
" "	99	-	6	-	-	Glos. - Avonmouth	98	100	82	100	100
" "	100	-	10	-	-	Gloucester	100	-	54	100	100
" "	100	-	24	-	-	Saul	100	-	2	100	100
Grumbein	100	-	32	100	100	Lancs. - Glazebury	100	-	75	100	100
La Plata	100	-	12	100	100	"	100	-	69	100	100
"	100	-	0	-	-	Lincs. - Boston	100	-	84	-	-
Mar del Plata	100	-	45	100	100	Northants. -					
Cord. - Isla Verde	100	-	1	100	100	Wellingborough	100	-	95	100	100

TABLE F4
(Cont'd)

S.oryzae (ii)

Locality	Mal	M+T	Lin	MeB	PH$_3$	Locality	Mal	M+T	Lin	MeB	PH$_3$
EUROPE AND NORTH AFRICA (Cont'd)						**AFRICA-SOUTH OF SAHARA**					
ENGLAND (Cont'd)						**BENIN**					
Northants. -						Cotonou	100	-	-	-	-
Wellingborough	100	-	82	100	100	**CENT. AFRICAN REP.**					
Worcs. - Defford	100	-	100	-	-	Bangui	91	100	29	-	-
GREECE						"	88	93	34	100	100
Lamja	100	-	1	-	-	"	99	100	48	100	100
Piraeus	100	-	16	-	-	"	100	-	-	100	100
"	100	-	0	100	100	**ETHIOPIA**					
"	100	-	-	100	100	Addis Ababa	100	-	100	100	100
LIBYA						" "	99	100	100	100	100
El-Marg	100	-	99	100	100	" "	100	-	100	-	-
Tobruk	100	-	69	-	-	" "	100	-	100	-	-
Tripoli	100	-	2	-	-	" "	100	-	-	-	-
"	100	-	23	-	-	" "	100	-	100	-	-
MOROCCO						" "	100	-	-	100	100
Meknes	100	-	8	-	-	Calitti	100	-	100	-	-
Rabat	100	-	-	-	-	Dabi	100	-	100	100	100
"	100	-	22	-	-	"	100	-	100	-	-
"	100	-	0	-	-	Dengago	100	-	57	100	100
"	100	-	26	-	-	Dire Dawa	100	-	89	100	100
"	100	-	11	100	100	Holetta	100	-	4	100	-
Sale	100	-	0	-	-	Nazret	100	-	100	100	100
"	100	-	0	-	-	"	100	-	78	-	-
"	100	-	14	100	100	"	100	-	98	-	-
POLAND						"	100	-	100	-	-
Gniezno	100	-	100	100	100	Saka	100	-	100	100	99
Poznan	100	-	100	100	-	Sebeta	100	-	100	-	-
PORTUGAL						Serbo	100	-	100	100	100
Ceiras	100	-	100	100	82	Soddu	100	-	0	-	-
Lisbon	100	-	98	100	100	Yenda	100	-	100	100	100
"	100	-	48	100	100	**GAMBIA**					
Queluz	-	-	-	99	100	Yundum	100	-	0	-	-
ROMANIA						**KENYA**					
Budesti	100	-	59	-	-	Bungoma	100	-	69	-	-
Calarasi	100	-	26	100	100	Mombasa	100	-	0	98	97
"	100	-	5	-	-	Nairobi	100	-	2	100	100
SCOTLAND						"	100	-	2	-	-
Edinburgh	100	-	38	-	-	"	100	-	0	100	100
Glasgow	100	-	15	100	100	Nakuru	100	-	5	-	-
New Abbey	100	-	100	100	100	Nunyuki	100	-	2	100	100
SPAIN						**MALAWI**					
Baena	-	-	-	100	99	Blantyre	100	-	65	-	-
YUGOSLAVIA						Limbe	-	-	-	100	100
Belgrade	100	-	99	100	100	Salina	-	-	-	-	100
Darda	100	-	-	-	-	**MOZAMBIQUE**					
"	100	-	-	-	-	Lourenco Marques	100	-	99	-	-
						" "	80	97	7	100	100
						" "	88	89	10	100	100

TABLE F4
(Cont'd)

S.oryzae (iii)

Locality	Mal	M+T	Lin	MeB	PH$_3$	Locality	Mal	M+T	Lin	MeB	PH$_3$
AFRICA-SOUTH OF SAHARA (Cont'd)						**WEST ASIA (Cont'd)**					
NIGERIA						**IRAN**					
Lagos	100	-	61	-	-	Teheran	100	-	100	100	-
"	100	-	100	-	-	**IRAQ**					
"	100	-	14	100	100	Abu-Ghraib	100	-	8	-	-
"	100	-	96	-	-	**ISRAEL**					
"	100	-	-	-	-	Kiryat - Gat	100	-	6	-	-
SENEGAL						Tel Aviv	100	-	3	100	100
Sefa	100	-	-	-	-	**NEPAL**					
"	100	-	-	-	-	Inner Terai	100	-	33	100	100
"	100	-	0	-	-	" "	100	-	2	100	100
SOMALIA						Lothar	100	-	100	100	100
Afgoi	100	-	-	100	100	Ranpur	56	100	2	99	100
"	100	-	26	100	100	**PAKISTAN**					
Mogadishu	100	-	25	-	-	Karachi	100	-	50	100	-
"	100	-	25	-	-	Lahore	96	98	64	100	100
"	100	-	56	100	-	"	-	-	20	-	-
"	100	-	31	100	100	**SAUDI ARABIA**					
"	100	-	-	100	100	Jeddah	100	-	78	100	100
"	100	-	24	100	100	"	100	-	78	100	99
SOUTH AFRICA						"	100	-	60	100	100
Pretoria	100	-	100	100	100	**SYRIA**					
"	100	-	12	100	100	Damascus	100	-	90	-	-
Settlers	100	-	87	100	100	"	100	-	100	-	-
SWAZILAND						**TURKEY**					
Luvenco	100	-	42	100	100	Sakarya	100	-	83	100	100
Malkerns	100	-	31	100	100	**YEMEN ARAB REP.**					
ZAMBIA						Hodeida	100	-	97	100	98
Chadiza	100	-	-	100	100	**EAST ASIA**					
WEST ASIA						**BURMA**					
AFGHANISTAN						Myaing	-	-	-	100	100
Kabul	100	-	85	100	-	Rangoon	100	-	69	-	-
CYPRUS						Sagaing	-	-	-	100	100
Athienou	100	-	61	-	-	Tatkon	100	-	1	-	-
Louroutzina	100	-	81	-	-	Thazi	-	-	-	100	100
"	100	-	44	-	-	**CHINA**					
Nicosia	100	-	-	-	-	Taiwan - Keelung	100	-	60	-	-
Pyrga	100	-	40	-	-	NanChiang	100	-	26	100	100
INDIA						Taipeh	100	-	7	100	100
Maharashtra-Bombay	100	-	38	100	90	TaoYuan	100	-	-	100	100
"	99	99	19	-	-						
"	100	-	37	100	100	**INDONESIA**					
"	100	-	34	100	100	Djakarta	100	-	62	-	-
Borivli	98	100	42	100	91	"	99	100	90	-	-
"	100	-	69	-	-	Parekedin	100	-	58	100	97
Punjab-Karnal	90	95	8	100	100	Petoju Jaharta	100	-	97	-	-
"	100	-	32	100	84	**JAPAN**					
U.P. - Delhi	100	-	32	100	100	Chiba	100	-	-	100	100
Faridabad	100	-	46	100	-	Yokohama	100	-	88	100	100
Mathura	100	-	11	100	100						
Naraina	100	-	18	100	100						

TABLE F4
(Cont'd)

S.oryzae (iv)

Locality	Mal	M+T	Lin	MeB	PH$_3$	Locality		Mal	M+T	Lin	MeB	PH$_3$
EAST ASIA (Cont'd)						AUSTRALASIA AND SOUTH PACIFIC (Cont'd)						
JAPAN (Cont'd)												
Yokohama	100	-	99	100	100	AUSTRALIA* (Cont'd)						
"	100	-	-	100	100							
"	-	-	-	100	100	NSW -	Homebush	90	-	72	-	-
KOREA REP.							Newcastle	47	37	51	100	100
Sosa	100	-	100	100	100		Peak Hill	100	-	100	-	-
PHILIPPINES							Yerong Creek	100	-	100	-	-
Laguna	100	-	38	100	100	Qd -	Brisbane	99	-	4	-	-
Pagsanjan	100	-	90	100	99		Dalby	92	-	79	-	-
"	100	-	100	-	-		Gindi	100	-	0	100	100
Pila	100	-	100	100	100		Jandowae	100	-	0	-	-
Zamboanga City	100	-	-	-	-		Wallumbilla	100	-	39	-	-
SINGAPORE							Warra	97	-	21	-	-
"	100	-	81	100	100		Yarranlea	100	-	80	-	-
"	100	-	-	100	100	S.A.-	Eyre Peninsula	75	-	100	-	-
"	-	-	-	100	100		Mile End	25	-	100	100	99
THAILAND							Moonta	99	-	100	-	-
Bangkok	100	-	100	-	-		Port Adelaide	70	-	100	-	-
Khon Kaen	-	-	-	-	100		" "	100	-	100	-	-
Rangsit	100	-	6	-	-		Salisbury	80	-	100	-	-
Thonburi	100	-	13	100	100		Strathalbyn	100	-	100	-	-
"	100	-	19	100	100	Tas.-	Launceston	100	-	100	-	-
U.S.S.R.						Vic.-	Echuca	90	-	64	-	-
							"	49	-	100	-	-
Georgiyevsk	-	-	88	-	-		"	46	-	100	100	100
Palazsaga	-	-	100	-	-		Melbourne	99	-	100	-	-
							Natya	100	-	100	-	-
AUSTRALASIA AND SOUTH PACIFIC							Newport	98	-	100	-	-
							Piangil	100	-	100	-	-
AUSTRALIA*						W.A.-	Cottesloe	0	-	100	-	-
							Geraldton	63	-	88	-	-
NSW - Barellan	97	-	100	-	-		Guildford	2	-	51	100	100
Barraba	100	-	38	-	-		Narrogin	100	-	100	-	-
Emerald Hill	100	-	13	-	-		North Fremantle	54	-	18	100	100
							Northam	100	-	13	-	-
							West Perth	33	-	97	-	-
						NEW GUINEA*						
						Bainyik		100	-	10	-	-
						Port Moresby		100	-	30	-	-
						Wewak		78	-	-	-	-

*These strains were tested in Australia using a discriminating concentration of 1.0% malathion and a 7 hr exposure period which would detect smaller increases in tolerance than would the discriminating concentration of 1.5% with a 6 hr exposure period used in the survey.

TABLE F5: SUMMARY OF WORLD OCCURRENCE OF RESISTANCE IN *SITOPHILUS ORYZAE* TO MALATHION, LINDANE, METHYL BROMIDE AND PHOSPHINE

	North America	Central America	South America	Europe,N.Afr.,USSR	Africa-S.of Sahara	Western Asia	Eastern Asia	Australasia	Totals	Totals (as %)
MALATHION										
No. countries sampled	2	6	4	12	13	12	8	2	59	-
No. with resistance	0	1	1	1	2	3	0	2	10	17%
No. of strains tested	5	15	23	53	61	35	26	39	257	-
No. resistant	0	1	4	1	4	4	0	19	33	13%
MALATHION + TPP										
No. countries sampled	2	6	4	12	13	12	8	-	57	-
No. with resistance	0	1	1	0	2	2	0	-	6	11%
No. of strains tested	5	15	23	53	61	35	26	-	218	-
No. resistant	0	1	3	0	3	2	0	-	9	4%
LINDANE										
No. countries sampled	2	6	4	13	11	12	8	2	58	-
No. with resistance	2	5	4	11	11	11	7	2	53	91%
No. of strains tested	5	12	23	50	51	35	21	38	235	-
No. resistant	2	10	22	41	34	32	16	19	176	75%
METHYL BROMIDE										
No. countries sampled	2	8	4	11	10	9	8	1	53	-
No. with resistance	0	0	0	0	1	0	0	0	1	2%
No. of strains tested	2	11	12	30	34	24	21	6	140	-
No. resistant	0	0	0	0	1	0	0	0	1	1%
PHOSPHINE										
No. countries sampled	2	8	4	11	10	7	8	1	51	-
No. with resistance	0	1	0	1	1	2	1	0	6	12%
No. of strains tested	2	10	13	29	33	20	22	6	135	-
No. resistant	0	1	0	1	1	4	1	0	8	6%

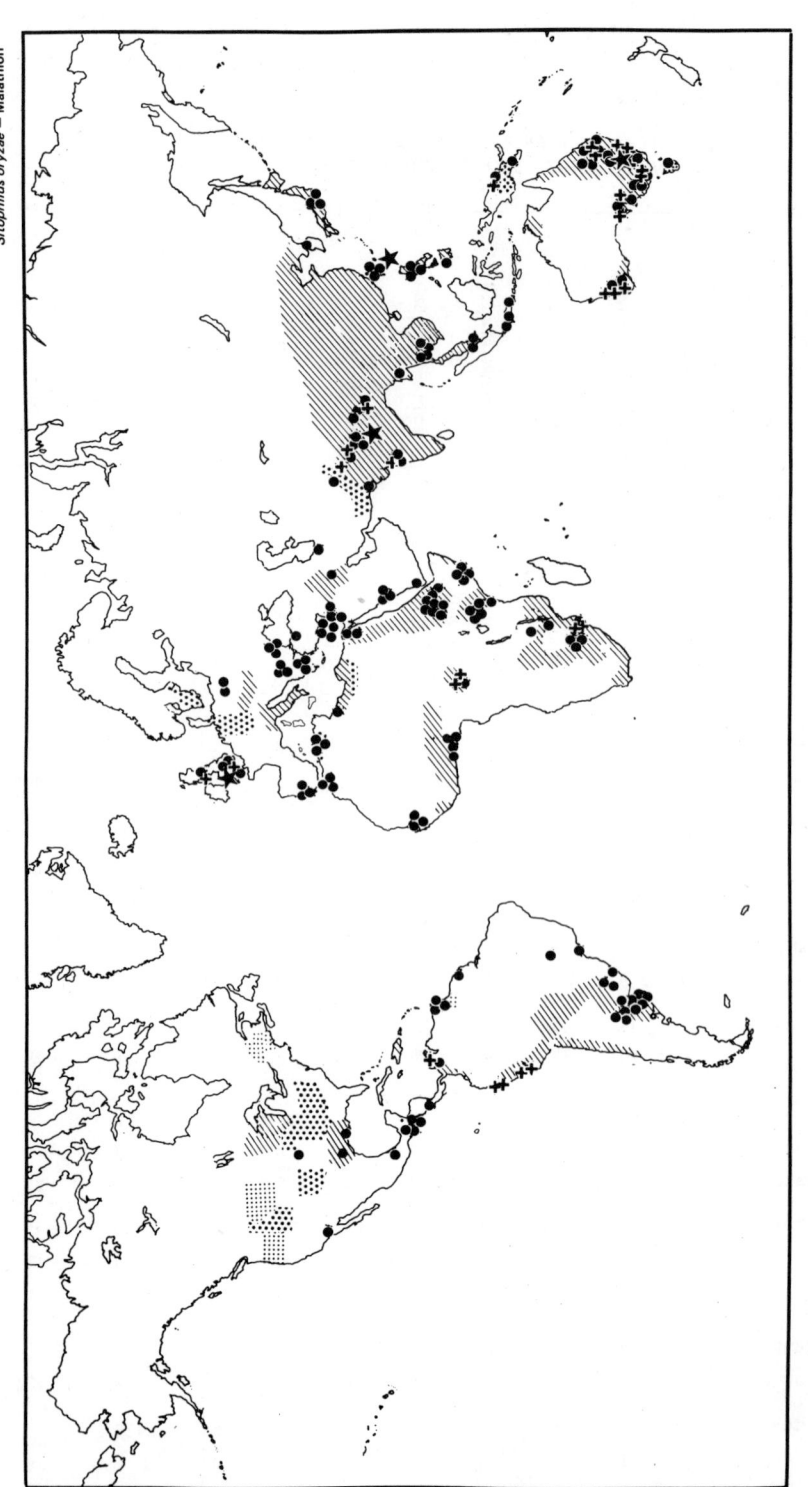

Fig. F 2 — The occurrence of malathion resistance in *Sitophilus oryzae*

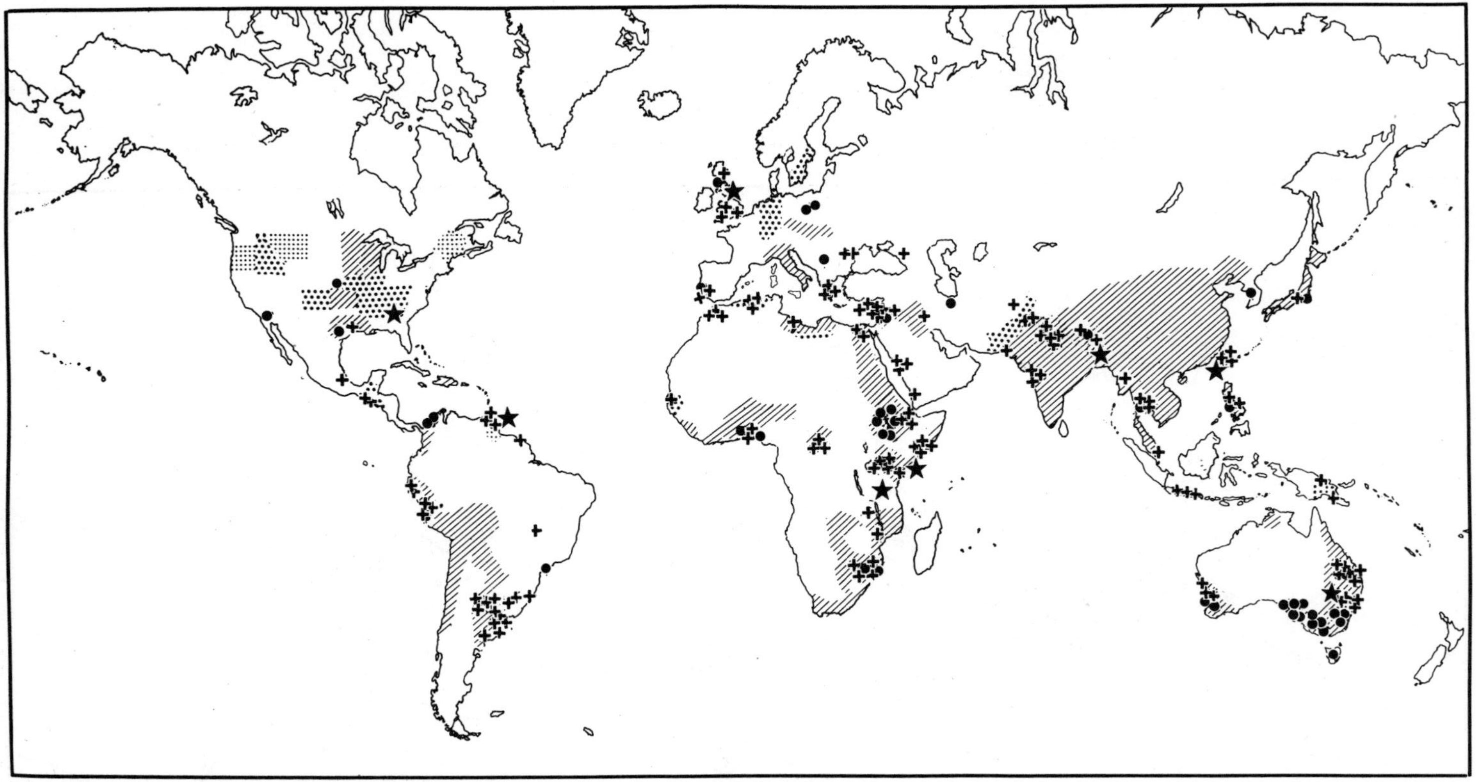

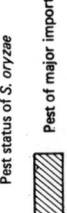

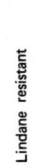

Fig. F3 – The occurrence of lindane resistance in *Sitophilus oryzae*

- ✚ Lindane resistant
- ● Lindane susceptible
- ★ Previously recorded resistance

Pest status of *S. oryzae*

- Pest of major importance
- Pest of moderate importance
- Pest of minor importance

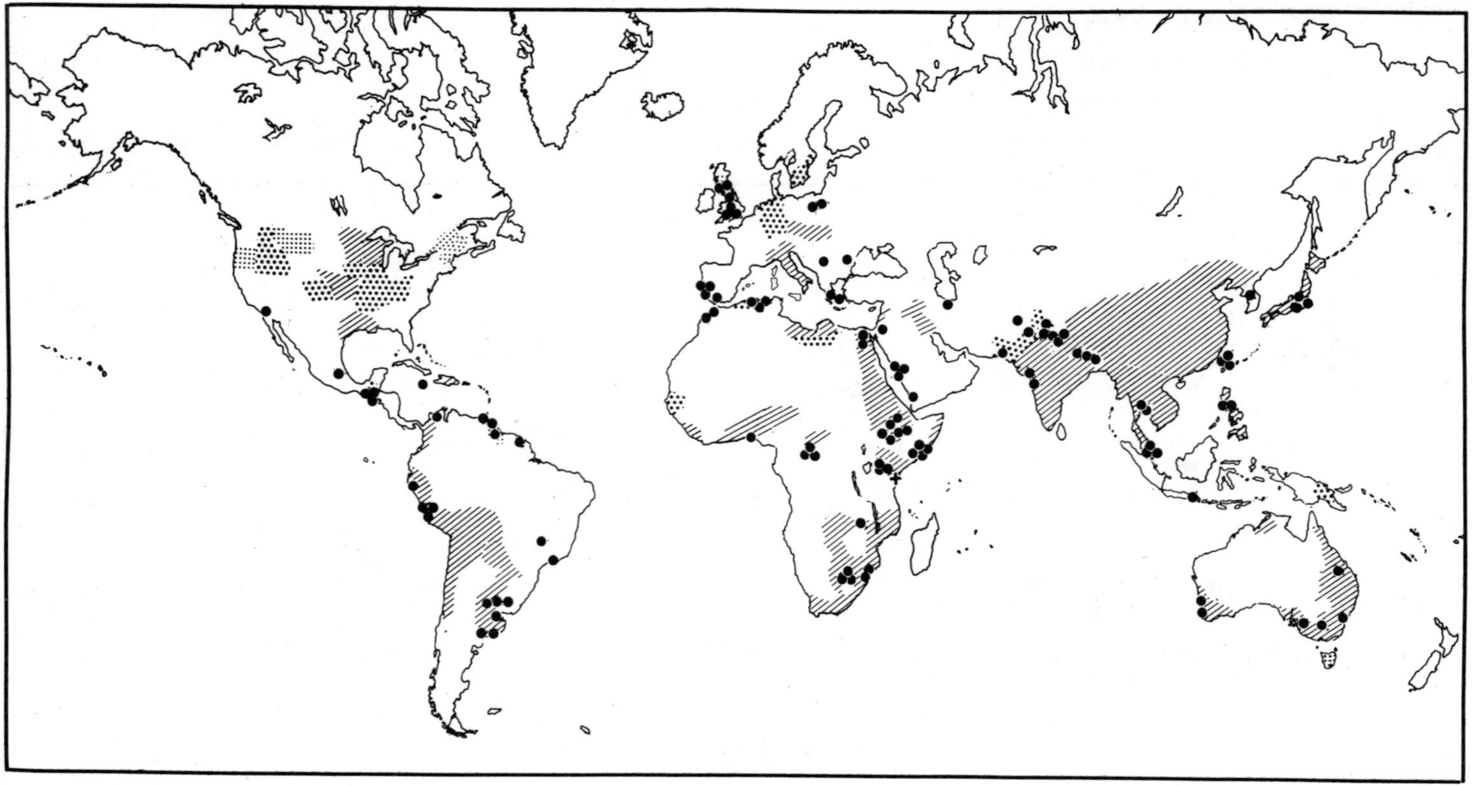

Fig. F 4 — The occurrence of methyl bromide resistance in *Sitophilus oryzae*

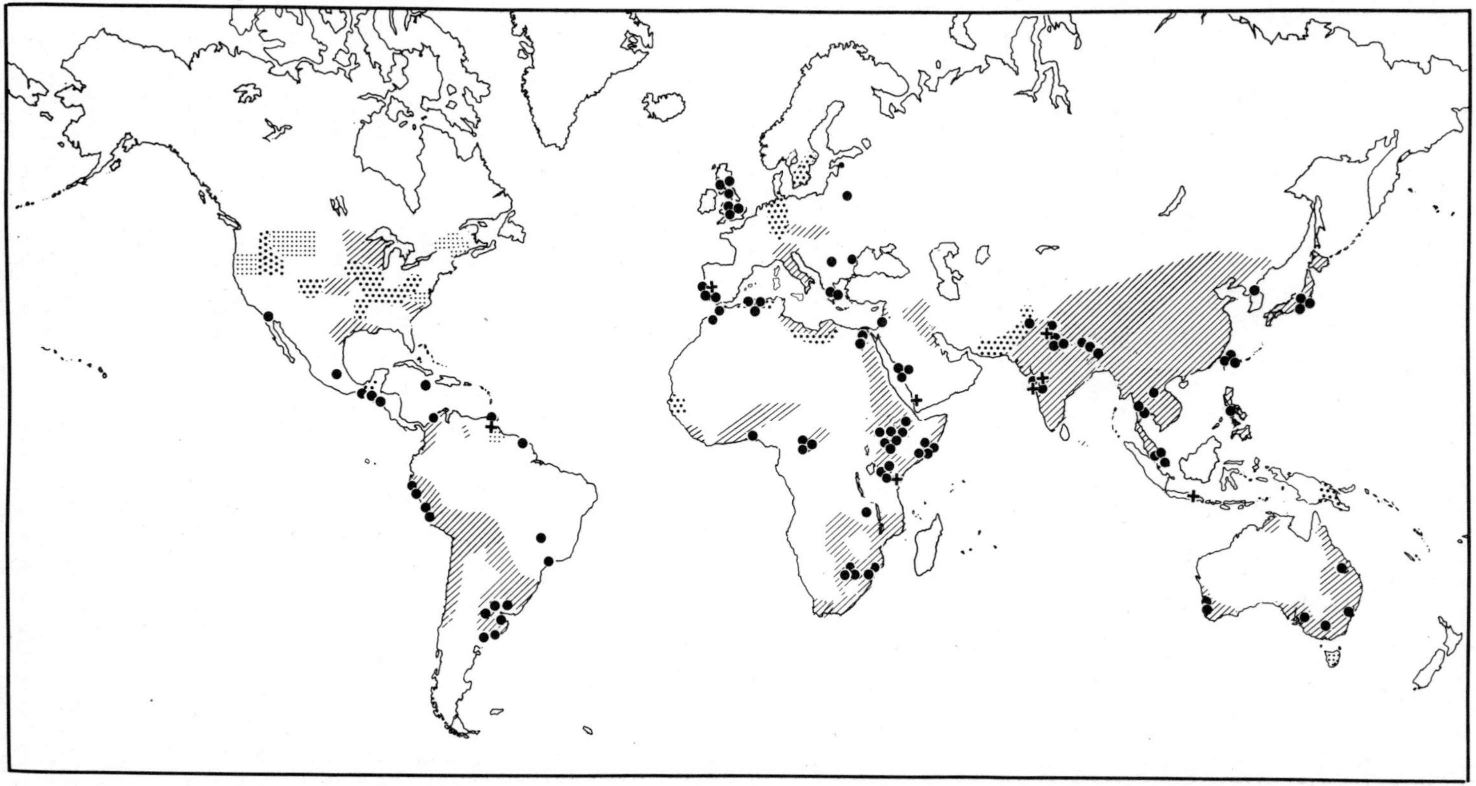

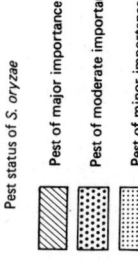

Fig. F 5 –The occurrence of phosphine resistance in *Sitophilus oryzae*

+ Phosphine resistant

● Phosphine susceptible

Pest status of *S. oryzae*

- Pest of major importance
- Pest of moderate importance
- Pest of minor importance

(F.4) *Sitophilus zeamais*

  Sitophilus zeamais is considerably less important than *S. oryzae* but complements this species in warm, moist areas. In tropical and subtropical zones, *S. zeamais*, like *S. oryzae*, has been subjected to intensive use of residual insecticides including DDT, particularly lindane and more recently malathion.

  There are few confirmed reports of resistance in *S. zeamais* and most concern lindane. Records for lindane resistance are available from Trinidad, Kenya, the Philippines and Brazil, and the resistance appears common in these countries. Nevertheless, lindane resistance does not seem to have been reported as frequently as in *S. oryzae*. DDT resistance has been reported from Brazil and the Philippines only, and despite local surveys in a number of countries, malathion resistance has been detected only in an imported strain in England. Similarly there have been no records of fumigant resistance.

  The confusion previously attendant with nomenclature of the *Sitophilus* species has created problems in attributing correctly some studies on the tolerance of these species. Nevertheless the species involved is usually reasonably clear from circumstantial evidence though a real possibility must exist that mixed species were involved.

(F.4.1) *Recorded insecticide resistances*

  Parkin *et al.* (1962a; Parkin 1965, 1967) at the Pest Infestation Laboratory, England using wheat treated with lindane dust and a 72 hr mortality assessment reported a x7.5 resistance in a strain of *S. zeamais* collected in Trinidad from rice imported from Thailand. A strain of similar origin from Kenya had a lindane resistance factor of x3.5 and a further strain from Kenya and one from southern Rhodesia, both collected from local maize, had resistance factors of x2 and x3.1 respectively (Parkin 1965). Subsequent collection in 1963 of three strains from standing maize in different localities in Kenya showed lindane resistance (x21) at Kitale where lindane was used in crib storage but not in other localities (Parkin and Forster 1966a, Wheatley 1967). Further collections from farms in Kenya where lindane was used extensively in crib storage had resistance factors of x32, x27 and x16 whereas a strain from a farm where lindane was rarely used showed only a x1.3 change in tolerance (Parkin and Forster 1968a). De Lima (1972) carried out a survey of lindane resistance in the main maize-growing areas of Kenya. This is one of the few surveys of stored grain pests within a country where the usage of pesticide and the occurrence and levels of resistance are related. He examined 40 samples from 23 sites using an impregnated filter paper technique with a 6 hr exposure period and responses measured as knockdown after a 24 hr recovery period. Twelve strains from the Nyanza and Western Provinces where lindane is widely used, had resistance factors ranging from x2 to x10 and two strains from the same locality had factors of x43 and x49; eight strains from the Rift Valley Province where lindane again was used widely, had factors from x3 to x29; 13 strains from the Central Province and Nairobi District where lindane use is low and intermittent, had low order factors of x2 to x5 with one from an experimental-seed store at x9 and one susceptible; while in 15 strains from the Eastern Province, factors were low (x1 to x8) on farms and in markets but higher (x11 to x29) on Agricultural Research Stations where lindane was used regularly. In Malawi, Pieterse and Schulten (1974) examined 11 strains of

S. zeamais collected from crib-stored maize during 1972 but were unable to detect either lindane or malathion resistance, though DDT, BHC and lindane had been used occasionally on farms and regularly for many years in central storage premises. Two strains, however, collected from imported seed maize were resistant to lindane though not to malathion. *S. zeamais* does not occur commonly in the United Kingdom and most infestations can be referred directly or indirectly to imported commodities. A survey of lindane resistance in 1973 revealed resistance in 6 of 6 strains examined reflecting the incidence of resistance in countries where the imports had originated (Dyte *et al.* 1975).

Laboratory selection of a reference strain in lindane-treated wheat at the Pest Infestation Laboratory resulted in a very slow development of resistance increasing from x1.8 in the 12th generation (Scott 1960) to a maximum of x12 in the 30th generation (Parkin and Green 1960, Scott 1961, Scott and Forster 1962, Parkin and Forster 1963a) after which the resistance level fell to approximately one half before the strain died out in the 46th generation (Parkin and Forster 1964a, 1965a, 1966a, Parkin 1965).

The lindane resistances detailed above were mostly of a low order. Champ and Cribb (1965a) were unable to detect lindane resistance in *S. zeamais* in Queensland and quoted a figure of x20 as the minimum resistance level necessary to be of significance in commercial practice. This was based on a 15.6 p.p.m. application rate of lindane in normal use in the smaller grains, twice this in maize and a minimum effective dosage of at most 0.8 p.p.m. for a susceptible strain.

In the Philippines, Morallo-Rejesus (1972) conducted a survey of DDT, lindane, carbaryl and malathion susceptibility using a modified impregnated filter paper technique and mortality after 24 hr continuous exposure. Nineteen strains were resistant to DDT with a maximum factor of x18 and all showed low level increases in tolerance to carbaryl to a maximum factor of x4 but were all susceptible to malathion. Subsequently (Morallo-Rejesus, personal communication) seven strains were shown resistant to lindane.

In Brazil at the Instituto Biologico de Sao Paulo, Mello (1972) using maize impregnated with insecticidal dusts, reported a > x25 and a x30 resistance to DDT and lindane respectively in a strain of *S. zeamais* (reported as *S. oryzae*) collected in 1969 from maize storages in Capinopolis in the state of Minas Gerais where control failures with DDT had been observed. This strain also showed a x2 tolerance to malathion, x7 to diazinon, x3 to tetrachlorvinphos and x2 to bromophos. At Porto Alegre, Corseuil and Vicenzi (1970) demonstrated lindane resistance from Rio Grande do Sul in a laboratory experiment evaluating dusts containing jodfenphos, DDT, malathion and synergised pyrethrins for protection of stored maize.

Dyte (1974, Dyte *et al.* 1975) has recorded malathion resistance from the United Kingdom in one of 6 samples collected during 1973. This is the only published record of malathion resistance in *S. zeamais*.

(F.4.2) *Recorded fumigant resistances*

Howe (1962a) listed the relative susceptibilities to methyl bromide of adults, larvae and pupae, and eggs and larvae of 11 strains of *Sitophilus zeamais* which originated in Jamaica, Kenya, Trinidad and the Transvaal giving factors ranging from x1.0 to x1.8. He attributed these changes to natural variation rather than resistance. Buckmire and Bennett (1969) at the Pest Infestation Laboratory, England, proposed a survey for fumigant resistance in stored product pests and Bennett (1970) reported small differences in response to phosphine of five strains of *Sitophilus zeamais* from unspecified localities around the world when compared with the normal laboratory strain.

(F.4.3) *Survey data*

Results from discriminating tests for malathion, lindane, methyl bromide and phosphine resistance in *Sitophilus zeamais* are given in Table F6 and Figures F6-F9 together with a summary of the occurrences in Table F7. The distribution maps (Figs. F6-F9) also show the classification of importance currently attributed to *S. zeamais* in the different countries.

(F.4.4) *Comments*

(F.4.4a) *Malathion resistance*

Malathion resistance was detected in six only of the 203 strains of *Sitophilus zeamais* tested, supplementing the single previously known (1974) record from the United Kingdom. The areas in which *S. zeamais* occurs and the principal grains concerned - maize and rice - are not subjected to the intensive use of malathion that often occurs in the drier areas where *S. oryzae* predominates. Nevertheless, as with *S. oryzae*, it appears that *S. zeamais* is not particularly prone to development of malathion resistance and probably less so than *S. oryzae*. A contributing factor to this difference between the species could be the greater mobility of *S. zeamais*. The high levels of flight activity resulting in extensive field infestation in maize and cross-infestation between storages, ensure dispersal and dilution of populations of *S. zeamais* and so reduce the effect of selection.

The two records of resistance from the Central American republics of El Salvador and Guatemala were from the Pacific coastal plain area where considerable quantities of maize are grown and stored in the huts of the local inhabitants. These huts and the maize stored in them have been sprayed for years with malathion and DDT in disease vector control programmes. Malathion is not used in El Salvador for control of *S. zeamais* but is recommended for control of phycitid moths and *Sitotroga cerealella* in central storages. In Guatemala, malathion is not used and control is based on fumigation and spraying with synergised pyrethrins. From the evidence available it seems reasonable to assume that the malathion resistances found on the coastal plain have been associated with use of malathion in disease vector control.

The background of the resistances present in the other records is obscure. The record from England presumably concerned an imported strain. The single occurrence from Australia was from a cereal seed store on a research station where malathion had not been used and which had contained seed from areas throughout Australia and overseas. The origin of the strain showing tolerance from New Guinea was also indeterminate - here the sample was collected from sweepings and residues in a wharf shed at Port Moresby. There is a record also from this port of a malathion resistant strain collected from maize residues on the deck of a ship unloading cargo from Japan. The two occurrences listed for China came from a rice-processing plant in Peking and a flour mill in Shanghai and may also concern importations. Because of its odour, malathion is used only in a small region in one province in China and is not produced - dichlorvos is used but there was no apparent resistance to this material. *S. zeamais* would be expected to move actively in world trade particularly in maize and rice and apart from the Central American records, the circumstances of the other occurrences were such as not to give any indications of the sources of the problem.

(F.4.4b) *Lindane resistance*

Lindane resistance was reported previously from the Caribbean, Brazil, East Africa and the Philippines. It was detected in 46 of the 48 countries sampled in the survey and 78% of the 186 samples concerned. The areas with low incidence were few - eastern U.S.A., Ethiopia and Korea, and none of these areas had a history of intensive lindane use. Lindane, however, has been used extensively for control of *S. zeamais* in many other parts of the world particularly with maize and less frequently rice in tropical and subtropical areas. This use has paralleled use of lindane in control of *S. oryzae*. From the limited information available, resistance in *S. zeamais* appears to have developed on a wide scale in *S. zeamais* somewhat later than in *S. oryzae*, possibly because of the higher dose rates frequently used in maize for *S. zeamais*. Nevertheless lindane resistance is now widespread and in most areas can now be assumed to be a normal attribute of *S. zeamais* as well as *S. oryzae* populations.

(F.4.4c) *Methyl bromide resistance*

Methyl bromide resistance has not been reported previously in *Sitophilus zeamais* though Howe's (1962a) report of differences up to x1.8 tolerance between strains from Jamaica, Kenya, the Transvaal and Trinidad, which he attributed to natural variation, must be qualified by the lack of variation in methyl bromide response between strains used during the survey. The mixtures of life-history stages used, however, could be a source of variation in Howes' tests. The changes in tolerance recorded in the survey were all of a lower order than those recorded by Howe. The maximum change was in a strain from a rice-processing plant in Peking, China where a 75% response was recorded in the discriminating test. This corresponds with an approximate x1.4 increase in tolerance to methyl bromide which is used widely in China. The strain was also resistant to malathion and lindane but its tolerance to phosphine which is also used widely was not checked. The other changes in tolerance observed during the survey were associated with marginal shifts in tolerance and are not considered confirmed records of resistance at this stage. Thus resistance to methyl bromide does not appear to be a problem of any consequence in *S. zeamais*.

(F.4.4d) *Phosphine resistance*

There have not been any records of changes in tolerance of *Sitophilus zeamais* to phosphine either in the laboratory or in the field. With the variation that appears generally inherent in phosphine response, and with the widespread use of phosphine in maize storage in the tropics and subtropics, often under unsatisfactory conditions, it seems likely that phosphine resistance, like malathion resistance, does not develop as readily in *S. zeamais* as in other species.

TABLE F6: RESPONSES OF REPRESENTATIVE WORLD STRAINS OF *SITOPHILUS ZEAMAIS* TO MALATHION, MALATHION + TPP, LINDANE, METHYL BROMIDE AND PHOSPHINE IN DISCRIMINATING TESTS FOR RESISTANCE

Locality	Mal	M+T	Lin	MeB	PH$_3$	Locality	Mal	M+T	Lin	MeB	PH$_3$
NORTH AMERICA						**CENTRAL AMERICA**					
MEXICO						**AND CARIBBEAN (Cont'd)**					
Chiapa	100	-	3	100	100	**JAMAICA (Cont'd)**					
Cortezar	100	-	6	-	-	Kingston	100	-	36	100	100
"	100	-	21	100	100	"	100	-	34	100	100
"	100	-	6	100	100	"	-	-	-	100	100
Mexico City	100	-	48	-	-	**NICARAGUA**					
" "	100	-	61	-	-	Managua	100	-	0	-	-
Poanas	100	-	17	100	100	Samulali	100	-	-	-	-
Vera Cruz	100	-	41	100	100	**TRINIDAD**					
U.S.A.						Port of Spain	100	-	35	-	-
Ark.-Fayetteville	100	-	98	100	100	" "	100	-	36	100	100
Cal.-Riverside	100	-	100	100	100	" "	100	-	100	100	100
La -Baton Rouge	100	-	94	-	-	" "	-	-	-	99	100
" "	100	-	78	100	100	" "	-	-	-	98	100
Va -Abingdon	100	-	100	100	99						
"	100	-	100	100	99	**SOUTH AMERICA**					
Franklin	100	-	100	100	100	**ARGENTINA**					
Nansemond	100	-	100	100	100	B.A.-Buenos Aires	100	-	34	-	-
Richmond	100	-	100	100	100	Pergamino	100	-	65	100	100
"	100	-	100	-	-	Corr.-Paseo de					
						los Libres	100	-	7	100	100
CENTRAL AMERICA						Ent. Rios-Diamente	100	-	52	-	-
AND CARIBBEAN											
ANTIGUA						**BRAZIL**					
Dunbars	100	-	12	100	100	Amaz. - Manaus	-	-	-	97	100
COLOMBIA						Goias- Anapolis	100	-	84	100	100
Barranquilla	100	-	100	-	-	G.B.-Rio de Janeiro	100	-	73	-	-
"	100	-	96	100	100	R.G. do Sul- Guaibo	100	-	100	-	-
EL SALVADOR						Porto Alegre	100	-	36	-	-
La Libertad	100	-	41	-	-	" "	100	-	100	100	100
"	100	-	95	-	-	" "	100	-	67	-	-
"	100	-	100	100	99	S.P.- Avare	100	-	89	100	100
"	94	100	26	-	-	Barretos	100	-	50	100	100
FRENCH GUIANA						"	100	-	55	100	100
Cayenne	100	-	96	100	100	Bauru	100	-	63	-	-
GUATEMALA						Itapetininga	100	-	100	-	-
Barcena	100	-	44	100	100	"	100	-	100	100	100
"	100	-	28	100	100	Jaguare	100	-	82	100	100
Escuintla	85	79	15	-	-	Ourinhos	100	-	65	100	100
Guatemala City	100	-	-	-	-	"	100	-	68	100	100
" "	100	-	77	100	100	Ribeirao Preto	100	-	80	-	-
" "	100	-	-	100	100	S.J.Rio Preto	100	-	60	100	100
						Sao Paulo	100	-	81	100	100
GUYANA						" "	100	-	100	100	100
Georgetown	99	95	19	-	-	**PERU**					
"	-	-	-	97	100	Callao	100	-	21	100	100
						Chiclayo	100	-	16	100	100
JAMAICA						**URUGUAY**					
Kingston	100	-	2	100	100	Rivera	100	-	0	100	100

TABLE F6
(Cont'd)

S.zeamais (ii)

Locality	Mal	M+T	Lin	MeB	PH$_3$	Locality	Mal	M+T	Lin	MeB	PH$_3$
EUROPE AND NORTH AFRICA						**AFRICA SOUTH OF SAHARA (Cont'd)**					
ENGLAND						MALAWI					
Lincs - Boston	100	-	84	-	-	Blantyre	100	-	90	100	-
Northants-Northampton	100	-	88	100	99	Limbe	100	-	81	-	-
Som. - Bridgewater	100	-	60	100	100	"	100	-	86	-	-
GREECE						Thucila	100	-	71	100	100
Piraeus	100	-	-	100	100						
LIBYA						MOZAMBIQUE					
Tobruk	100	-	69	-	-	Beira	100	-	53	-	-
Tripoli	100	-	23	-	-	Chibuto	100	-	32	100	100
MOROCCO						Lourenco Marques	100	-	19	100	100
Rabat	100	-	45	-	-	Quelmane	100	-	100	-	-
Sale	100	-	0	-	-	NIGERIA					
PORTUGAL						Lagos	100	-	92	-	-
Lisbon	100	-	91	-	-	"	100	-	-	-	-
"	100	-	48	100	100	RHODESIA					
"	100	-	0	-	-	Aspindale	100	-	-	96	100
Queluz	100	-	100	99	-	Causeway	100	-	33	100	100
SCOTLAND						SENEGAL					
Edinburgh	100	-	38	-	-	Sefa	100	-	-	-	-
YUGOSLAVIA						SOMALIA					
Darda	100	-	-	-	-	Afgoi	100	-	-	100	100
"	100	-	-	-	-	Mogadishu	100	-	-	100	100
AFRICA - SOUTH OF SAHARA						SOUTH AFRICA					
BENIN						Delmas	-	-	88	100	100
Cotonou	100	-	96	100	100	TANZANIA					
"	100	-	-	-	-	Dar-es-Salaam	100	-	84	-	-
CENT. AFRICAN REP.						" "	100	-	75	-	-
Bangui	100	-	90	-	-	" "	100	-	29	-	-
"	100	-	100	-	-	" "	100	-	100	100	100
"	100	-	78	-	-	ZAMBIA					
"	100	-	-	100	100	-	-	-	70	-	-
Km 22	100	-	100	-	-	Chadiza	100	-	-	100	100
ETHIOPIA						Kabwe	100	-	12	-	-
Addis Ababa	100	-	-	-	-	Mongu	100	-	40	100	100
" "	100	-	100	-	-						
" "	100	-	-	100	100	**WEST ASIA**					
Dabi	100	-	100	-	-						
Jiran	100	-	100	-	-	BAHRAIN	100	-	7	-	-
Saka	100	-	100	99	100	"	100	-	8	-	-
Serbo	100	-	100	100	100	"	100	-	20	-	-
GAMBIA						ISRAEL					
Yundum	100	-	0	-	-	Kiryat - Gat	100	-	6	-	-
KENYA						Tel Aviv	-	-	-	100	100
Nairobi	100	-	47	100	100	KUWAIT	100	-	11	-	-
Nakuru	100	-	46	-	-						

TABLE F6
(Cont'd)

S. zeamais (iii)

Locality	Mal	M+T	Lin	MeB	PH_3	Locality	Mal	M+T	Lin	MeB	PH_3
WEST ASIA (Cont'd)						**EAST ASIA (Cont'd)**					
TURKEY						PHILIPPINES (Cont'd)					
Carsamba	100	-	100	100	-	Echaque	100	-	62	100	100
Sakarya	100	-	83	100	100	Haringcanardan	100	-	9	-	-
						Ilagan	100	-	24	-	-
EAST ASIA						La Granja	100	-	82	100	100
BURMA						Laguna	100	-	92	100	100
Rangoon	100	-	1	-	-	Legaspi	100	-	58	-	-
"	100	-	84	-	-	Ligao	100	-	39	-	-
CHINA						Lipa City	100	-	54	-	-
Hopeh - Peking	95	-	5	75	-	Mandawe City	100	-	90	100	100
"	100	-	5	-	-	Manila	100	-	36	-	-
Kiangsu - Shanghai	76	-	0	-	-	Pagsanjan	100	-	100	-	-
Taiwan - Taipei	100	-	42	-	-	Pili	100	-	59	100	100
"	100	-	79	-	-	San Mateo	100	-	49	100	100
Tao Yuan	100	-	-	100	100	Santiago	100	-	56	-	-
						Sorsogen	100	-	90	-	-
INDONESIA						Talisay	100	-	72	-	-
Djakarta	100	-	74	-	-	Tanauan	100	-	83	-	-
"	100	-	38	-	-	Zamboanga City	100	-	59	100	100
						" "	100	-	-	-	-
JAPAN						SINGAPORE	100	-	16	100	100
Chiba	100	-	93	100	100	"	100	-	23	-	-
"	100	-	89	-	-	"	100	-	45	100	100
"	100	-	38	100	100						
..	100	-	100	100	100	**THAILAND**					
Hakodate	100	-	96	100	100	Bangkok	100	-	100	100	100
Obihiro	100	-	100	100	100	"	100	-	52	-	-
Sapporo	100	-	100	-	-	"	100	-	21	100	100
Tsuyama	100	-	100	-	-	"	100	-	4	100	100
Yokohama	100	-	25	99	100	"	100	-	2	100	100
"	100	-	32	100	100	Thonburi	100	-	88	100	100
"	100	-	69	100	100	"	100	-	37	-	-
"	100	-	89	-	-	"	100	-	94	100	100
"	100	-	-	100	100						
						U.S.S.R.					
KOREA REP.						Krasnodar	-	-	100	-	-
Inchon	100	-	100	100	100	"	100	-	100	-	-
Sosa	100	-	100	100	100	"	100	-	100	-	-
Suwon	100	-	98	-	-	"	-	-	100	-	-
"	100	-	96	100	99	Sagorvaki	-	-	100	-	-
MALAYSIA											
Alor Setar	100	-	75	97	100	**AUSTRALASIA AND**					
" "	100	-	100	100	100	**SOUTH PACIFIC**					
Penang	100	-	16	-	-	AUSTRALIA					
						ACT - Canberra	53	100	-	-	-
PHILIPPINES											
Bilar	100	-	81	100	100	NEW BRITAIN					
Bucay	100	-	85	-	-	Kerevat	100	-	32	-	-
Cagayan de Oro City	100	-	69	-	-						
Cebu City	100	-	79	100	100	NEW GUINEA					
College	-	-	-	100	100	Lae	100	-	11	-	-
Davao City	100	-	44	100	100	Port Moresby	93	-	11	-	-
" "	100	-	63	-	-	Wewak	100	-	-	-	-
Dumaquete City	100	-	-	-	-	"	100	-	70	-	-

TABLE F7: SUMMARY OF WORLD OCCURRENCE OF RESISTANCE IN *SITOPHILUS ZEAMAIS*, TO MALATHION, LINDANE, METHYL BROMIDE AND PHOSPHINE

	North America	Central America	South America	Europe,N.Afr.,USSR	Africa-S.of Sahara	Western Asia	Eastern Asia	Australasia	Totals	Totals (as %)
MALATHION										
No. countries* sampled	2	9	4	8	13	4	9	3	52	-
No. with resistance	0	2	0	0	0	0	1	2	5	10%
No. of strains tested	18	23	26	17	39	7	67	6	203	-
No. resistant	0	2	0	0	0	0	2	2	6	3%
MALATHION + TPP										
No. countries sampled	2	9	4	8	13	4	9	3	52	-
No. with resistance	0	2	0	0	0	0	0	0	2	4%
No. of strains tested	18	23	26	17	39	7	65	4	199	-
No. resistant	0	2	0	0	0	0	0	0	2	1%
LINDANE										
No. countries sampled	2	9	4	6	12	4	9	2	48	-
No. with resistance	2	9	4	5	11	4	9	2	46	96%
No. of strains tested	18	20	26	17	31	7	63	4	186	-
No. resistant	11	17	21	11	22	6	54	4	146	78%
METHYL BROMIDE										
No. countries sampled	2	8	4	3	11	2	7	-	37	-
No. with resistance	0	2	1	0	1	0	2	-	6	16%
No. of strains tested	13	17	18	5	18	3	35	-	109	-
No. resistant	0	2	1	0	1	0	2	-	6	6%
PHOSPHINE										
No. countries sampled	2	8	4	3	11	2	7	-	37	-
No. with resistance	0	0	0	0	0	0	0	-	0	0%
No. of strains tested	13	17	18	4	17	2	34	-	105	-
No. resistant	0	0	0	0	0	0	0	-	0	0%

Sitophilus zeamais – Malathion

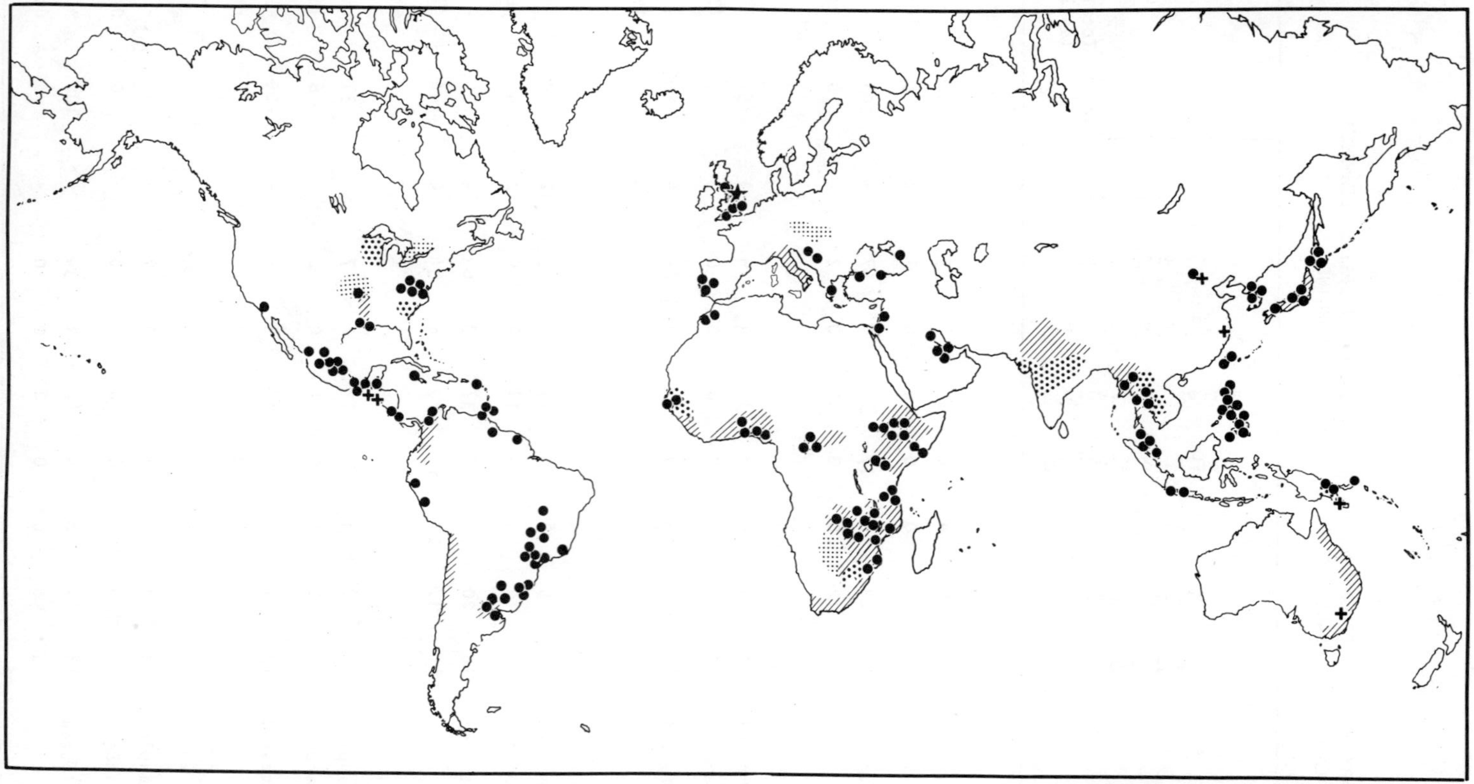

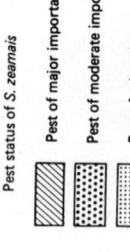

Fig. F 6 – The occurrence of malathion resistance in *Sitophilus zeamais*

+ Malathion resistant

● Malathion susceptible

★ Previously recorded resistance

Pest status of *S. zeamais*

- Pest of major importance
- Pest of moderate importance
- Pest of minor importance

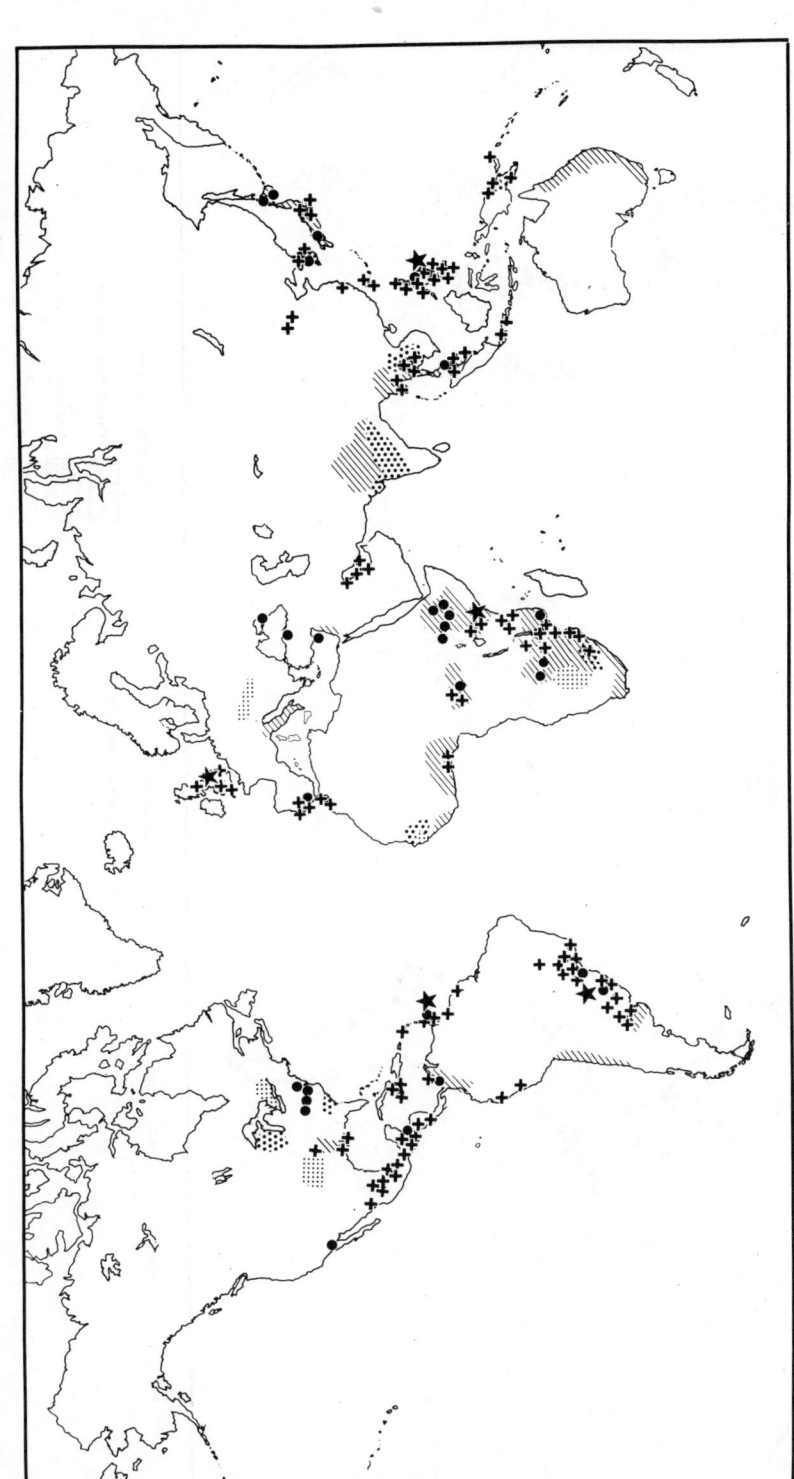

Fig. F 7 – The occurrence of lindane resistance in *Sitophilus zeamais*

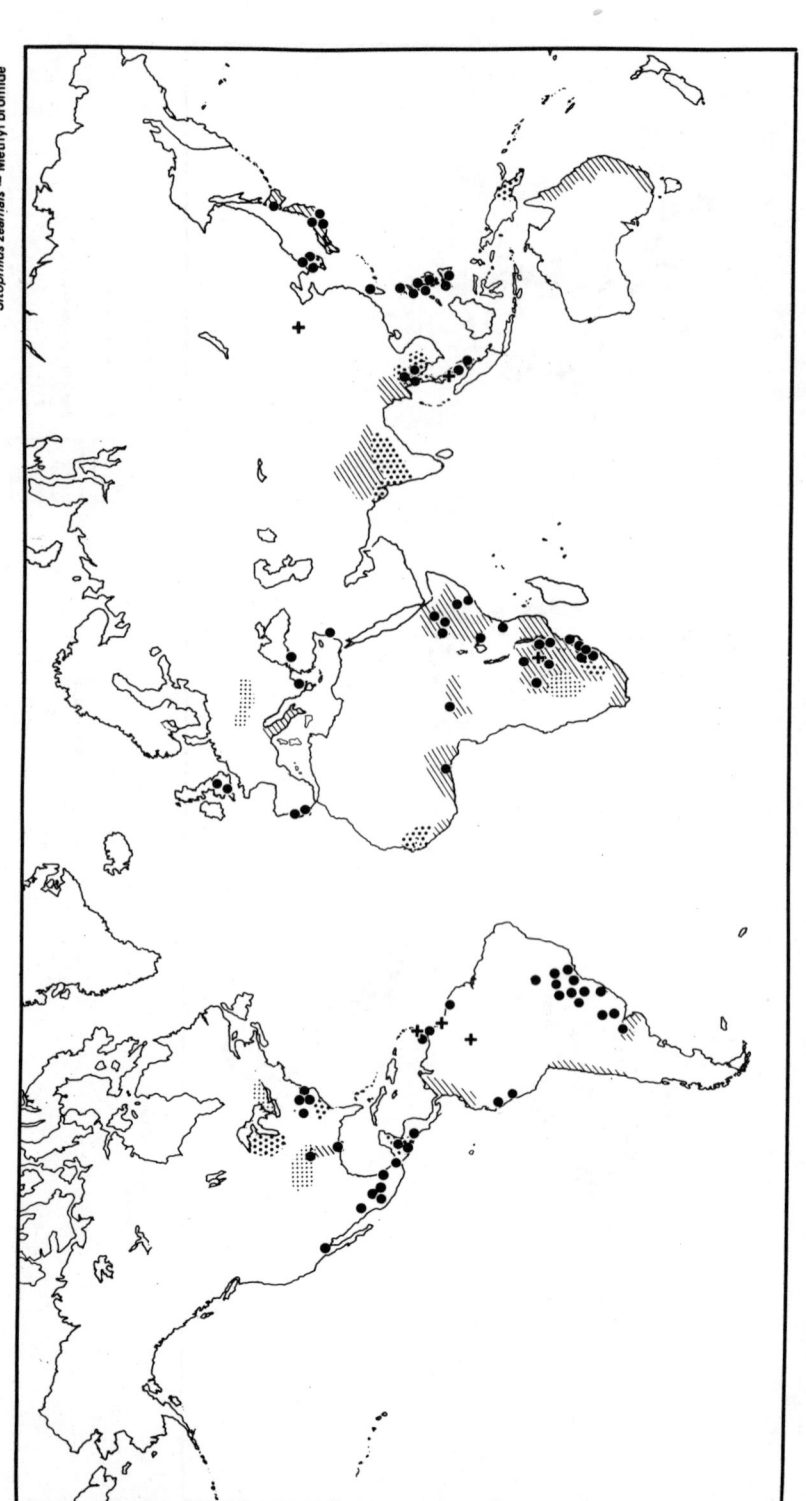

Fig. F8 – The occurrence of methyl bromide resistance in *Sitophilus zeamais*

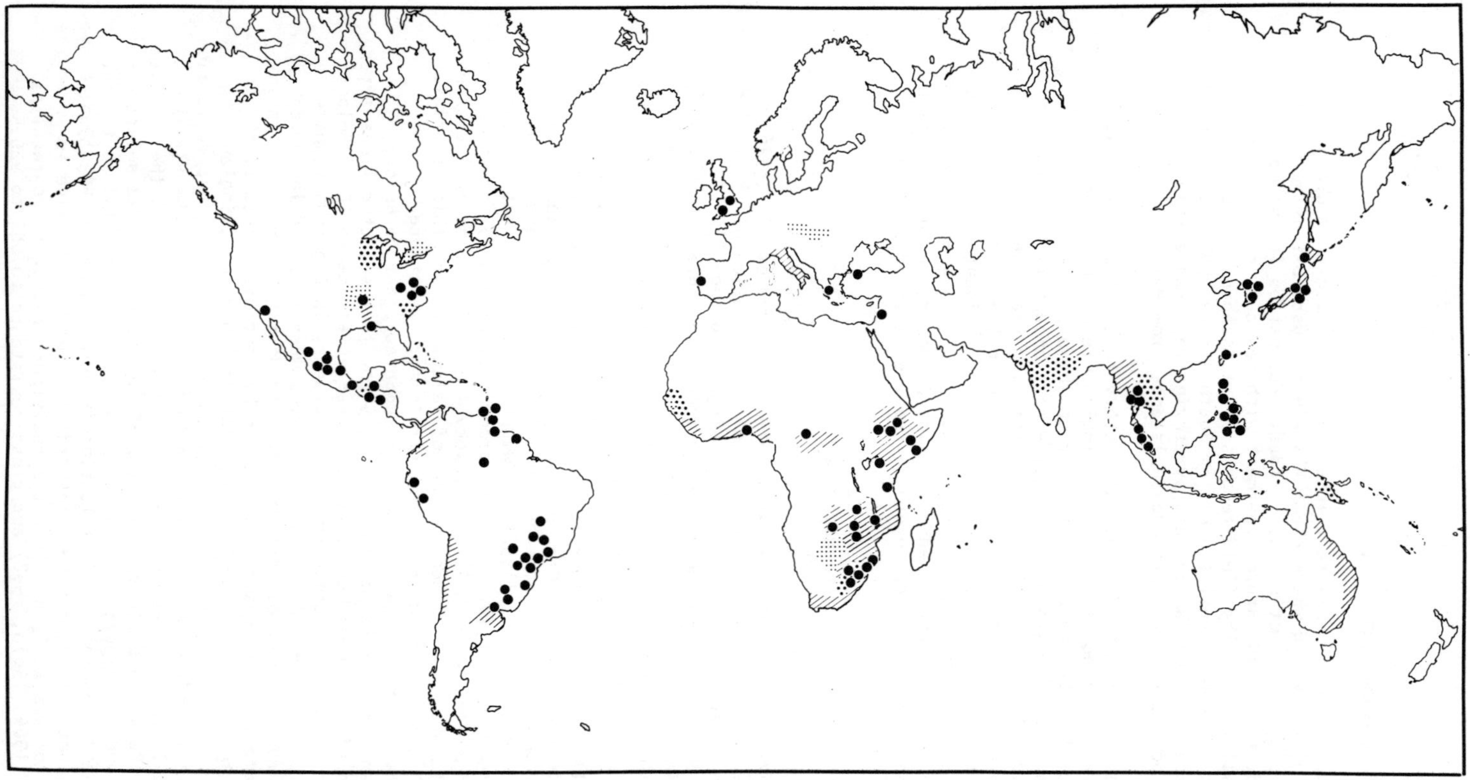

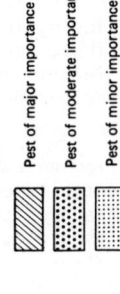
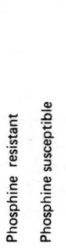

Fig. F.9 — The occurrence of phosphine resistance in *Sitophilus zeamais*

(F.5) *Sitophilus granarius*

Sitophilus granarius is restricted to the cool-to-cold grain growing areas of the world where chemicals are not used extensively for control and indeed are sometimes not needed. Fumigation is the major control method, though admixture treatment with residual chemicals is finding increasing use. The field records of resistance are few - single records of lindane resistance from South Africa, DDT and DDT + dieldrin resistances from Canada, and pyrethrins resistance from England and a possible methyl bromide resistance from Malta.

S. granarius is a traditional laboratory test animal resulting from an earlier centering of basic entomological research in the established areas of Europe and North America where *S. granarius* reaches its maximum pest status. A considerable amount is now known of its biology and behaviour and it is not surprising that there have been extensive laboratory studies to determine its potential for developing resistance to the more important insecticides and fumigants. Thus there have been selection programmes for resistance to DDT in Sweden, pyrethrins and mercury in England, synergised allethrin, propoxur, fenthion, methyl bromide and phosphine in Canada, hydrogen cyanide in the U.S.A., and malathion in France.

(F.5.1) *Recorded insecticide resistances*

Parkin (1967) reported lindane resistance in a strain of *S. granarius* from stored barley in South Africa. Lloyd et al. (1976a) using an impregnated filter-paper method in an examination of 18 strains collected from inland sites in the United Kingdom during 1970-1972, recorded from 9 strains a markedly lower response to lindane than from the susceptible laboratory strain. Ten of the strains showed responses to pyrethrum that were lower than expected as did all of 5 strains tested with DDT. During a further survey in 1973 in the United Kingdom, Dyte et al. (1975, 1976a; Lloyd et al. 1976a) recorded 2 of 42 strains resistant to lindane. Most other records of insecticide resistance are from laboratory-selected strains. Orr in Toronto (in Sevintuna and Musgrave 1961) reported a significant tolerance of lindane from repeated exposure to direct spraying, Mathlein (1952) in Sweden recorded a slight increase in tolerance to DDT and pyrethrins + piperonyl butoxide after seven generations of selection with DDT and Lakocy (1970) in Poznan an increase in tolerance to DDT after 12 generations of selection with this material. Blackith (1953), selecting a laboratory strain for resistance to paralysis by pyrethrins and using an impregnated paper technique, reported a mean increase in tolerance of x3.5 for four selected lines after 10 generations of selection but found no cross-resistance to DDT. Holborn (1957) using a strain collected from a farm granary in England, reported a x3.5 difference in susceptibility to pyrethrins when compared in dusted grain with a laboratory strain but did not find any change with pyrethrins + piperonyl butoxide. This strain was shown by laboratory selection pressure with pyrethrins at the Pest Infestation laboratory, England, to have potential for development of high pyrethrins resistance, the resistance increasing to > x5 in the third year (Anon.1958a), to x12 in the F_{18} (Anon. 1959b), to x18 in the F_{22} (Parkin and Lloyd 1960a,b), to x34 in the F_{25} from 18 selections (Parkin and Lloyd 1961), to x52 in the F_{29} from 22 selections (Parkin and Lloyd 1961,1962; Lloyd and Parkin 1963), and to x132 in the F_{44} from a total of 31 generations of selection (Parkin and Lloyd 1963,1964; Lloyd 1965). Subsequent selection with pyrethrins + piperonyl butoxide to which the resistance factor was x5.5, increased the resistance factor for pyrethrins to x138 (Lloyd 1966, 1967a, 1968) and cross-resistance to DDT to x29.5

(Lloyd 1967b, 1969a). The resistance to malathion and lindane was examined in a branch-line stock from the F_{25} when the pyrethrins resistance factor was x34 - this gave a factor of x5.7 for malathion and x5.5 for lindane (Lloyd and Parkin 1963). Rajak and Hewlett (1971) using unspecified material from this heavier than normal pyrethrins-resistant strain, recorded also a x2 tolerance to the fumigant phosphine. Selection of a Canadian strain with allethrin + piperonyl butoxide for 6 generations by Sevintuna and Musgrave (1961), gave only a x2 increase in tolerance though this is not inconsistent with early results of Lloyd. Kumar and Morrison (1964,1965) from the Macdonald College, Quebec using the Macdonald College Test Kit for detecting resistance, reported a x2 change in tolerance to DDT in a strain of *S. granarius* from Quebec and a x5 and a x2 change in tolerance to DDT and dieldrin respectively in a strain from Winnipeg. Subsequently they demonstrated tolerance to propoxur and to fenthion after 14 generations of selection with the respective materials (Kumar and Morrison 1967, Macdonald College 1967). The propoxur-resistant strain showed a x2.5 cross-resistance to DDT. Finally Coulon (personal communication) at the Laboratoire de Phytopharmacie at Versailles, selected *S. granarius* in malathion-treated wheat for 18 generations over three years but did not detect any change in tolerance to malathion.

The only field reports of malathion resistance have been from the United Kingdom and Australia. Dyte *et al.* (1975, 1976a; Lloyd *et al.* 1976a) using an FAO-type test recorded 3 of 43 strains resistant to malathion and to malathion synergised with triphenyl phosphate during a survey in 1973 of established infestations inland in the United Kingdom. In New South Wales, Australia, Greening *et al.* (1975), also using an impregnated paper test recorded x2.6 resistance from a strain collected in Sydney in 1971 from a flour and provender mill from which malathion-resistant strains of *Tribolium castaneum* and later *Sitophilus oryzae* and *Rhyzopertha dominica* were obtained. Fifteen further strains collected principally from farms from throughout New South Wales, were susceptible.

(F.5.2) *Recorded fumigant resistances*

Most reports of resistance to fumigants relate to laboratory-induced tolerances. There are exceptions to this. Blackith and Gorringe (1953) of the Imperial College Field Station England, observed a x350 tolerance to mercury vapour in eggs of *S. granarius* from a strain which had been unintentionally exposed to this material in an incubator for less than 10 generations. They then tested the susceptibility in the egg stage to mercury vapour of six populations of *S. granarius* from ships and warehouses in England and recorded a wide (x50) difference in tolerance between the most and the least susceptible strain. Howe (1962a) reported a x1.6 tolerance to methyl bromide of eggs and larvae of a strain collected from wheat in Malta.

Brudnaya *et al.* (1966) at the All-Union Institute of Grain and Grain Product Research in Moscow, carried out repeated fumigations of *S. granarius* with methyl bromide and chloropicrin and selected strains that were more tolerant to these materials and produced more progeny. Monro *et al.* (1961, Monro and Upitis 1956, Monro 1964) at the Science Service Laboratory, London Ont., selected for 27 generations with methyl bromide, a mixed strain of *S. granarius* collected from various sites locally, and raised the tolerance to x5.5 that of the original unselected strain at a relatively constant rate throughout the selection programme. Two further strains from Montreal and Guelph were selected also, again inducing methyl bromide resistance, while a branch-line stock from the 13th selection of the Montreal stock that had been reared for 26 generations without selection pressure, showed no reversal of resistance. The selected strains showed cross-resistance to a wide range of other fumigants including ethylene dibromide (x3), hydrogen cyanide (x2), and phosphine which reached a maximum level of x12.8. Further selection with methyl bromide raised the resistance level of the London strain progressively to x7 for methyl bromide after 40 generations (Monro 1963, 1964), reaching a maximum of x7.8 for methyl bromide after 50 generations of selection. A further 25 generations in culture without selection over a period of 4.3 years reduced the resistance factor to x5.5 (Bond and Upitis 1972). Subsequently a resistance level of x17 has been reported from this laboratory's selection programme with methyl bromide (Anon. 1974b). At the same time Lindgren and Vincent (1965, 1966a) at the University of California, Riverside, selected *S. granarius* for 25 generations with hydrogen cyanide increasing the tolerance to this material three times and to methyl bromide and ethylene dibromide very slightly.

Monro *et al.* (1972) established also a phosphine resistant strain of *S. granarius* by selecting survivors from a phosphine assay in which there was a marked change in slope above 90% mortality (Bond and Monro 1961). These survivors were selected 28 times during 41 generations of culture and when tested after a further nine generations without selection, survivors were found after 78 hr exposure, whereas complete mortality occurred in the parent stock after 18 hr exposure (Monro 1963, Monro *et al.* 1972). A check of cross-resistance at the 18th selection indicated no significant change in tolerance to methyl bromide or other fumigants with the exception of chloropicrin where a x2.7 increase was recorded. These selections and tests were made at comparatively high concentrations of phosphine (11-14 mg/l falling to 2mg/l after 78 hr). Selection was continued for 3 generations at a low concentration (0.61mg/l falling to 0.32mg/l) and when the selected strain was tested at low and high concentrations, the selected strain required at the low concentration an exposure three times as long as that given to the unselected strain to reach the $LC_{99.9}$, whereas at the high concentration the increase was in excess of 10 fold, and this level of response was reached in a slightly lesser time at the low concentration. Normal concentration-time relationships do not apply to phosphine responses of *S. granarius* over the range of concentrations considered here (Qureshi *et al.* 1965, Bond *et al.* 1969, Monro *et al.* 1972) and it appears that different types of response involving normal toxication and a form of narcosis are compounded at higher dose levels (Monro *et al.* 1972, Winks 1973) though at lower concentrations and with longer exposures, linear concentration-time relationships are obtained (Lindgren and Vincent 1966b). Nevertheless these studies have demonstrated the ability of *S. granarius* to develop a tolerance to phosphine.

(F.5.3) *Survey data*

Results from discriminating tests for malathion, lindane, methyl bromide and phosphine resistance in *Sitophilus granarius* are given in Table F7 and Figures F10-F13, together with a summary of the occurrences in Table F8. The distribution maps (Figs. F10-F13) also show the classification of importance currently attributed to *S. granarius* in different countries.

(F.5.4) *Comments*

(F.5.4a) *Malathion resistance*

Malathion resistance was diagnosed in four of the 28 countries sampled. These were Australia (4 states), Argentina, England and Greece.

Of the 14 strains showing resistance, eight were from Australia. The Australian strains as with *S. oryzae*, were tested at a more sensitive discriminating dose. In this instance the samples came from a cross-section of wheat, oats and barley storages and flour mills in New South Wales, Victoria and South Australia. There was a general association with malathion again both as an admixture treatment with grain and as a general fabric and disinfestation spray for storage buildings. The resistance levels involved were not high reaching a maximum level of approximately x3 (x6 at $LC_{99.9}$) in the field-collected strains. These levels are sufficiently high, however, to allow survival from direct contact with fresh deposits of malathion, and thus would offset any residual life of these deposits.

The record of malathion resistance from a single sample from wheat residues in a flour mill in Buenos Aires, Argentina can be associated also with extensive use of malathion comparable with that of the southern areas of Australia. The other records of malathion resistance were from England and Greece. They were marginal occurrences only. In England the three records concerned infestations in grain held at mills and were the only positive records from among 42 strains of *S. granarius* collected from various localities in England, Wales and Scotland. Malathion is used commonly in England for grain protection though the target species is often *Oryzaephilus surinamensis*. In Greece, malathion is used only by small producers, and not by the State authorities who prefer fumigants - the occurrence in Amfiali was in a strain collected from wheat stored locally but which had been imported from the U.S.A., while the record from Lamia was from material collected from bulk barley in a cooperative storage where phosphine only was used.

Malathion resistance is not a significant problem in *Sitophilus granarius* but an emerging problem is apparent, particularly in the warmer areas of its occurrence.

(F.5.4b) *Lindane resistance*

Lindane resistance had a similar distribution pattern to malathion resistance. Thirty-five strains of the 157 strains tested were resistant though this involved 50% of the 28 countries involved. The records were essentially an intensification of those for malathion at latitudes below approximately 40°N and S and extending into the warmer areas of occurrence of *Sitophilus granarius*. This corresponded with areas where there was some overlap with *S. oryzae* and lindane had been used regularly in control measures. Lindane resistance rarely occurred at higher latitudes such as in England, Scotland and Poland and then only around major trade centres. Thus, in the European and Mediterranean region, the resistance was present in Morocco but uncommon in Spain; it was common in Greece and Cyprus with occasional records only from Belgrade in Yugoslavia, Budesti in south-eastern Romania, and Sakarya from the north-east coast in Turkey. The resistance was not detected from inland Turkey or from the north-central zone of Iran, or from Afghanistan. Similarly, the resistance was not present in strains collected from high altitude areas such as Ethiopia. In the Americas, lindane resistances were reported from Argentina, Brazil and Chile in the south and southern Ontario in the north. In Australia, the resistance was present in most areas of occurrence of *S. granarius* but was less common in the southern areas of Victoria.

The levels of resistance generally have not been of a high order and reflect the limited use of lindane in control of *S. granarius*.

(F.5.4c) *Methyl bromide resistance*

There have not been any significant occurrences of methyl bromide resistance in the field despite the widespread use of methyl bromide for control of *Sitophilus granarius* and the induced tolerances that have been demonstrated in the laboratory. When considered in conjunction with the slow rate at which laboratory resistances increase, it appears that economic resistance to methyl bromide in *S. granarius* is not an immediate danger.

(F.5.4d) *Phosphine resistance*

Phosphine resistance has also been demonstrated in the laboratory and phosphine is being used much more widely today than previously for control of *Sitophilus granarius*. The eight records of change in tolerance have not been confirmed by further studies or associated with known phosphine usage and may have resulted from relatively non-specific low-level cross-tolerances. A possible exception is the strain from Buenos Aires which has retained an approximate x2 increase in tolerance over several generations of culture.

The increased use of phosphine and the demonstrated potential for resistance, indicate a considerably greater likelihood of resistance to phosphine than to methyl bromide. The restrictions imposed on *S. granarius* by its inability to fly increase the frequency of comparatively closed populations. This restricts the genetic pool available for the selection of resistance, but may result in repeated selection of the same populations and thus a higher probability of resistance if the capacity is present.

TABLE F8: RESPONSES OF REPRESENTATIVE WORLD STRAINS OF *SITOPHILUS GRANARIUS* TO MALATHION, MALATHION + TPP, LINDANE, METHYL BROMIDE AND PHOSPHINE IN DISCRIMINATING TESTS FOR RESISTANCE

Locality	Mal	M+T	Lin	MeB	PH_3	Locality	Mal	M+T	Lin	MeB	PH_3
NORTH AMERICA						**EUROPE AND NORTH AFRICA (Cont'd)**					
CANADA						**ENGLAND (Cont'd)**					
B.C.-Vancouver	100	-	100	100	100	Wilts.- Salisbury	100	-	100	-	-
"	-	-	-	100	100	Worcs.- Defford	100	-	100	100	100
Ont.-Middlesex	100	-	100	100	98	Yorks.- N.Pickering	100	-	100	100	100
"	100	-	100	100	100	" "	100	-	100	100	99
Thamesford	100	-	100	100	-						
Woodstock	100	-	76	100	-	**GERMANY F.R.**					
U.S.A.						Hamburg	100	-	100	100	100
Cal.-Riverside	100	-	100	-	-						
Ill.- Piatt Co.	100	-	100	-	-	**GREECE**					
Ore.-Silverton	100	-	100	100	100	Amfiali	98	100	99	-	-
Va -Marion	100	-	100	-	-	Lamia	95	93	2	100	-
						. "	100	-	1	100	100
SOUTH AMERICA						Piraeus	100	-	4	100	-
ARGENTINA						"	100	-	60	-	-
B.A.-Buenos Aires	89	85	95	100	97						
						MOROCCO					
BRAZIL						Casablanca	100	-	95	-	-
R.G. do Sul -Uruguiana	100	-	17	-	-	Safi	100	-	33	-	-
CHILE						Taza	100	-	13	-	-
Talagante	100	-	37	100	100						
						NETHERLANDS					
EUROPE AND NORTH AFRICA						Wageningen	100	-	100	100	100
ENGLAND						**POLAND**					
Berks.- Abingdon	100	-	100	100	100	Breganowo	100	-	100	100	100
Wantage	100	-	100	100	100	Dlon	100	-	100	-	-
"	100	-	100	-	-	Dobrojewo	100	-	100	100	100
Bucks.- Haddenham	100	-	100	100	100	Golonczewo	100	-	100	-	-
Corn. - Bodmin	100	-	100	-	-	Gulewo	100	-	100	100	100
Cumb.- Silloth	98	95	100	100	100	Hypoczka	100	-	100	-	-
Devon - Bideford	100	-	100	100	100	Kaweczyn	100	-	100	100	100
Sampford Peverell	100	-	100	100	100	Komorniki	100	-	100	100	100
Dorset - Puddletown	100	-	100	-	-	Krzyzanowo	100	-	100	-	-
Sturminster Newtown	94	98	100	100	100	Lacot	100	-	100	-	-
						Lakizewo	100	-	100	-	-
Essex - Braintree	100	-	100	100	100	Nagradowice	100	-	100	-	-
Glos.- Avonmouth	100	-	100	100	100	Paninewo	100	-	100	100	100
"	100	-	100	100	100	Pawlowice	100	-	100	100	100
Lancs.- Manchester	100	-	100	100	100	Pepowo	100	-	100	100	100
Leics.- Swithland	100	-	100	-	-	Poznan	100	-	100	-	-
Lincs.- Boston	100	-	100	-	-	"	100	-	100	100	96
Gainsborough	100	-	100	100	100	"	100	-	100	100	-
"	94	96	100	100	100	Stomowo	100	-	100	100	100
"	100	-	100	100	100	Topola	100	-	100	-	-
Som.- Sandford	100	-	90	100	100	Tarnowo Podgorne	100	-	100	-	-
Suff.- Felsham	100	-	100	100	100	Tuczepy	100	-	100	-	-
Framlingham	100	-	100	100	100	Wagrowiec	100	-	100	100	100
Haughley	100	-	100	-	-	Zieliniec	100	-	100	-	-
Saxmundham	100	-	100	-	-	Zodyn	100	-	100	-	-

TABLE F8
(Cont'd)   *S.granarius* (ii)

Locality	Mal	M+T	Lin	MeB	PH$_3$	Locality	Mal	M+T	Lin	MeB	PH$_3$
EUROPE AND NORTH AFRICA (Cont'd)						**EUROPE AND NORTH AFRICA (Cont'd)**					
PORTUGAL						YUGOSLAVIA (Cont'd)					
Lisbon	100	-	99	-	-	Belgrade	100	-	86	100	100
"	100	-	100	100	-	Darda	100	-	100	-	-
"	100	-	-	-	-						
Queluz	100	-	100	-	-	**AFRICA - SOUTH OF SAHARA**					
ROMANIA						ETHIOPIA					
Baza Oravita	100	-	100	100	100	Addis Ababa	100	-	-	-	-
Bucharest	100	-	100	100	-	" "	100	-	100	100	100
Budesti	100	-	94	100	-	" "	100	-	100	-	-
Calarasi	100	-	100	100	-	" "	100	-	100	-	-
"	100	-	100	-	-	" "	100	-	100	-	-
SCOTLAND						Nazret	100	-	-	-	-
Alness	100	-	100	100	100						
Berwick-on-Tweed	100	-	100	100	100	MOZAMBIQUE					
Campbeltown	100	-	100	100	100	Beira	100	-	100	-	-
Dumfries	100	-	100	100	100	SWAZILAND					
"	100	-	98	100	100	-	100	-	100	100	100
Dunbar	100	-	100	100	100						
Glasgow	100	-	100	100	99	**WEST ASIA**					
Kilmarnock	100	-	100	100	99	AFGHANISTAN					
Kirkaldy	100	-	100	100	100	Kabul	100	-	100	-	-
New Abbey	100	-	100	100	100	CYPRUS					
Old Meldrum	100	-	100	-	-	Athienou	100	-	32	100	-
SPAIN						Louroutzina	100	-	37	100	98
Alcala	100	-	100	-	-	Nicosia	100	-	12	100	99
"	100	-	100	100	99	IRAN					
"	100	-	100	100	99	Karadj	100	-	100	100	100
Aranda de Duero	100	-	100	100	-	Teheran	100	-	100	100	-
Baena	100	-	-	100	100	"	100	-	100	100	97
"	100	-	100	-	-	"	100	-	100	100	99
"	100	-	99	100	100	TURKEY					
Caceres	100	-	29	-	-	Ankara	100	-	100	100	98
Granada	100	-	100	100	98	"	100	-	100	100	-
"	100	-	100	100	100	"	100	-	100	-	-
"	100	-	100	100	100	"	100	-	100	-	-
Jaen	100	-	90	100	100	Sakarya	100	-	97	-	-
"	100	-	100	100	99	SAUDI ARABIA					
Madrid	100	-	42	-	-	Jeddah	100	-	100	100	100
Murcia	100	-	100	100	99						
St. Juan Puerto	100	-	100	100	100	**U.S.S.R.**					
Sevilla	100	-	100	100	100	Moscow	100	-	100	100	100
SWEDEN						"	100	-	100	100	98
-	100	-	100	100	100						
WALES						**AUSTRALASIA AND SOUTH PACIFIC**					
Aberystwyth	100	-	100	100	100						
Northop	100	-	100	100	100	AUSTRALIA*					
Pyle	100	-	100	-	-	NSW - Milbrulong	100	-	95	-	-
YUGOSLAVIA											
Belgrade	100	-	60	-	-						
"	100	-	100	100	100						

TABLE F8
(Cont'd)

S. granarius (iii)

Locality	Mal	M+T	Lin	MeB	PH_3	Locality	Mal	M+T	Lin	MeB	PH_3
AUSTRALASIA AND SOUTH PACIFIC (Cont'd)						**AUSTRALASIA AND SOUTH PACIFIC (Cont'd)**					
AUSTRALIA* (Cont'd)						AUSTRALIA* (Cont'd)					
NSW - Rozelle	90	-	72	100	100	Vic.- Newport	100	-	95	-	-
Yerong Creek	100	-	90	-	-	Raywood	100	-	100	-	-
S.A.- Edunda	97	78	77	-	-	Teddywaddy	99	97	97	-	-
Murray Bridge	67	28	94	-	-	W.A.- Watercarrin	97	77	82	100	100
Port Adelaide	100	-	80	100	100						
" "	100	-	100	-	-						
Port Lincoln	96	98	-	-	-						
Vic.- Ballarat	97	96	0	-	-						
Echuca	10	25	76	100	100						
"	39	38	76	-	-						
Glenroy	100	-	-	100	100						
Melbourne	100	-	100	-	-						
"	100	-	100	-	-						
"	99	75	100	-	-						

*These strains were tested in Australia using a discriminating concentration of 1.0% malathion and a 7 hr exposure period which would detect smaller increases in tolerance than would the discriminating concentration of 1.5% with a 6 hr exposure period used in the survey.

TABLE F9: SUMMARY OF WORLD OCCURRENCE OF RESISTANCE IN *SITOPHILUS GRANARIUS* TO MALATHION, LINDANE, METHYL BROMIDE AND PHOSPHINE

	North America	Central America	South America	Europe,N.Afr.,USSR	Africa-S.of Sahara	Western Asia	Eastern Asia	Australasia	Totals	Totals (as %)
MALATHION										
No. countries sampled	2	-	3	14	3	5	-	1	28	-
No. with resistance	0	-	1	2	0	0	-	1	4	14%
No. of strains tested	9	-	3	110	8	14	-	19	163	-
No. resistant	0	-	1	5	0	0	-	8	14	9%
MALATHION + TPP										
No. countries sampled	2	-	3	14	3	5	-	1	28	-
No. with resistance	0	-	1	2	0	0	-	1	4	14%
No. of strains tested	9	-	3	110	8	14	-	18	162	-
No. resistant	0	-	1	4	0	0	-	8	13	8%
LINDANE										
No. countries sampled	2	-	3	14	3	5	-	1	28	-
No. with resistance	1	-	3	7	0	2	-	1	14	50%
No. of strains tested	9	-	3	108	6	14	-	17	157	-
No. resistant	1	-	3	15	0	4	-	12	35	22%
METHYL BROMIDE										
No. countries sampled	2	-	2	13	2	4	-	1	24	-
No. with resistance	0	-	0	0	0	0	-	0	0	0%
No. of strains tested	7	-	2	72	2	10	-	5	98	-
No. resistant	0	-	0	0	0	0	-	0	0	0%
PHOSPHINE										
No. countries sampled	2	-	2	12	2	4	-	1	23	-
No. with resistance	1	-	1	3	0	3	-	0	8	35%
No. of strains tested	5	-	2	64	2	7	-	5	85	-
No. resistant	1	-	1	3	0	3	-	0	8	9%

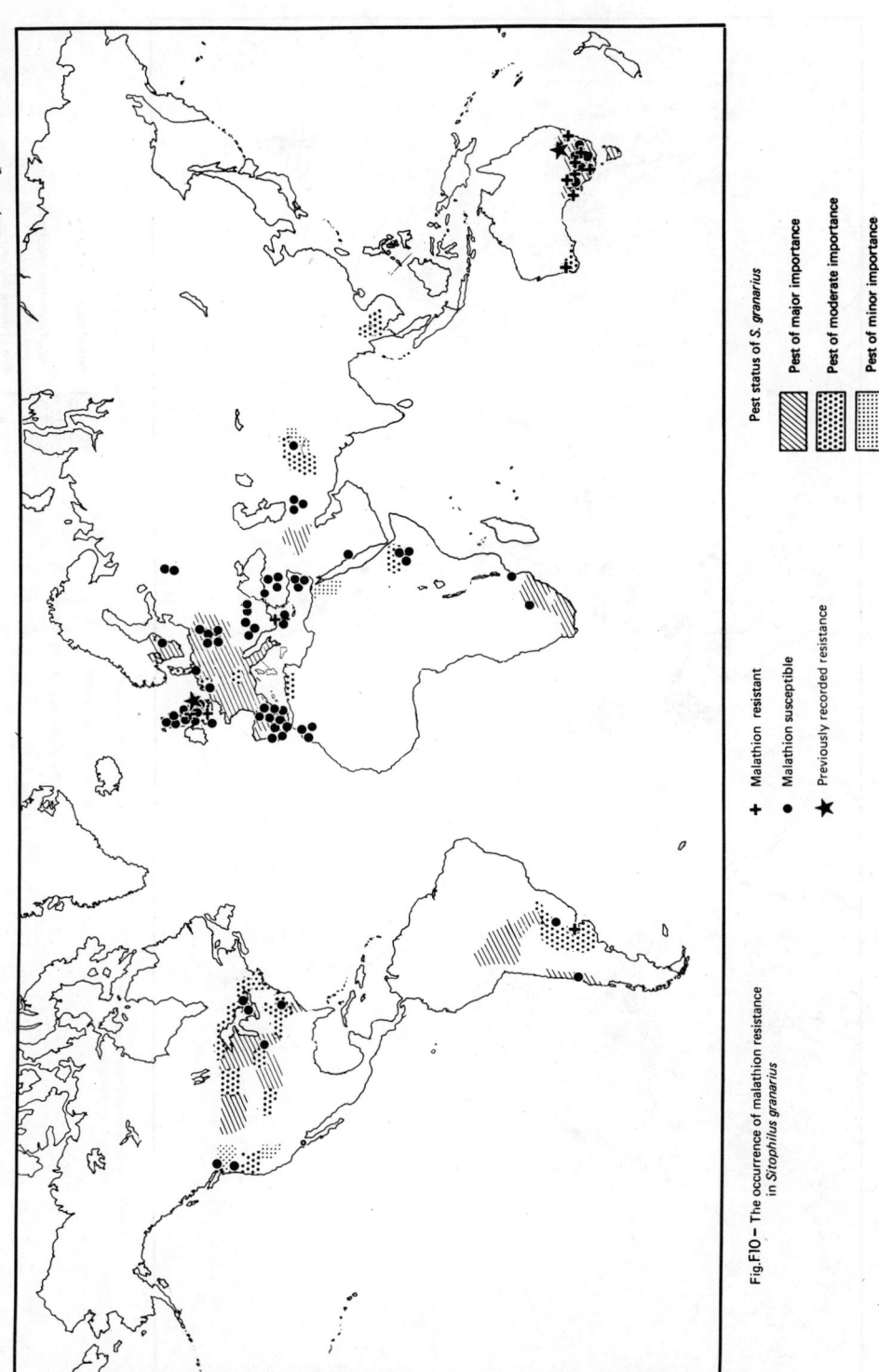

Fig. F10 — The occurrence of malathion resistance in *Sitophilus granarius*

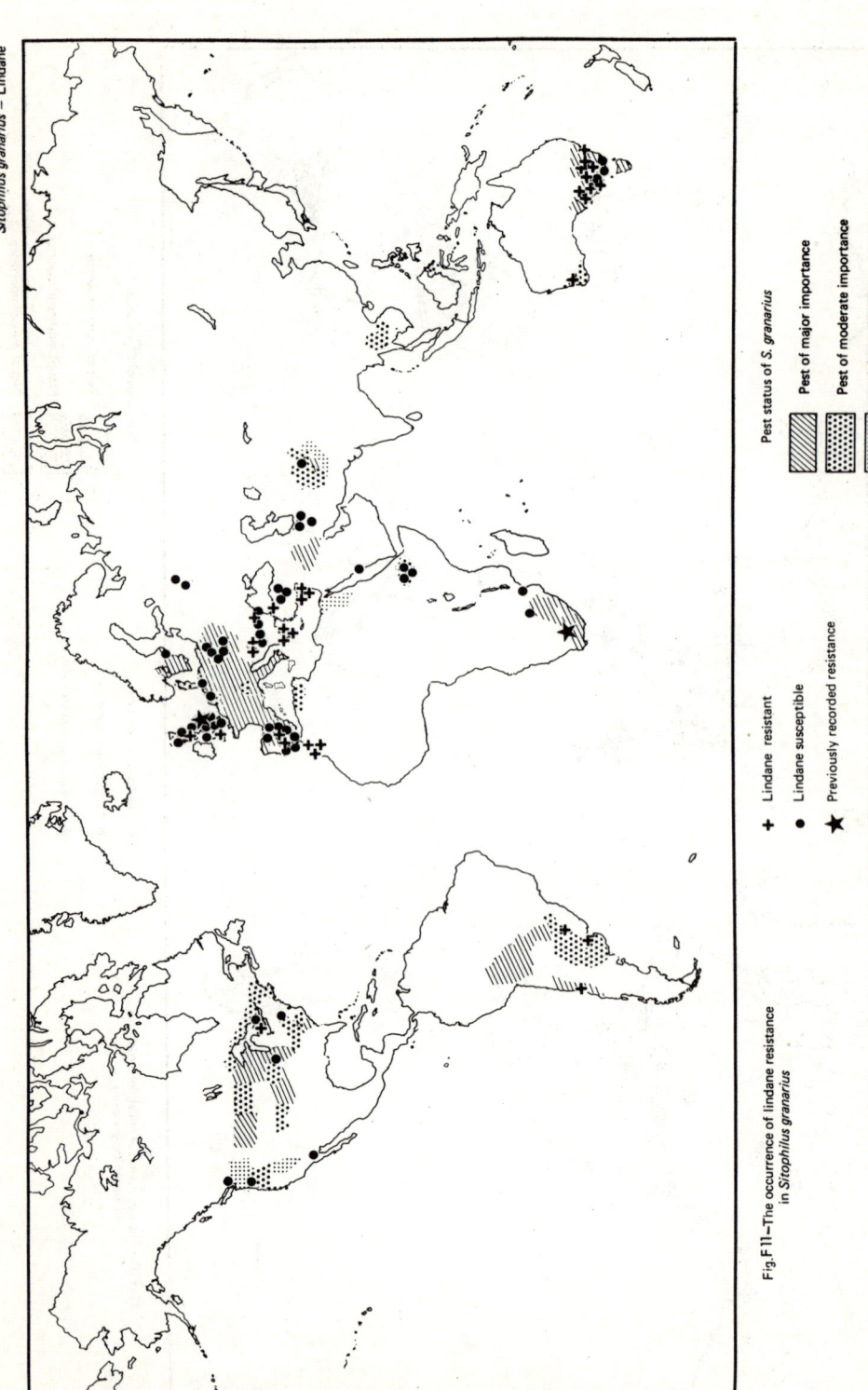

Fig. F11—The occurrence of lindane resistance in *Sitophilus granarius*

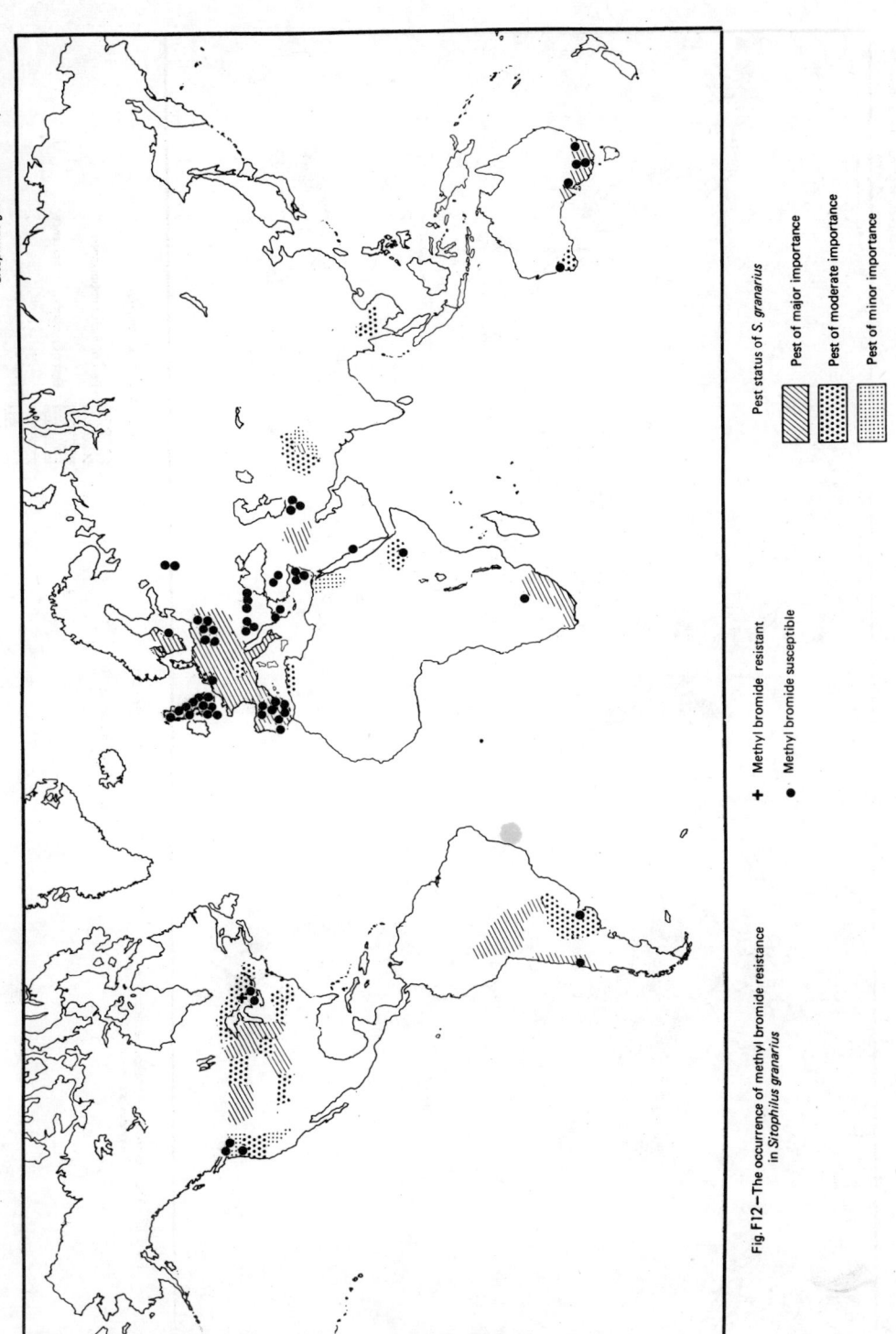

Fig. F12—The occurrence of methyl bromide resistance in *Sitophilus granarius*

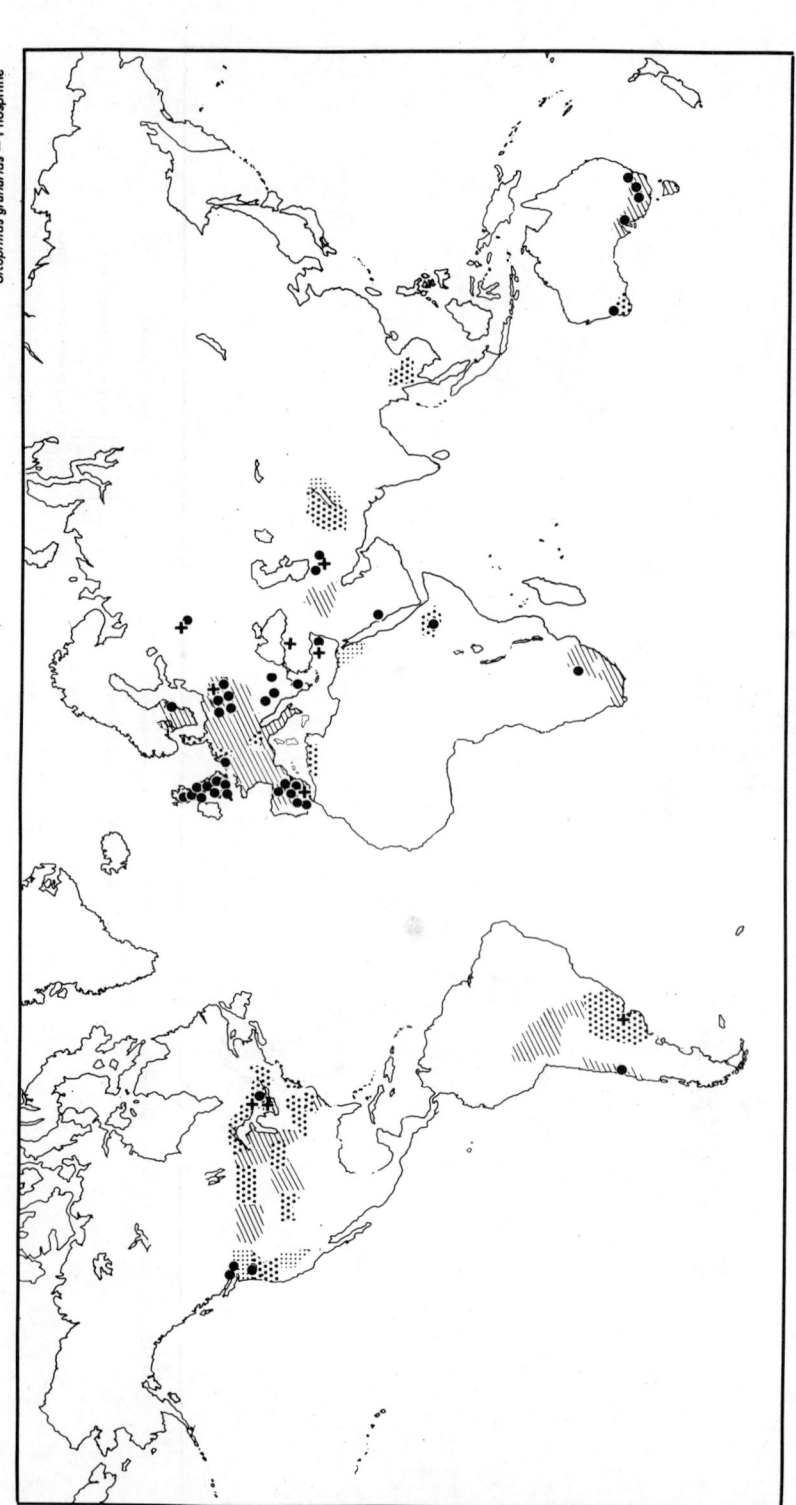

Fig. F13 – The occurrence of phosphine resistance in *Sitophilus granarius*

(F.6)   *Rhyzopertha dominica*

Rhyzopertha dominica ranks third in importance to *Sitophilus oryzae* and *Sitotroga cerealella* as a pest of stored grain but is of considerably less significance in practice. It reaches its greatest importance in the warm, dry areas where grain is left undisturbed for long periods in storage.

(F.6.1)   *Recorded insecticide resistances*

The only report of insecticide resistance in *Rhyzopertha dominica* before 1974 concerned interceptions at Avonmouth, England of three strains in maize ex South Africa, cocoa ex Ghana and wheat residues on a ship - all three strains were resistant to malathion and lindane (Dyte et al. 1973a). The strain from South Africa was not resistant to malathion synergised with triphenyl phosphate so resembling the malathion-specific type of resistance reported from *Tribolium castaneum* (Dyte and Wildey 1976, *vide* Section F14). The lack of records of resistance is a little surprising because this species is known to have characteristically flat dose-response lines with materials such as malathion (Champ et al. 1969).

In Australia, where malathion has been admixed with all wheat immediately after harvest, control of *R. dominica* has always been regarded as marginal. In March 1971 in New South Wales, Greening et al. (1975) collected a strain from the terminal elevators at Sydney which was diagnosed as resistant to malathion (x9) and dichlorvos (x10) using impregnated paper tests. Subsequently during 1971/1972 these authors detected malathion resistance in 17 grain storages in country centres of New South Wales (52 localities sampled) and dichlorvos resistance in 12 such storages (of 33 sampled). Both resistances were diagnosed also from the terminal elevators at Newcastle. In 1972/1973, malathion resistance was recorded from 16 of 25 wheat and barley storages in country centres and again at both terminal elevators. By 1973/1974, four only of 21 centres sampled yielded susceptible insects. The first records from farms of strains showing malathion resistance were made in 1973 and further records from provender merchants followed from widespread localities within the state. Similarly 8 further records of dichlorvos resistance were listed from bulk grain storages and a poultry farm. This worsening of the problem was accentuated by the collection from the Sydney terminal elevators in April 1973 of a strain with x65 resistance to malathion and x16 resistance to dichlorvos.

Dyte et al. (1975) in a survey of the pesticide resistance status of storage pests in the United Kingdom reported 2 of 8 strains of *R. dominica* resistant to lindane and 2 of 7 strains resistant to malathion. These resistances presumably were imported as *R. dominica* would not be expected to occur commonly in England.

(F.6.2)   *Recorded fumigant resistances*

Fumigation is the most widely used method of control of *Rhyzopertha dominica* on a world basis. This is a direct reflection of the greater importance of *R. dominica* in grain held in storage for a period, when fumigation is the most satisfactory and often the only practicable method of control.

There has been little work done on the responses of *R. dominica* to fumigants. In the context of resistance, Howe (1962a) was unable to detect any change in response to methyl bromide in a strain collected from maize in South Rhodesia.

(F.6.3) *Survey data*

Results from discriminating tests for malathion, lindane, methyl bromide and phosphine resistance in *Rhyzopertha dominica* are given in Table F9.and Figures F14-F17 together with a summary of the occurrences in Table F10. The distribution maps (Figs. F14-F17) also show the classification of importance currently attributed to *R. dominica* in different countries.

(F.6.4) *Comments*

(F.6.4a) *Malathion resistance*

Malathion resistance in *Rhyzopertha dominica* was detected in 23 of the 50 countries sampled, involving 50 of the 158 strains tested. The resistance appeared to be distributed throughout the general areas of occurrence of the species particularly in the warm, dry areas that favour *R. dominica*. Resistance was noticeably absent from some but certainly not all humid areas. Thus it was not recorded in strains from the Burma, Malaysia, Thailand area, and from East Africa from Kenya, through Tanzania and Malawi to the north-east of South Africa, although it was present in strains from the Philippines, Indonesia and West Africa. This is probably a reflection of the reduced pest status of *R. dominica* in the former groups of countries and the inter-related short storage periods that are necessarily a feature of humid climates. In Australia following the recording of malathion resistance in New South Wales in March 1971 (Greening *et al*. 1975, Section F.6.1), resistant strains were diagnosed from Queensland in June, South Australia in November of the same year, and from Northern Territory in 1973 and Victoria in 1974 (Champ unpublished data). It seems probable that resistance had been present for some time and the extra attention given to sampling *R. dominica* following Greening's record, resulted in the detection of the resistance from widely-distributed centres.

The levels of resistance encountered were generally not high - the usual level was of the order of x5 at the LC_{50}. However, this must be qualified by consideration of the flat dosage-mortality regression lines that are characteristic of *R. dominica*. These illustrate the difficulty in completely controlling susceptible strains with malathion and indicate that even low resistance factors, though allowing the general level of the population to be suppressed, will ensure significant survival from economically feasible dose rates.

All the occurrences listed with the exception of one strain from imported grain in a flour mill in Kuwait concerned resistances that were suppressible with triphenyl phosphate. The carboxyesterase system that is suppressed specifically by triphenyl phosphate, however, contributes significantly to normal tolerance of *R. dominica* to malathion (Fig. F1) and low-level resistances not specific to malathion may have been present in the strains. Cross-resistance tests on the strains listed from Australia indicate nevertheless, that these resistances were specific to malathion. Further investigations in Australia determined a non-specific type of malathion resistance that was not suppressible with triphenyl phosphate and involved a wide range of cross-resistances to other organophosphorus compounds (Section F.14). There had been an incipient problem with *R. dominica* since malathion was introduced in 1964 and by 1971/1972, *R. dominica* had become the predominant insect in infestations particularly in New South Wales, reflecting the widespread occurrence of malathion resistance. Dichlorvos and fenitrothion were introduced into

control programmes for treatment of the fabric of buildings following
detection of malathion resistance in *Tribolium castaneum* in 1968 and
Sitophilus oryzae in 1970. Dichlorvos also has been used widely as an
admixture treatment and the appearance of dichlorvos resistance was to
be expected. This type of organophosphorus-compound resistance is
confined largely to New South Wales but occurrences in Queensland have
become more frequent and a resistant strain has been diagnosed from
Western Australia. The infestations concerned in these outbreaks usually
have involved low populations of insects but the problem is compounded by
the innate variability of the species, the fact that populations are often
survivors from treatments, and the necessity for complete disinfestation
to meet nil tolerance requirements for export grain. The most serious
outbreaks have concerned situations where hygiene and inspections has
been unsatisfactory and reinfesting populations allowed to build up. The
levels of resistance to malathion in these outbreaks are now of a high order
(x240) and give rise to new concern as to the potential of the emerging
problem of organophosphorus-compound resistance in *R. dominica*.

(F.6.4b) *Lindane resistance*

Lindane resistance in *Rhyzopertha dominica* was more widespread
than malathion resistance occurring in 41 of the 51 countries sampled and
91 of the 137 strains tested. There was no apparent pattern in its
occurrence. Iraq was one of the few countries with a climate apparently
favourable for this species where resistances to all pesticides tested
including lindane were almost absent. *R. dominica* was not regarded as
a serious pest in Iraq except in the Baghdad and Basra areas. A contributing factor to this would be the low levels of infestation associated with
both underground and overground farm storage in Central and North Iraq
particularly.

Lindane is rarely used for control of *R. dominica* though there
would be contact indirectly in general use of lindane as a hygiene
treatment. Thus the occurrence of resistance is not of particular significance.

(F.6.4c) *Methyl bromide resistance*

The only record of a change in tolerance of *Rhyzopertha
dominica* to methyl bromide among the 93 samples tested from 45 countries
was in a strain collected from grain residues in silos at Piraeus in
Greece. The change, however, was not outside the probability limits of
a chance occurrence in a normally susceptible strain. The strain was
resistant to lindane and phosphine but not to malathion and this may have
conferred some low-level non-specific tolerance. Methyl bromide is used
in Greece only in chamber fumigation on a limited scale.

Methyl bromide has been widely, and in some countries intensively
used for control of *R. dominica*. The scale and often inefficiency of
many of these fumigations should have created conditions favourable for
development of resistance but none has been recorded. This indicates
that *R. dominica* may not be particularly prone to develop methyl
bromide resistance. Thus, for example, a strain collected in the Caribbean
area from a stack of bagged paddy that had been fumigated with methyl
bromide 13 times in 18 months and was infested, gave a normal response

with the discriminating test for resistance. There were numerous other less spectacular instances demonstrating the stability of normal methyl bromide tolerance levels in R. dominica.

(F.6.4d) *Phosphine resistance*

The situation with phosphine resistance in *Rhyzopertha dominica* is completely different from that with methyl bromide. Twenty-two strains of the 92 tested were shown to have abnormal tolerance to phosphine and many of these were shown to satisfy such criteria of true resistance as heritability and intensification of the resistance with selective breeding from survivors.

Eleven countries were involved. The occurrences, however, appeared as pockets of resistance rather than diffusely distributed throughout the world. In most instances, phosphine resistance was confirmed in other species collected from the same localities (specifically *Sitophilus oryzae* and *Tribolium castaneum*). These localities were particularly the Bombay and Delhi areas of India, Guyana, Argentina, Libya, Greece and the Central African Republic, with lesser though significant occurrences in Australia.

The highest levels of resistance were recorded from India. They concerned the Bombay area of Maharashtra State and to a lesser extent the area about Delhi. The resistance levels of adults from these strains at the LD_{50} approximated x7 in the strains from Borivli, x1.4 in the strain from Hapur and x1.9 from Kaili. Breeding from the survivors from discriminating tests with the Hapur strains has raised the resistance level to a level similar to that shown by the Borivli strains. Phosphine usage in the Bombay-Borivli area has been discussed with reference to *Sitophilus oryzae*. Here samples were taken from central storages containing wheat and rice infested with a mixture of species including R. dominica. These storages were subject to an extensive programme of insecticide and fumigant controls including fumigation 3 or 4 times a year at the rate of 2g phosphine/ton supplemented as necessary with surface treatment only with this material. The records from the Delhi area were from wheat in a trade shop at Hapur and from a farm store at Kaili nearby.

The patterns of pest occurrence in India have shown significant changes in recent years with R. dominica increasing its relative pest status. In the period from 1944 to 1950, safe storage periods for grain averaged about three months, increasing to six months during 1950 to 1960 with the introduction of the newer synthetic pesticides and more organised pest control programmes. Now, with further improvements including more efficient use of fumigants, the safe storage periods approximate three years ranging from a minimum of six months to a maximum of five years. Until 1960, *Trogoderma granarium* was the major pest, particularly in the northern areas but by 1965 its numbers had been reduced to very low levels and *Sitophilus oryzae* and *Rhyzopertha dominica* were the most important species in almost all cereals and particularly in wheat and rice. R. dominica also occasionally attacked paddy. Until 1967, fumigants were imported and the main emphasis for control was on surface treatment principally with BHC. At this time malathion was substituted for BHC and India commenced manufacturing fumigants which introduced a period of increasing use of these materials. Production in 1972 of aluminium phosphide tablets was of the order of 170 - 200 tons, most of which was used in India. The stage has now been reached with phosphine, as with malathion, that the levels of control previously obtained are not being realised (S.V. Pingale, personal communication). The reasons for these reduced effects have not been determined but the records of resistance in *R. dominica* as well as *Sitophilus oryzae* and *Tribolium castaneum* imply some contribution from resistance and show a prospect of considerably greater difficulty in maintaining the present expected safe storage periods for grain.

The only other record from Asia of phosphine resistance in *R. dominica* was from bagged wheat and barley in the port silos in Keelung in China (Taiwan). This was of a low order only. Phosphine is used in China (Taiwan) but usually only when pests are present in large numbers.

Phosphine resistance comparable with that from India, has been recorded from *R. dominica* in Guyana. Here a strain showing > x12 resistance in adults was collected from empty bags and wheat residues in an empty grain bin at a flour mill in Georgetown where phosphine was used to disinfest empty bins. It was stated that exposure periods were often as short as 24 hours from introduction of the aluminium phosphide tablets. All wheat used in the mill was imported from the U.S.A. Phosphine is used also in fumigation of paddy in Guyana though exposure periods are 4-5 days. There were other records of phosphine resistance at lower levels from the Caribbean area. In Jamaica, a resistant strain was collected from maize meal in a grain store in Kingston and another from paddy in a rice mill in Spanish Town where phosphine was used occasionally at the rate of 5g/ton for 72 hr. In Guadeloupe a strain collected from a research station at Petit Bourg also showed low-level resistance. There is active movement of infestable commodities around the Caribbean particularly in rice exports from Guyana, providing an effective dispersal mechanism for resistance.

In Argentina, phosphine resistance was detected in bagged wheat residues again stored under a flour mill. This mill in Buenos Aires had a high turnover rate of grain using 25000 tons a month with a storage capacity of 30000 tons. Grain was inspected on arrival and if storage for several months was anticipated, malathion dust was added (to which resistance was also detected), otherwise phosphine was used at the rate of 4 g/ton often for 2-3 days only.

The records of phosphine resistance from Libya were comparable with those from India and Guyana. These involved strains of *R. dominica* taken in Tripoli from poultry food containing crushed maize and wheat. Phosphine is used on a wide range of commodities in Libya usually at a dose rate in grain of 1-3 g/ton for not less than 3 days. Under some circumstances, the tablets are dropped in water to achieve a quick fumigation. Another record from the Mediterranean area was from maize and dust residues in the port silos at Piraeus in Greece. Phosphine is the principal fumigant used in Greece and is widely used in mills, silos and flat storage. Dose rates were 6 g/ton for silos, 10 g/ton for other storage and for transport, and 3-4 g/ton for adequately closed storages. When first introduced it was used at the rate of 20 g/ton on open storages and the rate was subsequently reduced to 8-10 g/ton with a minimum of 4 days exposure. It was usually not considered possible to cover the open storages during fumigation.

Phosphine resistance was also recorded from Bangui in the Central African Republic in *R. dominica* collected from an oil mill and from bagged wheat stacked in an open shed. The significance of the protracted periods involved in transporting grain to countries such as Central African Republic has been outlined when considering *Sitophilus oryzae*, another species in which phosphine resistance was found in this country which uses virtually no pesticides in the control of stored product pests. The wheat concerned in the above record had been shipped from Greece where phosphine resistance was known to occur, and had been fumigated with phosphine before leaving Greece.

In Australia where phosphine is widely used and fumigations have not always been carried out under optimal conditions, the records of changes in tolerance were of a low order and principally serve notice of an emerging problem. All states except Victoria were concerned. *R. dominica* is well adapted to the climate of many areas of Australia and with the spread of malathion and general organophosphorus compound resistance, it has assumed an increased importance. The progeny from survivors from discriminating tests on the sample from Cambooya, Queensland showed a very marked increase in tolerance which approached the levels of the strains from India.

A common feature of all these outbreaks of phosphine resistance is their association with use of phosphine under conditions that conflict with the accepted principles of fumigation and the known properties of phosphine. Phosphine is a very efficient fumigant but laboratory experience with stored grain insects generally indicates that they have little difficulty in developing resistance to it. *R. dominica*, though not investigated specifically, appears from the survey to be no exception and possibly more prone to development of phosphine resistance than most other species. There appears to be and have been widespread misuse of phosphine, particularly with respect to exposure times and gas tightness during fumigation. It is surprising that the occurrences of phosphine resistance in *R. dominica* are not more extensive though the frequency of 23% of strains tested in the survey is itself sufficiently serious, and a portent for the future of phosphine for control of *R. dominica*.

TABLE F10 RESPONSES OF REPRESENTATIVE WORLD STRAINS OF *RHYZOPERTHA DOMINICA* TO MALATHION, MALATHION + TPP, LINDANE, METHYL BROMIDE AND PHOSPHINE IN DISCRIMINATING TESTS FOR RESISTANCE

Locality	Mal	M+T	Lin	MeB	PH_3	Locality	Mal	M+T	Lin	MeB	PH_3
NORTH AMERICA						**EUROPE AND NORTH AFRICA**					
U.S.A.						**ALGERIA**					
Cal. - Riverside	93	100	100	-	-	Dar-es-Abeida	100	-	100	100	100
Kans.- McPherson	100	-	100	100	100	**BULGARIA**					
N.C. - Clinton	100	-	97	100	100	Kostinbrod	100	-	74	-	100
Farmville	100	-	98	100	100	**ENGLAND**					
CENTRAL AMERICA AND CARIBBEAN						Devon. - Barnstaple	100	-	100	-	100
ANTIGUA	100	-	83	100	100	Glos. - Avonmouth	43	100	23	-	-
COLOMBIA						"	93	100	57	-	-
Barranquilla	77	100	53	100	100	"	99	100	62	-	-
"	100	-	62	-	-	"	100	-	100	100	100
DOMINICAN REP.						Lancs. - Glazebury	100	-	100	100	100
San Cristobal	100	-	81	99	100		100	-	100	100	100
GUADELOUPE						Lincs. - Boston	58	100	100	100	100
Petit Bourg	72	100	97	-	-	**GREECE**					
" "	63	100	100	100	92	Lamia	100	-	4	-	-
GUYANA						"	100	-	0	100	100
E.C. Demerara	100	-	87	-	-	Piraeus	100	-	2	98	63
Georgetown	100	-	100	100	5	**LIBYA**					
JAMAICA						Gabal el Aghadar	100	-	30	100	100
Kingston	91	100	100	100	100	Tripoli	86	100	14	100	100
"	68	100	100	100	95	"	100	-	96	100	50
Spanish Town	97	100	100	100	98	"	100	-	41	-	-
"	92	100	84	-	-	"	-	-	100	100	9
TRINIDAD						**MOROCCO**					
Port of Spain	100	-	100	100	100	Ouedzem	100	-	100	100	100
VENEZUELA						**POLAND**					
La Morita	100	-	86	100	100	Poznan	100	-	100	-	-
"	92	100	84	100	100	**PORTUGAL**					
						Lisbon	32	100	100	100	100
SOUTH AMERICA						**SCOTLAND**					
ARGENTINA						Glasgow	100	-	30	100	100
B.A. - Buenos Aires	92	100	32	100	61	**YUGOSLAVIA**					
" "	91	100	64	100	84	Belgrade	-	-	17	100	100
" "	100	-	99	100	100	**AFRICA - SOUTH OF SAHARA**					
Mar del Plata	100	-	63	100	-						
Tres Arroyos	100	-	26	100	100	**CENT. AFRICAN REP.**					
Cord.- Isla Verde	100	-	63	-	-	Bangui	24	100	28	-	-
Sta Fe - Rosario	100	-	98	100	100	"	96	100	53	100	50
San Francisco	100	-	97	-	-	"	100	-	40	100	60
BRAZIL						**ETHIOPIA**					
Goias - Anapolis	97	100	94	100	100	Addis Ababa	-	-	-	100	100
PERU						**KENYA**					
Chiclayo	100	-	99	-	-	Mombasa	99	100	54	100	100

TABLE F10
(Cont'd)   *R.dominica (ii)*

Locality	Mal	M+T	Lin	MeB	PH$_3$	Locality	Mal	M+T	Lin	MeB	PH$_3$
AFRICA - SOUTH OF SAHARA (Cont'd)						**WEST ASIA (Cont'd)**					
MALAWI						INDIA (Cont'd)					
Blantyre	100	-	89	-	-	U.P. - Kaili	100	-	33	100	86
"	100	-	89	-	-	Meerut	93	100	84	-	-
Limbe	100	-	93	100	100	IRAN					
MOZAMBIQUE						Khorramshar	100	-	100	100	100
Beira	52	100	79	-	-	IRAQ					
Lourenco Marques	100	-	100	100	100	Abu-Ghraib	88	-	100	100	100
" "	-	-	-	100	98	Arbil	100	-	100	100	100
NIGERIA						Baghdad	100	-	100	100	100
Lagos	54	100	97	100	100	"	100	-	100	100	100
"	91	100	48	-	-	Diwanyah	100	-	100	100	100
"	100	-	100	100	100	Mosul	100	-	100	100	100
SENEGAL						Rifae	100	-	100	100	100
Bambey	100	-	6	-	-	Swaira	100	-	100	100	100
"	-	-	-	99	99	ISRAEL					
SOUTH AFRICA						Tel Aviv	100	-	2	-	-
Eloff	-	-	-	100	-	KUWAIT	4	96	100	100	100
Pretoria	100	-	20	-	-	NEPAL					
"	100	-	100	100	100	Inner Terai	100	-	83	100	100
Radium	100	-	30	-	-	PAKISTAN					
TANZANIA						Karachi	65	100	88	-	-
Dar-es-Salaam	100	-	40	-	-	"	51	100	100	-	-
" "	100	-	45	100	100	SAUDI ARABIA					
						Jeddah	78	100	61	100	99
WEST ASIA						"	85	100	64	100	100
BAHRAIN	82	100	81	100	100	"	62	100	100	100	100
"	93	100	86	100	100	"	100	-	67	-	-
"	100	-	100	100	100	"	100	-	100	100	100
"	45	100	73	100	100	TURKEY					
CYPRUS						Ankara	100	-	26	-	-
Nicosia	100	-	100	100	100	Kösk	-	-	-	100	100
Pyrga	100	-	91	-	-	Sakarya	100	-	74	100	100
INDIA						YEMEN ARAB REP.					
Maharashtra - Bombay	100	-	46	-	-	Hodeida	69	100	91	100	100
"	100	-	46	99	37						
"	100	-	7	-	-	**EAST ASIA**					
"	58	100	58	-	-	BURMA					
"	100	-	11	-	-	Rangoon	100	-	0	100	100
Borivli	92	100	89	100	3	"	100	-	21	-	-
"	86	100	41	-	-	"	100	-	56	100	100
"	100	-	39	100	4	CHINA					
Punjab - Karnal	100	-	23	-	-	Taiwan-Keelung	100	-	83	100	94
"	100	-	11	99	100						
Rajasthan - Kota	93	100	87	-	-	INDONESIA					
U.P. - Delhi	100	-	20	-	-	Djakarta	100	-	100	100	100
"	100	-	11	-	-	"	97	100	78	100	100
Hapur	100	-	23	100	93	"	88	100	81	-	-
"	100	-	11	-	-						

TABLE F10
(Cont'd)

R.dominica (iii)

Locality	Mal	M+T	Lin	MeB	PH_3	Locality	Mal	M+T	Lin	MeB	PH_3
EAST ASIA (Cont'd)						**AUSTRALASIA AND SOUTH PACIFIC (Cont'd)**					
JAPAN						AUSTRALIA (Cont'd)					
Omiya	100	-	100	100	100	NSW - Moree	84	100	-	-	-
Yokohama	99	100	78	-	-	Peak Hill	100	-	-	-	-
KOREA REP.						Trangie	87	100	-	-	-
Sosa	99	100	78	-	-	Willow Tree	100	-	-	-	-
MALAYSIA						Qd - Biloela	98	100	-	-	-
Alor Setar	100	-	88	100	100	Cambooya	73	100	100	100	81
" "	100	-	-	-	-	Clermont	100	-	-	-	-
PHILIPPINES						Gindie	100	-	-	-	-
Sorsogon	90	100	44	100	100	Hodgson	100	-	-	-	-
Tanauan	54	100	67	100	100	Macalister	100	-	-	-	-
THAILAND						Miles	100	-	-	100	100
Bangkok	100	-	-	-	-	S.A. - Adelaide	100	-	100	-	-
"	100	-	77	-	-	Red Hill	88	100	100	100	98
Khon Kaen	100	-	68	-	-	Vic. - Bridgewater	100	-	100	-	-
Rangsit	100	-	30	-	-	Bunnaloo	100	-	-	-	-
						Echuca	100	-	-	100	100
AUSTRALASIA AND SOUTH PACIFIC						Natya	100	-	100	-	-
						W.A. - Beacon	8	100	-	-	-
AUSTRALIA						Geraldton	100	-	-	-	-
NSW - Armatree	28	100	100	100	98	Guildford	100	-	-	-	-
Berrigan	100	-	-	-	100	"	99	100	-	-	-
Gular	37	100	-	-	-	Kwelkan	100	-	-	100	98
						Mingenew	100	-	-	-	-
						West Perth	100	-	-	-	-

TABLE F11:   SUMMARY OF WORLD OCCURRENCE OF RESISTANCE IN *RHYZOPERTHA DOMINICA* TO MALATHION, LINDANE, METHYL BROMIDE AND PHOSPHINE

	North America	Central America	South America	Europe,N.Afr.,USSR	Africa-S.of Sahara	Western Asia	Eastern Asia	Australasia	Totals	Totals (as %)
MALATHION										
No. countries sampled	1	8	3	9	8	12	8	1	50	-
No. with resistance	1	4	2	3	3	7	2	1	23	46%
No. of strains tested	4	15	10	21	18	45	18	27	158	-
No. resistant	1	8	3	5	5	16	4	8	50	32%
MALATHION + TPP										
No. countries sampled	1	8	3	9	8	12	8	1	50	-
No. with resistance	0	0	0	0	0	1	0	0	1	2%
No. of strains tested	4	15	10	21	18	45	18	27	158	-
No. resistant	0	0	0	0	0	1	0	0	1	1%
LINDANE										
No. countries sampled	1	8	3	10	8	12	8	1	51	-
No. with resistance	1	7	2	6	8	9	8	0	41	80%
No. of strains tested	4	15	10	23	18	45	16	6	137	-
No. resistant	2	9	8	13	15	30	14	0	91	66%
METHYL BROMIDE										
No. countries sampled	1	8	2	8	9	10	6	1	45	-
No. with resistance	0	0	0	1	0	0	0	0	1	2%
No. of strains tested	3	11	7	15	13	29	9	6	93	-
No. resistant	0	0	0	1	0	0	0	0	1	1%
PHOSPHINE										
No. countries sampled	1	8	2	9	9	10	6	1	46	-
No. with resistance	0	3	1	2	2	1	1	1	11	24%
No. of strains tested	3	11	6	17	12	29	9	7	94	-
No. resistant	0	4	2	3	3	5	1	4	22	23%

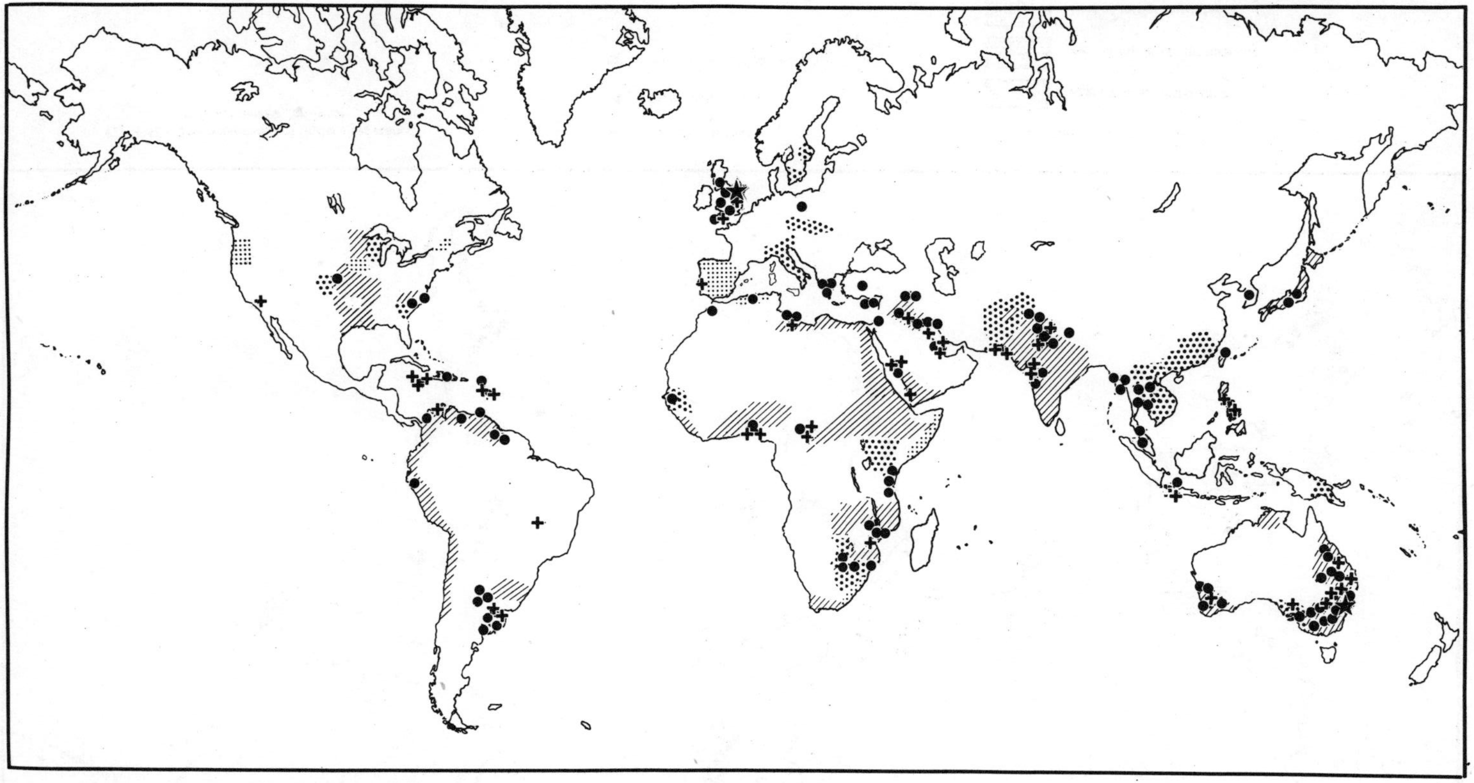

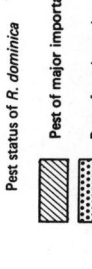

Fig. F14 – The occurrence of malathion resistance in *Rhyzopertha dominica*

+ Malathion resistant
● Malathion susceptible
★ Previously recorded resistance

Pest status of *R. dominica*

- Pest of major importance
- Pest of moderate importance
- Pest of minor importance

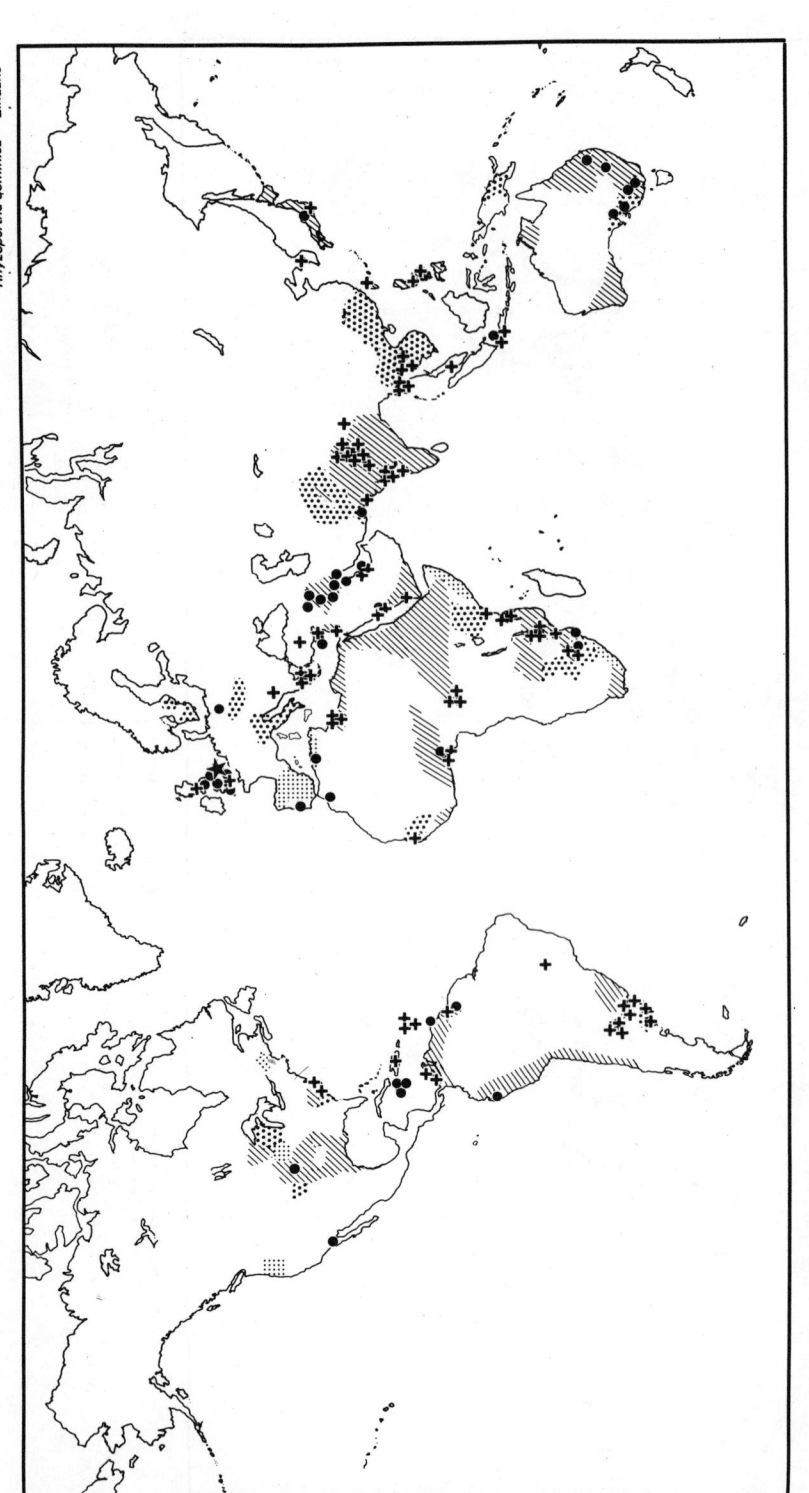

Fig. F.15 — The occurrence of lindane resistance in *Rhyzopertha dominica*

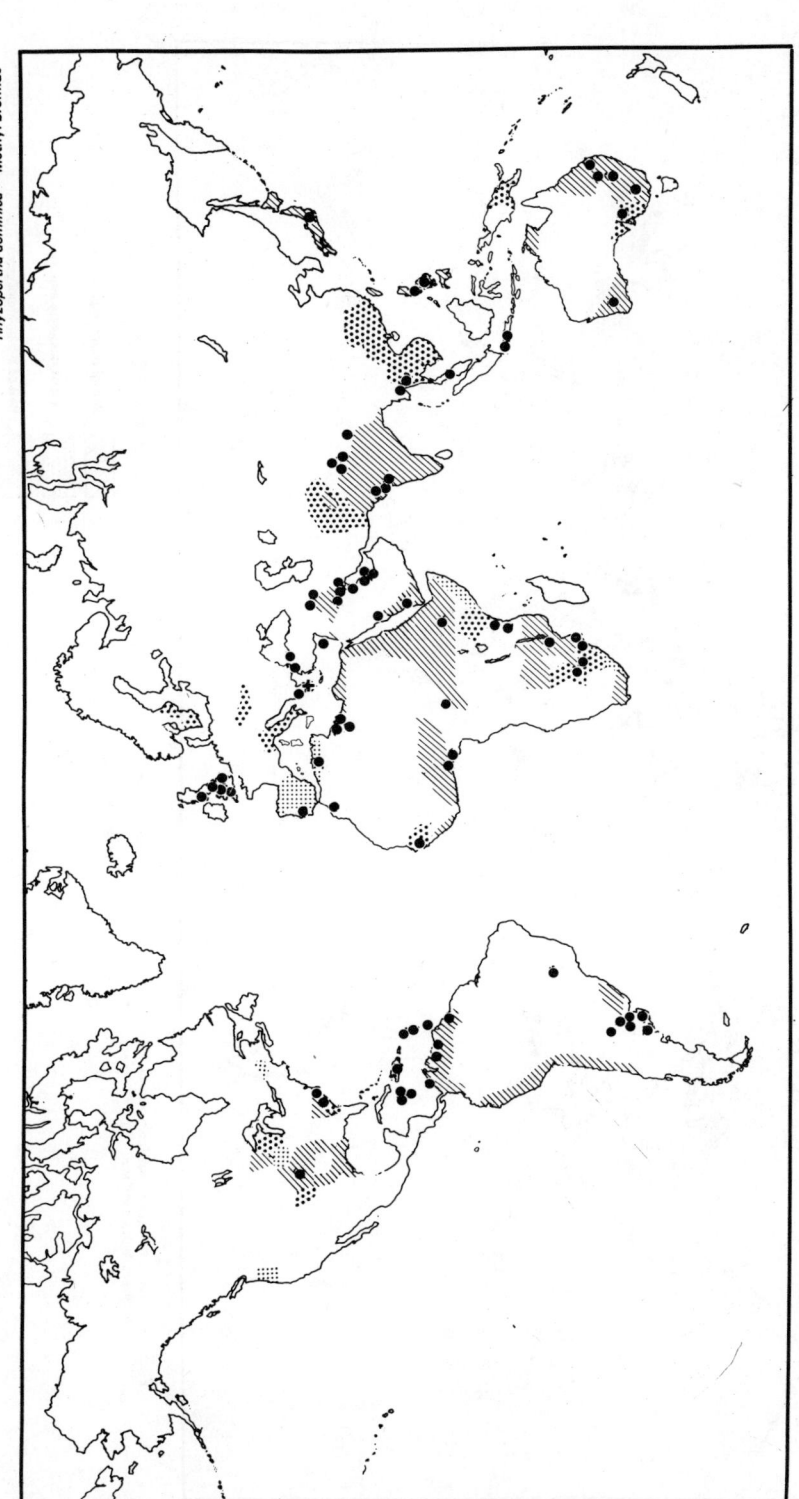

Fig. F16.–The occurrence of methyl bromide resistance in *Rhyzopertha dominica*

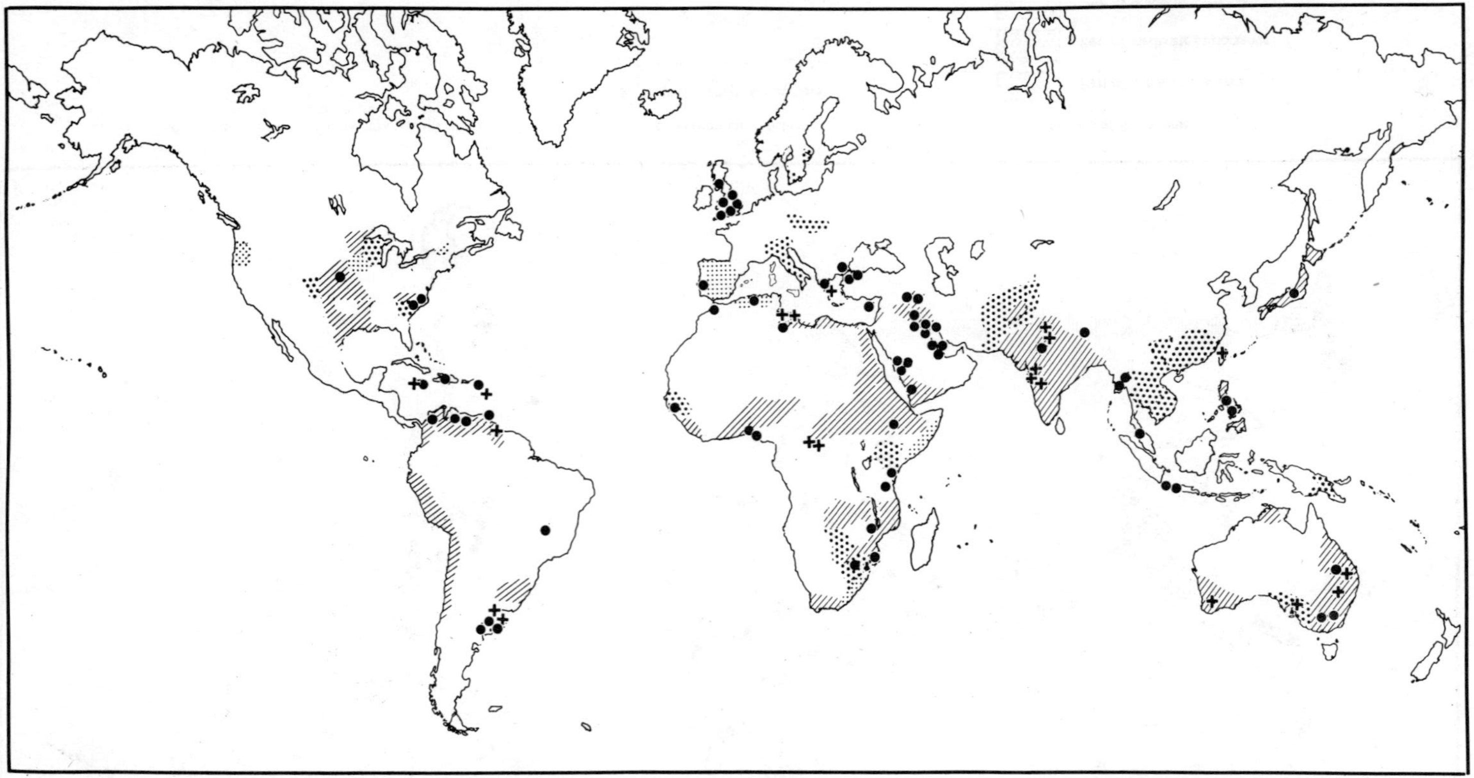

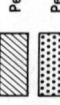

Fig. F17– The occurrence of phosphine resistance in *Rhyzopertha dominica*

+ Phosphine resistant

● Phosphine susceptible

Pest status of *R. dominica*

Pest of major importance

Pest of moderate importance

Pest of minor importance

(F.7) *Tribolium castaneum*

Tribolium castaneum is the commonest stored product insect in the world and it is not surprising that early reports of resistance concerned this species. It is endemic to tropical and sub-tropical areas but has become established in heated premises in many cool but not cold countries. Although *T. castaneum* is a pest of cereal grains, it does not cause damage comparable with other species and is often regarded as a nuisance rather than a source of direct loss. Its importance increases as a contaminant of milled cereal products and reaches considerably higher levels of importance with oil seeds such as peanuts. This gradation of importance is reflected in some measure by the intensity of use of chemicals for its control. However, this must be qualified by its common occurrence in mixed infestations with consequent exposure to pesticides used primarily against other species. However, the adults are active and conspicuous, and infestations are often obvious at lower population levels than for other species, thus increasing the frequency of applied control measures in some circumstances.

(F.7.1) *Recorded insecticide resistances*

T. castaneum is a classical tool for population ecology studies and for genetic research. As with *Sitophilus granarius* there has been a number of laboratory selection programmes aimed at intensifying insecticide resistances. These programmes have concerned DDT, lindane and malathion and with the available background of formal genetic data for *T. castaneum*, most investigations in stored grain insects on the genetic aspects of resistance have concerned this species. Field records of resistance are well-documented and cover most areas of the world and most insecticides.

Holborn (1957) at the Cooper Technical Bureau, England, using dust impregnated wheat and a 6 day exposure period recorded a strain of undefined origin that was x4 more tolerant to lindane and x3.5 more tolerant to pyrethrins than the most susceptible strain he had available. Despite widespread use of lindane and pyrethrins, subsequent reports of resistance to these materials are rather sparse, apparently because of limited examinations of strains for such resistances. The first field reports of lindane resistance were from Sierra Leone (x4) (Anon. 1958c) and Nigeria (Anon. 1958d). Subsequently Holborn (1960, in Busvine 1962b, Brown 1963) recorded low level BHC resistance in strains of *T. castaneum* from Nigeria and Sierra Leone and a x12 resistance from Kenya. There was a further report of a low level resistance to both lindane and dieldrin in a malathion-resistant strain from Nigeria (Parkin and Bright 1965). Then Speirs (1967) recorded lindane resistance in a strain of *T. castaneum* collected from peanuts in Georgia U.S.A., and Champ and Campbell-Brown (1970a,b) using knockdown response from exposure to impregnated filter paper reported lindane resistance in strains collected from a number of localities in Queensland and New South Wales in peanuts particularly, but also in wheat, barley, sorghum and a provender mill. The maximum resistance factor observed was x100.

Dyte and Blackman (1970a, Dyte 1970, Dyte *et al*. 1973c) also using an impregnated paper technique examined 18 strains from 11 countries for lindane resistance and provided the first extensive listing of occurrences which included England, The Gambia, Malawi, Malaysia, Senegal, the Seychelles and Zambia as well as further records from Australia and Kenya. Three strains only were susceptible including the only strain tested from Japan. A wide range of commodities was concerned. A further survey in 1973 by Dyte et al.(1975) within the United Kingdom found 24 of 28 strains resistant to lindane. Toppozada et al. (1969) in Egypt compared the tolerance to lindane treated wheat of *T. castaneum* collected from a storage in Alexandria Harbour with base-line data supplied by Parkin (vide Parkin 1965) and reported a x2 increase. In India, where lindane was commonly used to disinfest godowns, Bhatia et al. (1971) were unable to demonstrate lindane resistance in a strain of *T. castaneum* collected from rice and flour in Delhi though Bhatia and Pradhan (1971) at the Indian Agricultural Research Institute selected a lindane resistant strain during 10 generations of breeding in lindane-impregnated flour. Initial selections were made at 20 ppm increasing to 600 ppm after which resistance levels were measured in the impregnated medium and by spray-film and direct spraying methods. The increases in tolerance were x55, >x87 and x13 respectively when selected progeny were compared with the parental strain. When comparisons were made with a reference susceptible strain there was good correspondence except for the impregnated medium method where a factor of x90 was found. In this latter method responses were measured as percentage emergence of adults from a 4 day oviposition period when compared with untreated batches, and thus include the responses of all development stages and associated behavioural patterns. The resistance appeared to be a typical lindane-cyclodiene type with little cross-resistance except to carbaryl (>x12).

Pieterse et al. (1972) using an impregnated paper method surveyed the occurrence of lindane resistance in *T. castaneum* in warehouses and stores throughout Malawi where lindane had been used regularly for many years and found a widespread problem. All 17 strains collected were resistant to lindane complementing the earlier record of resistance in a single strain by Dyte and Blackman (1970a). Subsequently Pieterse and Schulten (1974) examined ten strains collected during 1971 from stored cob maize in cribs on small farms in different parts of Malawi. Both strains obtained from the Northern Region were resistant as were two of the six strains from the Central Region, and one of the two from the Southern Region.

In Canada, Bond (1973) using the FAO method for detecting resistance in *T. castaneum*, recorded lindane resistance in two strains collected from a flour mill where lindane was used irregularly, and from an adjacent feed mill where lindane was not used.

With pyrethrins which are widely used in pest control in foodstuffs, the first survey of susceptibility was carried out in the U.S.A. by Vincent and Lindgren (1967) who examined 13 strains from 7 States using a topical application technique and 7 day mortalities. They found changes in tolerance in single samples from New York (x5.7), Iowa (x2.6), Minnesota (x1.6) and Kansas (x1.6). In south-east U.S.A., synergised pyrethrins were in general use for controlling insect pests of peanuts before being replaced with malathion during 1959-1961. Speirs and Zettler (1969, Speirs 1967) using a similar technique and 5 day mortalities examined the resistance pattern of a field-collected

malathion-resistant strain from peanuts in Georgia which subsequently
had been selected with malathion and found a x13 resistance to pyrethrins.
In Queensland where pyrethrins were used only occasionally in peanuts,
Champ and Campbell-Brown (1970b) using an impregnated paper technique
and knockdown, found very low increases in tolerance to both synergised
(x1.2) and unsynergised pyrethrins (x1.4). Cichy (1971) at the Warsaw
Institute of Ecology using continuous exposure to filter papers
impregnated with water emulsions of insecticide and 72 hr mortalities,
selected with pyrethrins and piperonyl butoxide a laboratory strain
showing x8 resistance and similarly independent lines showing x28
resistance to demeton methyl and x4 to DDT.

Documented field occurrences of DDT resistances in *T. castaneum*
are few. Dyte and Blackman (1963, 1967b; Dyte 1967) reported an
approximate x2 tolerance in a strain from East London, South Africa
and confirmed the resistance by selecting this stock for 7 generations
in flour containing DDT. Concentrations were progressively increased
from 10 ppm to 37.5 ppm and they recorded a x12 resistance level
after the fourth generation. Three further generations of selection
allowed breeding in flour containing 60 ppm compared with a corresponding
20 ppm for a laboratory susceptible strain (Dyte *et al.* 1964). Selection
was stopped after the thirteenth generation which was bred in flour
containing 100 ppm DDT. The resistance level measured by a topical
application method approximated x49 whereas two similar lines initially
selected at 5 and 20 ppm had factors of x6 and x4 respectively (Dyte
and Blackman 1966b, 1967b). A branch line derived from seventh
generation adults was selected by topical treatment of unmated adults
and cultured in untreated flour. It reached a resistance factor of x60
in its thirteenth generation of selection (Dyte *et al.* 1965a)
and increased to x166 after 15 generations (Dyte and Blackman 1967a,b).
Apart from the ability of the resistant strain to breed in flour containing
100 ppm DDT, the observations that this strain would produce progeny in
maize or wheat treated with DDT dust at 10 ppm and in peanuts treated
with 30 ppm were particularly significant. These concentrations will
preclude breeding of susceptible strains.

A further programme for selection of DDT resistance in
T. castaneum was instituted by Bhatia and Pradhan (1968) using methods
similar to those they had used for selecting the lindane resistant strain
(Bhatia and Pradhan 1971) discussed above. A composite strain from eight
localities in Delhi, Punjab, Bihar and Uttar Pradesh where DDT was
commonly recommended for disinfestation of godowns, was selected in
wheat flour containing initially 7.5 ppm increasing to 250 ppm in the
eighth generation. Breeding then was continued in flour maintained at
this level of DDT. Resistance levels again were measured in the
impregnated medium, and by spray-film and direct spraying methods
giving resistance factors of x95, >x113 and x37 respectively when
compared with the parental strain. These factors were based on survivors
from the previous generation selection, and true resistance factors for
the latter two methods using progeny bred in insecticide-free medium were
x94 and x28 respectively. It was observed also that the resistance level
decreased on relaxation of selection pressure and that there was a strong
correlation between longer larval development periods and resistance.
The selected strain was tolerant to carbaryl (>x12), a wide range of
organophosphorus compounds including malathion (x13), and marginally so
to cyclodienes and pyrethrins (x2)(Bhatia and Pradhan 1970).

In the field, Bhatia *et al.* (1971) confirmed a x2.3 resistance
to DDT in a strain collected in 1970 from rice bags and floor debris in
a godown in Delhi, and Parkin and Bright (1965) recorded a low level of
resistance from a malathion-resistant strain from peanuts in Nigeria
where DDT had been used in the storage areas. In the U.S.A., Speirs
et al. (1971) reported x82 DDT-resistance from material collected in
1967 in floor sweepings of an old peanut shelter in Georgia - it had
been common practice to spray empty storages in the area with DDT
when they were empty. Dyte (1970, Dyte *et al.* 1973c) using a topical
application method and discriminating doses also reported DDT
resistance in Kenya, Malawi and Senegal.

Considerably more attention has been focussed on malathion
resistance in *T. castaneum* than in any other species. This was
primarily a result of the spectacular level of control it exercised
when first introduced to replace lindane and pyrethrins and the
equally spectacular build up of populations subsequently as a
manifestation of resistance. The realisation that almost complete
control of this ubiquitous pest was readily attainable with a material
of low cost and low mammalian toxicity, and that this advantage was
quickly lost through resistance, has increased the impact of this
resistance both in the current situation and as a warning of future
problems with other insecticides and pest species. As with most
pesticides, different mechanisms can contribute to malathion resistance
and so influence cross-resistance patterns. Experience has shown that
it is useful to distinguish between resistance that is specific to
malathion and closely related compounds, and resistance that includes
malathion with unrelated organophosphorus insecticides and different
classes of compounds.

Malathion came into general use in stored product pest control
in the late 1950's particularly following reports of its effectiveness by
Lindgren *et al.*(1954) and Parkin (1958). One intensive application was
in control of pests of peanuts and by 1961 resistance was recorded in a
strain of *T. castaneum* from the peanut growing area of Northern Nigeria
(Anon.1959a, Hayward 1962, Parkin *et al.*1962c). This strain was shown to
have a resistance factor of x52 and two further strains from bag stacks of
peanuts at Kano had resistances of the same order (Parkin and Forster
1963b, Parkin 1965, Parkin and Dyte 1967). Twelve more field strains from
Nigeria were tested subsequently. Three strains from peanuts arriving in
Kano from buying stations and two strains collected from these buying
stations, had high resistances similar to those of strains taken from the
bag stacks. Similar high resistances were reported from other strains
taken in Kano from cassava and cowpeas. A further strain taken in Kano
from Guinea corn and three strains taken in Ibadan from cowpeas, maize
and tobacco all showed moderate resistance ($\sim$x10-16), whereas a strain
taken in Samaru from sorghum had doubtful resistance (Parkin and Forster
1964b,1965c, Parkin 1965). The original strain was selected by Parkin
and Forster (1965c, 1966d) in whole wheat flour containing malathion and
though there was no clear increase in resistance after 4 generations, a
five-fold increase was apparent in the seventh generation. Independently
Dyte and Blackman (1966a) increased progressively the resistance factor of
a separate line of the same strain by selection from x70 to about x200 and
then to x455 (Dyte and Rowlands 1968) when measured by a topical application
method. An examination of the cross-resistance pattern of the unselected
strain did not show any change in tolerance to other organophosphorus
compounds or carbaryl or synergised pyrethrins (Parkin and Bright 1965).
This was confirmed with the selected strain by further cross-resistance
tests (Dyte and Blackman 1969), metabolic studies and tests for synergism
(Dyte *et al.* 1965b, Dyte and Rowlands 1968) which indicated that the
resistance was specific to malathion-type compounds and was probably one

of enhanced degradation by **carboxyesterases** which could be inhibited by
triphenyl phosphate. Subsequently FAO recommended incorporating triphenyl
phosphate in the test method for detecting resistance in *T. castaneum* to
determine whether or not this malathion-specific type of resistance was
present (Anon. 1970a).

Another area where peanuts were grown extensively and where
malathion resistance was detected early was the south-eastern United
States. Here malathion replaced synergised pyrethrins for pest control
in stored peanuts in 1961 and in stored citrus pulp animal feed in 1962.
In facilities where malathion was used for about 2 years, an apparent
decrease in its effectiveness against stored products was noted (Speirs
et al. 1964, 1967). They checked ten strains of *T. castaneum* for
resistance to malathion by topical application and noted resistance
factors to x11.3 that were directly related to the length of time
malathion had been used for insect control in the storage facility.
Subsequent cross-resistance studies including tests on two of the strains
above which had been selected with malathion to maximum resistance
factors of x17 and x19, showed resistances present to organophosphorus
compounds which were unrelated to malathion indicating a non-specific
type of malathion resistance (Speirs 1967, Speirs and Zettler 1969)
Vincent and Lindgren (1967) using a topical application method and 7
day mortality conducted a survey of the susceptibility of 11 strains of
T. castaneum collected from various flour and feed mills and warehouses
throughout the United States and recorded apparently significant increases
in tolerance from strains from Iowa (x6.4), New York (x3.2) and Kansas
(x2.3) but not from four strains from California, two from Tennessee and
single strains from Minnesota and Utah.

At the same laboratory but using a direct spray method and 24 hr
responses, Strong *et al.* (1969) supported Vincent and Lindgren's findings
on the absence of malathion resistance in California by an examination of
39 strains collected from widely separated locations in California during
1965-1967. By 1972, however, Zettler (personal communication) described
T. castaneum in the U.S.A. as being moderately resistant to malathion in
Georgia and slightly resistant in Florida and California, and recorded a
strain showing x66 resistance from Alabama. Further records of an unspecified type of malathion resistance in *T. castaneum* were made by
Zettler (1974c, 1975) from six strains collected from stored peanuts in
Georgia and Alabama and from nine strains collected from rice storage
facilities in Texas and Louisiana. All strains were resistant and the
maximum resistance factor was x109 from peanuts and x38 from rice.

The third major occurrence of malathion resistance in *T. castaneum*
associated with peanuts was in Australia where malathion at 44 ppm had come
into general use as a protectant for unshelled peanuts in 1963. In 1968
large numbers of *T. castaneum* reappeared in the silo-storage complex in
the Kingaroy peanut-growing area in Queensland and a series of strains
tested according to the FAO method gave resistance factors of up to a
maximum of x18 (Champ and Campbell-Brown 1970b, Champ 1969). The highest
malathion residue in peanuts with which living *T. castaneum* was associated
was 132 ppm (in shells). Further strains collected during 1968 from the
other peanut growing areas of Atherton and Gayndah were also resistant as
were numerous other strains collected during 1968-1969 from a wide range
of localities in Queensland from wheat, barley, sorghum, safflower and
provender mills and farm produce stores. The resistances were all of a
non-specific malathion type as indicated by discriminating tests with
triphenyl phosphate and confirmed for the original collections from peanuts

by cross-resistance studies on a wide range of organophosphorus
compounds and carbamates. Mortality responses of a susceptible and
two resistant strains were compared in malathion-treated wheat giving
resistance factors at 10 day assessments of x6.5 and x1.6 for strains
collected from peanuts and wheat which showed resistance factors of
x8.4 and x2 respectively using the FAO impregnated paper method.
Comparisons in malathion-impregnated whole wheat flour indicated
breeding at 10 ppm whereas reproduction in the susceptible strain was
severely inhibited above 1 ppm. These authors discussed the
significance of pest control in peanuts as a situation predisposing
to development of resistance. They also discussed the significance of
broken grain and cereal dust generally in creating a similar situation.

Greening (1970), using an impregnated paper method, surveyed
resistance to malathion in New South Wales in 1968/1969 and detected
malathion-specific resistance in three localities in the Sydney
metropolitan area and four country centres among 48 samples collected
from storages and grain transport facilities. By 1970/1971 malathion
resistance had been diagnosed from bulk grain storages at 6 country
centres, in rail transport and at the terminal elevators in Sydney.
Subsequent sampling indicated that by 1973/1974, 96% of samples collected
contained resistant insects and that resistance was present in most
infestations throughout the grain handling system including terminal
storages, provender mills and stores, and involving wheat, barley, maize,
oilseeds and pelleted stock-feed (Greening *et al.* 1975). Although
resistance was detected on stock and poultry farms north of Sydney in
1970, the first record from farm-stored grain in the grain-producing
areas was made from central New South Wales in 1971 (Greening 1969, 1973).
Few other occurrences on grain farms were noted by these authors until
1973/1974 when 10 farms sampled were found infested with malathion-
resistant insects. As with earlier studies most records concerned
malathion-specific resistance - exceptions were 9 localities in northern
New South Wales, one in central New South Wales and in the Sydney area
(Greening *et al. op. cit.*).

Champ and Campbell-Brown (1970b) reported a further occurrence
of malathion-specific resistance in a provender mill in Sydney. This
strain had a resistance factor of x11 which was increased by two
selections in malathion impregnated flour to x83 in the F_1 and x244 in the
F_2. These authors also examined numerous samples from outbreaks of
T. castaneum in Victoria and South Australia but all proved susceptible
to malathion. More recently, however, Amos *et al.* (1974) reported a
malathion-specific resistant strain from a flour mill at Echuca, Victoria
and a strain with non-specific resistance from Horsham, Victoria. Both
were collected in 1971.

In India, Bhatia *et al.* (1971) using a spray-film method and
4 day mortality responses, reported a x38 resistance to malathion from
a strain collected in Delhi in 1970 from rice bags and the floor of a
government godown where food stocks had been stored since 1967 and
which had received repeated surface treatment with malathion. Malathion
had been used in the godowns since 1964-1965 replacing lindane and BHC.
When introduced, malathion was sprayed at a concentration of 0.17% but
because of operator reports of ineffectiveness the concentration was
increased to 0.25% by 1967 and to 0.33-0.5% by 1972. These concentrations
are well below those normally recommended for surface treatment with
malathion (2.5%) and were introduced because of the relatively high
consumption of wholemeal flour milled from unwashed grain. Nevertheless
such concentrations would predispose to rapid resistance development.
Bhatia *et al. (op. cit.)* also referred without detail to strains from

Madras and Mysore which were malathion-resistant and Pasalu (1974, Pasalu
and Bhatia 1974a,b) described selection of a malathion-specific resistant
strain which showed x36 resistance at the time of collection from a godown
at Naraina, New Delhi and after six generations of malathion pressure was
found to have a x54 tolerance to malathion using an impregnated paper
method. The selected strain did not show any appreciable cross-resistance
to a range of other organophosphorus compounds and the resistance was
suppressed by synergising the malathion with triphenyl phosphate. Further
records from India were obtained by Dyte *et al.* (1973b) and also by
Rajak *et al.* (1973) who in a survey of malathion resistance in storage
insects reported resistance factors of x21 and x2 in a grain warehouse and
flour mill respectively in Meerut and x17 and x2 in similar facilities in
Hapur. Rajak used a method similar to the recommended FAO method but with
6 day end-point mortality responses.

On the African continent, Toppozada *et al.* (1969, Eldefrawi
1969) in Egypt compared the tolerance to malathion-treated wheat of
T. castaneum collected from Alexandria Harbour where malathion had not
been used previously, with base-line data supplied by Parkin (*vide*
Parkin 1965) and reported a x7.6 increase. In Malawi, where DDT, lindane
and malathion had been used widely over the previous 20 years and where
Dyte and Blackman (1970a) had reported previously a non-specific malathion
resistance from Balaka, Pieterse *et al.* (1972) using a FAO type method
reported non-specific malathion resistance in 11 of 17 strains of
T. castaneum collected from warehouses and stores throughout the country.
The resistance to malathion and triphenyl phosphate was confirmed by a
general cross-resistance to bromophos. Subsequently Pieterse and
Schulten (1974) examined ten strains collected during 1971 from cob maize
stored in cribs on small farms throughout Malawi but unlike in strains
from central premises, they were able to demonstrate significant malathion
resistance in one strain only of six taken from the Central Region of
the country. This strain responded to triphenyl phosphate suggesting
a malathion-specific resistance.

The provision by FAO of a recommended method for detecting
pesticide resistance in *T. castaneum* which in fact endorsed methods
already in use in several laboratories, facilitated surveys of the
resistance status of this species. Apart from the surveys in individual
countries listed above, Dyte and Blackman (1970a) summarised current
knowledge of the distribution of malathion resistance and supplemented
this with further records using FAO-type discriminating tests on material
provided by workers in other countries. Twenty-two strains were examined
of which four from Australia (1), The Gambia (2) and Zambia (1) were
susceptible, whereas seven strains from England (1), The Gambia (2),
Japan (1), Malaysia (1), Nigeria (1) and the Seychelles (1) showed
malathion-specific resistance. The remaining eleven from The Gambia (2),
Kenya (2), Malawi (1) and Senegal (6) showed a non-specific type of
malathion resistance which in most instances was associated with a
malathion-specific resistance in the same strain (Dyte and Blackman
1970a,b; Dyte 1970; Dyte *et al.* 1973b,c). Cross-resistance tests on
the strains from Malawi and Senegal confirmed the presence of carbaryl
resistance also (Dyte 1970, Dyte *et al.* 1973c). These occurrences were
supplemented by a survey in 1970 of infestations intercepted in cargoes
entering the United Kingdom (See Section F13). Eighteen of 20 samples
were malathion-resistant and a single inland control failure reported at
the same time proved to be due to resistance (Green 1975). In 1973 a
further survey was conducted at inland sites following increasing numbers
of control failures. Twenty-three occurrences of resistance (of 28)
of which 19 were of the malathion-specific type were reported from
England and Scotland indicating a considerable spread of resistant

strains on imported material (Dyte et al. 1975, Green op. cit.). These authors also demonstrated the survival of resistant T. castaneum after short exposures to wettable powder formulations of malathion (1600 mg/m^2, 2-24 hr) and fenitrothion (950 mg/m^2, 2-6 hr)(Green op. cit., et al. 1976).

In South America, Rai et al. (1972) in Guyana conducted experiments comparing pirimiphos methyl with malathion and listed dose levels for malathion with responses that can only be interpreted as a positive diagnosis of malathion resistance in T. castaneum in that country which is a focal point for rice shipments throughout the Caribbean.

In Canada, Bond (1973) using the FAO method during studies on fumigant resistance recorded non-specific malathion resistance associated with lindane resistance in two strains collected from a flour mill and an adjacent feed mill.

Resistance in T. castaneum to a synthetic juvenile hormone (cis/trans mixture of methyl 10, 11-epoxy-7-ethyl-3, 11-dimethyl-2, 6-trideca dienoate) has also been recorded which resulted in higher doses being required to produce responses equivalent to those observed in a susceptible strain including adult, larval and pupal mortality, and delay and prevention of pupation (Dyte 1972). The strain concerned had a wide range of other resistances present involving DDT, lindane, pyrethrins, 4 carbamates, 22 organophosphorus compounds and various other materials (vide Section F.14.7).

(F.7.2) *Recorded fumigant resistances*

Reports relating to fumigant resistance in T. castaneum are few. A programme of selection for methyl bromide resistance at the Pest Infestation Laboratory at Slough was discontinued after six generations of selection by culturing from survivors in samples fumigated as pupae had failed to produce any sign of a change in the tolerance level (Anon. 1960). Reynolds and Sylvester (1964) and Howe (1962a) in comparing the relative susceptibility of T. castaneum collected from The Gambia, Nigeria, South Rhodesia, The Transvaal and Trinidad with the Pest Infestation Laboratory stock cultures also could not detect any change in tolerance to methyl bromide. In the U.S.A., Lindgren and Vincent (1965, 1966a) surveyed the susceptibility to methyl bromide, ethylene dibromide and hydrogen cyanide of ten field-collected strains from flour and feed mills and warehouses in California (6), Minnesota (1), Tennessee (2) and Utah (1). The only increase in tolerance found was to hydrogen cyanide (x1.4) in the single strain from Minneapolis, Minnesota which came from a source with a known history of general fumigation with hydrogen cyanide and methyl bromide and regular spot fumigation with fumigant mixtures. Bhatia and Bansode (1971) were unable to detect any resistance to a range of seven fumigants in the laboratory-selected DDT-resistant strain of Bhatia and Pradhan (1968). In Australia Winks (1969) reported immediate increases in tolerance to phosphine from selection of a previously laboratory-slected lindane-resistant strain. First generation progeny showed a x5 increase in tolerance and F_2 progeny a x6.5 increase. In India, Kem (1975) selected a field-collected strain with phosphine for ten generations and obtained a strain with x11.9 resistance. This strain had no cross-resistance to five other fumigants, nor any significant increased tolerance to ten contact insecticides tested.

The only field report of resistance relates to the ethylene-dibromide component of an ethylene dibromide-methyl bromide

spot fumigant. Difficulty had been noted in controlling insects in the
machinery of a Canadian flour mill where the spot fumigant had been used
in a regular monthly sanitation programme for at least 6 years and where
before this an ethylene dibromide : ethylene dichloride : carbon
tetrachloride (1:1:3) mixture had been used for several years. Bond
(1973) showed that the strain concerned had a x2 tolerance to ethylene
dibromide and to the mixture, but no significant change in tolerance to
methyl bromide. He was also unable to detect any change in tolerance to
either material in a strain collected from an adjacent feed mill where
fumigants were not used. The resistant strain was selected twice with
ethylene dibromide under laboratory conditions increasing the tolerance
to x3 that of the unselected strain. A further 4 generations of selection
increased the tolerance to x4 (Anon. 1974b). This report is of considerable
significance as it demonstrates economic resistance to the halogenated
hydrocarbon fumigants in the field for the first time.

(F.7.3) *Survey data*

Results from discriminating tests for malathion, lindane,
methyl bromide and phosphine resistance in *Tribolium castaneum* are
given in Table F12 and Figures F18-F21, together with a summary of
the occurrences in Table F13. The distribution maps (Figs. F18-F21) also
show the classification of importance currently attributed to *T. castaneum*
in the different countries.

(F.7.4) *Comments*

(F.7.4a) *Malathion resistance*

Malathion resistance, already reported from most areas of the
world, was recorded from 75 of the 78 countries sampled involving 438
of the 504 strains tested. Single samples only were available from the
three countries, Finland, Netherlands and Romania, from which resistance
was not recorded and these do not provide grounds for assuming that
resistance is absent in these countries. All had climates which were
unsuitable for *T. castaneum* but there were numerous records of resistance
from countries with comparable winters presumably resulting from movement
of infested commodities in trade. It is a reasonable conclusion that
malathion-resistant strains of *T. castaneum* are present in all areas of
the world and from a consideration of the frequency of occurrence of the
resistance, the probability is high that any strain encountered will be
resistant. This probability decreases only in Europe, the eastern sea-
board of the U.S.A. and southern Australia apparently because of limited
use of insecticide and limited movement of infestable commodities
into these areas from the warm areas where the resistances appear
to have originated. The very low incidence of resistance in Burma
is exceptional but conforms broadly to the pattern of minimal use
of malathion together with a very limited import trade thus restricting
introduction of resistance strains from elsewhere.

Approximately half of the resistant strains was resistant also
to malathion synergised with triphenyl phosphate. These strains were
distributed throughout most of the warm areas of the world and though
also in cooler areas, the frequency of occurrence in these areas appeared
lower. In most strains resistant to the synergised malathion, however,
there was a response to the addition of triphenyl phosphate which ranged
from nil to almost complete. It appears that there is very little
if any response to triphenyl phosphate in susceptible strains of this
species. The data in Table F13 therefore indicate that the major

contribution to malathion resistance comes from the triphenyl phosphate-suppressible type. Strains in which non-TPP suppressible resistance played the major role were reported principally from areas which correspond with the early records of this type of resistance such as Australia, Malawi, south-east U.S.A. and Senegal. The areas where only the TPP-suppressible malathion-specific resistance was recorded or predominated were the mid-latitudes of the U.S.A., Europe north of the Mediterranean, a band through Africa from Nigeria and Benin through Central Africa to North Africa excepting Algeria and through the Arabian Peninsula to Iraq and Iran, the mainland area of south-east Asia through to Japan and Korea, and south-east Australia. This pattern of occurrence did not suggest any determinative factors responsible other than a less intensive use of pesticides generally in these areas. The occurrence of the different types of resistance does indicate that there remains a considerable proportion of the world where many organophosphorus compounds unrelated to malathion would remain effective against *T. castaneum* even where non-specific malathion resistance is present.

(F.7.4b) *Lindane resistance*

Lindane resistance was widespread as in *Sitophilus oryzae* again reflecting the extensive use of lindane throughout the world, the long standing of the resistance and the resultant increased opportunity for dispersal in trade. The frequency of diagnoses was higher than with *S. oryzae*, because of the wider range of infestable commodities and climatic conditions concerned and the dependently-related, common occurrence of *T. castaneum*. Resistance was recorded from 75 of the 76 countries sampled and 93% of the 497 samples involved compared with reports from 13 countries previously. The areas from which strains were collected that were not resistant, were principally North America, north-eastern Europe and south-eastern Australia, where use of lindane has been on a considerably reduced scale compared with elsewhere. In Australia, the absence of lindane resistance in the states of Tasmania, Victoria and South Australia and in southern New South Wales appears related particularly to the absence of compulsorily-treated sorghum and maize seed from this region. It seems reasonable to assume that it now would be normal to expect difficulty throughout the world in the control of *T. castaneum* with lindane.

(F.7.4c) *Methyl bromide resistance*

The response of geographically-isolated strains of *T. castaneum* to methyl bromide is remarkably uniform and survival in discriminating tests for resistance can usually be accepted as indicating a real change in tolerance even when a few individuals only are involved. Increases in frequency of these non-responding individuals by selective breeding from the survivors of discriminating doses provide unequivocal evidence of resistance and such was demonstrated in strains examined during the survey. Thus the progeny of survivors of a strain from Peking, China, which gave 89% response in the primary test, gave a 25% response (Winks, personal communication). Similarly a strain from Nazret in Ethiopia which gave a 96% response in the primary test, gave an 82% response when the progeny of the survivors was checked. Survivors were recorded in 8 strains of the 288 tested and although survival in all these strains was less than 11%, this represented up to an approximate x1.2 change in tolerance at the $LD_{99.9}$ which, taking into account minimum concentration-time products for 99.9% kill of immature stages, may be significant in control of *T. castaneum* when dose rates used have little margin over minimum recommended effective dose rates (e.g. data from Anon. 1974c, Tables 1 and 3) and if the particular resistances expressed in adults are present

in the immature stages. In most instances, however, dose levels with
methyl bromide are such that, assuming satisfactory distribution, control
would be achieved and the occurrences can be regarded more as early warn-
ings of a developing problem rather than as a serious existing problem.
The lack of success in limited laboratory selection programmes (Section
F.7.2) and the time methyl bromide has already been in use indicate that
development of widespread resistance will be slow. Nevertheless, poor
control over distribution and maintenance of concentration-time products
is often a feature of fumigation, and increases in failure rate in
fumigation due to resistance must now be considered a reality.

The only common feature among the strains was that most
were collected from premises with a considerable throughput of grain
of diverse origin, for example from a terminal elevator at Vancouver,
silos in Ethiopia and flour mills in Bahrain, Ethiopia, Peking
and Shanghai. All strains were malathion-and lindane-resistant and
those from China showed tolerance to phosphine also. The only
recorded association with methyl bromide concerned fumigation of
mill products but there was no obvious relationship with this, and
presumably the resistances were chance occurrences resulting from the
continual introduction of new strains into the residual infestations
in residues and grain offal in the premises concerned.

The only field report concerning methyl bromide resistance was
from Balaka, Malawi where survivors have been recorded 48 hrs after
treatment of infestations of *T. castaneum* with methyl bromide at the
rate of 32 g/m^3 (Smithyman, personal communication). It was
not possible to obtain for testing any of the insect material concerned
- the 5 strains tested from Limbe, Salina and Thucila were all methyl
bromide-susceptible.

(F.7.4d) *Phosphine resistance*

As with methyl bromide resistance, phosphine resistance was
unknown before the survey except for two reports from laboratory
selection (Section F.7.2). The ease with which tolerance was increased
in these selection programmes contrasts with the low incidence of
phosphine resistance found in the survey. Fifteen strains only of 267
tested were resistant involving 13 of the 69 countries sampled. Again
responses in discriminating tests were high - all greater than 88%
mortality - but because of the less steep regression lines resulting
from the greater variability in response of individuals to phosphine, these
survivals indicated tolerances to x3 normal. In most instances, selective
breeding from survivors of discriminating doses gave marked increases in
tolerance similar to those in the laboratory selections quoted above (Section)
F.7.2). Thus the response of a strain from Georgetown which without selection
was 94% in a discriminating test decreased to 77% after one selection and to 4%
after two selections with a resistance factor then approximating x6. Similarly
the response of strains from India and Nepal decreased from 88% and 92%
respectively in the initial test to 64% and 32% after single selection
of survivors with both resistance factors approximately x5.

A number of the occurrences of phosphine resistance in
T. castaneum were from localities where phosphine resistance was recorded
in *Rhyzopertha dominica* also, and where the strains concerned were taken
from the same sample, resistance was present in both species and in
Sitophilus oryzae also taken from the same sample. The localities
concerned were Georgetown (Guyana), Pireaus (Greece), Bangui (Central
African Republic) and Borivli (India) and the circumstances associated

with these occurrences are discussed in relation to *S. oryzae* (Section F.3.4d) and *R. dominica* (Section F.6.4d). The ubiquitous nature of *T. castaneum* and the ease with which selection apparently takes place, increases the probability that where conditions have resulted in development of phosphine resistance in other stored grain pests, resistance in *T. castaneum* can be expected. The conditions in the four localities listed above certainly predisposed to the occurrence of resistance either from importing a wide range of infested commodities and resistances from elsewhere, or providing marginal or inadequate dose levels in fumigation.

The association in the field with phosphine varied among the other strains showing resistance. The strain from Washington was a laboratory strain that was susceptible to malathion, lindane and methyl bromide and was unusual in that all other known instances of phosphine tolerance have been in strains carrying some form of resistance to other chemicals. The strains from Montserrat were from rice and stock-feed in a provender store and a distribution premises - Montserrat like the Central African Republic is an importer of cereal products and so of resistances. Other occurrences from Africa were in Malawi and Somalia. In Malawi phosphine has been used on a limited scale particularly for fumigation of maize in bags - the dose rate used was 1 g/m^3 for 5 days. In Somalia also, limited amounts of phosphine have been used again on maize - dose rates used for bulk grain at inloading were 4-6 tablets per ton for 3 days although one minimum exposure time of 24 hr was quoted. The strain from Nepal was collected from the Hetavra feed mill from maize imported from the U.S.A. Similarly the strain from Syria was collected from bagged wheat imported from Russia which was stored in the open under tarpaulins in Damascus - strains from nearby stacks of wheat from Australia and Canada were susceptible. Phosphine is used occasionally in Syria for fumigation of bagged grain at the rate of 1 g/sack under sheets for 36 hr. The remaining four strains came from flour mills in Peking and Shanghai in China and Cottesloe in Western Australia, and from a provender mill in Yokohama, Japan. As with methyl bromide resistant strains and the strains collected from situations where other phosphine-resistant species were present, many of the strains in this latter group were collected from premises with a considerable throughput of grain of diverse origin or from the imported commodities themselves. The data do not allow conclusions as to whether the resistances were imported or not, although this seems probable in some instances. Phosphine usage was not always in accord with accepted principles.

Phosphine resistance in *T. castaneum* appears still to be a problem which is only just beginning to develop. The resistance levels found in the survey in adults of field strains are low and would not allow adult survival under normal fumigation conditions. Relationships between tolerance changes in adults and immature stages have not been investigated as resistant strains have not been available. However, the natural tolerance of immature stages normally approaches levels that may allow survival in susceptible strains (Table F1).
It follows that if the strains with tolerant adults have immature stages with similar increases in tolerance, the latter would present a significant problem, even at these resistance levels.

TABLE F12   RESPONSES OF REPRESENTATIVE WORLD STRAINS OF *TRIBOLIUM CASTANEUM* TO MALATHION, MALATHION + TPP, LINDANE, METHYL BROMIDE AND PHOSPHINE IN DISCRIMINATING TESTS FOR RESISTANCE

Locality	Mal	M+T	Lin	MeB	PH$_3$	Locality	Mal	M+T	Lin	MeB	PH$_3$
NORTH AMERICA						**CENTRAL AMERICA AND CARIBBEAN (Cont'd)**					
CANADA											
B.C. - Vancouver	36	100	35	98	100	GUADELOUPE	19	100	14	-	-
"	100	-	100	-	-	**GUATEMALA**					
Man. - St. Boniface	97	100	100	100	100	Escuintla	6	63	0	100	100
Winnipeg	81	96	64	100	-	Guatemala City	9	77	0	-	-
Que. - Montreal	10	24	0	100	100	" "	11	84	3	-	-
"	1	0	0	100	100	**GUYANA**					
MEXICO						E.C. Demerara	0	78	0	100	100
Mexico City	42	90	13	-	-	Essoquibo	31	98	1	99	100
" "	53	99	0	100	100	Georgetown	47	97	0	-	-
U.S.A.						"	48	99	0	100	-
Ark. - Fayetteville	-	-	-	-	99	"	7	95	0	100	100
Cal. - Fresno	100	-	100	100	-	"	24	98	4	100	100
"	98	100	16	100	100	"	63	100	4	100	94
Riverside	98	100	96	100	100	"	31	97	2	100	100
Fla - Leesville	4	2	0	100	100	"	13	99	1	-	99
Kans.- McPherson	0	100	80	100	100	**JAMAICA**					
Mo.- St. Louis	78	48	41	95	100	Kingston	0	87	0	-	-
N.C. - Clinton	100	-	86	-	-	"	33	94	2	-	-
Wallace	100	-	61	100	100	"	24	100	5	100	100
N.M. - Clovis	17	100	75	100	100	"	9	92	0	100	100
S.C.- Ft Jackson	100	-	15	-	-	"	43	98	6	100	100
Tex. - Kelly	100	-	100	100	100	"	11	99	4	-	-
Va - Franklin	81	100	61	-	-						
Richmond	100	-	100	100	100						
"	100	-	81	100	100	Spanish Town	9	97	0	100	100
Wash.	100	-	100	100	94	" "	43	92	2	-	-
						MONTSERRAT					
CENTRAL AMERICA AND CARIBBEAN						Plymouth	0	100	4	100	98
						"	13	98	8	100	98
ANTIGUA						"	0	97	6	100	99
Dunbars	38	55	0	-	-	**NICARAGUA**					
"	87	97	11	100	100	La Calera	10	-	-	-	-
"	24	55	54	100	100	"	-	-	-	100	100
"	18	97	-	100	100	**ST. KITTS**	0	96	0	100	100
COLOMBIA						"	0	61	9	100	100
Barranquilla	57	100	0	-	-	"	0	98	11	-	-
"	95	98	9	-	-	**TRINIDAD**					
"	88	99	5	-	-	Curepe	33	76	5	100	100
"	30	98	5	-	-	Port of Spain	35	100	2	100	100
"	31	94	2	100	100	" "	10	94	2	-	100
EL SALVADOR						" "	4	90	3	99	100
La Libertad	4	38	0	100	100	" "	40	96	0	-	-
"	70	83	3	-	-	" "	100	-	0	-	-
"	58	94	0	-	-	" "	44	100	0	100	-
San Martin	10	55	0	-	-	" "	36	99	0	-	-
"	11	61	0	-	-	" "	32	70	1	-	-
San Salvador	9	73	0	-	-	" "	67	100	6	-	100
Soyapango	0	16	0	-	-	" "	100	-	0	100	100

TABLE F12
(Cont'd)   *T.castaneum (ii)*

Locality	Mal	M+T	Lin	MeB	PH$_3$	Locality	Mal	M+T	Lin	MeB	PH$_3$
CENTRAL AMERICA AND CARIBBEAN (Cont'd)						**EUROPE AND NORTH AFRICA** (Cont'd)					
TRINIDAD (Cont'd)						EGYPT					
Port of Spain	94	100	1	100	100	Behira	90	100	21	100	-
" "	100	-	5	100	100	Cairo	79	97	14	-	-
" "	100	-	16	100	100	Giza	98	100	56	-	-
VENEZUELA						Madkou	83	100	25	-	-
Guacara	29	96	0	100	-						
La Encrucijada	0	90	0	-	100	ENGLAND					
"	0	74	1	100	99	Camb. - Haddenham	83	100	30	-	-
La Morita	0	91	0	-	-	Devon - Exmouth	28	100	82	100	100
"	0	64	1	100	100	Barnstaple	34	100	12	96	100
Valencia	-	-	-	-	100	Essex-Battlesbridge	86	94	43	100	99
SOUTH AMERICA						Braintree	100	-	80	100	100
ARGENTINA						Glos. - Avonmouth	0	100	7	100	100
B.A. - Buenos Aires	1	50	2	100	100	"	52	98	100	100	100
" "	2	59	0	100	100	"	94	100	5	100	99
" "	90	99	18	100	100	"	92	100	20	100	100
" "	81	98	16	100	100	Bristol	20	100	100	100	100
Grumbein	71	99	12	100	-	Gloucester	100	100	100	100	100
La Plata	58	88	10	100	100	Saul	91	100	1	100	100
Necochea	76	84	1	-	-	Hants.- Botley	99	100	11	-	-
Pergamino	100	-	7	100	100	Lancs. - Glazebury	2	100	89	100	100
Tres Arroyos	93	98	3	100	-	London	20	100	92	100	100
" "	89	99	46	100	-	Northants. - Wellingborough	53	100	98	100	100
Ent. Rios - Diamente	61	97	-	-	-	Notts.-Gainsborough	3	92	0	-	-
BRAZIL						Oxon.- Adderbury	100	-	7	100	100
Amaz.- Manaus	27	82	37	100	-	Suff.- Sudbury	70	94	8	100	100
Goias- Anapolis	35	89	5	-	100	Yorks.-N. Pickering	73	100	0	100	100
"	63	82	10	-	-	Middlesborough	3	2	67	-	-
G.B. - Rio de Janeiro	3	31	0	-	100.						
R.G. do Sul - Guaibo	73	99	13	-	100	FINLAND					
S.P. - Barretos	16	85	0	100	100	Tikkurila	100	-	100	100	100
S.J. Rio Preto	9	73	17	100	100	GREECE					
Santos	9	84	3	100	-	Lamia	52	93	0	100	100
CHILE						"	82	94	0	100	99
Santiago	6	99	1	100	100	Piraeus	95	98	0	100	96
PERU						"	56	100	28	100	100
Callao	-	-	-	100	100	"	100	-	0	-	-
Chiclayo	72	98	13	-	-	"	100	-	6	100	100
"	69	99	3	-	-	Thessalonica	-	-	-	100	99
Lambayeque	20	46	0	100	100	LIBYA					
EUROPE AND NORTH AFRICA						Tripoli	7	100	0	99	100
ALGERIA						"	8	100	18	-	-
						"	97	100	55	-	-
						El Marj	58	100	7	-	-
Algiers	0	73	0	-	-	Merzek	47	100	2	100	100
Blida	0	72	0	100	100	MOROCCO					
Dar-es-Abeida	45	98	0	100	100	Casablanca	100	-	2	100	100
						Fes	100	-	0	-	-
BULGARIA.						Rabat	92	100	0	-	-
Burgas	100	-	3	-	-	"	68	100	30	-	-
"	83	100	81	100	100	"	100	100	2	-	-

TABLE F12
(Cont'd)

T.castaneum (iii)

Locality	Mal	M+T	Lin	MeB	PH$_3$	Locality	Mal	M+T	Lin	MeB	PH$_3$
EUROPE AND NORTH AFRICA (Cont'd)						**AFRICA - SOUTH OF SAHARA (Cont'd)**					
MOROCCO (Cont'd)						**CENT. AFRICAN REP.**					
Sale	83	95	0	-	-	Bangui	88	100	74	100	-
"	92	100	6	-	-	"	90	100	73	100	-
Taza	0	100	0	-	-	"	100	-	78	100	100
NETHERLANDS						"	2	97	6	100	98
Wageningen	100	100	84	100	100	"	14	100	6	100	100
POLAND						"	72	100	4	100	100
Poznan	85	100	18	100	100	"	30	100	11	100	100
Zieliniec	85	100	100	-	-	"	25	100	3	100	100
PORTUGAL						"	79	100	2	100	100
Lisbon	79	92	4	100	100	"	4	97	5	100	100
"	100	-	49	100	100	"	93	100	11	100	100
"	57	100	12	100	100	Km 22	36	100	0	100	-
ROMANIA						"	25	96	6	100	100
Baza Oravita	100	-	4	-	-	"	13	100	0	-	-
SCOTLAND						**CHAD**					
						Abeche	90	100	4	100	99
Cumbermauld	33	71	0	100	100	Biltine	-	-	-	100	100
Kirkaldy	42	100	12	100	100	Bongor	97	100	0	100	100
New Abbey	91	100	83	100	100						
" "	100	-	100	100	100	**ETHIOPIA**					
Glasgow	48	100	0	100	100	Addis Ababa	54	61	3	98	100
"	39	100	18	100	100	" "	87	98	46	100	100
	100	-	99	100	100	Nazret	100	-	54	100	100
SPAIN						"	100	-	76	-	-
Alcala	78	100	16	100	100	"	0	87	3	96	100
Baena	100	-	14	100	100	**GAMBIA**					
"	97	100	7	100	100	Denton Bridge	32	98	0	100	100
Granada	90	98	29	100	100	**KENYA**					
"	-	-	-	99	100	Kisumu	0	100	2	100	100
Huelva	98	99	0	100	100	Mombasa	41	100	19	100	100
Jaen	62	98	7	100	100	Nairobi	2	83	0	100	100
Lerma	-	-	-	100	100	"	2	93	0	100	100
Murcia	79	95	32	100	100	"	15	68	0	-	-
"	66	61	71	100	100	"	13	100	0	100	-
WALES						"	8	96	1	-	-
Cardiff	78	100	-	-	-	Nakuru	4	95	0	100	100
						"	1	100	0	100	100
YUGOSLAVIA											
Belgrade	100	-	3	100	100	**MALAWI**					
Beli Manastir	100	-	44	100	100	Blantyre	55	83	0	-	-
Darda	97	100	55	100	100	"	69	100	5	-	-
						Limbe	52	73	22	-	-
AFRICA - SOUTH OF SAHARA						"	75	98	9	100	100
						"	9	69	0	100	100
BENIN						"	48	100	14	-	-
Cotonou	17	100	0	100	100	"	29	100	7	100	100
"	33	100	0	100	100	Salina	37	39	6	100	95
"	52	100	21	100	100	Thucila	25	33	0	100	100

TABLE F12
(Cont'd)

T.castaneum (iv)

Locality	Mal	M+T	Lin	MeB	PH$_3$	Locality	Mal	M+T	Lin	MeB	PH$_3$
AFRICA - SOUTH OF SAHARA (Cont'd)						**AFRICA - SOUTH OF SAHARA (Cont'd)**					
MOZAMBIQUE						SOUTH AFRICA (cont'd)					
Beira	64	95	0	100	100	Eloff	32	83	1	100	100
"	52	100	0	100	-	Pretoria	89	100	13	-	-
"	40	91	5	100	100	"	5	100	16	100	100
"	31	95	10	100	100	"	13	96	7	100	99
"	33	97	1	100	100	Radium	38	100	0	100	99
Chibuto	42	100	0	100	100	Settlers	39	96	0	100	100
Lourenco Marques	16	100	17	100	-	TANZANIA					
"	62	100	11	100	100	Dar-es-Salaam	13	88	1	100	100
" "	49	100	17	100	100	" "	11	100	0	100	100
" "	10	27	1	100	99	" "	0	100	0	100	100
" "	79	100	12	100	100	" "	40	100	0	100	100
" "	6	98	2	100	100	" "	10	98	0	100	100
Umbeluzi	67	100	-	100	100	" "	12	100	0	-	-
NIGERIA						" "	21	100	0	100	100
Lagos	69	100	8	100	-	" "	40	100	12	-	-
"	4	100	18	100	100	UGANDA					
"	42	100	10	100	100		33	82	0	100	100
"	49	100	14	100	100	ZAMBIA					
"	56	100	12	100	100	Chipata	26	84	4	100	100
"	100	-	-	-	-	Kabwe	-	-	-	-	100
"	3	100	22	100	100	**WEST ASIA**					
"	37	100	27	100	100	AFGHANISTAN					
"	88	100	19	-	-	Kabul	0	100	40	-	-
RHODESIA						BAHRAIN	2	97	13	-	-
Aspindale	75	86	1	100	100	"	3	100	6	98	99
Salisbury	62	98	1	100	99	"	10	100	8	99	100
"	83	95	1	100	100	"	13	100	8	100	100
"	83	100	5	-	-	"	0	100	4	-	-
"	76	98	-	100	99	"	1	100	18	100	100
SENEGAL						CYPRUS					
Bambey	4	18	2	100	-	Athienou	-	-	-	100	99
"	14	17	23	100	-	Louroutzina	72	97	4	100	100
"	13	25	22	100	-	"	35	100	3	-	-
"	24	31	5	-	-	Nicosia	3	84	0	-	-
"	17	22	56	100	-	"	41	93	3	-	-
Nioro du Rip	0	10	0	100	100	"	1	89	9	-	-
" "	11	15	0	100	100	Pyrga	72	95	2	-	-
Sefa	23	24	0	100	100	INDIA					
SOMALIA						Maharashtra-Bombay	0	98	2	-	-
Afgoi	27	79	0	100	98	"	18	99	24	-	-
Mogadishu	66	100	4	100	100	"	1	98	2	-	-
"	19	100	1	100	100	"	40	98	16	-	-
"	68	100	3	100	99	"	5	100	5	-	-
"	69	95	10	100	100	"	2	100	15	-	-
"	88	96	11	100	100	Borivli	2	100	36	100	88
"	55	88	1	100	100	"	0	97	3	-	-
"	75	100	2	100	99	Rajasthan-Kota	22	63	24	100	-
SOUTH AFRICA						U.P. - Chamau	52	82	2	-	-
Delmas	100	-	2	100	100						

TABLE F12
(Cont'd)

T.castaneum (v)

Locality	Mal	M+T	Lin	MeB	PH$_3$	Locality	Mal	M+T	Lin	MeB	PH$_3$
WEST ASIA (Cont'd)						**WEST ASIA (Cont'd)**					
INDIA (Cont'd)						**SAUDI ARABIA**					
U.P. - Delhi	68	98	2	-	-	Jeddah	1	100	5	-	-
"	60	97	2	99	100	"	7	100	11	-	-
"	93	98	10	-	-	"	44	100	0	-	-
Hapur	53	98	6	-	-	"	40	100	12	-	-
"	3	99	6	-	-	"	7	100	8	-	-
Kaili	59	99	6	-	-	"	69	100	56	-	-
Mathura	65	93	15	-	-	Riyadh	1	100	14	100	99
Meerut	38	96	6	100	-	"	2	96	0	-	-
Naraina	1	98	4	-	-	"	17	100	2	-	-
IRAN						"	8	100	34	100	100
Ahwaz	0	98	0	100	99	**SYRIA**					
Khorramshahr	56	100	11	-	-	Damascus	39	87	36	100	100
"	44	100	17	100	100	"	28	80	68	100	97
Teheran	33	100	63	100	-	"	91	100	18	100	100
"	100	-	67	-	-	**TURKEY**					
"	0	100	7	99	100	Ankara	100	-	53	-	-
IRAQ						"	72	97	4	100	100
Abu Ghraib	69	100	23	100	100	Sakarya	100	-	12	100	100
" "	71	100	12	-	-	**YEMEN ARAB REP.**					
" "	0	100	2	-	-	Hodeida	3	100	5	-	-
" "	87	100	80	100	100	"	4	100	10	-	-
" "	0	100	14	100	100	"	11	100	79	-	-
Baghdad	3	98	9	-	-	"	57	100	1	-	-
"	20	100	10	-	-	"	15	100	3	100	100
"	12	100	11	-	-	"	24	100	0	100	100
"	14	100	14	100	99	"	25	100	15	100	100
"	2	100	100	-	-						
Basra	13	100	3	-	-	**EAST ASIA**					
Heath	53	100	4	-	-	**BURMA**					
Karbala	50	100	56	100	100	Kyauk-pa-Daung	100	-	89	-	-
Khalis	78	100	3	100	100	Monywa	100	-	88	-	-
Khanekean	0	98	3	100	100	Myaing	98	100	78	100	100
Rarneli	59	100	16	-	-	Myitha	100	-	88	-	-
Sulaymaniyaha	2	100	9	-	-	Nyauing-U	100	-	90	-	-
Swaira	87	100	10	100	100	Rangoon	100	-	63	100	100
Thekar	35	100	49	100	100	"	100	-	17	100	100
ISRAEL						"	100	-	62	-	-
Kiriyat-Gat	94	85	0	100	100	"	100	-	57	100	100
Tel Aviv	45	95	0	100	100	"	100	-	1	100	100
KUWAIT	6	96	5	-	-	"	99	100	79	100	100
"	0	99	0	-	-	Sagaing	100	-	52	100	100
"	0	92	45	100	100	Schwebo	100	-	74	-	-
"	5	100	92	100	100	Tewingtha	100	-	42	-	-
NEPAL						Thazi	100	-	85	100	100
Hetavra	0	93	-	100	92	Ywathagi	98	100	69	-	-
Inner Terai	83	100	5	-	-	**CHINA**					
						Hopeh - Peking	8	-	4	-	-
PAKISTAN						"	53	99	55	89	93
Karachi	7	92	20	-	-	"	46	-	92	-	-
"	20	90	34	-	99	Kiangsu - Shanghai	19	100	53	91	97
"	2	100	0	-	-	Taiwan - Keelung	20	100	25	-	-
"	0	88	0	-	-						

TABLE F12
(Cont'd)

T. castaneum (vi)

Locality	Mal	M+T	Lin	MeB	PH$_3$	Locality	Mal	M+T	Lin	MeB	PH$_3$
EAST ASIA (Cont'd)						**EAST ASIA (Cont'd)**					
CHINA (Cont'd)						**THAILAND (Cont'd)**					
Taiwan - Nan-Chiang	8	100	40	-	-	Bangkok	94	100	7	-	-
Taipei	6	100	6	-	-	"	100	-	97	100	100
"	44	91	9	100	100	Khon Kaen	95	100	96	100	100
Tao-Yuan	14	100	10	100	-	Rangsit	89	100	58	-	-
INDONESIA						Thonburi	99	100	74	-	-
Djakarta	65	90	86	100	100	"	100	-	91	-	-
"	69	95	44	-	-	"	100	-	88	-	-
"	71	100	64	100	-	"	95	100	64	-	-
"	55	99	43	-	-						
"	43	99	20	-	-	**AUSTRALASIA AND**					
"	55	98	43	100	-	**SOUTH PACIFIC**					
"	44	91	15	100	100	**AUSTRALIA**					
"	42	98	22	100	100	NSW - Bribbaree	0	100	100	-	-
						Erigolia	69	100	100	-	-
JAPAN						Lockhart	10	100	100	100	100
Chiba	44	100	39	100	-	Merah North	23	81	84	100	100
"	55	99	60	-	-	Moree	1	96	81	-	-
Tsuyama	-	-	73	100	100	Peak Hill	0	99	100	-	-
Yufiun	36	100	50	-	-	Sydney	0	100	99	-	-
Yokohama	33	100	21	100	100	Qd - Biloela	19	79	41	100	100
"	41	100	10	100	100	Clermont	0	99	71	-	-
"	13	98	0	100	91	Dalby	6	100	85	-	-
KOREA REP.						Hodgson	96	96	82	-	-
Pusan	58	100	62	-	-	Retro	5	32	17	-	-
Sosa	88	100	62	-	-	Rockhampton	98	98	57	-	-
						Springsure	0	91	94	-	-
MALAYSIA						S.A.- Bordertown	0	100	100	-	-
Alor Setar	0	100	1	-	-	Hamley Bridge	0	100	100	-	-
" "	1	99	3	-	-	Port Adelaide	100	-	100	-	-
" "	43	100	13	-	-	Laura	100	-	100	-	-
" "	22	100	16	-	-	Mannum	100	-	100	-	-
" "	30	100	16	-	-	Mile End	4	90	82	-	-
Batu Caves	0	100	3	-	-	Strathalbyn	100	-	100	-	-
" "	3	100	3	-	-	Tas.- Launceston	100	-	100	-	-
" "	0	98	0	-	-	Vic.- Boort	6	100	100	-	-
Kuala Lumpur	0	100	2	-	-	"	11	100	100	-	-
Penang	0	94	11	100	100	Geelong	0	100	100	-	-
PHILIPPINES						Glen Loth	0	99	100	-	-
Batangas	77	100	77	100	100	Kaniva	0	100	100	-	-
"	64	100	80	100	-	NorthMelbourne	28	100	100	-	-
College	64	100	39	-	-	Teddywaddy	0	100	100	-	-
Pagsanjan	60	100	50	100	100	W.A.- Corrigin	11	100	99	-	-
"	69	100	56	100	-	Cottesloe	0	46	64	100	96
Santa Cruz	88	100	37	100	-	Geraldton	0	99	77	100	100
SINGAPORE						"	0	99	35	-	-
"	28	100	16	-	-	Guildford	0	19	57	-	-
"	45	98	5	-	-	West Perth	0	34	36	-	-
"	91	100	47	-	-	" "	34	71	88	-	-
"	81	100	49	100	99	**BOUGAINVILLE**					
"	27	100	35	100	100	Kieta	71	92	29	-	-
"	63	100	46	-	-	"	38	99	25	-	-
"	45	97	22	-	-	**NEW BRITAIN**					
THAILAND						Kerevat	13	40	100	-	-
Bangkok	97	100	1	-	-	"	0	97	65	-	-
"	100	-	22	-	-						

TABLE F12
(Cont'd) $T.castaneum\ (vii)$

Locality	Mal	M+T	Lin	NeB	PH$_3$
AUSTRALASIA AND SOUTH PACIFIC (Cont'd)					
NEW GUINEA					
Lae	11	100	57	-	-
Madang	4	91	57	-	-
Port Moresby	10	100	70	-	-
" "	1	100	83	-	-
" "	8	96	80	-	-
" "	0	100	86	-	-
" "	8	99	70	-	-
" "	4	100	91	-	-
" "	0	62	76	-	-
Wewak	0	99	2	-	-

TABLE F13: SUMMARY OF WORLD OCCURRENCE OF RESISTANCE IN *TRIBOLIUM CASTANEUM* TO MALATHION, LINDANE, METHYL BROMIDE AND PHOSPHINE

	North America	Central America	South America	Europe,N.Afr.,USSR	Africa-S.of Sahara	Western Asia	Eastern Asia	Australasia	Totals	Totals (as %)
MALATHION										
No. countries sampled	3	12	4	16	16	14	9	4	78	-
No. with resistance	3	12	4	13	16	14	9	4	75	96%
No. of strains tested	23	63	23	76	103	93	74	50	505	-
No. resistant	14	59	22	57	97	90	55	45	439	87%
MALATHION + TPP										
No. countries sampled	3	11	4	16	16	14	11	4	79	-
No. with resistance	3	10	4	8	13	12	6	4	60	76%
No. of strains tested	23	62	23	76	103	93	74	50	504	-
No. resistant	7	49	22	20	47	39	16	26	226	45%
LINDANE										
No. countries sampled	3	11	4	15	16	14	9	4	76	-
No. with resistance	3	11	4	14	16	14	9	4	75	99%
No. of strains tested	23	61	22	75	100	91	75	50	497	-
No. resistant	17	61	22	69	100	90	75	29	463	93%
METHYL BROMIDE										
No. countries sampled	3	11	4	14	16	12	8	1	69	-
No. with resistance	2	0	0	1	1	1	1	0	6	9%
No. of strains tested	18	34	16	56	89	38	32	5	288	-
No. resistant	2	0	0	1	2	1	2	0	8	3%
PHOSPHINE										
No. countries sampled	3	11	4	13	16	13	8	1	69	-
No. with resistance	1	2	0	1	3	3	2	1	13	19%
No. of strains tested	17	36	14	55	79	36	25	5	267	-
No. resistant	1	3	0	1	3	3	3	1	15	6%

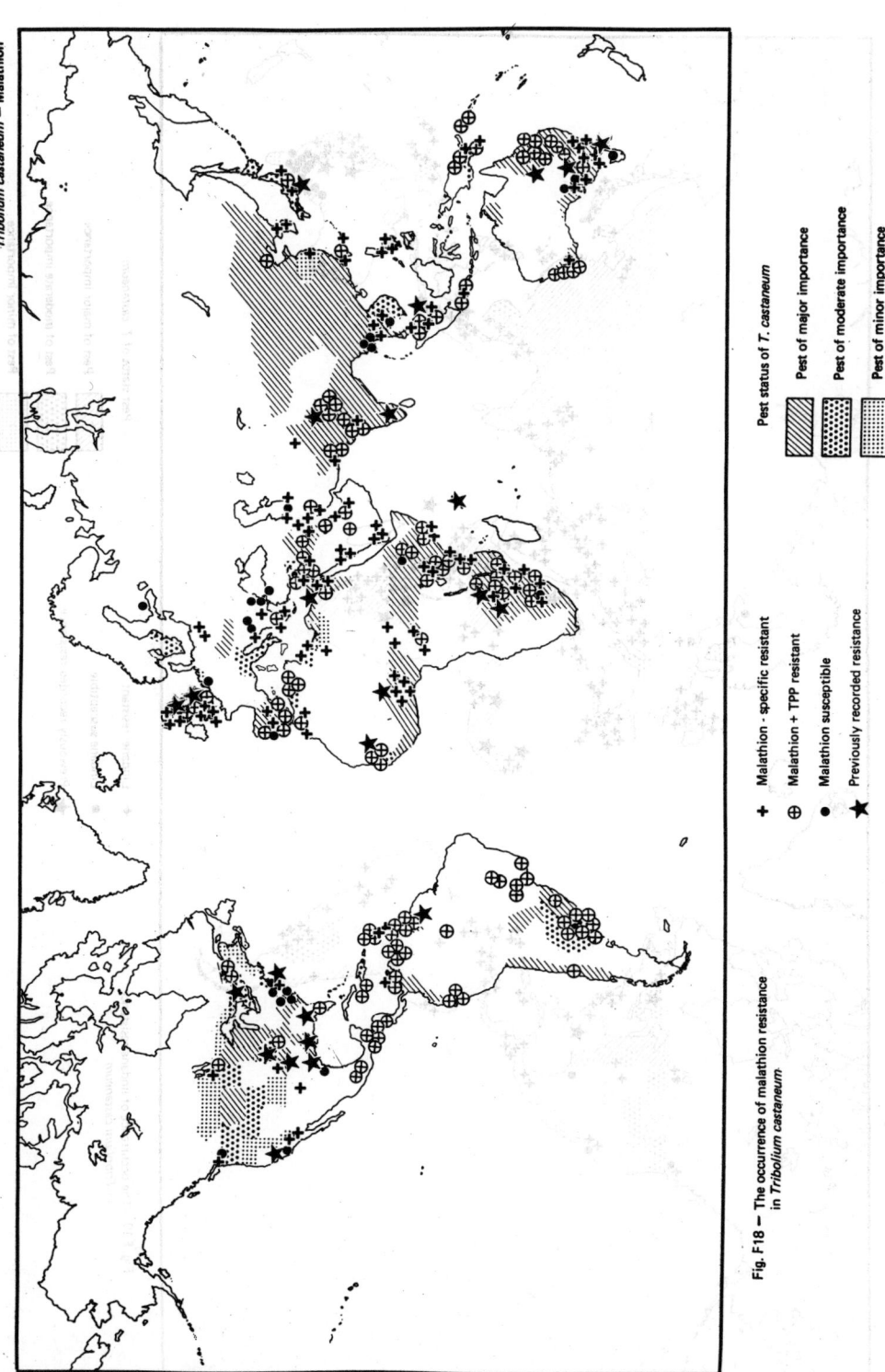

Fig. F18 — The occurrence of malathion resistance in *Tribolium castaneum*

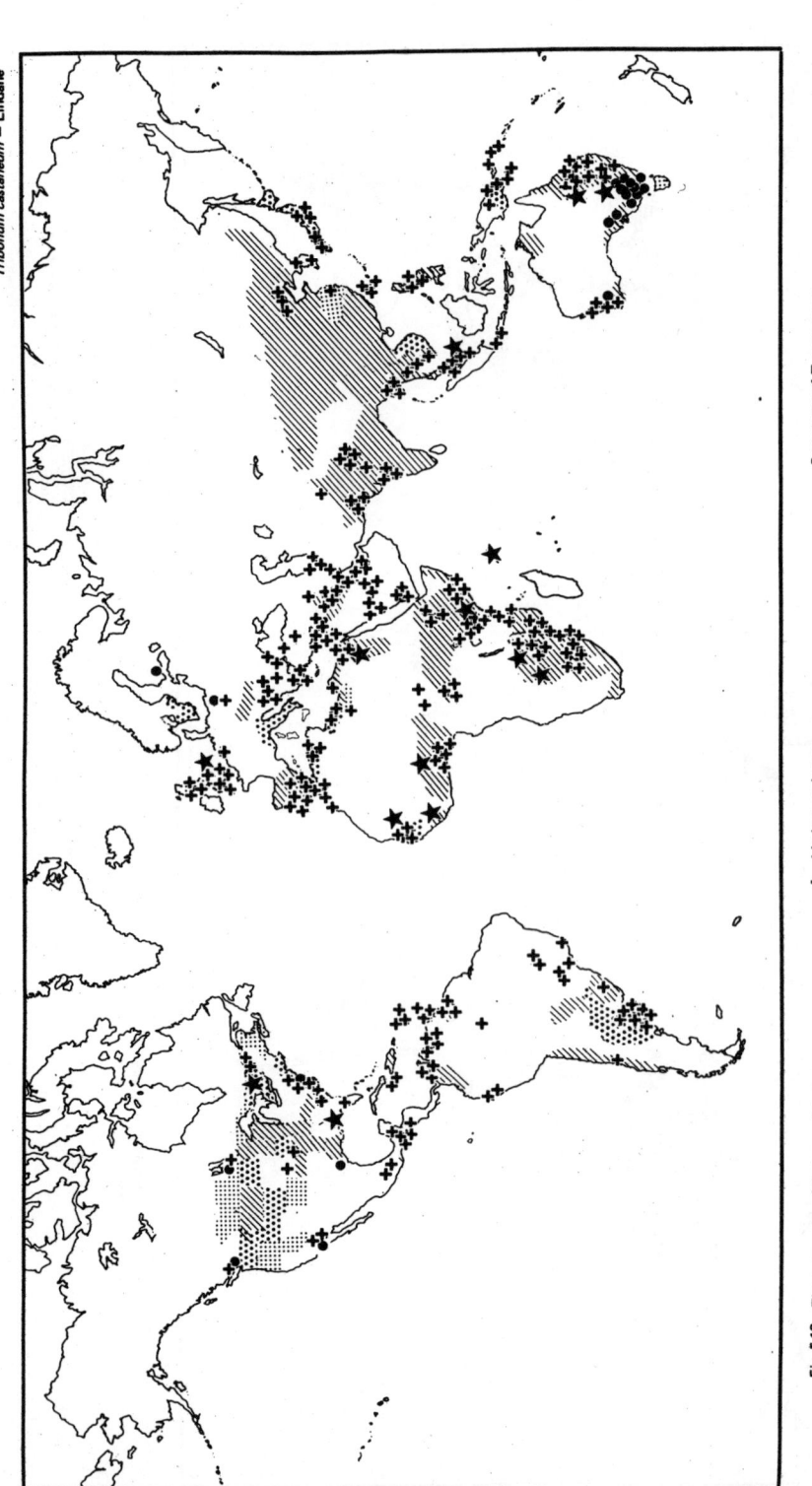

Fig. F 19 —The occurrence of lindane resistance in *Tribolium castaneum*

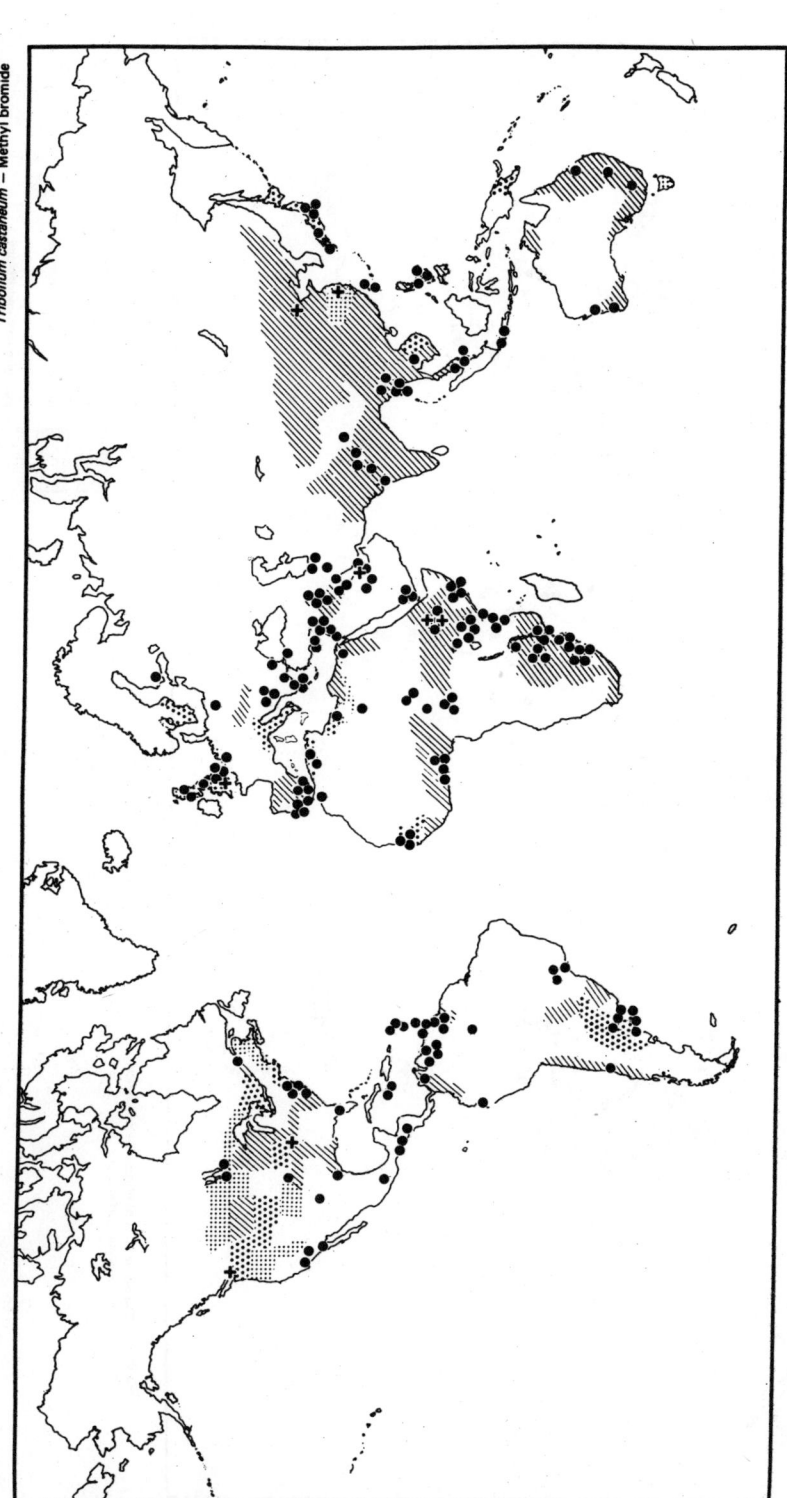

Fig. F 20 — The occurrence of methyl bromide resistance in *Tribolium castaneum*

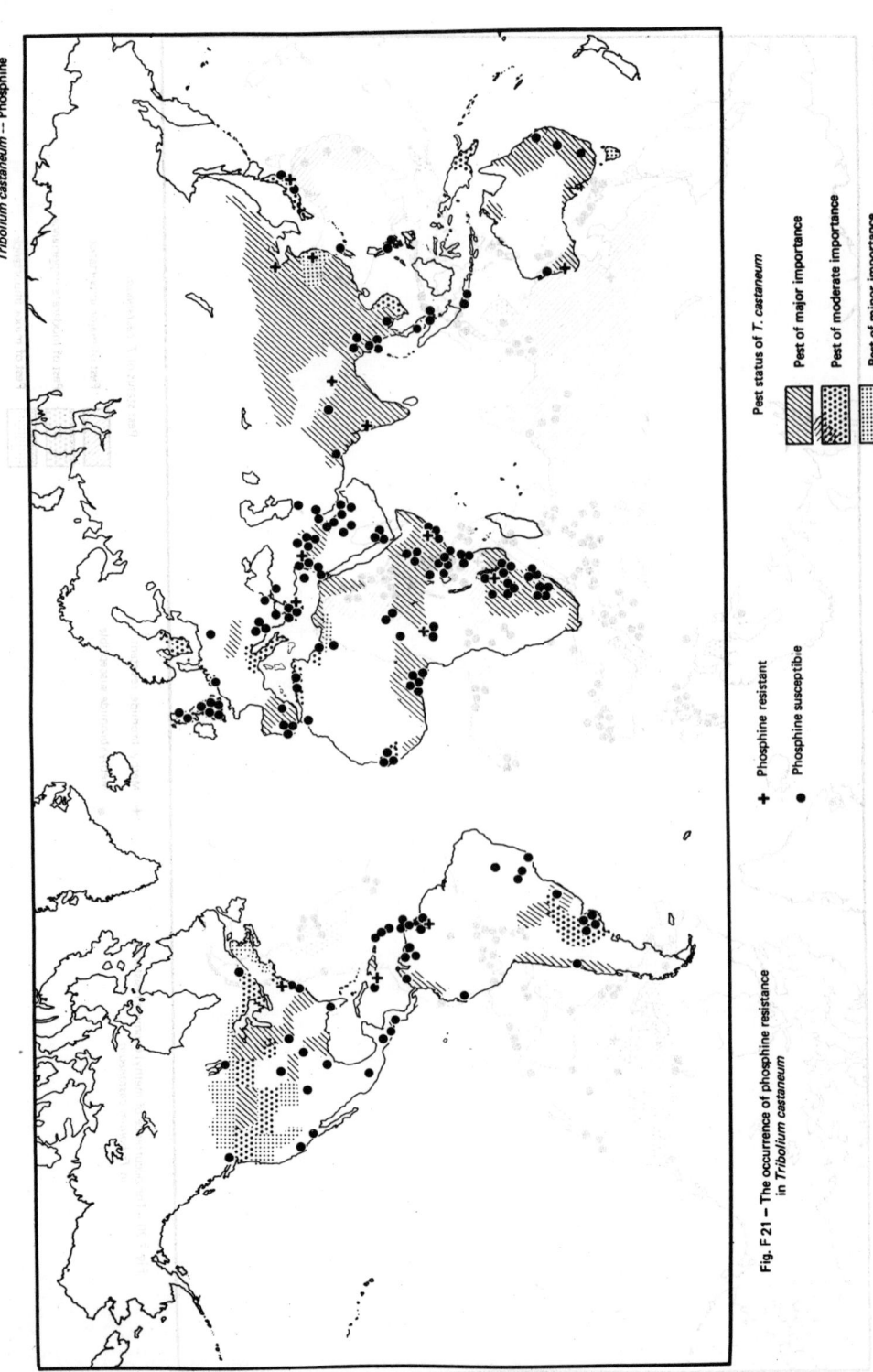

Fig. F 21 – The occurrence of phosphine resistance in *Tribolium castaneum*

(F.8) *Tribolium confusum*

Tribolium confusum is considerably less common than *T. castaneum* and of lesser importance though it tends to replace it in colder areas. It is of particular significance in flour mills where infestations become established in mill machinery. *T. confusum* is usually subjected to considerably less exposure to pesticides than is *T. castaneum* and a high proportion of this exposure would concern fumigants.

(F.8.1) *Recorded insecticide resistances*

Maeda (1958) in Japan working with DDT-impregnated media selected DDT-resistant strains of *T. confusum* that had a x2-x8 tolerance after 5-12 generations of selection. In Canada, Kumar and Morrison (1965) using the Macdonald College Test Kit reported a x25 resistance to DDT in a strain from Macdonald College, Quebec. A survey in the United States of the susceptibility of a series of strains of *T. confusum* to pyrethrins and malathion was carried out by Vincent and Lindgren (1967) in association with a fumigant susceptibility survey. Strains from Kentucky (1), Minnesota (4) and New York (1) were tested using a topical application method and 7 day mortalities but the maximum change in tolerance to either material of x1.3 was acceptably within limits of normal variation. Similarly Strong *et al.* (1969) using a direct spray method and 24 hr responses did not report any significant increase in tolerance to malathion in 8 strains collected from widely separated areas of California during 1965-1967.

In Africa, Pieterse and Schulten (1974) examined 5 strains collected during 1971 from crib-stored maize throughout Malawi but were unable to detect any lindane or malathion resistance using a discriminating test based on an impregnated filter paper method. Dyte and Blackman (1973), however, reported without detail a strain of *T. confusum* from Senegal which was resistant to malathion but not to malathion synergised with triphenyl phosphate, suggesting a malathion-specific type of resistance as in *T. castaneum*.

In the United Kingdom, Dyte *et al.* (1973a, Blackman *et al.* 1976) tested five strains from Worcestershire (2), Warwickshire (2) and Yorkshire collected during 1970-1972 and found all were resistant to malathion but not to lindane. During the summer of 1972, 10 samples from provender and flour mills in the West Midlands Region were tested and all proved susceptible to both malathion and lindane (Blackman *et al.* 1976). A further survey in 1973 showed 5 of 23 strains resistant to lindane and 12 of 24 strains resistant to malathion (Dyte *et al.* (1975, 1976a; Blackman *et al.* 1976). In addition to these records from the United Kingdom, response data without detail have been given for a malathion-resistant strain from Poland (Dyte *et al.* 1975, Blackman *et al.* 1976) and further records reported of malathion resistance from Botswana, Egypt and Russia, and lindane resistance from Sweden and Senegal (Blackman *et al.* 1976). In Australia, Greening *et al.* (1975) reported malathion resistance in *T. confusum* collected in 1971 from three flour mills in Sydney and one in southern New South Wales, and from stock-feed pellets collected in 1972 from stables near Sydney. Samples collected from Newcastle and many country centres were susceptible. Amos *et al.* (1974) reported a strain resistant to both malathion and malathion synergised by triphenyl phosphate which was collected at Murtoa, Victoria also in 1971.

(F.8.2) *Recorded fumigant resistances*

Gough (1939) at the Imperial College of Science and Technology Field Station at Slough, in an examination of factors affecting the tolerance of *T. confusum* to hydrogen cyanide found that the progeny of the more tolerant individuals selected over seven generations were more tolerant (x1.7) than the progeny of the more susceptible individuals and that this increase was heritable. At the same laboratory (then the Pest Infestation Laboratory), Hope and Phillips (1957; Anon. 1958b, 1959c) reported a heritable tolerance (x3) to methyl formate after 35 selections with this material. Monro et al. (1961) subjected *T. confusum* to a regime of selection with methyl bromide similar to that used successfully with *Sitophilus granarius* (vide Monro et al. 1961, Monro 1963) but the increase in tolerance was of a low order only and the programme was discontinued (Monro 1964).

At Riverside, California, Lindgren and Vincent (1965, 1966a) surveyed the susceptibility to methyl bromide, ethylene dibromide and hydrogen cyanide of seven field-collected strains from flour and feed mills and warehouses in California (2), Iowa (1), Kentucky (1), Minnestoa (2) and New York (1). The only changes in tolerance found were to ethylene dibromide (x1.8) and hydrogen cyanide (x1.4) in the strain from Buffalo, New York which had been fumigated monthly with a methyl bromide : ethylene dibromide (70:30) mixture. This resistance resembled the ethylene dibromide resistance recorded by Bond (1973), in *T. castaneum* from Canada.

(F.8.3) *Survey data*

Results from discriminating tests for malathion, lindane, methyl bromide and phosphine resistance in *Tribolium confusum* are given in Table F14 and Figures F22-F25, together with a summary of the occurrences in Table F15. The distribution maps (Figs. F22-F25) also show the classification of importance currently attributed to *T. confusum* in the different countries.

(F.8.4) *Comments*

(F.8.4a) *Malathion resistance*

Malathion resistance, previously known only from Australia, Poland, Senegal and the United Kingdom was diagnosed in 27 of the 33 countries sampled involving 78 of 122 strains tested. The resistant strains appeared to be distributed generally throughout the areas of occurrence of *T. confusum*. As with *T. castaneum* approximately half of the resistant strains were resistant also to malathion synergised with triphenyl phosphate, although these strains were confined to 56% of countries showing malathion resistance, compared with 80% in *T. castaneum*.

Resistance in Australia, Canada and Central and South America usually was not suppressible with triphenyl phosphate indicating a non-specific malathion type of resistance, whereas resistance in the U.S.A., the United Kingdom, Europe and North Africa, and western Asia was predominantly suppressible indicating a malathion-specific type - elsewhere both types were present or samples numbers were insufficient to indicate trends. Most of the strains were collected in grain and flour residues from flour and provender mills and stores for milled grain particularly flour. There was no obvious association with malathion usage and occurrence probably represented interception of resistant strains derived from infested commodities moving normally in trade.

The most extensive sampling of *T. confusum* has been from
Australia where, except in tropical areas, *T. confusum* is the major
beetle pest of flour mills. Flour mills, by regulation must be licensed
to mill grain for the extensive export trade and so are inspected
regularly to ensure adequate standards of hygiene. Insecticides and
fumigants are used regularly to maintain these standards and malathion,
in addition to being present in all grain from admixture when delivered
from farms, is used regularly for surface treatment in mills as a 2.5%
spray at the rate of 1 gal/1000 s.ft. every 6-8 weeks. Most occurrences
of *T. confusum* in these mills and elsewhere are monitored for resistance
and although malathion resistance levels usually are not high, resistance
appears to be present in most areas. Exceptions appear to be farms and
storages away from milling premises or the channels of distribution of
their products, supporting observations of Greening *et al.* (1975) for
New South Wales. Both non-specific and malathion-specific resistance
have been recorded in all mainland states except the Northern Territory,
and the maximum resistance levels found have been x18 for malathion-
specific resistance in a strain from a flour mill at Port Adelaide and x5
for non-specific resistance in a strain from a flour mill at Laura also in
South Australia (Champ, unpublished data).

Malathion resistance is an economic problem in control of
T. confusum although this appears to be restricted particularly to flour
mills and similar premises in the cooler countries. The resistances do
not present the same threat to adequate control as do resistances in
other storage pests because the circumstances usually allow alternative
control measures such as improved hygiene and use of spot fumigants to
be used to advantage.

(F.8.4b) *Lindane resistance*

Lindane resistance was considerably less common in *T. confusum*
than in *T. castaneum* occurring in 41% of the 94 samples tested compared
with 93% of 497 samples tested in the latter species. Similarly only
23 countries of 34 sampled were involved compared with 75 of 76 sampled
for *T. castaneum*. This reflects the more limited distribution of
T. confusum and particularly its restriction to cooler climates where
lindane has not been used as extensively. Lindane resistance was also
less common than malathion resistance and this appears to be a direct
result of the introduction of higher standards of hygiene throughout the
world at about the same time as malathion was introduced. Infestations
in milling and related premises were tolerated at much higher levels and
control programmes were much less intensive when lindane was the major
residual pesticide available. Changes in consumer demand to freedom from
pest infestations in the 1960-1965 period, required an intensification of
pest suppression in mills and storage, and the availability of the relatively
non-toxic and commercially-acceptable malathion enabled this to become a
reality. Malathion has been used much more extensively and intensively
than lindane in areas where *T. confusum* occurs as a significant pest
and this provides a reasonable explanation for the less common occurrence
of lindane resistance.

There is little data available on resistance levels in *T. confusum*
but it seems probable that these would be reasonably high in strains which
showed little or no response in the discriminating dose tests such as those
from Jamaica, Greece, Morocco, Gambia, Mozambique and Turkey. Interpretation
of the resistances in terms of current lindane usage would be pointless as it
is probable that the resistances are of comparatively long standing, and
obvious associations such as between the extensive use of lindane in Africa

and the highest world incidence of lindane resistance in that area, are probably as much as can be deduced. Irrespective it is unlikely that the resistance has any great significance today as a wide range of alternative controls are in use in most situations where *T. confusum* has pest potential.

(F.8.4c) *Methyl bromide resistance*

Methyl bromide resistance in *T. confusum* was surprisingly common occurring in 25 of the 92 strains examined and involving 15 of the 30 countries sampled. The other seven species gave a combined total of 17 only of 802 strains examined involving 13 of the 83 countries concerned. The resistance in *T. confusum* was heritable but showed little change in tolerance from selective breeding from the survivors of treatment at the discriminating dose. Thus strains from Piraeus, Greece, and Teheran gave 94% and 87% response initially and the progeny of survivors 91% each after several generations of selection. Resistance levels ranged from x1.1 to x1.3 at the LD_{50} and x1.2 to x1.7 at the $LD_{99.9}$. *T. confusum* is slightly more tolerant of methyl bromide than *T. castaneum* and resistance levels need not be as high for equivalent loss of control - this may be offset by the more frequent occurrence of *T. confusum* in cooler areas and the greater relative increase in dose of methyl bromide necessary although temperature effects on the resistance are unknown.

The samples from the flour mill at St Marys', Ontario illustrate again the gradual increase in tolerance of *T. confusum* as discussed earlier. The first sample was collected in December 1969 when ethylene dibromide/ethylene dichloride/carbon tetrachloride was used as a spot fumigant and methyl bromide tolerance was x1.2 ($LD_{99.9}$, x1.1). Further sampling of outbreaks at the time of the survey (May 1972) when ethylene dibromide (70) and methyl bromide (30) had replaced EDB/EDC/CT mixture, showed a slight increase in tolerance of methyl bromide to x1.3 ($LD_{99.9}$, x1.2) paralleling increases in tolerance of malathion, lindane and phosphine.

The association with methyl bromide in the field of other resistances recorded in the survey is less clear. The strains from Finland were from a bread factory and a provender mill where controls were based on methyl bromide fumigation. In China, *T. confusum* is not regarded as a serious problem in flour mills and control stresses hygiene with use of chloropicrin or phosphine as necessary. Methyl bromide is used, however, to fumigate both grain and processed grain. The strain from Yokohama was from a provender mill with a turnover of 4.4×10^5 m.t. where methyl bromide was used occasionally in the silos. In the Argentine, the records were from flour mills where hydrogen cyanide and not methyl bromide was used. Data on fumigant use was not available for other occurrences such as from the U.S.A. and the United Kingdom although it is known that in countries such as Cyprus and Iran methyl bromide is used regularly. In Greece where several records of resistance were made from port silos at Piraeus and local flour mills where spillage and residues were a serious problem, the limited controls used were based on phosphine. The only other fumigant used was ethylene dichloride (3)/carbon tetrachloride (1) whilst methyl bromide was not used although its introduction into widescale use was under consideration.

The overall impression is of some association between *T. confusum* and methyl bromide usage but not at levels which would impose high selection pressure for resistance. The prevalence of resistance in *T. confusum* implies that resistance to methyl bromide may develop more readily than in other species. However, until more is known of the characteristics of the resistance and its relationships with other materials such as phosphine, hydrogen cyanide and pesticides generally (Section E.14.8) particularly as resistance levels are still marginal, interpretation of the circumstances of the various occurrences will be difficult.

(F.8.4d) *Phosphine resistance*

Changes in tolerance of phosphine were recorded in more strains of *T. confusum* than with any other of the species considered. Thus 28 strains of 92 tested were resistant involving 14 of the 30 countries sampled. This frequency of occurrence, however, approximated the situation with phosphine resistance in *Rhyzopertha dominica*, and more particularly with methyl bromide resistance in *T. confusum*. Although occurrence of methyl bromide and phosphine resistances in single strains of *T. confusum* appeared independent (Section F.14.8b), the occurrence of both resistances in particular countries appeared not to be independent. Thus similarities could be expected in the conditions under which the resistances had appeared.

The strains from the flour mill at St. Marys', Ontario showed an increase in tolerance of phosphine as well as malathion, lindane and methyl bromide although there was no record of use of phosphine in the mill as there was for the other three materials. Apart from any considerations of cross-tolerance, phosphine is now used so widely around the world that it is probable that most populations of *T. confusum* in central storage facilities such as those sampled, would come either in direct contact with phosphine or be diluted in the course of normal trade movements by individuals that have had such contact. The higher frequency of resistance in some countries can be associated with more intensive use of phosphine in those countries, for example Jamaica, the Argentine, Greece, the Middle East countries and Australia. The most tolerant strain collected was from a batch of CSMA fly-rearing medium in Louisiana, U.S.A. The resistance factor approximated x3. Most strains had lower factors and although these factors were marginal in terms of allowing significant survival at normal dose rates, selection to higher levels occurred readily. Thus a strain from Piraeus in Greece showed x7 resistance at the $LD_{99.9}$ after a single selection. It is apparent that a minor problem exists in control of *T. confusum* with phosphine although this has barely reached identifiable proportions.

TABLE F14: RESPONSES OF REPRESENTATIVE WORLD STRAINS OF *TRIBOLIUM CONFUSUM* TO MALATHION, MALATHION + TPP, LINDANE, METHYL BROMIDE AND PHOSPHINE IN DISCRIMINATING TESTS FOR RESISTANCE

Locality	Mal	M+T	Lin	MeB	PH$_3$	Locality	Mal	M+T	Lin	MeB	PH$_3$
NORTH AMERICA						**EUROPE AND NORTH AFRICA (Cont'd)**					
CANADA											
Alb. - Calgary	62	83	49	100	100	**ENGLAND**					
Ont. - St. Marys	58	69	100	99	97	Glos. - Avonmouth	90	100	86	99	100
" "	0	1	71	96	87	"	94	100	100	100	100
Woodstock	34	69	94	100	98	"	100	-	100	100	100
U.S.A.						Gloucester	100	-	95	98	98
Ark. - Fayetteville	92	100	100	100	97	Hants. - Botley	100	-	100	97	100
Cal. - Fresno	80	100	100	100	100	Lancs. - Manchester	100	-	100	100	99
"	98	100	100	100	100	Lincs.-Gainsborough	63	100	100	100	100
Riverside	42	91	100	100	100	"	96	100	100	100	100
Fla - Ft. Meade	80	100	100	100	100	Northants-Wellingbor.	96	100	97	85	95
Ill. - Piatt Co.	99	100	100	-	-	Northumb.-Newcastle	60	100	100	100	99
La - Baton Rouge	99	100	100	-	-	Staff. - Walsall	93	100	100	98	99
" "	100	-	100	98	73	Suff. - Bures	100	-	100	99	98
Mo - St. Louis	-	-	-	95	100	Saxmundham	100	-	100	100	98
N.M. - Clovis	96	99	-	98	100	**FINLAND**					
Ore. - Grant's Pass	97	100	98	100	100	Tikkurila	-	-	-	100	100
Tex. - Kelly	45	100	100	100	100	"	100	-	100	98	98
						"	98	100	100	99	100
CENTRAL AMERICA AND CARIBBEAN						"	100	-	100	100	99
						"	100	-	100	100	100
JAMAICA						**GERMANY F.R.**					
Kingston	43	99	1	100	98	Hamburg	96	100	100	96	98
NICARAGUA						**GREECE**					
Samulali	53	85	100	-	-	Pireaus	35	36	42	93	100
						"	31	79	14	94	98
SOUTH AMERICA						"	88	100	1	99	98
ARGENTINA						"	94	89	38	100	95
B.A. - Buenos Aires	52	96	100	97	99	"	93	100	1	-	-
" "	10	91	13	100	98	"	100	-	100	94	98
Junin	10	39	92	100	97	**MOROCCO**					
BRAZIL						Fes	100	-	0	99	100
S.P. - Bauru	10	37	92	-	-	Taza	-	-	-	100	100
CHILE						**NETHERLANDS**					
La Monte	76	96	100	-	-	Wageningen	100	-	100	-	-
						Woimerveer	100	-	100	99	99
EUROPE AND NORTH AFRICA						**SCOTLAND**					
						Campbeltown	100	-	100	100	100
EGYPT						"	100	-	100	100	100
Cairo	-	-	-	100	100	Glasgow	83	100	100	100	100
Giza	95	100	51	100	100	Inverness	100	-	100	100	100
"	-	-	-	100	100	Lanark	97	100	100	100	100
ENGLAND						Leith	100	-	100	100	100
BUCKS. - Haddenham	100	-	96	100	100						
Cambr. - Cambridge	100	-	100	100	100						
Dorset - Sturminster Newton	88	100	99	95	100	**SPAIN**					
						Granada	100	-	100	91	98

TABLE F14
(Cont'd)

T.confusum (ii)

Locality	Mal	M+T	Lin	MeB	PH$_3$	Locality	Mal	M+T	Lin	MeB	PH$_3$
EUROPE AND NORTH AFRICA (Cont'd)						**WEST ASIA (Cont'd)**					
WALES						TURKEY					
Cardiff	78	100	-	-	-	Ankara	100	-	100	100	100
Glamorgan	80	100	100	100	100	"	100	-	99	100	100
						Sakarya	100	-	5	100	100
AFRICA - SOUTH OF SAHARA						**EAST ASIA**					
CENT. AFRICAN REP.						CHINA					
Bangui	42	94	89	100	100	Hopeh-Peking	55	92	100	99	100
ETHIOPIA						"	65	-	-	-	-
Addis Ababa	59	76	100	-	-	Kiangsu-Shanghai	85	-	-	-	-
" "	81	88	100	98	100	"	32	88	100	89	100
" "	100	-	94	100	100	JAPAN					
" "	100	-	100	-	-	Tsuyama	100	-	66	-	-
Nazret	100	-	96	100	100	Yokohama	32	97	10	100	100
						"	98	100	97	95	97
GAMBIA						"	-	-	-	100	100
Denton Bridge	100	-	0	100	100	KOREA REP.					
KENYA						Suwon	74	100	19	100	99
Nairobi	100	-	97	-	-	**AUSTRALASIA AND SOUTH PACIFIC**					
MALAWI						AUSTRALIA					
Blantyre	-	-	45	100	100	NSW - Newtown	29	58	-	100	96
MOZAMBIQUE						"	95	96	-	100	-
Lourenco Marques	59	96	2	100	100	Pyrmont	95	95	-	-	-
NIGERIA						Qd - Dalby	90	93	-	-	100
Lagos	100	-	-	-	-	Warwick	100	-	-	-	-
"	80	-	19	100	100	"	94	83	-	-	-
SOMALIA						S.A.- Jamestown	100	-	-	-	-
Mogadishu	91	100	26	100	100	Laura	8	14	93	100	98
						Mannum	90	87	-	-	-
WEST ASIA						Port Adelaide	6	100	-	-	-
CYPRUS						" "	8	100	100	100	98
Zyyi	65	100	100	95	97	Mile End	91	90	-	-	-
IRAN						Salisbury	100	-	-	-	-
Teheran	94	100	90	87	98	Vic.- Ballarat	82	78	-	-	-
IRAQ						Charlton	92	96	-	-	-
Abu-Ghraib	49	86	100	97	100	Irymple	100	-	-	-	-
Arbil	1	100	100	100	97	Maryborough	91	94	-	-	-
Heath	100	-	100	93	99	Melbourne	100	-	100	100	100
"	79	100	24	-	-	Murtoa	70	71	-	100	93
Karbala	1	100	89	100	98	North Melbourne	98	100	-	-	-
Khalis	79	100	50	-	-	W.A.- Cottesloe	65	72	-	100	99
Rarneli	100	-	100	98	100	"	90	99	-	-	-
						Geraldton	99	100	-	-	-
SYRIA						"	100	-	-	-	-
						"	100	-	-	-	-
						Northam	98	93	-	-	-
Damascus	62	100	100	100	99	Watercarrin	100	-	-	-	-

TABLE F15: SUMMARY OF WORLD OCCURRENCE OF RESISTANCE IN *TRIBOLIUM CONFUSUM* TO MALATHION, LINDANE, METHYL BROMIDE AND PHOSPHINE

	North America	Central America	South America	Europe,N.Afr.,USSR	Africa-S.of Sahara	Western Asia	Eastern Asia	Australasia	Totals	Totals (as %)
MALATHION										
No. countries sampled	2	2	3	10	7	5	3	1	33	-
No. with resistance	2	2	3	7	5	4	3	1	27	82%
No. of strains tested	15	2	5	40	12	13	8	27	122	-
No. resistant	12	2	5	20	6	8	7	18	78	64%
MALATHION + TPP										
No. countries sampled	2	2	3	10	7	5	3	1	33	-
No. with resistance	2	2	3	1	3	1	2	1	15	45%
No. of strains tested	15	2	5	40	11	13	6	27	119	-
No. resistant	6	2	5	3	4	1	3	14	38	32%
LINDANE										
No. countries sampled	2	2	3	10	8	5	3	1	34	-
No. with resistance	2	1	2	4	8	3	2	1	23	68%
No. of strains tested	14	2	5	39	12	13	6	3	94	-
No. resistant	4	1	3	11	9	5	4	1	38	40%
METHYL BROMIDE										
No. countries sampled	2	1	1	10	7	5	3	1	30	-
No. with resistance	2	0	1	5	1	3	2	1	15	50%
No. of strains tested	14	1	3	41	9	11	6	7	92	-
No. resistant	4	0	1	11	1	5	2	1	25	27%
PHOSPHINE										
No. countries sampled	2	1	1	10	7	5	3	1	30	-
No. with resistance	2	1	1	5	0	3	1	1	14	47%
No. of strains tested	14	1	3	41	9	11	6	7	92	-
No. resistant	5	1	2	11	0	4	1	4	28	30%

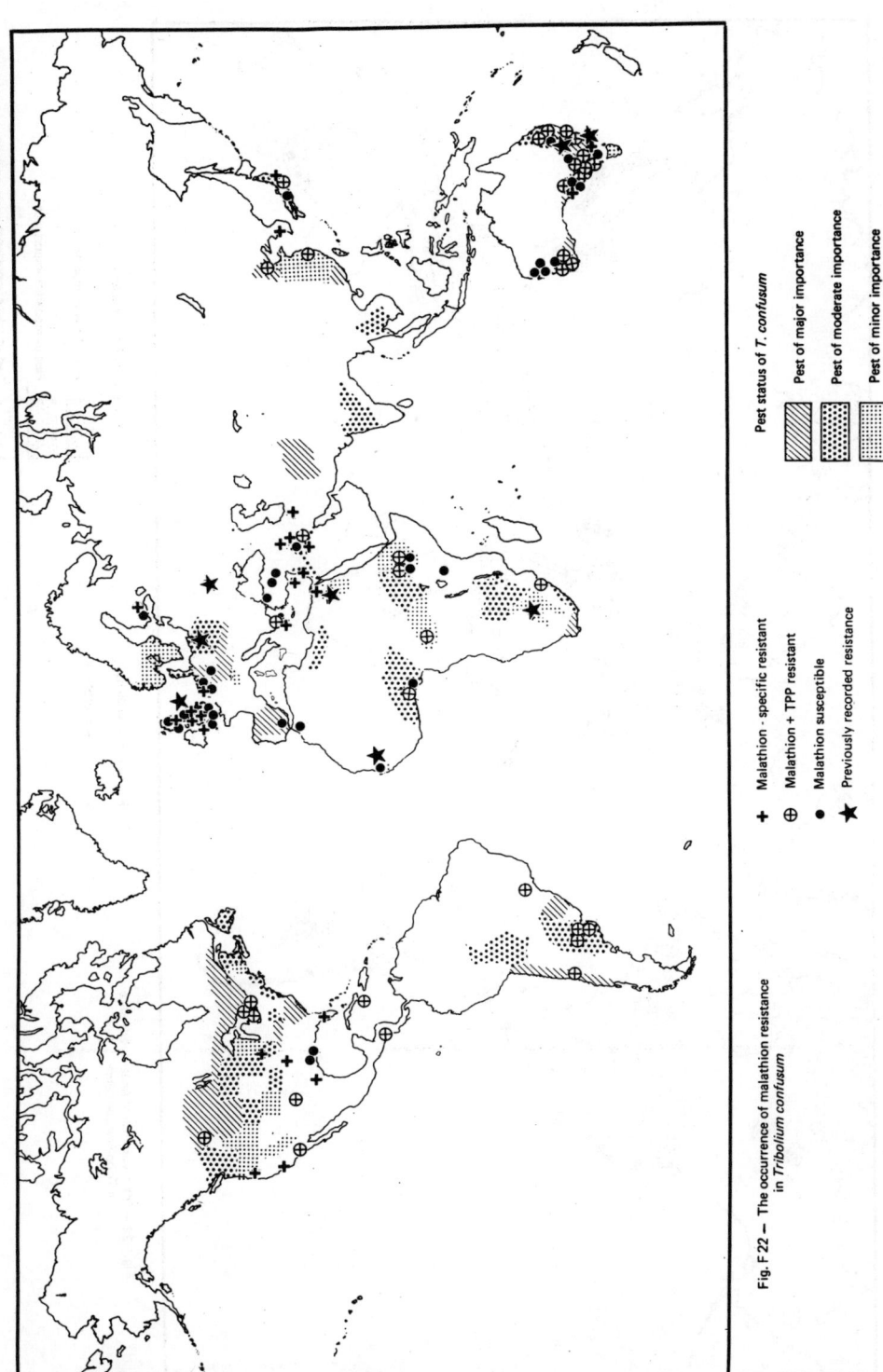

Fig. F 22 — The occurrence of malathion resistance in *Tribolium confusum*

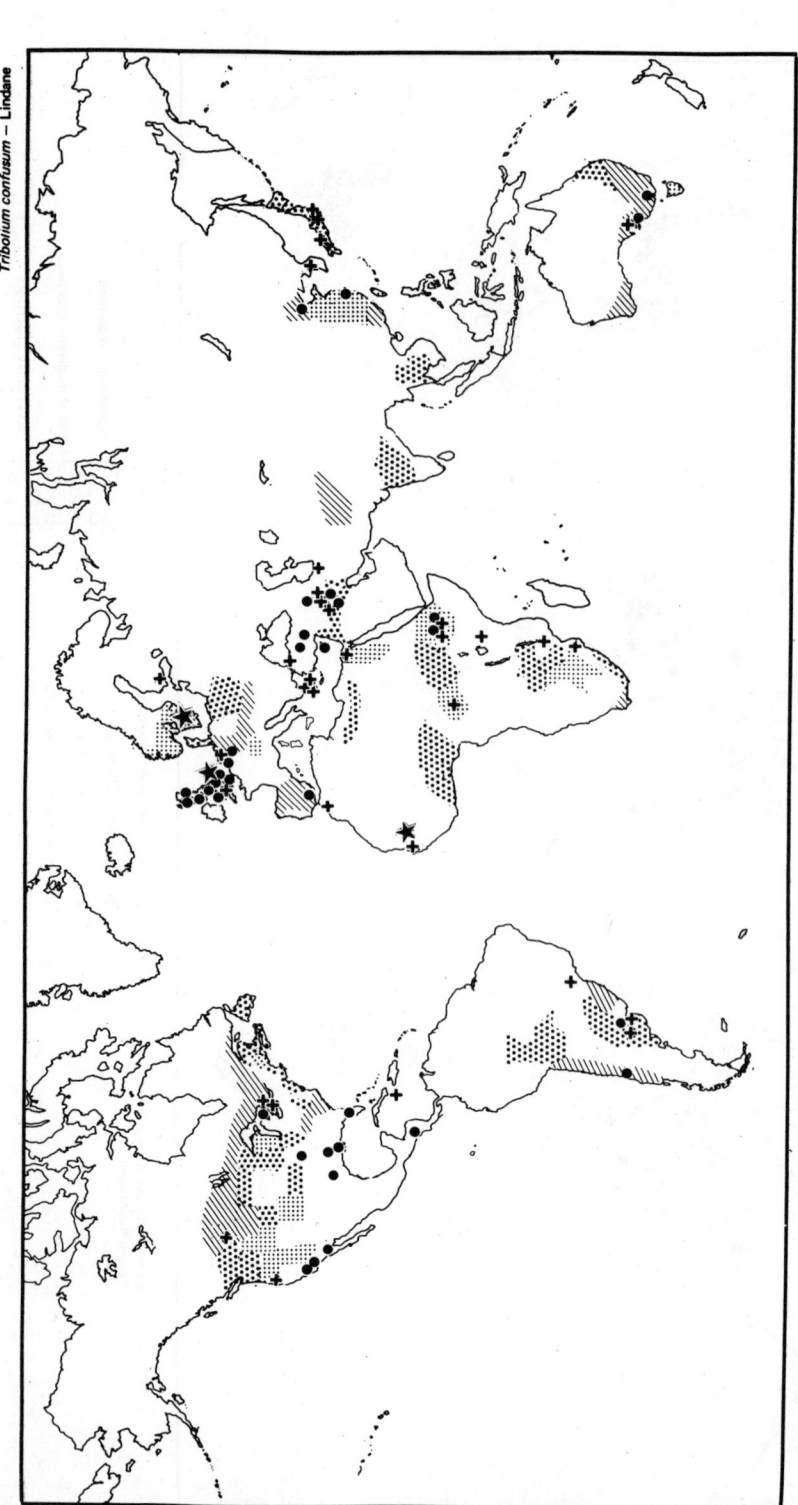

Fig. F 23 — The occurrence of lindane resistance in *Tribolium confusum*

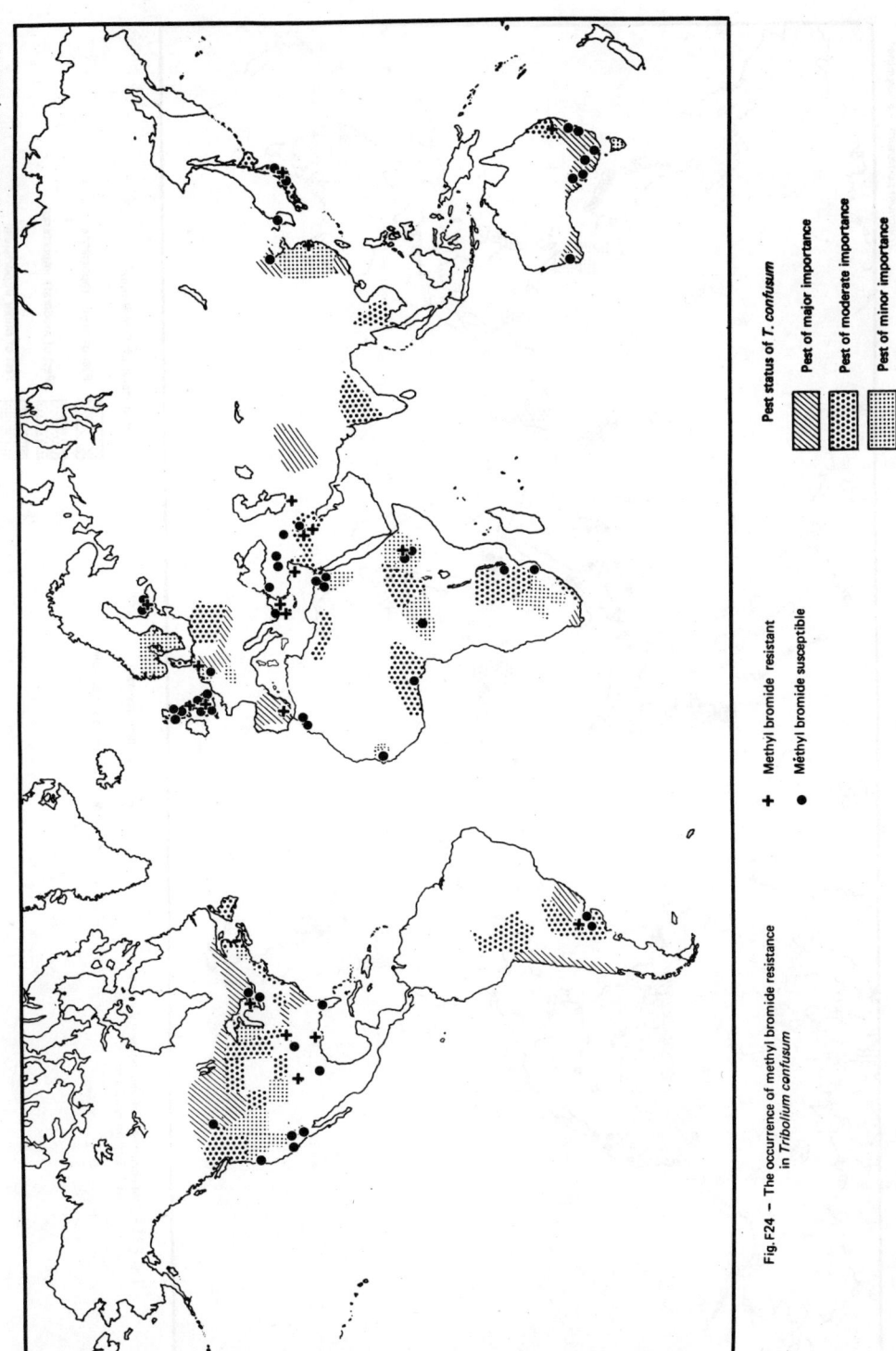

Fig. F24 – The occurrence of methyl bromide resistance in *Tribolium confusum*

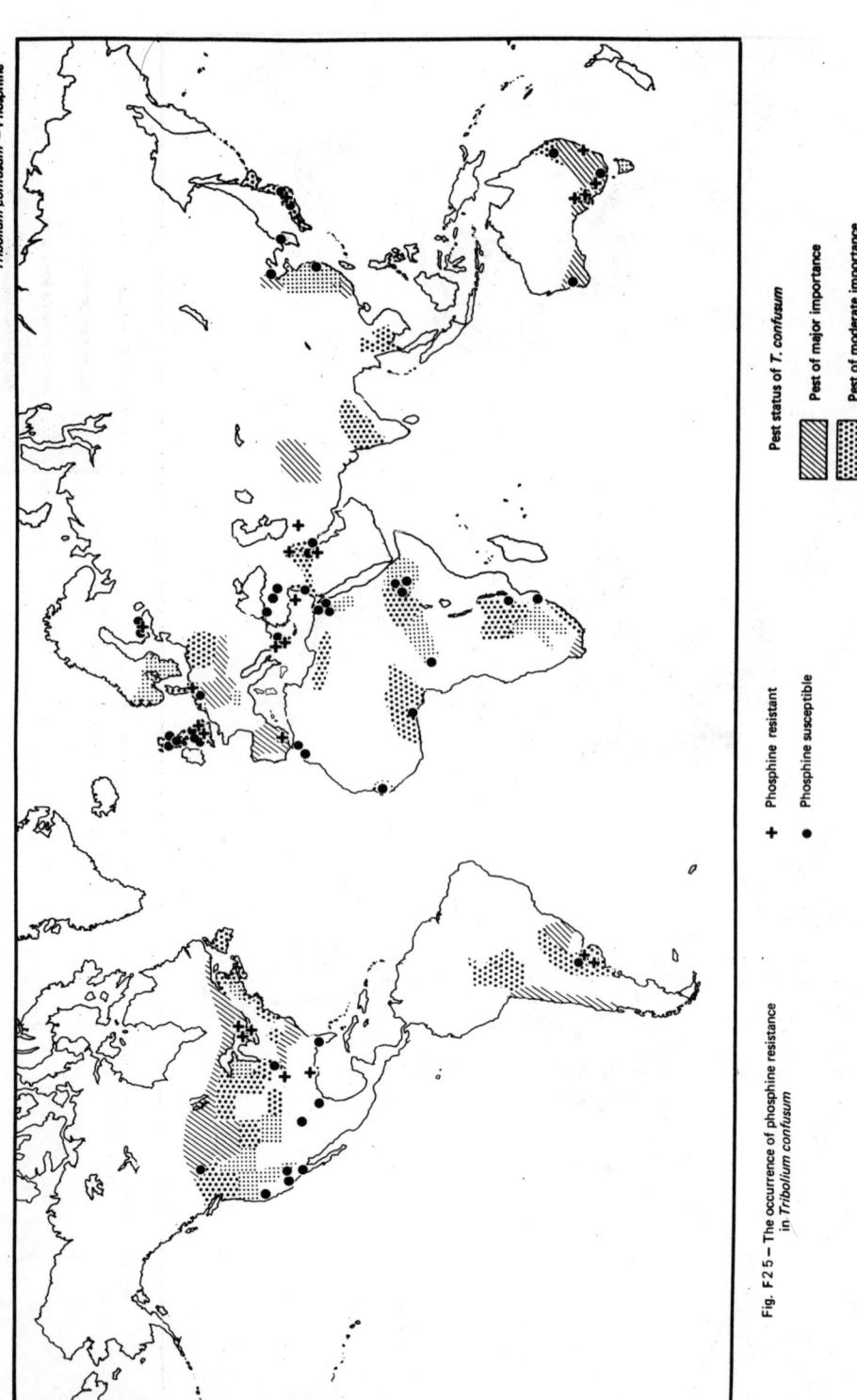

Fig. F25 – The occurrence of phosphine resistance in *Tribolium confusum*

(F.9)   *Oryzaephilus surinamensis*

Oryzaephilus surinamensis is the most widely distributed stored grain pest in the world thriving in areas with cold winters such as Western Europe through to the tropics. Adults are long-lived and often present in large numbers but the immature stages are more sensitive to environmental stresses than many other stored grain pest species and appear to provide a weak link in the field-selection of resistances. Thus despite the widespread occurrence of *O. surinamensis* in most climatic zones of the world and the widespread use of insecticides in the warmer areas and fumigants in the cooler areas, records of resistance are few and resistance does not appear to develop with the same ease as for example with the common *Tribolium castaneum*.

(F.9.1)   *Recorded insecticide resistances*

In general, *O. surinamensis* is comparatively susceptible to most pesticides except lindane. All reports of resistance to insecticide have originated from the Pest Infestation Control Laboratory, Slough. The early records concern lindane. Parkin *et al*. (1962b; Parkin 1965, 1967) reported strains resistant to lindane dusted grain from a farm near Canterbury, England (x2) and from maize in Kenya (x2.7). These tolerances were low and probably would not have significantly contributed to an immediate breakdown of control. Similar tolerances to lindane were reported by Parkin and Forster (1966c; Parkin 1969) in two strains from East Pakistan (Bangladesh) and subsequently both strains were demonstrated as susceptible to malathion (Parkin and Forster 1967b). More recently Dyte *et al*. (1975, Green 1975) reported that in a survey in the United Kingdom in 1973, 30 of 39 strains tested were lindane-resistant. This is the only extensive survey carried out for lindane resistance in *O. surinamensis* and is indicative of the situation that probably prevails elsewhere.

O. surinamensis is regarded as the most serious pest of stored grain in the United Kingdom (Green 1971) and considerable attention has been paid to monitoring the malathion-resistance status of infestations in imported commodities and in local premises to enable countermeasures where required to prolong the useful life of malathion for farm use. In 1966, Wilkin (1967) using a filter paper method with malathion dissolved in butyl phthalate (Wilkin 1966), surveyed the susceptibility of 16 strains of *O. surinamensis* collected from farms in England and could not demonstrate any abnormal change in tolerance to malathion. A strain of *O. surinamensis*, however, collected from barley in Hampshire in 1965 on a farm where malathion and dichlorvos had been used, was selected with malathion and was shown to be resistant to malathion and malathion synergised with triphenyl phosphate (Dyte *et al*. 1973a,d). Dyte *et al*. (1973a, 1976e) using an impregnated filter paper method and discriminating doses, examined a further 22 strains of *O. surinamensis* collected during 1971-1972 from 12 counties in England and 3 in Scotland but again were unable to demonstrate malathion resistance in the field though lindane resistance was present mostly at low levels in 13 of the 14 samples tested with this material. A survey in 1973 again did not reveal any malathion resistance in 52 strains tested (Dyte *et al*. 1975, 1976a,e) and in 1974, samples from 151 inland sites consisting mainly of farms and provender mills but including a miscellany of other sites, revealed only one small infestation of resistant *O. surinamensis* in a warehouse in the London area known to have traded in imported commodities from a ship subsequently shown to be infested with a resistant strain (Green 1975). The measures taken presumably eradicated this infestation. Dyte *et al*. (1976f) also recorded malathion resistance in *O. surinamensis* obtained from dried fruit in Greece. A further strain of unspecified overseas origin that was bred from the survivors of a single exposure to a dose of malathion that killed all susceptible insects, showed resistance of x10 and x7 to malathion alone and synergised with triphenyl phosphate and a wide range of cross-resistances to other organophosphorus compounds, suggesting that a malathion-specific resistance of the carboxyesterase type as in *T. castaneum* was not solely responsible for the increase in tolerance (Tyler and Binns 1973). Similar tests were carried out with 8 malathion-resistant strains collected during this survey from 7 other countries using a more restricted range of alternative insecticides (Dyte et al. 1976e). The level of malathion resistance in these strains ranged from x3 to x37, and all were resistant to malathion synergised with triphenyl phosphate and cross-resistant to fenitrothion and bromophos. The levels of cross-resistance to other organophosphorus compounds varied considerably.

Dyte and Wilkin (1963) observed a small number of unaffected individuals in tests in which adult *O. surinamensis* were exposed for 98 hr to filter paper impregnated with carbaryl in butyl phthalate. The parent stock failed to produce progeny in food medium containing 25 ppm carbaryl (Parkin et al.(1963) give 1 ppm carbaryl as sufficient to suppress breeding in *O. surinamensis*) whereas a culture established in part from these unaffected individuals and reared in carbaryl-containing medium was able after six generations to breed in a medium containing 30 ppm carbaryl. Further selection raised the tolerance for breeding to 80 ppm (Dyte and Wilkin 1964) and then 120 ppm (Dyte and Wilkin 1965).

(F.9.2) *Recorded fumigant resistances*

There is very little information available on any aspects of the toxicity of fumigants to *O. surinamensis* and most of this is confined to the adult stage with the exception of Bang and Telford's (1966) data on larval responses and more recently Vincent and Lindgren's (1972b) examination of the relative tolerance of all life history stages to ethyl formate and phosphine. The only recorded instance of fumigant resistance is from Italy where Dal Monte (1969), of the Ministry of Agriculture, Rome, reported resistance to a mixture of carbon tetrachloride and ethylene dichloride in all stages of a strain collected from stored grain in Puglia. Howe (1962a) was unable to detect methyl bromide resistance in adults collected from rice in Trinidad.

(F.9.3) *Survey data*

Results from discriminating tests for malathion, lindane, methyl bromide and phosphine resistance in *Oryzaephilus surinamensis* are given in Table F16 and Figures F26-F29, together with a summary of the occurrences in Table F17. The distribution maps (Figs. F26-F29) also show the classification of importance currently attributed to *O. surinamensis* in the different countries.

(F.9.4) *Comments*

(F.9.4a) *Malathion resistance*

The introduction of malathion into stored grain pest control reduced very significantly the pest status of *O. surinamensis*. Whereas previously, heavy infestations had been a characteristic of stored cereals particularly in bulk, it became difficult to find any insects at all where control programmes were based on malathion. Despite the gradual reappearance of serious problems with the other major stored cereal pests as they acquired resistance to malathion, *O. surinamensis* remained

noticeably absent. Inevitably resistance appeared but in Australia for
example, where resistance is monitored closely and all unexplained pest
outbreaks given particular attention, there have been only limited outbreaks
localised in 3 areas that can be attributed to resistance from the first
recording at Kununoppin in Western Australia in June 1971 up to 1974.
Although further outbreaks are now apparent, *O. surinamensis* remains a pest
of little consequence. By contrast resistance appeared in *Rhyzopertha dominica*
about the same time but this species consequently became a significant and
limiting factor in grain production and export. The concern expressed
in the United Kingdom about the possibility of introduction of resistant
strains (Section F.9.1) indicates effectiveness of current control measures
with malathion and the failure to detect resistance in inland surveys,
supports the suggestion that resistance does not appear readily in this species
- particularly taking into account local conditions that would predispose to
development of malathion resistance in the species such as the low
temperatures and high grain moisture contents that may be involved in
storage, and the types of storage which often provide harbourage for
residual populations in the fabric of the structures.

Fourteen strains only of the 183 tested in the survey were resistant
to malathion involving 10 countries of the 53 sampled. All 14 malathion-
resistant strains were resistant also to malathion synergised with triphenyl phosphate. Resistances were confined to nine areas around the
world - the Mississippi River basin of Central U.S.A., Central America,
a single occurrence in Rio de Janeiro, the Western Mediterranean, Bombay,
eastern Asia, and in Australia, south-west Western Australia, south-west
Victoria and south-east Queensland.

The resistant strains from the U.S.A. were all collected from
army commissary stores in flour, and in the sample from Illinois, cornmeal
impregnated with anticoagulant rodent bait. Malathion may have been used
in the Illinois store but insecticides were not used in the other stores.
The strain from El Salvador was collected from maize and flour. Resistance
was suspected also in a strain from Barranquilla, Colombia collected among
damaged sacks of dried milk and wheat imported from the U.S.A. in a food
packaging plant using a spray containing a malathion/DDVP mixture (0.8%,
1.6%). The strain from El Salvador was collected from maize and flour and
that from Guadeloupe from rice. The sample from Rio de Janeiro was taken
also from rice moderately infested with a range of species, which had been
stored for some time in a warehouse where malathion was used as a low-volume
spray every 3-6 months. The resistance levels in the strains from South America
were low - to x4 using an impregnated paper test (Dyte and Forster, unpublished
data). The records from the Western Mediterranean were from Cyprus, Greece
and Israel. The sample from Cyprus was from wheat from Louroutzina being
delivered to central storage in Nicosia. In Cyprus malathion is used in
both farm and central storage for admixture and for surface and structural
treatment, whereas in Israel, malathion is used only for disinfestation of
structures. The strain from Greece was from a moderate infestation in a
currant store where malathion had been used in previous years. The records
from Japan and China (Taiwan) were from provender mills. In Japan, malathion
usage was limited by the government in 1972 whereas in China (Taiwan) malathion
was used as a wettable powder for treatment of empty structures and as a dust for
monthly surface treatment of bulk grain in store. Resistance levels in
these strains from Asia were higher than in those from South America - to x7.
The three strains from the Bombay area of India were from rice husks and
wheat in the Bombay storages and rice at Borivli where a range of insecticide
and fumigant resistances including malathion resistance was detected also in
Sitophilus oryzae, *Rhyzopertha dominica* and *Tribolium castaneum*. The

resistance levels in these strains were of the order of x20-x40 again using the impregnated paper test and were the highest recorded for *O. surinamensis* collected in the survey - malathion usage in India has been discussed in Section F.7.1. The Australian records were derived from one strain from Western Australia included in the survey and from various sources of unpublished data from local surveys for resistance. Rimes (personal communication) has recorded two further occurrences of resistance from Western Australia in central storages containing barley and wheat, and from a survey in October, November 1972, three occurrences from a total of 381 produce stores inspected and nine occurrences from 526 cereal-producing farms inspected. In Victoria, Williams (personal communication) detected resistance in a sample from wheat and three samples of oats collected during a survey of oat storages. In Queensland Bengston (personal communication) has recorded resistance from wheat, sorghum and a produce store in three localities in south-east Queensland. All these occurrences can be associated with the extensive use of malathion that has taken place for the past ten years in the grain industry in Australia. Nevertheless malathion resistance does not appear to have significantly increased the minor pest status of *O. surinamensis* in recent years although the above records may be indicative of such a change. The oat industry in Victoria is exceptional in that it has had a continuing problem with *O. surinamensis* although the vast majority of samples tested have been susceptible.

(F.9.4b) *Lindane resistance*

As with most of the other major stored grain pests the incidence of lindane resistance in *O. surinamensis* was high. All malathion-resistant strains were lindane-resistant and in addition these strains were the most lindane-tolerant of collections from particular areas. Resistance was detected in 46 of the 52 countries sampled involving 112 of the 152 strains involved. In those countries where resistance was not detected, single samples only were involved except in Romania where of the two samples concerned, one was suspected as resistant. The movement of *O. surinamensis* in international trade, the long history of association with lindane and effectiveness of control with malathion, make any detailed analysis in terms of other factors affecting occurrence inappropriate. The frequency of occurrence of resistance in the survey (74%) corresponded closely with the frequency recorded by Dyte *et al.* (1975) in the United Kingdom (77%) supporting the suggestion that there could be a 3 out of 4 expectation of resistance throughout the world effectively invalidating control of *O. surinamensis* with this material.

(F.9.4c) *Methyl bromide and phosphine resistances*

A change in tolerance of methyl bromide and phosphine was detected in a single strain only of approximately 60 samples examined from 23 countries. This strain which was susceptible to malathion, was collected from a heavy infestation in dog food at an army medical laboratory in Maryland, U.S.A. Detail of pesticide usage is not available. The increase in tolerance of both materials was low and although the strain was not examined in detail, repeat tests on unselected progeny and progeny from survivors in discriminating tests provided further evidence of a real change -

Methyl bromide - P_1 95%, unselected progeny 99%, selected progeny 97%

Phosphine - P_1 94%, selected progeny 79%.

These occurrences identify a potential problem only in *O. surinamensis*.

TABLE F16: RESPONSES OF REPRESENTATIVE WORLD STRAINS OF *ORYZAEPHILUS SURINAMENSIS* TO MALATHION, MALATHION + TPP, LINDANE, METHYL BROMIDE AND PHOSPHINE IN DISCRIMINATING TESTS FOR RESISTANCE

Locality	Mal	M+T	Lin	MeB	PH$_3$	Locality	Mal	M+T	Lin	MeB	PH$_3$
NORTH AMERICA						**SOUTH AMERICA (Cont'd)**					
CANADA						**CHILE**					
Alb. - Medicine Hat	100	-	100	-	-	Platina	100	-	93	-	-
MEXICO						**PERU**					
Mexico City	100	-	92	100	99	Callao	100	-	100	-	-
U.S.A.											
Ark. - Fayetteville	100	-	99	100	100	**EUROPE AND NORTH AFRICA**					
Cal. - Fresno	100	-	75	100	100						
"	100	-	72	100	99	**ALGERIA**					
"	100	-	82	-	-	Blida	100	-	100	100	-
Riverside	100	-	81	-	-	"	100	-	23	100	100
Ill.- Granite City	97	88	58	-	-						
Kans.- McPherson	100	-	98	100	100	**BULGARIA**					
La. Ft. Polk	86	45	35	100	100	Kostinbrod	100	-	97	-	-
Md. - Ft. Meade	100	-	-	95	94	**ENGLAND**					
New Mex. - Clovis	100	-	0	-	-	Beds. - Sterington	100	-	-	-	-
Ore. - Grant's Pass	100	-	100	-	-	Berks.- Newbury	100	-	98	-	-
S.C. - Ft. Jackson	100	-	100	100	100	Wallingford	100	-	81	100	100
" "	100	-	40	100	100	Wantage	100	-	98	100	100
Tenn.- Memphis	33	-	-	-	-	Bucks.- Highwycombe	100	-	80	100	100
Tex. - Kelly	100	-	-	-	-	C'Wall- Bodmin	100	-	-	-	-
Va - Marion	100	-	100	-	-	Stithians	100	-	91	100	100
Richmond	100	-	90	-	-	Devon - Bideford	100	-	93	-	-
						Sampford Peverell	100	-	88	-	-
CENTRAL AMERICA AND CARIBBEAN						" "	100	-	94	100	100
						Dorset- Blandford Forum	100	-	88	100	100
COLOMBIA						Maiden Newton	100	-	-	-	-
Barranquilla	99	97	13	-	-	Puddletown	100	-	95	100	-
"	100	-	97	-	-	Sturminster Newton	100	-	92	100	100
EL SALVADOR						Essex - Braintree	100	-	100	-	-
San Martin	98	97	29	-	-	Hatfield Peverell	100	-	100	-	-
FRENCH GUIANA						Glos. - Avonmouth	100	-	85	100	100
Cayenne	100	-	-	-	-	Wotton-Under-Edge	100	-	-	-	-
GUADELOUPE						Hants.BrownCandover	100	-	93	100	100
Petit Bourg	57	59	63	-	-	" "	100	-	73	99	100
GUYANA						Winchester	100	-	45	100	100
Georgetown	100	-	28	-	-	Kent - Bexley	100	-	-	-	-
JAMAICA						Maidstone	100	-	92	-	-
Kingston	100	-	56	-	-	Lancs.-Liverpool	100	-	97	-	-
TRINIDAD						Preston	100	-	100	-	-
Port of Spain	100	-	100	-	-	Lincs.-Sleaford	100	-	99	100	100
						London-Tilbury	100	-	97	-	-
SOUTH AMERICA						Notts.-Toxford	100	-	94	-	-
ARGENTINA						Oxon. -Glympton	100	-	97	100	100
Cord. - Guatimozin	100	-	88	-	-	Salop.-Hadnall	100	-	58	100	100
BRAZIL						"	100	-	100	100	100
Amaz. - Manaus	100	-	-	-	-	"	100	-	93	-	-
G.B. - Rio de Janeiro	94	82	41	-	-	Som. - Bath	100	-	-	-	-
						"	100	-	-	-	-

TABLE F16
(Cont'd)

O. surinamensis (ii)

Locality	Mal	M+T	Lin	MeB	PH$_3$	Locality	Mal	M+T	Lin	MeB	PH$_3$
EUROPE AND NORTH AFRICA (Cont'd)						**EUROPE AND NORTH AFRICA (Cont'd)**					
ENGLAND (Cont'd)						**SPAIN (Cont'd)**					
Blue Anchor	100	-	-	-	-	Baena	100	-	98	100	100
Hinton St. George	100	-	-	-	-	Granada	100	-	87	100	100
Suff.- Felsham	100	-	100	100	100	Jaen	100	-	76	100	100
Suss.- Pulborough	100	-	89	-	-	"	100	-	99	100	99
Woodcote	100	-	49	-	-	Lerma	100	-	100	-	-
Warw.- Sutton Coldfield	100	-	100	100	100	Madrid	100	-	95	-	-
Wilts.-Westbury	100	-	-	-	-	Murcia	100	-	50	100	99
Yorks.-Boroughbridge	100	-	74	-	-	"	100	-	95	99	100
"	100	-	58	-	-	San Juan del Puerto	100	-	95	100	100
Northallerton	100	-	99	-	-	**WALES**					
Stokesley	100	-	100	-	-	Abergavenny	100	-	95	-	-
GREECE						Northrop	100	-	2	100	100
Amfiali	100	-	51	-	-						
Lamia	100	-	60	-	-	**AFRICA - SOUTH OF SAHARA**					
"	100	-	71	100	100						
Patras	99	96	-	-	-	**ETHIOPIA**					
Thessalonica	100	-	79	-	-	Addis Ababa	100	-	95	-	-
MOROCCO						" "	100	-	100	-	-
Rabat	100	-	100	-	-	**MALAWI**					
"	100	-	89	100	100	Blantyre	100	-	62	-	-
Sale	100	-	98	-	-	**MOZAMBIQUE**					
NETHERLANDS						Beira	100	-	-	100	100
Wageningen	100	-	97	100	100	"	100	-	64	-	-
POLAND						"	100	-	47	-	-
Poznan	100	-	2	-	-	Lourenco Marques	100	-	100	-	-
PORTUGAL						**NIGERIA**					
Ceiras	100	-	100	-	-	Lagos	100	-	98	100	100
Lisbon	100	-	100	-	-	"	100	-	100	-	-
"	100	-	94	-	-	**RHODESIA**					
Queluz	100	-	99	100	100	Aspindale	100	-	100	-	-
"	100	-	100	-	-	Salisbury	100	-	35	-	-
ROMANIA						"	100	-	0	-	-
Bucharest	100	-	100	-	-	**SOMALIA**					
Budesti	100	-	99	-	-	Mogadishu	100	-	8	100	100
SCOTLAND						**SOUTH AFRICA**					
Dumfries	100	-	100	100	100	Delmas	100	-	53	-	-
Dunbar	100	-	100	100	100	Pretoria	100	-	23	-	-
Dundee	100	-	98	-	-	**ZAMBIA**					
Glasgow	100	-	-	100	100	Livingstone	-	-	46	-	-
"	100	-	90	100	100						
Inverness	100	-	98	100	100	**WEST ASIA**					
Kilmarnock	100	-	100	-	-	**BAHRAIN**	100	-	36	-	-
New Abbey	100	-	89	-	-	**CYPRUS**					
" "	100	-	93	-	-	Athienou	100	-	15	-	-
Pencaitland	100	-	93	100	100	Louroutzina	96	79	37	100	100
SPAIN						Nicosia	100	96	42	-	-
Alcala	100	-	100	-	-						
Aranda de Duero	100	-	95	-	-						

TABLE F16
(Cont'd)

O. surinamensis (iii)

Locality	Mal	M+T	Lin	MeB	PH_3	Locality	Mal	M+T	Lin	MeB	PH_3
WEST ASIA (Cont'd)						**EAST ASIA (Cont'd)**					
INDIA						HONG KONG	100	-	100	100	100
Maharashtra-Bombay	2	3	0	-	-						
"	86	30	8	-	100	INDONESIA					
Borivli	6	7	4	-	-	Djakarta	100	-	1	-	-
						"	100	-	95	100	100
IRAN						"	100	-	100	-	-
Khorramshahr	100	-	93	-	-						
						JAPAN					
IRAQ						Yokohama	86	66	19	-	-
Baghdad	100	-	43	100	100						
Rarneli	100	-	43	-	-	KOREA REP.					
						Suwon	100	-	100	-	-
ISRAEL											
Tel Aviv	77	-	-	-	-	THAILAND					
						Bangkok	100	-	100	-	-
NEPAL						"	100	-	75	-	-
-	100	-	83	-	-	U.S.S.R.					
						Baku	-	-	89	-	-
PAKISTAN						"	-	-	96	-	-
Karachi	100	-	58	-	-						
"	100	-	86	100	100	**AUSTRALASIA AND SOUTH PACIFIC**					
SAUDI ARABIA											
Jeddah	100	-	-	-	-	AUSTRALIA					
"	100	-	-	100	-	NSW - Milbrulong	100	-	-	-	-
						Yerong Creek	100	-	-	-	-
SYRIA						Qd - Warwick	100	-	-	-	-
Damascus	100	-	92	-	-	S.A.- Mile End	100	-	-	-	-
						Port Adelaide	100	-	-	-	-
TURKEY						" "	100	-	-	-	-
Carsamba	100	-	100	100	100	Tas.-Launceston	100	-	-	-	-
Sakarya	100	-	53	100	100	Vic.-Melbourne	100	-	100	-	-
						"	100	-	-	-	-
YEMEN ARAB REP.						Mildura	100	-	-	-	-
Hodeida	100	-	56	-	-	Raywood	100	-	-	-	-
						W.A.-Kununoppin	1	2	6	-	-
EAST ASIA						Watercarrin	100	-	-	-	-
						Yearlering	100	-	-	-	-
BURMA											
Rangoon	100	-	-	-	-						
"	100	-	90	100	100						
CHINA											
Hopeh-Peking	100	-	77	-	-						
"	100	-	83	100	100						
"	100	-	74	-	-						
Taiwan-Taipei	52	19	4	-	-						

TABLE F17: SUMMARY OF WORLD OCCURRENCE OF RESISTANCE IN *ORYZAEPHILUS SURINAMENSIS* TO MALATHION, LINDANE, METHYL BROMIDE AND PHOSPHINE

	North America	Central America	South America	Europe,N.Afr.,USSR	Africa-S.of Sahara	Western Asia	Eastern Asia	Australasia	Totals	Totals (as %)
MALATHION										
No. countries sampled	3	7	4	12	7	12	7	1	53	-
No. with resistance	1	2	1	0	0	3	2	1	10	19%
No. of strains tested	19	8	5	88	15	20	14	14	183	-
No. resistant	3	2	1	0	0	5	2	1	14	8%
MALATHION + TPP										
No. countries sampled	3	7	4	12	7	11	7	1	52	-
No. with resistance	1	2	1	1	0	2	2	1	10	19%
No. of strains tested	18	8	5	88	15	19	14	14	181	-
No. resistant	2	2	1	1	0	5	2	1	14	8%
LINDANE										
No. countries sampled	3	6	4	13	8	10	7	1	52	-
No. with resistance	2	5	3	12	8	10	5	1	46	88%
No. of strains tested	16	7	4	78	15	17	13	2	152	-
No. resistant	11	6	3	55	11	16	9	1	112	74%
METHYL BROMIDE										
No. countries sampled	2	-	-	9	3	5	4	-	23	-
No. with resistance	1	-	-	0	0	0	0	-	1	4%
No. of strains tested	9	-	-	38	3	6	4	-	60	-
No. resistant	1	-	-	0	0	0	0	-	1	2%
PHOSPHINE										
No. countries sampled	2	-	-	9	3	5	4	-	23	-
No. with resistance	1	-	-	0	0	0	0	-	1	4%
No. of strains tested	9	-	-	36	3	6	4	-	58	-
No. resistant	1	-	-	0	0	0	0	-	1	2%

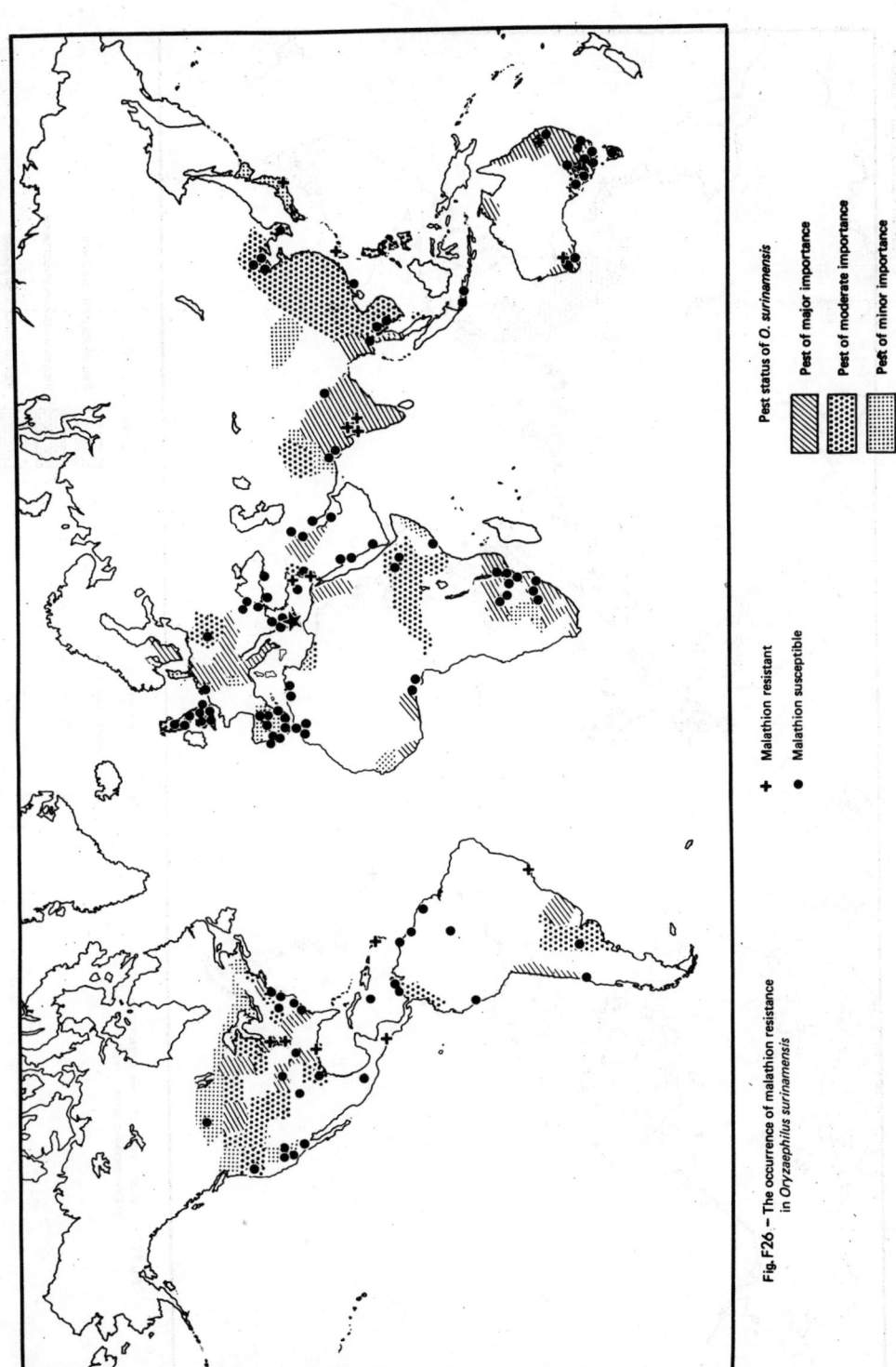

Fig. F26 — The occurrence of malathion resistance in *Oryzaephilus surinamensis*

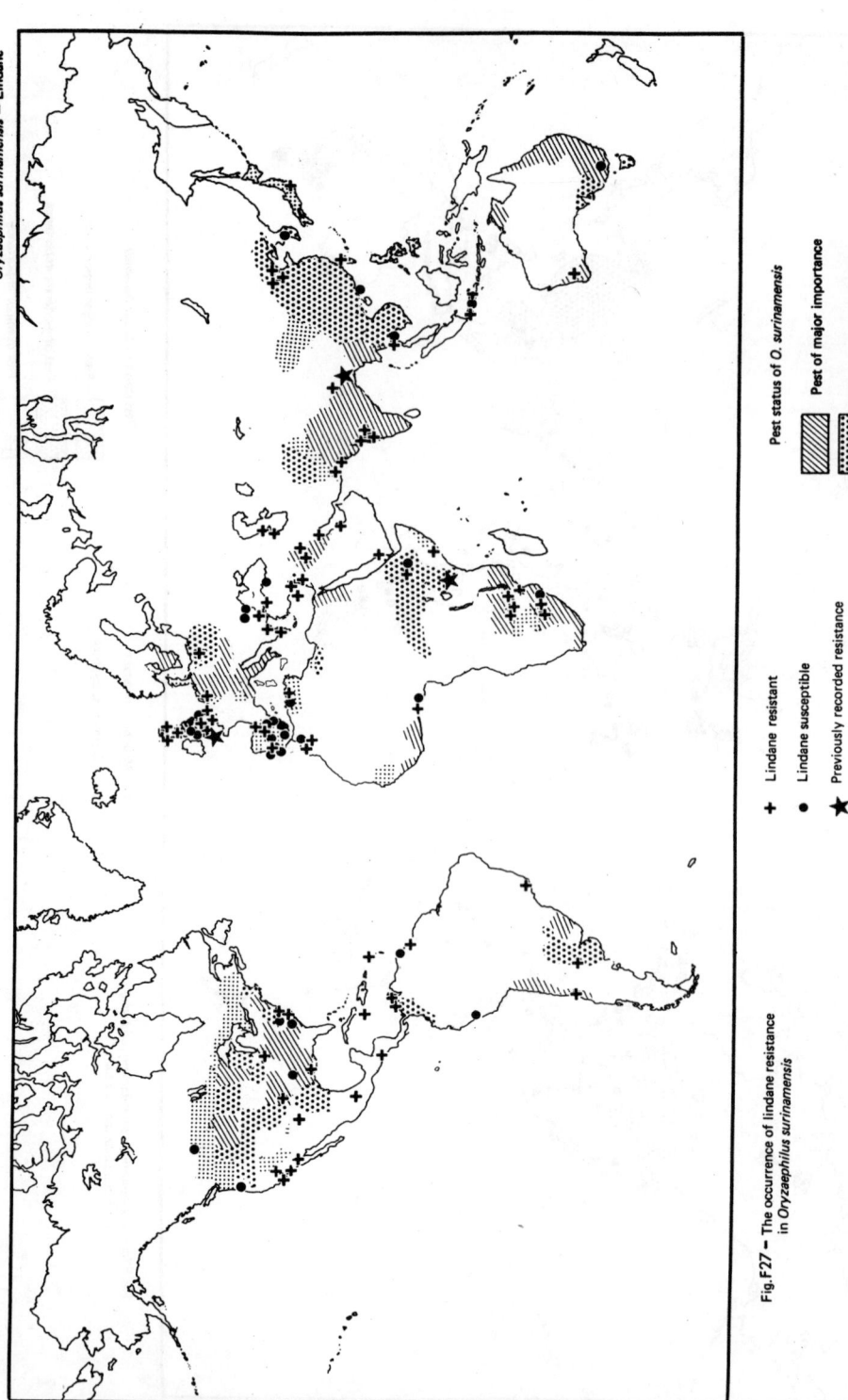

Fig.F27 – The occurrence of lindane resistance in *Oryzaephilus surinamensis*

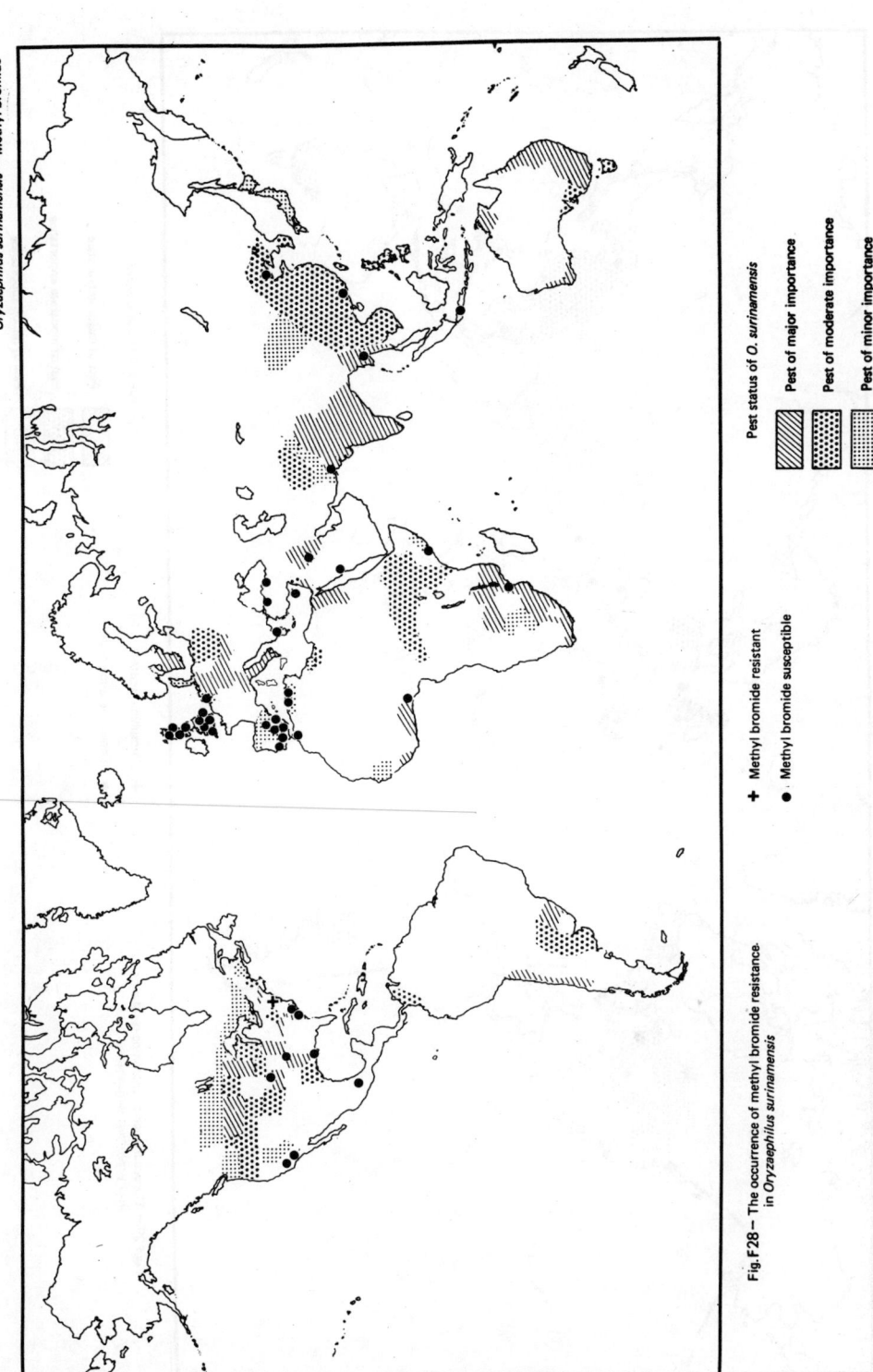

Fig. F28 — The occurrence of methyl bromide resistance in *Oryzaephilus surinamensis*

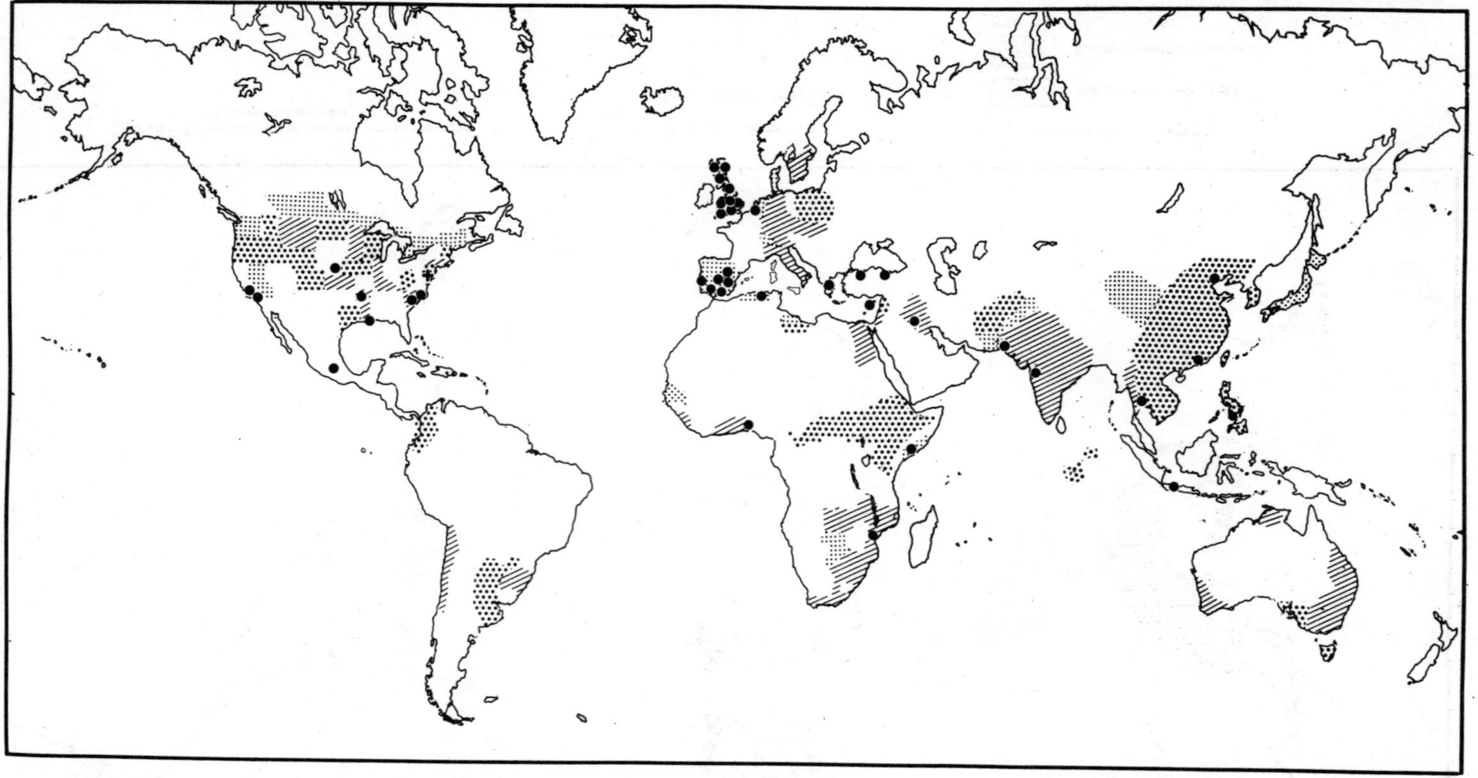

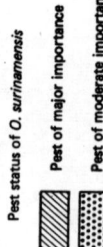

Fig.F29 – The occurrence of phosphine resistance in *Oryzaephilus surinamensis*

(F.10) *Oryzaephilus mercator*

Oryzaephilus mercator is the least important of the stored grain pests considered here. It does assume greater importance, however, in oilseeds in many parts of the world and has been reported as the dominant *Oryzaephilus* species in West Africa.

(F.10.1) *Recorded insecticide resistances*

Dyte and Blackman (1964, 1965b; Green 1967) confirmed malathion resistance in *O. mercator* from The Gambia by selection of a suspect resistant strain found by Green et al. (1964) during a field trial breeding on malathion-treated peanuts containing a 1 ppm malathion residue. The selected strain was able to breed with difficulty at concentrations of 17 ppm malathion in a medium containing rolled oats, wheat flour and yeast. Parkin and Forster (1966c, 1967b) however, were not able to demonstrate either lindane or malathion resistance in two strains from East Pakistan. Subsequently Dyte and Forster (1973a,b;1976) using a FAO-type test with discriminating doses based on 24 hr knockdown responses, reported further strains from The Gambia (2) and from Senegal (6) and Swaziland (1) resistant to malathion and also to malathion synergised with triphenyl phosphate and to lindane. A strain from flour in Sardinia gave similar responses and another strain ex peanuts from Botswana was malathion susceptible and lindane resistant. Malathion susceptible strains were recorded also from England and Spain - the strain from Spain was also lindane susceptible (Dyte et al. 1973a).
The suggestion of a non-specific type of resistance in the triphenyl phosphate tests was supported by cross-resistance data obtained from 48 hr mortality data on a strain (originally from Bambey, Senegal) derived from 6 generations of selection which increased the resistance factor for malathion from x1.9 to x4 when measured with the 24 hr knockdown response. High level resistance to bromophos, jodfenphos and tetrachlorvinphos was present but only low levels of resistance to fenitrothion (x2) and diazinon (x4). These authors (Dyte and Forster 1973b), were unable to detect lindane resistance in the strain from Swaziland using a graded-dose response test based on 24 hr knockdown, but demonstrated a x7.6 resistance when a 48 hr mortality response was used. The discriminating dose test, however, did indicate a suspect resistance (99%). The malathion selected strain from Senegal showed an unmeasurable (>> x100) lindane resistance using the 24 hr knockdown response and >x89 resistance using the 48 hr mortality response.

Subsequently Dyte et al. (1975, Dyte and Forster 1976) surveyed the incidence of lindane and malathion resistance in the United Kingdom in 1973 and reported 3 of 7 strains resistant to lindane and none resistant to malathion.

(F.10.2) *Recorded fumigant resistances*

There have not been any records of fumigant resistance in *O. mercator* and the only reference to tests being carried out was Howe (1962a) who was unable to detect methyl bromide resistance in adults collected from peanuts in The Gambia.

(F.10.3) *Survey data*

Results from discriminating tests for malathion, lindane, methyl bromide and phosphine resistance in *Oryzaephilus mercator* are given in Table F18 and Figures F30-F33 together with a summary of the occurrence in Table F19. The distribution maps (Figs. F30-F33) also show the classification of importance currently attributed to *O. mercator* in the different countries.

(F.10.4) *Comments*

(F.10.4a) *Malathion resistance*

Malathion resistance, already recorded from The Gambia, Senegal and Swaziland in Africa was detected in single strains from a further 3 countries in North and Central America. The strains involved were resistant also to malathion synergised with triphenyl phosphate and to lindane. The strain from Modesto, California was collected in 1968 from almonds in an almond handling warehouse and maintained subsequently in the U.S.D.A. Stored Products Insects Laboratory at Fresno. Both samples from the Caribbean were collected in port warehouses in imported paddy and rice. The paddy sampled at Kingston, Jamaica was lightly infested with *O. mercator* only and had been imported from Guyana. A malathion-lindane spray mixture was in use in this storage. The rice in Port-of-Spain, Trinidad was moderately infested with a range of beetle pests including *Sitophilus* and *Tribolium* species and apparently had a similar origin - insecticides were not used. Collections in rice in export storages in Guyana yielded malathion-susceptible *O. surinamensis* only but it does appear that resistant *O. mercator* was moving about the Caribbean in trade and that rice, of which Guyana was the main source, was implicated. The problem did not appear to be of consequence, however, and although overall world sampling was limited, malathion resistance in *O. mercator* was not a factor of significance in stored grain pest control and its importance as pointed out by Dyte and Forster (1973b) was apparently confined to an intensification of the existing problem of malathion resistance in *Tribolium castaneum* in West Africa particularly in peanuts.

(F.104b) *Lindane resistance*

Lindane resistance was less common in *O. mercator* than in *O. surinamensis* occurring in 50% only of the 34 samples tested. Eleven countries of the 18 sampled were involved and these did not conform to any recognisable pattern. The high resistance levels reported,> x89 for example by Dyte and Forster (1973b), and an apparent even chance that resistance will be present in an infestation of *O. mercator*, severely restrict the usefulness of lindane in control programmes and indicate the necessity for resistance monitoring of infestations if this material is to be used.

(F.10.4c) *Methyl bromide and phosphine resistances*

Fourteen strains from 8 countries were monitored for methyl bromide resistance and 13 strains from 10 countries for phosphine resistance. The only change in tolerance detected was in a strain from Sagaing, Burma on which repeat tests with methyl bromide showed survival of a single individual. Further tests were not carried out and the evidence is insufficient to consider this strain as showing resistance. Thus as with *O. surinamensis* fumigant resistance is not currently an identifiable problem.

TABLE F18 RESPONSES OF REPRESENTATIVE WORLD STRAINS OF *ORYZAEPHILUS MERCATOR* TO MALATHION, MALATHION + TPP, LINDANE, METHYL BROMIDE AND PHOSPHINE IN DISCRIMINATING TESTS FOR RESISTANCE

Locality	Mal	M+T	Lin	MeB	PH$_3$	Locality	Mal	M+T	Lin	MeB	PH$_3$
NORTH AMERICA						**AFRICA - SOUTH OF SAHARA**					
U.S.A.											
Cal. - Fresno	100	-	97	100	100	CENT. AFRICAN REP.					
Modesto	84	91	55	100	100	Bangui	100	-	100	-	-
Riverside	100	-	100	100	-	KENYA					
Tex. - Kelly	100	-	100	100	-	Mombasa	100	-	76	-	-
CENTRAL AMERICA AND CARIBBEAN						GAMBIA					
						Denton Bridge	100	-	94	100	100
ANTIGUA	100	-	0	-	-	MOZAMBIQUE					
JAMAICA						Beira	100	-	79	-	-
Kingston	93	99	36	-	-	Lourenco Marques	100	-	73	-	-
						" "	100	-	78	-	-
TRINIDAD						" "	100	-	100	-	-
Port of Spain	88	58	30	-	-	NIGERIA					
						Lagos	100	-	100	-	-
SOUTH AMERICA						"	100	-	99	100	100
BRAZIL						RHODESIA					
Amaz.- Manaus	100	-	-	-	100	Aspindale	100	-	100	-	-
PERU											
Callao	100	-	100	-	-	**WEST ASIA**					
EUROPE AND NORTH AFRICA						SAUDI ARABIA					
						Jeddah	100	-	100	100	100
ENGLAND						"	100	-	-	-	-
Cumb.- Carlisle	100	-	100	-	-						
Som. - Keynsham	100	-	86	100	100	**EAST ASIA**					
Staffs. - Uttoxeter	100	-	100	100	100	BURMA					
POLAND						Rangoon	100	-	7	100	100
Poznan	100	-	93	-	-	"	100	-	88	-	-
						"	100	-	88	-	-
PORTUGAL						"	100	-	0	-	-
Lisbon	100	-	100	-	-	Sagaing	100	-	100	99	100
SCOTLAND											
Drumchapel	100	-	100	100	100	THAILAND					
Glasgow	100	-	73	100	100	Thonburi	100	-	100	-	-
Kilmarnock	100	-	100	-	-	"	100	-	100	100	100

TABLE F19: SUMMARY OF WORLD OCCURRENCE OF RESISTANCE IN *ORYZAEPHILUS MERCATOR* TO MALATHION, LINDANE, METHYL BROMIDE AND PHOSPHINE

	North America	Central America	South America	Europe, N.Afr., USSR	Africa-S. of Sahara	Western Asia	Eastern Asia	Australasia	Totals	Totals (as %)
MALATHION										
No. countries sampled	1	3	2	4	6	1	2	-	19	-
No. with resistance	1	2	0	0	0	0	0	-	3	16%
No. of strains tested	4	3	2	8	10	2	7	-	36	-
No. resistant	1	2	0	0	0	0	0	-	3	8%
MALATHION + TPP										
No. countries sampled	1	3	2	4	6	1	2	-	19	-
No. with resistance	1	2	0	0	0	0	0	-	3	16%
No. of strains tested	4	3	2	8	10	2	7	-	36	-
No. resistant	1	2	0	0	0	0	0	-	3	8%
LINDANE										
No. countries sampled	1	3	1	4	6	1	2	-	18	-
No. with resistance	1	3	0	3	3	0	1	-	11	61%
No. of strains tested	4	3	1	8	10	1	7	-	34	-
No. resistant	2	3	0	3	5	0	4	-	17	50%
METHYL BROMIDE										
No. countries sampled	1	-	-	2	2	1	2	-	8	-
No. with resistance	0	-	-	0	0	0	0	-	0	0%
No. of strains tested	4	-	-	4	2	1	3	-	14	-
No. resistant	0	-	-	0	0	0	0	-	0	0%
PHOSPHINE										
No. countries sampled	1	-	1	2	2	1	3	-	10	-
No. with resistance	0	-	0	0	0	0	0	-	0	0%
No. of strains tested	2	-	1	4	2	1	3	-	13	-
No. resistant	0	-	0	0	0	0	0	-	0	0%

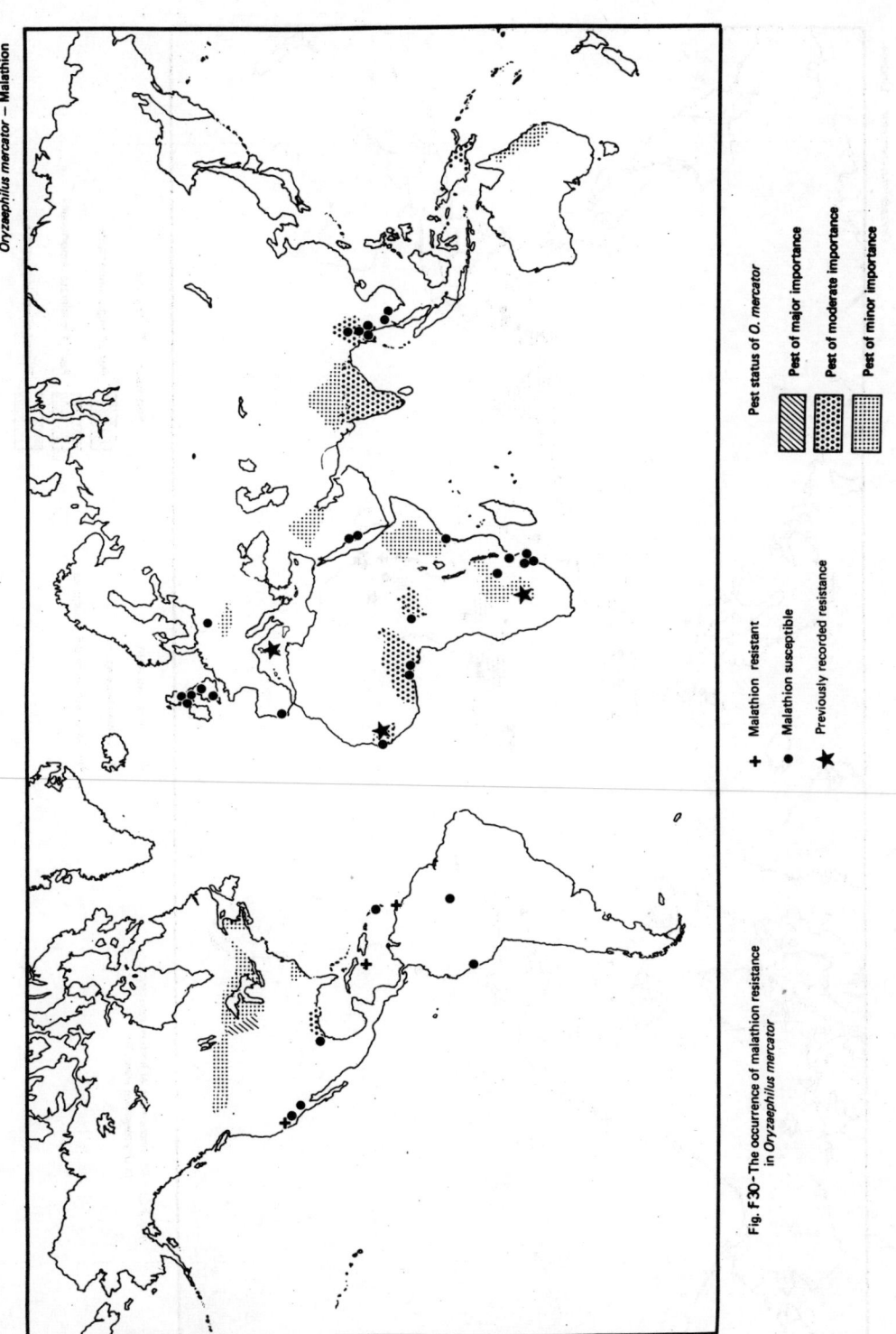

Fig. F 30 - The occurrence of malathion resistance in *Oryzaephilus mercator*

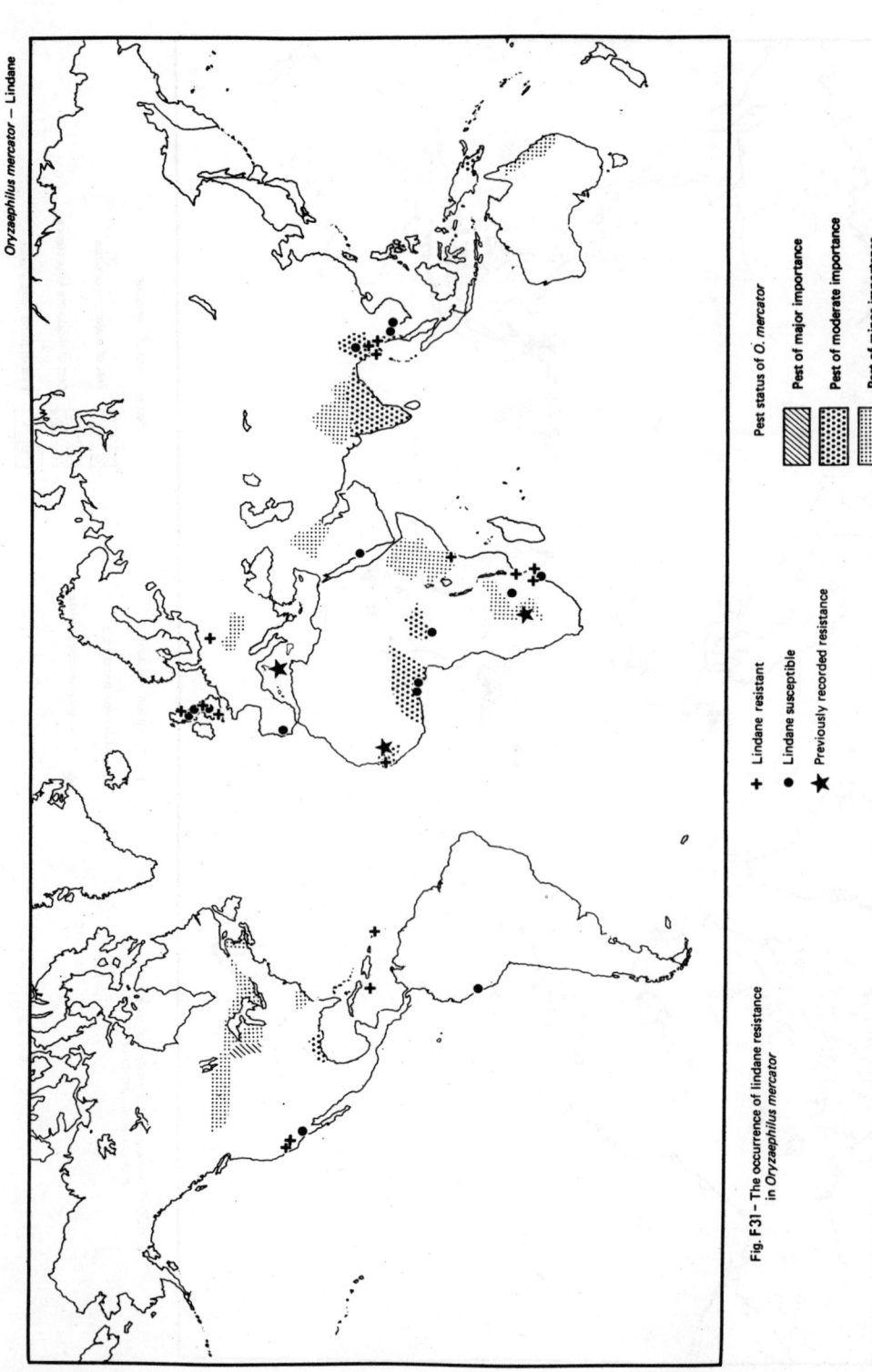

Fig. F31 – The occurrence of lindane resistance in *Oryzaephilus mercator*

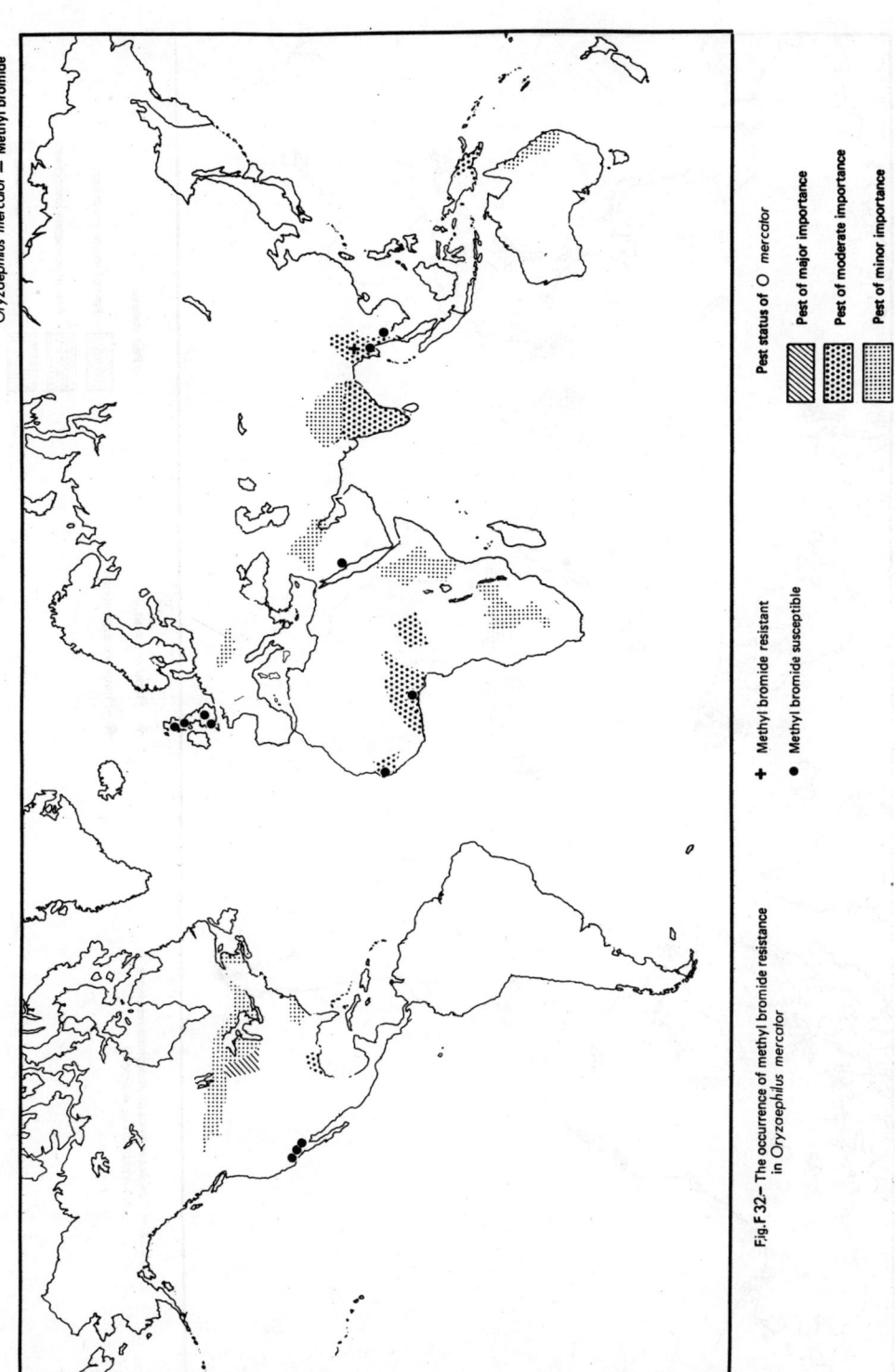

Fig.F 32.— The occurrence of methyl bromide resistance in *Oryzaephilus mercator*

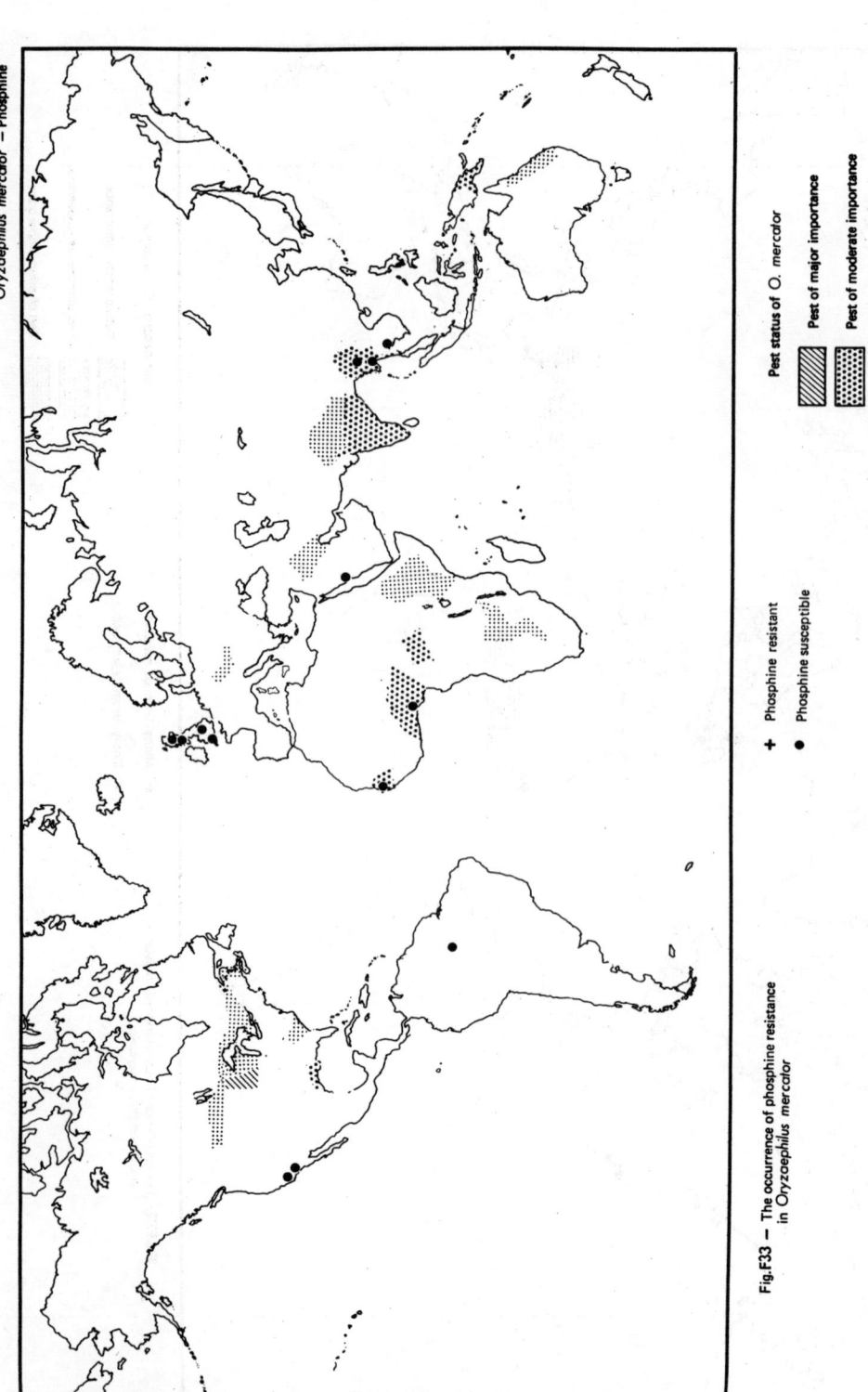

Fig.F33 — The occurrence of phosphine resistance in *Oryzaephilus mercator*

(F.11)   *Other Coleoptera*

(F.11.1)   *Latheticus oryzae*

Latheticus oryzae is a cosmopolitan species that co-habits with other beetles such as the *Tribolium* species in which resistance is prevalent. It is probable that in countries such as India where it is regarded as a pest of major importance and lindane and malathion resistances are present in all other major beetle pests, resistances would be revealed by systematic monitoring. There is a single report of malathion resistance in a listing of resistances by Dyte *et al.* (1975; from Blackman and Peckover 1976) referring to one strain of a series from 5 countries that was collected during this survey and examined by these authors.

(F.11.2)   *Tenebroides mauritanicus*

Tenebroides mauritanicus is mainly a pest of flour mills in the cooler countries and of residues of grain in stores and provender mills in the warmer countries. There are no reports of pesticide resistance occurring in the field. In the laboratory, Monro *et al.* (1961) subjected *T. mauritanicus* to a regime of selection with methyl bromide similar to that used successfully with *Sitophilus granarius* (Monro *et al.* 1961, Monro 1963) but the increase in tolerance was of a low order only and the programme was discontinued (Monro 1964).

(F.11.3)   *Trogoderma granarium*

Trogoderma granarium is usually a major pest only when storage conditions are unsatisfactory and control methods inadequate. Larvae may be long-lived and control is made difficult by their sheltering in cracks their resistance to starvation and other stress and by their entering a diapause under these conditions. There has not been any systematic examination of strains for resistance and the only report concerns malathion resistance in a strain from Tunisia (Yana 1967).

(F.11.4)   *Dermestes maculatus*

Dermestes maculatus is the most common dermestid beetle in animal products in warmer areas and occurs in commodities such as fishmeal and animal skins. Cross-infestation into grain is not uncommon when infestable commodities are stored nearby as for example from infested hides in ships cargoes or from dried fish in market places in developing countries. Lindane is widely used to control outbreaks and dieldrin and lindane resistance has been suspected in populations from South Africa and Australia - laboratory tests, however, failed to confirm the presence of resistance (Shuttleworth and Galloway 1961; Shaw and Lloyd 1967, 1969). Subsequently Dyte *et al.* (1975; Tyler and Binns 1976) have reported lindane and malathion resistance from a fellmonger's premises in Yorkshire in the United Kingdom.

Shaw and Lloyd (1967, 1969; Shaw *et al.* 1968) selected an Australian strain of *D. maculatus* for lindane resistance by culturing in a lindane-impregnated medium based on fishmeal. Each selection was made from the fittest and most tolerant of three lines established from the previous generation in high, medium and low concentrations of lindane. After 11 generations of selection, tolerance had increased approximately x10 enabling the selected line to breed in food containing 220 ppm of lindane. A further 3 generations of selection increased the tolerance more rapidly and allowed breeding in medium containing 450 ppm of lindane. This generation (F_{14}) was compared with a susceptible wild strain from South Africa by topical treatment

of adults and gave a resistance factor of x26 based on mortality assessed
16 days after treatment. A further generation was reared in medium
containing 550 ppm lindane.

Laboratory examinations of tolerance levels of *D. maculatus*
have been few and there has not been any extensive monitoring for resistance.
The positive response to selection of the single strain by Shaw and Lloyd
(*op. cit.*) may have been fortuitous but on probability may be taken to
reflect a species predisposition to resistance development. Interaction of
such a characteristic with the known widespread use of residual pesticides
for control of *D. maculatus* can be expected to result in field occurrences
of resistance and it appears that monitoring for resistance would reveal
these changes in tolerance.

(F.11.5) *Bruchidae*

The Bruchidae contains many species of economic significance as
pests of legumes. They do not attack cereals but are frequently present
as contaminants from cross-infestation in storages, particularly in market
places and merchants storages. There has not been any systematic examination
for resistances in the major pest species of *Callosobruchus* and *Acanthoscelides*
although lindane has been used very extensively as admixture with legumes for
their control. Fumigants have also been used widely and there has been an
unconfirmed report of resistance to methyl bromide from Malawi where
fumigation of bruchids at the rate of 2 lb/1000 cu.ft for 24 hr gave a poor
kill and increasing the dose to 3 lb for 36 hr still allowed survival although
phosphine gave complete kills.

The only confirmed report of resistance is from *Caryedon serratus*,
the groundnut seed borer. Parkin and Forster (1968c; Anon. 1969b) recorded
x15 and x11 dieldrin resistance in two strains from The Gambia following an
observation that peanuts were being successfully attacked after experimental
treatment with a fungicidal seed dressing containing dieldrin. This resistance
presumably was a cross-resistance resulting from lindane dust treatment of
nut-in-shell peanuts which had been used since 1956 for control of this pest.
Dyte and Forster (1969, 1970a; Anon. 1969b) using topically applied doses
of lindane, confirmed resistance levels up to about x3 in these strains and a
further 7 strains obtained from various parts of The Gambia where the lindane
treatment was used.

(F.12)   *Lepidoptera*

(F.12.1)   *Ephestia cautella*

Ephestia cautella is a common pest of a wide range of cereal and cereal products in tropical and warm areas. Pyrethrins and latterly malathion and dichlorvos have been used extensively in control programmes. Although synergists have not been necessary for phycitid control with pyrethrins, much of the exposure has been to synergised material aimed at a complex of pests. Malathion has never been regarded as a satisfactory material for control of *E. cautella*.

In common with most Lepidoptera and indeed most holometabolous insects, there are considerable increases in the tolerance to pesticides through the larval stages of *E. cautella*, particularly in the last larval stage where the marked increase in size during the feeding phase is accompanied by a marked increase in tolerance which continues to increase after feeding ceases and weight decreases to pupation. Many field applications of insecticides are made when there is a significant proportion of the large and conspicuous final-instar larvae in the population and often when these larvae are migrating from the commodity in search of pupation sites. The natural tolerance of these individuals prevents high immediate kills and probably effective control of the infestation. This has resulted in many claims for resistance in the phycitids which more correctly should be related to differences in susceptibility between the different stages of development. Nevertheless the partial kills of populations provide a resistance-selecting mechanism.

Parasites operate much more successfully in Lepidoptera than in Coleoptera particularly in the late season peaks of pest occurrence which characterise the Lepidoptera. When pests become resistant their parasites may still succumb to the pesticide treatment. This means that much bigger pest populations can develop in treated commodities than in untreated ones where parasites continue to have some effect.

The first reports of resistance in *E. cautella* came from the U.S.A. from peanut and citrus pulp storages in Georgia and Florida where parallel resistances to synergised pyrethrins and malathion were developing in *Tribolium castaneum*. Synergised pyrethrins were used to control insects in these industries before 1959 but after 1961 they were largely replaced by malathion. Brown (1961) and Lloyd and Parkin (1963) recorded personal communications from L.S. Henderson of the U.S. Department of Agriculture concerning resistance to pyrethrins in Florida and Georgia, and subsequently Speirs (1967) reported without detail malathion resistance from stored peanuts in Georgia and citrus pulp in Florida. Zettler (personal communication 1972) described *E. cautella* as moderately resistant to malathion in Georgia and slightly resistant in Florida in the period up to 1971. Subsequently Zettler *et al.* (1973) monitored strains of *E. cautella* collected during 1971 and 1972 from peanut- and grain-storage facilities in Florida, Georgia and Alabama using a topical-dosing method on 15-18 mg final-instar larvae and 5-day mortalities. Median lethal dosages only were reported indicating x3.3 and x2.5 increases in tolerance of synergised pyrethrins in strains from Florida and maximum increases for malathion of x1.7 in strains (2) from Georgia, x6.5 (3 strains) from Florida, and x7.2 in a single strain from Alabama. The slope of the dosage-mortality line for the susceptible reference strain and synergised pyrethrins was steeper than those of the test strains indicating that higher resistance factors could be expected at the $LD_{99.9}$ but with malathion all test strains had slopes steeper than that of the reference strain (line slope - 1.8) (Zettler, personal communication 1972). The changes in tolerance of malathion at the $LD_{99.9}$ were thus reduced to less than x1 in the strains from Georgia and one strain from Florida, x3.5 and x1.8 in the other strains from Florida and x2.3 in the strain from Alabama. All LD_{50}'s were considerably less than the $LD_{99.9}$ of the reference strain (LD_{50}'s test strains, 10-20 mg/g; $LD_{99.9}$ reference strain, 155 mg/g). These data involved extrapolation as the maximum dose used was of the order of 65 mg/g. The line slopes suggest that malathion is not an effective toxicant for E. cautella and it seems probable that many of the reports of resistance relate to a higher frequency of the more tolerant individuals within the normal range of tolerances.

Joubert and de Beer (1968) in South Africa reported breakdown in control of E. cautella 20 weeks after treatment with 10 ppm malathion in a maize storage trial in North Transvaal. They described resistance as a serious threat to use of malathion and referred to infestations of resistant strains in other centres in 1965-66. The levels of malathion and its residues quoted and implied, suggest that normal range of tolerance of E. cautella to malathion would account for the lack of control. There is little information available on the levels of malathion required in grain to suppress E. cautella populations. In India, Godavaribai et al. (1962) reported that a concentration of 16 ppm in wheat flour was necessary to retard growth and multiplication, although some eggs and larvae were able to complete development to the adult stage in medium containing 80 ppm malathion. In northern Australia, extensive use of a 10-20 ppm admixture of malathion with wheat grain was never regarded as adequate since its introduction in 1963-65 - Winks and Bailey (1965) cited laboratory experiments in Queensland in 1962 in which E. cautella was bred from wheat containing up to 26.5 ppm malathion.

The ability of E. cautella to develop resistance thus has not been demonstrated in the field unequivocally. In the laboratory, Hashimoto and Fukami (1964) of the Tokyo Agricultural Chemical Inspection Station selected final-instar larvae at an 80% mortality level using films of methyl parathion in petri dishes and obtained an increase in tolerance after the sixth generation of selection which rose to x7 after 16 generations of selection. The selected strain showed cross-resistance to malathion although it was significant that the susceptible reference strain showed considerable natural tolerance to malathion. This positive response to selection in a single strain indicates a high probability of resistance developing in the field following regular exposure to selecting doses of insecticide.

(F.12.2) *Plodia interpunctella*

Plodia interpunctella is a more widely-distributed species and occurs in considerably cooler areas than are tolerated by *Ephestia cautella*. It tends to be associated with milled products rather than grain and will attack a wide range of stored commodities other than cereals. Changes in tolerance of pesticides during the development stages are similar to those in E. cautella and there is the added complication of a distinct diapause in some populations of P. interpunctella. As with E. cautella there has been extensive use of pyrethrins, malathion and dichlorvos and many of these treatments have been made under conditions where partial kills only would be achieved.

The first report of resistance was to malathion in peanuts from Georgia (Speirs 1967) although it has become evident that pyrethrins resistance may have been present earlier (Speirs and Zettler 1969, Zettler et al. 1973). Synergised pyrethrins were used in the U.S.A. to control insects in stored products until approximately 1959 when malathion came into extensive use replacing pyrethrins. Failures in control with pyrethrins had been common because treatments were expensive and were not applied as frequently as necessary, insufficient quantities were used and applications were not always made at the correct time (La Hue 1969, 1971), but there was no suggestion of resistance to pyrethrins being present in *P. interpunctella* until Zettler et al. (1973) reported last stage larvae of a single strain from Florida that showed x2.5 resistance at the LD_{50} to topically-applied synergised pyrethrins (at $LD_{99.9}$, x8).

In 1969 La Hue (1969) in Manhattan, Kansas, recorded a malathion-resistant strain of *P. interpunctella* from maize that had been surface-treated with malathion and contained a surface residue of 81 ppm malathion. This strain was able to breed in maize containing 40 ppm malathion but was controlled by dichlorvos. In California, Armstrong and Soderstrom (1975) at the USDA Fresno Stored Products Insects Laboratory, compared a series of strains collected from almonds and dried fruit by rearing them from egg to adult stage in malathion-impregnated almond nutmeat. The laboratory-reference strain which had been in culture since 1967 showed an LD_{50} of about 1.8 ppm whereas strains collected from culled figs in Fresno and packaged prunes in San Jose had extrapolated LD_{50}'s of 0.2 ppm or less, and strains from oil-grade almond stock and almond hulls from Sacramento, Modesto and Chico had LD_{50}'s of about 95, 53 and 15 ppm respectively. Further tests indicated no resistance to diazinon which was used widely in California, or to pirimiphos methyl (Armstrong, personal communication 1972). The malathion resistance reported by Speirs (1967) from Georgia, was not detailed but Zettler (personal communication) at the USDA Stored Products Laboratory at Savannah, Georgia, has summarised Speirs' assessment of the malathion resistance status of *P. interpunctella* in the USA up to 1971 as highly resistant in Iowa, Kansas and South Carolina, moderately resistant in Georgia and Illinois and slightly resistant in California. Brower (1973a) in a separate study at the same laboratory of the gamma-radiation sensitivity of malathion-resistant strains, recorded x828 malathion resistance in the F_{19} of the strain from shelled corn in Iowa and >x850 in the F_1 of a strain from a cottonseed oil mill in Alabama. In 1971-1972 Zettler et al. (1973) checked strains from peanut- and grain-storage facilities in Georgia, Florida, Alabama, Kansas and Illinois and recorded resistances approximating x200 in all instances. The test data were based on 120 hr responses of 15-18 mg final-instar larvae to topically-applied malathion. Zettler (1974a) also checked the tolerance to pirimiphos-methyl of five strains from Georgia and Illinois that showed similar levels of resistance but as with Armstrong (op. cit.), was unable to detect any significant change in tolerance of this material. He subsequently (Zettler 1974b) suggested after esterase studies that the resistance was not specific to malathion.

It is most probable that these resistances are not confined to the United States and that a systematic search in areas where pyrethrins and malathion have been used regularly, would reveal similar resistances.

(F.12.3) *Ephestia elutella*

Ephestia elutella is a pest of grain and cereal products in cool temperate areas and is found only in warm temperature areas in commodities such as stored tobacco. Pyrethrins have been used for control though not to the same extent as in control of *E. cautella* or *P. interpunctella*.

Resistance has not been reported from *E. elutella*. The listings of pyrethrum resistance in Lindgren and Vincent (1965, 1966a) from Brown (1961) should be referred to *E. cautella*

(F.12.4) *Sitotroga cerealella*

Sitotroga cerealella is a widely distributed species but is essentially a field pest attacking standing grain and particularly un-threshed grain left to dry in the field. It is most important as a pest of rice, maize and sorghum and though the introduction of combine harvesting and threshing has reduced losses considerably, it remains of considerable significance in subsistence farming. Insecticides are not used to any great extent but of those that have been used, lindane has found widest use particularly in developing countries where *S. cerealella* reaches its maximum pest status. As with many other species there has not been a systematic examination of tolerance levels in *S. cerealella* and the only record of resistance is to lindane from stored paddy occurring throughout Sri Lanka (Anon. 1967, Fernando 1967).

Kot (1970) has used the term "partial resistance" to describe the phenotype expression of segregating resistant populations and attributes changes in slope of dosage-mortality lines for demeton methyl against *S. cerealella* to this cause. Although this may be so, the evidence from the slope of the lines provided is not sufficient to justify a conclusion of demeton-methyl resistance in this species.

(F.12.5) *Tineidae*

There is a wide range of species of Tineidae found associated with grain particularly in grain residues. The only records of resistance, however, relate to *Tineola bisselliella* and *Tinea pellionella* which are better known as pests of animal fibres but occasionally are taken from the stored grain habitat in circumstances where hygiene is poor and infestations are of long-standing.

Kühne and Becker (1965) selected *T. bisselliella* for 20 generations with dieldrin and mitin-FF and increased the tolerance levels to x70 and x2 respectively. The resistance was selected further and endrin and chlordane resistance was reported associated with the dieldrin resistance (Kühne 1967). Dieldrin has been the most commonly used material in moth-proofing of animal fibre textiles but there is now increasing evidence that this process which used doses of dieldrin nominally in considerable excess (500 ppm, 100 fold excess) of that required for complete protection, can no longer be relied on for this protection. Odeneal (1961) reported without detail, resistance to chlorinated hydrocarbons in clothes moths and fur beetles from 5 states in the U.S.A. In eastern Australia, there have been field reports of dieldrin resistance in *T. pellionella* involving mature larvae surviving in wool containing 1000 ppm dieldrin and breeding in wool containing 100 ppm dieldrin (Hoskinson, personal communication). There is also field evidence to suggest that tolerance to moth-proofing treatment is increasing in *Attagenus megatoma* (Dermestidae) also.

(F.13) *Movement of resistant strains in world trade*

The movement of infested parcels of grain or grain products in trade has provided an effective means of ensuring the widest-possible geographic spread of storage pests since cereals were first harvested. The continued introduction of the major pest species into unsuitable climates outside their normal distribution, ensures also that all these species may be found in all parts of the world notwithstanding that infestations of some species may be somewhat transient under local conditions, for example *Sitophilus granarius* in the tropics. This dispersal mechanism concerns interchange of strains as well as species and the wide range of commodities that attract the major pests creates problems in preventing the spread of acquired strain characteristics such as pesticide resistance. This is particularly so within countries where movement of commodities to and from farms, seed and animal feed processing plants, other mills and central storage provides considerable opportunity for the dissemination and introduction of resistance into most pest populations. Once established in an area, the spread of resistance into neighbouring areas is inevitable unless rigorous restrictions are placed on movement of infestable commodities. Usually, however, the resistance has spread before it has been detected and controls can be implemented. A similar situation prevails in international movement of grain and grain products but because the range of commodities and their movements are more limited and more effective controls can be introduced to prevent movement of infested commodities, there is greater opportunity to prevent the dispersal to other regions of new resistance as they emerge.

(F.13.1) *Movement of insects in international trade*

The extent of the world cereal trade and its association with pests of major importance is outlined in Section D. Because of the bulk and relatively low value of cereals, international transport is almost exclusively by sea although rail and road transport may become significant in limited areas. The infestations which occur in ships originate either in the commodity before loading or from cross-infestation from residues of previous cargoes or other infested cargoes being carried. It is often difficult, however to determine the precise origin of particular infestations. The origin, extent and effects of infestations in cargo ships have been described extensively by Freeman (1948a,b, 1950, 1957, 1958, 1960, 1962, 1968a,b,e,f, 1971, 1973 1974) and Monro (1951, 1969). The magnitude of the problem has been established for example by specific data covering imports to England particularly (Aitken 1964,1975; Anon. 1965; Freeman *op. cit.*, 1965, 1968c,d, *et al.*1973; Freeman and Heape 1969; Howe and Freeman 1955; Hurlock 1963, 1964; Matthews 1958; Turtle 1961), to Canada (Monro *op. cit.*), Portugal (Cabral and Gouveia 1960, Gouveia 1967) and to Japan (Kiritani *et al.* 1959). Although there appears to be a steady decline in the severity of the problem (Freeman and Heape 1969, Freeman *et al.* 1973, Freeman 1974) resulting from consumer demand and the associated introduction of compulsory inspection and fumigation of export commodities in the countries of origin, and the extensive and spectacular control exercised by malathion and the newer pesticides, the problem remains serious.

Of particular concern are the residual infestations in ships as these provide a continuing source of infestation by populations containing the various species and strains which have been carried previously in infested cargoes (Freeman 1960, *et al.* 1973; Freeman and Heape 1969; Hurlock 1961a,b; Turtle and Freeman 1960). Less importantly but of significance also are the resident populations in the food storage and handling areas of the ship (e.g. Evans and Porter 1965). The extent of these residual infestations has been highlighted by the introduction of compulsory inspection and treatment of ships' holds in exporting countries to supplement pre-loading measures as for example in Australia, Canada, Nigera and the U.S.A. where export grain and other commodities can be loaded only into ships that have been certified by inspection to be free of infestation.

(F.13.2) *Occurrence of resistant insects in ships and their cargoes*

Exporting countries depend on pesticides to meet the demand of trade for freedom from insects in export commodities. The extensive use of pesticides in these countries and the resultant prevalence of pesticide resistance necessarily must result in a high frequency of infestations in cargoes that contain resistant individuals. These contributions to the residual infestations in ships would be intensified by selection pressure from the pesticide residues from protectant-treated cargoes and from pesticide use in disinfestation of the ships.

Attention was drawn to the problem of the movement of resistant strains of grain insects in world trade during the FAO Symposium on Resistance of Agriculture Pests to Pesticides in Rome in September 1969. There is now considerable data available supporting this. Dyte (1970, *et al.* 1973b,c; Dyte and Blackman 1970a,c; Rowlands *et al.* 1973) tested 14 strains of *Tribolium castaneum* from cargoes being unloaded in English and Scottish ports and found 12 lindane-resistance strains in cargoes that included maize, rice bran, cottonseed expeller cake and peanut products that had been loaded in Burma, India, Kenya, Nigeria, The Sudan and Tanzania. Similarly of 18 strains tested for malathion resistance, 16 were found to be resistant involving all the countries listed above and The Gambia and Senegal - with the exception of single strains from Kenya and Senegal, all strains showed a malathion-specific type of resistance. A recent indication of the extent of the problem in imported cargoes has been the report by Green (1975) "that virtually all *T. castaneum* now being imported in Liverpool docks are malathion-resistant." Green (*op. cit.*) also referred without detail to malathion-resistant strains of *Oryzaephilus surinamensis* in imports to England from India and Cyprus and commented that the resistance in strains from Indian cargoes were of such a high level that the insects would not be controlled in bulk grain with permitted levels of malathion. The threat of introduction of malathion resistance into local populations of *O. surinamensis* is regarded sufficiently seriously in the United Kingdom to warrant special attention to any infested cargo originating in a country where resistance is known to occur. Champ and Campbell-Brown (1970b) reported interceptions in Australian ports of *T. castaneum* in cargoes of peanuts from India which were both malathion (x57)- and lindane (x36)- resistant and supplemented this with data from strains collected from residual infestations found during preloading inspections of ships chartered to carry wheat for export. All these strains were resistant to malathion and lindane and resistance factors up to x148 and x112 based on exposure to deposits on filter paper were recorded for non-specific and malathion-specific resistances respectively, and >x100 for lindane resistance. These authors did not find any susceptible strains, suggesting as above that the frequency of occurrence of resistance in infestations is high and emphasising the seriousness of the problem.

All available data concerning the incidence of resistance in infestation in cargoes and in resident populations in ships are summarised in Tables F20 and F21. These summaries include unpublished data from the Pest Infestation Control Laboratory, England particularly relating to interceptions of infestations in cargoes, and from the CSIRO Stored Grain Research Laboratory, Australia from material collected during pre-loading inspections of ships under contract to carry export wheat from Australia. The proportions of strains showing resistance were similar with cargoes and residual infestations approximating 70% with malathion and 90% with lindane. These frequencies were considerably higher than the 40% and 70% of strains resistant to malathion and lindane respectively that were recorded in the survey for samples of all species when grouped similarly. The greatest contribution came from *Tribolium castaneum* both in terms of frequency of occurrence of the species in infestations and of proportions of strains showing resistance. In residual infestations in grain-carrying ships, *all T. castaneum* strains were malathion-resistant and 98% were lindane-resistant. An almost parallel situation existed in the strains collected from infested cargoes. Although the number of strains involved was small, the high incidence of resistance in *Rhyzopertha dominica* is particularly significant. It is clear that resistant strains are being dispersed widely in infested commodities in world trade and that inspection procedures and recommendations for control of infestation must take into account the probability that resistant strains are concerned.

Malathion resistance was first detected in *T. castaneum* in the early 1960's and now can almost be considered a normal attribute of a *T. castaneum* population. The frequency of this resistance in other species is still comparatively low and although the much later emergence of detected resistance in these species (e.g. *S. oryzae*, 1969; *R. dominica*, 1971) indicates behaviour and a genetic constitution less conducive to development of resistance, the demonstrated movement of resistant strains in trade is providing a mechanism allowing a rapid rate of spread and further more widespread and intensive occurrences throughout the world. The same mechanism will allow the rapid distribution of new resistances as they develop, for example as is probably happening now with fumigant resistances.

TABLE F20: RESISTANCE TESTS ON INSECTS FROM CARGOES BEING UNLOADED

NOTE: Cross-infestation between commodities on ships may occur so that the country of origin of the cargo may not be the country of origin of the infestation.

Country of Origin of Cargo	Country or State where Intercepted	Malathion			Lindane	
		No. of strains	No. resistant Malathion	No. resistant Mal+TPP	No. of strains	No. resist.
Sitophilus oryzae						
Unknown	New Guinea	1	0	-	1	1
Sitophilus zeamais						
Australia	New Guinea	1	0	0	-	-
Japan	"	1	1	0	1	1
Unknown	"	1	0	0	1	1
"	England	1	0	-	1	1
Sub-total *S. zeamais*		4	1	0	3	3
Sitophilus granarius						
France	England	1	0	-	1	0
Rhyzopertha dominica						
U.S.A.	China	1	0	0	1	1
Ghana	England	1	1	-	1	1
S. Africa	England	1	1	-	1	1
Unknown	England	2	2	-	2	2
Sub-total *R. dominica*		5	4	0	5	5
Tribolium castaneum						
Australia	England	1	0	0	1	0
"	New Guinea	2	2	1	2	2
Brazil	W. Germany	1	1	1	1	1
"	England	1	1	1	1	1
Burma	"	1	1	0	1	1
Egypt	"	1	0	0	1	0
Gambia	"	1	1	0	-	-
India	"	4	4	0	4	4
"	Queensland	1	1	1	1	1
"	Victoria	1	1	1	-	-
Japan	New Guinea	1	1	1	1	1
Kenya	England	1	1	0	1	1
"	Scotland	1	1	1	1	1
Nigeria	England	6	6	0	3	2
Senegal	"	1	1	1	1	1
Sudan	"	1	1	0	1	1
"	W. Germany	1	1	0	-	-
Tanzania	England	1	1	0	1	1
Thailand	Scotland	1	0	0	1	1
Unknown	England	1	1	1	1	1
"	New Guinea	6	6	2	6	6
Sub-total *T. castaneum*		35	32	11	29	26
Oryzaephilus surinamensis						
Spain	England	1	0	-	-	-
Thailand	"	1	0	-	1	1
Sub-total *O. surinamensis*		2	0	-	1	1
Oryzaephilus mercator						
East Africa	Scotland	1	0	-	1	1
Grand total (all species)		49	37	11	41	37
ditto as per cent		100	76%	28%	100	90%

TABLE F21: **RESISTANCE TESTS ON INSECTS FROM RESIDUAL INFESTATIONS IN SHIPS**

Country or State where Intercepted	Malathion			Lindane	
	No. of Strains	Malathion	Mal+TPP	No. of Strains	No. resistant
Sitophilus oryzae					
S. Australia	7	1	-	6	3
Victoria	1	1	-	1	1
W. Australia	3	1	-	2	2
Sub-total *S. oryzae*	11	3	-	9	6
Sitophilus zeamais					
New Guinea	3	1	1	3	3
Queensland	1	0	-	1	1
S. Australia	1	0	-	1	1
Victoria	2	0	-	2	2
W. Australia	1	0	-	1	1
Sub-total *S. zeamais*	8	1	1	8	8
Rhyzopertha dominica					
S. Australia	3	1	0	-	-
Victoria	3	1	0	-	-
W. Australia	1	1	0	-	-
Sub-total *R. dominica*	7	3	0	-	-
Tribolium castaneum					
New Guinea	9	9	9	9	9
New South Wales	14	14	2	4	4
S. Australia	11	11	6	11	11
Victoria	12	12	8	12	12
W. Australia	5	5	2	5	4
Sub-total *T. castaneum*	51	51	27	41	40
Tribolium confusum					
Victoria	2	2	1	-	-
W. Australia	1	0	-	-	-
Sub-total *T. confusum*	3	2	1	-	-
Oryzaephilus surinamensis					
S. Australia	1	0	-	-	-
Victoria	3	0	-	-	-
Sub-total *O. surinamensis*	4	0	-	-	-
Grand total (all species)	84	60	29	58	54
Ditto as per cent	100	71	-	100	93

(F.14) *Multiple resistances and cross-resistances*

(F.14.1) *Introduction and definitions*

Strains resistant to one toxicant may or may not be resistant to others. When there is resistance to more than one compound it is useful to distinguish true cross-resistance, where a single resistance mechanism gives protection against several compounds, from multiple resistance which occurs when a particular strain has acquired independent mechanisms for dealing with several insecticides. Studies of cross-resistance and multiple resistance are thus closely connected with work on resistance mechanisms. It is therefore appropriate to consider our knowledge of resistance mechanisms in storage pests in the present section.

The identification of resistance mechanisms may be difficult in practice but is worth attempting because when the mechanism can be identified we can confidently predict the characteristics of a resistant strain from our previous knowledge of other strains with that particular mechanism. In particular it may help in selection of effective alternative control measures, or at least in the avoidance of alternative chemicals which are of little value. More studies of cross-resistance and multiple resistance have been undertaken with insecticides than with fumigants. It is therefore convenient to consider the main classes of insecticide separately but to discuss the different fumigants together.

(F.14.2) *DDT*

DDT was used in early laboratory selections for resistance in stored-product insects but subsequently it has become less popular with regulatory authorities, and interest has waned. Our knowledge of the mechanisms of DDT resistance in storage pests is correspondingly rather limited, and is best interpreted in relation to more extensive studies of DDT resistance in other insects. In the latter, two main detoxication mechanisms have been demonstrated in resistant strains. First, the conversion of DDT to DDE by the enzyme DDT-dehydrochlorinase, and second, the oxidation of DDT to various products including dicofol. Associated with each of these processes is a fairly characteristic pattern of cross-resistance to DDT analogues and inhibition by synergists, but it is important to realise that details vary from species to species, and that in general few non-Diptera have been extensively studied.

Both types of mechanism may contribute to the natural tolerance level of susceptible strains. For instance, among storage pests DDE is formed by susceptible *Sitophilus granarius*, *Tribolium castaneum* and *Plodia interpunctella* (Rowlands and Lloyd 1969b, Hoskins and Witt 1958). The same is probably true of *Tribolium destructor* since in this species DDT is synergized by DMC, which is an inhibitor of DDT-dehydrochlorinase (Dyte and Daly 1969). Unfortunately, the natural substrate of the enzyme in these susceptible insects is unknown. Mechanisms other than detoxication are involved in establishing the natural susceptibility levels of some species, including some very tolerant to DDT for example *Trogoderma granarium* (Gupta et al. 1971).

Bhatia and Pradhan (1970) showed that a laboratory-selected DDT-resistant strain of *Tribolium castaneum* had negligible cross-resistance to other organochlorine insecticides, though the strain had some resistance to organophosphorus compounds and to carbaryl (Table F23). In another DDT-

TABLE F22: RESISTANCE PATTERN OF *TRIBOLIUM CASTANEUM* (STRAIN CTC 12)

Strain CTC 12 from Kingaroy, Qd., Australia (unselected field strain)

dimethyl, P=S compounds					
malathion	x18	x8		x4	-
fenitrothion	x4	x4		x6	-
bromophos	x1	x6	Pyrethrins	-	x15
jodfenphos	x1	x4	" + PB (1:5)	x1.5	x5
phenthoate	-	x7	Bioallethrin	-	x16
pirimiphos-methyl	-	x6	" + PB (1:5)	-	x71
cyanophos	x3	-	Resmethrin	-	x2.2
Abate	>x9	-	" + PB (1:5)	-	x0.9
chloropyriphos-methyl	x4		Bioresmethrin	x4.3	x3.8
			" + PB (1:5)	x6.4	x4.5
dimethyl, P=0 compounds			Cismethrin	x0.7	-
malaoxon	-	x61	" + PB (1:5)	x1.5	-
bromoxon	-	x16	Tetramethrin	-	x9
fenitroxon	-	x10	" + PB (1:5)	x3.1	x97
dicrotophos	-	x11	Kikuthrin	-	x120
tetrachlorvinphos	>x227	x55	" + PB (1:5)	-	x27
dichlorvos	x3	-	Prothrin	-	x25
			" + PB (1:5)	-	x27
diethyl, P=S compounds			Phenothrin	-	x6
pirimiphos-ethyl	-	x4	" + PB (1:5)	-	x20
diazinon	x11	x10			
phoxim	x10	-	References	Carter	Lloyd
chlorpyriphos	x5	-		*et al.*	*et al.*
Hoe 2910	x6	-		(1975)	(1976b)
diethyl, P=0 compounds					
diazinon	-	x47	Method	Dust on	Topical
tetrachlorvinphos ethyl	-	>x20		wheat	end
				6-day	point
phosphonates				mortal.	mortal.
K 37	-	x3			
cyanolate	-	x37			
			JH mimics		
carbaryl	x13	-	JH-1	+	
arprocarb	x25	-	Altozar	+	
promecarb	x3	-	Altosid	+	
			DMF	+	
lindane	x36	-	Bowers' 2B	+	
aldrin	-	x3			
DDT	x2	-	Piperonyl butoxide	+	
			Tributyltin acetate	+	
References	Champ &	Dyte	PH-60-40	+	
	Campbell-	(1974			
	Brown	&			
	(1970b)	unpublished	References	Dyte *et al.* (1976c,e)	
		data)		Forster, unpublished	
				data; Carter (1975)	
Method	Impregnated	Topical			
	paper	9-day	Method	Admixture with food.	
	2.5-24 hr KD	mortal.		Failure and delay of	
				metamorphosis or	
				reduction in progeny	

Note + = resistance present but not measured by conventional methods

TABLE F23: RESISTANCE PATTERNS OF *TRIBOLIUM CASTANEUM* (other strains)

Country Locality Strain Selected with	India - - DDT	India - - lindane	India - - malathion	Nigeria Kano - malathion	Australia N.S.W. (CTC 97) malathion	Australia Qd. (CTC 95) malathion	USA Georgia (GP-4) malathion	Malaysia Kuala Lumpur -
malathion	x3	x1	x54	x263	x260	x254	x19	x32
malathion + TPP	-	-	-	x1	x2	x4	-	-
pirimiphos methyl	-	-	-	-	x1	x1	-	-
jodfenphos	-	-	x1	x1	x1	x3	-	-
bromophos	-	-	-	x1	x1	x1	-	-
bromoxon	-	-	-	-	x2	x5	-	-
fenitrothion	-	-	x1	x1	x1	x1	-	x1.6
fenitroxon	-	-	-	-	x1	x5	-	-
diazinon	x9	x1	x2	-	x1	x2	x8	-
diazoxon	-	-	-	-	x1	x2	-	-
tetrachlorvinphos	-	-	-	-	x1	x6	-	-
cyanophos	-	-	-	-	x1	x2	-	-
dichlorvos	-	-	x1	-	x1	x2	-	-
maloaxon	-	-	-	-	-	-	-	-
phenthoate	-	-	-	x19	-	-	-	-
malathion-ethyl	-	-	-	x25	-	-	-	-
acethion	-	-	-	x10	-	-	-	-
mevinphos	x8	-	-	x2	-	-	-	-
phoxim	-	-	-	x1	-	-	x5	-
chlorphoxim	-	-	-	x1	-	-	x20	-
surecide	-	-	-	-	-	-	-	-
dicrotophos	-	-	-	x1	-	-	-	-
dimethoate	-	-	-	x1	-	-	-	-
parathion	x9	x2	x2	-	-	-	-	-
chlorpyriphos-Me	-	-	-	-	x1	x2	-	-
fospirate	-	-	-	-	x4	x4	-	-
methacrifos	-	-	-	-	x1	x1	-	-
disulfoton	-	-	x3	-	-	-	-	-
phorate	-	-	x2	-	-	-	-	-
phosphamidon	-	-	x1	-	-	-	-	-
carbaryl	>x12	>x12	x2	-	x3	x1	-	-
lindane	x1	>x86	x1	-	x3	x240	-	+
dieldrin	x2	>x291	-	-	-	-	-	-
aldrin	x1	>x100	-	-	-	-	-	-
endrin	x2	>x165	-	-	-	-	-	-
heptachlor	x3	>x232	-	-	-	-	-	-
chlordane	x2	>x34	-	-	-	-	-	-
toxaphene	x1	>x9	-	-	-	-	-	-
endosulphan	x2	>x7	-	-	-	-	-	-
DDT	x94	x1	x1	-	x2	x3	-	-
pyrethrins	x2	x2	x5	x1	x1	x1	x13	x0.7
" PB (1 : 10)	-	-	-	-	x1	x1	-	-
" PB (1 : 5)	-	-	-	-	-	-	-	-
bioresmethrin	-	-	-	-	-	-	-	x1.5
" PB (1 : 5)	-	-	-	-	x2	x2	-	x2.9
cismethrin	-	-	-	-	-	-	-	x1.6
" PB (1 : 5)	-	-	-	-	-	-	-	x1.3
tetramethrin + PB (1 : 5)	-	-	-	-	-	-	-	x0.6
fenoxythrin	-	-	-	-	x2	x1	-	x9.8
Reference	Bhatia & Pradhan (1970)(1972)		Pasalu & Bhatia (1974b)	Dyte & Blackman (1972)	Champ & Turner (Unpublished)		Speirs & Zettler (1969)	Carter et al. (1975)
Method	Films on glass 4-day mortality		Impregnated paper 24 hr. mortality	Topical 9-day mortality	Impregnated paper 2.5, 5 or 24 hr KD		Topical 5-day mortality	Dusted wheat 6 day mortality

Note (1) The strains from Australia, Nigeria and USA were malathion-resistant field strains subsequently selected with malathion in the laboratory
(2) Another malathion-resistant (x18) strain (GP-1) from Georgia, USA, was also resistant to phoxim (x10) and diazinon (x11).
(3) See Parkin and Bright (1965), *Pest Infest. Res.* 1964 for more data on another strain from Kano (Nigeria).

resistant strain of the same species from South Africa, DDT could be synergised by DMC, WARF-antiresistant and piperonyl butoxide though none of these three compounds were synergists for DDT in the susceptible strain (Dyte, unpublished data). DMC and WARF-antiresistant both inhibit the dehydrochlorination of DDT in other insects, but piperonyl butoxide is usually regarded as an inhibitor of oxidases, particularly mixed function oxidases. It seems likely, therefore, that both types of detoxication are involved in the South African strain. DDT resistance has also been found in the non-specific organophosphorus resistant strain (CTC 12) (Table F22) of *T. castaneum* Australia (Champ and Campbell-Brown 1970b) and in strains from Kenya, Malawi and Senegal (Dyte 1970). Tests of synergists for DDT in these four strains showed that both WARF-antiresistant and piperonyl butoxide were effective in the Kenyan strain which was, in this respect, comparable to the strain from South Africa. However, both compounds were ineffective DDT synergists in the strains from Australia, Malawi and Senegal, and these may well represent a different type of DDT resistance.

Rowlands and Lloyd (1969b) have made a preliminary study of the metabolism of DDT in the resistant strain from South Africa and in a susceptible strain. The main metabolite in both strains was DDE but more was produced by the resistant than the susceptible insects. Other metabolites (e.g. DDA, DBP) which could be formed from DDE were also present in both strains but there were no dicofol-type products. These metabolic studies thus support the view that enhanced DDT-dehydrochlorination to DDE is involved in this strain. However, they do not explain the synergism by piperonyl butoxide and since DMC synergism does not completely overcome the DDT resistance, it is possible that other mechanisms may play a part.

Resistance to DDT has been reported in a strain of *Sitophilus granarius* selected with pyrethrins (Lloyd and Parkin 1963). When the resistance to pyrethrins was x33 the cross-resistance to DDT was x14. But after further selection with pyrethrins the resistance to pyrethrins rose to x138 and that to DDT to 29.5 (Lloyd 1969a). The increase in the level of resistance to both DDT and pyrethrins after selection with the latter insecticide suggests that a common mechanism may be involved in the resistance to both compounds, particularly as the level of resistance to lindane, dieldrin and aldrin remained virtually unchanged during this selection experiment. Lloyd (1969a) has also studied the cross-resistance of this strain to a number of DDT analogues (Table F24). He found that the cross-resistance to DDT (x29.5) extended also to compounds which cannot be dehydrochlorinated such as bulan (x22) and prolan (x14) or which are difficult to dehydrochlorinate like ortho-chloro DDT. There was also a cross-resistance to deutero-DDT (x30), iso-DDT, fluoro-DDT, methoxychlor and perthane and Lloyd concluded that the main mechanism involved was probably oxidative detoxication.

This interpretation was supported by his (1969b) studies with synergists which showed that compounds known to inhibit the oxidative detoxication of DDT and pyrethroids, like piperonyl butoxide, sesamex and SKF 525A, synergised DDT against the resistant but not the susceptible weevils. By contrast, two of three compounds known to inhibit the dehydrochlorination of DDT to DDE in other insects, namely DMC and piperonyl cyclonene, failed to synergise DDT against either strain. There were, however, some anomalies. Thus WARF-antiresistant which is best known as an inhibitor of dehydrochlorination was an effective synergist in the resistant beetles - this and the cross-resistance to *iso*-DDT and fluoro-DDT indicate that more than one mechanism may be present.

TABLE F24:   RESISTANCE PATTERNS OF *SITOPHILUS* SPP.

Species Country Locality Strain Selected with	*SITOPHILUS ORYZAE* (1)			*SITOPHILUS GRANARIUS*		
	Australia Guildford,W.A. (CSO 231) -	Australia Wondai, Qd. (LR2) -	India - - Malathion	Australia Echuca, Vic. (CSG 12) -	England (R-Py) (2) pyrethrins	
malathion	x24	-	x49	x4	x6	-
malathion + TPP	x13	-	Not suppressed	x1	-	-
fenchlorphos	-	x4	-	-	-	-
pirimiphos methyl	x5	-	-	x2	-	-
jodfenphos	x1	-	x9	x21	-	-
bromophos	x1	-	-	x1	-	-
bromoxon	x8	-	-	x1	-	-
fenitrothion	x6	-	x39	x1	-	-
fenitroxon	x5	-	-	x1	-	-
diazinon	x10	-	x18	x2	-	-
diazoxon	x7	-	-	x2	-	-
tetrachlorvinphos	x7	-	-	x2	-	-
cyanophos	x5	-	-	x2	-	-
dichlorvos	x6	-	x7	x1	-	-
chlorpyriphos methyl	x2	-	-	x1	-	-
fospirate	x3	-	-	x1	-	-
methacrifos	x1	-	-	x1	-	-
phosphamidon	-	-	x6	-	-	-
parathion	-	-	x33	-	-	-
methyl parathion	-	-	x21	-	-	-
endosulfan	-	-	x1	-	-	-
disulfoton	-	-	x7	-	-	-
carbaryl	x3	-	x5	x1	x3	-
lindane	x12	x28	x3	x1	x6	x7
dieldrin	-	>x1000	x4	-	x2	x2
aldrin	-	-	-	-	-	x3
endrin	-	-	x3	-	-	-
isodrin	-	-	x1	-	-	-
DDT	x2	x14	x11	x1	x15	x30
deutero DDT	-	-	-	-	-	x30
Bulan	-	-	-	-	-	x22
Prolan	-	-	-	-	-	x14
fluoro DDT	-	-	-	-	-	x37
pyrethrins	x7	-	x4	x1	x34	x148
" + PB (1 : 10)	x4	-	-	x1	x6	x8
allethrin	-	-	-	-	x40	x28
" + PB (1 : 10)	-	-	-	-	x7	x14
bioallethrin	-	-	-	-	-	x38
" + PB (1 : 10)	-	-	-	-	-	x9
resmethrin	-	-	-	-	-	x207
" + PB (1 : 10)	-	-	-	-	-	x8
bioresmethrin	-	-	-	-	-	x79
" + PB (1 : 10)	x4	-	-	x2	-	x8
tetramethrin	-	-	-	-	-	>x10
" + PB (1 : 10)	-	-	-	-	-	x8
fenoxythrin	x3	-	-	x1	-	-
dinoseb	-	-	-	-	x6	-
References	Champ & Turner (unpublished)	Champ & Cribb (1965a)	Bansode (1974)	Champ & Turner (unpublished)	Lloyd & Parkin (1963)	Lloyd (1969a, 1973)
Method	Impregnated paper 5-24 hr KD	Impregnated paper 24 hr KD	-	Impregnated paper 5-24 hr KD	Topical End point mortality	Topical End Point mortality

Note (1) Ten other field strains of *S. oryzae* from Australia which were malathion resistant (x2 to x8) were also resistant to fenitrothion and dichlorvos.
      (2) The R-Py strain of *S. granarius* was also resistant to methoxychlor, dicofol, perthane, *o*-chloro-DDT, and iso-DDT, but factors of resistance could not be measured.

Rowlands and Lloyd (1968;1969a,b) have studied the metabolism of DDT in these pyrethroid-resistant weevils. They found that dicofol was a major metabolite. The resistant beetles also produced lesser amounts of DDD and FW152 (the dicofol analogue formed from DDD) but only small amounts of DDE and DDA together with a number of products resulting from further degradation. By contrast the susceptible beetles yielded no dicofol but mainly DDE and trace amounts of DDD and DDA. Moreover, metabolism was less rapid than in the resistant strain. There were no studies on the metabolism of synergised DDT but the qualitative differences in the metabolites identified from the two strains accord well with those expected on the basis of the cross-resistance and synergism studies. There is therefore fairly good evidence that this strain is resistant to DDT largely because of an oxidative detoxication process yielding dicofol.

Champ (1967) has shown that in *Sitophilus oryzae* DDT resistance is controlled mainly by one or more sex-linked, semi-dominant, genes with other modifying factors also present. No studies on the biochemistry of resistance to DDT have been made, but DDT and lindane resistance in this species can both be suppressed almost completely by sesamex and can confer a slight tolerance to malathion, and Champ (1968) suggests that some of the minor modifying genetic factors may affect susceptibility to unrelated compounds. *S. oryzae* has a neo-XY sex chromosome system and 10 pairs of autosomes. This system appears to have been derived from a more primitive XY + 11 pairs of autosomes as found in *S. granarius*. Champ (1967) has pointed out that as a result the genetic equivalent of the sex-linked DDT resistance of *S. oryzae* might well be autosomally inherited in *S. granarius*. The mechanism conferring pyrethrins and DDT resistance in *S. granarius* is a possible equivalent since Lloyd and Parkin (1963) found that reciprocal crosses between resistant and susceptible strains gave F_1 beetles of similar and intermediate pyrethrins resistance. These authors therefore tentatively suggested that the genetic basis of pyrethrins resistance was largely autosomal and semi-dominant. It would be of considerable interest to examine the resistance to pyrethrins and synergised DDT in the DDT-resistant strain of *S. oryzae* so that the resistance of these two related species could be compared in greater detail.

(F.14.3) *Lindane and cyclodienes*

A form of lindane resistance involving a cross-resistance to cyclodiene insecticides is widely known in insects of medical importance. The mechanism concerned is unusual in that it does not appear to involve enhanced detoxication. Cross-resistance to a cyclodiene insecticide such as dieldrin in a lindane-resistant storage pest may well indicate a comparable mechanism is involved, even though dieldrin may not be a potential alternative for practical use.

In fact, there have been relatively few studies of cross-resistance in lindane-resistant strains. A cross-resistance to dieldrin, has, however, been demonstrated in a laboratory-selected lindane-resistant strain of *T. castaneum* (Bhatia and Pradhan 1972) and also in lindane-resistant field strains of *S. oryzae* (Champ and Cribb 1965a), *Caryedon serratus* (Dyte and Forster 1969) and the potato tuber moth, *Phthorimaea operculella* (Champ and Shepherd 1965). In contrast to this, the laboratory-selected dieldrin-resistant strain of *Tineola bisselliella* had a very weak cross-resistance to lindane though it was highly cross-resistant to other cyclodienes (Kühne 1967).

In *T. castaneum* a strain highly resistant to DDT had no cross-resistance to lindane (Bhatia and Pradhan 1970) and a lindane-resistant strain from Zambia had no cross-resistance to DDT (Dyte 1970). Thus, resistance to these two organchlorine insecticides can depend on different mechanisms. In the case of lindane resistance the mechanisms involved have been little studied. In an Australian strain of *T. castaneum*, Champ and Campbell-Brown (1969) found that a complex, multifactorial, type of inheritance was present which involved all seven linkage groups for which they had genetic markers available (there are ten pairs of chromosomes in this species but genetic mapping is too incomplete to assume perfect concordance between chromosomes and linkage groups, *vide* Section F15). They concluded that the biochemical mechanisms associated with lindane resistance in this strain were probably equally complex.

The metabolism of lindane has been studied in *T. castaneum* by Rowlands *et al.* (1973). They used a lindane-resistant strain from Zambia which was not cross-resistant to DDT, malathion or carbaryl, and compared it with a susceptible strain. Metabolism was rapid and the susceptible strain metabolised 85% of the dose applied within 5 hours. Lindane metabolism in *T. castaneum* contrasted with that reported in non-storage insects in that only small quantities of water-soluble metabolites were produced. The main metabolite was 2,4-dichlorophenol, but pentachlorophenol was also produced and lesser quantities of several other metabolites were identified. Metabolism in the resistant strain yielded the same metabolites as in the susceptibles but quantitative comparisons were difficult because the susceptibles excreted intact lindane more slowly, and metabolised it more rapidly. The more rapid excretion of unchanged lindane in the resistant strain could be a cause or an effect of resistance.

With all 8 beetles studied during the present survey strains resistant to lindane and not to malathion were found and this is also the case with *Dermestes maculatus* (Tyler and Binns 1976). It is apparent therefore that mechanisms giving resistance to lindane but not to malathion exist in these nine species. There is however, some circumstantial evidence indicating that at least in the two species of *Oryzaephilus* a second mechanism governing resistance to both malathion and lindane may also occur (Dyte *et al.* 1975). The incidence of lindane resistance in each of these species is about 60-70%. We might therefore expect that about two-thirds of the malathion-resistant strains of *Oryzaephilus* would also be lindane resistant, and about one-third lindane susceptible. In fact, all of the malathion-resistant strains so far tested have also been lindane resistant and strains resistant to malathion but not lindane have yet to be found. It is possible therefore that in both of these species resistance to malathion can confer a true cross-resistance to lindane.

(F.14.4) *Pyrethroids*

Resistance to pyrethroids is now known in six species of storage pests. In each of these species the field strains involved are also resistant to one or more non-pyrethroid insecticides and it is difficult to determine whether multiple resistance or true cross-resistance is involved.

The first report was in *Sitophilus granarius* collected from an English farm granary in 1953 (Holborn 1957). Synergised pyrethrins had been used in the granary but at that time malathion had not been introduced for stored product use in the United Kingdom. This resistance was thus not the result of selection with malathion. All other field occurrences of pyrethroid resistance have involved strains which were also resistant to malathion and other organophosphorus compounds (OPs). They were detected after both synergised pyrethrins and malathion had been used, and may involve multiple resistance or true cross-resistance.

The two moths in which pyrethroid resistance is known have been much less well studied than the beetles. Zettler et al. (1973) described a relatively low level of resistance to synergised pyrethrins (x2.5) in a strain of *Plodia interpunctella* with a high (over x200) resistance to malathion. Malathion-resistant strains of this species in the USA can be controlled by dichlorvos (La Hue 1969) and have no cross-resistance to pirimiphos-methyl (Zettler 1974a) but these alternative insecticides were tested against strains in which the pyrethroid resistance was unknown. With *Ephestia cautella* two strains with relatively low levels of resistance (x3.3, x2.5) to synergised pyrethrins had low resistance (x2.2, x6.5) to malathion (Zettler et al. 1973) but the susceptibility to alternative insecticides has yet to be studied.

Among the beetles, resistance to pyrethroids has been demonstrated in *S. oryzae*, *T. castaneum* and *O. surinamensis* as well as *S. granarius* and it may also occur in *T. confusum*. It has been most studied in the strain of *S. granarius* mentioned above and in the CTC 12 non-specific resistant strain of *T. castaneum*.

In *S. granarius* the cross resistance was first studied when the resistance to pyrethrins was x34 and that to DDT x15 (Table F24). The strain then had a relatively high resistance to allethrin (x40), but to pyrethrins or allethrin synergised by piperonyl butoxide (PB) the resistance was only 6 to 7-fold. The strain also had a resistance up to x6 to the unrelated compounds malathion, carbaryl, lindane and dinoseb.

Subsequently, when the resistance level of the strain had increased to x148 the cross-resistance to pyrethroids and DDT analogues was studied in more detail. The results with DDT analogues have been discussed above. Those with pyrethroids revealed cross-resistance levels ranging from x28 to x207 for unsynergised compounds (Table F24), no LD_{50} for tetramethrin being obtainable on the resistant strain (Lloyd 1973). However, when the synergist PB was used all the cross-resistance levels dropped dramatically to x8 to x14. All this work involved end point mortalities after topical application. Lloyd and Williams (1973) subsequently showed that when knockdown was measured on impregnated papers the resistance factor for unsynergised pyrethrins was much reduced (from x148 to x18) but the resistance levels for pyrethrins + PB, DDT, malathion and lindane were little affected. Even so PB still reduces the pyrethrins resistance by a factor of three, and it is clear that in this strain the pyrethroid resistance mechanism is strongly inhibited by PB.

Lloyd et al. (1976b) have recently made comparable studies on the CTC 12 strain of *T. castaneum*. In this strain resistance levels to other pyrethroids ranged from x2 and x120 for unsynergised and from x1 to x97 for synergised pyrethroids. With some compounds, for example pyrethrins and kikuthrin, PB reduced the resistance level but with others, for example bioallethrin and tetramethrin, it considerably increased the resistance factor.

Rowlands and Lloyd (1976) have studied the metabolism of labelled pyrethrin I in susceptible and pyrethroid-resistant strains of *S. granarius*. They found the same eight metabolites in adults of both strains. In the susceptible beetles the two main metabolites were demethyl pyrethrin II resulting from oxidation of the isobutenyl moiety of the chrysanthemic acid side chain, and a conjugate resulting from terminal hydroxylation in the pentadienyl side chain of the alcohol. The resistant insects produced from 2 to 3 times as much of these two metabolites as did the susceptibles. When pyrethrin I was applied with PB, oxidative attack was inhibited in both strains and hydrolysis of the ester was the main metabolic pathway. This hydrolytic attack was at a similar rate in both strains, and not much more rapid than that found in the susceptible strain in the absence of PB.

These workers (Rowlands and Lloyd 1976) also studied the metabolism of tetramethrin in the two strains of *S. granarius*. In both, oxidation of the isobutenyl moiety (giving demethyl tetramethrin) was more pronounced than with pyrethrin I, so much so that hydrolysis of the ester was of negligible importance. Metabolism was faster in the resistant strain. PB strongly inhibited oxidative attack in both strains and virtually eliminated it during the first 20 hours after treatment.

Lloyd *et al*. (1976b) have made a preliminary study of the metabolism of tetramethrin in susceptible and resistant (CTC 12) *T. castaneum*. They found that large quantities of an unidentified metabolite were excreted by both strains. This was not demethyl tetramethrin nor a simple phthalimido derivative. The metabolite appeared to have lost the aromatic ring and C-N moiety of tetramethrin. It appeared to be a chrysanthemic ester containing an epoxide group in the alcohol part of the molecule. The ester linkage appeared to be intact so if this compound proves to be the main metabolite, and re-esterification had not occurred, ester hydrolysis would appear to be unimportant in detoxication. There was some evidence that intact tetramethrin was absorbed and excreted more readily in the resistant than the susceptible insects but this may well have been a result rather than a cause of the resistance.

Pyrethrin synergists act mainly by inhibiting detoxication but they also affect the penetration of insecticides into the insect, and may affect translocation from tissue to tissue after penetration (Casida 1970, Yamamoto 1973). In pyrethroid detoxication, oxidative attack on the acid moiety is important and this can be inhibited by synergists such as PB. Oxidative attack on the alcohol may also be important in some compounds but in others for example allethrin, it may not occur. This second pathway may also be inhibited by PB. Finally, the ester linkage may be hydrolysed. This third pathway is probably of little importance in allethrin metabolism and even less so with the pyrethrins, but with the synthetic primary alcohols like tetramethrin and resmethrin, it is of greater significance (Casida 1973), and the same is true of phenothrin at least in mammals (Miyamoto *et al*. 1974). Esterase hydrolysis is little affected by synergists like PB which inhibit mixed function oxidases, but is more readily inhibited by esterase inhibitors (Jao and Casida 1974a). Recent work indicates that in several insects, esterases hydrolysed bioresmethrin more readily than cismethrin and correspondingly the (+)-trans isomer more readily than the (+)-cis isomer of tetramethrin (Jao and Casida 1974b). The effectiveness of esterase inhibitors as synergists for the two isomers varies accordingly (Jao and Casida 1974a).

In *S. granarius* the resistance appears to involve mainly the first two (oxidative) pathways, and the DDT resistance of this strain which appears to be a true cross-resistance, also involves oxidative detoxication. However, although PB much reduces the resistance it does not totally overcome it. This may reflect reduced penetration in the presence of PB or an accessory mechanism of resistance. The latter might be responsible for the resistance to unrelated insecticides for example malathion and lindane in this strain. Information is lacking on the effect of PB on this strain's resistance to these other compounds.

The interpretation of the effects of PB on pyrethroids in the CTC 12 strain of *T. castaneum* is more difficult. It reduced the resistance to some compounds but increased it in others. The fact that this strain is resistant to and degrades PB (as discussed below) is a complicating factor. It is, however, clear that a mechanism different from that occurring in the R-py strain of *S. granarius* is involved. It is interesting that Carter *et al.* (1975) found that the resistance to bioresmethrin was higher than that to cismethrin regardless of the presence of PB and that in Lloyd *et al.*'s (1976b) study the resistance to bioresmethrin was greater than that to resmethrin (which is a mixture of bioresmethrin and cismethrin). This suggests that the resistance mechanism may be less well able to deal with (+)-cis than (+)-trans isomers, whereas in *S. granarius* the opposite appears to be the case at least in the absence of PB.

Much less is known of pyrethroid resistance in other beetles. In *S. oryzae* selection with pyrethrins + PB yielded a resistant strain (Cichy 1971). This strain also responded to selection for DDT resistance and OP resistance but whether one or several mechanisms were involved is not known. A field strain of the same species from Australia which was resistant to several OP insecticides was also resistant to pyrethrins (+ or - PB) and bioresmethrin + PB (Table F24). Subsequent work by Carter *et al.* (1975) failed to confirm this though they found some tolerance to synergised tetramethrin in this strain and at the time of their study the level of malathion resistance appeared to have dropped considerably.

An OP-resistant field strain of *O. surinamensis* from Australia was also resistant to pyrethroids (Table F26) and Ruth Forster (unpublished data) has found that other OP-resistant strains of this species which were detected in the present survey were also resistant to pyrethroids. In the Australian strain (COS 11) the resistance to bioresmethrin (x8) was greater than that to pyrethrins (x3) (Table F26). A comparable situation may well exist in OP-resistant field strains of *S. granarius* and *T. confusum* which have a low resistance (x3 and x2) to bioresmethrin but none detected to pyrethrins (Table F24 and F25). This contrasts with the laboratory-selected (R-Py) strain of *S. granarius* and the CTC 12 strain of *T. castaneum* in both of which the level of resistance to bioresmethrin is much less than that to pyrethrins.

These results suggest that a third mechanism of pyrethroid resistance may be involved in these field strains. A pyrethroid resistance dealing more effectively with bioresmethrin than pyrethrins might be expected if enhanced hydrolysis of the ester linkage were involved. As yet, however, there is no direct evidence of a detoxication mechanism of this type. Further studies of these strains are obviously desirable. In particular, we need to know whether the detoxication system which can hydrolyse phosphorus esters can also hydrolyse the ester linkage in pyrethroids; this is particularly important because this ester link seems to be more vulnerable to hydrolysis in the modern synthetics like bioresmethrin, tetramethrin and phenothrin than in the natural pyrethroids.

TABLE F25: RESISTANT PATTERNS OF MALATHION-RESISTANT *RHYZOPERTHA DOMINICA* AND *TRIBOLIUM CONFUSUM*

Species	*RHYZOPERTHA DOMINICA*		*TRIBOLIUM CONFUSUM*	
Country Locality Strain	Australia Armatree, N.S.W. (CRD 37)	Australia Sydney, N.S.W. (CRD 118)	Australia Port Adelaide, S.A. (CTCO 16)	Australia Laura, S.A. (CTCO 37)
malathion	x5	>x100	x18	x5
malathion + TPP	x1	x10	x1	x3
pirimiphos methyl	x1	x1	x1	x4
jodfenphos	x1	x16	x3	x2
bromophos	x1	x15	x2	x2
bromoxon	x1	x30	x1	x12
fenitrothion	x1	x3	x1	x3
fenitroxon	x1	x6	x1	x6
diazinon	x1	x3	x3	x17
diazoxon	x1	-	x1	x3
tetrachlorvinphos	x1	x5	x2	x4
cyanophos	x1	>x100	x1	x4
dichlorvos	x1	x3	x1	x15
chlorpyriphos methyl	x1	x33	x1	x3
fospirate	x1	x43	x1	x1
methacrifos	x2	x13	x1	x2
carbaryl	x1	x1	x4	x9
lindane	x1	x1	x1	x2
DDT	x1	x1	-	-
pyrethrins	x1	x1	x1	x1
" + PB (1 : 10)	x1	x1	x1	x1
bioresmethrin	x1	x1	x1	x2
fenoxythrin	x1	x1	x2	x2
Reference (all strains)	Champ and Turner (unpublished)			
Method (all strains)	Impregnated paper, 2.5-24 hr KD			

TABLE F26: RESISTANCE PATTERNS OF *ORYZAEPHILUS* SPP., *EPHESTIA CAUTELLA* AND *PLODIA INTERPUNCTELLA*

Species	*ORYZAEPHILUS SURINAMENSIS*		*O. MERCATOR*	*EPHESTIA CAUTELLA*	*PLODIA INTERPUNCTELLA*
Country	Israel	Australia	Senegal	Japan	U.S.A.
Locality	Tel Aviv	Kununoppin, W.A.	Bambey	-	Ga, Ill
Strain	-	(COS 11)	(SR)	methyl parathion	-
Selected with	malathion	-	malathion		
malathion	x10	x9	x7	x3	x206
malathion + TPP	x7	x5	x8	-	-
pirimiphos methyl	x8	x8	-	-	x1
jodfenphos	x35	x9	x28	-	-
bromophos	x19	x9	x158	-	-
bromoxon	-	x5	-	-	-
fenitrothion	x11	x20	x2	x3	-
fenitroxon	-	x65	-	-	-
diazinon	x7	x11	x4	-	-
diazoxon	-	x8	-	-	-
tetrachlorvinphos	x7	x12	x64	-	-
cyanophos	x3	x1	-	-	-
dichlorvos	-	x2	-	-	-
chlorthion	x2	-	-	-	-
dicapthon	x2	-	-	-	-
phenothoate	x7	-	-	-	-
phoxim	x6	-	-	-	-
parathion	-	-	-	x1	-
EPN	-	-	-	x1	-
fenthion	-	-	-	x2	-
paraoxon methyl	-	-	-	x2	-
parathion methyl	-	-	-	x7	-
chlorpyriphos methyl	-	x1	-	-	-
fospirate	-	x2	-	-	-
methacrifos	-	x1	-	-	-
carbaryl	-	x12	-	-	-
lindane	-	x>20	x89	-	-
DDT	-	x2	-	-	-
pyrethrins	-	x3	-	-	-
" + PB (1 : 10)	-	x4	-	-	-
bioresmethrin	-	x8	-	-	-
fenoxythrin	-	x5	-	-	-
References	Tyler & Binns (Unpublished)	Champ & Turner (Unpublished)	Dyte & Forster (1973b)	Hashimoto & Fukami (1964)	Zettler (1974a)
	Impregnated paper 48 hr mort.	Impregnated paper 2.5-24 hr KD	Impregnated paper 48 hr mort.	Dry film on glass 24 hr mort.	Topical 120 hr mort.

Note (1) The strains from Israel and Senegal were field strains showing malathion resistance and subsequently selected in the laboratory before cross-tolerance studies.

(2) Eight other malathion-resistant strains of *O. surinamensis* obtained during the survey were also resistant to malathion + TPP, and to fenitrothion, bromophos, and jodfenphos.

(F.14.5) *Organophosphorus compounds*

With one possible exception, all known cases of resistance to organophosphorus (OP) compounds in storage pests involve resistance to malathion. This reflects both the wide use of malathion for practical control and possibly the neglect of other OP compounds in resistance monitoring. Cross-resistance to OP compounds in storage pests can therefore be regarded as the extent to which malathion resistance confers resistance to other insecticides.

The exception is the laboratory strain of *S. oryzae* which was selected for resistance to demeton-methyl by Cichy (1971). This strain responded rapidly to selection with this insecticide as well as to selection with DDT or synergised pyrethrins but its resistance to malathion was not studied.

Malathion resistance has been most studied in *Tribolium castaneum*. In this species at least two types occur which have been referred to as "malathion-specific" and "non specific". Champ and Smith (1972) have reported that the major genes involved in these two types of resistance are in the same linkage group. Dyte and Rowlands (1968) studied the metabolism and synergism of malathion in a strain of *T. castaneum* from Nigeria. They found that the malathion resistance (x39 before, and x450 after laboratory selection) could be completely overcome by the synergist triphenyl phosphate (TPP) so that there was no cross-resistance to malathion plus TPP even in the highly selected malathion-resistant strain. They also found that metabolism of malathion was more rapid in the resistant beetles and that these produced more malathion monoacid and malathion diacid than did the susceptible insects. Moreover, higher levels of malaoxon (the toxic metabolite of malathion) were produced in the susceptible beetles, though the levels of other metabolites were similar in the two strains. Malathion monoacid and malathion diacid are detoxication products produced by hydrolysis of the carboxyesters of malathion. When the beetles were treated with malathion synergised by TPP the levels of these metabolites were reduced to about the same level in both strains and there were no striking differences in the levels of other metabolites produced. It is probable that the resistance mechanism in this strain is one of enhanced degradation by carboxyesterases not only because carboxyesterase products are produced more rapidly by the resistant beetles but because TPP both inhibits this degradative pathway and restores the toxicity of malathion thus completely overcoming the resistance.

This type of resistance confers a cross resistance only to malathion analogues (e.g. Kano strain in Table F23). Most such analogues with the exception of phenthoate are laboratory chemicals rather than insecticides with practical potential. In effect, therefore, all insecticides except phenthoate remain fully effective against the strains of the type. They are, therefore, called malathion-specific resistant strains.

Strains of *T. castaneum* which are resistant to several other OP insecticides besides malathion are also known (Tables F22 and F23). Among these strains with non-specific OP resistance that most extensively studied is strain CTC12 from Australia first reported by Champ and Campbell-Brown (1970b). This strain is resistant to malathion synergised by TPP and a wide diversity of OP structures including dimethyl-, diethyl-, and dipropyl-compounds, phosphates, phosphonates and phosphorthioates and compounds with either aromatic or aliphatic leaving groups. It is now known to be resistant to 25 organophosphorus compounds, 4 carbamates, 3 organochlorines, 8 pyrethroids, 5 juvenile hormone mimics, 1 organo-tin compound, the substituted urea PH-60-40 which inhibits cuticle formation, and the synergist piperonyl butoxide (Table F22).

A feature of the resistance to OP compounds is that resistance levels to phosphates (P = O compounds) are higher than the levels of resistance to the analogous phosphorothioates (P = S compounds). Few phosphates except dichlorvos and tetrachlorvinphos are important in control of stored product pests, but this feature may help with the recognition of this type of resistance in other strains and species and also is significant with regard to the mechanism of resistance postulated.

The metabolism of ^{14}C-labelled malathion has been studied in the CTC12 strain of *T. castaneum* (Dyte et al. 1970). Adult beetles metabolised malathion at the same rate as did susceptible beetles. Moreover, the various detoxication products of malathion were apparently produced at similar rates in the two strains. However, the level of the toxic metabolite malaoxon found in homogenates of the resistant strain was lower than in the susceptible beetles, and the resistant strain produced two metabolites of malaoxon i.e. desmethyl malaoxon and methyl thiolophosphate, which were not found in the susceptible beetles. Desmethyl malathion was not detected in beetles of either strain. It appears, therefore, that this resistant strain is able to desmethylate malaoxon but not malathion.

The metabolism of ^{14}C-labelled tetrachlorvinphos in this strain and in a susceptible strain was also studied by these workers. Susceptible beetles metabolised this compound rather slowly, the main metabolite being desmethyl tetrachlorvinphos with small quantities of non-phosphorus containing metabolites derived from the leaving group. In the resistant beetles metabolism was much more rapid and much more desmethyl tetrachlorvinphos was produced, though the levels of the other metabolites were similar to those in the susceptible strain. These studies suggest that this strain is able to detoxify phosphates (P = O compounds) but not phosphorothioates (P = S compounds) by de-alkylation. The fact that there was a qualitative difference in the metabolism produced in the two strains from malathion, and that this mechanism could account for the higher levels of cross resistance to phosphates support this view. However, this must remain a tentative interpretation because as yet a synergist overcoming the resistance has not been found and no *in vitro* studies have been reported.

In the absence of a diagnostic synergist this second type of resistance in *T. castaneum* is probably best recognised by the contrasting resistance levels to phosphates and phosphorothioates. Thus, the strain CTC 95 (Table F23) appears to be one whose malathion resistance is dependent mainly on malathion-specific resistance as indicated by the great reduction (from x254 to x4) in the level of malathion resistance produced by the use of TPP. However, this synergist does not completely overcome the resistance in this strain so there is reason to suspect a second mechanism. The highest OP cross-resistance levels are to tetrachlorvinphos (x6), bromoxon (x5) and fenitroxon (x5) all of which are phosphates, whereas to bromophos, and fenitrothion (analogous phosphorthioates) no cross-resistance was detected. This indicates that the second mechanism of OP resistance is comparable to that of strain CTC12. The high resistance to lindane in this strain is probably due to an independent mechanism so strain CTC95 probably shows multiple resistance involving three distinct mechanisms.

The cross-resistance spectra of other malathion-resistant strains of *T. castaneum* has been less well studied (Table F23). As yet there is no reason to postulate more than two mechanisms of OP resistance but more would not be unlikely.

Malathion resistance in other species has received less attention. Because TPP has proved useful in the detection of malathion-specific resistance in *T. castaneum* it has been used in tests on malathion-resistant strains of other species. Strains in which the malathion resistance can be overcome by TPP, and which have little or no cross-resistance to other OPs have been detected in *R. dominica* and *T. confusum* (Table F25). In both species other strains which are resistant to malathion synergised by TPP and also to other OP compounds are known, and by analogy with *T. castaneum* it appears that specific and non-specific OP resistances can also be distinguished in these species by this means. Because of the lack of detailed investigation, tests involving TPP synergism in these species cannot be interpreted with the same confidence as in *T. castaneum*, but the overall picture appears to be comparable.

Problems might arise because carboxyester hydrolysis may contribute to the natural susceptibility level of some species in which case TPP would synergise malathion to some degree in all strains. This appears to be the case in *Dermestes maculatus*, though not *D. lardarius* (Dyte et al. 1966) and also *Rhyzopertha dominica*. In monitoring tests it means that some TPP synergism would be expected in strains with non-specific OP resistance so that these would not be recognised as such unless their resistance level was fairly high. This problem can be met by using a lower discriminating dose in tests with malathion + TPP. In *R. dominica* a reduction of about 5-fold has been suggested (Dyte et al. 1975, also Fig. F1). Has this been done in the present survey more strains of this species might have survived a discriminating dose of malathion + TPP. It is also possible that TPP may antagonise the penetration of malathion into some insects. This might lead to false indications of non-specific resistance and would necessitate raising the malathion dose in discriminating tests using TPP. Finally, it is possible that TPP may prove to be a less specific inhibitor of carboxyesterases in some species.

While the malathion-specific resistances of *T. castaneum*, *T. confusum* and *R. dominica* appear to be comparable, the non-specific resistances of these three species and those found in *O. surinamensis*, *O. mercator*, *S. oryzae* and *S. granarius* do not appear to be very similar according to the cross-resistance data available (Tables F22 to F26). Some of these strains are resistant to pyrethroids but others like the CRD 118 strain of *R. dominica* are not. At present we do not know how much this may be due to multiple resistance and how much is true cross-resistance. We are also not yet in a position to assess whether more than one type of non-specific OP resistance occurs in any one species.

Little has been published on the cross-resistance spectra of OP-resistant moths. A strain of *Ephestia cautella* selected with parathion-methyl developed a low level of resistance to this compound and to four other dimethoxy-OPs including the P = O analogue paraoxon-methyl (Table F26). However, there was no resistance to parathion (the diethoxy analogue) and scarcely any to EPN (x1.4) suggesting this resistance might be limited to dimethoxy OP compounds. In the USA two malathion-resistant strains of this species were also resistant to synergised pyrethrins, but these may well be cases of multiple resistance.

Malathion-resistant strains of *Plodia interpunctella* in the USA can be controlled by dichlorvos (La Hue 1969) and have no cross-resistance to pirimiphos-methyl (Zettler 1974a). Zettler (1974b) has compared the esterases in larval homogenates of a malathion-resistant strain of this species with those of a malathion-susceptible strain. Differences were found but it was difficult to distinguish those which might be associated with the resistance

from incidental strain differences. Similar levels of acetyl
cholinesterase activity were found in the two strains but the resistant
strain showed less carboxylesterase and butyrylcholinesterase activity
than did the susceptible strain. Zettler tentatively concluded that
the resistance was unlikely to be malathion-specific.

(F.14.6) *Juvenile hormone mimics*

Compounds which mimic the effects of insect juvenile hormone
(JH) are receiving attention as a new type of insect control chemical.
It was hoped that resistance to these compounds would not develop, but
resistance to JH analogues has now been reported in insecticide-resistant
strains of beetles, moths and flies so these compounds can no longer be
regarded as an answer to the resistance problem.

Resistance to JH mimics was first demonstrated in the CTC 12
strain of *Tribolium castaneum* (Dyte 1972). The compound used was a
synthetic *cis/trans* mixture of the first JH to be isolated from the
cecropia silk moth. The test method involved adding the compounds to
flour in which 14-day-old larvae were placed and thereafter examining
them twice weekly. By increasing the dose of JH mimic, adult mortality,
pupal mortality, delay in pupation, prevention of pupation and larval
mortality were produced in both strains. A larger dose was required for
each of these effects in the resistant strain, but, because of variation
in the average age at which these effects occur, mortality at any
particular time was not a simple function of dosage. Thus, whereas in
both strains 10 ppm gave 100% mortality, at 30 ppm some insects seemed
to be protected from pupal death by their failure to pupate. Subsequently
four further JH mimics have been studied and resistance was found to all
of them in the CTC12 strain (Dyte *et al*. 1976b, Table F22).

Resistance to JH analogues has also been reported in the house
fly and tobacco budworm and in each instance, as in *T. castaneum*, a strain
resistant to a diversity of traditional pesticides was involved. Thus, it
seems that some types of non-specific resistance to traditional insecticides
can involve a true cross-resistance to the pesticides being developed from
juvenile hormone mimics.

The metabolism of two JH mimics has been compared in the
susceptible and CTC 12 resistant strains of *T. castaneum* by Dyte
et al. (1976c). Published studies on other insects have shown that
the main pathways involve either esterase hydrolysis yielding the acid
derivative, or hydration of the epoxide yielding a diol. If both these
processes should occur successively, the acid-diol derivative is produced.
Whereas these pathways are known to occur with the three rather similar
natural JHs so far isolated from insects, some of the synthetic analogues
with hormone activity may lack the epoxide group or may have the methoxy
ester modified or replaced. The compounds to which *T. castaneum* is
resistant include several different types of modification of the JH
molecule. The synthetic JH has both epoxide and methyl ester; Bowers'
compound 2b had the epoxide but the carboxyester is replaced by a
methylenedioxyphenyl moiety; DMF lacks the epoxide but has the methyl
ester; and Altosar and Altosid both lack the epoxide and have the
methoxy ester replaced by ethoxy- or isopropoxy-esters respectively.

Studies using radio-active JH showed that in adults the diol,
acid and acid-diol derivatives were all produced in both susceptible and
resistant strains. Both produced about 1 to 2 times as much diol as acid

and a relatively small amount of acid-diol. Although the metabolism appeared to be qualitatively similar in both strains, it was more rapid in the resistant insects - the half lives for JH were 8 hours and 3 hours respectively.

Bowers' compound 2b was also studied after labelling in the methylene group. This compound was metabolised to the diol derivative and to the catechol produced by attack on the methylene of the methylenedioxyphenyl group. Metabolism was very much more rapid in the resistant strain. In the susceptibles, the main product was the diol with very little catechol produced. In the resistant beetles, however, the main pathway was that leading to the production of the catechol derivative.

These results, and the differing structures of the JH mimics to which the CTC 12 strain is resistant indicate that the resistance is not associated with enhanced detoxication by one particular pathway. This is perhaps not surprising in view of the diversity of traditional insecticides to which the strain is resistant.

Resistance to JH mimics does not appear to have been looked for in other species. This will become necessary if compounds of this type are introduced for the practical control of storage pests. It would appear to be particularly worth-while if a resistant strain should appear which shares with the CTC 12 strain a higher level of resistance to phosphates than to phosphorothioates.

(F.14.7) *Other non-gaseous chemicals*

Carbamates have been tested against relatively few resistant strains of storage pests but Champ and Campbell-Brown (1970b) showed that the CTC 12 strain of *T. castaneum* was resistant to three carbamates (Table F22). Using a discriminating dose test they also showed that resistance to carbaryl in this species was quite common in Queensland, Australia. However, some strains which were resistant to malathion synergised with TPP were susceptible to carbaryl and so it seems likely that the carbaryl resistance of the CTC 12 strain is a case of multiple resistance rather than true cross resistance. Resistance to carbaryl was also reported in this species by Bhatia and Pradhan (1970,1972) in strains selected with DDT or lindane (Table F23), and in *S. granarius* by Lloyd and Parkin (1963) in a strain selected with pyrethrins (Table F24).

A propoxur-resistant strain of *S. granarius* produced by selection with propoxur was found to be somewhat resistant to two other carbamates and DDT (Kumar and Morrison 1967). The propoxur resistance of this strain was overcome by the synergist PB so it appears that oxidative detoxication was involved.

The CTC 12 strain of *T. castaneum* has also been found to be resistant to tributyltin chloride (Forster, unpublished data) and to 1-(4-chlorophenyl)-3-(2,6 difluorobenzoyl)-urea (PH-60-40), a compound which represents a new class of insecticides acting by the inhibition of cuticle formation (Carter 1975). However, another strain of *T. castaneum* from Kuala Lumpur which was resistant to malathion, lindane, and synergised tetramethrin (Table F23), was slightly more sensitive to PH-60-40 than the susceptible strain tested, and this compound was fully effective against an OP-resistant strain of *S. oryzae* (CSO 231 in Table F24) (Carter 1975).

The CTC 12 strain of *T. castaneum* is also resistant to
piperonyl butoxide (Dyte et al. 1976d). This was first suspected
from metabolic studies with C^{14}-labelled PB which showed that the CTC 12
strain degraded PB much more rapidly than did the susceptibles with
predominant attack at the methylene group. In susceptibles and in a
malathion-specific resistant strain (Kano), the main attack was at the
side chain ether linkage. In view of this result it seemed probable
that the CTC 12 strain could be resistant to the synergist PB, but since
the strain is resistant to all the insecticides which might be expected
to be synergised by PB it was difficult to demonstrate PB resistance.
However, PB is known to affect the development of insects even in the
absence of any insecticide and this effect was therefore compared in the
susceptible and resistant strains. When PB was added to the food both
strains showed increases in the larval and pupal periods. With 5% PB,
pupation was almost totally suppressed in the susceptible strain whereas
in the resistant strain some pupae and adults were produced. At this
concentration mortality was 52% in the susceptibles and 12% in the
resistant. At 7.5% PB, all the susceptible larvae were dead by the fifth
week but in the resistant strain 26% were still alive at the twelfth week
although none of them had pupated. Overall, the resistant strain was less
effected by the treatments than the susceptible strain at each dose level
and thus the resistance to PB was demonstrated. This resistance to PB needs
to be considered when the effects of PB as a synergist on this strain are
considered.

(F.14.8) *Fumigants*

(F.14.8a) *Recorded fumigant cross-resistances*

Cross-resistance to other fumigants has been studied in several
fumigant-resistant strains produced by laboratory selection. The results
are summarised in Table F27. Lindgren (in Lindgren and Vincent 1965) showed
that a strain of *Sitophilus oryzae* with a 2.5-fold resistance to hydrogen
cyanide had little increased tolerance to either ethylene dibromide or
methyl bromide, and that the same was true of a strain of *S. granarius*
with a 3-fold resistance to hydrogen cyanide. The phosphine-resistant
strain of *S. granarius* produced by Monro et al. (1972) and the phosphine-
resistant strain of *T. castaneum* selected by Kem (1975) also had relatively
specific resistances in that they both lacked any notable cross-tolerance
to other fumigants (Table F27). However, in the case of *S. granarius* there
was a 2.7-fold resistance to chloropicrin. The level of phosphine resistance
in this strain varied according to whether a high or low dose was used and
this made difficult the interpretation of crosses designed to study the
inheritance of phosphine resistance. It is of interest that although this
strain was produced by selection on the adult stage, the resistance produced
occurs in adults and pupae, though not in other immature stages (Anon. 1974b).

In contrast to these relatively specific fumigant resistances two
strains of *Sitophilus granarius* selected by Monro et al. (1961) with methyl
bromide both developed resistance not only to this fumigant but also to
several others including phosphine and acrylonitrile. Further studies
have been made on one of these strains (strain LWA) by Upitis et al.(1973).
Continued selection with methyl bromide produced a strain with a 7.8-fold
resistance after 44 selections. This resistance level did not increase
during a further six selections. Throughout the selection programme the
slope of the log-dose probit-mortality lines remained parallel to those of
the unselected strain. Crosses between susceptible and selected beetles
gave F_1 and F_2 hybrids which were intermediate in tolerance between the two
parental strains, and back crosses of F_1 beetles mated to either parental
strain gave insects with a tolerance level intermediate between that of the
F_1 and the parental strain used. The heterogeneity and slope of the response
line for the F_2 was similar to that of the F_1. Although the LWA strain was

TABLE F27: RESISTANCE PATTERNS INVOLVING FUMIGANTS

Species	SITOPHILUS ORYZAE	TRIBOLIUM CASTANEUM		TRIBOLIUM CONFUSUM
Country	U.S.A.	India	Canada	U.S.A.
Strain	-	-	Field	Field
Selected with	HCN	Phosphine	nil	MeBr/EDB
phosphine	-	x12	-	-
methyl bromide	x1	x1	x1	x1
ethylene dibromide	x1	x1	x2	x2
ethylene dichloride	-	x1	-	-
HCN	x2	-	-	x1
carbon bisulphide	-	x1	-	-
carbon tetrachloride	-	x1	-	-
References	Lindgren & Vincent (1965)	Kem (1975)	Bond (1973)	Lindgren & Vincent (1965)

Species	SITOPHILUS GRANARIUS				
Country	Canada	Canada	Canada	U.S.A.	England
Strain	-	LWA	GCA	-	R-Py
Selected with	PH_3	Me Br	Me Br	HNC	pyrethrins
phosphine	x3	x6	x13	-	x2
methyl bromide	x1	x6	x2	x1	x1
acrylonitrile	x1	x5	x3	-	-
ethylene dibromide	x1	x3	x2	x1	-
ethylene oxide	x1	x5	x2	-	-
HCN	x1	x2	x1	x3	-
chloropicrin	x3	x2	x2	-	-
References	Monro et al. (1972)	Monro et al. (1961)	Monro et al. (1961)	Lindgren & Vincent (1965)	Lloyd & Parkin (1963) Rajak & Hewlett (1971)

Note: All strains except the Canadian *T. castaneum* and the *T. confusum* were selected for resistance in the laboratory.

produced by selection of the adult stage with methyl bromide, the eggs, pupae and larval instars were also tolerant to this fumigant, though not as tolerant as were the adults. The selected strain also had a longer life cycle, 44-48 days compared with 34-36 days in the susceptible strain (Upitis *et al.* 1973).

Ellis and Morrison (1967) and Ellis (1972a) using different test methods have confirmed the cross-resistance to ethylene dibromide in the LWA strain or some derivatives from it. (Ellis 1972a obtained his insects from Monro but he refers to the strain as LMR and there are differences in the number of selections it had reputedly undergone.) Ellis (1972a,b,c) has studied resistance to ethylene dibromide in this methyl bromide-selected strain. Both he (Ellis 1972a) and Upitis *et al.* (1973) found that the mean body weight of the resistant beetles was greater and their mean respiratory rate much lower than in the susceptible strain. There was also a small change in the relative susceptibility to methyl bromide of the sexes. In the susceptible strain the females were more tolerant than the males, but this was reversed in the resistant strain (Upitis *et al.* 1973). Ellis (1972a) found a comparable change in the relative susceptibility of the two sexes to ethylene dibromide but the difference was not significant.

Ellis (1972a) showed that only about 25% of the tolerance to ethylene dibromide in the resistant beetles could be attributed to the weight difference between the two strains. However, metabolic rate per unit mass generally decreases with increasing body weight and the lower oxygen consumption of the resistant insects was entirely a function of their greater weight. He also showed (Ellis 1972c) that the resistant beetles contained more lipid because of their greater weight, but considered that inert storage of ethylene dibromide in lipid was not a factor contributing to the resistance.

Studies with radio-labelled fumigant showed that the resistant beetles had a slower rate of uptake and a faster rate of metabolism of ethylene dibromide than did susceptible beetles of the same weight. Ellis (1972b) considers that this indicates a biochemical mechanism of resistance to ethylene dibromide (and by analogy, to methyl bromide). However, there is no evidence that the enhanced metabolism is a cause rather than a symptom of the resistance. The apparently polygenic inheritance and the wide sprectum of cross-resistance to unrelated fumigants would suggest a complex of several non-specific mechanisms in this strain among which increased weight and reduced penetration may be contributing factors.

(F.14.8b) *Survey data*

The results obtained during the present survey may give some indications regarding the nature of fumigant cross-resistance, though there are obvious limitations in using discriminating dose data for this purpose. Moreover, few resistance levels were measured so vigour tolerance is not precluded in all cases. The fumigant-resistant strains found were either resistant to phosphine, methyl bromide, or both of these compounds. There is obviously no direct evidence to indicate whether the strains resistant to both fumigants were instances of multiple resistance or true cross resistance.

The highest incidence of fumigant resistance was in *Tribolium confusum*. In this species 25 of 92 strains tested were resistant to methyl bromide and 28 resistant to phosphine. On a chance basis we might expect that 8 of the strains would be resistant to both these fumigants. In fact, 12 strains of *T. confusum* were found to be resistant to both fumigants.

Similar calculations can be made for the other species examined but since in these the incidence of fumigant resistance was generally much lower than in *T. confusum* they are best considered together. No phosphine resistance was found in *Sitophilus zeamais* or *Oryzaephilus mercator*, and no methyl bromide resistance in *S. granarius*, so there are four other species in which strains resistant to each of the two fumigants could occur (*S. oryzae*, *T. castaneum*, *R. dominica* and *O. surinamensis*). Among these four species 2% of the strains tested were resistant to methyl bromide and under 9% were resistant to phosphine. Thus we might expect that not more than 0.2% of the strains would be resistant to both fumigants on a random basis, that is only 1 strain among over 500 tested. In fact, in these four species 5 strains resistant to both of these fumigants were found.

In each case methyl bromide resistance and phosphine resistance occurred together more frequently than might be expected, though the tendency is not statistically significant. A possible explanation is that a single mechanism conferring resistance to both compounds, that is a generalised fumigant resistance, may have evolved comparable perhaps to that produced by Monro *et al*. (1961) by laboratory selection with methyl bromide. However, several other explanations are possible and these strains may merely be examples of multiple resistance. For instance, it seems likely that insects which have been exposed to one fumigant may have a greater chance of being exposed to others because the incidence of fumigations as a control technique is not random but depends very much on the availability of facilities and trained personnel. Moreover, even among fumigant-resistant strains those resistant to both fumigants were uncommon. Apart from 12 strains of *T. confusum*, only 2 strains of *T. castaneum* and single strains of *S. oryzae*, *R. dominica* and *O. surinamensis* were involved. Most fumigant resistant strains were either resistant to phosphine and not methyl bromide, or resistant to methyl bromide and not to phosphine.

Resistance to phosphine but not methyl bromide occurred in 7 strains of *S. oryzae*, 8 strains of *S. granarius*, 21 strains of *R. dominica*, 13 strains of *T. castaneum* and 16 strains of *T. confusum*. In each of these species except *T. castaneum* the phosphine-resistant strain included some which were susceptible to malathion and some susceptible to lindane, so there is no reason to associate resistance to either of these contact insecticides with phosphine resistance. In *T. castaneum* the incidence of malathion resistance and of lindane resistance is so high that it is not surprising that none of the phosphine-resistant strains of this species were susceptible to either of these insecticides.

Fewer strains were resistant to methyl bromide but not to phosphine, but 5 strains of *S. zeamais*, 6 strains of *T. castaneum*, and 13 strains of *T. confusum* showed this pattern of fumigant resistance. Some of the methyl bromide-resistant strains of the species other than *T. castaneum* were susceptible to malathion, and in the case of *T. confusum* some of them were susceptible to lindane. As noted above, the incidence of resistance to both insecticides is very high in *T. castaneum* and the same is true of lindane resistance in *S. zeamais*. There is no reason, therefore, to associate methyl bromide resistance with resistance to either malathion or lindane.

(F.14.8c) *Types of fumigant resistance*

Despite the paucity of our knowledge on fumigant resistance at least six different types can be recognised on the basis of existing information. These are listed in Table F28.

The six groups appear to be mutually exclusive except that some of the phosphine-resistant strains listed under type 6 are only provisionally placed there as their susceptibility to hydrogen cyanide, or both hydrogen cyanide and ethylene dibromide is unknown.

All measurements of resistance levels are dependent on the methods used, so qualitative rather than quantitative differences have been utilised to distinguish the groups. Even so, future work may show that some of these differences are artificial because the detection and measurement of fumigant resistance presents special problems. For example, the concentration x time relationship may not be constant over the full dosage range (e.g. Bell and Glanville 1973b) and narcosis may occur at high dose levels (Lindgren 1938). These difficulties appear to be particularly important with phosphine *(vide* Section F.1.3b). The table omits strains in which multiple resistance is suspected and the examples from the present survey are limited to instances where five or more strains of a species are involved.

Among the pests of growing crops, resistance to the fumigant hydrogen cyanide has been reported in one aphid and three scale insects (Brown 1958). Extensive studies appear to have been made only with the California red scale *Aonidiella aurantii*. Resistance to hydrogen cyanide in this species is associated with resistance to methyl bromide, ethylene oxide and hydrogen sulphide (Quayle 1938, Yust and Shelden 1952) but not to ethylene dibromide (Lindgren and Gerhardt 1947). This resistant strain can only be fitted into type 2 in Table F28, but its susceptibility to phosphine is not known.

Some of the different types of fumigant resistance tabulated may reflect differences in methodology rather than different fumigant resistance mechanisms. However, it also seems likely that as our knowledge of fumigant resistance increases some of the types will need repeated subdivision to accommodate the different mechanisms revealed. Future work is likely to considerably modify the tentative groupings of Table F28.

(F.14.8d) *Relationship with insecticide resistance*

Bond (1973) has shown that a field strain of *T. castaneum* from Canada which was resistant to ethylene dibromide had no cross-resistance to methyl bromide. This strain was resistant to lindane and malathion as was a fumigant-susceptible strain collected in nearby premises. The insecticide resistance of the strain would therefore appear to be due to multiple resistance. This ethylene dibromide-resistant strain has responded to selection in the laboratory and now has a 4-fold resistance to this fumigant (Anon. 1974b). Kem (1975) reported that his selected strain of *T. castaneum* with a x12 resistance to phosphine lacked any definite cross-resistance to DDT, carbaryl, pyrethrins or the seven organophosphorus compounds which he tested.

The lack of evidence for true cross-resistance to insecticides in fumigant-resistant strains is complemented by a virtual absence of evidence that insecticide resistance can confer true cross-resistance to fumigants. Bhatia and Bansode (1971) found no cross-resistance to seven fumigants in

TABLE F28: TENTATIVE CLASSIFICATION OF THE DIFFERENT TYPES OF FUMIGANT RESISTANCE KNOWN IN STORAGE PESTS

Type	Resistant to	Example	Lab. or Field	Ref.
1	MeBr, PH_3, EDB and others	S. granarius	Lab. (2 strains)	1
2	MeBr but not PH_3 (in adults)	S. zeamais	Field (6 strains)	2
		T. castaneum	Field (6 strains)	2
		T. confusum	Field (13 strains)	2
3.	MeBr but not HCN (only in mature larvae)	P. interpunctella	Field	3,4
4.	EDB but not MeBr	T. castaneum	Field	5
		T. confusum	Field	6
5.	HCN but not MeBr or EDB	S. oryzae	Lab	6
		S. granarius	Lab	6
6 (a)	PH_3 but not MeBr, EDB or HCN	S. granarius	Lab	7
(b)	PH_3 but not MeBr or EDB	T. castaneum	Lab	8
(c)	PH_3 but not MeBr	S. oryzae	Field (7 strains)	2
		S. granarius	Field (8 strains)	2
		R. dominica	Field (21 strains)	2
		T. castaneum	Field (13 strains)	2
		T. confusum	Field (16 strains)	2

References
1. Monro et al. (1961)
2. FAO Survey
3. Sardesai (1972)
4. Prevett (1971)
5. Bond (1973)
6. Lindgren & Vincent (1965)
7. Monro et al. (1972)
8. Kem (1975)

a highly DDT-resistant strain of *T. castaneum* (Table F27) and Ellis and
Morrison (1967) found no evidence of ethylene dibromide resistance in a
DDT-resistant strain of *S. granarius*. Bond (1973) found that a lindane-
resistant strain of *T. castaneum* was fully susceptible to both ethylene
dibromide and methyl bromide. Many instances of multiple resistance
involving both fumigants and insecticides were found in the present
survey but the only evidence which may indicate a true cross-resistance
between insecticides and fumigants appears to be the report by Rajak
and Hewlett (1971) of a low cross-resistance to phosphine in adults
of a pyrethroid-resistant strain of *S. granarius*. This strain was about
30% heavier than the susceptible beetles with which it was compared, but
earlier studies on the same strain had given no evidence of resistance to
methyl bromide (Lloyd and Parkin 1963, Howe and Hole 1967).

Of particular interest are the data of Bansode (1974) who
recorded a x2.6 tolerance of methyl bromide and x1.7 of phosphine in a
strain of *Sitophilus oryzae* that had been selected with malathion to a
resistance level of x49. The resistance was not suppressed with triphenyl
phosphate. This strain had a wide range of tolerances to other organophosphorus compounds, a carbamate, cyclodienes, DDT and pyrethrins. Only
a preliminary report on this study is available as yet, but if comparisons
at the $LD_{99.9}$ level give a similar picture this strain may conform essentially
with a type 1 resistance (Table F28). Upitis *et al*. (1973) working with
S. granarius demonstrated a complex polyfactorial inheritance of a methyl
bromide resistance also conforming with type 1 and it is feasible that
there are common non-specific factors involved in these tolerances to methyl
bromide, phosphine, malathion and other insecticides.

(F.14.8e) *Lepidoptera*

There appears to be no direct evidence for the existence of
fumigant-resistant strains of the moths infesting stored products, but a
resistance mechanism must occur in at least one species. Sardesai (1972)
has demonstrated that diapausing larvae of *Plodia interpunctella* are more
tolerant of methyl bromide than are non-diapausing larvae of this species,
though the two types of larvae are equally susceptible to hydrogen cyanide.
This cannot be dismissed as merely a difference between different
developmental states since some strains of *P. interpunctella* undergo diapause
in conditions where others do not (Prevett 1971). Strain differences in the
ability to diapause must therefore reflect strain differences in susceptibility
to methyl bromide. Moreover, selection with methyl bromide may result in an
increase in the incidence of diapause, and climatic conditions favouring
diapause will also favour methyl bromide resistance; always providing, of
course, that the environment permits diapause to be expressed. Diapausing
larvae of other storage moths are known to be less susceptible to fumigants
than non-diapausing larvae, for example to phosphine in *Ephestia elutella*
(Bell and Glanville 1973b). However, the evidence for both diapausing and
non-diapausing strains occurring naturally in species other than
P. interpunctella is lacking or equivocal, so in these species diapause has
yet to be demonstrated as a mechanism of resistance.

(F.14.9) *Cross-resistance to non-chemical measures*

Insects vary in their response to a wide range of physical
features in their environment. Intra-specific variation in susceptibility
to physical control measures is therefore to be expected and strains
resistant to irradiation or oxygen depletion might well arise.

Several storage pests for example *Sitophilus oryzae*, *Tribolium castaneum*, *Callosobruchus maculatus* and *Plodia interpunctella* have been selected in the laboratory with sublethal doses of gamma-irradiation for 25-30 generations without radio-resistant strains being produced (*vide* Section F16). However, an inherited increase in tolerance to ionizing irradiation has been demonstrated in Diptera, so radio resistance remains a possibility for storage pests. There is, however, no reason to suppose that insecticide or fumigant resistant strains would be cross-tolerant to irradiation, though actual tests of this possibility appear to be relatively few.

With oxygen depletion, which in practice can result from air-tight storage, or flushing with nitrogen etc., the situation is more complex. Bailey (1965) has shown that diapausing larvae of *Trogoderma granarium* are unusually resistant to low oxygen tension and this is probably associated with their respiration rate. Diapause in *T. granarium* is at least partly genetic (Nair and Desai 1973) so inherited strain differences in the ability to diapause will reflect genetic differences in the ability of different strains to resist oxygen depletion. The same may well be true of other storage pests with diapause and non-diapause strains, which are known to have a reduced oxygen requirement during diapause for example *Plodia interpunctella* (Sardesai 1968).

Jay and Pearman (1971) found that in laboratory experiments a strain of *T. castaneum* with a 7-fold resistance to malathion was rather more susceptible to atmospheres with reduced oxygen and enhanced carbon dioxide concentrations than was a susceptible strain. There was thus no evidence of cross-resistance, but the enhanced susceptibility could have been an incidental strain difference rather than a feature of the malathion resistance.

Another instance of enhanced susceptibility was reported in the fumigant resistant California red scales mentioned above. These were found to be more susceptible to anoxia in an atmosphere of almost pure nitrogen than were susceptible scales (Yust and Sheldon 1952). In this case a common mechanism for fumigant resistance and enhanced susceptibility to anoxia seems more probable particularly as a selected strain with higher HCN resistance had somewhat higher susceptibility to anoxia. As noticed above these scales were resistant to several fumigants including methyl bromide. It will obviously be of interest to determine whether any strains of storage pests which are resistant to several fumigants have an enhanced susceptibility to anoxia. If this should prove to be the case it would contribute to our understanding of the resistance mechanism and the negative correlation might be turned to practical use.

(F.15) *The inheritance of resistance in stored grain insects*

A knowledge of the mode of inheritance of resistance is basic to an understanding of the development and stability of resistance in populations and the tolerance levels they achieve. Studies are few and the extent of these studies is limited by a lack of marker genes in most species. Significant sources of information available on formal genetics of stored product insects are the *Tribolium* information Bulletin and the monograph by its Editor A. Sokoloff, "Genetics of *Tribolium*" (1966a), Robinson's (1971)"Lepidoptera Genetics", and the series of publications by S.G. Smith on Coleoptera cytogenetics culminating in a listing of the karyotypes of 83 stored product beetles (Smith and Brower 1974). The following summary concerns the information available on the mode of inheritance of resistance in stored grain pests and outlines the mutants and linkage information available for the stored product pests generally.

(F.15.1) *Sitophilus spp.*

Lindane resistance is common in the three species *Sitophilus oryzae, S. zeamais* and *S. granarius*. Malathion resistance is as yet localised and there are now indications of an emerging phosphine-resistance problem.

Sitophilus granarius has 11 pairs of autosomes and an XY sex-determining mechanism whereas the derivative species *S. oryzae* and *S. zeamais* have 10 pairs of autosomes and neo-XY systems (Smith 1952a, Smith and Brower 1974). Thus as with the *Tribolium* spp. many linkage groups are involved and genetic studies correspondingly complex. Mutants in the stored product weevils are few although a range of body-colour and eye-colour mutants have been described from another Curculionid, the cotton boll weevil *Anthonomus grandis* (Bartlett 1967). Campbell-Brown and Champ (1971) described a fused antennae (*fa*) strain of *Sitophilus oryzae* with a single-factor, semi-dominant autosomal mode of inheritance. Champ (1967) referred without description to a semi-dominant black body-colour mutant of *S. oryzae* that was lethal in the homozygous state and subsequently (unpublished data) isolated a red-eyed autosomal mutant of this species and an apparently similar mutant in *S. granarius*. An undescribed pearl-eye mutant also in *S. granarius* has been reported in Stock Lists of Mutants from the Pest Infestation Laboratory, Slough (*Tribolium Inf. Bull. 9*). Other heritable variations affecting size and colouration appear common in Curculionidae but have not been definitely isolated. Thus in summary little is known of the formal genetics of this group of Curculionidae.

Although DDT resistance in insects is usually autosomal, Champ (1967) reported two sex-linked partially-dominant factors from two different strains of *S. oryzae*. The resistance factors were allelic or resulted from association with modifying factors and the resistances were subsequently shown to be sesamex-suppressible (unpublished data). Lloyd and Shaw (1968) reported a partially-dominant DDT resistance from *S. granarius* which was partially-suppressed by sesamex (Lloyd 1969b). This resistance however, was autosomal and homology may exist between the autosomes concerned with this resistance in the more primitive *S. granarius* and the neo-XY chromosomes of *S. oryzae* as has been suggested between the linkage groups of the prothoraxless mutant (IX) of *T. castaneum* and the sex-linked prothoraxless-like mutant of *T. confusum* and the closely-linked homeotic mutants labiopedia in both species (Dawson 1968).

Lindane resistance in *Sitophilus oryzae* has been shown to be inherited at least in some strains by a limited number of factors (Champ, unpublished data). Strains have been isolated with resistance inherited as a single, semi-dominant, autosomal factor as well as strains from the same population which have two segregating, major controlling factors.

The only data for fumigants and stored product insects concerns methyl bromide and *Sitophilus granarius*. Upitis et al. (1973) demonstrated polyfactorial inheritance of resistance to this material in a strain of *S. granarius* showing a x7.8 tolerance after 45 generations of selection with methyl bromide. This interpretation was based on results from crosses between susceptible and selected strains which yielded F_1 and F_2 progenies and test crosses which did not segregate and had intermediate tolerance with no change in slope of the dose-response lines.

(F.15.2) *Rhyzopertha dominica*

Lindane resistance is also common in *Rhyzopertha dominica*. Two types of malathion resistance have been recorded - resistance specific to malathion is the general type whereas at least in Australia a resistance occurs which is not suppressible with triphenyl phosphate and which confers tolerance to dichlorvos and a wide range of other organophosphorus compounds. Phosphine resistance is also an emerging problem.

The karyotype of *R. dominica* is 8AA+XY (Smith and Brower 1974) but there is only one mutant available, black (*b*) body colour, a semi-dominant autosomal, fully penetrant but apparently showing some lethal effects in the homozygous state (Dyte et al. 1965c, Champ and Genn 1971). The latter authors also described a fused-antenna mutant that had poor penetrance.

The malathion-specific resistance appears to be partially dominant and linked with *b* (Champ, unpublished data). Examinations of the other resistances have not been made.

(F.15.3) *Tribolium castaneum*

Malathion and lindane resistances in *Tribolium castaneum* are present in all areas of the world and there is now difficulty in finding susceptible strains. Both specific and non-specific malathion resistances occur. In addition there are indications of methyl bromide and phosphine resistance problems.

Tribolium castaneum, in addition to being the commonest pest of stored products in the world, is a widely-used tool for studies on population ecology and genetics. The karyotype is 9AA+XY conforming with the primitive condition of the *Tribolium* group (Smith 1952b,c). It is the only stored product species for which extensive linkage data is available (Sokoloff 1966a, et al. 1966; Dawson 1971, 1972a,b; Dewees 1967, Schmitz and Englert 1967) though of the nine pairs of autosomes seven only in addition to the sex-determining chromosomes, have sufficient linkage data for establishing map positions. The remaining autosomes, VI and X have effectively single genes only and these have limited usefulness. Microphthalmic (*Mo*, VI) behaves as a dominant with recessive lethal effects, and shows variability in expression on outcrossing to the extent that mutants are difficult to distinguish from wild-type beetles. Abbreviated appendages (*aa*, X) is a recessive of good penetrance but of variable expression on outcrossing. Moreover, prothoraxless (*ptl*, IX) is a semi-dominant of variable expression with poor penetrance on outcrossing and is lethal in the homozygous state. Thus incomplete analyses only of the different factors contributing to resistance are possible.

Erdman (1966, 1970) has reported a correlation between DDT resistance and sooty (s, autosomal recessive, II) with indications of maternal effects. As discussed earlier (Section Fl.1g), Dyte et al. (1968a) found that DDT-tolerance was correlated with body colour in a range of *Tribolium* species and that the darker the body colour the more tolerant the species was to DDT though dark body colour mutants of *T. castaneum* including sooty were little different from the normal reference stock in response to DDT.

Some linkage relationships of lindane resistance in a laboratory-intensified resistant strain were established by Champ and Campbell-Brown (1969). Their examination was restricted to determination of dominant and partially-dominant factors in linkage groups II, IV, V, VII and VIII and all types of resistance factors in linkage group III. Responses were measured as knockdown after 72 hr exposure to lindane-impregnated filter papers and effects of each linkage group and their interactions were analysed by the methods outlined by Tsukamoto (1964). All linkage groups tested I, II, III, V, VII and VIII were concerned with the expression of lindane resistance and dominant or partially dominant factors were demonstrated in linkage groups II, IV, V and VIII together with a recessive factor in linkage group III. These data suggested a multifactorial type of resistance.

These authors using similar techniques subsequently (unpublished data) examined the contributions of linkage groups I, II, III, IV, V, VI VII and VIII to resistance to knockdown by malathion in two strains one of which showed a triphenyl phosphate (TPP) suppressible type of malathion resistance and the other a non-suppressible type. The strain carrying the TPP-suppressible resistance did not segregate any non-TPP responding phenotypes although the resistance level could be increased by selection whereas in the other strain which showed resistance to malathion + TPP, a low response to triphenyl phosphate was always present which may have been related to the activity of TPP in the normal susceptible state of *T. castaneum*. Linkage groups I, VI and VIII were concerned with expression of the resistance in both strains with the major contribution in both strains coming from a partially-dominant factor in linkage group VI which produced dosage-mortality responses in test-crossing that were typical of those expected with single factor inheritance. The cross-over frequencies with microphthalmic (linkage group VI) of the major factors controlling both types of malathion resistance were 19 for the strain showing the specific type and 18 for the strain showing the non-specific type suggesting a similar locus or alleles. Pasalu (1974) at the Indian Agricultural Research Institute studied the inheritance of a triphenyl phosphate-suppressible malathion resistance by reciprocal mass-crossing between susceptible and resistant strains and determining susceptibility levels and segregation of progenies by exposure to impregnated filter paper. From the segregations of F_1, F_2 and test-cross progenies he concluded that this malathion resistance was inherited as a single major autosomal, partially dominant factor.

(F.15.4) *Tribolium confusum*

Resistances occur that are similar to those in *T. castaneum*. There is considerably less linkage information available for *Tribolium confusum* than for *T. castaneum* (Sokoloff 1966a). *T. confusum* has the less primitive neo-XY system with 8 pairs of autosomes (Smith 1952b,c).

Linkage relationships suitable for mapping have been established for the sex chromosomes (linkage group I) and autosomal linkage groups II, III and V. A number of other mutants also are available which segregate independently of these groupings but in effect any analyses of the linkage relationships of resistance in *T. confusum* would be very incomplete.

The formal genetics of the malathion-specific and non-specific resistances, or of the lindane resistance present in *T. confusum* have not been established. This information when established will provide a valuable opportunity to seek homologies between resistances in the two species.

(F.15.5) *Other Tribolium species*

Resistance to pesticides has not been reported in the species of *Tribolium* other than *T. castaneum* and *T. confusum*. These other species are of considerably lesser importance but a knowledge of their genetics will assist in an ultimate understanding of the inheritance of resistance in the *Tribolium* group.

The mutants described for these other species are similar to corresponding mutants in *T. castaneum* and *T. confusum*.

Tribolium madens has the (9AA+XY) karyotype similar to *T. castaneum* but with supernumerary autosomes present (Shaw in Halstead 1969). Four autosomal recessive mutants all of variable expression have been reported - fused antennal segments (*fas-1*), split elytra (*spl*), the incompletely penetrant creased abdominal segments (*cas*) (Sokoloff 1963), and the poorly penetrant bent tibia (*btt*) (Sokoloff 1964a).

In *Tribolium destructor* (8AA+neoXY) much of the autosomal element of the neo-XY system as in *T. confusum* has been eliminated (Smith 1952b,c). The three mutants described are the autosomal recessives *cas* and *btt* (Sokoloff 1964a) again incompletely and poorly penetrant respectively, and *spl* (Sokoloff 1966b).

The mutants *cas, fas* and *spl* have been reported also from *Tribolium brevicornis* and *cas* and sternites incomplete (*sti*), an autosomal recessive of good penetrance but variable expression from *Tribolium anaphe* (Sokoloff 1964a,1971).

(F.15.6) *Other Tenebrionidae*

Latheticus oryzae - Malathion but not lindane resistance has been reported from *L. oryzae* and the prevalence of resistance in cohabiting species such as *T. castaneum* indicates the probability of other resistances developing. Ten mutants are known and five of these have been placed in two of the ten possible linkage groups (9AA+XY, Smith and Brower 1974). The completely penetrant red eye (*r*) and the semi-lethal truncated elytra (*te*) are sex-linked recessives (Sokoloff and Shrode 1960). The marker identifying linkage group II is the recessive pearl (*p*) which has good penetrance and uniform expression, and is epistatic to red (*r*) (Sokoloff 1959). It is linked with the recessives creased abdominal sternites (*cas*) which has variable expression and incomplete penetrance, and brown body (*bwb*) (Sokoloff 1964a) whose viability is reduced. Unassigned autosomal recessives are tucked elytra (*tke*), fused antennal segments-1 (*fas-1*) (Sokoloff 1965) and elongate elytra (*ele*) (Sokoloff and Hoy 1965) - all have variable expression and are incompletely penetrant.

Other mutants known are incomplete meso-metathoracic suture (*ims*) (Hoy 1966b), a recessive with incomplete penetrance, and droopy elytra (*dre*) presumably a phenodeviant (Sokoloff 1965).

Gnathocerus cornutus - There have been unconfirmed reports of pesticide resistance in *G. cornutus* but these probably can be related to poor hygiene. Resistance, however, remains a strong possibility as with *L. oryzae*. Two autosomal recessive mutants, pearl eye (*p*) (Dyte and Blackman 1961) and light ocular diaphragm (*lod*) (Sokoloff and Ho 1963), are known which are not linked and have been assigned provisionally as markers of linkage groups II and III of the ten possible (9AA+XY, Snow 1962).

Tenebrio molitor - Resistance has not been reported from *T. molitor*. The probability of resistance developing is not high as this species occurs principally under conditions of poor hygiene and in damp habitats where exposure to pesticides and so selection, is not intensive. The karyotype is 9AA+XY (Smith 1952c). A number of genes have been isolated that affected larval body colour (Arendsen Hein 1920) and eye colour, and produced head and eye, and appendage abnormalities. Three interacting recessive genes determined changes in eye colour - the sex-linked yellow (*h*), and the autosomal red (*f*) and the epistatic flesh-coloured (*g*) (Ferwerda 1928). The mutant pearl (Dyte *et al.* 1965d) is thought to be a recurrence of *g*. The factor *g* was linked with a dominant semi-lethal gene *B* controlling a head and eye abnormality (Ferwerda 1928) in a provisional linkage group II. Other mutants expressed as appendage abnormalities have been antennae compressed (*Atc*), the incompletely penetrant antennae and tarsi reduced with one joint (*At 10/10*), and the completely penetrant tarsi abnormal (*Ta*) - all have variable expression (Arendsen Hein 1924).

(F.15.7) *Silvanidae*

Oryzaephilus surinamensis is probably the most widely distributed storage insect. Lindane resistance is common but malathion resistance is spasmodic and does not appear to develop readily. Pearl eye is an autosomal recessive mutant (Blackman 1966) and *bicornis* and *small* are strains collected directly from the field (Aitken 1966) - their genetics is complex (Ashman and Higgs 1968). The karyotype is 8AA+XY (Robertson 1959).

Ahasverus advena is a minor pest of grain in many areas but rarely reaches major pest status. Dyte (1963) listed a black mutant but its mode of inheritance is unknown. The karyotype is 8AA+XY. (Smith and Brower 1974).

(F.15.8) *Bruchidae*

Resistance has been definitely reported only in *Caryedon serratus* (8AA+XY) in which species no marker mutations are known. However, other species may well prove to have resistant strains when they have been more extensively studied, and the members of this family are particularly suitable for genetic study. The sex chromosome complement appears to vary in some species (see Smith and Brower 1974).

Acanthoscelides obtectus - An autosomal recessive red (*r*) lacking black pigment in the cuticle was described by Skaife (1925). The karyotype is 9AA+XY or XO.

Callosobruchus chinensis - Skaife (1925) showed that at least two semi-dominant factors chic (C) and intense (I) were involved in determining the black elytral markings in this species. More recently, a recessive black mutation ($sC-b$) has been reported by Kashiwagi and Utida (1972). Strains with 9AA and either XO, XY or XYY have been reported.

Callosobruchus maculatus - Breitenbecher (1921) described an autosomal multiple allelic series controlling elytral and body colour. Red (R) was the top dominant of the series, with black (R^b) recessive to red but dominant to white (R^w) which in turn was dominant to the tan colour of the wild type. These three dominant mutants produced obvious effects in females but only very slight effects in males. He subsequently reported two further autosomal dominants, red elytral spots (S) which was expressed only in females and was not allelic with the R locus (Breitenbecher 1923) and macula (M) affecting elytral colour and which was only expressed in males (Breitenbecher 1925a). He also (1925b,c) described two autosomal recessives, apterous (which can be symbolized as a) and piebald (p) both of which are expressed only in females. The karyotype is 9AA+XO or XY.

Zabrotes subfasciatus - Preliminary genetic studies on the elytral markings have been made by Seeliger (1943). The karyotype is 12AA+XY.

(F.15.9) *Cryptolestes spp.*

The *Cryptolestes* species are recorded commonly associated with resistant strains of other species in pesticide-treated commodities. Resistance has not been demonstrated unequivocally but it is probable that if systematic monitoring was carried out, pesticide resistance would be detected.

C. pusillus has a 8AA+XY karyotype (Robertson 1959). A black-coloured form with some of the characteristics of an autosomal recessive mutant (Lefkovitch 1963) was later (Lefkovitch and Currie 1967) considered to be part of a complex of characters differentiating a distinct subspecies, *C. pusillus fuscus*.

C. turcicus also has a 8AA+XY karyotype (Smith and Brower 1974). Six mutants have been isolated but the modes of inheritance of two only have been studied. Red eye (r) is a sex-linked recessive (Dyte *et al.* 1965c) and crooked antennae (cka), an autosomal recessive of variable expression and incomplete penetrance (Sokoloff 1965). Runty (rty) (Sokoloff 1965) and deformed prothorax (dfp) (Hoy 1966a), an incompletely penetrant semi-lethal have been described, and tiny and pink have appeared only in the 1969 and 1971 Stock Lists from the California State College (*Tribolium Inf. Bull.* 11,14).

(F.15.10) *Dermestidae.*

Trogoderma granarium is principally a pest of unsatisfactory types of storage under conditions where effective use of chemicals is difficult. There is a single record of malathion resistance. The only mutant reported is pearl eye (p) an autosomal recessive (Reynolds and Sylvester 1962).

Dermestes maculatus is the commonest *Dermestes* species. Lindane resistance has been recorded associated with widespread use of lindane and dieldrin. Twenty-two mutants have been reported and there are nine possible linkage groups (8AA+XY, John and Shaw 1967). Two strains with 1 and 2 supernumerary y chromosomes (2y, 3y chromosomes) have been reported although that with 3y chromosomes was unstable at meiosis (Shaw 1968). Pearl eye (p) is a sex-linked recessive (Shaw 1966) and fuscous (fu), a semi-dominant body colour mutant, has been assigned as a marker for linkage group 11. Of the remaining unassigned mutants, those affecting eye colour are the autosomal recessives white eye (w) (Philip 1940), rufous (ru) (Shaw and Welch 1966), and red eye, pink eye and muddy eye whose genetics have not been determined (Shaw and Welch 1966, Shaw 1967a). Other mutants affecting body and wing colour are white abdomen, an autosomal dominant (Dyte et al.1968b), the poorly penetrant white tip and light wing, both autosomal recessives (Philip 1940), and platinum (pla) and copper (co) whose mode of inheritance has not been worked out (Dyte and Binns 1971). Light wing (l)(Dyte et al. 1965d) may be a recurrence of Philip's light wing. Mutants affecting the antennae are deformed antennae (def) an autosomal recessive of variable expression (Shaw and Welch 1967), and three whose genetics have not been studied, light antennae (Shaw and Welch 1966), dark antennae and double antennae (Stock Lists, Pest Infestation Laboratory, Tribolium Inf. Bull. 11). Mutants causing general morphological abnormalities are the autosomal recessives creased sternites (Shaw and Welch 1966) and second sex pit which has sex-limited expression (Philip 1940), the undetermined second sex pit (ssp) of Dyte et al. (1965d) which may be a recurrence of Philip's mutant sexpitless (Shaw 1967a), dented pronotum (dp) (Dyte et al. 1969), short elytra (sh) (Dyte et al. 1965d) and ridged elytra (re) (Shaw and Welch 1966).

Dermestes frischii has a karyotype similar to D. maculatus 88AA+XY, John and Shaw 1967) and Shaw (1967b) has reported a strain of this species also with a supernumerary chromosome (2y chromosomes). A creased sternites mutant has been listed also from the Pest Infestation Laboratory (Tribolium Inf. Bull. 9,10,11,13).

(F.15.11) Other Coleoptera

Carpophilus dimidiatus (Nitidulidae) occurs widely in damp grain residues, standing maize, dried fruit and other habitats in which exposure to pesticides is minimal. Resistance has not been recorded and the only mutant reported is pearl eye, an autosomal recessive (Amos and Scott 1965).

Lasioderma serricorne (Anobiidae) is a pest of a wide range of commodities that reaches its maximum pest status in stored tobacco where there is considerable exposure to pesticides often under conditions allowing selection for resistance. Resistance, however, has not been reported. Coffelt and Vick (1973) have described a black body colour mutant which has a semi-dominant autosomal mode of inheritance.

(F.15.12) Ephestia cautella

Resistances to malathion and synergised pyrethrins have been reported from the U.S.A. and Hashimoto and Fukami (1964) have selected in the laboratory a strain resistant to methyl-parathion.

As with most Lepidoptera, visible mutants are few. Hagstrum (1974) has described black (bl) and brown (br) wing-colour and white (we) and yellow (ye) eye-colour mutants. All were autosomal recessives. The only other mutant described has been a low temperature conditional lethal mutant which produces females only at 20°C and equal numbers of each sex at 30°C (Takahashi et al. 1968, Takahashi and Kuwahara 1970).

(F.15.13) *Ephestia elutella*

There are no reports of pesticide resistance from *E. elutella*. The only mutants known are an undescribed red-eyed form (Childs, personal communication) and an autosomal recessive white eye (Prickett and Bell, unpublished data).

(F.15.14) *Ephestia kuehniella*

There have not been any reports of pesticide resistance in *Ephestia kuehniella* although there has been considerable pressure placed on it in pest control programmes in milling premises.

Considerable attention has been paid to the formal genetics of this species commencing with the early studies of Whiting (1919) on body and wing colour mutants and a later proliferation of research by Kühn and others following the description by Kühn and Henke (1930) of the rotäugig eye-colour mutant (a) and its pleiotropic effects, and the subsequent explanations of the biochemical mechanisms involved. Robinson (1971) has listed 46 genes of which 8 controlled eye colour, 27 wing colour and patterns, 2 wing morphology, 3 protein synthesis and the remaining 6 incidental characters such as chromosome behaviour, embryo development, scale morphology, and colour of testes and of malpighian tubules. Hanser (1955) has described a scaleless (glasflüglige) mutant (gl) also and Cotter (1974) has since described the autosomal dominant ala nigra (an) which affects wing colour. Much of the work related to these genes has concerned biological and biochemical interpretations of their expression and their pleiotropic effects. There have been limited attempts only to establish linkage data for *E. kuehniella* presumably because of the high chromosome number (30) and the spread of the work over some 30 years which probably has precluded simultaneous availability of many of the strains. In addition many of the genotypes of similar types of mutant and their crosses would be phenotypically difficult to distinguish. Data on linkage have been summarised by Robinson (1971) and Cotter (1974). The first cases demonstrated were between schwartzschuppig (b), a black wing colour and biochemica (bch), an eye colour mutant (Kühn and Berg 1956), and between the sex-linked Dunkles Feld (df) wing pattern mutant and Schwartz's lethal (ls), a mutant affecting embryonic development. Subsequently Cotter (1974) quoting unpublished work of Yearick - Nupponen has reported that there appears to be another linkage group comprising An (ala nigra), Us (ungegliedertes Symmetriesystem), ml (musterlos), and Ch (charcoal) independent of the groups characterised by b, a (red eye), wa (white eye) and sa (short antennae). Twenty one other instances of apparently independent assortment of gene pairs have been reported principally involving one of the pair as an eye-colour gene or involving wing colour and wing pattern combinations (*vide* Robinson 1971).

(F.15.15) *Plodia interpunctella*

This widely distributed species has been exposed to a wide range of insecticides in many areas of the world. There have been many unconfirmed reports of resistance and authenticated cases of malathion and pyrethrins resistance from the U.S.A.

The known mutants of *Plodia interpunctella* are few and as with *Ephestia kuehniella* they principally concern wing colour and pattern, and eye colour. The recessive Gelbfärbung (g) is characterised by a replacement of the reddish wing and body colour with yellow, and the independently-inherited dominant Ringzeichnung (R) is expressed as a complete or partial loss of the black pattern of the forewings and helle Randlinie (*he*) lightens black markings on the forewings and general body pigmentation (Schwartz 1953). Other less well established mutants described by the same author affect the intensity of pigmentation of the wings and body. Brower (1972c) described a sex-linked dominant mutant, melanic (*M*) affecting forewings, and an autosomal recessive scaleless (*scl*) wing mutant which is inherited more regularly than the clear wing (*cw*) of Takahashi *et al.* (1973). The eye colour mutants are the recessives white (*w*) (Smith 1956), red (rotaugig, *ra*, Almeida 1958a) and black (Almeida 1958b). Recently Wehrmaker (1975a,b) has described flavosignata (*fs*), a non-autonomous autosomal recessive which produces yellow eyes and also affects the pigmentation of the wing scales so that the normally-reddish scales are yellow.

The *ra* genes of *P. interpunctella* and *E. kuehniella* are not homologous as shown by accumulation of kynurenine rather than tryptophane as in the latter species, and by parabiotic experiments with mutant and wild-type pupae - black, however, has been likened to biochemica in *E. kuehniella* (Almeida *op. cit.*, Kühn and Almeida 1961). By reciprocal intergeneric feeding tests, Wehrmaker (1975b) showed that *fs* in *Plodia* is homologous to *a* in *E. kuehniella*.

(F.15.16) *Sitotroga cerealella*

The only report of resistance in *Sitotroga cerealella* is to lindane from Sri Lanka.

Volkova (1940) described a "shortened-wings" mutant without detail of its inheritance and an autosomal "folded-wings" mutant in which other aspects of development, fertility and egg viability appeared normal. The only other mutant reported has been an autosomal recessive red eye (*r*) (Mills and Bell 1972). This eye mutant resembled the autosomal recessive red eye (*r*) described from another gelechiid, *Phthorimaea operculella*, by Champ and Shepherd (1971).

(F.15.17) *Comments*

A summary of mutants described from stored product insects is given in Tables F29 and F30. Many of the strains are now lost but if the example of *T. castaneum* and *T. confusum* is considered where the number of mutants reported from both species increased from 7 in 1959 to more than 150 in 1966 (Lerner 1966) and to more than 260 currently, it is apparent that concerted searches probably would reveal recurrences of the lost mutants in addition to much new material.

The known information on the mode of inheritance of resistances in stored-product insects is listed in Table F31. The data is sparse and significant progress will be delayed until there are considerably more mutants described and linkage relationships obtained. When this information is available, there does appear considerable scope for establishing homologies between biochemical mechanisms contributing to the resistance syndrome in stored product insects. It seems likely that many resistances, whether apparently showing single factor inheritance or otherwise, will have other contributing factors operating to enhance the resistance and cross-resistance.

TABLE F 29:   MARKER GENES IN STORED PRODUCT COLEOPTERA

In beetles females are XX and males are XY or XO. Sex-linked genes on the X-chromosome form linkage group I. Other (autosomal) linkage groups are numbered II, III, IV etc.
A = Autosomal genes untested for linkage.   U = Undetermined inheritance

A. obtectus	A - *r*	
A. advena	U - *b*	
C. chinensis	A - *θ, I, sC-b*	
C. maculatus	A - *R, R^b, R^w, S, M, a, p*	
C. dimidiatus	A - *p*	
C. turcicus	I - *r* U - *dfp, rty*	A - *cka*
D. frischii	U - *cas*	
D. maculatus	I - *p* A - *cas, def, l, ru, ssp, w, Wa, wt* U - *co, da, dba, dp, la, m, pk, pla, r, re, sh, spl*	II - *fu*
G. cornutus	II - *p*	III - *lod*
L. serricorne	A - *b*	
L. oryzae	I - *r, te* A - *dre, ele, fas-1, tke*	II - *p, bwb, cas* U - *ims*
O. surinamensis	A - *p*	
R. dominica	A - *b, mal*	
S. granarius	A - *r, ddt*	U - *p*
S. oryzae	I - *ddt-1, ddt-2*	A - *fa, b, r, lind-1, lind-2*
T. molitor	I - *h* U - *Atc, At10/10, p, Ta*	A - *B, f, g*
T. anaphe	A - *sti*	U - *cas*
T. brevicornis	U - *cas, fas, spl*	
T. castaneum	I - *bb, pd, pd-1, dpm, dve, l$_1$, l$_2$, l$_3$, l$_4$, l$_5$, l$_6$, l$_7$, M^r, ma, ma^D, ma-1, mi, pok, pte, py, r, r^D, rHo, rph, r^S, r-1, r-2, rg, rs, ser, sp, te, te-2, ti* II - *p, p^{Pk}, p^S, i, i^D, pg, Ren* III - *au, b, b^S, bcd, b^D, bt, b-1, b^S-1, lod, lod^D, msg, mt, sc* IV - *Be, ctp, dfl, eju, fas-2, h, ims, ju, my, rju, s, Spa, sti, w* V - *j, j^E, j^K, fas-3, fas-3a, fas-3b, m, m^D, mc, mceg, mc-1, spl, ru* VI - *Mo, Mo-1, mal-1, mal-2* VII - *c, c^s, ble, Fta, ju-7, Sa, Sa-1, Sa-2, Sa-3, Sa-4, sa, sa-1, sa-2, sa-3, sa3* VIII - *ap, ap^D, ap^S, elb, sh, sh$^{H\&D}$, sh^S, sq* IX - *lp, mas, ppas, ptl, ptl^D, ptl-1* X - *aa* A - *aas, akb, apt, atf, bal, Bamp, bd, bet, bf, bj, bl, bra, bt, btt, by, cas, Chr, cos, cspl, cye, d, da, da-1, dep, Df, dff, dft, dim, dre, dt, elb-1, ele, em, eta, eu, fas-1, fas-4, fas-5, fas-6, gdf, gl, imp, jac, knp, la, lmg, mgt, Mu, pe, pec, ph, re, ro, rps, rt, rta, rue, sb, sct, slk, smp, spce, sss, sta, stu, tro, ty, u, umb, vgl, wa, we* U - *Npp, pdp, scp, ohs*	

TABLE F29(contd.):   MARKER GENES IN STORED PRODUCT COLEOPTERA

T. confusum	I - $St, aer, apt, dep, cru, es, es^{lt}, l, lp, pas, ptll, ma, r, r^u, tet$
	II - $p, p^H, p^S, p^{H-1}, cas, cas-1, dpe, dpe-1, e_\ell, fas-2$
	III - $b, b^z, b-1, b-2, b-3, b-4, msg, rus, sp$
	IV - thu, thu^S V - $e, e^{L\&H}, ble$
	VI - dj A - $bdl, btf, btt, dep, dl, dre, ele, em,$
	$ems, ems^{DT}, fas-1, imp, ims, knp, lgl, lod, mag, nd, pel, poc, psl,$
	$rby, rju, ro, sc, sep, sh, sti, stl, stt, ty, umb, we, wgl, wspl-1$
	U - apt, dfl, Re, twa
T. destructor	A - cas, btt U - spl
T. madens	A - $btt, cas, fas-1, spl$
T. granarium	A - p

TABLE F30:   MARKER GENES IN STORED PRODUCT LEPIDOPTERA

In moths males are XX and females XY or XO. Sex-linked genes on the X-chromosome form linkage group I. Other (autosomal) linkage groups are numbered II, III, IV etc.
A = autosomal genes untested for linkage. U = undetermined inheritance

E. cautella	A - bl, br, we, ye
E. elutella	A - r, w
E. kuehniella	I - df, ls
	II - b, bch
	III - Us, An, ml, Ch
	IV - a, a^k
	V - wa
	VI - sa
	A & U - Eye colour: $alb, br, M-a, t$. Wing colour: $bl, dia, f,$ $\overline{fa}, (\overline{h}), he, (m)$. Wing pattern: $d, df-2, dz, Hu, (q), qu, (r), (rb),$ $S, (Sb), (Sc), sh, Sy, Syb, u$. Wing morphology: $gl, kfl, (p)$. Chromosome behaviour: (Mo). Protein synthesis: p^0, p^3, p^4. Testes colour: rt. Malpighian tubule colour: (om). Scale morphology: vd. Scale pattern: Sr.
P. interpunctella	I - M
	A & U - black, $cw, fs, g, he, R, ra, scl, w$
S. cerealella	A - fw, r U - sw

TABLE F31: THE PRINCIPAL MODES OF INHERITANCE OF RESISTANCE TO PESTICIDES KNOWN FROM STORED PRODUCT PESTS

Species	Karyotype	Pesticide	Type of Inheritance	Dominance	Linkages	Reference
S. oryzae	10AA+neoXY	DDT	Single factor (+ alleles ?)	Partial	Sex-linked	Champ 1967
		Lindane	Single factor (+alleles ?)	Partial	Autosomal	Champ, unpublished
		Pyrethrins	Multifactorial	Partial	Autosomal	"
S. granarius	11AA+XY	Pyrethrins	Single factor	Partial	Autosomal	Lloyd & Shaw 1968
		Methyl bromide	Multifactorial	–	Autosomal	Upitis *et al.* 1973
R. dominica	8AA+XY	Malathion	Single factor	Partial	b	Champ, unpublished
T. castaneum	9AA+XY	DDT	–	Recessive	$IV + m$	Erdman 1970
		Lindane	Multifactorial	Partial	$II, IV, V,$ $VIII+I(=X)$	Champ and Campbell-Brown 1969
				Recessive	III	"
		Malathion (specific)	Single factor	Partial	$VI+I(=X)$ VIII	Champ and Smith, unpublished
		Malathion (non-specific)	Single factor	Partial	$VI+I(=X)$ VIII	Champ and Smith, unpublished

In 1940, Serebrovsky suggested use of chromosomal translocations as a basis for reducing the reproductive potential of *Musca domestica* and *Sitophilus granarius*. Subsequently there has been considerable interest in suppressing insect populations by the introduction of these and other chromosomal aberrations or mutations. Examples include the use of recessive or conditional lethal mutations and non-lethal but otherwise deleterious mutations involving sex-distorting mechanisms, or inducing sterility through cytoplasmic incompatibility of different geographic races, or less promisingly through interspecific incompatibility from artifically-induced mating from use of pheromones or other normal-behaviour disrupting agents. Applications in stored-product entomology have been few. Sokoloff (1964b, 1966a) has listed numerous lethal and semi-lethal mutations in *T. castaneum* and *T. confusum* (Section F.15.3,4) including a dominant synthetic lethal in *T. castaneum* formed by the combination of *Fta* and *Sa* (Linkage group VII). Takahashi et al. (1968, Takahashi and Kuwahara 1970) described a temperature-conditional mutant in *Ephestia cautella* (Section F.15.12) and Shaw (1967b, 1968) has reported supernumerary y chromosomes in *Dermestes frischii* and *D. maculatus* (Section F.15.10) which cause significant distortions in sex ratios in these species. Although the movement of stored-product insects in commerce reduces the likelihood of the genetic isolation that will result in extensive cytoplasmic incompatibility between races from different localities, trade channels are sufficiently well-defined to allow a reasonable possibility of its occurrence. Ahmed et al. (in press) have reported infertility in eggs from crosses of strains of *Ephestia cautella* from the U.S.A. and Iraq that has been tentatively attributed to this cause.

Replacement of insecticide-resistant strains (which are usually less fertile) with susceptible strains by introduction of genes contributing to susceptibility has been suggested. The population to be controlled is swamped with susceptible insects, preferably males only, at a time when commodity damage would be minimal, and pesticides used when the susceptibility genes are integrated into the population (Wool 1971, 1975; Brower 1975a). The possibilities, however, for this and the other more sophisticated methods of genetic control of stored product pests appear limited in the light of present knowledge and the complex of species usually involved directly or ready to move into the ecological niche created by the removal of individual species. In principle also, the addition of further insects to an infestation in a consumable commodity does not appeal to those storing or handling a commodity and particularly not to buyers. Genetic engineering is, as yet, a dream in most stored product species with the possible exception of *Tribolium castaneum* and *Dermestes maculatus*. Nevertheless it has been pointed out rightly by Brower (1975a) and Wool (1975) who have recently reviewed the potential for genetic control of stored product insect populations, that developmental work should be continued and the methods evaluated objectively as a practical proposition. The methods currently available may not be adequate but the next generation of techniques may well be useful and irrespective there will be considerable benefits accruing to our basic knowledge of the genetics of the stored product group of insects.

(F.16) *Ionising radiations*

The use of ionising radiation has been proposed as a counter to pesticide resistance involving physical control methods that should operate independently of changing levels of pesticide tolerance and should induce a predictably stable response from exposed individuals. The relevant aspects are thus whether insects can develop resistance to these radiations and whether there is any relationship between tolerance to pesticides and tolerance to radiations.

Effects of ionising radiations on stored product insects have been studied extensively both in relation to sterile male release techniques and more particularly, general disinfestation of grain and similar products. Most attention in sterile-male release investigations has centred on the phycitid Lepidoptera which, *prima facie*, appear suitable subjects for this technique. Bull and Wond (1963) have suggested use of the method for *Ephestia kuehniella* and Amuh (1971) for *Ephestia cautella*, while Ahmed et al. (1972) and Ashrafi et al. (1972) have discussed the applicability to the group generally of a technique based on inherited partial sterility, which is essentially an extension of the sterile-male release method in which the lethal effect is manifested largely by induction of multiple translocations in succeeding generations. Although the sterile-male release method has been considered for other stored product species (Cornwell et al. 1966, Pradhan et al. 1971), the prospects, as for genetic control, do not appear good particularly among the Coleoptera because of the potential increase in contamination of commodities with the insects, and because of their dispersal characteristics. The use of chemosterilants to produce sterile males has also received attention but the risks of contamination present severe difficulties for practical applications in stored products.

Radiation resistance has been reported from *Aedes aegypti* (Terzian and Stahler 1966, Stahler 1971), *Drosophila melanogaster* (Ogaki and Nakashima-Tanaka 1966, Parsons et al. 1969, Nothel 1970), and *D. nebulosa* (Marques 1973). In stored product insects selection for resistance to gamma irradiation has not resulted in any change in sensitivity to acute doses of irradiation despite exposure to substerilising doses for two generations with *Sitophilus granarius* (Cornwell and Morris 1959), for seven generations with *Plodia interpunctella* (Hossain et al. 1972a), and *Callosobruchus maculatus* (Hossain et al. 1972b) and for six generations with *Sitophilus oryzae* and *Tribolium castaneum* (Brower et al. 1973). Brower (1974h,a,c,d) also reported similar findings for *S. oryzae*, *T. castaneum*, *P. interpunctella* and *C. maculatus* after 25, 25, 30 and 30 generations respectively of selection, though finding that the selected strains showed decreases in reproductive capacity. These decreases in genetic fitness agreed with results of Bartlett et al. (1966) for *T. castaneum*, Ashrafi et al. (1972) for *P. interpunctella* and *Callosobruchus maculatus* (Hossain et al., op. cit.), but conflicted with increases in fitness reported for *T. castaneum* by Erdman (1966) and for *T. confusum* by Crenshaw (1965).

As with pesticides, individual species have characteristic levels of tolerance of radiation. These levels have been shown to vary among geographically-isolated strains of *T. castaneum* and *T. confusum* (Soliman 1973) and strains of these species reared under different environmental conditions particularly of temperature (Ducoff and Bosma 1967, Yang and Ducoff 1969). The rate of recovery from radiation-induced damage was also temperature-dependent (Yang and Ducoff 1971). In addition to the intrinsic variability in response of individuals to exposure to irradiation, differences in sensitivity occur between the developmental

stages. These have been demonstrated for example for all stages of
Trogoderma granarium (Nair and Rahalkar 1963) including the influence of
diapause (Rahalkar and Nair 1968), *T. glabrum* (Tilton et al. 1966),
T. inclusum and *T. variabile* (Brower and Tilton 1972a), *Gibbium psylloides*
Brower and Scott 1972), *Sitophilus zeamais* and *S. granarius* (Brown et al.
(1972), *Oryzaephilus surinamensis* and *O. mercator* (Brower and Tilton 1972b),
Tenebroides mauritanicus and *Cryptolestes pusillus*(Brower and Mahany 1973),
Tribolium castaneum and *T. confusum* (Erdman 1962), *T. castaneum* and
T. madens (Brower and Tilton 1973), *Cathartus quadricollis* (Brower 1974e),
Gnathocerus maxillosus (Brower 1974f), *Latheticus oryzae* (Brower 1975b), and
Ephestia cautella (Cogburn et al. 1973) while Chawla et al. (1973) have
compared the sensitivity of larvae, pupae and adults of *Oryzaephilus
surinamensis* and Huque (1971) has compared the sensitivity of eggs and
first stage and fully-grown larvae of *Corcyra cephalonica*. Sensitivity to
radiation may change also with age within the developmental stage.
Differences with age have been reported from eggs of *T. castaneum* (Brown
and Davis 1973), *T. confusum* (Vereecke and Pelerents 1969), *Tenebrio
molitor* (Po-Chedley 1969, Brower 1972a), *Gibbium psylloides* (Brower 1972b),
Plodia interpunctella (Brower 1974g), and *Ephestia cautella* (Gonen and
Fishbain 1974), larvae of *T. molitor* (Menhinik and Crossley 1968 though not
by Jayaraman and Ducoff 1970), pupae of *T. castaneum* and *T. confusum*
(Soliman 1972, 1973) and pupae and adults of *T. confusum* (Ducoff and
Bosma 1966 and Ducoff 1967 respectively). Differences between males
and females in sensitivity to radiation also occur, for example *Trogoderma
granarium* (Carney 1959), *T. inclusum* and *T. variabile* (Brower and Tilton
1972a), *Tenebrio molitor* and *T. obscurus* (Brower 1973b), *Oryzaephilus
surinamensis* (Brower and Tilton 1972b), *Tribolium castaneum* and *T. madens*
(Brower and Tilton 1973), *Cathartus quadricollis* (Brower 1974e),
Latheticus oryzae (Brower 1975b) and *Ephestia cautella* (Cogburn
et al. 1973) where females are more sensitive than males, while responses
are similar in *Gibbium psylloides* (Brower and Scott 1972) and in
Gnathocerus maxillosus (Brower 1974f).

Irradiation techniques are rarely used at present for disinfestation
of commodities infestable with stored grain pest species. However, should
developments allow extensive commercial use of the method which from the
trend of present work presumably would involve sterilising rather than acute
exposures, some form of resistance would seem inevitable as an expression of
the inherent variability between individuals and as evidenced by the presence
of mechanisms which can confer differences in sensitivity between stages of
development, sexes and with age, and as radiation resistance is known already
from non-storage species. Nevertheless, the extensive selection work of
Brower, Tilton and Hossain of the U.S. Department of Agriculture at Savannah,
indicates that resistance may not appear readily. The limited pool of
genetic variability necessarily available in their laboratory
investigations, however, must be taken into consideration as a restriction
on the probability of their selecting resistance to exposure to radiation.

Though *prima facie* the presence of chemical insecticide resistance
would not be expected to result in cross-resistance to irradiation, it remains
possible that there are protective mechanisms associated with non-specific
pesticide resistances that may incidentally lessen damage resulting from
radiation exposure. Nevertheless, the known radiation protective mechanisms
such as changes in sulphydryl activity appear unlikely to be associated with
chemical pesticide resistance other than fortuitously. Pesticide resistance
is commonly associated with changes in development rates again fortuitously,

though sometimes determinatively when, for example, a laboratory culturing method is superimposed that selectively favours early or late developing individuals that have linkages between rate of development and resistance. When there are differences between strains in rates of development, particularly reproductive development, or when rates of such development show marked increase in variability within populations, these may be expressed as changes in sensitivity to substerilising doses of irradiation at particular times of treatment. In laboratory work particularly, there may thus be indirectly correlated association between pesticide resistance and radiation sensitivity and selection may occur and frequencies increased within the population of the more radiation tolerant individuals.

Brower (1974b) found that gamma irradiation (2 krad) of adults of unspecified age of a susceptible strain of *T. castaneum* and one resistant to malathion and DDT did not result in any proportionate differences in number of progeny from a 14 day oviposition period. Similarly, he showed that the presence of malathion resistance did not affect progeny production in 2 strains of *Plodia interpunctella* (Brower 1973a), whereas Erdman (1966, 1970) in Washington found that a DDT-tolerant strain of *T. castaneum* was more sensitive than a susceptible strain and Lüers (1963) reported that a strain of *Drosophila melanogaster*, resistant to DDT and hydrogen cyanide was less sensitive to x-rays than a susceptible strain. These differences in sensitivity may well represent minor strain differences not connected with the resistance to chemicals in the strains used.

G.   THE GENERAL PROBLEM OF RESISTANCE IN STORED CEREAL INSECTS

The survey was instituted as a direct result of the concern about the emerging pesticide resistance problem in cereal grain storage which was expressed by delegates from a number of countries during the 1969 FAO Symposium on Pesticide Resistance in Pests of Agricultural Importance. The basic objective was to define in general terms the extent and perspective of the problem.

Insecticide resistance was known to be prevalent and increasing in many of the storage pest species. Although this resistance was of considerable economic significance and constituted arreal threat to safe storage of cereals, fumigation in many instances provided an effective answer to the problem. Resistance as a problem of fumigation of stored products in the field was virtually unknown.

During 1972-1973, 61 countries were visited to collect strains of the major stored grain pests for pesticide resistance monitoring. A total of 799 samples was collected and these, together with strains from a further 25 countries were cultured at the Pest Infestation Control Laboratory, Slough, before testing for resistance to malathion, lindane, methyl bromide and phosphine. A total of 1687 strains of 8 species were tested.

(G.1)   *Summary of resistance status of individual species*

The main findings of the survey with respect to individual species are summarised in Tables G1 and G2 - the latter includes earlier records also.

(G.1.1)   *Sitophilus oryzae*

S. oryzae is the most important storage pest of raw cereal grains in the world and any occurrence of resistance to pesticides used in its control is a matter of considerable consequence and can have serious economic implications.

Malathion resistance, already known from Australia, England, India and China (Taiwan), was detected from a further 7 countries in South America, Africa and Asia. Thirty-three of the 257 strains tested showed increases in tolerance which were generally of a low order (to x8). The resistance appeared to involve cross-resistance to a wide range of compounds and taking into account the major role of this species as a storage pest in most warm cereal producing areas of the world, and the wide usage of malathion for its control, the resistance must be regarded as a significant though still limited problem of increasing importance.

Lindane resistance was much more widespread occurring in 53 of the 58 countries sampled and 75% of the 235 strains tested compared with records from 7 countries previously. It is apparent that an economic problem has existed with lindane resistance for at least 15 years and the general occurrence of the resistance has been offset to a large degree by replacement with materials such as malathion.

TABLE G1 : SUMMARY OF WORLD OCCURRENCE OF RESISTANCE TO MALATHION, LINDANE, METHYL BROMIDE AND PHOSPHINE IN THE MAJOR PESTS OF STORED CEREALS.

	Sitophilus oryzae	*Sitophilus zeamais*	*Sitophilus granarius*	*Rhyzopertha dominica*	*Tribolium castaneum*	*Tribolium confusum*	*Oryzaephilus surinamensis*	*Oryzaephilus mercator*	Totals	Totals (as %)
MALATHION										
No. countries sampled	59	52	28	50	78	33	53	19	86	-
No. with resistance	10	5	4	23	75	27	10	3	78	92%
No. of strains tested	257	203	163	158	505	122	183	36	1627	-
No. resistant	33	6	14	50	439	78	14	3	637	39%
MALATHION + TPP										
No. countries sampled	57	52	28	50	79	33	52	19	86	-
No. with resistance	6	2	4	1	60	15	10	3	60	71%
No. of strains tested	218	199	162	158	504	119	181	36	1575	-
No. resistant	9	2	13	1	226	38	14	3	306	19%
LINDANE										
No. countries sampled	58	48	28	51	76	34	52	18	86	-
No. with resistance	53	46	14	41	75	23	46	11	82	96%
No. of strains tested	235	186	157	137	497	94	152	34	1492	-
No. resistant	176	146	35	91	463	38	112	17	1078	72%
METHYL BROMIDE										
No. countries sampled	53	37	24	45	69	30	23	8	83	-
No. with resistance	1	6	0	1	6	15	1	0	21	26%
No. of strains tested	140	109	98	93	288	92	60	14	894	-
No. resistant	1	6	0	1	8	25	1	0	42	5%
PHOSPHINE										
No. countries sampled	51	37	23	46	69	30	23	10	82	-
No. with resistance	6	0	8	11	13	14	1	0	33	41%
No. of strains tested	135	105	85	94	267	92	58	13	849	-
No. resistant	8	0	8	22	15	28	1	0	82	10%

TABLE G2 : KNOWN OCCURRENCE OF PESTICIDE RESISTANCE IN PESTS OF STORED PRODUCTS (vide Section F3-F12)

	Organochlorine compounds			Organophosphorus compounds			Carbamates	Pyrethroids	Juvenile hormones	Fumigants				
	DDT	Lindane/HCH	Dieldrin	Malathion	Dichlorvos	Other				Methyl bromide	Ethylene dibromide	HCN	Phosphine	Other
COLEOPTERA														
Sitophilus oryzae	F	F	F	F	F		F	F				L	F	
Sitophilus zeamais		F		F						F				
Sitophilus granarius	F	F	L	F	F		L	F		L	L	L	F	L
Rhyzopertha dominica		F		F	F									
Trogoderma granarium				F										
Tribolium castaneum	F	F	F	F	F		F	F	F	F	F	F	F	
Tribolium confusum	F	F		F			F			F	F	F	F	L¹
Latheticus oryzae				F										
Oryzaephilus surinamensis		F		F	F		F	F						
Oryzaephilus mercator		F		F	F									
Dermestes maculatus		F		F										
Tenebroides mauritanicus										L				
Caryedon serratus	F	F												
LEPIDOPTERA														
Sitotroga cerealella		F												
Ephestia cautella				F	L		F							
Plodia interpunctella				F			F							
Tineola bisselliella		L²												
Tinea pellionella		F												
Phthorimaea operculella	F	F	F											
ACARINA														
*Acarus siro**		F												
*Glycyphagus destructor***		F												

¹ Methyl formate   L - Laboratory   * Wilkin (1973, *et al.* 1976a,)
² Endrin, chlordane   F - Field   ** Wilkin *et al.* (1976c)

Methyl bromide resistance was not demonstrated conclusively from
S. oryzae whereas increases in tolerance up to x2.5 were recorded for
phosphine. These changes in tolerance were recorded in 8 strains from
6 countries and although resistance factors in excess of those recorded
will be necessary for a general breakdown of control, an emerging problem
with phosphine resistance is apparent.

(G.1.2) *Sitophilus zeamais*

S. zeamais is of less overall importance than *S. oryzae* but has
particular significance as a pest of maize and rice in developing
countries where physical loss or damage to the commodity may be of greater
consequence than reduction in the value of the grain for trading purposes.

Malathion resistance was detected in only 6 of the 203 strains
tested supplementing a single previously-known record from the United
Kingdom. The problem is as yet very restricted but presumably will
increase.

Lindane resistance, previously reported from the Caribbean, Brazil,
East Africa and the Philippines, was detected in 46 of the 48 countries
sampled and 146 of the 186 strains concerned. Lindane resistance in
S. zeamais appears to have developed later than in *S. oryzae* but nevertheless
is widespread and as in *S. oryzae*, can be assumed now to be a normal
attribute of *S. zeamais* as well as *S. oryzae* populations.

Changes in tolerance of methyl bromide were found in 6 strains from
6 countries. The maximum increase was approximately x1.4. The changes
were marginal and thus resistance does not appear currently to be a problem
of any consequence in *S. zeamais*.

There were no records of change in tolerance of *S. zeamais* to
phosphine either in the laboratory or in the field and it appears that
phosphine resistance like malathion resistance does not develop readily in
this species

(G.1.3) *Sitophilus granarius*

S. granarius is principally a pest of wheat, barley and related cool
temperate grain crops. Although this concerns particularly the major
producers of Eurasia such as France, other European countries and the
U.S.S.R., residual chemicals are not used extensively and fumigation is the
major control method.

Malathion resistance, previously known only from Australia and
England, was detected in 14 strains only of 163 tested involving two
additional countries, Argentina and Greece. Eight of these resistant
strains came from Australia. Resistance levels were not high reaching
a maximum level of approximately x3 which nevertheless would have been
sufficient to allow survival from contact with fresh deposits of malathion.
Although malathion resistance is not presently a significant problem in
S. granarius, an emerging problem is apparent particularly in the warmer
areas of its occurrence.

Lindane resistance was more widespread but had a distribution pattern
similar to malathion resistance occurring mainly in the warmer areas where

there was overlap with *S. oryzae* and lindane was used regularly in control programmes. Thirty-five of the 157 strains tested were resistant involving 14 of the 28 countries sampled. The levels of resistance were generally low and reflect the general unimportance of lindane in control of *S. granarius*.

Methyl bromide resistance was not detected and 8 strains only of 85 tested showed increases in tolerance to phosphine. The changes in tolerance were small and although high-level resistances have been demonstrated in the laboratory, it appears that economic resistance to these fumigants in *S. granarius* is not an immediate problem.

(G.1.4) *Rhyzopertha dominica*

R. dominica reaches its maximum importance in warm, dry areas particularly when grain is left undisturbed in storage for long periods. It is considered the third most important pest of raw cereal storage after *Sitophilus oryzae* and *Sitotroga cerealella* and it is with *R. dominica* that the most serious problems of pesticide resistance are emerging.

Malathion resistance was diagnosed in 23 of the 50 countries sampled and 50 of the 158 strains tested compared with previous records from Australia and England only. The levels of resistance encountered have generally not been high (to x5) but recently strains have been detected with resistance factors to malathion of x150 and x16 resistance to dichlorvos giving rise for new concern as to the potential of organophosphorus resistance in *R. dominica*.

Lindane is a particularly effective material against *R. dominica* but is not used commonly in control programmes against this species. Although resistance was detected in 91 of the 137 strains tested involving 41 of the 51 countries samples, it was not considered of particular significance.

A change in tolerance of methyl bromide was detected in one strain only of 93 tested whereas phosphine resistance was found in 22 strains of 94 tested involving 11 countries in most of which phosphine-resistant strains of other species were collected from the same localities. Resistance levels to phosphine reached a maximum of x12. It appears that there will be in the near future, if there is not at present, difficulty in controlling *R. dominica* with phosphine.

(G.1.5) *Tribolium castaneum*

T. castaneum is the commonest stored product insect in the world. Although essentially a warm climate species, it moves very actively in world trade and will survive in colder countries in heated premises. It causes minor damage only to raw cereal grains but infestations can be heavy when dockage is present and it is the primary pest of milled products.

Both malathion and lindane resistance are widespread and were found in 87% and 93% respectively of the approximately 500 strains tested with these materials. It is reasonable now to accept that these resistances will be present in the species wherever it is found and control programmes must be designed accordingly. This does not imply that either material will immediately fail to control infestations of *T. castaneum*, rather that a high probability of resistance exists and difficulties in control may occur frequently. The malathion resistance in approximately half the strains was specific to malathion-type compounds and this type of resistance appeared to be a major component also of resistance in strains showing a non-specific type of resistance. The distribution of these different types of resistance indicates that there remains a considerable part of the world where many organophosphorus compounds unrelated to malathion will remain effective against *T. castaneum* even when non-specific malathion resistance is present.

Changes in tolerance to methyl bromide were found in 8 strains of the 288 tested. The factors for change in tolerance were low (x1.2). However, if the remarkable uniformity of response to methyl bromide of geographically-isolated strains of *T. castaneum* is taken into account together with the small margins between dose rates used and minimum effective dose rates for immature stages, such levels of change of tolerance approach significance in the control of *T. castaneum*. There is already a documented occurrence of control breakdown with ethylene dibromide involving a x2 change in tolerance.

Phosphine tolerance in *T. castaneum* is increased readily in the laboratory whereas in the survey, 15 strains only of 267 tested showed increases in tolerance the highest being x3 normal. The mobility and interchange in strains of *T. castaneum* together with less intensive fumigation pressure from the acceptance of light infestations, probably have delayed expression of phosphine resistance. However, the ubiquitous nature of *T. castaneum* and a general increase in use of phosphine, indicate that an economic problem with resistance is inevitable.

(G.1.6) *Tribolium confusum*

T. confusum tends to replace *T. castaneum* in cooler areas and is usually exposed to considerably less pressure from pesticides. It is, however, less mobile and confined more to milling premises where regular spot treatments are carried out particularly with fumigants.

Compared with *T. castaneum*, malathion and lindane resistances were less common (64% and 40% of 122 and 94 strains tested respectively) reflecting the more limited use of these materials. Again approximately half the malathion-resistant strains showed a malathion-specific type of resistance. There is undoubtedly an economic problem in control of *T. confusum* with malathion but this appears to be restricted particularly to flour mills and similar premises in cooler countries. It is unlikely that the lindane resistance will have any particular significance since lindane is not used extensively in the areas where *T. confusum* occurs and a range of alternative controls is available.

T. confusum showed the highest frequencies of methyl bromide and phosphine resistances of all the species tested. It is problematical whether this reflected greater potential in this species for expression of the resistances, the restricted mobility of the species, or a tendency to frequent particular habitats where varied strains of insects accumulate and fumigants are used regularly and frequently (e.g. in flour mills which in effect are foci for grain movements from widely dispersed areas). Nevertheless 27% and 30% respectively of the 92 strains tested showed resistance to methyl bromide and phosphine. Changes in tolerance of methyl bromide were low (to x1.3 at the LD_{50} and 1.7 at the $LD_{99.9}$) but higher with phosphine (to x7 in survivors from discriminating tests). It is likely that a local problem exists in control of *T. confusum* with fumigants and that this will probably increase in the foreseeable future.

(G.1.7) *Oryzaephilus surinamensis*

O. surinamensis has a wider distribution than *T. castaneum* but is considerably less common. It does not appear to develop resistance to pesticides with the same ease as other species and although lindane resistance is prevalent (74% of 152 strains tested involving 88% of countries sampled) there are few occurrences of other resistances. Malathion resistance, known previously from unpublished data from Australia and Israel, was detected from 14 strains only of 183 tested involving 10 countries. This limited occurrence, although involving resistance levels up to x40 normal, has not created widespread problems as yet but it is a matter of considerable concern to those cool temperate countries such as the United Kingdom where *O. surinamensis* is the most significant pest of cereal grain storage. Lindane has been a major insecticide for controlling *O. surinamensis* but as there is general susceptibility to organophosphorus compounds, these have tended to replace lindane.

Single strains only showed increases in tolerance to methyl bromide and phosphine.

(G.1.8) *Oryzaephilus mercator*

O. mercator is not a major pest of cereal storage as reflected by the few samples available for testing in the survey. Malathion resistance, already reported from Africa, was found in North America and the Caribbean. The major significance of this resistance appears to be in commodities other than cereal grains, as for example in peanuts in West Africa. Lindane resistance was found in half of the strains tested.

Changes in tolerance of methyl bromide and phosphine were not detected.

(G.1.9) *Other species*

Although the monitoring aspects of the survey were confined to the species listed above, concern must be expressed at the emergence of malathion and pyrethrin resistance in the phycitid moths. These pests, although not approaching the importance of the beetle pests, are a major consideration in pest control programmes and require commensurate attention.

Sitotroga cerealella ranks very high as a pest particularly in developing countries. As insecticides usually are not used intensively for its control and in many circumstances infestations do not persist in storage, resistance already reported for lindane, does not appear to be a problem. A careful examination of its pest status and pesticide tolerance levels are needed before the significance of the resistance can be assessed.

Reports of resistances in other grain-infesting species are few and the localised occurrences are not a matter of immediate concern. However, a close watch should be maintained for any change in the pest status of species resulting from resistance.

(G.2) *The current status of the pesticides concerned in the survey*

The number of strains from each country and the types of situation concerned have been extremely varied. The information obtained on the occurrence of resistance is essentially qualitative, that is, the survey indicates only those countries where resistance occurs and does not define the situation within countries. More extensive sampling would reveal resistances in many countries that have not yet been detected and may change the pattern in those countries where it has been recorded. Nevertheless certain general patterns are evident.

(G.2.1) *Lindane*

Lindane resistance was found to occur in nearly all countries. It was found in all 8 species and in 82 of the 85 countries sampled involving 1077 of the 1490 strains tested with this material. It was most prevalent in *Tribolium castaneum*, the commonest pest in many parts of the world, occurring in 93% of the 497 strains tested and all but one of the 76 countries sampled. This resistance was prevalent also in the other warm-climate species - *Sitophilus zeamais* 79% of strains tested, *S. oryzae* 74%, *Oryzaephilus surinamensis* 74% and *Rhyzopertha dominica* 66% - reflecting the more intensive use of lindane in warm areas. Such use would involve *S. zeamais* and maize and rice particularly, but to a much lesser extent *R. dominica* against which lindane is not used as frequently in the drier areas where it is a significant pest such as through the Middle East. Moreover, normal strains of *R. dominica* are considerably more susceptible to lindane than any of the other species (Table F2) thus giving a greater margin between applied dosages and minimum lethal dosages which would also retard development of resistance. The frequency of occurrence of resistance in *O. surinamensis* would be influenced by its wider distribution which extends into cool areas where lindane use would be minimal. Lindane resistance was less common also in the cool-climate species *T. confusum* and *S. granarius*. With *T. confusum* whose distribution reaches into the warm areas the frequency of occurrence of this resistance was 41% whereas with *S. granarius* which cannot become established in warm areas the frequency dropped to 22%.

Lindane was introduced into agriculture and stored product pest control in the immediate post-war years replacing DDT. It was used very widely, often carelessly, to disinfest infested grain and grain residues by admixture and surface treatment as well as for residual application to structures and for prevention of reinfestation. The long residual life of lindane and a frequent association with conditions of poor hygiene, allowed extensive selection to occur as residue levels decayed slowly and imposed a range of selection pressures which would have included extended periods at pressures optimum for selection of resistance in the species concerned. These circumstances have prevailed in many areas particularly where lindane usage included admixture with feed and food grains, and in seed dressings which are still in use today. The extensive movement of these resistances through trade channels is now well documented both locally and internationally (Sections F.3.1, F.7.1, F13).

It is not known to what extent selection for DDT resistance predisposed to lindane resistance and it was not until the late 1950's that lindane resistance began to emerge as a significant problem. The presence of lindane resistance does appear in some circumstances to have some relation to changes in tolerance of organophosphorus compounds (Section F.14.3). This can complicate the interpretation of results from discriminating tests with field-collected material if doses are set close to reference $LC_{99.9}$

levels. With *Sitophilus oryzae*, Champ (1967) found that increasing levels of DDT and lindane resistances were associated with an increase in low-level tolerance to malathion. It has also been observed that there is a correlation between survival in discriminating tests for malathion resistance in field-collected material and level of lindane resistance and that this tolerance is not carried through to the F_1 progeny of field strains.

Despite the occurrence of lindane resistance in practically every country, its active movement through trade channels and the high probability that strains encountered would be resistant, lindane is still used widely in stored product pest control. It has many advantages - activity against a wide a wide range of beetle and moth pests at relatively low dose rates, comparatively low mammalian toxicity, extreme persistence under a wide range of adverse conditions including alkaline surfaces and the presence of heavy metals, a low but significant vapour pressure combined with repellency to some species, and minor effects only on seed germination. There has not been any replacement pesticide developed with similar properties although pirimiphos-methyl, an organophosphorus compound with a considerably lower mammalian toxicity does have many of the essential features of lindane. Pirimiphos-methyl is toxic at economic dose levels to a wide range of storage pests including some strains resistant to malathion and some Lepidoptera, is very stable in admixture with grain and has a vapour effect but like other organophosphorus compounds its persistence is reduced on alkaline surfaces and it is not compatible with fungicides containing heavy metals.

(G.2.2) *Malathion*

Whereas lindane is still used widely, other chemicals particularly malathion have supplanted it in many applications including usages where direct contact with foodstuffs may occur. The occurrence of lindane resistance has thus created difficulties but in general terms these could not be regarded as serious restrictions in practical pest control as long as malathion remained effective. The industries have become dependent, however, on malathion to maintain the low tolerances of insect infestation required by markets today and the emergence of resistance to this material has much more serious consequences.

Malathion resistance was found in the survey to be almost as widespread as lindane resistance, occurring in 78 of the 86 countries samples and involving 637 of 1627 strains tested with this material. This resistance was present in all species tested but was most prevalent in the *Tribolium* spp. and *Rhyzopertha dominica*. It was found in 439 of the 505 strains of *T. castaneum* tested and was detected in all but three of the 78 countries sampled. Moreover, these three countries, which were all in Europe, were represented by only single samples of this species. Approximately half of these malathion-resistant strains of *T. castaneum* were resistant to malathion synergised with triphenyl phosphate (TPP) indicating a wide range of tolerances to other organophosphorus compounds. The situation with *T. confusum* resembled that with *T. castaneum* but frequencies of occurrence of the resistances were lower. Again resistance to TPP-synergised malathion occurred in approximately half of the resistant strains although this involved proportionately fewer countries. The ubiquity of *T. castaneum* and its active movement in infested commodities in commerce would account for these occurrences. Fortunately *T. castaneum* in terms of actual loss of foodstuffs particularly grain, is not a serious pest. The major losses occur through contamination by the physical presence of the insects, the resultant adverse consumer reaction and the cost of measures to prevent this.

The situation with the other species is potentially more serious. Malathion resistance was found in 50 of 158 strains of *R. dominica* involving half of the 50 countries concerned. In all but one instance, the resistance was suppressed with TPP indicating for this species that there was not any high level resistance to other organophosphorus compounds. The now documented occurrence of dichlorvos resistance in Australia (Sections F.6.1, F.6.4a, Table F25) gives notice of considerably greater problems in control programmes. *R. dominica* usually exhibits considerable variation between individuals in responses to organophosphorus compounds, a factor which has played a major role in expediting expression of resistance in this species. The cross-resistances shown by these dichlorvos-resistant strains indicate that most organophosphorus compounds can be discounted for use against them (Table F25, Fig. G1). Resistance to pirimiphos-methyl was not present but this material was not particularly effective against normal *R. dominica* and did not prevent development of progeny at normal application rates. Malathion resistances in *Sitophilus* spp. and *Oryzaephilus* spp. did not respond to TPP, and moreover, where tests have been made, all strains have shown a wide range of apparent cross-resistances to other organophosphorus compounds (Tables F24, F26, Fig. G1). The frequencies of occurrence in these two genera were comparatively low involving 10 countries each for *S. oryzae* and *O. surinamensis*, 5 for *S. zeamais*, 4 for *S. granarius* and 3 for *O. mercator*. The importance of these species in terms of potential loss or damage to cereals (Section D) indicates, however, that these low frequencies of malathion resistance can be interpreted only as a portent of a worsening problem not only with malathion but also with many of the replacement chemicals introduced. It must be recognised also that in field infestations, most of the pests occur as mixtures of species and that the presence of resistance in any one species can become the operative factor in necessitating modified pest control programmes.

(G.2.3) *Methyl bromide*

The emergence of resistance to fumigants under practical conditions is another matter of serious concern. Forty-two strains (5%) of the 894 tested from all species showed increases in tolerance of methyl bromide involving 21 (26%) of the 83 countries from which samples were taken; these increases in tolerance were generally of a low order but reflected a situation developing in practical fumigation in which the levels of tolerance accepted as normal for particular species are no longer applicable. All species except *S. granarius* and *O. mercator* were concerned although the increases in tolerance were recorded for single strains only of *S. oryzae*, *R. dominica* and *O. surinamensis*. Most records (60% of total) were from *T. confusum* which although a widely-distributed species, is essentially a pest of flour mills and similar establishments in cool countries.

It is now over 40 years since Le Goupil (1932) discovered the insecticidal properties of methyl bromide and during much of this time, the material has been used extensively for a wide range of fumigant applications under a wide range of conditions, many of which would probably have been conducive to development of resistance. Strains carrying a high level of resistance have been demonstrated in the laboratory for *S. granarius* (Section F.5.2) although resistance in this species was not detected in the

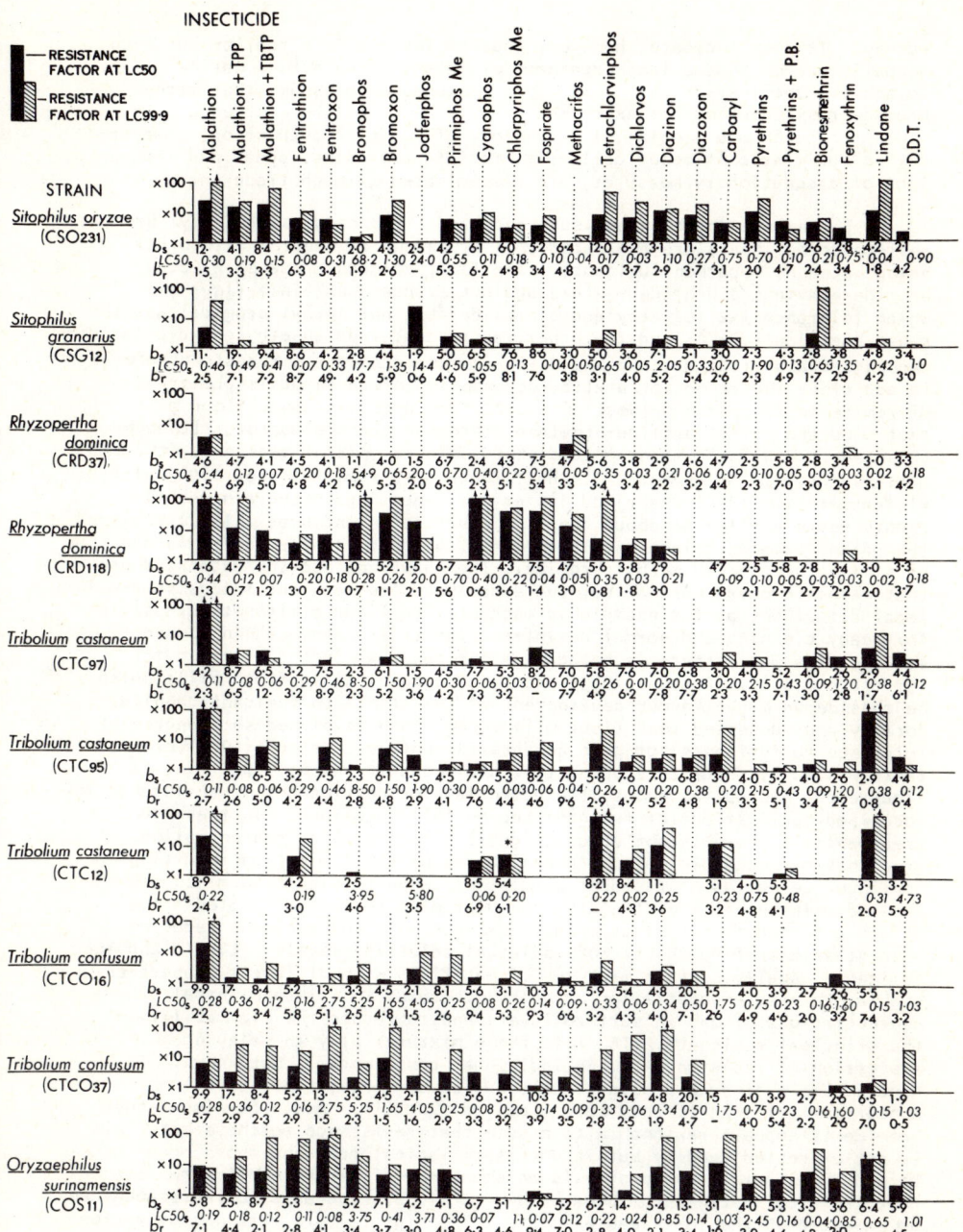

FIG. G1 : The resistance patterns of representative strains carrying the organophosphorus-compound resistances known to be present in stored grain beetles in Australia. (Champ and Turner unpublished data).

survey. It does indicate, however, that resistance is a reality but with
a qualification that a long programme of intensive selection was necessary
to achieve the resistance. Thus it is reasonable to assume that methyl
bromide resistance is not expressed readily and that the emergence of a
widespread economic problem will be slow. The latter assumption, however,
may be optimistic if genetically-fit, resistant strains appear and through
lack of precautionary measures, are disseminated through trade channels.

There have not been any substantiated field reports of methyl bromide
resistance but there are two records of breakdown of control with other
halogenated hydrocarbon fumigants. The first concerned use of a methyl
bromide-ethylene dibromide mixture against *T. castaneum* in a flour mill in
which tolerance (x2) of ethylene dibromide (but not methyl bromide) was the
cause (Section F.7.2), and the other report concerned an ethylene dichloride-
carbon tetrachloride mixture and *O. surinamensis* (Section F.9.2). There is
in addition, the report of a x1.8 increase in tolerance of ethylene
dibromide in *T. confusum* (Section F.8.2) from duct work in a flour mill
that used the methyl bromide-ethylene dibromide mixture monthly although
this report did not comment on the status of the control measure. Ethylene
dibromide became of commercial significance as a fumigant about 1945 although
Niefert *et al.* (1925) described its effectiveness against grain weevils
twenty years earlier at about the time when Cotton and Roark (1927) realised
the potential of ethylene dichloride as a fumigant. These materials are thus
of similar vintage to methyl bromide. They have not had the extensive use
that methyl bromide has had but there has been frequent and intensive use in
local situations as for example in machinery in flour mills not accessible
for ready cleaning and normal disinfestation treatments, and in wooden
storage bins where there is harbourage for residual infestations in the
structure itself. Cotton (1963) described conditions in flour mills which
he considered might favour development of resistance to fumigants applied
locally and suggested that flour mill insects had developed some degree of
tolerance to fumigants but gave no data in support. Griggs (1964) claimed that
there was no evidence to support this claim and it was not until the
occurrences listed above that the potential for development of resistance in
localised applications of fumigant became fully apparent. In these spot
treatments if there is not adequate gas-tight enclosure, concentration
gradients may occur in the peripheral areas of the fumigation resulting in
continued selection of resident populations when routine treatments are
applied. In the case of the ethylene dibromide resistance quoted above,
it was presumed that convection air currents were concerned in reducing
concentrations of fumigants to sub-lethal selecting levels. Liquid and spot
fumigants usually contain mixtures of materials with different properties
such as to ensure penetration of bulks of grain, to maintain lethal
concentrations of gas in surface layers and to reduce the flammability of
the principal toxicants. The role these mixtures play in delaying
expression of resistance to the individual components is not known. It is
conceivable that such a delay occurs but equally resistance to all may
appear earlier than if they were used individually. Again as with methyl
bromide it appears reasonable to assume that resistance to these materials
is not expressed readily but it does seem logical from the above occurrences
to add the rider to the prognosis on emergence of methyl bromide resistance
that isolated instances of breakdown directly attributable to resistance can
be expected in the foreseeable future. The high failure rate in fumigation
that currently exists due to inadequate techniques would tend to obscure
correct identification of causes and it will not be until adequate monitoring
of otherwise unexplainable fumigant failures is carried out, that a true
picture will emerge. Resistance is a common explanation of these failures
but field reports are equivocal - in this context a considerable number of
tests have been carried out on strains of different species that have been
collected after fumigations that failed for no other apparent reason, but in
every instance, the responses of the strains have been normal.

(G.2.4) *Phosphine*

The problem with phosphine resistance is more immediate and acute. There is little doubt that insects develop tolerance to phosphine readily and the manner of use of phosphine is such that inadequate techniques of fumigation are often employed. This misuse has been accentuated by recommendations for use that have been made which imply that a substantial degree of gas-tightness is not necessary - for example that sheeting may not be required for bulks of grain. Eighty-two strains (10%) of 849 tested from all species showed increases in tolerance involving 33 (41%) of the 82 countries sampled. Changes in tolerance were not detected in *S. zeamais* or *O. mercator* and there was a single record only from *O. surinamensis*. As with methyl bromide, the highest frequency of occurrence was from *T. confusum* (28 of 91 samples, 34% of total records) but in this instance, *R. dominica* had a comparable frequency (22 of 94 samples, 27% of total records). The levels of resistance recorded were up to x10-x12 in *R. dominica*. In most instances resistances were confirmed by breeding from survivors from discriminating tests and it is probable that although there are no substantiated field reports of resistance, these are occurring but as with methyl bromide, would be obscured by the comparatively high rate of failures of fumigations from other causes.

The occurrence of the higher levels of phosphine resistance appeared to be limited to a comparatively few localities from which in most instances resistance was recorded in a number of species often taken in the same sample. The principal areas concerned are shown in Table G3. There were common features in all these occurrences relating to modes of use of phosphine in the localities concerned or in the places of origin of imported grain (Sections F.3.4d, F.5.4d, F.6.4d, F.7.4.d, F.8.4d). These observations can be extended to include to a lesser extent malathion resistance also.

TABLE G3: OCCURRENCE OF PHOSPHINE RESISTANCE IN DIFFERENT SPECIES COLLECTED FROM THE SAME AREA
(Data given are percentage responses in discriminating tests)

Locality	S.oryzae	S.granarius	R.dominica	T.castaneum	T.confusum
Georgetown (Guyana)	91	-	5	94	-
Buenos Aires (Argentina)	100	97	61	100	98
Piraeus (Greece)	100	100	63	96	95
Tripoli (Libya)	100	-	9	100	-
Bangui (C.A.R.)	100	-	50	98	100
Bombay-Borivli (India)	90	-	3	88	-
Delhi area (India)	84	-	86	100	-

The occurrence and implications of the resistances summarised here are considered in detail in the relevant Sections (F3-F9).

H. REFERENCES

ADKISSON, P.L. (1957). - The relative susceptibility of the life history stages of the rice weevil to certain fumigants. *J. econ. Ent.* 50:761-764.

AHMED, M.S.H., AL-HAKKAH, Z.S. and AL-SAQUR, A.M. (1972). - Inherited sterility in the fig moth, *Cadra (Ephestia) cautella* Walker. In "Peaceful Uses of Atomic Energy". Int. Atom. Energy Agency, 4th int. Conf., Geneva, 1971. (I.A.E.A.: Vienna). p. 383.

AHMED, M.S.H., AL-HASSANY, I.A., LAMMONZA, S.R. and OUDA, N.A. (in press). - Crossing experiments with two *Ephestia cautella* strains. *J. econ. Ent.*

AITKEN, A.D. (1964). - Some insects of interest recorded on stored products during 1962. *Entomologist's mon. Mag.* 99:88,89.

AITKEN, A.D. (1966). - A strain of small *Oryzaephilus surinamensis* (L.) (Coleoptera, Silvanidae) from the Far East. *J. stored Prod. Res.* 2:45-55.

AITKEN, A.D. (1975). - Insect Travellers. Vol. 1. Coleoptera. *U.K. Min. Agric. Fish.& Food, Tech. Bull.* 31. (HMSO: London).

ALMEIDA, F.F. de (1958a). - Parabiosen zwischen verschiedenen Genotypen von *Plodia interpunctella* und *Ephestia kühniella*. *Z. Naturforsch.* 13b:617-621.

ALMEIDA, F.F. de (1958b). - Über die fluoreszierenden Stoffe in den Augen dreier Genotypen von *Plodia interpunctella*. *Z. Naturforsch.* 13b:687-691.

AMOS, T.G. and SCOTT, J.S. (1965). - Pearl eye in *Carpophilus dimidiatus* F. (Coleoptera, Nitidulidae). *Amer. Nat.* 99:421-423.

AMOS, T.G., WILLIAMS, P., DU GUESCLIN, P.B. and SCHWARZ, M. (1974). - Compounds related to juvenile hormone: activity of selected terpenoids on *Tribolium castaneum* and *T. confusum*. *J. econ. Ent.* 67:474-476.

AMUH, I.K.A. (1971). - Potentialities for application of the sterile-male technique to the control of the cocoa moth, *Cadra cautella* Walk. In "Application of Induced Sterility for Control of Lepidopterous Populations". Int. Atom. Energy Agency, Proc. Panel, Vienna, 1970. (I.A.E.A.: Vienna). p. 7.

ANON. (1958a). - Increased resistance of granary weevil to insecticides. *Pest Infest. Res.* 1957:23.

ANON. (1958b). - Variation in insect resistance. *Pest Infest. Res.* 1957:24.

ANON. (1958c). - Resistance of strains of *Tribolium castaneum* to lindane. *Pest Infest. Res.* 1957:47.

ANON. (1958d). - Groundnuts. *Rep. W. Afr. stored Prod. Res. Unit* 1957:17-32.

ANON. (1959a). - VIII Groundnuts. Infestation of stored groundnuts by *T. castaneum*. *Rep. W. Afr. stored Prod. Res. Unit* 1958:21,22.

ANON. (1959b). - The increased resistance of insects to insecticides. The grain weevil and pyrethrum. *Pest Infest. Res.* 1958:25.

ANON. (1959c). - Variation in insect resistance. *Pest Infest. Res.* 1958:25,26.

ANON. (1960). - Fumigants. *Pest Infest. Res.* 1959:30.

ANON. (1961). - Insect prevention and control in farmers stock peanuts. *U.S. Dep. Agric.* AMS:453.

ANON. (1965). - Appendix 1. *Infest. Control* 1962-64:77-87.

ANON. (1967). - Methods for the detection and measurement of resistance in agricultural pests. *Rep. 1st Session FAO Working Party of Experts on Resistance of Pests to Pesticides, Meeting Rep.* PL/1965/18. (FAO: Rome). 106 pp.

ANON. (1968). - International collaborative program for the development of standardised tests for resistance in important pests of agriculture. Item 5. *Rep. 3rd Session FAO Working Party of Experts on Resistance of Pests to Pesticides, Meeting Rep.* PL/1967/M/8. (FAO: Rome). pp. 9-11.

ANON. (1969a). - Recommended methods for the detection and measurement of resistance of agricultural pests to pesticides. 1. General principles. *Plant Prot. Bull. FAO* 17:76-82.

ANON. (1969b). - World survey for resistance. Item 1/10/8. *Rep. 4th Session FAO Working Party of Experts on Resistance of Pests to Pesticides, Meeting Rep.* PL/1968/M/10. (FAO: Rome). p. 22.

ANON. (1970a). - Recommended methods for the detection and measurement of resistance of agricultural pests to pesticides. Tentative method for adults of the red flour beetle, *Tribolium castaneum* (Herbst). FAO Method No. 6. *Pl. Prot. Bull. FAO* 18:107-113.

ANON. (1970b). - Pest resistance to pesticides in agriculture. Importance, recognition and countermeasures. AGP: CP/26. (FAO: Rome).

ANON. (1974a). - Recommended methods for the detection and measurement of resistance of agricultural pests to pesticides. Tentative method for adults of some major beetle pests of stored cereals with malathion or lindane. FAO Method No. 15. *Pl. Prot. Bull. FAO* 22:127-137.

ANON. (1974b). - Fumigants - Mode of action, use and residue analysis. *Res. Inst. London, Ont., Res. Branch Rep.* 1973, p. 186.

ANON. (1974c). - Fumigation with methyl bromide under gas proof sheets. Ministry of Agriculture, Fisheries and Food, Slough, England.

ANON. (1975). - Recommended methods for the detection and measurement of resistance of agricultural pests to pesticides. Tentative method for adults of some major beetle pests of stored cereals with methyl bromide and phosphine. FAO Method No. 16. *Pl. Prot. Bull. FAO* 23:12-25.

ARENDSEN HEIN, S.A. (1920). - Studies on variation in the mealworm *Tenebrio molitor* L. I. Biological and genetical notes on *Tenebrio molitor* L. *J. Genet.* 10:227-263.

ARENDSEN HEIN, S.A. (1924). - Studies on variation in the mealworm *Tenebrio molitor* L. II. Variations in tarsi and antennae. *J. Genet.* 14:1-38.

ARMSTRONG, J.W. and SODERSTROM, E.L. (1975). - Malathion resistance in some populations of the Indian meal moth infesting dried fruits and tree nuts in California. *J. econ. Ent.* 68:505-507.

ASHMAN, F. and HIGGS, G.A. (1968). - A horned strain of *Oryzaephilus surinamensis* (L.) (Coleoptera, Silvanidae) from the Mediterranean Region. *J. stored Prod. Res.* 4:203-211.

ASHRAFI, S.H., TILTON, E.W. and BROWER, J.H. (1972). - Inheritance of radiation-induced partial sterility in the Indian meal moth, *Plodia interpunctella*. *J. econ. Ent.* 65:1265-1268.

BAILEY, S.W. (1965). - Air-tight storage of grain; its effect on insect pests. - IV *Rhyzopertha dominica* (F.) and some other Coleoptera that infest stored grain. *J. stored Prod. Res.* 1:25-33.

BANG, Y.H. and TELFORD, H.S. (1966). - Effect of sublethal doses of fumigants on stored-grain insects. *Washington agric. Exp Sta., Tech. Bull.* 50.

BANSODE, P.C. (1974). - Studies on the development of resistance to malathion in *Sitophilus oryzae* (L.) *Ent. Newsl. (Indian agric. Res. Inst.)* 4:8.

BARKER, P.S. (1967). - Susceptibility of eggs and young adults of *Cryptolestes ferrugineus* and *C. turcicus* to methyl bromide. *J. econ. Ent.* 60:1434-1436.

BARKER, P.S. (1969). - Susceptibility of eggs and young adults of *Cryptolestes ferrugineus* and *C. turcicus* to hydrogen phosphide. *J. econ. Ent.* 62:363-365.

BARTLETT, A.C. (1967). - Genetic markers in the boll weevil. *J. Hered.* 58:159-163.

BARTLETT, A.C., BELL, A.E. and ANDERSON, V.L. (1966). - Changes in quantitative traits of *Tribolium* under irradiation and selection. *Genetics* 54:699-713.

BASKARAN, P. and MOOKHERJEE, P.B. (1971). - Effect of food on the susceptibility of *Cadra cautella* Walker and *Trogoderma granarium* Everts to Phosphine. *Indian J. Ent.* 33:23-29.

BELL, C.H. and GLANVILLE, V (1970). - Toxicity of phosphine to insects. Toxicity to moths. *Pest Infest. Res.* 1969:55-57.

BELL, C.H. and GLANVILLE, V. (1973a). - Provision and fumigation of diapausing moth larvae. *Pest Infest. Control* 1968-70:144,145.

BELL, C.H. and GLANVILLE, V. (1973b). - The effect of concentration and exposure in tests with methyl bromide and phosphine on diapausing larvae of *Ephestia elutella* (Hübner) (Lepidoptera, Pyralidae). *J. stored Prod. Res.* 9:165-170.

BELL, C.H. and WALKER, D.J. (1973). - Diapause induction in *Ephestia elutella* (Hübner) and *Plodia interpunctella* (Hübner) (Lepidoptera, Pyralide) with a dawn-dusk lighting system. *J. stored Prod. Res.* 9:149-158.

BENNETT, R.G. (1969). - The influence of age and concentration of fumigant on the susceptibility of pupae of *Tribolium castaneum* (Herbst) (Coleoptera, Tenebrionidae) to methyl bromide. *J. stored Prod. Res.* 5:119-126.

BENNETT, R.G. (1970). - Variation in susceptibility to fumigants between different populations. *Pest Infest. Res.* 1969:58,59.

BENNETT, R.G., DUFFIN, P. and GOODSHIP, G. (1969). - Toxicity of methyl bromide to pupae of *Tribolium castaneum*. *Pest Infest. Res.* 1968:63.

BERTAND, G., BROCQ-ROUSSEU and DASSONVILLE (1919). - Influence de la temperature et d'autres agents physiques sur le pourvoir insecticide de la chloropicrine. *Dokl. Akad. Nauk. SSSR.* 169:1059-1061

BHATIA, S.K. and BANSODE, P.C. (1971). - Studies on resistance to insecticides in *Tribolium castaneum* (Herbst). IV. Susceptibility of p,p'-DDT-resistant strains to some fumigants. *Indian J. Ent.* 33:45-49.

BHATIA, S.K. and PRADHAN, S. (1968). - Studies on resistance to insecticides in *Tribolium castaneum* (Herbst). I. Selection of a strain resistant to p,p'-DDT and its biological characteristics. *Indian J. Ent.* 30:13-32.

BHATIA, S.K. and PRADHAN, S. (1970). - Studies on resistance to insecticides in *Tribolium castaneum* (Herbst). II. Cross-resistance characteristics of the p,p'-DDT-resistant strain. *Indian J. Ent.* 32:32-38.

BHATIA, S.K. and PRADHAN, S. (1971). - Studies on resistance to insecticides in *Tribolium castaneum* (Herbst). III. Selection of a strain resistant to lindane and its biological characteristics. *J. stored Prod. Res.* 16:331-337.

BHATIA, S.K. and PRADHAN, S. (1972). - Studies on resistance to insecticides in *Tribolium castaneum* (Herbst). V. Cross-resistance characteristics of a lindane-resistant strain. *J. stored Prod. Res.* 8:89-93.

BHATIA, S.K., YADAV, T.D. and MOOKHERJEE, P.B. (1971). - Malathion resistance in *T. castaneum* in India. *J. stored Prod. Res.* 7:227-230.

BLACKITH, R.E. (1953). - Bioassay systems for the pyrethrins. V. Experiments with a resistant strain of *Calandra granaria*. L. *Ann. appl. Biol.* 40:113-120.

BLACKITH, R.E. and GORRINGE, B.S. (1953). - Response of pests to fumigation. I. Toxicity of mercury vapour to the eggs of *Calandra granaria* (L). *Bull. ent. Res.* 44:217-224.

BLACKMAN, D.G. (1966). - A pearl-eye mutation in *Oryzaephilus surinamensis* (L.) (Coleoptera, Silvanidae). *J. stored Prod. Res.* 2:167-169.

BLACKMAN, D.G. and PECKOVER, J.I. (1976). - Incidence of resistance in the United Kingdom. Other beetles. The long-headed flour beetle. *Pest Infest. Control* 1971-73:84.

BLACKMAN, D.G., SIMPSON, W.J. and BROMILOW, C. (1976). - Incidence of resistance in the United Kingdom. Other beetles. The confused flour beetle. *Pest Infest. Control* 1971-73:83.

BOND, E.J. (1973). - Increased tolerance to ethylene dibromide in a field population of *Tribolium castaneum* (Herbst). *J. stored Prod. Res.* 9:61-63.

BOND, E.J. and MONRO, H.A.U. (1961). - The toxicity of various fumigants to the Cadelle, *Tenebroides mauritanicus*. *J. econ. Ent.* 54:451-454.

BOND, E.J., ROBINSON, J.R. and BUCKLAND, C.T. (1969). - The toxic action of phosphine. Absorption and symptoms of poisoning in insects. *J. stored Prod. Res.* 5:289-298.

BOND, E.J. and UPITIS, E. (1972). - Persistence of tolerance to methyl bromide in *Sitophilus granarius* after cessation of selection. *J. stored Prod. Res.* 8:221,222.

BREITENBECHER, J.K. (1921). - The genetic evidence of a multiple (triple) allelomorph system in *Bruchus* and its relation to sex-limited inheritance. *Genetics* 6:65-90.

BREITENBECHER, J.K. (1923). - A red-spotted sex-limited mutation in *Bruchus*. *Amer. Nat.* 51:59-65.

BREITENBECHER, J.K. (1925a). - The inheritance of a macula mutation concerned with elytral spotting and latent traits in the male of *Bruchus*. *Biol. Bull. mar. biol. Lab., Woods Hole* 49:265-274.

BREITENBECHER, J.K. (1925b). - An apterous mutation in *Bruchus*. *Biol. Bull. mar. biol. Lab., Woods Hole* 48:166-170

BREITENBECHER, J.K. (1925c). - The inheritance of sex-limited bilateral asymmetry in *Bruchus*. *Genetics* 10:261-277.

BROWER, J.H. (1972a). - Combined effects of egg age and radiation dosage on egg hatch of *Tenebrio molitor* (Coleoptera, Tenebrionidae). *Can. Ent.* 104:141-144.

BROWER, J.H. (1972b). - Interaction of age and radiation dosage on hatch of *Gibbium psylloides* eggs (Coleoptera, Ptinidae). *Ann. ent. Soc. Am.* 65:1237,1238.

BROWER, J.H. (1972c). - "Scaleless" and "Melanic", two undescribed mutations in *Plodia interpunctella* (Lepidoptera, Phycitidae). *J. Kans. ent. Soc.* 45:421-426.

BROWER, J.H. (1973a). - Gamma-radiation sensitivity of malathion-resistant strains of the Indian meal moth. *J. econ. Ent.* 66:461,462.

BROWER, J.H. (1973b). - Sensitivity of *Tenebrio molitor* and *T. obscurus* to gamma irradiation. *J. econ. Ent.* 66:1175-1179.

BROWER, J.H. (1974a). - Radio resistance of the red flour beetle, *Tribolium castaneum* (Coleoptera: Tenebrionidae) exposed to sublethal gamma irradiation for 25 generations. *Can. Ent.* 106:241-246.

BROWER, J.H. (1974b). - Radio sensitivity of an insecticide-resistant strain of *Tribolium castaneum* (Herbst). *J. stored Prod. Res.* 10:129-131.

BROWER, J.H. (1974c). - Lack of radio-resistance in *Plodia interpunctella* (Lepidoptera: Phycitidae) exposed to sublethal gamma irradiation for 30 generations. *Radiation Res.* 57:73-79.

BROWER, J.H. (1974d). - Inability of populations of *Callosobruchus maculatus* to develop tolerance to exposures of acute gamma irradiation. *Ann. ent. Soc. Am.* 67:287-291.

BROWER, J.H. (1974e). - Radio sensitivity of the square necked grain beetle, *Cathartus quadricollis* (Guérin-Ménéville) (Coleoptera: Cucujidae). *J. Kans. ent. Soc.* 47:254-259.

BROWER, J.H. (1974f). - Radio sensitivity of the slenderhorned flour beetle, *Gnathocerus maxillosus* (Coleoptera: Tenebrionidae). *Fla Ent.* 57:91-95.

BROWER, J.H. (1974g). - Age as a factor in determining radio-sensitivity of eggs of *Plodia interpunctella*. *Environ. Entomol.* 3:945,946.

BROWER, J.H. (1974h). - Fitness and radio-resistance of the rice weevil, *Sitophilus oryzae* (Coleoptera: Curculionidae), after 25 generations of substerilizing irradiation. *J. Kans. ent. Soc.* 47:437-444.

BROWER, J.H. (1975a). - Potential for genetic control of stored product insect populations. *Proc. 1st int. wking Conf. stored-Prod. Ent., Savannah,* 1974, pp. 167-180.

BROWER, J.H. (1975b). - Radio-sensitivity of the longheaded flour beetle. *J. econ. Ent.* 68:220-222.

BROWER, H.J., HOSSAIN, M.M. and TILTON, E.W. (1973). - Radiation sensitivity of successively irradiated generations of *Tribolium castaneum* (Herbst) (Coleoptera, Tenebrionidae) and *Sitophilus oryzae* (L.) (Coleoptera, Curculionidae). *J. stored Prod. Res.* 9:43-49.

BROWER, J.H. and MAHANY, P.G. (1973). - Gamma radiation sensitivity of the cadelle, *Tenebroides mauritanicus* (Coleoptera: Ostomidae) and the flat grain beetle, *Cryptolestes pusillus* (Coleoptera: Cucujidae). *J. Georgia ent. Soc.* 8:174-184.

BROWER, J.H. and SCOTT, H.C. (1972). - Gamma radiation sensitivity of the spider beetle, *Gibbium psylloides* (Coleoptera, Ptinidae). *Can. Ent.* 104:1551-1556.

BROWER, J.H. and TILTON, E.W. (1972a). - Gamma-radiation effects on *Trogoderma inclusum* and *T. variable*. *J. econ. Ent.* 65:250-254.

BROWER, J.H. and TILTON, E.W. (1972b). - Comparative gamma radiation sensitivity of the sawtoothed grain beetle and the merchant grain beetle. *Environ. Entomol.* 1:735-738.

BROWER, J.H. and TILTON, E.W. (1973). - Comparative gamma radiation sensitivity of *Tribolium madens* (Charpentier) and *T. castaneum* (Herbst). *J. stored Prod. Res.* $\underline{9}$:93-100.

BROWN, A.W.A. (1958). - The spread of insecticide resistance in pest species. *Adv. Pest Contr. Res.* $\underline{2}$:351-414.

BROWN, A.W.A. (1961). - The challenge of insecticide resistance. *Bull. ent. Soc. Am.* $\underline{7}$:6-19.

BROWN, A.W.A. (1963). - Insect Resistance. Part I. Nature and prevalence of resistance. *Fm Chem.* $\underline{126}$:22-24,26,28.

BROWN, G.A., BROWER, J.H. and TILTON, E.W. (1972). - Gamma radiation effects on *Sitophilus zeamais* and *S. granarius*. *J. econ. Ent.* $\underline{65}$:203-205.

BROWN, G.A. and DAVIS, R. (1973). - Sensitivity of red flour beetle eggs to gamma radiation as influenced by treatment age and dose rate. *J. Georgia ent. Soc.* $\underline{8}$:153-157.

BROWN, W.B. (1954). - Fumigation with methyl bromide under gas-proof sheets. *Bull. Pest Infest. Res.* No. I.

BROWN, W.B. (1959). - Fumigation with methyl bromide under gas-proof sheets. *Bull. Pest Infest. Res.* No. I, 2nd Ed.

BROWN, W.B., HOLE, B.D. and GOODSHIP, P.G. (1969). - Toxicity of phosphine to insects. *Pest Infest. Res.* $\underline{1968}$:59-62.

BRUDNAYA, A.A., CHUDINOVA, A.N. and ANOSKINA, N.L. (1966). - Effect of sublethal doses of toxic chemicals on granary pests. *Ent. Obozr.* $\underline{45}$:83-94.

BUCKMIRE, K.U. and BENNETT, R.G. (1969). - Variation in susceptibility to fumigants between different populations. *Pest Infest. Res.* $\underline{1968}$:63.

BULL, J.O. and WOND, T. (1963). - Control of the Mediterranean flour moth, *Anagasta kuehniella* Zell., by sterile male release. II. Susceptibility to gamma radiation. *United Kingdom Atomic Energy Authority Rep. 5, A.E.R.E.* - R3967, p. 42.

BURGES, H.D. (1960). - Studies on the dermestid, *T. granarium* Everts - IV. Feeding, growth and respiration with particular reference to diapause. *J. Insect Physiol.* $\underline{5}$:317-334.

BURGES, H.D. (1962). - Diapause, pest status and control of khapra beetle, *Trogoderma granarium* Everts. *Ann. appl. Biol.* $\underline{50}$:611-617.

BUSVINE, J.R. (1942). - Specificity of fumigants. *Nature, Lond.* $\underline{150}$:208.

BUSVINE, J.R. (1962a). - Die Dosis in Insektizidprüfungen. *Verhl. XI Kongr. Ent.* $\underline{2}$:592-594.

BUSVINE, J.R. (1962b). - Insecticide resistance among pests of stored products. *Proc. 11th int. Congr. Ent., Vienna, 1960.* $\underline{3}$:220-221.

BUSVINE, J.R. (1968a). - Detection and measurement of insecticide resistance in arthropods of agricultural and veterinary importance. *Wld Rev. Pest Control* $\underline{7}$:27-41.

BUSVINE, J.R. (1968b). - Design of tests for detecting and measuring insecticide resistance. *Soc. Chem. Ind. Monograph* 29:18-34.

BUSVINE, J.R. (1971). - A critical review of the techniques for testing insecticides. 2nd ed. (Cwlth Agric. Bur.: Slough).

BUSVINE, J.R. (1973). - Progress in research on resistance to arthropods. *Pestic. Sci.* 4:491-499.

CABRAL, A.L. and GOUVEIA, A.J.S. (1960). - Condicoes fitossanitarias de produtos ultramarinos em armazens do Porto de Lisboa (Alcantara - Norte). *Estud. Ens. e Doc.* 68, 119 pp.

CAMPBELL-BROWN, M.J. and CHAMP, B.R. (1971). - A fused-antenna mutant of *Sitophilus oryzae* (L.) (Coleoptera, Curculionidae). *J. stored Prod. Res.* 7:217-220.

CARNEY, G.E. (1959). - Differential response of male and female adults of *Trogoderma granarium* Everts towards sterilizing doses of gamma-radiation. *Nature, Lond.* 183:338-339.

CARTER, S.W. (1975). - Laboratory evaluation of three novel insecticides inhibiting cuticle formation against some susceptible and resistant stored products beetles. *J. stored Prod. Res.* 11:187-193.

CARTER, S.W., CHADWICK, P.R. and WICKHAM, J.C. (1975). - Comparative observations on the activity of pyrethroids against some susceptible and resistant stored product beetles. *J. stored Prod. Res.* 11:135-142.

CASIDA, J.E. (1970). - Mixed-function oxidase involvement in the biochemistry of insecticide synergists. *J. agr. Fd Chem.* 18:753-772.

CASIDA, J.E. (1973). - Biochemistry of the pyrethrins. *In* "Pyrethrum the natural insecticide". ed. by J.E. Casida. (Academic Press: N. York, London). pp. 101-120.

CHAMP, B.R. (1967). - The inheritance of DDT resistance in *Sitophilus oryzae* (L.) (Coleoptera, Curculionidae) in Queensland. *J. stored Prod. Res.* 3:321-334.

CHAMP, B.R. (1968). - A test method for detecting insecticide resistance in *Sitophilus oryzae* (L.) (Coleoptera, Curculionidae). *J. stored Prod. Res.* 4:175-178.

CHAMP, B.R. (1969). - World survey for resistance. Items 7/10/5,6. *Rep. 4th Session FAO Working Party of Experts on Resistance of Pests to Pesticides, Meeting Rep.* PL/1968/M/10. (FAO: Rome). pp. 37,38.

CHAMP, B.R. and CAMPBELL-BROWN, M.J. (1969). - Genetics of lindane resistance in *Tribolium castaneum* (Herbst) (Coleoptera, Tenebrionidae). *J. stored Prod. Res.* 5:399-406.

CHAMP, B.R. and CAMPBELL-BROWN, M.J. (1970a). - Insecticide resistance in Australian *Tribolium castaneum* (Herbst). I. A test method for detecting insecticide resistance. *J. stored Prod. Res.* 6:53-70.

CHAMP, B.R. and CAMPBELL-BROWN, M.J. (1970b). - Insecticide resistance in Australian *Tribolium castaneum* (Herbst) (Coleoptera, Tenebrionidae). II. Malathion resistance in eastern Australia. *J. stored Prod. Res.* 6:111-131.

CHAMP, B.R. and CRIBB, J.N. (1963). - Lindane resistance of *Sitophilus sasakii* (Tak.) in Queensland. *Grain Stor. Newsl. (FAO, Rome)* 5(4):2.

CHAMP, B.R. and CRIBB, J.N. (1965a). - Lindane resistance in *Sitophilus oryzae* (L.) and *Sitophilus zeamais* Motsch. (Coleoptera, Curculionidae) in Queensland. *J. stored Prod. Res.* 1:9-24.

CHAMP, B.R. and CRIBB, J.N. (1965b). - An investigation of peanut storage pests in Queensland. 2. Insecticide treatment of bulk nut-in-shell peanuts in silos. *Qd. J. agric. anim. Sci.* 22:241-257.

CHAMP, B.R. and GENN, B.G. (1971). - Some mutants of stored-products Coleoptera. *J. stored Prod. Res.* 7:293-298.

CHAMP, B.R. and SHEPHERD, R.C.H. (1965). - Insecticide resistance in *Phthorimaea operculella* (Zell.): comparative response to DDT, DDD, endrin, dieldrin, isobenzan, lindane, and azinphos-ethyl. *Qd J. agric. Anim. Sci.* 22:511-513.

CHAMP, B.R. and SHEPHERD, R.C.H. (1971). - A red-eyed mutant of *Phthorimaea operculella* (Zeller) (Lepidoptera, Gelechiidae). *J. stored Prod. Res.* 7:221-226.

CHAMP, B.R. and SMITH, M.J. (1972). - Genetics of resistance in stored product insects. *Div. Ent. Ann. Rep. 1971-72, CSIRO, Canberra, Australia,* p. 151.

CHAMP, B.R., STEELE, R.W., GENN, B.G. and ELMS, K.D. (1969). - A comparison of malathion, diazinon, fenitrothion and dichlorvos for control of *Sitophilus oryzae* (L.) and *Rhyzopertha dominica* (F.) in wheat. *J. stored Prod. Res.* 5:21-48.

CHATTERJI, S.M. (1955). - Effect of nutrition and starvation on the susceptibility of *Corcyra cephalonica* Stainton to carbon disulphide. *Curr. Sci.* 24:206,207.

CHAWLA, S.S., PERRON, J.M., HUOT, L. and CORRIVAULT, G.W. (1973) - Sensitivity of *Oryzaephilus surinamensis* to Cesium-137. II. Irradiation of larvae, pupae, and adults. *J. econ. Ent.* 66:605-608.

CHILDS, D.P., OVERBY, J.E., COX, E.L. and NIFFENEGGER, D. (1973). - Toxicity of phosphine at various concentrations and temperatures to the cigarette beetle. *Rep. U.S. Dep. Agric., Agric. Res. Serv.* No. ARS-S-16, 8 pp.

CICHY, D. (1969). - The influence of some ecological factors on the susceptibility of *Tribolium castaneum* (Herbst) (Coleoptera, Tenebrionidae) to pyrethrin. *Ekol. pol. (A)* 17:159-166.

CICHY, D. (1971). - The role of some ecological factors in the development of pesticide resistance in *Sitophilus oryzae* (L.) and *Tribolium castaneum* (Herbst). *Ekol. pol. (A)* 19:563-616.

COFFELT, J.A. and VICK, K.W. (1973). - A black mutation of *Lasioderma serricorne* (F.) (Coleoptera, Anobiidae). *J. stored Prod. Res.* 9:65-70.

COGBURN, R.R., TILTON, E.W. and BROWER, J.H. (1973). - Almond moth: gamma radiation effects on life stages. *J. econ. Ent.* 66:745-751.

COLLINS, W.E. and KING, H.L. (1953). - A modified technique for laboratory evaluation of contact insecticides. *J. econ. Ent.* 46:51-53.

COOPER, L.M. and BENGSTON, M. (1974). - Tobacco beetle and its control. *Aust. Tob. Grow. Bull.* 21:36-40.

CORNWELL, P.B., BULL, J.O. and PENDLEBURY, J.B. (1966). - Control of weevil populations *(Sitophilus granarius* (L.)) with sterilizing and substerilizing doses of gamma radiation. *In* "The Entomology of Radiation Disinfestation of Grain". (Pergamon Press: London). pp. 71-95.

CORNWELL, P.B. and MORRIS, J.A. (1959). - Susceptibility of the grain and rice weevils, *Calandra granarius* L. and *C. oryzae* L., to gamma radiation. *United Kingdom Atomic Energy Authority, A.E.R.E.* - R3065, 32 pp.

CORSEUIL, E. and VICENZI, M.L. (1970). - Acao de alguns inseticidas na protecao do milho armazenado. *Revta Esc. Agron. Vet., Porto Alegre* 10:9-14.

COTTER, W.B. (1974). - Studies of melanism in *Ephestia kühniella* Z. I. The genetic and physiological characteristics of the mutant, ala nigra (An). *J. Hered.* 65:213-216.

COTTON, R.T. (1932). - The relation of respiratory metabolism of insects to their susceptibility to fumigants. *J. econ. Ent.* 25:1088-1103.

COTTON, R.T. (1963). - Rotating fumigants can overcome resistance of hardy insect pests. *NWest Miller* 268(11):32.

COTTON, R.T. and ROARK, R.S. (1927). - Ethylene dichloride-carbon tetrachloride mixture; a new non-burnable non-explosive fumigant. *J. econ. Ent.* 20:636-639.

COVENEY, R.D. and CORBAN, P.A. (1970). - Simple field test for organo-phosphorus resistance. *Trop. Stored Prod. Centre, Rep.* 1965-1966. (HMSO: London). p.24.

CRAUFURD-BENSON, H.J. (1938). - An improved method for testing liquid contact insecticides in the laboratory. *Bull. ent. Res.* 29:41-56.

CRENSHAW, J.W. (1965). - Radiation-induced increases in fitness in the flour beetle *Tribolium confusum*. *Science* 149:426,427.

DAL MONTE, G. (1969). - World survey for resistance. Item 5/10/5. *Rep. 4th Session FAO Working Party of Experts on Resistance of Pests to Pesticides, Meeting Rep.* PL/1968/M/10. (FAO: Rome). p.31.

DAVEY, P.M. and AMOS, T.G. (1960). - Lindane dusted paddy stored in Trinidad and resistance of the new species *Sitophilus sasakii*. *Pest Infest. Res.* 1959:56.

DAWSON, P.S. (1968). - Genetic evidence for an hypothesis concerning evolution in *Tribolium*. *J. Heredity* **59**:188-190.

DAWSON, P.S. (1971). - The blob mutant in *Tribolium castaneum*. *Can. J. Genet. Cyt.* **13**:801-810.

DAWSON, P.S. (1972a). - Linkage group IV of *Tribolium castaneum*. *Can. J. Genet. Cyt.* **14**:675-680.

DAWSON, P.S. (1972b). - Sex and crossing over in linkage group IV of *Tribolium castaneum*. *Genetics* **72**:525-530.

DE LIMA, C.P.F. (1972). - Lindane resistance in field strains of *Sitophilus zeamais* in Kenya. *J. stored Prod. Res.* **8**:167-175.

DEVARAJ URS, K.C. and MOOKHERJEE, P.B. (1966). - Effect of oilseed food on the biology of *Tribolium castaneum* Herbst, *Trogoderma granarium* Everts and *Corcyra cephalonica* Staint. and their susceptibility to pyrethrins. *Indian J. Ent.* **28**:234-240.

DEVARAJ URS, K.C. and MOOKHERJEE, P.B. (1967). - Effect of oilseed food on the susceptibility of three species of storage insects to methyl bromide. *Indian J. Ent.* **29**:297,298.

DEWEES, A.A. (1967). - Sex differences in recombination values for linkage group V of *Tribolium castaneum*. *Tribolium Inf. Bull.* **10**:89,90.

DIERICK, G.F.E.M. (1942). - Dip test method:Ovicides. *Diss. Univ. Amsterdam*, 117 pp.

DUCOFF, H.S. (1967). - Changes with age in the radio-sensitivity of adult flour beetles. *Radiat. Res.* **31**:612.

DUCOFF, H.S. and BOSMA, G.C. (1966). - Influence of pupal age on sensitivity to radiation. *Biol. Bull.* **130**:151-156.

DUCOFF, H.S. and BOSMA, G.C. (1967). - Acute lethality after x-irradiation of *Tribolium confusum* adults. *Ent. exp. appl.* **10**:153-165.

DYTE, C.E. (1963). - Section on stocks. *Tribolium Inf. Bull.* **6**:16-20.

DYTE, C.E. (1967) - Summary of cases of resistance to pesticides in agricultural pests. Item 1/10/5. *Rep. 1st Session FAO Working Party of Experts on Resistance of Pests to Pesticides, Meeting Rep.* PL/1965/18. (FAO: Rome). p.30.

DYTE, C.E. (1970). - Insecticide resistance in stored-product insects with special reference to *Tribolium castaneum*. *Trop. stored Prod. Inf.* **20**:13-18.

DYTE, C.E. (1972). - Resistance to synthetic juvenile hormone in a strain of the flour beetle *Tribolium castaneum*. *Nature, Lond.* **238**:48,49.

DYTE, C.E. (1974). - Problems arising from insecticide resistance in storage pests. *Europ. Pl. Prot. Org. Bull.* **4**:275-289.

DYTE, C.E. and BINNS, T. (1971). - Section on "New Mutants". *Tribolium Inf. Bull.* **14**:65.

DYTE, C.E. and BLACKMAN, D.G. (1961). - A pearl-eyed mutation in *Gnathocerus cornutus* (Fab.) (Coleoptera, Tenebrionidae), with notes on similar variants in other beetles. *Proc. R. ent. Soc. Lond. (A)* 36:168-172.

DYTE, C.E. and BLACKMAN, D.G. (1963). - Increased resistance of stored-product insects to insecticides. The rust-red flour beetle and DDT. *Pest Infest. Res.* 1962:38.

DYTE, C.E. and BLACKMAN, D.G. (1964). - Increased resistance of stored-product insects to insecticides. The merchant grain beetle and malathion. *Pest Infest. Res.* 1963:31.

DYTE, C.E. and BLACKMAN, D.G. (1965a). - A recessive allele of the black mutation in *Tribolium castaneum* (Herbst). *Entomologist* 98:78,79.

DYTE, C.E. and BLACKMAN, D.G. (1965b). - Increased resistance of stored-product insects to insecticides. The merchant grain beetle and malathion. *Pest Infest. Res.* 1964:34.

DYTE, C.E. and BLACKMAN, D.G. (1966a). - Increased resistance of stored-product insects to insecticides. The rust-red flour beetle and malathion. *Pest Infest. Res.* 1965:40.

DYTE, C.E. and BLACKMAN, D.G. (1966b). - Increased resistance of stored-product insects to insecticides. The rust-red flour beetle and DDT. *Pest Infest. Res.* 1965:41.

DYTE, C.E. and BLACKMAN, D.G. (1967a). - Increased resistance of stored-product insects to insecticides. The rust-red flour beetle and DDT. *Pest Infest. Res.* 1966:38.

DYTE, C.E. and BLACKMAN, D.G. (1967b). - Selection of a DDT-resistant strain of *Tribolium castaneum* (Herbst) (Coleoptera, Tenebrionidae). *J. stored Prod. Res.* 2:211-228.

DYTE, C.E. and BLACKMAN, D.G. (1969). - Insecticide resistance. Malathion resistance in rust-red flour beetles. *Pest Infest. Res.* 1968:43,44.

DYTE, C.E. and BLACKMAN, D.G. (1970a). - The spread of insecticide resistance in *Tribolium castaneum* (Herbst) (Coleoptera, Tenebrionidae) *J. stored Prod. Res.* 6:255-261.

DYTE, C.E. and BLACKMAN, D.G. (1970b). - Insecticide resistance. The spread of resistance in rust-red flour beetles. *Pest Infest. Res.* 1969:42,43.

DYTE, C.E. and BLACKMAN, D.G. (1970c). - Insecticide resistance. Resistant rust-red flour beetles in imported cargoes. *Pest Infest. Res.* 1969:43.

DYTE, C.E. and BLACKMAN, D.G. (1972). - Laboratory evaluation of organo-phosphorus insecticides against susceptible and malathion-resistant strains of *Tribolium castaneum* (Herbst) (Coleoptera, Tenebrionidae) *J. stored Prod. Res.* 8:103-109.

DYTE, C.E. and BLACKMAN, D.G. (1973). - Insecticide resistance. Malathion resistance in the confused flour beetle. *Pest Infest. Control* 1968-70:120.

DYTE, C.E., BLACKMAN, D.G. and BINNS, T. (1969). - Section on "New Mutants". *Tribolium Inf. Bull.* 11:65.

DYTE, C.E., BLACKMAN, D.G., COGAN, P.M., FORSTER, R., PECKOVER, J. *et al.* (1976a). - Worldwide survey of resistance in grain pests. *Pest Infest. Control* 1971-73:70,71.

DYTE, C.E., BLACKMAN, D.G. and FORSTER, R. (1973a). - Insecticide resistance to malathion and lindane in stored-product insects in the United Kingdom. *U.K. Min. Agric. Fish. & Food, Pest Infest. Control Lab., Tech. Inf. Circ.* 1973/18. 5 pp.

DYTE, C.E., BLACKMAN, D.G. and MULCAHY, F.H. (1964). - Increased resistance of stored-product insects to insecticides. The rust-red flour beetle and DDT. *Pest Infest. Res.* 1963:31.

DYTE, C.E., BLACKMAN, D.G. and MULCAHY, F.H. (1965a). - Increased resistance of stored-product insects to insecticides. The rust-red flour beetle and DDT. *Pest Infest. Res.* 1964:34.

DYTE, C.E., BLACKMAN, D.G. and PECKOVER, J.I. (1973b). - Insecticide resistance. Malathion-specific resistance in the rust-red flour beetle. *Pest Infest. Control* 1968-70:118,119.

DYTE, C.E. and DALY, J.A. (1969). - Selectivity and synergism of insecticides. Contrasting susceptibilities to DDT in different flour beetles. *Pest Infest. Res.* 1968:44,45.

DYTE, C.E. and DALY, J.A. (1970). - Selectivity and synergism of insecticides. Contrasting susceptibilities to DDT in different beetles. *Pest Infest. Res.* 1969:45,46.

DYTE, C.E., DALY, J.A. and READE, A. (1968a). - Selectivity and synergism of insecticides. Contrasting susceptibilities to DDT in different flour beetles. *Pest Infest. Res.* 1967:37.

DYTE, C.E., ELLIS, V.J. and LLOYD, C.J. (1966). - Studies on the contrasting susceptibilities of the larvae of two hide beetles (*Dermestes* spp. Coleoptera, Dermestidae) to malathion. *J. stored Prod. Res.* 1:223-234.

DYTE, C.E. and FORSTER, R. (1969). - Insecticide resistance. Lindane-dieldrin resistance in the groundnut borer. *Pest Infest. Res.* 1968:43.

DYTE, C.E. and FORSTER, R. (1970a). - Insecticide resistance. Lindane resistance in the groundnut borer. *Pest Infest. Res.* 1969:43.

DYTE, C.E. and FORSTER, R. (1970b). - Insecticide resistance. Lindane resistance in the maize weevil. *Pest Infest. Res.* 1969:43.

DYTE, C.E. and FORSTER, R. (1973a). - Insecticide resistance. Malathion resistance in the merchant grain beetle. *Pest Infest. Control* 1968-70:120,121.

DYTE, C.E. and FORSTER, R. (1973b). - Studies on insecticide resistance in *Oryzaephilus mercator* (Fauv.) (Coleoptera, Silvanidae). *J. stored Prod. Res.* 9:159-164.

DYTE, C.E. and FORSTER, R. (1976). - Incidence of resistance in the United Kingdom. Other beetles. The merchant grain beetle. *Pest Infest. Control* 1971-73:83,84.

DYTE, C.E., FORSTER, R. and AGGARWAL, S. (1976b). - The rust-red flour beetle. Resistance to juvenile hormone mimics. *Pest Infest. Control* 1971-73:79,80.

DYTE, C.E., GREEN, A.A. and PINNIGER, D.B. (1975). - Some consequences of the development of insecticide resistance in stored-product insects. *Proc. 1st int. wking Conf. stored-Prod. Ent., Savannah*, 1974, pp. 261-271.

DYTE, C.E., LLOYD, C.J. and BLACKMAN, D.G. (1966). - Synergists for DDT in DDT-resistant beetles. *Pest Infest. Res.* 1965:33.

DYTE, C.E. and ROWLANDS, D.G. (1968). - The metabolism and synergism of malathion in resistant and susceptible strains of *Tribolium castaneum* (Herbst) (Coleoptera, Tenebrionidae). *J. stored Prod. Res.* 4:157-173.

DYTE, C.E., ROWLANDS, D.G. and BLACKMAN, D.G. (1965b). - Synergism of organophosphorus insecticides. Selective synergism of malathion in resistant flour beetles. *Pest Infest. Res.* 1964:28.

DYTE, C.E., ROWLANDS, D.G., DALEY, J.A. and BLACKMAN, D.G. (1970). - Insecticide resistance. A new type of resistance in rust-red flour beetles. *Pest Infest. Res.* 1969:41,42.

DYTE, C.E., ROWLANDS, D.G., DALEY, J.A., BLACKMAN, D.G., PECKOVER, J.I. and FIELD, J. (1973c). - Insecticide resistance. Non-specific resistance in the rust-red flour beetles. *Pest Infest. Control* 1968-70:118,119.

DYTE, C.E., ROWLANDS, D.G. and EDWARDS, J.P. (1976c). - Incidence of resistance in the United Kingdom. The rust-red flour beetle. Metabolism of juvenile hormone mimics. *Pest Infest. Control* 1971-73:80,81.

DYTE, C.E., ROWLANDS, D.G. and EDWARDS, J.P. (1976d). - Incidence of resistance in the United Kingdom. The rust-red flour beetle. Resistance to piperonyl butoxide. *Pest Infest. Control* 1971-73:82,83.

DYTE, C.E., SHAW, D.D. and BLACKMAN, D.G. (1965c). - Genetic studies on insecticide resistance. *Pest Infest. Res.* 1964:30.

DYTE, C.E., SHAW, D.D. and BLACKMAN, D.G. (1965d). - Section on "New Mutants". *Tribolium Inf. Bull.* 8:41,42.

DYTE, C.E., SHAW, D.D., BLACKMAN, D.G., BINNS, T.J. and KING, M.E. (1968b). - Genetics. Miscellaneous studies. *Pest Infest. Res.* 1967:36,37.

DYTE, C.E., TYLER, P.S., BLACKMAN, D.G., FORSTER, R., COGAN, P.M., BINNS, T.J. and PECKOVER, J. (1976e). - Incidence of resistance in the United Kingdom. The saw-toothed grain beetle. Monitoring for resistance in the United Kingdom. *Pest Infest. Control* 1971-73:71-73.

DYTE, C.E., TYLER, P.S., BLACKMAN, D.G., FORSTER, R., COGAN, P.M., BINNS, T.J. and PECKOVER, J. (1976f). - Incidence of resistance in the United Kingdom. The saw-toothed grain beetle. Cross-resistance studies. *Pest Infest. Control* 1971-73:74.

DYTE, C.E. and WILDEY, K.B. (1976). - Incidence of resistance in the United Kingdom. Other beetles. The lesser grain borer. *Pest Infest. Control* 1971-73:85.

DYTE, C.E. and WILKIN, D.R. (1963). - Increased resistance of stored-product insects to insecticides. The saw-toothed grain beetle and carbaryl. *Pest Infest. Res.* 1962:38.

DYTE, C.E. and WILKIN, D.R. (1964). - Increased resistance of stored-product insects to insecticides. The saw-toothed grain beetle and carbaryl. *Pest Infest. Res.* 1963:31.

DYTE, C.E. and WILKIN, D.R. (1965). - Increased resistance of stored-product insects to insecticides. The saw-toothed grain beetle and carbaryl. *Pest Infest. Res.* 1964:34.

DYTE, C.E., WILKIN, D.R. and FORSTER, R. (1973d). - Insecticide resistance. Malathion resistance in the saw-toothed grain beetle. *Pest Infest. Control* 1968-70:121.

ELDERFRAWI, M.E. (1969). - World survey of resistance. Item 5/10/6. *Rep. 4th Session FAO Working Party of Experts on Resistance of Pests to Pesticides, Meeting Rep.* PL/1968/M/10. (FAO: Rome). p. 31.

ELLIS, C.R. (1972a). - Susceptibility of two strains of *Sitophilus granarius* to 1,2-dibromoethane. 1. Effect of weight-dependent respiration and fumigant uptake on strain susceptibility. *J. econ. Ent.* 65:42-47.

ELLIS, C.R. (1972b). - Susceptibility of two strains of *Sitophilus granarius* to 1,2-dibromoethane. 2. Strain difference in uptake and metabolism of radio-labelled fumigant. *J. econ. Ent.* 65:1573-1577.

ELLIS, C.R. (1972c). - Susceptibility of two strains of *Sitophilus granarius* to 1,2-dibromoethane. 3. Effect of total lipid on strain susceptibility. *J. econ. Ent.* 65:1579-1582.

ELLIS, C.R. and MORRISON, F.O. (1967). - Small chamber tests of ethylene dibromide and ethylene dichloride on adult grain-infesting Coleoptera. *Can. J. Zool.* 45:435-448.

ERDMAN, H.E. (1962). - Comparative x-ray sensitivity of *Tribolium confusum* and *T. castaneum* (Coleoptera, Tenebrionidae) at different developmental stages during their life cycle. *Nature, Lond.* 195:1218.

ERDMAN, H.E. (1966). - Modification of fitness in species and strains of flour beetles due to x-ray and DDT. *Ecology* 47:1066-1072.

ERDMAN, H.E. (1970). - Effects of x-radiation and the insecticide DDT on mortality and reproduction of flour beetles, *Tribolium confusum* and *T. castaneum*, with a genetic interpretation for DDT resistance. *Ann. ent. Soc. Am.* 63:191-197.

EVANS, B.R. and PORTER, J.E. (1965). - The incidence, importance and control of insects found in stored food and food-handling areas of ships. *J. econ. Ent.* 58:479-481.

FERNANDO, H.E. (1967). - Summary of cases of resistance to pesticides in agricultural pests. Item 7/16/21. *Rep. 1st Session FAO Working Party of Experts on Resistance of Pests to Pesticides, Meeting Rep.* PL/1965/18. (FAO: Rome). p. 78.

FERWERDA, F.B. (1928). - Genetische Studien am Mehlkäfer, *Tenebrio molitor* L. *Genetica* 11:1-110.

FISK, F.W. and SHEPARD, H.H. (1938). - Laboratory studies of methyl bromide as an insect fumigant. *J. econ. Ent.* 31:79-84.

FREEMAN, J.A. (1948a). - World foci of infestation and principal channels of dissemination to other points, with suggestions for detection and standard of inspection. *FAO agric. Stud.* 2:15-34.

FREEMAN, J.A. (1948b). - Stored product pests: a survey of the principal entomological problems in the United Kingdom. *Ann. appl. Biol.* 35:294-301.

FREEMAN, J.A. (1950). - Methods of spread of stored products insects and origin of infestation in stored products. *Proc. 8th int. Congr. Ent., Stockholm,* 1948, pp. 815-825.

FREEMAN, J.A. (1957). - Infestation of grain in international trade. A review of problems and control methods. *J. Sci. Fd Agric.* 11:623-629.

FREEMAN, J.A. (1958). - The control of infestation in stored products moving in international trade. *Proc. 10th int. Congr. Ent., Montreal,* 1956. 4:5-16.

FREEMAN, J.A. (1960). - Stored products pests and international trade. *Rep. 7th Cwlth Ent. Conf., London,* 1960, pp. 61-68.

FREEMAN, J.A. (1962). - Inherent vice and infestation of cargo. *J. ins. Inst. London* 50:97-108.

FREEMAN, J.A. (1965). - On the infestation of rice and rice products imported into Britain. *Proc. 12th int. Congr. Ent, London,* 1964, pp. 632-634.

FREEMAN, J.A. (1968a). - Problems of infestation of commodities carried by sea with special reference to imports into Great Britain. *Rep. int. Conf. Prot. Stored Products, Lisbon,* 1967, *EPPO Paris Publ. Ser.* A No. 46E, pp. 15-30.

FREEMAN, J.A. (1968b). - Implications for infestation control of the use of freight containers in international trade. *Rep. int. Conf. Prot. Stored Products, Lisbon,* 1967, *EPPO Paris Publ. Ser.* A No. 46E, pp. 87-89.

FREEMAN, J.A. (1968c). - The control of infestation in Greek currants. *Rep. int. Conf. Prot. Stored Products, Lisbon,* 1967, *EPPO Paris Publ. Ser.* A No. 46E, pp. 103-106.

FREEMAN, J.A. (1968d). - On the infestation of almonds, with special reference to those imported into Great Britain from Mediterranean countries. *Rep. int. Conf. Prot. Stored Products, Lisbon,* 1967, *EPPO Paris Publ. Ser.* A No. 46E, pp. 121-124.

FREEMAN, J.A. (1968e). - Legislative and commercial measures for control of infestation in agricultural products. *Rep. int. Conf. Prot. Stored Products, Lisbon,* 1967, *EPPO Paris Publ. Ser.* A No. 46E, pp. 155-157.

FREEMAN, J.A. (1968f). - Problems in the carriage of infested commodities in freight containers. *Proc. int. Container Symposium, London,* 1968, pp. 79-81.

FREEMAN, J.A. (1971). - Prevention and control of infestation in conventional cargo ships. *Proc. 3rd British Pest Control Conf., Jersey,* 1971, pp. 109-117.

FREEMAN, J.A. (1973). - Infestation and control of pests of stored grain in international trade. *In* "Grain Storage - Part of a system". (ed. by R.N. Sinha and W.E. Muir). (Avi Publ. Co. Inc.: Westport, Conn.). Chap. 5, pp. 99-136.

FREEMAN, J.A. (1974). - A review of changes in the pattern of infestation in international trade. *Europ. Pl. Prot. Org. Bull.* $\underline{4}$:251-273.

FREEMAN, J.A. and HEAPE, R.J. (1969). - Infestation problems of imports. *Infest. Control* $\underline{1965\text{-}67}$:86-112.

FREEMAN, J.A., SMITH, K.G. and HART, J.J. (1973). - Infestation problems of imports. *Pest Infest. Control* $\underline{1968\text{-}70}$:148-172.

GEOGHIOU, G.P. and GIDDEN, F.E. (1965). - Contact toxicity of insecticide deposits on filter paper to adult mosquitos. *Mosquito News* $\underline{25}$:204-208.

GODAVARIBAI, S., KRISHNAMURTHY, K. and MAJUMDER, S.K. (1962). - Bacterial spores with malathion for controlling *Ephestia cautella*. *Pest Technol.* $\underline{4}$:155-158.

GODAVARIBAI, S., KRISHNAMURTHY, K. and MUJUMDER, S.K. (1964). - Malathion for stored product insect control. *Int. Pest Control* $\underline{6}$:9,10.

GODDEN, D. and HOWE, R.W. (1965). - The susceptibility of the developmental stages of *Tribolium castaneum* to methyl bromide. *Tribolium Inf. Bull.* $\underline{8}$:76.

GODDEN, D., HOWE, R.W. and GOODSHIP, G. (1965). - Experimental verification of dosage required for high percentage kills. *Pest Infest. Res.* $\underline{1964}$:48,49.

GONEN, M. and FISHBAIN, I. (1974). - Sensitivity of different age eggs of the tropical warehouse moth *Ephestia cautella* (Lepidoptera, Phycitidae), to gamma radiation. *Phytoparasitica* $\underline{2}$:47-49.

GOOS, A. (1961). - Biologiczne badania toksycznosci insektycydow na wotku zbozowym (*Calandra granaria* L.) (Materiaty do metodyki badan biologicznej oceny srodkow ochrony roslin cz II). Poznan.pp. 121-147.

GORE, K.S. (1958). - Laboratory experiments with some organic phosphorus insecticides as wheat protectants. Ph.D. Thesis, Cornell University, Ithaca, N.Y.

GOUGH, H.C. (1939). - Factors affecting the resistance of the flour beetle *Tribolium confusum* Duv., to hydrogen cyanide. *Ann. appl. Biol.* $\underline{26}$:533-571.

GOUVEIA, A.J.S. (1967). - Quelques aspects des conditions phytosanitaires des products d'outremer au port de Lisbonne. *Garcia de Orto* $\underline{15}$:409-418.

GOVINDAN, M. and CUTKOMP, L.K. (1961). - Carbon disulphide effectiveness on flour beetles in relation to temperature and exposure time. *J. econ. Ent.* $\underline{54}$:1121-1127.

GREEN, A.A. (1967). - Summary of cases of resistance to pesticides in agricultural pests. Item 1/10/2. *Rep. 1st Session FAO Working Party of Experts on Resistance of Pests to Pesticides, Meeting Rep.* PL/1965/18. (FAO: Rome). p. 30.

GREEN, A.A. (1971). - The protection of cereals stored on farms. *Proc. 3rd Brit. Pest Control Conf., St. Helier,* 5th Sess. Pap. No. 13, 6 pp.

GREEN, A.A. (1975). - Resistance as a problem in the control of insects and mites infesting stored products in Britain. *Proc. 4th Brit. Pest Control Conf., St. Helier,* 7th Sess. Pap. No. 14, 6 pp.

GREEN, A.A., DARLEY, M.J., PINNIGER, D.B., BRADGATE, N.C. et al. (1976). - Incidence of resistance in the United Kingdom. The rust-red flour beetle. Resistance in field populations. *Pest Infest. Control* **1971-73**:78,79.

GREEN, A.A., KANE, M.J., TYLER, P.S., ROWLANDS, D.G. and PATTINSON, I. (1964). - Control of insects infesting groundnuts. Admixture of malathion with groundnuts: field trial. *Pest Infest. Res.* **1963**:24.

GREENING, H.G. (1969). - Grain insects in farm machinery and storages. *Agric. Gaz. N.S.W.* **80**:554.

GREENING, H.G. (1970). - Malathion resistance in the rust-red flour beetle *Tribolium castaneum* Herbst in New South Wales. *J. Aust. ent. Soc.* **9**:160-162.

GREENING, H.G. (1973). - Grain insects in farm machinery and storages. *Agric. Gaz. N.S.W.* **84**:216.

GREENING, H.G., WALLBANK, B.E. and ATTIA, F.I. (1975). - Resistance to malathion and dichlorvos in stored product insects in New South Wales. *Proc. 1st int. wking Conf. Stored-Prod. Ent., Savannah,* 1974, pp. 607-617.

GRIGGS, R. (1964). - What about insect resistance in flour mill pests? Millers should examine their fumigation techniques before deciding to change chemicals. *NWest Miller* **270**:14,17,18,20.

GUPTA, B., AGARWAL, H.C. and PILLAI, M.K.K. (1971). - Distribution, excretion and metabolism of ^{14}C-DDT in the larva and adult *Trogoderma granarium* in relation to toxicity. *Pestic. Biochem. Physiol.* **1**:180-187.

HAGSTRUM, D.W. (1974). - Four non-allelic autosomal recessive eye- and wing-colour mutants of *Cadra cautella* (Lepidoptera, Phycitidae). *J. Georgia ent. Soc.* **9**:88-90.

HAGSTRUM, D.W. and SHARP, J.E. (1975). - Locomotor behaviour of diapausing and non-diapausing larvae of *Cadra cautella* (Walker) (Lepidoptera, Pyralidae). *Proc. 1st int. wking. Conf. Stored-Prod. Ent., Savannah,* 1974, pp. 618-622.

HALSTEAD, D.G.H. (1969). - A new species of *Tribolium* from North America previously confused with *Tribolium madens* (Charp.) (Coleoptera, Tenebrionidae). *J. stored Prod. Res.* **4**:259-304.

HANSER, G. (1955). - Eine neue glasflüglige Mutante (*gl*) von *Ephestia kühniella*. *Z. Naturforsch.* **10b**:161-166.

HASHIMOTO, Y. and FUKAMI, J. (1964). - Resistance to insecticides in almond moth, *Ephestia cautella* Walker. I. Development of methyl-parathion resistance. *Jap. J. appl. Ent. Zool.* **8**:62-67.

HAYWARD, L.A.W. (1962). - Groundnuts. *Rep. W. Afr. stored Prod. Res. Unit* **1961**;12-15.

HEWLETT, P.S. (1974). - Time from dosage to death in beetles, *Tribolium castaneum*, treated with pyrethrins or DDT, and its bearing on dose-mortality relations. *J. stored Prod. Res.* 10:27-41.

HINDS, W.E. and TURNER, W.E. (1910). - Carbon disulphide for the rice-weevil in corn. *J. econ. Ent.* 3:47-57.

HITCHEN, J.M. and WOOD, R.J. (1974). - A genetical investigation of resistance to "knockdown" and "kill" by DDT in adults of the mosquito *Aedes aegypti* L. *Can. J. Genet. Cytol.* 16:177-182.

HOLBORN, J.M. (1957). - The susceptibility to insecticides of laboratory cultures of an insect species. *J. Sci. Fd Agric.* 8:182-188.

HOPE, J.A. and PHILLIPS, R. (1957). - Variation in insect resistance. *Pest Infest. Res.* 1956:30,31.

HORSFALL, W.R. (1934). - Some effects of ethylene oxide on the various stages of the bean weevil and the confused flour beetle. *J. econ. Ent.* 27:405-409.

HOSKINS, W.M. and WITT, J.M. (1958). - Type of DDT metabolism as illustrated in several insect species. *Proc. 10th int. Congr. Ent., Montreal, 1956.* 2:151-156.

HOSSAIN, M.M., BROWER, J.H. and TILTON, E.W. (1972a). - Radiation sensitivity of successively irradiated generations of the Indian meal moth. *J. econ. Ent.* 65:673-676.

HOSSAIN, M.M., BROWER, J.H. and TILTON, E.W. (1972b). - Sensitivity to an acute gamma radiation exposure of successively irradiated generations of the cowpea weevil. *J. econ. Ent.* 66:1566-1568.

HOWE, R.W. (1962a). - The entomological problems of assessing the success of a fumigation of stored produce. *Proc. 11th int. Congr. Ent., Vienna, 1960.* 2:288-290.

HOWE, R.W. (1962b). - The influence of diapause on the status as pests of insects found in houses and warehouses. *Ann. appl. Biol.* 50:611-617.

HOWE, R.W. (1967). - The influence of age of the parents on some characteristics of the offspring of insects bred in the laboratory. *J. stored Prod. Res.* 3:371-385.

HOWE, R.W. (1970). - Susceptibility of the developmental stages of the grain weevil. *Pest Infest. Res.* 1969:57.

HOWE, R.W. (1973). - The susceptibility of the immature and adult stages of *Sitophilus granarius* to phosphine. *J. stored Prod. Res.* 8:241-262.

HOWE, R.W. and FREEMAN, J.A. (1955). - Insect infestation of West African produce imported into Britain. *Bull. ent. Res.* 46:643-668.

HOWE, R.W. and HOLE, B.D. (1966). - The susceptibility of the developmental stages of *Sitophilus granarius* (L.) (Coleoptera, Curculionidae) to methyl bromide. *J. stored Prod. Res.* 2:13-26.

HOWE, R.W. and HOLE, B.D. (1967). - Predicting the dosage of fumigant needed to eradicate insect pests from stored products. *J. appl. Ecol.* 4:337-351.

HOWE, R.W., HOLE, B.D., WELCH, J.J. and GOODSHIP, G. (1965). - Experimental verification of dosages required for high percentage kills. *Pest Infest. Res.* 1964:48.

HOWE, R.W. and WAGNER, C.M. (1969). - Susceptibility of the developmental stages of *Sitophilus granarius* to phosphine. *Pest Infest. Res.* 1968:62,63.

HOY, M.A. (1966a). - Section of "New Mutants". *Tribolium Inf. Bull.* 9:57.

HOY, M.A. (1966b). - Section of "New Mutants". *Tribolium Inf. Bull.* 9:59.

HUQUE, H. (1971). - Effect of gamma radiation on various developmental stages of *Corcyra cephalonica*. *Int. J. appl. Radiat. Isotopes.* 22:439.

HURLOCK, E.T. (1961a). - Occurrence of *Gibbium psylloides* de Czen (Col., Ptinidae) in ships' holds. *Entomologist's mon Mag.* 96:129,130.

HURLOCK, E.T. (1961b). - Persistence of khapra beetle in ships' holds. *Pest. Technol.* 3:144-146.

HURLOCK, E.T. (1963). - The infestation of Canadian produce inspected in the United Kingdom ports between 1953 and 1959. *Can. Ent.* 95:1263-1284.

HURLOCK, E.T. (1964). - Infestation of foodstuffs from the United States of America inspected in the United Kingdom between 1953 and 1961. *Bull. ent. Res.* 55:173-192.

IORDANOU, N.T. and WATTERS, F.L. (1969). - Temperature effects on the toxicity of five insecticides against five species of stored-product insects. *J. econ. Ent.* 62:130-135.

JAO, .L.T. and CASIDA, J.E. (1974a). - Esterase inhibitors as synergists for (+)-trans-chrysanthemate insecticide chemicals. *Pestic. Biochem. Physiol.* 4:456-464.

JAO, L.T. and CASIDA, J.E. (1974b). - Insect pyrethroid-hydrolysing esterases. *Pestic. Biochem. Physiol.* 4:465-472.

JAY, E.G. and PEARMAN, G.C. (1971). - Susceptibility of two species of *Tribolium* (Coleoptera: Tenebrionidae) to alterations of atmospheric gas concentrations. *J. stored Prod. Res.* 7:181-186.

JAYARAMAN, S. and DUCOFF, H.S. (1970). - Effects of x-irradiation at the larval stage on *Tenebrio molitor* (L.) *Biol. Bull.* 43:270.

JOHN, B. and SHAW, D.D. (1967). - Karyotype variation in dermestid beetles. *Chromosoma* 20:371-385.

JOUBERT, P.C. and de BEER, P.R. (1968). - The toxicity of contact insecticides to seed infesting insects. No. 6. Tests with bromophos on maize. *Dept. Agric. Techn. Serv., S. Africa, Tech. Comm.* No. 84, 9 pp.

KAMEL, A.H. and FAM, E.Z. (1962). - The effect of carbon disulphide on the building up of resistant strains of *Sitophilus oryzae* L. (Coleoptera, Curculionidae). *Bull. Soc. Ent. Egypt* 46:285-290.

KASHIWAGI, M. and UTIDA, S. (1972). - A new mutant in *Callosobruchus chinensis* L. (Coleoptera: Bruchidae). *Appl. Ent. Zool.* 7:95-96.

KEM, T.R. (1975). - Studies on the development of resistance to phosphine in *Tribolium castaneum* (Herbst). *Ent. Newsl. (Indian Agric. Res. Inst.)* **5**:6,7.

KENAGA, E.E. (1957). - Some biological, chemical and physical properties of sulphuryl fluoride as an insecticidal fumigant. *J. econ. Ent.* **50**:1-6.

KIRITANI, K., MURAMATSU, T. and YOSHIMURA, S. (1959). - Fauna of storage pests of south-east Asian produce imported into Japan. *Osaka Pl Prot.* **7**:184-207.

KOT, J. (1970). - The phenomenon of partial resistance to insecticides in some arthropds. *Ekol. pol.* **18**:351-359.

KROHNE, H.E. and LINDGREN, D.L. (1958). - Susceptibility of life stages of *Sitophilus oryzae* to various fumigants. *J. econ. Ent.* **51**:157,158.

KUENEN, D.J. (1958). - Influence of sublethal doses of DDT upon the multiplication rate of *Sitophilus granarius* (Coleopt. Curculionidae). *Ent. exp. appl.* **1**:147-152.

KUHN, A. and ALMEIDA, F.F. de (1961). - Fluoreszierende Stoffe und Ommochrome bei Genotypen von *Plodia interpunctella*. *Z. Vererbungslehre* **92**:126-132.

KUHN, A. and BERG, B. (1956). - Zur Genetischen Analyse der Mutation *biochemica* von *Ephestia kühniella*. *Z. Vererbungslehre* **87**:25-35.

KUHN, A. and HENKE, K. (1930). - Eine Mutation der Augenfarbe und der Entwicklungsgeschwindigkeit bei der Mehlmotte *Ephestia kühniella* Z. *Wilhelm Roux's Arch. Entw. Mech. Org.* **122**:204-212.

KUHNE, H. (1967). - Co-resistenz und Resistenz-Vererbung eines gegen Dieldrin resistenten Laboratorium Stammes der Kleidermotte *(Tineola bisselliella* (Hum.), Lep.). *Zool. Beitr.* **13**:397-407.

KUHNE, H. and BECKER, G. (1965). - Zuchtung giftresistenter Kleidermotten *(Tineola bisselliella* (Hum.), Lep.). *Z. angew Ent.* **56**:61-89.

KUMAR, V. and MORRISON, F.O. (1963). - The susceptibility levels of certain stored product pest populations to chemicals used for their control. *Phytoprotection* **44**:101-105.

KUMAR, V. and MORRISON, F.O. (1964). - Macdonald College test kits for testing the susceptibility of stored-product insect pests to residual insecticides. *Can. Ent.* **96**:122.

KUMAR, V. and MORRISON, F.O. (1965). - Recording the susceptibility levels of current stored product pest populations to current insecticides. *Proc. 12th int. Congr. Ent., London*, 1964, pp. 656-657.

KUMAR, V. and MORRISON, F.O. (1967). - Carbamate and phosphate resistance in adult granary weevils. *J. econ. Ent.* **60**:1430-1434.

LA HUE, D.W. (1969). - Control of malathion-resistant Indian meal moths *Plodia interpunctella* (Hübner) with dichlorvos resin strips. *Proc. N. Centr. Branch ent. Soc. Amer.* **24**:117-119.

LA HUE, D.W. (1971). - Controlling the Indian meal moth in shell corn with dichlorvos PVC resin strips. *U.S. Dep. Agric., agric. Res. Serv.* ARS 51-42.

LAKOCY, A. (1970). - Actual problems of the resistance of agricultural pests to pesticides in Poland. *Bull. Inst. Ochrony Roslin* **47**:89-103.

LEFKOVITCH, L.P. (1963). - Differing status of colour forms in *Cryptolestes* Gangl. (Cucujidae). *Tribolium Inf. Bull.* **7**:45,46.

LEFKOVITCH, L.P. and CURRIE, J.E. (1967). - Some morphological, biological and genetical differences between *Cryptolestes pusillus fuscus* spp. n. and *C. pusillus pusillus* (Schönherr) (Coleoptera, Cucujidae). *J. stored Prod. Res.* **3**:311-320.

LE GOUPIL (1932). - Les propriétés insecticides du bromure de méthyle. *Rev. path. végétale entomol. agr. France* **19**:169-172.

LERNER, I.M. (1966). - Foreword to "The Genetics of Tribolium and Related Species" by A. Sokoloff. (Academic Press: New York).

LIN, T. (1973). - Determination of rice weevil (*Sitophilus oryzae* Linne) resistance to insecticides and the development of substitute chemicals. *Taiwan Agric. Quarterly* **8**:115-123.

LINDGREN, D.L. (1935). - The respiration of insects in relation to the heating and fumigation of grain. *Minn. agric. Exp. Sta., Tech. Bull.* 109.

LINDGREN, D.L. (1938). - The stupefaction of red scale, *Aonidiella aurantii* by hydrocyanic acid. *Hilgardia* **11**:213-225.

LINDGREN, D.L. and GERHARDT, P.D. (1947). - The response of California red scale to fumigation with ethylene dibromide and ethylene dibromide-HCN. *J. econ. Ent.* **40**:680-682.

LINDGREN, D.L., KROHNE, H.E. and VINCENT, L.E. (1954). - Malathion and chlorthion for control of insects infesting stored grain. *J. econ. Ent.* **47**:705,706.

LINDGREN, D.L. and VINCENT, L.E. (1965). - The susceptibility of laboratory-reared and field-collected cultures of *Tribolium confusum* and *T. castaneum* to ethylene dibromide, hydrocyanic acid and methyl bromide. *J. econ. Ent.* **58**:551-555.

LINDGREN, D.L. and VINCENT, L.E. (1966a). - Development of resistance in stored-product insects to insecticides. *Cereal Sci. Today* **11**:12-14,26.

LINDGREN, D.L. and VINCENT, L.E. (1966b). - Relative toxicity of hydrogen phosphide to various stored-product insects. *J. stored Prod. Res.* **2**:141-146.

LINDGREN, D.L., VINCENT, L.E. and KROHNE, H.E. (1955). - The 'Khapra beetle' - *Trogoderma granarium* Everts. *Hilgardia* **24**:1-36.

LINDGREN, D.L., VINCENT, L.E. and STRONG, R.G. (1958). - Studies on hydrogen phosphide as a fumigant. *J. econ. Ent.* **51**:900-903.

LLOYD, C.J. (1965). - Increased resistance of stored-product insects to insecticides. The grain weevil and pyrethrum. *Pest Infest. Res.* 1964:32,33.

LLOYD, C.J. (1966). - Increased resistance of stored-product insects to insecticides. The grain weevil and pyrethrum. *Pest Infest. Res.* 1965:39,40.

LLOYD, C.J. (1967a). - Increased resistance of stored-product insects to insecticides. The grain weevil and pyrethrum. *Pest Infest. Res.* 1966:37.

LLOYD, C.J. (1967b). - Summary of cases of resistance to pesticides in agricultural pests. Item 8/10/1. *Rep. 1st Session FAO Working Party of Experts on Resistance of Pests to Pesticides, Meeting Rep.* PL/1965/18. (FAO: Rome). p. 86.

LLOYD, C.J. (1968). - Increased resistance of stored-product insects to insecticides. The grain weevil and pyrethrum. *Pest Infest. Res.* 1967:33,34.

LLOYD, C.J. (1969a). - Studies on the cross-tolerance to DDT-related compounds of a pyrethrin-resistant strain of *Sitophilus granarius* (L.) (Coleoptera, Curculionidae). *J. stored Prod. Res.* 5:337-356.

LLOYD, C.J. (1969b). - The synergism of DDT, deutero-DDT and methoxychlor in a pyrethrin-resistant strain of *Sitophilus granarius* (L.) (Coleoptera, Curculionidae). *J. stored Prod. Res.* 5:357-363.

LLOYD, C.J. (1973). - The toxicity of pyrethrins and five synthetic pyrethroids, to *Tribolium castaneum* (Herbst), and susceptible and pyrethrin-resistant *Sitophilus granarius* (L.) *J. stored Prod. Res.* 9:77-92.

LLOYD, C.J., BLACKMAN, D.G., CHAPMAN, P.A. and RUCZKOWSKI, G. (1976a). - Incidence of resistance in the United Kingdom. The grain weevil. Monitoring for resistance in the United Kingdom. *Pest Infest. Res.* 1971-73:74,75.

LLOYD, C.J. and PARKIN, E.A. (1963). - Further studies on a pyrethrin-resistant strain of the granary weevil, *Sitophilus granarius* (L.). *J. Sci. Fd Agric.* 14:655-663.

LLOYD, C.J., ROWLANDS, D.G. and RUCZKOWSKI, G. (1976b). - Incidence of resistance in the United Kingdom. The rust-red flour beetle. Resistance to synthetic pyrethroids. *Pest Infest. Control* 1971-73:81,82.

LLOYD, C.J. and SHAW, D.D. (1968). - Genetics. Pyrethrin resistance in grain weevils. *Pest Infest. Res.* 1967:36.

LLOYD, C.J., STEPHENS, A. and AITKEN, R.B. (1973). - Resistance monitoring in field strains of the grain weevil *Sitophilus granarius* L. *U.K. Min. Agric. Fish.& Food, Pest Infest. Control Lab., Res. Rep. No. 8.*

LLOYD, C.J. and WILLIAMS, V. (1973). - Detection of resistance in grain weevils. *Pest Infest. Control* 1968-70:122,123.

LOSCHIAVO, S.R. (1955). - Mortalities of males and females of *Tribolium confusum* Duv. (Coleoptera, Tenebrionidae) exposed to residual deposits of p,p'-DDT. *Can. Ent.* 37:407-410.

LOSCHIAVO, S.R. (1960). - Effects of low doses of ethylene dibromide on some stages of the confused flour beetle *Tribolium confusum*. *J. econ. Ent.* 53:762-767.

LUERS, H. (1963). - Reactions to x-rays of a normal and a HCN-unsusceptible stock of *Drosophila melanogaster*. *Int. J. Radiat. Biol.* 6:380.

McDOUGALL, W.A. (1964). - Insecticides fail on two pests. *Qd agric. J.* 90:24.

MacDONALD COLLEGE, (1967). - Summary of cases of resistance to pesticides in agricultural pests. Item 6/10/63. *Rep. 1st Session FAO Working Party of Experts on Resistance of Pests to Pesticides, Meeting Rep.* PL/1965/18. (FAO: Rome). p.55.

MAEDA, O. (1958). - Development of DDT resistance in the flour beetle *Tribolium confusum* Duv. *Botyu-Kagaku* 23:66-74.

MARQUES, E.K. (1973). - The development of radio-resistance in irradiated *Drosophila nebulosa* populations (Dipt., Drosophilidae). *Mutat. Res.* 17:57-72.

MATHLEIN, R. (1952). - Undersökningar över uppkomst av DDT-resistens hos kornvivel, *Calandra granaria* L. *Meddn St. Växsk. Anst.* No. 62, 20 pp.

MATTHEWS, G.A. (1958). - A preliminary investigation into the insect infestation of cargoes from West Africa. *Ann. appl. Biol.* 46:259-263.

MELLO, E.J.R. (1972). - Constatacao de resistencia ao DDT e lindane em *Sitophilus oryzae* (L.) em milho armazenado, na localidade de Capinipolis, Minas Gerais. *Rep. Instituto Biologico de Sao Paulo, Brazil*.

MENHINICK, E.F. and CROSSLEY, D.A. (1968). - A comparison of radiation profiles of *Acheta domesticus* and *Tenebrio molitor*. *Ann. ent. Soc. Am.* 61:1359-1365.

MILANI, R. (1963). - Genetical aspects of insecticide resistance. *Bull. Wld Hlth Org.* 29(Suppl):77-87.

MILLS, R.B. and BELL, K.O. (1972). - A red-eyed mutation in *Sitotroga cerealella* (Oliver) (Lepidoptera, Gelechiidae). *J. stored Prod. Res.* 8:235-236.

MIYAMOTO, J., SUZUKI, T. and NAKAE, C. (1974). - Metabolism of phenothrin or 3-phenoxybenzyl *d-trans*-chrysanthemumate in mammals. *Pestic. Biochem. Physiol.* 4:438-450.

MONRO, H.A.U. (1951). - Insect pests in cargo ships. *Can. Dep. Agric. Publ.* 855.

MONRO, H.A.U. (1963). - Fumigants - ancient and modern. *Pestic. Prog.* 1:133-137.

MONRO, H.A.U. (1964). - Insect resistance to fumigants. *Pest Control* 32:11-13,26.

MONRO, H.A.U. (1969). - Insect pests in cargo ships (Revised). *Can. Dep. Agric. Publ.* 855.

MONRO, H.A.U., MUSGRAVE, A.J. and UPITIS, E. (1961). - Induced tolerance of stored product beetles to methyl bromide. *Ann. appl. Biol.* **49**:373-377.

MONRO, H.A.U. and UPITIS, E. (1956). - Selection of populations of the granary weevil *Sitophilus granarius* (L.) more resistant to methyl bromide fumigation. *Can. Ent.* **88**:37-40.

MONRO, H.A.U., UPITIS, E. and BOND, E.J. (1972). - Resistance of a laboratory strain of *S. granarius* (L.) (Coleoptera, Cuculionidae) to phosphine. *J. stored Prod. Res.* **8**:199-207.

MOOKHERJEE, P.B., SHARMA, G.C. and TULI, S. (1964). - Effect of post-treatment temperature on the susceptibility of the khapra *(Trogoderma granarium* Everts) grubs to pyrethrins. *Indian J. Ent.* **26**:215-220.

MORALLO-REJESUS, B. (1972). - Survey of Philippine populations of rice weevil (*Sitophilus oryzae* (L.)) for resistance to insecticides. *3rd Ann. Intensified Corn Prod. Conf., U.P. College of Agriculture, Philippines,* 1972.

MORI, T., IKEGAMI, Y. and TATEYA, A. (1969). - Susceptibility of the pupae of rice weevil *Sitophilus zeamais* Motschulsky, to hydrogen phosphide. *Res. Bull. Plant Prot. Serv. Japan* **7**:67-70.

MORI, T. and KAWAMOTO, N. (1968). - Studies on the properties and effects of fumigant, aluminium phosphide. *Res. Bull. Plant Prot. Serv. Japan* **3**:23-35.

MORIARTY, F. (1969). - The sublethal effects of synthetic insecticides on insects. *Biol. Rev.* **44**:321-357.

MOSTAFA, S.A.S., KAMEL, A.H., EL-NAHAL, A.K.M. and EL-BOROLLOSY, F.M. (1972). - Toxicity of carbon bisulphide and methyl bromide to the eggs of four stored product pests. *J. stored Prod. Res.* **8**:193-198.

MUKERJEA, T.D. (1953). - The relationship between the stage of development and susceptibility to DDT and the pyrethrins of *Diataraxia oleracea* (L.), *Tenebrio molitor* L. and *Periplaneta americana* (L.). *Bull. ent. Res.* **44**:121-161.

MURTHY, K.S.R.K. and SRIVASTAVA, B.P. (1971). - Effect of food on the susceptibility of pulse beetle *Callosobruchus maculatus* Fab. (Bruchidae: Coleoptera) to different fumigants. *Indian J. Ent.* **33**:148-151.

MUTHU, M. (1973). - Some aspects of phosphine as a fumigant. *In.* "Fumigation and Gaseous Pasteurization", (ed. by S.K. Majumder and J.S. Venugopal).(Acad. Pest Control Sci.:Mysore). pp. 21-36.

MUTHU, M. and PINGALE, S.V. (1956). - Extent of variation caused by certain factors in the toxicity of ethylene dichloride-carbon tetrachloride mixtures to insects. *Indian J. Ent.* **17**:193-200.

NAIR, K.S.S. and DESAI, A.K. (1973). - Studies on the isolation of diapause and non-diapause strains of *Trogoderma granarium* Everts (Coleoptera, Dermestidae). *J. stored Prod. Res.* **9**:181-188.

NAIR, K.K. and RAHALKAR, D.W. (1963). - Studies on the effects of gamma radiation on the different developmental stages of the khapra beetle, *Trogoderma granarium* Everts. *In* "Radiation and Radioisotopes applied to Insects of Agricultural Importance". (Int. Atom. Energy Agency: Vienna). pp. 465-477.

NAKAKITA, H., SAITO, T. and IYATOMI, K. (1974). - Effect of phosphine on the respiration of adult *Sitophilus zeamais* Motsch. (Coleoptera, Curculionidae). *J. stored Prod. Res.* 10:87-92.

NEWSOM, L.D. (1967). - Summary of cases of resistance to pesticides in agricultural pests. Item 6/10/64. *Rep. 1st Session FAO Working Party of Experts on Resistance of Pests to Pesticides, Meeting Rep.* PL/1965/18. (FAO: Rome), p. 55.

NIEFERT, I.E., COOK, F.C., ROARK, R.C., TONKIN, W.H., BACK, E.A. and COTTON, R.T. (1925). - Fumigation against grain weevils with various volatile organic compounds. *U.S. Dep. Agric. Bull.* 893.

NOTHEL, H. (1970). - Investigations on radio-sensitive and radio-resistant populations of *Drosophila melanogaster*. I. Decreased radio-sensitivity in stage 7-oocytes of the irradiated population RO1. *Mutat. Res.* 10:463-474.

ODENEAL, J.F. (1961). - Insect resistance survey II. *Soap and Chem Spec.* 37:103-107.

OGAKI, M. and NAKASHIMA-TANAKA, E. (1966). - Inheritance of radio resistance in *Drosophila*. *Mutat. Res.* 3:438-443.

OUTRAM, I. (1966). - Studies on the sorption and toxicity of sulphuryl fluoride in certain species. Ph.D. Thesis, University of London.

OUTRAM, I. (1967). - Factors affecting the resistance of insect eggs to sulphuryl fluoride - I. The uptake of sulphuryl-^{35}S fluoride by insect eggs. *J. stored Prod. Res.* 3:255-260.

ÖZER, M. (1961). - Phostoxin' in degisik doz, muddet ve isida *Calandra granaria* L. ve *Calandra oryzae* L. nin biyolojik safhalarina karsi toksik etkisi. *Koruma* 2:19-35.

PARKIN, E.A. (1944). - Control of the granary weevil with finely ground mineral dusts. *Ann. appl. Biol.* 31:84-88.

PARKIN, E.A. (1958). - A provisional assessment of malathion for stored-product insect control. *J. Sci. Fd Agric.* 9:370-375.

PARKIN, E.A. (1965). - The onset of insecticide resistance among field populations of stored product insects. *J. stored Prod. Res.* 1:3-8.

PARKIN, E.A. (1967). - Summary of cases of resistance to pesticides in agricultural pests. Items 1/10/1,3; 2/10/4,5. *Rep. 1st Session FAO Working Party of Experts on Resistance of Pests to Pesticides, Meeting Rep.* PL/1965/18. (FAO: Rome). pp. 30,32.

PARKIN, E.A. (1969). - World survey for resistance. Items 4/10/2-5. *Rep. 4th Session FAO Working Party of Experts on Resistance of Pests to Pesticides, Meeting Rep.* PL/1968/M/10. (FAO: Rome). p. 29.

PARKIN, E.A. and BALL, G.M. (1962). - The age of beetles and their susceptibility to insecticides. *Pest Infest. Res.* 1961:35.

PARKIN, E.A. and BRIGHT, P.E. (1965). - Increased resistance of stored product insects to insecticides. Cross-resistance of malathion-resistant flour beetles. *Pest Infest. Res.* 1964:34.

PARKIN, E.A. and DYTE, C.E. (1967). - Summary of cases of resistance to pesticides in agricultural pests. Item 1/10/6. *Rep. 1st Session FAO Working Party of Experts on Resistance of Pests to Pesticides, Meeting Rep.* PL/1965/18. (FAO: Rome). p. 30.

PARKIN, E.A. and FORSTER, R. (1963a). - Increased resistance of stored-product insects to insecticides. The rice weevil and lindane. *Pest Infest. Res.* 1962:37,38.

PARKIN, E.A. and FORSTER, R. (1963b). - Increased resistance of stored-product insects to insecticides. The rust-red flour beetle and malathion. *Pest Infest. Res.* 1962:38,39.

PARKIN, E.A. and FORSTER, R. (1964a). - Increased resistance of stored-product insects to insecticides. The rice weevil and lindane. *Pest Infest. Res.* 1963:31.

PARKIN, E.A. and FORSTER, R. (1964b). - Increased resistance of stored-product insects to insecticides. The rust-red flour beetle and malathion. *Pest Infest. Res.* 1963:32.

PARKIN, E.A. and FORSTER, R. (1965a). - Increased resistance of stored-product insects to insecticides. The maize weevil and lindane. *Pest Infest. Res.* 1964:33.

PARKIN, E.A. and FORSTER, R. (1965b). - Increased resistance of stored-product insects to insecticides. The rice weevil and lindane. *Pest Infest. Res.* 1964:33.

PARKIN, E.A. and FORSTER, R. (1965c). Increased resistance of stored-product insects to insecticides. The rust-red flour beetle and malathion. *Pest Infest. Res.* 1964:33,34.

PARKIN, E.A. and FORSTER, R. (1965d). - Tests of newer insecticides. Dichlorvos vapour. *Pest Infest. Res.* 1964:36.

PARKIN, E.A. and FORSTER, R. (1966a). - Increased resistance of stored-product insects to insecticides. The maize weevil and lindane. *Pest Infest. Res.* 1965:40.

PARKIN, E.A. and FORSTER, R. (1966b). - Increased resistance of stored-product insects to insecticides. The rice weevil and lindane. *Pest Infest. Res.* 1965:40.

PARKIN, E.A. and FORSTER, R. (1966c). - Increased resistance of stored-product insects to insecticides. Saw-toothed grain beetle and lindane. *Pest Infest, Res.* 1965:40.

PARKIN, E.A. and FORSTER, R. (1966d). - Increased resistance of stored-product insects to insecticides. The rust-red flour beetle and malathion. *Pest Infest. Res.* 1965:40.

PARKIN, E.A. and FORSTER, R. (1967a). - Increased resistance of stored-product insects to insecticides. The rice weevil and malathion. *Pest Infest. Res.* 1966:37.

PARKIN, E.A. and FORSTER, R. (1967b). - Increased resistance of stored-product insects to insecticides. Saw-toothed grain beetle and malathion. *Pest Infest. Res.* 1966:37.

PARKIN, E.A. and FORSTER, R. (1968a). - Increased resistance of stored-product insects to insecticides. The maize weevil and lindane. *Pest Infest. Res.* 1967:34.

PARKIN, E.A. and FORSTER, R. (1968b). - Increased resistance of stored-product insects to insecticides. The rice weevil and bromophos. *Pest Infest. Res.* 1967:34.

PARKIN, E.A. and FORSTER, R. (1968c). - Increased resistance of stored-product insects to insecticides. The groundnut borer and dieldrin, *Pest Infest. Res.* 1967:34.

PARKIN, E.A. and GREEN, A.A. (1960). - Some practical problems in the use of contact insecticides against stored-product insects. *Rep. 7th Cwlth. Ent. Conf., London,* 1960, pp. 68-72.

PARKIN, E.A. and LLOYD, C.J. (1960a). - Selection of a pyrethrum-resistant strain of the grain weevil, *Calandra granaria* L. *J. Sci. Fd Agric.* 11:471-477.

PARKIN, E.A. and LLOYD, C.J. (1960b). - Increased resistance of insects to contact insecticides. The grain weevil and pyrethrum. *Pest Infest. Res.* 1959:25.

PARKIN, E.A. and LLOYD, C.J. (1961). - Increased resistance of insects to contact insecticides. The grain weevil and pyrethrum. *Pest Infest. Res.* 1960:36.

PARKIN, E.A. and LLOYD, C.J. (1962). - Increased resistance of stored-product insects to insecticides. The grain weevil and pyrethrum. *Pest Infest. Res.* 1961:33,34.

PARKIN, E.A. and LLOYD, C.J. (1963). - Increased resistance of stored-product insects to insecticides. The grain weevil and pyrethrum. *Pest Infest. Res.* 1962:37.

PARKIN, E.A. and LLOYD, C.J. (1964). - Increased resistance of stored-product insects to insecticides. The grain weevil and pyrethrum. *Pest Infest. Res.* 1963:31.

PARKIN, E.A., SCOTT, E.I.C. and FORSTER, R. (1962a). - Increased resistance of stored-product insects to insecticides. The resistance of field strains of beetles. (a) *Sitophilus zeamais* and *S. oryzae*. *Pest Infest. Res.* 1961:34.

PARKIN, E.A., SCOTT, E.I.C. and FORSTER, R. (1962b). - Increased resistance of stored-product insects to insecticides. The resistance of field strains of beetles. (b) *Oryzaephilus surinamensis*. *Pest Infest. Res.* 1961:34.

PARKIN, E.A., SCOTT, E.I.C. and FORSTER, R. (1962c). - Increased resistance of stored-product insects to insecticides. The resistance of field strains of beetles. (c) *Tribolium castaneum*. *Pest Infest. Res.* 1961:34,35.

PARKIN, E.A., WARMAN, A. and FORSTER, R. (1963). - Carbaryl and fenchlorphos dusts. *Pest Infest. Res.* 1962:40.

PARSONS, P.A., MacBEAN, I.I. and LEE, B.T.O. (1969). - Polymorphism in natural populations for genes controlling radio-resistance in *Drosophila*. *Genetics.* 61:211-218.

PASALU, I.C. (1974). - Studies on the inheritance of resistance to malathion in *Tribolium castaneum* (Herbst). *Ent. Newsl. (Indian Agric. Res. Inst.)* 4:7.

PASALU, I.C. and BHATIA, S.K. (1974a). - Laboratory evaluation of some insecticides against malathion-resistant and susceptible strains of *Tribolium castaneum* (Herbst). *Bull. Grain Technol. India* 12:175-179.

PASALU, I.C. and BHATIA, S.K. (1974b). - Specific nature of malathion resistance in *Tribolium castaneum* in India. *Bull. Grain Technol. India* 12:229.

PEDERSEN, J.R. (1960). - Susceptibility of certain stages of the rice weevil to a methallyl chloride fumigant formulation in wheat of various moistures. *J. econ. Ent.* 53:288-291.

PENNELL, J.T., MISKUS, R. and CRAIG, R. (1964). - The use of gas chromatography for the quantitative determinations of micro-amounts of insecticides picked up by mosquitos. *Bull. Wld Hlth Org.* 30:91-95.

PETERS, G. (1938). - Die biologisch-chemische Eignungsprüfung gasförmig wirkender Insektizide. *Anz. P. Schädlingsk* 14:116-122.

PETERS, G. and GANTNER, W. (1935). - Zur Frage der Abtötung des Kornkäfers mit Blausäure. *Z. angew. Ent.* 21:547-559.

PHILIP, V. (1940). - A genetical analysis of three small populations of *Dermestes vulpinus* F. (Coleoptera). *Proc. Indian Acad. Sci.* 12:133-171.

PIETERSE, A.H. and SCHULTEN, G.G.M. (1974). - Investigations on insecticide resistance in *Tribolium castaneum, T. confusum* and *S. zeamais* in small maize cribs in Malawi. *Trop. Agric. Trin.* 51:63-67.

PIETERSE, A.H., SCHULTEN, G.G.M. and KUIJKEN, W. (1972). - A study of insecticide resistance in *Tribolium castaneum* (Herbst) (Coleoptera, Tenebrionidae) in Malawi (Central Africa). *J. stored Prod. Res.* 8:183-191.

PINNIGER, D.B. (1975). - The behaviour of insects in the presence of insecticides: the effect of fenitrothion and malathion on resistant and susceptible strains of *Tribolium castaneum* Herbst. *Proc. 1st int. wking Conf. Stored Prod. Ent., Savannah,* 1974, pp. 301-308.

PO-CHEDLEY, D.S. (1969). - Radio-sensitivity and water content for yellow mealworm embryos. *J. econ. Ent.* 62:1505.

POTTER, C. (1938). - The use of protective films of insecticide in the control of indoor insects with special reference to *Plodia interpunctella* Hb. and *Ephestia elutella* Hb. *Ann. appl. Biol.* 25:836-854.

POTTER, C. and GILLHAM, E.M. (1946). - Effects of atmospheric environment before and after treatment, on the toxicity to insects of contact poisons. I. *Ann. appl. Biol.* 33:142-159.

PRADHAN, S. (1949a). - Studies on the toxicity of insecticide films. II. Effect of temperature on the toxicity of DDT films. *Bull. ent. Res.* 40:239-265.

PRADHAN, S. (1949b). - Studies on the toxicity of insecticide films. III. Effect of relative humidity on the toxicity of films. *Bull. ent. Res.* 40:431-444.

PRADHAN, S., CHATTERJI, S.M., SETHI, G.R., BHAMBURKAR, M.W. and PRASAD, H. (1971). - Feasibility of controlling stored grain pests by the sterile-male technique. In "Sterility Principle for Insect Control or Eradication". Int. Atom. Energy Agency, Symposium, Athens, 1970. (I.A.E.A.: Vienna). p. 365.

PRADHAN, S. and GOVINDAN, M. (1954). - Effect of temperature on the degree of susceptibility of insects to fumigation. Indian J. Ent. 16:115-136.

PREVETT, P.F. (1971). - Some laboratory observations on the development of two African strains of Plodia interpunctella (Hübn.) (Lepidoptera, Phyticidae), with particular reference to the incidence of diapause. J. stored Prod. Res. 7:253-260.

PUNJ, G.K. (1970). - The effect of nutrition on the susceptibility of larvae of Trogoderma granarium Everts (Coleoptera, Dermestidae) to certain fumigants. J. stored Prod. Res. 6:181-185.

PUNJ, G.K. (1971). - The effect of certain artificial diets on the susceptibility to ethylene oxide of larvae of Trogoderma granarium Everts (Coleoptera, Dermestidae). J. stored Prod. Res. 6:347-349.

PUNJ, G.K. and GIRISH, G.K. (1968). - Effect of nutrition and starvation on the susceptibility of adults of Tribolium castaneum (Herbst) to carbon disulphide. Bull. Grain Technol. India 6:187-191.

PUNJ, G.K. and GIRISH, G.K. (1969). - Relative toxicity of certain fumigants to Trogoderma granarium Everts (Coleoptera, Dermestidae). J. stored Prod. Res. 4:339-342.

QUALE, H.J. (1938). - The development of resistance to hydrocyanic acid in certain scale insects. Hilgardia 11:183-210.

QURESHI, A.H., BOND, E.J. and MONRO, H.A.U. (1965). - Toxicity of hydrogen phosphide to the granary weevil, Sitophilus granarius, and other insects. J. econ. Ent. 58:324-331.

RAHALKAR, D.N. and NAIR, R.R. (1968). - Influence of diapause on the radio-sensitivity of khapra beetle larvae. In "Isotopes and Radiation in Entomology" (Int. Atom. Energy Agency: Vienna). pp. 149-154.

RAI, B.K., BULLOCK, O.J.W. and SLADE, P. (1972). - Admixture of pirimiphos methyl with paddy to control storage pests. Unpublished report, Central Agriculture Station, Mon Repos, Guyana.

RAJAK, R.L., GHATE, M. and KRISHNAMURTHY, K. (1973). - Bioassay technique for resistance to malathion of stored product insects. Int. Pest Control 15:11-16.

RAJAK, R.L. and HEWLETT, P.S. (1971). - Effects of some synergists on the insecticide potency of phosphine. J. stored Prod. Res. 7:15-19.

RATTAN LAL and ATTRI, B.S. (1967). - Effect of food on the susceptibility of Trogoderma granarium Everts to insecticides. Indian J. Ent. 29:329-338.

RATTAN LAL and SINGH, Y.P. (1966). - Effect of food on the susceptibility of Tribolium castaneum Herbst to contact insecticides. Indian J. Ent. 28:89-93.

REYNOLDS, E.M., BENNETT, R.G., DUFFIN, P. and GOODSHIP, G. (1968). - Toxicity of methyl bromide to insects: concentration-time relationships. *Pest Infest. Res.* 1967:57,58.

REYNOLDS, E.M., ROBINSON, J.M. and HOWELLS, C. (1967). - The effect on *Sitophilus granarius* (L.) (Coleoptera, Curculionidae) of exposure to low concentrations of phosphine. *J. stored Prod. Res.* 2:177-186.

REYNOLDS, E.M. and SYLVESTER, N.K. (1962). - The inheritance of a pearl-eyed mutation in *Trogoderma granarium* Everts (Coleoptera, Dermestidae). *Entomologists mon. Mag.* 97:221-224.

REYNOLDS, E.M. and SYLVESTER, N.K. (1964). - The toxicity of methyl bromide to six populations of *Tribolium castaneum*. *Tribolium Inf. Bull.* 7:70,71.

REYNOLDS, E.M., SYLVESTER, N.K., WELCH, J. and GOODSHIP, G. (1964). - Toxicity of fumigants to insects. The toxicity of sulphuryl fluoride. *Pest Infest. Res.* 1963:42-44.

RICHARDSON, H.H. (1943). - Toxicity of derris, nicotine and other insecticides to eggs of the housefly and the Angoumois grain moth. *J. econ. Ent.* 36:729-731.

ROBERTSON, J.G. (1959). - Cytological affinities in *Cryptolestes* and *Oryzaephilus* (Coleoptera, Cucujidae). *Can. J. Genet. Cytol.* 1:78-83.

ROBINSON, R. (1971). - "Lepidoptera genetics". (Pergamon Press: New York).

ROWLANDS, D.G. (1967). - The metabolism of contact insecticides in stored grains. *Residue Rev.* 17:105-177.

ROWLANDS, D.G., DALY, J.A. and BLACKMAN, D.G. (1973). - Lindane resistance in the rust-red flour beetle. *Pest Infest. Control* 1968-70:119.

ROWLANDS, D.G. and LLOYD, C.J. (1968). - Studies on pyrethrin-resistant grain weevils: penetration and metabolism of DDT. *Pest Infest. Res.* 1967:35,36.

ROWLANDS, D.G. and LLOYD, C.J. (1969a). - Detoxication in pyrethrum-resistant weevils. *Pest Infest. Res.* 1968:42.

ROWLANDS, D.G. and LLOYD, C.J. (1969b). - DDT metabolism in susceptible and pyrethrin-resistant *Sitophilus granarius* (L.). *J. stored Prod. Res.* 5:413-415.

ROWLANDS, D.G. and LLOYD, C.J. (1976). - Incidence of resistance in the United Kingdom. The grain weevil. Metabolism and synergism of pyrethroids in susceptible and resistant strains. *Pest Infest. Control* 1971-73:75-77.

SARDESAI, J.B. (1968). - The evolution, bionomics, and physiology of diapause in *Plodia interpunctella* (Hübner) and its relationship to the action of some fumigants. Ph.D. Thesis, Univ. Calif., Riverside, 222 pp. Univ. Microfilms Inc.

SARDESAI, J.B. (1972). - Response of diapausing and non-diapausing larvae of *Plodia interpunctella* to hydrogen cyanide and methyl bromide. *J. econ. Ent.* 65:1562-1565.

SATO, K., HIGUCHI, Y. and SUWANAI, M. (1973). - Studies on the characteristics of action of fumigants - I. The 50 per cent knock-down dose of hydrogen phosphide to the azuki bean weevil, *Callosobruchus chinensis* L., calculated from the uptake amounts of oxygen by the weevil. *Botyu-Kagaku* 38:22-25.

SCHMITZ, T.H. and ENGLERT, D.C. (1967). - The mottled mutation in *Tribolium castaneum*. *Can. J. Genet. Cytol.* 9:335-341.

SCHWARTZ, V. (1953). - Zur Phänogenese der Flügelzeichnung von *Plodia interpunctella*. *Z. Vererbungslehre* 85:51-96.

SCOTT, E.I.C. (1960). - Increased resistance of insects to contact insecticides. The rice weevil and lindane. *Pest Infest. Res.* 1959:25.

SCOTT, E.I.C. (1961). - Increased resistance of insects to contact insecticides. The rice weevil and lindane. *Pest Infest. Res.* 1960:37.

SCOTT, E.I.C. and FORSTER, R. (1962). - Increased resistance of stored-product insects to insecticides. The rice weevil and lindane. *Pest Infest. Res.* 1961:34.

SEELIGER, R. (1943). - Genetische Untersuchungen an dem Flügelmuster des Bohnenkäfers *Zabrotes subfasciatus* Boh. (Coleopt. Bruchidae). *Z. Vererbungslehre* 81:196-251.

SEREBROVSKY, A.S. (1940). - On the possibility of a new method for the control of insect pests. *Zoologicheskii Zhurnal* 19:618.

SEVINTUNA, C. and MUSGRAVE, A.J. (1961). - Observations on males and females of *Sitophilus granarius* (L.) the granary weevil, GC Strain, exposed for 6 generations to allethrin and piperonyl butoxide. *Can. Ent.* 93:545-552.

SHAW, D.D. (1966). - The inheritance of two mutations 'pearl' and 'fuscous' in *Dermestes maculatus* de Geer (Coleoptera, Dermestidae). *J. stored Prod. Res.* 1:261-265.

SHAW, D.D. (1967a). - Section on "New Mutants". *Tribolium Inf. Bull.* 10:65,66.

SHAW, D.D. (1967b). - Sex chromosome variation in *D. maculatus* and *D. frischii*. *Tribolium Inf. Bull.* 10:134,135.

SHAW, D.D. (1968). - Selection for supernumerary Y-chromosomes in *Dermestes maculatus* (Coleoptera, Dermestidae). *Can. J. Genet. Cytol.* 10:54-62.

SHAW, D.D. and LLOYD, C.J. (1967). - Increased resistance of stored-product insects to insecticides. The leather beetle and lindane. *Pest Infest. Res.* 1966:37,38.

SHAW, D.D. and LLOYD, C.J. (1969). - Selection for lindane resistance in *Dermestes maculatus* de Geer (Coleoptera, Dermestidae). *J. stored Prod. Res.* 5:69-72.

SHAW, D.D., LLOYD, C.J. and BINNS, T.J. (1968). - Increased resistance of stored-product insects to insecticides. The leather beetle and lindane. *Pest Infest. Res.* 1967:34.

SHAW, D.D. and WELCH, J.J. (1966). - Section of "New Mutants". *Tribolium Inf. Bull.* 9:58.

SHAW, D.D. and WELCH, J.J. (1967). - Formal genetics. The leather beetle. *Pest Infest. Res.* 1966:40.

SHEPARD, H.H., LINDGREN, D.L. and THOMAS, E.L. (1937). - The relative toxicity of insect fumigants. *Minn. agric. Exp. Sta., Tech. Bull.* 120, 23pp.

SHUTTLEWORTH, S.G. and GALLOWAY, A.C. (1961). - Insecticide for controlling dieldrin resistant Dermestidae (Coleoptera) on skins. *Res. Bull. Leath. Inds. Res. Inst. Grahamstown* 262:1-5.

SKAIFE, S.H. (1925). - On variation and heredity in the Bruchidae. *Trans. R. Soc. S. Afr.* 12:221-242.

SLOAN, W.J.S. (1967). - Summary of cases of resistance to pesticides in agricultural pests. Item 7/10/4. *Rep. 1st Session FAO Working Party of Experts on Resistance of Pests to Pesticides, Meeting Rep.* PL/1965/18. (FAO: Rome). p. 73.

SMITH, L.B. (1956). - A mutant in the Indian meal moth *Plodia interpunctella* (Hbn.) (Lepidoptera, Phyticidae) with a white eye. *Ann. Rep. ent. Soc. Ontario* 87:40-42.

SMITH, S.G. (1952a). - The cytology of *Sitophilus (=Calandra) oryzae* (L.), *S. granarius* (L.), and some other Rhynchophora (Coleoptera). *Cytologia* 17:50-70.

SMITH, S.G. (1952b). - The evolution of heterochromatin in the genus *Tribolium* (Tenebrionidae, Coleoptera). *Chromosoma* 4:585-610.

SMITH, S.G. (1952c). - The cytology of some tenebrionid beetles (Coleoptera). *J. Morph.* 91:325-364.

SMITH, S.G. and BROWER, J.H. (1974). - Chromosome numbers of stored-product Coleoptera. *J. Kans. ent. Soc.* 47:317-328.

SNOW, R. (1962). - The chromosome number of *Gnathocerus cornutus*. *Tribolium Inf. Bull.* 5:39.

SOKOLOFF, A. (1959). - The genetics of "pearl" in *Latheticus oryzae* Waterh. *Can. J. Genet. Cytol.* 1:183-188.

SOKOLOFF, A. (1963). - Section on "New Mutants". *Tribolium Inf. Bull.* 6:23-32.

SOKOLOFF, A. (1964a). - Section on "New Mutants". *Tribolium Inf. Bull.* 7:31-43.

SOKOLOFF, A. (1964b). - A dominant synthetic lethal in *Tribolium castaneum* Herbst. *Am. Nat.* 98:127,128.

SOKOLOFF, A. (1965). - Section on "New Mutants". *Tribolium Inf. Bull.* 8:43,44.

SOKOLOFF, A. (1966a). - The genetics of *Tribolium* and related species. (Academic Press: New York).

SOKOLOFF, A. (1966b). - Section of "Stock Lists". *Tribolium Inf. Bull.* 9:10.

SOKOLOFF, A. (1971). - Section on "Stock Lists". *Tribolium Inf. Bull.* 14:24.

SOKOLOFF, A. and HO, F.K. (1963). - Section of "New Mutants". *Tribolium Inf. Bull.* 6:23-32.

SOKOLOFF, A. and HOY, M.A. (1965). - Possible genetic basis for prothetely in *Tribolium castaneum* and *Latheticus oryzae*. Tribolium Inf. Bull. <u>8</u>:150,151.

SOKOLOFF, A., HOY, M.A. and JOHNSON, G.R. (1966). - Linkage studies in *Tribolium castaneum*. XI. The map position of "platinum eye" and "spatulate". J. stored Prod. Res. <u>1</u>:225-260.

SOKOLOFF, A. and SHRODE, R.R. (1960). - Linkage studies on *Latheticus oryzae* Waterh. I. Recombination between "red" and "truncated elytra". Can. J. Genet. Cytol. <u>2</u>:418-428.

SOLIMAN, M.H. (1972). - Age at pupation and population effects on survival of adults of the flour beetle *Tribolium* after x-irradiation. Int. J. Radiat. Biol. <u>22</u>:425-430.

SOLIMAN, M.H. (1973). - "Geographic" variation in sensitivity to x-irradiation in *Tribolium*. I. Differences in adult emergence due to species, strain and age at pupation. Radiat. Res. <u>54</u>:510-520.

SPEIRS, R.D. (1967). - Summary of cases of resistance to pesticides in agricultural pests. Items 6/10/65-72, 6/16/7-9. *Rep. 1st Session FAO Working Party of Experts on Resistance of Pests to Pesticides*, Meeting Rep. PL/1965/18. (FAO: Rome). pp. 55, 56, 63, 66.

SPEIRS, R.D., BOLES, H.P., LANG, J.H. and REDLINGER, L.M. (1964). - Malathion resistance in red flour beetles collected from stored products. Bull. ent. Soc. Am. <u>10</u>:157,176.

SPEIRS, R.D., REDLINGER, L.M. and BOLES, H.P. (1967). - Malathion resistance in red flour beetles. J. econ. Ent. <u>60</u>:1373,1374.

SPEIRS, R.D., REDLINGER, L.M. and JONES, R. (1971). - DDT-resistant red flour beetles from a Georgia peanut sheller. J. econ. Ent. <u>64</u>:1328,1329.

SPEIRS, R.D. and ZETTLER, J.L. (1969). - Toxicity of three organophosphorus compounds and pyrethrins to malathion-resistant *Tribolium castaneum* (Herbst) (Coleoptera, Tenebrionidae). J. stored Prod. Res. <u>4</u>:279-283.

STAHLER, N. (1971). - Changes in survival and tolerance levels in a radiation-resistant strain of *Aedes aegypti* (Diptera, Culicidae), during 90 generations. Ann. ent. Soc. Am. <u>64</u>:1247-1249.

STRONG, R.G., PARTIDA, G.J. and ARCHER, T.L. (1969). - Genetic plasticity of stored-product insects in relation to insecticides: comparative susceptibility of confused and red flour beetles from various areas of California to malathion. J. econ. Ent. <u>62</u>:470-474.

SUN, Y.P. (1947). - An analysis of some important factors affecting the results of fumigation tests on insects. Minn. agric. Exp. Sta., Tech. Bull. 177, 104pp.

SURTEES, G. (1966). - Locomotory behaviour of pyrethrum-resistant and susceptible strains of grain weevil, *Sitophilus granarius* (L.) (Coleoptera, Curculionidae). Anim. Behaviour <u>14</u>:201-203.

SYLVESTER, N.K. (1964). - Studies on the toxic effect of certain fumigants to insect eggs and their variation in resistance with stages of embryonic development. M.Sc Thesis, Reading University.

SYLVESTER, N.K. and GOODSHIP, G. (1960). - Susceptibility of *Tribolium confusum* eggs to fumigation. *Pest Infest. Res.* 1959:34.

SYLVESTER, N.K., GOODSHIP, G. and WELCH, J. (1963). - The toxicity of fumigants to insects. Toxicity of fumigants to insect eggs. *Pest Infest. Res.* 1962:50-52

TAKAHASHI, F., KITAMURA, C., KUWAHARA, Y. and FUKAMI, N. (1968). - Studies on sex pheromones of Pyralididae. II. Mass rearing of virgin females of the almond moth *Cadra cautella* Walker (Phycitinae). *Botyu-Kagaku* 33:163-168.

TAKAHASHI, F. and KUWAHARA, Y. (1970). - Studies on sex pheromones of Pyralididae. III. The inherittance of the abnormal sex ratio condition in a strain of the almond moth, *Cadra cautella* Walker (Phycitinae). *Botyu-Kagaku* 35:11-21.

TAKAHASHI, F., SAKURAI, M. and KUWAHARA, Y. (1973). - A clear wing mutant in *Plodia interpunctella* Hübner (Lepidoptera: Pyralidae) - the mode of its inheritance and modification of its expression by temperature. *Kontyû* 41:256-261.

TEOTIA, T.P.S. and PANDEY, K.K. (1967). - The influence of temperature and humidity on the contact toxicity of some insecticide deposits to *Tribolium castaneum* Herbst. *Indian J. Ent.* 5:155-160.

TERZIAN, L.A. and STAHLER, N. (1966). - A selected strain of *Aedes aegypti* resistant to gamma radiation. *Radiat. Res.* 28:643-646.

TIELECKE, H. (1960). - Eine Analyse unterschiedlicher Resistenz männlicher und weiblicher Kornkäfer gegen synthetische Kontaktinsektizide. *Proc. 11th int. Congr. Ent., Vienna,* 1960. 2:657-662.

TILTON, E.W., BURKHOLDER, W.E. and COGBURN, R.R. (1966). - Effects of gamma radiation on *Trogoderma glabrum* and *Attagenus piceus*. *J. econ. Ent.* 59:944-948.

TOPPOZADA, A., ISMAIL, F.I. and ELDEFRAWI, M.E. (1969). - Susceptibility of local strains of *Sitophilus oryzae* (L.) and *Tribolium castaneum* (Herbst) to insecticides. *J. stored Prod. Res.* 5:393-397.

TSUKAMOTO, M. (1964). - Methods for the linkage-group determination of insecticide resistance factors in the housefly. *Botyu-Kagaku* 29:51-59.

TURTLE, E.E. (1961). - The control of infestation in commodities in ports in the United Kingdom. *Sanitarian* 2(70):145-147.

TURTLE, E.E. and FREEMAN, J.A. (1960). - Residual insect infestations and their control in holds of ships. *Proc. 4th int. Congr. Crop. Prot.,* Hamburg, 1957. 2:1729-1732.

TYLER, P.S. and BINNS. T.J. (1973). - Laboratory evaluation of insecticides against susceptible and malathion-resistant strains of *O. surinamensis* (L.) (Coleoptera, Silvanidae). *Min. Agric. Fish. & Food, Pest Infest. Control Lab., Res. Rep.* No. 9.

TYLER, P.S. and BINNS, T.J. (1976). - Incidence of resistance in the United Kingdom. Other beetles. The hide beetle. *Pest Infest. Control* 1971-73:84,85.

UPITIS, E., MONRO, H.A.U. and BOND, E.J. (1973). - Some aspects of inheritance of tolerance to methyl bromide by *Sitophilus granarius* (L.) *J. stored Prod. Res.* 9:13-17.

VEREECKE, A. and PELERENTS, C. (1969). - Sensitivity to gamma radiation of *Tribolium confusum* eggs at various developmental stages. *Ent. exp. Appl.* 12:62-66.

VINCENT, L.E. and LINDGREN, D.L. (1967). - Susceptibility of laboratory and field collected cultures of the confused flour beetle and red flour beetle to malathion and pyrethrins. *J. econ. Ent.* 60:1763-1764.

VINCENT, L.E. and LINDGREN, D.L. (1972a). - Toxicity of phosphine to the life stages of four species of dermestids. *J. econ. Ent.* 65:1429-1431.

VINCENT, L.E. and LINDGREN, D.L. (1972b). - Hydrogen phosphide and ethyl formate: fumigation of insects infesting dates and other dried fruits. *J. econ. Ent.* 65:1667-1669.

VOLKOVA, K.V. (1940). - Genetics of the grain moth. *C.R. Acad. Sci.USSR.* 26:604-606.

WEHRMAKER, A. (1975a). - A non-autonomous mutant of eye color and scale color in *Plodia interpunctella* (Lepidoptera). *Naturwissenschaften* 62:440.

WEHRMAKER, A. (1975b). - A simple feeding method for demonstrating non-autonomous development of mutant traits in Lepidoptera. *Verh. Dfsch. Zool. Ges.* 67:138-142.

WHEATLEY, P.E. (1967). - Summary of cases of resistance to pesticides in agricultural pests. Item 1/10/4. *Rep. 1st Session FAO Working Party of Experts on Resistance of Pests to Pesticides, Meeting Rep.* PL/1965/18. (FAO: Rome). p. 130.

WHITING, P.W. (1919). - Genetic studies on the Mediterranean flour moth, *Ephestia kühniella* Zeller. *J. exp. Zool.* 28:413-445.

WHITNEY, W.K. and WALKDEN. H.H. (1961). - Concentrations of methyl bromide lethal to insects in grain. *U.S. Dep. Agric., Mktg Res. Rep.* 511, 25pp.

WILKIN, D.R. (1966). - Insects infesting farm stores. Tolerance to insecticides of farm strains of *O. surinamensis*. *Pest Infest. Res.* 1965:24.

WILKIN, D.R. (1967). - Insects infesting farm stores. Tolerance to insecticides of farm strains of *O. surinamensis*. *Pest Infest. Res.* 1966:25.

WILKIN, D.R. (1973). - Resistance to lindane in *Acarus siro* from an English cheese store. *J. stored Prod. Res.* 9:101-104.

WILKIN, D.R., PRAGNELL, L. *et al.* (1976a). - Incidence of resistance in the United Kingdom. Mites. Mites from cheese stores. *Pest Infest. Control* 1971-73:85,86.

WILKIN, D.R., PRAGNELL, L. *et al.* (1976b). - Incidence of resistance in the United Kingdom. Mites. *Acarus siro* from farm stored grain. *Pest Infest. Control* 1971-73:86.

WILKIN, D.R., PRAGNELL, L. et al. (1976c). - Incidence of resistance in the United Kingdom. Mites. *Glycyphagus destructor*. *Pest Infest. Control* 1971-73:86.

WINKS, R.G. (1969). - Resistance to the fumigant phosphine in a strain of *Tribolium castaneum* (Herbst). *Insect Toxicol. Inf. Serv.* 12:178.

WINKS, R.G. (1971). - The inhibitory effect of phosphine on reproduction of *Tribolium castaneum* (Herbst). M.Sc. Thesis, University of Queensland, 145 pp.

WINKS, R.G. (1973). - Some aspects of the response of *Tribolium castaneum* (Herbst) to phosphine. Ph.D. Thesis, University of London, 214 pp.

WINKS, R.G. and BAILEY, S.W. (1965). - Treatment and storage of export wheat in Australia. *Trop. stored Prod. Inf.* 11:431-438.

WOOL, D. (1971). - A new approach to insecticide resistance. Ecological-genetic speculations. *Bull. ent. Soc. Am.* 17:133-135.

WOOL, D. (1975). - Genetic control of insecticide-resistance in stored product insects: prospects and preliminary investigations. *Proc. 1st int. wking Conf. Stored-Prod. Ent., Savannah, 1974*, pp.310-316.

YAMAMOTO, I. (1973). - Mode of action of synergists in enhancing the insecticidal activity of pyrethrum and pyrethroids. In "Pyrethrum the natural insecticide", ed. by J.E. Casida. (Academic Press; N. York and London). pp. 195-210.

YANA, A. (1967). - Note sur l'efficacité d'un insecticide de la série des organophosphorés: le sumithion (fenitrothion) sur quelques souches Tunisiennes d'insectes des grains. *Doc. tech. Inst. Nat. Rech. Agron. Tunisie* 25:1-15.

YANG, T.C.H. and DUCOFF, H.S. (1969). - Radio sensitivity studies of x-irradiated *Tribolium castaneum* larvae. *Radiat. Res.* 39:643-654.

YANG, T.C.H. and DUCOFF, H.S. (1971). - Recovery studies of x-irradiated *Tribolium castaneum* (flour beetle) larvae. *Radiat. Res.* 46:290-300.

YUST, H.R. and SHELDEN, F.F. (1952). - A study of the physiology of resistance to hydrocyanic acid in the California red scale. *Ann. ent. Soc. Am.* 45:220-228.

ZETTLER, J.L. (1974a). - PP511: toxicity to malathion-resistant strains of the Indian meal moth. *J. econ. Ent.* 67:450,451.

ZETTLER, J.L. (1974b). - Esterases in a malathion-susceptible and a malathion-resistant strain of *Plodia interpunctella* (Lepidoptera: Phycitidae). *J. Georgia ent. Soc.* 9:207-213.

ZETTLER, J.L. (1974c). - Malathion resistance in *Tribolium castaneum* collected from stored peanuts. *J. econ. Ent.* 67:339-340.

ZETTLER, J.L. (1975). - Malathion resistance in strains of *Tribolium castaneum* collected from rice in the U.S.A. *J. stored Prod. Res.* 11:115-117.

ZETTLER, J.L. and LeCATO, G.L. (1974). - Sublethal doses of malathion and dichlorvos: effects on fecundity of the black carpet beetles. *J. econ. Ent.* 67:19-21.

"best fit," making it impossible to distinguish the quality of fits resulting from any of a number of pairs or trios of values for the correlated parameters. Davies and Ku (1977) have provided an example of this problem in an application of the Richards equation, pointing out that the parameters m, $M(0)$, and K must be considered as a set of correlated parameters rather than independently estimable. This difficulty makes it virtually impossible to interpret estimated values of m in the Richard's equation. The similar shapes of von Bertalanffy ($m = \frac{2}{3}$), Gompertz ($m = 1$) and logistic ($m = 2$) equations suggest that each may provide a reasonable fit to most growth data and estimates of m should have broad confidence limits.

4. Statistical Inference about Growth Parameters

When using linearized data, statistical inference is based on linear regression with the condition that the residuals about the regression line are independently and identically distributed. This condition may often restrict the use of data that are close to zero or to the upper asymptote of the growth curve.

NLIN techniques produce an asymptotic standard error for each parameter estimate, which is based on the assumption that other parameters are known rather than estimated. When parameters are correlated, as in most growth equations, this assumption is not reasonable and the asymptotic standard errors are generally too small. More conservative estimates of standard errors can be calculated in this case (e.g., Schoener and Schoener, 1978). The jackknife technique is a promising approach (Mosteller and Tukey, 1977).

The statistical application of standard errors to hypothesis testing varies with the intended comparison. The data of growth are organized into a hierarchy from the single data point or record to the individual and finally to one or more levels of samples. In testing the hypothesis that growth data of two individuals do not differ significantly, one may use the standard errors for growth parameters estimated separately for each individual. In this case, the individual is the sample and the within sample variation is based on the deviations of each record from the growth curve. Difficulties may be encountered when deviations are autocorrelated, indicating that the growth model does not describe the trend of the data perfectly.

When one wishes to test the hypothesis that two samples of individuals differ, perhaps in a comparison of species, seasons, or locations, the appropriate within-sample variation is that based on individuals, not on records analyzed irrespective of individual. Different types of data are handled differently for such comparisons. In cross-sectional data (see Section V,B) each record represents a different individual and variation among records is therefore identical to variation among individuals. In the case of longitudinal data, variation among samples must be compared to variation among individuals. The most straight-forward approach to such comparisons is to base analyses on estimated parameters of equations fitted to the growth data for each individual separately. Mixed longitudinal data present the greatest difficulty because there may be too few records per individual to estimate growth parameters for each separately. One approach to this problem, short of throwing out repeated data for individuals, is to fit a common growth curve to the entire data set, calculate the average residual for each individual, and perform an analysis of variance based on these values. The jackknife technique may also be used to calculate within sample estimates of variance of parameters for mixed longitudinal data.

5. Increment Analysis

Most data of growth are expressed as size or mass at the end of successive intervals of time. Because growth builds upon previous growth, deviations of data from fitted curves are usually autocorrelated and thus the data do not strictly conform to the assumptions that underlie statistical inference about regression. In order to achieve independence of data, some authors have calculated increments of growth as the difference between measurements at successive times, $M(t + 1) - M(t)$, as the basis for curve fitting (e.g., Ross, 1980).

All growth curves may be expressed in increment form by taking the difference between expressions for $M(t + 1)$ and $M(t)$. For the logistic model [Eq. (6)], the resulting equation is

$$M(t + 1) - M(t) = \frac{[1 - \exp(-K)][M(\infty) - M(t)][M(t)]}{M(t) + [M(\infty) - M(t)][\exp(-K)]} \qquad (17)$$

Increment equations have no parameter for initial size $M(0)$ or age at inflection I. $M(\infty)$ and K may be determined by NLIN techniques (e.g., Schoener and Schoener, 1978; Ross, 1980).

Another feature of increment equations is that time or age does not enter as a variable. As a result, analysis of increments can be applied to interval data on individuals of unknown age to estimate K, $M(\infty)$, and m if appropriate. Such analyses are therefore suitable for capture–recapture data (Fabens, 1965; Schoener and Schoener, 1978). If intervals are longer than 1 day or are unequal, increments can be divided by interval length to obtain increase per day (Schoener and Schoener, 1978), or the growth model can be made to incorporate interval length (Fabens, 1965).

Ricklefs and White (1975) have proposed a similar technique that uses interval data to estimate relative ages of chicks but does not rely on a particular mathematical model of growth. The technique was designed for application to growth of seabirds, which develop over long periods on inaccessible islands but which may be visited for brief periods. It was shown that data collected at the beginning and end of a period of 1 week permitted a reasonable estimate of the growth curve provided that nestlings representing all stages of development were present. Such data could be used directly as a basis for fitting growth equations. Ricklefs and White (1975, 1978) also suggested that analysis of increments can be used to compare the growth performance of samples of birds with respect to season, year, locality, or other variables.

E. RELATIVE GROWTH

Organs having different specific growth rates change in relative proportion to each other during the course of development. Huxley (1924, 1932) provided the first mathematical description of such relationships. For organs with sizes or masses x and y having specific growth rates a and b, Huxley showed that the growth of one with respect to the other followed the equation $y = $ constant $x^{b/a}$. Substituting k for b/a and incorporating a constant, we get the familiar equation $y = cx^k$ for allometric growth (Huxley and Teissier, 1936). When the specific growth rates are equal, k is one and the organs grow in direct proportion to each other according to $y = cx$. Under these conditions, growth is said to be isomorphic. When k is not equal to one, growth is heterogonic. When k exceeds one, y increases in relative proportion to x; when k is less than one, it decreases. Values of k are usually estimated by linear regression of the logarithm of y upon the logarithm of x, according to the model $\log y = \log c + k \log x$ (see Cock, 1966).

In spite of the general application of allometric analysis to describe relative growth, data usually do not fit a single regression throughout the duration of development, nor would they be expected to. Reeve and Huxley (1945) pointed out that the sum of a number of equations of the form $y = bx^a$ is not a similar equation if the as are unequal. In addition, two organs obeying sigmoid growth do not obey allometric growth except under restrictive conditions (Kavanaugh and Richards, 1942), although it is generally possible to transform data so as to make sigmoid growth processes comparable in an allometric analysis (e.g., Robertson, 1908).

Two or more allometric plots can be compared by standard analysis of covariance to determine whether values of k differ significantly. If the slopes of the regression do differ, the intercepts (log b) are not interpretable, and one can meaningfully compare elevations of two allometric curves only if an a priori decision is made about the point at which the comparisons are to be made. Cock (1966) suggested on statistical grounds that the estimation of elevation be based upon estimated values of y at the mean value of x, but I can imagine instances in which one might want to compare elevations at a particular development stage, such as hatching.

F. Longitudinal Correlations

Because growth is a process of accumulation, size at a particular age depends upon the past history of growth. Correlations and regressions of sizes across age in longitudinal data may indicate the contributions of growth performance at a particular time to size at some time in the future. Schifferli (1973) used this technique to determine the effect of egg weight on subsequent growth in Great Tits. Similar studies on domestic fowl and other species will be discussed in a later section.

Correlations need not involve measurements of the same character. Crosscorrelations between two measurements, such as weight and subsequent feather growth, have been used to test hypotheses about the pattern of variation in development processes. Correlations may also be based upon increments in size as an index to the relationship of growth performance during different periods. Autocorrelations with lags of one to a few days can be used in species with long development periods to test whether growth performance is correlated from one day to the next in a periodic fashion.

G. Multivariate Analyses

Because development involves the simultaneous change of many characteristics, multivariate analyses may be required to provide truly comprehensive descriptions of growth patterns. O'Connor's (1978b) study of growth in three passerines is the only one to have attempted such an approach thus far. O'Connor applied principal components analysis (PCA) based on correlation matrices calculated for several sets of data. In one, components of variation in mass, winglength, and primary feather length were calculated for each age group from hatching to fledging. The first component (PCI) was related to overall size of the chick, and the second (PCII) generally contrasted weight against wing and primary length (i.e., heavy, short-winged versus light, long-winged for nestlings of a given age). O'Connor described how the amounts of variance explained by PCI and PCII change during development. As one might expect, as mass levels off. PCI explains less of the total variance and PCII more. In a second analysis, O'Connor treated the masses at each age as the variables in a PCA. In this analysis, PCI generally consisted of strong factor loadings from masses at earlier ages and PCII from masses at later ages, although the data for House Martins resulted in a more complicated pattern of factor loadings. O'Connor performed two other PCAs based on additional indices of growth and development (e.g., peak mass, age at fledging, day 0 mass) in an attempt to extract the few key variables among the many intercorrelated ones available that adequately describe the pattern of growth. O'Connor (1978b) concluded that "a knowledge of nestling weight and one other measure—winglength, age, or length of primary—was sufficient to describe the development status of any nestling [p. 171]."

VI. Genetics of Growth

The evolutionary interpretation of patterns in life histories, including development (Lack, 1968), is based on the assumption that differences between species are inherited and have resulted from the selective effects of different environments. However, in spite of the central importance of the genetic premise, there have been few demonstrations of the genetic basis for variation in life-history traits within natural populations, let alone between populations and species.

Poultry scientists have more practical reason to study the genetics underlying body size, growth rate, and traits of commercial importance and, indeed, most information on avian genetics is found in the poultry literature. To be sure, the experiments required to determine mode of inheritance are more practical with domesticated than wild populations, but intensive research programs on especially suitable natural populations are now beginning to provide opportunities for genetic analysis.

A. COMPONENTS OF PHENOTYPIC VARIANCE

Because the data of growth are, for the most part measurements of such continuous variables as mass and length, the estimation of genetic variation involves the analysis of variance and covariance and is generally referred to as "quantitative genetics" (Falconer, 1960).

The variance (V) of a trait is a measure of the dispersion of observations about the mean and is estimated by the formula

$$V = \frac{1}{n-1}\left[\sum_{i=1}^{n} X_i^2 - \left(\sum_{i=1}^{n} X_i\right)^2 / n\right]$$

where X_i is the measurement for individual i and n the number of individuals in the sample. The variance of a sample of measurements, the phenotypic variance V_P is the sum of contributions of genetic factors (the genetic variance V_G) and environmental factors (the environmental variance V_E). The genetic variance is further subdivided into the additive variance V_A attributed to the expression alleles in homozygous form, the dominance variance V_D arising from the interaction of alleles in heterozygous form, and the interaction variance V_I comprising the influences of different genes on the expression of alleles at a particular locus. The environmental variance is also subdivided into a general component V_{Eg} arising from permanent effects of the environment, and a special component V_{Es} arising from temporary or localized circumstances. Hence the phenotypic variance is defined in terms of its components by

$$\underbrace{V_P}_{\text{phenotypic}} = \underbrace{V_A + V_D + V_I}_{\text{genetic}} + \underbrace{V_{Eg} + V_{Es}}_{\text{environmental}}$$

Because the variance of a measurement increases more or less in direct relationship to the square of the mean, variability of measurements is more meaningfully compared among species by calculating coefficients of variation (the standard deviation divided by the mean). In a sample of 251 nestlings from early broods of European Starlings, coefficients of variation (CVs) in asymptotes [$M(\infty)$] and growth-rate constants (K) of fitted logistic equations were 10 and 8% of the mean (Ricklefs and Peters, 1979). In three other studies of temperate-zone songbirds, CVs of $M(\infty)$ were 4–9%, and those of K were 6–8% (Ricklefs, 1976). O'Connor (1977) tabulated variation in several indices of development for three passerines (Table VI). The data indicate that development in House Martins is considerably more variable than it is in House Sparrows and Blue Tits. In addition, in House Martins wing length (CV at fledging = 5%) is apparently a better index of developmental age than mass (CV = 18%). In domestic fowl, coefficients of variation

TABLE VI

Coefficients of Variation (%) for Indexes of Development and Growth in Three Species of Passerines[a]

Trait	Blue Tit	House Martin	House Sparrow
Nestling period	3	17	9
Peak mass	6	8	6
Fledging mass	6	18	9
Fledging wing length	4	5	6
Fledging tarsus length	4	2	3

[a] From O'Connor (1977).

1. AVIAN POSTNATAL DEVELOPMENT

in body mass at 8 weeks of age were 15% (Siegel, 1962) and 9% (Thomas et al., 1958). From this brief survey, it appears that phenotypic variation in most measures of development is likely to lie between 5 and 15%.

B. The Estimation of Variance Components

From the standpoint of potential for evolutionary response, the variance due to the additive effects of genes (V_A) is the most important component of V_P. The rate of evolutionary response to selection is directly proportional to the ratio V_A/V_P, which is known as the heritability of a trait (h^2). Because V_P can be measured directly, the challenge of quantitative genetics is to estimate the additive genetic component V_A. Genetic models of continuous traits based on the additive effects of many genes suggest that V_A can be estimated from the phenotypic correlations among relatives. Estimates of heritabilities derived in this manner by and large are consistent with the phenotypic response to selection (see Falconer, 1960).

Without going into the details of the genetic models, estimates of V_A can be based on the phenotypic covariance among relatives, as outlined in Table VII, and estimates of h^2 can be based on the statistics of regression and correlation. The most straightforward methods of estimating V_A and h^2 are from the resemblances of parents and their offspring and among half sibs. Half-sib comparisons are normally accomplished in breeding programs by mating one male (sire) to several females (dams) thus producing half sibs

TABLE VII

Estimation of Additive Genetic Variance (V_A) and Heritability (h^2) from Phenotypic Resemblance between Relatives[a,b]

Relatives	Covariance	Regression (b) or correlation (r)
Offspring and one parent	½V_A	$b = ½h^2$
Offspring and mid-parent	½V_A	$b = h^2$
Half sibs	¼V_A	$r = ¼h^2$
Full sibs	½V_A + ¼V_D + V_{Ec}	—

[a]From Falconer (1960).
[b]Note that b is the slope of the regression of offspring measurement upon parent measurement; r is the coefficient of correlation between the two. The covariance among siblings is equal to the variance among groups of siblings (see, for example, Sokal and Rohlf, 1969, for details of calculations of variance components).

having their father's genotype in common. Dam effects in half-sib comparisons can be produced by mating one female successively to different males.

Estimates of heritabilities in birds are restricted almost entirely to domestic fowl and other galliforms. Some of these values pertinent to development and life-history traits, extracted from Kinney (1969), are presented in Table VIII, in which several patterns are apparent. First, heritabilities of size and mass traits are generally higher than those of life-history traits. Second, estimates based on different methods of analysis are in general agreement for most traits. Because dam effects are based on full-sib correlations in poultry breeding experiments, the correlations include dominance effects and the effects of full sibs being reared in a common environment (V_{E_c}). Because

TABLE VIII

Estimates of Heritabilities of Certain Traits in Domestic Fowl[a]

Trait	Basis for estimate[a]				
	Sire	Dam	S + D	b_{OP}	R
Body mass					
4 weeks (♀)	0.31	0.95	0.50	—	—
8 weeks (♀)	0.39	0.61	0.42	0.36	0.27
Pullets (heavy strains)	0.53	0.56	0.51	0.44	0.41
Mature (♀, heavy strains)	0.49	0.52	0.56	0.59	—
Mass gain					
Various ages	0.53	0.68	—	—	—
Length					
Keel	0.47	—	0.48	0.54	—
Shank	—	—	0.46	0.45	—
Feathering (various areas and ages)	—	—	0.40	—	—
Eggs					
Mature mass (heavy strains)	0.58	0.54	0.58	0.46	—
Albumen/yolk ratio	0.56	0.31	0.21	—	—
Life-history traits					
Age at sexual maturity	0.39	0.35	0.37	—	—
Long-term production	0.22	0.30	0.27	—	—
Rate of production (long term)	0.15	0.64	0.31	—	—
Hatchability (% fertile eggs)	0.14	—	—	—	—
Early mortality (0–10 weeks)	0.02	0.27	0.04	—	—

[a] Sire, half-sib correlations with common father; Dam, usually full-sib correlations among progeny of each mother; S + D, sire and dam effects combined; b_{OP}, slope of regression of parent–offspring correlation; R, realized heritability based on response to selection.

chicks are raised under controlled conditions independent of the care of their mother, the only component of variance due to common environment results from maternal effects, probably expressed in the composition of eggs. These effects are reflected in the difference between sire and dam estimates of h^2. In Table VIII, these are particularly marked only for traits of young chicks (4-week mass, 8-week mass, mass gain, early mortality), as might be expected, and long-term production, which could have a sex-linked genetic component. In natural populations the component of covariance among sibs due to common environment is likely to be very large because of the effects of parental care on growth performance.

Several studies involving natural populations of sedentary birds have presented the opportunity to calculate the correlation between parent and offspring as an estimate of heritability in several measurements of body size (Table IX). The heritabilities indicate considerable discrepancies among species and may, in part, be confounded by correlations between the measurements of parents and by genotype environemnt correlations, both of which tend to inflate h^2 (Boag and Grant, 1978; Smith and Dohndt, 1980).

Few values for the heritability of rate of growth have been estimated either in domestic birds or wild populations. The heritabilities of weight at a particular age and of gain between one age and the next include components of variation due to both growth rate K and final size $M(\infty)$. Heritabilities of several measures of growth performance were obtained in the study of Thomas et al. (1958) on broiler strains of New Hampshire chickens (Table X). Values of h^2 were uniformly high (0.4–0.8) for mass, mass gain, and feed consumption. For feed conversion, however, h^2 was significantly different

TABLE IX

HERITABILITIES IN SEVERAL MEASUREMENTS OF BODY SIZE IN NATURAL POPULATIONS BASED ON PARENT–OFFSPRING REGRESSION AND CORRELATION

Trait	Great Tit[a]	Song Sparrow[b]	Medium Ground Finch[c]
Mass	0.59	0.04	0.84
Wing length	—	0.14	0.53
Tarsus length	0.76	0.32	0.43
Beak length	—	0.33	0.62
Beak depth	—	0.51	0.82
Beak width	—	0.50	0.95

[a]Parus major, from Van Noordwijk et al. (1980) and Garnett (1981).
[b]Melospiza melodia, from Smith and Zach (1979).
[c]Geospiza fortis, from Boag and Grant (1978).

TABLE X

ESTIMATES OF HERITABILITIES OF VARIOUS MEASURES
OF GROWTH PERFORMANCE IN A BROILER STRAIN
OF NEW HAMPSHIRE CHICKENS[a,b]

Trait	Age (weeks)			
	4	6	8	10
Mass	0.66	0.51	0.64	0.56
Mass gain		0.65	0.52	0.42
Feed consumption		0.73	0.76	0.60
Feed conversion[c]		0.48	0.03	0.11

[a] Heritabilities based on sire component of variance (half-sib covariance).
[b] From Thomas et al. (1958).
[c] Mass of food per unit of mass gain.

from zero only for 4–6 weeks, suggesting that there is little genetic variation for aspects of digestion, assimilation, and metabolism that contribute to feed conversion. Their results also suggest that genetic components of variance for mass gain, feed consumption, and probably mass, are highly correlated and perhaps are different expressions of the same trait. A change in either $M(\infty)$ or K will alter $M(x)$, $[M(x + t) - M(x)]$, and, presumably, food consumption during the early part of the growth curve.

To date the only attempt to determine the heritable components of variation in growth parameters in a natural population is that of Ricklefs and Peters (1981) in a study of European Starlings. They combined full-sib analyses of variance in $M(\infty)$ and K with experiments in which chicks were switched among broods to reduce the effects of the common environment. In unmanipulated broods, the covariance among full sibs, which estimated $\frac{1}{2}V_A + \frac{1}{4}V_D + V_{Ec}$, amounted to 62 and 31% of the total variance in $M(\infty)$ in two sets of observations, and to 34 and 42% of the variance in K. When chicks were switched to foster parents at hatching so that the common environment of full sibs extended only to effects of egg quality and incubation, the covariance among full sibs was 14% of the total for $M(\infty)$ and 18% for K. Covariance among foster sibs, resulting from the effects of care during the nestling period, accounted for 28 and 27% of the variances, respectively. When eggs were switched just after laying, so that V_{Ec} included only egg-quality effects, the covariance among full sibs was not significantly different from zero, whereas that among foster sibs was 26% of the total variance in $M(\infty)$ and 77% of that in K. Therefore, in one experiment at least, there was no indication of heritable variation in either $M(\infty)$ or K.

C. Response to Selection

Selection applied to measures of body size usually elicits a marked response that is detectable within one or a few generations. The most useful measure of the rate of selection response is the realized heritability, which is the ratio of the deviation of the progeny of selected parents from the mean of their parents' generation, divided by the deviation of the parents from their own generation mean. For example, in Siegel's (1962) selection for 8-week body mass in White Plymouth Rock chickens, mean values for females in the unselected population were 715 (±117 SD) g. Female parents selected in the first generation for high mass had 8-week values averaging 878 g (1.4 SD above the mean), and their female offspring had masses averaging 775 g. Thus the selection differential was 878 − 715 = 163 g and the response differential was 775 − 715 = 60 g, yielding a realized heritability of $60/163$ = 37%. After four generations of selection 8-week mass had increased to 951 g, an increase of 33%. Such responses are not unusual. In six generations of selection, Verghese and Nordskog (1968) achieved a 50% increase and 20% decrease in mass of leghorn pullets. Such experiments reveal substantial additive genetic variance for body size.

Breeding programs for domestic animals usually involve very strong selection in order to achieve rapid results. If phenotypic variation were normally distributed, a selection differential of 1.4 SD would represent about 20% of the population; a selection differential of 2 SD would correspond to about 5% of the population. As well as being unrealistic, strong selection requires maintaining a large population in order to obtain enough selected individuals. In addition, fecundity must at least equal the inverse of the selected percentage so that the population does not decline. In the study by Verghese and Nordskog (1968), selection differentials were about 1 SD (40% of the population), which seems to be typical of such programs.

Selection for body mass at a particular age confounds responses in $M(\infty)$ and K. Although these parameters have not been resolved in any selection program, Marks (1978a) has examined the effects of 37 generations of selection for 4-week body mass of Japanese Quail on parameters of fitted logistic equations. Comparing females of the P (selected) and C (control) lines fed on a diet containing 28% protein, $M(\infty)$ increased 115% from 122 to 262 g and K increased 35% from 0.113 to 0.153 days^{-1}. The greater response in $M(\infty)$ is not surprising because by 4 weeks Japanese Quail attain about 80% of $M(\infty)$. After 29 generations of selection, egg mass had increased by 22% and neonate mass by 27% (Marks, 1975).

Response to selection does not continue indefinitely, either because genetic variation is eroded by constant selection or because correlated responses of other phenotypic traits weigh against the selection program. Response to selection for 4-week body mass in Japanese Quail began to level off after 25–30 generations at between 125 and 150% increase over control values. Realized heritabilities declined from 49% during generations 1–10 to 34% (11–20), 24% (21–30), and 14% (31–40) in the P line (Marks, 1978b). Long-term strong selection normally causes reduced viability (Verghese and Nordskog, 1968). After 40 generations of selection, fertility of Japanese Quail eggs had dropped from 88 to 51% and hatchability from 61 to 45% (Marks, 1979).

D. GENETIC COVARIANCE AND CORRELATED RESPONSES TO SELECTION

The correlation between inherited components of variation in two traits can be estimated by techniques analogous to those used to estimate heritabilities (Falconer, 1960). For measurements of mass at various ages, both phenotypic and genetic correlations are generally high. For example, in the study by Martin et al. (1953) of selection on Rhode Island Red broilers, 3-week mass exhibited phenotypic correlations of 83% with 6-week mass, 74% with 9-week, and 68% with 12-week mass. Corresponding genetic correlations were similar (91, 83, and 70%), even though the heritabilities of the masses at each age were only 27–31%. Such strong correlations suggest that masses at different ages during the development period are under the control of a small number of genetic factors that influence underlying growth processes.

Genetic correlations between size and other life-history traits are generally low, as shown in Table XI. The predicted response of unselected traits to selection on correlated traits is proportional to the square roots of the heritabilities of the two traits and their genetic correlation. Hence if the heritabilities of two traits are similar and they are perfectly correlated, both will respond equally rapidly to selection on either. Empirical results are in general agreement (e.g., Siegel, 1963; Marks, 1979). Responses to selection of uncorrelated traits may, however, be difficult to predict from the characteristics of unselected lines. Selection for body and egg mass in poultry produces asymmetrical responses in the unselected character which are not predicted by theory (Verghese and Nordskog, 1968). Attempts to select for egg and body mass simultaneously revealed little response in a low-body-mass–high-egg-mass experiment, as one might expect from the difficulty of getting a large egg out of a small hen, and marked response only in body mass in a high-body-mass–low-egg-mass experiment (Nordskog et al., 1974).

Such antagonistic selection on the measurements of development have presumably resulted in the tremendous variety of development patterns in

TABLE XI

Genetic[a] and Phenotypic[b] Correlations (%) between Size and Other Life-History Traits in Domestic Fowl[c]

Trait	A	B	C	D	E	F
A, 8-Week mass	—	68	52	37	−10	NA[d]
B, Pullet mass	56	—	91	40	3	6
C, Mature mass	43	43	—	36	6	3
D, Mature egg mass	21	17	23	—	10	−4
E, Age at first egg	7	−17	12	−1	—	−23
F, Annual egg production	NA	NA	11	−3	−20	—

[a] Upper right.
[b] Lower left.
[c] From data in Kinney (1969).
[d] NA, Not available.

birds, but the literature provides almost no clues to the genetic bases of such changes or to the kinds of selection required to produce them (Cock, 1966). The analysis by Ricklefs and Peters (1981) of starling growth curves revealed no significant phenotypic correlation between $M(\infty)$ and K in the experiments as a whole or within broods. In one experiment, the two parameters exhibited significant positive covariance among foster sibs, indicating a similar response of $M(\infty)$ and K to some factor, perhaps nutrition, but not among full sibs. Selection for 4-week body weight in quail revealed positively correlated responses in $M(\infty)$ and K. Among species, $M(\infty)$ and K are negatively correlated (see following discussion).

VII. Susceptibility of Growth to Environmental Factors

Under natural conditions, chicks are sometimes underfed, poorly nourished, or poorly cared for in a variety of other ways. The effects of such mistreatment range from achieving independence at a relatively low mass, to retardation of development, permanent stunting, and death (Ricklefs, 1968b, 1976). Although such effects are pervasive there have been few systematic studies of the response of growth and development to experimentally controlled factors in natural populations. In particular, one would like to know the extent to which growth abnormalities are reversible under restored suitable conditions and whether development processes respond to poor conditions in such a way as to reduce their impact on survival and adult condition. One might imagine that reduced growth rate and metabolism, achieved, perhaps, through facultative hypothermia or altered hormone balance, could reduce the impact of restricted energy or protein supply.

Observations on natural populations have revealed evidence both for and against such adaptive responses. The development rate of European Swifts (*Apus apus*) appears to be sensitive to feeding conditions, resulting in fledging periods ranging from 35 to 56 days (Lack, 1956). In shearwaters (*Puffinus* spp.), however, rate of feather growth and length of the fledging period vary little over a wide range of body mass of the chick (Richdale, 1963; Harris, 1966). Most experimental studies have been done with domestic fowl and, to a lesser extent, with waterfowl. A few of these are summarized here to provide an indication of the development response of one group of birds.

A. Force-Feeding

The food intake of precocial chicks might be regulated at one of three different levels according to (1) the demands of growing tissues, (2) the maximum rates of digestion and assimilation, or (3) the ability of the chick to find and consume food. One way to test the second type of limitation is to force-feed chicks over and above their voluntary rate of consumption in an *ad libitum* trial and determine whether the chick is able to metabolize the additional food. Experiments of this type were performed by Nir *et al.* (1974, 1978) on chicks during the early part of the development period (14–29 days and 3–21 days). Diet was mixed with water and introduced through a tube directly into the crop until it was fully distended.

The results of these experiments, summarized here only for light breeds (New Hampshire and White Leghorn cross), are as follows. Force-feeding did not increase food consumption until the beginning of the second week of the experiment, after which intake increased to a maximum of 70% over controls. During the 18-day experimental period, food intake of force-fed chicks was 43% greater than that of controls. Overfed chicks gained substantially more mass than did controls (3.6- and 7.1-fold increases versus 1.8- and 5.7-fold increases in the two studies). Overfeeding stimulated skeletal growth only slightly (shank length 54 versus 52 mm). At the end of the experiment, the proportions of water, protein, and ash in the lean carcass did not differ between treatment and control. About 40% of the difference in body masses at the end of the experiment was lipid; the remainder represented tissue growth, primarily by the viscera (especially the crop, duodenum, and liver). The viscera of experimental chicks had nearly double the mass of that of control chicks at the end of the experiment; the remainder of the carcass had 19% more mass in the experimental chicks.

Experiments with force-feeding suggest that the ability to utilize additional food is associated with hypertrophy of visceral organs of storage (crop and liver) and digestion. Once metabolized energy and nutrients increase, growth of other tissues appears to be stimulated. The early overfeeding regime did not affect the size of individuals when fully grown, however, and the response of chicks of heavy breeds to the same feeding program was much less pronounced. The results suggest that *ad libitum* levels of food consumption by small chicks may be near the maximum that can be effectively handled by the digestive process without hypertrophy of visceral organs (to 19 versus 12% of total body mass at 21 days in the experiment of Nir *et al.*, 1978).

B. INTERMITTENT FEEDING

Poultry scientists have applied various programs of intermittent feeding in attempting to increase feed efficiency. Their experiments provide insight into the ability of growing chickens to recover from periodic fasts of various lengths. Experimental feeding programs include denial of food for a portion of each day, denial of food on alternate days, and altering the light cycle to reduce the feeding period. The latter program confounds feeding regime with daylength effects.

Nir and Nitsan (1979) deprived chicks of a light-bodied strain of chicken of food on alternate days from 15 to 46 days of age. The deprived birds about doubled their food intake on the repletion day and their rate of growth was depressed only slightly compared to controls. As in the aforementioned overfed birds, the size of the crop increased greatly, to more than twice the mass of that of control chicks at 21 days of age, and 1.7 times the mass at 31 days of age. Other digestive organs did not exhibit marked increases in relative size and so it appears that chicks accommodate chronic intermittent feeding by increasing the storage capacity of the crop. In chicks placed on an intermittent feeding regime, after 24 hours of depletion the contents of the small intestine were one-third those of replete controls compared to complete emptying of the gut after 24 hours of food deprivation in previously unrestricted chicks. These results suggest that an enlarged crop can store more food than a chick is able to digest in 1 day.

Yule and Fuelling (1979) deprived broiler chicks, grown under 23-hour photoperiods, of food for various periods between 28 and 56 days of age. Compared to control individuals whose mass averaged 2.16 kg on day 56, performance was unaffected by restricting chicks to 8 hours of feeding per day (mass = 2.17 kg) but decreased by restriction to alternate days (1.94 kg) and to 1 of every 3 days (1.50 kg). The food conversion ration (food intake/growth) for the four treatments increased from 2.24 and 2.24 to 2.32 and 2.47, indicating that growth was inhibited slightly more than food intake by intermittent feeding. Auckland (1978) reared turkeys from 2 to 16 weeks on photoperiods of 23, 14, and 8 hours day^{-1}, thus effectively restricting food consumption. He also divided the treatments into an *ad libitum* group and a second group given the same amount of food at a slightly restrictive level to isolate the effects of photoperiod from those of food quantity. There were no significant differences with respect to photoperiod in either group for body mass, food intake, or food conversion efficiency. Taken together, the results of these studies indicate that although daily food deprivation is easily compensated by increased feeding rate, chronic deprivation for periods of 1 out of 2 days or 2 out of 3 days cannot be fully compensated.

C. Drastically Restricted Food or Nutrient Intake

A remarkable phenomenon of poultry growth is the effect of severe food or nutrient restriction. Chickens maintained on protein, amino acid, or energy-deficient diets at levels just sufficient to fill maintenance requirements, can be maintained at a physiological age of about 10 days for months (McCance, 1960; Dickerson and McCance, 1960). Return to an unrestricted, nutritional diet restores growth and development to a normal rate with little subsequent effect on adult body size or egg production (McRoberts, 1965). Similar effects have been reported for Mallards initially maintained on seed mixtures (11–14% protein) in which mass increase to 14 days was 1.6-fold compared to 9.3-fold for ducklings fed on invertebrates. When diets of the two groups were switched at 14 days, the body mass of previously restricted chicks increased rapidly whereas that of the control leveled off (Street, 1978).

Results of an experiment on the effects of protein restriction (9% versus 28%) on the growth and age at sexual maturity of Japanese Quail (Morse and Vohra, 1971) are shown in Fig. 4. The low protein diet resulted in only very slow growth and virtually no developmental change. When a normal diet was restored, growth and development rates returned to normal; onset of sexual maturity was delayed by approximately the period of restricted diet.

Experiments on fowl indicate considerable flexibility of growth in response to diet without significant long-term effects when good conditions are restored. This responsiveness makes good sense for a self-feeding chick with a long development period during which feeding conditions might be expected to vary. The extent to which altricial and semiprecocial species are similarly responsive has not been explored experimentally. Although under-

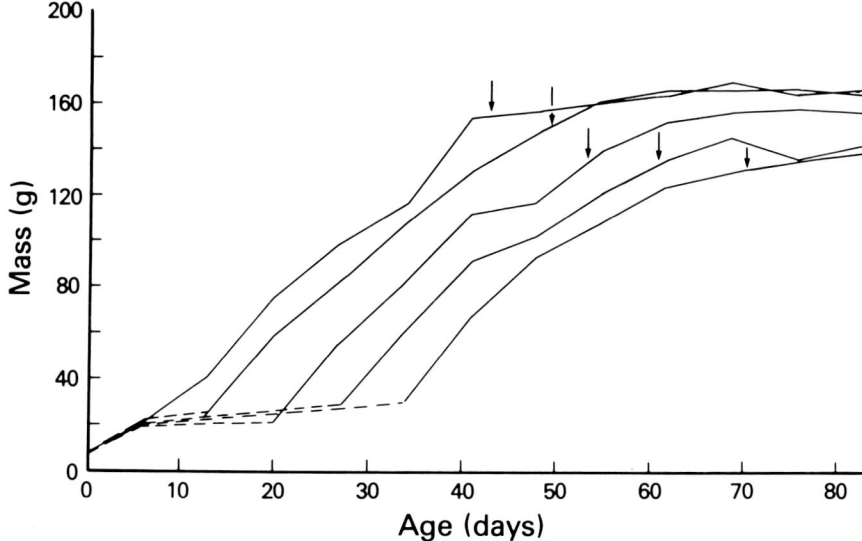

FIG. 4. Increase in body mass of Japanese Quail chicks reared on poult starter (28%,—) and a protein-deficient diet (9%,---). Age by which 50% of individuals in each treatment had achieved sexual maturity is indicated by an arrow. (After data in Morse and Vohra, 1971.)

nourishment frequently leads to death, anecdotal observations of many species reveal retarded growth or development coupled with long-term persistence in the nest (e.g., Ricklefs, 1976).

D. CHRONIC FOOD DEPRIVATION

Experiments on prolonged fasting are understandably uncommon, and there is therefore little to be said about the responses of birds, especially chicks, to starvation. Because such conditions in nature would normally result from abandonment with death virtually guaranteed, there can be little selection for metabolic compensation. However, the chicks of some species, especially pelagic seabirds, may undergo long fasts fairly regularly during periods of poor feeding.

Baldwin and Kendeigh (1932) determined that the survival time of starved House Wren nestlings depended upon age and ambient temperature. At 37°C, survival time increased from about 10 hours at hatching to 4 days at 15 days of age, probably as a result of increasing lipid reserves (Ricklefs, 1967b; O'Connor, 1978c). At 19°C, survival decreased from 35 hours at hatching to

17 hours at 7–11 days as thermogenic responses consumed energy more rapidly, and then increased to 25 hours at 15 days of age. In small chicks, death at 19°C was probably caused by hypothermia as less than 2% of initial mass was lost. At 37°C, two chicks 0 and 2 days of age had lost only 15% of their initial mass by the time of death. An 11-day-old chick lost 43.2% at the same temperature.

Koskimies (1948) compared the responses of adult and nestling (4–5 weeks old) European Swifts to starvation. Swifts apparently interrupt development in response to food restriction in nature (Lack, 1956). Fasted at 24°C, four chicks survived an average of 9.6 days during which time they lost 53% of their initial mass (49.2 g). Adults (42.2 g) survived only 4.6 days and their mass decreased 38%. If the greater mass of the chicks were mostly lipid, it would account for the difference in survival and mass loss. Koskimies also found reversible hypothermia accompanied by decreased metabolite rate after about 6 days of starvation in the nestlings. If this represented an adaptive response, it came only after severe effects of food deprivation were expressed. Records from the literature indicate examples of swift nestlings surviving 13 and 21 days of starvation.

E. Cold Exposure

Most chicks depend on their parents for some contribution to homeothermy during the early part of the growth period. At this time, they are vulnerable to cold stress which in turn may affect their growth and subsequent development. Aulie and Steen (1976) found that depriving Willow Ptarmigan of heat lamps for 20 minutes each hour (ambient temperatures fell from >30°C to 16–20°C) had no effect on growth of the body and several organs through 3 weeks of age. In a similar experiment with bantam chicks, intermittent cold exposure stimulated growth in body mass and an increase in the maximum mass-specific metabolic rate under cold stress (Aulie, 1976b, 1977; Aulie and Grav, 1979). In contrast, acute cold stress (10°C for 4 hours) of newly hatched broiler chicks severely retarded growth during the subsequent 2-week period (Sagher, 1975).

In precocial chicks with partially developed homeothermy at hatching, mild chronic cold stress increases the expenditure of energy for temperature regulation. The effect of such treatment on growth has been studied by Osbaldiston (1966) and Kleiber and Dougherty (1934). Osbaldiston (1966) reared broiler chickens either alone or in large groups at six temperatures ranging from 13 to 27°C. In group-reared chicks, which huddle together at low temperature, the relative growth rate during the first week varied from

152% per week at 27°C to 168% at 21°C, but only 119% at 13°C. Temperature had no consistent effect on growth during subsequent age intervals. The lower early growth rate of chicks in the colder environments was associated with decreased food intake rather than increased metabolism, suggesting that behavioral thermoregulation (huddling) competed for feeding time.

Birds raised singly at temperatures of 32, 24, 16, and 7°C exhibited markedly decreased growth rates at low temperature, which persisted through 3 weeks of age. By 3 weeks, chicks reared at 32°C weighed about three times as much as chicks reared at 7°C. Weight-specific food intake increased with lowered ambient temperature to the point that chicks kept at 7°C consumed three times as much food relative to body mass during the first 2 weeks as did chicks at 32°C. Hence the total food intake of the groups was similar. Osbaldiston estimated that the percentage of metabolizable energy allocated to temperature regulation increased from 40% at 32°C to 43.7% at 24°C, 49% at 16°C, and 57% at 7°C. Although body temperature was not reported and may have been a confounding factor, Osbaldiston's results indicate first that chicks can increase food consumption greatly in response to the metabolic demands of cold stress and second that the response is not completely compensating. The apparent preference of group-reared birds for maintaining high temperatures rather than feeding suggests that the singly reared chicks may have suffered from hypothermia when reared at low temperature.

Kleiber and Dougherty (1934) reared White Leghorn chicks in groups at temperatures of 21, 27, 32, and 38–40°C between ages 6 and 15 days. Growth was depressed at the highest temperatures in association with decreased food intake. Comparing chicks reared at 21 and 32°C, average body mass during the experiment was similar (74.2 versus 71.7 g), but food intake and metabolized energy (42.3 versus 35.0 kcal day^{-1}) were higher in the cold-stressed group. However, because food intake did not fully compensate the increased heat production under cold stress (35.4 versus 23.2 kcal day^{-1}), the net energy available for production was lower (6.9 versus 11.8 kcal day^{-1}). As a result, although the cold-stressed chicks gained more mass (4.9 versus 4.4 g day^{-1}), and accumulated more protein (1.10 versus 0.97 g day^{-1}), they stored virtually no fat (0.06 versus 0.44 g day^{-1}). These results are in agreement with Osbaldiston's (1966) study in that food consumption can be increased voluntarily in response to cold stress but not sufficiently to compensate for the heat expenditure of temperature regulation. Unfortunately it was not determined whether the results were produced by competition of growth and homeothermy for limited energy or the secondary effects on growth of slight hypothermia. Similar experiments with older fully homeothermic chicks would be informative.

VIII. Intraspecific Variation in Postnatal Growth

Experimental studies have revealed that postnatal growth can be influenced by variation in diet quality, quantity of food, temporal pattern of food availability, and temperature. In natural populations, variability in these conditions may be brought about relative to locality, habitat, season, and weather. In addition to these external factors, growth rate is also affected by such variables as brood size, which mediates the effects of food limitation and perhaps temperature as well (Royama, 1966; Mertens, 1969; O'Connor, 1975a), egg size and composition, position in the hatching sequence, age and experience of parents (e.g., DeSteven, 1978; Ollason and Dunnet, 1978; Harvey et al., 1979; Crawford, 1980), and individual variability in the quality of parental care, which is expressed as variance among broods.

Variability in rate of postnatal growth has been used frequently as an index to the suitability of the environment for breeding and to test statistically the effects of more or less controlled variables, such as brood size and age of parent, on reproductive performance. Relatively little effort has been put into determining the mechanisms by which these factors act on the development process. Only more recently have studies attempted to analyze intrapopulation variability as a phenomenon of interest in its own right (e.g., Ricklefs and Peters, 1979; Ross, 1980).

A. Locality, Year, and Season

Although postnatal growth of many species of birds has been measured in several localities, over several years, or at different times of the year, the influence of these factors on growth rate is difficult to determine statistically because of the limited scope of such studies. When measurements in each of several localities are conducted during a single but different year, it is impossible to separate the effects of year and locality, as well as between-locality differences in habitat. Similarly, the systematic effects of season related to daylength and temperature changes cannot be discerned during a single year because annual variation in weather cannot be accounted for.

A sufficiently large body of data on growth of the European Starling has accumulated for Ricklefs and Peters (1979) to determine the effects of locality and season. On the basis of the variance in $M(\infty)$ and K within a population in Pennsylvania, they conservatively estimated that least significant differences are 6.6 g for $M(\infty)$ and 0.034 day^{-1} for K. Among localities, $M(\infty)$ varied between 75 and 84 g for early (first) broods and between 62 and 83 g for late broods. Hence locality effects (Pennsylvania, New York, British Columbia, Czechoslovakia, Scotland) were significant and in three of four localities

1. AVIAN POSTNATAL DEVELOPMENT

having data for second broods, values of $M(\infty)$ decreased as the season progressed. Values of K varied between 0.37 and 0.42 for early broods and between 0.36 and 0.37 day^{-1} for late broods. Hence locality effects were significant for early but not for late broods; chicks in late broods grew significantly slower than those in early broods only in Pennsylvania and New York. The study allows no particular generalization other than cautioning against generalizing from limited studies. For example, tests of the significance of seasonal effects revealed no change in $M(\infty)$ or K in Scotland, decrease in both $M(\infty)$ and K in New York (one season only), decrease in $M(\infty)$ but not K in Czechoslovakia, and vice versa in Pennsylvania.

A comparison of fledging weights of Rhinoceros Auklets (*Cerorhinca monocerata*) in the Puget Sound–Vancouver Island area (Summers and Drent, 1979) revealed significant variation (339–521 g). In 1969, growth performance was poor at two localities; in 1974, it was very poor at one and excellent at a second. The average of a 4-year study at one locality was intermediate. A study on Razor-billed Auks (*Alca torda*; Lloyd, 1979) revealed significant differences, of the order of 20%, among average fledging masses during 3 years, and a similar systematic decline in fledging mass during the season.

In a 3-year study of the Ipswich Sparrow (*Passerculus sandwichensis princeps*), Ross (1980) detected significant effects of year, month, habitat, and brood size, as well as several interactions among these, upon measures of nestling growth. Ross's thorough analysis of variance based on a sample of 917 individuals might serve as a useful model for future studies of this kind.

Anecdotal accounts of variation in growth parameters with respect to locality, habitat, year, and season are reported in most thorough accounts of the breeding biology of species, but no purpose would be served by recounting them here. Studies of spatial and temporal variation have not progressed in scope and depth of analysis sufficiently to reveal generalizations concerning environmental effects on growth, or to serve as indices to environmental conditions. The potential nonetheless is there and has been realized in a few studies.

B. Weather and Food Supply

For species whose food supply or method of foraging is sensitive to weather, one would expect variation in growth performance to be correlated with weather variables. This has been shown in a number of studies, particularly those of aerially feeding birds (e.g., Lack and Lack, 1951; E. K. Dunn, 1975). Moreover, Bryant (1975) and Van Balen (1973) have demonstrated parallels between growth rate and direct measures of food abundance.

The relationship between weather or food supply and nestling growth has been studied systematically in few species and, in many cases, feeding conditions have been inferred from growth performance, particularly in seabirds (e.g., Nelson, 1978), without independent measurement of feeding or food supply. Properly done, this type of study may serve two purposes: first, to estimate the sensitivity of feeding and chick growth to variation in particular ecological conditions and, second, to provide a relevant, albeit indirect, indicator of the environment. Of course, interpreting growth variation observed in the second type of study depends upon generalizing results of the first type, which seems premature at this point for many types of birds.

The relationship between growth rate and food supply is probably best understood for swallows and martins, largely through the work of Bryant (1975, 1978a,b). In the second of these papers, Bryant tested the relationship between mean nestling mass in broods of various ages and brood size, time of season, and food abundance. A multiple regression–partial correlation analysis revealed that time of season was important during the early phases of growth, later-hatched chicks being lighter, and that food abundance was positively related to mass during the latter part of the growth period. Variation in mean nestling wing length was not significantly related to any of the variables.

C. Brood Size

A negative effect of brood size upon nestling growth has been widely reported (see Klomp, 1970) and generally accepted as evidence that parents have limited capacity to provide food for their broods. On the premise that chicks in larger broods compete intensively for food, several authors have considered the effects of such competition and the mechanisms by which parents and young control it (Bryant, 1978b; O'Connor, 1978a; Howe, 1976, 1978). Under conditions of variable and unpredictable food supplies, parents may commence incubation before the last egg is laid, apparently so that the nestlings hatch asynchronously and a size hierarchy forms within the brood. Parents can then discriminate between nestlings on the basis of size and selectively starve the smallest chick or chicks if there is insufficient food (Lack, 1954), a phenomenon called brood reduction by Ricklefs (1965). O'Connor (1978a) expanded concepts of the adaptive value of brood reduction by introducing inclusive fitness to our thinking about it. Howe (1976, 1978) suggested that the detrimental effects of competition on the last-hatched nestling in broods of Common Grackles (*Quiscalus quiscula*) could be offset by the trend of increasing egg size and provisioning observed with sequence of laying in that species. Further discussion of brood reduction may be found in Clark and Wilson (1981), Hahn (1981), and Richter (1982).

Bryant (1978a,b) has determined factors responsible for differences in masses of nestling House Martins within broods. During the early part of the growth period, most of the variation in nestling masses could be attributed to the degree of asynchrony in hatching, that is, age differences. Later, brood size is the only significant predictor of differences in nestling masses, presumably because of the greater effects of competition on later-hatched chicks in large broods. The relationship of body mass to brood size in general must be related to the consequences of brood size upon nestling nutrition, but body temperature may also play a role whose importance has not been investigated.

D. Egg Size

In domestic fowl, large eggs produce large chicks (Halbersleben and Mussehl, 1922) which generally have greater survival, grow more rapidly, and attain larger size (Upp, 1928; Wiley, 1950; Skoglund et al., 1952; Merritt and Gowe, 1965; Blyth et al., 1965). Similar relationships have been reported for nondomesticated species (e.g., Parsons, 1970; Schifferli, 1973; O'Connor, 1975b, 1979; Nisbet, 1978; Ankney, 1979, Boersma et al., 1980). Data of growth are difficult to interpret in certain cases because growth rate and final size are not usually distinguished, egg size and final size have strong genetic as well as phenotypic correlations (see Section IV,D), and the consequences of egg size for chick size and quality are not well understood.

That large eggs produce large neonates has been demonstrated for several species (e.g., Herring Gull, Parsons, 1970; Great Tit, Schifferli, 1973; Common Grackle, Howe, 1976; Brown-headed Cowbird, Nolan and Thompson, 1978). Larger size can be accomplished in three ways: more advanced stage of growth accompanied by a longer incubation period, larger size at a comparable stage of growth, and greater reserves of energy, water, or nutrients in yolk, fat bodies, and tissues. Any or all of these effects may be present and they may vary among species. The relative proportions of yolk and albumen in eggs vary with respect to egg size in some species. The relationships between the log of yolk mass and the log egg mass have slopes of 1.14 (± 0.15 SE) in Japanese Quail, indicating relative constancy of proportions, but 0.80 (± 0.11 SE) in the Laughing Gull (*Larus atricilla*) and 0.31 (± 0.18 SE) in the European Starling (Ricklefs et al., 1978). Presumably, these changes in proportion of yolk have some bearing on the quality of neonates hatched from eggs of different sizes. Ricklefs et al. (1978) determined that quail neonates hatched from large eggs contained relatively large quantities of lipid in body fat depots (but not yolk), and had relatively large legs. Gull neonates from large eggs contained significantly higher levels of water and relatively heavier integuments than smaller chicks. Absolute amounts of lipid did not differ. These patterns suggested that quail embryos respond to egg provisioning through the storage of lipids and that gull embryos may adjust their time of hatching, as indicated by Parsons (1972). The different responses make sense in that quail feed themselves and rely on stored lipids during the first few days after hatching, whereas gull chicks are fed by their parents and compete among themselves.

In the domestic fowl, differences in the neonatal mass of chicks related to egg size appear to be compensated for by variation in growth rate after hatching. Small chicks appear not to be disadvantaged, even when reared together with large chicks, under domestic conditions (Skoglund et al., 1952). For example, neonates produced by eggs averaging 71 and 43 g (a 1.65-fold range) varied in mass over a 1.75-fold range. By 4 weeks this range had been reduced to 1.28, and then to 1.18 by 8 weeks and 1.12 by 12 weeks.

Among passerines, Howe (1976) found that mass differences at hatching within broods of Common Grackles were detectable at 12 days of age. The most detailed studies of the relationship between egg size and growth are those of Schifferli (1973) on Great Tits and O'Connor (1975b) on Blue Tits, House Sparrows, and House Martins. In Schifferli's study, growth curves of chicks from heavy and light eggs were parallel but slightly displaced with respect to age. Ratios of mass on one day to that of the previous day were nearly identical for nestlings from small and large eggs, hence the effect of egg size appeared to be restricted to $M(0)$ and, to a lesser extent, $M(\infty)$, avoiding K altogether. Because O'Connor estimated $M(0)$ by mass at a fixed time on the day of hatching, his estimates include age variation due to variation in time of hatching. O'Connor's figure portraying growth curves of House Sparrows having high and low initial masses suggests that they are essentially superimposable, although displaced by about 1 day. Because $M(0)$ or egg size was not measured, it is difficult to interpret O'Connor's findings.

E. Brood Effects

Variation among broods unrelated to year, season, climate, habitat, brood size, and other similar factors, reflects aspects of the inherent quality of parents. Much variation in growth and development clearly can be attributed to differences between nests whose origin may be traced to genotype, age and experience of parent, attributes of nest sites or breeding territories not often measured, or phenotypic variation in the quality of parental care of unknown basis. Even though such variation may be extensive, it has been systematically analyzed for growth parameters only for the European Starling (Ricklefs and Peters, 1979, 1981). They found that among early (first) nests, 12% of the variation in asymptote [$M(\infty)$] occurred among years, whereas 64% occurred among broods within years and 24% occurred within broods. The comparable figures for growth rate (K) were 4, 49, and 47%. Experiments in which eggs and neonates were switched among nests (Ricklefs and Peters, 1981) indicated that little of the among-broods variation had a genetic basis.

Similarity within broods reflects the common environment, but competition among siblings may order chicks into a hierarchy of mass gain and thereby reduce within-brood similarity. Although we could speculate on factors responsible for variance in growth and development among broods and on the specific behavior and ecological conditions that affect the level of this variance, observational and experimental studies have yet to produce a substantial phenomenology as a basis for further work.

IX. Variation among Species

A. Patterns of Interspecific Variation

Comparisons of growth rates among species of birds have been reviewed by Lack (1968), Ricklefs (1968b, 1973, 1976, 1979b), Case (1978), O'Connor (1978c), Dunn (1979a), and Drent and Daan (1980). Most of the variation in growth rate K over the entire spectrum of the class Aves has been related to adult body size $M(\infty)$, position in the precocial–altricial spectrum, geographical locality (especially tropical versus temperate zone), ecological status (especially pelagic versus inshore among marine species), nest mortality, and feeding rates. O'Connor (1978c) has considered the significance of variation in growth pattern among passerine birds on a finer scale. Comparisons among widely disparate species are usually based upon simple indices of rate of growth in body mass. Over the broad range of birds, rate of increase in mass is generally related to other attributes of development and is inversely related to the length of the development period. Comparisons based on a single abstraction of development tend to obscure a great deal of additional variation of immense biological significance, such as relative age of fledging, pattern of lipid deposition, relative growth of organs, and so forth. However, for comparisons over several orders of magnitude, growth in body mass is a useful and readily available measure of development.

Growth rates (K of Gompertz equations) vary from a high of 0.40–0.45 in certain small species of passerines (Ricklefs, 1968b) to 0.01–0.02 in certain large seabirds and cursorial land birds (Ricklefs, 1973). Other species not included in the analysis might expand this range, but at present it appears that the development periods of birds extend over at least a 30-fold range.

Growth parameters of fitted equations vary in inverse relationship to adult body mass. Ricklefs (1968b) found that among altricial land birds (primarily passerines and raptors), growth rate (logistic equation) and asymptote (g) were related by the equation $K = 1.11[M(\infty)]^{-0.28}$. Therefore, the length of the development period varies as the 0.28 power of body mass, and the absolute rate of growth as the 0.72 power of body mass. For precocial, cursorial birds, primarily galliforms, Ricklefs (1973) estimated the relationship between K and $M(\infty)$ to have a logarithmic slope of -0.36; precocial and altricial birds were not, however, compared by a proper analysis of variance.

The relationship for all birds together ($n = 165$) had a slope of -0.340 (± 0.026 SE), hence growth rate scales inversely with the cube root of body weight (Ricklefs, 1979b). The coefficient of determination ($r^2 = 0.58$) indicated that 42% of the variation in growth rate is unrelated to adult mass; variation about the regression line spans a nearly 10-fold range.

Much of the variation about the regression of growth rate upon body size is related to position on the precocial–altricial spectrum (Ricklefs, 1973). Measuring deviations from the K versus $M(\infty)$ regression in $\log_{10}$ units, 30 species of altricials and semialtricials (excluding pelagic seabirds, see following) had an average value of 0.160 with a standard deviation of 0.089; 21 species of precocial 2 and precocial 3 birds averaged -0.284 ± 0.140 SD. Hence the means of the two groups were separated by 0.44 log units, a 2.8-fold difference. There is also considerable variation among groups of precocial species compared on the same scale, ranging from galliforms (-0.355 ± 0.133 SD, $n = 8$) to shorebirds (Charadriidae and Scolopacidae; -0.184 ± 0.123 SD, $n = 6$) and ducks (-0.081 ± 0.167 SD, $n = 5$). Two species of rails (precocial 4) were similar to galliforms (0.318 ± 0.028 SD), but semiprecocial gulls and terns, excluding pelagic species, appear to grow nearly as rapidly as altricials ($+0.105 \pm 0.054$ SD, $n = 11$). Therefore, although there is a strong correlation between growth rate and development pattern, the condition of the neonate itself is not sufficient to account for this pattern.

Pelagic seabirds, all of which rear a single chick, grow slowly regardless of the condition of the neonate (Ricklefs, 1968b, 1973, 1982; Werschkul and Jackson, 1979). Ricklefs (1982) compared growth curves from 43 studies of altricial and semialtricial terrestrial and nearshore species with those from 29 studies of pelagic seabirds, including altricial, semialtricial, and semiprecocial species. The slopes of the K upon $M(\infty)$ regressions (logarithmic) did not differ (common slope = -0.321 ± 0.029 SE). The regression line for the pelagic species was displaced $0.25 \log_{10}$ units (a factor of 0.56) below that of the others ($P < 0.0001$).

Among small passerine birds [$M(\infty)$ less than 100 g], Ricklefs (1976) found that tropical species grow about 23% more slowly on average than temperate-zone and arctic species. Differences between related species in the two regions were less, suggesting the possibility of a taxonomic component to the comparison. Most of the tropical data came from Panama and Trinidad; the growth rates of nine species in the Amazon basis (Manaus, Brazil) did not differ significantly from the temperate-zone sample, but were significantly more rapid than those of species from Panama and Trinidad (Oniki and Ricklefs, 1981). The two tropical samples of species differed taxonomically, thus making it difficult to draw any general conclusions concerning geographical patterns in growth rate.

Lack (1968), Cody (1973), and Case (1978) have suggested that growth rates are closely related to mortality rate and feeding habit such that many of the correlations outlined earlier are based upon ecological factors. Hence the slow growth of pelagic seabirds is related to the low nestling mortality rates associated with breeding on remote islands and the difficulty of finding food (also expressed in the single-chick broods of such species). Case (1978) additionally pointed out that chicks of slowly growing species of birds are either self-feeders or are fed large meals at infrequent intervals.

Although mortality undoubtedly exerts strong selection for rapid growth, Ricklefs (1969a) failed to find a relationship between daily nest mortality rates and growth rates of temperate-zone passerine birds. Furthermore, the more slowly growing tropical species are exposed to almost twice as much mortality as nestlings of temperate-zone species (Ricklefs, 1969b).

Interpreting correlations that do exist between growth rate and other factors is made difficult by the fact that the independent variables are generally highly correlated among themselves. This problem may be circumvented in part by partial correlation analysis in which correlation with variables other than the two of interest are accounted for statistically, but data are often inadequate and such studies have yet to be undertaken. One may also choose species so as to control for certain unwanted effects, but in doing so one often reduces the sample variation in growth rate and runs the additional risk of becoming anecdotal. These problems must be kept in mind when nonexperimental data are brought to bear upon the hypotheses about growth rate outlined in the next section.

B. Hypotheses Concerning Growth Rates

Explanations for variation in growth rate may be placed in three categories. The first includes those that relate variation in growth rate directly to factors in the environment, the primary ones being predation on nestlings and chicks, availability of food, nutritional quality of the diet, and competition among siblings. Such hypotheses generally imply that growth rate is evolutionarily flexible independently of the overall pattern of development and that it is the primary target of selective forces. A second category of hypotheses concerns the level of parental reproductive effort, optimized in accordance with the demography of the population, which is then reflected in development rate. The effect of reproductive effort on growth is analogous to that of variation in food availability, although parental effort rather than nestling growth is the primary target of selection. The third category includes hypotheses concerning the limitation of growth rate by internal constraints imposed by nestling anatomy and physiology, that is, by the rate at which chicks either process energy and nutrients or convert them to biomass. According to such hypotheses, variation in growth rate follows upon variation in anatomy and physiology, especially position on the precocial–altricial spectrum, and therefore is an indirect consequence of selection of the development pattern of the chick.

1. Optimization with Respect to Mortality and Fecundity

Lack (1968) regarded observed growth rate as a balance between mortality factors selecting for more rapid growth and food limitation selecting for slower growth. When food supply is limited, slow growth reduces food requirement and enables parents to rear more offspring. Weighing against this benefit is the longer exposure of the vulnerable nestlings to mortality factors. All else being equal, one would expect growth rate to vary in direct proportion to mortality rate. Lack (1968) and Case (1978) found data consistent with this prediction, Ricklefs (1969a) did not (Section IX,A).

Food requirement does not increase in direct proportion to growth rate because maintenance energy presumably is not altered and the most rapid growth takes place while the chick is only partly grown and therefore uses less than its maximum maintenance requirement (for a contrasting analysis see Drent and Daan, 1980). Ricklefs (1969a) constructed a simple model of maximum energy requiremet E_{max} based on the logistic equation. The value of E_{max} depended on two quantities, the mass-specific maintenance energy requirement H and the product of growth rate K and the energy density of tissues G. When GK exceeded H, these were related by $E_{max} = M(\infty)(H +$

$GK)^2/4GK$; for values of K such that GK was less than H, $E_{max} = HM(\infty)$. Values of G and H for many birds are such that GK may usually exceed H, as illustrated by the following calculations. Energy densities of adult birds are on the order of 8 kJ g^{-1}. For altricials, logistic growth rate K is related to adult mass by the expression $K = 1.11M(\infty)^{-0.28}$ (Ricklefs, 1968b). For adult passerines, H(kJ g^{-1} day^{-1}) = 3.64 $M(\infty)^{-0.28}$ (Lasiewski and Dawson, 1967). Because G is more or less independent of $M(\infty)$, the relationship between H and GK is independent of $M(\infty)$, with $GK > H$ [8.9 $M(\infty)^{-0.28}$ versus 3.64 $M(\infty)^{-0.28}$]. Putting these values into the expression for maximum energy requirement, we find $E_{max} = 4.42\ M(\infty)^{0.72}$, that is, approximately 20% above H. At this point, increasing K by 10% results in an increase in E_{max} of about 4%.

In the model proposed by Ricklefs (1969a), such energetic consequences of rapid growth were balanced by increased nestling survival and increased number of broods per year made possible by the shorter nestling period. These factors appeared to offset changes in energy requirement so completely that the optimum growth rate was judged to be the fastest achievable no matter the energy cost. The predicted result was a brood of a single rapidly growing chick, contrary to the general circumstance of several offspring per brood. The model was weakened somewhat by several simplifying assumptions (e.g., values of G and H constant over the growth period), but more by the assumption that number of broods increases in direct relationship to K. This certainly is not applicable to single-brooded species and the relationship of the duration of stages other than the nestling period to variation in K is unknown. Case (1978) suggested modifications to the model that result in optimized rather than maximized values of K. I have formulated models that incorporate the assumptions mentioned above as explicit expressions. These more complicated models also predict optimum values for growth rate within the ranges observed in natural populations. Although modeling holds much promise for understanding variation in growth rate among species, the present models are useful primarily because they indicate to us how little we understand about the relationship of an adaptation, such as growth rate, to evolutionary fitness.

2. Sibling Competition

Werschkul and Jackson (1979) suggested that competition for food within broods exerts strong selection for rapid growth because larger chicks are more likely to be fed. The consequences of asynchronous hatching for late-hatched chicks reveals the potential force of sibling competition. Werschkul and Jackson suggested that evolutionary expression of sibling competition appears in the slow growth of species that rear a single chick (B1) for which there is no sibling competition, compared to those with larger broods (B2+). Ricklefs (1982) pointed out that because B1 species are virtually all pelagic seabirds, it is difficult to separate the effects of sibling competition and those of other ecological factors held in common by B1 species. Ricklefs (1982) also tested the relationship between growth rate and the effects of sibling competition, estimated by within-brood mortality rate, in B2+ species of temperate-zone passerines, but found no correlation. Although Werschkul and Jackson's hypothesis is an attractive one, it is virtually impossible to evaluate properly at present.

3. Food Availability and Growth Rate

Lack (1968) and Case (1978) have both argued that growth rate of some species is adjusted to the amount of food that parents can provide, which in turn is related to the availability of food in the environment. This argument is most clearly applied to pelagic seabirds and other B1 species because, after brood size is reduced to one, further energy savings are possible only by reducing the requirement of each chick. The energy savings of slow growth may not be substantial, however. In the simple model presented in (Section IX,B,1), the maximum energy requirement of rapidly growing altricial species was only about 20% greater than the minimum possible with slow growth. Ricklefs and White (1981) confirmed this when they estimated that doubling the growth rate of the slowly growing Sooty Tern (B1) would increase the maximum energy requirement by 20%. Such a change would increase the total amount of energy gathered by the parents by only 5%. For Leach's Storm-Petrel, doubling the chick's rate of growth (excluding lipid deposition) would increase the maximum energy requirement by less than 5% (Ricklefs et al., 1980b). Drent and Daan (1980) have suggested that energy requirement is directly proportional to growth rate, but their analysis is based on differences between species greatly differing in adult size and their result is a fortuitous consequence of the similar allometric scaling of growth rate and metabolism rate and their mistaken assumption that a change in growth rate causes a proportional change in metabolizable energy requirement. Presumably the maintenance component of metabolizable energy is independent of growth rate.

For B2+ species, reducing growth rate may allow a larger brood to be reared although this benefit is opposed by increased nest mortality and is difficult to evaluate. Ricklefs (1968c) suggested that because brood size changes by integers it adjusts energy requirement to food supply only coarsely; birds may use growth rate to achieve finer adjustment of energy requirement. This model predicts that among species with smaller clutches one would expect to find a lower average growth rate and, importantly, greater variation in growth rate (Ricklefs, 1968c; Fretwell *et al.*, 1974). Ricklefs (1976) employed data for tropical passerines to test this hypothesis with negative result.

4. Diet Quality and Growth Rate

Certain foods do not contain sufficient nutrients relative to metabolizable energy to support rapid growth (see Section IV,B). That nutrition is responsible for variation in growth rate in nature is doubtful because of the parent's ability to select nutritional items for the diet and because of the relative uniformity of growth rate among species with vastly different diets. Although protein deficiency in the oily diets of petrels has been suggested as a cause of their slow growth, many other pelagic seabirds with high-protein (fish) diets also grow slowly. Nutrients may, however, limit growth in some tropical frugivores that do not supplement the nestling diet with insects (e.g., Bearded Bellbird and Oilbird). In general, diet limitation on growth rate still needs to be evaluated by suitable nutritional analyses on a case-by-case basis.

5. Adult Reproductive Effort

The amount of food delivered to nestlings depends on its availability in the environment and the effort that adults expend to gather it. If growth rate is sensitive to rate of food provisioning, it may be argued that variation in reproductive effort rather than food availability is responsible for some variation in growth rate. Theoretical arguments indicate that reproductive effort should vary inversely with the expectation of future reproduction. There is little to be gained by risking a large amount of future reproduction through increasing effort on behalf of the present brood (Williams, 1966; Gadgil and Bossert, 1970). Goodman (1974) and Dunn (1979a) have used this argument to relate the great longevity (i.e., future reproduction) and low fecundity (i.e., reproductive effort) in pelagic seabirds. Indeed lifespan and fecundity are inversely related over all birds as a group, but Ricklefs (1977b) cautioned that this pattern may derive as easily from density dependence as from demographic optimization, and he presented evidence that minimized the relative importance of the latter factor.

6. Rate of Assimilation

Growth requires energy and nutrients and competes with other avenues of expenditure when either is limiting. Inasmuch as survival is more critical than growth, the energy needs of temperature regulation and activity must be satisfied before energy is allocated to growth. If the amount of assimilated energy were limited by the size or degree of functioning of the alimentary tract and its associated organs, growth rate would be limited below the maximum rate of assimilation by allocation of resources to maintenance activities. The alimentary organs of precocial species may be relatively smaller than those of altricials and certainly a relatively greater proportion of energy is allocated to maintenance, particularly in self-feeding precocials (levels 1 to 3, and partly 4, of Nice, 1962).

Data on growth rates are consistent with the energy limitation–allocation hypothesis. Among precocial species growth rate increases from galliforms to shorebirds, waterfowl, and semiprecocials (e.g., gulls and terns). This sequence parallels my subjective impression of the relative availability of food to each group, hence the expenditure of energy to obtain food. At the same time, the sequence may parallel the rate of food gathering in the absence of an upper limit to rate of assimilation.

Experiments on intermittent food deprivation in fowl revealed that starved chicks can greatly increase their rates of food intake during periods of repletion, but long-term experiments are usually accompanied by enlargement of storage organs in the forepart of the alimentary tract. Most experience with hand-feeding birds suggests that chicks are easily sated and that there is therefore an upper limit to rate of assimilation. It remains to be determined whether maximum rates of assimilation and the relative allocation of resources differ between rapidly growing and slowly growing species.

7. Allocation of Tissue

Difficulties with most of the hypotheses listed earlier led Ricklefs (1979a,b) to suggest that growth rate is determined by a balance between mature and embryonic functions of tissues. The idea has deep roots in the literature of embryology (e.g., Schmalhausen, 1930) but its application had been restricted to the slowing of the growth rate of the individual during the development period. If growth and mature function (e.g., muscle contraction) were mutually exclusive functions of a tissue, it would stand to reason that growth should slow as the individual matures and that precocial chicks would grow more slowly than altricial chicks. Where this hypothesis has been tested by anatomical and physiological data, it has been shown to be adequate. The slowing of growth rate with age is associated with increasing functional capacity for temperature regulation and mobility and with changes in the composition of tissues consistent with differentiation and maturing function (see Section III,B,1). In general, differences among species in precocity of development are associated with expected differences in rate of growth. Ricklefs (1979b) has focused attention on skeletal muscle as the tissue most responsible for these differences because it is most clearly associated with functional differences between altricial and precocial birds, namely, mobility and development of temperature regulation. Both types of species have equal requirements for food processing, blood circulation and, to some extent, neural function, hence one would not expect nor does one find such differences in visceral organs (Section II,B,2).

Certain semiprecocial species, such as gulls and terns, some alcids, and, to a lesser degree, waterfowl present a difficulty for the tissue-maturation hypothesis because of their relatively rapid rates of growth. Ricklefs (1979a) has shown, however that the legs of Common Terns grow as slowly as do those of Japanese Quail, as expected by the theory. That the tern chick as a whole grows more rapidly is related to the fact that its legs decrease in size relative to the body as a whole throughout development. Legs make up 15.8% of the mass of the neonate but only 4.5% of the adult, increasing by a factor of 3.1 between hatching and fledging. In the quail, the legs make up 18.1% of the mass of the neonate and 16.5% of the adult, increasing 17.6 times between hatching and adulthood. The difference between the growth increments is almost exactly what would be required to allow the tern to complete its body growth in only 40% of the time taken by the quail. Ricklefs (1979b) showed that growth rate among precocial birds in general is inversely related to the relative size of the legs of the adults. This suggests that growth rate may be an indirect consequence of adaptations of the adults for foraging and other aspects of their ways of life.

Slow growth of pelagic seabirds may be more difficult to explain by the tissue-allocation hypothesis, but the slow growth of procellariiforms is certainly consistent with their precocious development of temperature regulation (Ricklefs *et al.*, 1980a), and there is some indication that slowly growing tropical terns may develop more precocially than temperate-zone counterparts (Ricklefs and White, 1981). A similar explanation for slow growth in pelagic pelecaniforms with altricial development is doubtful, however.

I have argued that most of the variation in rate of growth among species is related to precocity of development and adult body proportions rather than to factors directly influencing growth rate, such as juvenile mortality and food quality and availability. If these ideas turn out to be correct, then one must accept a view of adaptation in which separate traits of the individual are interrelated through a complex and nonobvious set of constraints that determine the course of evolutionary adaptation.

ACKNOWLEDGMENTS

I am grateful to Joseph B. Williams and Larry Clark for helpful comments on the manuscript. The author's research on patterns of development in birds has been generously supported by the National Science Foundation, the U.S. Antarctic Research Program, and the National Geographic Society. The preparation of this chapter was supported in part by NSF DEB80-21732.

REFERENCES

Ankney, C. D. (1980). Egg weight, survival, and growth of Lesser Snow Goose goslings. *J. Wildl. Manage.* **44,** 174–182.
Annison, E. F. (1971). Lipid and acetate metabolism. *In* "Physiology and Biochemistry of the Domestic Fowl" (D. J. Bell, and B. M. Freeman, eds.), pp. 321–337. Academic Press, New York.
Assenmacher, I. (1973). The peripheral endocrine glands. *In* "Avian Biology" (D. S. Farner, and J. R. King, eds.), Vol. 3, pp. 183–286. Academic Press, New York.
Auckland, J. N. (1978). The effect of photoperiod on performance of female turkeys from two to 16 weeks of age. *Br. Poult. Sci.* **19,** 691–693.
Aulie, A. (1976a). The pectoral muscles and the development of thermoregulation in chicks of willow ptarmigan. *Comp. Biochem. Physiol. A* **53,** 343–346.
Aulie, A. (1976b). The shivering pattern in an arctic (willow ptarmigan) and a tropical bird (bantam hen). *Comp. Biochem. Physiol. A* **53,** 347–350.
Aulie, A. (1977). The effect of intermittent cold exposure on the thermoregulatory capacity of bantam chicks (*Gallus domesticus*). *Comp. Biochem. Physiol. A* **56,** 545–549.
Aulie, A., and Grav, H. J. (1979). Effect of cold acclimation on the oxidative capacity of skeletal muscle and liver in young bantams. *Comp. Biochem. Physiol. A* **62,** 335–338.
Aulie, A., and Steen, J. B. (1976). Thermoregulation and muscular development in cold exposed willow ptarmigan chicks (*Lagopus lagopus* L.). *Comp. Biochem Physiol. A* **55,** 291–295.
Austin, G. T., and Ricklefs, R. E. (1977). Growth and development of the Rufous-winged Sparrow (*Aimophila carpalis*). *Condor* **79,** 37–50.
Baldwin, S. P., and Kendeigh, S. C. (1932). Physiology of the temperature of birds. *Sci. Publ. Cleveland Mus. Nat. Hist.* **3,** 1–196.
Baldwin, S. P., Oberholser, H. C., and Worley, L. G. (1931). Measurements of birds. *Sci. Publ. Cleveland Mus. Nat. Hist.* **2,** 1–165.
Balmer, R. T., and Strobusch, A. D. (1977). Critical size of newborn homeotherms. *J. Appl. Physiol.* **42,** 571–577.
Bartholomew, G. A. (1966). The role of behavior in the temperature regulation of the Masked Booby. *Condor* **68,** 523–535.
Bell, D. J., and Freeman, B. M. (eds.) (1971). "Physiology and Biochemistry of the Domestic Fowl," Vol. 1. Academic Press, New York.
Bellairs, R. (1960). Development of birds. *In* "Biology and Comparative Physiology of Birds" (A. J. Marshall, ed.), Vol. 1, pp. 127–188. Academic Press, New York.
Bergtold, W. H. (1913). A study of the House Finch. *Auk* **30,** 40–73.
Bernstein, M. H. (1973). Development of thermoregulation in Painted Quail *Excalfactoria chinensis*. *Comp. Biochem. Physiol. A* **44,** 355–366.
Bilby, L. W., and Widdowson, E. M. (1971). Chemical composition of growth in nestling blackbirds and thrushes. *Br. J. Nutr.* **25,** 127–134.
Blivaiss, B. B. (1947). Development of secondary sexual characters in thyroidectomized Brown Leghorn hens. *J. Exp. Zool.* **104,** 267–305.
Blyth, J. S. S., Pun, C. F., and Sang, J. H. (1965). Survey of line crosses in Brown Leghorn flock. II. Relations of hatched chick weight to egg weight in inbred lines and their crosses. *Br. Poult. Sci.* **6,** 217–223.
Boag, P. T., and Grant, P. R. (1978). Heritability of external morphology in Darwin's Finches. *Nature (London)* **274,** 793–794.

Boersma, P. D., Wheelwright, N. T., Nerini, M. K., and Wheelwright, E. S. (1980). The breeding biology of the Fork-tailed Storm-Petrel (*Oceanodroma furcata*). *Auk* **97**, 268–282.

Böni, A. (1942). Ueber den Entwicklung der Temperaturregulation bei verschiedenen Nesthockern. *Arch. Suisses Ornithol.* **2**, 1–56.

Boomgardt, J., and Baker, D. H. (1973). Effect of dietary energy concentration on sulfur amino acid requirements and body composition of young chicks. *J. Anim. Sci.* **36**, 307–311.

Boorman, K. N., and Lewis, D. (1971). Protein metabolism. *In* "Physiology and Biochemistry of the Domestic Fowl" (D. J. Bell, and B. M. Freeman, eds.), pp. 339–372. Academic Press, New York.

Breitenbach, R. P., and Baskett, T. S. (1967). Ontogeny of thermoregulation in the Mourning Dove. *Physiol. Zool.* **40**, 207–217.

Brody, S. (1937). Relativity of physiologic time and physiologic weight. *Growth* **1**, 60–67.

Brody, S. (1945). "Bioenergetics and Growth." Reinhold, New York.

Bryant, D. M. (1975). Breeding biology of House Martins *Delichon urbica* in relation to aerial insect abundance. *Ibis* **117**, 180–216.

Bryant, D. M. (1978a). Environmental influences on growth and survival of nestling House Martins *Delichon urbica*. *Ibis* **120**, 271–283.

Bryant, D. M. (1978b). Establishment of weight hierarchies in the broods of House Martins *Delichon urbica*. *Ibis* **120**, 16–26.

Bryant, D. M., and Gardiner, A. (1979). Energetics of growth in House Martins (*Delichon urbica*). *J. Zool.* **189**, 275–304.

Burger, J. (1980). The transition to independence and postfledging parental care in seabirds. *In* "Behavior of Marine Animals" (J. Burger, B. L. Olla, and H. E. Winn, eds.), Vol. 4, pp. 367–447. Plenum, New York.

Butler, E. J. (1971). The role of trace elements in metabolic processes. *In* "Physiology and Biochemistry of the Domestic Fowl" (D. J. Bell, and B. Freeman, eds.), pp. 398–426. Academic Press, New York.

Calder, W. A., and King, J. R. (1974). Thermal and caloric relations of birds. *In* "Avian Biology" (D. S. Farner, and J. R. King, eds.), Vol. 4, pp. 259–413. Academic Press, New York.

Carey, C., Rahn, H., and Parisi, P. (1980). Calories, water, lipid and yolk in avian eggs. *Condor* **82**, 335–343.

Case, T. J. (1978). On the evolution and adaptive significance of postnatal growth rates in the terrestrial vertebrates. *Q. Rev. Biol.* **55**, 243–282.

Clark, L., and Balda, R. P. (1981). The development of effective endothermy and homeothermy by nestling Piñon Jays. *Auk* **98**, 615–619.

Clark, A. B., and Wilson, D. S. (1981). Avian breeding adaptations: hatching asynchrony, brood reduction, and nest failure. *Q. Rev. Biol.* **56**, 253–277.

Coates, M. E. (1971). The role of vitamins in metabolic processes. *In* "Physiology and Biochemistry of the Domestic Fowl" (D. J. Bell, and B. M. Freeman, eds.), pp. 373–396. Academic Press, New York.

Cock, A. G. (1966). Genetical aspects of growth and form in animals. *Q. Rev. Biol.* **41**, 131–190.

Cody, M. L. (1973). Coexistence, coevolution and convergent evolution in seabird communities. *Ecology* **54**, 31–44.

Cramp, S. (ed.) (1977). "Handbook of the Birds of Europe, the Middle East, and North Africa," Vol. 1. Oxford Univ. Press, Oxford.

Crawford, R. D. (1980). Effects of age on reproduction in American Coots. *J. Wildl. Manage.* **44**, 183–189.

Csapo, A., and Herrmann, H. (1951). Quantitative changes in contractile proteins of chick skeletal muscle during and after embryonic development. *Am. J. Physiol.* **165**, 701–710.

Davies, O. L., and Ku, J. Y. (1977). Re-examination of the fitting of the Richards growth function. *Biometrics* **33**, 546–547.

Dawson, W. R., and Allen, J. M. (1960). Thryoid activity in nestling Vesper Sparrows. *Condor* **62**, 403–405.

De Steven, D. (1978). The influence of age on the breeding biology of the Tree Swallow (*Iridoprocne bicolor*). *Ibis* **120**, 516–523.

Dickerson, J. W. T. (1960). The effect of growth on the composition of avian muscle. *Biochem. J.* **75**, 33–37.

Dickerson, J. W. T., and McCance, R. A. (1960). Severe undernutrition in growing and adult animals. III. Avian skeletal muscle. *Br. J. Nutr.* **14**, 331–338.

Drent, R. H., and Daan, S. (1980). The prudent parent: energetic adjustments in avian breeding. *Ardea* **68**, 225–252.

Dunn, E. H. (1973). "Energy Allocation of Nestling Double-Crested Cormorants." Ph.D. Thesis, Univ. of Michigan, Ann Arbor.

Dunn, E. H. (1975a). The timing of endothermy in the development of altricial birds. *Condor* **77**, 288–293.

Dunn, E. H. (1975b). Growth, body components and energy content of nestling Double-crested Cormorants. *Condor* **77**, 431–438.

Dunn, E. H. (1976). The relationship between brood size and age of effective homeothermy in nestling House Wrens. *Wilson Bull.* **88**, 478–482.

Dunn, E. H. (1979a). Time-energy use and life history strategies of northern seabirds. *Res. Rep. U.S. Fish Wildl. Serv.*, No. 11, pp. 141–166.

Dunn, E. H. (1979b). Age of effective homeothermy in nestling Tree Swallows according to brood size. *Wilson Bull.* **91**, 455–457.

Dunn, E. H. (1980). On the variability of energy allocation of nestling birds. *Auk* **97**, 19–27.

Dunn, E. K. (1975). The role of environmental factors in the growth of tern chicks. *J. Anim. Ecol.* **44**, 743–754.

Fabens, A. J. (1965). Properties and fitting of the von Bertalanffy growth curve. *Growth* **29**, 265–289.

Falconer, D. S. (1960). "Introduction to Quantitative Genetics." Ronald, New York.

Falconer, I. R. (1971). The thryoid glands. In "Physiology and Biochemistry of the Domestic Fowl" (D. J. Bell, and B. M. Freeman, eds.), Vol. 1, pp. 459–472. Academic Press, New York.

Fisher, H. (1972). The nutrition of birds. In "Avian Biology" (D. S. Farner, and J. R. King, eds.), Vol. 2, pp. 431–469. Academic Press, New York.

Fjeldså, J. (1977). "Guide to the Young of European Precocial Birds." Skarv Nature Publ., Tisvildeleje, Denmark.

Foster, M. S. (1978). Total frugivory in tropical passerines: a reappraisal. *Trop. Ecol.* **19**, 131–154.

Freeman, B. M. (1971). Body temperature and thermoregulation. In "Physiology and Biochemistry of the Domestic Fowl" (D. J. Bell, and B. M. Freeman, eds.), Vol. 2, pp. 1115–1141. Academic Press, New York and London.

Fretwell, S., Bowen, D. E., and Hespenheide, H. A. (1974). Growth rates of young passerines and the flexibility of clutch size. *Ecology* **55**, 907–909.

Gadgil, M., and Bossert, W. H. (1970). Life historical consequences of natural selection. *Am. Nat.* **104**, 1–24.

Garnett, M. C. (1981). Body size, its heritability and influence on juvenile survival among Great Tits, *Parus major. Ibis* **123**, 31–41.

Glick, B. (1960). The effect of bovine growth hormone, DCA, and cortisone on weight of bursa of Fabricius, adrenal glands, heart and body weight of young chickens. *Poult. Sci.* **39**, 1527–1533.

Gompertz, B. (1825). On the nature of the function expressive of the law of human mortality, and on a new mode of determining the value of life contingencies. *Phil. Trans. R. Soc. London* **115**, 515–585.

Goodman, D. (1974). Natural selection and a cost ceiling on reproductive effort. *Am. Nat.* **108**, 247–268.

Hahn, D. C. (1981). Asynchronous hatching in the Laughing Gull: cutting losses and reducing rivalry. *Anim. Behav.* **29**, 421–427.

Halbersleben, D. L., and Mussehl, F. E. (1922). The relation of egg weight to chick weight at hatching. *Poult. Sci.* **1**, 143–144.

Harris, M. P. (1966). Breeding biology of the Manx Shearwater *Puffinus puffinus*. *Ibis* **108**, 17–33.

Hartree, A. S., and Cunningham, F. J. (1971). The pituitary gland. In "Physiology and Biochemistry of the Domestic Fowl" (D. J. Bell, and B. M. Freeman, eds.), Vol. 1, pp. 427–257. Academic Press, New York.

Harvey, P. H., Greenwood, P. J., Perrins, C. M., and Martin, A. R. (1979). Breeding success of Great Tits *Parus major* in relation to age of male and female parent. *Ibis* **121**, 217–219.

Hayashida, T. (1969). Relatedness of pituitary growth hormone from various vertebrate classes. *Nature (London)* **222**, 294–295.

Hazelwood, R. L., and Hazelwood, B. S. (1961). Effects of avian and rat pituitary extracts on tibial growth and blood composition. *Proc. Soc. Exp. Biol. Med.* **108**, 10–12.

Herrmann, H. (1952). Studies of muscle development. *Ann. N.Y. Acad. Sci.* **55**, 99–108.

Horton, E. W. (1971). Prostaglandins. In "Physiology and Biochemistry of the Domestic Fowl" (D. J. Bell, and B. M. Freeman, eds.), Vol. 2, pp. 589–601. Academic Press, London.

Houston, D. C. (1978). The effect of food quality on breeding strategy in Griffon vultures. *J. Zool.* **186**, 175–184.

Howarth, B., Jr., and Marks, H. L. (1973). Thyroidal ^{131}I uptake of Japanese Quail in response to three different dietary goitrogens. *Poult. Sci.* **52**, 326–331.

Howe, H. F. (1976). Egg size, hatching synchrony, sex, and brood reduction in the Common Grackle. *Ecology* **57**, 1195–1207.

Howe, H. F. (1978). Initial investment, clutch size, and brood reduction in the Common Grackle (*Quiscalus quiscula* L.). *Ecology* **59**, 1109–1122.

Huxley, J. S. (1924). Constant differential growth ratios and their significance. *Nature (London)* **114**, 895–896.

Huxley, J. S. (1932). "Problems of Relative Growth." Methuen, London.

Huxley, J. S., and Teissier, G. (1936). Terminology of relative growth. *Nature (London)* **137**, 780–781.

Kaufman, L. (1927). Récherches sur la croissance du corps et des organes du Pigeon. *Biol. Gen.* **3**, 105–128.

Kaufman, L. (1930). Innere und ässere Wachstums-faktoren. Untersuchungen an Hühnern und Tauben. *Wilhelm Roux Arch. Entwicklungsmech. Org.* **122**, 395–431.

Kaufman, L. (1962). "Metamorphosis" in pigeons. *Acta Biol. Cracov. Ser. Zool.* **5**, 317–326.

Kaufman, L., and Nowotna, A. (1934). Heterogonisches Wachstum und chemische Zusammensetzung der Leber bei Tauben. *Pflügers Arch. Gesamte Physiol. Menschen Tiere* **235**, 247–255.

Kavanaugh, A. J., and Richards, O. W. (1942). Mathematical analysis of the relative growth of organisms. *Proc. Rochester Acad. Sci.* **8**, 150–174.

Kendeigh, S. C. (1939). The relation of metabolism to the development of temperature regulation in birds. *J. Exp. Zool.* **82**, 419–438.

Kendeigh, S. C., and Baldwin, S. P. (1928). Development of temperature control in nestling House Wrens. *Am. Nat.* **62,** 249–278.

Kendeigh, S. C., Dol'nik, V. R., and Gavrilov, V. M. (1977). Avian energetics. *In* "Granivorous Birds in Ecosystems" (J. Pinowski, and S. C. Kendeigh, eds.), pp. 127–204. Cambridge Univ. Press, Cambridge.

King, A. S. (1966). Structural and functional aspects of the avian lungs and air sacs. *Int. Rev. Gen. Exp. Zool.* **2,** 171–267.

King, D. B. (1969). Effect of hypophysectomy of young cockerels, with particular reference to body growth, liver weight, and liver glycogen level. *Gen. Comp. Endocrinol.* **12,** 242–255.

King, D. B., and King, C. R. (1973). Thyroidal influence on early muscle growth of chickens. *Gen. Comp. Endocrinol.* **21,** 517–529.

King, J. R., and Farner, D. S. (1961). Energy metabolism, thermoregulation and body temperature. *In* "Biology and Comparative Physiology of Birds" (A. J. Marshall, ed.), Vol. 2, pp. 215–288. Academic Press, New York.

Kinney, T. B. (1969). A summary of reported estimates of heritabilities and of genetic and phenotypic correlations for traits of chickens. *U.S. Dep. Agric. Agric. Handb.*, No. 363.

Kirkpatrick, C. M. (1944). Body weights and organ measurements in relation to age and season in Ring-necked Pheasants. *Anat. Rec.* **89,** 175–194.

Kleiber, M., and Dougherty, J. E. (1934). The influence of environmental temperature on the utilization of food energy in baby chicks. *J. Gen. Physiol.* **17,** 701–726.

Klomp, H. (1970). The determination of clutch size in birds, a review. *Ardea* **58,** 1–124.

Koskimies, J. (1948). On temperature regulation and metabolism in the Swift, *Micropus a. apus* L., during fasting. *Experientia* **4,** 274–276.

Koskimies, J. (1962). Ontogeny of thermoregulation and energy metabolism in some gallinaceous birds. *Trans. Congr. Int. Union Game Biol., 5th, Bologna*, 149–160.

Koskimies, J., and Lahti, L. (1964). Cold-hardiness of the newly hatched young in relation to ecology and distribution in ten species of European ducks. *Auk* **81,** 281–307.

Kuhlmann, F. (1909). Some preliminary observations on the development of instincts and habits of young birds. *Psych. Rev. (Mon. Suppl.)* **11,** 49–85.

Lack, D. (1954). "The Natural Regulation of Animal Numbers." Oxford Univ. Press, London.

Lack, D. (1956). "Swifts in a Tower." Methuen, London.

Lack, D. (1968). "Ecological Adaptations for Breeding in Birds." Methuen, London.

Lack, D., and Lack, E. (1951). The breeding biology of the Swift *Apus apus. Ibis* **98,** 606–619.

Laird, A. K., Tyler, S. A., and Barton, A. D. (1965). Dynamics of normal growth. *Growth* **29,** 233–248.

Lasiewski, R. C, and Dawson, W. R. (1967). A re-examination of the relation between standard metabolic rate and body weight in birds. *Condor* **69,** 13–23.

Latimer, H. B. (1924). Postnatal growth of the body, systems and organs of the single-comb white leghorn chicken. *J. Agric. Res.* **29,** 363–399.

Latimer, H. B. (1925a). The relative postnatal growth of the systems and organs of the chicken. *Anat. Rec.* **31,** 233–253.

Latimer, H. B. (1925b). The postnatal growth of the central nervous system of the chicken. *J. Comp. Neurol.* **38,** 251–297.

Latimer, H. B. (1927). Postnatal growth of the chicken skeleton. *Am. J. Anat.* **40,** 1–57.

Lerner, I. M. (1937). Relative growth and hereditary size limitation in the domestic fowl. *Hilgardia* **10,** 511–560.

Libby, D. A., Meites, J., and Schaible, J. (1955). Growth hormone effects in chickens. *Poult. Sci.* **34,** 1329–1331.

Lloyd, C. S. (1979). Factors affecting breeding of Razorbills *Alca torda* on Skokholm. *Ibis* **121**, 165–175.

Lucas, A. M., and Stettenheim, P. R. (1972). Avian anatomy. Integument, parts I and II. *U.S. Dep. Agric. Agric. Handb.*, No. 362.

Marks, H. L. (1971). Selection for four-week body weight in Japanese Quail under two nutritional environments. *Poult. Sci.* **50**, 931–937.

Marks, H. L. (1975). Relationship of embryonic development to egg weight, hatch weight, and growth in Japanese Quail. *Poult. Sci.* **54**, 1257–1262.

Marks, H. L. (1978a). Growth curve changes associated with long-term selection for body weight in Japanese Quail. *Growth* **42**, 129–140.

Marks, H. L. (1978b). Long term selection for four-week body weight in Japanese Quail under different nutritional environments. *Theor. Appl. Genet.* **52**, 105–111.

Marks, H. L. (1979). Changes in unselected traits accompanying long-term selection for four-week body weight in Japanese Quail. *Poult. Sci.* **58**, 269–274.

Marshall, A. J. (ed.) (1960–1961). "Biology and Comparative Physiology of Birds," Vols. 1 and 2. Academic Press, New York.

Martin, G. A., Glazener, E. W., and Blow, W. L. (1953). Efficiency of selection for broiler growth at various ages. *Poult. Sci.* **32**, 716–720.

McCance, R. A. (1960). Severe undernutrition in growing and adult animals. I. Production and general effects. *Br. J. Nutr.* **14**, 59–73.

McNabb, F. M. A., and McNabb, R. A. (1977a). Skin and plumage changes during the development of thermoregulatory ability in Japanese Quail chicks. *Comp. Biochem. Physiol. A* **58**, 163–166.

McNabb, F. M. A., and McNabb, R. A. (1977b). Thyroid development in precocial and altricial avian embryos. *Auk* **94**, 736–742.

McRoberts, M. R. (1965). Growth retardation of day-old chickens and physiological effects at maturity. *J. Nutr.* **87**, 31–40.

Meier, A. H., and Ferrell, B. R. (1978). Avian endocrinology. *In* "Chemical Zoology" (A. H. Brush, ed.), Vol. 10, pp. 213–271. Academic Press, New York.

Merritt, E. S., and Gowe, R. O. (1965). Postembryonic growth in relation to egg weight. *Poult. Sci.* **44**, 477–486.

Mertens, J. A. L. (1969). The influence of brood size on the energy metabolism and water loss of nestling Great Tits *Parus major major*. *Ibis* **111**, 11–17.

Millward, D. J., and Garlick, P. J. (1976). The energy cost of growth. *Proc. Nutr. Soc.* **35**, 339–349.

Morse, K., and Vohra, P. (1971). The effect of early growth retardation of *Coturnix* (Japanese Quail) on their sexual maturity. *Poult. Sci.* **50**, 283–284.

Morton, E. S. (1973). On the evolutionary advantages and disadvantages of fruit eating in tropical birds. *Am. Nat.* **107**, 8–22.

Moss, F. P. (1968a). The relationship between the dimensions of fibres and the number of nuclei during normal growth of skeletal muscle in the domestic fowl. *Am. J. Anat.* **122**, 555–564.

Moss, F. P. (1968b). The relationship between the dimensions of the fibres and the number of nuclei during restricted, degrowth and compensatory growth of skeletal muscle. *Am. J. Anat.* **122**, 565–572.

Mosteller, F., and Tukey, J. W. (1977). "Data Analysis and Regression." Addison-Wesley, Reading, Massachusetts.

Moudgal, N. R., and Li, C. H. (1961). Immunochemical studies of bovine and ovine pituitary growth hormone. *Arch. Biochem. Biophys.* **93**, 122–127.

Nair, K. R. (1954). The fitting of growth curves. *In* "Statistics and Mathematics in Biology" (O. Kempthorne, T. A. Bancroft, J. W. Gower, and J. L. Lush, eds.), pp. 119–132. Iowa State Univ. Press, Ames.

Nalbandov, A. V., and Card, L. E. (1942). Results of hypophysectomy in growing chickens. *Poult. Sci.* **21**, 474.

Neff, M. (1972). Untersuchungen über das embryonale und postembryonale Organwachstum bei Vogelarten mit verschiedenen Ontogenesemodus. *Rev. Suisse Zool.* **79**, 1471–1597.

Nelson, J. B. (1966). The behaviour of the young Gannet. *Br. Birds* **59**, 393–419.

Nelson, J. B. (1978). "The Sulidae." Oxford Univ. Press, Oxford.

Nice, M. M. (1943). Studies in the life history of the Song Sparrow. II. The behavior of the Song Sparrow and other passerines. *Trans. Linn. Soc. N.Y.* **6**, 1–328.

Nice, M. M. (1962). Development of behavior in precocial birds. *Trans. Linn. Soc. N.Y.* **8**, 1–211.

Nir, I., and Nitsan, Z. (1979). Metabolic and anatomical adaptations of light-bodied chicks to intermittent feeding. *Br. Poult. Sci.* **20**, 61–71.

Nir, I., Shapira, N., Nistan, Z., and Dror, Y. (1974). Force-feeding effects on growth, carcass and blood composition in the young chick. *Br. J. Nutr.* **32**, 229–239.

Nir, I., Nitsan, Z., Dror, Y., and Shapira, N. (1978). Influence of overfeeding on growth, obesity and intestinal tract in young chicks of light and heavy breeds. *Br. J. Nutr.* **39**, 27–35.

Nisbet, I. C. T. (1978). Dependence of fledging success on egg size, parental performance and egg-composition among Common and Roseate terns, *Sterna hirundo* and *S. dougallii*. *Ibis* **120**, 207–215.

Nolan, V., Jr., and Thompson, C. F. (1978). Egg volume as a predictor of hatchling weight in the Brown-headed Cowbird. *Wilson Bull.* **90**, 353–358.

Nordskog, A. W., Tolman, H. S., Casey, D. W., and Lin, C. Y. (1974). Selection in small populations of chickens. *Poult. Sci.* **53**, 1188–1219.

O'Connor, R. J. (1975a). The influence of brood size upon metabolic rate and body temperature in nestling Blue Tits *Parus caeruleus* and House Sparrows *Passer domesticus*. *J. Zool.* **175**, 391–403.

O'Connor, R. J. (1975b). Initial size and subsequent growth in passerine nestlings. *Bird-Banding* **46**, 329–340.

O'Connor, R. J. (1975c). Growth and metabolism in nestling passerines. *Symp. Zool. Soc. London* **35**, 277–306.

O'Connor, R. J. (1975d). Nestling thermolysis and developmental change in body temperature. *Comp. Biochem. Physiol. A* **52**, 419–422.

O'Connor, R. J. (1977). Differential growth and body composition in altricial passerines. *Ibis* **119**, 147–166.

O'Connor, R. J. (1978a). Brood reduction in birds: selection for fratricide, infanticide and suicide? *Anim. Behav.* **26**, 79–96.

O'Connor, R. J. (1978b). Structure in avian growth patterns: a multivariate study of passerine development. *J. Zool.* **185**, 147–172.

O'Connor, R. J. (1978c). Growth strategies of nestling passerines. *Living Bird* **16**, 209–238.

O'Connor, R. J. (1979). Egg weights and brood reduction in the European Swift (*Apus apus*). *Condor* **81**, 133–145.

Odum, E. P. (1942). Muscle tremors and the development of temperature regulation in birds. *Am. J. Physiol.* **136**, 618–622.

Ollason, J. C., and Dunnet, G. M. (1978). Age, experience and other factors affecting breeding success of the fulmar (*Fulmarus glacialis*) in Orkney. *J. Anim. Ecol.* **47**, 961–976.

Oniki, Y., and Ricklefs, R. E. (1981). More growth rates of birds in the humid New World Tropics. *Ibis* **123**, 349–354.
Osbaldiston, G. W. (1966). The response of the immature chicken to ambient temperature. In "Physiology of the Domestic Fowl" (C. Horton-Smith, and E. C. Amoroso, eds.), pp. 228–234. Oliver & Boyd, Edinburgh and London.
Parsons, J. (1970). Relationship between egg size and post-hatching chick mortality in the Herring Gull (*Larus argentatus*). *Nature (London)* **228**, 1221–1222.
Parsons, J. (1972). Egg size, laying data and incubation period in the Herring Gull. *Ibis* **114**, 536–541.
Pearce, J., and Brown, W. O. (1971). Carbohydrate metabolism. In "Physiology and Biochemistry of the Domestic Fowl" (D. J. Bell, and B. M. Freeman, eds.), pp. 295–320. Academic Press, New York.
Pembrey, M. S. (1895). The effect of variation in external temperature upon the output of carbonic acid and the temperature of young animals. *J. Physiol. (London)* **18**, 363–379.
Pembrey, M. S., Gordon, M. H., and Warren, R. (1895). On the response of the chick, before and after hatching, to change of external temperature. *J. Physiol. (London)* **17**, 331–348.
Portmann, A. (1935). Die Ontogenese der Vögel als Evolutionsproblem. *Acta Biotheor.* **1**, 59–90.
Portmann, A. (1938). Beiträge zur Kenntnis der postembryonalen Entwicklung der Vögel. I. Vergleichende Untersuchungen über die Ontogenese der Hühner und Sperlingsvögel. *Rev. Suisse Zool.* **45**, 273–348.
Portmann, A. (1950). Le developpement postembryonaire. In "Traité de Zoologie" (P.-P. Grasse, ed.), Vol. 15, pp. 521–535. Masson, Paris.
Portmann, A. (1955). Die postembryonale Entwicklung der Vögel als Evolutionsproblem. *Proc. Int. Ornithol. Congr. 11th,* pp. 138–151.
Pruitt, K. M., DeMuth, R. E., and Turner, M. E., Jr. (1979). Practical application of generic growth theory and the significance of the growth curve parameters. *Growth* **43**, 19–35.
Pütter, A. (1920). Wachstumsähnlichkeiten. *Pfluegers Arch. Gesamte Physiol. Menschen Tiere* **180**, 298–340.
Raheja, K. L., and Snedecor, J. G. (1970). Comparison of subnormal multiple doses of L-thryoxine and L-triiodothyronine in propylthiouracil-fed and radiothyroidectomized chicks (*Gallus domesticus*). *Comp. Biochem. Physiol.* **37**, 555–563.
Rahn, H., Paganelli, C. V., and Ar, A. (1975). Relation of avian egg weight to body weight. *Auk* **92**, 750–765.
Reeve, E. C. R., and Huxley, J. S. (1945). Some problems in the study of allometric growth. In "Essays on Growth and Form" (W. E. le gros Clark and P. B. Medawar, eds.), pp. 121–156. Oxford Univ. Press, London.
Richards, F. J. (1959). A flexible growth function for empirical use. *J. Exp. Bot.* **10**, 290–300.
Richdale, L. E. (1963). Biology of the Sooty Shearwater *Puffinus griseus*. *Proc. Zool. Soc. London* **141**, 1–117.
Richter, W. (1982). Hatching asynchrony: the nest failure hypothesis and brood reduction. *Amer. Nat.* **120**, 828–832.
Ricker, W. E. (1979). Growth rates and models. *Fish Physiol.* **8**, 679–743.
Ricklefs, R. E. (1965). Brood reduction in the Curve-billed Thrasher. *Condor* **67**, 505–510.
Ricklefs, R. E. (1967a). A graphical method of fitting equations to growth curves. *Ecology* **48**, 978–983.
Ricklefs, R. E. (1967b). Relative growth, body constituents and energy content of nestling Barn Swallows and Red-winged Blackbirds. *Auk* **84**, 560–570.

Ricklefs, R. E. (1968a). Weight recession in nestling birds. *Auk* **85**, 30–35.
Ricklefs, R. E. (1968b). Patterns of growth in birds. *Ibis* **110**, 419–451.
Ricklefs, R. E. (1968c). On the limitation of brood size in passerine birds by the ability of adults to nourish their young. *Proc. Natl. Acad. Sci. USA* **61**, 847–851.
Ricklefs, R. E. (1969a). Preliminary models for growth rates of altricial birds. *Ecology* **50**, 1031–1039.
Ricklefs, R. E. (1969b). An analysis of nesting mortality in birds. *Smithson. Contrib. Zool.* **9**, 1–48.
Ricklefs, R. E. (1973). Patterns of growth in birds. II. Growth rate and mode of development. *Ibis* **115**, 177–201.
Ricklefs, R. E. (1974). Energetics of reproduction in birds. *Publ. Nuttall Ornithol. Club* **15**, 152–292.
Ricklefs, R. E. (1975). Patterns of growth in birds. II. Growth and development of the Cactus Wren. *Condor* **77**, 34–45.
Ricklefs, R. E. (1976). Growth rates of birds in the humid New World tropics. *Ibis* **118**, 179–207.
Ricklefs, R. E. (1977a). Composition of eggs of several bird species. *Auk* **94**, 350–356.
Ricklefs, R. E. (1977b). On the evolution of reproductive strategies of birds: reproductive effort. *Am. Nat.* **111**, 453–478.
Ricklefs, R. E. (1979a). Patterns of growth in birds. V. A comparative study of development in the Starling, Common Tern, and Japanese Quail. *Auk* **96**, 10–30.
Ricklefs, R. E. (1979b). Adaptation, constraint, and compromise in avian postnatal development. *Biol. Rev. (Cambridge Philos. Soc.)* **54**, 269–290.
Ricklefs, R. E. (1982). Some considerations on sibling competition and avian growth rates. *Auk* **98**, 141–147.
Ricklefs, R. E., and Hainsworth, F. R. (1968). Temperature regulation in nestling Cactus Wrens: development of homeothermy. *Condor* **70**, 121–127.
Ricklefs, R. E., and Peters, S. (1979). Intraspecific variation in the growth rate of nestling Starlings (*Sturnus vulgaris*). *Bird-Banding* **50**, 338–348.
Ricklefs, R. E., and Peters, S. (1981). Parental components of variance in growth rate and body size of nestling European Starlings (*Sturnus vulgaris*) in eastern Pennsylvania. *Auk* **98**, 39–48.
Ricklefs, R. E., and White, S. C. (1975). A method for constructing nestling growth curves from brief visits to seabird colonies. *Bird-Banding* **46**, 135–140.
Ricklefs, R. E., and White, S. C. (1978). Growth rate of the Brown Noddy (*Anous stolidus*) on the Dry Tortugas, June 1972. *Bird-Banding* **49**, 301–312.
Ricklefs, R. E., and White, S. C. (1981). Growth and energetics of chicks of the Sooty Tern (*Sterna fuscata*) and Common Tern (*S. hirundo*). *Auk* **98**, 361–378.
Ricklefs, R. E., Hahn, D. C., and Montevecchi, W. A. (1978). The relationship between egg size and chick size in the Laughing Gull and Japanese Quail. *Auk* **95**, 135–144.
Ricklefs, R. E., White, S. C., and Cullen, J. (1980a). Postnatal development of Leach's Storm-Petrel. *Auk* **97**, 768–781.
Ricklefs, R. E., White, S. C., and Cullen, J. (1980b). Energetics of postnatal growth in Leach's Storm-Petrel. *Auk* **97**, 566–575.
Robertson, T. B. (1908). On the normal rate of growth of an individual and its biochemical significance. *Wilhelm Roux Arch. Entwicklungsmech. Org.* **25**, 581–614.
Ross, H. A. (1980). Growth of nestling Ipswich Sparrows in relation to season, habitat, brood size, and parental age. *Auk* **97**, 721–732.
Royama, T. (1966). Factors governing feeding rate, food requirement and brood size of nestling Great Tits *Parus major*. *Ibis* **108**, 313–347.

Sagher, R. M. (1975). The effect of cold stress on muscle growth in young chicks. *Growth* **39**, 281–288.
Salt, G. W. (1964). Respiratory evaporation in birds. *Biol. Rev. (Cambridge Philos. Soc.)* **39**, 113–136.
Schiess, L. R. (1963). Die postembryonale Ausbildung der Körperproportionen bei Vögeln. *Rev. Suisse Zool.* **70**, 689–740.
Schifferli, L. (1973). The effect of egg weight on the subsequent growth of nestling Great Tits *Parus major*. *Ibis* **115**, 549–558.
Schmalhausen, I. (1930). Das Wachstumsgesetz als Gesetz der progressiven Differenzierung. *Wilhelm Roux Arch. Entwicklungsmech. Org.* **123**, 153–178.
Schmekel, L. (1961). Datem über des Gewicht des Vögeldottersackes von Schlüpftag bis zum Schwinden. *Rev. Suisse Zool.* **68**, 103–110.
Schoener, T. W., and Schoener, A. (1978). Estimating and interpreting body-size growth in some *Anolis* lizards. *Copeia* **1978**, 390–405.
Scott, M. L. (1973). Nutrition in reproduction—direct effects and predictive functions. *In* "Breeding Biology of Birds" (D. S. Farner, ed.), pp. 46–68. Natl. Acad. Sci., Washington, D.C.
Scott, M. L., Nesheim, M. C., and Young, R. J. (1969). "Nutrition of the Chicken." Scott, Ithaca, New York.
Seastedt, T. R., and MacLean, S. F., Jr. (1977). Calcium supplements in the diet of nestling Lapland Longspurs *Calcarius lapponicus* near Barrow, Alaska. *Ibis* **119**, 531–533.
Shilov, I. A. (1973). "Heat Regulation in Birds. An Ecological–Physiological Outline." Amerind, New Delhi.
Siegel, P. B. (1962). Selection for body weight at eight weeks of age. I. Short term response and heritabilities. *Poult. Sci.* **41**, 954–962.
Siegel, P. B. (1963). Selection for body weight at eight weeks of age. II. Correlated responses of feathering, body weights, and reproductive characteristics. *Poult. Sci.* **42**, 896–905.
Singh, A., Reineke, E. P., and Ringer, R. K. (1968). Influence of thyroid status of the chick on growth and metabolism, with observations on several parameters of thyroid function. *Poult. Sci.* **47**, 212–219.
Skoglund, W. C., Seegar, K. C., and Ringrose, A. T. (1952). Growth of broiler chicks hatched from various sized eggs when reared in competition with each other. *Poult. Sci.* **31**, 796–799.
Skutch, A. F. (1976). "Parent Birds and Their Young." Univ. of Texas Press, Austin.
Smith, J. N. M., and Dohndt, A. A. (1980). Experimental confirmation of heritable morphological variation in a natural population of Song Sparrows. *Evolution* **34**, 1155–1158.
Smith, J. N. M., and Zach, R. (1979). Heritability of some morphological characters in a Song Sparrow population. *Evolution* **33**, 460–467.
Snow, B. K. (1970). A field study of the Bearded Bellbird in Trinidad. *Ibis* **112**, 299–329.
Snow, D. W. (1960–1961). The natural history of the oilbird, *Steatornis caripensis*, in Trinidad, W. I., Parts 1 and 2. *Zoologica (N.Y.)* **46**, 27–47; **47**, 199–221.
Sokal, R. R., and Rohlf, J. (1969). "Biometry." Freeman, San Francisco, California.
Solomon, J., and Greep, R. O. (1959). The growth hormone content of several vertebrate pituitaries. *Endocrinology* **65**, 334–335.
Spiers, E. D., McNabb, R. A., and NcNabb, F. M. A. (1974). The development of thermoregulatory ability, heat seeking activities, and thyroid function in hatching Japanese Quail (*Coturnix coturnix japonica*). *J. Comp. Physiol.* **89**, 159–174.
Stanwood, C. J. (1913). The Olive-backed Thrush (*Hylocichla ustulata swainsonii*) at his summer home. *Wilson Bull.* **25**, 118–137.

Street, M. (1978). The role of insects in the diet of Mallard ducklings—an experimental approach. *Wildfowl* **29**, 93–100.
Streich, G., and Swetosarow, E. (1937). Die Entwicklung der Proportionalität im Wachstumsprozess der Vögel. *Zool. Jahr. Abt. Allg. Zool. Physiol. Tiere* **58**, 113–126.
Sturkie, P. D. (ed.) (1976). "Avian Physiology," 3rd ed. Springer-Verlag, New York.
Summers, K. R., and Drent, R. H. (1979). Breeding biology and twinning experiments of Rhinoceros Auklets on Cleland Island, British Columbia. *Murrelet* **60**, 16–22.
Sumner, E. L., Jr. (1929). Comparative studies on the growth of young raptors. *Condor* **31**, 85–111.
Sutter, E. (1943). Über das embryonale und postembryonale Hirnwachstum bei Hühnern und Sperlingsvögeln. *Mem. Soc. Helv. Sci. Nat.* **75**, 1–110.
Sutter, E. (1951). Growth and differentiation of the brain of nidifugous and nidicolous birds. *Proc. Int. Ornithol. Congr. 10th*, pp. 636–644.
Tanabe, Y. (1965). Relation of thyroxine secretion rate to age and growth rate in the cockerel. *Poult. Sci.* **44**, 591–595.
Thomas, C. H., Blow, W. L., Cockeram, C. C., and Glazener, E. W. (1958). The heritability of body weight, gain, consumption and feed conversion in broilers. *Poult. Sci.* **37**, 862–869.
Tixier-Vidal, A., and Follett, B. K. (1973). The adrenohypophysis. *In* "Avian Biology" (D. S. Farner, and J. R. King, eds.), Vol. 3, pp. 109–182. Academic Press, New York.
Upp, C. W. (1928). Egg weight, day old chick weight, and rate of growth in Single Comb Rhode Island Red chicks. *Poult. Sci.* **7**, 151–155.
Van Balen, J. H. (1973). A comparative study of the breeding ecology of the Great Tit *Parus major* in different habitats. *Ardea* **61**, 1–93.
Vandeputte-Poma, J. (1968). Quelques données sur la composition du "lait de pigeon." *Z. Vgl. Physiol.* **58**, 356–363.
Vandeputte-Poma, J. (1980). Feeding, growth and metabolism of the pigeon, *Columba livia domestica*. Duration and role of crop milk feeding. *J. Comp. Physiol. B* **135**, 97–99.
Van Noordwijk, A. J., Van Balen, J. H., and Scharloo, W. (1980). Heritability of ecologically important traits in the Great Tit. *Ardea* **68**, 193–203.
Verghese, M. W., and Nordskog, A. W. (1968). Correlated responses in reproductive fitness to selection in chickens. *Genet. Res.* **11**, 221–238.
Voitkevich, A. A. (1966). "The Feathers and Plumage of Birds." October House, New York.
von Bertalanffy, L. (1938). A quantitative theory of organic growth. *Hum. Biol.* **10**, 181–213.
von Bertalanffy, L. (1957). Quantitative laws in metabolism and growth. *Q. Rev. Biol.* **32**, 217–231.
Wekstein, D. R., and Zolman, J. F. (1969). Ontogeny of heat production in chicks. *Fed. Proc. Fed. Am. Soc. Exp. Biol.* **28**, 1023–1028.
Werschkul, D. B., and Jackson, J. A. (1979). Sibling competition and avian growth rates. *Ibis* **121**, 97–102.
West, G. C. (1965). Shivering and heat production in wild birds. *Physiol. Zool.* **38**, 111–120.
White, G. C., and Ratti, J. T. (1977). Estimation and testing of parameters in Richards growth model for Western Grebes. *Growth* **41**, 315–323.
White, S. C. (1974). "Ecological Aspects of Growth and Nutrition in Tropical Fruit-Eating Birds." Ph.D. Dissertation, Univ. of Pennsylvania, Philadelphia.
Wiley, W. H. (1950). The influence of egg weight on the pre-hatching and post-hatching growth rate in the fowl. *Poult. Sci.* **29**, 570–574, 595–605.
Williams, A. J., Siegfried, W. R., and Cooper, J. (1982). Egg composition and hatchling precocity in seabirds. *Ibis* **124**, 456–470.
Williams, G. C. (1966). Natural selection, the costs of reproduction, and a refinement of Lack's principle. *Am. Nat.* **100**, 687–690.

Winchester, C. F., and Davis, G. K. (1952). Influence of thyroxine on growth of chicks. *Poult. Sci.* **31**, 31–34.

Yarbrough, C. G. (1970). The development of endothermy in nestling Gray-crowned Rosy Finches, *Leucosticte tephrocotis griseonucha*. *Comp. Biochem. Physiol.* **34**, 917–925.

Yule, W. J., and Fuelling, D. E. (1979). Effect of different patterns of food restriction from different ages on growth and efficiency of broilers. *Br. Poult. Sci.* **20**, 273–279.

Chapter 2

THE ONTOGENY OF AVIAN BEHAVIOR

Susan M. Smith

I.	Introduction	85
II.	Behavior of Embryos: Prehatching Behavior	90
	A. Early Motor Activity	91
	B. Development of Sensory Capacities	94
III.	Behavior at Hatching	99
IV.	Posthatching Behavior	103
	A. Imprinting	103
	B. Songs and Calls	114
	C. Food and Feeding	124
	D. Migration and Orientation	135
	E. Play	140
V.	Concluding Remarks	144
	References	145

I. Introduction

The study of behavioral ontogeny is the study of how behavior changes in the developing individual. It is an extremely diverse field, encompassing a wide range of topics and approaches.

Most current studies are relatively short laboratory investigations of behavioral changes in hand-reared birds. For example, Bateson and Jaeckel (1976) investigated how length of prior experience affects the preference of newly hatched domestic chicks (*Gallus gallus*) for a familiar, rather than an unfamiliar, conspicuous visual stimulus (either a red, flashing, rotating light or a yellow, flashing, rotating light). In general, they found that the longer the prior experience with a particular color, the greater the chicks' preference for it over the unfamiliar light in subsequent tests. This general pattern had one exception, however: if the prior experience was 30 minutes long, chicks showed less preference subsequently than if the prior experience was either 15 or 45 minutes. That is, red-trained chicks preferred red to yellow, flashing lights progressively more if they had been trained with red, flashing lights for 15, 45, and 60 minutes; but if trained for 30 minutes, their preference for red over yellow was less. Bateson and Jaeckel (1976) attributed this rather puzzling result to the chicks' preferences for slight novelty, which were superimposed on their growing preference for familiarity. In this sort of study, factors affecting one particular type of behavior, in this case the chicks' approach to a conspicuous visual stimulus, are chosen and studied in detail.

A rarer approach involves relatively longer term investigations of behavioral changes, and often includes both laboratory and field studies. One of the classical examples is Nice's (1943) work on young Song Sparrows (*Melospiza melodia*). In addition to intensive observations on 11 hand-reared individuals (most of which were followed for several months, and one for 4 years) superb data on the development of young Song Sparrows in the wild were obtained for comparative purposes. Nice's study remains a landmark in integrative investigation in behavioral ontogeny.

The diversity of conceptual and methodological approaches in behavioral ontogeny appears to be approximately equal to the number of people working in the area. While recognizing that no chapter has enough room to analyze or even list all such approaches, it may be useful to look at three broad categories: physiological–morphological, psychological, and ethological. The first deals with changes in the structure and function of a developing bird's anatomy: what the bird has to behave with, so to speak. Included here are the studies by Ricklefs on patterns of growth in birds (e.g., Ricklefs, 1968, 1969, 1973, 1979; Ricklefs and White, 1981). Other physiological studies include investigations of the development of temperature regulation by young birds (e.g., Bartholomew and Dawson, 1954; Dawson and Bennett, 1981; Sherry, 1981), or the development of oxygen consumption by young birds (e.g., Birchard and Kilgore, 1980). Although these studies provide invaluable background information for understanding behavioral ontogeny, they will not be considered further here.

Both of the other approaches, that is, psychological and ethological, have been involved in most of the work covered by this chapter. The differences between the two approaches are often subtle, and are best explained as ones of emphasis. The aforementioned study of Bateson and Jaeckel (1976) is typical of the psychological approach. The chicks they used were hatched in a dark, still-air incubator, then transferred to another incubator and kept in individual compartments in constant darkness and constant temperature except during the experiments. Before the experiment, the chicks were individually exposed to constant white light for 30 to 60 minutes; after the experiments they were returned to their dark, single compartments. Although it is clear that all factors are meticulously controlled in such an arrangement it is equally clear that any resemblance to the normal environment of Red Junglefowl chicks (wild *Gallus gallus*) is strictly accidental! This psychological approach often involves studying how far one can distort the natural aspects of processes involved in the ontogeny of a behavior pattern while still generating its main characteristics in the laboratory. I doubt that anyone would disagree that a young bird's external environment must have an important influence on its behavioral development, and investigators who use this approach must beware of the behavioral artifacts such unnatural environments must produce.

In contrast, the ethological approach concerns studying ontogeny in a context of natural events that occur in the process of development (Hess, 1973), either including field studies, or else involving knowledge of what the natural environment is for the species being investigated. Although this approach can avoid some of the aforementioned behavioral artifacts, it nevertheless often means that many factors cannot be precisely controlled.

Each approach therefore has its own pitfalls. But each also has important contributions to make; both approaches are necessary for a complete understanding of the natural processes involved in behavioral ontogeny.

So much for approaches. What are some of the problems with which behavioral ontogeny is concerned? Any discussion of this must begin with a consideration of the Great Dichotomy Debate, which continues to be argued whereever ontogeny researchers gather. Many terms have been used to describe the dichotomy: nature–nurture, instinctive–intelligent, and innate–learned are just a few of them. Several prominent investigators have stated repeatedly that such dichotomies are inherently useless, false, and misleading (e.g., Schneirla, 1952, 1966; Hinde, 1968, 1970). Schneirla felt that research contered around this sort of dichotomy led to "blind alley" projects and suggested (1966) an alternative dichotomy of *maturation* (tissue growth and differentiation) and *experience* (those stimuli from the developmental medium), but warned that even here these two are, in his words, separable only as abstractions.

Hinde (1970) raised three main objections to these dichotomies. First, he felt that they could lead to neglecting effects of the environment other than those produced by learning (e.g., effects of temperature during early embryology on later behavior). Second, Hinde argued that "innate behavior" is generally defined solely in negative terms (e.g., unlearned). Third, Hinde rejected such dichotomies because they tend to deal with classification of units of behavior, rather than with the processes on which development depends.

I shall consider these in reverse order. Although an understanding of processes involved in behavioral development is clearly important, study of the development of individual behavior patterns is also valid so long as it is viewed within the context of the biology of the whole developing organism.

Thus innate–learned dichotomies should not be rejected on these grounds alone. Hinde's second objection, that innate behavior is defined only in negative terms, is harder to refute. Still, although it is simplest to define innate behavior in this way, it is nevertheless possible to refer to "behavior patterns that a young individual can perform as soon as it reaches a particular age in a somewhat restricted developing environment"—which no editor in his right mind would accept when a shorter (but negative) definition is available. Finally, there is Hinde's objection that innate–learned dichotomies tend to neglect environmental effects other than those produced by learning. Insofar as they do neglect such effects, the dichotomies are indeed misleading. Some early studies assumed that it was possible to classify every behavior pattern as either entirely learned or entirely innate, the basic idea being that no behavior pattern that was innate could be in any way modified by experience, nor could any learned behavior pattern have any genetic component. This extreme form of the dichotomy is indeed false and misleading, and should be rejected. It is now well documented that every embryo must have been in the correct environment during development in order for it to behave normally after hatching (e.g., Kuo, 1967). If needing this environment is defined as needing previous experience, then no behavior pattern can be classified as entirely innate, and the dichotomy breaks down.

Even though such an extreme form of the innate–learned dichotomy is clearly false, rejecting the whole idea may lead to throwing out the baby with the bathwater. I feel, along with Burghardt and Bekoff (1978), that such dichotomies are not necessarily useless or dangerous. However, it may be more useful to consider "innate" and "learned" as opposite ends of a continuum, with various combinations in between.

If we are to accept the notion of such a continuum, it must be possible to provide examples of each end: of both primarily learned behavior and primarily innate behavior. Curio *et al.*, (1978), in an ingenious experiment, allowed "tutor" European Blackbirds (*Turdus merula*) to mob a stuffed owl in an aviary, while "observer" blackbirds watched. However, these observers were prevented from seeing the stuffed owl; instead, they could see a stuffed Australian honeyeater. Later, when given the opportunity, the observer blackbirds mobbed the stuffed honeyeater. This misdirection of mobbing toward the honeyeater is a clear example of learned behavior (see also Vieth *et al.*, 1980).

Brown-headed Cowbirds (*Molothrus ater*) are brood parasites; no young cowbird can learn what species it is by who feeds it. West *et al.* (1981) hand-reared male and female cowbirds in isolation from 2 days after hatching until their first breeding season. Their naive females responded to male cowbird songs (and to cowbird songs only) by adopting copulatory postures. Both their recognition of the male song and their ability to perform the precopulatory display are "behavior patterns that young individuals can perform as soon as they reach the appropriate age in a somewhat restricted developing environment"—that is, innate. Similarly, naive young Turquoise-browed Motmots (*Eumomota superciliosa*) and Great Kiskadees (*Pitangus sulphuratus*) strongly avoid the first yellow and red ring pattern they encounter, even though they readily approached an "innocuous" pattern of green and blue, or white and green rings, and also a pattern of yellow and red stripes. Apparently these two neotropical predators possess an innate aversion to a generalized coral snake pattern: potentially lethal prey (Smith, 1975, 1977a).

In general, however, it is usually impossible to term the development of a given behavior pattern as either entirely innate or entirely learned; it is instead the result of the interplay between genetic and environmental factors. Even in Curio *et al.*'s "cultural transmission" of appropriate mobbing stimulus for blackbirds, mentioned earlier, there was evidence of genetic constraints on what stimuli could be learned. Thus when observer blackbirds were allowed to see a multicolored plastic detergent bottle rather than a honeyeater during mobbing by the tutors, the observers did learn to mob the bottle, but with much less intensity than shown toward the honeyeater. As Marler (1975) has pointed out, if a given species relies on a complex set of learned behavior patterns as basic to its ecology and behavior, it would be surprising if the form, timing, and direction of this learning were left only to chance. Hence much, if not most, learning takes place within genetic contraints, just as most innate behavior requires at least some practice. (An exception here would be young Turquoise-browed Motmots' and Great Kiskadees' aversion to the generalized coral snake pattern. A bite from a coral snake would be lethal to a young motmot or kiskadee; under such circumstances there could be no opportunity for a young bird to hone its recognition skills, because any mistake would probably be fatal.) Faulty performance of most other behavior patterns, however, does not have fatal consequences, so opportunities for refining skills are usual.

This interplay between genetic and experiential factors is pointed out nicely by Hailman's (1969) title, *How an instinct is learned*. Many other examples of learning within genetic constraints are provided by the development of song in birds (e.g., Thorpe, 1958; Marler, 1970); this will be considered in greater detail later (Section IV,B). Thus referring back to our model of the continuum from innate to learned, it is safe to say that the development of most behavior falls somewhere between the two extremes.

Finally, if we are to talk about genetic factors in development, something should be said about the genetic basis of behavior. It is well established that genetic differences have strong effects on the behavior of individual birds.

These genetic–behavioral effects can be both interspecific and intraspecific. Dilger (1960, 1962) worked on interspecific genetic effects in nest building in the African parrot genus *Agapornis*. Whereas *A. roseicollis* typically carries nesting material tucked into the lower back and rump feathers, *A. fischeri* generally carries it in its bill. When Dilger (1960) studied the development of carrying behavior in hybrids between these two species, he found that they showed behavioral elements from both parents.

A cogent example of within-species genetic effects on behavior is provided by West *et al.* (1981). They studied male Brown-headed Cowbirds, hand-reared in isolation, from two populations: one from Maryland, and the other from Texas. At sexual maturity, the males from Texas sang a song that contained an element not found in the Maryland males' songs. Moreover, females from the two populations that were hand-reared in isolation consistently responded more often to male songs from their own geographic area. Because these birds were reared under identical conditions, these differences in behavior must reflect genetic differences between the two populations. Similarly, Kovach (1978, 1980) has shown that genetically based behavioral differences can be selected in the Japanese Quail. Using selection through 14 generations, Kovach succeeded in obtaining a "blue line" of quail that chose to approach blue in preference to red, and a "red line" that preferred red to blue.

Using various hybrids and back crosses, Kovach (1980) demonstrated four to eight segregating units of inheritance involved in this color choice, with a small directional dominance of factors responsible for the blue preference. Such carefully executed genetic studies of behavior in birds are very rare. As Kovach (1980) stated, these kinds of data point the way to studying the mediation of gene effects, environment effects, and gene–environment interactions in the ontogeny of such visually guided behavior patterns. More such studies are needed.

II. Behavior of Embryos: Prehatching Behavior

A surprising number of people tend to ignore the prehatching behavior of birds, apparently feeling that true behavior begins only after hatching. Any complete study of behavioral ontogeny, however, should consider all of the behavior of the individual throughout its entire developmental history and thus must include the prehatching period as well.

By far the best-studied species with respect to prehatching activity are the domestic chicken (Kuo, 1932a,b,c,d,e; Hamburger, 1963; Alconero, 1965; Oppenheim, 1974) and duck (Gottlieb and Kuo, 1965; Oppenheim, 1970;

Gottlieb, 1976a). Indeed, there is far more information on prehatching behavioral development in precocial birds (see also Nice, 1962) than for altricial species, although some comparative data exist for later stages (e.g., Oppenheim, 1972a). Accordingly, much of the discussion in this section will concentrate on chicks and ducklings.

A. Early Motor Activity

Kuo (1932b) divided the postitional changes of the developing chick into seven stages: stage 1 (orientation), stage 2 (flexion and torsion), stage 3 (lying at the large end), stage 4 (fixation of position), stage 5 (changes in the positional relationship between the yolk sac and the embryo), stage 6 (turning of the body to lie lengthwise in the egg), and stage 7 (protrusion of the neck). Kuo's pioneering work is superbly detailed and remains the classical account of chick development. Stages 1–3 take approximately 9 days, stage 4 is usually achieved on about day 11, stages 5–6 from day 11 to day 15, and stage 7 from day 16 until hatching (approximately day 20).

Kuo claimed that each of the seven stages is necessary and critical in order for hatching to be achieved; failure to pass through any given stage would result in death in the shell. Most of Kuo's work consisted of describing and experimenting with the positioning of the embryo and extraembryonic membranes within the shell, and less attention was paid to the embryo's movements themselves.

More recently, workers such as Oppenheim (1974) have studied the early embryonic movements of chicks in considerable detail. Oppenheim (1974) has divided the prehatching period into three, rather than seven, stages; the early phase (days 3 to 9), the middle phase (days 9 to 16), and the final or hatching phase (day 17 to hatching).

In domestic chicks, the earliest motor activity begins on about day 3 or 4, with the appearance of very slight unilateral contractions of the neck muscles. Most of these first contractions are too weak to cause the head to turn. Early movements are infrequent, but gradually increase in number (Hamburger, 1963; Foelix and Oppenheim, 1973).

Almost immediately, these movements change to rather stereotyped, undulating body movements, caused by contractions alternating first on one side, then on the other. This results in the embryo's moving in almost a swimming fashion. Initially such movements are only in the neck, but later they spread down the trunk and by day 5 include the whole body. The first active limb movements appear on day 6.

By the end of the early phase, these movements have a regular pattern: a burst of several contractions, a pause, then a new burst. Throughout this early period these movements steadily increase in rate. Thus the early phase involves four main features: active muscle contractions spreading from the neck along the trunk and then to the limbs, stereotyped "swimming" movements, a regular periodicity of movements, and a steady increase in rate of movements.

The middle phase in chicks lasts from day 9 to day 16. During this phase, the movements of the chick change from the earlier stereotyped undulations to jerky, almost convulsive, apparently uncoordinated movements. At the same time the movements gradually become aperiodic. Also during this stage more and more parts of the body become independently active (e.g., individual toes, eyelids, and beak). Finally, during the middle phase the rate of activity rises to a peak at about day 11 (i.e., near the midpoint of the incubation period), and then sharply decreases again (Fig. 1).

Hamburger and Oppenheim (1967) have distinguished three types of activity that can be seen in chicks in the middle and final phases. Both type-I and type-II activity occur in the middle phase, and all three are performed in the final phase. Type-I activity consists of irregular, low-amplitude movements, usually jerky and uncoordinated, that may involve just one part of the body, several, or all parts at once. Type-II activity consists mostly of "sudden startles," involving an almost synchronous activation of the entire neuromuscular system.

The final phase in chicks (day 17 to hatching) is marked by the appearance of type-III activity, in addition to the continuation of the other two types. Type-III activity is stereotyped, involving a vigorous wriggle of the whole body, including a powerful lifting of the shoulder region and the coordinated

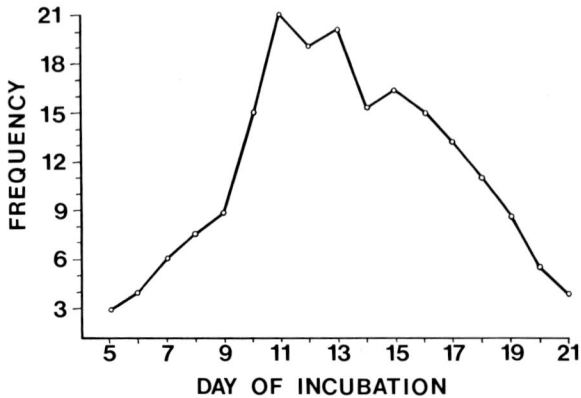

FIG. 1. Frequency (per minute) of all observable embryonic movements of the domestic chick during the incubation period. (Modified from Oppenheim, 1974.)

flapping of both wings. Bouts of this coordinated type-III activity may last for several minutes. Between such bouts, chicks perform the more uncoordinated type-I and type-II movements.

Early in the final phase about 90% of all activity is type I or II, but within a couple of hours before hatching there is a sudden upswing of type-III activity, and by hatching this activity type comprises more than 90% of all observed activity (Oppenheim, 1974). For details of hatching behavior, see Section III.

The same general pattern of development is found in ducks as well. Adjusting for the relatively longer incubation in ducks, Kuo's seven stages can be found (Gottlieb and Kuo, 1965), and Oppenheim's three phases can also be determined (e.g., Foelix and Oppenheim, 1973).

As mentioned earlier, remarkably little information is available on early developmental behavior of altricial birds, especially for any period much before hatching. However, Harth (1971) and Foelix and Oppenheim (1973) have given some data on early behavioral development before hatching in the domestic pigeon, which is usually classified as an altricial species. Although they differ in certain details from chicks and ducklings, most of the basic patterns in squabs are remarkably similar to those of chicks and ducklings. Pigeons, for example, also show a peak in activity (average movements per unit time) approximately halfway through their incubation period, after which their motor activity declines until hatching.

So far I have been discussing early motor activity without considering its causes. In early work, Kuo (1932a,b,c,d,e) assumed that the embryo's activity was caused or else stimulated by sensory input. However, the elegant work of Hamburger and colleagues has now demonstrated that this activity is not greatly affected either by transection of the spinal cord or by removal of sensory input (Hamburger, 1963, 1973; Hamburger *et al.*, 1965, 1966; Decker and Hamburger, 1967; Hamburger and Oppenheim, 1967; Hamburger and Narayanan, 1969). They worked with chick embryos in which a segment of neural tissue in the cervical spinal cord was removed at 2 days of incubation. Even in these spinal chick embryos, they have shown that spontaneous motor activity continues, although beyond the age of 8 days of incubation spinal chicks and those with various areas of the brain ablated show 10–20% less activity than do intact chicks. Thus early motor activity is truly spontaneous (see also Oppenheim, 1972b).

These early embryonic movements are apparently mediated by the nervous system (neurogenic) right from the start, rather than being due at any stage simply to the actions of the muscle fibers themselves (myogenic). There are three lines of evidence for this. Kuo (1939) and later Hamburger (1963) found that all movements in the chick embryo are blocked by curare (a neuromuscular blocking agent). Similarly, transplantation experiments by

Alconero (1965) show that excised somites were capable of spontaneous contractions only when they included neural tissue that had actually innervated the muscle. Finally, Ripley and Provine (1972) clearly showed that even the earliest movements in the chick embryo are closely correlated with bioelectric activity in both the spinal cord and the peripheral motor nerves.

B. Development of Sensory Capacities

In chicks, and probably in most other birds, early development involves a state of "motor primacy," that is, the sensory nerves and muscles do not differentiate and start to function until several days after the motor system does. Thus, although the first motor activity begins at day 3 in chicks, external stimuli of any kind will not elicit any response before the sixth or seventh day (Oppenheim, 1974).

Gottlieb (1968a, 1971a) and Vince (1974) have reviewed the timing of the onset of function in the various sensory modalities in birds. They reported that the first to appear is nonvisual photic sensitivity, followed by tactile (or cutaneous), vestibular, proprioceptive, auditory, and visual sensitivity, in that order. Although rather few species of birds have been studied from the point of view of prehatching chronology of sensory modality onset, it is noteworthy that the same sequence has also been reported in mammals (Gottlieb, 1971a). This certainly suggests that this pattern of development may have been present in their reptilian ancestors, and (more to the point for this chapter) that it may well be common to all birds.

Bursian (1964) found that domestic chicks apparently respond to light at day 3 of incubation, well before the eyes are functional. The chicks showed this nonvisual photic sensitivity by modifying their rate of embryonic motility in response to unnaturally strong light. From day 3 to day 7, chicks responded to bright light by moving more, but on the eighth and ninth day they became inhibited by strong light. Heaton and Harth (1974a) tested pigeon embryos on days 11 through 17 of incubation, and found clear evidence of light sensitivity when the visual system is known to be nonfunctional. They concluded that embryonic pigeons also have some form of dermal light sensitivity.

Apparently this dermal sensitivity may be lost at approximately the time that the visual system becomes functional. Heaton and Harth (1974a) also tested chicks at 19 days of incubation, and found that at this age the head had to be exposed to the light in order for any response to occur.

Tactile (cutaneous) sensitivity is probably the first modality to become functional with respect to sensorimotor reflex activity. Using loops of baby hair touched lightly to the oral region, various workers have shown tactile sensitivity to appear by the end of the first third of incubation in both precocial species (chicks and ducklings) and altricial species (pigeons; Gottlieb, 1968a). Sensitivity first appears in the oral region, but within a day or two the rest of the body also responds to tactile stimuli (Vince, 1974). This onset of sensitivity is synchronous with the establishment of the cutaneous reflex arc (Visintini and Levi-Montalcini, 1939).

The onset of vestibular sensitivity has been studied by rotating developing eggs on a disk. This was first studied by Visintini and Levi-Montalcini (1939), who reported onset of vestibular sensitivity in domestic chicks at day 8 of incubation. They found that this response (postrotational head shaking, or nystagmus) coincided with the completion of the appropriate neural structures. However, this finding has apparently not been confirmed by Gottlieb and Oppenheim (Impekoven, 1976). Similarly, Kuo (1932a) did not find active movements in response to egg rotation until after the piercing of the membranes just before hatching, and Kovach (1970b) found postural righting reflexes not appearing until about 30 hours before hatching. Apparently, then, vestibular sensitivity may not appear until quite late in incubation (see also Oppenheim, 1973). If this is true, then the chronological sequence mentioned previously may have to be corrected.

According to Gottlieb's (1968a) sequence, proprioception is the next sensory modality to become functional. Again, Visintini and Levi-Montalcini (1939) reported the onset of this response to be correlated in domestic chicks with the establishment of the appropriate nervous connections (on about the tenth day of incubation).

The next sense to become functional is hearing. Cochlear microphonic effects of stimulation using pure tones of low frequencies have been found in chicks at 12 days of incubation (Vanzulli and Garcia-Austt, 1963). Moreover, the first indication of evoked responses at the level of the cochlear nuclei appeared on day 11 of incubation. On subsequent days both the threshold sensitivity of the response and the frequency range over which it could be recorded improved. These findings were confirmed by Saunders *et al.* (1973) and by Saunders (1974), who also demonstrated that the onset of auditory function was clearly related to the histogenesis of the acoustic ganglion and cochlear nuclei between days 11 and 13 of incubation in domestic chicks. Konishi (1973), working with duck embryos, found the onset of auditory sensitivity to be later than in chicks: at about 70% of incubation, compared with about 60% in chicks. As with chicks, duck embryos showed improvement in threshold and frequency range with time.

Gottlieb (1965a,b, 1966, 1968b, 1971a, 1974) studied in great detail the ontogeny of auditory discrimination in duck embryos. He worked primarily with two species: Mallards (*Anas platyrhynchos*) from both wild and domestic stocks and Wood Ducks (*Aix sponsa*). In particular, Gottlieb examined prehatching responses to maternal vocalizations in these species. Mallard embryos respond on the day before hatching to maternal calls of several species by increase in heart rate. However, they also respond by increase in their own vocalizations and their bill clapping; and here they discriminate, this increase occurring only in response to maternal calls of their own species. Gottlieb therefore used bill clapping as an index of response in further tests. With this technique Gottlieb has found that Mallard embryos can differentiate between maternal calls of their own and other species up to 5 days before hatching (incubation period is 26 days). Later work showed that this ability to discriminate can be accelerated or retarded and even partially destroyed by varying the amount of auditory stimulation received by the embryo (Gottlieb, 1975a,b, 1976b, 1978, 1979, 1980a).

One of the last senses to mature in avian embryos is vision. Even in precocial birds the structural maturation of this system requires most of the incubation period; in altricial species the maturation process is usually not complete until well after hatching (Nice, 1943, 1962; Heaton and Harth, 1974b). In domestic chicks, neuroanatomical and biochemical evidence (reviewed by Gottlieb, 1968a, 1971a) suggests that differentiation in both the retina and the optic lobes is essentially complete by day 18 (2 days before hatching). This finding has been confirmed by Wada et al. (1979).

Heaton and Harth (1974b) took the neurally mediated pupillary reflex as an index of visual sensitivity, and compared the onset of visual sensitivity in six species: the duck, Bobwhite Quail (*Colinus virginianus*), Japanese Quail, domestic chicks, pigeons, and Common Grackles (*Quiscalus quiscula*). Converting to proportions to allow comparisons, they found that the visual system of ducks becomes functional after 67% of the incubation period has elapsed; both Bobwhite (65 to 67%) and Japanese Quail (66 to 72%) showed similar development patterns. Rather surprisingly, the precocial domestic chicks (85%) and the so-called altricial pigeon (88%) showed very similar development patterns, but the Common Grackles showed the more typical altricial pattern, their eyes becoming sensitive only 8 days after hatching (Fig. 2).

These data are particularly interesting when Heaton and Harth's data on normal light levels are considered. They have shown that in all six species the egg shell is too thick to let in enough light to elicit a pupillary reflex. Thus precocial species, especially the ducks and quail, develop sensitivity long before they would be able to use it normally. However, once the shell is broken, the eyelid plus the nictitating membrane, even when both are shut, together let in enough light to elicit a pupillary response; hence the first pupillary reflex could be produced naturally at or soon after hatching in precocial species. It is interesting that truly altricial species such as grackles achieve visual sensitivity so much later than this.

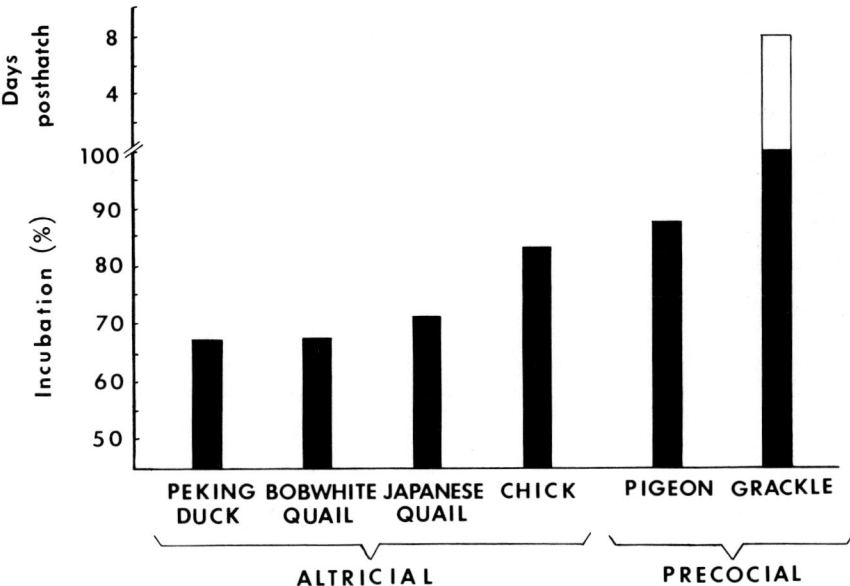

FIG. 2. Onset of the pupillary reflex in six avian species. ■, Prehatching; □, posthatching. (After Heaton and Harth, 1974b. Copyright 1974 by the American Psychological Association. Adapted by permission of the publisher and author.)

As for color sensitivity, Sedláček (1972) demonstrated that domestic chick embryos could discriminate only blue and red from other parts of the spectrum at 18 days of incubation, although Wada et al. (1979) have found that illumination of eggs at this stage by strong light of various colors could influence subsequent color preferences by newly hatched chicks. Under normal conditions, both chicks and ducklings can respond selectively to a greater range of colors by hatching (Hess, 1956; Kear, 1964; Oppenheim, 1968a,b). Kovach (1971, 1980) has similar data on color discrimination in Japanese Quail.

Little data exist on the timing of development of other sensory modalities in birds. Oppenheim and Levin (1975) examined the development of thermal sensitivity in domestic chick embryos. They found that physiological responses (change in heart rate and duration of amnion contractions) to temperature changes can appear as early as day 6 of incubation, but behavioral responses (changes in rate of neuromuscular activity) were not seen until day 15 of incubation.

Virtually nothing is known about the development of chemical senses (taste and smell) in birds. At least for smell, however, it is known that the nostrils are obstructed until a few hours before hatching in a number of avian species (Vince, 1974, 1977), so it is unlikely that functioning of this sense would begin much before hatching.

Continued controversy exists over the potential effects of sensory input during embryonic growth on later behavioral development. For example, during middle to late incubation there is a period during which a developing domestic chick's head rests on its thorax, and is passively lifted and dropped with each heart beat (Kuo, 1932a). Schneirla (1952) and Lehrman (1953) claimed that this early experience might have a strong influence on the later development of pecking movements in chicks and other young birds. Lorenz (1961) and others have strongly challenged this suggestion. Lorenz pointed out that the intraegg environment is more similar between species than are the behavioral characters supposedly derived from them, that is, all species as embryos receive passive head movements derived from heart beats, but some will peck on hatching, others will dabble, and still others are capable only of gaping.

The particular issue of pecking-response development remains unresolved. As for the more general question of what (if any) influence early sensory experience exerts on later behavioral development, at best it can be said that some evidence supports a definite influence, whereas other evidence does not.

The development of proprioception, it has been suggested, (e.g., Kuo, 1932b) may depend on, and be the result of, self-stimulation from the embryo's natural movements. This, however, is apparently not the case; when Hamburger *et al.* (1966) removed the spinal sensory ganglia associated with tactile and proprioceptive input from the legs in domestic chick embryos, they found no change in either the amount or the periodicity of leg motility between 8 and 16 days of incubation. Similar results were obtained by Provine (1972, 1973).

Provine's later work also suggests that early sensory stimulation and, indeed, motor "practice" are unnecessary for the normal development of wing flapping and flight in domestic chicks. Working now with hatchling chicks, Provine (1981) found that chicks immobilized with elastic bandages from day 1 until just before testing on day 13 showed not only normal flapping rates but also achieved flight distances equal to those of unbandaged (control) chicks.

Gottlieb, however, has clearly demonstrated that early sensory experience can play a highly important although often subtle role in later behavioral development. Working primarily with ducklings, Gottlieb performed a series of experiments to study exactly how normally occurring exposure to embryonic vocalizations might contribute to later ability to recognize their own species' maternal call. Gottlieb (1980a) stated that sensory experience can affect the development of perceptual preference in birds in three main ways: maintenance, facilitation, and induction. So far, Gottlieb has found examples of the first two and has been working toward demonstrating the third. In order to develop their normal preference for the maternal call at its species-typical rate (4 notes/second), ducklings must be exposed to their species-typical embryonic peeping at that rate (Gottlieb, 1980a). Exposure to other rates of peeping will neither maintain nor modify this normal preference. This is an example of Gottlieb's *maintenance* effect.

Gottlieb's *facilitation* effect has also been found in ducklings (Gottlieb, 1979). Ducklings with normal auditory experience developed discriminatory ability 4 days before hatching, whereas those without auditory experience did not attain this ability until 2 days after hatching. Therefore, even though embryonic experience is not necessary for development, it nevertheless plays a facilitative role. Gottlieb has since been working on inductive effects.

Until more recently, the role of early embryonic experince on later behavioral development had not received as much attention as it might have. This area should be a very productive one in the future.

III. Behavior at Hatching

All birds must hatch, but not all birds are at the same stage of embryonic development when they do hatch. The fascinating aspect of this is that behavior patterns associated with hatching are remarkably similar in a wide variety of altricial and precocial species (Oppenheim, 1972, 1973; Vince, 1974). This section will include events leading up to the point of so-called *climax*, as well as the climax, or hatching, itself. Once again, more information is available for domestic chicks and ducklings than for other species; I shall therefore begin this discussion with an account of prehatching and hatching in domestic chicks and then consider comparative data on other species.

Oppenheim (1972a) and others have referred to six stages leading up to escape from the eggshell in birds. Briefly, these are: pretuck I, pretuck II, tucking, membrane penetration, pipping, and climax. At approximately day 16 of incubation, domestic chicks assume the position known as pretuck I. The embryo lies on its left side, its neck arched under the air space and its beak and anterior head buried in the yolk between the legs. At this stage there is a fair amount of Hamburger and Oppenheim's (1967) type-I and type-II activity, but this is before the appearance of their type-III activity.

The short pretuck-II stage coincides with the first sporadic appearances of type-III activity. This behavior eventually results in a change of embryonic position; the embryo lifts its head and beak out of the yolk and tucks them under its right wing. This is known as *tucking*. During the next few hours, the entire embryo shifts position slightly, so that the beak and right shoulder protrude into the air space but remain covered by the membranes (Brooks, 1978). Very shortly afterward, the beak actually penetrates the membranes and reaches the air space; lung ventilation is begun at this time [although Oppenheim (1972a) has reported a few embryonic chicks beginning respiratory movements somewhat prior to membrane penetration]. Initially, respiration is irregular and sporadic, but gradually the chicks perform stronger, more regular breathing. Bill clapping becomes noticeable at this stage; this is also the time when domestic chicks begin to vocalize.

The first cracks in the eggshell usually appear on the upper surface of the blunt end of the egg. The movements involved in cracking the shell are termed *pipping*. Typically, the first cracks are made many hours prior to hatching. Pipping movements consist of strong lifting or back thrusts of the head and beak against the shell. These movements serve to force the embryo's egg tooth against the inside of the shell; they increase in frequency during the last hours before hatching.

During these last hours the pipping hole becomes enlarged. Then, in a rather sharp transition, the actual hatching, or climax, behavior begins. Hamburger and Oppenheim (1967) have listed three motor patterns associated with this behavior: a vigorous oblique back thrust of the beak against the shell, the simultaneous pressure of the shoulder and thoracic area against the shell, and the pressure of the tarsal joints against the shell at the narrow end of the egg, adding force that is directed up to the blunt end. Soon after they begin, these movements acquire a strong rotary component; this is always counterclockwise in chicks, and, interestingly enough, also in every other species studied so far (Oppenheim, 1972a; Vince, 1974; Impekoven, 1976). The strong thrusts of the head along with this turning component result in a circular crack, the eggshell being progressively chipped around its circumference. When this crack extends about two-thirds of the way around, the chick is able to push the shell cap off, and so emerges.

With every few modifications this same basic pattern is shown by the vast majority of bird species studied, regardless of whether they are precocial or altricial or to what order they belong (Oppenheim, 1972a; Vince, 1974). Every species studied by Oppenheim passed through the same basic six stages just described. Some slight variability occurs; for example, in chicks and ducklings, the timing of the first crack in the eggshell occurs after membrane penetration, whereas in other species such as various charadriiforms (gulls, terns, and Black-tailed Godwit [*Limosa limosa*]) it happens during the tucking stage (Lind, 1961; Oppenheim, 1972a; Impekoven, 1976). Perhaps only the onset of vocalizations shows any definite pattern of variability; embryos of precocial and semiprecocial species begin to call at the time of membrane penetration, whereas altricial species do not call until they emerge from the egg (Oppenheim, 1972a; Vince, 1974; Impekoven, 1976).

Bakhuis (1974) and colleagues (Bakhuis and van de Nes, 1979; Bakuis and Bour, 1980) have studied hatching in chick embryos that have been transferred into clear glass "eggshells." They have distinguished a *cracking* phase and a *turning* phase. In cracking, all four limbs are extended, bracing the body firmly against the eggshell, while the beak (plus egg tooth) cracks the shell. This alternates with the turning phase, during which the right leg flexes as the body relaxes away from the shell. The flexing of the knee pulls the body around slightly, resulting in the typical counterclockwise rotation. Once the crack is sufficiently extended, the pressure exerted during the cracking phase pushes the shell cap off, allowing the chick to emerge.

At one time it was thought that the chick was asleep at the point of climax; this was referred to as *paradoxical sleep* (Bakhuis and van de Nes, 1979, and included references). However, the later work by Bakhuis and colleagues has demonstrated that climax is always associated with wakefulness, not sleep.

Vince and colleagues have worked for some time now on factors that accelerate hatching in chicks. In Japanese Quail embryos, Vince (1966, 1969) showed that a fascinating sequence of events happens when late embryos are given 2 to 3 days of continuous, regular clicking (simulating the clicking produced by a certain type of prehatching breathing). Although at first the quail embryos show no measurable response (to the observer), embryos so stimulated hatch much sooner than unstimulated controls. More recently, Woolf et al. (1976) found that similar acceleration can be induced even when the duration of clicking is reduced to about 2 hours.

The same technique also accelerates hatching in domestic chicks (Vince et al., 1976). Moreover, these effects are apparently the result of acceleration of specific activities involved with bringing the embryo into position for hatching. Oppenheim (1973) suggested that such effects might be the result of increased secretion of thyroxine, but Ockleford and Vince (1980) tested this and found no acceleration effects from thyroxine.

In a systematic attempt to determine the cause of such effects, Vince (1979) used six different indices to measure acceleration in Japanese Quail chicks. These indices were onset of lung ventilation, yolk-sac withdrawal, vocalization, pipping, clicking, and membrane penetration. Experimental chicks received continuous clicks (3/second) starting at about 3 days before hatching (day 14 of incubation). Vince's experimental chicks began breathing 9 hours before the controls did and hatched 23 hours before the controls.

Indeed, the experimentals were earlier than the controls in all of the indices. However, each index seemed to be affected at a rate different from that of any of the others. These results suggest that no single mechanism underlies the effects of accelerating stimuli. Vince (1979) suggested that continuous accelerating stimulation may affect avian development in two main stages: first the onset of breathing is advanced, then there is a pause during which development proceeds normally, after which there seems to be a second period of responsiveness. Vince and colleagues have continued to investigate this subject.

Finally, the question arises, Are there any species in which hatching behavior is significantly different? According to Oppenheim (1972a), the answer must be yes: the Australian moundbuilders (Megapodiidae). These birds are famous for laying their eggs in mounds of decaying vegetation or sand and using the heat produced naturally from rotting plant material or sun-warmed sand to incubate the eggs. Baltin (1969), working with the Australian Brushturkey (*Alectura lathami*), found several differences between megapode hatching processes and those of most other birds. For example, artificially incubated megapode eggs fail to hatch if they are exposed to light. Also, unlike those of other birds, the eggs of megapodes have a movable air space that is always at the highest point of the egg, regardless of whether this is the blunt end, pointed end, or even one side. This is apparently because the air space is actually under, rather than between, the two shell membranes. Nevertheless, the air space is usually at the blunt end in nature, because megapodes lay their eggs pointed end down. Baltin (1969) reported that cracks appear at numerous points on the shell during hatching, and that the legs emerge first, followed quite a while later by the head. Instead of simply removing a shell cap, megapodes break the entire shell into small pieces during hatching. Given this unusual pattern of emergence, it is perhaps not surprising that the egg tooth, which is present at the midpoint of incubation, is lost before hatching occurs (Clark, 1960, 1964). Indeed, based on such anatomical features as the vestigial egg tooth and vestigial down feathers, as well as various behavioral characteristics of newly hatched megapodes, Clark (1960, 1964) argued that megapodes undergo considerably more development before hatching than do other members of the order Galliformes. However, even though Oppenheim (1972a) concluded that megapodes are "markedly different" from other birds at hatching, many of the behavior patterns involved with hatching in megapodes seem very similar to those of other birds (Baltin, 1969; Impekoven, 1976). It seems safe to say, at least for the birds so far studied, that the general processes involved in hatching are remarkably similar throughout the entire class.

IV. Posthatching Behavior

The vast majority of studies of avian behavioral ontogeny deal with behavioral development after hatching. Young birds at this stage are more accessible and are at least somewhat closer to attaining typical adult behavior patterns.

Ontogeny studies have been published on every conceivable kind of behavior. Five areas of posthatching behavior have been selected for consideration here in some detail.

A. IMPRINTING

1. General Characteristics

In a seminal paper that soon became a milestone in the study of behavior, Lorenz (1935) turned our attention to a fascinating phenomenon: imprinting. Lorenz's original description concerned the reactionships of young birds to their companions; Lorenz pointed out that most birds do not recognize their own species instinctively; instead they must learn this by their early experience with fellow members of their species.

Lorenz felt that the process by which these first social bonds were learned was unique and called it *Prägung* or stamping in (as in coins); this has been translated into English as "*imprinting.*" The results of this learning process have been familiar for centuries; newly hatched birds of many species, such as domestic goslings, ducklings, quail, and others normally form a relatively rapid social attachment either to the biologically appropriate conspecific or (in its absence) to some other object.

Lorenz's paper provided the impetus for an enormous number of studies related to imprinting; any truly adequate coverage of these would require an entire book rather than one small section of a chapter. As Immelmann (1975) pointed out, the "classical" cases of imprinting in birds all relate directly or indirectly to the problem of species recognition; therefore, most of my discussion will be of imprinting in this context. For more general accounts, see Bateson (1966, 1976), Hess (1973), Immelmann (1975), and Hess and Petrovich (1977).

Too often, people define imprinting simply as "the following response." This is clearly incorrect; rather, imprinting is the process by which young birds may acquire a following response—or various other kinds of behavior patterns. Unfortunately, this confusion still pervades some of the thinking with respect to how imprinting actually works.

Imprinting with respect to choice of companion can be broken into two main categories: filial imprinting and sexual imprinting. Filial imprinting refers to the process by which a newly hatched bird learns the characteristics of its parents; it results in the following response of newly hatched precocial birds to parent–objects. In contrast, sexual imprinting is the process by which young birds acquire preferences that may be used in subsequent pair formation. Simply because of the time involved, studies of filial imprinting (which can be very short) far outnumber studies of sexual imprinting (which often must be considerably longer).

Much has been written concerning Lorenz's claim that imprinting is a unique form of learning (e.g., Bateson, 1966, 1971; Hinde, 1970; Hess, 1973; Hess and Petrovich, 1977). Both Hinde and Bateson have felt strongly that it is not fundamentally different from other forms of learning; Hess has apparently felt that it is. It is true that the basic neurochemical processes associated with imprinting probably do not differ from those in any other kinds of learning. Nevertheless, imprinting does have certain characteristics that make it easily distinguishable from the other forms of learning that occur in a normally developing bird.

Lorenz listed four criteria that he felt characterized imprinting and made it unique:

1. It can occur only during a restricted time period: the critical period.
2. It is irreversible, that is, once learned, it cannot be forgotten.
3. It involves generalization: the learning of supraindividual, species-specific characters.
4. It may occur before the time at which the appropriate reaction itself first occurs (e.g., mate characteristics may be imprinted long before an animal is sexually mature).

In briefly considering these characteristics, it is important to bear in mind that Lorenz's original discussion of the imprinting process was made in the context of sexual, rather than filial, imprinting. It is now generally felt that only the first two characteristics apply to all forms of imprinting.

The restriction to a critical period is perhaps the most important characteristic of imprinting. Some semantic argument over the term has occured (e.g., Sluckin, 1964, Bateson, 1966 and others have felt that the term *critical* period implies an all-or-nothing effect; they and others have adopted the substitute term *sensitive* period, although this does not seem to alter the basic concept in any major way. The fact remains that imprinting is restricted to a particular time period, regardless of what it is called, and that for any given genetic strain raised under specified conditions, this time period can often be characterized very precisely (e.g., Gallagher, 1977).

A second important characteristic of imprinting has been referred to as *irreversibility*, although perhaps the term *persistence* is a more accurate translation of Lorenz's original meaning (Immelmann, 1972, 1975). In terms of social bonding, this characteristic applies primarily to sexual rather than filial imprinting. Experimental evidence for both of these main characteristics will be discussed later.

In considering experimental studies of imprinting, it may be helpful to divide the studies into broad categories on the basis of the sort of question being asked. A large number of studies have explored the types of objects on which young birds can imprint. Others have studied how external conditions affect this imprinting. Still others have focused on who imprints, regarding both intersexual and interspecific differences, whereas others have been aimed at the critical periods—when they occur naturally, and what factors can modify them. Finally, certain studies are aimed at detecting the underlying mechanisms of imprinting. I shall briefly consider a few examples to illustrate some of these basic experimental approaches.

2. *What Makes an Effective Surrogate?*

That very young precocial birds will imprint on an incredible variety of unnatural objects has been repeatedly verified; the range of objects that have been offered as surrogate parent–objects is limited only by the ingenuity of the investigators working on this aspect of imprinting. Precocial species such as newly hatched ducks and geese will follow virtually any moving object they see; indeed, the early work of Fabricius and Boyd (1954) showed that ducklings will follow objects as diverse as a walking person or a moving matchbox.

Not all such unnatural potential surrogate companions are equally effective. There is some evidence that, all else being equal, round objects are better than objects with other shapes (Hess, 1959; Schaefer and Hess, 1959; Smith, 1962). Within certain limits, size is also an important variable, although the optimal size is not always that of the natural parent or siblings; for example Hinde *et al.* (1956) found that young European Coots (*Fulica atra*) and Common Gallinules (*Gallinula chloropus*) were more likely to become imprinted on a person carrying a large conspicuous canvas hide over his shoulder than on a life-sized model of a bird. Hinde and others concluded that the effectiveness of the potential surrogate increases, at least to a certain point, with its conspicuousness. Bateson (1964) found that day-old domestic chicks prefer to approach a conspicuous static pattern rather than two less conspicuous patterns; similarly, Bateson and Reese (1969) found that chicks prefer to approach a more conspicuous flashing color (orange) rather than a less conspicuous one (green).

Klopfer and Hailman (1964) found a more subtle effect of conspicuousness in ducklings. They used two model types as surrogates: plain (a flat, white, life-sized duck model), and "striking" (a basically yellow decoy with symmetrically arranged patches and stripes of red, green, blue, and brown). Both model types elicited following equally well; moreover, birds trained on these two models showed no difference in the amount of subsequent following during testing. There was nevertheless some difference in effectiveness between the plain and the striking models: ducklings trained on the striking model were strongly inclined to follow it, whereas many trained on the plain model switched over to follow the striking model during testing (Klopfer and Hailman, 1964). Nevertheless, beyond a certain point, a stimulus may actually be too conspicuous, causing a young bird to flee rather than approach (e.g., Weidmann, 1958).

Another important variable is stimulus complexity. Nicki and Rogers (1975) offered day-old domestic chicks various stimuli consisting of painted checkerboard-like matrices with half of the pattern black and half white, but differing in the size of the blocks that made up the pattern. Using patterns that contained 4, 36, 100, 400, and 900 bits of information, they found that young chicks showed a strong preference for the simplest (2×2 block) pattern offered; with increasing stimulus complexity the number of approaches per minute dropped sharply, then gradually increased, only to fall away again at the 900-bit level. Thus, disregarding for the moment the strong preference for the 2×2 pattern, these data do show that up to a point stimulus effectiveness increases with complexity. As for the strong preference for the simplest pattern, I wonder whether this might be explained by the fact that the four contrasting blocks most closely matched the size of the chicks themselves, and thus the size of potential companions.

Still another factor to be considered here is movement. Smith (1962), Moltz (1963), and many others have shown that domestic chicks become imprinted to stationary as well as moving objects; this is true for most other species of birds as well. Nevertheless, slow movement away from the young bird may enhance an object's effectiveness (Bateson, 1966).

Sounds can also be very important. Unlike altricial birds, precocial young usually begin calling before hatching. Newly hatched individuals of various precocial species are attracted by these calls (Collias, 1952; Gottlieb, 1966) as well as the "maternal calls" of their own parents (Smith and Bird, 1964; Gottlieb, 1966, 1968a, 1979, 1980b). Consequently, for precocial species, both the calls of siblings and of parents are potentially important in imprinting. Indeed, calls may actually be more effective than visual stimuli for imprinting in certain species. Gottlieb (1966) showed that newly hatched domestic chicks approached the auditory, rather than the visual stimulus when given a choice between recordings of their maternal call and the sight of a stuffed hen. Similarly, Storey and Shapiro (1979) found that naive ducklings aged 1 to 6 days preferred a localized maternal call over a silent live female duck in all test ages.

Both Gottlieb (1966) and Storey and Shapiro (1979) also found that the appropriate auditory and visual stimuli, when combined, are more effective than either auditory or visual stimuli alone. In other words, within certain limits, the more natural a stimulus, the more likely it is to be an effective companion surrogate. This was tested by Beaver et al. (1976), who showed that a live surrogate was far more effective than an inanimate stimulus in eliciting imprinting in Japanese Quail.

An important technique in the study of imprinting is cross-fostering. A variety of animals have been used in this kind of experiment. For example, Hindman (1981) did a series of experiments with domestic chicks that were given a live companion—either another chick or a guinea pig. The test procedure involved living with the original companion for 3 weeks, living with the other type for 3 weeks, then returning to the original for another 3 weeks, so that the 9-week-old chicks had spent a total of 6 weeks with the original companion and 3 with the other kind before testing. Control chicks spent 9 weeks in isolation. Amazingly enough, when given a choice, the experimental chicks chose to associate with their original companion regardless of which it was, that is, chicks raised with 6 weeks' experience of living with guinea pigs preferred them to live chicks. This is particularly interesting considering that they had had 3 weeks' experinece with live chicks prior to testing.

Although this study perhaps pertains more to filial imprinting, the majority of cross-fostering experiments have related directly to sexual imprinting; they will be considered toward the end of this section.

3. Sensitive Periods

A plethora of papers exists on various aspects of sensitive periods (for a review see Bateson, 1979). One interesting point emerging from such studies is that filial imprinting and sexual imprinting may be different processes involving separate sensitive periods, with that for filial imprinting coming before that for sexual imprinting (Schutz, 1965; Gallagher, 1976, 1977). Bateson (1979) has pointed out that not all phenomena termed sensitive periods are necessarily the same.

Efforts to describe precisely the limits of sensitive periods have produced some conflicting results. Some studies have been able to document beautifully defined sensitive periods (e.g., Hess, 1959; Gottlieb, 1961; Gallagher, 1977). Others, such as R. T. Brown's (1975) study of ducklings, have not. But this should not be too surprising, considering that these studies are often on domesticated species in which strains can vary markedly from one another

(Vince, 1974); also, all were done under variously unnatural conditions. In spite of such difficulties, it appears that for many species (or strains), the sensitive period for filial imprinting occurs soon after hatching (regardless of whether this is best measured in hours posthatching, e.g., Hess, 1959, or, in terms of developmental age, Gottlieb, 1961; see also Landsberg, 1976). This filial sensitive period generally peaks rather quickly, then wanes, whereas the data of Schutz (1965), Vidal (1976), and Gallagher (1977) all suggest that the sensitive period for sexual imprinting starts somewhat later and lasts considerably longer.

At least three theories have been proposed to explain what triggers the onset of sensitive periods. Clearly, imprinting can occur only after the relevant senses are sufficiently developed; Paulson (1965) and others (see Bateson, 1979) have suggested that the onset of sensitive periods may be related to changes in the efficiency of the visual system and their effects on neural connections in relevant underlying systems (see also Riesen, 1975). Bateson and his colleagues have examined this subject extensively, and have shown that after exposure to constant white light for approximately 0.5 hour, day-old domestic chicks approach a conspicuous visual stimulus more readily than do those kept in the dark up to the time of testing (Bateson et al., 1972; Bateson and Seaburne-May, 1973). Cherfas (1977, 1978) has obtained similar results. Also, the social preferences of light-exposed chicks are restricted more rapidly to the familiar conspicuous object than are those of dark-reared chicks (Bateson and Wainwright, 1972). They have concluded that exposure to light probably enhances visual development generally (Vauclair and Bateson, 1975) and, through this, the onset of the sensitive period.

Some investigators have suggested that the onset of sensitive periods may be simply the result of a general increase in the level of "arousal" (Kovach, 1970a; Fischer, 1970; Martin and Schutz, 1974). Indeed, some have explained Bateson's effect of white light in this way. This, however, does not appear to be correct. Exposure to stimuli in sensory modalities other than vision does not have similar effects; rather, such stimuli actually make chicks less responsive to conspicuous visual stimuli (Graves and Siegel, 1968; Bateson and Seaburne-May, 1973).

Bischof (1979) has created a model of imprinting involving changes in synaptic connections, based in part on data suggesting that there are considerably higher protein turnover rates in the brain during sensitive periods than at other times. Finally, certain researchers have suggested that the onset of sensitive periods is triggered by changes in hormone levels, particularly corticosterones (Martin, 1975, 1978; Weiss et al., 1977).

Again, there is considerable controversy about what factors are most important in the termination of sensitive periods. One factor that is frequently postulated is fear. Although there may be some evidence that fear of unfamiliar objects plays some role in the termination of the filial sensitive period (e.g., Landsberg and Weiss, 1976), there is considerable evidence that sexual sensitive periods last well past the point when fear responses begin (e.g., Gallagher, 1977). Bateson (1979) felt that sensitive periods are "self-terminating," in a process that depends entirely on the influence of external events. Whether or not any endogenous mechanisms are also involved in the termination of either filial or sexual sensitive periods (and I suspect that they are), there can be no question that external events can strongly influence these processes.

4. Irreversibility

Lorenz's second major characteristic of imprinting is the irreversibility of imprinted information. The term irreversibility has stimulated much argument; unfortunately, perhaps, because persistence is probably the more accurate translation of Lorenz's actual meaning (Immelmann, 1975).

Salzen and Meyer (1968) investigated filial imprinting in domestic chicks, and showed that although newly hatched chicks would form an attachment to a particular object during their first 3 days after hatching, this could be superseded by experience with a second type of object during the next 6 days. This, they felt, proved reversibility, thus showing Lorenz's concept to be false. However, they made no tests for the persistence of attachment of their chicks to either of the two object types. Immelmann (1975) has suggested that irreversibility with respect to filial (as opposed to sexual) imprinting may in any case be irrelevant, because the following response naturally disappears with age. Certainly, Moltz and Rosenblum's (1958) claim that Lorenz's concept was false simply because ducklings followed less and less as they became more independent seems somewhat absurd. They made no attempt to look for subsequent sexual preferences.

5. The Role of Imprinting in the Development of Species Recognition

Lorenz originally described imprinting as the process by which birds learn what are their appropriate companions. The idea that imprinting plays a vital role in most species' self-recognition has been accepted without uestion for a long time (Lorenz, 1935; Gottlieb, 1971b; Hess and Petrovich, 1977). However, there is strong evidence that argues against this conclusion.

Much of this evidence comes from cross-fostering experiments. Among the earliest of these is the study by Warriner et al. (1963) of mate choice in black-plumaged and white-plumaged domestic pigeons. Young of both colors were raised in the laboratory; half of each type was raised with black parents, and the other half with white parents. No young pigeon was allowed to see any birds other than its "parents" during raising. When they were old enough to mate, groups of sixteen were placed together in aviaries. The males showed strong preferences for courting females of the same color as their former parents, but the females (which were not tested separately) did not show any such preferences (Warriner et al., 1963). At first glance, these data seem to support the idea that imprinting is important in species recognition, at least for males.

Similar intraspecific cross-fostering experiments have been done with white and normally colored Japanese Quail by Gallagher (1976, 1977). The results clearly showed that early experience plays a significant role in later sexual preference of male Japanese Quail. At the interspecific level, Immelmann (1969, 1972) obtained similar results in cross-fostering experiments using males of three estrildid finches: Zebra Finches (*Poephila guttata*), Bengalese Finches (*Lonchura striata*), and Common Silverbills (*Lonchura malabarica*). Again, once at reproductive age, these young males showed very strong subsequent preferences for courting members of their fostering species, regardless of which species that was.

Incredibly enough, however, neither Immelmann nor Gallagher paid any attention whatsoever to the behavior of females, even though accepted theory clearly states that in virtually all animals, where any actual selection is made, it is the females, not the males, that select their mates (e.g., Orians, 1969b; Emlen and Oring, 1977). Payne and Payne (1977), in just one of the many well-documented field studies involving mate selection, reported that female Village Indigobirds (*Vidua chalybeata*) visited many males in an area, then subsequently returned and mated with just a few. Whether or not a male prefers to court any particular morph in the laboratory may have little to do with actual mate choice in the wild. With this in mind, the data of Walter (1973) on intraspecific cross-fostering in Zebra Finches are exceptionally interesting. Using white and wild-type morphs, Walter found, as had Immelmann, that mate selection by male Zebra Finches is clearly determined by their early experience. However, Walter also investigated female behavior and found a fascinating thing: all but one female tested preferred normally colored males to white males, regardless of their previous experience. Walter concluded that female Zebra Finches do not imprint.

More recently, Sonnemann and Sjölander (1977) have claimed to have evidence that female Zebra Finches can be sexually imprinted. They performed cross-fostering experiments between Zebra Finches and Bengalese Finches; their females were later scored by how much time they spent sitting beside caged male Bengalese Finches as compared with caged male Zebra Finches. Although the cross-fostered females did spend more time by Bengalese males than did the controls, they still showed a "marked interest" in males of their own species. I strongly suspect that what Sonnemann and

Sjölander measured was not sexual preference at all, but rather simply the social preference of their female Zebra Finches. Because Bengalese Finches were familiar to cross-fostered females but not to the controls, it is not surprising that the experimentals spent more time with them than the controls did. This does not necessarily say anything about which of the two the females would choose to mate with. (Sonnemann and Sjölander assumed males select the mates.) It is unfortunate that the distinction between social and sexual preferences of females was not made in the experimental design; more data are needed to resolve this difficulty. Nevertheless, the data of Walter and the Sonnemann and Sjölander data support the idea that female Zebra Finches are considerably more resistant to sexual imprinting than are males.

Female resistance to sexual imprinting occurs in several duck species as well. Schutz (1965) did interspecific cross-fostering experiments with several species of ducks and reported that, although male mate choice (under laboratory conditions) was strongly influenced by early experience, female mate choice was considerably less so. More recently, Klint (1975) studied sexual imprinting in mallards and concluded that, because female Mallards have such a low imprinting potential, imprinting could play no role in females' later recognition of potential mates. When one stops to consider the natural history of the species involved (something not done frequently enough by many who study imprinting), this female resistance to imprinting turns out to be rather fortunate. Male and female Mallards look very different, and only females care for the young. Given the fact of female mate choice, young females that became sexually imprinted on their brown-mottled mother would leave rather fewer genes to the next generation than would any sisters they had who resisted imprinting and used other means to choose green-headed males! Indeed, in any sexually dimorphic species in which only the females care for the young, given female mate choice, sexual imprinting may play little role in subsequent pairing in nature.

Unlike ducks, geese show little if any sexual dimorphism. Cooch and colleagues, in their pioneering work on the color phases of *Anser caerulescens*, the Lesser Snow and Blue Geese, interpreted some of their observed selective matings as having been influenced by imprinting (Cooch and Beardmore, 1959; Cooke and Cooch, 1968). Unfortunately, they also assumed that males choose the mates, and consequently some of their explanations were somewhat convoluted. In this species, biparental care is the rule, so "hybrid" female offspring would have the chance to imprint on both parental types. Examining their data with the understanding that females exert mate choice would clarify some of Cooch's problems of interpretation.

In species with true sexual monomorphism females do have the opportunity to learn future mate characteristics from the individual that gives them parental care. Therefore it is not surprising that early experience plays a strong role in later mate choice by females of both gulls (Harris, 1970) and doves (Brosset, 1971). This correlation between color dimorphism and female resistance to imprinting was noted as early as 1965 by Schutz.

If females, rather than males, are the ones that select their mates, why do males become sexually imprinted at all? (As we have seen, that they do is well documented.) First of all, even though females exert final choice, it is still important for males to be able to distinguish appropriate potential mates. It is clearly disadvantageous to males to waste time and energy courting females of the wrong species. A second potential role of sexual imprinting, which might be applicable to both sexes, is the ingenious theory proposed by Bateson (1978). Bateson suggested that the most important function of imprinting may actually be to learn the characteristics of the next of kin, so that birds can subsequently avoid inbreeding via mate selection. The idea is that a bird imprints precisely on its next of kin, then subsequently chooses a slightly different individual to mate with. This is a fascinating theory that warrents further study. Ironically, Bateson reported solely on the behavior of the males in the study, including neither mention nor consideration of the behavior of females involved in the experiments. However, regardless of whether males chose unfamiliar rather than familiar females, or whether females rejected familiar males and accepted unfamiliar ones, the outcome of the experiments supports Bateson's theory.

6. *Other Kinds of Imprinting*

Finally, some mention should be made of other sorts of abilities that may be acquired by processes similar to imprinting. Immelmann (1975) has reviewed some of these in detail. This discussion will be restricted to four main examples.

Young Loggerhead Shrikes (*Lanius ludovicianus*) apparently learn how to impale or wedge prey by a form of imprinting. At about 21 days after hatching, these altricial birds begin taking food, turning sideways, and placing it on their perches (*dabbing* behavior). Within approximately 36 hours, a pulling component appears, and the young shrikes turn sideways, reach out, then pull the food along the perch toward them (*dragging* behavior). At first, there is no particular orientation to this behavior; both dabbing and dragging are performed as frequently on dowel perches as on thorn branches (Smith, 1972). However, when a piece of food being dragged happens to catch on something, the shrike begins to concentrate its dragging along that part of the branch; it thus quickly learns to direct its dragging to appropriate places (thorns or forks). However, experience with such places must occur within a certain period; shrikes that had personal experience with suitable impaling locations between 20 and 40 days posthatching all learned to impale normally, but those raised on only smooth dowel perches until 75 days posthatching did not learn how to impale, even if they were allowed to see older shrikes in an adjacent cage successfully impaling on thorn branches during their sensitive period. There is also evidence that this learning involves irreversibility. Wemmer (1969) showed that young shrikes preferred the kind of impaling device they were reared with, even when it was less efficient.

Löhrl (1959) has demonstrated a clear case of imprinting in the learning of the home area by Collared Flycatchers (*Ficedula albicollis*). As with many other European migrants, Collared Flycatchers return next spring to the area where they lived just after fledging. Löhrl was able to show that it is the period immediately before fall migration that is critical, and that approximately 2 weeks' exposure at this time is sufficient for imprinting to occur. Similar locality imprinting has been shown in Short-tailed Shearwaters (*Puffinus tenuirostris*). Serventy (1967) found that this bird's return to its natal island is not genetically based; rather, it is based on a clear-cut form of imprinting, the critical period being that time during which the young begin their nocturnal emergences from the nesting burrows. Finally, it is well known that wintering birds return year after year to the same location (Diamond and Smith, 1973; Ely, 1973; Smith and Stiles, 1979). Ralph and Mewaldt (1975) have shown that such location fidelity also involves a form of imprinting, at least for two sparrow (*Zonotrichia*) species, in which the critical period for imprinting was midwinter.

The more general concept of habitat preferences may also involve a form of imprinting, at least in certain species. One of the first studies to show this in birds was done by Klopfer (1963). Working with hand-reared Chipping Sparrows (*Spizella passerina*), Klopfer showed that 2- to 4-month-old *Kaspar Hauser* birds reared without sight of any foliage showed the same preference for pine over oak leaves that wild-trapped adults did. However, isolates that were reared with oak (but not pine) foliage showed a decreased preference for pine, thus showing an influence of early experience. Herring Gulls (*Larus argentatus*) may also learn general habitat characteristics through a form of imprinting. Drost (1958) took a number of gulls from coastal areas of the North Sea, and raised them in a zoo setting in the interior of West Germany. These birds departed in their first autumn, having been banded for identification. Three to fours years later, when they were sexually mature, several of these gulls returned to the very spot where they had been raised (a simple case of home-area imprinting); but others located in other inland areas, suggesting a strong learning (and perhaps imprinting) of the general characteristics of inland habitat. Although Immelmann (1975) has proposed such cases as examples of imprinting, no direct eivdence of the existence or extent of any sensitive period has been documented so far.

A fourth possible type of imprinting involves the establishment of host preferences in social parasites. Individual females of many Old World social parasites are highly specific in their choice of host. For example, each female European Cuckoo (*Cuculus canorus*) lays her eggs in nests of only one host species (Lack, 1963), and it is tempting to infer that that species is the one she was reared with (e.g., Lack, 1968). Similar individual specificity of host preferences has been found in other parasitic cuckoos as well (Friedmann, 1968; Liversidge, 1970).

The best-documented cases of host imprinting are among the viduine finches (subfamily Viduinae of the Estrildidae). These parasites are extremely specialized, each species or subspecies parasitizing only one host species. Young females apparently become imprinted on the host species during their period of parental care and, indeed, only come into the reproductive condition after seeing reproductive activities of the host species (Payne, 1973, 1977).

Two other sorts of skill are acquired only during a critical period, with a high degree of irreversibility or persistence also being well documented. One of these is song learning (Nottebohm, 1969); the other is learning how to use the stars in celestial orientation (Emlen, 1972). Each of these will be covered in greater depth in its own section later.

B. Songs and Calls

1. Introduction

When considering the development of vocal behavior in birds, it may be useful to begin by attempting to distinguish between *songs* and *calls*. In general, songs are usually considered to be the most complex vocalizations a given species produces, and often include relatively pure tones. The most common uses of songs are for advertising, either for mates or in territory proclamation and defense. In contrast, calls are usually shorter than songs, often considerably less complex, and may contain harsh or grating notes rather than pure tones. Calls serve a wide variety of functions, and a given species may have a different and readily distinguishable call for each function. Even so, it is not always easy to say whether a particular vocalization is a true song or a call. Bertram (1970) has described several loud vocalizations given by the Hill Myna (*Gracula religiosa*) that could be classified functionally as either calls or songs.

Before turning to the vast literature on the development of true songs, I shall discuss briefly a few aspects of call development. There is considerable evidence that, unlike songs, calls can and often do develop in the absence of previous auditory experience. This is so for many nonpasserines (Lanyon, 1960; Lade and Thorpe, 1964) and also for passerines (Messmer and Messmer, 1956; Marler, 1956; Lanyon, 1957, 1979). Nevertheless, even though in many species calls can develop independently of example, this does not necessarily imply that early sensory input is not important; Gottlieb and co-workers have shown that auditory experience enhances call development in ducklings, and Nottebohm (1967) has found that Chaffinches (*Fringilla coelebs*) deafened at an early age produce clearly abnormal calls (see also Nottebohm, 1972b).

There have been rather few long-term studies of how adult calls develop in either nonpasserines or passerines, but these seem to indicate that adult calls develop by progressive modification (either elaboration or restriction) of the chick calls. This has been found in gulls (Moynihan, 1962), domestic chicks (Andrew, 1963), motmots (Smith, 1977b), and Chaffinches (Wilkinson, 1980). Far more studies involve only a very short time period, usually close to hatching (e.g., Scoville and Gottlieb, 1980).

An interesting area of investigation concerns the development of call recognition in birds. Such studies have been carried out on both passerines and nonpasserines and at various ages in a bird's development. Many such studies involve parent–young recognition (e.g., Beer, 1969, 1970a,b, 1972; Evans, 1973, 1977, 1979, 1980). Beer (1979) reviewed the literature on nonpasserines and pointed out that several earlier studies fail to distinguish between the parents' ability to recognize their own chicks and the chicks' ability to distinguish the calls of their parents. However, Ingold (1973) has clearly shown that adult Razor-billed Auks (*Alca torda*) can recognize their own chicks individually by calls alone. Similarly in passerines, Beecher *et al.* (1981) have found that adult Bank Swallows (*Riparia riparia*), can recognize their own offspring by calls once the young are at least 17 days old. They showed that the basis for this recognition is what they term the *signature call* of the chicks, which first appears when the young are 15–17 days posthatching.

Individual differences in call notes and the development and modification of such differences can be important later in life as well. In a fascinating study, Mundinger (1970) showed that in various cardueline finches females recognize the flight calls of their mates. Moreover, the flight calls of mated pairs are virtually identical, indicating that learning and modification by a sort of vocal mimicry occurs well after these finches have reached reproductive age. Here, then, is a kind of call that does involve some learning that can occur at any time and thus involves no sensitive period. The latter is very important in typical song learning (see beyond).

The vast majority of papers on avian vocal development concern song development in passerines. Because this process typically involves both a sensitive period and some degree of irreversibility, it is clearly related to imprinting.

How much learning is involved in passerine song development? Most reviews (e.g., J. L. Brown, 1975) indicate that there is a complete range from wholly innate to almost entirely learned songs, depending on the species. This has been strongly challenged by Kroodsma (1977a, 1978), who has felt that exposure to normal adult songs is necessary for young birds of most species studied, if song development is to proceed in truly normal pathways. Certainly, the often quoted example of Song Sparrows' ability to develop normal songs even when reared in acoustic isolation (Mulligan, 1966) can now be seen to be incorrect (Kroodsma, 1977b). However, I would argue that there remain some passerines in which learning plays at best a slight role in song development, such as brood parasites like the Brown-headed Cowbird. Although much has recently been made of the influence of experience on the eventual pattern of this species' song (King et al., 1980; West et al., 1981), the fact remains that this experience is not necessary in order for male cowbirds raised in acoustic isolation to sing perfectly recognizable species-specific (and indeed population-specific) songs. In fact, songs of males raised in acoustic isolation were actually more attractive to female cowbirds than were songs of wild-caught adults. The basis of this attraction is a particular element rarely given by wild males. West et al. (1981) were able to show that songs that include this element are more likely to invoke attack from older males than are songs that do not; hence most wild males learn to eliminate this element, simply to survive socially. I suggest that this simple modification, which probably does not even occur in every male's song, is very different from the acquisition of major structural aspects of song that is the usual case in song learning. Because songs of males raised in acoustic isolation are identical in every other aspect to those of wild-caught males, and because these songs are even more attractive to females than the socially modified songs, I feel that, in essence, cowbird songs are innate. Such exceptions aside, however, most oscines require considerable learning in order to acquire normal songs; suboscine songs may be innate (D. E. Kroodsma, personal communication).

2. Sensitive Periods

Most passerines need precise auditory experience for normal song development. Moreover, this experience must occur during a restricted time period: the sensitive period. Certain details of the timing and extent of sensitive periods are known for several species, for example, White-crowned Sparrows (*Zonotrichia leucophrys*), Song Sparrows, Zebra Finches, and

Long-billed Marsh-Wrens (*Cistothorus palustris*). This period varies markedly from species to species. In Zebra Finches, it is relatively brief and early in life; young males by day 66 have learned the song elements and by day 80 have acquired the proper sequence and timing. At this point song learning is complete (Immelmann, 1969; Price, 1979). Similarly, young male Song Sparrows, have a peak of sensitivity between 5 and 10 weeks of age (Mulligan, 1966). In White-crowned Sparrows, the sensitive period is even earlier. Marler (1970) found that a White-crowned Sparrow sang a good copy of a training song provided that he heard it between 8 and 50 days after hatching. Songs heard before 8 days or after 50 days had little effect. However, young male White-crowned Sparrows do not normally sing until their first spring, several months after the end of their critical period. This indicates a well-developed auditory memory. For a review of other species, see Marler and Mundinger (1971).

Most such studies tend to report on when hand-reared birds kept in contolled conditions are able to learn carefully presented training songs, and report this as *the* sensitive period for the species in question. Kroodsma and Pickert (1981), working with Long-billed Marsh-Wrens, have shown the danger of this sort of assumption. Male marsh-wrens usually stop singing relatively early in the breeding season, but second clutches are common. The latest broods thus hatch predictably after adult males have stopped singing for the year. How, then, do late-hatched males learn marsh-wren songs? In an elegant set of experiments, Kroodsma and Pickert demonstrated, at least in marsh-wrens, that the sensitive period is significantly affected by photoperiod. Eighteen nestling males were taken in early July and divided into two groups: one given a June photoperiod, and the other an August photoperiod. Later tests showed that the sensitive period for the August group was far longer than that for the June group. Several of the August males were able to learn songs in the following spring, but none of the June males did. But these birds had all been hatched at the same time! Clearly, the limits on sensitive periods are not as rigidly set as many people have assumed.

Another factor known to affect the length of the sensitive period, at least in some species, is hormone levels. Nottebohm (1969) castrated a juvenile male Chaffinch before it was old enough to come into song and then waited until it was 2 years old before implanting it testosterone. Normally, Chaffinches are incapable of any song learning after their tenth month (Thorpe, 1958). However, Nottebohm's castrated male, on receiving the testosterone, learned to sing the song of a tutor when 2 years old. Nevertheless, gonadal androgens are not essential for normal song development in all species. For example, male Zebra Finches castrated at 9–17 days posthatching all developed perfectly normal songs, although they sang less than intact males (Arnold, 1975).

Finally, in certain species the sensitive period never ends, and the birds can keep on learning new vocal patterns throughout their lives. This is clearly the case in vocal mimics such as Common Mockingbirds (*Mimus polyglottos*).

3. Steps in Song Learning

Various authors have described different numbers of steps in the development of song in passerines. Nice (1943) described six stages, but most of the later reviews follow the three stages described by Nottebohm (1970). Nottebohm felt that the early food-begging calls given by nestlings gradually change into the long, rambling type of vocalization referred to as the *subsong*. This is a loose aggregation of notes, usually given at low volume by young birds after feeding. It is generally thought to be a sort of vocal practice with no communicatory significance (Nottebohm, 1975). Juvenile subsong is said to resemble roughly the normal adult song in length, pitch, and tonal quality (J. L. Brown, 1975). However, my own experience with young birds, both in the field and the laboratory, suggests that this resemblance can be very rough indeed; for example, young wild Black-capped Chickadees (*Parus atricapillus*) often sing subsongs that are many times longer than adult songs. The age at which young birds perform subsongs varies with the species. Most temperate-zone and tropical birds begin subsong when fledglings, but a few species begin subsong when almost 1 year old (Nottebohm, 1970). For those beginning in late summer, subsong often stops during the winter, then begins again early in the spring, at which time it grows in complexity. Gradually, elements of the typical adult song appear within the subsong. This combination of elements makes up the intermediate stage, known as *rehearsed song* (Lanyon, 1960) or *plastic song* (Nottebohm, 1970). Then, gradually, the last of the subsong elements are lost, the adult elements become louder, and the elements are at last put together in one or more rigidly stereotyped patterns. This process is known as *crystallization,* and the final product(s) are referred to as *full song*. This general pattern seems to hold for the vast majority of songbirds, and at least subsong has been found in one nonpasserine, the Orange-winged Parrot (*Amazona amazonica*) by Nottebohm (1975). However, such stages do not occur widely among nonpasserines, and they are apparently absent in most galliforms, anseriforms, and columbiforms (but see Sparling, 1979).

4. Experiments with Deaf Birds

One experimental method of studying vocal ontogeny is to remove both cochleas, thus rendering a bird chronically deaf. This can be done at various stages in a bird's development, and thus the contribution of auditory feedback in song development can be determined (Konishi and Nottebohm, 1969). Apparently all of the nonpasserines treated in this way develop all of their vocalizations normally. Konishi (1963) showed that domestic chicks deafened on the first day after hatching developed all adult calls normally. Similarly, Ring Doves deafened 5 days after hatching also develop normal adult vocalizations (Nottebohm and Nottebohm, 1971). Schleidt (1964) reported that Turkeys (*Meleagris gallopavo*) deafened soon after hatching also developed all normal calls.

Among passerines, most of the work involving deafening has centered on song rather than call development. However, it is clear that passerines differ in how deafening affects the development of calls. For instance, Common Cardinals (*Cardinalis cardinalis*) that were deafened as juveniles develop perfectly normal calls (Dittus and Lemon, 1970), but Chaffinches that have been similarly treated produce only abnormal calls (Nottebohm, 1967, 1972a).

With regard to song development, the effects of deafening also vary with the species. Apparently, nobody has studied the effects of deafening on song development on any social parasite, such as the Brown-headed Cowbird. However, canaries are also reported to develop normal song when hand-reared from an early nestling stage (Metfessel, 1935; Poulsen, 1959). However, deafening experiments have shown that canaries must have auditory feedback in order to develop normal song. Early work by Marler *et al.* (1973) showed that canaries deafened at an early age produced many abnormal vocalizations, including loud screeches and stretches of silent "song" during which the birds maintained a singing posture with open bill and pulsating throat but emitted no sound. More recently, however, Marler and Waser (1977) found that deafening affects canaries' song development less than it does that of certain sparrows. The songs of deaf canaries, although abnormal, contained more species-specific features than did the songs of similarly treated sparrows.

In contrast to canaries, most passerines depend to a great extent on learning for their song development. It is thus not surprising that deafening early in development has marked effects on how these birds eventually sing. Nevertheless, all species so far treated in this way have managed to develop some sort of vocalizations, and these have usually contained at least some broad characteristics of the species' normal song, such as typical length or frequency range.

An interesting set of experiments has been done using deafening at various ages in developing White-crowned Sparrows (Konishi, 1965). White-crowns' normal sensitive period is approximately 8–50 days posthatching (Konishi, 1965; Marler, 1970), but these sparrows normally do not start to sing until their first spring, several months later. Konishi deafened some males after exposure to the normal song during their sensitive period, but before they had started to sing. The birds all produced highly abnormal songs. In contrast, if young males are deafened soon after they begin to sing naturally, the song after deafening is virtually unchanged. Hence, auditory feedback appears to be necessary in order for White-crowned Sparrows to use their auditory memory in song development, but is unnecessary in song maintenance after crystallization, even though the normal white-crown song (regardless of dialect) is moderately complex. Konishi (1963, 1965) concluded that song maintenance after crystallization therefore depends on some nonauditory memory. Two possibilities exist: proprioceptive memory and motor memory (J. L. Brown, 1975). So far, there is no experimental evidence as to which is working in this case.

Nottebohm (1968) investigated the effects of deafening young male Chaffinches at various stages of their song development. Nottebohm deafened the young birds at five different ages: 3, 7–8, 10, 11 months, and adult. As with White-crowned Sparrows, Chaffinches deafened as adults continued to sing normal songs, indicating song maintenance through some nonauditory memory. For those deafened prior to crystallization, the later the deafening, the closer to full song. Chaffinches deafened while in plastic song showed some deterioration of song patterns developed up to that time, but they still sang something closer to full song than those deafened at 3 months of age (i.e., in the subsong stage).

Finally, however, not all birds deafened as adults maintain normal song. Dittus and Lemon (1970) found that cardinals deafened well after song crystallization showed gradual deterioration of song. The significance of this difference in memory function remains unclear.

Interpretation of these experiments with deafened birds requires some caution. Clearly, deafening prevents any auditory feedback, as well as any access to auditory models. But deafening may also have other effects. For example, deafening may affect hormone and, in particular, androgen levels. It is known, for example, that deaf Chaffinches sing far less than normal Chaffinches and that singing rate can be increased by implanting testosterone (Nottebohm, 1975). Although deafened young Chaffinches, even when implanted with testosterone, still developed abnormal songs, the possibility remains that deafening in other species may have more effects than simple removal of auditory feedback.

5. Sources and How to Choose Them

As we have seen, most passerines require some learning for normal song development. But how does a naive young bird select the correct song to learn? Anyone who has taken an early morning walk in late May or June knows that young birds are exposed to a bewildering array of bird songs to choose from, but instances of birds learning the "wrong" song in nature are extremely rare (Baptista and Wells, 1975).

Thorpe (1958) was one of the first to show that juvenile songbirds often selectively imitate only songs of conspecifics. Thorpe offered young Chaffinches songs of several different species in an attempt to teach them the wrong song, and found they were highly resistant to learning such songs, even from closely related species. Similar results were obtained for White-crowned Sparrows by Marler (1970).

Konishi (1970) suggested that a possible basis for this selectivity might be the upper frequency range of auditory sensitivity. Konishi showed that different passerines have different upper sensitivity ranges and that their song ranges vary accordingly. However, although this may be one factor in song selectivity, it is not a sufficient explanation for all cases. For example, both White-crowned and Song sparrows, which breed sympatrically, share the same frequency band, but they do not imitate each other's songs (Marler and Tamura, 1964).

Another factor that clearly is extremely important in some species is the effect of social bonds. Various species learn songs selectively from individuals with whom they are bonded. One such bond is that between parents and offspring. European Bullfinches (*Pyrrhula pyrrhula*) learn only the sounds made by their father (Nicolai, 1959); the same is true of Bengalese Finches (Dietrich, 1980). Indeed, certain species (unlike Thorpe's Chaffinches), if raised with a tutor of a different species, can actually be induced to learn that other species' song. Lanyon (1957, 1960) has shown this for meadowlarks (*Sturnella* spp.), and Immelmann (1969) reported similar behavior in Zebra Finches.

Baptista and Morton (1981) have reported that a White-crowned Sparrow, hand-reared from a nestling, learned a song remarkably similar to that of an estrildrid, the Red Avadavat or Strawberry Finch (*Amadava amadava*) (Fig. 3). The young bird had been allowed to interact socially with both other White-crowned Sparrows and with Strawberry Finches. This is strikingly different from the resistance to wrong song learning reported for White-crowned Sparrows by Marler (1970). Baptista and Morton felt that this difference may be one of technique. Marler used only tape-recorded songs, whereas Baptista and Morton allowed social interactions between tutors and pupils. L. F. Baptista (personal communication) has since taught other White-crowned Sparrows to sing Strawberry Finch songs, using this live-tutor technique. Clearly, social bonds can greatly affect a young bird's choice of song template in many species.

Learning from socially bonded conspecific rivals has long been known to be important for many species. Such learning has been documented in

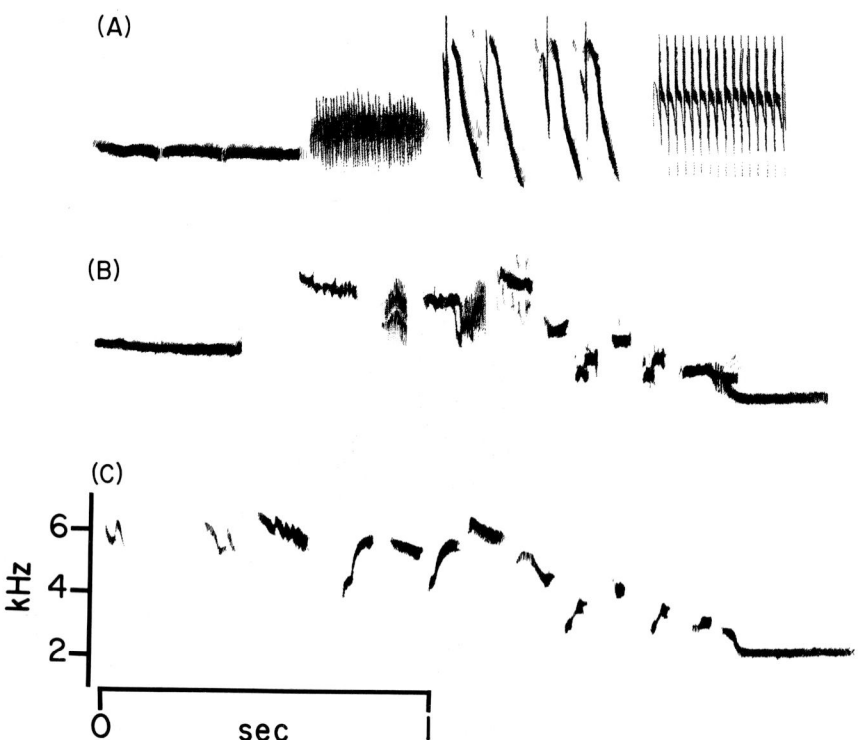

FIG. 3. (A) Tracing of sound spectrogram of White-crowned Sparrow song from Lompoc, California. (B) White-crowned Sparrow's imitation of *Amandava amandava* song. (C) Song of male *Amandava amandava*. (After Baptista and Morton, 1981.)

Chaffinches (Thorpe, 1958), Indigo Buntings (*Passerina cyanea*), (Rice and Thompson, 1968), and White-crowned Sparrows (Marler and Tamura, 1964), to name just a few. Depending on the relative timing of such learning and of dispersal, this kind of learning can lead to the formation of local dialects. Local song dialects have been described in many avian taxa (e.g., see Marler and Mundinger, 1971; Baptista, 1974, 1975, 1977; Jenkins, 1978). Space limitations do not permit discussing the enormous literature on dialects in bird song; for reviews see Lemon (1975) and Baptista and King (1980).

Another sort of social bond that can affect song learning is that between mates. Convergence in calls of finches (Mundinger, 1970) has already been mentioned; similar convergence in songs between members of a pair has also been reported for several species that duet (e.g., Thorpe and North, 1965; Diamond, 1972).

Finally, a few species normally learn phrases from other species. One such group includes the more-or-less generalized mimics, such as lyrebirds (Menuridae), mockingbirds (*Mimus* sp.), certain starlings (Sturnidae) and others (Bertram, 1970). A special case involves song learning in certain social parasites, especially the highly specialized ones. Several viduine finches learn to sing the songs of their host species (Payne, 1973; Payne and Payne, 1977). Such specific vocal mimicry is thought to play an important role in the functioning of the particular host–parasite systems.

Clearly, then, social bonding has been shown to be of great importance in learning selectivity in many species. Nevertheless, social bonding alone is not sufficient to explain wholly how a young bird chooses what to learn. Marler (1967, 1970) demonstrated that juvenile White-crowned Sparrows, given taped songs of conspecifics, Harris's Sparrows (*Zonotrichia querula*, congeneric with white-crowns), and Song Sparrows, learned only the correct songs. Moreover, three naive white-crowns that were allowed to hear only taped Song Sparrow songs sang simple songs similar to those of birds raised in acoustic isolation, thus showing no evidence of learning from the tutor tapes. These sparrows, therefore, were selecting correctly in the complete absence of social bonding.

More recently, a technique involving mixtures of taped song elements has produced some exciting results. Swamp Sparrows (*Melospiza georgiana*) and Song Sparrows sing songs similar in length but very different in temporal organization. Marler and Peters (1977) made artificial "songs" on tapes by splicing together an array of Swamp Sparrow-like and Song Sparrow-like phrases and temporal patterns. When these artificial songs were played to young male Swamp Sparrows, the young birds learned only those songs made up of Swamp Sparrow syllables; moreover, they did so regardless of whether the temporal pattern involved was that of a Swamp Sparrow or that of a Song Sparrow (see also Peters *et al.*, 1980). Similarly, young Long-billed Marsh-Wrens, when presented with spliced tapes containing elements of Long-billed and congeneric Sedge Wrens (*Cistothorus platensis*), again preferentially learned the songs that contained Marsh-Wren trills (Kroodsma, 1978).

In 1970 Marler postulated that White-crowned Sparrows and other songbirds possess an "innate auditory template" that allows them to learn selectively the appropriate song. This has been questioned several times in the literature. Nottebohm (1972b, 1975), while not necessarily disagreeing with the theory, pointed out that there was no indication of exactly how such a template might operate. However, at least part of the answer has been supplied. Dooling and Searcy (1980) have shown that naive young Swamp Sparrows, during their sensitive period, give a significantly greater cardiac orienting response to conspecific songs than to songs of their congener, the Song Sparrow. This response occurred in the decelerative phase of the cardiac response; hence their naive Swamp Sparrows showed greater cardiac deceleration to conspecific vocalizations than to Song Sparrow vocalizations. This discrimination was evident even with the young birds' very first exposure to the proper song. These data are clear support for Marler's innate auditory template theory. Even though Baptista and Morton (1981) have shown that social bonding can supersede the innate template, this does nothing to undermine Marler's theory, because selectivity persists in the absence of social bonding.

6. Mechanisms: Hypoglossal Dominance

Some information is available on the neural basis for song learning in birds. In many species, such as the Chaffinch, the left and right sound-producing membranes of the syrinx can be controlled independently. In the Chaffinch, after song crystallization, cutting the left hypoglossal nerve to the syrinx results in loss of most of the components of the normal vocal repertoire (both songs and calls), whereas cutting the right hypoglossal nerve has essentially no effect on the bird's vocalizations (Nottebohm, 1975). However, when only one side (either left or right) is cut before the bird has started to develop its song, normal song will develop under the control of whichever side is intact (Nottebohm et al., 1979). Thus each hypoglossal nerve, at least prior to crystallization, is equally capable of vocalization control, but under normal conditions the left one is clearly dominant. For a more detailed consideration of brain pathways involved in vocal learning in birds, see Nottebohm (1980).

C. Food and Feeding

Basically, one can divide food-related problems that are faced by young birds into two major categories: problems associated with obtaining food from the parent and problems associated with finding, choosing, catching, and handling food on their own.

1. Obtaining Food from the Parents

All young birds must grow rapidly. Anyone who has hand-reared birds of any kind is familiar with the urgency of their need for food. Nevertheless, young birds do not necessarily beg for food indiscriminately. One can therefore ask, How does a young bird select when and from whom to beg?

Tinbergen and Kuenen (1973) did one of the first studies on this subject. They pointed out that in any truly altricial species, the nestlings' eyes do not open until several days after hatching, so vision can play little role in eliciting begging from newly hatched nestlings. However, altricial birds of this age lie quietly in the nest until an adult arrives, at which point they immediately respond with vigorous gaping. What cues elicit this behavior? Working with European Blackbirds and Song Thrushes (*Turdus philomelos*), Tinbergen and Kuenen demonstrated that the mechanical stimulus of jarring the nest, caused by the weight of the arriving parent, is what the young nestlings initially respond to. Later, after their eyes open, visual cues become important. To be effective, an object must be larger than 3 mm across, it must be above eye level, and, preferably, it should move toward the nestlings.

Mechanical cues are effective for altricial species that nest in trees, but they cannot work for ground nesters or those that nest in holes. For many such species, early begging is elicited by parental vocalizations (Vince, 1974).

Although relatively little research has been done on early begging behavior in altricial species, there is an enormous body of literature on food begging by more precocial species. Again, Tinbergen and colleagues did some of the earliest work on this subject. Tinbergen and Perdeck's (1950) investigation of the begging behavior of young Herring Gull chicks is a classic that is referred to in virtually all texts on animal behavior. Using a series of painted cardboard models of a parent's head, they varied the color and pattern of the models' "bills." Adult Herring Gulls have yellow bills with a red spot toward the end of the lower mandible, and normal begging behavior involves pecking at this red spot by the young chicks. Using the cardboard models, Tinbergen and Perdeck showed that the mandibular spot was the most important aspect of the parent's head for eliciting begging. The effectiveness of the spot was partly due to contrast (gray bills with either black or white spots elicited more pecks than did plain bills), but the red color itself was also important. Thus a plain red bill was more attractive than a plain bill of any other color, and a red spot was more effective than a spot of any other color. Indeed, having discovered that redness and contrast were so effective in stimulating Herring Gull chicks' begging, Tinbergen and Perdeck went on to construct the now-famous model that actually elicited more pecks than a real parent's head did: a simple narrow rod, painted red with three white rings near the tip (i.e., redder and with more contrast than the actual bill).

Not all gulls have yellow bills with contrasting spots, but the food-begging behavior by chicks of most *Larus* species and, indeed, also of various terns (*Sterna* spp.) is quite similar (Hailman, 1967). Three such species with solid red bills—Black-headed Gulls (*Larus ridibundus;* Weidmann and Weidmann, 1958), Laughing Gulls (*L. atricilla;* Hailman, 1967), and Arctic Terns (*Sterna paradisaea;* Weidmann, 1961)—all show strong preferences for red bills as chicks. Interestingly, the same is also true for Black-legged Kittiwakes (*Rissa tridactyla*), even though parents have plain yellow bills. Adults, however, do have red inside the mouth (Hailman, 1967).

Nevertheless, not all larid species prefer red bills as chicks. The chicks of Sandwich Terns (*Sterna sandvicensis*) and Sooty Terns (*S. fuscata*) respond more strongly to black bills than to red ones (Weidmann, 1961; Cullen, 1962). Not surprisingly, both of these species have essentially black bills as adults. The Sooty tern's is solid black, and the sandwich tern's is black with a small yellow tip. Nevertheless, in each of these species red bills were second best to black in attracting chicks' pecks.

Nyström (1973) reported that 2-day-old Herring Gull chicks showed a far slower extinction rate for pecking red strips than for pecking either white or blue strips. Nyström suggested that the longer extinction for red explained the results indicating preference for red in larids. However, it is perhaps more likely that it is the gulls' preference for red that causes the slower extinction rate for this color.

Hörlyk and Lind (1978) have found, rather surprisingly, that newly hatched European Oystercatchers (*Haematopus ostralegus*) show no color preferences at all, even though the adults' red bill seems so striking to human eyes. Using models of adult bills, they found that both shape and contrast are influential in eliciting pecking responses, but color appeared to be unimportant.

When larid chicks peck at their parents' bills, they normally turn their heads so that the chicks's bill clasps that of the parent. Hailman (1967) found that this clasping component of begging apparently needs personal experience of seeing the parent's bills, because gull chicks hand-reared in the dark do not develop this behavior.

2. Obtaining Food on Their Own

a. Introduction. Sooner or later, every bird must develop the ability to obtain its own food. The timing of this varies from many weeks after hatching in certain altricial birds, especially large raptors, to virtually immediately upon hatching in many precocial species such as galliforms.

There seem to be a large number of studies on very young precocial birds in which questions relating to food getting and those relating to imprinting overlap. Sluckin (1964) and others have pointed out that newly hatched chicks tend to peck at the stimuli that elicit approach. This has also been stressed by Strobel and MacDonald (1974). Hence, a great many papers concern stimuli that elicit pecking by young precocial birds, but discuss the results mostly or entirely in the context of imprinting. A reexamination of these data within the context of the development of feeding behavior would be interesting.

b. Color Choices. Many such studies involve color preferences in newly hatched precocial chicks. Kear (1964, 1966) tested chicks in an arena in which they were offered spots of various colors and found some interesting interspecific differences in color choice. Like Hailman (1967) and others, Kear found that gull chicks prefer red. Rather surprisingly, so do both Common Gallinule and European Coot chicks. Gallinules have red bills as adults, but coot bills are stark white. (Weidmann, 1965, using a different technique, found that coot chicks prefer yellow and white rods to red ones.) In contrast to these species, Kear found that Ring-necked Pheasant chicks (*Phasianus colchicus*) and several species of duckling, all show a strong preference for green spots. This latter result is perhaps not surprising because chicks of these species feed themselves immediately after hatching, primarily on vegetation. Oppenheim (1968a) studied the preference for green in newly hatched Mallard ducklings in greater detail and he found that contrast is important. The ducklings were attracted to green most strongly when it contrasted with the background. They persisted in this preference when incubated in total darkness, and even when a window was made in the eggshell and the embryos' eyes were illuminated with white or yellow light for many hours before hatching. Hence, experience appears to play no role in the development of this color preference in Mallards. In contrast to pheasants and ducklings, domestic chicks tend to avoid green in favor of either blue or red (Kovach, 1971; Fischer *et al.*, 1975). Moreover, unlike ducklings, domestic chicks' color preferences do appear to be affected by early experience. Wada *et al.* (1979) found that illuminating embryos with green light between day 18 and hatching shifted the hatchlings' color preferences from red to green. Kovach (1978, 1980) found that Japanese Quail chicks, similar to domestic chicks, also choose red or blue over green, and in this case there is a clear-cut genetic basis for such preferences.

c. Parental Influence. Even in precocial species that start to feed themselves immediately, after hatching parental behavior can have a strong and important influence on feeding by the chicks. In an excellent field study, Stokes (1971) found that mother Red Junglefowl normally adopt a particular stance pointing to available food for the chicks. Moreover, after the young are old enough to wander away on their own, a mother still indicates food sources, now by exaggerated and apparently stereotyped pecking movements.

At the same time these visual cues are available, hens also emit various vocalizations, such as feeding calls and distress calls. Snapp (1969) demonstrated that chicks of domestic fowl that could hear maternal feeding calls but could not see an adult hen did not feed, but rather exhibited distress behavior. Snapp concluded that visual cues are more important here than auditory ones. In a somewhat different approach, Cowan and Evans (1974) investigated the behavior of older domestic chicks and their ability to distinguish calls of their own parents from those of other hens. They showed that chicks selectively pecked at food in response to their own parents' calls.

Finally, Gochfeld (1980) has found that the consistency with which fish are presented (head first and tail dangling left to right) to newly hatched chicks of Common Terns (*Sterna hirundo*) may enhance the development of the chicks' ability to handle and manipulate food items.

d. Social Facilitation. Another factor that can affect early feeding by precocial chicks is social facilitation. Again, both visual cues (the sight of another bird eating) or auditory ones (such as a pencil tapping) can enhance pecking in hatchlings of several species (Tolman, 1967a,b). For reviews of the effects of social facilitation on pecking in young birds, see Tolman (1968), Wilson (1968), and Strobel and MacDonald (1974).

e. Curiosity. Curiosity is clearly an important factor in the early pecking behavior of both altricial and precocial birds. It is virtually universal for young birds early in their feeding development to begin pecking at various objects, usually those contrasting with the background, regardless of whether these are edible or inedible. Barraud (1961) has referred to this behavior as *exploratory pecking*. At first, this curiosity is apparently very general, but later it may become more specific, directing a bird's pecks to a particular region of a potential food item (see later). Vince (1960) studied curiosity levels in young Great Tits (*Parus major*) and showed that young birds' curiosity level is high up to a certain age, but that beyond this the level wanes noticeably. Similar results have been found for many other species (Smith, 1973, 1974, 1978).

f. Development of Food Recognition. How do birds come to recognize appropriate food items? In normal development young birds show curiosity and exploratory pecking toward a wide variety of items, both suitable and unsuitable, and then gradually come to concentrate their pecks on the edible items. From this sort of information one might conclude that food recognition is usually learned by experience, but this is not always the case.

In a series of experiments, Hogan (1973a,b, 1975) and Hogan-Warburg and Hogan (1981) studied the development of food preferences in domestic chicks. Their basic observation was that newly hatched chicks peck indiscriminately at sand and food, whereas by 3 days of age chicks' pecks are directed more at food. Sand, of course, is also important for chicks to ingest, because they need it for their gizzard. It would be interesting to know how chicks discriminate between something biologically useful (either food or sand) and something not at all useful to them. In any case, the authors concluded that various learning processes are essential in the development of food recognition by domestic chicks, but that these are involved in an intricate interplay with genetic predispositions.

Kear (1962) investigated the development of seed preferences in various species of finches and found that at first they showed only a slight degree of preference between seed types, but that this grew stronger with age. Kear went on to show that these preferences appear to be related to the kind of seed that the birds can husk most efficiently, giving them the most kernel weight per unit time. Because the different species had bills of different sizes, they came to prefer different kinds of seeds (Kear, 1962).

In contrast, a large number of species seem to have fairly strong genetic predispositions for particular prey types. Rabinowitch (1968), investigating Herring and Ring-billed Gull (*Larus delawarensis*) chicks, found that, although familiarity was potentially important, when forced to choose between unfamiliar foods the young gulls preferred worms to pink catfood, and pink catfood to green catfood. Rabinowitch (1969) later showed that young Zebra Finches also have consistent preferences, even among unfamiliar food types. Ligon and Martin (1974) found that hand-reared young Piñon Jays (*Gymnorhinus cyanocephalus*) prefer piñon seeds to other objects and concluded that this specialist has an inborn recognition of its normal food. Similarly, both Loggerhead Shrikes (Smith, 1973) and American Kestrels (*Falco sparverius;* Mueller, 1974) have an apparently inborn recognition of mice. In these two species, young birds given regular experience with live mice gradually developed the ability to kill them by a certain age; but naive birds of the same age were also capable of attacking their first mouse in the same manner. When naive shrikes of the appropriate age were given a stuffed mouse, they approached with evident interest and apparent caution, and, so long as the stuffed mouse was upright, directed their pecks to the back of the mouse's neck. However, the cotton-filled mouse sometimes fell over during the shrikes' approach. As soon as it was on its side or back, the birds directed their pecks to any projecting extremity, such as the feet, ears, or tail of the mouse. Loggerhead Shrikes thus seem to have a very precise innate recognition for the conformity that makes up "mouse," but only so long as this is in the correct orientation. As soon as the same stimulus is overturned, the shrikes apprently no longer recognized it as a mouse.

Busbee (1976), working on a more southern population of Loggerhead Shrikes, found that his birds apparently needed some experience before being able to kill mice and concluded that this attack behavior is not innate. I suspect the main problem here is semantic. I have shown that young shrikes, once at the appropriate age, need no prior experience to be able to recognize mice as food and to direct their pecks to the back of the neck. I therefore concluded that this attack behavior is innate. That these skills improve with practice, considering the complexity (and danger) of the task, is hardly surprising.

In an interesting study, Davies and Green (1976) found that hand-reared Reed Warblers (*Acrocephalus scirpaceus*), given regular experience with live flies, showed gradual improvement in flycatching with age. Nevertheless, naive warblers given their first live flies at the age when experienced warblers could catch flies showed skill equal to that of the experienced warblers.

In summarizing the theories concerning prey recognition in birds, Mueller (1974) mentioned three hypotheses. One is that young birds learn appropriate food types from their parents. Another is that food recognition is based primarily on learning associated with personal experience. The third is that food recognition is largely independent of the bird's prior experience and is thus primarily genetically based. Clearly, different hypotheses could be correct for different species of birds. Some precocial species may well acquire their food preferences from a combination of parental direction and personal experience. However, at least in certain species, parental teaching alone apparently plays little, if any, role in early food preferences. Young American Kestrels, for example, are usually fed small mammals and birds, but they prey almost exclusively on large insects when they first become independent (Mueller, 1974). Loggerhead Shrikes show similar diet differences. Learning by personal experience alone seems to play a greater role in manipulation (see later) than in actual recognition of food. It would be interesting to see whether hand-reared finches of the species Kear (1962) used, if kept naive until the age at which it was reported that they had marked seed preferences, also showed such preferences. Nevertheless, in a wide variety of birds, there is strong evidence that the development of food recognition has a strong genetic basis.

g. *Development of Food-Handling Techniques.* In contrast with the development of food recognition, there seems to be a far stronger role of personal learning involved in the development of food-handling techniques. Perhaps the simplest such technique is just picking up food items. One of the most thorough studies of how this develops is that of Cruze (1935), done on domestic chicks. If chicks are given their first chance to peck at small objects at different ages (1 to 5 days after hatching), having been kept in the dark and hand-fed beforehand, the accuracy of their pecks can be shown to improve with age. The pecks of a naive 5-day-old chick are considerably more accurate than those of a naive 2-day-old chick. This improvement cannot be due to learning, because all chicks were equally inexperienced. Nevertheless, learning also has an effect, because Cruze found that experienced chicks pecked more accurately than inexperienced chicks at every age tested.

Davies and Green (1976) found that naive hand-reared Reed Warblers, although just as capable as experienced warblers at catching flies, are significantly poorer at handling them once the flies are caught. Although the naive birds steadily improved with time, they did not achieve the skill of the experienced warblers. Davies and Green suggested that a sort of critical period, in which personal experience is necessary, may be involved.

One rather specialized feeding technique involves holding food with the feet (found commonly in raptors, as well as in passerines such as titmice, corvids, and shrikes). Hinde and Tinbergen (1958) studied the development of foot use by young titmice and other species in some detail. They found that young Chaffinches do not learn to use feet in feeding, even when raised by titmice foster parents. This difference is probably inherited, because adult Chaffinches do not normally use their feet in this way.

Vince (1961a,b) discussed a modification of this sort of behavior in a series of papers on string pulling in birds. Titmice, which normally use their feet to hold down food, were presented with a food item suspended by a string from a perch. They had little difficulty in obtaining the food simply by pulling up loops of string with their bill and holding them under the foot while reaching down to get the next loop; eventually, the food arrived within reach. Vince also found that species that do not use their feet in feeding can develop their own form of string pulling. Greenfinches (*Carduelis chloris*) often draw the string up slowly in their bills with the mandibulating movements normally used for opening seeds (Vince, 1961b).

As mentioned earlier, the development of still another food-handling technique, impaling and wedging, requires personal experience during a critical period (Wemmer, 1969; Smith, 1972) in Loggerhead Shrikes. In an interesting comparative study, Lorenz and von St. Paul (1968) studied the development of this technique in three European shrikes: Red-backed (*Lanius collurio*), Woodchat (*L. senator*), and Great Gray or Northern (*L. excubitor*). Neither Red-backed nor Woodchat Shrikes needed any personal experience before being able to direct dragging behavior to thorns (true impaling), but both needed considerable experience before learning to direct dragging toward forks in branches (wedging). In contrast, *L. excubitor* young needed no experience before directing dragging to forks, but they had to learn by experience to impale on thorns. Lorenz and von St. Paul related these differences to the natural history of the three species.

h. The Problem of Large Prey. Any bird that normally eats animal food must encounter a wide size range of potential prey. Smaller items present relatively few problems to the would-be predator, which can probably just pick them up and swallow them. The larger the potential prey, however, the more problems it may give the predator, but at the same time larger prey yield more energy intake per unit of energy expended in catching them. Thus it is to the advantage of many birds to select and deal with prey that are as large as possible.

In such prey, often certain parts such as the neck or head of small vertebrates are more vulnerable than others. One can therefore ask, What factor(s) do naive birds use to direct their pecks along the length of a large potential prey item?

Hand-reared young of six altricial species were used to study this question: Turquoise-browed Motmots, Great Kiskadees, Blue Jays (*Cyanocitta cristata*), Black-capped Chickadees, Gray Catbirds (*Dumetella carolinensis*), and Loggerhead Shrikes (Smith, 1973, 1974, 1976, 1978). The young birds, tested when they were at the age when fledglings in the field could catch live prey, were given appropriately sized wooden dowel models bearing various cues such as eyespots, neck constriction, pointed "tail," and motion (Fig. 4). Many cues elicited very similar responses from all six species, but others elicited some very interesting differences.

The ends of a model were more attractive than the middle. All six species directed significantly more than 67% of their pecks to the two end thirds of the plain, unmarked model. This makes sense, because most terrestrial animals have their most vulnerable body parts near one end of their total body length. The six species also all pecked significantly more at the cue-bearing end of any model with either eyespots or neck constrictions. This, however, was probably simple curiosity (pecking at the different end) rather than any inborn recognition of eyes or neck, because models that bore both cues at once did not elicit a more highly directed response.

Not all species pecked simply at the different end of every model. An exception was the kiskadees' response to the model with a pointed tail. Here the different third was the pointed end, but the kiskadees gave a highly

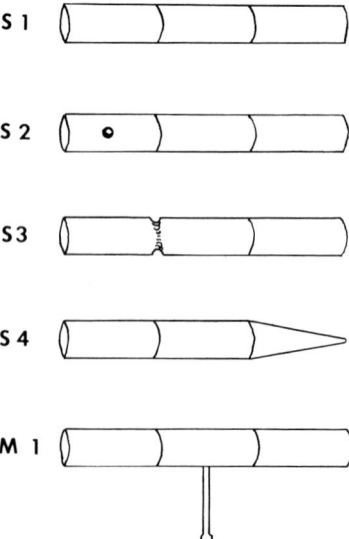

FIG. 4. Some wooden dowel models showing major cues offered to young shrikes, jays, chickadees, catbirds, motmots, and kiskadees. Actual sizes of the models varied with the species. In addition to the plain model (S 1), the cues shown are eyespot (S 2), neck constriction (S 3), pointed tail (S 4), and motion (M 1). Motion was achieved by inserting the nail into a diagonal slot in the cage floor, grasping it from below, and moving it along the diagonal track.

significant response to the blunt end instead. Yet it would seem to be easier for kiskadees to pick up the model by the sharpened end. Clearly, this response is not due to curiosity; it suggests that kiskadees (which are known to eat lizards) possess an innate recognition of "tail" in large prey. Because many lizards can shed their tails when caught, a kiskadee catching a lizard by the head (blunt end) would be able to eat the whole thing, whereas a kiskadee catching a lizard by its tail would probably get to eat only the tail.

When presented with moving models, three species (shrikes, motmots, and kiskadees) directed most of their pecks to the leading third of the model; the other three species did not. This clearly reflects the dietary differences of the six species. The three that regularly eat vertebrate prey all pecked most at the "head" end; the others did not. Because all six species were raised in the same manner, these differences seem to be genetically based.

Mueller (1974) presented hand-reared American Kestrels with various potential prey objects, including a tissue ball, a tissue "mouse," and a stuffed mouse. Of nine kestrels tested, only one attacked a tissue ball, two attacked the tissue mouse, and two (the same two) attacked the stuffed mouse. Mueller did not report on where the initial pecks were directed along the length of these models.

i. Warning Coloration and Noxious Prey. Many noxious species, both vertebrate (e.g., certain snakes and salamanders) and invertebrate (such as many insects), are conspicuously colored, usually with red, yellow, orange, black, or some combination of these. It is well documented that adult birds in the wild tend to avoid such prey (Jones, 1932, 1934; Morrell and Turner, 1970). How do wild birds develop such aversions?

One hypothesis is that naive birds attack any possible prey, regardless of color or pattern, then learn by experience to avoid those with aposematic (warning) coloration. This seems to be the case for many birds tested. For example, Alcock (1973) found that naive Red-winged Blackbirds (*Agelaius phoeniceus*) attacked the first aposematic stinkbugs they ever saw about as readily as they had attacked mealworms. Presumably, if the stinkbugs tasted bad enough, the blackbirds would learn to avoid attacking them. Various studies have investigated this form of learning more closely. Hale and Green (1979) found an interesting difference in the ability of very young domestic chicks to learn when consequences are bad as opposed to when consequences are good. Working with two age groups (½ day old, 2½ days old) they found that when the consequences were positive, only the older chicks showed learning, but when the consequences were negative both groups learned. Hence, avoidance learning seems to appear earlier than acceptance learning. Moreover, Gittleman *et al.* (1980) found, looking simply at negative consequences, that 4-day-old domestic chicks learned to avoid conspicuous prey more readily than cryptic prey. Thus, at least in this species, young chicks are particularly capable of learning to reject prey types, and they do so even more readily if the prey is conspicuously colored.

Another hypothesis suggests that, in addition to learning, young birds may possess a genetically based general aversion to novel stimuli. This theory, proposed by Coppinger (1970), is widely accepted (e.g., Turner, 1977), but there is very little evidence to support it. Coppinger's work and other studies that seemed to support novelty rejection are based on young birds raised on severely restricted diets, very unlike what such birds normally encounter in the wild. Studies of young birds raised on at least moderately varied diets give no evidence of novelty rejection. Alcock's Red-winged Blackbirds (mentioned earlier) attacked their first stinkbugs with no hesitation. Similarly, when raised on moderately varied diets, young kiskadees, House Sparrows (*Passer domesticus*) and Blue Jays all readily attacked the first bright colors and contrasting patterns they ever saw (Smith, 1978, 1980). Thus novelty rejection alone cannot explain such birds' subsequent rejection of conspicuously colored prey.

This does not mean that novelty has no effect on a bird's attack behavior. Shettleworth (1972) found that young domestic chicks learned to reject novel stimuli more rapidly than they learned to reject familiar ones.

Not all rejection of noxious prey must be learned by naive young birds. At least some species apparently possess inborn recognition of potentially dangerous prey. Certain species that regularly eat noxious insects show apparently innate patterns of treatment that render such prey harmless. Fry (1969) found that naive young Red-throated Bee-eaters (*Merops bulocki*) deal with live honeybees in an apparently innate sequence of stereotyped behavior consisting of two distinct actions. First, the bird beats the honeybee sharply against the perch; then it rapidly shifts is grip and wipes the abdomen against the perch, thereby devenoming it. Birkhead (1974) reported similar unlearned behavior toward wasps in other wasp-eating birds.

Although innate recognition in the previously mentioned species is accompanied by specialized handling, allowing the bird to eat the noxious prey, other species show simple innate avoidance of such prey. Davies and Green (1976) found that hand-reared Reed Warblers would not attack live wasps, even though they readily caught and ate live flies minutes later. Yet the same warblers had attacked and eaten live wasp mimics (hoverflies) earlier! When one wasp was devenomed and painted blue, a young Reed Warbler attacked it. Thus Reed Warblers apparently possess a precise recognition of, and aversion to, "wasp."

Two species of neotropical predators, Turquoise-browed Motmots and Great Kiskadees, also possess an innate aversion to a potentially lethal type of prey, the coral snake (Smith, 1975, 1977a). Early experiments with motmots showed that these birds would not attack a wooden dowel model painted with a pattern of yellow and red rings, even though they readily attacked red, yellow, and even yellow- and red-striped models. Hand-reared young kiskadees also avoided a yellow- and red-ring model; moreover, when given a model painted with a true coral snake pattern (red, yellow, and black rings), they gave alarm calls and other evidence of fear in addition to not attacking the model.

Yellow and red are common aposematic colors and ring patterns are among the most common as well. It remained possible, therefore, that the kiskadees' and motmots' response to the yellow and red rings might have been nothing more than an innate aversion to a generalized pattern of warning coloration. If this were so, then other, nonpredatory birds, especially generalists that often eat large insects, might also show such an innate aversion. However, young Blue Jays, Red-winged Blackbirds, and House Sparrows raised and tested in the same manner showed no hesitation in attacking both yellow- and red-ring patterns and true coral snake patterns. Clearly, therefore, the motmots' and kiskadees' aversion is a specialization enabling them to avoid attacking the lethal coral snake.

j. Later Development. So far, the discussion has concerned young birds usually only a few days to a few weeks old. However, in at least some species development of food recognition and handling continues for months or even years.

This gradual development is particularly evident in birds that prey on fish. Orians (1969a) documented that immature Brown Pelicans (*Pelecanus occidentalis*) have significantly poorer foraging success when diving for fish than do adult pelicans. Similar data were obtained for Little Blue Herons (*Egretta caerulea*) by Recher and Recher (1969). More recently, Quinney and Smith (1980) found that, as with the other two species, adult Great Blue Herons (*Ardea herodias*) show far greater foraging efficiency than do juveniles. They divided foraging behavior into two major categories: *strikes*, and *probes*. Adults were successful in 62% of their strikes, as compared with a 33% success rate for immatures. It may be significant that neither age group was even seen to swallow after a probe. Moreover, adults rarely probed, although juveniles did so fairly regularly. Quinney and Smith suggested that probing may be a form of orientation learning and not actual foraging at all.

D. MIGRATION AND ORIENTATION

Almost three-quarters of the bird species breeding in Canada and the northern United States migrate south for the winter. This phenomenon of migration has fascinated people for centuries. How do young birds, just a few weeks old, determine when and where to go during their first migratory trip? This is not necessarily taught to them by older, experienced birds. Early observations of certain shorebirds revealed that the adults had already departed by the time the young were ready to leave, so the young had to make their way south on their own. Nor is this ability restricted to shorebirds, because experiments with hand-reared young of species that normally migrate in mixed-age flocks have confirmed this ability in a wide variety of species.

It is now well known that adult birds are able to obtain directional information from a large number of different sources, such as celestial cues (sun and stars), the earth's magnetic field, and perhaps even such seemingly implausible sources as infrasound, polarized light, and variations in the strength of gravity (Keeton, 1979). Under optimal conditions, these cues may provide redundant information, and the bird may pay attention to only one source, for example, the sun's position. However, when information from that source is blocked, as on a heavily overcast day, the bird is able to switch to an alternate source of information (e.g., magnetic field) and continue on its way (Emlen, 1975a,b; Keeton, 1979). This so-called hierarchy of cues is thus highly adaptive.

How do inexperienced young birds develop the ability to use these cues? Preliminary observations seemed to suggest that this ability is entirely innate. In a famous set of experiments, Sauer and Sauer (1960) found that hand-reared young European warblers could orient correctly under autumn "skies" in a planetarium. More recently, Wiltschko and Gwinner (1974) reported evidence for an innate magnetic compass in Garden Warblers (*Sylvia borin*).

It now appears, however, that any conclusion that such abilities are entirely innate may be an oversimplification. Although there is undoubtedly a strong genetic base for the acquisition of such abilities, there is often a learned component as well. In an elegant set of experiments, Emlen (1969, 1970, 1972, 1975b) documented this for star-compass orientation in Indigo Buntings. In early tests, naive hand-reared buntings that were allowed to see normal skies from mid-August to mid-September all oriented south when placed under a stationary planetarium sky. However, another group of naive buntings that were exposed to normal skies only after migration had started (mid-September to mid-October) showed essentially random orientation when tested in the same manner. This suggested two things to Emlen: first, that indigo buntings do not appear to have a genetic recognition of specific stellar cues and, second, that young birds must have personal experience in seeing natural skies some time before normal migration in order to acquire the ability to use stellar cues in orientation. Later experiments confirmed both of these conclusions. Buntings that have not seen the night sky until after the first migration season has begun never learn to use the star compass, no matter how often they see the sky thereafter. Because learning can take place only within a critical period, this suggests that the ability to use stellar cues is acquired, at least in Indigo Buntings, by a sort of imprinting. It is interesting to note that there can be no reinforcement, in the usual sense of the term, involved here, because the birds were confined to aviaries throughout this sensitive period (Emlen, 1972).

Actually, what the young buntings respond to initially is the apparent rotation of the starry sky during the night. Soon, however, buntings allowed to see this rotation during their sensitive period learn star patterns that indicate where the axis of rotation is. Thereafter, they can orient successfully simply by seeing the position of a few key stars. Moreover, each individual bunting selects and learns its own key stars. Hence, although one bunting might need the entire area within 35° of Polaris in order to orient correctly, another might need to see only a much smaller sector of the night sky (Emlen, 1975a).

Emlen raised one group of buntings under a planetarium sky rotating around an incorrect axis—that through the star Betelgeuse. As expected, these birds learned to use star patterns appropriate to the new axis and thus oriented incorrectly when shown the normal sky. Moreover, they continued this mistaken orientation a year later, even after 6 full months under the normal sky. Again, this clear irreversibility supports the idea that imprinting is involved.

As Emlen (1972, 1975a) pointed out, this lack of an inborn star map is actually adaptive. Over evolutionary time, stellar cues themselves are not reliable, because the earth's axis wobbles very slowly. The current "spring stars" will, therefore, become autumn stars 13,000 years from now, and Vega will eventually become the new pole star (Emlen, 1972). For this reason, it is better to have the flexibility of learning from celestial rotation.

Keeton and colleagues, working with domestic pigeons, have shown similar subtle learning involved in the acquisition by young birds of the ability to use directional cues. Experienced pigeons are capable of orienting correctly using either the sun or magnetic cues alone (Keeton, 1969, 1971). In contrast, inexperienced pigeons apparently need both at once. Thus young first-flight pigeons, if released under totally overcast skies, usually departed randomly (Keeton and Gobert, 1970). Moreover, first-flight pigeons that carried bar magnets attached to their backs also departed randomly, even on sunny days (Keeton, 1971). The previously mentioned results are from young pigeons raised under normal conditions of having seen natural skies and having experienced the regular geomagnetic field. However, when young pigeons are raised with no chance to see the sun, they are found to orient perfectly even if their first homing flight is under totally overcast skies.

Keeton (1979) interpreted this as follows. Having never viewed the sun, these birds had not incorporated the sun into their navigational system and thus had no problem orienting when it was missing. Keeton did not speculate on what cues such birds used for their orientation, but a likely candidate is the geomagnetic field. These results are similar to those obtained by Wiltschko and Gwinner (1974), who hand-reared Garden Warblers with no opportunity to see either the sun or stars and found that the birds could orient correctly using simply the normal geomagnetic field. Apparently, the nonvisual, geomagnetic orientation system of both pigeons and Garden Warblers can mature independently, without requiring any learning processes based on experience with celestial cues.

Wiltschko and Gwinner (1974) concluded from these data that Garden Warblers possess an innate magnetic compass. Although they may be right, I would be more convinced of this if they had hand-reared Garden Warblers under conditions in which they could not perceive the geomagnetic field. If such birds could later orient correctly using geomagnetic cues, then they do indeed possess an innate magnetic compass. Alternatively, they might, like Emlen's buntings, need personal experience during a critical period in order to make subsequent use of the cue for orientation.

In an interesting paper, Wiltschko et al. (1976) showed that the coupling of times, directions, and sun azimuths is not inherited by young pigeons; it must instead be learned. They hand-reared young pigeons under a permanently 6-hour-slow, clock-shifted photoperiod and found that their birds had no difficulty in orienting normally. Their pigeons had thus apparently learned that the sun rises in the south and sets in the north. Hence, the sun compass requires personal experience in order to be calibrated.

Finally, learning may also be involved in determining the relative rank of the various sources of information that make up a bird's hierarchy of cues (Keeton, 1979). Hand-reared birds of different species seem to form different hierarchies, although it is sometimes difficult to know whether these differences are really species specific or simply reflect differing investigative techniques. Thus for domestic pigeons and also for indigo buntings celestial cues appear to have higher rank than geomagnetic cues, but for Ring-billed Gulls geomagnetic cues apparently rank above information from sun or stars (Southern, 1978).

So far I have focused primarily on the question, What does a naive bird use in order to find its way south on its first migratory trip? However, one can also ask, Why is south the direction it chooses? Perhaps the most ambitious theory proposed to explain this is that of Gwinner (1972, 1974, 1977), who argued strongly that the basis for bird migration is an endogenous circannual rhythm. He felt that migratory birds possess a genetically based fall and spring "program" that determines the timing and direction of the bird's migratory trips. In early work with hand-reared Willow Warblers (*Phylloscopus trochilus*), Gwinner (1972) reported that the young caged birds in their first autumn showed the greatest amount of migratory restlessness, or *Zugunruhe*, on the dates when they would have been crossing the Sahara Desert in the wild. Gwinner concluded that the amount of migratory restlessness is correlated with the length of the migratory route normally taken on that date. Similar data have since been obtained for other European warblers as well (e.g., Gwinner, 1977). Some of Gwinner's later work has yielded some quite amazing results. Gwinner hand-reared 59 Garden Warblers and kept them in a constant 12-hour photoperiod during their first fall. These birds were tested for *Zugunruhe* regularly throughout that fall in standard round cages. The birds were not allowed to see any celestial cues, but they were exposed to the earth's normal magnetic field. During August and September, these warblers showed a significant southwesterly directional preference, whereas from October to December the birds preferred a south-by-southeasterly direction. These two directional preferences, which differ significantly from each other, correspond to the normal migratory path for this species: first southwest across Europe from Germany to Spain, then across Gibraltar and back southeast across Africa (Fig. 5). Yet these birds had

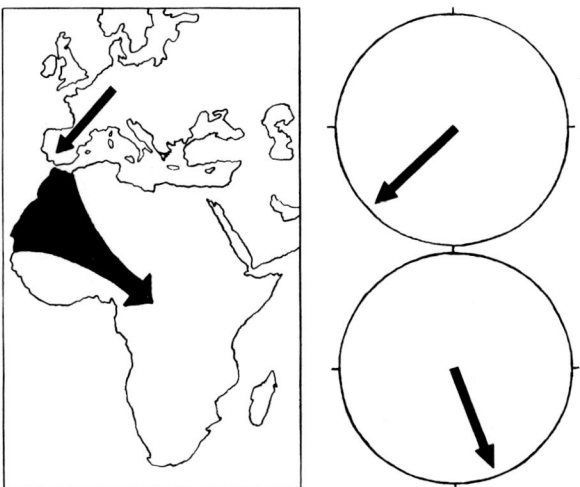

FIG. 5. Changes in migratory direction of Garden Warblers on fall migration (left). Average direction of migratory restlessness of Garden Warblers (right), tested in circular orientation cages early (August–September, upper diagram, $n = 233$) and later (October–December, lower diagram, $n = 172$) in the migratory period. (Numbers refer to the number of individual test nights during which the birds exhibited migratory restlessness.) The mean direction preferred in August–September and that preferred in October–December are significantly different. (After Gwinner, 1977. With permission, from the *Annual Review of Ecology and Systematics*, Volume 8. Copyright 1977 by Annual Reviews Inc.)

been kept under a constant regime of 12 hours light and 12 hours dark throughout the fall. These data certainly suggest some long-term timing device that is, at least in Garden Warblers, apparently coupled to a very precise set of genetically based directional preferences. More data on other species with natural "dog-legs" in their normal migratory routes are needed to test this interesting hypothesis.

Berthold and Querner (1981) have found evidence that lends support to Gwinner's theory. Earlier work had shown that the correlation between amount of *Zugunruhe* and normal length of migratory route holds good for a number of European warbler species. Berthold and Querner worked with various populations of the Blackcap (*Sylvia atricapilla*), a species that ranges from northern Europe south to the Cape Verde Islands, off the West African coast. Blackcaps from the northern populations are strongly migratory, whereas those from Africa are only partially migratory. When nestlings from four populations (Finnish, German, French, and African) were hand-reared and tested for *Zugunruhe* in the fall, the Finnish birds showed the most (longest) restlessness, with progressively less in the German, French, and African birds. *Zugunruhe* lasted about 50 days in the African birds and more than 150 days in the Finnish ones. Berthold and Querner than did a cross-breeding experiment using German and African Blackcaps. The resulting 32 interpopulational hybrids showed intermediate migratory restlessness, and an intermediate precentage of birds displayed restlessness compared with that of the two parental types. Because the amount of *Zugunruhe* is thus genetically based and because it is strongly correlated with migratory behavior, Berthold and Querner concluded that the migratory behavior of Blackcaps is itself also genetically based, through an endogenous program controlled by or linked to a circannual rhythm.

Not all species' migratory behavior seems to fit Gwinner's hypothesis. For instance, Southern (1969, 1972, 1976) found that Ring-billed Gulls, even chicks 2–20 days after hatching show a consistent southeast preference when placed in a circular cage 8 feet in diameter. This is well before the age and/or date when migration normally starts for this species. Moore (1975) obtained similar results with young Herring Gulls. Clearly, many more data are needed to determine the generality of Gwinner's hypothesis.

E. PLAY

1. General Characteristics

Play must be one of the most difficult kinds of activities to distinguish and define. But there is no question that genuine play is widespread in young birds (Ficken, 1977; Fagen, 1981). In an excellent review of avian play,

Ficken (1977) listed three main characteristics of play that distinguish it from other, nonplay activites. First, play often involves incomplete or reordered sequences, or even incomplete movements, such as actions of prey capture not preceded by normal stalking. It may also involve mixtures of normally separate activities. Next, play often includes exaggeration and repetition of motor patterns. Finally, play behavior can often be elicited by a far wider range of stimuli than nonplay behavior.

Even with these characteristics in mind, however, there remains some difficulty in interpretation, especially when observing birds in captivity. Ficken (1977) warned that behavior that appears to meet the criteria for play may actually be simple vacuum activities, redirection, or responses to suboptimal stimuli associated with captivity. For instance, when hand-reared Turquoise-browed Motmots looked over their shoulder, then whirled around in an evident attempt to catch their own tails, I interpreted it as true play (Smith, 1977b), but field studies of young birds would be necessary to confirm this conclusion.

2. Play with Objects

This is perhaps the most common sort of play seen in young birds, being commonly observed both in captivity and in the field. Generally, such play involves movements later associated with feeding. Young Northern Gannets (*Sula bassana*) frequently play with nest material, tossing up sticks and feathers and catching them (Nelson, 1966). Similarly, young Ascension Frigatebirds (*Fregata aquila*) catch feathers and strands of seaweed from each other in midair (Stonehouse and Stonehouse, 1963). Such picking up of inedible objects, flying with them, then dropping and retrieving them is also common in many species of gulls and terns (e.g., see Ashmole and Tovar, 1968; Delius, 1973). B. Beck (unpublished, 1976), in Herring and Great Black-backed Gulls (*Larus argentatus* and *L. marinus*), found six clear differences between play dropping and predatory dropping of objects. Play dropping often occurs on soft substrates, it involves minimal inspection and manipulation after the drop, it often includes midair catching and redropping, it often occurs in strong winds, and, finally, it occurs randomly with respect to tide time. Interestingly, Beck also found that even though Great Black-backed Gulls do not perform predatory dropping of edible objects at any age, young birds nevertheless do play drop.

Play with inedible objects is very well known in young predatory birds. All sorts of manipulations have been reported: pouncing repeatedly; carrying objects up, dropping them and catching them with the feet; tossing and retrieving; flinging, grasping, shaking, and biting (e.g., Sumner, 1929, 1934; Hoffmeister and Setzer, 1947; Cade, 1953; Parker, 1975). These behavior patterns occur in young birds in the field as well as in captivity.

Parrots are notorious for such play (although the majority of observations have been made on captives). Similar behavior occurs in young wild passerines of several species as well. Wild young Loggerhead Shrikes frequently pounce on twigs and other inanimate objects, fly up to bushes, and perform impaling movements with them (Smith, 1972). Garden Warblers repeatedly pick up, drop, and catch small stones (Sauer, 1956). Swallows sometimes play with feathers (J. P. Hailman cited in Ficken, 1977). However, the most extensive and complex play in passerines is apparently done by crows and jays. Most of the observations of play in this family have been of members of the genus *Corvus*. Many of these observations include tossing objects in the air and catching them, transferring objects from the beak to the feet and back again, and/or various balancing acts. Thorpe (1966) watched a captive Common Raven (*Corvus corax*) repeatedly throwing a rubber ball up in the air and catching it, often lying on its back and passing the ball back and forth from its beak to its feet. Gwinner (1966) reported similar manipulations in several *Corvus* species and commented on the "inventive" nature of corvid play.

Play with objects is probably the easiest kind to interpret. Certainly, there can be little doubt that such manipulations are potentially important in the development of skills used in both foraging and feeding.

3. *Locomotory Play*

This type of play involves repetitions of acts of locomotion that apparently lack any immediate purpose. For example, Roberts (1934) reported that a group of Common Eiders (*Somateria mollissima*) repeatedly rode down a rapids, only to dash back upstream to do it again. Incredibly, a similar pattern has been seen in an Anna's Hummingbird (*Calypte anna*), which repeatedly floated down a small stream of water flowing from a hose (Stoner, 1947). I have seen a group of Swallow-tailed Kites (*Elanoides forficatus*) in Costa Rica repeatedly wheeling up in the air at one end of a lake, then dashing down along the surface of the water, their breast feathers actually touching the water in some cases, before returning to repeat the process (S. M. Smith, unpublished field notes). Many of the spectacular aerial maneuvers performed by corvids (Gwinner, 1966) may be considered a form of locomotory play.

This kind of play is more difficult to interpret from an adaptive point of view. Although aerial maneuverability probably does need practice and is clearly useful, there seems to be no obvious advantage to a hummingbird to hone its water-floating skills. Perhaps the best explanation is that of Beach (1945), who said that one important function of play is to enhance muscular development and coordination in young animals, because the resulting good physical condition will be important later in their lives.

4. Social Play

Social play involves two or more birds, usually siblings, but sometimes parents and offspring or groups of unrelated juveniles. Such play is widespread among bird species. Groups of young Great Frigatebirds (*Fregata minor*) play follow the leader over rock pools (Gibson-Hill, 1947); somewhat similar behavior has been reported in small groups of young European Buzzards (*Buteo buteo*) by Weir and Picozzi (1975).

Mock battles are common. Typically such encounters do not include any actual agonistic displays, and there is frequent role reversal, in which the chaser becomes the chased. Aerial mock battles have been seen between young Peregrine Falcons (*Falco peregrinus*) by Parker (1975) and in ravens by Gwinner (1966) and Fagen (1981). Gwinner reported that strong wind apparently stimulates this kind of play. Mock battles on the ground had been reported in Fischer's Turacos (*Tauraco fischeri*), involving birds too young to fly (Moreau, 1938). Similarly, young Barn Owls (*Tyto alba*) engage in vigorous running, pushing, and wrestling matches with siblings (Trollope, 1971). This kind of social intaction is very widespread among both nestlings and fledglings of a large number of avian species (Nice, 1943; Löhrl, 1975; Fagen, 1981).

Young birds will sometimes engage in apparent sexual play. Young Song Sparrows have been seen hopping on each other's back and pecking (Nice, 1943). Kruijt (1964) found that young Red Junglefowl would direct sexual behavior toward an inanimate object, such as a feather. Many reports of such misdirected sexual behavior exist for hand-reared birds in captivity, but these are difficult to interpret becuase they may be simply redirected activities (Ficken, 1977).

Social play is clearly important to developing birds. In many social mammals, social play is necessary if an animal is to be able to form normal social relationships in later life (Beach, 1945). There is some evidence that this is also true in birds. Boag and Alway (1980) found that social interactions within the brood had an important influence on subsequent dominance rank in both Red Grouse (*Lagopus lagopus scoticus*) and Japanese Quail; similar effects occur in Spruce Grouse (*Canachites canadensis*) as well (Alway and Boag, 1979).

5. The Problem of Subsong

Opinions differ as to whether subsong in passerines should be consided as play. Certainly, it does share some of play's characteristics (Thorpe, 1966; Ficken, 1977). However, the work of Nottebohm (1970, 1972a, 1975) has shown that subsong is a distinct stage in song development. Young birds that do give subsong are incapable of singing true full songs, and adults that have attained full song rarely (if ever) give subsongs. It therefore seems best to interpret this as simply a regular stage in vocal development, not true play.

V. Concluding Remarks

Much research still needs to be done in many aspects of behavioral ontogeny. For example, we have a pressing need for careful studies on the early prehatching behavior of truly altricial species such as passerines. We also need more data on hatching behavior itself, especially of altricial species. The technique developed by Bakhuis and colleagues, involving the use of clear glass "eggshells," promises to be especially productive.

For posthatching studies in general, we need more carefully planned field studies of wild nestling and fledgling behavior. Data from such studies will allow experiements with hand-reared birds to be designed to ask more biologically relevant questions.

Within the more specialized subject of imprinting, there are two aspects in particular in which future research should yield exceptionally interesting and important results. One concerns sexual imprinting in females, which is only beginning to receive the attention it deserves. We need many more data here, especially with respect to whether or not the species has sexual color dimorphism and whether or not both parents care for the young. The second potentially fascinating aspect of imprinting concerns the effects of various environmental factors on the length and general functioning of sensitive periods, such as the study by Kroodsma and Pickert (1981) on the effects of photoperiod on song learning by marsh-wrens. Similar environmental effects may well be found in other forms of imprinting as well.

Among studies of song learning, two techniques in particular promise to be highly productive. One involves the experimental use of living tutors and the manipulation of social interactions between pupil and tutor, such as Baptista and colleagues have been doing. The other is tape splicing and the consequent ability to create pseudosongs of various syllables, rhythms, patterns and lengths while using actual vocalizations as building blocks, as done by Marler and colleagues. The use of these two techniques, especially in conjunction, should improve our understanding of song development enormously.

So many aspects of the ontogeny of foraging and feeding need more work that it is hard to pinpoint just a few of them. The development of specialized food-handling techniques, such as flycatching (as studied in Reed Warblers by Davies and Green, 1976), merits far more attention. Coppinger's theory of novelty rejection needs more investigation. We are just beginning to find out how the young of some species develop the ability to deal with potentially dangerous prey, such as wasps.

As for the ontogeny of migration, two aspects in particular seem to be obvious areas for future research. One involves looking for the necessity of imprinting, such as Emlen (1972) has shown for the development of the star compass, in the ability to use other cues such as sun compass or magnetic field. The other is the development of the so-called hierarchy of cues. Will birds raised under various environmental conditions develop differently ordered hierarchies?

Play in birds, although relatively little studied, is both widespread and potentially of great importance in the development of a bird's behavior. We need many more careful field studies of play in wild birds. For more detailed reviews of this important topic, see Thompson (1964), Thorpe (1966), Ficken (1977), and Fagen (1981).

Behavioral ontogeny is a fascinating and important area of avian biology. It is, as I hope this chapter has shown, a subject in which much exciting research is currently in progress and in which we can expect new techniques to bring fascinating new insights in the future.

REFERENCES

Alcock, J. (1973). The feeding response of hand-reared Red-winged Blackbirds (*Agelaius phoeniceus*) to a stinkbug (*Euschistus conspersus*). Am. Midl. Nat. **89**, 307–313.

Alconero, B. B. (1965). The nature of the earliest spontaneous activity of the chick embryo. J. Embryol. Exp. Morphol. **13**, 255–266.

Alway, J. H., and Boag, D. A. (1979). Behaviour of captive Spruce Grouse at the time broods break up and juveniles disperse. Can. J. Zool. **57**, 1311–1317.

Andrew, R. J. (1963). Effects of testosterone on the behavior of the domestic chick. J. Comp. Physiol. Psychol. **56**, 933–940.

Arnold, A. P. (1975). The effects of castration on song development in Zebra Finches (*Poephila guttata*). J. Exp. Zool. **191**, 261–277.

Ashmole, N. P., and Tovar, H. (1968). Prolonged parental care in Royal Terns and other birds. Auk **85**, 90–100.

Bakhuis, W. L. (1974). Observations on hatching movements in the chick (*Gallus domesticus*). J. Comp. Physiol. Psychol. **87**, 997–1003.

Bakhuis, W. L., and Bour, H. L. M. G. (1980). The behavioural state during climax (hatching) in the domestic fowl (*Gallus domesticus*). Behaviour **73**, 77–105.

Bakhuis, W. L., and van de Nes, J. C. M. (1979). The causal organization of climax behaviour in the domestic fowl (*G. domesticus*). Behaviour **70**, 185–230.

Baltin, S. (1969). Zur Biologie und Ethologie des Telegalla Huhns (*Alectura lathami:* Gray) unter besonderer Berücksichtigung des Verhaltens während der Brutperiod. Z. Tierpsychol. **6**, 524–572.

Baptista, L. F. (1974). The effects of songs of wintering White-crowned Sparrows on song development in sedentary populations of the species. Z. Tierpsychol. **34**, 147–171.

Baptista, L. F. (1975). Song dialects and demes in sedentary populations of the White-crowned Sparrow (*Zonotrichia leucophrys nuttalli*). Univ. Calif. Berkeley Publ. Zool. **105**, 1–52.

Baptista, L. F. (1977). Geographic variation in song and dialects of the Puget Sound White-crowned Sparrows. *Condor* **79**, 356–370.
Baptista, L. F., and King, J. R. (1980). Geographical variation in song and song dialects of montane White-crowned Sparrows. *Condor* **82**, 267–284.
Baptista, L. F., and Morton, M. L. (1981). Interspecific song acquisition by a White-crowned Sparrow. *Auk* **98**, 383–385.
Baptista, L. F., and Wells, H. (1975). Additional evidence of song-misprinting in the White-crowned Sparrow. *Bird-Banding* **46**, 269–272.
Barraud, E. M. (1961). The development of behaviour in some young passerines. *Bird Study* **8**, 111–118.
Bartholomew, G. A., and Dawson, W. R. (1954). Temperature regulation in young pelicans, herons, and gulls. *Ecology* **35**, 466–472.
Bateson, P. P. G. (1964). Relation between conspicuousness of stimuli and their effectiveness in the imprinting situation. *J. Comp. Physiol. Psychol.* **58**, 407–411.
Bateson, P. P. G. (1966). The characteristics and context of imprinting. *Biol. Rev. (Cambridge Philos. Soc.)* **41**, 177–220.
Bateson, P. P. G. (1971). Imprinting. *In* "Ontogeny of Vertebrate Behavior" (H. Moltz, ed.), pp. 369–378. Academic Press, New York.
Bateson, P. P. G. (1976). Specificity and the origins of behavior. *In* "Advances in the Study of Behavior" (J. Rosenblatt, R. A. Hinde, and C. Beer, eds.), Vol. 6, pp. 1–20. Academic Press, New York.
Bateson, P. P. G. (1978). Sexual imprinting and optimal outbreeding. *Nature (London)* **273**, 659–660.
Bateson, P. P. G. (1979). How do sensitive periods arise and what are they for? *Anim. Behav.* **27**, 470–486.
Bateson, P. P. G., and Jaeckel, J. B. (1976). Chicks' preferences for familiar and novel conspicuous objects after different periods of exposure. *Anim. Behav.* **24**, 386–390.
Bateson, P. P. G., and Reese, E. P. (1969). The reinforcing properties of conspicuous stimuli in the imprinting situation. *Anim. Behav.* **17**, 692–699.
Bateson, P. P. G., and Seaburne-May, G. (1973). Effects of prior exposure to light on chicks' behaviour in the imprinting situation. *Anim. Behav.* **21**, 720–725.
Bateson, P. P. G., and Wainwright, A. A. P. (1972). The effects of prior exposure to light on the imprinting process in domestic chicks. *Behaviour* **42**, 274–290.
Bateson, P. P. G., Horn, G., and Rose, S. P. R. (1972). Effects of early experience on regional incorporation of precursors into RNA and protein in the chick brain. *Brain Res.* **39**, 449–465.
Beach, F. (1945). Current concepts of play in animals. *Am. Nat.* **79**, 523–541.
Beaver, P. W., Shrout, P. E., and Hess, E. (1976). Relative effectiveness of an inanimate stimulus and a live surrogate during imprinting in Japanese quail, *Coturnix coturnix japonica*. *Anim. Learn. Behav.* **4**, 193–196.
Beck, B. (1976). Predatory shell dropping by Herring Gulls. Paper presented at Annual Meeting of Animal Behavior Society, Boulder, Colorado.
Beecher, M. D., Beecher, I. M., and Hahn, S. (1981). Parent-offspring recognition in Bank Swallows (*Riparia riparia*). II. Development and acoustic basis. *Anim. Behav.* **29**, 95–101.
Beer, C. G. (1969). Laughing Gull chicks: recognition of their parents' voices. *Science (Washington, D.C.)* **166**, 1030–1032.
Beer, C. G. (1970a). On the responses of Laughing Gull chicks (*Larus atricilla*) to the calls of adults. I. Recognition of the voices of the parents. *Anim. Behav.* **18**, 652–660.
Beer, C. G. (1970b). On the responses of Laughing Gull chicks (*Larus atricilla*) to the calls of adults. II. Age changes and responses to different types of call. *Anim. Behav.* **18**, 661–677.

Beer, C. G. (1972). Individual recognition of voice and its development in birds. *Proc. Int. Ornithol. Congr. 15th*, pp 339–356.
Beer, C. G. (1979). Vocal communication between Laughing Gull parents and chicks. *Behaviour* **70**, 118–146.
Berthold, P., and Querner, U. (1981). Genetic basis of migratory behavior in European warblers. *Science (Washington, D.C.)* **212**, 77–79.
Bertram, B. (1970). The vocal behaviour of the Indian Hill Mynah, Gracula religiosa. *Anim. Behav. Monogr.* **3**, 81–192.
Birchard, G. F., and Kilgore, D. L., Jr. (1980). Ontogeny of oxygen consumption by embryos of two species of swallows (Hirundinidae). *Condor* **82**, 402–405.
Birkhead, T. R. (1974). Predation by birds on social wasps. *Br. Birds* **67**, 221–229.
Bischof, H.-J. (1979). A model of imprinting evolved from neurophysiological concepts. *Z. Tierpsychol.* **51**, 126–139.
Boag, D. A., and Alway, J. H. (1980). Effect of social environment within the brood on dominance rank in gallinaceous birds (Tetraonidae and Phasianidae). *Can. J. Zool.* **58**, 44–49.
Brooks, W. S. (1978). Avian prehatching behavior: functional aspects of the tucking pattern. *Condor* **80**, 442–444.
Brosset, A. (1971). "L'imprinting" chez les Colombidés—étude des modifications comportementales au cours de vieillisement. *Z. Tierpsychol.* **29**, 279–300.
Brosset, A. (1973). Étude comparative de l'ontogenèse des comportements chez les rapaces Accipitridés et Falconidés. *Z. Tierpsychol.* **32**, 386–417.
Brown, J. L. (1975). "The Evolution of Behavior." Norton, New York.
Brown, R. T. (1975). Following and visual imprinting in ducklings across a wide age range. *Dev. Psychobiol.* **8**, 27–33.
Burghardt, G., and Bekoff, M. (1978). "The Development of Behavior." Garland, New York.
Bursian, A. V. (1964). The influence of light on the spontaneous movements of chick embryos. *Bull. Exp. Biol. Med. (Engl. Transl. of Byull. Eksp. Biol. Med.)* **58**, 7–11.
Busbee, E. L. (1976). The ontogeny of cricket killing and mouse killing in Loggerhead Shrikes (Lanius ludovicianus L.). *Condor* **78**, 357–365.
Cade, T. J. (1953). Behavior of young Gyrfalcon. *Wilson Bull.* **65**, 26–31.
Cherfas, J. J. (1977). Prior exposure to light improves avoidance learning in day-old chicks. *Anim. Behav.* **25**, 732–735.
Cherfas, J. J. (1978). Simultaneous colour discrimination in chicks is improved by brief exposure to light. *Anim. Behav.* **26**, 1–5.
Clark, G. A., Jr. (1960). Notes on the embryology and evolution of the megapodes (Aves: Galliformes). *Postilla (Yale Peabody Mus.)* **45**, 1–7.
Clark, G. A. Jr. (1964). Ontogeny and evolution in the megapodes (Aves: Galliformes). *Postilla (Yale Peabody Mus.)* **78**, 1–37.
Collias, N. E. (1952). The development of social behavior in birds. *Auk* **69**, 127–159.
Cooch, F. G., and Beardmore, J. A. (1959). Assortative mating and reciprocal differences in the Blue-Snow Goose complex. *Nature (London)* **183**, 1833–1834.
Cooke, F., and Cooch, F. G. (1968). The genetics of polymorphism in the goose Anser caerulescens. *Evolution* **22**, 289–300.
Coppinger, R. P. (1970). The effect of experience and novelty on avian feeding behavior with reference to the evolution of warning coloration in butterflies. II. Reactions of naive birds to novel insects. *Am. Nat.* **104**, 323–335.
Cowan, P. J., and Evans, R. M. (1974). Calls of different individual hens and the parental control of feeding behavior in young Gallus gallus. *J. Exp. Zool.* **188**, 353–360.
Cruze, W. W. (1935). Maturation and learning in chicks. *J. Comp. Psychol.* **19**, 371–409.

Cullen, J. M. (1962). The pecking response of young Wideawake Terns, *Sterna fuscata*. *Ibis* **103**, 162–170.
Curio, E., Ernst, U., and Vieth, W. (1978). The adaptive significance of avian mobbing. II. Cultural transmission of enemy recognition in Blackbirds: effectiveness and some constraints. *Z. Tierpsychol.* **48**, 184–202.
Davies, N. B., and Green, R. E. (1976). The development and ecological significance of feeding techniques in the Reed Warbler (*Acrocephalus scirpaceus*). *Anim. Behav.* **24**, 213–229.
Dawson, W. R., and Bennett, A. F. (1981). Field and laboratory studies of the thermal relations of hatchling Western Gulls. *Physiol. Zool.* **54**, 155–164.
Decker, J. D., and Hamburger, V. (1967). The influence of different brain regions on periodic motility of the chick embryo. *J. Exp. Zool.* **165**, 371–384.
Delius, J. D. (1973). Agonistic behaviour of juvenile gulls, a neuroethological study. *Anim. Behav.* **21**, 236–246.
Diamond, A. W., and Smith, R. W. (1973). Returns and survival of banded warblers wintering in Jamaica. *Bird-Banding* **44**, 221–224.
Diamond, J. M. (1972). Further examples of dual singing by southwest Pacific birds. *Auk* **89**, 180–183.
Dietrich, K. (1980). Vorbildwahl in der Gesangentwicklung beim Japanischen Möwchen (*Lonchura striata* var. *domestica* Estrildidae). *Z. Tierpsychol.* **52**, 57–76.
Dilger, W. C. (1960). The comparative ethology of the African parrot genus *Agapornis*. *Z. Tierpsychol.* **17**, 649–685.
Dilger, W. C. (1962). The behavior of lovebirds. *Sci. Am.* **206**, 88–98.
Dittus, W. P. J., and Lemon, R. E. (1970). Auditory feedback in the singing of Cardinals. *Ibis* **112**, 544–548.
Dooling, R., and Searcy, M. (1980). Early perceptual selectivity in the Swamp Sparrow. *Dev. Psychobiol.* **13**, 499–506.
Drost, R. (1958). Über die Ansiedlung von Jung ins Binnenland verfrachteten Silbermöwen (*Larus argentatus*). *Vogelwarte* **19**, 169–173.
Ely, C. A. (1973). Returns of North American birds to their wintering grounds in southern Mexico. *Bird-Banding* **44**, 228–229.
Emlen, S. T. (1969). The development of migratory orientation in young Indigo Buntings. *Living Bird* **8**, 113–126.
Emlen, S. T. (1970). Celestial rotation: its importance in the development of migratory orientation. *Science (Washington, D.C.)* **170**, 1198–1201.
Emlen, S.T (1972). The ontogenic development of orientation capabilities. *NASA Spec. Publ.* **SP-262**, 191–210.
Emlen, S. T. (1975a). Migration: orientation and navigation. *In* "Avian Biology" (D. S. Farner, and J. R. King, eds.), Vol. 5, pp. 129–219. Academic Press, New York.
Emlen, S. T. (1975b). The stellar-orientation system of a migratory bird. *Sci. Am.* Aug., 102–111.
Emlen, S. T., and Oring, L. W. (1977). Ecology, sexual selection, and the evolution of mating systems. *Science (Washington, D.C.)* **197**, 215–223.
Evans, R. M. (1973). Differential responsiveness of young Ring-billed Gulls and Herring Gulls to adult vocalizations of their own and other species. *Can. J. Zool.* **51**, 759–770.
Evans, R. M. (1977). Auditory discrimination-learning in young Ring-billed Gulls (*Larus delawarensis*). *Anim. Behav.* **25**, 140–146.
Evans, R. M. (1979). Responsiveness of young Herring Gulls to stimuli from their own and other species: effects of training with food. *Can. J. Zool.* **57**, 1452–1457.
Evans, R. M. (1980). Development of individual call recognition in young Ring-billed Gulls (*Larus delawarensis*): an effect of feeding. *Anim. Behav.* **28**, 60–67.

Fabricius, E., and Boyd, H. (1954). Experiments on the following-reaction of ducklings. *Rep. Wildfowl Trust* **1952–3**, 84–89.
Fagen, R. (1981). "Animal Play Behavior." Oxford Univ. Press, New York.
Ficken, M. S. (1977). Avian play. *Auk* **94**, 573–582.
Fischer, G. J. (1970). Arousal and impairment: temperature effects on following during imprinting. *J. Comp. Physiol. Psychol.* **73**, 412–420.
Fischer, G. J., Morris, G. L., and Ruhsam, J. P. (1975). Color pecking preferences in white leghorn chicks. *J. Comp. Physiol. Psychol.* **88**, 402–406.
Foelix, R. F., and Oppenheim, R. W. (1973). Synaptogenesis in the avian embryo: ultrastructure and possible behavioral correlates. *In* "Studies in Development of Behavior and the Nervous System" (G. Gottlieb, ed.), Vol. 1, pp. 103–139. Academic Press, New York.
Friedmann, H. (1968). The evolutionary history of the avian genus *Chrysococcyx*. *Bull. U.S. Natl. Mus.*, No. 265.
Fry, C. H. (1969). The recognition and treatment of venomous and non-venomous insects by small bee-eaters. *Ibis* **111**, 23–29.
Gallagher, J. E. (1976). Sexual imprinting: effects of various regimens of social experience on mate preference in Japanese quail *Coturnix coturnix japonica*. *Behaviour* **57**, 91–114.
Gallagher, J. E. (1977). Sexual imprinting: a sensitive period in Japanese quail (*Coturnix coturnix japonica*). *J. Comp. Physiol. Psychol.* **91**, 72–78.
Gibson-Hill, C. A. (1947). Notes on the birds of Christmas Island. *Bull. Raffles Mus. Singapore* **18**, 87–165.
Gittleman, J. L., Harvey, P. H., and Greenwood, P. J. (1980). The evolution of conspicuous coloration: some experiments in bad taste. *Anim. Behav.* **28**, 897–899.
Gochfeld, M. (1980). Learning to eat by young Common Terns: consistency of presentation as an early cue. *Proc. Colonial Waterbird Group* **3**, 108–118.
Gottlieb, G. (1961). Developmental age as a baseline for determination of the critical period in imprinting. *J. Comp. Physiol. Psychol.* **54**, 422–427.
Gottlieb, G. (1965a). Imprinting in relation to parental and species identification by avian neonates. *J. Comp. Physiol. Psychol.* **59**, 345–356.
Gottlieb, G. (1965b). Prenatal auditory sensitivity in chickens and ducks. *Science (Washington, D.C.)* **147**, 1596–1598.
Gottlieb, G. (1966). Species identification by avian neonates: contributory effect of perinatal auditory stimulation. *Anim. Behav.* **14**, 282–290.
Gottlieb, G. (1968a). Prenatal behavior of birds. *Q. Rev. Biol.* **43**, 148–174.
Gottlieb, G. (1968b). Prenatal development of vocal abilities in birds. *In* "Animal Behavior in Laboratory and Field" (A. W. Stokes, ed.), pp. 157–160. Freeman, San Francisco.
Gottlieb, G. (1971a). Ontogenesis of sensory function in birds and mammals. *In* "The Biopsychology of Development" (E. Tobach, L. R. Aronson, and E. Shaw, eds.), pp. 67–128. Academic Press, New York.
Gottlieb, G. (1971b). "Development of Species Recognition in birds." Univ. of Chicago Press, Chicago, Illinois.
Gottlieb, G. (1974). On the acoustic basis of species identification in Wood Ducklings (*Aix sponsa*). *J. Comp. Physiol. Psychol.* **87**, 1038–1048.
Gottlieb, G. (1975a). Development of species identification in ducklings. I. Nature of perceptual deficit caused by embryonic deprivation. *J. Comp. Physiol. Psychol.* **89**, 387–399.
Gottlieb, G. (1975b). Development of species identification in ducklings. II. Experiential prevention of perceptual deficit caused by embryonic auditory deprivation. *J. Comp. Physiol. Psychol.* **89**, 675–684.
Gottlieb, G. (1976a). Conceptions of prenatal development: behavioral embryology. *Psychol. Rev.* **83**, 215–234.

Gottlieb, G. (1976b). The roles of experience in the development of behavior and the nervous system. In "Neural and Behavioral Specificity" (G. Gottlieb, ed.), Vol. 3, pp. 25–54. Academic Press, New York.

Gottlieb, G. (1978). Development of species identification in ducklings. IV. Change in species-specific perception caused by auditory deprivation. *J. Comp. Physiol. Psychol.* **92**, 375–387.

Gottlieb, G. (1979). Development of species identification in ducklings. V. Perceptual differentiation in the embryo. *J. Comp. Physiol. Psychol.* **93**, 831–845.

Gottlieb, G. (1980a). Development of species identification in ducklings. VI. Specific embryonic experience required to maintain species-typical perception in Peking ducklings. *J. Comp. Physiol. Psychol.* **94**, 579–587.

Gottlieb, G. (1980b). Development of species identification in ducklings. VII. Highly specific early experience fosters species-specific perception in Wood Ducklings. *J. Comp. Physiol. Psychol.* **94**, 1019–1027.

Gottlieb, G., and Kuo, Z.-Y. (1965). Development of behavior in the duck embryo. *J. Comp. Physiol. Psychol.* **59**, 183–188.

Graves, H. B., and Siegel, P. B. (1968). Prior experience and the approach response in domestic chicks. *Anim. Behav.* **16**, 18–23.

Gwinner, E. (1966). Über einige Bewegungsspiele des Kolkraken (*Corvus corax* L.). *Z. Tierpsychol.* **23**, 28–36.

Gwinner, E. (1972). Endogenous timing factors in bird migration. *NASA Spec. Publ.* **SP-262**, 321–338.

Gwinner, E. (1974). Endogenous temporal control of migratory restlessness in warblers. *Naturwissenschaften* **61**, 405.

Gwinner, E. (1977). Circannual rhythms in bird migration. *Annu. Rev. Ecol. Syst.* **8**, 381–405.

Hailman, J. P. (1967). The ontogeny of an instinct. *Behaviour (Suppl.)* **15**, 1–159.

Hailman, J. P. (1969). How an instinct is learned. *Sci. Am.* **221**, 98–106.

Hale, C., and Green, L. (1979). Effect of initial pecking consequences on subsequent pecking in young chicks. *J. Comp. Physiol. Psychol.* **93**, 730–735.

Hamburger, V. (1963). Some aspects of the embryology of behavior. *Q. Rev. Biol.* **38**, 342–365.

Hamburger, V. (1973). Anatomical and physiological basis of embryonic motility in birds and mammals. In "Studies on the Development of Behavior and the Nervous System" (G. Gottlieb, ed.), Vol. 1, pp. 52–76. Academic Press, New York.

Hamburger, V., and Narayanan, C. H. (1969). Effects of the deafferentiation of the trigeminal area on the motility of the chick embryo. *J. Exp. Zool.* **170**, 411–426.

Hamburger, V., and Oppenheim, R, (1967). Prehatching motility and hatching behavior in the chick. *J. Exp. Zool.* **166**, 171–204.

Hamburger, V., Balaban, M., Oppenheim, R., and Wenger, E. (1965). Periodic motility of normal and spinal chick embryos between 8 and 17 days of incubation. *J. Exp. Zool.* **159**, 1–14.

Hamburger, V., Wenger, E., and Oppenheim, R. (1966). Motility in the absence of sensory input. *J. Exp. Zool.* **162**, 133–160.

Harris, M. P. (1970). Abnormal migration and hybridization of *Larus argentatus* and *L. fuscus* after interspecific fostering experiments. *Ibis* **112**, 488–498.

Heaton, M. B., and Harth, M. S. (1974a). Non-visual light responsiveness in the pigeon: developmental and comparative considerations. *J. Exp. Zool.* **188**, 251–264.

Heaton, M. B., and Harth, M. S. (1974b). Developing visual function in the pigeon embryo with comparative reference to other avian species. *J. Comp. Physiol. Psychol.* **86**, 151–156.

Hess, E. H. (1956). Natural preferences of chicks and ducklings for objects of different colours. *Psychol. Rep.* **2**, 477–483.

Hess, E. H. (1959). Imprinting. *Science (Washington, D.C.)* **130,** 133–141.
Hess, E. H. (1973). "Imprinting. Early Experience and the Developmental Psychobiology of Attachment." Van Nostrand, New York.
Hess, E. H., and Petrovich, S. B. (1977). "Imprinting." Dowden, Hutchinson & Ross, Stroudsburg, Pennsylvania.
Hinde, R. A. (1968). Dichotomies in the study of development. *In* "Genetic and Environmental Influences on Behaviour" (J. M. Thoday, and A. S. Parkes, eds.). Oliver & Boyd, Edinburgh.
Hinde, R. A. (1970). "Animal Behaviour: A Synthesis of Ethology and Comparative Psychology." McGraw-Hill, New York.
Hinde, R. A., and Tinbergen, N. (1958). The comparative study of species-specific behavior. *In* "Behavior and Evolution" (A. Roe, and G. G. Simpson, eds.), pp. 251–268. Yale Univ. Press, New Haven, Connecticut.
Hinde, R. A., Thorpe, W. H., and Vince, M. A. (1956). The following response of young Coots and Moorhens. *Behaviour* **9,** 214–242.
Hindman, J. L. (1981). Attachment in chicks: effects of companion species on social preferences. *Dev. Psychobiol.* **14,** 13–18.
Hoffmeister, D. F., and Setzer, H. W. (1947). The postnatal development of two broods of Great Horned Owls (*Bubo virginianus*). *Univ. Kans. Publ. Nat. Hist.* **1,** 157–173.
Hogan, J. A. (1973a). The development of food recognition in young chicks. I. Maturation and nutrition. *J. Comp. Physiol. Psychol.* **83,** 355–366.
Hogan, J. A. (1973b). The development of food recognition in young chicks. II. Learned associations over long delays. *J. Comp. Physiol. Psychol.* **83,** 367–373.
Hogan, J. A. (1975). Development of food recognition in young chicks. III. Discrimination. *J. Comp. Physiol. Psychol.* **89,** 95–104.
Hogan-Warburg, A. J., and Hogan, J. A. (1981). Feeding strategies in the development of food recognition in young chicks. *Anim. Behav.* **29,** 143–154.
Hörlyk, N.-O., and Lind, H. (1978). Pecking response of artificially hatched Oystercatcher *Haematopus ostralegus* young. *Ornis Scand.* **9,** 138–145.
Immelmann, K. (1969). Song development in the Zebra Finch and other estrildid finches. *In* "Bird Vocalizations" (R. A. Hinde, ed.), pp. 61–74. Cambridge Univ. Press, London.
Immelmann, K. (1972). Sexual and other long-term aspects of imprinting in birds and other species. *Adv. Study Behav.* **4,** 147–174.
Immelmann, K. (1975). Ecological significance of imprinting and early learning. *Annu. Rev. Ecol. Syst.* **6,** 15–37.
Impekoven, M. (1976). Prenatal parent-young interactions in birds and their long-term effects *In* "Advances in the Study of Behavior" (J. Rosenblatt, R. A. Hinde, and C. Beer, eds.), Vol. 7, pp. 201–253. Academic Press, New York.
Ingold, P. (1973). Zur lautlichen Beziehung des Elters zu seinem Küken bei Tordalken (*Alca torda*). *Behaviour* **45,** 155–190.
Jenkins, P. F. (1978). Cultural transmission of song patterns and dialect development in a free-living bird population. *Anim. Behav.* **26,** 50–78.
Jones, F. M. (1932). Insect coloration and the relative acceptability of insects to birds. *Symp. R. Entomol. Soc. London* **80,** 345–386.
Jones, F. M. (1934). Further experiments on coloration and relative acceptability of insects to birds. *Symp. R. Entomol. Soc. London* **82,** 443–453.
Kear, J. (1962). Food selection in finches with special reference to interspecific differences. *Proc. Zool. Soc. London* **138,** 163–204.
Kear, J. (1964). Colour preference in young Anatidae. *Ibis* **106,** 361–369.

Kear, J. (1966). The pecking response of young Coots *Fulica atra* and Moorhens *Gallinula chloropus*. *Ibis* **108**, 118–122.

Keeton, W. T. (1969). Orientation by pigeons. Is the sun necessary? *Science (Washington, D.C.)* **165**, 922–928.

Keeton, W. T. (1971). Magnets interfere with pigeon homing. *Proc. Natl. Acad. Sci. USA* **68**, 102–106.

Keeton, W. T. (1979). Avian orientation and navigation: a brief overview. *Br. Birds* **72**, 451–470.

Keeton, W. T., and Gobert, A. (1970). Orientation by untrained pigeons requires the sun. *Proc. Natl. Acad. Sci. USA* **65**, 853–856.

King, A. P., West, M. J., and Eastzer, D. H. (1980). Song structure and song development as potential contributors to reproductive isolation in cowbirds (*Molothrus ater*). *J. Comp. Physiol. Psychol.* **94**, 1028–1039.

Klint, T. (1975). Sexual imprinting in the context of species recognition in female Mallards. *Z. Tierpsychol.* **38**, 385–392.

Klopfer, P. H. (1963). Behavioral aspects of habitat selection: the role of early experience. *Wilson Bull.* **75**, 15–22.

Klopfer, P. H., and Hailman, J. P. (1964). Basic parameters of following and imprinting in precocial birds. *Z. Tierpsychol.* **21**, 755–762.

Konishi, M. (1963). The role of auditory feedback in the vocal behavior of the domestic fowl. *Z. Tierpsychol.* **20**, 349–367.

Konishi, M. (1965). The role of auditory feedback in the control of vocalization in the White-crowned Sparrow. *Z. Tierpsychol.* **22**, 770–783.

Konishi, M. (1970). Comparative neurophysiological studies of hearing and vocalizations in songbirds. *Z. Vl. Physiol.* **66**, 247–272.

Konishi, M. (1973). Development of auditory neuronal responses in avian embryos. *Proc. Natl. Acad. Sci. USA* **70**, 1795–1798.

Konishi, M., and Nottebohm, F. (1969). Experimental studies in the ontogeny of avian vocalizations. *In* "Bird Vocalizations" (R. A. Hinde, ed.), pp. 29–48. Cambridge Univ. Press, London.

Kovach, J. K. (1970a). Critical period or optimal arousal? Early approach behavior as a function of stimulus, age, and breed variables in chicks. *Dev. Psychobiol.* **3**, 73–77.

Kovach, J. K. (1970b). Development and mechanisms of behavior in the chick embryo during the last five days of incubation. *J. Comp. Physiol. Psychol.* **73**, 392–406.

Kovach, J. K. (1971). Effectiveness of different colours in the elicitation and development of approach behaviour in chicks. *Behaviour* **38**, 154–168.

Kovach, J. K. (1978). Color preferences in quail chicks: generalizations of the effects of genetic selection. *Behaviour* **65**, 263–269.

Kovach, J. K. (1980). Mendelian units of inheritance control color preferences in quail chicks (*Coturnix coturnix japonica*). *Science (Washington, D.C.)* **207**, 549–551.

Kroodsma, D. E. (1977a). Correlates of song organization among North American wrens. *Am. Nat.* **111**, 995–1008.

Kroodsma, D. E. (1977b). A reevaluation of song development in the Song Sparrow. *Anim. Behav.* **25**, 390–399.

Kroodsma, D. E. (1978). Aspects of learning in the ontogeny of bird song: where, from whom, when, how many, which, and how accurately? *In* "Ontogeny of Behavior" (G. Burghardt, and M. Bekoff, eds.), pp. 215–230. Garland, New York.

Kroodsma, D. E., and Pickert, R. (1981). Environmentally dependent sensitive periods for avian vocal learning. *Nature (London)* **288**, 477–479.

Kruijt, J. (1964). Ontogeny of social behaviour in Burmese red junglefowl. *Behaviour (Suppl.)* **12**, 1–201.
Kuo, Z.-Y. (1932a). Ontogeny of embryonic behavior in Aves. I. The chronology and general nature of the behavior of the chick embryo. *J. Exp. Zool.* **61**, 395–430.
Kuo, Z.-Y. (1932b). Ontogeny of embryonic behavior in Aves. II. The mechanical factors in the various stages leading to hatching. *J. Exp. Zool.* **62**, 453–487.
Kuo, Z.-Y. (1932c). Ontogeny of embryonic behavior in Aves. III. The structural and environmental factors in embryonic behavior. *J. Comp. Psychol.* **13**, 245–271.
Kuo, Z.-Y. (1932d). Ontogeny of embryonic behavior in Aves. IV. The influence of embryonic movements upon the behavior after hatching. *J. Comp. Psychol.* **14**, 109–122.
Kuo, Z.-Y. (1932e). Ontogeny of embryonic behavior in Aves. V. The reflex concept in the light of embryonic behavior in birds. *Psychol. Rev.* **39**, 499–515.
Kuo, Z.-Y. (1939). Studies in the physiology of the embryonic nervous system. I. Effect of curare on motor activity of the chick embryo. *J. Exp. Zool.* **68**, 371–386.
Kuo, Z.-Y. (1967). "The Dynamics of Behavior Development." Random House, New York.
Lack, D. (1963). Cuckoo hosts in England. *Bird Study* **10**, 185–201.
Lack, D. (1968). "Ecological Adaptations for Breeding in Birds." Methuen, London.
Lade, B. I., and Thorpe, W. H. (1964). Dove songs as innately coded patterns of specific behaviour. *Nature (London)* **202**, 366–368.
Landsberg, J.-W. (1976). Posthatch age and developmental age as a baseline for determination of the sensitive period for imprinting. *J. Comp. Physiol. Psychol.* **90**, 47–52.
Landsberg, J.-W., and Weiss, J. (1976). Stress and increase of the corticosterone level prevent imprinting in ducklings. *Behaviour* **57**, 173–189.
Lanyon, W. E. (1957). The comparative biology of the meadowlarks (*Sturnella*) in Wisconsin. *Publ. Nuttall Ornithol. Club* **1**, 1–67.
Lanyon, W. E. (1960). The ontogeny of vocalization in birds. *In* "Animal Sounds and Communication" (W. E. Lanyon, and W. N. Tavolga, eds.), pp. 321–347. Amer. Inst. of Biol. Sci., Washington, D.C.
Lanyon, W. E. (1979). Development of song in the Wood Thrush (*Hylocichla mustelina*), with notes on a technique for hand–rearing passerines from the egg. *Am. Mus. Novitates*, No. 2666.
Lehrman, D. S. (1953). A critique of Konrad Lorenz's theory of instinctive behavior. *Q. Rev. Biol.* **28**, 337–363.
Lemon, R. E. (1975). How birds develop song dialects. *Condor* **77**, 385–406.
Ligon, J. D., and Martin, D. J. (1974). Piñon seed assessment by the Piñon Jay, *Gymnorhinus cyanocephalus*. *Anim. Behav.* **22**, 421–429.
Lind, H. (1961). Studies on the behavior of the Black-tailed Godwit (*Limosa limosa*). *Medd. Naturfredningsradets Reservatudvalg (Copenhagen)* **66**, 1–157.
Liversidge, R. (1970). The biology of the Jacobine Cuckoo (*Clamator jacobinus*). *Ostrich (Suppl.)* **8**, 117–137.
Löhrl, H. (1959). Zur Frage des Zeitpunktes einer Prägung auf die Heimatregion beim Halsbandschnäpper (*Ficedula albicollis*). *J. Ornithol.* **100**, 132–140.
Löhrl, H. (1975). Droh- und Schreckverhalten nestjunger oder flügger Vögel. *Vogelwelt* **96**, 64–68.
Lorenz, K. (1935). Der Kumpan in der Umwelt des Vogels. *J. Ornithol.* **83**, 137–213, 389–413.
Lorenz, K. (1961). Phylogenetische Anpassung und adaptive Modifikation des Verhaltens. *Z. Tierpsychol.* **18**, 139–187.
Lorenz, K., and von St. Paul, U. (1968). Die Entwicklung des Spiessens und Klemmens bei den drei Würtgerarten *Lanius collurio, L. senator,* und *L. excubitor*. *J. Ornithol.* **109**, 137–156.

Marler, P. (1956). The voice of the Chaffinch and its function as a language. *Ibis* **98**, 231–261.
Marler, P. (1967). Comparative study of song development in sparrows. *Proc. Int. Ornithol. Congr. 14th*, pp. 231–244.
Marler, P. (1970). A comparative approach to vocal learning: song development in White-crowned Sparrows. *J. Comp. Physiol. Psychol. Monogr.* **71**, 1–25.
Marler, P. (1975). On strategies of behavioural development. *In* "Function and Evolution in Behaviour" (G. P. Baerends, C. Beer, and A. Manning, eds.), pp. 254–275. Methuen, London.
Marler, P., and Mundinger, P. (1971). Vocal learning in birds. *In* "The Ontogeny of Vertebrate Behavior" (H. Moltz, ed.), pp. 389–450. Academic Press, New York.
Marler, P., and Peters, S. (1977). Selective vocal learning in a sparrow. *Science (Washington, D.C.)* **198**, 519–521.
Marler, P., and Tamura, M. (1964). Culturally transmitted patterns of vocal behaviour in sparrows. *Science (Washington, D.C.)* **146**, 1483–1486.
Marler, P. and Waser, M. S. (1977). Role of auditory feedback in canary song development. *J. Comp. Physiol. Psychol.* **91**, 8–16.
Marler, P., Konishi, M., Lutjen, A., and Waser, M. S. (1973). Effects of continuous noise on avian hearing and vocal development. *Proc. Nat. Acad. Sci. USA* **70**, 1393–1396.
Martin, J. T. (1975). Hormonal influences in the evolution and onotogeny of imprinting behaviour in the duck. *In* "Progress in Brain Research" (W. H. Gispen, T. B. Van Wimersma Greidanus, B. Bohus, and D. de Wied, eds.), Vol. 42, pp. 357–366. Elsevier, Amsterdam.
Martin, J. T. (1978). Embryonic pituitary adrenal axis, behavior development and domestication in birds. *Am. Zool.* **18**, 489–499.
Martin, J. T., and Schutz, F. (1974). Arousal and temporal factors in imprinting in Mallards. *Dev. Psychobiol.* **7**, 69–78.
Messmer, E., and Messmer, I. (1956). Die Entwicklung der Lautäusserungen und einiger Verhaltensweisen der Amsel (*Turdus merula merula* L.) unter natürlichen Bedingungen und nach Einzelaufzucht in schalldichten Bäumen. *Z. Tierpsychol.* **13**, 341–441.
Metfessel, M. (1935). Roller canary song produced without learning from external sources. *Science (Washington, D.C.)* **81**, 470.
Moltz, H. (1963). Imprinting: an epigenetic approach. *Psychol. Rev.* **70**, 123–138.
Moltz, H., and Rosenblum, L. A. (1958). Imprinting and associative learning: the stability of the following response in Peking ducks (*Anas platyrhynchos*). *J. Comp. Physiol. Psychol.* **51**, 580–583.
Moore, F. R. (1975). Influence of solar and geomagnetic stimuli on the migratory orientation of Herring Gull chicks. *Auk* **92**, 655–661.
Moreau, R. E. (1938). A contribution to the biology of the Musophagiformes, the so-called plantain-eaters. *Ibis* **2**, 639–671.
Morrell, G. M., and Turner, J. R. G. (1970). Experiments on mimicry. I. The response of wild birds to artificial prey. *Behaviour* **36**, 116–130.
Moynihan, M. (1962). Hostile and sexual behavior patterns of South American and Pacific Laridae. *Behaviour (Suppl.)* **8**, 1–365.
Mueller, H. C. (1974). The development of prey recognition and predatory behaviour in the American Kestrel *Falco sparverius*. *Behaviour* **49**, 313–324.
Mulligan, J. A. (1966). Singing behavior and its development in the Song Sparrow, *Melospiza melodia*. *Univ. Calif. Berkeley Publ. Zool.* **81**, 1–76.
Mundinger, P. C. (1970). Vocal imitation and individual recognition of finch calls. *Science (Washington, D.C.)* **168**, 480–482.
Nelson, J. B. (1966). The behaviour of the young Gannet. *Br. Birds* **59**, 393–419.

Nice, M. M. (1943). "Studies in the Life History of the Song Sparrow," Vol. II. Dover, New York.
Nice, M. M. (1962). Development of behavior in precocial birds. *Trans. Linn. Soc. N.Y.* **8**, 1–211.
Nicki, R. M., and Rogers, J. A. (1975). Approach and following behavior of 24-hour-old chicks as a function of stimulus complexity. *Anim. Behav.* **23**, 116–123.
Nicolai, J. (1959). Familientradition in der Gesangsentwicklung des Gimpels (*Pyrrhula pyrrhula* L.). *J. Ornithol.* **100**, 39–46.
Nottebohm, F. (1967). The role of sensory feedback in the development of avian vocalizations. *Proc. Int. Ornithol. Congr. 14th*, pp. 265–280.
Nottebohm, F. (1968). Auditory experience and song development in the Chaffinch *Fringilla coelebs*: ontogeny of a complex motor pattern. *Ibis* **110**, 549–568.
Nottebohm, F. (1969). The 'critical period' for song-learning in birds. *Ibis* **111**, 386–387.
Nottebohm, F. (1970). Ontogeny of bird song. *Science (Washington, D.C.)* **167**, 950–956.
Nottebohm, F. (1972a). The origins of vocal learning. *Am. Nat.* **106**, 116–140.
Nottebohm, F. (1972b). Neural lateralization of vocal control in a passerine bird. II. Subsong, calls, and a theory of vocal learning. *J. Exp. Zool.* **179**, 35–49.
Nottebohm, F. (1975). Vocal behavior in birds. *In* "Avian Biology" (D. S. Farner, and J. R. King, eds.), Vol. 5, pp. 287–332. Academic Press, New York.
Nottebohm, F. (1980). Brain pathways for vocal learning in birds: a review of the first 10 years. *In* "Progress in Psychobiology and Physiological Psychology" (J. M. Sprague, and A. N. Epstein, eds.), Vol. 9 pp. 85–124. Academic Press, New York.
Nottebohm, F., and Nottebohm, M. E. (1971). Vocalizations and breeding behaviour of surgically deafened Ring Doves (*Streptopelia risoria*). *Anim. Behav.* **19**, 311–327.
Nottebohm, F., Manning, E., and Nottebohm, M. E. (1979). Reversal of hypoglossal dominance in Canaries following unilateral syringeal denervation. *J. Comp. Physiol.* **134**, 227–240.
Nyström, M. (1973). Extinction, disinhibition and spontaneous recovery of the pecking response in young Herring Gulls. *Behaviour* **45**, 271–281.
Ockleford, E. M., and Vince, M. A. (1980). Effects of thyroxine on prehatching developmental rate and behavior in the chick. *J. Comp. Physiol. Psychol.* **94**, 280–288.
Oppenheim, R. W. (1968a) Color preferences in the pecking response of newly hatched ducks (*Anas platyrhynchos*). *J. Comp. Physiol. Psychol. Monogr. (Suppl.)* **66**, 1–17.
Oppenheim, R. W. (1968b). Light responsivity in chick and duck embryos just prior to hatching. *Anim. Behav.* **16**, 276–280.
Oppenheim, R. W. (1970). Some aspects of embryonic behavior in the duck (*Anas platyrhynchos*). *Anim. Behav.* **18**, 335–352.
Oppenheim, R. W. (1972a). Prehatching and hatching behaviour in birds: a comparative study of altricial and precocial species. *Anim. Behav.* **20**, 644–655.
Oppenheim, R. W. (1972b). The embryology of behavior in birds: a critical review of the role of sensory stimulation in embryonic movement. *Proc. Int. Ornithol. Congr. 15th*, pp. 283–302.
Oppenheim, R. W. (1973). Prehatching and hatching behavior: a comparative and physiological consideration. *In* "Studies on the Development of Behavior and the Nervous System" (G. Gottlieb, ed.), Vol. 1, pp. 163–244. Academic Press, New York.
Oppenheim, R. W. (1974). The ontogeny of behavior in the chick embryo. *In* "Advances in the Study of Behavior" (D. S. Lehrman, J. S. Rosenblatt, R. A. Hinde, and E. Shaw, eds.), Vol. 5, pp. 133–171. Academic Press, New York.
Oppenheim, R. W., and Levin, H. L. (1975). Short-term changes in incubation temperature:

behavioral and physiological effects in the chick embryo from 6 to 20 days. *Dev. Psychobiol.* **8**, 103–115.
Orians, G. H. (1969a). Age and hunting success in the Brown Pelican (*Pelecanus occidentalis*). *Anim. Behav.* **17**, 316–319.
Orians, G. H. (1969b). On the evolution of mating systems in birds and mammals. *Am. Nat.* **103**, 589–603.
Parker, A. (1975). Young male peregrines passing vegetation fragments to each other. *Br. Birds* **68**, 242–243.
Paulson, G. W. (1965). Maturation of evoked responses in the duckling. *Exp. Neurol.* **11**, 324–333.
Payne, R. B. (1973). Behavior, mimetic songs and song dialects, and relationships of the parasitic Indigobirds (*Vidua*) of Africa. *Ornithol. Monogr.* **11**, 1–333.
Payne, R. B. (1977). The ecology of brood parasitism in birds. *Annu. Rev. Ecol. Syst.* **8**, 1–28.
Payne, R. B., and Payne, K. (1977). Social organization and mating success in local song populations of Village Indigobirds, *Vidua chalybeata*. *Z. Tierpsychol.* **45**, 113–173.
Peters, S. S., Searcy, W. A., and Marler, P. (1980). Species song discrimination in choice experiments with territorial male Swamp and Song Sparrows. *Anim. Behav.* **28**, 393–404.
Poulsen, H. (1953). A study of incubation responses and some other behaviour patterns in birds. *Vidensk. Medd. Dansk Naturh. Foren.* **115**, 1–131.
Poulson, H. (1959). Song learning in the Domestic Canary. *Z. Tierpsychol.* **16**, 173–178.
Price, P. H. (1979). Developmental determinants of structure in Zebra Finch song. *J. Comp. Physiol. Psychol.* **93**, 260–277.
Provine, R. R. (1972). Ontogeny of bioelectric activity in the spinal cord of the chick embryo and its behavioral implications. *Brain Res.* **41**, 365–378.
Provine, R. R. (1973). Neurophysiological aspects of behavioral development in the chick embryo. *In* "Studies on the Development of Behavior and the Nervous System" (G. Gottlieb, ed.), Vol. 1, pp. 77–102. Academic Press, New York.
Provine, R. R. (1981). Development of wing-flapping and flight in normal and flap-deprived domestic chicks. *Dev. Psychobiol.* **14**, 279–291.
Quinney, T. E., and Smith, P. C. (1980). Comparative foraging behaviour and efficiency of adult and juvenile Great Blue Herons. *Can. J. Zool.* **58**, 1168–1173.
Rabinowitch, V. E. (1968). The role of experience in the development of food preferences in gull chicks. *Anim. Behav.* **16**, 425–428.
Rabinowitch, V. E. (1969). The role of experience in the development and retention of seed preferences in Zebra Finches. *Behaviour* **33**, 222–236.
Ralph, C. J., and Mewaldt, L. R. (1975). Timing of site fixation upon the wintering grounds in sparrows. *Auk* **92**, 698–705.
Recher, H. F., and Recher, J. A. (1969). Comparative foraging efficiency of adult and immature Little Blue Herons (*Florida caerulea*). *Anim. Behav.* **17**, 320–322.
Rice, J. O., and Thompson, W. L. (1968). Song development in the Indigo Bunting. *Anim. Behav.* **16**, 462–469.
Ricklefs, R. E. (1968). Patterns of growth in birds. *Ibis* **110**, 419–451.
Ricklefs, R. E. (1969). Preliminary models for growth rates of altricial birds. *Ecology* **50**, 1031–1039.
Ricklefs, R. E. (1973). Patterns of growth in birds. II. Growth rate and mode of development. *Ibis* **115**, 177–201.
Ricklefs, R. E. (1979). Patterns of growth in birds. V. A comparative study of development in the Starling, Common Tern, and Japanese Quail. *Auk* **96**, 10–30.
Ricklefs, R. E., and White, S. C. (1981). Growth and energetics of chicks of the Sooty Tern (*Sterna fuscata*) and Common Tern (*S. hirundo*). *Auk* **98**, 361–378.

Riesen, A. H. (1975). "The Developmental Neuropsychology of Sensory Deprivation." Academic Press, New York.
Ripley, K. L., and Provine, R. R. (1972). Neural correlates of embryonic motility in the chick. Brain Res. 45, 127–134.
Roberts, B. (1934). Notes on the birds of central and south-east Iceland with special reference to food habits. Ibis 13, 239–264.
Salzen, E. A., and Meyer, C. C. (1968). Reversibility of imprinting. J. Comp. Physiol. Psychol. 66, 269–275.
Sauer, E. G. F., and Sauer, E. M. (1960). Star navigation of nocturnal migrating birds. The 1958 planetarium experiments. Cold Spring Harbor Symp. Quant. Biol. 25, 463–473.
Sauer, F. (1956). Über das Verhalten junger Gartengrasmücken Sylvia borin. J. Ornithol. 97, 156–189.
Saunders, J. C. (1974). The development of auditory evoked responses in the chick embryo. Minerva Otorinolaringol. 24, 221–229.
Saunders, J. C., Coles, R. B., and Gates, G. R. (1973). The development of auditory evoked responses in the cochlea and cochlear nuclei of the chick. Brain Res. 63, 59–74.
Schaefer, H. H., and Hess, E. H. (1959). Color preferences in imprinting objects. Z. Tierpsychol. 16, 161–172.
Schleidt, W. M. (1964). Über die Spontaneität von Erbkoordinationen. Z. Tierpsychol. 21, 235–256.
Schneirla, T. C. (1952). A consideration of some conceptual trends in comparative psychology. Psych. Bull. 49, 559–597.
Schneirla, T. C. (1966). Behavioral development and comparative psychology. Q. Rev. Biol. 41, 283–302.
Schutz, F. (1965). Sexuelle Prägung bei Anatiden. Z. Tierpsychol. 22, 50–103.
Scoville, R., and Gottlieb, G. (1980). Development of vocal behaviour in Peking ducklings. Anim. Behav. 28, 1095–1109.
Sedláček, J. (1972). Development of the optic afferent system in chick embryos. Adv. Psychobiol. 1, 129–170.
Serventy, D. L. (1967). Aspects of the population ecology of the Short-tailed Shearwater, Puffinus tenuirostris. Proc. Int. Ornithol. Congr. 14th, pp. 165–190.
Sherry, D. F. (1981). Parental care and the development of thermoregulation in red junglefowl. Behaviour 76, 250–279.
Shettleworth, S. J. (1972). The role of novelty in learned avoidance of unpalatable 'prey' by domestic chicks (Gallus gallus). Anim. Behav. 20, 268–273.
Sluckin, W. (1964). "Imprinting and Early Learning." Methuen, London.
Smith, F. V. (1962). Perceptual aspects of imprinting. Symp. Zool. Soc. London 8, 171–191.
Smith, F. V., and Bird, M. W. (1964). The sustained approach of the domestic chick to coloured stimuli. Anim. Behav. 12, 60–63.
Smith, S. M. (1972). The ontogeny of impaling behaviour in the Loggerhead Shrike, Lanius ludovicianus L. Behaviour 42, 232–247.
Smith, S. M. (1973). A study of prey-attack behaviour in young Loggerhead Shrikes, Lanius ludovicianus L. Behaviour 44, 113–141.
Smith, S. M. (1974). Factors directing prey-attack by the young of three passerine species. Living Bird 12, 55–67.
Smith, S. M. (1975). Innate recognition of coral snake pattern by a possible avian predator. Science (Washington, D.C.) 187, 759–760.
Smith, S. M. (1976). Predatory behaviour of young Turquoise-browed Motmots, Eumomota superciliosa. Behaviour 61, 309–320.

Smith, S. M. (1977a). Coral snake recognition and stimulus generalisation by naive Great Kiskadees (Aves: Tyrannidae). *Nature (London)* **265**, 535–536.
Smith, S. M. (1977b). The behavior and vocalizations of young Turquoise-browed Motmots. *Biotropica* **9**, 127–130.
Smith, S. M. (1978). Predatory behaviour of young Great Kiskadees (*Pitangus sulphuratus*). *Anim. Behav.* **26**, 988–995.
Smith, S. M. (1980). Responses of naive temperate birds to warning coloration. *Am. Midl. Nat.* **103**, 346–352.
Smith, S. M., and Stiles, F. G. (1979). Banding studies of migrant shorebirds in northwestern Costa Rica. *Stud. Avian Biol.* **2**, 41–47.
Snapp, B. D. (1969). Recognition of maternal calls by parentally naive *Gallus gallus* chicks. *Anim. Behav.* **17**, 440–445.
Sonnemann, P., and Sjölander, S. (1977). Effects of cross-fostering on the sexual imprinting of the female Zebra Finch *Taeniopygia guttata*. *Z. Tierpsychol.* **45**, 337–348.
Southern, W. E. (1969). Orientation behavior of Ring-billed Gull chicks and fledglings. *Condor* **71**, 418–425.
Southern, W. E. (1972). Magnets disrupt the orientation of juvenile Ring-billed Gulls. *BioScience* **22**, 476–479.
Southern, W. E. (1976). Migrational orientation in Ring-billed Gull chicks. *Auk* **93**, 78–85.
Southern, W. E. (1978). Orientation responses of Ring-billed Gull Chicks: a reevaluation. *In* "Animal Migration, Navigation, and Homing" (K. Schmidt-Koenig, and W. T. Keeton, eds.), pp. 311–317. Springer-Verlag, New York.
Sparling, D. W. (1979). Evidence for vocal learning in prairie grouse. *Wilson Bull.* **91**, 618–621.
Stokes, A. W. (1971). Parental and courtship feeding in red jungle fowl. *Auk* **88**, 21–29.
Stonehouse, B., and Stonehouse, S. (1963). The frigate bird *Fregata aquila* of Ascension Island. *Ibis* **103B**, 409–422.
Stoner, E. A. (1947). Anna hummingbird at play. *Condor* **49**, 36.
Storey, A. E., and Shapiro, A. J. (1979). Development of preferences in white Peking ducklings for stimuli in the natural post-hatch environment. *Anim. Behav.* **27**, 411–416.
Strobel, M. G., and MacDonald, G. E. (1974). Induction of eating in newly hatched chicks. *J. Comp. Physiol. Psychol.* **86**, 493–502.
Sumner, E. L., Jr. (1929). Notes on the growth and behavior of young Golden Eagles. *Auk* **46**, 161–169.
Sumner, E. L., Jr. (1934). The behavior of some young raptorial birds. *Univ. Calif. Berkeley Publ. Zool* **40**, 331–361.
Thompson, A. L. (1964). "A New Dictionary of Birds." McGraw-Hill, New York.
Thorpe, W. H. (1958). The learning of song patterns by birds with especial reference to the song of the Chaffinch, *Fringilla coelebs*. *Ibis* **100**, 535–570.
Thorpe, W. H. (1966). Ritualization in ontogeny. I. Animal play. *Phil. Trans. R. Soc. London Ser. B* **451**, 311–319.
Thorpe, W. H., and North, M. E.W. (1965). Origin and significance of the power of vocal imitation: with especial reference to the antiphonal singing of birds. *Nature (London)* **208**, 219–222.
Tinbergen, N., and Kuenen, D. J. (1973). The releasing and directing stimulus situations of the gaping response in young Blackbirds and thrushes (*Turdus m. merula* L. and *T. e. ericetorum* Turton). *In* "The Animal in Its World" Vol. 2, pp. 17–51. Harvard Univ. Press, Cambridge, Massachusetts.
Tinbergen, N., and Perdeck, A. C. (1950). On the stimulus situation releasing the begging response in the newly hatched Herring Gull chick (*Larus argentatus* Pont.). *Behaviour* **3**, 1–39.

Tolman, C. W. (1967a). The feeding behaviour of domestic chicks as a function of rate of pecking by a surrogate companion. *Behaviour* **29**, 57–62.
Tolman, C. W. (1967b). The effect of tapping sounds on feeding behaviour of domestic chicks. *Anim. Behav.* **15**, 145–148.
Tolman, C. W. (1968). The role of the companion in social facilitation of animal behaviour. *In* "Social Facilitation and Imitative Behaviour" (E. C. Simmel, R. A. Hoppe, and G. A. Milton, eds.). Allyn & Bacon, Boston.
Trollope, J. (1971). Some aspects of behavior and reproduction in captive Barn Owls (*Tyto alba alba*). *Avic. Mag.* **77**, 117–125.
Turner, J. R. G. (1977). Butterfly mimicry: the genetical evolution of an adaptation. *Evol. Biol.* **10**, 163–206.
Vanzulli, A., and Garcia-Austt, E. (1963). Development of cochlear microphonic potentials in the chick embryo. *Acta Neurol. Latinoam.* **9**, 19–23.
Vauclair, J., and Bateson, P. P. G. (1975). Prior exposure to light and pecking accuracy in chicks. *Behaviour* **52**, 196–201.
Vidal, J. M. (1976). L'empreinte chez les animaux. *Recherche* **63**, 24–35.
Vieth, W., Curio, E., and Ernst, U. (1980). The adaptive significance of avian mobbing. III. Cultural transmission of enemy recognition in Blackbirds: cross-species tutoring and properties of learning. *Anim. Behav.* **28**, 1217–1229.
Vince, M. A. (1960). Developmental changes in responsiveness in the Great Tit (*Parus major*). *Behaviour* **15**, 219–243.
Vince, M. A. (1961a). Developmental changes in learning capacity. *In* "Current Problems in Animal Behaviour" (W. H. Thorpe, and O. L. Zangwill, eds.), pp. 225–247. Cambridge Univ. Press, Cambridge.
Vince, M. A. (1961b). "String-pulling" in birds. III. The successful response in Greenfinches and Canaries. *Behaviour* **17**, 103–129.
Vince, M. A. (1966). Artificial acceleration of hatching in quail embryos. *Anim. Behav.* **14**, 389–394.
Vince, M. A. (1969). Embryonic communication, respiration, and the synchronization of hatching. *In* "Bird Vocalizations" (R. A. Hinde, ed.), pp. 233–260. Cambridge Univ. Press, London.
Vince, M. A. (1974). Development of the avian embryo, Part 1: Behaviour. *In* "Development of the Avian Embryo, A Behavioural and Physiological Study" (B. M. Freeman, and M. A. Vince, eds.), pp. 3–116. Chapman & Hall, London.
Vince, M. A. (1977). Taste sensitivity in the embryo of the domestic fowl. *Anim. Behav.* **25**, 797–805.
Vince, M. A. (1979). Effects of accelerating stimulation on different indices of development in Japanese quail embryos. *J. Exp. Zool.* **208**, 201–220.
Vince, M. A., Reader, M., and Tolhurst, B. (1976). Effects of stimulation on embryonic activity in the chick. *J. Comp. Physiol. Psychol.* **90**, 221–230.
Visintini, F., and Levi-Montalcini, R. (1939). Relazione tra differenciazione strutturale dei centri delle vie nervose nell'embryone di pollo. *Schweiz. Arch. Neurol. Psychiatr.* **43**, 1–45.
Wada, M., Goto, J., Nishiyama, H., and Nobukuni, K. (1979). Colour exposure of incubating eggs and colour preference of chicks. *Anim. Behav.* **27**, 359–364.
Walter, M. J. (1973). Effects of parental colouration on the mate preference of offspring in the Zebra Finch, *Taeniopygia guttata castanotis* Gould. *Behaviour* **46**, 154–173.
Warriner, C. C., Lemmon, W. B., and Ray, T. S. (1963). Early experience as a variable in mate selection. *Anim. Behav.* **11**, 221–224.

Weidmann, R., and Weidmann, U. (1958). An analysis of the stimulus situation releasing food-begging in the Black-headed Gull. *Anim. Behav.* **6**, 114.

Weidmann, U. (1958). Verhaltensstudien an der Stockente (*Anas platyrhynchos* L.), II. *Z. Tierpsychol.* **15**, 277–300.

Weidmann, U. (1961). The stimuli eliciting begging in gulls and terns. *Anim. Behav.* **9**, 115–116.

Weidmann, U. (1965). "Colour preference" and pecking response in young Moorhens, *Gallinula chloropus*, and Coots, *Fulica atra*. *Ibis* **107**, 108–110.

Weir, D., and Picozzi, N. (1975). Aspects of social behaviour in the Buzzard. *Br. Birds* **68**, 125–141.

Weiss, J., Köhler, W., and Landsberg, J.-W. (1977). Increase of the corticosterone level in ducklings during the sensitive period of the following response. *Dev. Psychobiol.* **10**, 59–64.

Wemmer, C. (1969). Impaling behavior of the Loggerhead Shrike, *Lanius ludovicianus*. *Z. Tierpsychol.* **26**, 208–224.

West, M. J., King, A. P., and Eastzer, D. H. (1981). The cowbird: reflections on development from an unlikely source. *Am. Sci.* **69**, 56–66.

Wilkinson, R. (1980). Calls of nestling Chaffinches *Fringilla coelebs:* the use of two sound sources. *Z. Tierpsychol.* **54**, 346–356.

Wilson, G. F. (1968). Early experience and facilitation of feeding in domestic chicks. *J. Comp. Physiol. Psychol.* **66**, 800–802.

Wiltschko, W., and Gwinner, E. (1974). Evidence for an innate magnetic compass in Garden Warblers. *Naturwissenschaften* **61**, 406.

Wiltschko, W., Wiltschko, R., and Keeton, W. T. (1976). Effects of a 'permanent' clock-shift on the orientation of young homing pigeons. *Behav. Ecol. Sociobiol.* **1**, 229–243.

Woolf, N. K., Bixby, J. L., and Capranica, R. R. (1976). Prenatal experience and avian development: brief auditory stimulation accelerates the hatching of Japanese quail. *Science (Washington, D.C.)* **194**, 959–960.

Chapter 3

AVIAN ECOLOGICAL ENERGETICS

Glenn E. Walsberg

I.	Introduction	161
	A. Scope	161
	B. Limitations of the Data	162
	C. Special Characteristics of Birds Pertinent to Their Ecological Energetics.	163
II.	Energetics of Some Prominent Activities and Events in the Avian Life Cycle	164
	A. Introduction	164
	B. Energy Allocation to Resource Defense	165
	C. Gonadal Growth and Gametogenesis	167
	D. Incubation	178
	E. Care of the Young	184
	F. Molt	187
III.	Total Energy Expenditure of Free-Living Birds	191
	A. Body Mass and Energy Expenditure	191
	B. Relationship between Total Energy Expenditure and Basal and Thermostatic Requirements	195
	C. Seasonal and Sexual Variation in Energy Expenditure and Acquisition...	201
VI.	Concluding Comments	211
	References	212

I. Introduction

A. SCOPE

Energy plays two significant roles in biological systems. It is a resource necessary for the propagation and maintenance of the highly ordered physiochemical systems that comprise living structures. It is also an important characteristic of an organism's physical environment, because random kinetic-energy content is a prime determinant of a substance's temperature and thus of the rate and stability of biochemical reactions that form the basis for all life. Thermal ecology, the analysis of kinetic-energy transfer between organism and environment and associated regulation of body temperature, will not be dealt with in detail in this chapter. There currently exists a sound knowledge of the thermoregulatory physiology of birds (or at least of resting birds); this is the subject of an earlier chapter in this treatise (Calder and King, 1974). To be ecologically useful, however, this knowledge must be integrated with an understanding of the thermal stresses placed upon birds occupying their natural environments. Although this is a rapidly expanding field, data on the thermal relationships of free-living birds are not sufficiently abundant to warrant generalizations. Thus this chapter will be restricted to consideration of the acquisition and allocation of chemical potential energy as a vital resource. This will include some aspects of thermal ecology, because thermoregulation requires use of potential energy.

Two additional considerations have guided the development of this chapter. The first is the necessity to emphasize an evolutionary approach to understand the origin of biological patterns. This requires that central consideration be given to the energy relationships of individuals, the units upon which natural selection operates. Other levels of potential interest, such as the energetics of ecosystems, consequently are deemphasized.

It is also necessary that an animal's energy relationships be examined in their natural context. Organisms exist only by continuous or nearly continuous exchange of mass and energy with their environment and thus cannot be meaningfully separated from that environment. Therefore, results from studies of birds held in the laboratory or aviaries are artifacts in the simplest sense because they represent the interaction of an animal with an artificial environment. Such studies have direct ecological relevance only to the degree that the artificial environment is reasonably similar to the animal's natural environment. This is commonly assumed, but rarely demonstrated.

Primary questions when evaluating energy as a vital resource are how pressures related to energy acquisition and allocation affect an organism's contribution to succeeding generations and how such pressures have thus influenced observed biological patterns. These are exquisitely difficult problems to approach, and few workers have attempted to explore directly the consequences of energy relationships for a bird's evolutionary fitness. To examine such questions will ultimately require an understanding of the energetic consequences of activities and characteristics important for birds in nature. Although the data base for such evaluations has expanded rapidly in recent years, there are few strong areas in ecological energetics.

B. Limitations of the Data

Data used for ecological analyses range from estimates of the total energy expenditure of free-living birds to laboratory studies of particular avenues of energy exchange. The former has the advantage of ecological realism, and should reflect important effects of factors such as the behavioral flexibility exhibited by birds in nature, although the accuracy and reliability of such analyses is often known only poorly. For example, daily energy expenditure is commonly estimated using time budgets of the bird's activities combined with measures or estimates of power-consumption rates characteristic of different activities. Extensive tests or sensitivity analyses of such techniques are rare. Some analyses have indicated that particular versions of the time-budget method are acceptably accurate (e.g., Utter and LeFebvre, 1973; Koplin et al., 1980; Ettinger and King, 1980), whereas analyses of other versions have found substantial error (Weathers and Nagy, 1980). Given such uncertainties, it is prudent to restrict attention generally to substantial patterns that appear repeatedly in spite of the inaccuracies of a particular technique.

Laboratory analyses should produce results that are usually more accurate and reliable than studies of free-living birds, but their relevance to birds in nature may be questionable. The value of such analyses usually lies in quantifying the energetic status of a bird engaged in a specific activity, although perhaps in an unnatural context. In this chapter, extensive use will be made of both laboratory studies and analyses of free-living birds. A third source of data will be relied upon less heavily; that is, analyses of the energy balance of birds held in captivity but not under carefully controlled conditions. Such compromises between laboratory examinations and studies of free-living birds often produce results that lack the benefits of either alternative: the rigorous control of variables possible in the laboratory or the behavioral realism characteristic of field studies.

C. Special Characteristics of Birds Pertinent to Their Ecological Energetics

Two characteristics of birds are particularly important when placing their energy relationships in an ecological context; these are that birds are endothermic homeotherms and that they exhibit determinant growth. The shift from ectothermy to continuous endothermy has produced high rates of power consumption compared to capacities for energy storage. For example, Calder (1974) has summarized values for maximum fat and body mass in 19 passerine species. Assuming that fat contains 40 kJ/g (Kleiber, 1961) and estimating total daily energy expenditure in free-living birds using Eq. (8) (derived in Section III), Calder's Eq. (45) predicts that a 25-gram bird can typically store, at a maximum, energy sufficient for only two days of normal activity. Unlike many ectotherms, birds generally cannot effectively store energy over major fractions of the annual cycle. Exceptions may be hibernating species and those that store large amounts of energy outside of the body. For example, Vander Wall and Balda (1977) have estimated that the seeds stored during the autumn by Clark's Nutcracker (*Nucifraga columbiana*) represent 2.2 to 3.3 times the animal's energy requirement for the fall and winter.

In addition, birds exhibit determinant growth; they usually reach approximate adult mass early in life and prior to becoming independent of their parents. Adult birds therefore do not allocate energy to somatic production with the exception of maintenance functions such as molt. This again contrasts with typical ectothermic vertebrates, which may exhibit somatic growth throughout their lives.

Birds thus have fewer avenues of energy allocation open to them than do many ectothermic animals. To a major degree, birds are restricted to allocating energy only to activity, somatic maintenance, immediate reproduction, or short-term storage. Other taxa often have additional avenues open to them, such as storing energy until the next annual cycle when it may be invested in reproduction (Tinkle and Hadley, 1973, 1975). More so than in many other groups, an annual cycle in birds is energetically separate from previous or subsequent cycles. Because it represents a discrete, repeated event in avian life history, the annual cycle is a particularly appropriate unit for ecological analyses. Two other time scales are also appropriate for energetic analyses. The small size of most birds dictates a short time scale for change in body temperature, which may occur over periods of minutes (Southwick, 1973; Calder and King, 1974; Veghte, 1975; Torre-Bueno, 1976). This defines the appropriate scale for analyses of thermostatic physiology and behavior. Other important types of behavior, such as foraging and social interactions, typically occur in 24-hour cycles. Daily cycles thus are most useful if the primary focus of an analysis deals with such phenomena.

II. Energetics of Some Prominent Activities and Events in the Avian Life Cycle

A. INTRODUCTION

This section examines the energetics of prominent avian activities, such as resource defense and reproduction. It deals primarily with those activities isolated from others in which a free-living bird may be engaged. For example, this section's consideration of incubation deals with the specific act of egg thermoregulation and not with the total energy budget of a bird during this phase of the annual cycle, in which it is engaged in egg thermoregulation as well as other important activities. The net effect of all activities that characterize a particular phase of the annual cycle is examined in Section III,C.

B. Energy Allocation to Resource Defense

Table I summarizes estimates of the fractional allocation of total daily energy expenditure (E_{TOT}) to territorial defense. Data for breeding and non-breeding birds are segregated because substantially different resources (e.g., access to mates) are defended in addition to food, which is defended by all species listed. Only a small fraction of daily energy expenditure is devoted to advertisement and territorial maintenance; the average is 11%, and most values fall below 8% of E_{TOT}. This suggests that the major costs of such defense are not energetic but rather other avenues of expenditure, such as time allocation or exposure to predation, or that the major energetic costs of such resource defense often are not quantified by workers. Most studies have focused on birds maintaining previously established territories. Energy

TABLE I

Energy Allocation to Territorial Defense and Advertisement as a Percentage of Total Daily Energy Expenditure

Species	Energy expenditure (percentage of E_{TOT})	Reference
Nonbreeding birds		
Oreotrochilus estella[a]	33.1	Carpenter (1976)
Calypte anna[b]	3.6	Pearson (1954); Stiles (1971); Ewald and Carpenter (1978)
Vestiaria coccinea[c]	2.5	Carpenter and MacMillen (1976a)
Nectarinia reichenowi	7.4	Gill and Wolf (1975)
Phainopepla nitens[d]	7.1	Walsberg (1977)
Breeding birds		
Actitis macularia[e]	0.9–11.7	Maxson and Oring (1980)
Oreotrochilus estella[f]	16.2	Carpenter (1976)
Calypte anna	3.4	Pearson (1954)

[a] Mean of values for males and females.
[b] Median of available values.
[c] Mean value.
[d] Mean value for birds in Sonoran Desert.
[e] Minimum and maximum values represent females during laying period and brooding period respectively.
[f] Incubating female.

costs of establishing territories may be substantially higher than simply maintaining them, but this point has received little attention. In addition, the mode and intensity of resource defense are likely to vary with changes in resource availability (Carpenter and MacMillen, 1976a; Ewald and Carpenter, 1978; Frost and Frost, 1980). For example, Ewald and Carpenter (1978) examined the energetic consequences of such variation by manipulating the food supply in territories defended by Anna's Hummingbirds (*Calypte anna*). As energy was made less available on the territory, territorial birds became less vigorous in resource defense. Fewer intruders were chased and intruders were pursued for shorter distances. Consequently, energy expenditure devoted to such chasing declined from about 4% of total daily expenditure when rich territories were defended to about 0.3% of total expenditure when food was absent in the territory.

Data describing the energetic consequences of major shifts in patterns of resource defense are available for three species. In nonbreeding Andean Hillstar hummingbirds (*Oreotrochilus estella*), individuals wintering in patches of native vegetation (*chuquiraga*) are nonterritorial (Carpenter, 1976). Individuals wintering in patches of introduced eucalyptus are territorial and exhibit a total daily expenditure elevated 46% over the nonterritorial birds (Carpenter, 1976). About two-thirds of this increase can be attributed to active territorial defense, with additional costs associated with changes in foraging patterns (e.g., increased time spent in flight and flycatching).

The Phainopepla (*Phainopepla nitens*) breeds in both its winter and summer range and exhibits a dramatic shift in territorial behavior between the two breeding periods (Walsberg, 1977). In their first breeding season in the Sonoran Desert, Phainopeplas defend large territories ($\bar{x} = 0.38$ ha), within which they conduct almost all foraging. In the second breeding period in the coastal chaparral and oak woodlands of southern California, much smaller territories are defended, typically consisting simply of a single tree in which the nest is placed. Food is usually not defended and is typically collected more than 50 m from the nest. Because of this increased distance to foraging sites, time spent in flight is increased two- to fivefold over comparable phases of reproduction in the Sonoran Desert, and daily energy expenditure is estimated to increase 20–27%.

Finally, in a Hawaiian honeycreeper, the Iiwi (*Vestiaria coccinea*), Carpenter and MacMillen (1976a) have compared the energy expenditure of territorial and nonterritorial individuals during the nonbreeding season. E_{TOT} is 17% greater in territorial individuals than in nonterritorial birds. Less than one-third of this increase can be directly attributed to the energy costs of active defense. Most of the increase in energy expenditure is due to subtle shifts in patterns of foraging that may serve as advertisement by increasing the conspicuousness of the territorial birds.

C. Gonadal Growth and Gametogenesis

1. The Male

In summary, these studies confirm the expectation that qualitative shifts in territorial behavior can have major effects on energy expenditure. Perhaps unexpectedly, they also indicate that such shifts in energy expenditure often may not be associated with active expulsion of intruders, which commonly accounts for less than 10% of total daily energy expenditure. Rather, major energetic shifts may be associated with more subtle phenomena, such as changes in foraging techniques and energy allocation to territorial advertisement.

Estimates of the total energy required for seasonal testicular growth and the maximum rate of such expenditure are summarized for seven species in Table II. Testicular recrudescence requires energy equivalent to only about 7–40% of a single day's basal expenditure. This energy cost is distributed over a 10- to 30-day growth period, and maximum rates of expenditure equal 2% or less of basal metabolism (BMR). These low rates indicate that the energy cost of testicular growth is negligible. Few data are available with

TABLE II
Energy Requirements of Testicular Recrudescence[a]

Species	Body mass (g)	Testes[b] g	Testes[b] %BMR[d]	Maximum growth rate[c] g/day	Maximum growth rate[c] %BMR[d]	Reference
ianus colchicus	1250	7.90	17.2	0.153	0.45	Greeley and Meyer (1953)
toris chukar	650	2.04	7.3	0.031	0.14	Mackie and Buechner (1963)
nortyx californicus	180	1.04	9.5	0.018	0.22	Anthony (1970)
mba palumbus	500	2.33	13.4	0.022	0.10	Ljunggren (1969)
nus vulgaris	76	3.88	40.3	0.191	1.97	Bissonnette and Chapnick (1930)
aius phoeniceus	61	1.32	16.9	0.050	0.64	Wright and Wright (1944)
otrichia leucophrys	27	0.61	14.0	0.035	1.06	King et al. (1966)

[a] Energy requirements calculated assuming a 75% net efficiency of synthesis and that energy density of testes equals 8.0 kJ/g (Ricklefs, 1974). Where values for histologically fixed tissues were reported by authors, fresh mass was estimated using the equation $Y = 0.364 + 1.388X$ (Kern, 1970, cited in King,). Basal metabolism was calculated using equations of Aschoff and Pohl (1970) for birds in the rest phase of their daily cycle.
[b] Represents difference between seasonal maximum and minimum testes mass or energy content.
[c] Values calculated from author's data for organ volumes assuming tissue density = 1 g/cm^3.
[d] Value represents energy requirements of testicular growth as a percentage of one day's basal metabolic expenditure.

which to estimate energy requirements of sperm and semen production. Using the data of Arscott and Parker (1963), Ricklefs (1974) calculated that daily semen production in male chickens accounts for about 0.8% of BMR. Additional data obviously are required, but this avenue of expenditure also appears to be insignificant.

2. *Ovarian and Oviducal Growth*

At their greatest size, the oviduct and functional but nonovulating ovary typically account for 5–15% of body mass, and the estimated energy content of these structures is a nearly linear function of body mass. (See legend of Fig. 1 for references and method of estimating energy content.) A least-squares regression of available data (logarithmically transformed) yields the equation

$$\ln E_o = \ln 0.655 + 0.938 M_B \tag{1}$$

where E_o is energy content of the ovary and oviduct (kJ) and M_B is body mass (g); $S_{yx} = 0.381$, $S_b = 0.0833$, and $r^2 = 0.962$. This equation indicates that

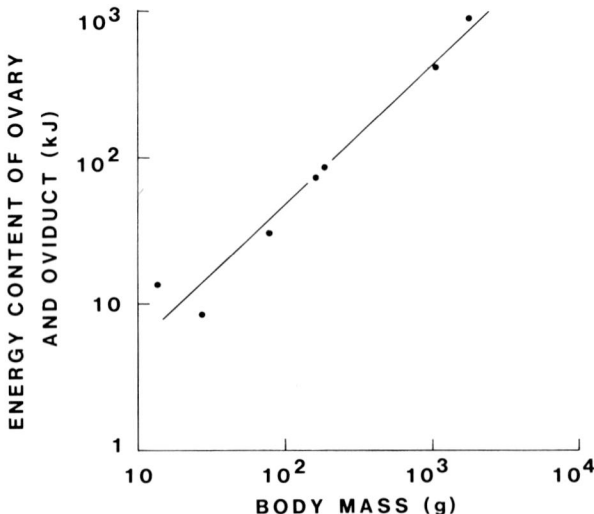

FIG. 1. Energy content of ovary and oviduct as a function of body mass. Values calculated assuming energy density of ovary equals 7.83 kJ/g and that of oviducts equals 6.36 kJ/g (Ricklefs, 1974). $E_o = 0.655 M_B^{0.938}$. Species, in order of increasing body mass, are *Riparia riparia* (Petersen, 1955), *Zonotrichia leucophrys* (King *et al.*, 1966), *Sturnus vulgaris* (Ricklefs, 1974), *Lophortyx gambelii* (Raitt and Ohmart, 1966), *Lophortyx californicus* (Lewin, 1963), *Anas platyrhynchos* (Johnson, 1961), and *Gallus gallus* (Romanoff and Romanoff, 1949).

the energy content of these organs is equal to 33% of one day's basal metabolic expenditure for a 10-g passerine bird and 89% of one day's basal expenditure for a 1-kg passerine bird (BMR was estimated for the inactive phase of the daily cycle using the equation of Aschoff and Pohl, 1970.) Available data indicate that the rapid phase of ovarian and oviducal growth may last 7–30 days (Petersen, 1955; Lewin, 1963; King et al., 1966; Raitt and Ohmart, 1966). If costs are distributed evenly over this period, then the average expenditure devoted to this growth is equivalent to 3% of BMR in *Lophortyx* spp., 9% of BMR in *Riparia*, and 2% of BMR in *Zonotrichia*. Unless future studies reveal a particularly rapid rate of ovarian and oviducal growth in larger species, this avenue of energy expenditure seems minor compared to other avenues during reproduction.

3. Energy Requirements of Egg Synthesis

The energetics of egg production may be usefully evaluated with regard to both the total energy requirements of synthesis and the maximum rate at which energy is required for egg production. The total energy cost of egg production is a product of egg mass and energy density, clutch size, and the energetic efficiency of biosynthesis. Egg mass (M_E) is an allometric function of body mass (M_B) (Rahn et al., 1975):

$$M_E = 0.277 M_B^{0.77} \qquad (2)$$

Thus the mass of a single egg represents 16% of body mass in a typical 10-g bird and 6% of body mass in a typical 1-kg bird. This relationship, however, subsumes many avian taxa. When data within a particular family or order are examined, the analysis by Rahn et al. (1975) indicates that egg mass usually scales approximately with $M_B^{0.67}$. There is substantial variation around these allometric averages. For example, the Mallard (*Anas platyrhynchos*) and the Brown Kiwi (*Apteryx australis*) are of similar size, but the Mallard egg is 4% of body mass, whereas the kiwi's egg is about 20% of body mass (Rahn et al., 1975). Entire taxa may be characterized by unusually large or small eggs. For example, birds of the order Procellariiformes produce eggs that are about twice the mass of that predicted by the allometric equation for all species (Rahn et al., 1975).

The energy density of eggs varies substantially with composition, because yolk tends to be more energy-dense than albumin. Whole-egg energy density ranges from 4.3 to 11.6 kJ/g (Table III). Eggs of precocial species average 7.76 ± 1.54 kJ/g, or 62% higher than the value for altricial species (4.78 ± 0.540 kJ/g); analysis of variance indicates that this difference is statistically significant ($P < 0.001$).

TABLE III

EGG ENERGY CONTENT AND APPROXIMATE LENGTH OF THE RAPID FOLLICULAR GROWTH PHASE[a]

	Body mass (g)	Egg mass (g)	Rapid follicular growth phase (days)	Egg energy content			References
				kJ/g	kJ/egg	egg/BMR	
Casuarius casuarius	—	623	—	6.86	4275	—	Carey *et al.* (1980)
Apteryx australis	2098	350	7½	11.45	4014	11.50	Calder and Rowe (1977), Calder *et al.* (1978)
Podilymbus podiceps	343	19.7	—	5.60	110	0.78	Hartmann (1955), Carey *et al.* (1980)
Pelecanus occidentalis	3510	92.1	—	5.74	528	0.68	Lawrence and Schreiber (1974)
Sula bassana	3150	118	—	3.76	442	0.62	Ricklefs and Montevecchi (1979)
Oceanodroma leucorhoa	43.5	10.2	—	7.61	77.6	2.52	Carey *et al.* (1980), Ricklefs *et al.* (1980)
Bubulcus ibis	375	26.9	—	4.92	132	0.88	Siegfried (1969), Carey *et al.* (1980)
Casmerodius albus	888	88.5	—	4.86	236	0.84	Hartmann (1955), Carey *et al.* (1980)
Egretta tricolor	342	27.5	—	4.89	135	0.96	Hartmann (1955), Carey *et al.* (1980)
Eudocimus albus	881	49.2	—	4.72	233	0.83	Kushlan (1977), Carey *et al.* (1980)
Anser anser	3250	200	—	7.70	1537	2.10	Romanoff and Romanoff (1949)
Branta canadensis	1890	197	7	7.55	1487	3.62	Raveling (1979), Carey *et al.* (1980)
Dendrocygna autumnalis	—	41.0	—	11.08	454	—	Cain (1976)
Anas platyrhynchos	1870	79.9	6–7	7.33	586	1.20	Ricklefs (1977)
Aix sponsa	665	42.6	7	8.67	369	1.62	Drobney (1980)
Gallus gallus	2500	58.0	7–8	6.95	402	0.67	Romanoff and Romanoff (1949)
Phasianus colchicus	1304	31.2	—	7.14	233	0.60	Nelson and Martin (1953), Carey *et al.* (1980)
Colinus virginianus	198	8.7	—	7.80	67.9	0.72	Amadon (1943), Case and Robel (1974)
Coturnix coturnix	97	9.9	—	6.87	68.0	1.22	Ricklefs (1977)
Meleagris gallopavo	4200	65.2	—	6.91	451	0.51	Romanoff and Romanoff (1949)
Rallus limicola	65	10.8	—	5.82	62.8	1.52	Poole (1938), Carey *et al.* (1980)
Porzana carolina	49.8	8.7	—	5.27	45.9	1.35	Connell *et al.* (1960), Carey *et al.* (1980)
Arenaria interpres	97.1	17.1	5–6	—	—	—	Schönwetter (1960–1978), Johnson (1979), Roudybush *et al.* (1979)
Actitis macularia	42.5	9.1	—	7.17	65.2	2.15	Carey *et al.* (1980), Maxson and Oring (1980)

Species							References
Calidris alpina	46.5	10.0	—	7.13	71.3	2.20	Maclean and Holmes (1971), Norton (1973)
Calidris bairdii	41.9	12.3	—	7.61	93.6	3.12	Jehl (1970), Carey et al. (1980)
Calidris mauri	27	6.7	5–8	—	—	—	Schönwetter (1960–1978), Roudybush et al. (1979)
Limosa lapponica	260	37	6–12	—	—	—	Schönwetter (1960–1978), Roudybush et al. (1979)
Phalaropus fulicaria	—	—	4–5	—	—	—	Roudybush et al. (1979)
Phalaropus lobatus	—	—	6–7	—	—	—	Roudybush et al. (1979)
Larus hyperboreus	1600	118	12	—	—	—	Roudybush et al. (1979)
Larus marinus	—	111	13	—	—	—	Drent (1970), Roudybush et al. (1979)
Larus glaucescens	1198	98	12	—	—	—	Baltz and Morejohn (1977), Morgan et al. (1978), Roudybush et al. (1979)
Larus occidentalis	1137	86.8	10–11	6.12	531	1.57	Baltz and Morejohn (1977), Roudybush et al. (1979), Carey et al. (1980)
Larus argentatus	894	94.0	11–13	7.02	576	2.03	Drent (1970), Roudybush et al. (1979), Carey et al. (1980)
Larus delawarensis	432	—	12	—	—	—	Stewart and Skinner (1967), Roudybush et al. (1979)
Larus atricilla	306	44.2	—	6.52	288	—	Hartmann (1955), Schreiber and Lawrence (1976)
Larus canus	383	51	5–8	—	—	—	Schönwetter (1960–1978), Roudybush et al. (1979)
Rissa tridactyla	400	52	9	—	—	—	Schönwetter (1960–1978), Roudybush et al. (1979)
Xema sabini	169	—	7–8	—	—	—	Johnston (1963), Roudybush et al. (1979)
Sterna paradisaea	110	18	6	—	—	—	Schönwetter (1960–1978), Roudybush et al. (1979)
Sterna albifrons	—	9.8	—	6.07	59.5	—	Carey et al. (1980)
Sterna maxima	492	70.2	—	6.14	431	2.36	Stewart and Skinner (1967), Carey et al. (1980)
Sterna sandvicensis	—	34.6	—	6.43	222	—	Carey et al. (1980)
Rynchops nigra	—	26.6	—	6.49	162	—	Carey et al. (1980)
Uria aalge	966	113	12–18	—	—	—	Lack (1968), Baltz and Morejohn (1977), Roudybush et al. (1979)
Cepphus grylle	431	52	8	—	—	—	Lack (1968), Roudybush et al. (1979)
Cepphus columba	484	55	10	—	—	—	Lack (1968), Bedard (1976), Roudybush et al. (1979)
Ptychoramphus aleuticus	164	30	8	—	—	—	Baltz and Morejohn (1977), Roudybush et al. (1979)
Lunda cirrhata	779	—	12–13	—	—	—	Spring (1971), Roudybush et al. (1979)
Columba livia	330	17.4	5–8	4.51	78.5	0.55	Bartelmez (1912), Carey et al. (1980)

(continued)

TABLE III Continued

	Body mass (g)	Egg mass (g)	Rapid follicular growth phase (days)	Egg energy content			References
				kJ/g	kJ/egg	egg/BMR	
Zenaida macroura	91.4	6.41	—	5.22	33.5	0.63	Ricklefs (1977), G. E. Walsberg (unpublished)
Streptopelia "risoria"	150	9.2	5–7	5.21	47.9	0.63	Cuthbert (1945)
Colaptes auratus	131	8.8	—	3.31	29.1	0.42	Baldwin and Kendeigh (1938), Carey et al. (1980)
Sayornis phoebe	19.2	2.5	—	5.14	12.9	0.76	Baldwin and Kendeigh (1938), Carey et al. (1980)
Corvus monedula	250	15.2	5	—	—	—	Stieve (1919)
Pica pica	162	7.2	—	3.62	26.1	0.20	Carey et al. (1980), Mugaas and King (1981)
Parus major	18.5	1.6	3–4	4.62	7.7	0.29	Kluijver (1951), J. A. L. Mertens (unpublished data cited in Carey et al. 1980)
Cistothorus palustris	12.0	1.1	—	4.82	5.3	0.27	Hartmann (1955), Kale (1965)
Turdus migratorius	74.9	6.7	—	4.48	30.1	0.41	Stewart (1937), Carey et al. (1980)
Catharus guttatus	30.0	6.5	—	3.89	25.3	0.67	Holmes et al. (1979), Carey et al. (1980)
Sturnus vulgaris	76.0	7.2	—	4.31	31.2	0.42	Johnson (1972), Ricklefs (1974)
Dendroica petechia	9.6	1.7	—	4.61	7.8	0.47	Baldwin and Kendeigh (1938), Carey et al. (1980)
Passer domesticus	26.0	2.8	—	4.60	12.88	0.38	Tangl (1903), Kendeigh (1973)
Passer montanus	22.1	2.9	—	4.27	12.38	0.43	Pinowski (1967)
Xanthocephalus xanthocephalus	—	4.6	—	4.15	19.1	—	Carey et al. (1980)
Agelaius phoeniceus	41.9	4.5	—	4.16	18.7	0.39	Brenner (1964), Carey et al. (1980)
Agelaius tricolor	46.0	3.7	3–4	—	—	—	Payne (1969)
Euphagus carolinus	56.1	6.8	—	4.77	30.1	0.51	Hartmann (1955), Carey et al. (1980)
Euphagus cyanocephalus	—	4.9	—	4.34	21.3	—	Carey et al. (1980)
Quiscalus quiscula	101	6.8	—	4.69	31.9	0.35	Amadon (1943), Carey et al. (1980)
Molothrus ater	36.2	2.9	—	3.66	10.6	0.25	Lustick (1970), Carey et al. (1980)
Carpodacus mexicanus	19.3	2.4	—	4.94	11.9	0.43	Poulson and Bartholomew (1962), Carey et al. (1980)
Poephila guttata	11.8	1.0	—	5.15	5.15	0.27	El-Wailly (1966)
Egretta thula	—	22.5	—	4.96	112	—	Carey et al. (1980)
Zonotrichia leucophrys	27.0	2.4	4	—	—	—	Kern 1970 cited in King, 1973)
Melospiza melodia	21.0	2.9	4	3.45	10.1	—	Carey et al. (1980)

[a] Represents energy content of egg as fraction of one day's basal metabolism. Basal metabolic rate (BMR) for rest phase of daily cycle calculated using

Table III summarizes data on total energy content of eggs, which ranges from 38 to 1352% of one day's basal expenditure ($\bar{x} \pm SD = 119 \pm 197\%$). The data are too few to be conclusive, but eggs of precocial species appear to average about twice as high in energy content compared to adult BMR as the eggs of altricial species ($53 \pm 13\%$ of BMR in altricial species, $121 \pm 57\%$ in precocial species). (These statistics exclude the exceptional value for the Brown Kiwi. Analysis of variance indicates that the difference is statistically significant; $P < 0.02$). The energy requirement for producing an entire clutch is quite variable, partially due to its dependence on clutch size (Table IV), and ranges at least from 1.5 to 25 times a single day's basal expenditure.

Another value of potential significance is the peak daily cost of egg synthesis. This is a function of clutch size, egg energy content, the interval between the laying of successive eggs, and the length of the rapid phase of follicular growth. The latter represents the length of time over which the bulk of the ovum is synthesized. Using an average when multiple values are available for a species, least-squares regression of logarithmically transformed data from Table III indicates that the length of the rapid growth phase (T_{RG}) is a function of egg mass:

$$\ln T_{RG} = \ln 2.96 + 0.278 \ln M_E \qquad (3)$$

where T_{RG} is in days, M_E is egg mass in grams, $r^2 = 0.810$, $S_{yx} = 0.7593$, and $S_b = 0.03734$. Thus the length of the rapid follicular growth phase is proportional to $M_E^{0.28}$ and varies from 3–4 days in small songbirds to 10–18 days in oceanic birds laying large eggs (Fig. 2). The relationship between egg mass and the average rate of yolk production per day (R; g/day) should be proportional to egg mass divided by the length of rapid growth phase:

$$R = M_E/T_{RG} \propto M_E/M_E^{0.28} = M_E^{0.72} \qquad (4)$$

Because egg mass is proportional to (body mass)$^{0.77}$,

$$R \propto (M_B^{0.77})^{0.72} = M_B^{0.54} \qquad (5)$$

This suggests that the average rate of ovarian growth increases more slowly with increases in body mass than do many other physiological rate functions, such as BMR, which commonly scale as $M_B^{0.7-0.8}$ (Calder, 1974). Thus the rate at which energy is used in egg synthesis tends to be a larger fraction of BMR in smaller species.

The time course of egg synthesis, estimates of biosynthetic efficiency, and egg energy content must be integrated to estimate the peak energy requirements of egg production. Ricklefs (1974) estimated that these peak daily

TABLE IV

Estimated Energy Cost of Egg Synthesis for Various Species[a]

Species	Rapid follicular growth period (days)	Clutch size	Laying interval (days)	Cost of entire clutch kJ	Cost of entire clutch %BMR[b]	Peak daily expenditure kJ	Peak daily expenditure %BMR
Apteryx australis	7.5	1	—	4014	1150	798	229
Pelecanus occidentalis[c]	10	3	2	2124	628	277	82
Anseriformes							
Branta canadensis[d]	7	5	1	9913	1128	956	109
Anas platyrhynchos[d]	6	8	1	3936	808	492	101
Aix sponsa[e]	7	12	1	5754	2523	479	210
Galliformes							
Gallus gallus	8	10	1	5360	890	536	89
Coturnix coturnix	6	10	1	901	1620	90.1	162
Meleagris gallopavo	9	10	1	7840	1060	784	106
Charadriiformes							
Larus atricilla	8	3	2	1129	875	137	106
Larus argentatus[e]	10	3	2	768	272	226	80
Larus occidentalis	10	3	2	2124	628.	277	82
Columbiformes							
Columba livia[d]	6	2	1	209	147	35.2	25
Zenaida macroura[c]	5	2	1	89.3	168	21.0	40
Streptopelia "risoria"	5	2	1	119	156	40.0	53
Passeriformes							
Sturnus vulgaris[c]	5	5	1	208	281	40.6	55
Carpodacus mexicanus[c,d]	4	4	1	34.5	127	15.3	56
Melospiza melodia[d]	4	4	1	53.9	189	13.5	46

[a] Data from sources cited in Table III, except as noted.
[b] Represents energy expenditure as a percentage of one day's basal metabolism. Basal metabolic rate (BMR) for rest phase of daily cycle calculated using equations of Aschoff and Pohl (1970).
[c] Length of rapid follicular growth period estimated using Eq. (3).
[d] Ratio of yolk energy content to total egg energy content assumed to equal that of domestic goose for *Branta*, that of *Larus atricilla* for *L. argentatus*, that of *Zenaida* for *Columba*, and that of *Sturnus* for *Carpodacus* and *Melospiza*.
[e] Computations of Drobney (1980).

energy requirements were about 39% of BMR in hawks and owls, 45% of BMR in passerines, 126% of BMR in Galliformes, 140% of BMR in shorebirds, 170% of BMR in Laridae, and 180% of BMR in ducks. These generalizations tend to be confirmed by estimates by other authors and those I have made for various species using a model similar to that derived by King (1973) (Table IV). For these estimates, I have assumed biosynthetic efficiency is 75% and that the follicle follows a logistic growth curve during the rapid growth phase. [See King (1973) for discussion of this assumption.] The entire

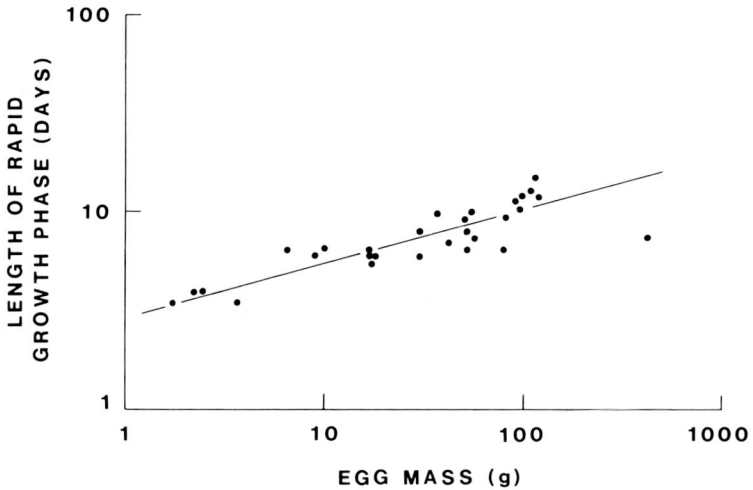

Fig. 2. Length of rapid follicular growth phase as a function of body mass. $T_{RG} = 2.94\ M_E^{0.274}$. Data from Table III.

egg white is assumed to be produced in the 24-hour period following ovulation. This latter assumption contrasts with the model of King (1973), who prorated costs through the follicular growth period. Results suggest a considerable range in peak energy requirements from relatively low values (37–55% of BMR) in some passerine species to very high values (160–216% of BMR) in ducks and the kiwi (Table IV). The very high value of 210% of BMR for the Wood Duck (*Aix sponsa*) presents one of the most direct estimates available (Drobney, 1980), and suggests that for at least some species egg synthesis may be the most important energetic event during reproduction. This is most likely to be true for precocial forms such as ducks, in which parental energy investment in the young after hatching may be comparatively small.

4. Sources of Energy for Egg Synthesis

An animal may make energy available for any process, including egg production, by increasing dietary intake, reducing allocation to other activities, or using internal energy stores. Unfortunately, no study has compared the relative importance of these three avenues in a free-living species during egg synthesis.

Several workers have observed increases in food consumption by laying females. For example, the following increases in time spent foraging by the laying female compared to the male during the same period have been described: 175% in *Anas platyrhynchos* (Dwyer et al., 1979), 64% in *Anas clypeata* (Afton, 1979), 57% in *Actitis macularia* (Maxson and Oring, 1980), 45% in *Calidris pusilla* (Ashkenazie and Safriel, 1979), and 11% in *Zonotrichia leucophrys* (Hubbard, 1978).

A number of studies have also documented reductions in the female's locomotor activity during laying. In the Willow Flycatcher (*Empidonax traillii*), the female allocates 13% less energy to activity during the laying period then does the male (Ettinger and King, 1980). This reduction appears to compensate for synthetic costs with the result that the female's total daily expenditure averages 5% less during the laying period than the male's. In other species, such as the Black-billed Magpie (*Pica pica*), the female's activity is reduced during egg synthesis compared to that of the male, but this reduction compensates only partially for increases due to egg production (Mugaas and King, 1981). Finally, in some species no decrease, or even an increase, in the female's activity has been observed (e.g., *Empidonax minimus, Vireo olivaceus*) (Holmes et al., 1979).

Use of energy stores in meeting demands of egg production is suggested by changes in female mass that have been documented in many species (e.g., Weller, 1957; Harris, 1970; Korschgen, 1977; Ankney and MacInnes, 1978; Raveling, 1979; Drobney, 1980). Few studies, however, have quantified both energy supply and demand in a manner yielding reliable estimates of the importance of this energy storage. Most data deal with waterfowl. In a detailed study of Wood Ducks, which lay large, energy-rich clutches, Drobney (1980) has demonstrated that during laying females deplete fat stores that contain energy equal to about 88% of the requirements of egg formation. Females also forage extensively during this period and increase the fraction of animal material compared to plant material in their diet, apparently as a mechanism to supply protein necessary for ovogenesis. Drobney (1980) has suggested that fat storage in this species not only supplies energy needed for egg formation, but also ameliorates pressure on the female's foraging behavior so that she can devote more time to feeding on invertebrates that have higher protein concentrations but may be less energetically profitable to acquire. Krapu (1981) has suggested a similar role for fat reserves in Mallards. In this species, the female loses 25% of her body mass in fat during the prelaying, laying, and incubation periods; most mass loss occurs during laying and early incubation. Fat storage apparently supplies energy necessary for egg formation as well as supplying a portion of the female's general energy requirement. This allows the female to forage extensively for relatively scarce aquatic invertebrates to meet protein requirements of egg formation. The importance of lipid reserves for egg production in this species is suggested by the correlation between clutch size and lipid reserves held by females. When mean values for six 10-day periods from

April to June are compared, the size of new clutches and the lipid reserves held by female Mallards are closely correlated [$r^2 = 0.967$; calculated from data in Krapu's (1981) Table 2]. Male Mallards also lose fat reserves during reproduction, although the loss tends to occur earlier than in females. This correlates with the male's intense activity early in the breeding season, when it is establishing an activity center (Krapu, 1981).

Extensive use of body reserves during reproduction has been documented in several larger anseriform species (Hanson, 1962; Barry, 1962; Krapu, 1974; Ankney, 1977; Ankney and MacInnes, 1978; Raveling, 1979), although its function appears to differ somewhat from that described for the Wood Duck and Mallard. In a number of these species, body reserves are apparently a major source of both the energy and protein requirements of egg synthesis. For example, both sexes of Canada Geese (*Branta canadensis*) lose weight after arrival on their arctic nesting grounds (Raveling, 1979). Females lose about 19% of their body mass in fat and males lose about 17%. That both sexes lose weight and that the female's 360-g fat loss is approximately six times the lipid content of an entire five-egg clutch suggests that egg synthesis in itself may not be the major cause of weight loss. This is difficult to evaluate without a complete energy budget. In contrast to Wood Ducks and Mallards, Raveling (1979) noted that female Canada Geese also lose an amount of protein from the body approximately equivalent to that incorporated in the clutch, and suggested that egg production in this case may be limited by protein reserves in the body. A similar role for nutrient reserves in clutch formation is also indicated in Lesser Snow Geese (*Anser caerulescens*; Ankney and MacInnes, 1978). In this species, energy and protein reserves also function importantly to maintain the female through the incubation period (Ankney and MacInnes, 1978). The reliance of Snow Geese and Canada Geese upon protein reserves for egg synthesis contrasts with the pattern observed in the ducks studied by Drobney (1980) and Krapu (1981), in which protein is supplied by intensive foraging that is facilitated by reliance upon fat stores. This difference may reflect adaptations to food scarcity at the time of egg laying in arctic-nesting geese (Krapu, 1981).

Data for nonanseriform species are sparse. The Red-billed Quelea (*Quelea quelea*) and Brown-headed Cowbird (*Molothrus ater*) are the only extensively studied passerine species (Jones and Ward, 1976; Ankney and Scott, 1980). In *Quelea*, substantial amount of both protein and fat are lost from the body during laying. The protein removed from the breast muscles is equal to 21% of the content of the clutch. In contrast, the loss in lipids is equal to 1.7 times that included in the eggs. This suggests that lipids may serve not only as an energy and material source for egg synthesis per se, but that they may also serve to reduce the rate at which energy must be acquired in foraging and thus facilitate a behavioral shift in which the bird specializes on protein-rich food such as insects. Fat storage therefore may allow behavioral flexibility that compensates for the bird's limited ability to store protein. In contrast, female cowbirds do not utilize fat or protein reserves during laying. Nutrients required for egg production are apparently obtained directly from the diet (Ankney and Scott, 1980).

Taken together, these studies demonstrate the complex interplay of behavioral and physiological adaptations that may be involved in meeting the nutrient requirements of ovogenesis. These phenomena deserve more extensive attention, but in addition these clearly fruitful approaches to an integrated analysis of the acquisition and allocation of a variety of resources (not simply energy) could be usefully applied to other portions of the annual cycle.

D. INCUBATION

1. Introduction

The energetics of avian incubation has been controversial for a variety of technical and philosophical reasons. One source of this controversy is that incubation energetics are commonly analyzed at three different levels, each involving different sets of questions:

1. How much heat does the adult supply to the eggs? This question deals solely with heat flow to the eggs, and does not consider other avenues of energy transfer from or to the adult.
2. What is the net change in the adult's heat balance associated with the specific act of egg thermoregulation? This change subsumes heat transfered to the eggs and a variety of other factors affecting the adult's heat balance as it incubates, such as nest insulation and microclimate.
3. What changes occur in the adult's energy budget during this phase of the annual cycle compared to other phases? Such analyses include the energetic effects of all of the bird's daily activities (e.g., foraging, social behavior) and not simply egg thermoregulation. This question is addressed in Section III,C.

A major difficulty in incubation energetics has been designing studies such that the questions addressed are directly relevant to ecological phenomena. Heat loss to the eggs (question 1) may be important to the adult's energy balance, but to yield ecologically relevant information it must be placed in the context of the adult's total energy budget (question 2). Unless heat loss through the brood patch is so intense that to maintain appropriate egg temperatures would exceed the adult's transfer capability, heat loss is important only to the degree that it affects the parent's general heat balance. In addition to transfer through the brood patch, this heat balance is determined by factors such as adult heat production, plumage insulation, nest microclimate and insulation, embryonic heat production, egg mass, and the degree of egg cooling during inattentive periods (Walsberg and King, 1978a). Not unexpectedly, it is difficult to analyze adult energy balance and form broad generalizations regarding incubation energetics.

2. Estimates of the Energetic Consequences of Incubation

The most direct estimates of power consumption during incubation have been made for cavity-nesting species using respiratory-gas analyses (Biebach, 1977, 1979; Gessaman and Findell, 1979; Vleck, 1981). These studies employed the nest cavity as a metabolic chamber in an open-flow gas-analysis system. Air is typically drawn in the entrance to the nest box, through the cavity in which the adult is incubating, and withdrawn at the rear of the nest chamber to be passed to the gas-analysis system. The technique thus is applicable only to those species that nest in cavities. In addition, this technique may importantly modify the nest microclimate and, consequently, the adult's power consumption. Presence of an adult should normally increase air temperature and humidity within the nest cavity. Flushing air through the cavity counteracts this effect and probably increases heat loss from the adult. For example, flow rates in Vleck's (1981) study were sufficient to replace completely the air within the cavity every 1.4–2.0 minutes. Workers have not accounted for the increases in energy expenditure such effects may produce. An additional difficulty in these analyses is that authors commonly compare the power consumption of incubating birds with that of nonincubating adults roosting in the nest. Nonbreeding adults of most species, including those studied by Biebach (1977, 1979), Gessaman and Findell (1979), and Vleck (1981), do not normally roost in the nest. Thus such comparisons have limited ecological relevance.

Vleck (1981) measured the oxygen consumption of Zebra Finches (*Poephila guttata*) that built nests within small metal cans. The metabolic rate of adults continuously incubating four-egg clutches was 20–23% above that of nonincubating individuals resting within the nest. Vleck also explored the effect of egg temperature on the adult's metabolic rate by changing the temperature of the clutch. A hollow glass egg was placed in the clutch, and the temperature of this egg was controlled by circulating water through it. By cooling this egg's surface temperature as much as 20°C, Vleck demonstrated that the adult is capable of at least doubling its metabolic rate in response to low egg temperatures.

Vleck's (1981) comparisons assume that the energy cost of incubation is properly defined as the elevation in metabolic rate observed in an incubating adult compared to that of a nonincubating adult resting within the nest. Ecological inferences based upon Vleck's data may be changed, however, if one compares the energy metabolism of an incubating adult with that of a nonbreeding individual in the more typical condition in which it does not exploit the nest's insulation. Vleck (1981) described the relationship between air temperature within the nest (T_N) and that in the environmental chamber within which the nest was placed (T_C) as: $T_N = 2.98 + 0.93T_C$ [Vleck's Eq. (1)]. As expected, air temperature within the nest is greater than that outside the nest. This elevation may importantly affect the power consumption during incubation. Using Calder's (1964) data and this equation to equate nest and environmental chamber temperatures allows calculation of the metabolic rate of a Zebra Finch perching in an environmental chamber. Such calculations indicate that the metabolic rate of an incubating bird is 3–21% *below* that of a bird perching outside the nest in the environmental chamber over ranges of chamber and nest temperatures of 10–28 and 7.5–26.9°C, respectively. This suggests that incubation requires an increase in energy expenditure compared to a nonincubating bird roosting in the nest, but a decrease compared to the normal condition of a nonincubating bird resting outside the nest. This clearly illustrates the necessity of framing such comparisons in an ecologically appropriate manner.

Biebach (1977, 1979) compared the nocturnal oxygen consumption of European Starlings (*Sturnus vulgaris*) incubating in nest boxes with that of nonbreeding adults roosting within the boxes. Metabolic rate in the two groups did not differ significantly when air temperature within the nest box equalled 10–22°C. Power consumption averaged 25–30% higher in incubating than nonincubating adults when air temperatures equalled −10 to +10°C. Biebach's data do not allow comparison of the power consumption of incubating birds with that of nonincubating birds perching outside of the nest. In a manner similar to Vleck (1981), Biebach (1979) placed artificial eggs in the clutch and controlled their temperature by pumping water through them from a reservoir. Artificial egg cooling produced increases of up to threefold in adult resting metabolism.

Gessaman and Findell (1979) presented data describing nocturnal carbon dioxide production for three free-living American Kestrels (*Falco sparverius*) incubating in nest boxes during periods in which air temperatures within the nest box ranged from 4 to 16°C. Although there is substantial variance in the data, average values for two individuals (one male, one female) indicate no important difference between the power consumption of incubating and nonincubating individuals in the nest box. The third individual exhibited metabolic rates up to 43% above average values for nonbreeding birds resting in the nest box at the same air temperatures.

In addition to these respiratory-gas measurements, several workers have analyzed the energetic consequences of incubation using biophysical models (Kendeigh, 1963; Mertens, 1977; Walsberg and King, 1978a,b). Kendeigh (1963) was the first worker to examine systematically the thermobiology of natural incubation. Kendeigh's equation estimates net heat loss from the eggs, which under steady-state conditions should equal heat supplied by the parents to the clutch. Kendeigh (1963) and subsequent users of this equation have assumed that such heat flow is directly additive to parental metabolism (e.g., Haftorn, 1978). These computations predict that energy expenditure during incubation is elevated above that of a resting, nonincubating individual by an increment equal to about 15–40% of BMR [e.g., elevation equals approximately 40% of BMR in the House Wren (*Troglodytes aedon*; Kendeigh, 1963), approximately 30% of BMR in the Zebra Finch (El-Wailly, 1966), and approximately 30% of BMR in the White-crowned Sparrow (Hubbard, 1978)]. Such estimates, however, rest upon the assumption that heat flow to the eggs is an energy burden that is completely additive to parental metabolism. King (1973) initially challenged this viewpoint by noting that all or a large fraction of this heat flux may be supplied by waste heat that is normally lost through other avenues.

Difficulties in use of the Kendeigh formula were also demonstrated by the analyses of Walsberg and King (1978a,b). They estimated the energetic consequences of incubation for three passerine species (*Zonotrichia leucophrys, Empidonax traillii, Agelaius phoeniceus*) using a biophysical model that subsumed physical and physiological characteristics of the adult and eggs, as well as physical properties of the nest and nest microclimate. These analyses indicated that energy expenditure during incubation in these species is typically 15–18% lower than that of a bird perching quietly outside of the nest. This reduction is produced primarily by the insulation of the nest, reducing heat loss over much of the adult's surface, and secondarily by placement of the nest in a sheltered microclimate that reduces thermal stress on the adult. The importance of this reduction in heat loss due to nest insulation had generally been ignored in previous studies, but the analyses by Walsberg and King indicate that this heat conservation can be so large that the effect of heat loss from the adult to the eggs is overwhelmed. If the use of the nest's insulation is characteristic of both incubating and nonbreeding birds, conclusions may be reversed. For example, I have used the model and data of Walsberg and King (1978a,b) to calculate the power consumption of nonincubating adults roosting in the nest rather than in their normal microclimate. This computation predicts that incubation would entail a 6–10% increase in resting energy expenditure *if* the nonbreeding adult normally took advantage of the nest's insulation. Again, this change indicates the critical importance of phrasing such comparisons in an ecologically appropriate manner.

Mertens (1977) also developed a biophysical model that incorporated qualities of the nest, adult, and eggs and used this model to estimate power consumption by incubating Great Tits (*Parus major*). Mertens's model predicts progressive increases of power consumption by an incubating bird as clutch size is increased, compared with power consumption of a nonincubating bird roosting in the nestbox. Unfortunately, Mertens did not provide data describing climatic conditions for this species' nesting period. Thus it is difficult to set the predictions of his model in context. Assuming, however, that ambient temperature normally averages about 15°C in the Netherlands during May allows one to estimate from Mertens's (1977) Figure 4 that incubation requires a power consumption equal to 1.4–1.5 times BMR for clutches of 5–10 eggs. Mertens did not estimate the power consumption of an individual roosting outside of the nest, although this could drop to basal levels, as the lower critical temperature of this species is 8°C (Mertens, 1977).

Two techniques in addition to analyses of respiratory gases or biophysical models have been used to analyze incubation energetics. In a pioneering study, El-Wailly (1966) attempted to quantify incubation costs of captive Zebra Finches using a food-balance technique. El-Wailly computed energy expenditure by measuring food intake in the diet and subtracting excretory energy loss. Changes in energy stored within the body were assumed to be insignificant. El-Wailly compared the power consumption of nonbreeding birds with that of incubating individuals, and concluded that supplying heat to the eggs entailed an increase in energy expenditure equivalent to 0–34% of BMR over an air temperature range of 14.5–34.4°C. However, such estimates depend upon the assumption that the occurrence of incubation behavior is the only important difference affecting energy balance between incubating and nonincubating birds. This assumption is probably seriously violated, because most species exhibit dramatic changes in activity with the onset of incubation. Because variables such as energy allocation to activity were not quantified or controlled by El-Wailly, his data cannot be used reliably to estimate changes in adult metabolism attributable to egg thermoregulation.

Finally, Mertens (1980) has estimated the energy requirements of incubation in Great Tits by measuring heat flux through the walls of the nest box. Mertens implanted 35 heat-flux transducers in the walls of a nest box used by a Great Tit incubating nine eggs. Net heat loss through the walls was taken as equal to the female's heat production. For the single bird studied, these values indicate that nocturnal heat production during incubation equaled two to four times BMR. In contrast, Mertens's values indicate that a nonincubating adult roosting outside of the nest would require only basal expenditure. These are the highest empirically derived values reported for power consumption during incubation. The values should be underestimates, because heat loss by evaporation and convection through the entrance hole in the nest box was not quantified. Although relatively high values might be expected, as this species produces the largest clutch of any bird yet studied, it is also notable that these values are substantially higher than Mertens's other estimates for the same species and setting. For example, data generated from the heat-flux measurements are 1.6–2.2 times higher than those produced by Mertens's biophysical model and 2–4 times higher than those based upon measurements of oxygen consumption (Mertens, 1977). Mertens (1977) estimated that his values for oxygen consumption are about 13% too low due to loss of exhalent air from the nest box, but an error of this size still cannot account for the disparity of these values and those calculated from heat-flux data.

3. Conclusions

In spite of considerable effort by workers, it is difficult to generalize regarding the energetic consequences of incubation for free-living birds. Variation in technical accuracy has undoubtedly produced part of the variation in estimated incubation costs. Such costs, however, should also be sensitive to differences in microclimate, clutch mass, nest construction, and adult physiology and behavior. Current data are not adequate to examine the importance of such effects. Such biological variation may partially explain the disparity between the results of Mertens's (1977, 1980) studies of Great Tits, which incubate very large clutches, and results obtained by other workers studying species that incubate smaller clutches. Most studies of incubation energetics have dealt with birds incubating 3–5 eggs. The most nearly complete and direct analyses of energy expenditure during incubation, those of Biebach (1977, 1979) and Vleck (1981), indicate that power consumption during incubation is increased about 20–30% over that of nonbreeding birds roosting in the nest cavity. Other analyses commonly indicate a similar or smaller increase, if any. Although these values are probably overestimates because the nest is continuously flushed with cool, dry air, the nest is still functioning as insulation to an unknown degree. If one accepts that the nest and adult are a functional unit characteristic of incubation, then a more useful comparison is that of the power consumption of an incubating bird to that of one in its normal microclimate outside of the nest. Such a comparison should produce values for incubating and nonincubating birds that are more similar than those reported by Biebach (1977, 1979) and Vleck (1981). This is supported by the analyses of Walsberg and King (1978a,b) and by my manipulations of Vleck's (1981) data described above, which predict a decrease in resting metabolism in an incubating bird compared to that of a bird outside of the nest in its normal microclimate.

E. CARE OF THE YOUNG

1. Nestling Energy Requirements

A number of workers have estimated the energy requirements of nidicolous young (Table V, Fig. 3), although rarely in conjunction with measurement of parental energy requirements. These estimates are typically derived from measurements of the nestling's food intake and excretory loss of energy (e.g., Dunn, 1975a, 1976) or by combining estimates of the nestling's resting metabolic rate with measurements of energy investment in the growth of new tissues (e.g., Ricklefs et al., 1980). The latter estimates do not account for factors such as nestling activity and are probably underestimates (Dunn, 1980).

Average energy expenditure per nestling ($\bar{E}_N$; kJ/day) and the peak daily energy expenditure per nestling during the nestling period (E_N^*; kJ/day) are both functions of adult body mass (M_B; g) (Fig. 3):

$$\ln(\bar{E}_N) = \ln 14.05 + 0.440 \ln(M_B) \qquad (6)$$

or

$$\ln(E_N^*) = \ln 13.40 + 0.528 \ln(M_B) \qquad (7)$$

These equations are computed using least-squares regression and logarithmically transformed data from Table V. In Eq. (6), $n = 10$, $r^2 = 0.913$, $S_{yx} = 1.608$, and $S_b = 0.1616$. In Eq. (7), $n = 9$, $r^2 = 0.9232$, $S_{yx} = 1.035$, and $S_b = 0.1245$.

Because adult energy expenditure scales with body mass (M_B) to the power of 0.6 [Eq. (8)], whereas nestling requirements scale with parental body mass to the power of 0.4–0.5, these equations suggest that the proportional increase in rates of parental food acquisition associated with feeding a single nestling should decrease with an increase in body mass. Although current data are quite sparse, these factors suggest that feeding typical broods may be a greater burden on small-bodied than large-bodied species.

The energy requirements of typical broods of various species are summarized in Table 5 and compared to estimates of parental expenditure. Most of the values for parental expenditure are only crude estimates based on the allometric relationship between body mass and daily energy expenditure [Eq. (8)]. It is apparent, however, that even if both parents contribute equally to care of the young, a parent must increase two- to threefold the amount of energy it acquires while foraging. An increase in food acquisition of three- to fivefold may be required if only a single parent feeds the young.

TABLE V

ESTIMATES OF NESTLING ENERGY EXPENDITURE FOR TYPICAL BROOD SIZES

Species	Adult mass (g)	Brood size	Peak expenditure		Average expenditure		Reference
			kJ	Brood/adult[a]	kJ	Brood/adult[a]	
Oceanodroma leucorhoa[b]	45	1	97	0.71	80	0.59	Ricklefs et al. (1980)
Phalacrocorax auritus	2047	3	1494	1.10	956	0.70	Dunn (1975a,b)
Eudocimus albus	881	1+	—	—	238	0.35	Kushlan (1977)
Larus argentatus	1000	3	2107	2.38	1225	1.39	Dunn (1976)
Cepphus columba	484	2	860	1.51	490	0.89	Koelink (1972)
Delichon urbica[c]	18	4	184	1.07	109	0.64	Bryant and Gardiner (1979)
Lanius collurio[b]	25	5	285	2.97	189	1.97	Diehl and Myrcha (1973)
Sturnus vulgaris	78	4	544	2.88	416	2.20	Westerterp (1973)
Passer domesticus	24	4	300	3.23	216	2.32	Blem (1975)
Zonotrichia leucophrys	22	4	408	4.37	298	3.19	Hubbard (1978)

[a]Ratio of brood expenditure to that of a single adult. Adult expenditure estimated using Eq. (8), except values for Zonotrichia, Eudocimus, and Delichon, which are author's estimates.
[b]Estimate of nestling expenditure accounts only for growth and maintenance metabolism. Components such as activity are not included, and this value is probably an underestimate.
[c]Calculated assuming that nestling RQ = 0.75.

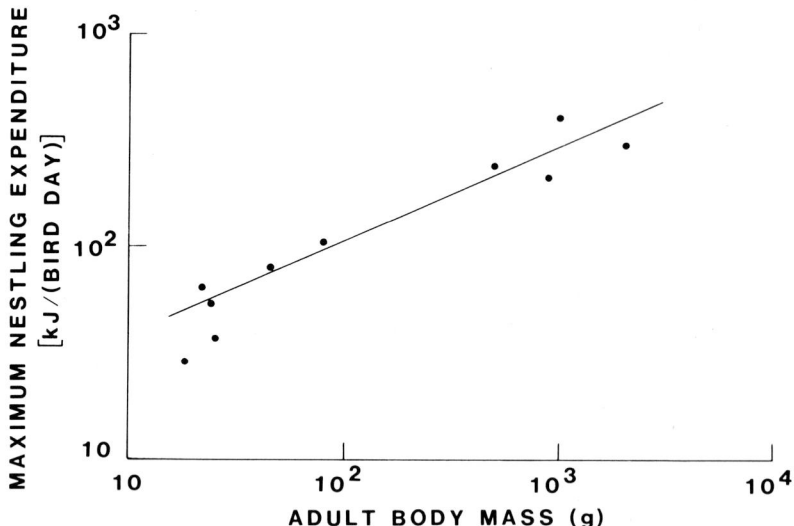

FIG. 3. Peak nestling energy expenditure as a function of adult body mass. $E_N^* = 13.40\, M_B^{0.528}$. Data from Table V.

These data also support the supposition that the energy requirements of feeding a typical brood require a greater proportional increase in parental foraging for small species compared to larger forms (e.g., average expenditure by a *Zonotrichia* brood is about three times that of a single parent, whereas average expenditure by a *Phalacrocorax* brood is only about 0.7 times that of a single parent).

2. Correlates of Parental Energy Expenditure with Brood Size

The relationship between parental energy expenditure and reproductive output as reflected by brood size has received little attention, especially considering the potential ecological significance of these relationships and the extensive attention that has been devoted to the determination of reproductive output in general. Only two studies have analyzed the energetic consequences of variation in brood size in free-living birds.

Estimating energy expenditure from time budgets, I compared parental energy expenditure midway through nestling growth in Phainopeplas raising broods of two or three young (Walsberg, 1978). In this species, both parents feed the young a diet of small fruits and of insects caught in flight. Pairs feeding three young spend 42% more time flycatching and 16% more time foraging for fruit than do individuals raising two young. However, this increased activity is estimated to result in only a 7% increase in parental energy expenditure.

Hails and Bryant (1979) used doubly labeled water to quantify the energy expenditure of House Martins (*Delichon urbica*) raising broods of various number and mass. Their data are not segregated by brood size and age, but rather reported only as a function of brood mass. This may produce substantial variation in results, since the mass-specific energy requirement of nestlings usually varies with age. In this species, parental foraging rate is correlated with *metabolic brood mass* [= (brood mass)$^{0.66}$], with the response of males more sensitive to brood mass and more closely correlated ($r^2 = 0.79$) than that of females ($r^2 = 0.48$). Daily energy expenditure in males is correlated significantly with both brood mass and number. Female expenditure is not correlated significantly with brood mass, but there is a significant, although loose, correlation ($r^2 = 0.29$) with brood size such that an increase in brood size of 33% from the typical brood of three young to one of four young entails an average increase in the female's daily expenditure of 18%.

F. Molt

Seasonal replacement of feathers apparently represents the major event of somatic production in the annual cycle of adult birds. Analysis of the energy cost of any productive process is difficult because it involves segregation of that cost from other components of the energy budget that may be changing simultaneously, such as activity metabolism. In addition, accurate analyses of the energetics of production require concurrent, and often technically difficult, estimates of the rate of synthetic processes. Thus molt energetics remains a confusing issue, although substantial efforts have been devoted to its analysis.

The energetics of molt has been approached using three techniques, each with intrinsic disadvantages. Measurements of oxygen consumption by resting birds are potentially the most accurate. However, most such measurements have been made during only a portion of the day/night cycle, and have explicitly or implicitly assumed that feather growth proceeds at a constant rate through the 24-hour cycle (e.g., Lustick, 1970; Chilgren, 1975). There is little information with which to evaluate this assumption. On the basis of unpublished data, Newton (1968) stated that the rate of feather growth is the same by day as by night in Bullfinches (*Pyrrhula pyrrhula*). Wood (1950) analyzed *growth bars* in feathers and concluded that in most species examined, growth proceeds at similar rates day and night. There were multiple exceptions, however.

Several authors have compared the oxygen consumption at rest in molting and nonmolting birds. In thermal neutrality, oxygen consumption is increased 9–46% (Table VI). Oxygen consumption not only increases during molt, but in carefully studied species the increase appears to be correlated with the intensity of molt (Gavrilov, 1974; Gavrilov and Dolnik, 1974; Chilgren, 1975). Typically, standard metabolic rate below thermal neutrality is also elevated 10–40%. In some species this increase is partially associated with increases in whole-body thermal conductance (including evaporative heat loss) during molt (Gavrilov and Dolnik, 1974; Chilgren, 1975). This is not apparent in at least one species, *Molothrus ater* (Lustick, 1970). In those species that do demonstrate increases in thermal conductance during molt, King's (1981) analysis indicates that such shifts may be largely due to increases in evaporative water loss. Thus King (1981) concluded from a variety of studies that there are as yet no data clearly indicating a decrease in plumage insulation.

Using data from Chilgren (1975) describing changes in nocturnal oxygen consumption, King (1981) estimated that the energy cost of molt in White-crowned Sparrows was about 833 kJ, or 505 kJ per gram of feather replaced. Using similar data for Chaffinches (*Fringilla coelebs*), Dolnik and Gavrilov (1979) estimated that 322 kJ, or 230 kJ/g feather, was expended. This value is substantially lower than those derived from measurements of metabolizable energy intake. The authors suggest that the disparity is due to decreased rates of feather synthesis at night, when O_2 consumption was measured.

Various workers have analyzed molt energetics using measurements of metabolizable energy intake by caged birds, which has the advantage of

TABLE VI

INCREASE IN BASAL METABOLISM DURING MOLT COMPARED TO PREMOLT LEVELS[a]

Species	Average percentage increase	Reference
Gallus gallus[b]	46	Perek and Sulman (1945)
Molothrus ater	13	Lustick (1970)
Fringilla coelebs	12	Gavrilov (1974)
Zonotrichia leucophrys[c]	26–34	Chilgren (1975)
Passer domesticus	9	Gavrilov (1974)

[a]Basal metabolism measured at night in *Fringilla*, *Passer*, and *Zonotrichia* and during the day in *Molothrus*. Authors did not specify time of measurement for *Gallus*.

[b]Premolting birds were hens laying eggs.

[c]Range of values for different phases of molt. Maximum occurs in middle of molting cycle.

encompassing entire 24-hour cycles. A critical disadvantage, however, is the lack of control or measurement of other portions of the energy budget, such as locomotor activity (King, 1981). The activity of caged birds is often greatly reduced during molt (Eyster, 1954; Chilgren, 1975). Only in the unlikely event that all factors such as activity are essentially constant throughout the premolting, molting, and postmolting periods will measurements of metabolizable-energy intake yield accurate estimates of the energy cost of feather production. It is to be expected, therefore, that highly variable results would be obtained using this technique. Indeed, this technique has yielded sharply differing results even within the same species [e.g., analyses of *Passer domesticus* by Davis (1955), Blackmore (1969), and Gavrilov (1974)]. In a variety of studies using caged birds, metabolizable-energy intake is statistically indistinguishable between premolt and molt periods [*P. domesticus* (Davis, 1955), *Lagopus lagopus* (West, 1968), *Zonotrichia leucophrys* (Chilgren, 1975)]. In contrast, Gavrilov (1974) and Gavrilov and Dolnik (1974) found that metabolizable-energy intake increased during molt in *P. domesticus* and *Fringilla coelebs*, and that the increase paralleled the intensity of molt. By subtracting the metabolizable-energy intake characteristic of nonmolting birds from that of molting birds, Gavrilov (1974), Gavrilov and Dolnik (1974), and Dolnik and Gavrilov (1979) estimated the energy cost of molt as 586 kJ total or 419 kJ/g feather in adult *F. coelebs*, 363 kJ total or 451 kJ/g feather in juvenile *F. coelebs*, 1491 kJ total or 785 kJ/g feather in adult *P. domesticus*, and 1545 kJ total or 813 kJ/g feather in juvenile *P. domesticus*. For adults, these increases in energy expenditure over the entire molting period are equal to an average of 30% of BMR in *F. coelebs* and 43% of BMR in *P. domesticus*. This increase is similar to that observed in resting oxygen consumption (Table VI).

A third technique used by Dolnik and Gavrilov to estimate energy intake required for feather synthesis involves analyzing the amino acid requirements of molt (Gavrilov, 1974; Gavrilov and Dolnik, 1974; Dolnik and Gavrilov, 1979). Sulfur-linked amino acids (cystine and cysteine) are less concentrated in foods eaten by birds than in feathers (e.g., dry feather keratin contains 6.8–8.2% cystine and cysteine, whereas plant proteins contain 0–2.9% cystine and cysteine; Ward and Lundgren, 1954; Newton, 1968). Using knowledge of the cystine and cysteine content of the feathers produced during molt and the concentration of these amino acids and energy in the diet, Gavrilov (1974) and Gavrilov and Dolnik (1974) calculated the food intake required to supply these amino acids. The energy content of that food is assumed to equal the net energy cost of molt [see Kendeigh et al., 1977; Eq. (5.63)].

King (1981) noted two unstated but crucial assumptions implicit in such calculations. The calculation assumes that the diet is the sole source of cystine or cysteine. However, these are not truly essential amino acids and are synthesized by birds from methionine (Boorman and Lewis, 1971). This apparently invalidates the technique of Gavrilov and Dolnik and Eq. (5.63) of Kendeigh et al. (1977).

In addition, these authors' calculations assume that the increment in energy intake associated with acquiring these amino acids is not available for use in maintenance functions such as regulatory thermogenesis (i.e., the increased heat production during molt is additive to thermostatic requirements). Indeed, Gavrilov and Dolnik (1974) believed that the bird must metabolize increased amounts of food to acquire necessary amino acids, and that the heat produced by this metabolism increases the probability of heat stress.

Finally, this technique also assumes that the amino acid requirement for feather formation cannot be supplied by normal dietary intake even if transamination does not occur. Rather, these authors apparently assumed that increased food intake is required and that cystine and cysteine in the food normally consumed by nonmolting birds is not available for feather formation. For example, Gavrilov (1974) calculated that a Chaffinch must ingest 58.3 g of food to acquire the amino acids necessary for feather formation. This is equivalent to a metabolizable energy intake of 588 kJ or 23% of this species' BMR over the 70-day period of molt (Gavrilov, 1974). Thus, these amino acids apparently could be supplied in the normal diet without increased food intake. In contrast, Gavrilov (1974) assumed that the bird must increase its food intake by an amount equal to 23% of BMR, apparently ignoring the amino acid content of the rest of the bird's food intake.

In view of such technical problems, it is remarkable that Gavrilov's (1974), Gavrilov and Dolnik's (1974), and Dolnik and Gavrilov's (1979) estimates of the energy cost of molt in *Passer domesticus* and *Fringilla coelebs* calculated using this method agree very closely with estimates based upon measurement of metabolizable-energy intake. For example, Gavrilov (1974) presented eight sets of data for different ages and races of *F. coelebs* and *P. domesticus*. The estimated energy costs of molt based upon the amino acid technique represent an increase in expenditure equal to 16–20% of BMR, and differ by an average of only 0.7% (range: 0–1.6%) from estimates based upon metabolizable energy intake.

Thus it is striking (and currently inexplicable) that the three major techniques for estimating the energetic consequences of molt have yielded generally similar results. From analyses of metabolizable-energy intake or cystine requirements of feather growth in House Sparrows and Chaffinches, Gavrilov (1974), Gavrilov and Dolnik (1974), and Dolnik and Gavrilov (1979) concluded that approximately 420–460 kJ were expended per gram of feather produced. Analyzing Chilgren's (1975) data for O_2 consumption by

White-crowned Sparrows, King (1981) estimated that molt required 505 kJ/g. Averaged over the entire molting period, these expenditures are typically equal to 20–40% of BMR.

These values are approximately 20 times higher than can be accounted for by the potential-energy content of the plumage (heat of combustion of keratin = 22.5 kJ/g; King and Farner, 1961). If any or all of the estimates currently available are accurate, then molt must involve energetically expensive processes other than simply keratin synthesis. Molt entails a variety of additional physiological modifications (e.g., approximate 20% increase in blood volume in *Zonotricha leucophrys*; Chilgren and DeGraw, 1977), but such processes have not been shown to be capable of producing such large increases in power consumption. One source for such an increase could be an elevated thermostatic requirement due to disruption of plumage insulation by molt. As noted above, there is little evidence to support this, particularly since elevated rates of power consumption persist throughout the thermal neutral zone.

Current data thus indicate that for unknown reasons feather replacement requires a substantial increase in energy expenditure in some species in captivity. The increase is large enough to significantly affect the energy relationship of free-living birds, although few studies of molting birds in nature are available (e.g., Ankney, 1979). Particularly needed are analyses of free-living species that produce a complete energy budget correlated with measurements of molt intensity.

III. Total Energy Expenditure of Free-Living Birds

A. Body Mass and Energy Expenditure

Table VII and Fig. 4 summarize estimates of total daily energy expenditure (E_{TOT}) in free-living birds. On the basis of average values when multiple estimates are available for a species in the same phase of the annual cycle and median values when estimates for different phases are available, a least-squares regression of the logarithmically transformed data for E_{TOT} (kJ/day) on body mass (M_B in grams) yields the relationship

$$\ln(E_{TOT}) = \ln 13.05 + 0.6052 \ln(M_B) \qquad (8)$$

where $n = 42$, $r^2 = 0.981$, $S_{yx} = 0.4150$, and $S_b = 0.0199$. Birds that forage in flight might be expected to exhibit relatively high rates of energy expenditure. Considering only these species yields the equation

TABLE VII

Estimates of Total Daily Energy Expenditure in Free-Living Birds[a]

	Body mass (g)	Energy expenditure (kJ/day)	Reference
Aptenodytes forsteri	25,200	6373	Le Maho et al. (1976)
Eudocimus albus	881	689	Kushlan (1977)
Branta leucopsis	1,550	920	Ebbinge et al. (1975)
Anas rubripes	1,000	586	Wooley and Owen (1978)
Anas strepera	955	553–687	Dwyer (1975)
Oxyura maccoa	761	738	Siegfried et al. (1976)
Elanus caeruleus	243	412	Tarboton (1978)
Elanus leucurus	331	474	Koplin et al. (1980)
Buteo regalis	1237(♂)–1983(♀)	1110–1566	Wakely (1978)
Falco sparverius	119	173	Koplin et al. (1980)
Lagopus lagopus	500	864	Moss (1973, 1977)
Lagopus leucurus	360	439	Moss (1973)
Lagopus mutus	420	419	Moss (1973)
Actitis macularia	42.5	46.7–153.7	Maxson and Oring (1980)
Calidris pusilla	27.0	142–218	Ashkenazie and Safriel (1979)
Asio flammeus	406	787	Graber (1962)
Asio otus	252	665	Graber (1962)
Aegolius acadicus	96	247	Graber (1962)
Calypte anna	4.8	22–34	Pearson (1954), Stiles (1971), Calder (1975)
Stellula calliope	3.2	31	Calder (1971)
Colibri coruscans	8.3	49	Hainsworth (1977)
Oreotrochilus estella	8.1	36–102	Carpenter (1976)
Empidonax traillii	12.6	51.7–62.2	Ettinger and King (1980)
Empidonax minimus	10.0	48.7–71.0	Holmes et al. (1979)
Delichon urbica	17.8	84.4–91.6	Hails and Bryant (1979)
Progne subis	49.0	139–227	Utter and LeFebvre (1973)
Petrochelidon pyrrhonota	24.6	106–134	Withers (1977)
Pica pica	162(♀)–183(♂)	232–299	Mugaas and King (1981)
Nucifraga caryocatactes	158	238	Andreev (1977)
Mimus polyglottos	46.9	73–176	Utter (1971)
Anthus spinoletta	22.0	50.0	Gibb (1956)
Calcarius lapponicus	27.0	105–167	Custer (1974)
Phainopepla nitens	24.0	58.5–89.1	Walsberg (1977, 1978)
Nectarinia famosa	13.5	71.0	Gill and Wolf (1975)
Nectarinia olivacea	12.1	55.1	Frost and Frost (1980)
Nectarinia reichenowi	15.0	68.2	Gill and Wolf (1975)
Vireo olivaceus	17.0	52.9–77.7	Holmes et al. (1979)
Vestiaria coccinea	16.9	65.6	Carpenter and MacMillen (1976)
Dendroica caerulescens	9.4	37.7–46.3	Holmes et al. (1979)
Spiza americana	35.0	100	Schartz and Zimmerman (1971)
Zonotrichia leucophrys	21.8	89.6–107.7	Hubbard (1978)

[a] Maximum and minimum are given if values are available for different sexes or portions of the annual cycle. Average value is given if multiple estimates are available for the same sex or phase of the annual cycle.

3. AVIAN ECOLOGICAL ENERGETICS

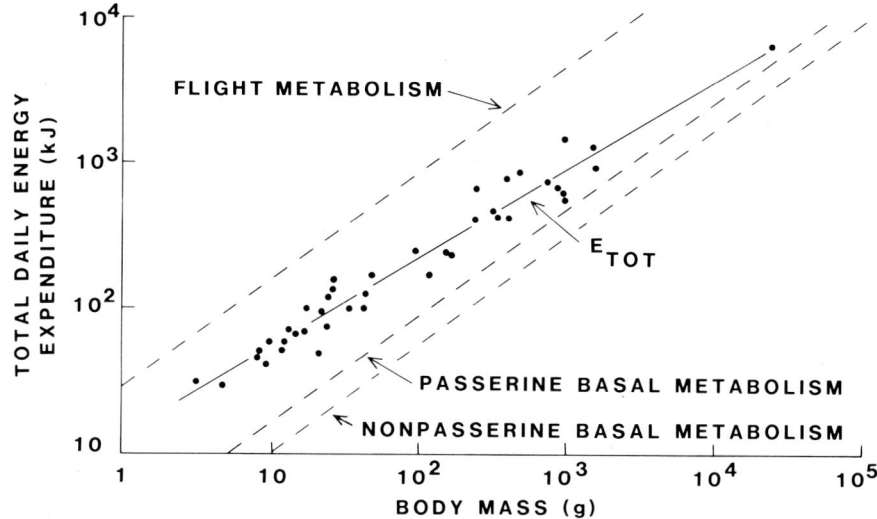

FIG. 4. Total daily energy expenditure as a function of body mass. Lines representing basal expenditure were calculated using equations of Aschoff and Phol (1970) and that representing expenditure during flight was calculated using the equation of Hart and Berger (1972). Other data from Table VII.

$$\ln(E_{TOT}) = \ln 14.17 + 0.607 \ln(M_B) \quad (9)$$

where $n = 11$, $r^2 = 0.978$, $S_{yx} = 0.3088$, and $S_b = 0.0305$. Considering data for all other species yields the relationship

$$\ln(E_{TOT}) = \ln 12.84 + 0.610 \ln(M_B) \quad (10)$$

where $n = 31$, $r^2 = 0.956$, $S_{xy} = 0.4257$, and $S_b = 0.0236$. None of these three equations is statistically distinguishable from any other ($P > 0.20$), and the predicted differences in energy expenditure are slight. For example, Eq. (9) predicts a level of expenditure about 10% above that predicted by Eq. (10). It is obvious, however, that species of similar mass may differ greatly in daily energy expenditure. For example, estimates of E_{TOT} for 21- to 25-g species range from 50 to 134 kJ/day. It is not yet possible to segregate the degree to which such variation reflects biological variation rather than errors in estimation.

It is particularly notable that E_{TOT} scales with $M_B^{0.61}$, because other physiological rate functions usually scale with $M_B^{0.7-0.8}$ (e.g., BMR, energy expenditure in flight; see Calder, 1974). Because metabolic rate in at least two major avenues of energy expenditure, flight and basal metabolism, scales with $(\text{mass})^{0.72-0.74}$, total daily energy expenditure might be expected to exhibit a similar relationship if birds of different size tend to allocate similar amounts of time to these different intensities of activity. This does not occur, however, because smaller species tend to be more intensely active than larger forms. Figure 5 shows the relationship between time spent in powered flight and body mass. Using median data when multiple values are available for a species and calculating a least-squares regression using logarithmically transformed data yields the relationship

$$\ln(T_F) = \ln 44.26 - 0.603 \ln(M_B) \qquad (11)$$

where T_F is the percentage of the active day spent in powered flight (i.e., excluding soaring or extensive gliding), $n = 29$, $r^2 = 0.501$, $S_b = 0.1074$, and $S_{yx} = 0.9012$. Although only a loose correlation, this does indicate a tendency for smaller species to allocate more time to this energetically expensive activity [e.g., Eq. (4) predicts that the average 10-g species spends 11% of its active day in flight, whereas the average 100-g species is in powered flight only 3% of the day]. In addition, other important physiological rate functions

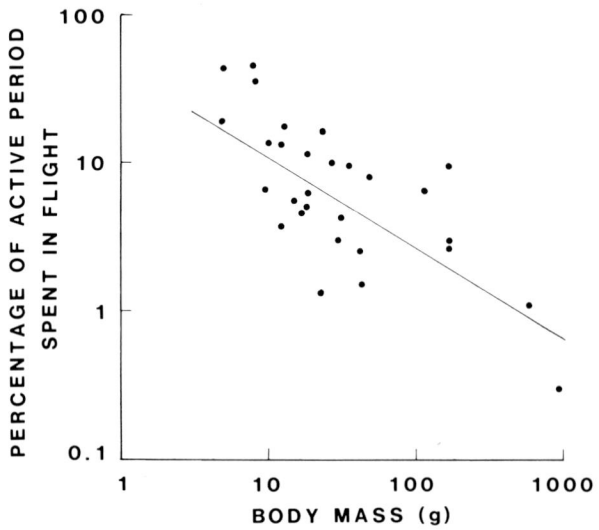

FIG. 5. Relationship between body mass and function of active day spent in powered flight. $T_F = 44.3 M_B^{-0.603}$. Median value is plotted if multiple values are available for a single species. Sources of data: Ashkenazie and Safriel (1979), Austin (1978), Boyd (1976), Boyden (1978), Carpenter (1976), Custer (1974), Dwyer (1975), Ettinger and King (1980), Frost and Frost (1980), Gill and Wolf (1975), Greenlaw (1969), Hainsworth (1977), Holmes et al. (1979), Hubbard (1978), Koplin et al. (1980), Maxson and Oring (1980), Mugaas and King (1981), Murdock (1975), Pearson (1954), Pinkowski (1979), Schartz and Zimmerman (1971), Siegfried et al. (1976), Stiles (1971), Tarboton (1978), Utter and LeFebvre (1973), Verbeek (1964, 1972), Walsberg (1977, 1978), Wiens (1969), and Wolf (1975).

may scale with $M_B^{0.4-0.6}$. For example, rates of egg synthesis apparently vary in proportion to $M_B^{0.5}$ [Eq. (5)], and nestling energy expenditure (an index of the rate at which adults must acquire energy for their young) apparently scales with $M_B^{0.4-0.6}$ [Eqs. (6) and (7)].

The result of this increased activity and mass-specific rate functions in smaller forms is that daily energy expenditure is typically equal to a higher multiple BMR than is characteristic of larger species. Using Eq. (8) to calculate power consumption and calculating BMR from the equation of Aschoff and Pohl (1970) for passerine species during the inactive phase of their daily cycle predicts that daily energy expenditure will average 3.1 times BMR for a 10-g bird and 1.8 times BMR for a 1000-g bird.

B. Relationship between Total Energy Expenditure and Maintenance Requirements

1. Allocation of Energy to Maintenance Metabolism

Energy allocated to maintenance metabolism (= basal and thermostatic demands) may be considered an obligatory or nearly obligatory expenditure, because this avenue is relatively insensitive to behavioral modifications except through microclimatic selection (Ettinger and King, 1980). Other expenditures of energy, such as activity, are more sensitive to behavioral shifts. This distinction is important, because allocation to obligatory demands may determine the amount of energy available for elective expenditures such as reproduction.

Table VIII summarizes data on fractional allocation to obligatory and elective demands. Typically, maintenance metabolism accounts for 40–60% of total expenditure. The minimum reported maintenance expenditure is 28% of E_{TOT} in female White-crowned Sparrows during egg synthesis. This reflects an increase in total expenditure during this period due to the elective allocation to egg synthesis with a relatively constant amount of energy devoted to maintenance functions. The maximum fractional allocation to maintenance expenditure is the 85% value reported by Mugaas and King (1981) for female Black-billed Magpies during the egg-laying period. As in White-crowned Sparrows, there is a substantial allocation to egg synthesis, but total elective expenditure is reduced at this time due to the female's reduced activity. The amount of energy devoted to maintenance functions is relatively stable.

Sufficient data are available to describe variation in these fractional allocations over the annual cycle in two species and through different phases of reproduction in six additional species (Table VIII). Over the annual cycle in Black-billed Magpies, the fraction of total expenditure devoted to mainte-

TABLE VIII

Phasic Variation in Total Daily Energy Expenditure (E_{TOT}) and Fractional Allocation to Obligatory Demands (E_{OBL})

Phase of annual cycle	Males		Females		Reference
	E_{TOT} (kJ/day)	E_{OBL}/E_{TOT}	E_{TOT} (kJ/day)	E_{OBL}/E_{TOT}	
Actitis macularia					Maxson and Oring (1980)
Egg laying	60	—	111	—	
Incubation	47	—	72	—	
Brooding	66	—	—	—	
Calidris pusilla					Ashkenazie and Safriel (1979)
Egg laying	150	0.69	218	0.46	
Incubation	161	0.57	174	0.60	
Nestlings (early)	159	0.59	168	0.62	
Nestlings (late)	146	0.59	—	—	
Oreotrochilus estella					Carpenter (1976)
Incubation and brooding	—	—	77	—	
Late nestlings	—	—	102	—	
Winter	36	—	—	—	
Empidonax traillii					Ettinger and King (1980)
Nest construction and egg laying	61	0.43	58	0.47	
Incubation	54	0.40	52	0.44	
Nestlings	60	0.45	54	0.41	
Empidonax minimus					Holmes et al. (1979)
Nest construction and egg laying	53	0.48	61	0.42	
Incubation	54	0.47	49	0.52	
Nestlings	64	0.39	69	0.37	
Care of fledged young	68	0.44	71	0.42	
Petrochelidon pyrrhonota					Withers (1977)
Nest construction	134	—	134	—	
Incubation	106	—	106	—	
Nestlings	111	—	111	—	
Pica pica					Mugaas and King (1981)

196

Nestlings	315	0.41	235	0.55
Nonbreeding				
July	258	0.58	228	0.58
August	244	0.61	217	0.61
September	255	0.58	227	0.58
October	271	0.53	241	0.53
November	264	0.58	298	0.58
December	298	0.51	266	0.51
Calcarius lapponicus					Custer (1974)
Egg laying	170	—	223	—
Incubation	159	—	151	—
Nestlings	155	—	137	—
Phainopepla nitens					Walsberg (1977)
Incubation (March)	55	0.74	56	0.73
Nestlings (April)	72	0.53	74	0.47
Incubation (June)	70	0.41	70	0.46
Nestlings (July)	86	0.67	89	0.59
Winter	60		60	0.80
Vireo olivaceus					Holmes et al. (1979)
Nest construction and egg laying	58	0.44	78	0.33
Incubation	68	0.38	53	0.41
Nestlings	61	0.42	69	0.37
Care of fledged young	57	0.52	64	0.47
Dendroica caerulescens					Holmes et al. (1979)
Nest construction	38	0.43	40	0.41
Egg laying	40	0.41	46	0.35
Incubation	40	0.41	39	0.42
Nestlings	41	0.39	44	0.37
Care of fledged young	41	0.39	41	0.40
Zonotrichia leucophrys					Hubbard (1978)
Courtship	108	0.56	104	0.55
Nest construction	96	0.54	98	0.54
Egg laying	90	0.52	104	0.30
Incubation	86	0.51	98	0.59
Nestlings	93	0.52	94	0.42

nance metabolism is rather stable. Except for the female during egg synthesis, maintenance expenditure accounts for 48–68% of total expenditure. Most variation in fractional allocation to obligatory requirements is due to changes in total expenditure associated with variation in the activity increment. For example, although maintenance expenditure ranges only from 141 to 158 kJ/day (12% variation) in males over the annual cycle, total daily expenditure ranges from 233 to 315 kJ/day (35% variation). Variation in obligatory requirements is thus small compared to that in elective expenditure. Similar patterns are observed in the Phainopepla (Table VIII). Over the annual cycle, maintenance metabolism accounts for 39–80% of total expenditure. This reflects, however, a relatively stable amount of energy devoted to maintenance metabolism in the face of a more variable total expenditure due to variation in elective components. In the male, for example, maintenance expenditure ranges from 35 to 41 kJ/day (17% variation), whereas expenditure devoted to activity varies 360% (14–52 kJ/day) and total expenditure varies 56% (55–86 kJ/day).

For only three other species, and only during reproduction, has allocation to maintenance metabolism been analyzed using methods sensitive to phasic variation in the thermal environment (Table IX). In *Calidris pusilla*, elective expenditure during reproduction varies substantially more than does obligatory expenditure. In contrast to *Calidris*, *Pica*, or *Phainopepla*, most phasic variation in total expenditure by *Empidonax traillii* and *Zonotrichia leucophrys* is due to shifts in obligatory rather than elective components (Table IX). For example, in *E. traillii* maintenance expenditure varies 36%, whereas total expenditure varies only 13% in either sex. In *Z. leucophrys* allocation

TABLE IX

RATIO OF SEASONAL VARIATION (V) IN ALLOCATION TO ELECTIVE AND OBLIGATORY COMPONENTS OF THE DAILY ENERGY BUDGET[a]

Species	$V_{ELECTIVE}/V_{OBLIGATORY}$		Reference
	Males	Females	
Calidris pusilla[b]	1.4	13.3	Ashkenazie and Safriel (1979)
Empidonax traillii[b]	0.5	0.5	Ettinger and King (1980)
Pica pica[c]	8.5	7.7	Mugaas and King (1981)
Phainopepla nitens[c]	6.9	6.8	Walsberg (1977)
Zonotrichia leucophrys[b]	0.3	1.8	Hubbard (1978)

[a] Variation in each component is calculated as seasonal maximum minus seasonal minimum. Data are used only for species in which estimates of thermostatic demands were sensitive to variation in ambient temperature.

[b] Variation during reproductive period.

[c] Variation over annual cycle.

to maintenance functions during the breeding period varies 210%, whereas E_{TOT} varies only about 11% in either sex.

In summary, these limited data indicate that maintenance metabolism subsumes a large fraction of total expenditure in a free-living bird; commonly as much energy is allocated to these functions as to all other activities combined. In some species, variation in maintenance metabolism may be important in determining variation in total expenditure. This clearly demonstrates the necessity to continue expansion of our knowledge, which is currently primitive, of the thermostatic relationships of birds in their naturally selected microclimates.

2. Interaction of Activity and Productive Metabolism with Thermoregulation

The relationship between heat produced as a by-product of activity and an endotherm's thermostatic requirement has been the subject of much discussion and confusion. Much of this discussion has centered on the question of whether this waste heat of activity or production substitutes for the resting thermostatic requirement of endotherms at low temperature. This is an important question in ecological energetics. It may determine, for example, the extent to which an active bird is stressed by low environmental temperatures or, conversely, the extent to which activity such as foraging entails an increase in power consumption over the resting level. If heat produced as a by-product of activity substitutes for the thermostatic requirement of a resting bird, then it is possible that there is little or no net energetic cost to such activity for birds in cold climates. In this section I shall briefly review knowledge of this phenomenon in birds and present an alternative viewpoint that may aid in revitalizing consideration of the interaction of cold- and exercise-induced thermogenesis.

Investigations of birds engaged in nonflight activity at low temperatures indicate that there may be a partial or complete substitution of the waste heat of activity for thermostatic demands (Pohl, 1969; Pohl and West, 1973; Paladino, 1979). For Common Redpolls (*Carduelis flammea*; Pohl and West, 1973) and White-crowned Sparrows (Paladino, 1979), the authors concluded that partial or complete substitution of exercise heat for the thermostatic requirement occurred at low temperatures (−45 to −30°C in *Carduelis*, −10 to 0°C in *Zonotrichia*). Results collected at more moderate temperatures indicated an addition of exercise thermogenesis to cold-induced thermogenesis.

Measurements of flight metabolism have produced mixed results. Budgerigars (*Melopsittacus undulatus*; Tucker, 1968) flying in a wind tunnel exhibit a metabolic rate that is independent of ambient temperature between 20 and 37°C. For a hummingbird, the Glittering-throated Emerald (*Amazilia fimbriata*), Berger and Hart (1972) concluded that exercise heat production partially substituted for the thermostatic increment. Power consumption during hovering increases with decreasing temperature at a rate equal to roughly one-half that expected if flight metabolism were simply additive to the resting thermostatic requirement.

Such results are initially confusing, because it is not obvious why heat produced as a by-product of activity does not substitute fully for the thermostatic increment. One possible explanation is that thermostatic costs are increased during exercise due to disruption of the body insulation. This has not been clearly demonstrated, and in at least some studies the relationship between O_2 consumption and ambient temperature indicate that thermal conductance is similar in resting and active birds (Pohl and West, 1973).

Many analyses of this problem have assumed implicitly that birds are rigid homeotherms; that is, that they regulate body temperature (T_B) around a single set point. This apparently is not a valid assumption, because birds regulate T_B at a variety of set points depending upon activity and physiological condition. In flight, there is an elective evaluation of T_B in at least some species. For example, Torre-Bueno (1976) demonstrated that T_B in European Starlings during flight is elevated to 43–44°C and is independent of ambient temperature over a range of 0–28°C. The elevation appears to be elective because insulation is apparently adjusted to maintain a high temperature (i.e., the increase in core temperature does not simply reflect inability to dissipate the heat load generated by flight). There is also evidence that T_B is electively increased during nonflight locomotion (Taylor *et al.*, 1971; Paladino, 1979). Paladino (1979) monitored core body temperature of White-crowned Sparrows resting and hopping on a treadmill. Body temperature increased from an average of 42.9°C at rest to 43.8°C at the maximum hopping speed of 12.8 cm/second. This elevation of T_B was independent of ambient temperature from 0 to 25°C, although the increase did not occur at $T_A = -10°C$. The elevation appears to be voluntary, because whole-body thermal resistance (largely a function of plumage insulation) increased with declining T_A in exercised birds. This indicates that the birds increased the effectiveness of their insulation when exposed to low temperatures. If they refrained from such adjustment, T_B would have decreased to values nearer resting levels.

There is also evidence hinting at elevations of body temperature during intense synthetic activities such as molt. Newton (1968), Lustick (1970), and Dolnik and Gavrilov (1979) noted that T_B may be elevated about 0.5–1.0°C during molt. (Dolnik and Gavrilov indicate that T_B is elevated during early phases of molt, but late in the molting period is depressed about 0.5°C below premolting levels.) Such increases apparently occur in spite of a stable or slightly increased whole-body thermal conductance. Molting birds are also characterized by increased metabolic rates, which may partially reflect an adaptation allowing elevated body temperatures. The data are sparse, however, and further measurements are required before definite statements can be made.

Finally, it is notable that most species studied exhibit a regulated, circadian cycle of resting body temperature (Calder and King, 1974). Resting T_B is normally 1–3°C lower during the inactive phase of the daily cycle than during the active phase.

Taken together, these data indicate that body temperature is regulated at varying levels and suggest that the set point is correlated with intensity of activity. The varying degrees of elevation largely appear to be elective and regulated. That is, the higher T_B characteristic of active birds is not necessarily a passive result of a heat load produced by exertion. What advantages might such variation in T_B confer? One can speculate, for example, that such elevations during activity may exploit the temperature dependence of physiological rate functions. Assuming $Q_{10} = 2.5$, a 2–4°C increase in tissue temperature during activity should produce a 20–44% increase in power production per unit muscle mass. This could substantially increase locomotor capabilities. A 1°C elevation of T_B during molt could increase rates of synthesis about 10%, thus shortening this presumably vulnerable period for the bird (see Section II,F for discussion of molt energetics). In a sense, these speculations are an extension of the view that in the evolution of high body temperatures endotherms have sacrificed energy economy for increased rates of sustained activity (Bennett and Ruben, 1979).

Such considerations suggest that the *additive* or *partially substitutive* nature of activity metabolism to the thermostatic requirement may be an adaptation associated with the elective maintenance of elevated body temperatures. One reason for the difficulty in understanding why the waste heat of activity does not always effectively substitute for the thermostatic heat production of a resting bird may be that active and resting birds are thermoregulating at different levels. The thermostatic requirement of active birds may thus be underestimated. Clearly, more data are needed, but perhaps a more useful viewpoint from which to investigate this problem is to address the question, What are the advantages of expending energy to maintain a higher body temperature during locomotor or intense synthetic activity?

C. Seasonal and Sexual Variation in Energy Acquisition and Expenditure

1. Variation in Energy Expenditure between Reproductive Phases

Data describing phasic variation in energy expenditure during reproduction are summarized in Table VIII. Authors have commonly divided reproduction into periods of nest construction, egg laying, incubation, and care of the nestling. The postfledging period has largely been ignored, which is unfortunate, as this may be a period of high juvenile mortality and energy requirements. In addition, foraging patterns of parents feeding fledged young may change substantially, because the fledgling's mobility may make the parent's return to a central nest site unnecessary.

In all species for which data are available, female parental expenditure is 15–85% greater than male expenditure during the egg-laying period ($\bar{x} = 32\%$; calculated from data in Table VIII). During incubation, the parent that takes the major role in nest attendance averages 3–50% lower in energy expenditure than its mate ($\bar{x} = 18\%$). Sexual differences in daily energy expenditure decrease during the nestling period and average about 10%. In most species, parental energy expenditure differs only 0–13%, with *Pica pica* the exception. In this species, the male takes the major role in feeding the young and his mass-specific daily energy expenditure averages 19% greater than the female's. For the species in Table VIII, male parental expenditure typically peaks during nest construction or nestling care, associated with the intense locomotor activity characteristic of these periods.

The female's expenditure typically peaks during egg synthesis. The least expensive phase differs widely in females, but typically occurs during incubation in the male (Table VIII). This reflects the great reduction in activity characteristic of a bird attending the nest, which commonly overwhelms any additional thermostatic expenditure required to warm the eggs (Walsberg and King, 1978b).

2. Parental Investment of Energy in Reproduction

Theories of parental investment (*sensu* Trivers, 1972) are a popular and heuristically useful approach to analyzing the selective pressures responsible for the diverse reproductive patterns observed in nature (e.g., Howe, 1977, 1978; Alcock, 1979; Lill, 1979). It is, however, a difficult body of theory to test critically. Parental investment is generally defined as any investment made by a parent in one reproduction at the expense of its ability to invest in future reproductions (Trivers, 1972). Thus such investment may include energy or time allocation, exposure to predators, or a host of other factors. To test parental investment theory critically, the aspects of such allocation that constitute actual parental investment must be determined (i.e., that allocation that occurs at the expense of the parent's ability to invest in future reproductions) and various modes of investment must be equated to a common currency. No investigator has yet overcome these difficulties. Many authors (e.g., Trivers, 1972; Gladstone, 1979) have suggested that energy may be used as a currency to estimate parental investment. Implementation of such suggestions requires determination of (1) the amount of energy expenditure devoted to reproduction, (2) the extent to which the parent has the option of allocating such energy to future reproductions, and (3) the extent to which energy represents a limiting resource for reproduction. For example, it is likely that a substantial fraction of the energy devoted to reproduction may not be correctly classified as parental investment because the parent bird does not possess the option of allocating that energy to a future reproduction. As noted previously, small birds store small amounts of energy compared to their high rates of consumption. For example, for the five small (<25 g) altricial species in Table VIII for which data are available describing female energy expenditure from egg laying to fledging, a female that elected to reduce her expenditure 25% during a reproductive event and store that energy to invest in a subsequent reproduction would be required to store fat equal to an average of 72% of normal body mass (range = 56–85%; calculated assuming fat = 40 kJ/g). Authors frequently assume that egg production is an important mode of investment, but in these five species the clutch contents represent an average of only 2% of the energy expended during reproduction. Thus changes in allocation to egg production in such species will negligibly affect the female's energy balance at the initiation of a subsequent reproduction. This suggests that, in altricial species, only changes in energy allocation that occur relatively late in the breeding cycle (i.e., care of the young) are likely to affect importantly the adult's ability to invest in subsequent reproductions.

Additional difficulties that arise in applying parental investment theory include determination of the fraction of adult energy expenditure during the reproductive period that can be classified as devoted to reproduction. For example, defense of a foraging territory that subsequently is held through multiple reproductions may represent simultaneous investment in present and future reproductions. Obligatory components of the adult's energy budget (i.e., basal and thermostatic demands) should also be excluded from estimates of parental investment in young.

The reliability of predictions of avian behavior made using parental investment theory and the assumption that energy is an appropriate currency for quantifying such investment have not been adequately evaluated by any worker. This could be accomplished by quantifying patterns of such investment in a variety of bird species and correlating such patterns with patterns of parental behavior. Table X summarizes estimates of parental energy investment in five monogamous passerine species. In light of the complications discussed earlier, parental investment is estimated only crudely as the elective components of the adult's energy budget during the latter (nestling) phase of reproduction. However, these patterns of investment do not change importantly if elective expenditure during the entire period from egg synthesis to fledging is included in the estimate. In most species listed in Table X, males and females contribute nearly equal fractions of total investment. Notable exceptions are the Red-eyed Vireo (*Vireo olivaceus*), in which the male contributes only a small fraction of the total investment, and the Black-billed Magpie, in which the male contributes nearly twice as much as the females.

A prediction from current theory in sexual selection is that the sex that invests more, and which consequently stands to lose more in a mismating, should be more discriminating in choice of mates than the sex that invests less. In addition, adaptations that ensure paternity should be most conspicuous in those species in which the male contributes a large fraction of total parental investment (i.e., adaptations that increase the likelihood that the young in which the male invests are actually his offspring); Trivers, 1972; Barash, 1976; Burley, 1977; Alcock, 1979). Unfortunately, of the species listed in Table X, sufficient behavioral data are available only for the Black-billed Magpie. The behavior of this species, however, agrees with that predicted by the male's large energetic contribution to reproduction. Birkhead (1979) reported that the male acts to ensure his paternity of future offspring by guarding his mate against copulations with other males. Such guarding behavior is more intense during the female's fertile period than at other times. Thus these data are consistent with parental investment theory, but comparative data for other species with contrasting patterns of investment are lacking.

TABLE X

PARENTAL INVESTMENT (PI) OF ENERGY DURING THE NESTLING PERIOD[a]

	Male PI (kJ/day)	Female PI (kJ/day)	Male PI/total PI[b]
Empidonax traillii	33	31	0.52
Empidonax minimus	21	27	0.44
Progne subis	106	143	0.43
Pica pica	186	106	0.61
Phainopepla nitens	34	37	0.48
Vireo olivaceus	9	19	0.32
Dendroica caerulescens	11	14	0.44
Zonotrichia leucophrys	45	55	0.45

[a]Data from references cited in Table VIII. Energy investment is assumed to equal elective components of the daily energy budget. See text for discussion.

[b]Total PI equals the sum of male and female investment.

3. Variation in Rates of Energy Acquisition

To this point, only variation in rates of energy expenditure has been considered. Obviously, parental energetics during reproduction may be affected importantly by the necessity to feed the young as well as by restrictions in foraging due to allocation of time to duties such as brooding. The net effects of such time restrictions, parental energy expenditure, and nestling energy expenditure have rarely been examined. Direct studies of foraging rates (e.g., Morse, 1968) during reproduction would be very useful, but are sparse. For the few species for which appropriate data are available, I have estimated variations in foraging rates by computing the average rate at which parents must acquire energy during their active period (E_{FOR}; Table XI). These estimates are derived using authors' estimates for adult and nestling energy expenditure, but with no attempt to estimate assimilation efficiency. The adults are assumed to remain in energy balance and to have their entire daily active period available for foraging except for time devoted

TABLE XI

ESTIMATES OF AVERAGE RATES OF ENERGY ACQUISITION DURING FORAGING[a]

Phase of annual cycle	Sex	Length of daily active period (hours)	Time available for foraging (hours)	Energy supplied to young (kJ/day)	Average foraging rate (kJ/hour)
Actitis macularia[b]					
Egg laying	Male	17.0	13.1	0	4.6
	Female	17.0	16.3	0	6.8
Incubation	Male	17.0	4.1	0	11.4
	Female	17.0	12.9	0	5.6
Brooding	Male	17.0	15.5	0	4.2
Empidonax traillii					
Nest construction and egg laying	Male	18.0	18.0	0	3.4
	Female	18.0	9.7	0	6.0
Incubation	Male	18.0	18.0	0	3.0
	Female	18.0	4.6	0	11.2
Nestlings	Male	18.0	18.0	25[c]	4.7
	Female	18.0	13.1	25[c]	6.0
Pica pica					
Egg laying	Male	13.2	13.2	0	17.7
	Female	12.8	11.5	0	22.4
Incubation	Male	15.1	14.9	0	18.5
	Female	11.5	1.5	0	109.2
Nestlings	Male	16.0	15.1	589[d]	59.9
	Female	16.0	13.9	124[d]	26.0

(*continued*)

TABLE XI *Continued*

Phase of annual cycle	Sex	Length of daily active period (hours)	Time available for foraging (hours)	Energy supplied to young (kJ/day)	Average foraging rate (kJ/hour)
July	Male	16.6	16.6	0	15.5
	Female	16.6	16.6	0	13.7
August	Male	15.2	15.2	0	16.1
	Female	15.2	15.2	0	14.3
September	Male	13.2	13.2	0	19.3
	Female	13.2	13.2	0	17.2
October	Male	11.5	11.5	0	23.6
	Female	11.5	11.5	0	21.0
November	Male	9.8	9.8	0	30.4
	Female	9.8	9.8	0	26.9
December	Male	8.9	8.9	0	33.5
	Female	8.9	8.9	0	30.0
Phainopepla nitens					
Incubation	Male	11.8	6.5	0	8.5
(March)	Female	11.8	5.6	0	10.0
Nestlings	Male	13.0	12.0	31[c]	8.6
(April)	Female	13.0	11.8	31[c]	8.9
Incubation	Male	14.5	8.2	0	8.5
(June)	Female	14.5	7.8	0	9.0
Nestlings	Male	14.6	13.5	47[c]	9.9
(July)	Female	14.6	13.4	47[c]	10.2
Winter	Male	10.3	10.3	0	5.8
	Female	10.3	10.3	0	5.8
Zonotrichia leucophrys					
Nest construction	Male	16.7	16.7	0	5.7
	Female	16.7	10.9	0	9.0
Egg laying	Male	16.7	16.7	0	5.4
	Female	15.8	8.9	0	11.7
Incubation	Male	16.7	16.7	0	5.2
	Female	15.4	4.4	0	22.2
Nestlings	Male	16.4	14.8	149[d]	16.4
	Female	15.3	10.0	149[d]	24.3

[a] Adult daily energy expenditure is listed in Table VII. Other data are from references cited in Table VIII. See text for discussion of assumptions and methods of calculations.
[b] Young are precocial and are not fed by adults.
[c] Maximum energy requirements of brood during development.
[d] Average of values given by Hubbard (1978).

to nest attendance or construction. This is probably an overestimate of available foraging time. Because parental and nestling energy requirements are underestimates (i.e., assimilation inefficiency is not accounted for), the computed values for required acquisition rates are also underestimates, but even these crude approximations should lend insight into relative shifts in foraging rates.

In the Willow Flycatcher the female's average energy acquisition rate is persistently higher than the male's, associated primarily with the female's extensive nest attendance and consequent reduction in time available for foraging (Table XI). E_{FOR} peaks during incubation for the female at a value 2.2 times her average rate while feeding young. In contrast, the male's E_{FOR} peaks during the nestling period at a value 1.7 times the nadir, which occurs during incubation. In the White-crowned Sparrow, the female's required acquisition rate is also persistently higher than the male's, again associated primarily with her more extensive nest attendance and associated reduction in foraging time. The female E_{FOR} peaks during the nestling period at a value slightly above that during incubation and 2.7 times the value during the period of nest construction. The female Black-billed Magpie also has higher required foraging rates than the male, except during the nestling period. Again, the peak value for the female occurs when feeding the young, whereas the male's peak occurs during incubation. The Spotted Sandpiper is the only precocial species for which necessary data are available. In this species, E_{FOR} is higher in the female than the male during egg production, but the male's required foraging rate is greater than the female's during incubation. This is largely due to the male's reduction in available foraging time associated with his primary role in incubation. The male's value peaks during incubation, with the nadir occurring during the brooding period. The value for E_{FOR} in the female peaks during egg laying.

In summary, the average rate at which a parent bird must acquire energy appears to differ greatly between reproductive phases; the value for a particular sex and species typically varies two- to threefold. In the male, E_{FOR} usually peaks during the nestling period, associated with the additional energy requirements of the young, and reaches its lowest value during incubation, due to decreased locomotor activity. The female's rate is commonly greater than the male's. Incubation and the nestling period are apparently periods in which the female's average foraging rate is very high compared to the egg-laying period.

These values suggest at which phase during reproduction the parent's foraging abilities are most likely to limit performance of vital functions, but in themselves cannot be used to demonstrate that such limits occur and are biologically significant. Such an evaluation requires currently unavailable information describing changes in energy availability in the environment and the degree to which an individual's time budget normally contains noncritical components such as apparent "loafing" time. It is possible that even during the nestling period, the parent normally has substantially more than sufficient time to forage and is not energetically limited in this sense. In the Willow Flycatcher, for example, loafing behavior persists through all phases of reproduction, and Ettinger and King (1980) suggested that a reduction in this component could allow the parents to raise more than their normal number of young in typical years. These authors suggest that clutch size has been fixed by other selection pressures, such as the advantage gained by maintaining a loafing component in the time budget that would allow an increase in foraging time and thus successful reproduction in nontypical years characterized by food shortage. Unfortunately, analyses such as Ettinger and King's are rare and the ecological significance of such energy relationships is difficult to evaluate.

4. Glimpses of Annual Cycles in Energetics

Comparative analyses of annual cycles in energy relationship might provide valuable insight into the factors that resulted in the evolution of the wide variety of life histories exhibited by birds. It is thus particularly unfortunate that most authors have examined only the breeding season. Few have considered nonreproductive portions of the annual cycle which, for example, could be important periods of mortality (Fretwell, 1972). There currently exist only two reasonably extensive studies of the annual energetic cycles of free-living birds.

Mugaas and King's (1981) analysis of the annual cycle of energy relationships in Black-billed Magpies (*Pica pica*) is the most nearly complete such analysis. The population studied resides in agricultural areas and Palouse prairie remnants of eastern Washington. This species flocks during the winter and forms breeding pairs in spring and early summer. The male has virtually no role in incubation, but does feed the female at the nest. The female spends long periods on the nest during incubation and consequently is quite inactive. Both sexes feed the young. Over the annual cycle, the male reaches his maximum daily energy expenditure during the nestling period, whereas the female's maximum occurs in early winter (Table VIII). Minimum levels of E_{TOT} occur during the egg-laying period for the male and during incubation for the female. Daily energy expenditure does not vary greatly over the annual cycle; maximum levels are about 35% above minima in males and 37% above minima in females if the exceptionally inactive incubation period is excluded (Table VIII). As noted perviously, most of this variation is due to changes in activity patterns (e.g., foraging) rather than to changes in the thermostatic increment. In the male, allocation to these obligatory thermostatic demands varies only about 19%, whereas allocation to elective components varies about 100%. (Variation is computed as percentage elevation of the maximum value compared to the minimum.) In the female, obligatory expenditures vary about 19%, whereas elective expenditures vary 544%.

Average foraging rates required for maintenance of energy balance (E_{FOR}) vary much more over the annual cycle than does total energy expenditure (Table XI). E_{FOR} varies 108% in the male and 120% in the female. Peak values occur in winter and are 70% (males) or 27% (females) higher than the peak values occurring in the breeding season. Annual variation in required foraging rates is produced primarily by changes in photoperiod rather than energy expenditure; daily energy expenditure varies 35% over the annual cycle, whereas daylength varies 87%.

I have analyzed the annual energetic cycle of the Phainopepla, which occupies arid and semiarid habitats in Mexico and the southwestern United States (Walsberg, 1977, 1978; Table VIII). The populations studied occupy the Sonoran Desert from approximately November through April, where breeding occurs in March and April. Phainopeplas largely desert this area in spring and appear in coastal chaparral and oak woodland habitats, where breeding occurs in June and July. Sexes share nearly equally in care of the eggs and young. Energy expenditure is similar in both sexes. Highest levels occur during the nestling periods and are 20–30% above that during incubation. For comparable stages of breeding, power consumption is about 20–30% higher in the summer nesting in the coastal chaparral than the spring nesting in the desert. These shifts reflect major changes in territorial and foraging behavior. In the Sonoran Desert, Phainopeplas defend territories within which almost all food is collected and the nest is constructed. In the chaparral habitat, Phainopeplas commonly nest in loose colonies and defend only a small area around the nest. Most foraging is accomplished at distances 50–500 m from the nest. This requires many energetically expensive foraging flights per day and a consequent increase in energy expenditure. Daily energy expenditure varies a total of 57% over the annual cycle, with almost all variation associated with these behavioral shifts. The basal and thermostatic demand is estimated to vary only 15%, whereas maximum allocation to elective components of the energy budget is 250% greater than the minimum such allocation.

Average foraging rates required for energy balance vary only 27%, only about one-half the range observed in rates of energy expenditure (Table XI). This reflects a tendency for rates of energy expenditure and time available for foraging (determined largely by photoperiod) to follow similar seasonal patterns. For example, the winter nonbreeding period and the early spring incubation period in the Sonoran Desert are characterized by both short daylengths and low levels of energy expenditure. Thus, in spite of varying rates of energy expenditure, the Phainopepla must acquire energy at approximately the same rate throughout the annual cycle.

In addition to these studies of *Pica* and *Phainopepla*, Carpenter (1976) presented data on portions of the annual cycle of *Oreotrochilus estella* (Table VIII). This large hummingbird inhabits *altiplano* regions of the Andes and faces an exceptionally rigorous habitat. Carpenter's data describe breeding female *Oreotrochilus* and males during the winter. The female's annual energetic cycle thus can be estimated by assuming that nonbreeding males and females do not differ importantly in power consumption. Use of these data and this assumption indicates that the female's daily energy expenditure peaks during the nestling period. The lowest rate of expenditure occurs in winter, associated with reduced activity and the use of nocturnal torpor. Daily energy expenditure during the nestling period is 2.9 times the winter value; this is the largest annual range in energy expenditure yet described. Unlike the Black-billed Magpie or Phainopepla, most annual variation of E_{TOT} in *O. estella* is produced by changes in the basal and thermostatic requirements, reflecting this species' extensive use of torpor during winter.

Although current information does not allow broad generalizations, it is notable that in the species for which appropriate data are available, winter is not a period of relatively intense energy expenditure as is commonly assumed. Indeed, energy expenditure during the winter may be relatively low, although required rates of energy acquisition may be quite high because of the decrease in time available for foraging due to shortened daylengths.

It is also notable that in both the Black-billed Magpie and the Phainopepla major shifts in power consumption are primarily produced by variation in activity patterns rather than by changes in the thermostatic increment. The obligatory basal and thermostatic requirement is large but relatively stable. In *Pica*, this stability reflects physiological capabilities that produce a wide zone of thermal neutrality (5–35°C; (Stevenson, 1971, cited in Mugaas and King, 1981). The thermal neutral zone of *Phainopepla* is relatively narrow (30–43°C; Walsberg, 1977), but its seasonal migration between the Sonoran Desert and coastal areas produces a relatively constant thermal regime over the annual cycle. This suggests that seasonal migration in this species may function importantly as a form of behavioral thermoregulation. This possible role of migration deserves more extensive and rigorous attention than it has previously received.

IV. Concluding Comments

Avian ecological energetics is a relatively new and rapidly developing discipline that in many respects is only on the edge of addressing its most important questions. The roots of ecological energetics lie in ecology and environmental physiology. In the latter discipline, basal and thermoregulatory metabolism has been a major focus. Available evidence suggests that this avenue of energy expenditure will also be a subject of major importance for the future development of ecological energetics. Data complied in Section III,B indicate that approximately one-half of a bird's total energy allocation is devoted to such maintenance requirements. With notable exceptions (e.g., Mugaas and King, 1981), such estimates are often of uncertain accuracy due to the sparse data base describing the thermal relationships of birds occupying their naturally selected microclimates. Thus there is an important need for data describing the thermal stresses imposed upon birds in nature, as well as continuing explorations of other aspects of basal and thermoregulatory expenditure, such as the relationship between thermoregulatory metabolism and synthetic or locomotor activity. The large fraction of a free-living bird's energy budget that is apparently devoted to basal and thermostatic demands makes such investigations imperative.

Behavioral ecology is the second major root of ecological energetics and it is clear that knowledge of behavioral phenomena is critical to a comprehension of the energetic constraints acting upon animals. This is exemplified by the insights gained by behavioral studies of the significance of weight loss by incubating birds. Adult birds commonly lose weight during incubation, which has been taken as evidence of energy stress associated with egg-temperature regulation (e.g., Folk *et al.*, 1966). By manipulating food availability during incubation in Red Junglefowl (*Gallus gallus*), Sherry *et al.* (1980) demonstrated that weight loss does not reflect partial starvation, but rather a change in the level at which body weight is regulated. This regulated loss of weight and reduction of food requirements allows the incubating female to resolve conflicts in time allocation between foraging and nest attendance. Thus the significance of the hen's change in energy balance is linked intimately with behavioral constraints on time allocation.

A second example of the importance to ecological energetics of simultaneous behavioral and physiological analyses is suggested by the contrast observed in Section III,C between phasic variation in rates of energy expenditure (primarily a physiological variable) and required rates of energy acquisition during foraging (which subsume both the animal's time and energy budgets). In a variety of species, power consumption appears to exhibit substantially less seasonal variation than does the average foraging rate. The more extreme variation in foraging rates suggests that problems of energy acquisition may be more likely to affect fitness than simply rates of energy expenditure. Such constraints can be understood only through combined analyses of both the animal's time allocation and its energy budget. Measurement of purely physiological variables, such as daily energy expenditure, are of limited ecological value unless correlated with knowledge of behavioral patterns and adaptations. The strength of ecological energetics lies in its integration of the two vigorous fields of physiology and ecology. Future advances in this discipline will depend importantly upon the vigor with which more extensive integrations are pursued.

ACKNOWLEDGMENTS

The analysis presented in this chapter was supported in part by National Science Foundation Grant DEB 80-04266 and an Arizona State University Faculty Grant-in-Aid.

I thank J. R. King, C. M. Vleck, and W. W. Weathers for making manuscripts available prior to publication and K. A. Voss-Roberts for assistance during preparation of this review. In addition, I am indebted to J. Alcock, J. R. Hazel, J. R. King, and R. L. Rutowski for stimulating discussions.

REFERENCES

Afton, A. D. (1979). The time budget of breeding Northern Shovelers. *Wilson Bull.* **91**, 42–49.
Alcock, J. (1979). "Animal Behavior: An Evolutionary Approach," 2nd ed. Sinauer, Sunderland, Massachusetts.
Amadon, D. (1943). Bird weights and egg weights. *Auk* **60**, 221–334.
Andreev, A. (1977). Bioenergetics of the Nutcracker (*Nucifraga caryocatactes*) under extremely low temperatures. *Zool. Zh.* **56**, 1578–1581 (in Russian).
Ankney, C. D. (1977). The use of nutrient reserves by breeding male Lesser Snow Geese, *Chen caerulescens caerulescens*. *Can. J. Zool.* **55**, 1984–1987.
Ankney, C. D. (1979). Does the wing molt cause nutritional stress in Lesser Snow Geese? *Auk* **96**, 68–72.
Ankney, C. D., and MacInnes, D. D. (1978). Nutrient reserves and reproductive performance of female Lesser Snow Geese. *Auk* **95**, 459–471.
Ankney, C. D., and Scott, D. M. (1980). Changes in nutrient reserves and diet of breeding Brown-headed Cowbirds. *Auk* **97**, 684–696.
Anthony, R. (1970). Ecology and reproduction of California Quail in southeastern Washington. *Condor* **72**, 267–287.
Arscott, G. H., and Parker, J. E. (1963). Dietary protein and fertility of male chickens. *J. Nutr.* **80**, 311–314.
Aschoff, J., and Pohl, H. (1970). Der Ruheumsatz von Vögeln als Funktion der Tageszeit und der Körpergrösse. *J. Ornithol.* **111**, 38–47.
Ashkenazie, S., and Safriel, U. N. (1979). Time-energy budget of the Semipalmated Sandpiper *Calidris pusilla* at Barrow, Alaska. *Ecology* **60**, 783–799.
Austin, G. T. (1978). Daily time budget of the postnesting Verdin. *Auk* **95**, 247–251.
Baldwin, S. P., and Kendeigh, S. C. (1938). Variations in the weight of birds. *Auk* **55**, 416–467.
Baltz, D. M., and Morejohn, G. V. (1977). Food habits and niche overlap of seabirds wintering on Monterey Bay, California. *Auk* **94**, 526–543.

Barash, D. P. (1976). Male response to apparent female adultery in the Mountain Bluebird (*Sialia currocoides*): an evolutionary interpretation. *Am. Nat.* **110**, 1097–1101.
Barry, T. W. (1962). Effect of late seasons on Atlantic Brant reproduction. *J. Wildl. Manage.* **26**, 19–26.
Bartelmez, G. W. (1912). The bilaterity of the pigeon's egg. A study in egg organization from the first growth of the oocyte to the beginning of cleavage. *J. Morphol.* **23**, 269–328.
Bedard, J. (1976). Coexistence, coevolution and convergent evolution in seabird communities: a comment. *Ecology* **57**, 177–184.
Bennett, A. F., and Ruben, J. A. (1979). Endothermy and activity in vertebrates. *Science (Washington, D.C.)* **206**, 649–654.
Berger, M., and Hart, J. S. (1972). Die Atmung beim Kolibri *Amazilia fimbriata* während des Schwirrfluges bei verschiedenen Umgebungstemperaturen. *J. Comp. Physiol.* **81**, 363–380.
Biebach, H. (1977). Der Energieaufwand für das Brüten beim Star. *Naturwissenschaften* **64**, 343.
Biebach, H. (1979). Energetik des Brütens beim Star (*Sturnus vulgaris*). *J. Ornithol.* **120**, 121–138.
Birkhead, T. R. (1979). Mate guarding in the magpie *Pica pica*. *Anim. Behav.* **27**, 866–874.
Bissonnette, T. H., and Chapnick, M. H. (1930). Studies on the sexual cycle in birds. II. The normal progressive changes in the testis from November to May in the European Starling (*Sturnus vulgaris*), an introduced, nonmigratory bird. *Am. J. Anat.* **45**, 307–343.
Blackmore, F. H. (1969). The effect of temperaure, photoperiod and molt on the energy requirements of the House Sparrow, *Passer domesticus*. *Comp. Biochem. Physiol.* **30**, 433–444.
Blem, C. R. (1975). Energetics of nestling House Sparrows *Passer domesticus*. *Comp. Biochem. Physiol. A* **52**, 305–312.
Boorman, K. N., and Lewis, D. (1971). Protein metabolism. *In* "Physiology and Biochemistry of the Fowl" (D. J. Bell, and B. M. Freeman, eds.), Vol. 1, pp. 338–372. Academic Press, New York and London.
Boyd, R. (1976). "Behavioral Biology and Energy Expenditure in a Horned Lark Population." Ph.D. Thesis, Colorado State Univ., Fort Collins.
Boyden, T. C. (1978). Territorial defense against hummingbirds and insects by tropical hummingbirds. *Condor* **80**, 216–221.
Brenner, F. J. (1964). Growth, fat deposition and development of endothermy in nestling Redwinged Blackbirds. *J. Sci. Lab. Denison Univ.* **46**, 81–89.
Bryant, D. M, and Gardiner, A. (1979). Energetics of growth in House Martins (*Delichon urbica*). *J. Zool.* **189**, 275–304.
Burley, N. (1977). Parental investment, mate choice, and mate quality. *Proc. Natl. Acad. Sci. USA* **74**, 3476–3479.
Cain, B. W. (1976). Energetics of growth for Black-bellied Tree Ducks. *Condor* **78**, 124–128.
Calder, W. A. (1964). Gaseous metabolism and water relations of the Zebra Finch, *Taeniopygia castanotis*. *Physiol. Zool.* **37**, 400–413.
Calder, W. A. (1971). Temperature relationships and nesting of the Calliope Hummingbird. *Condor* **73**, 314–321.
Calder, W. A. (1974). Consequences of body size for avian energetics. *In* "Avian Energetics" (R. A. Paynter, ed.). Nuttall Ornithol. Club, Cambridge, Massachusetts.
Calder, W. A. (1975). Daylength and the hummingbirds' use of time. *Auk* **92**, 81–97.
Calder, W. A., and King, J. R. (1974). Thermal and caloric relations of birds. *In* "Avian Biology" (D. S. Farner, and J. R. King, eds.), Vol. 4, pp. 259–413. Academic Press, New York.

Calder, W. A., and Rowe, B. (1977). Body mass changes and energetics of the Kiwi's egg cycle. *Notornis* **24,** 129–135.

Calder, W. A., III, Parr, C. R., and Karl, D. P. (1978). Energy content of eggs of the Brown Kiwi (*Apteryx australis*); an extreme in avian evolution. *Comp. Biochem. Physiol. A* **60,** 177–179.

Carey, C., Rahn, H., and Parisi, P. (1980). Calories, water, lipid, and yolk in avian eggs. *Condor* **82,** 335–343.

Carpenter, F. L. (1976). Ecology and evolution of an Andean hummingbird (*Oreotrochilus estella*). *Univ. Calif. Berkeley Publ. Zool.* **106,** 1–74.

Carpenter, F. L., and MacMillen, R. E. (1976a). Energetic cost of feeding territories in an Hawaiian honeycreeper. *Oecologia* **26,** 213–223.

Carpenter, F. L., and MacMillen, R. E. (1976b). Threshold model of feeding territoriality and test with a Hawaiian honeycreeper. *Science (Washington, D.C.)* **194,** 639–642.

Case, R. M., and Robel, R. J. (1974). Bioenergetics of the Bobwhite. *J. Wildl. Manage.* **38,** 638–652.

Chilgren, J. D. (1975). "Dynamics and Bioenergetics of Post-Nuptial Molt in Captive White-Crowned Sparrows (*Zonotrichia leucophrys gambelii*)." Ph.D. Thesis, Washington State Univ., Pullman.

Chilgren, J. D., and DeGraw, W. A. (1977). Some blood characteristics of White-crowned Sparrows during molt. *Auk* **94,** 169–171.

Connell, C. E., Odum, E. P., and Kale, H. (1960). Fat-free weights of birds. *Auk* **77,** 1–9.

Custer, T. W. (1974). "Population Ecology and Bioenergetics of the Lapland Longspur (*Calcarius lapponicus*) near Barrow, Alaska." Ph.D. Thesis, Univ. of California, Berkeley.

Cuthbert, N. L. (1945). The ovarian cycle of the Ring Dove (*Streptopelia risoria* L.). *J. Morphol.* **77,** 351–377.

Davis, E. A., Jr. (1955). Seasonal changes in the energy balance of the English Sparrow. *Auk* **72,** 385–411.

Diehl, B., and Myrcha, A. (1973). Bioenergetics of nestling Red-backed Shrikes (*Lanius collurio*). *Condor* **75,** 259–264.

Dolnik, V. R., and Gavrilov, V. M. (1979). Bioenergetics of molt in the Chaffinch (*Fringilla coelebs*). *Auk* **96,** 253–264.

Drent, R. H. (1970). Functional aspects of incubation in the Herring Gull. *Behaviour* **17** (Suppl.), 1–132.

Drobney, R. D. (1980). Reproductive bioenergetics of Wood Ducks. *Auk* **97,** 480–490.

Dunn, E. H. (1975a). Caloric intake of nestling Double-crested Cormorants. *Auk* **92,** 553–565.

Dunn, E. H. (1975b). Growth, body components and energy content of nestling Double-crested Cormorants. *Condor* **77,** 431–438.

Dunn, E. H. (1976). The development of endothermy and existence energy expenditure in Herring Gull chicks. *Condor* **78,** 493–498.

Dunn, E. H. (1980). On the variability in energy allocation of nestling birds. *Auk* **97,** 19–27.

Dwyer, T. J. (1975). Time budget of breeding Gadwalls. *Wilson Bull.* **87,** 335–343.

Dwyer, T. J., Krapu, G. L., and Janke, D. M. (1979). Use of prairie pothole habitat by breeding Mallards. *J. Wildl. Manage.* **43,** 526–531.

Ebbinge, B., Canters, K., and Drent, R. (1975). Foraging routines and estimated daily food intake in Barnacle Geese wintering in the northern Netherlands. *Wildfowl* **26,** 5–19.

El-Wailly, A. J. (1966). Energy requirements for egg-laying and incubation in the Zebra Finch, *Taeniopygia castanotis*. *Condor* **68,** 582–594.

Ettinger, A. O., and King, J. R. (1980). Time and energy budgets of the Willow Flycatcher (*Empidonax traillii*) during the breeding season. *Auk* **97,** 535–546.

Ewald, P. W., and Carpenter, F. L. (1978). Territorial responses to energy manipulations in the Anna Hummingbird. *Oecologia* **31,** 277–292.
Eyster, M. B. (1954). Quantitative measurement of the influence of photoperiod, temperature, and season on the activity of captive songbirds. *Ecol. Monogr.* **24,** 1–28.
Folk, C. Hudec, K., and Toufar, J. (1966). The weight of the Mallard, *Anas platyrhynchos*, and its changes in the course of the year. *Zool. Listy* **15,** 249–260.
Fretwell, S. D. (1972). Populations in a seasonal environment. *Monogr. Popul. Biol.*, No. 5. Princeton Univ. Press.
Frost, S. K., and Frost, P. G. H. (1980). Territoriality and changes in resource use by sunbirds at *Leonotis leonurus* (Labiatae). *Oecologia* **45,** 109–116.
Gavrilov, V. M. (1974). Metabolism of molting birds. *Zool. Zh.* **53,** 1363–1375 (in Russian).
Gavrilov, V. M., and Dolnik, V. R. (1974). Bioenergetics and regulation of the postnuptial and postjuvenal moult in Chaffinches (*Fringilla coelebs coelebs* L.). *Tr. Zool. Inst. Akad. Nauk SSSR* **55,** 14–61.
Gessaman, J. A., and Findell, P. R. (1979). Energy cost of incubation in the American Kestrel. *Comp. Biochem. Physiol. A* **63,** 57–62.
Gibb, J. (1956). Food, feeding habits, and territory of the Rock Pipit, *Anthus spinoletta. Ibis* **98,** 506–530.
Gill, F. B., and Wolf, L. L. (1975) Economics of feeding territoriality in the Golden-winged Sunbird. *Ecology* **56,** 333–345.
Gladstone, D. E. (1979). Promiscuity in monogamous colonial birds. *Am. Nat.* **114,** 545–557.
Graber, R. R. (1962). Food and oxygen consumption in three species of owls (Strigidae). *Condor* **64,** 473–487.
Greeley, F., and Meyer, R. K. (1953). Seasonal variation in testis stimulating activity of male pheasant pituitary glands. *Auk* **70,** 350–358.
Greenlaw, J. S. (1969). "The Importance of Food in the Breeding System of the Rufous-Sided Towee, *Pipilo erythrophthalmus* (L.)." Ph.D. Thesis, Rutgers Univ., New Brunswick, New Jersey.
Haftorn, S. (1978). Energetics of incubation by the Goldcrest *Regulus regulus* in relation to ambient air temperature and the geographical distribution of the species. *Ornis Scand.* **9,** 22–30.
Hails, C. J., and Bryant, D. M. (1979). Reproductive energetics of a free-living bird. *J. Anim. Ecol.* **48,** 471–482.
Hainsworth, F. R. (1977). Foraging efficiency and parental care in *Colibri coruscans. Condor* **79,** 69–75.
Hanson, H. C. (1962). The dynamics of condition factors in Canada Geese and their relation to seasonal stresses. *Tech. Pap. Arct. Inst. North Am.*, No. 12.
Harris, J. H., Jr. (1970). Evidence of stress response in breeding Blue-winged Teal. *J. Wildl. Manage.* **34,** 747–755.
Hart, J. S., and Berger, M. (1972). Energetics, water economy and temperature regulation during flight. *Proc. Int. Ornithol. Congr. 15th*, pp. 189–199.
Hartmann, F. A. (1955). Heart weight in birds. *Condor* **57,** 221–238.
Holmes, R. T., Black, C. P., and Sherry, T. W. (1979). Comparative population bioenergetics of three insectivorous passerines in a deciduous forest. *Condor* **81,** 9–20.
Howe, H. F. (1977). Sex-ratio adjustment in the Common Grackle. *Science (Washington, D.C.)* **198,** 744–746.
Howe, H. F. (1978). Initial investment, clutch size, and brood reduction in the Common Grackle (*Quiscalus quiscula* L.). *Ecology* **59,** 1109–1122.
Hubbard, J. D. (1978). "Breeding Biology and Reproductive Energetics of Mountain White-Crowned Sparrows in Colorado." Ph.D. Dissertation, Univ. of Colorado, Boulder.

Jehl, J. R., Jr. (1970). Sexual selection for size differences in two species of sandpipers. *Evolution* **24**, 311–319.
Johnson, O. W. (1961). Reproductive cycle of the Mallard Duck. *Condor* **63**, 351–364.
Johnson, O. W. (1979). Biology of shorebirds summering on Enewetak Atoll. *Stud. Avian Biol.* **2**, 193–205.
Johnson, S. R. (1972). "Thermal Adaptation in North American Sturnidae." Ph.D. Thesis, Univ. of British Columbia, Vancouver.
Johnston, D. W. (1963). Heart weights of some Alaskan birds. *Wilson Bull.* **75**, 435–446.
Jones, P. J., and Ward, P. (1976). The level of reserve protein as the proximate factor controlling the timing of breeding and clutch-size in the Red-billed Quelea *Quelea quelea*. *Ibis* **118**, 547–573.
Kale, H. W. (1965). Ecology and bioenergetics of the Long-billed Marsh Wren (*Telmatodytes palustris griseus* Brewster) in Georgia salt marshes. *Publ. Nuttall Ornithol.* **Club, No. 5.**
Kendeigh, S. C. (1963). Thermodynamics of incubation in the House Wren, *Troglodytes aedon*. *Proc. Int. Ornithol. Congr., 13th*, pp. 884–904.
Kendeigh, S. C. (1973). Monthly variation in the energy budget of the House Sparrow throughout the year. *In* "Productivity, Population Dynamics, and Systematics of Granivorous Birds" (S. C. Kendeigh, and J. Pinowski, eds.), pp. 17–44. Polish Sci. Publ., Warsaw.
Kendeigh, S. C., Dolnik, V. R., and Gavrilov, V. M. (1977). Avian energetics. *In* "Granivorous Birds in Ecosystems" (J. Pinowski, and S. C. Kendeigh, eds.), pp. 127–204. Cambridge Univ. Press, London.
Kern, M. D. (1970). "Annual and Steroid-Induced Changes in the Reproductive System of the Female White-Crowned Sparrow, *Zonotrichia leucophrys gambelii*." Ph.D. Thesis, Washington State Univ., Pullman.
King, J. R. (1973). Energetics of reproduction in birds. *In* "Breeding Biology of Birds" (D. S. Farner, ed.), pp. 78–107. Natl. Acad. Sci., Washington, D.C.
King, J. R. (1981). Energetics of avian molt. *Proc. Int. Ornithol. Congr. 17th*, pp. 312–317.
King, J. R., and Farner, D. S. (1961). Energy metabolism, thermoregulation, and body temperature. *In* "Biology and Comparative Physiology of Birds" (A. J. Marshall, ed.), Vol. 2, 215–288. Academic Press, New York.
King, J. R., Follett, B. J., Farner, D. S., and Morton, M. L. (1966). Annual gonadal cycles and pituitary gonadotropins in *Zonotrichia leucophrys gambelii*. *Condor* **68**, 476–487.
Kleiber, M. (1961). "The Fire of Life." Wiley, New York.
Kluijver, H. N. (1951). The population ecology of the Great Tit, *Parus m. major* L. *Ardea* **39**, 1–135.
Koelink, A. F. (1972). "Bioenergetics of Growth in the Pigeon Guillemot, *Cepphus columba*." M.S. Thesis, Univ. of British Columbia, Vancouver.
Koplin, J. R., Collopy, M. W., Bammann, A. R., and Levenson, H. (1980). Energetics of two wintering raptors. *Auk* **97**, 795–806.
Korschgen, C. E. (1977). Breeding stress of female Eiders in Maine. *J. Wildl. Manage.* **41**, 360–373.
Krapu, G. L. (1974). Feeding ecology of Pintail hens during reproduction. *Auk* **91**, 243–483.
Krapu, G. L. (1981). The role of nutrient reserves in Mallard reproduction. *Auk* **98**, 29–38.
Kushlan, J. A. (1977). Population energetics of the American White Ibis. *Auk* **94**, 114–122.
Lack, D. (1968). "Ecological Adaptations for Breeding in Birds." Methuen, London.
Lawrence, J. M., and Schreiber, R. W.(1974). Organic material and calories in the egg of the Brown Pelican, *Pelecanus occidentalis*. *Comp. Biochem. Physiol. A* **47**, 435–440.
Le Maho, Y. (1977). The Emperor Penguin: a strategy to live and breed in the cold. *Am. Sci.* **65**, 680–693.

Le Maho, Y., Delditte, P., and Chatonnet, J. (1976). Thermoregulation in fasting Emperor Penguins under natural conditions. *Am. J. Physiol.* **231,** 913–922.

Lewin, V. (1963). Reproduction and development of young in a population of California Quail. *Condor* **65,** 249–278.

Lill, A. (1979). An assessment of male parental investment and pair bonding in the polygamous Superb Lyrebird. *Auk* **96,** 489–498.

Ljunggren, L. (1969). Seasonal studies of Wood Pigeon populations. II. Gonads, crop glands, adrenals and the hypothalamo–hypophysial system. *Viltrevy* **6,** 41–126.

Lustick, S. (1970). Energy requirements of molt in cowbirds. *Auk* **87,** 742–746.

Mackie, R. J., and Buechner, H. K. (1963). The reproductive cycle of the Chukar. *J. Wildl. Manage.* **27,** 246–260.

Maclean, S. F., Jr., and Holmes, R. T. (1971). Bill lengths, wintering areas, and taxonomy of North American dunlins, *Calidris alpina*. *Auk* **88,** 893–901.

Maxson, S. J., and Oring, L. W. (1980). Breeding season time and energy budgets of the polyandrous Spotted Sandpiper. *Behaviour* **74,** 200–263.

Mertens, J. A. L. (1977). The energy requirements for incubation in Great Tits, *Parus major* L. *Ardea* **65,** 184–196.

Mertens, J. A. L. (1980). The energy requirements for incubation in Great Tits and other bird species. *Ardea* **68,** 185–192.

Morgan, K. K., Paganelli, C. V., and Rahn, H. (1978). Egg water loss and nest humidity during incubation in two Alaskan Gulls. *Condor* **80,** 272–275.

Morse, D. H. (1968). A quantitative study of the foraging of spruce-woods warblers. *Ecology* **49,** 779–784.

Moss, R. (1973). The digestibility and intake of winter foods by wild Ptarmigan in Alaska. *Condor* **75,** 293–300.

Moss, R. (1977). The digestion of heather by Red Grouse during the spring. *Condor* **79,** 471–477.

Mugaas, J. N., and King, J. R. (1981). The annual variation in daily energy expenditure of the Black-billed Magpie: a study of thermal and behavioral energetics. *Stud. Avian Biol.* **5,** 1–78.

Murdock, L. C. (1975). "Physiology and Bioenergetics of the American Coot, *Fulica americana.*" M.S. Thesis, California State Univ., Fullerton.

Myrcha, A., Pinowski, J., and Tomek, T. (1972). Energy balance of nestlings of Tree Sparrows, *Passer m. montanus*, and House sparrows, *Passer d. domesticus*. *In* "Productivity, Population Dynamics and Systematics of Granivorous Birds" (S. G. Kendeigh, and J. Pinowski, eds.), pp. 59–83. Polish Sci. Publ., Warsaw.

Nelson, A. L., and Martin, A. C. (1953). Gamebird weights. *J. Wildl. Manage.* **17,** 36–42.

Newton, I. (1978). The temperatures, weights and body composition of molting Bullfinches. *Condor* **70,** 323–332.

Norton, D. W. (1973). "Ecological Energetics of Calidrine Sandpipers Breeding in Northern Alaska." Ph.D. Thesis, Univ. of Alaska, Fairbanks.

Paladino, F. V. (1979). "Energetics of Terrestrial Locomotion in White-Crowned Sparrows (*Zonotrichia leucophrys gambelii*)." Ph.D. Thesis, Washington State Univ., Pullman.

Payne, R. B. (1969). Breeding seasons and reproductive physiology of Tricolored Blackbirds and Redwinged Blackbirds. *Univ. Calif. Berkeley Publ. Zool.* **90,** 1–115.

Pearson, O. P. (1954). The daily energy requirements of a wild Anna Hummingbird. *Condor* **56,** 317–322.

Perek, M., and Sulman, F. (1945). The basal metabolic rate in molting and laying hens. *Endocrinology (Baltimore)* **36,** 240–243.

Petersen, A. (1955). The breeding cycle in the Bank Swallow. *Wilson Bull.* **67**, 235–286.
Pinkowski, B. C. (1979). Time budget and incubation rhythm of the Eastern Bluebird. *Am. Midl. Nat.* **101**, 427–433.
Pinowski, J. (1967). Estimation of the biomass produced by a Tree Sparrow [*Passer m. montanus* (L.)] population during the breeding season. *In* "Secondary Productivity of Terrestrial Ecosystems" (K. Petrusewicz, ed.), Vol. 1, pp. 357–367. Polish Sci. Publ., Warsaw.
Pohl, H. (1969). Some factors influencing the metabolic response to cold in birds. *Fed. Proc. Fed. Am. Soc. Exp. Biol.* **28**, 1059–1064.
Pohl, H., and West, G. C. (1973). Daily and seasonal variation in metabolic response to cold during rest and forced exercise in the Common Redpoll. *Comp. Biochem. Physiol. A* **45**, 851–867.
Poole, E. L. (1938). Weights and wing areas in North American birds. *Auk* **55**, 511–517.
Poulson, T. L., and Bartholomew, G. A. (1962). Salt utilization in the House Finch. *Condor* **64**, 245–252.
Rahn, H., Paganelli, C. V., and Ar, A. (1975). Relation of avian egg weight to body weight. *Auk* **92**, 750–765.
Raitt, R. J., and Ohmart, R. D. (1966). Annual cycle of reproduction and molt in Gambel Quail of the Rio Grande Valley, southern New Mexico. *Condor* **67**, 541–561.
Raveling, D. G. (1978). The timing of egg laying by northern geese. *Auk* **95**, 294–303.
Raveling, D. G. (1979). The annual cycle of body composition of Canada geese with special reference to control of reproduction. *Auk* **96**, 234–252.
Ricklefs, R. E. (1974). Energetics of reproduction in birds. *In* "Avian Energetics" (R. A. Paynter, ed.). Nuttall Ornith. Club. Cambridge.
Ricklefs, R. E. (1977). Composition of eggs of several bird species. *Auk* **94**, 350–356.
Ricklefs, R. E., and Montevecchi, W. A. (1979). Size, organic composition and energy content of North American Gannet *Morus bassanus* eggs. *Comp. Biochem. Physiol. A* **64**, 161–165.
Ricklefs, R. E., White, S. C., and Cullen, J. (1980). Energetics of postnatal growth in Leach's Storm-Petrel. *Auk* **97**, 566–575.
Romanoff, A. L., and Romanoff, A. J. (1949). "The Avian Egg." Wiley, New York.
Roudybush, T. E., Grau, C. R., Petersen, M. R., Ainley, D. G., Hirsch, K. V., Gilman, A. P., and Patten, S. M. (1979). Yolk formation in some Charadriiform birds. *Condor* **81**, 293–298.
Schartz, R. L., and Zimmerman, J. L. (1971). The time and energy budget of the male Dickcissel (*Spiza americana*). *Condor* **73**, 65–76.
Schönwetter, M. (1960–1978). "Handbuch der Oologie" (W. Meise, ed.), Lief. 1–24. Akademie-Verlag, Berlin.
Schreiber, R. W., and Lawrence, J. M. (1976). Organic material and calories in Laughing Gull eggs. *Auk* **93**, 46–52.
Sherry, D. F., Mrosovsky, N., and Hogan, J. A. (1980). Weight loss and anorexia during incubation in birds. *J. Comp. Physiol. Psychol.* **94**, 89–98.
Siegfried, W. R. (1969). Energy metabolism of the Cattle Egret. *Zool. Afr.* **4**, 265–273.
Siegfried, W. R., Burger, A. E., and Frost, P. G. H. (1976). Energy requirements for breeding in the Maccoa Duck. *Ardea* **64**, 171–190.
Southwick, E. E. (1973). Remote sensing of body temperature in a captive 26-g bird. *Condor* **75**, 464–466.
Spring, L. (1971). A comparison of functional and morphological adaptations in the Common Murre (*Uria aalge*) and Thick-billed Murre (*Uria lomvia*). *Condor* **73**, 1–27.
Stevenson, R. E. (1971). "Temperature Acclimatization in the Black-Billed Magpie (*Pica pica hudsonia*, Sabine)." Ph.D. Thesis, Montana State Univ., Bozeman.

Stewart, P. A. (1937). A preliminary list of bird weights. *Auk* **54**, 324–332.
Stewart, P. A., and Skinner, R. W. (1967). Weights of birds from Alabama and North Carolina. *Wilson Bull.* **79**, 37–42.
Stieve, H. (1919). Die Entwicklung des Eierstockseises der Dohl (*Colaeus monedula*). Ein Beitrag zur Frage nach den physiologischerweise im Ovar strattfinden Rückbildungsvorgangen. *Arch. Mikrosk. Anat. Abt. II* **92**, 137–288.
Stiles, F. G. (1971). Time, energy, and territoriality of the Anna Hummingbird (*Calypte anna*). *Science (Washington, D.C.)* **173**, 818–821.
Tangl, F. (1903). Beiträge zur Engergetik der Ontogenese. I. Mitteilung. Die Entwicklungsarbeit im Vogelei. *Pfluegers Arch. Gesamte Physiol. Menschen Tiere* **93**, 303–398.
Tarboton, W. R. (1978). Hunting and the energy budget of the Black-shouldered Kite. *Condor* **80**, 88–91.
Taylor, C. R., Dmi'el, R., Fedak, M., and Schmidt-Nielsen, K. (1971). Energetic cost of running and heat balance in a large bird, the Rhea. *Am. J. Physiol.* **221**, 597–601.
Tinkle, D. W., and Hadley, N. F. (1973). Reproductive effort and winter activity in the vivparous montane lizard *Sceloporus jarrovi*. *Copeia* **1973**, 272–277.
Tinkle, D. W., and Hadley, N. F. (1975). Lizard reproductive effort: caloric estimates and comments on its evolution. *Ecology* **56**, 427–434.
Torre-Bueno, J. R. (1976). Temperature regulation and heat dissipation during flight in birds. *J. Exp. Biol.* **65**, 471–482.
Trivers, R. L. (1972). Parental investment and sexual selection. *In* "Sexual Selection and the Descent of Man, 1871–1971" (B. Campbell, ed.), pp. 136–179. Aldine, Chicago, Illinois.
Tucker, V. A. (1968). Respiratory exchange and evaporative water loss in the flying budgerigar. *J. Exp. Biol.* **48**, 67–87.
Utter, J. M. (1971). "Daily Energy Expenditure of Free-Living Purple Martins (*Progne subis*) and Mockingbirds (*Mimus polyglottos*) with Comparison of Two Northern Populations of Mockingbirds." Ph.D. Thesis, Rutgers Univ., New Brunswick, New Jersey.
Utter, J. M., and LeFebvre, E. A. (1970). Energy expenditure for free flight by the Purple Martin (*Progne subis*). *Comp. Biochem. Physiol.* **35**, 713–719.
Utter, J. M., and LeFebvre, E. A. (1973). Daily energy expenditure of Purple Martins (*Progne subis*) during the breeding season: estimates using D_2O^{18} and time budget methods. *Ecology* **54**, 597–604.
Vander Wall, S. B., and Balda, R. P. (1977). Coadaptations of the Clark's Nutcracker and the pinyon pine for efficient seed harvest and dispersal. *Ecol. Monogr.* **47**, 89–111.
Veghte, J. H. (1975). "Thermal Exchange between the Raven (*Corvus corax*) and Its Environment." Ph.D. Thesis, Univ. of Michigan, Ann Arbor.
Verbeek, N. A. M. (1964). A time and energy budget study of the Brewer Blackbird. *Condor* **66**, 70–74.
Verbeek, N. A. M. (1972). Daily and annual time budget of the Yellow-billed Magpie. *Auk* **89**, 567–582.
Vleck, C. M. (1981). Energetic cost of incubation in the Zebra Finch. *Condor* **83**, 229–237.
Wakely, J. S. (1978). Activity budgets, energy expenditures, and energy intakes of nesting Ferruginous Hawks. *Auk* **95**, 667–676.
Walsberg, G. E. (1977). Ecology and energetics of contrasting social systems in *Phainopepla nitens* (Aves: Ptilogonatidae). *Univ. Calif. Berkeley Publ. Zool.* **108**, 1–63.
Walsberg, G. E. (1978). Brood size and the use of time and energy by the Phainopepla. *Ecology* **59**, 147–153.
Walsberg, G. E., and King, J. R. (1978a). The heat budget of incubating Mountain White-crowned Sparrows (*Zonotrichia leucophrys oriantha*) in Oregon. *Physiol. Zool.* **51**, 92–103.
Walsberg, G. E., and King, J. R. (1978b). The energetic consequences of incubation for two passerine species. *Auk* **95**, 644–655.

Ward, W. H., and Lundgren, H. P. (1954). The formation, composition, and properties of the keratins. *Adv. Protein Chem.* **9**, 243–297.

Weathers, W. W., and Nagy, K. A. (1980). Simultaneous doubly labeled water (^{3}HH^{18}O) and time budget estimates of daily energy expenditure in *Phainopepla nitens*. *Auk* **97**, 861–867.

Weller, M. W. (1957). Growth, weights, and plumages of the Redhead (*Aythya americana*). *Wilson Bull.* **69**, 5–38.

West, G. C. (1968). Bioenergetics of captive Willow Ptarmigan under natural conditions. *Ecology* **49**, 1035–1045.

Westerterp, K. (1973). The energy budget of the nestling Starling *Sturnus vulgaris*, a field study. *Ardea* **61**, 137–158.

Wiens, J. A. (1969). An approach to the study of ecological relationships among grassland birds. *Ornithol. Monogr.* **8**, 1–93.

Withers, P. C. (1977). Energetic aspects of reproduction by the Cliff Swallow. *Auk* **94**, 718–725.

Wolf, L. L. (1975). Energy intake and expenditures in a nectar-feeding sunbird. *Ecology* **56**, 92–104.

Wood, H. B. (1950). Growth bars in feathers. *Auk* **67**, 486–491.

Wooley, J. B., and Owen, R. B. (1978). Energy costs of activity and daily energy expenditure in the Black Duck. *J. Wildl. Manage.* **42**, 739–745.

Wright, P. L., and Wright, M. H. (1944). The reproductive cycle of the male Red-winged Blackbird. *Condor* **46**, 46–59.

Chapter 4

HORMONAL CORRELATES OF BEHAVIOR

Jacques Balthazart

I.	Introduction	221
II.	The Classical Approach	223
	A. Correlational Studies	223
	B. Experimental Manipulations	227
III.	Attempts toward a Synthesis	260
	A. Dose–Response Relationships	260
	B. Latency of Action	264
	C. Specificity of Action of Gonadal Steroids: The Behavioral Level	268
	D. Mode of Action	272
IV.	Measuring Hormones and Their Fate	288
	A. Methods for Measuring Plasma Steroids	289
	B. The "Good" Steroids Are Present in Birds	290
	C. Covariations of Plasma Steroids and Behavior	291
	D. Do the Doses of Hormones Usually Injected Produce Physiological Levels of Steroids?	300
	E. Effect of Behavior on Hormones	302
V.	Brain Mechanisms	304
	A. The Need for Studies of Brain Mechanisms	304
	B. Brain Metabolism of Steroids	314
	C. Steroid Receptors	330
VI.	Conclusions	334
	References	335

I. Introduction

This chapter discusses types of correlations that can be established between behavioral events in birds and changes in their endocrine system. Owing to limitations of the available material and also to my own interest, the focus will be on reproductive behavior and the hormones directly involved in reproduction, that is, the gonadal steroids (androgens, estrogens, and progestins) and some of the pituitary hormones (gonadotropins and prolactin).

It is obvious that in most instances a correlation between a behavioral and an endocrine signal would be completely uninteresting if not presented along with a causal interpretation. We shall therefore focus on two types of situations and correlations according to their causal meaning: changes in endocrine physiology that induce modifications in the overt behavior of a bird and performance of behavioral activities that directly affect any aspect of the endocrine system.

The first type of situation (hormone induces behavior) has been studied far more than the second. This is largely due to technical and methodological limitations that will be described later.

Several criteria must be met before we can claim that a given hormone is responsible for the occurrence of a specific behavior. (1) Removal of the secretory gland should abolish the behavior. (2) Replacement of the hormone either by a graft of an endocrine gland or by hormone therapy should restore the behavior. (3) Variations of hormone concentration and behavior should be correlated. Additional criteria would also have to be added to ensure that the hormone acts physiologically in the intact animal. For example, the hormone injections that experimentally activate the behavior must produce changes in the endocrine milieu of the same amplitude as spontaneously observed in intact birds. Also, experimentally induced phase shifts in the endocrine cycles (daily, annual) must produce similar shifts in behavior (see details in Hiroshige, 1974).

The first two criteria can generally be met by uncomplicated techniques, and consequently many data are available. Since Berthold (1849) demonstrated that castration abolishes and testicular grafts restore sexual behavior in the cockerel, many studies have shown by endocrinectomy and subsequent hormone therapy that behavioral events must depend in some way on endocrine secretions (see Section II). Validation of this conclusion by fulfillment of the other criteria had to wait until more recently. Correlations originally established by histological studies of the endocrine glands proved in some cases to be confusing because there is frequently a discrepancy between the appearance of a gland and its real secretory activity. The establishment of precise correlations required the availability of sensitive methods for hormone assays (see Section IV,A), and these became available only recently.

This chapter reviews the literature on hormones and behavior in birds, following the historical and technical approach that was used to gain this knowledge. Results obtained by gonadectomy and hormone replacement are summarized first, and an attempt is made to extract some of the features common to these studies. Hormone assays are then described and their contribution to the study of behavior analyzed. The last section focuses on brain mechanisms of hormone action and reviews a number of unsolved problems and future trends in this field of research.

II. The Classical Approach

A. CORRELATIONAL STUDIES

1. Hormonal Correlates in Juvenile Birds: Sexual Maturation

At hatching, young birds have very small, largely nonfunctional gonads. The ratio of testicular or ovarian weight to body weight is low, and histological studies reveal that Leydig cells in the male are not differentiated and no free sperm is present in the lumen of the seminiferous tubules. In females, ovarian follicles are small and the interstitial cells are undifferentiated.

After a variable period of time [from 5–7 weeks in the Japanese Quail to 2–3 years in gulls or even 9–10 years in the Royal Albatross (*Diomedea epomophora*; Richdale, 1952, cited in Silver *et al.*, 1979)], the gonads enter a phase of rapid growth [e.g., in the male quail (Ottinger and Brinkley, 1979a), cockerels (de Reviers, 1971; Sharp *et al.*, 1977), and Zebra Finches (*Poephila guttata*; Sossinka, 1975)] and become functional, that is, they start secreting larger amounts of steroids and producing sperm or eggs (for a review see Abs, 1976).

The quiescent state of the gonad in young birds is generally associated with an absence of many species-typical behavior patterns: chicks do not crow, young passerines do not sing, sexual and courtship behavior are not observed or appear only with low frequencies (e.g., copulatory responses in chicks; Andrew, 1966; Vidal, 1971; see Section IV,C,1 for possible explanations of precocial behavior), and little aggressiveness is observed.

Several lines of evidence suggest that gonadal immaturity and behavioral quiescence are causally related.

1. The development of behavior generally parallels the differentiation and growth of the gonads. Sossinka (1975) showed that the onset of singing and courtship in the male Zebra Finch coincides with the period of fast maturation of the testes. Similarly, crowing, mating attempts, and cloacal contacts first occur in young male Japanese Quail at the age (32–36 days posthatch) when testes weight and plasma testosterone sharply increase (Ottinger and Brinkley, 1978, 1979a,b).

2. Castration before puberty interferes with the development of adult behavior, which is frequently completely abolished. Female and male Japanese Quail gonadectomized on the day of hatching show no sexual behavior when tested in adulthood with sexually active partners (R. E. Hutchison, 1978). Castration of young pigeons delays (~20 days) the "breaking of the voice," a change in vocalization that is very probably testosterone dependent (Abs, 1980). Testosterone injections indeed advance the maturation of the voice in pigeons (Abs, 1980) as well as in Mallards (*Anas platyrhynchos*;

Etienne, 1964) and Gadwalls (*A. strepera;* Schommer, 1978). However, the effects of castration on song development are not always as clear. Arnold (1975a), for example, reported that castration of male Zebra Finches at the age of 9–17 days does not prevent song development and only slightly decreases the singing rate.

3. Injection of gonadal steroids into immature birds induces precocious adult behavior in most instances. Injections or implants of testosterone or one of its esters stimulate copulatory responses in young chicks (e.g., Hamilton, 1938; Noble and Zitrin, 1942; Wood-Gush, 1963; Andrew, 1975a; Balthazart and Hendrick, 1978a; Young and Rogers, 1978; Balthazart and Hirschberg, 1979a; Benoff, 1979), domestic ducklings (Balthazart and Stevens, 1975a, 1976; Balthazart, 1978a; Balthazart and Hendrick, 1979a), and young turkeys (Schein and Hale, 1959). The same hormone also induces crowing in young chicks (Hamilton, 1938; Noble and Zitrin, 1942; Marler *et al.*, 1962; Andrew, 1975b) and Japanese Quail (Guyomarc'h, 1974). [For a review of testosterone effects on avian vocalizations see Andrew, (1969)]. Its effects on strutting displays in turkey poults have been discussed by Schein and Hale (1959) and Schleidt (1970). Testosterone advances the onset of singing in male Zebra Finches (shift from 34–37 days to 18–19 days posthatch; Sossinka *et al.*, 1975); increases aggressiveness in chicks (Andrew, 1975b; Young and Rogers, 1978), Herring Gulls (*Larus argentatus;* Boss, 1943), and domestic ducklings (Balthazart and Stevens, 1975a, 1976); and brings on occurrences of various courtship displays in several species of ducks including domestic ducks, Mallards, Northern Pintails (*Anas acuta*), and Redheads (*Aythya americana*) (Phillips and McKinney, 1962; Etienne, 1964; Balthazart and Hendrick, 1979a).

In juvenile female birds, estradiol similarly induces an early development of adult sexual responses. In female domestic ducklings, for example, the daily injection of 0.25 mg estradiol benzoate (EB) from Days 4 to 17 posthatch increases the sexual receptivity, as revealed by a higher incidence of squatting posture in the presence of testosterone-injected male ducklings (Balthazart and Stevens, 1976).

Several questions are raised by these experiments in which steroids were injected into juvenile animals. Why are high doses of hormones required to induce precocious adult behavior (0.1 mg testosterone propionate (TP) per kilogram body weight elicits strutting in adult turkeys, whereas 15 mg/kg only induces weak responses in the turkey poult; Schleidt, 1970)? Are the imperfect behavior patterns frequently observed in these experiments a result of inadequate hormone levels, do they reflect the need for some kind of learning, or are they produced by the incomplete development of the nervous system? Tentative answers to those questions are presented in the second part of this chapter, after the discussion of results concerning hormone assays in the plasma and studies of brain metabolism of steroids.

2. Seasonal Breeding

Most species of birds live in an environment subject to regular cyclic variations of both short (diurnal cycle) and long (seasonal changes) periodicities. Evolution has favored the species and individuals that are capable of producing their young at the most favorable time of the year, which implies that animals must be able to anticipate the arrival of the most suitable season for breeding. Hormonal mechanisms regulate and synchronize all aspects of reproduction, from the maturation of sperm and eggs to the expression of the species-specific behavior patterns that are employed to defend a territory, court a female, and ensure the fertilization of the eggs. The activity of the pituitary and the gonads, which produce the most important hormones in this context, is itself controlled by the environment (mainly the photoperiod) and also in numerous species by endogenous rhythms, which persist in constant conditions with periods close to 12 months. Several reviews of these control mechanisms are available (Lofts et al., 1970; Immelmann, 1971; Follett, 1973, 1978; Gwinner, 1973, 1975a; Farner, 1975; Follett and Davies, 1975; Lofts, 1975; Murton and Westwood, 1977; Farner and Wingfield, 1980; Follett and Robinson, 1980), but their detailed analysis is outside the scope of the present chapter, which will focus only on the hormonal mechanisms that control reproductive behavior.

It has long been known that the seasonal onset of reproductive behavior in birds is associated with an increase in the size of the gonads. In males, seasonal changes in testes size and histology have been correlated with reproductive behavior in a number of species from different orders, including the Great Tit (*Parus major;* Silverin, 1978a), Pied Flycatcher (*Ficedula hypoleuca;* Silverin, 1975, 1978b), White-crowned Sparrow (*Zonotrichia leucophrys;* King et al., 1966; Wingfield and Farner, 1980), European Starling (*Sturnus vulgaris;* Temple, 1974), Mallard (Höhn, 1947, 1960), Rook (*Corvus frugilegus;* Lincoln et al., 1980), Common Eider (*Somateria mollissima;* Gorman, 1974), feral pigeons (*Columba livia;* Murton and Westwood, 1975) and Wood Pigeons (*Columba palumbus;* Lofts et al., 1966). An example of this type of study is given in Fig. 1, which summarizes important changes in size and histology of the testes and the relationship between these changes and behavior during the reproductive cycle of the Pied Flycatcher. When birds arrive in Sweden in early May, their testes are rapidly growing. This tendency continues throughout the nest-building phase of the cycle and culminates around the time of egg laying. Toward the end of the incubation period, the gonads regress rapidly, marking the onset

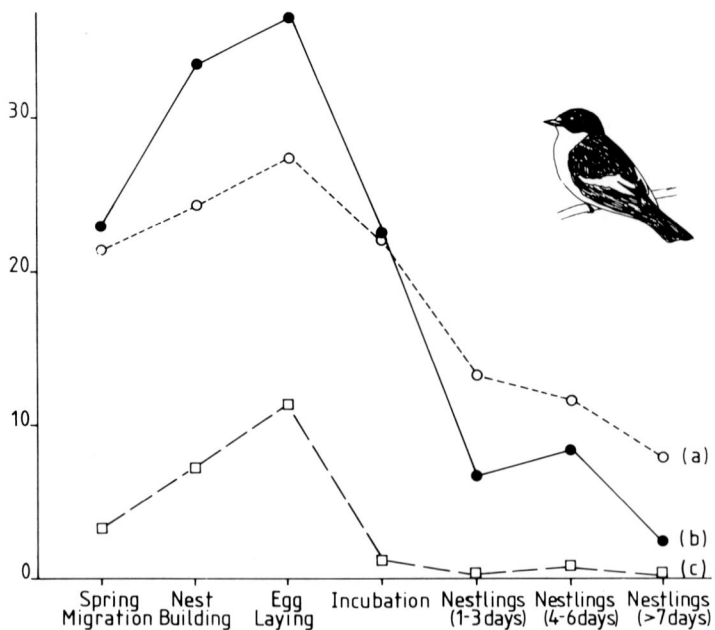

FIG. 1. Changes in testicular histology of the male Pied Flycatcher during the breeding season. (a) Diameter of the seminiferous tubules ($\mu m \times 0.1$), (b) Paired testes weight (mg), (c) Index number of Leydig cells ($\times 0.1$). (Drawn from data in Silverin, 1975.)

of the refractory period. Histological criteria of testicular activity closely parallel the changes in gonadal weight.

Similar information is available for female birds. Seasonal changes in the sizes of the largest ovarian follicle, the ovary, or the oviduct are correlated with variations in the reproductive behavior [e.g., Mallard (Höhn, 1947), Northern Pintail (Phillips and Van Tienhoven, 1962), White-crowned Sparrow (Wingfield and Farner, 1980), and Great Tits (Silverin, 1978a)].

In domestic or laboratory birds, there is also ample evidence that the onset of reproductive behavior, when birds are shifted from short to long photoperiod or simply exposed to congeners of the opposite sex, is associated with an increase in gonadal size or activity [e.g., Japanese Quail (Sachs, 1969), Ring Dove (Lehrman, 1964a; Cheng, 1974, 1979), and Canary (Hutchison, et al., 1968)]. The correlation between gonadal stage and reproductive behavior is also demonstrated by data from Murton and Westwood (1975), who showed that in the male feral pigeon there is throughout the year an inverse relationship between the size of the testes in free-living birds and the latency to show the bowing display when a male is paired with a female in captivity.

4. HORMONAL CORRELATES OF BEHAVIOR

B. Experimental Manipulations

Overwhelming evidence now supports the conclusion that the changes in gonadal condition cause the seasonal variations of reproductive behavior in birds. With very few exceptions (see Section II,B,7), these behavior patterns disappear or occur with much lower frequencies in gonadectomized animals, and replacement therapy with synthetic steroids usually restores the behavior, although in some instances precastration levels of activity cannot be reinstated (e.g., J. B. Hutchison, 1978). These experiments permit identification of the active steroid(s) among all those that are produced by the gonad (Idler, 1972); however, they cannot prove by themselves that a specific steroid is actually responsible for the occurrence of a given behavior in physiological conditions (metabolites of the steroid could be the active compounds, see Section V,B). Controls must also be made by measuring the plasma levels of hormone following the injections to ensure that physiological concentrations have been induced and thus that the restoration of behavior does not result from a pharmacological effect of the steroid.

With this type of experiments, it has nevertheless been possible to demonstrate that many aspects of bird behavior are controlled by the endocrine secretions of the gonads. This lends further support to the correlation just described. The results of this type of experiment are reviewed in Section II,B, using a functional classification for the behavioral activities (Hinde, 1973). A special case will be made, however, for the Ring Dove and the Canary, in which this type of analysis has been most extensively performed.

1. Aggression, Establishment of Hierarchies, and Territorial Defense

In birds, the male is generally much more aggressive than the female (Collias, 1944), and it is thus not surprising that most of the research available on the hormonal basis of agonistic behavior concerns male birds. Among the notable exceptions to this rule are the phalaropes (*Phalaropus*), in which females have a brighter plumage and are more active than males both in aggressive behavior and in behavior related to pair formation. There is evidence that this unusual situation correlates with a changed hormonal status as, in Wilson's Phalarope (*P. tricolor*), the ovarian content of testosterone exceeds that of the testes ((Höhn and Cheng, 1967), whereas the contrary is generally observed [e.g., in domestic fowl, Mallards, and Red-winged Blackbirds (*Agelaius phoeniceus*)]. This probably explains the brighter plumage and the aggressiveness of female phalaropes, as Johns (1964) demonstrated that injections of TP induce plumage of the female nuptial type in dull juveniles or adults in nonbreeding plumage.

Before being able to court and eventually mate with a female, male birds of most species have to establish and defend a territory or at least select a female and prevent other males from approaching her. In many species, these behavioral activities are under the control of androgens. Although correlations have been detected in field studies between the physiological state of the testes and/or the plasma levels of androgens on one hand and aggressive behavior and establishment of a territory on the other hand (e.g., Temple, 1974; Silverin, 1975, 1978a,b; Wingfield and Farner, 1978, 1980), few experimental manipulations seem to have been done to prove that androgens are a causal factor in the establishment of a territory. Boss (1943), however, has mentioned that young male Herring Gulls injected with androgens display the same aggressive reactions as those seen in adult males during the breeding season. Noble and Wurm (1940a) reported the occurrence of active territory defense in Black-crowned Night-Herons (*Nycticorax nycticorax*) which had been treated with TP. Watson and Moss (1971) implanted androgens in two male Red Grouse (*Lagopus lagopus scoticus*), and the two birds almost doubled their territory size. Also, Silverin (1980b) showed that treatment with long-acting testosterone can prolong territorial behavior in the Pied Flycatcher.

Numerous laboratory experiments, however, have established the role of androgens in the control of aggression and thus possibly of territorial defense. As early as 1939, Allee *et al.* showed that injections of TP increase the social status in intact White Leghorn hens.

Capons are known for their lack of aggressiveness (Goodale, 1913), but injections of testosterone rapidly stimulate their aggressive behavior (Davis and Domm, 1943). Similarly, castration of Japanese Quail markedly decreases aggressiveness, which is restored by treatment with testosterone (Selinger and Bermant, 1967; Tsutsui and Ishii, 1981). In addition, the transfer of male quail from short days (where they are functionally castrated; Sachs, 1969; Adkins and Adler, 1972) to long days stimulates their testicular activity, increases their plasma testosterone levels (Follett, 1976; Follett and Maung, 1978), and at the same time increases their aggressiveness (Balthazart *et al.*, 1979).

A decreased aggressiveness following castration and its restoration by treatments with testosterone or one of its esters has been observed in many species belonging to at least three orders other than Galliformes. Examples are indeed found in the Anseriformes [wild Mallards (Etienne and Fischer, 1964) and domestic ducks (Deviche, 1979a)], Columbiformes [Ring Dove (Hutchison, 1970a; Cheng and Lehrman, 1975; Adkins, 1981; Saad *et al.*, personal communication, 1981)] and Passeriformes [Zebra Finch (Arnold, 1975b)].

A few experiments seem, however, to contradict this general conclusion. Vowles and Harwood (1966) studied the aggressive and defensive behavior in Ring Doves during the breeding cycle and after intramuscular injections of various hormones into intact birds. They concluded that prolactin and progesterone, but not testosterone, increase defensive behavior toward a predator. Various explanations have been proposed (see Vowles and Harwood, 1966; Silver *et al.*, 1979) to explain the discrepancy between this conclusion and results of similar experiments that assign testosterone an important role in the control of agonistic behavior of doves (e.g., Bennett, 1940; Hutchison, 1970a). One interesting suggestion is that the hormonal stimulation may interact with situational variables; for example, testosterone could increase aggressiveness toward a conspecific of the opposite sex, although it would have negligible effects in a nest-defense situation. The same idea was suggested by Crook and Butterfield (1968), who showed that testosterone administration to low-ranking members of a hierarchy of intact male Red-billed Quelea (*Quelea quelea*) produces no significant change in their ability to win encounters in individual distance infringement, although it does improve their success in fights over material used to build nests. There is a correlation between incidence of aggressive encounters and shifts in plumage coloration that are thought to be determined by luteinizing hormone (LH; Butterfield and Crook, 1968). Crook and Butterfield (1968) injected LH preparations into intact birds, and their results suggest that aggression during encounters in the absence of nest material could be controlled by the direct action of LH (i.e., without any involvement of gonadal steroids). A similar claim has been made for the European Starling, whose fighting ability is increased rather than decreased by castration and is not affected by injections of testosterone (Davis, 1957). This possibility will be analyzed in greater detail in a section devoted to the behavioral role of gonadotropins in birds (see Section II,B,7).

As stated earlier, female birds are generally less aggressive than their male conspecifics. This is probably related mainly to the lower androgen production of the ovary compared to the testes (Höhn and Cheng, 1967), as testosterone treatment markedly enhances the frequency of aggressive behavior or improves the status in the hierarchy in Ring Doves (Bennett, 1940), female chicks (Hamilton and Golden, 1939), and adult hens (Allee *et al.*, 1939).

It must be pointed out here that if, for reasons of convenience in this presentation, the effects of hormones on both aggressive behavior and the position in a hierarchy have been mixed, these two types of effects must be sharply distinguished, as they are not necessarily controlled by the same causal factor. A very aggressive bird can eventually be low in the hierarchy due to interactions with other variables such as ontogenetic development, morphology, or previous conditioning (e.g., Lumia, 1972; Rohwer and Rohwer, 1978). The phenotypic difference between males and females at the level of aggressive behavior may also be enhanced by the presence of high estrogen levels in the females. Estrogen administration does tend to lower aggressiveness [in hens (Allee and Collias, 1940), although in some studies no effect could be detected, for example, in the Canary (Shoemaker, 1939)].

2. *Courtship and Sexual Behavior*

Sexual behavior can easily be defined as all behavior patterns directly involved or preceding the process of mating. As far as we know, this group of activities is controlled in all species of birds by the same type of hormones: androgens in the males, estrogens and eventually progestins in the female. Courtship immediately appears as a more complex group of activities. In the ethological literature, it is described as the result of conflicting tendencies that promote the simultaneous occurrence of three categories of behavior: attacking, fleeing, and mating (Hinde, 1966). There is, in addition, little certainty that those activities we call courtship in different species (e.g., social displays in ducks and bowing in the pigeon) are truly homologous. This complexity is reflected at the level of hormonal control. Many experiments have revealed a preeminent role of androgens in the control of these activities, but claims have frequently been made that other hormones (e.g., gonadotropins) could also be involved. Available evidence is reviewed here and in Section II,B,7, which is devoted to this problem.

a. Male Sexual Behavior. Historically, the first experiment on the effects of hormones was devoted to the sexual behavior of birds. In 1849, Berthold showed that the endocrine secretions of the testes control the comb growth and the sexual behavior of chickens. Since that time, the demonstration that castration abolishes, whereas testosterone treatments restore, sexual behavior has been confirmed in cockerels (Davis and Domm, 1941, 1943; Domm *et al.*, 1942) and extended to numerous species such as the Japanese Quail (Beach and Inman, 1965; Adkins and Nock, 1976a; Adkins, 1977; Adkins *et al.*, 1980), the wild or domestic Mallard (Phillips and McKinney, 1962; Etienne and Fischer, 1964; Deviche, 1979a), two species of gulls (Noble and Wurm, 1940b; Boss, 1943), the Zebra Finch (Pröve, 1974; Arnold, 1975b), the Ring Dove (Cheng and Lehrman, 1975; Adkins-Regan, 1981), and the pigeon (Erpino, 1969).

It is thus likely that testosterone stimulates sexual behavior in every species of bird. Additional evidence for this has been researched using antiandrogenic drugs—mainly cyproterone acetate (CA). Adkins and Mason (1974) and Silver (1977) demonstrated that injections of CA block sexual behavior in intact male Japanese Quail and bow cooing and nest soliciting in Ring Doves. Although it was generally acknowledged that the effects of CA on behavior were much less pronounced than the effects on morphology (e.g., on the cloacal gland in quail), we do not believe that results of these experiments can be taken as evidence that testosterone is involved in the control of sexual behavior for the following reasons:

1. In addition to its antiandrogenic activity (competition for binding with androgen receptors; Attardi and Ohno, 1976), CA also possess some antigonadotropic properties, that is; it can decrease gonadotropin secretion (see Neumann and Steinbeck, 1974). This probably results in a decrease of all testicular secretions, including testosterone, and it cannot be ascertained that the behavioral deficit is related to specific blockage of testosterone effects in the brain.

2. There is good evidence that the behaviors that are blocked by CA injection (e.g., copulation in quail or nest soliciting in doves) are induced by estradiol injections and that their occurrence after testosterone injection is mediated through the aromatization of testosterone in the brain [in quail (Adkins and Nock, 1976b; Adkins et al., 1980); in doves (Cheng and Lehrman, 1975; J. B. Hutchison, 1978; Silver et al., 1979; Adkins-Regan, 1981)]. It is likely that CA does not compete for the estrogen receptor in birds, as already demonstrated in mammals (Attardi and Ohno, 1976), so that a blockage of behavior by CA is hardly explained by specific antiandrogenic or antiestrogenic properties of the drug.

3. The effects of CA on sexual behavior in mammals are generally weak or even do not exist (Zucker, 1966; Steinbeck et al., 1967; Beach and Westbrook, 1968; Whalen and Edwards, 1969), whereas their action on peripheral sex organs is extremely clear (reviewed in Neumann and Steinbeck, 1974).

4. In addition, McEwen et al. (1979) have suggested that CA could partially interfere with the aromatization of testosterone in the rat brain, whereas Luttge et al. (1975) claim that it could act as an antiestrogen.

Although the first of these objections is partially met by the work of Balthazart (1978a), who showed in immature ducks that CA blocks the sexual behavior induced by exogenous testosterone [which rules out the interpretation involving only a decrease of the testicular output of testosterone (antigonadotropic effect of CA)], it remains that the precise mechanism of action of CA cannot be proven by the available data. It is clear, however, that testosterone (eventually, through its metabolites; see Section V,B) exerts a major control on this type of activity.

b. Courtship. A more cautious conclusion has to be formulated when considering the mechanism that controls courtship in male birds. Although castration abolishes courtship in all species studied (to the best of our knowledge), difficulties have sometimes been encountered when trying to re-establish the behavior by androgen therapy. Testosterone injections have sometimes failed to stimulate courtship behavior in immature animals or in nonbreeding birds.

Castration abolishes strutting in quail (Beach and Inman, 1965) and waltzing in cockerels (Domm *et al.*, 1942; Davis and Domm, 1943; McCollom *et al.*, 1971), and these courtship behaviors are restored by androgen injections. Similarly, in the Zebra Finch, Pröve (1974) and Arnold (1975b) have shown that courtship and courtship songs directed toward females are dramatically reduced following castration, but readily reappear following TP injections.

In the Ring Dove both the bowing display directed toward females and a nest-oriented display, wing flipping (eventually associated with the nest coo—the pattern is also called nest soliciting), are suppressed by castration and restored by daily injections of TP (Hutchison, 1970a; Cheng and Lehrman, 1975; Saad *et al.*, personal communication, 1981). Estradiol benzoate also activates nest soliciting in males (Hutchison, 1970a) and is actually active at a lower dose (Cheng and Lehrman, 1975). (The problem of hormonal specificity will be reviewed in detail in Section III,C).

In the closely related pigeon, however, discordant results have been obtained. Erpino (1969) showed that castration suppresses or reduces bowing and displacement preening in males and that these are restored in birds implanted with pellets of TP. By injecting intact pigeons with exogenous hormones, Murton *et al.* (1969) reached slightly different conclusions. In their study, males given testosterone showed an increase in the frequency of the bowing display and later began nest cooing, thus producing a reproductive cycle closely resembling that seen in control birds. However, they also observed that injection of mammalian follicle-stimulating hormone (FSH) increased the sexuoaggressive components of the bowing and produced much driving (chasing) behavior. On that basis, they suggested that parts of the courtship displays in pigeons could be under the control of FSH. This claim is indirectly supported by a number of other experiments (e.g., Murton and Westwood, 1975), and it could provide satisfactory interpretations for a number of results in Ring Dove experiments (e.g., castration does not immediately eliminate chasing and nest soliciting; Hutchison, 1970b). More direct explanations involving only steroids are also available.

In ducks, the same type of equivocal results have been obtained. In a number of studies, castration has been shown to abolish social displays [Wild Mallard (Phillips and McKinney, 1962; Etienne and Fischer, 1964), domestic ducks (Deviche, 1979a)]. Various types of displays have also been shown to be stimulated by TP injections in castrated ducks [wild Mallard (Phillips and McKinney, 1962; Etienne and Fischer, 1964), domestic duck (Deviche, 1979a)], in intact ducklings [Mallard, Northern Pintail, and Redhead (Phillips and McKinney, 1962; domestic ducks (Balthazart and Hendrick, 1979a)], or in subadult and adult nonbreeding Gadwalls (Schommer, 1977) and Common Eiders (Gorman, 1974). In contrast, we failed in several experiments to stimulate social displays in young ducklings or immature domestic ducks (Balthazart and Stevens, 1975a, 1976; Deviche and Balthazart, 1976; Balthazart, 1978a). In one experiment, TP injections even inhibited social displays in adult intact males (Balthazart and Deviche, 1977). On the basis of a positive correlation between the annual variations in plasma FSH and in social-behavior frequency (Balthazart and Hendrick, 1976), it was suggested that FSH together with androgens could play a role in the control of ducks' displays. However, no conclusive evidence for this notion had been collected (Deviche and Balthazart, 1976; Balthazart and Deviche, 1977; see Section II,B,7).

If experiments in which androgen treatments failed to stimulate courtship can be taken as evidence for a control by other hormones, it must nevertheless be clear that the failure can always be related to inadequate testing situations. The social releasers of courtship behavior are largely unknown, and consequently it can never be ascertained that some of them were not missing during the tests (Balthazart and Stevens, 1976). Schein and Hale (1959) also demonstrated that early social experience or the nature of the eliciting stimuli can markedly influence the behavior induced by androgen injections in turkeys. Previous social and sexual experience could also be important, as suggested by Erpino (1969), to explain why his naive (i.e., unfamiliar with their mates) pigeons completely lost their sexual behavior following castration, whereas as many as 65% of the pigeons (in established pairs) castrated by Carpenter (1933a,b) retained some copulatory behavior indefinitely. The behavioral responses following hormone treatment thus need to be analyzed relative to the particular testing situation and the previous experience of the animals (Lehrman, 1964b).

c. Female Sexual Behavior. Much less research has been devoted to the hormonal control of sexual behavior in female birds than in males. This somehow contrasts with the situation in mammals where the lordosis response of the female, which can easily be elicited (e.g., with a palpation of the flanks by the hand of the experimenter) and is easily quantifiable, has been a commonly used model in the study of hormones and behavior (e.g., see reviews in Komisaruk, 1978; Feder, 1978; Lisk, 1978).

All the available data, however, clearly support the conclusion that sexual receptivity in female birds is controlled by the endocrine secretions of the ovary (mainly estrogen, presumably in synergism with progesterone). As early as 1913, Goodale noted that ovariectomized hens shown "no sex instincts." Davis and Domm (1943) confirmed that bilaterally ovariectomized hens do not squat in response to the male's approach. This is corrected by estrogen treatments. Administration of estrogen in bilaterally ovariectomized poulards induces squatting (Allee and Collias, 1940; Davis and Domm, 1941, 1943). Further research in Japanese Quail confirms this conclusion. Female quail that are functionally castrated by exposure to a short photoperiod cease egg laying within about 2 weeks (median 16 days) and show a complete loss of sexual receptivity (Adkins and Adler, 1972). Estrogen injection at a dosage equal to 0.05–0.1 mg/day was highly effective in restoring sexual receptivity in these females. Progesterone alone was not effective in this respect (doses up to 5 mg), but there was some suggestion in that study that progesterone plus estrogen was more effective than estrogen alone in stimulating squatting. This was confirmed by Noble (1972), who demonstrated that female Japanese Quail housed in continuous darkness squat in 87% of the tests if they are treated with EB (200 µg/day), and in 100% of the tests if they receive, in addition, progesterone at a daily dose of 200 µg. Control oil-injected females never squat in these conditions. Adkins and Nock (1976a), however, subsequently showed that, in ovariectomized females or in females exposed to short days, 25 µg EB per day is sufficient to restore all aspects of sexual receptivity to the levels observed in intact females raised in long days. Noble and Wurm (1940b) also showed that administration of estrogen in gonadectomized female Laughing Gulls (*Larus atricilla*) stimulates sexual receptivity and food begging, whereas these patterns are absent in uninjected birds.

Removal of the left ovary (the one that is normally functional in birds) (Lofts and Murton, 1973; Murton and Westwood, 1977) in female doves leads rapidly to the disappearance of the female courtship behaviors (wing flipping and nest cooing) that are shown by intact birds in response to the male courtship. However, after about 6–10 weeks, wing flipping generally reappears, sometimes accompanied by all other characteristics of mature females. It can be shown that this recovery is due to a regeneration of ovarian follicles on the right side and that their surgical ablation will then lead to a complete suppression of the behavior (Cheng, 1973a).

Estrogen alone at the daily dose of 50 µg/animal fully restores the behavior of the females (wing flipping plus nest coo) with the exception of nest building (Cheng, 1973b). Progesterone alone does not reliably restore any of these behavior patterns, but in synergism with estrogens it stimulates nest cooing and nest-oriented behavior (Cheng and Silver, 1975).

In the Ring Dove, Cheng (1973a, 1979) distinguished two types of squatting according to the context in which they are observed. In relatively early stages of the cycle (up to stage 5), the females crouch only in response to a prolonged courtship of the male, including self-preening near the female, circling, displacement preening, and courtship feeding. Later (stage 6), she crouches immediately after being introduced to the male, even before the latter starts to court (proceptive sexual crouch). The hormonal basis of these two types of crouch are somewhat different but they are both restored in the ovariectomized dove by injection of EB alone at a daily dose of 50 μg (Cheng, 1973b). The treatment with EB plus progesterone only slightly improves the sexual responsiveness of the females (Cheng and Silver, 1975). Surprisingly, the daily injection of 100 μg EB, in spite of the fact that it increased oviducal weight more than the treatment with 50 μg EB, had almost no stimulating effect on sexual behavior. It was speculated that the failure of large doses of estrogen to induce female courtship could be due to the negative-feedback effect on the hypothalamo–hypophyseal axis, resulting in too low levels of gonadotropins (LH or FSH) or of gonadotropin-releasing hormones (e.g., LHRH; for more details see Cheng, 1979). Experimental evidence collected later supports this interpretation (Cheng, 1977; see also Section II,B,7).

In summary, it seems warranted to conclude that the endocrine activity of the ovary largely controls the sexual behavior of female birds. Estrogens are the major hormones involved in this control.

In female mammals there is good evidence that in some species estrogen alone will elicit lordosis (e.g., rabbit and cat), whereas in others the sequential effects of estradiol and progesterone are absolutely required (e.g., guinea pig), and estrogens alone even at very high dosage will have no consistent effect (see Feder et al., 1979a). Large doses of estradiol (≥25 μg) have been used in most experiment dealing with birds, and to our knowledge no experiment has shown an activation of sexual behavior *sensu stricto* in female birds by small doses of estrogen combined with progesterone. Such a synergism is clearly acting in the control of nest soliciting in Ring Doves (Cheng, 1977). Adrenal progesterone is probably still available in ovariectomized birds. It is thus impossible, on the basis of the available data, to decide whether exogenous progesterone acted only as a complementary activator in the experiments described earlier or whether its presence was absolutely required for the full occurrence of female sexual behavior in physiological conditions.

Various aspects of the hormonal regulation of courtship and mating, such as induction of male behavior in females by testosterone injections, specificity of steroid action, or possible role of steroid metabolsim, will be considered in more detail in Sections III,C, V,A,4, and VB.

3. Nest Building

There is a wide variation among species in the type of nest (e.g., Drent, 1975), the amount of building behavior, and the amount of cooperation between sexes in this activity (Lehrman, 1961). The amount of nest building varies from nothing in those species that either lay their eggs on the ground (e.g., Japanese Quail) or drop them in the nests of other species (cowbirds, Old World cuckoos) to the construction of very elaborate nests that very effectively protect the eggs and nestlings (see examples in Drent, 1975). The female can build alone [e.g., ducks (McKinney, 1969) and Canary (Hinde, 1958)], build in cooperation with her mate (e.g., Ring Dove; Lehrman, 1964a), or, exceptionally, the male builds alone (e.g., Red-billed Quelea; Morel and Bourlière, 1956)]. Nest building can last from a few hours prior to oviposition (e.g., hen; Wood-Gush, 1975) to several days (e.g., Canary; Hinde, 1958, 1965). In some species it continues after egg laying [e.g., Ring Dove (White, 1975c) and ducks (McKinney, 1969)].

Hormonal correlates of nest-building behavior have been studied in only a small number of species but, in these, in great detail. The female Japanese Quail, a favorite animal in studies of behavioral endocrinology, takes almost no care of her eggs; she simply drops them on the floor. In open-range conditions, laid eggs are, however, in a small depression, and some nest construction is eventually observed (Stevens, 1961; Rothsstein, 1967). So far as we know, the hormonal basis of this behavior (or lack of it) has not been investigated.

In many strains of domestic hens, nest-building activities and care of the eggs are very rudimentary, but hens of some modern strains when placed in adequate conditions (deep-litter floor pens) engage in relatively long periods of nest building just before oviposition (1–4 hours) and in this way build a rough nest that is used for successive eggs within a sequence (Wood-Gush, 1975). When kept in pens equipped with trap nests, the hens are restless before laying and give a special call, which can be induced in immature birds by EB injections (Wood-Gush and Gilbert, 1969). The hen then examines different nests, enters one, and sits. This behavior is hormonally controlled by the postovulatory follicle, as demonstrated elegantly by Wood-Gush and Gilbert (1964, 1970), and it can be reproduced in ovariectomized hens by injections of estrogen and progesterone (Wood-Gush and Gilbert, 1973). In domestic hens nest-building behavior can be independent of the cock (Wood-Gush, 1975), although according to McBride *et al.* (1969) cocks of feral domestic fowl will eventually escort the female to the potential nest site when she gives the nesting call.

Various studies suggest that androgens could be involved in the control of nest building in male and female birds (when both participate). For example, injections of TP (but not estrogens) in intact or castrated Black-crowned Night-Herons of both sexes induce nest-building behavior (Noble and Wurm, 1940a), and testosterone treatment stimulates some intact female Budgerigars (*Melopsittacus undulatus*) to perform nest-box oriented behavior (Brockway, 1969). Estradiol is, however, more active than testosterone in this respect.

The detailed studies of the Ring Dove clearly show that it is necessary to differentiate between the hormonal stimuli that are active in males and females. Furthermore, it is crucial to consider the interaction between the sexes in the control of nesting behavior.

Ring Dove. In the Ring Dove the male usually selects the nest material and transports it to the nest site, whereas the female normally stays in the nest and builds with the material carried by the male. These activities are observed mainly during the late morning or early afternoon (Martinez-Vargas and Erickson, 1973; Martinez-Vargas, 1974; Erickson and Martinez-Vargas, 1975). In early experiments, Lehrmann (1955, 1958a, 1965) demonstrated that the readiness to build a nest and to incubate is not present immediately when a male and a female are mated but develops gradually during the breeding cycle.

This observation, together with the finding that these behavior patterns can be induced by exogenous hormones (Lehrman, 1958b), were taken as evidence for a strong hormonal dependency of the behaviors. More recently, however, it was discovered that an intact, untreated male dove will immediately perform active nest building if paired with a hormone-treated female (Martinez-Vargas and Erickson, 1973). This does not mean that hormones play no role in promoting nest-building behavior in the males. Sexually active males are physiologically ready to gather nest material when introduced to a female that has been previously treated with estradiol and progesterone (Martinez-Vargas and Erickson, 1973), but castrated males will not respond in this way even to treated females, which proves that in addition to its control by social determinants the nest-building behavior of the male is hormonally controlled. It has been shown, in addition, that injections of TP or EB strongly activate nest-material gathering in castrated male doves (Martinez-Vargas, 1974).

Additional studies demonstrate that testosterone implanted in minute amounts in the anterior hypothalamus (the preoptic region or the ventral neostriatum intermediate) is also effective in this respect (Erickson and Hutchison, 1977). Nest building is essentially confined to the late hours of the morning in intact male doves, and Erickson and Hutchison (1977) found that the birds that received effective brain implants of TP exhibit this same daily rhythm of behavior. This suggests that the diurnal changes in nest building are not directly dependent upon similar changes in testicular secretion. This, however, does not mean that plasma testosterone levels are stable throughout the day, and additional data actually confirm that testosterone in Ring Doves fluctuates greatly during the diurnal cycle (Balthazart *et al.*,

1981a). In contrast to the male, the female doves do not seem ready to perform nest-building behavior immediately after they are paired to active males. The female's ovary must be active, that is, it must secrete estrogen and progesterone if she is to build a nest.

A female dove when paired with an active male never starts building until several days after pairing (Lehrman, 1958a, 1964a, 1965). If intact females are treated with diethylstilbestrol while in isolation cages, nest building is immediately observed in some birds when they are paired with an active male (Lehrman, 1958b). On that basis, Lehrman suggested that the courtship behavior of the male stimulates estrogen secretion in the female, which in turn induces the nest-building activity. This conclusion is only partly supported by more recent experiments using ovariectomized doves. Lehrman *et al.* (1961) and Erickson (1970) directly demonstrated that the male's courtship stimulates ovarian activity in the female, but it was also proved that estrogens alone are ineffective in eliciting nest-building behavior in ovariectomized female doves (Cheng, 1973b). A combined estrogen plus progesterone treatment is, however, very effective for eliciting nest-building activity in ovariectomized female doves (Fig. 2). It also facilitates this behavior in their untreated mates (Cheng and Silver, 1975). The combined treatment with diethylstilbestrol and progesterone also activates the building behavior in intact female doves (Martinez-Vargas and Erickson, 1973). In this experiment more males engaged in nest building when paired with hormone-injected females than when paired with untreated ones. The hormonal condition of the female thus not only affects its behavior but also determines to a large extent the behavior of its mate. If testosterone in the male plays some role in the control of nest building (see earlier), social determinants (e.g., the female's behavior) are also very important.

Additional stimuli are also provided by the nest itself. White (1975a,b,c) has shown that the condition of the nest effectively determines the level of nest-building activity displayed by the birds. If, for example, nests are destroyed every day by the experimenter (empty group), birds remove more nest material from the dispensers supplied in their cage, and more of this material will be placed in the nest bowl (mainly by females). Conversely, if full nests (i.e., nest bowls containing a completed nest) are provided (full group), very little building activity is observed.

Surprisingly, the condition of the nest also seems to affect the sex roles during building. In a normal pair of doves (normal plus empty groups in White, 1975a,b), the male usually collects and transports the nest material to the female which, using tucking movements, inserts it into the nest. When White (1975b) provided full nests to the birds, these roles disappeared. In these pairs, both males and females engaged with similar (low) frequencies in all categories, from collecting the material outside the nest to the actual

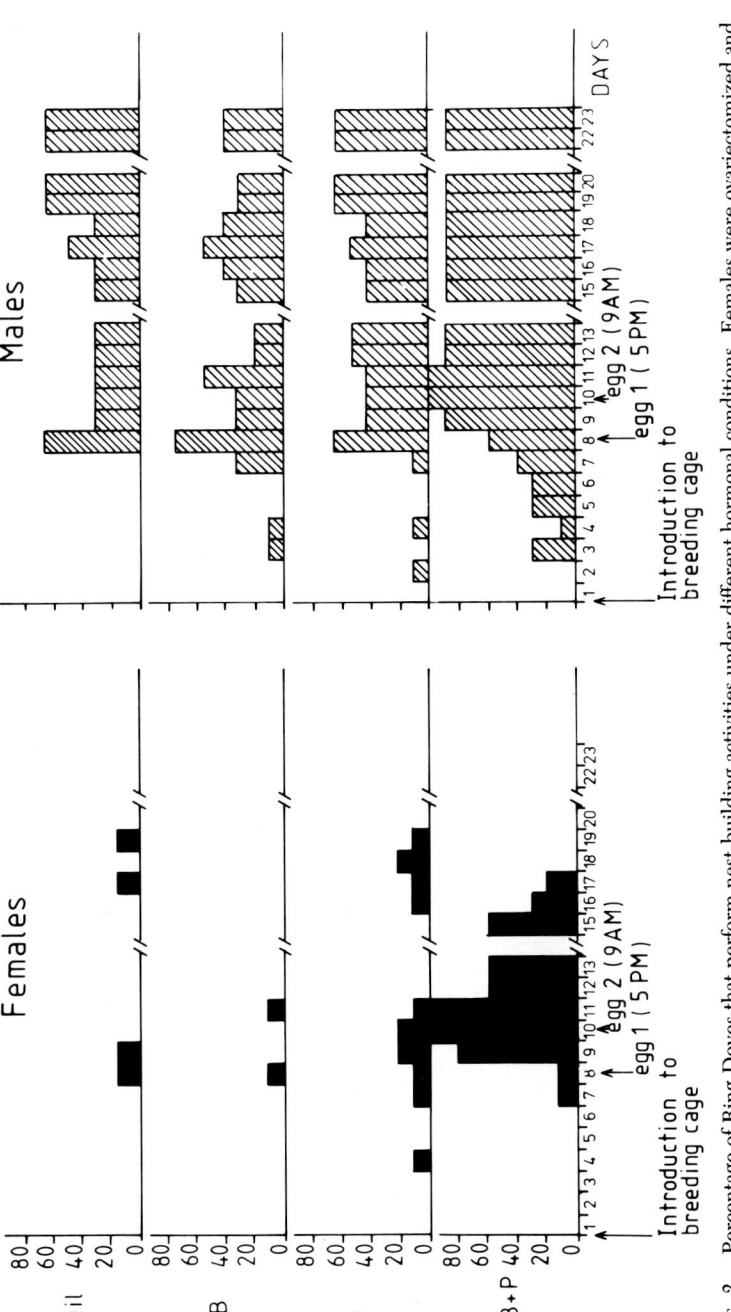

FIG. 2. Percentage of Ring Doves that perform nest-building activities under different hormonal conditions. Females were ovariectomized and injected for several days with either oil, estradiol benzoate (EB; 50 μg/day), progesterone (P; 100 μg/day) or received the combined treatment (EB + P). Only the EB + P treatment markedly increases building behavior in females. It also facilitates building activity in their untreated mates (right). (Redrawn from Cheng and Silver, 1975. Copyright 1975 by the American Psychological Association. Adapted by permission of the publisher and author.)

building (placing material into the nest bowl). Such changes in behavioral sex roles controlled by the nest condition could be hormonally mediated. Cheng and Balthazart (1982) showed that the nest condition may alter the plasma levels of gonadotropins in the doves. No conclusive data are available yet.

Canary. In the Canary, also, considerable amounts of data demonstrate that, in addition to hormonal factors, the nest condition plays an important role in the control of nest building (see reviews by Hinde, 1965; Hinde and Steel, 1966). In this species the female usually does most of the building. She gathers suitable nest material, carries it to a nest site, and then forms it into a nest structure by a number of stereotyped movements (Hinde, 1965). The outside of the nest is usually built with grass and the inside lined with feathers. The male makes no real contribution to the building; only occasionally will he be seen picking up material (Hinde, 1958). During that period, however, he actively courts the female, feeds her occasionally, and copulates with her.

Both nest-building and copulatory behavior show an exact relationship with the day on which the first egg will be laid: they culminate 2 or 3 days before laying, which suggests that they are related to a particular endocrine state of the animals (Hinde, 1965). Injections of EB into intact females during the winter (when they normally show no building behavior) produce active building in some individuals, which supports the notion of an endocrine control of the behavior. It must be noticed, however, that large doses (0.5 mg thrice weekly) of estrogen have to be injected to facilitate the behavior and that these doses are toxic (lethal) for some birds (Warren and Hinde, 1959). No improvement or inhibition of the effect of estrogens was detected in this study when progesterone was injected simultaneously.

Hinde and co-workers have beautifully demonstrated how these hormonal stimuli interact with stimuli provided by the nest in controlling the nest building. By manipulating hormone levels and nest condition, they were able to elaborate a model that explains the behavioral changes throughout the reproductive cycle of the female Canary. In response to increased day length in spring (Bell and Hinde, 1963) and to stimuli from the male (Warren and Hinde, 1961a,b; Hinde, 1965), the female's ovary starts secreting estrogens. This hormone induces the female to build a nest with grass. Estrogen also begins to increase sensitivity of the brood patch (Steel and Hinde, 1963; Hinde *et al.*, 1963). Increased sensitivity to the nest cup then contributes to a reduction of nest building (Hinde, 1958, 1965) and to the selection of feathers instead of grass as material (Hinde and Steel, 1962). Additional ("secondary") hormones (progesterone and prolactin) secreted under the influence of estrogens or in response to the stimulation of the nest cup would also contribute some of these changes (e.g., increase in brood-patch sensitivity; Steel and Hinde, 1963).

Very small doses of estrogen (0.05 mg thrice weekly) induce complete defeathering and stage 2 vascularity of the brood patch. Stage 3 vascularity, which is normally observed a few days before egg laying, requires a bit more estrogen (between 0.05 and 0.2 mg), but this is still much less than the doses required to induce the building behavior in only some birds and which proves to be lethal in others (0.5 mg; Hinde, 1965). As brood-patch vascularization and nest-building behavior occur simultaneously during the breeding cycle of intact birds, it is difficult to conclude that estradiol alone controls the building behavior in physiological conditions in the female Canary.

More recent research has clarified this problem. It was first hypothesized that estrogen normally acts with another hormone to facilitate nest building during the reproductive cycle, as is the case in the Ring Dove (see earlier). In several experiments, however, it proved impossible, as in the earlier study (Warren and Hinde, 1959), to establish the existence of a synergism between progesterone (0.125 or 0.25 mg thrice weekly) and estradiol in the control of nest building (Hinde and Steel, 1978). Testosterone propionate (2.0 mg thrice weekly), alone or in combination with EB, tends to suppress gathering of material in birds kept on normal winter day lengths.

The effect of a mammalian gonadotropin, pregnant mare serum gonadotrophin (PMSG), was also tested and in the course of that study it appeared that, although PMSG induced a considerable reproductive development and in some case egg laying in intact females during the winter months, the effect on nest building varied with the season. The facilitation of behavior was slight between November and January but much more important in September–October or in February–March (Steel and Hinde, 1966). In a subsequent study (Steel and Hinde, 1972a) it was demonstrated that this differential effect of PMSG is related to the short day length during the winter months. During a similar experiment carried out in March it was shown that PMSG activates more nest building in canaries kept under a long photoperiod (20L:4D) than under a short one (9.5L:13.5D). As the photoperiod was not influencing the speed of the PMSG-induced ovarian development (large doses were used, which swamped differences in endogenous levels), it was reasoned that it was probably altering the effectiveness of the endogenous estrogens produced by the ovary under the influence of PMSG.

In a number of experiments it was established that the photoperiod actually modifies the behavioral response to exogenous estrogens. It was confirmed that under a short photoperiod (normal winter day length: 8–10L:16–14D) a small dose of EB (50 μg thrice weekly) only slightly increased the gathering of nest material in ovariectomized canaries; in long days (20L:4D) the same treatment stimulated both gathering and placing material in the nest (Steel and Hinde, 1972b). The critical day length for this facilitation by long days of the estrogenic effect seems to be close to 12 hours of light (Hinde and Steel, 1978). Comparable results were also obtained in males, which were used partly because they can be more easily gonadectomized completely (Hinde and Steel, 1975). Estrogen-treated castrated males kept on short days (8L:16D) show little nest-building behavior [which confirms results of Warren and Hinde (1959), who used larger doses—500 instead of 50 μg thrice weekly], but the same treatment elicits vigorous building behavior in birds kept under 16L:8D (Fig. 3).

This effect of long days on the estrogen-induced behavior could also be obtained in other experiments using skeleton photoperiods, which mimic long days by a change in the distribution but not the total duration of light received each day (for a more detailed description of the procedure and its

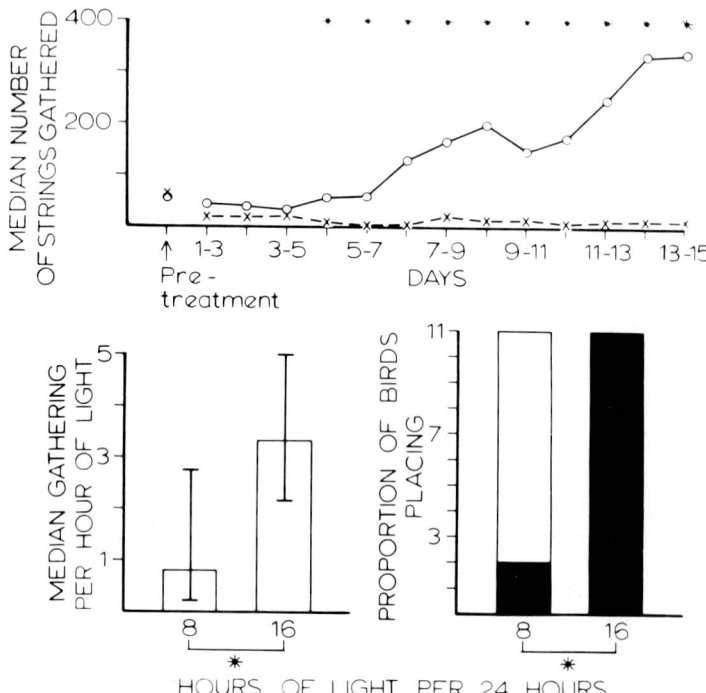

FIG. 3. Nest-building activity in estrogen-treated, castrated male Canaries kept either on long or short days. (Upper panel) Nest material gathered over the experimental period (15 days plus 3 days before injections) for birds on long (16L:8D; O—O) and short (8L:16D; X--X) days. (Lower left panel) Median and interquartile range of the amounts of material gathered per hour of light over the 15-day period. (Lower right panel) Number of birds (■) placing material in nest as a proportion of the total group. *, $p < .01$ (one-tailed tests). (From Hinde and Steel, 1978.)

theoretical backgrounds see Follett, 1973; Murton and Westwood, 1977). Castrated male canaries kept under a 14-hour skeleton long day involving 9 hours of light (e.g., 8L:5D:1L:10D) were compared with birds kept in a continuous-light short day (9L:15D). The latency for placing criterion were significantly different for these two groups of birds when they were estrogen treated (Steel and Hinde, 1976). This experiment together with other data reviewed in Hinde and Steel (1978) thus support the notion that the variable effect of estrogen on nest building is mediated by a circadian mechanism similar (at least functionally) to that implicated in the control by the photoperiod of gonadotropin secretions (e.g., Follett, 1973, 1978; Farner, 1975). Possible alternative mechanisms for this effect such as direct effect of gonadotropins or of luteinizing hormone–releasing hormone (LHRH) have been discussed by Hinde and Steel (1978) and are reviewed in Section II,B,7. No critical mechanism has been identified, although several lines of research are likely to be rewarding, especially with the availability of new physiological and biochemical techniques (see Sections IV and V).

The situation in the Canary is even more complicated by the fact that the effect of exogenous estrogen on nest building is also influenced by stimuli from the male (Fig. 4). Under a photoperiod of 12L:12D, exposure of ovariectomized, estrogen-treated females to the song of males (by means of tape-recorded songs) markedly increases the gathering of nest material (Hinde and Steel, 1976). There is, however, an interaction of this effect with the effect of photoperiod, and the action of male song cannot be demonstrated under shorter or longer day lengths.

Budgerigar. Experimental data demonstrating complex interactions between hormones, environment, and nest-building behavior are also available for a third species, the Budgerigar. Although this species lives in an environment very different from that of the Canary (tropical latitude, nesting after the irregular rains), its reproductive behavior is similarly controlled by hormones (estrogen accelerates nest-box entry in the female) having important interactions with external factors such as photoperiod, male vocalizations, or condition of the nest (for a review of these data see Hinde and Steel, 1978).

The studies of nest-building behavior in the Ring Dove and the Canary probably represent the best example of the interaction between hormonal and environmental stimuli in the control of behavior. These data perfectly illustrate the notion that gonadal steroids are generally necessary but far from sufficient for the normal expression of the behavior. Some external stimuli not only affect the production of gonadal steroids but also control their effectiveness in the induction of specific behavior patterns. Hormones thus have to be considered as a part of a network of interactions rather than as primary stimuli. This point will be made even more evident when we

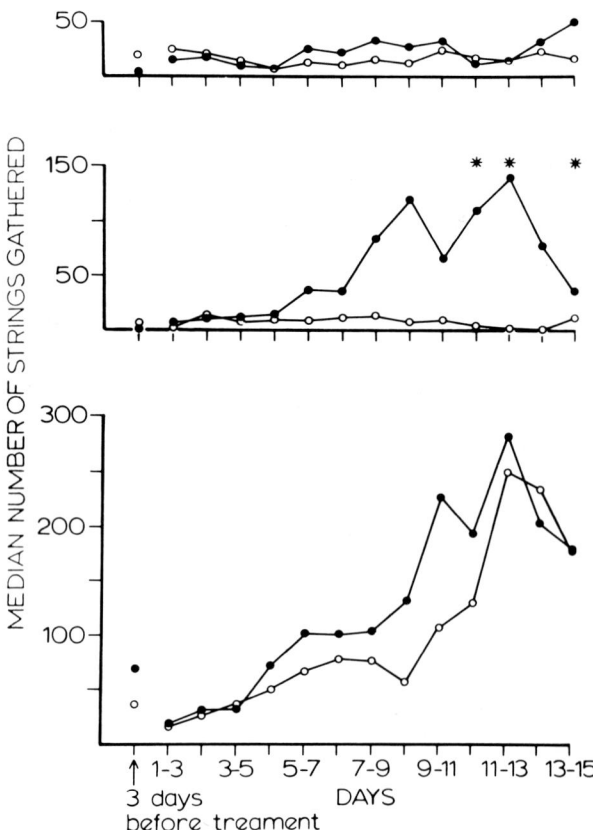

FIG. 4. Interaction between the photoperiod and social stimuli in the control of estrogen-induced nest-building activity in ovariectomized female Canaries. Birds were kept under three different photoperiods (8L:16D, upper panel; 12L:12D, middle panel; 14L:10D, lower panel) and were exposed (●) or not exposed (○) to tape-recorded male songs. The graph shows the median amounts of nest material gathered over successive 3-day periods. Asterisks indicate points where the difference between exposed and nonexposed groups is significant at $p < .05$. (From Hinde and Steel, 1976.)

review the effect of behavior on the endocrine physiology of birds (see Section IV,E).

4. Incubation

Incubation of the eggs (usually) marks the end of the courtship period in birds. In some species it is associated in the male with a collapse of the testes [e.g., Pied Flycatcher (Silverin, 1975), Short-tailed Shearwater (*Puffinus tenuirostris;* Marshall and Serventy, 1956), pigeon (Champy and Colle, 1919)], which has frequently been interpreted as the sign of a decreased androgen output. In other species that show the ability to raise several broods within a reproductive season or to recycle when their nest is destroyed, such a decrease in testicular activity is usually not seen [e.g., *Zonotrichia leucophrys pugetensis* (Wingfield and Farner, 1980), Ring Dove (Silver and Barbière, 1977)]. In many species the female incubates alone while being fed by the male, but in some cases both male and female share the duty of incubation. The hormonal bases of incubation have been analyzed in a limited number of species, but especially intensively in the Ring Dove.

The hormonal control of incubation behavior has been controversial. This has impelled many investigators in this field to design clever experiments and to produce very detailed and critical reviews of the literature. This makes the subject a very good example of how experimental data can be discussed and evaluated. This controversy proceeded historically in two steps. During a first period the laboratories of Riddle and of Lehrman claimed that prolactin or progesterone, respectively, were the critical hormonal stimuli involved in the initiation of incubation. In the late 1960s, it seemed that it was definitively established that progesterone was of paramount importance in the control of incubation. More recently, on the basis of studies involving radioimmunoassay, gonadectomy, and functional adrenalectomy, the role of progesterone has been denied, and it has been claimed that the effects of exogenous progesterone previously demonstrated by Lehrman (1958b, 1965) were obtained only by pharmacological mechanisms. We shall try to summarize all this history in the following pages.

a. The Case of the Ring Dove. Because Riddle *et al.* (1935) found that prolactin influences the brooding behavior of hens, this hormone was thought to control most aspects of parental behavior in birds. In 1944, Riddle and Lahr demonstrated that if Ring Doves are implanted with progesterone and then placed in unisexual pairs in a cage containing a nest, they will quickly (3–7 days!) begin to incubate if eggs are placed in the nest. After 14 days of this progesterone-induced incubation, the authors measured the crop weight in these birds [the growth of this organ is a good biological assay for prolactin (Riddle, 1937)] and found that it had increased. They concluded from these data that progesterone had induced incubation by stimulating the release of endogenous prolactin from the bird's hypophysis.

This interpretation was disputed by Lehrman (1958b) on the basis that participation in incubation itself stimulates prolactin secretion (Patel, 1936) and that there was no evidence of an increased prolactin secretion at the *beginning* of the progesterone-induced incubation. This resulted in an active dispute (Riddle, 1963; Lehrman, 1963) in which Lehrman and co-workers provided, with the techniques available at that time, elegant demonstration of a prominent role of progesterone in the induction of incubation. They showed that pairs of Ring Doves that are not ready to sit on eggs exhibit incubation behavior if they have been injected with progesterone for 7 days before the test (Lehrman, 1958b). Prolactin in their experiments was effective only in maintaining (Lehrman and Brody, 1964) but not in inducing incubation behavior (Lehrman and Brody, 1961). Even in large doses that caused maximal crop growth, prolactin induced incubation in only 40% of the birds and thus proved significantly less effective than progesterone in this respect.

Lehrman (1958a) and Lehrman *et al.* (1961) also demonstrated that birds that are normally not ready to incubate will do so if before the test they have been kept in heterosexual pairs for 1 week and that this effect was enhanced

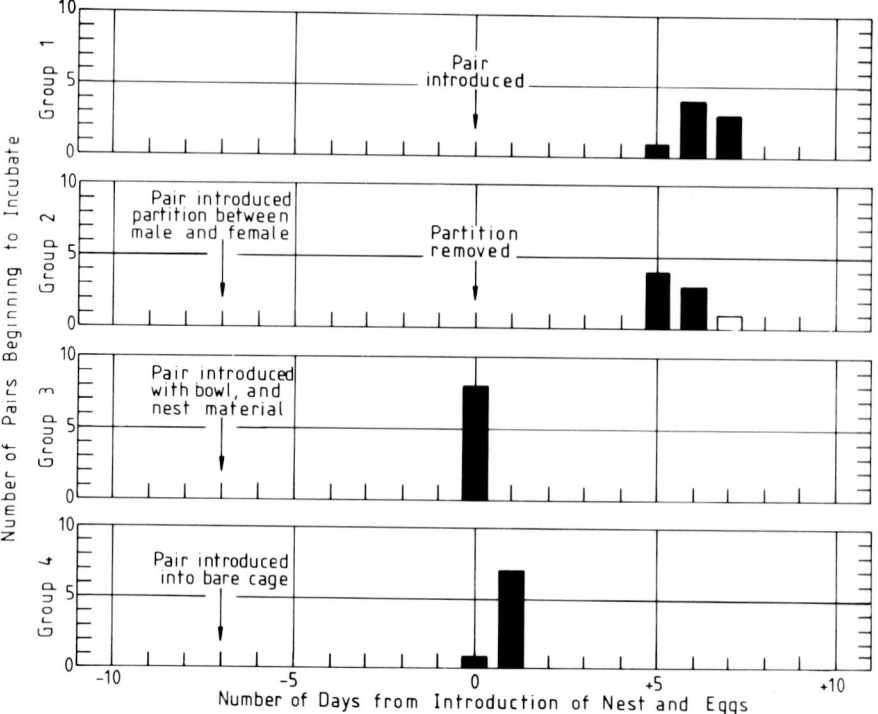

Fig. 5. Effect of mate and nesting material on the latencies of incubation behavior in Ring Doves. Birds that are introduced to a cage containing a nest and eggs start incubating after 5–6 days (group 1; standard). If they are first exposed to their mate for 7 days, incubation starts after 0–1 day (group 4; effect of mate alone) and simultaneous exposure to mate and nest material is even more effective (group 3; effect of mate and nesting material). This does not result from a habituation to the test cage (group 2; effect of habituation to cage). (Redrawn from Lehrman, 1958a. Copyright 1958 by the American Psychological Association. Adapted by permission of the publisher.)

by the simultaneous presence of nest material (Fig. 5). They concluded that, during the reproductive cycle of intact birds, the courtship of the mate and the participation in nest building stimulates the secretion of progesterone in both sexes, which in turn induces the birds to sit on the eggs. This original conclusion had to be refined when the same type of experiment was redone using gonadectomized birds. Stern and Lehrman (1969) and Stern (1974) demonstrated that progesterone alone is not very effective in inducing incubation in castrated males. However, priming with TP or EB restores the effectiveness of progesterone. Similarly, injections of progesterone alone in gonadectomized females elicit little sitting behavior, but the combined (sequential) treatment with estrogens and progesterone is very effective in this respect (Cheng and Silver, 1975).

At that point, the hypothesis of Lehrman concerning the role of progesterone in the induction of incubation could still be retained, with the modification that progesterone has to synergize with gonadal hormones in both sexes to facilitate the behavior. This view is challenged, however, as far as the male is concerned, by the following observations:

1. By radioimmunoassay, it could be demonstrated that there is no change in plasma progesterone throughout the reproductive cycle of intact male doves (Silver et al., 1974). If progesterone shows a sharp rise in the female around the time of egg laying and thus of the start of incubation, in males a baseline value (± 1.2 ng/ml) is maintained throughout the cycle.

2. When, during a normal breeding cycle, experienced intact male doves are injected with dexamethasone (Dex, an ACTH blocker that should deplete the circulating progesterone) starting on the day when a first egg is discovered in the nest, they show a normal initiation of sitting in their own cages although Dex tends to disrupt the initiation of incubation when males are tested alone in a novel cage (Silver and Buntin, 1973).

3. Gonadal steroids do not seem to be obligatory stimuli for the initiation of sitting because:
   a. A fair proportion of castrated males (60%) will sit on eggs during a test situation if they have been prepared before by exposure to a female plus a nest bowl and nest material for a period of 7 to 10 days (Silver et al., 1973).
   b. Castrated males given TP before being paired with an intact female but not during the courtship period will incubate normally when the female lays eggs (Silver and Feder, 1973).
   c. Males of the closely related pigeon will incubate if they are castrated after mating (Riddle and Dikshorn, 1932; Patel, 1936).

These results seem difficult to reconcile with the original concept of Lehrman or with its modified version, which proposed that hormonal stimuli (progesterone plus gonadal steroids) are essential in the transition from courtship to incubation in the dove. It has consequently been suggested that the effect of progesterone on the incubation in the male may be pharmacological (Silver et al., 1979). Parts of the contradictory data have been criticized on methodological grounds. It has been pointed out, for example, that Dex was injected too late (when incubation was already initiated) in the experiment of Silver and Buntin (1973), so that its conclusion concerned the maintenance rather than the initiation of sitting (Cheng, 1975, 1979). There is also no proof that Dex effectively suppresses progesterone in doves at the dosage used in these experiments. The contribution of the adrenal glands and the testes in the production of testosterone or progesterone is also unknown. Additional data relevant to domestic hens and cocks (Tanabe et al., 1979) suggest that the role of the adrenal glands is far from negligible, making the results of castration experiments difficult to interpret in this context.

In conclusion, despite intensive and very exacting research carried out during more than 30 years, the factors underlying the induction of incubation in the male Ring Dove remain unclear. A group of experiments support the notion first presented by Lehrman (1958b, 1965) that a true endogenous change must occur in a dove before it starts incubating eggs. The original experiments did not show whether this change had to occur in both male and female or in the female only, as is the case for nest-building behavior (see earlier; males will build immediately provided they are paired with a female that is in the adequate situation). Silver et al. (1973) gave this proof by establishing that a male freshly introduced to an incubating female will not begin to incubate immediately.

Another group of results suggest that endogenous changes are not necessary (e.g., experienced castrated birds still start to incubate if placed in adequate conditions) and stress the importance of situational cues such as the presence of a nest and the behavior of a female partner. This view is also supported by the finding that even during the incubation phase of the cycle, a male is still ready to court a new female introduced in its cage after its mate and nest have been removed.

Whether all these mechanisms, which can be experimentally demonstrated, actually play a role in the control of the reproductive cycle will have to be determined in future experiments. It is possible that birds in some experiments were submitted to pharmacological hormone treatments that activated nonphysiological mechanisms. Similarly, parts of the experiments manipulating environmental cues may have induced incubation through superstimulations usually never encountered by the birds. It nevertheless remains true that plasma progesterone does not seem to increase in male doves around the time when they start to incubate (Silver et al., 1974). How baseline levels of one hormone could facilitate one behavior at a given time will be discussed in the section devoted to change in sensitivity to hormones in the brain and their biochemical basis (Section V,C).

b. Other Species. Other than doves, little is known about the hormonal mechanism that induces incubation in birds. Whatever the causes of incubation, there is a good deal of evidence suggesting that this period of the cycle is associated with a decrease of gonadal activity probably induced by prolactin. Three types of more or less direct correlational evidence prove that testicular and ovarian activity are diminished during incubation.

1. Song and overt sexual activities cease in many species of birds more or less at the time when incubation begins (review by Eisner, 1960). According to Eisner (1960), it seems that the decrease or suppression of singing occurs only in those species in which the male incubates [e.g., European Starling (Kluijver, 1933), Black-headed Grosbeak (*Pheucticus melanocephalus;* Weston, 1947)] whereas song is maintained in species where the female incubates alone [e.g., Willow Warbler (*Phylloscopus trochilus;* Kuusisto, 1941), Red-eyed Vireo (*Vireo olivaceus;* Lawrence, 1953)] or only in unmated males in the species from the first group (Burkitt, 1919; Kluijver, 1933). Unfortunately, a number of exceptions make it necessary to consider this conclusion with much caution. This caution is also reinforced by the finding of Silver and Barbière (1977) that even during incubation male doves will immediately court a new female if their own female and their nest are removed. This suggests that their testicular activity is not decreased during incubation (supported by the finding that testicular weight does not change systematically throughout the cycle; Silver and Barbière, 1977) and that courtship behavior can be suppressed by situational as well as hormonal control during the incubation period.

2. Secondary sexual characteristics, which are known as good indicators of gonadal activity, change during incubation. Collias (1950) noted that the comb, which is known as a typical androgen-dependent organ, shrinks in broody hens. Similarly, Noble and Wurm (1940a) reported that the colors of unfeathered areas of Black-crowned Night-Herons, which are also androgen dependent, fade after the egg-laying period.

3. Finally, in a large number of species there is direct evidence that gonads of both sexes are regressed during incubation (see references in Eisner, 1960; Silver et al., 1979). Detailed histological analysis confirms that the regressed state of the gonad is associated with a decrease in the secretory activity of its endocrine parts (e.g., Silverin, 1975, 1978a,b; for a detailed review of gonadal histophysiology see also Lofts and Murton, 1973), and

more recent studies have shown that the plasma levels of testosterone and estradiol in males and females, respectively, are low when birds of various species are incubating (reviewed by Wingfield and Farner, 1980).

There is also little doubt that incubation of the eggs is associated in birds with an increased secretion of prolactin. This is demonstrated by histological studies of the pituitary gland (Legait, 1959; for a review of studies using light and electron microscopy see Tixier-Vidal and Follett, 1973) and bioassay of pituitary prolactin content by the increase in pigeons and doves of the weight of the crop sac, which is a typical prolactin-dependent organ (Riddle et al., 1932, 1933; Bates and Riddle, 1936; Lahr and Riddle, 1938; Lehrman, 1955; Silver and Barbière, 1977). Radioimmunoassays (RIAs) or radioreceptor assays have confirmed this notion (e.g., March and McKeown, 1973; Scanes et al., 1977; Buntin and Forsyth, 1979; Sharp et al., 1979; Goldsmith and Williams, 1980; El Halawani et al., 1980; Burke and Dennison, 1980).

Two groups of questions have been asked concerning the high prolactin secretion by incubating birds: (1) Does it cause decreased gonadal activity and (2) does it play a role in the initiation or maintenance of the behavior, or, conversely, is it a consequence of the act of sitting on the eggs? The first of these questions cannot, at the present time, receive a general answer. An antigonadal effect of prolactin has been established in a large number of species belonging to different orders: Passeriformes (Bailey, 1950; Lofts and Marshall, 1956; Thapliyal and Saxena, 1964), Galliformes (Bates et al., 1935; Breneman, 1942; Nalbandov, 1945), and Columbiformes (Riddle and Bates, 1933; Bates et al., 1937); but in other species no depression of gonadal size or activity or inhibition of photoperiod-induced testicular growth has been detected following prolactin injections (Laws and Farner, 1960; Meier and Dusseau, 1968; Hammer, 1968; Jones, 1969; March and McKeown, 1973; Silverin, 1980a). It has been proposed that prolactin would have an antigonadal effect in nonmigrants and weak migrants but would not affect strong migrants. These discordant data might also be explained by differences in the injection procedures from one experiment to another: there is indeed a pronounced daily rhythm in the response of tissues to prolactin [e.g., crop sac (Meier et al., 1971) fat deposition (Meier and Davis, 1967)]. A detailed discussion of this problem can be found in Silverin (1980a) or in De Vlaming (1979).

The availability of RIAs for prolactin in a number of bird species [chicken (Scanes et al., 1976), turkey (Burke and Papkoff, 1980), and Mallard (Goldsmith and Williams, 1980; using the heterologous assay of McNeilly et al., 1978)] has allowed important progress in the question of whether prolactin is a cause or a consequence of incubation. The view of Riddle and collaborators that prolactin causes incubation behavior (e.g., Riddle, 1963) has been challenged by Lehrman (1963), who suggested that the hormone is involved only in the maintenance of sitting (Lehrman and Brody, 1961, 1964). Studies involving manipulations of behavior and repetitive assays of plasma prolactin even suggest that the hyperprolactinemia observed during the incubation period is caused by the act of sitting on the eggs. In the turkey, El Halawani *et al.* (1980) showed that serum prolactin levels of broody hens drop markedly within a day of nest deprivation and increase again when the nest becomes available. Furthermore, at that time, birds resume nesting within 5 minutes, but serum prolactin increases only gradually.

It must finally be mentioned that a complete set of data on the hormonal correlates of incubation can be obtained by analyzing the literature on the control of the incubation patch (brood-patch development). This structure, which permits an increased heat transfer from the parents to the eggs, develops in most species around the time when incubation starts. Its morphological characteristics, the phenology of its appearance, its distribution among orders of birds and between males and females, have been reviewed by Eisner (1960) and Drent (1975). The hormonal control of brood-patch development has also been discussed by Drent (1975) and Silver *et al.* (1979) (with special reference to the role of androgens). The broadly accepted pattern is that prolactin synergizes with gonadal steroids (estrogens mainly but also progestins and androgens) to control the defeathering, vascularization, epidermal thickening, and increase in tactile sensitivity in the patch area. These data generally fit well with what was summarized in the preceding pages concerning the endocrine status of the birds during and just before the incubation period (see the aforementioned reviews for further details).

5. Care of the Young

With the exception of the Ring Dove, little information is available concerning the hormonal correlates of behavior associated with the care of young birds. Parental behavior takes very different forms according to species. In precocial birds (e.g., galliforms and anseriforms) the parents and generally the female only lead the young chicks, call them to food, and eventually warm them. In altricial species (e.g., passeriforms and columbiforms) the parents have to feed the young and warm them and, in some species, also clean the nest by removal of the feces. It is thus unlikely that the conclusions related to the Ring Dove apply to a wide range of species. In addition, indirect evidence suggests that the dependence of parental behavior on hormonal stimuli is weaker than for the other behavioral groups discussed so far. According to Eisner (1960), who listed a number of examples, parental behavior can be elicited without hormonal treatment in a variety of circumstances, which suggests that it does not depend on a very precise hormonal condition.

a. Ring Dove. In doves during the parental phase, the plasma levels of LH, FSH, testosterone, estradiol, and progesterone are low (Silver *et al.*, 1974, 1980; Cheng and Follett, 1976; Feder *et al.*, 1977; Peczely and Pethes, 1979; Cheng and Balthazart, 1982). The circulating levels of prolactin are high and progressively decline toward the end of the brooding phase (Goldsmith *et al.*, 1981). By a radioreceptor assay, Buntin and Forsyth (1979) detected less prolactin in the pituitary gland of incubating pigeons than in laying birds, which they interpreted as a sign of increased release in the plasma. Pituitary prolactin concentrations were at their highest values immediately after the hatching of the young, when the crop sac shows its maximal growth (Lehrman, 1955; Silver and Barbière, 1977).

Relative to parental behavior, the Columbidae are peculiar in that both parents feed their young by regurgitation of a crop "milk," which consists essentially of desquamated cells from the crop epithelium. Lehrman (1955) demonstrated that this behavior is in part controlled by prolactin. This hormone was already known to cause growth of the crop sac and production of the crop milk (Riddle *et al.*, 1932, 1933). Lehrman showed that experienced doves (i.e., those that have already bred) will readily feed squabs placed in their cage if they have been previously injected for 7 days with prolactin (10 of 12 birds), whereas none of the control birds fed squabs in these conditions. This effect depends, however, upon the previous breeding experience of the parents, as the same hormonal treatment is without effect in naive birds (i.e., birds with no previous breeding experience). Lehrman (1955) also suggested, in contradiction to Riddle (1935), that this effect of prolactin is not primarily through an action on some brain center but more directly related to its effect on the crop. He showed, indeed, that if experienced birds are injected, as in the previous experiment, with prolactin and if in addition their crop is anesthetized, these birds will feed squabs significantly less often than birds that received the anesthetic elsewhere to control for systemic effects of the injection.

It was later demonstrated that the secretion of prolactin is stimulated in the adult doves by their squabs. If young Ring Dove squabs are introduced to experienced foster parents that have already laid eggs, they will significantly accelerate the crop growth within a 4-day period (Hansen, 1966), this effect being modulated by the previous experience of the parents, the age of the squabs that are introduced (Hansen, 1971a,b), and the quality of the parent–young interaction (Buntin *et al.*, 1977).

Lehrman (1955) had originally proposed that feeding behavior is a learned response that is reinforced by the release in the tension of the crop, which occurs when the crop milk is regurgitated to the young. This accounted for the fact that prolactin injections do not elicit feeding in inexperienced doves, but it was invalidated by later experiments showing that feeding responses can be induced prior to crop engorgement in inexperienced doves (Klinghammer and Hess, 1964; Hansen, 1966).

In a later study, Lott and Comerford (1968) proposed a model that accounts for all previously reported data. Considering that feeding behavior follows a period of incubation of the eggs supposedly facilitated by progesterone (Lehrman, 1958b), they suggested that brooding of the young is controlled by progesterone, possibly acting together with the changes in the crop sac induced by prolactin. They were able to show that if squabs are introduced to inexperienced parents after seven daily injections of hormones, birds that were injected with progesterone brooded more often than those injected with prolactin or with control solutions. Prolactin did not improve the effect of progesterone on brooding. The combined treatment with prolactin and progesterone was, however, much more effective in inducing feeding of the young than the injection of progesterone alone, prolactin alone, or control solutions. These data demonstrate that it is possible to establish parental behavior in inexperienced Ring Doves by hormone treatments. The presence of crop milk is not an absolute requirement, as some progesterone-injected birds (2 of 20) also fed the introduced squabs. Even in control birds one single parent (of 18) fed, which suggests that exogenous hormones are not absolutely essential for the occurrence of parental behavior. This is reinforced by unpublished data of R. Saad and R. Silver (cited in Silver *et al.*, 1979), who showed that castrated males proceeding through a reproductive cycle will fed squabs introduced to the breeding cage.

It thus seems that hormones can markedly increase the likelihood of parental behavior (brooding and feeding in doves), but this effect interacts strongly with experimental and situational factors. Experienced doves feed when stimulated by prolactin alone (Lehrman, 1955), but inexperienced ones require progesterone in addition (Lott and Comerford, 1968). Once established, the behavior seems to become relatively independent of its hormonal inducing stimuli.

b. Other Species. It is difficult to evaluate how far the foregoing conclusions relevant to doves can be extended to other birds species. Feeding young in other altricial birds mainly involves collecting food rather than producing and regurgitating crop milk. There is thus little chance that prolactin is so closely involved in feeding behavior of other species. Furthermore, in precocial species, parents only lead their young and call them for food, these activities being eventually interrupted by limited periods of brooding. The hormonal basis of these activities is unclear.

According to Riddle *et al.* (1935), prolactin injections induce hens to become broody, but this is only if the hens have active ovaries and are laying eggs. Egg-laying hens also become broody by simply exposure to chicks

(Collias, 1946, 1950). These data were generally taken as evidence that prolactin (injected or secreted in response to chicks) acts only in birds primed with sex steroids.

Radioimmunological data partly disagree with the earlier view that prolactin controls parental behavior. If plasma prolactin increases in a number of species around the time of egg laying [Ruffed Grouse (*Bonasa umbellus*; Etches *et al.*, 1979)] and during incubation [Ruffed Grouse (Etches *et al.*, 1979), bantam hens (Sharp *et al.*, 1979), turkey (Burke and Dennison, 1980)], other data show that it is back to baseline levels during the period of parental care [Mallards (Goldsmith and Williams, 1980)]. Older data suggest, furthermore, that pituitary prolactin is not necessarily high during the parental phase (hen; Saeki and Tanabe, 1955) and that broodiness toward chicks and pituitary prolactin content can vary independently (Saeki and Tanabe, 1955; Nakajo and Tanaka, 1956; Breitenbach and Meyer, 1959). It was also speculated that prolactin facilitates parental behavior through its antigonadotropic effect. By decreasing the secretion of gonadal steroid it would reduce aggressive and sexual behavior, which would, if present, interfere with the care of the young (Lehrman, 1955; Nalbandov *et al.*, 1945). Eisner (1960) reviewed experimental evidence showing that loss of the gonads slightly facilitates parental behavior but concluded that this is certainly not the sole explanation. In addition, it has since appeared in that the antigonadotropic role of prolactin is not firmly established in birds (see Section II,B,4 and reviews by De Vlaming, 1979; Silverin, 1980a).

Considering the most recent data, it is impossible to assign conclusively a role to prolactin in the control of parental behavior, and the proposed mechanism of action for this hormone also seem dubious. It is likely that the availability of prolactin radioimmunoassay in many bird species will soon permit important progress in the understanding of the control of parental behavior.

6. Other Behaviors

Several types of nonreproductive behavior are also known to be influenced by gonadal hormones. In most instances, however, it has been established only that exogenous hormones alter these behaviors, but few data are available to support the notion that these changes are physiological and occur in intact animals. Most of these hormonal effects on nonreproductive behavior concern androgens, and the relevant literature has been reviewed by Silver *et al.* (1979). We shall thus only mention the most important studies.

a. Persistence, Attention, and Emotional Responses. Andrew and collaborators have established that testosterone affects perceptual and/or attentional mechanisms in the chick (Andrew, 1972a,b; Andrew and Rogers,

1972). In their studies, administration of testosterone to young male chicks increased their persistence of search for a given stimulus and in a given area. This type of effect can be obtained in a variety of conditions using different stimuli and may concern many behavioral responses (Andrew, 1975c; Andrew and de Lanerolle, 1975; see also Archer, 1973a,b,c).

The literature on this subject has been extensively reviewed by Andrew (1972a, 1975d, 1978). Although very large doses of hormones were used, there is reason to believe that this effect of testosterone is physiological because adult cocks are more persistent than hens, and antiandrogen injections decrease the persistence of cocks as does a neonatal castration (Rogers, 1974; see also Andrew, 1972a, 1978; and critical papers by Cummins *et al.*, 1974; Andrew and Archer, 1977).

The steroid specificity of this type of effect is not known. Young and Rogers (1978) showed only that 5α-dihydrotestosterone decreased the persistence of response to a given stimulus, whereas in the same study testosterone increased this persistence and 5α-androstenedione did not affect it significantly. Andrew (1975b) has suggested that general effects such as increased persistence of attention to a particular type of stimulus or changes in perceptual mechanisms could play a role in many of the testosterone effects on behavior, including reproductive behavior. This line of research should certainly be pursued.

b. Imprinting. Steroids could alter behavior of young altricial birds toward an imprinting stimulus and also control the expression of the imprinted responses. Testosterone injections decrease a chick's tendency to approach and stay near an intermittent light source (James, 1962). They also inhibit the "following" reaction of ducklings (Millikan, 1972). Balthazart and De Rycker (1979) demonstrated that such an effect can also be induced by low doses (50–250 μg) of TP in chicks and that it is accompanied by a number of behavioral changes (e.g., decreased frequency of twitter calls, increase of peeping). These effects are androgen dependent, and are induced by testosterone and 5α-DHT. Estradiol has the opposite effect.

It is difficult to evaluate whether the experiments reveal physiological mechanisms. Those doses of TP (50–250 μg) that depress "following" seem low because they do not depress plasma LH or stimulate comb growth (Puts, 1978; Balthazart and Puts, 1979). They nevertheless produce plasma levels of testosterone two to four times above the physiological ranges during the first hours after the injection (J. Balthazart, unpublished).

Whether testosterone is responsible for sex differences in the behavior of chicks during imprinting sessions (their endocrine status at that time is very different: see Sections IV,C,1 and V,A,4) or controls in some way the end of the sensitive phase (plasma testosterone seems to increase in males after 40 hours of life, which could explain the decrease of the following tendency with age; J. Balthazart, unpublished results) similarly remains questionable. Schutz (1975) also suggested that testosterone can reveal latent sexual imprinting in female Mallards. Schutz reared female Mallards in the exclusive company of Muscovy ducklings (*Cairina moschata*). When adult, these females were given free choice between Muscovy and Mallard mates, and all paired with Mallards. This choice could be reversed by testosterone injections. Schutz explained these data by saying that testosterone transformed the female's releasing mechanism for partner choice to that of the male, and in this way revealed a latent sexual imprinting. Other interpretations, however, seem possible, and the biological meaning of this effect remains obscure.

Hormones of the pituitary–adrenal axis (corticosterone and ACTH) also seem to be involved more or less directly in the regulation of the imprinting process. Martin (1978a) has reviewed this subject.

ACTH and corticosterone injections exert opposite effects on the approach behavior of ducklings toward an imprinting model (respectively increase and inhibit the behavior). That this effect is physiological is supported by the finding that corticosterone acts in very low doses (500 ng), that injection of an antiserum to corticosterone (which suppress the effects of endogenous corticosterone) has the reversed effects of a corticosterone injection, and that plasma corticosterone is negatively correlated to "following" time in intact ducklings (Martin, 1978b). Landsberg and Weiss (1977) also showed that if ducklings are exposed to stress during their imprinting period they cannot be imprinted. Their corticosterone plasma level is then much elevated. Martin (1975, 1978a) suggested that pituitary–adrenal hormones could control two independent aspects of imprinting. On one hand, they could limit the period of time during which imprinting can take place. On the other hand, they could explain some behavioral differences between wild and domestic Mallards. (Wild birds show a higher pituitary–adrenal activity and also a greater avoidance of any novel stimulus. During imprinting sessions they approach a moving imprinting model less readily than the domestic birds.)

c. Activity Rhythms. Rhythms of circadian activity may depend partly on gonadal hormones. Gwinner (1974) showed that, under the influence of testosterone, the circadian rhythm of locomotor activity in the European Starling splits into two components that run with different circadian frequencies. This suggests that testosterone controls the mutual coupling of different circadian oscillators related to locomotor activity. This result was extended in a later study (Gwinner, 1975b) in which many relationships could be established between two measures of activity [daily activity time (α) and period of the activity rhythms (τ)] and endocrine condition. Daily activity time and possibly τ are controlled by testosterone. They are affected by castration and testosterone injections and in intact animals are correlated with testes size. Together with mammalian data (see references in Gwinner, 1975b), these studies suggest that seasonal variations in activity such as observed during the reproductive season or in the migration period are controlled by the testicular secretions.

d. Miscellaneous. There is also evidence in birds that testosterone may modulate some learning processes, as suggested by studies of song plasticity in passerines (Nottebohm, 1969, 1975; Güttinger, 1979). Testosterone also affects the learning of a passive avoidance task by young chicks (Clifton et al., 1979).

Gonadal steroids affect body weight in birds. Generally, estrogens increase it whereas androgens may decrease or increase it, probably according to the dose injected and the physiological stage of the animals (e.g., Hazelwood, 1972; Pietras and Wenzel, 1974; Balthazart and Hendrick, 1979a). The vernal premigratory fattening observed in many species is also controlled to a large extent by hormones (for a review see Wingfield and Farner, 1980). It is difficult to know in most cases whether these effects result from changes in eating behavior or in the digestive and anabolic metabolism. Much more information is available in mammals (for a review see Wade, 1976).

7. Nonsteroid Hormones

In addition to gonadal steroids, a number of hormones, mostly secreted by the anterior lobe of the pituitary, have been proposed as specific activators of behaviors. We have already discussed the possible role of prolactin in the incubation and brooding behavior, so no further mention will be needed here (see Sections II,B,4 and 5).

ACTH in mammals affects a number of behavioral processes, including grooming (Jolles et al., 1979; Gispen and Isaacson, 1981), avoidance behavior (Bohus and De Wied, 1966; Wimersma Greidanus et al., 1974), or sexual excitement (Bertolini et al., 1968, 1969). Much less is known in birds. Injections of ACTH are known to facilitate imprinting behavior in ducklings (Martin, 1975, 1978a,b; Section II,B,6). When given intraventricularly, ACTH also affects a number of comfort behavior patterns generally considered to be displacement activities (Delius et al., 1976; Deviche and Delius, 1981). In one study, systemic ACTH injections were also known to increase sexual behavior (Deviche, 1976), but no confirmation of this effect was obtained in later studies (Deviche, 1979a,b).

In mammals, LHRH is clearly involved in the control of female sexual behavior, as demonstrated by the works of Moss and McCann (1973, 1975), Pfaff (1973), or Foreman and Moss (1977). These studies actually offer the best evidence for a specific and physiological action of a hypothalamic peptide hormone on the central nervous system (Moss and Dudley, 1981). Only one study has demonstrated a behavioral role of LHRH in birds. Cheng (1977) showed that, in synergism with doses of EB that are inactive by themselves, LHRH activates female-type behavior such as nest soliciting and squatting in Ring Doves. In contrast, McDonald (1979) failed to modify courtship and nest-building behavior in males of the same species.

Gonadotropins. In studies on avian behavior, a fair amount of interest has been devoted to the possibility that gonadotropins (LH and FSH) exert a direct influence (i.e., not through their effect on steroid synthesis) on behavior. Detailed reviews of this subject can be found in Balthazart (1978c) or Deviche (1982).

A number of studies *indirectly* suggest a possible role of gonadotropins in the control of behavior:

1. Hypothalamic TP implants induce courtship behavior in castrated male Ring Doves (Hutchison, 1971), but their effectiveness decreases in long-term castrates (Hutchison, 1974a; see Section V,A,3). Such implants are, however, slightly more effective in long-term castrates maintained in long days (Hutchison, 1974b), which could be an effect of high gonadotropic activity in such animals. Other interpretations are nevertheless possible (Hutchison, 1975).
2. In the Red-billed Quelea, there is a correlation (Butterfield and Crook, 1968) between some types of aggressive interactions and plumage characteristics that are thought to be controlled by LH (Witschi, 1954).
3. Castration does not abolish song in the European Starling (Davis, 1957), and castrated males are generally dominant over intact birds (Mathewson, 1961). Similarly, Collias et al. (1961) showed that castrated Village Weaverbirds (*Ploceus cucullatus*) continue to sing, display, and hold a territory.
4. In domestic ducks, there is a significant correlation in the annual cycle between changes in plasma FSH and social displays, but the latter are not related to plasma testosterone (Balthazart and Hendrick, 1976; see also Section IV,C).

Such indirect correlation data suggested that LH or FSH could be behaviorally important in birds, and consequently injection experiments were undertaken.

Mathewson (1961) showed that injections of mammalian LH in the subordinate bird of a pair of European Starlings generally results in a reversal of dominance. Crook and Butterfield (1968) injected LH into Red-billed Queleas, and this induced some increase in social status of low-ranking birds in the hierarchy, but only in a nonreproductive context (absence of nest material). In the same species, Lazarus and Crook (1973) later observed that LH injections increase the frequency of agonistic behavior in gonadectomized females. Murton et al. (1969) also claimed that LH injections increase the probability that attacking will follow bowing display in the pigeon.

A behavioral role for FSH has been suggested for two species of birds, the pigeon and the duck. In the pigeon, FSH injections increased the sexual-aggressive components of the bowing display (Murton et al., 1969). This result was extended in a latter study during which FSH injections into estradiol-primed intact male pigeons were shown to enhance the aggressive components of the courtship (Murton and Westwood, 1975; for reviews see also Murton and Westwood, 1977; Lofts and Murton, 1973). It was thus reasoned that a progressive decrease in plasma FSH should be responsible for the transition from aggressive to nest-oriented and sexual courtship during the cycle. Additional assay data in male Ring Doves (Cheng and Balthazart, 1982) support this idea.

In ducks, the suggestion that gonadotropins may control behavior is supported by a number of facts:

1. In very young birds, testosterone injections often fails to induce social display although they activate sexual behavior (Balthazart and Stevens, 1975a, 1976).
2. Injections of TP into intact adult ducks in the springtime decrease rather than activate their social displays (Balthazart and Deviche, 1977), and it was shown that this was concomitant with a drop in plasma FSH concentrations (Balthazart et al., 1977a).
3. There is a correlation between plasma FSH and social displays in the course of the annual cycle (Balthazart and Hendrick, 1976) and during the daily changes in activity (Balthazart and Hendrick, 1979b).

Many injection experiments were performed in an attempt to explain these observations, but the injection of mammalian FSH induced only minor behavioral changes, and, furthermore, few reproducible effects were detected (e.g., Deviche and Balthazart, 1976; Balthazart and Deviche, 1977; Balthazart and Hendrick, 1979a).

Most of the experiments mentioned in this section suffer from major drawbacks. Many were performed on noncastrated animals, hence effects induced through stimulation of steroid synthesis cannot be excluded. Furthermore, many used inadequate methodologies (real controls frequently lacking), pharmacological doses of hormones were injected, and these were of mammalian origin so that their activity in birds is questionable (for critical review see Arnold, 1975b). Although suggestive evidence points to a possible role of LH or FSH in the control of bird behavior, alternative explanations are always possible. In the absence of any conclusive data on this problem, this question should thus be considered open.

III. Attempts toward a Synthesis

A. Dose–Response Relationships

The variations in the hormone preparations, the solvents used, and the injection schedules make it very difficult to draw any general rules on dose relationships in the experiments on hormones and behavior in birds. Most of the studies carried out until recently presented behavioral results obtained after injection of one single dose of steroid, so that most attempts to synthesize have to rely on comparisons between different experiments.

No statement can be made about experiments using subcutaneous implantation of hormone pellets (e.g., Sachs, 1969; R. E. Hutchison, 1978). These cannot be compared with other studies in which hormones were periodically injected as oil solutions, and usually the studies cannot even be compared with each other, as the diffusion rates of the hormone from the implant are generally unknown. (The total weight is frequently reported, but this has little physiological significance.)

If we restrict our analysis to those studies that used the daily injection of steroids in oil solution (most frequently done), some conclusions can be reached for two steroid preparations that are esters of naturally occuring compounds: TP and EB (see Balthazart, 1978b,c). The injection of daily doses of 1 to 10 mg TP/kg body weight [30–50 mg/kg in the male Japanese Quail (Adkins and and Adler, 1972), but lower doses would probably work; see later] generally restores aggressive, sexual, and courtship behavior in castrated birds. In a number of studies, however, it was found impossible to restore the behavior of castrated animals to the precastration levels (e.g., Hutchison, 1970a, 1976; Pröve, 1974). The origin of this discrepancy is unclear. It is probably not due to inadequacy of the doses, which very probably induce plasma levels of hormone equal or superior to those observed in intact animals (e.g., Sossinka et al., 1980). In females EB restores sexual and courtship behavior after ovariectomy (or functional castration by exposure to short days; e.g., Adkins and Adler, 1972; Adkins and Nock, 1976a,b) if injected at daily doses in the range 0.25–5 mg/kg.

When TP and EB induce the same behavior pattern in males, generally EB is active at lower doses or more active at the same dose. In the Japanese

Quail, for example, 5 mg TP/day for 10 days induced copulation in 5 of 10 males, whereas 0.05 mg EB/day activated the same behavior in 7 of 10 (Adkins and Adler, 1972). In the Ring Dove the daily injection of 300 µg EB stimulated nest soliciting more effectively than TP at the same dose (Hutchison, 1970a; see also Cheng and Lehrman, 1975).

A number of more recent studies have shown that smaller doses of steroids can in some conditions be effective. In Ring Doves, for example, as little as 5 µg EB per animal per day (i.e., 0.03 mg/kg) significantly improved the responsiveness of ovariectomized females to the courtship of the male, and maximum effects were obtained with 50 µg (Cheng, 1973b).

Low doses of hormones are, however, not always active, and a number of studies involving detailed dose responses show that unexpectedly high amounts of testosterone or estradiol have to be injected in order to be behaviorally active, especially when unesterified steroids are used (which decreases the half-life of the injected compounds). Deviche and Schumacher (1982), for example, studied the behavioral response of castrated male Japanese Quail to various doses of testosterone ranging from 0.1 to 10 mg per animal per day. Little or no sexual behavior was observed in birds injected with doses below 1 mg, but most animals receiving 5 mg/day of testosterone copulated and crowed during behavioral tests. The minimal effective dose is thus between 1 and 5 mg/animal (6–30 mg/kg).

The amounts of hormone required are even larger if the purpose of the experiment is to induce adult behavior in juvenile birds. The relative insensitivity of immature birds to testosterone had already been noted by Schleidt (1970) and is also known in mammals (e.g., Södersten et al., 1977). It is confirmed by the dose–response curve established for the chick using a slightly different procedure, (single injection of a long-acting testosterone ester, testosterone enanthate, TE). Andrew (1972a, 1975a,b,c) demonstrated that high doses of testosterone are required to activate sexual behavior in domestic chicks. Males injected with 0, 5, 12.5, or 25 mg of TE had, in this test, copulatory scores of 4.05, 6.30, 7.20, and 7.13. It must be noticed that 12.5 mg induced a larger response than 5 mg, which is thus not sufficient to obtain a maximal effect. Although these doses of long-acting androgen cannot be directly compared with doses for daily TP injections, they correspond fairly well with cumulative doses used in similar studies [e.g., Collias, (1950), 0.7 mg TP/day for 20 days, i.e., 14 mg; Gardner and Fisher, (1968), 0.1–0.5 mg/day for 24 days, i.e., 2.4 to 12 mg]. These effective doses are several orders of magnitude above the physiological levels (plasma levels of testosterone in birds range from 0.1 to 10 ng/ml). This is partly related to the schedule of injections, which always fails to establish a relatively constant circulating level of steroid (e.g., Balthazart, 1978a; in mammals, Smith et al., 1977b; Södersten et al., 1980). This problem is partly corrected by the use of silastic implants, which make it possible to establish (in gonadectomized animals) constant circulating levels of steroids in the physiological range (Turek et al., 1976; Desjardins and Turek, 1977; for a review see Smith et al., 1977b). We have studied the effectiveness of testosterone and TP silastic implants in inducing copulatory responses in a hand-test situation (J. Balthazart and D. Hirschberg, unpublished results). Both types of implant were active in this respect, although TP implants were more potent than those of testosterone (20 mm TP is about as effective as 100 mm of testosterone). This mode of hormone administration is much more efficient than daily injections (Fig. 6).

In Japanese Quail, a 50-mm testosterone silastic implant that released 0.15 mg of hormone per day effectively restored copulation and induced a large growth of the cloacal gland (androgen-dependent structure; Adkins et al., 1980), but if the hormone was administered in the form of daily injections, 1 mg/day of testosterone had the same effect as the 0.15 mg of testosterone released from the implant (Adkins, 1977). Our results in the young chick tend to corroborate this result. A 50-mm testosterone implant that released about 0.15 mg/day of testosterone (Balthazart and Hirschberg, 1979b) effectively stimulated sexual behavior in young chicks (Balthazart and Hirschberg, 1979a; compare with doses of TP in Collias, 1950; Gardner and Fisher, 1968). There are too few systematic data, however, to permit detailed comparison of biological efficiencies of a given hormone administered in different forms. Figure 7, nevertheless, clearly demonstrates the importance of the techniques of administration.

Even with silastic implants, it is uncertain whether the hormone doses generally used (microgram or, more frequently, milligram range) mimic the circulating levels observed in untreated birds during the reproductive season. Plasma concentrations of testosterone measured in mature birds by RIAs or gas chromatography are generally in the range 1–10 ng/ml [e.g., duck 0.03–2 ng/ml (Garnier, 1971; Jallageas et al., 1974), 1–6 ng/ml (Balthazart and Hendrick, 1976); Ring Dove, 0.2 to 6 ng/ml (Hutchison and Katongole, 1975; Feder et al., 1977); chicken, 0.8–7.8 ng/ml (Furr and Thomas, 1970); White-crowned Sparrow, 0.5–4 ng/ml (Wingfield and Farner, 1978)].

Jallageas and Assenmacher (1970, 1973) demonstrated that the daily injection of 5 mg of TP in castrated ducks (i.e., 1.5–2.5 mg/kg) produces a testosteronemia corresponding to normal levels during the reproductive period (1.8 ng/ml), whereas lower doses (1 mg/day) induce much lower levels (0.4 ng/ml). The enormous difference between injected doses (5 mg) and circulating levels [~2 ng/ml, or 2.5×10^6 times less, whereas blood volume is only 150–200 ml (Jallageas and Attal, 1968)] cannot be explained completely, but it certainly relates to the catabolism of the hormone at the

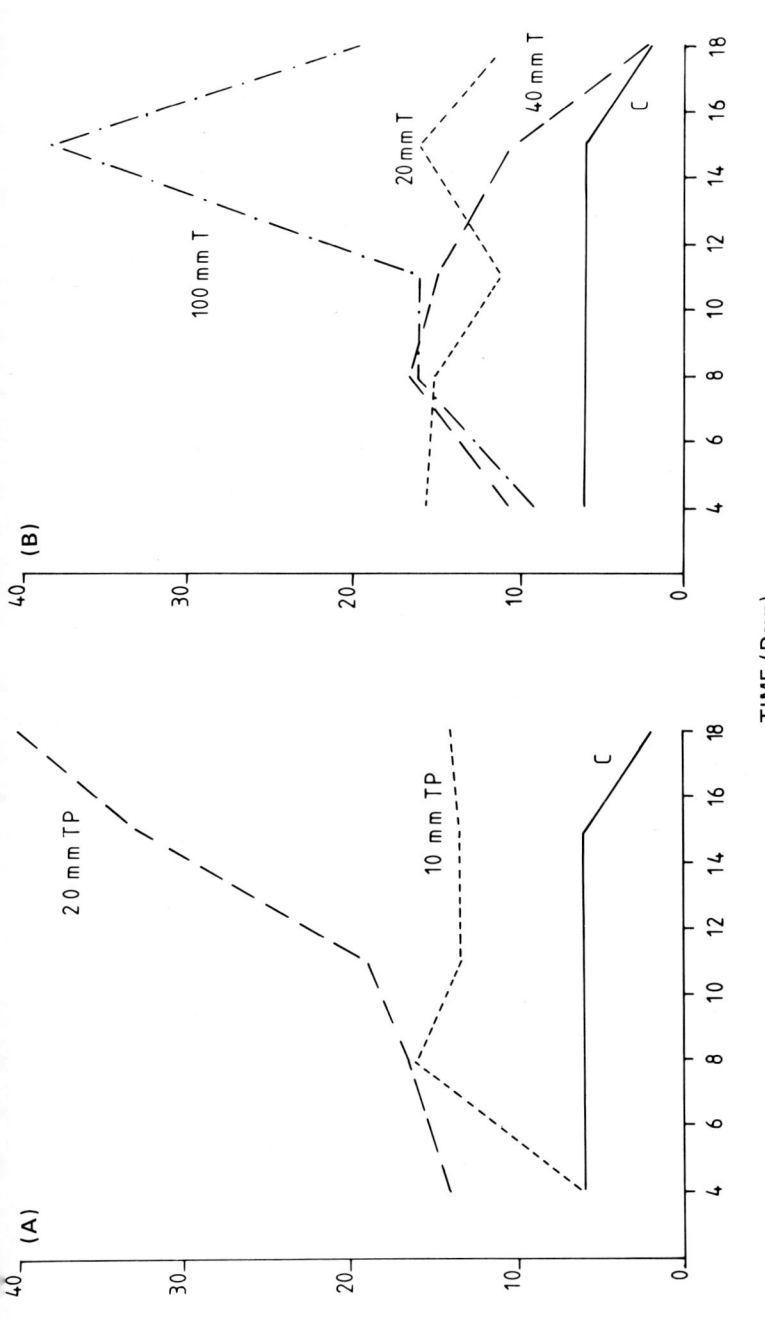

FIG. 6. Median copulation scores young domestic chicks implanted at 4 days of age with silastic implants (1.02 mm I.D.; 2.16 mm O.D.) of various lengths (mm) containing either testosterone propionate (TP; panel A) or testosterone (T; panel B) or left empty as control (C). Abscissae: days from implantation. Ordinates: behavioral scores in responses to the hand test. Points were given for each behavior pattern shown by the bird. (See Balthazart and Hirschberg, 1979a, for the details of the scoring system; J. Balthazart and D. Hirschberg, unpublished data.)

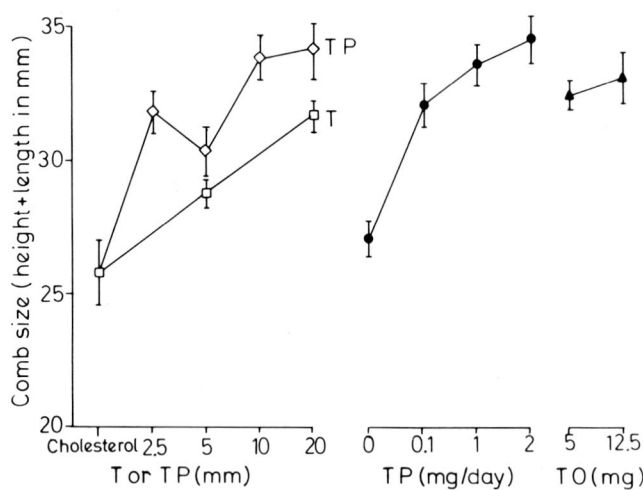

FIG. 7. Effect of various treatments with testosterone or its esters on comb growth in chicks. Four-day-old chicks were either implanted with silastic capsules (1.02 mm I.D.; 2.16 mm O.D.) containing testosterone (T) or testosterone propionate (TP), or received a daily injection of TP for 14 days, or a single injection of testosterone oenanthate (TO). Comb size was measured after 14 days. TP implants are more effective than T implants. 1 mg of TP daily (total, 14 mg) has more or less the same effect as a 20-mm TP implant that contained at the beginning 13.05 ± 1.06 mg but released much less than 1 mg during the 14-day period. (J. Balthazart and D. Hirschberg, unpublished data.)

injection site, to the very short half-life of testosterone in birds (14 minutes according to Jallageas and Assenmacher, 1973), and also, possibly, to the rapid catabolism endured by the hormone in the target cells (essentially 5β-reduction; see Steimer and Hutchison, 1981; Section V,B).

The first of these two points finds confirmation in the experiments showing that minute amounts of hormones placed directly in the hypothalamus qualitatively restore sexual and courtship behavior in castrated birds (quantitative differences, in contrast, are frequently observed). Forty micrograms of TP implanted in the preoptic area or in the anterior hypothalamus of the Ring Dove induce the same behavioral effects as the daily injection of 300 µg TP for 15 days (Hutchison, 1971). This shows that very little of the peripherally injected hormone has to reach the brain to be behaviorally effective.

B. Latency of Action

Despite important progress in recent years, the detailed mechanisms by which steroids affect nerve cells and behavior remain largely unknown. The

4. HORMONAL CORRELATES OF BEHAVIOR

latency between the injection of a hormone and the occurrence of the behavior is an interesting characteristic that invites further research. It has been established that a tracer dose of steroid injected in the general circulation reaches the brain rapidly but probably leaves it within a few hours. Very small doses, however, remain for longer periods (up to 24 hours), and they could be behaviorally important (Dudley et al., 1979; Blaustein et al., 1979). Meyer (1973) showed by autoradiography in chicks that labeled testosterone is already found in the brain 30 minutes after injection, but the number of labeled cells in the hypothalamus decreases significantly between 30 minutes and 2 to 3 hours postinjection. This agrees with autoradiographic data obtained in the Chaffinch (*Fringilla coelebs*) by Zigmond et al. (1973) and with direct measurements of steroid concentrations in the brain of doves (Zigmond et al., 1972). With rare exceptions (Vowles and Harwood, 1966; Clifton et al., 1979; Clifton and Andrew, 1981), behavioral effects of steroids are never observed within such short delays. Even if larger doses of TP (an ester with longer half-life than testosterone) are injected, they remain in the body fluid (plasma) for longer periods but occurrence of the behavior is nevertheless not synchronized with changes in the plasma level of the hormone (Fig. 8; Sossinka et al., 1980).

Behavioral effects of systemic steroid injections or implants are generally not observed before 2 to 3 days after the beginning of the treatment. Sachs

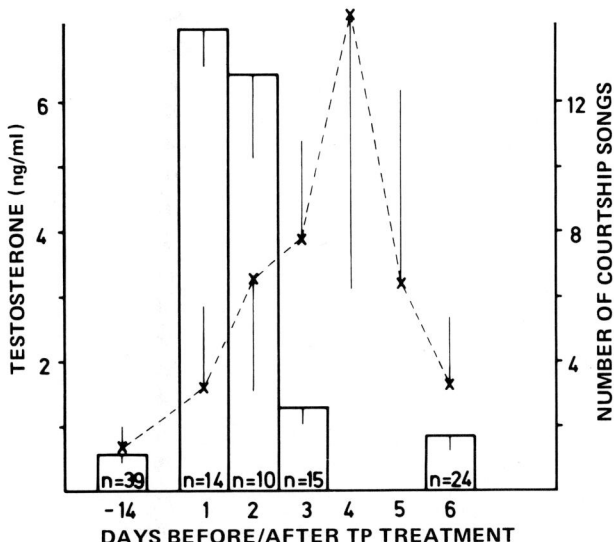

FIG. 8. Temporal relationships between frequency of courtship songs and plasma level of testosterone after testosterone injections in castrated male Zebra Finches. Birds received 20 μg of TP/g body weight on days 1 and 0 subcutaneously. Columns, plasma testosterone ($\bar{x} \pm$ SD); interrupted lines, number of songs per 30 minutes. (From Sossinka et al., 1980.)

(1969) showed that TP injections activate sexual behavior in male Japanese Quail maintained under short days (i.e., functionally castrated), but the effect occurs only after 1 to 5 days, depending on the individual or the strain of birds. Hutchison (1971) presented statistical comparisons of the latencies observed between the initiation of various steroid treatments and the restoration of several courtship displays in castrated male Ring Doves (Table I). Hutchison's data clearly show that the latency (several days) for the restoration of behavior is not due to

1. The time required for the hormone to reach the brain from its peripherial site of injection. [The bowing display is the only exception to this rule, but comparison with results obtained for the other two behavior patterns suggests that this is due to a dose–response problem rather than to a difference in latency. (This interpretation is also supported by experimental data in Hutchison, 1970b.)
2. The time needed for the hydrolysis of TP into testosterone
3. The time required to aromatize testosterone into estradiol [For some behavior patterns there is now good evidence that the effect of testosterone is actually due at the biochemical level to the action of estradiol formed by aromatization of testosterone in the target cells (see Section V,B).]

As in Table I, latencies of 2–5 days seem to be the rule in this type of experiment (see J. B. Hutchison, 1978, p. 301). A number of studies show,

TABLE I

Latency (Median and Ranges) for the Restoration of Three Courtship Behaviors in Castrated Male Ring Doves[a,b]

Treatment	Latency (days)		
	Chasing	Bowing	Nest soliciting
Intramuscular injection of TP (300 μg/day)	2.5 (0–3)	5[c] (3–11)	1.5 (0–3)
TP implanted in PM + HAM area	2 (0–6)	1.5[c] (0–2)	3.5 (2–4)
T implanted in PM + HAM area	1.5 (0–2)	—	2 (0–2)
E_2 implanted in PM + HAM area	3 (2–9)	—	2 (1–4)

[a]Adapted from Hutchison (1971).
[b]TP, Testosterone propionate; PM + HAM, preoptic–anterior hypothalamic region; T, testosterone; E_2, estradiol.
[c]$p < .002$. Comparison of the results labeled with superscript c.

furthermore, that different behavior patterns in a given species are not activated after similar latencies. Balthazart and Stevens (1976) showed, for example, that TP injections stimulate sexual behavior in young ducklings but that the different patterns do not emerge simultaneously. Grasping of the partner's neck feathers appears after 2 days (median), mounting attempts after 3.5, and mounting after 6 days. Whether this sequence is explained by hormonal mechanisms or by experiential factors (need for some motor learning) is not clear in this study, but similar data are available for adult experienced birds in which learning factors are probably minimized. In doves, for example, nest soliciting, bowing, and chasing are not restored simultaneously by TP injections (Hutchison, 1970b); and, interestingly, they also disappear at different times after castration. In Zebra Finches, undirected song and courtship song disappear after castration and are reactivated at different rates by testosterone injections (Sossinka et al., 1980). The available data are inadequate to show whether these differential changes in the timing of several patterns in a species are related to dose problems (different behavior patterns would require different hormone levels, and these would only build up in the plasma after several injections) or are true latency differences.

Similarly, there is no obvious way to explain the latency of at least 1 day between the injection of a steroid and the activation of reproductive behavior. The resolution of this problem will probably have to wait until we better understand the biochemical mechanisms involved in this activation. If the behavioral effects of steroids are produced by the same intracellular mechanisms as their effects on peripheral tissues (e.g., uterotropic effect of estradiol, effects of androgens on the rat prostate, or of estrogens and progesterone on the chick oviduct; see Section V,C), explanations will have to be sought in the rate of synthesis of new mRNA and of new proteins or in a later event in the chain of biochemical interactions between the steroid and its target cell. Evidence has accumulated suggesting that the effects of estradiol and progesterone on the lordosis response in female rodents is mediated by such interactions with cytoplasmic and nuclear receptors (Blaustein and Feder, 1979a,b, 1980; Moguilewsky and Raynaud, 1977, 1979; Parsons et al., 1980). In birds, a few data also support the notion that effects of steroids in the brain are mediated by interactions with specific intracellular receptors (Kawashima et al., 1978, 1979a,b, 1980; Balthazart et al., 1980c). However, there is no definite proof of this concept yet (Feder et al., 1979b), and alternative mechanisms such as binding with membrane receptors (Teyler et al., 1980; Giorgi, 1980) that would induce faster effects could also be present. This, for example, would take into account the finding of Vowles and Harwood (1966) that a single injection of 0.4 mg TP or EB definitively changes the defensive behavior of Ring Doves within a 30-minute period.

C. Specificity of the Action of Gonadal Steroids: The Behavioral Level

It was recognized very early that testosterone secreted by the testes determines the behavioral characteristics of the male (mounting and copulation), whereas estradiol was thought to be responsible for the sexual receptivity of the female (eventually in synergism with progesterone). This view is, however, an oversimplification. As early as 1961, Young pointed out that sometimes androgens stimulate female behavior whereas estrogens stimulate male behavior. This was substantiated by numerous other studies (e.g., Pfaff, 1970). More recently, it was discovered that testosterone is actively metabolized in its target organs. In mammals, this conversion mainly leads to the formation of estrogens and 5α-reduced androgens (e.g., Selmanoff et al., 1977; Jouan and Samperez, 1980). In birds, large amounts of 5β-reduced androgens are produced in addition (see Section V,B). Testosterone can thus in some instances only be considered a prohormone (see Adkins, 1981). The metabolism of androgen in the brain and its biological significance, especially with respect to behavior, will be considered later in this chapter (see Section V,B). The question discussed here concerns more specifically the validity or usefulness of the concept of hormonal specificity at the behavioral level. Is there any reason to retain this notion when some behavior patterns can be activated by both TP and EB in castrated animals? Detailed studies of this problem are now available.

As early as 1949, Guhl demonstrated that copulatory movements can be activated in castrated male chickens by estrogen injections. However, it was only more recently that data became available that permit a direct comparison of the effects of estrogens and androgens in male and female birds. Adkins and Adler (1972) compared the effect of EB and TP in Japanese Quail of both sexes that were functionally castrated by exposure to short days (Adkins, 1973; Adkins and Nock, 1976a; see Table II). Testosterone propionate and EB at the dosages used in this experiment significantly activated the homotypical behavior patterns in males and females, respectively (i.e., cloacal-contact movements and strut in the former, squat in the latter). In addition, however, EB stimulated cloacal-contact movements as well as sexual receptivity in males. On the basis of this study, male birds thus appear more bisexual than females. It is relatively easy to stimulate female behavior in males by EB injections, but TP injections in females activate little or no male behavior. This strikingly contrasts with the situation in mammals, in which the female appears more bisexual than the male (Young, 1961; Aren-Engelbrektsson et al., 1970; Pfaff, 1970; see Section V,A,4 for possible mechanisms explaining this difference).

It is also of interest that EB acted in this experiment at much smaller doses

TABLE II

Effects of Testosterone Propionate (TP) or Estradiol Benzoate (EB) Injections in Male and Female Japanese Quail[a,b]

Treatments	Behavior patterns					
	Cloacal-contact movements (CCM)		Squat (receptivity)		Strut	
	Males	Females	Males	Females	Males	Females
TP (5 mg/day)	$\frac{5}{10}$	$\frac{9}{9}$	$\frac{0}{10}$	$\frac{9}{5}$	$\frac{5}{9}$	$\frac{1}{9}$
EB (0.05 mg/day)	$\frac{7}{10}$	$\frac{0}{8}$	$\frac{7}{10}$	$\frac{4}{8}$	$\frac{1}{10}$	$\frac{0}{8}$

[a]Reconstructed from data in Adkins and Adler (1972).
[b]The ratios are the number of birds that showed the behavior (numerator) with respect to the number of birds tested (denominator).

than TP, although no precise conclusion could be drawn because of the lack of a detailed dose–response curve (at least for some hormone × sex combinations). These data thus suggest that the notion of steroid specificity had to be considered, in the best case, a relative one. By detailed quantitative experiments, Cheng and Lehrman (1975), using the Ring Dove as an experimental model, have adequately quantified the limits and usefulness of the concept. According to these authors, much confusion arose in the literature because ambiguous terms such as masculine or feminine behavior were used. Parts of the behavioral repertoire of males indeed overlap with that of females and vice versa. Södersten (1972, 1974), for example, showed that occasionally female rats will exhibit mounting without any hormone treatment. Cheng and Lehrman (1975) proposed to classify the Ring Dove behaviors into three categories:

1. Male specific: that is, those that appear *only* in males, such as bow cooing, hop charging, mounting, and copulation;
2. Female specific: that is, those that appear only in females, such as squatting;
3. Isomorphic (as opposed to the dimorphic considered above): that is, those that appear in both males and females, such as wing flipping, nest soliciting, and nest building.

In gonadectomized male and female Ring Doves, they tested the effects of TP (200, 400, or 800 µg/day) or EB (50, 100, or 200 µg/day) injections on the activation of these three types of behavior. Injections and observations were carried on until 80% of the birds in each group showed a given behavior pattern. Minimal effective doses were then computed for each sex–hormone– behavior combination by multiplying the daily dose injected by the number of injections required to activate the behavior [for simplification only two latencies (10 and 20 days) were considered]. Results of this experiment are summarized in Table III.

Table III reveals several features of specificity. For instance, more TP is needed in females than in males to activate male-specific behavior, and more EB is required in males than in females to activate squatting. However, it also appears that the minimum dose of EB required to induce homotypical behavior in females (500 μg, i.e., 50 μg × 10 days) is much lower than the dose of TP that activates homotypical behavior in males (2000 μg, i.e., 200 μg × 10 days). There is a ratio of 1:4 between the effective baselines for TP and EB, which does not permit direct comparison of doses for these two hormones (see also Section III,A). To adjust for this difference, Cheng and Lehrman (1975) proposed computing for each behavior–hormone–sex an index of effectiveness equal to the dose of hormone required to induce homotypical behavior in homotypical sex (i.e., 2000 μg for TP and 500 μg for EB) divided by the dose of hormone required to activate a given behavior in a specific sex (e.g., for the activation of bow coo in males the effectiveness of TP is 2000/4000 = 0.5). These indexes are given in brackets in Table III.

TABLE III

Doses of Testosterone Propionate (TP) and Estradiol Benzoate (EB) Required to Activate Various Aspects of Reproductive Behavior in Male and Female Gonadectomized Ring Doves[a]

	Behavior patterns			
	Male specific		Female specific	Isomorphic nest solicitation
Treatment	Bow coo	Hop charge	Squat	
TP				
Male	4,000[c] (0.5)	2000[b] (1.0)	No[d] (0)	2000[b] (1.0)
Female	16,000[c] (0.125)	4000[b] (0.5)	No (0)	8000[c] (0.25)
EB				
Male	No (0)	1000[b] (0.5)	2000[b] (0.25)	2000[c] (0.25)
Female	No (0)	No (0)	500[b] (1.0)	500[b] (1.0)

[a] After Cheng and Lehrman (1975).

[b,c] Doses shown in the table are total amounts in micrograms injected per animal over a period of 10 (superscript b) or 20 (superscript c) days that activate the behavior in 80% of the birds. The indexes of effectiveness are also given in brackets (see text for their calculation).

[d] No, corresponding behavior pattern cannot be activated by this hormone in this sex.

Their analysis permits distinction of three types of specificity which also interact to some extent.

1. Hormone specificity. Male-specific behavior is readily induced in males and females by TP but not by EB, whereas the reverse is true for female-specific behavior (EB active but not TP).

2. Sex specificity. Male behaviors are more easily induced by TP in males than in females, whereas female-specific behaviors are more easily induced by EB in females than in males. Hop charging, a male-specific pattern, is induced by both TP and EB but more readily in males than females (in females EB is not active at all).

3. Behavior-pattern specificity. Different behavior patterns require different doses of hormones for their activation. For example, hop charging and nest soliciting require less TP than bowing in both males and females, and the EB threshold for wing flipping is lower than for nest soliciting in males and females (data not shown in Table III). For the isomorphic behaviors, nest soliciting (and also wing flipping, not shown here) there is, in addition to the aforementioned effects, an interaction between hormone and sex specificity. These behaviors are indeed activated more readily by TP in males but more readily by EB in females.

Although these data are probably an oversimplification of the real world (e.g., the evaluation of threshold doses by the multiplication of daily doses by number of injections is not desirable because it will confuse potentially independent problems such as dose and latency), they clearly support the notion that steroid hormones affect reproductive behavior in a specific way. More specifically, "male" and "female" hormones (i.e., androgens and estrogens) can be distinguished by their effects both at the quantitative and qualitative levels.

Problems will arise, however, when trying to compare results from different species. In Japanese Quail, for example, both TP and EB activate cloacal-contact movements in males but are ineffective in females (Adkins and Adler, 1972; Table II). This differs from results in the Ring Dove (Table III), which are probably best explained by interspecific variation in the interactions between sex and hormone specificity. The classification of a behavior pattern as dimorphic and isomorphic will also vary from one species to another. It was already mentioned that mounting is occasionally observed in female rats (Södersten, 1972, 1974), but Cheng and Lehrman (1975) have claimed that female doves do not display mounting-like behavior in appropriate conditions (even if injected with large doses of TP). They are nevertheless capable of mounting other birds during agonistic encounters (M. F. Cheng, personal communication). This is probably not a phylogenetic difference, as homosexual mounting is occasionally observed in pairs of female domestic ducks (author's observation).

We shall see later that the notion of specificity of the gonadal steroids evolved considerably when the biochemical mechanisms underlying behavior became the subject of intensive study. The metabolism of steroids in the brain tends to blur the distinction between androgens and estrogens, as they can be transformed one into the other; the receptor specificity tends to restore it at another level (see Sections V,B and V,C).

D. Mode of Action

Although it is frequently assumed that steroid hormones activate reproductive behavior by specific actions on discrete areas in the central nervous system, numerous studies both in birds and mammals show that this is not the only mode of action. Six different mechanisms by which hormones may influence behavior are listed by Hinde (1970). Most of these mechanisms can be illustrated by studies involving birds.

1. Effect of Hormones on the Development of the Nervous System

During pre- and perinatal life, steroids exert a profound influence on the development of the nervous system, which results in permanent modifications of behavior (e.g., sexual differentiation). A clear distinction must be made between these so-called organizational effects of steroids, which are characterized by a critical period and a high degree of irreversibility, and the so-called activational effects observed in adult animals. The organizational effects of steroids have long been considered to reflect direct actions of the hormones on the growth of neural connections. Experimental evidence for this has been obtained only relatively recently (e.g., Toran-Allerand, 1976, 1978; Toran-Allerand et al., 1980). The data concerning this mode of action will be discussed later (Section V,A,4). Here we shall consider only the activational effect of hormones.

2. Peripheral Effects

Hormones can influence behavior by acting on peripheral organs (i.e., not on the brain) in three basically different ways: they can either change the sensory input to the brain, affect the development of effectors of the behavior (e.g., specific muscles), or alter some external structures that serve as social signals. These three types of action, which finally result in behavioral

4. HORMONAL CORRELATES OF BEHAVIOR

changes, can be identified in birds. Some mammalian examples have been analyzed in depth and are also presented here.

a. Changes in Sensory Inputs. The copulation of rats is activated by some central actions of androgens, but it is well known that ejaculation depends on stimulations received by the penis during a series of intromissions. In addition to their action at the brain level, androgens increase the density of sensitive cornified papillae of the glans penis, which indirectly affects the ejaculatory patterns (Beach and Levinson, 1950). According to Larsson *et al.* (1973), these two actions of androgens can be dissociated by injections of 5α-DHT, which reproduce the effects on the penis papillae, and injections of estradiol, which reproduce the central effects of testosterone. This interpretation has, however, been questioned (Yahr, 1979; Södersten and Gustafsson, 1980).

Similar peripheral effects of hormones are known in birds. The nest-building behavior of the Canary is controlled mainly by estrogen (see Section II,B,3). In this species, the building behavior is also influenced by the stimuli that the female receives from the nest cup. The characteristics of the nest (size and texture) affect, for example, the amount of building or the type of nest material that is selected. The effectiveness of these stimuli is itself affected by the state of the brood patch. During the building period, the female loses feathers from her breast, where the skin becomes vascular and more sensitive to tactile stimulation (for a review see Hinde, 1965). These changes in brood-patch sensitivity are controlled mainly by estrogen, with some synergism by progesterone or prolactin (Steel and Hinde, 1963; Hinde, 1965). Estrogens thus influence nest-building behavior partly through their action on the brood patch.

There is also evidence that the pituitary hormone prolactin controls the feeding of the young in Ring Doves in part by its peripheral effects. Lehrman (1955) showed that prolactin injections activate feeding behavior by adult doves, but if afferent stimulation from the crop sac is removed by application of a local anesthesic to the crop, the proportion of adults that feed squabs is reduced. Although this conclusion must be partly reevaluated in the light of more recent criticism (Klinghammer and Hess, 1964) and experimental data (Lott and Comerford, 1968), it is still probably true that part of the effect of prolactin on feeding in doves is mediated by a change in the sensory inputs from the crop sac to the brain (see Section II,B,5).

As far as we know, no such peripheral effect of hormones is known in the control of sexual behavior of birds, although indirect evidence suggests that penile sensory inputs in ducks could play a role in copulation. Balthazart and Deviche (1977) demonstrated that EB injections in domestic ducks increase the relative frequency of precopulatory displays (grasping the female's neck feathers, mounting) but do not affect the frequency of copulation. A similar differential stimulation has previously been reported in several studies of rats (Davidson, 1969; Södersten, 1973), and the failure to copulate has been related to inappropriate penile development (Larsson et al., 1973; see before: DHT but not EB stimulates the development of sensitive papillae on the penis). As EB has little effect on penile weight in ducks, whereas TP increases it markedly (Balthazart and Hendrick, 1979a), the failure of EB to induce copulation could reflect only the lack of penile development (and thus inadequacy of sensory inputs to the brain), whereas central activation of the behavior would be adequate. Alternative explanations are, however, possible, and the critical experiments such as combined treatments with EB + DHT or treatment with TP in ducks with deafferentated penis have yet to be performed, so no conclusion can be put forward.

b. *Changes in Social Signals.* Many species of birds exhibit an important sexual dimorphism in their plumage and in a number of integumentary derivatives and skin appendages such as beak, comb, wattles, cloacal gland, and penis (for a review see Stettenheim, 1972). These structures are generally under hormonal control. Feather loss and replacement is controlled by hormones (Payne, 1972) and, furthermore, the type of feathers (female or male type) grown at a molt is steroid dependent. Under the influence of estrogens, birds generally develop a female-type plumage, whereas in the absence of this hormonal stimulus a male-type plumage develops [male plumage is "anhormonal"; e.g., Japanese Quail (R. E. Hutchison, 1978), Mallard (Caridoit, 1938)]. Some exceptions probably exist, as Johns (1964) reported that phalaropes of two species, if injected with testosterone (but not estradiol), form a plumage of the female type. These species are among the few birds in which the female wears the bright nuptial plumage and takes the active role in courtship, whereas males have a duller plumage and are much less aggressive (see Höhn and Cheng, 1967; Section II,B,1). In chickens, comb size is controlled by androgens (Gallagher and Koch, 1935; Boris et al., 1970), and the bill color in several species is also under hormonal control (Witschi, 1961).

Any body-surface structure is likely to be used as a social signal during interactions between the sexes, although experimental proof for this is uncommon (see, however, Klint, 1980 for an experimental analysis of the role of the male plumage in mate selection). Some experimental details are available on the role of plumage pattern in Harris's Sparrow (*Zonotrichia querula*). In this species, the amount of black feathering on the head and breast is very variable among individuals. This partly correlates with sex and age, but also with dominance status (Rohwer, 1975; Rohwer and Rohwer, 1978). Treatment of Harris's Sparrows with testosterone fails to increase their social status unless the feathers in the head and breast area are simultaneously dyed in black to mimic the appearance of high-ranking birds. The nuptial plumage of birds is generally controlled by sex steroids and/or LH (Witschi, 1961; Lofts and Murton, 1973; Wingfield and Farner, 1980). It is likely that the amount of black feathering in the winter plumage of Harris's Sparrow is also controlled by these hormones, but experimental data confirming this view are still lacking. Apparently, there is no species of bird in which definitive proof has been established that a given feature in external appearance is hormonally controlled and at the same time plays a role in the control of behavior.

In mammals, changes in olfactory as well as visual stimuli can be important. Michael and Saayman (1968) demonstrated, for example, that estradiol administered intravaginally in rhesus monkeys, increases the tendency of the male to mount the female, probably through a change in olfactory stimuli produced by the female. No such effect is known in birds although Jacob *et al.* (1979) demonstrated that the chemical composition of the uropygial secretion of female ducks changes during the reproductive period. Contrary to widespread belief, at least some birds are quite sensitive to odors (Wenzel, 1971, 1973), and in ducks olfactory stimuli may play a role in the control of reproductive behavior (Balthazart and Schoffeniels, 1978). It is thus conceivable that hormones may also affect avian behavior through olfactory stimuli.

c. Changes in Effectors. The performance of a given behavior ultimately depends on the activation of functional muscles (or structures) by central outputs. Hormones can also act at this level by promoting the growth of specialized structures specifically involved in behavior. For example, vocalizations in birds are generated by the air flow through the syrinx. This is controlled by specialized muscles (Nottebohm, 1975; Brackenbury, 1980) that adjust the tension of membranes in the syrinx. The weight, and thus, probably, the power of some of these muscles are controlled by steroid hormones. In Japanese Quail, for example, the weight of the sternotrachealis muscles increases by more than 20% in castrated birds injected daily with 1 mg of testosterone (Deviche *et al.*, 1982a). The same effect can be obtained by injecting testosterone into immature quail (J. Balthazart, unpublished results); in immature as well as in adult castrated quail, 5α-DHT completely mimics the effects of testosterone on the weight of the sternotrachealis muscles (J. Balthazart, unpublished; Deviche and Schumacher, 1982).

Similarly, testosterone increases the weight of the syrinx and affects its cholinergic enzymes in Zebra Finches and Canaries (Luine *et al.*, 1980). Castration also decreases the syringeal weight by 25% and decreases the level of cholinergic activity in the Zebra Finch (Harding *et al.*, 1980). Testosterone and 5α-DHT restore these variables to control level or above, whereas estradiol has very little effect in this respect. Through these morphological and biochemical changes in the syrinx, androgens may affect the quality of the vocalization in birds while controlling their frequency at the central level (see later; Lieberburg and Nottebohm, 1979).

The androgen-dependent growth of the cloacal gland in male quail and of the penis in male ducks also probably affects sexual behavior in these species (at least improves the sperm transfer during the copulation), although no direct evidence for this is available.

3. Central Effects

The best-documented effects of hormones on behavior are those mediated by a central action at the brain level. It is, however, relatively recently that it became possible to localize with some precision the brain areas involved in the effects of steroids on behavior. This has been achieved by the combination of two different techniques:

1. The autoradiography of diffusible, water-soluble compounds (for reviews see Stumpf, 1971; Stumpf and Sar, 1975);
2. The stereotaxic implantation of small amounts of steroids in the brain, originally developed by Lisk (1962), which allowed the identification of the brain centers (labeled by autoradiography) that actually play a role in the control of behavior.

Autoradiographic studies have now been performed after the injection of tritiated testosterone (T), estradiol (E_2) or progesterone (P) in a wide variety of avian species: domestic fowl [adult T (Barfield et al., 1978), chick T (Meyer, 1973; Meyer et al., 1976), chicken T, E_2, P (Wood–Gush et al., 1977), chick embryo E_2 (Martinez-Vargas et al., 1975a)], domestic duck [P, corticosterone (Rhees et al., 1972)], Ring Dove [T, E_2 (Kim et al., 1978; Martinez-Vargas et al., 1974, 1975b, 1976)], Zebra Finch [T (Arnold et al., 1976)], Chaffinch [T (Zigmond et al., 1973)] and Brambling [T (Lücke and Haase, 1980)]. Lücke and Haase (1980) have summarized the details of these and other studies.

In connection with stereotaxic implantation of steroids, autoradiography has identified a small number of areas that seem to be of paramount importance in the endocrine control of behavior. These deserve detailed consideration.

a. Preoptic Area, Anterior and Posterior Hypothalamus. There is enough agreement among autoradiographic studies in birds to accept that the preoptic–hypothalamic continuum is one of the major sites of androgen and estrogen binding within the brain. Testosterone or its metabolites accumulate, among others, in the nucleus preopticus medialis (POM), the nucleus paraventricularis magnocellularis (PVM), nucleus lateralis and medialis hypothalami posterioris (PLH and PMH) of most species studied so far. These results are consistent with the finding that testosterone implants positioned in the preoptic and anterior hypothalamic regions are very effective in inducing copulatory behavior and/or sexual display in the domestic fowl (Barfield, 1965, 1969) and Ring Doves (Hutchison, 1967, 1970b, 1971; Barfield 1971a). There is a very good correspondence between the areas that accumulate radioactivity after injection of [^{3}H]testosterone and are sensitive to testosterone implants, as illustrated by Fig. 9.

All the sites in the preoptic–anterior hypothalamus area that are correlated with a clear activation of sexual behavior when they receive a testosterone implant also accumulate radioactivity, as shown by the autoradiographs. The reverse, however, is not true. Other sites of accumulation do not seem to be related to the sex behavior (e.g., in posterior hypothalamus, the PMH or PVM). It appears that accumulation of testosterone in the ventromedial and tuberal hypothalamus is related instead to the regulation of gonadotropin secretion. In ducks implants of testosterone in the ventromedial hypothalamus inhibit light-induced gonadal growth (Gogan, 1968). In Tree Sparrows (*Spizella arborea*) TP implants in the tuberal hypothalamus cause testicular regression (Cusick and Wilson, 1972). Electrical or surgical lesions in the basal tuberal hypothalamus also disrupt gonadotropic function in male birds (Sharp and Follett, 1969; Ravona *et al.*, 1973; Davies and Follett, 1975a, 1980). A role of the anterior hypothalamus in the control of gonadotropin secretions in birds is not, however, to be neglected (Davies and Follett, 1975b,c; Davies, 1980).

Despite the overall agreement concerning the location of the sites involved in behavior control, species differences in the neuroendocrine organization of behavior are likely to exist. In domestic fowl, for example, TP implants in the preoptic area–anterior hypothalamus activate sexual behavior (Barfield, 1964, 1969) but not aggression or crowing, whereas courtship is only moderately facilitated (Phillips and Barfield, 1977). In contrast, in Ring Doves TP implants in the same area activate both courtship and aggression (Barfield, 1971a; Hutchison, 1971).

In fowl only a partially successful localization of the site of action of androgens on courtship and aggressive behavior has been achieved (Barfield, 1969). Some aggressive behavior was displayed by birds bearing TP implants in the lateral forebrain in and below the paleostriatum (Barfield, 1965), and "waltzing" could be activated in one bird by combining implants in the lateral forebrain and the preoptic area. This suggests that waltzing may depend on a simulateneous activation of the two independent behavioral systems responsible for aggressive and sexual behavior, which is consistent with the classical ethological view that displays result of an interaction between different behavioral systems (or motivations) (cf. Hinde, 1970). However, too few animals were used in this study, the implants were too large, and the experiments have not been replicated. Definitive conclusions are not possible.

It seems that the organization of the psychoneuroendocrine system takes place early during the ontogeny, most of it during embryonic life, as the uptake of [^{3}H]testosterone is very similar in the very young chick (Meyer, 1973; Meyer et al., 1976) and in the adult fowl (Barfield et al., 1978). In the chick embryo, estrogen-binding sites are detected by autoradiography as early as day 10 of incubation in areas such as the POM and parts of the posterior hypothalamus (Martinez-Vargas et al., 1975a). There is, however, a progressive development of the estrogen–neuron system during embryonic life, going from no specific binding on Day 6 of incubation to the appearance of numerous areas labelled by [^{3}H]estradiol in the telencephalon, diencephalon, and mesencephalon of the newly hatched chick (Martinez-Vargas et al., 1975a; Kim et al., 1978). Although no direct comparison can be established with the adult fowl (no autoradiographs following [^{3}H]estradiol injection are available), the pattern of labeled neurons observed in the newly hatched chick closely resembles that described in the adult male and female Ring Dove (Martinez-Vargas et al., 1976).

The demonstration of estrogen-binding cells in the 10-day-old chick embryo correlates well with the finding that estrogen given on days 10–13 to male bird embryos strongly affects their brain differentiation and demasculinizes their adult sexual behavior (in quail, Adkins, 1975, 1979; Whitsett et al., 1977; in chicks, Wilson and Glick, 1970; see Section V,A,4 for the problem of sexual differentiation of the brain). Furthermore, the preoptic region of young chicks is sensitive to testosterone implants, which elicit strong copulatory responses (Gardner and Fisher, 1968; Crawford and Glick, 1975). Also, electrical lesions in this area interfere with the expression of sexual behavior in this species (Haynes and Glick, 1974). This fits in well with the presence of androgen- and estrogen- (testosterone can be aromatized in E_2, which would bind to these sites; Callard et al., 1978a,b) binding sites in this area (Meyer, 1973; Martinez-Vargas et al., 1975a,b).

In females, much less information is available about the central sites of steroid action. By autoradiography, the sites of estradiol uptake have been determined in adult Ring Doves (Martinez-Vargas et al., 1976) and in the chick embryo (Martinez-Vargas et al., 1975a). The distribution of binding in the Ring Dove is consistent with the information available in mammals (e.g., rat, mouse, guinea pig, and squirrel monkey). Estradiol cells are found in four major areas: the preopticostrial region, the basal hypothalamus, the amygdaloid region (nucleus taeniae and nearby portions of archistriatum), and parts of the midbrain. The significance of estrogen uptake in these cleus paraventricularis magnocellularis (PVM), nucleus lateralis and medialis hypothalami posterioris (PLH and PMH) of most species studied so far. These results are consistent with the finding that testosterone implants positioned in the preoptic and anterior hypothalamic regions are very effective in inducing copulatory behavior and/or sexual display in the domestic fowl (Barfield, 1965, 1969) and Ring Doves (Hutchison, 1967, 1970b, 1971; Barfield 1971a). There is a very good correspondence between the areas that accumulate radioactivity after injection of [^{3}H]testosterone and are sensitive to testosterone implants, as illustrated by Fig. 9.

All the sites in the preoptic–anterior hypothalamus area that are correlated with a clear activation of sexual behavior when they receive a testosterone implant also accumulate radioactivity, as shown by the autoradiographs. The reverse, however, is not true. Other sites of accumulation do not seem to be related to the sex behavior (e.g., in posterior hypothalamus, the PMH or PVM). It appears that accumulation of testosterone in the ventromedial and tuberal hypothalamus is related instead to the regulation of gonadotropin secretion. In ducks implants of testosterone in the ventromedial hypothalamus inhibit light-induced gonadal growth (Gogan, 1968). In Tree Sparrows (*Spizella arborea*) TP implants in the tuberal hypothalamus cause testicular regression (Cusick and Wilson, 1972). Electrical or surgical lesions in the basal tuberal hypothalamus also disrupt gonadotropic function in male birds (Sharp and Follett, 1969; Ravona *et al.*, 1973; Davies and Follett, 1975a, 1980). A role of the anterior hypothalamus in the control of gonadotropin secretions in birds is not, however, to be neglected (Davies and Follett, 1975b,c; Davies, 1980).

Despite the overall agreement concerning the location of the sites involved in behavior control, species differences in the neuroendocrine organization of behavior are likely to exist. In domestic fowl, for example, TP implants in the preoptic area–anterior hypothalamus activate sexual behavior (Barfield, 1964, 1969) but not aggression or crowing, whereas courtship is only moderately facilitated (Phillips and Barfield, 1977). In contrast, in Ring Doves TP implants in the same area activate both courtship and aggression (Barfield, 1971a; Hutchison, 1971).

In fowl only a partially successful localization of the site of action of androgens on courtship and aggressive behavior has been achieved (Barfield, 1969). Some aggressive behavior was displayed by birds bearing TP implants in the lateral forebrain in and below the paleostriatum (Barfield, 1965), and "waltzing" could be activated in one bird by combining implants in the lateral forebrain and the preoptic area. This suggests that waltzing may depend on a simulatenous activation of the two independent behavioral systems responsible for aggressive and sexual behavior, which is consistent with the classical ethological view that displays result of an interaction be-

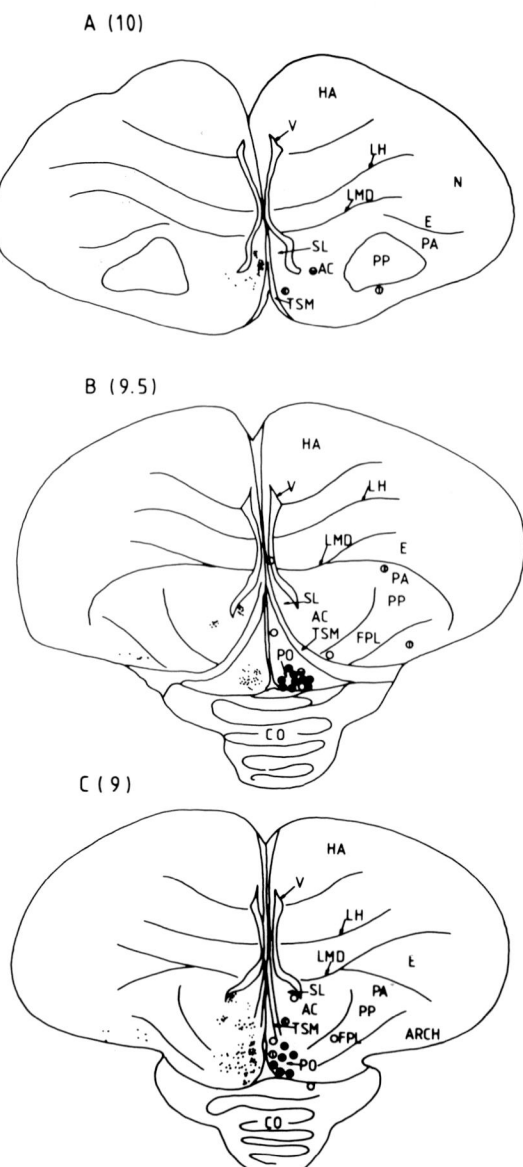

Fig. 9. Survey of transverse sections through the forebrain and midbrain of the domestic fowl showing correspondence between principal sites of radioactive-hormone concentration (small black dots on left side) and sites where testosterone-priopionate implants activate sexual behavior (large dots on right side). The anterior–posterior stereotaxic coordinates of sections A to F are given in parentheses. Coordinates followed by an asterisk correspond to sections in which implant positions were extrapolated from adjacent sections 0.5 mm away (these always circumscribe nonresponsive sites). Symbols: single implants: ●, copulate; ◐, mount only; ○, no sex behavior; bilateral implants: ◑, copulate; ⊘, no sex behavior. Abbreviations: AC, nucleus accumbens; ARCH, archistriatum; CO, chiasma opticum; E, ectostriatum; FPL, fas-

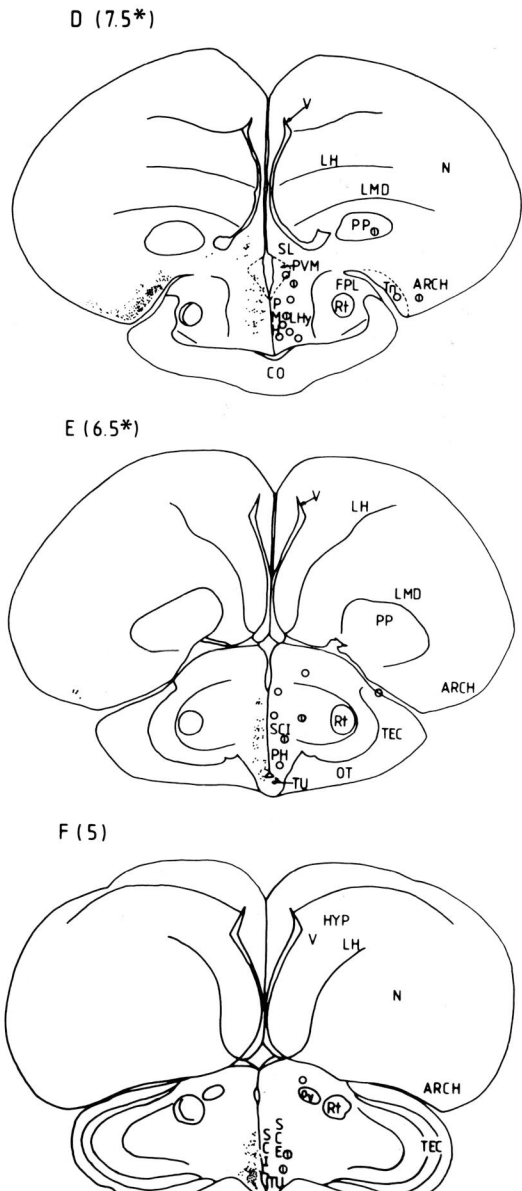

ciculus prosencephali lateralis; HA, hyperstriatum accessorium; HYP, hyperstriatum; LH, lamina hyperstriatica; LHy, nucleus lateralis hypothalami; LMD, lamina medullaris dorsalis; N, neostriatum; OT, tractus opticus; OV, nucleus ovoidalis; PA, paleostriatum augmentatum; PH, posterior hypothalamus; PMH, nucleus medialis hypothalami posterioris; PO, preoptic area; PP, paleostriatum primitivum; PVM, nucleus paraventricularis magnocellularis; Rt, nucleus rotundus; SCE, stratum cellulare externum; SCI, stratum cellulare internum; SL, nucleus septalis lateralis; TEC, tectum; Tn, nucleus taeniae; TSM, tractus septomesencephalicus; TU, nucleus tuberis; V, ventricle. (Redrawn from Barfield, 1969, and Barfield *et al.*, 1978.)

tween different behavioral systems (or motivations) (cf. Hinde, 1970). However, too few animals were used in this study, the implants were too large, and the experiments have not been replicated. Definitive conclusions are not possible.

It seems that the organization of the psychoneuroendocrine system takes place early during the ontogeny, most of it during embryonic life, as the uptake of [³H]testosterone is very similar in the very young chick (Meyer, 1973; Meyer et al., 1976) and in the adult fowl (Barfield et al., 1978). In the chick embryo, estrogen-binding sites are detected by autoradiography as early as day 10 of incubation in areas such as the POM and parts of the posterior hypothalamus (Martinez-Vargas et al., 1975a). There is, however, a progressive development of the estrogen–neuron system during embryonic life, going from no specific binding on Day 6 of incubation to the appearance of numerous areas labelled by [³H]estradiol in the telencephalon, diencephalon, and mesencephalon of the newly hatched chick (Martinez-Vargas et al., 1975a; Kim et al., 1978). Although no direct comparison can be established with the adult fowl (no autoradiographs following [³H]estradiol injection are available), the pattern of labeled neurons observed in the newly hatched chick closely resembles that described in the adult male and female Ring Dove (Martinez-Vargas et al., 1976).

The demonstration of estrogen-binding cells in the 10-day-old chick embryo correlates well with the finding that estrogen given on days 10–13 to male bird embryos strongly affects their brain differentiation and demasculinizes their adult sexual behavior (in quail, Adkins, 1975, 1979; Whitsett et al., 1977; in chicks, Wilson and Glick, 1970; see Section V,A,4 for the problem of sexual differentiation of the brain). Furthermore, the preoptic region of young chicks is sensitive to testosterone implants, which elicit strong copulatory responses (Gardner and Fisher, 1968; Crawford and Glick, 1975). Also, electrical lesions in this area interfere with the expression of sexual behavior in this species (Haynes and Glick, 1974). This fits in well with the presence of androgen- and estrogen- (testosterone can be aromatized in E_2, which would bind to these sites; Callard et al., 1978a,b) binding sites in this area (Meyer, 1973; Martinez-Vargas et al., 1975a,b).

In females, much less information is available about the central sites of steroid action. By autoradiography, the sites of estradiol uptake have been determined in adult Ring Doves (Martinez-Vargas et al., 1976) and in the chick embryo (Martinez-Vargas et al., 1975a). The distribution of binding in the Ring Dove is consistent with the information available in mammals (e.g., rat, mouse, guinea pig, and squirrel monkey). Estradiol cells are found in four major areas: the preopticostrial region, the basal hypothalamus, the amygdaloid region (nucleus taeniae and nearby portions of archistriatum), and parts of the midbrain. The significance of estrogen uptake in these regions is, however, hard to establish because of the lack of implant studies in females. It is known that electrolytic or surgical lesions of the preoptic–basal hypothalamus region disrupt egg laying in the hen and Japanese Quail (Ralph and Fraps, 1959; Ralph, 1959; Egge and Chiasson, 1963; Davies, 1980) and that estrogen implants in these areas interrupt egg laying in the quail (Stetson, 1972), which suggests that estrogen binding in these sites plays a role in the neuroendocrine control of ovulation.

We know of only one study that has been devoted to the behavioral effects of estradiol implants in the brain of a female bird. Gibson and Cheng (1979) placed unilateral (30-gauge) EB implants throughout the hypothalamic regions of ovariectomized Ring Doves from an area rostral to the preoptic area to sites posterior to the PMH. Implants in the PMH were the most effective in stimulating sexual crouches and nest soliciting, but some responses were also observed in females bearing implants in the nucleus accumbens or preoptic–anterior hypothalamus area. In addition, the same study showed that lesions of the PMH area significantly decrease the courtship behavior displayed by ovariectomized doves in response to EB injections, whereas lesions in the anterior–medial hypothalamus or anterior to the preoptic area had no effect. These data are thus consistent with the notion that PMH is a very important site in the mediation of courtship in the female Ring Dove.

The PMH accumulates [^{3}H]estradiol in autoradiographic studies (Martinez-Vargas *et al.*, 1976), and it has been suggested that it is analogous to the mammalian ventromedial hypothalamus (VMH). Although numerous discrepancies exist in the mammalian literature (for a review see Gibson and Cheng, 1979), and the preoptic area is frequently referred to as "the probable site of action of estrogen in facilitating lordosis behavior [Gorski, 1976]," the most recent experiments using very small implants [30-gauge cannula filled with a mixture of estrogen: cholesterol in a 1:250 ratio (this reducing the possibility of leakage)] clearly point to the importance of the VMH in mammals (Rubin and Barfield, 1980; see also Davis *et al.*, 1979). This indirectly supports the conclusion that, in doves, the PMH (analogous to mammalian VMH) is the crucial area in the mediation by estradiol of courtship; generalization to other bird species appears to be premature because no other studies are available. The present data, however, point to a difference, in birds as in mammals (McEwen, 1981), in the neuroanatomical sites involved in the control of male and female behavior (male behavior controlled by testosterone action in preoptic area–anterior hypothalamus, female behavior controlled by E_2 effects in PMH).

b. The Midbrain: The Nucleus Intercollicularis and the Control of Vocalizations. Autoradiographic studies have also drawn attention to a structure in the tectum of the mesencephalon, the nucleus intercollicularis (ICo). In

1973, Zigmond et al. showed that this anatomically heterologous structure (Karten, 1967) accumulated testosterone and/or its Δ^4-reduced metabolites in a song bird, the Chaffinch (Fig. 10). This has since been confirmed in several other species: the chick (Meyer et al., 1976), the adult fowl (Barfield et al., 1978), the Zebra Finch (Arnold et al., 1976), the Brambling (Lücke and Haase, 1980), and the Ring Dove (Martinez-Vargas et al., 1976). In the latter species it was also shown that the ICo accumulates radioactivity after a peripheral injection of tritiated estradiol (Martinez-Vargas et al., 1976). In several cases the ICo was one of the most heavily labeled areas in the brain (e.g., Ring Dove, Zebra Finch).

It had already been suggested that the ICo was involved in the control of vocal behavior in a number of avian species (e.g., Popa and Popa, 1933; Brown, 1965; Potash, 1970). The autoradiographic findings helped renew interest in this subject, and the works of A. P. Arnold and F. Nottebohm (Arnold et al., 1976; Arnold, 1980; Nottebohm, 1975; Nottebohm and Arnold, 1976; Nottebohm et al., 1976) have led to a detailed model of the central control of vocalization in two oscines, the Canary and the Zebra Finch. Numerous other studies showed that low-threshold electrical stimulation of the ICo elicits vocalizations, whereas bilateral electrolytic lesions of the ICo produce muting in a large number of species (see references in

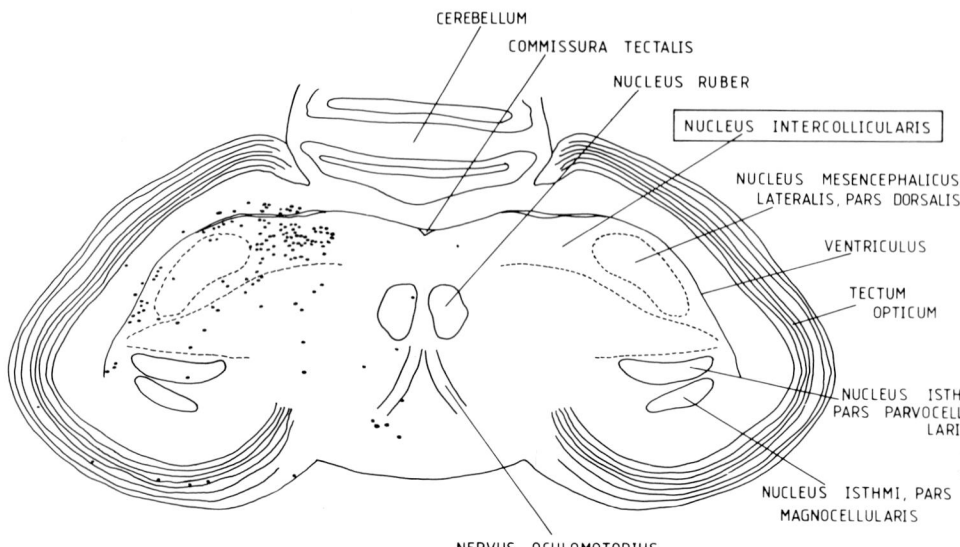

FIG. 10. Distribution of labeled cells in a section through the midbrain of a Chaffinch one hour after injection of [^{3}H]testosterone. Dots represent cells containing at least five times the density of grains found between cell bodies (background). (Modified from Zigmond et al., 1973.)

4. HORMONAL CORRELATES OF BEHAVIOR

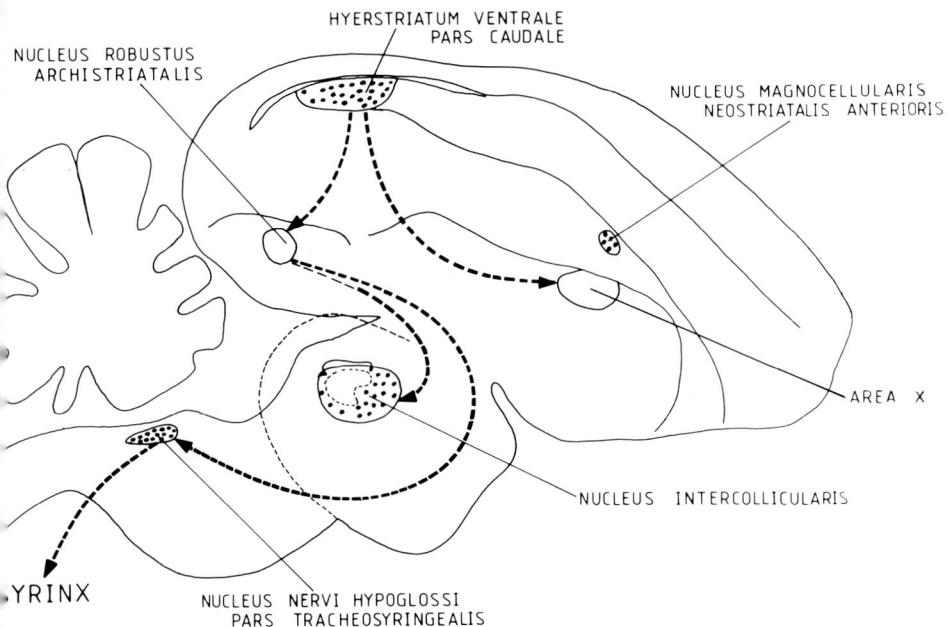

FIG. 11. Parasagittal reconstruction of the Canary brain showing relationships between nuclei involved in song control and nuclei that accumulate radioactivity after injection of tritiated testosterone. Lesion of the hyperstriatum ventrale pars caudale (HVc) and of the nucleus robustus archistriatalis (RA) disrupts singing in the Canary. These nuclei project to other nuclei such as area X, the nucleus intercollicularis (ICo), and the nucleus nervi hypoglossi pars tracheosyringealis (n XII ts). Several of these nuclei (HVc, ICo, n XII ts) show high densities of labeled cells in autoradiographic studies (black dots on the graph). Fibers link the nucleus magnocellularis neostriatalis anterior (MAN) and area X in the Zebra Finch. As MAN also accumulates testosterone it could be involved in song control. (Modified from Arnold et al., 1976 and Nottebohm et al., 1976.) Additional data also show that the RA accumulates radioactivity and that there are connections from MAN to HVc and RA and from ICo to n XII ts (Arnold, 1980).

Cohen and Cheng, 1981). The central mechanisms that control avian vocalization are now among the best understood systems of the neuroendocrine control of behavior.

In the Canary, five brain structures are known to be involved in the control of vocal behavior (Nottebohm et al., 1976). Their localization and relationships are summarized in Fig. 11. Three of these structures (HVc, ICo, and nXII ts) accumulate radioactivity after injection of [^{3}H]testosterone in the Zebra Finch and Chaffinch, which suggests that testosterone acts at several levels to control singing behavior in passerine birds (in addition to the peripheral control of syringeal musculature mentioned earlier). Several of these structures (HVc, RA, and nXII ts) are sexually dimorphic (Nottebohm and Arnold, 1976), and their development is under the control of steroid hormones (Nottebohm, 1980; Gurney and Konishi, 1980; Arnold, 1980).

In Ring Doves also, the ICo region accumulates radioactivity after injection of [^{3}H]estradiol in females (Martinez-Vargas et al., 1976) and of [^{3}H]testosterone in males (Martinez-Vargas et al., 1974). More experiments have confirmed the role of steroids acting in this area to control the vocal behavior of doves. Cohen and Cheng (1981) showed that estradiol implants in the ICo restore the nest coo in ovariectomized doves, whereas TP implants have the same effect in males (Cohen, 1980). Lesions of the ICo lead to a reduction of nest cooing in both sexes (Cohen, 1980; Cohen and Cheng, 1981). The ICo area is thus clearly involved in song control in at least two orders of birds (Passeriformes and Columbiformes), but perhaps this cannot be generalized to all avian species. Phillips and Barfield (1977) failed to elicit crowing in fowl by implanting them with testosterone in this area, although other vocalizations are altered by electric stimulation of the ICo in chicks (Andrew, 1975d). Whether this relates to the special nature of the vocalization studied (crowing is very probably not homologous to song in passerines or to nest coo in doves) or represents a true species difference cannot be evaluated with the available evidence.

Interspecific morphological differences in the brain areas involved in the control of vocalization have also been detected. Two telencephalic structures, the magnocellularis neostriatalis anterioris (MAN) and the hyperstriatum ventrale caudal (HVc), which are present in oscine birds and accumulate steroids (Arnold et al., 1976; Nottebohm et al., 1976), seem to be absent in two other species (chicken and pigeon) for which brain atlases have been published (Van Tienhoven and Juhasz, 1962; Karten and Hodos, 1967). Experiments in progress could, however, lead to a revision of this idea, as Ring Doves seem to have a nucleus equivalent to the HVc (J. Cohen, personal communication). It must be remembered, however, that the ICo in the fowl and chick accumulates radioactivity after an intravenous injection of [^{3}H]testosterone (Meyer, 1973; Barfield et al., 1978). The role of this steroid uptake is not yet clear. Meyer et al. (1976) suggested that it could be related to the increase in the persistence in attending to objects that is observed after systemic injection of testosterone (Andrew, 1972a,b, 1978).

Andrew (1975d) noted that electrical stimulation of the ICo induces calling and alerting in chicks, whereas lesions of the ICo produce a number of behavioral modifications (e.g., muteness, impaired visual targeting behavior, including loss of scanning behavior and absence of exploratory pecking at moving targets). Andrew and de Lanerolle (1975) suggested that chicks with ICo lesions treat visual objects as if they were not conspicuous, but no implant study is available to ascertain that these behavioral changes are related to the testosterone uptake by the ICo region. Further research is needed to clarify this subject.

c. *Telencephalic Structures.* The uptake of radioactivity has been detected by autoradiography in several telencephalic structures. The hyperstriatum ventrale (HV), especially its caudal part (HVc) concentrates testosterone in the Zebra Finch (Arnold et al., 1976), in the Brambling (Lücke and Haase, 1980), and in the domestic hen (Wood–Gush et al., 1977). In the hen, it has also been shown that the HV accumulates radioactivity following injection of [^{3}H]estradiol and [^{3}H]progesterone (Wood–Gush et al., 1977). The HV in ducks also accumulates corticosterone and progesterone (Rhees et al., 1972), although the localization of labeled cells does not exactly fit that found in the hen (for details on differences see Wood–Gush et al., 1977).

The significance of steroid receptors in the HV has not been worked out. Hyperstriatal lesions in the hen make them less reactive to frightening stimuli (Gentle et al., 1978) and in general terms affect numerous aspects of escape responses, responses to predators, and attacking behavior (Gentle et al., 1978). The telencephalon of birds is also involved in various aspects of reproduction (Beach, 1951; Juhasz and van Tienhoven, 1964; Phillips, 1964; Åkerman, 1966a,b) but it is uncertain that the binding of steroids in the HV is related to these effects.

There are good reasons to believe that steroid binding in HVc is related to the control of vocalizations at least in songbirds (see earlier) and, in addition, indirect evidence suggests that hormonal effects on the HVc could take part in the control of the imprinting process (restriction of the social preferences to a visual object; Horn, 1979) in young precocial birds. Since the mid-1970s, G. Horn, P. Bateson and co-workers have clearly demonstrated that filial imprinting in chicks can be related to biochemical (e.g., incorporation of radioactive uracil) and even morphological (modification of the length of the synaptic apposition zone) changes that are restricted to the HV (Bateson et al., 1972; Horn et al., 1979; Bradley et al., 1981; Bradley and Horn, 1981). Lesions and stimulation experiments, combined with refined experimental procedures for behavioral testing and training, largely support the notion that the changes observed in the HV are specific for the imprinting process (Bateson et al., 1975, 1978; Horn, 1979; McCabe et al., 1979). On the other hand, we have shown that androgen injections interfere with the behavior of chicks during sessions of imprinting on an artificial model. Small doses of testosterone or DHT (50–250 μg) essentially decrease the "following" reaction and the frequency of twitter calls but strongly increase the frequency of peeping (Balthazart and De Rycker, 1979; Balthazart and Puts, 1979; see also Section II,B,6,b). Whether these actions are results of a localized effect on the hyperstriatum remains, however, to be determined.

The nucleus taeniae in the archistriatium (considered homologous with the mammalian amygdala; Zeier and Karten, 1971) also accumulates testosterone and/or its metabolites in the fowl (Barfield *et al.*, 1978), the Zebra Finch (Arnold *et al.*, 1976), the Ring Dove (Martinez–Vargas *et al.*, 1974), and the Brambling (Lücke and Haase, 1980). Although this area in birds is considered to be involved in the control of agonistic behavior (e.g., Åkerman, 1966b; Maley, 1969; Phillips and Youngren, 1971), TP implants in the archistriatum did not increase aggressive behavior in two different species (fowl; Barfield, 1969; dove; Barfield, 1971a). The role of testosterone binding in this area is thus unclear.

The same conclusion must, unfortunately, be reached for other telencephalic areas retaining steroids. The septum, at least in some of its parts, accumulates testosterone in the fowl (Barfield *et al.*, 1978), the Zebra Finch (Arnold *et al.*, 1976), the Chaffinch (Zigmond *et al.*, 1973), and the Brambling (Lücke and Haase, 1980). However, TP implants in the septum of the fowl do not activate courtship or copulation (Barfield, 1969). The binding of corticosterone and progesterone in this area similarly cannot be assigned a definite biological role (see Rhees *et al.*, 1972).

There is, finally, a distinct nucleus in the neostriatum of songbirds (e.g., Canary, Zebra Finch) that is apparently absent in the chicken or pigeon (Stokes *et al.*, 1974; Arnold *et al.*, 1976) and strongly accumulates testosterone in the Zebra Finch (Arnold *et al.*, 1976). Considering that this nucleus magnocellularis neostriatalis anterioris projects to an area involved in song control (area X of Nottebohm *et al.*, 1976), it is possible that this uptake is related to androgenic modulation of song.

d. General Comments. Although the preceding pages provide only a schematic summary of the available data, they clearly suggest a number of conclusions about the cephalic effects of hormones on behavior. A limited number of species have been studied by autoradiography and almost no replication of data is available. It is thus difficult (sometimes impossible) to evaluate the extent of interspecific differences.

When considering the biological implications of the hormone bindings that have been detected, conclusions remain generally tentative (with a few exceptions: e.g., binding of testosterone in the preoptic area) due to the lack of detailed implant studies. Most of the data on the central control of behavior by hormones are concerned with sexual and courtship behavior. Other aspects of reproductive behavior have generally been ignored, except for one study by Erickson and Hutchison (1977), who showed that nest-material collection by Ring Doves is activated by testosterone acting in the preoptic–anterior hypothalamus area and, secondarily, on the ventral neostriatum intermediale.

The largest part of the data on the hormone–brain–behavior interactions concerns testosterone. A more limited amount of information is available for estradiol, and even less for progesterone. It is known, however, that the peripheral effects of progesterone on the reproductive behavior of male doves [essentially induction of incubation (but see Section II,B,4 for discussion of this effect) and inhibition of androgen-induced behavior such as a bow cooing and nest calling; Erickson et al., 1967] can be reproduced by chronic implantation of progesterone in the preoptic and anterior hypothalamus area and in the lateral forebrain system (Komisaruk, 1967). Similarly, progesterone is antiandrogenic in chicks when injected systemically. Precocial testosterone-induced copulation is likewise suppressed by implants of progesterone in the periventricular areas of the preoptic–hypothalamic continuum (Meyer, 1972; Meyer and Angelo, 1975), where uptake of [^{3}H]progesterone has been demonstrated autoradiographically in at least one bird, the domestic duck (Rhees et al., 1972). The biological meaning of these effects cannot, however, be evaluated yet. Even in these cases, in which the most advanced knowledge has been gained by the combined use of autoradiography and brain implantation techniques, a number of problems remain unsolved because of methodological limitations and inadequacies.

It is likely that the conventional implant technique, which has been used in most avian studies so far, tends to overestimate the size of the areas responsive to the hormone. Hormone diffusion from such implants can extend as far as 2 mm from the site of implantation (Palka et al., 1966) and thus provides hormonal stimulation over wide areas. This explains, for example, why large implants of TP (22 gauge) or TP pellets activate courtship behavior in doves whatever their localization in the hypothalamus or in the archi- and paleostriatum, whereas smaller implants (27-gauge) are active only if positioned in the anterior hypothalamus and preoptic area (Barfield, 1971a).

Although leakage of hormones from brain implants into the systemic circulation is limited as judged by effects on morphological structures (comb in fowl, oviduct in female doves), it is not negligible. In castrated rats, Smith et al. (1977a) found that intracranial implants (200-μg pellets) resulted in clearly detectable circulating levels of testosterone. Furthermore, the amount of leakage into the general circulation clearly depended, in their study, on the position of the implant in the brain (e.g., implants in the median eminence resulted in plasma levels twice as high as those in the preoptic area) (for detailed discussion of these problems see also Smith et al., 1977b).

These facts can potentially confuse the neuroanatomical localizations obtained so far in all avian studies. These difficulties can, however, be overcome. There has been a trend toward the use of smaller implants (30-gauge). Furthermore, Rubin and Barfield (1980) showed that such implants can be filled with hormone diluted with cholesterol and still remain behaviorally effective [a dilution of estradiol with cholesterol (1:250) still primes estrous responsiveness in the female rat]. Using [^{3}H]estradiol, Davis *et al.* (1979) showed that such diluted implants limit diffusion in the brain (e.g., the contralateral hypothalamus is not affected), and almost no radioactivity is recovered in peripheral target organs for estrogens (e.g., uterus or pituitary). The use of small implants containing diluted hormones thus permits a more refined localization of the brain nuclei responsive to hormones, but avian studies of this type are unfortunately not available.

A final problem with autoradiographic studies is that a control for the chemical nature of the steroid concentrated in cell nuclei is generally lacking. Testosterone can, for example, be converted into estrogens or into 5α-reduced metabolites (see Section V,B,1). Both are suspected of having specific receptors in the avian brain and both are capable of activating some aspects of behavior. When tritiated testosterone is injected and an autoradiographic analysis of the brain performed, it is thus critical to identify the nature of the recovered radioactivity.

Competition with unlabeled steroids can partially answer this question (e.g., injection of an excess of unlabeled estradiol should prevent the concentration of tritiated estradiol derived by aromatization from testosterone but should not affect testosterone uptake). In some studies, biochemical controls have also been performed. Zigmond *et al.* (1973) showed, for example, by extraction and chromatography that 93% of the radioactivity retained in brain cell nuclear fractions after injection of [^{3}H]testosterone consists of testosterone or its 5α- and 5β-reduced metabolites. A study by Sheridan (1981) of the neonatal rat has furthermore provided a method that allows one to demonstrate by autoradiography (i.e., mapping all neuroanatomical localizations) if and where unaromatized androgens are concentrated by the brain. It has been established that the C-1β and C-2β hydrogens are lost during aromatization of testosterone to estradiol whereas the C-1α and C-2α are retained. Taking advantage of this stereospecificity of aromatization, it is thus possible to compare the brain of animals injected with [^{3}H]testosterone labeled in the α or β positions of carbons 1 and 2. The application of this technique to birds should aid the resolution if several questions about the control of reproductive behavior in the future.

IV. Measuring Hormones and Their Fate

The foregoing analysis of the central effects of hormones has brought us closer to the last problems that will be treated here. What are the biochemical mechanisms involved in the facilitation of behavior by hormones, and What is the biological relevance of the experimental effects observed in the castration and hormone-replacement experiments?

The solution of these problems required first that the experimenters be able to measure repeatedly the circulating levels of steroids (and other hormones) in birds. This became possible in the early 1970s and permitted the resolution questions such as:

1. Is the hormone that we inject into an animal the same one secreted by its corresponding endocrine gland? It is now evident, for example, that circulating androgens differ from one class of vertebrates to another. Testosterone is the main androgen in mammals, but this role is probably played by 11-cetotestosterone in fishes (Ozon, 1972) and possibly by 5α-DHT in amphibians (Adkins, 1981).

2. Is there a correlation between variations (in time, between individuals, between strains) of behavior and of plasma concentrations of steroids? The demonstration that a given steroid activates a specific behavior requires not only that injection of this compound restore the behavior in gonadectomized animals but also that spontaneous variations of the behavior observed in intact animals be correlated with changes in the plasma levels of that steroid.

When these data started to accumulate in the mid-1970s (avian but even more mammalian studies), it became evident that the control of behavior involved not only hormonal changes in the plasma, but also variation of cephalic sensitivity to hormones. In several instances changes in sensitivity to hormone according to sex, age, strain, or environmental conditions had been demonstrated, and they did not seem to find their explanation in the analysis of the plasma levels of hormones. The availability of tritiated steroids with high specific activity made it feasible to undertake the analysis of the metabolism of steroids and to quantify the binding of these steroids in the central nervous system. Results of these studies will now be reviewed.

A. Methods for Measuring Plasma Steroids

During the 1970s, research methodology in the field of hormone–behavior interactions changed substantially, mainly as a result of the appearance of biochemical techniques that made it possible to trace the fate of steroids from their secretion by endocrine glands to their interactions with target organs. Before 1970, steroids in biological fluids could be assayed by only three methods: biological assays, colorimetric and fluorimetric methods, and gas chromatography. All these techniques suffered from the same drawbacks: they were time-consuming and insensitive (at best in the microgram–nanogram range), and it was thus almost impossible to assay repeatedly one hormone in a same animal. In the early 1970s, RIAs for each class of gonadal steroids allowed detailed analysis of the relationships between plasma hormone levels and behavior [e.g., testosterone (Furuyama et al., 1970; Ismail et al., 1972), estradiol (Abraham, 1969; Niswender and Midgley, 1970), and progesterone (Abraham et al., 1971)]. In the first half of the 1970s, the first RIAs for avian gonadotropins were also developed [LH (Follett et al., 1972) and FSH (Croix et al., 1974; Follett, 1976)].

Radioimmunoassay is based on a competition between a labeled and an unlabeled antigen (Ag) for specific antibodies (Ab) according to the reaction

$$Ag^* + Ab \rightleftarrows Ag^*Ab + AgAb$$
$$+$$
$$Ag$$

If the amounts of antibody and of labeled antigen (hormone in this case) remain constant, any increase in the amount of unlabeled antigen (hormone) will result in a decrease of the antibody-bound radioactivity. To perform an assay, a standard curve is first established that shows the decrease of the antibody-bound radioactivity as a function of the increasing amounts of unlabeled hormone. The binding of radioactivity is then determined in identical conditions but using plasma or any other biological fluid instead of the known amounts of hormone. This binding is then read on the standard curve, which gives the amount of hormone present in the sample.

The conditions of the assays, the details on their practical realization, and their limitations can be found in numerous papers and textbooks (e.g., Jaffe and Behrman, 1974). Let us simply say here that RIA is generally very sensitive (in the 10^{-12}-g range) and reasonably specific. Cross-reaction with interfering compounds can usually be minimized by chromatographic purification of the sample (e.g., Wingfield and Farner, 1975), and antibodies of increasing specificity are now being developed by several laboratories. The problems related to the validation of a specific assay in a given species are beyond the scope of this chapter, but it must be stressed that the greatest care should be exercised at this level (e.g., Cekan, 1979).

B. The "Good" Steroids Are Present in Birds

The advent of RIA for steroid hormones first permitted a verification that the steroids that had been used for several decades in endocrinological studies of birds are actually present in the plasma of the species under study and thus that the effects observed in earlier experiments following injections of steroids are likely to have a physiological meaning. Testosterone (T), estradiol (E_2), and progesterone (P) have now been detected in the plasma of a wide variety of species [e.g., T (Balthazart, 1976; Balthazart and Hendrick, 1976; Feder et al., 1977; Harding and Follett, 1979; Ottinger and Brinkley, 1978; Pröve, 1978), E_2 (Korenbrot et al., 1974; Wingfield and Farner, 1975), and P (Silver et al., 1974; see also later)], and it seems likely that they are present in all bird species.

The detection by RIA of a steroid is not by itself really conclusive evidence of its presence because of the possible cross-reaction of related compounds in the immunological reaction. Even if it has been shown that the antibody used in a given assay does not react with a series of related compounds, cross-reaction with an unknown compound is still possible. The major steroids have, however, been definitely identified in the gonads and adrenals of several bird species. These identifications, which require large amounts of hormones, use criteria such as isopolarity of the steroid and some of its derivatives in several chromatographic systems, recrystallization at constant specific activity, constant isotope ratios, and various physical methods (e.g., nuclear magnetic resonance, infrared absorption spectra) (for methodology see Sandor and Idler, 1972; for examples see Fevold and Eik-Nes, 1962, 1963; Connell et al., 1966; Fevold and Pfeiffer, 1968; Nakamura and Tanabe, 1972; Galli et al., 1973; Kime, 1980).

Although birds (at least some species) produce steroids that are unknown in mammals (e.g., 11-cetotestosterone in the ovary of Japanese Quail and testes of duck, also tentatively identified in the plasma of ducks; Colombo and Colombo-Belvedere, 1978; Colombo-Belvedere et al., 1981; D. Garnier and P. Deviche, unpublished), it can be concluded with reasonable certainty that they also produce significant amounts the three steroids that have been mainly studied in mammals (T, E_2, P). This at least justified the study of their biological properties in birds.

C. COVARIATIONS OF PLASMA STEROIDS AND BEHAVIOR

1. Ontogeny

Using RIA, it has been possible to confirm and refine the conclusion, originally reached by histological study of the gonad (see Section II,A), that there is a close correlation between the onset of reproductive behavior in young birds and the increase of gonadal activity reflected in the plasma by the concentrations of testosterone. In the male Japanese Quail, for example, plasma testosterone concentrations remain low until 21 days of age and then rise rapidly to reach adult values around day 40 (Ottinger and Brinkley, 1978, 1979a; Hirano et al., 1978). This period of the quail's life corresponds precisely to the time when the first crowings, mating attempts, and cloacal contacts the observed (respective onsets of 32.6, 34.3, and 36.9 days; Ottinger and Brinkley, 1978, 1979b).

These studies of the quail are apparently the only ones that relate explicitly the plasma levels of a steroid to behavior during ontogeny. In other species, however, changes in plasma steroids during maturation have been measured and can be compared (more or less directly) with the available knowledge of behavior. In the developing cockerel, for example, a clear-cut increase of plasma testosterone can be detected after the twentieth week of life (Sharp et al., 1977). It is known that the onset of puberty with the accompanying behavioral manifestations in fowl kept in similar conditions occurs between 16 and 24 weeks of age. Ontogenetic aspects of endocrine functions have, however, scarcely been studied.

Radioimmunoassay studies have also renewed interest in the problem of "infantile sexuality" in birds. It has been known for a long time that many birds show incomplete courtship or nest-building behavior long before their gonads are in breeding condition (Curio, 1960). Young Mallards, for example, perform a number of their adult social displays (e.g., grunt whistle or head up–tail up) when only about 60 days old (Kaltenhäuser, 1971). Also, precocial strutting displays have been observed in turkeys of both sexes (Schleidt, 1970). Andrew (1966) even demonstrated that strong copulatory responses can be elicited from 2-day-old chicks simply by stimulating them with repeated thrusts of the experimenter's hand toward the chick's bill. Hinde (1970) speculated that these juvenile manifestations of adult behavior could be hormone independent, noting that the lordosis responses of newborn guinea pigs form parts of the excretory pattern that do not depend on ovarian hormones (Beach, 1966). Hinde (1970), however, recognized that some hormonal basis for this type of precocious behavior had not been ruled out.

Additional data show that significant amounts of steroids are present in the plasma of very young birds even though the gonads do not look secretory when the birds hatch. In chicks, testosterone is first present, in approximately equal amounts in the blood of both sexes, on Day 5.5 of incubation. Later in incubation, plasma testosterone rises in females but much more so in males, and significant sex differences are observed from Day 7.5 to 17.5. In support of this, Guichard et al. (1977, 1980) have demonstrated that, *in vitro*, gonads of male and female chick and quail embryos produce different amounts of various steroids including estrone, estradiol, and testosterone. These sex differences were also observed *in vivo* by Tanabe et al. (1979).

Their data also suggest that the relatively high testosteronemia in male chick embryos could depend on adrenal rather than testicular production of the hormone (testosterone content of the adrenal pairs is 100-fold higher than that of the testes at 17 days of incubation). This fits in well with the observation by Ottinger and Bakst (1981) that interstitial cells in testes of quail embryos retain a fibroblast-like appearance until the first few weeks posthatch.

Whatever their origin, steroid hormones are present in the plasma of very young birds, and in some cases they show a peak in concentration around the time of hatching (Balthazart and Stevens, 1975b; Ottinger and Bakst, 1981). There is thus a possibility that "juvenile" sexuality has some hormonal basis, a notion reinforced by the observation that testosterone injections activate, among others, copulatory responses in young chicks (Andrew, 1975a) and ducklings (Balthazart and Stevens, 1975a, 1976) and strutting displays in turkey poults (Schleidt, 1970). The higher dose of hormone required as well as its long latency of action suggest, however, that the receptor structures for testosterone do not react exactly as in adult animals (Schleidt, 1970). This could be because of differences in testosterone metabolism or in receptor concentrations between adult and young or because of some difference in functional maturation. The question of whether endogenous steroids play a role in the control of juvenile sexuality thus remains open.

2. Reproductive Cycle and Annual Variations

Variations of the plasma levels of gonadal and pituitary hormones during the reproductive cycle have been studied in a number of avian species. However, detailed information is available for only three species: the Ring Dove, the White-crowned Sparrow, and the domestic duck. In the Ring Dove, both male and female, quantitative analyses of behavior at different stages of the reproductive cycle are available together with assays of plasma LH (Cheng and Follett, 1976; Silver *et al.*, 1980; Cheng and Balthazart, 1982), FSH (Cheng and Balthazart, 1982), testosterone and DHT (Feder *et al.*, 1977), estradiol (Korenbrot *et al.*, 1974), and progesterone (Silver *et al.*, 1974). These data are summarized in Fig. 12.

These assay data support a large number of the conclusions that had been drawn on the basis of experiments in which exogenous hormones had been injected into intact or gonadectomized animals (most of these originally formulated by Lehrman, e.g., 1958b, 1961, 1964a,b, 1965) such as:

1. All those steroids that are supposed to play a role in the control of Ring Dove behavior (T, E_2, and P) are actually detected in the plasma of both sexes (except E_2 in males);

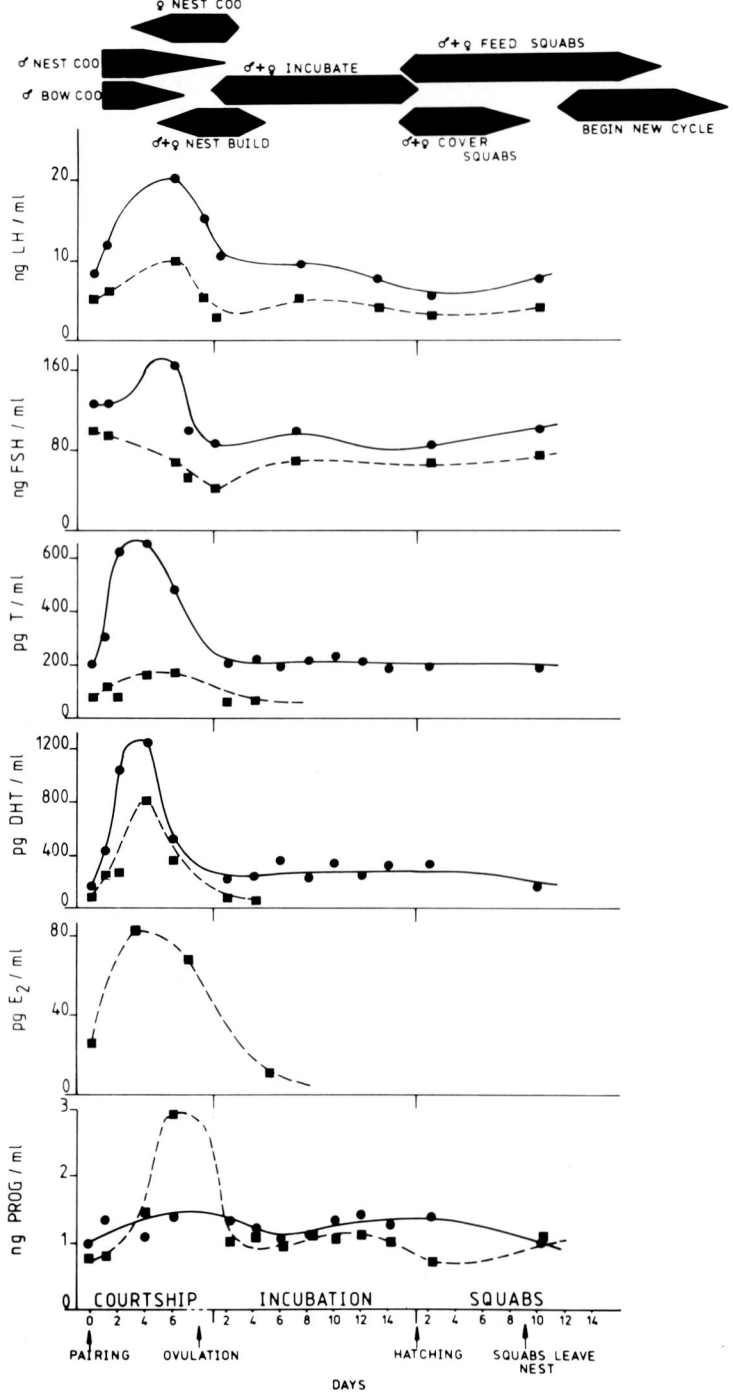

2. A few days after pairing estradiol concentration rises in the plasma of females, which induces them to nest coo (Cheng, 1973b);
3. Slightly later (courtship day 5 in Silver *et al.*, 1974) plasma progesterone reaches a peak and, in combination with estradiol, activates nest building (see Cheng and Silver, 1975).

A number of the mechanisms revealed by experiments using injections of exogenous hormones are thus likely to play a role during the cycle of intact doves. The RIA data raise, however, a number of questions that have so far found no answer (see the original papers for more detailed discussions such as:

1. What is the biological meaning of the sharp increase of plasma testosterone (and dihydrotestosterone) observed in male doves within 24 hours after pairing? It was originally speculated that this would activate the bowing display and possibly nest cooing, but considering that males exhibit these displays *immediately* after being paired and that males castrated shortly before being paired with an intact female proceed through a complete reproductive cycle (R. Saad and R. Silver, unpublished, cited in Silver *et al.*, 1979) this interpretation becomes dubious.

2. What is the relevance for the control of a normal cycle of the effects of exogenous progesterone on incubation behavior in males? As plasma progesterone does not change significantly during the reproductive cycle in males, it is difficult to conceive how this hormone could induce the incubation behavior, as suggested by earlier injection experiments (see Section II,B,4 for detailed discussion of this problem).

One problem in all the dove data is that they were collected by different experimenters on different groups of birds using slightly different experimental procedures (essentially sampling times). There is thus no way to compare precisely the results for different hormones and for the behavior (only schematical relationship can be presented, as in Fig. 12). A study including measures of several hormones and behavioral quantification on the same birds would be most welcome. It must also be noticed that plasma

---

FIG. 12. Changes in plasma LH, FSH, testosterone (T), dihydrotestosterone (DHT), estradiol (E_2), and progesterone (PROG) during the reproductive cycle of the Ring Dove. LH values are expressed in ng chicken LH-AE_1, FSH values in rat FSH-RP_1. Hormone plasma levels are reported for males (●) and females (■) separately when available. The main behavioral events are shown schematically at the top. The coincidence in time of data points for different hormones is only approximate due to different experimental schedules used in the various studies. [Drawn from data in Cheng and Follett, 1976 (LH); Silver *et al.*, 1980 (LH); Cheng and Balthazart, 1982 (LH, FSH); Feder *et al.*, 1977 (T, DHT); Korenbrot *et al.*, 1974 (E_2); and Silver *et al.*, 1974 (PROG).]

levels of testosterone, progesterone, estrone, and estradiol have been measured in another species of dove (the Collared Dove, *Streptopelia decaocto*), and the results generally fit in well with what is known in the Ring Dove (see Peczely and Pethes, 1979).

Detailed descriptions of the variations of several plasma steroids during reproduction are also available for the White-crowned Sparrow (summarized in Wingfield and Farner, 1978, 1980). No detailed recording of behavior is, however, available in these data (the primary focus of these experiments was the endocrinology of reproduction), and the activity of the birds can only be inferred from their stage in the reproductive cycle. The same conclusion holds true for the study by Donham (1979), who described very precisely the variations of plasma LH, testosterone, DHT, estrone, estradiol, and progesterone in male and female Mallards during the annual cycle but does not give details on the behavior of his experimental birds (only their stage in the cycle; e.g., laying, incubating, is supplied).

In the domestic duck, quantitative recordings of several aspects of reproductive behavior were performed by Balthazart and Hendrick (1976) in a group of birds that were at the same time bled and their plasma assayed for LH, FSH, and testosterone (see Fig. 13). The results showed a good correlation ($r = 0.78$, $p < .001$) between the annual variations of plasma testosterone concentrations and the frequency of sexual behavior. Surprisingly, no significant relationship could be established between testosterone and the social displays (grunt whistle, head up–tail up, etc.) in the course of the annual cycle ($r = .36$, NS), but the changes in this group of social activities were positively correlated with the concentration of FSH in the plasma ($r = .58$, $p < .05$). This raised the hypothesis that gonadotropins could play a direct role in the control of courtship in ducks. However, despite several experiments in which testosterone or FSH were injected in ducks of different ages and their effects on behavior compared (Deviche and Balthazart, 1976; Balthazart and Deviche, 1977; Balthazart and Hendrick, 1979a), it is impossible thus far to ascertain that gonadotropins have a direct role in the activation of these displays, although all available data are consistent with this notion (see Balthazart, 1977; Silver *et al.*, 1979).

Information on the changes of plasma testosterone during the reproductive cycle and their relationship with social behavior are also available for a number of other species, including the European Starling, (Temple, 1974), the Red-winged Blackbird (Kerlan and Jaffe, 1974), the Rook (Lincoln *et al.*, 1980), the Cape Cormorant, (Berry *et al.*, 1979) and the Black-faced (Lesser) Sheathbill (Burger and Millar, 1980) (for a review of this kind of data, including assays of LH, see Murton and Westwood, 1977).

In females of several species detailed accounts of the variations of plasma hormones (mainly LH, estradiol, and progesterone) during the ovulatory

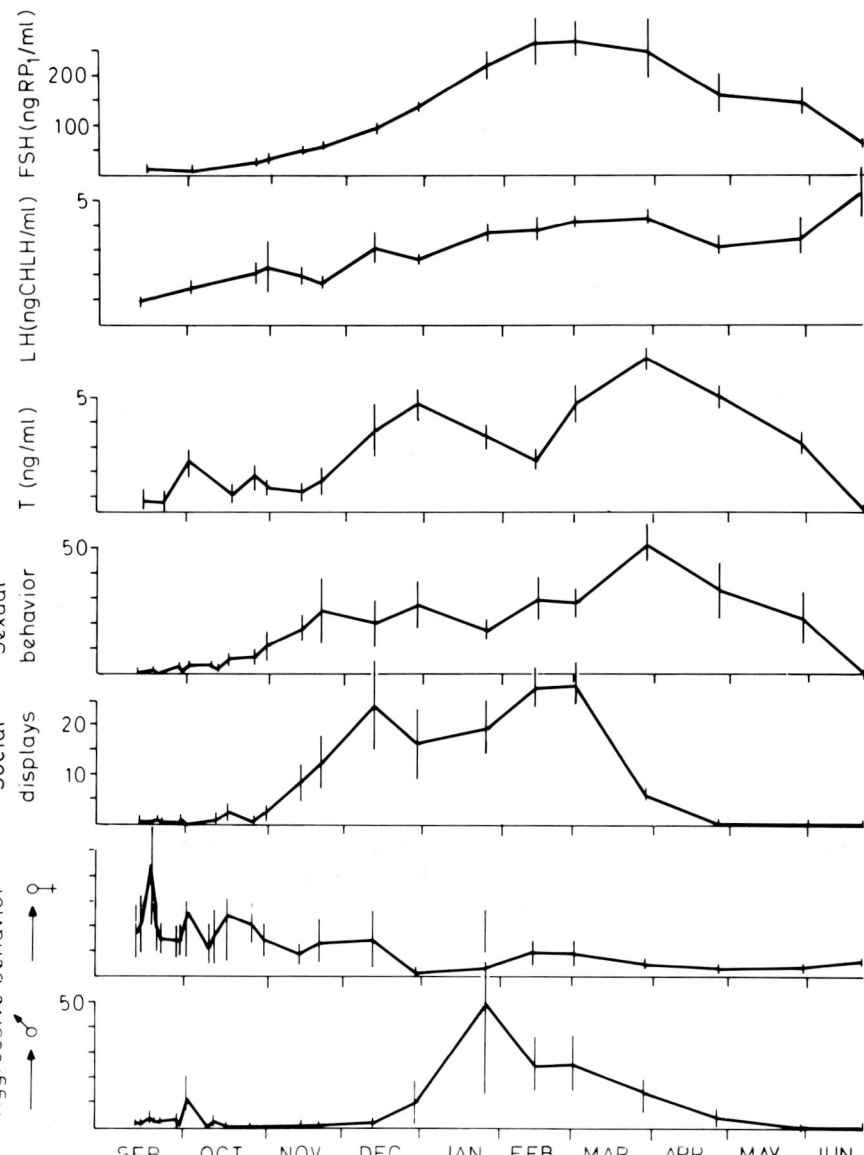

FIG. 13. Annual variations of plasma FSH, LH, and testosterone (T) and of reproductive behavior in a group of five male domestic ducks. Data shown are means ± SEM. (Modified from Balthazart and Hendrick, 1976.)

cycle are also available [e.g., hen (Johnson and van Tienhoven, 1980) and duck (Tanabe et al., 1980), Japanese Quail (Doi et al., 1980)]. These assays are not accompanied by descriptions of behavior. They help to understand the control of ovulation but reveal little about the control of behavior. Changes in behavior (e.g., receptivity) of females during their ovulatory cycle are in fact poorly described and understood.

3. Daily and Ultradian Variations

The availability of RIA has also permitted the study of shorter periodicities (daily, or less) in endocrine physiology that could not be detected by histological techniques and the relation of them to behavior. Changes in LH and/or FSH plasma concentrations during the day have been studied in several species [Japanese Quail (Gledhill and Follett, 1976), domestic fowl (Scanes et al., 1978), duck (Balthazart et al., 1977b), and Ring Dove (Balthazart et al., 1981a)]. Although these studies generally failed to demonstrate reproducible patterns of change during the day, with the exception of the fowl (Scanes et al., 1978) and the quail during only the first day of photostimulation (Follett et al., 1977; Wada, 1979), they showed that the pituitary secretions change sometimes by more than 100% within a few hours. It was thus useful to investigate whether plasma steroids followed these short-term changes of LH and FSH and to ask whether these possible variations in plasma steroids were related to behavior.

Daily variations in plasma testosterone have now been demonstrated in a few birds [e.g., fowl (Schanbacher et al., 1974), Japanese Quail (Ottinger and Brinkley, 1979a), and Ring Dove (Balthazart et al., 1981a)]. In quail, plasma testosterone varies diurnally, with high concentrations being observed in the morning and late afternoon, the lowest point in the mid afternoon (~5 PM (Ottinger and Brinkley, 1979a). This seems related to the daily distribution of crowing, an androgen-dependent vocalization that is heard with higher frequencies in the morning and evening than near noon (Guyomarc'h and Thibout, 1969). Detailed correlations of this type are available, especially in ducks. In this species, quantitative recordings of behavior were performed several times from dawn to dusk on groups of birds that were also submitted to repeated blood sampling. The samples collected in this way were later assayed for LH, FSH, and testosterone (Balthazart, 1976; Balthazart and Hendrick, 1979b). In two different groups of ducks that were observed and bled in December and March, respectively, the diurnal variations of plasma testosterone (and FSH) were correlated with variations in social displays and sexual behavior (Fig. 14). This occurred despite the fact that the patterns of variation were completely different in the two groups.

This suggests that short-term variations in these androgen-dependent behaviors could be controlled by the rapid variations in plasma testosterone. It

4. HORMONAL CORRELATES OF BEHAVIOR

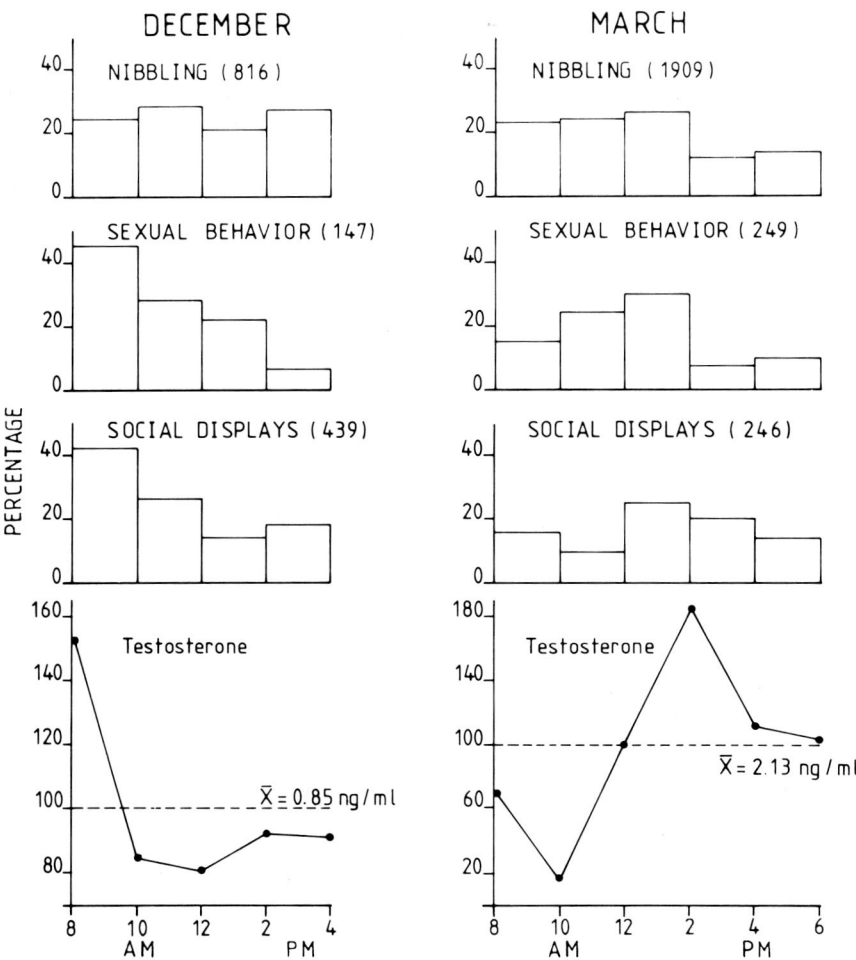

FIG. 14. Daily variations of behavior and plasma testosterone in two different groups of 10 male domestic ducks observed in December and March. Means of the 10 individual values are used for each variable. Behavior frequencies during each 2-hour period are presented as percentages of the total frequencies observed (in brackets). Plasma testosterone levels are expressed as percentages of the daily means represented by the dotted line. There is a correlation between sexual behavior or social displays and plasma testosterone but not between testosterone and a nonsocial behavior such as nibbling the feathers. (Drawn from data in Balthazart et al., 1979.)

is, however, still unclear whether the changes of hormone levels can cause behavioral changes within hours (which would be rather fast compared to other known latencies of action for steroids; see Section III,B) rather than the converse (occurrence of behavior changing the plasma hormone levels;

see Section IV,E). The critical experiments in which either behavior or hormones would be deliberately manipulated have not been performed so far.

Ultradian periodicities (<24 hours) of plasma hormones are also known in birds. Luteinizing hormone, for example, is secreted in a pulsatile fashion in the domestic fowl (Wilson and Sharp, 1975), the Japanese Quail (Gledhill and Follett, 1976), and the domestic duck (Balthazart *et al.*, 1980a). On the basis of mammalian data (e.g., Bartke *et al.*, 1973; Rowe *et al.*, 1975; Lincoln *et al.*, 1977), it is also likely that plasma testosterone shows cyclic variations with periods much less than 24 hours (in the range of 0.5 to 3 hours). This has actually been shown in roosters (Wilson *et al.*, 1979).

On the other hand, it is known that many behavior patterns in birds appear during the day in a cyclic fashion, with periods of activity regularly alternating with periods of rest [e.g., crowing in Japanese Quail (Guyomarc'h and Thibout, 1969) and social display in ducks (Raitasuo, 1964)]. It could theoretically be expected that some relationship exists between these periodicities of endocrine and behavioral variables (Wilke, 1977), but experimental data are not yet available.

D. Do the Doses of Hormones Usually Injected Produce Physiological Levels of Steroids?

The previously described correlations greatly improved our understanding of the hormonal basis of behavior. Sexual behavior, for example, is induced in immature or castrated ducks by testosterone injections (Balthazart and Stevens, 1976; Deviche, 1979a,b), and the annual variation of sexual behavior frequencies is strongly correlated with that of plasma testosterone (Balthazart and Hendrick, 1976; Fig. 13, this chapter). This supports the notion that testosterone actually activates sexual behavior in intact ducks. However, proof that the effects of injections mimic physiological events also requires that the amounts of injected hormone induce physiological levels of the compound in the blood (Mills, 1966, cited in Hiroshige, 1974). Few assays permit a test of this assertion for studies of avian behavior. When data are available, they throw some doubts on the validity of many of the experiments performed until recently.

In most experiments on avian behavior, large doses of hormones have been injected, resulting in supraphysiological plasma levels of steroids. In the Ring Dove, an injection of 100 μg of progesterone (a dose commonly used; Lehrman, 1958b; Erickson *et al.*, 1967; Stern and Lehrman, 1969; Stern, 1974; Cheng and Silver, 1975) produces plasma levels too high to measure 1 hour after injection and which remain higher than the physiological values even after 24 hours (4.86 ng/ml compared to 1–2 ng/ml in males and females during the reproductive cycle; see Fig. 12; Silver *et al.*, 1974). Similarly, estradiol plasma levels in castrated male doves injected with 200 μg of EB (as in many experiments e.g., Stern, 1974; Cheng and Lehrman, 1975) are in the nanogram range, 10 to 100 times the endogenous concentrations measured in breeding females (Korenbrot *et al.*, 1974).

Many experiments with testosterone do not give a better picture. The doses of TP that were used by Balthazart and Stevens (1975a, 1976) to activate sexual behavior in duckling (5 mg TP per animal per day) produce plasma testosterone levels that are 2–5 times higher than physiological values in the few hours after the injection (Balthazart and Stevens, 1975b; Balthazart and Hendrick, 1979a). Values as high as 100 ng/ml (at least 10 times higher than the physiological range) have even been detected in the hour following such injections (Deviche *et al.*, 1979).

It must be recognized that as long as hormone therapies were given in the form of daily injections, there was no way to replace the steroids of gonadectomized animals in a physiological way. The injected compounds are quickly metabolized, and it is thus necessary to give rather large amounts of hormone to obtain behavioral activation. We described, for example, the changes in plasma testosterone in immature ducks following a single injection of 10 mg of TP (Balthazart, 1978a). The data clearly show that during the first hours after injection the birds are subjected to supraphysiological levels of hormone which could activate, possibly in a pharmacological way, brain mechanisms involved in behavior control. Later, plasma levels of testosterone fall below concentrations observed in sexually active males in spring and are thus probably inactive. When dose–response curves are available, they demonstrate that, if a schedule of daily injections is used, large doses of steroids (in the milligram range) must be injected to activate behavior in birds (see Section III,A).

Progress has been made by using silastic implants of steroids that release hormones at constant rates for periods as long as 1–2 months (Stratton *et al.*, 1973; Turek *et al.*, 1976; Desjardins and Turek, 1977) and maintain plasma concentrations in the physiological range (Smith *et al.*, 1977b). A number of experiments on avian behavior using this mode of hormone administration have now been performed (e.g., Balthazart and Hirschberg, 1979a; Nottebohm, 1980; Gurney and Konishi, 1980) with in some cases control over the rate of release from the implants (Balthazart and Hirschberg, 1979b; Adkins *et al.*, 1980) or over the plasma levels of steroids following implantation (Balthazart *et al.*, 1979). In several selected cases, it can now be ascertained that therapies with physiological levels of steroids restore sexual behavior in castrated animals [e.g., in quail (Adkins *et al.*, 1980; Balthazart *et* al., 1983)]. Despite their considerable advantages, the silastic implants suffer, however, from one important drawback; they do not release different steroids at the same rate (Balthazart and Hirschberg, 1979b; Adkins et al., 1980), so that their usefulness in studies comparing the effects of different steroids on behavior is questionable.

E. Effect of Behavior on Hormones

Radioimmunoassays in behavioral studies on birds have also shown that plasma hormones (gonadotropins, gonadal steroids) are not only controlled by environmental factors, mainly the photoperiod (see reviews by Immelmann, 1971; Follett, 1973, 1978; Farner, 1975; Farner and Wingfield, 1980; Follett and Robinson, 1980; Wingfield and Farner, 1980) but that the preceding history, the social experience, the social context, and the behavioral activity of a bird modify its endocrine physiology. Almost by definition, stress increases corticosterone plasma levels [e.g., fowl (Scanes et al., 1980), duck (Harvey et al., 1980)], and generally the simple fact of bleeding a bird repeatedly suffices to affect its corticosterone levels (e.g., Harvey et al., 1980). Surprisingly, this type of manipulation can also disturb the functioning of other endocrine systems. Repeated bleeding, for example, decreases plasma LH and testosterone in the fowl (Wilson and Sharp, 1975; Wilson et al., 1979), and there is some indication that it modifies plasma LH and FSH in the Ring Dove (Balthazart et al., 1981a,b). It is also known that the way animals are bled can strongly modify the concentrations of the hormones that are measured in the samples (e.g., in the rat; Döhler et al., 1977).

In the Zebra Finch, housing conditions have a very marked influence on plasma testosterone titers. Males caught from an aviary and kept for one night in a cage have, for example, much lower plasma testosterone concentrations than birds kept for a long period in visual isolation from other birds in the same cage (Pröve, 1978; Sossinka et al., 1980). Wingfield and Farner (1980) have also demonstrated that captive White-crowned Sparrows have much lower testosterone levels in their plasma than birds sampled in the field.

External perturbations in the reproductive cycle of birds also definitely affect their endocrine physiology. The best-documented example here is the change in pituitary activity following clutch destruction or removal. In 1976, Donham et al. reported that if eggs are removed from incubating female Mallards, plasma LH of the birds, which was very low, increased within one day to levels similar to those of laying females. This finding has since been confirmed in at least two species: the White-crowned Sparrow (Wingfield and Farner, 1979; changes in plasma testosterone, DHT, E_1, E_2, and corticosterone are also described in this study) and the Ring Dove (Silver et al., 1980). Several experiments also strongly suggest that performance of a given behavior can alter endocrine functions of the performer or its social partner. This idea had already been raised by studies on the Ring Doves performed by Lehrman and co-workers.

Lehrman et al. (1961) originally demonstrated that the presence of a male in the cage of a female Ring Dove causes a gradual increase in gonadotropin secretion of the female, as reflected by oviducal growth and increased frequency of ovulation. Barfield (1971b) demonstrated later that this response is dependent on the time of exposure to the male and on the amount of courtship exhibited by the male. Friedman (1977) also showed how auditory and visual courtship stimuli interact to produce the reproductive development in the female.

It is only recently that the endocrine changes associated with courtship in doves have been accurately measured and that it could be proved that LH secretion actually increases in a female Ring Dove when she is paired with a male (Cheng and Follett, 1976; Cheng and Balthazart, 1982). It was also found that similar changes (increase in plasma LH and testosterone right after pairing) also occur in males (Feder et al., 1977; Silver et al., 1980; Cheng and Balthazart, 1982; O'Connell et al., 1981).

In mammals, there is now abundant literature showing that plasma levels of LH, testosterone, and other hormones in the male are drastically affected by copulation or simple exposure to females (e.g., Katongole et al., 1971; Fox et al., 1972; Purvis and Haynes, 1974; Illius et al., 1976; Gray et al., 1976; Moss et al., 1977; see also references in Harding and Follett, 1979). Few data of this type are available for birds. In a study on the Rook, Lincoln et al. (1980), however, showed that a male shot immediately after copulation had plasma levels of testosterone and LH much higher than the mean of the corresponding population. Pröve (1978) also showed that male Zebra Finches paired with a female have high plasma testosterone levels compared to males of the same age without contacts with conspecifics. Similarly, male White-crowned Sparrows caged with an estradiol-implanted female (which consequently gives numerous courtship postures) have higher plasma levels of LH and testosterone than males caged alone or caged with a cholesterol-implanted female (Wingfield and Farner, 1980. So far as we know, only one field study of this type has been reported. Harding and Follett (1979) showed that concentrations of hormones (testosterone, DHT, and LH) in the plasma of male Red-winged Blackbirds caught during aggressive encounters are different from those in males caught while they were foraging and had not engaged recently in aggressive interactions.

Numerous effects of behavioral interactions on reproductive physiology are known in birds (for reviews, see Rosenblatt, 1978; Sossinka et al., 1980). In few studies, however, have the actual plasma levels of hormones been measured. Another problem with most of these studies is that the specificity of the stimuli that cause the endocrine changes is generally poorly defined. Even if general controls (for example, showing that the rise in plasma androgen in male doves observed after pairing with a female is not induced by exposure to a new cage) are available, they usually do not suffice to identify the active stimuli. The behavior of the female (which type?), the performance of the male (what behavior patterns?), some interactions between birds, or stress factors could still be the cause of the androgen rise.

One interesting attempt toward a precise characterization of a behavioral stimulus acting on endocrine physiology has been reported. Cohen and Cheng (1979) demonstrated that female Ring Doves with bilateral hypoglossal nerve sections fail to show the follicular growth usually observed when they are exposed to males. Considering that these lesioned females only show a reduction in the vocalization and wing-flipping components of nest coos but no other change in courtship behavior and that the behavior of the stimulus males does not seem to be modified by the lesion of the females, these data suggest that the female's performance of nest coo may stimulate her own follicular growth, probably through an increased LH and/or FSH secretion. If this interpretation is validated by hormone assays (lacking in this experiment) and further behavioral experiments, it would represent a nearly unique case showing effects of performing one behavior on the physiology of the performer. In most studies performed so far, only the effect of behavior on the endocrine status of another individual has been described (Cheng, 1979).

V. Brain Mechanisms

A. THE NEED FOR STUDIES OF BRAIN MECHANISMS

Although much remains to be discovered, the results thus far of investigations of the interactions of hormones and behavior in birds clearly highlight one fact: the behavioral activity of the animals is not controlled only by changes in plasma levels of steroids. Several lines of evidence suggest that endocrine or neural responses in the target organs play a critical role. We shall now review some of these data.

1. Individual Differences

It is common knowledge that marked individual differences in behavior (e.g., sexual or aggressive) are generally observed within a group of animals.

These do not seem to be controlled by differences in the plasma levels of steroids. It has long been known that in male mammals the individual level of sexual activity before castration is correlated with that shown after the surgery when animals are submitted to a standard hormone therapy (e.g., in guinea pig; Grunt and Young, 1953). Similar data are available for birds.

Deviche (1979a) showed that the amount of sexual behavior and social display exhibited by a castrated male duck injected with TP is significantly correlated with its precastration activity. A similar result was reported by Arnold (1975b), who studied Zebra Finch behavior. Hutchison (1970a, 1971) further demonstrated that *qualitative* individual differences in male Ring Dove courtship are maintained following castration and androgen therapy. Males that before castration display only the aggressive components of courtship (chasing, bowing) in the absence of nest-oriented courtship displays, retain this particularity when treated with testosterone after castration. These data support the notion that individual differences in plasma steroids are not responsible for behavioral differences. When birds are submitted to a standard hormone therapy and have presumably similar plasma levels of steroids, the behavioral differences are maintained. This generally agrees with results of RIAs. In domestic ducks, we have shown repeatedly that there is no significant correlation between the individual levels of sociosexual behavior and the plasma concentrations of testosterone, LH, or FSH (Balthazart, 1977; Balthazart et al., 1977a; Balthazart and Hendrick, 1978b). In particular, no correlation could be found within a group of 30 ducks between the levels of testosterone in the plasma and the frequency of sexual behavior or social displays (Fig. 15).

Additional studies of the duck have revealed that this discrepancy is not due to factors such as:

1. Rapid variations of hormone plasma levels [We showed that the individual differences are stable throughout a day despite rapid variations within one animal (e.g., Balthazart et al., 1977b).
2. Rapid fluctuations in behavior [Several studies demonstrated that the differences observed are stable for at least 15 days. The Kendall coefficients of concordance between the behavior of 30 male ducks observed on four occasions were 0.69 and 0.80 for social display and sexual behavior, respectively ($p < .001$ in both cases).]

Furthermore, correlations between behavior and plasma hormones are not higher when animals are observed from dawn to dusk for several consecutive days, and hormone assays are performed for several samples (up to six) collected at different hours within a day. Analyzing single behavior patterns (e.g., copulations) instead of groups of activities (e.g., sexual behavior) does not improve the correlation coefficients (Fig. 16). We also showed that there is no correlation between the androgen plasma levels and any

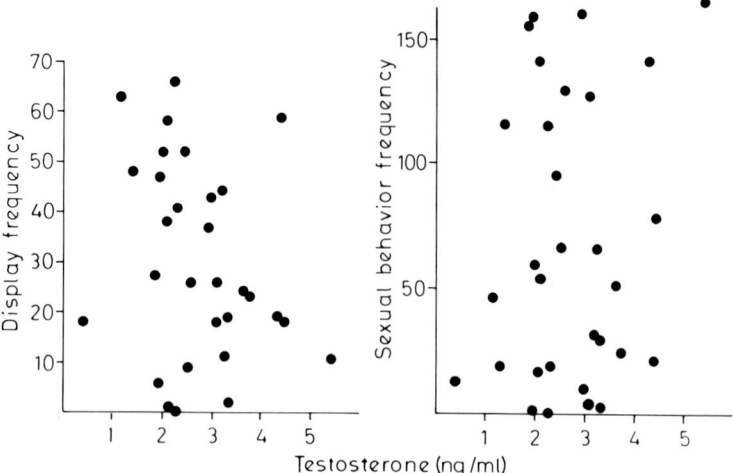

FIG. 15. Relationships between plasma testosterone and social displays (left panel) or sexual behavior (right panel) in a group of 30 male domestic ducks. Behavior of each bird was observed during 16 periods of 15 minutes in the presence of two females. A blood sample was then collected and assayed for androgens. There is no significant correlation (Spearman rank-correlation coefficient r: left, $-.28$; right, $.05$) between behavior and plasma testosterone. (Drawn from data in Balthazart et al., 1977a.)

measure of social or sexual behavior even in birds that are injected daily with a standard dose of TP (Balthazart, 1978a).

A similar conclusion has been reached for the Japanese Quail. In this species, there is no relationship between plasma testosterone and aggressive or courtship behavior (Balthazart et al., 1979). Ishii and Tsutsui (1981) confirmed this conclusion for aggressive behavior and also showed that replacement therapy with either constant or variable doses of TP in a group of castrated males did not change their relative aggressiveness. The generality of this conclusion is challenged, however, by the observation by Pröve (1978) that the amount of undirected songs and courtship songs given by an individual in a group of male Zebra Finches, is correlated with the amount of testosterone measured in a sample of its plasma collected immediately after the behavioral test. It could be, however, that this correlation reveals the effect on plasma androgen of performing a behavior rather than the reverse (see Section IV,E).

The absence of correlation between individual measures of plasma hormone and behavior has also been reported in several mammlian studies (e.g., rat; Damassa et al., 1977). Damassa et al. (1977) also showed that silastic implants of testosterone that produce plasma testosterone levels less than 10% of the normal values can maintain sexual behavior of castrated rats

4. HORMONAL CORRELATES OF BEHAVIOR 307

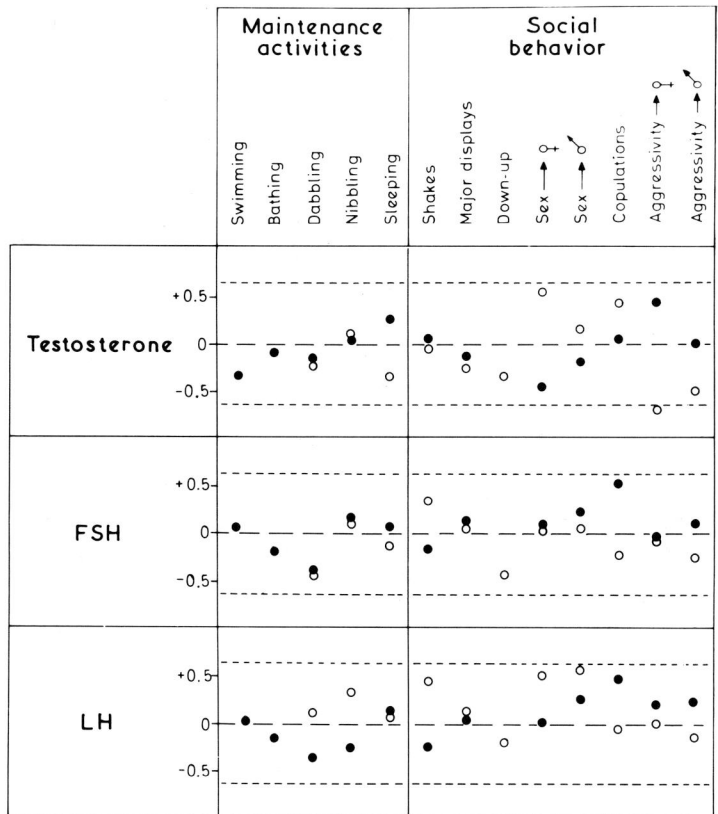

FIG. 16. Relationships between individual variations in several types of social and maintenance behavior and plasma levels of testoserone, FSH, or LH in groups of 10 male domestic ducks. Two different groups of birds were studied in December (○) or in March (●). Their behavior was recorded for 52 hours within 20 days (○) or 80 hours within 9 days (●) generally from dawn to dusk. Individual values for hormone assays are the means of six different samples for each bird collected on a single day, every second hour from 8 AM to 6 PM. Correlations between individual measures for hormones and behavior were computed by the Spearman rank correlation coefficient r and are reported in the figure (vertical axis, correlation level; horizontal axis, the different behavior patterns). Interrupted lines show the critical correlation coefficients for $2p = .05$ ($r = .632$). Points outside these lines represent significant correlations between behavior and hormones (one exception: aggress toward females is negatively correlated with testosterone in the December study). (Drawn from data in J. Balthazart and J. C. Hendrick, 1979b, and unpublished data.)

at a nearly normal level. They consequently suggested that the lack of correlation between behavior and plasma hormones results from the fact that the androgen requirement for the activation of behavior is less than the amount normally present in the blood. The control of behavior would then take place at other levels, for example, in the use of testosterone by the brain cells.

2. Strain Differences

Differences in behavioral performances between strains of birds from a same species have been frequently reported. In the chicken and Japanese Quail, such differences in sexual behavior have been artificially selected, resulting in genetic lines that differ significantly in their adult mating frequencies [e.g., chicken (Siegel, 1965, 1972), quail (Sefton and Siegel, 1975)]. Although these low- and high-mating lines of birds eventually differ in their plasma testosterone titers (e.g., in chicken; Benoff et al., 1978a; the high-mating line has more testosterone than the low mating line), it can be ascertained that this is not the only cause of the behavioral difference between the lines. Treatment even with high doses of testosterone of birds in low- and high-mating lines always fail to abolish the behavioral differences between the lines [chicken (McCollom et al., 1971), quail (Cunningham et al., 1977)]. Even the precocious sexual behavior induced by hand-thrust tests (Andrew, 1975a) in chicks during their first weeks of life is differentially activated by testosterone in the different lines of chickens (Benoff, 1979). In addition to its effects on the gonadal production of steroids, artificial selection for mating ability has thus altered the neural mechanisms mediating sexual behavior.

3. Changes in Sensitivity to Steroids

Although it has been demonstrated in many studies (see Section IV,C) that changes of behavior in time correspond fairly well with variations in the plasma steroids that are supposed to induce the behavior, discrepancies have been identified. These are frequently interpreted as resulting from changes in responsiveness of the brain to steroids. This subject has been reviewed extensively by J. B. Hutchison (1975, 1976, 1978) and by Hinde and Steel (1978); only a few examples will be described here.

There is in doves a decreased sensitivity to androgens following a prolonged androgen deficit. To demonstrate this, Hutchison (1969, 1974a) castrated male Ring Doves that had been selected as being behaviorally homogenous. Then these doves received intrahypothalamic TP implants 15, 30, or 90 days after castration. The implants, which were all adequately positioned, restored effectively behavior in the 15-day castrates but were somewhat less effective and completely ineffective in the 30- and 90-day castrates, respectively (Fig. 17).

This decrease in sensitivity to androgens can, however, be reversed at least partially by systemic or intrahypothalamic treatment with testosterone

4. HORMONAL CORRELATES OF BEHAVIOR

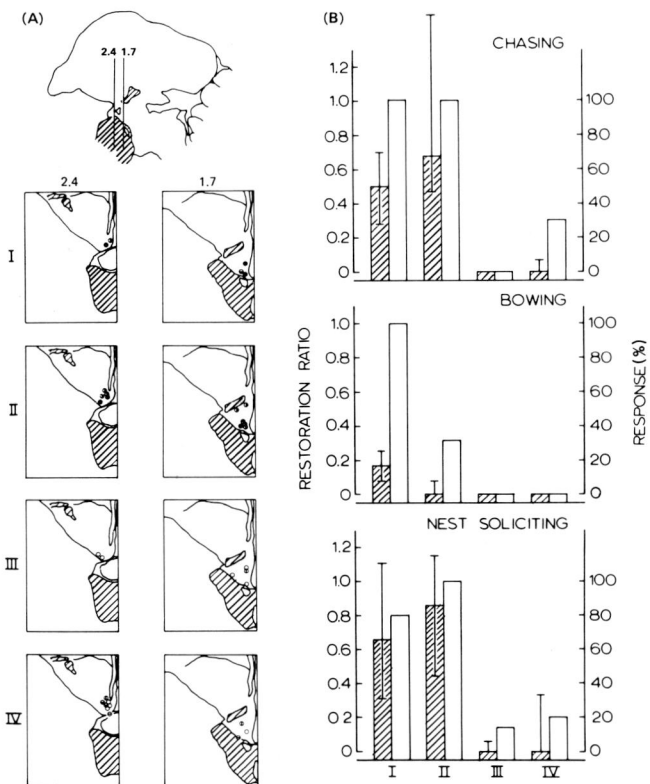

FIG. 17. Behavioral effects of intrahypothalamic implants of testosterone propionate in castrated male Ring Doves maintained on a 8.5L:15.5D photoperiod and implanted 15 (group I), 30 (group II), or 90 (group III) days after castration and in males maintained on a 13L:11D photoperiod implanted 90 days after castration (group IV). (A) Position of the implants and the responses they activated. ●, chasing, bowing, and nest soliciting; ◐, chasing and nest soliciting; ◑, chasing and bowing; ①, chasing; ⊖, nest soliciting; ○, no response. (B) Comparison of precastration to postimplantation courtship expressed by the ratio postimplantation peak duration: precastration peak duration (crosshatched bars; medians and ranges are shown). Open bars represent the percentage of birds that displayed at least 5 seconds of a courtship pattern. (From Hutchison, 1976.)

(J. B. Hutchison, 1976, 1978; see, however, in these papers the methodological problems linked with such experiments). Diminished sensitivity to androgens is also influenced by environmental conditions. Long-term castrates (90 days) that have been exposed to a long photoperiod (13L:11D) retain a slightly better androgen sensitivity than birds kept in short days (8.5L:15.5D) (Hutchison, 1974b, 1976; see Fig. 17).

Effects of the photoperiod on the brain is sensitivity to androgens have been identified in several other studies. McDonald and Liley (1978) studied the effects of photoperiod on reproductive behavior in androgen-injected castrated male Ring Doves. They showed that males exposed to long days exhibit consistently higher levels of nest building than short-day males. This different responsiveness to androgens does not seem to be related to photoperiod-induced differences in luteinizing hormone releasing factor (LHRH) (nor, probably, to differences in gonadotropin secretion) because injections of large doses of LHRH do not increase the amount of nest building induced by the androgens (McDonald, 1979). Also, in Canaries and Budgerigars, long days enhance the effectiveness of estradiol in inducing nest-building behavior (Hinde et al., 1974; Gosney and Hinde, 1975), and gonadotropins are probably not directly involved in this effect (Hinde and Steel, 1975, 1978; see also Section II,B,3).

Other environmental factors such as exposure to vocalizations of conspecifics are known to affect steroid-induced behavior [in Ring Doves (J. B. Hutchison, 1978), in Canaries (Hinde and Steel, 1976; see also Section II,B,3), and in Budgerigars (Steel et al., 1977)]. It has already been pointed out by J. B. Hutchison (1978) that the aforementioned changes in responsiveness to steroids do not need to be controlled by biochemical mechanisms within the brain. Nonhormonal effects such as visual isolation or "lack of use" of the courtship displays could, for example, explain the decreased responsiveness to androgens in long-term castrated Ring Doves. It is, however, most likely that the changes in sensitivity to steroids relate at least in part to changes in the cellular action of testosterone. This leads naturally to the research strategy exposed in the last section of this chapter.

4. Sex Differences and Sexual Differentiation

In some species of birds (e.g., many ducks, galliforms, and passerines) there is marked dimorphism in secondary sexual characteristics and in various aspects of behavior. In the remaining species, which are more numerous, sexual dimorphism is less conspicuous and is often limited to behavior directly related to copulation. Sexually dimorphic structures and behavior patterns of birds are largely controlled by steroid hormones, as shown by all the data reviewed in this chapter. However, some of the behavioral differences between sexes in birds are not controlled only by their adult endocrine status and consequently cannot be abolished by gonadectomy followed by an adequate hormone therapy. For example, intact adult female Japanese Quail do not crow or copulate (Adkins and Adler, 1972). If they are ovariectomized or functionally ovariectomized by exposure to short days (Adkins, 1973), testosterone injections induce them to crow but not to copulate, even if large doses of hormones are used (Adkins and Adler, 1972). Similarly, testosterone injections facilitate copulatory movements in young male chicks but not in females (Andrew, 1975a).

Such experiments demonstrate that the behavioral dimorphism of adult birds does reflect not only the adult endocrine status (because the dimorphism is not abolished by treatment with adequate hormones) but also is based on more stable differences. Furthermore, the experiments of Andrew (1975a) confirm that these differences are already organized in very young animals. Wilson and Glick (1970) and Adkins (1975, 1976, 1979) have demonstrated that this differentiation is induced by steroids and occurs during the second third of the embryonic life (before Day 11 in the Japanese Quail, before Day 13 in the chicken). The potential to display male copulatory behavior in quail and chicken is lost (demasculinization) if the embryo is exposed at the adequate time to estrogenic hormones. As little as 1 µg of EB given to male quail embryos significantly reduces their masculine behavior (Adkins, 1979).

This process spontaneously occurs in female embryos under the influence of the estrogens secreted by the ovaries (Guichard et al., 1980; Woods and Brazzill, 1981) but can be blocked by treatment of female embryos with the antiestrogen CI-628 (Adkins, 1978). Testosterone can also induce this demasculinization process (quail; Adkins, 1975; chicken; Wilson and Glick, 1970), probably through its aromatization within the brain (Callard et al., 1978a,b; Adkins-Regan et al., 1982). The nonaromatizable androgen, dihydrotestosterone propionate, has no demasculinizing effect (Adkins, 1978) even when injected in large doses. Considering the available mammalian data, it would, however, be premature to consider that testosterone and 5α-reduced androgens play no role in sexual differentiation (for a review see MacLusky and Naftolin, 1981). Testosterone is much less active than estradiol in the differentiation process [minimum effective doses in quail: 1 µg of EB, 100 µg of TP (Adkins, 1979)], and this probably explains why males are not demasculinized by their own endocrine secretions (Whitsett et al., 1977).

The differentiation revealed by these injection experiments is likely to be a physiological rather than pharmacological one. It is induced by low doses of hormones (1 µg of EB), which compare reasonably with the estimates of estrogen production by embryonic gonads (about 1 ng/gonad per day produced in vitro by chicken ovaries; Guichard et al., 1977, 1980) and with concentrations of estradiol in the plasma (~1 ng/ml; Woods and Brazzill, 1981) and in the allantoic fluid (50 ng/ml; Ozon, 1965). It is also known that embryonic gonads secrete androgens and estrogens in a sexually dimorphic manner (Guichard et al., 1977, 1980), and these sex differences are found in the plasma concentration of the steroids (e.g., Woods et al., 1975; Tanabe et al., 1979; Woods and Brazzill, 1981). The endocrine signals that are supposed to induce the differentiation are thus actually present at the right time. The finding that antiestrogens prevent the demasculinization of females without affecting the copulatory behavior of males (Adkins, 1976) also supports the conclusion that the effects of injected estradiol on sex differentiation are physiological.

It is still doubtful whether the demasculinizing effects of estradiol on sexual behavior are strictly limited to a critical period of the embryonic life. R. E. Hutchison (1978) has demonstrated that if genetic female Japanese Quail are ovariectomized neonatally and treated with TP as adults, a significant proportion (57%) display male sexual behavior patterns. We achieved the same result by ovariectomizing young females at 1 week of age (Schumacher and Balthazart, 1983). It thus seems that the period of sensitivity during which estrogen suppresses male-type behavior is not confined to embryonic life. Exact delineation of this period remains to be achieved.

A surprising conclusion arising from the studies of avian differentiation mentioned so far is that birds (at least quail and chickens) and mammals seem to differentiate in completely opposite ways (for reviews see Adkins, 1978; MacLusky and Naftolin, 1981). In mammals, exposure to androgens during the pre- or neonatal period results in a permanent defeminization of the embryos, whereas the absence of androgens at this time results in the feminine pattern of responsiveness to hormones. Females appear thus as a "neutral" sex from the hormonal point of view. In the absence of hormones, a female-like behavioral phenotype develops. In physiological conditions, male embryos are supposed to be defeminized by the secretions of their testes (paradoxically, it is mainly through aromatization by the brain of testosterone into estradiol that the effect is produced; i.e., males are defeminized by a "female" hormone), whereas the females would be protected at least in some species (MacLusky and Naftolin, 1981) from the potential defeminizing effects of their endogenous estradiol by a circulating feto-neonatal estrogen-binding protein (FEBP). This binds estradiol (but not testosterone) and prevents its action at the intracellular level (review by Adkins, 1978; Toran-Allerand, 1978; McEwen, 1978; Harlan et al., 1979; Donovan, 1980; MacLusky and Naftolin, 1981).

In mammals, females appear to be more bisexual than males. If stimulated by adequate hormones (testosterone for male copulatory patterns, estradiol and progesterone for female receptivity), gonadectomized females readily perform both male and female types of behavior, whereas males show female receptivity only very exceptionally. In contrast, male quail and chickens are more bisexual than females (see Section III,C). The male seems to be the neutral sex and estradiol demasculinizes the female embryos (Fig. 18).

This difference may result from the fact that the female is the homogametic sex in mammals whereas in birds it is the male (Adkins and Adler, 1972;

4. HORMONAL CORRELATES OF BEHAVIOR

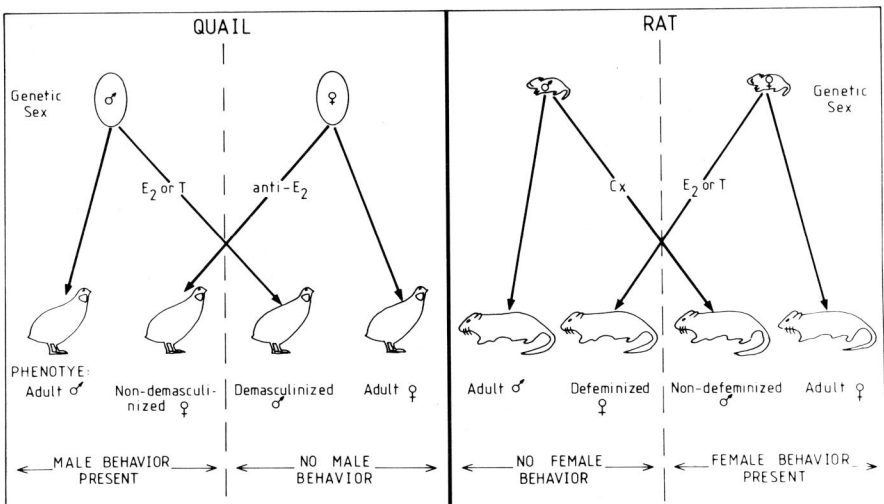

FIG. 18. Schematical presentation of the major processes that occur during sexual differentiation of the brain in Japanese Quail and rats. The presence of estradiol (E_2) during embryogenesis in quail leads to a loss of male copulatory behavior (demasculinization). Estradiol, if present during the neonatal life of rats, produces animals that do not easily exhibit lordosis in response to estrogens and progesterone when adult (defeminization). These effects can be blocked by neonatal castration (Cx) rats or injection of an antiestrogen (anti-E_2) in quail eggs. Testosterone (T) mimics the effects of E_2, probably through its aromatization in the brain.

Adkins, 1978). We should not, however, conclude prematurely that this represents consistent difference between birds and mammals. Gurney and Konishi (1980) demonstrated that estradiol can masculinize the song system in female Zebra Finches. In this species, the male sings under the influence of testosterone whereas the female does not, even if exposed to androgens. This dimorphism corresponds to morphological sex differences in the neural substrates that mediate the behavior [e.g., larger HVc and nucleus robustus archistriatalis (RA) in males]. If females are treated neonatally with estradiol, then either testosterone or 5α-DHT may induce them to sing when an adult. These hormonal treatments also suppress the neuroanatomical sex dimorphism. In the Zebra Finch, the neonatal differentiation of adult behavior thus operates according to a pattern similar although *not identical* to the mammalian one. Females are the "neutral" sex, and neonatal E_2 *masculinizes* their song system (in mammals neonatal E_2 mainly *defeminizes* sexual behavior).

Whether this represents a species difference or is related to the kind of behavior that was studied (song in Zebra Finch, mating in quail and chicken) remains unknown. It must be stressed in addition that all models developed in this section on differentiation are only schematical representations of the mainstream of the data and do not take into account a number of experimental facts. The concept of a neutral sex is probably deceptive, as suggested by the finding of Toran-Allerand (1976, 1978) that in mammals both male and female patterns of neural differentiation require active induction by steroids. It is not only the loss of bisexuality that must be hormone dependent (demasculinization in birds, defeminization in mammals), but also the acquisition of a given behavioral repertoire (e.g., masculinization), which probably has some hormonal basis largely unknown so far. In the male rat, for example, it seems that the presence of hormones in the neonatal period is necessary for the development of the full ejaculatory pattern in the adult, but whether this effect is due to estradiol or another steroid is still a matter of dispute (Södersten and Hansen, 1978; Södersten, 1979).

Besides their intrinsic interest, studies of sexual differentiation demand research on the brain mechanisms involved in hormonal activation of behavior. Until we know in more detail how the behavior is activated by steroids in the adult brain, it will obviously be difficult to search for the ontogeny and sexual differentiation of this mechanism. The differentiating effects of hormones could be morphological (Toran-Allerand, 1976, 1978) but also could concern the brain's metabolism of steroids, the concentration of steroid receptors, or any postranscriptional event in the cellular action of the hormones (MacLusky and Naftolin, 1981). These possibilities have begun to be analyzed in mammals but few data are available for birds.

B. Brain Metabolism of Steroids

The absence of relationships between the behavior of an animal and hormone titers in the plasma, as just described (Section V,A), may in some instances reflect only failures of methodology. Rapid changes, either spontaneous (see Section IV,C,3) or stress induced (see Section IV,E), may obscure some correlations. Furthermore, the concentrations of steroids measured in blood samples after extraction by an organic solvent (e.g., diethyl ether or dichloromethane) do not take into account the fact that a part of the steroid is bound to plasma proteins and is consequently not readily available for intracellular actions (e.g., Martin, 1980).

Taken together, the studies reported in Section V,A nevertheless reveal the need for studies of the encephalic mechanisms of hormone action. This is furthermore the logical direction of the research as soon as it is established that behavior is activated in many instances by central effects of steroids (see Section III,D).

In peripheral target organs, such as the chick oviduct, the interaction of a steroid with responsive cells follows the general process depicted in Fig. 19 (for reviews see Jensen and Desombre, 1973; King, 1976; Schrader and O'Malley, 1980; Katzenellenbogen, 1980). The systemic steroid is eventually bound to a plasma protein. The free steroid may enter the cells, probably by passive diffusion (for counterarguments see Giorgi, 1980). It is then eventually metabolized into a number of compounds that may or may not have specific cytosolic receptors. The hormone–receptor complex is then translocated to the nucleus where it interacts with DNA, induces the transcription of new RNA (or at least increases its synthesis), which will then be translated into new proteins that change cell functions. It is generally believed that a mechanism such as this also explains the activation of behavior by steroids in the central nervous system. Good evidence for this has been obtained in a number of mammalian studies (e.g., Blaustein and Feder, 1979a,b, 1980;

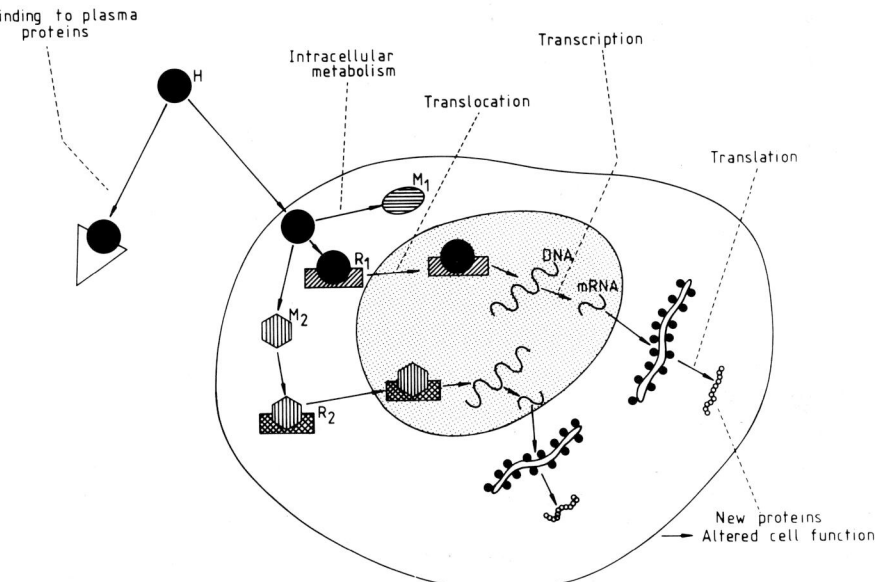

FIG. 19. Cellular mechanisms of steroid hormone action. The circulating hormone (H) can either bind to plasma proteins or enter a cell. There it is eventually transformed into several metabolites (M_1, M_2) that bind (M_2) or do not bind (M_1) to specific cytoplasmic receptors (R_2). The hormone itself can bind to cytoplasmic receptor (R_1). The hormone–receptor complexes are then translocated to the cell nucleus and attach themselves to acceptor sites on the chromosome containing the genes (DNA). This elicits synthesis of new messenger RNA (transcription) which leave the nucleus and migrate to the endoplasmic reticulum, where new proteins are synthesized (translation). These new proteins (e.g., enzymes) may then change cellular functioning.

Moguilewsky and Raynaud, 1977, 1979; Parsons *et al.*, 1980). There may also be other modes of action, although not yet positively identified (see Giorgi, 1980; Teyler *et al.*, 1980).

In birds, apparently, no data are available that relate behavior to steroid-binding plasma proteins, to membrane effects of steroids, or to mechanisms that control the permeability of cell membranes to steroids. The metabolism of steroid hormones and the concentration of steroid receptors in the brain of birds have, however, been described and measured. In a number of cases meaningful relationships with behavior can be established.

1. Testosterone

a. Metabolic Pathways. Many studies have been devoted to the study of the metabolism of testosterone by the brain in birds. With few exceptions (e.g., Horst and Paulke, 1977; Steimer and Hutchison, 1981) all these experiments were performed *in vitro*. They used, however, a wide variety of species so that we can be relatively confident about the general character of the metabolic pathways (Fig. 20).

The transformation of androgens (testosterone and androstenedione) into estrogens by aromatization has been detected in the brain of chickens (Callard *et al.*, 1978a) and Ring Doves (Steimer and Hutchison, 1980) but will very probably be found in every species because it is widely distributed among vertebrates (Callard *et al.*, 1978a,b). In mammals, aromatase has been detected in all species studied (e.g., Naftolin, *et al.*, 1975) but the enzymatic activity was in each case very low (conversion of androstenedione into estrone generally around or below 1% under the conditions used by Naftolin *et al.*, 1975).

In contrast, Steimer and Hutchison (1980) measured an aromatization of testosterone into estradiol by the hypothalamic area of the Ring Dove of which the magnitude was much above levels typical of mammals. Although it is not easy to compare quantitatively these results with the mammalian data because of different incubation conditions, it can be said that in the studies by Steimer and Hutchison (1980, 1981) using testosterone as a substrate, similar amounts of estradiol and of 5α-reduced androgens were formed. This ratio is never found in mammals, in which 5α-reduction is much more active than aromatization (Selmanoff *et al.*, 1977). Whether this high aromatization rate is a widespread characteristic of birds is not yet possible to evaluate. As in mammals, it seems that the greatest aromatizing activity in birds is located in the anterior hypothalamus–preoptic area complex (Callard *et al.*, 1978a; Steimer and Hutchison, 1980) and in a forebrain sample including the nucleus taeniae (Callard *et al.*, 1978a).

The 5α-reduction *in vitro* of Δ^4-C_{19}-androgens (testosterone, androstenedione) into 5α-DHT (5α-androstan-17β-ol-3-one, or 5α-DHT) and 5α-an-

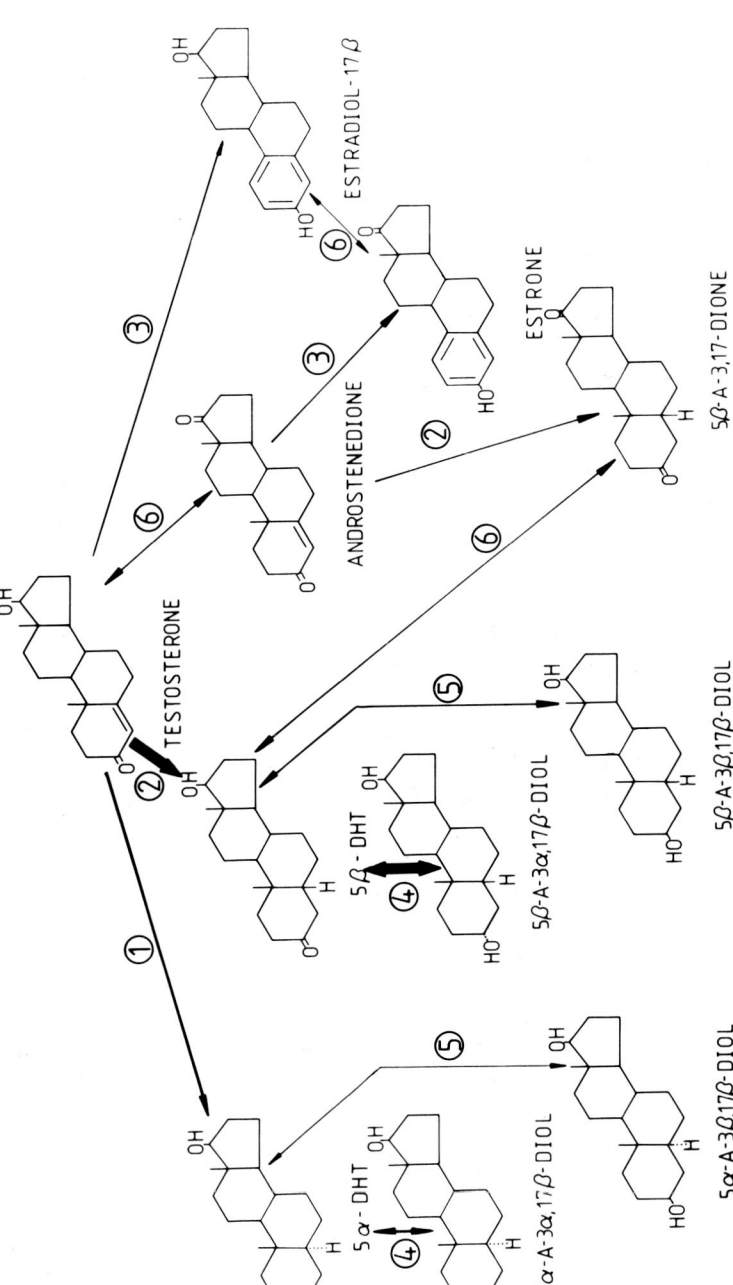

FIG. 20. Metabolic pathways followed by testosterone in its target organs as revealed by *in vitro* studies. Widths of arrows between molecules are roughly proportional to the corresponding enzymatic activities. Enzymatic systems involved are noted by encircled numbers using the following code: 1, 5α-reductase; 2, 5β-reductase; 3, aromatase; 4, 3α-hydroxysteroid dehydrogenase; 5, 3β-hydroxysteroid dehydrogenase; 6, 17β-hydroxysteroid dehydrogenase. 5α-DHT, 5α-dihydrotestosterone; 5β-DHT, 5β-dihydrotestosterone; A, androstane (as diol or dione).

drostane-3α,17β-diol (5α,3α-diol) was first demonstrated in the anterior pituitary and the hypothalamus by Jaffe (1969) and Kniewald *et al.* (1971). This process also occurs in the brain and/or pituitary gland of birds and has been identified in a wide range of species such as the chicken (Nakamura and Tanabe, 1974), the European Starling (Massa *et al.*, 1977), the Japanese Quail (Davies *et al.*, 1980), the Ring Dove (Steimer and Hutchison, 1981), and the domestic duck (Willems, 1978). The 5α-DHT formed by the irreversible reduction of testosterone can further be transformed reversibly under the influence of a 3α-hydroxysteroid dehydrogenase (3α-HSDH) into the 5α,3α-diol. As in mammals (e.g., Jouan and Samperez, 1980), a 3β-hydroxysteroid dehydrogenase (3β-HSDH) is probably also present in the brain of birds but has not yet been demonstrated mainly because the substrate for this enzyme (5α-DHT) is formed too slowly, so that the formation of other 5α-reduced androgens might be too slow to allow their detection by methods commonly used (Massa *et al.*, 1977; 5α,3α-diol is itself usually in concentrations too low to be detected and positively identified).

In contrast to mammals, the 5α-reduction in birds is not the main reductive pathway followed by testosterone in the brain. A very active 5β-reductase activity was first demonstrated by Nakamura and Tanabe (1974) in the brain of the chicken and later confirmed in all other species studied (see references above). In the bird brain, the 5β-reductase activity is usually 10–20 times more active than the 5α-reductase. It leads to compounds [5β-DHT or 5β-androstan-17β-ol-3-one; 5β-androstane-3α,17β-diol (5β,3α-diol); 5β-androstane-3β,17β-diol (5β,3β-diol) and 5β-androstane-3,17-dione (5β-dione)] that have a physical structure extremely different from that of 5α-reduced androgens (rings A/B are trans in the α form and cis in the β form) (Giorgi, 1980) and consequently their biological properties are also very different. The subcellular distributions of these enzymatic activities in the avian brain are mainly unknown, although Steimer and Hutchison (1981) mentioned that 5β-reductase is essentially nonnuclear whereas 5α-reduction is equally distributed between nuclear and nonnuclear fractions.

In the chicken comb, 5β-reductase activity is found mainly in the cytosol fraction, whereas the 5α-reductase is essentially microsomial. The 3α- and 3β-HSDH activities are located in the cytosol and the mitochondrial and microsomial fractions depending upon the substrates (5α- or 5β-DHT) that are used (Mori *et al.*, 1974). These localizations usually correspond to those reported for the mammalian brain (e.g., Jouan and Samperez, 1980) when comparisons are possible [there is no 5β-reduction of testosterone in the brain of mammals except in the golden hamster, *Mesocricetus auratus* (Callard *et al.*, 1978b)].

The precursor–product relationships presented in Fig. 20 remain in some of their parts hypothetical in birds. There is, for example, no definitive proof of the irreversibility of 5α- and 5β-reduction and of aromatization. Most pathways are, however, consistent with mammalian data and with results of time-course experiments described by Steimer and Hutchison (1981). These also presented the proof that 5β-reduced androgens cannot be formed by isomerization of the corresponding 5α-reduced steroids. 5β-Reductase in the dove brain is clearly $NADPH_2$ dependent (Steimer and Hutchison, 1981), but the cofactor requirements of the steroid-metabolizing enzymes in the bird brain have not been systematically investigated (for data concerning the rooster comb see Mori et al., 1974). Few data are available concerning the specificity of these enzymes. The relative affinities of 3α- and 3β-HSDH for 5α- and 5β-DHT are, for example, unknown and it is not known whether one or several forms of each enzyme are present as in the rat kidney (e.g., Ghraf et al., 1979). That at least different enzymes are acting on 5α- and 5β-DHT is, however, suggested by the different subcellular localization of hydrogen transfer on 5α- and on 5β-DHT in the chicken comb (Mori et al., 1974).

b. Changes in Enzymatic Activities and Their Control. Research on testosterone metabolism in the avian brain has just started and, generally, has not yet gone beyond the stage of descriptive studies. In a number of cases, however, major changes in enzymatic activities according to the physiological stage of animals have been detected and show how the intracellular metabolism of testosterone could be important in the fine adjustment of behavior.

In the European Starling, the *in vitro* conversion of testosterone into 5β-DHT and 5β,3α-diol by the pituitary gland and the hyperstriatum significantly increases when the birds become photorefractory in May (Bottoni and Massa, 1981). We have obtained a similar result in the domestic duck, which was shown to have a much higher ratio of 5β- versus 5α-reductase activity in the pituitary after the postnuptial molt in July than at the peak of the breeding season in March (J. Willems and J. Balthazart, unpublished). It is not known at present how these changes are controlled.

Slightly more information is available for the Japanese Quail, in which the effects on testosterone metabolism of photoperiod, castration, and hormone replacement by various steroids have been studied. In these experiments, tissues [hyperstriatum, (HYS), hypothalamus (HYP), pituitary gland (PIT), and pieces of the cloacal gland (CG), a typical androgen-dependent structure] were dissected just after killing the birds and were incubated *in vitro* with [^{14}C]testosterone. The metabolites produced were then extracted, chromatographed, and quantified by liquid scintillation. Figure 21 shows, for example, the accumulation of five major matabolites observed by this method in adult male Quail raised in long days. The tissue specificity of

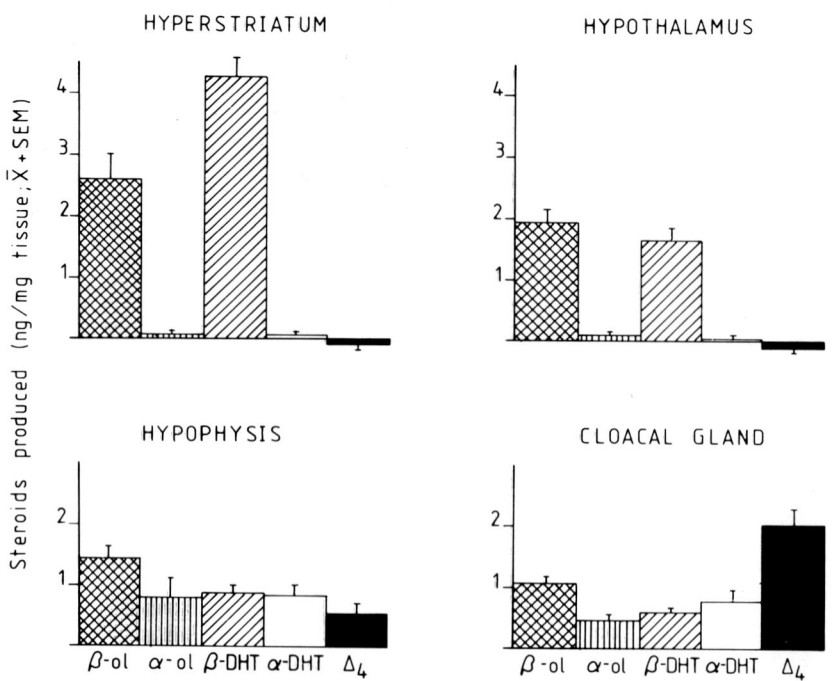

FIG. 21. *In vitro* metabolism of testosterone in four target organs of male Japanese Quail. Pieces of tissue from intact adult birds were incubated for 3 hours at 41°C with [^{14}C]testosterone (20,000 dpm/mg tissue). The metabolites produced were then extracted, chromatographed, and quantified by scintillation counting. The five compounds shown account for 50 to 75% of the metabolized radioactivity. β-ol, 5β-Androstane-3α,17β-diol; α-ol, 5α-androstane-3α,17β-diol; β-DHT, 5β-dihydrotestosterone; α-DHT, 5α-dihydrotestosterone; Δ_4, androstenedione. (Data from J. Balthazart and M. Schumacher, unpublished.)

metabolism is obvious here. For instance, there is the very high 5β-reduction in nervous tissues (HYP and even more in HYS) in comparison to the pituitary and cloacal gland, the high 5α-reductase activity in the pituitary and cloacal gland, and also the very high production of androstenedione in the cloacal gland.

This pattern of metabolism generally appears very constant across different studies (see Davies *et al.*, 1980; Massa *et al.*, 1980; Balthazart *et al.*, 1979, 1980b; Deviche *et al.*, 1982b), although moderate quantitative differences have been observed (e.g., more or less androstenedione produced according to the experiment). In two of the target organs for androgens (PIT and CG) transfer from short to long photoperiods markedly affects this pattern of metabolism. In both structures, it has, for example, been shown repeatedly that long photoperiods enhance the conversion of testosterone into 5α-reduced androgens [for PIT (Davies *et al.*, 1980; Deviche *et al.*, 1982b), for CG (Massa *et al.*, 1980; Balthazart *et al.*, 1979; Deviche *et al.*, 1982b)].

It is likely that changes in the amount of 5α-reduction by the cloacal gland are controlled by the circulating levels of testosterone. The increase in 5α-reduction observed after transfer in long days when plasma testosterone is known to increase (Follett, 1976; Follett and Maung, 1978) can also be obtained by treating castrated birds with testosterone (Balthazart *et al.*, 1979; J. Balthazart, L. Bottoni and R. Massa, unpublished results; Balthazart *et al.*, 1983). These changes are at best only indirectly related to mechanisms controlling behavior (see Section III,D). In this context, it would be of major importance to demonstrate similar changes in testosterone metabolism by the hypothalamus, which is the main site of action of androgen on reproductive behavior. Most studies carried out so far, however, agree that androgen metabolism by the hypothalamus does not seem to be altered (markedly) by the photoperiod, castration, or treatments with steroids such as testosterone, 5α-DHT, or estradiol (Davies *et al.*, 1980; Balthazart *et al.*, 1979, 1980b; Deviche *et al.*, 1982b; Balthazart *et al.*, 1983). An exception to this rule is that on one occasion, an increased production of 5β,3α-diol was observed in male quail after transfer to long days (Davies *et al.*, 1980), and the reversed effect (decreased accumulation of 5β,3α-diol) was observed in birds treated systematically with testosterone (Davies *et al.*, 1980; Balthazart *et al.*, 1983).

It could be that important changes in testosterone metabolism take place in some specific hypothalamic nuclei under the experimental conditions considered above but that these are not detected because the hypothalamus, unlike the cloacal gland or hypophysis, is a very heterogeneous sample and changes in a specific area are diluted in nonresponsive tissues. The finding that castration of doves increases, whereas TP injections decrease, 5β-reduction in the preoptic area but not in the neighboring area basalis (Steimer and Hutchison, 1981) supports this interpretation. More studies of testosterone metabolism in limited parts of the hypothalamus obtained by microdissection are urgently needed.

It has also been demonstrated that the conversion of testosterone to 5β-reduced metabolites in neuroendocrine tissues (both hypothalamus and hyperstriatum) dramatically decreases during the ontogeny in the cockerel (from the first to the twentieth week of life; Massa and Sharp, 1981) and in male and female quail (from the first to the fifth week of life; J. Balthazart, unpublished results). Castration of the cockerels at either 4 or 12 weeks of age did not markedly alter this process, and its control remains poorly understood.

In conclusion, the intracellular transformation of testosterone into other androgens is clearly dependent on the physiological stage of birds, but the mechanisms by which these controls are exerted remain generally unknown. A control by systemic testosterone appears likely in some cases. For example, the cloacal gland 5α-reductase increases after transfer to long days (Massa et al., 1980; Balthazart et al., 1979), decreases after castration (Massa et al., 1980; Balthazart et al., 1983); and is restored by testosterone therapy (J. Balthazart, L. Bottoni and R. Massa, unpublished results; Balthazart et al., 1983). In other cases, the photoperiod acts without involving plasma testosterone, but rather by some unidentified mechanisms—perhaps a direct action of the gonadotropins [e.g., the pituitary 5α-reductase, which is increased by long photoperiods (Davies et al., 1980; Deviche et al., 1981b) and further stimulated by castration but decreased by testosterone treatment (Davies et al., 1980)]. Similar controls of steroid metabolizing enzymes by gonadotropins have already been demonstrated in the gonads (e.g., Verhoeven, 1980).

Few quantitative data are available about the aromatization of androgens in birds. It is not known whether, as in mammals, the reaction is sexually dimorphic, with males having a much greater hypothalamic aromatase activity than females (e.g., Naftolin et al., 1975).

c. Biological Properties of Metabolites. In order to evaluate the importance of the testosterone metabolism in its target organs, it is obviously necessary to know the biological properties of the metabolites that are produced. The steroids derived from testosterone control three groups of variables related to the reproduction: the development of secondary sexual characteristics (e.g., rooster comb), the plasma levels of LH and FSH (by a feedback action on the hypothalamus and the pituitary gland), and sexual behavior.

Secondary Sexual Characteristics. Two target organs for androgens have been used as models to study the action of testosterone metabolites. They are the cloacal gland of Japanese Quail and the comb of chickens. Results of the two groups of studies generally coincide. Androstenedione promotes the growth of both structures [quail (Adkins, 1977; Massa et al., 1980), chick (Balthazart and Hirschberg, 1979a)], but it is not known whether it is by a direct action or through its conversion into testosterone (see Fig. 20).

Generally, 5α-reduced androgens (5α-DHT; 5α,3α-diol; 5α-androstane-3,17-dione) strongly stimulate cloacal-gland growth (Adkins, 1977; Massa et al., 1980) and growth of the chicken comb (Young and Rogers, 1978; Balthazart and Hirschberg, 1979a). It must be observed that 5α-androstane-3β,17β-diol produced no change in cloacal-gland size when given in a 10-mm silastic implant (Massa et al., 1980).

5β-Reduced C_{19}-steroids are usually reported as having no effect on the growth of secondary sexual characteristics (Adkins, 1977; Massa et al., 1980). In three of five experiments on young chicks, we found, however, that 5β-DHT slightly although significantly increased comb size (Balthazart and Hirschberg, 1979a; Balthazart et al., 1981b). A similar result has also been reported by Deviche et al. (1982a) but this time for the quail cloacal gland. Its size is consistently increased by about 25% following the daily injection of 2 mg of 5β-DHT. In the same study, the weight of the sternotracheal muscles (another target for androgens; e.g., Lieberburg and Nottebohm, 1979) was also regularly increased to 110–120% of control levels by the 5β-DHT injections. Whether this weak androgenic effect is explained by a cross-reaction of 5β-DHT for binding with the testosterone receptor (Lieberburg and Nottebohm, 1979) or by some other process (e.g., competition between the injected 5β-DHT and low levels of endogenous androgens for the enzymes that catabolize testosterone, resulting in the buildup of significant concentrations of testosterone in some target organs) remains to be determined.

Estradiol has no effect either on the cloacal gland (Adkins and Adler, 1972; Adkins and Pniewski, 1978; Balthazart et al., 1980b) or on chick comb growth (Young and Rogers, 1978; Balthazart and Hirschberg, 1979a). Actually, a decrease in comb size was observed in these studies, probably resulting from the feedback effects of estradiol on plasma gonadotropins and endogenous androgen.

Plasma Gonadotropins. The few detailed studies available thus far suggest that the controls of plasma gonadotropins are similar to those of secondary sexual characters, with the exception that estradiol is here very active. In the Japanese Quail, α-reduced androgens, androstenedione, and estradiol strongly depress plasma LH whereas 5β-reduced androgens have no effect (Davies et al., 1980). It seems that FSH may be controlled in a slightly different way. In the single experiment performed so far, testosterone and estradiol decreased the plasma level of FSH but 5α-DHT was inactive in this respect (Davies et al., 1980). For LH, similar trends were observed in young chicks (testosterone; 5α-DHT; 5α,3α-diol and androstenedione decrease plasma LH, whereas 5β-DHT and 5α,3α-diol have no effect) but these generally failed to be significant probably because too small doses of hormones were administered (Balthazart and Hirschberg, 1979a). In a study of castrated Ring Doves, Saad et al. (personal communication, 1981) found that testosterone and androstenedione markedly depress plasma LH, 5α-DHT only slightly decreased it (not significant), whereas 5β-DHT had absolutely no effect.

Reproductive Behavior. It is in the Japanese Quail that the most advanced knowledge of the behavioral effects of testosterone metabolites has been gained. The work of Adkins and collaborators has shown that all effects of testosterone on reproductive behavior can be mimicked by some of its metabolites. It has been known for long that estradiol activates copulation in males of this species (Adkins and Adler, 1972) just as testosterone does. Estradiol is even effective in much smaller doses than testosterone (50 μg of EB have the same effect as 5 mg of TP). Recent data confirm that the aromatization of testosterone into estradiol is a critical step in the androgen activation of copulation. The nonaromatizable androgens (e.g., 5α-DHT) indeed do not activate copulation whereas the aromatizable androstenedione does (Adkins, 1977; Adkins et al. 1980). Furthermore, the antiestrogen nitromifene citrate (CI-628) blocks testosterone-induced copulation in quail (Adkins and Nock, 1976b). Finally, Adkins et al. (1980) showed that the aromatase inhibitor ATD (1,4,6-androstatrien-3,17-dione) blocks testosterone-induced copulation but has no effect on estradiol-induced copulation. Taken together these data clearly show that in order to activate copulation in quail a part of the testosterone reaching the brain has to be aromatized into estradiol.

Other sexual behaviors such as crowing and strutting seem, on the contrary, to depend on 5α-reduced androgens. They are indeed very efficiently elicited by 5α-DHT, which in some experiments even proved more active than testosterone (Adkins and Pniewski, 1978; Adkins et al., 1980). These data suggest that crowing and strutting do not depend on aromatization and are instead controlled by steroids produced by the 5α-reduction pathway. It is impossible, however, to evaluate whether 5α-reduction is *necessary* for the activation of these patterns of if testosterone itself is the active hormone. No experiment with 5α-reductase inhibitors has been performed so far and no good inhibitor seems to be available. Most show little activity, are not very specific, or are toxic (for studies using such inhibitors but performed in mammals see Dupuy et al., 1977, 1978; Kao and Weisz, 1979; Blohm et al., 1980).

5β-Dihydrotestosterone, which is the metabolite of testosterone produced in the largest amounts by the quail brain (Balthazart et al., 1979; Davies et al., 1980), seems devoid of any behavioral activity in this species. When injected at a daily dose of 1 mg for 15 consecutive days, it did not activate copulation or strutting in castrated males (Adkins, 1977), whereas testosterone was active in these conditions.

The results of experiments with Ring Doves generally fit in well with the quail data, although they cannot be directly compared because other behavior patterns not homologous with those observed in quail were studied. Copulation, for example has generally not been studied in doves because of technical problems (a male dove will not mount a sexually receptive female unless the latter is familiar; Cheng and Lehrman, 1975). In castrated Ring

Doves, nest soliciting and nest cooing are elicted by estradiol as well as by testosterone (Cheng and Lehrman, 1975; Adkins, 1981), which suggests that aromatization could be critical in the control of these activities. The nonaromatizable androgen 5α-DHT (or its esters), although much less potent than testosterone, is weakly effective in restoring this courtship pattern (J. B. Hutchison, 1976, 1978; Adkins-Regan, 1981), but no study with aromatase inhibitors or with antiestrogens is available and no firm conclusion can be drawn.

The behavioral actions of 5α-DHT are difficult to evaluate in the Ring Dove. Even for behavior patterns such as chasing (hop charging) and bow cooing that are stimulated by testosterone but not estradiol, reports on 5α-DHT action are contradictory and responses range from very strong activation to no effect at all (J. B. Hutchison, 1976, 1978; Adkins-Regan, 1981; Saad *et al.*, personal communication, 1981). The discrepancies could result from the use of free 5α-DHT in some studies and of some of its esters (acetate, propionate) in others, as suggested by the experiments of Adkins-Regan (1981). A similar difference between the behavioral actions of 5α-DHT and its propionate derivative seems to be the rule in mammals (Yahr, 1979). 5β-DHT, which is produced in very large amounts during *in vitro* incubations of dove brain homogenates with radioactive testosterone, seems to be behaviorally inactive when given systematically (Saad *et al.*, personal communication, 1981) or even when implanted into the anterior hypothalamus–preoptic area (Steimer and Hutchison, 1981).

The quail and Ring Dove data thus suggest that 5β-reduction of testosterone in the bird brain represents a pathway of testosterone inactivation that depresses the effectiveness of the hormone. This interpretation is also supported by the following facts reported by Steimer and Hutchison (1981):

1. 5β-Reduced androgens are not detected in hypothalamic cell nuclei after the intramuscular injection of tritiated testosterone, which suggests that they are rapidly eliminated from the brain cells.
2. 5β-Diols are a major product of testosterone metabolism in the liver, which is known as a site of androgen inactivation.
3. The brain areas that are typical target structures for androgens (preoptic area, anterior ventromedial and posterior hypothalamus) show less 5β-reduction than adjacent regions not directly involved in androgen action.

This interpretation is, however, challenged by our finding that, in young domestic chicks, 5β-DHT strongly activates the copulatory responses elicited by hand-thrust tests (Balthazart and Hirschberg, 1979a; Balthazart *et al.*, 1981b). During five independent experiments, it was demonstrated that 5β-DHT or 5β,3α-diol are very effective in eliciting these copulatory responses. It had been speculated that 5β-DHT is not androgenic by itself but that it is competing with the low levels of endogenous testosterone for the catabolic enzymes, resulting in the build-up of significant concentrations of testosterone in the brain. The fact that 5β-DHT remains effective in castrated chicks argues against this possibility (Balthazart et al., 1981c) although some interaction with adrenal steroids (Tanabe et al., 1979) has not been ruled out. Some possible reasons for the discrepancies between effects of 5β-DHT in chicks on one hand and in quail and doves on another hand (age of the birds, nature of the behavioral test, interaction with other steroids) are discussed by Balthazart et al. (1981c).

The effectiveness of 5β-DHT in inducing copulatory responses in chicks appears even more surprising if we consider that the same behavior can also be activated, provided sufficient doses are injected, by testosterone as well as estradiol and 5α-DHT (Young and Rogers, 1978). To the best of our knowledge, none of these steroids can be transformed one into the other (Fig. 20), and there is no known receptor that would bind them all.

Studies with specific inhibitors of 5β-reductase (not known at present) and of the specificity of androgen receptors would be needed to resolve the aforementioned problems. If the 5β-reduction of testosterone were finally shown to be involved in behavior control, it would then remain to demonstrate whether 5β-reduced androgens have specific intracellular receptors in the brain as in the chick's liver (Lane et al., 1975; Adigweme and Abul-Hajj, 1981) or if they act at the membrane level like 5β-reduced progestins in the rat uterus (Kubli-Garfias et al., 1979). Ontogenetic changes in receptors for 5β-androgens could also be expected because these compounds exert a strong control on porphyrin synthesis by induction of the enzyme δ-aminolevulinic acid synthetase during embryonic life. This role probably becomes only subsidiary in the adult (e.g., Irving et al., 1976). Cross-reaction of 5β-DHT with testosterone for binding to an androgen receptor, such as demonstrated by Lieberburg and Nottebohm (1979) in the Zebra Finch syrinx, could also explain some of the 5β-DHT effects.

d. *Who Cares about Testosterone Metabolism?* Although studies of testosterone metabolism by the avian brain began only quite recently, they already provide or at least suggest answers to a number of questions about behavior and its endocrine control. First, the description of the metabolic pathways followed by testosterone (Fig. 20) gives sense to the observations of the specificity of steroids in the control of behavior, already described in Section III,C. The paradoxical findings that some behavior patterns in birds are elicited by androgens as well as estrogens (e.g., copulation in quail, nest soliciting in doves) or by testosterone and by 5α-DHT (e.g., crowing in quail) become clear as soon as we know that the avian brain can transform testosterone into estradiol or 5α-DHT. It is also interesting to note that generally the neuroanatomical localization of the metabolic activities conforms with the sites in the nervous system implicated in the control of behavior. Aromatase activity is, for example, mainly located in the hypothalamus–preoptic area (in chicks; Callard et al., 1978a), where estrogen implants are maximally effective for stimulating courtship behavior (e.g., dove; Hutchison, 1970a,b; 1971). The steroid specificities described earlier (Section III,C) can then be reinterpreted in terms of rates of conversion into metabolites and in availability for the brain of steroid molecules injected systematically (half-lives for different steroid could differ markedly; e.g., Tenniswood et al., 1981).

It should, however, be kept in mind that a number of notions on the specificity of steroids were presented here only in a simplified form. For example, the idea that testosterone activates copulation in quail *only* through its aromatization and that crowing depends on 5α-reduced metabolites ignores a number of experimental facts, such as:

1. No proof of the impossibility to aromatize 5α-DHT in the avian brain has been presented yet. [5α-DHT is aromatized in the embryonic chicken ovary (Weniger and Zeis, 1982).]

2. In our hand 5α-DHT activated sexual behavior in about 50% of the castrated quail tested (two independent experiments; M. Schumacher and J. Balthazart, unpublished; Deviche and Schumacher, 1982). This disagrees with Adkins's results (Adkins, 1977; Adkins and Pniewski, 1978) but it must be pointed out that we used free 5α-DHT, whereas in most experiments Adkins and collaborators used esters (propionate or benzoate), and similar differences between behavioral effectiveness of 5α-DHT and its esters have been reported in mammals (Yahr, 1979).

3. The enzyme inhibitors and hormone antagonists sometimes have side effects, and results obtained must consequently be interpreted with caution. In female rats, for example, the antiestrogen CI-628 antagonizes the effects of testosterone and 6α-fluorotestosterone equally well, although the latter is probably not aromatizable (Yahr and Gerling, 1978). The antiandrogen cyproterone acetate seems to interfere to some extent with the process of aromatization in the rat brain (McEwen et al., 1979).

Injection of the aromatase inhibitor ATD was also shown to maintain precastration levels of male sexual behavior in rats (Landau, 1980). Similar problems have been encountered in mammalian studies (Yahr and Gerling, 1978; Yahr, 1979; Södersten and Gustafsson, 1980). The involvement of aromatization in the testosterone-induced copulation of quail must consequently be considered as a theory rather than as a fact in order not to sterilize future research (which will require more attention to the biochemical process of behavior activation).

Testosterone metabolism could also answer some of the questions raised by individual differences in behavior (see Section V,A,1). We have studied the changes in reproductive behavior, plasma gonadtropins and testosterone, and cloacal-gland size in a group of 20 male quail after transfer to long days. These changes were then compared to the individual measures of testosterone metabolism within target organs such as the cloacal gland, the pituitary, the hypothalamus, and the hyperstriatum (Balthazart et al., 1979). It appeared that the individual measures of aggressive behavior were postively correlated with hyperstriatal conversion of testosterone into androstenedione and negatively correlated with hypothalamic conversion of testosterone into 5β-DHT. In addition, the frequency of courtship displays (struttings) was positively correlated with the hypothalamic 17β-hydroxysteroid dehydrogenase activity (production of androstenedione). In contrast, no correlation between these behavioral variables and plasma testosterone was observed, which confirms studies reported earlier (Section V,A,1).

In rats, Dessi-Fulghieri et al. (1976) have already found a significant correlation between individual measures of aggression and the rate of production of androstenedione in the brain, but they found no relationship between behavior and plasma testosterone. These studies suggest that testosterone metabolism in target organs may be of critical importance in the control of some individual differences. Other experiments would be needed, however, to permit a reliable generalization of this idea.

We already presented data supporting the notion that very young birds are less sensitive to testosterone than adults (see Section II,A,1; III,A; IV,C,1). This difference may be explicable by differences in the intracellular metabolism of testosterone. A dramatic decrease in hypothalamic 5β-reductase activity has indeed been demonstrated in the maturing cockerel (Massa and Sharp, 1981) and in the young male and female Japanese Quail (J. Balthazart, unpublished). If we accept the viewpoint that 5β-reductase represents an inactivation shunt of testosterone (see V,B,1,c), the ontogenetic decrease of this enzymatic activity then results in a progressive potentiation of the behavioral effects of testosterone. This would adequately explain the increased sensitivity to testosterone of older birds. The final proof of this interpretation must, however, await more direct experimental facts, such as could be obtained by the use of specific inhibitors of 5β-reduction. Unfortunately, no such compound is known at present except for estradiol, which inhibits 5β-reductase activity in the pig liver (Van Doorn and Clark, 1973), in the quail pituitary (Balthazart et al., 1980b), and in the dove hypothalamus (Steimer and Hutchison, 1981). Estradiol is, however, not well suited for this type of study because it is behaviorally active by itself.

It must also be mentioned that the long-term androgen deficit induced by castration of Ring Doves coincides with an increase of 5β-reduction of testosterone (i.e., probably greater inactivation of the hormone) in the preoptic area, but not in the adjacent parolfactory area; Steimer and Hutchison, 1981). This effect can be reversed by treatment with TP. This could represent the biochemical basis of the decreased sensitivity to androgen previously described in long-term castrates (J. B. Hutchison, 1974a,b 1976, 1978; cf. Section V,A,3).

The aforementioned examples of studies still in progress have been cited only to illustrate a new research strategy which has become possible thanks to the availability of radioactive steroids with high specific activity. Within a few years a review of the literature should contain much more definite results.

2. Estradiol and Progesterone

Very little information is available concerning the brain's metabolism of estrogens and progestins in birds and its involvement in the control of behavior. Estradiol-17β ($E_2β$), estrone (E_1), and estradiol-17α ($E_2α$) occur in brain homogenates incubated with testosterone or androstenedione, which shows that these steroids, as in mammals, can be transformed one into the other in birds (Callard et al., 1978a; Steimer and Hutchison, 1980). In mammals, estrogen metabolism is, however, far less active than the metabolism of testosterone (Zigmond, 1975). In doves, the equilibrium between $E_2α$ and $E_2β$ seems to favor the β-isomer (Steimer and Hutchison, 1980). In mammals E_1 and $E_2β$ can be further transformed into catechol-estrogens (2-hydroxyestrone and 2-hydroxyestradiol-17β) that have significant effects both on reproductive behavior and on gonadotropin secretion. These actions imply binding to estrogen receptors and/or interference with the metabolism of the brain catecholamines (for reviews see Parvizi and Ellendorff, 1980; Reddy et al., 1981; and Jellinck, et al., 1981). Sexual differences of estrogen metabolism have been observed at several levels in the rat brain (Reddy et al., 1981) and could play some role in sexual dimorphism of behavior.

As yet, there are no quantitative data relating estrogen metabolism in the avian brain to behavioral effects. More can be said about progesterone metabolism. Sharp and Massa (1980) have demonstrated that in the hyperstriatum, hypothalamus, and pituitary gland of laying hens progesterone is actively converted into 5β-pregnan-3,20-dione (5β-DHP), 5β-pregnan-3α-ol-20-one (5β,3α-ol), and to a lesser extent into 5α-pregnane-3,20-dione (5α-DHP). A conversion to 5α-DHP had already been reported by Nakamura and Tanabe (1974). It is likely that the enzymes 5α- and 5β-reductase, which reduce progesterone, are the same as those acting on testosterone (Zigmond, 1975). Their neuroanatomical localization appears to be the same (e.g., more 5β-reductase in hyperstriatum than in hypothalamus) and progesterone interferes with the *in vitro* 5α-reduction of testosterone in mammals (Voigt *et al.*, 1970). This action probably reflects a competition of the two hormones for the same enzymes. Furthermore, Stern (1972) demonstrated that progesterone inhibits *in vivo* the accumulation in the hypothalamus of doves of a testosterone metabolite that has the same chromatographic mobility as 5α-DHT (according to recent data this would be a pool of 5α- and 5β-DHT).

These data, although very limited, suggest two ways in which progesterone metabolism could be relevant to the control behavior. It is known that 5β-reduced progesterone metabolites exert a marked depressant action on the central nervous system of mammals (Gyermek *et al.*, 1967; Holzbauer, 1971). Their presence in large amounts in the brain of birds could thus regulate directly or indirectly their behavior. On the other hand, the likely competition of testosterone and progesterone for the 5α- and 5β-reductase activities could be an explanation for the antiandrogenic effects of progesterone. It is indeed known that, progesterone inhibits androgen-dependent behavior such as bow cooing (Erickson *et al.*, 1967; Hutchison, 1976) in Ring Doves, and that this effect is central because it can be obtained by implanting progesterone in the preoptic area and anterior hypothalamus (Komisaruk, 1967). Similarly in young chicks, progesterone suppresses by a central action testosterone-induced precocial copulation (Meyer, 1972; Meyer and Angelo, 1975).

It is possible that, in the doses used in these experiments, progesterone significantly reduced the production of 5α-DHT in the brain of treated animals as a consequence of its competition with testosterone for the 5α-reductase. If 5α-DHT plays some role in the activation of the behaviors under study (see Section V,B,1,*c*), the competition hypothesis then adequately explains at the molecular level the behavioral observations. For the time being this represents only a working hypothesis rather than experimental facts. It is, however, easily testable.

C. Steroid Receptors

Interaction of a steroid with a specific cytoplasmic receptor and the translocation of the steroid–receptor complex to the cell nucleus constitutes the last endocrine event leading to the production of a behavioral effect (see Fig. 19; see reviews by McEwen, 1980, 1981; McEwen *et al.*, 1980). Further steps in the activation of behavior include changes in the transcription and translation of DNA, but once initiated these reactions no longer require the presence of steroid molecules.

Indirect estimations of the binding sites for steroids can be obtained *in vivo* by autoradiography or uptake studies. Autoradiography can supply very precise neuroanatomical localization of the binding but usually no quantitative evaluation (see Section III,D,3 for results obtained in birds). By measuring the retention of radioactivity in whole homogenates or in purified nuclear fractions from the brain of birds that have been injected *in vivo* with labeled steroids (e.g., Stern, 1972; Benoff *et al.*, 1978b; J. B. Hutchison, 1978) such quantitative data can be obtained. The estimation *in vivo* of binding in uptake studies, however, is confused by a number of other processes, such as peripheral catabolism and intracellular metabolism of the hormone, which usually prevent accurate measurement of the total number of binding sites.

Since the mid-1970s measurement has become possible in brain tissues by *in vitro* methods using labeled steroids with very high specific activity. Cytoplasmic (free) and nuclear (occupied) receptors can be quantified in this way (e.g., Clark and Peck, 1977).

Studies of the brain of mammals have demonstrated the presence of receptors with high affinity ($K_d = 10^{-8}$–10^{-10} M) limited capacity (fmol/mg protein), and high specificity for all classes of steroids (androgens, estrogens, progestins, and corticosteroids). Reviews of this subject and of the involvement of these receptors in the control of behavior are available (e.g., King, 1976; McEwen, 1980, 1981; Balthazart, 1982) but are beyond the scope of this chapter.

In birds, despite the fact that the chick oviduct was one of the earliest models for hormone–receptor interactions (for a review see Schrader and O'Malley, 1980), little is known about brain receptors for steroids. Cytosolic receptors for androgens have been characterized in peripheral organs [comb and wattles of cocks (Dubé and Tremblay, 1974), syrinx of songbirds (Lieberburg and Nottebohm, 1979)], for estrogens [chick and hen oviduct (Harrison and Toft, 1975; Kon *et al.*, 1980)], and for progestins [chick oviduct (Sherman *et al.*, 1970; Schrader and O'Malley, 1972, 1980; Gschwendt, 1980)], but only progestin receptors have been characterized so far in the brains of only two species: the hen (Kawashima *et al.*, 1978) and the Ring Dove (Balthazart *et al.*, 1980c).

Kawashima *et al.* (1978) demonstrated the presence of a progestin binder in the cytosol of the anterior pituitary, the hypothalamus, and the oviduct magnum of laying hens. The dissociation constant of the binder is similar in the three tissues (1–3 × 10^{-10} M) although it could be a bit lower in the pituitary than in the hypothalamus (Kawashima *et al.*, 1979b). The nature of these two receptors could thus be slightly different. Kawashima *et al.* (1979a) also demonstrated that the concentrations of cytoplasmic progestin receptors in the hypothalamus and pituitary change during the ovulatory cycle and show peaks at 18 and 8 hours before ovulation. This is confirmed in a later study (Kawashima *et al.*, 1980) showing, in addition, that the decrease in cytosol receptor after the peaks at 18 and 8 hours is associated with an increase in nuclear receptors. This suggests that progesterone at these times is translocated to the nucleus. These changes are probably related to the ovulatory process because they are not observed in nonlaying hens (Kawashima et al., 1979b). It is likely that they are triggered by the variations of plasma progesterone, estradiol, and testosterone during the cycle. Important changes in cytoplasmic progestin receptors are found after injection of these hormones (Kawashima et al., 1979b). Estradiol injections, especially, increase the cytosol concentration of progestin receptors, which probably sensitizes the brain to circulatory progesterone and could play a critical role in the control of ovulation. A similar mechanism seems to operate for the activation of lordosis in female mammals (Blaustein and Feder, 1979a,b, 1980; Moguilewsky and Raynaud, 1977, 1979). The relevance to the hen's behavior of the changes in receptor concentrations remains, however, to be investigated.

We have presented (Balthazart et al., 1980c) results of receptor assays that could help resolve the problem concerning the hormonal basis of incubation in doves (cf. Section II,B,4). In this species, injections of progesterone facilitate incubation behavior (Lehrman, 1958b), but this effect is dependent on the simultaneous presence of gonadal steroids (Stern and Lehrman, 1969; Stern, 1974). However, it was shown more recently that plasma progesterone concentration does not change in male doves during the reproductive cycle (Silver et al., 1974). This raised, together with other data discussed previously (see Section II,B,4), two groups of questions:

1. Does progesterone play a physiological role in the induction of incubation in male doves?
2. What is the biochemical basis for the synergism between sex steroids (estradiol and testosterone) and exogenous progesterone in the induction of incubation?

It is known that estradiol can increase the concentration of progestin receptors in the mammalian hypothalamus (Blaustein and Feder, 1979a; Moguilewsky and Raynaud, 1979), and this suggested a mechanism involving changes in progesterone binding to explain its action on incubation. By an *in vitro* assay using the synthetic progestin R 5020 (17α,21-dimethyl-19-nor-pregna-4,9-diene-3,20-dione, i.e., promegestone) as a ligand, we demonstrated the presence of a progestin binder in the brain of castrated doves. It has a high affinity ($K_d \approx 4 \times 10^{-10}$ M), a limited capacity ($\sim$ 100 fmol/mg protein) and is progestin specific. Only progesterone and R 5020 are strong competitors. The cytosolic binding observed in the hypothalamus is significantly reduced (about 50%) if birds are given a 100-μg progesterone injection 4 hours before sacrifice, which suggests that the receptor can be translocated to the nucleus. The neuroanatomical localization of this binding is also correlated with its potential biological significance. It is more concentrated in the preoptic area and hypothalamus [a site of action for progesterone in the induction of incubation (Komisaruk, 1967)] than in the other areas that were studied.

Interestingly, we found that testosterone priming [200 µg of TP daily for 7 days; a treatment which, according to Stern and Lehrman (1969), restores the facilitation effect of progesterone on incubation in castrated males] increases the hypothalamic concentration of progestin receptors but is without effect in the hyperstriatum or in the midbrain. A similar but more pronounced effect is observed following priming for 7 days with EB (50 µg/day per animal), but treatment with either hormone (TP or EB) for only 2 days does not significantly affect he progestin-receptor concentration.

It is interesting to correlate these assay data with a number of behavioral observations. It then appears likely that the synergism between exogenous progesterone and testosterone in the activation of incubation is related to the androgenic induction of hypothalamic progestin receptors. Whether this synergism actually plays a role in the reproductive cycle of intact male doves is more difficult to evaluate. When a male dove is paired with a female, its plasma testosterone concentration quickly rises to reach a peak about 3 days later (Feder et al., 1977; see Section IV,C,2 and Fig. 12). After a latency of 6–10 days, the female lays eggs and both sexes engage in incubation. It is conceivable that the increased androgen levels observed during the courtship phase of the cycle (between pairing and egg laying) could be sufficient to increase the hypothalamic concentration of progestin receptors in the males. This would result in an increased sensitivity of the brain to progesterone by means of an enhanced translocation to the cell nuclei. In physiological conditions, progesterone would thus control behavior not by a change in plasma concentration but by causing an increased sensitivity to a constant signal. The time course of the concentration increase of the receptor after testosterone priming in castrated birds is consistent with this model. Receptor concentration is increased after 7 days of priming, which is close to the normal egg-laying latency, but no or little effect is observed after only 2 days. Direct evidence supporting this hypothesis, such as could be obtained by measuring changes in nuclear progestin receptors during the cycle, is not yet available.

The aforementioned study is only one example of how changes in steroid-receptor concentration could play a role in the control of behavior. This mechanism is well known in mammals but so far very little can be said about birds. Even the presence of steroid receptors in birds has not been generally demonstrated. Their specificity is consequently largely unknown. It should not be assumed a priori that these specificities will be the same as in mammals. For example, in the hypothalamus of the hen and Ring Dove, corticosterone competes somehow for binding with the progestin receptor (Kawashima et al., 1978; Balthazart et al., 1980c), whereas this cross-reaction does not exist in the guinea pig (Blaustein and Feder, 1979a) or in the rat (MacLusky and McEwen, 1980). Similarly, the specificity of hormone antagonists (e.g., antiandrogens such as CA or antiestrogens such as mitromifene citrate, CI-628) may be different in birds and in mammals. Until we know more about receptors in the bird brain, the results of experiments using these drugs and data on steroid specificity should be interpreted carefully.

VI. Conclusions

Many data have accumulated, mainly since 1960, that demonstrate how hormones (chiefly steroids) control various aspects of avian behavior and how this interacts with a number of independent factors such as endocrine status during ontogeny, photoperiod, or behavior of a congener. Some of the most detailed models describing these interactions have resulted from studies of birds. Investigations of the Ring Dove provide a unique case in which the interaction of environmental and hormonal stimuli is quite well understood. Our knowledge on the neuroendocrine control of vocalizations in songbirds is also one of the best-known models of brain–hormone–behavior integration.

Studies of bird behavior, however, lag behind mammalian work in a number of aspects:

1. In many cases little effort has been made to use physiological doses of hormones in replacement therapy, thus the physiological meaning of the results remains questionable.
2. Few assay data are available, and consequently in most cases we know only that one hormone if injected activates a given behavior, but we have no proof of its real presence as a "hormonal correlate of behavior."
3. Studies of the intracellular events of hormonal action are conspicuously missing.

In the late 1970s, reliable techniques became available for measuring steroids in biological fluids, for quantifying their metabolism, and for characterizing their receptors. These are the tools that will lead to answers to many questions concerning subjects such as steroid specificity, sexual differentiation, and changes in sensitivity to steroids. It can also be reasonably hoped that soon studies on the posttranscriptional events related to steroid action in the brain will become possible. The interaction between hormones and neurotransmitters (e.g., Meyerson et al., 1979) will then be of special interest. These studies will renew interest in "old-fashioned" experiments using castration–hormone replacement and will help to design them in a more physiological way. A decision as to whether one hormone or another is the hormonal correlate or even the "causal" stimulus for one behavior can no longer rely on the simple comparison of dose–response curves for the two components. Factors such as half-life of the compounds in the blood, their intracellular metabolism, or the receptor occupancy they induce must also be taken into account. This is one of the directions of future research.

Finally, it must be stressed that very different reproductive strategies have evolved in birds, but our knowledge of the endocrine control of reproduction is based on only a few species. It is now time to examine other species and to learn how different strategies (oportunistic versus seasonal breeding; monogamous versus polygamous species; transient versus long-term pair bonds) correlate with different endocrine mechanisms. These studies will frequently have to be done in the field, which is now possible thanks to the brilliant work by Wingfield (Wingfield and Farner, 1978, 1979, 1980; Wingfield et al., 1980). It is also likely that many studies will demonstrate the importance of behavior in the control of endocrine physiology. Hormonal correlates of behavior will undoubtedly be an exciting subject of research for the next decade.

ACKNOWLEDGMENTS

I am indebted to Professor E. Schoffeniels for his consistent interest in my work. Aspects of my research presented in this chapter were supported by a grant (NBR 2.4544.76) from the Fonds de la Recherche Fondamentale Collective attributed to Professor E. Schoffeniels and a grant from the Fonds National de la Recherche Scientifique to myself. I sincerely thank Dr. Mei-Fang Cheng of Rutgers University for reading the manuscript and making numerous improvements. I also thank M. Dunon and M. Gillet for carefully typing the manuscript and J. Evard for the preparation of the photographs.

REFERENCES

Abraham, G. E. (1969). Solid-phase radioimmunoassay of estradiol-17β. *J. Clin. Endocrinol.* **23**, 866–870.
Abraham, G. E., Swerdloffer, R., Tulchinsky, D., and Odell, W. D. (1971). Radioimmunoassay of plasma progesterone. *J. Clin. Endocrinol.* **32**, 619–624.
Abs, M. (1976). Physiology of juvenile development and sexual maturation in non-passerine birds. A review. *Arch. Gefluegelkd* **40**, 153–167.
Abs, M. (1980). Zur Bioakustik des Stimmbrucks bei Vögeln. *Zool. Jahrb. Physiol.* **84**, 289–382.
Adigweme, A. I., and Abul-Hajj, Y. J. (1981). Macromolecular binding of 5β-H steroids in chick embryo liver. *J. Steroid Biochem.* **14**, 91–100.
Adkins, E. K. (1973). Functional castration of the female Japanese quail. *Physiol. Behav.* **10**, 619–621.
Adkins, E. K. (1975). Hormonal basis of sexual differentiation in the Japanese quail. *J. Comp. Physiol. Psychol.* **89**, 61–71.

Adkins, E. K. (1976). Embryonic exposure to an antiestrogen masculinizes behavior of female quail. *Physiol. Behav.* **17**, 357–359.

Adkins, E. K. (1977). Effects of diverse androgens on the sexual behavior and morphology of castrated male quail. *Horm. Behav.* **8**, 201–207.

Adkins, E. K. (1978). Sex steroids and the differentiation of avian reproductive behavior. *Am. Zool.* **18**, 501–509.

Adkins, E. K. (1979). Effect of embryonic treatment with estradiol or testosterone on sexual differentiation of the quail brain. *Neuroendocrinology* **29**, 178–185.

Adkins, E. K. (1981). Hormone specificity, androgen metabolism and social behavior. *Am. Zool.* **21**, 257–272.

Adkins, E. K., and Adler, N. T. (1972). Hormonal control of behavior in the Japanese quail. *J. Comp. Physiol. Psychol.* **81**, 27–36.

Adkins, E. K., and Mason, P. (1974). Effects of cyproterone acetate in male Japanese quail. *Horm. Behav.* **5**, 1–6.

Adkins, E. K., and Nock, B. L. (1976a). Behavioral responses to sex steroids of gonadectomized and sexually regressed quail. *J. Endocrinol.* **68**, 49–55.

Adkins, E. K., and Nock, B. L. (1976b). The effects of antiestrogen CI-628 on sexual behavior activated by androgen or estrogen in quail. *Horm. Behav.* **7**, 417–429.

Adkins, E. K., and Pniewski, E. E. (1978). Control of reproductive behavior by sex steroids in male quail. *J. Comp. Physiol. Psychol.* **92**, 1169–1178.

Adkins, E. K., Boop, J. J., Koutnik, D. L., Morris, J. B., and Pniewski, E. E. (1980). Further evidence that androgen aromatization is essential for the activation of copulation in male quail. *Physiol. Behav.* **24**, 441–446.

Adkins-Regan, E. (1981). Effect of sex steroids on the reproductive behavior of castrated male ring doves (*Streptopelia* sp.). *Physiol. Behav.* **26**, 561–566.

Adkins-Regan, E., Pickett, P., and Koutnick, D. (1982). Seasonal differentiation in quail: conversion of androgen to estrogen mediates testosterone-induced desmasculinization of copulation but not other male characteristics. *Horm. Behav.* **16**, 259–278.

Åkerman, B. (1966a). Behavioural effects of electrical stimulation in the forebrain of the pigeon. I. Reproductive behaviour. *Behaviour* **26**, 328–338.

Åkerman, B. (1966b). Behavioural effects of electrical stimulation in the forebrain of the pigeon. II. Protective behaviour. *Behaviour* **26**, 339–350.

Allee, W. C., and Collias, N. (1940). The influence of estradiol on the social organization of flocks of hens. *Endocrinology (Baltimore)* **27**, 87–94.

Allee, W. C., Collias, N., and Lutherman, C. Z. (1939). Modification of the social order in flocks of hens by the injection of testosterone propionate. *Physiol. Zool.* **12**, 412–440.

Andrew, R. J. (1966). Precocious adult behaviour in the young chick. *Anim. Behav.* **14**, 485–500.

Andrew, R. J. (1969). The effects of testosterone on avian vocalizations. *In* "Bird Vocalizations" (R. A. Hinde, ed.), pp. 97–130. Cambridge Univ. Press, London and New York.

Andrew, R. J. (1972a). Recognition processes and behaviour, with special reference to effects of testosterone on persistence. *Adv. Study Behav.* **4**, 175–208.

Andrew, R. J. (1972b). Changes in search behaviour in male and female chicks, following different doses of testosterone. *Anim. Behav.* **20**, 741–750.

Andrew, R. J. (1975a). Effects of testosterone on the behaviour of the domestic chick. I. Effects present in males but not in females. *Anim. Behav.* **23**, 139–155.

Andrew, R. J. (1975b). Effects of testosterone on the behaviour of the domestic chick. II. Effects present in both sexes. *Anim. Behav.* **23**, 156–168.

Andrew, R. J. (1975c). Effects of testosterone on the calling of the domestic chick in a strange environment. *Anim. Behav.* **23**, 169–178.

Andrew, R. J. (1975d). Midbrain mechanisms of calling and their relation to emotional states. *In*

"Neural and Endocrine Aspects of Behaviour in Birds" (P. Wright, P. G. Caryl, and D. M. Vowles, eds.), pp. 275–304. Elsevier, Amsterdam.
Andrew, R. J. (1978). Increased persistence of attention produced by testosterone, and its implications for the study of sexual behaviour. In "Biological Determinants of Sexual Behaviour" (J. B. Hutchison, ed.), pp. 255–275. Wiley, Chichester.
Andrew, R. J., and Archer, J. (1977). Testosterone and chick behavior. Horm. Behav. **8**, 120–123.
Andrew, R. J., and de Lanerolle, N. (1975). The effects of muting lesions on emotional behaviour and behaviour normally associated with calling. Brain Behav. Evol. **10**, 377–399.
Andrew, R. J., and Roger, L. J. (1972). Testosterone, search behavior and persistence. Nature (London) **237**, 343–346.
Archer, J. (1973a). The influence of testosterone on chick behavior in novel environments. Behav. Biol. **8**, 93–108.
Archer, J. (1973b). Effects of testosterone on immobility responses in the young male chick. Behav. Biol. **8**, 551–556.
Archer, J. (1973c). A further analysis of responses to a novel environment by testosterone-treated chicks. Behav. Biol. **9**, 389–396.
Aren-Engelbrektsson, B., Larsson, K., Södersten, P., and Wilhelmsson, M. (1970). The female lordosis pattern induced in male rats by estrogen. Horm. Behav. **1**, 181–188.
Arnold, A. P. (1975a). The effects of castration on song development in Zebra finches (Poephila guttata). J. Exp. Zool. **191**, 261–278.
Arnold, A. P. (1975b). The effects of castration and androgen replacement on song, courtship and aggression in Zebra finches (Poephila guttata). J. Exp. Zool. **191**, 309–326.
Arnold, A. P. (1980). Sexual differences in the brain. Am. Sci. **68**, 165–173.
Arnold, A. P., Nottebohm, F., and Pfaff, D. W. (1976). Hormone concentrating cells in vocal control and other areas of the brain of the Zebra finch (Poephila guttata). J. Comp. Neurol. **165**, 487–512.
Attardi, B., and Ohno, S. (1976). Androgen and estrogen receptors in the developing mouse brain. Endocrinology (Baltimore) **99**, 1279–1290.
Bailey, R. E. (1950). Inhibition with prolactin of light-induced gonad increase in White-crowned Sparrows. Condor **52**, 247–251.
Balthazart, J. (1976). Daily variations of behavioural activities and of plasma testosterone levels in the domestic duck, Anas platyrhynchos. J. Zool. **180**, 155–173.
Balthazart, J. (1977). "Le Contrôle hormonal du comportement chez le canard domestique mâle (Anas platyrhynchos L.)." Doctoral Thesis, Univ. of Liège, Belgium.
Balthazart, J. (1978a). Behavioural and physiological effects of testosterone propionate and cyproterone acetate in immature male domestic ducks, Anas platyrhynchos. Z. Tierpsychol. **47**, 410–421.
Balthazart, J. (1978b). Le contrôle hormonal du comportement reproducteur chez l'oiseau mâle. Première partie. Rev. Quest. Sci. **149**, 311–332.
Balthazart, J. (1978c). Le contrôle hormonal du comportement reproducteur chez l'oiseau mâle. Deuxième partie. Rev. Quest. Sci. **150**, 95–117.
Balthazart, J. (1982). Steroids receptors and behavior. In "Aspects of the Endocrinology of Birds: Practical and Theoretical Implications" (C. G. Scanes, M. A. Ottinger, A. D. Kenny, J. Balthazart, J. Cronshaw, and I. Chester Jones, eds.). Texas Tech. Univ. Press, Lubbock.
Balthazart, J., and De Rycker, C. (1979). Effects of androgens and oestrogens on the behaviour of chicks in an imprinting situation. Z. Tierpsychol. **49**, 55–64.
Balthazart, J., and Deviche, P. (1977). Effects of exogenous hormones on the reproductive behaviour of adult male domestic ducks. I. Behavioural effects of intramuscular injections. Behav. Processes **2**, 129–146.
Balthazart, J., and Hendrick, J. C. (1976). Annual variation in reproductive behavior, testoster-

one, and plasma FSH levels in the Rouen duck, *Anas platyrhynchos*. *Gen. Comp. Endocrinol.* **28**, 171–183.

Balthazart, J., and Hendrick, J. C. (1978a). Steroidal control of plasma luteinizing hormone, comb growth and sexual behaviour in male chicks. *J. Endocrinol.* **77**, 149–150.

Balthazart, J., and Hendrick, J. C. (1978b). Short and long-term variations of pituitary–gonadal hormones in the plasma of male ducks and their relationships with behaviour. *Comp. Endocrinol. Proc. Int. Symp. 8th*, p. 189.

Balthazart, J., and Hendrick, J. C. (1979a). Effects of exogenous gonadotropic and steroid hormones on the social behaviour and gonadal maturation of male domestic ducklings, *Anas platyrhynchos* L. *Arch. Int. Physiol. Biochim.* **87**, 741–761.

Balthazart, J., and Hendrick, J. C. (1979b). Relationships between the daily variations of social behavior and of plasma FSH, LH and testosterone levels in the domestic duck, *Anas. platyrhynchos* L. *Behav. Processes* **4**, 107–128.

Balthazart, J., and Hirschberg, D. (1979a). Testosterone metabolism and sexual behavior in the chick. *Horm. Behav.* **12**, 253–263.

Balthazart, J., and Hirschberg, D. (1979b). Differential release of various androgens from silastic implants. *IRCS Med. Sci. Libr. Compend.* **7**, 123.

Balthazart, J., and Puts, C. (1979). Androgens control the behavior of chicks in an imprinting situation. *Abstr. Annu. Meet. Anim. Behav. Soc.*, June 1979, New Orleans.

Balthazart, J., and Schoffeniels, E. (1978). Pheromones are involved in the control of sexual behaviour in birds. *Naturwissenschaften* **66**, 55–56.

Balthazart, J., and Stevens, M. (1975a). Effects of testosterone propionate on the social behaviour of groups of male domestic ducklings, *Anas platyrhynchos* L. *Anim. Behav.* **23**, 926–931.

Balthazart, J., and Stevens, M. (1975b). Plasma testosterone levels in very young domestic ducklings. *IRCS Med. Sci. Libr. Compend.* **3**, 345.

Balthazart, J., and Stevens, M. (1976). Social behaviour of testosterone injected male ducklings in the presence of oestrogen-treated females. *Behaviour* **57**, 288–306.

Balthazart, J., Deviche, P., and Hendrick, J. C. (1977a). Effects of exogenous hormones on the reproductive behaviour of adult male domestic ducks. II. Correlation with morphology and hormone plasma levels. *Behav. Processes* **2**, 147–161.

Balthazart, J., Hendrick, J. C., and Deviche, P. (1977b). Diurnal variations of plasma gonadotrophins in male domestic ducks during the sexual cycle. *Gen. Comp. Endocrinol.* **32**, 376–389.

Balthazart, J., Massa, R., and Negri-Cesi, P. (1979). Photoperiodic control of testosterone metabolism, plasma gonadotrophins, cloacal gland growth and reproductive behaviour in the Japanese quail. *Gen. Comp. Endocrinol.* **39**, 222–235.

Balthazart, J., Willems, J., and Hendrick, J. C. (1980a). Changes in pituitary responsiveness to luteinizing hormone releasing hormone during an annual cycle in the domestic duck, *Anas platyrhynchos* L. *J. Exp. Zool.* **211**, 113–123.

Balthazart, J., Bottoni, L., and Massa, R. (1980b). Effects of sex steroids on testosterone metabolism, plasma gonadotropins, cloacal gland growth and reproductive behaviour in the Japanese quail. *Boll. Zool.* **47**, 185–192.

Balthazart, J., Blaustein, J. D., Cheng, M. F., and Feder, H. H. (1980c). Hormones modulate the concentration of cytoplasmic progestin receptors in the brain of male ring doves (*Streptopelia risoria*). *J. Endocrinol.* **86**, 251–261.

Balthazart, J., Reboulleau, C., and Cheng, M. F. (1981a). Diurnal variations of plasma FSH, LH and testosterone in male ring doves kept under different photoperiods. *Gen. Comp. Endocrinol.* **44**, 202–206.

Balthazart, J., Malacarne, G., and Deviche, P. (1981b). Stimulatory effects of 5β-dihydrotestosterone on the sexual behavior in the domestic chick. *Horm. Behav.* **15**, 246–258.

Balthazart, J., Balthazart-Raze, C., and Cheng, M. F. (1981c). Hormonal control of the gonadal regression and recovery observed in short days in male and female doves. *J. Endocrinol.* **89**, 79–89.

Balthazart, J., Schumacher, M., and Ottinger, M. A. (1983). Seasonal differences in the Japanese quail: behavior, morphology and intracellular metabolism of testosterone. *Gen. Comp. Endocrinol.* **51**, in press.

Barfield, R. J. (1964). Induction of copulatory behavior by intracranial placement of androgen in capons. *Am. Zool.* **4**, 301.

Barfield, R. J. (1965). Induction of aggressive and courtship behavior by intracerebral implants of androgen in capons. *Am. Zool.* **5**, 203.

Barfield, R. J. (1969). Activation of copulatory behavior by androgen implanted into the preoptic area of the male fowl. *Horm. Behav.* **1**, 37–52.

Barfield, R. J. (1971a). Activation of sexual and aggressive behavior by androgen implanted into the male ring dove brain. *Endocrinology (Baltimore)* **89**, 1470–1476.

Barfield, R. J. (1971b). Gonadotrophic hormone secretion in the female ring dove in response to visual and auditory stimulation by the male. *J. Endocrinol.* **49**, 305–310.

Barfield, R. J., Ronay, G., and Pfaff, D. W. (1978). Autoradiographic localization of androgen-concentrating cells in the brain of the male domestic fowl. *Neuroendocrinology* **26**, 297–311.

Bartke, A., Steele, R. E., Musto, N., and Caldwell, B. W. (1973). Fluctuations in plasma testosterone levels in adult male rats and mice. *Endocrinology (Baltimore)* **92**, 1223–1228.

Bates, R. W., and Riddle, O. (1936). Effect of route of administration on the bioassay of prolactin. *Proc. Soc. Exp. Biol. Med.* **34**, 847–849.

Bates, R. W., Lahr, E. L., and Riddle, O. (1935). The gross action of prolactin and FSH on the mature ovary and sex accessories of the fowl. *Am. J. Physiol.* **111**, 361–368.

Bates, R. W., Riddle, O., and Lahr, R. L. (1937). Mechanism of the anti-gonad action of prolactin in adult pigeons. *Am. J. Physiol.* **119**, 610–614.

Bateson, P. P. G., Horn, G., and Rose, S. P. R. (1972). Effects of early experience on regional incorporation of precursors into RNA and protein in the chick brain. *Brain Res.* **39**, 449–465.

Bateson, P. P. G., Horn, G., and Rose, S. P. R. (1975). Imprinting: correlations between behaviour and incorporation of (^{14}C) uracil into chick brain. *Brain Res.* **84**, 207–220.

Bateson, P. P. G., Horn, G., and McCabe, B. J. (1978). Imprinting; the effect of partial ablation of the medial hyperstriatum ventrale of the chick. *J. Physiol. (London)* **285**, 23P.

Beach, F. A. (1951). Effects of forebrain injury upon mating behaviour in male pigeons. *Behaviour* **4**, 36–59.

Beach, F. A. (1966). Ontogeny of "coitus-related" reflexes in the female guinea pig. *Proc. Natl. Acad. Sci. USA* **56**, 526–533.

Beach, F. A., and Inman, N. G. (1965). Effects of castration and androgen replacement on mating in male quail. *Proc. Natl. Acad. Sci. USA* **54**, 1426–1431.

Beach, F. A., and Levinson, G. (1950). Effects of androgen on the glans penis and mating behavior of castrated male rats. *J. Exp. Zool.* **114**, 159–168.

Beach, F. A., and Westbrook, W. H. (1968). Morphological and behavioral effects of an "antiandrogen" in male rats. *J. Endocrinol.* **42**, 379–382.

Bell, R. Q., and Hinde, R. A. (1963). Brood patch sensitivity of female canaries brought into reproductive condition in winter. *Anim. Behav.* **11**, 561–565.

Bennett, M. A. (1940). The social hierarchy in ring doves. II. The effect of treatment with testosterone propionate. *Ecology* **21**, 148–165.

Benoff, F. H. (1979). Testosterone induced precocious sexual behavior in chickens differing in adult mating frequency. *Behav. Processes* **4**, 35–41.

Benoff, F. H., Siegel, P. B., and Van Krey, H. P. (1978a). Testosterone determination in lines of chickens selected for differential mating frequency. *Horm. Behav.* **10**, 246–250.

Benoff, F. H., Siegel, P. B., and Van Krey, H. P. (1978b). Hypothalamic uptake of (^{3}H)-testosterone in chickens differing in mating frequency. *Physiol. Behav.* **20,** 803–805.
Berry, H. H., Millar, R. P., and Louw, G. N. (1979). Environmental cues influencing the breeding biology and circulating levels of various hormones and triglycerides in the cape cormorant. *Comp. Biochem. Physiol. A.* **62** 879–884.
Berthold, A. D. (1849). Transplantation der Hoden. *Arch. Anat. Physiol.* **16,** 42–46.
Bertolini, A., Vergoni, W., Gessa, G. L., and Ferrari, W. (1968). Induction of sexual excitement with intraventricular ACTH: permissive role of testosterone in the male rabbit. *Life Sci.* **7,** 1203–1206.
Bertolini, A., Vergoni, W., Gessa, G. L., and Ferrari, W. (1969). Induction of sexual excitement by the action of adrenocorticotropic hormone in brain. *Nature (London)* **221,** 667–669.
Blaustein, J. D., and Feder, H. H. (1979a). Cytoplasmic progestin-receptors in the guinea-pig brain: characteristics and relationship to the induction of sexual behaviour. *Brain Res.* **169,** 481–497.
Blaustein, J. D., and Feder, H. H. (1979b). Cytoplasmic progestin receptors in the female guinea pig brain and their relationship to refractoriness in expression of female sexual behaviour. *Brain Res.* **177,** 489–498.
Blaustein, J. D., and Feder, H. H. (1980). Nuclear progestin receptors in guinea pig brain measured by an *in vitro* exchange assay after hormonal treatments that affect lordosis. *Endocrinology (Baltimore)* **106,** 1061–1069.
Blaustein, J. D., Dudley, S. D., Gray, J. M., Roy, E. J., and Wade, G. N. (1979). Long-term retention of estradiol by brain cell nuclei and female rat sexual behavior. *Brain Res.* **173,** 355–359.
Blohm, T. R., Metcalf, B. W., Laughlin, M. E., Sjoerdsma, A., and Schatzman, G. L. (1980). Inhibition of testosterone 5α-reductase by a proposed enzyme-activated, active site-directed inhibitor. *Biochem. Biophys. Res. Commun.* **95,** 273–280.
Bohus, B., and De Wied, D. (1966). Inhibitory and facilitatory effect of two related peptides on extinction of avoidance behavior. *Science (Washington, D.C.)* **153,** 318–320.
Boris, A., Cox, D. C., and Hurley, J. F. (1970). Comparison of the effects of six antiandrogens on the chick comb stimulation by testosterone. *Proc. Soc. Exp. Biol. Med.* **134,** 985–987.
Boss, E. R. (1943). Hormonal determination of adult characters and sex behavior in herring gulls *(Larus argentatus). J. Exp. Zool.* **94,** 181–203.
Bottoni, L., and Massa, R. (1981). Seasonal changes in testosterone metabolism in the pituitary gland and central nervous system of the European starling *(Sturnus vulgaris). Gen. Comp. Endocrinol.* **43,** 532–536.
Brackenbury, J. (1980). Respiration and production of sounds by birds. *Biol. Rev. (Cambridge Philos. Soc.)* **55,** 363–378.
Bradley, P., and Horn, G. (1981). Imprinting. A study of cholinergic sites in parts of the chick brain. *Exp. Brain Res.* **41,** 121–123.
Bradley, P., Horn, G., and Bateson, P. (1981). Imprinting. An electron microscopic study of chick hyperstriatum ventrale. *Exp. Brain Res.* **41,** 115–120.
Breitenbach, R. P., and Meyer, R. K. (1959). Pituitary prolactin levels in laying, incubating and brooding pheasants *(Phasianus colchicus). Proc. Soc. Exp. Biol. Med.* **101,** 16–19.
Breneman, W. R. (1942). Action of prolactin and estrone on weights of reproductive organs and viscera of the cockerel. *Endocrinology (Baltimore)* **30,** 610–614.
Brockway, B. F. (1969). Hormonal and experiential factors influencing the nestbox oriented behaviour of budgerigars *(Melopsittacus undulatus). Behaviour* **35,** 1–26.
Brown, J. L. (1965). Vocalizations evoked from the optic lobe of a songbird. *Science (Washington, D.C.)* **149,** 1002–1003.

Buntin, J. D., and Forsyth, I. A. (1979). Measurements of pituitary prolactin levels in breeding piegons by crop sac radioreceptor assay. *Gen. Comp. Endocrinol.* **37**, 57–63.
Buntin, J. D., Cheng, M. F., and Hansen, E. W. (1977). Effect of parental feeding activity on squab-induced crop sac growth in ring doves *(Streptopelia risoria). Horm. Behav.* **8**, 297–309.
Burger, A. E., and Millar, R. P. (1980). Seasonal changes of sexual and territorial behaviour and plasma testosterone levels in male lesser sheathbills *(Chionis minor). Z. Tierpsychol.* **52**, 397–406.
Burke, W. H., and Dennison, P. T. (1980). Prolactin and luteinizing hormone levels in female turkeys *(Meleagris gallopavo)* during a photo-induced reproductive cycle and broodiness. *Gen. Comp. Endocrinol.* **41**, 92–100.
Burke, W. H., and Papkoff, H. (1980). Purification of turkey prolactin and the development of a homologous radioimmunoassay for its measurement. *Gen. Comp. Endocrinol.* **40**, 297–307.
Burkitt, J. P. (1919). Relation of song to the nesting of birds. *Ir. Nat.* **28**, 97–101.
Butterfield, P. A., and Crook, J. H. (1968). The annual cycle of nest building and agonistic behaviour in captive *Quelea quelea* with reference to endocrine factors. *Anim. Behav.* **16**, 308–317.
Callard, G. V., Petro, Z., and Ryan, K. J. (1978a). Phylogenetic distribution of aromatase and other androgen-converting enzymes in the central nervous system. *Endocrinology (Baltimore)* **103**, 2283–2290.
Callard, G. V., Petro, Z., and Ryan, K. J. (1978b). Conversion of androgen to estrogen and other steroids in the vertebrate brain. *Am. Zool.* **18**, 511–523.
Caridoit, F. (1938). Recherches expérimentales sur les rapports entre testicules, plumages d'éclipses et mues chez le canard sauvage. *Sta. Zool. Wimereux* **13**, 47–67.
Carpenter, C. R. (1933a). Psychobiological studies of social behavior in Aves. I. The effect of complete and incomplete gonadectomy on the primary sexual activity of the male pigeon. *J. Comp. Psychol.* **16**, 25–57.
Carpenter, C. R. (1933b). Psychobiological studies of social behavior in Aves. II. The effect of complete and incomplete gonadectomy on secondary sexual activity, with histological studies. *J. Comp. Psychol.* **16**, 59–96.
Cekan, S. Z. (1979). On the assessment of validity of steroid radioimmunoassays. *J. Steroid Biochem.* **11**, 1629–1634.
Champy, C., and Colle, P. (1919). Sur une corrélation entre la glande du jabot du pigeon et les glandes génitales. *C. R. Seances Soc. Biol. Ses Fil.* **82**, 818–819.
Cheng, M. F. (1973a). Effect of ovariectomy on the reproductive behavior of female ring doves *(Streptopelia risoria). J. Comp. Physiol. Psychol.* **83**, 221–233.
Cheng, M. F. (1973b). Effect of estrogen on behavior of ovariectomized ring doves *(Streptopelia risoria). J. Comp. Physiol. Psychol.* **83**, 234–239.
Cheng, M. F. (1974). Ovarian development in the female ring dove in response to stimulation by intact and castrated male ring doves. *J. Endocrinol.* **63**, 43–53.
Cheng, M. F. (1975). Induction of incubation behaviour in male ring doves *(Streptopelia risoria)*: a behavioural analysis. *J. Reprod. Fertil.* **42**, 267–276.
Cheng, M. F. (1977). Role of gonadotrophin releasing hormones in the reproductive behaviour of female ring doves *(Streptopelia risoria). J. Endocrinol.* **74**, 37–45.
Cheng, M. F. (1979). Progress and prospects in ring dove research: a personal view. *Adv. Study Behav.* **9**, 97–129.
Cheng, M. F., and Balthazart, J. (1982). The role of nest-building activity in the gonadotrophin secretions and the reproductive successes of ring doves *(Streptopelia risoria). J. Comp. Physiol. Psychol.* **96**, 307–324.

Cheng, M. F., and Follett, B. K. (1976). Plasma luteinizing hormone during the breeding cycle of the female ring dove. *Horm. Behav.* **7**, 199–205.

Cheng, M. F., and Lehrman, D. S. (1975). Gonadal hormone specificity in the sexual behavior of ring doves. *Psychoneuroendocrinology* **1**, 95–102.

Cheng, M. F., and Silver, R. (1975). Estrogen-progesterone regulation of nest building and incubation behavior in ovariectomized ring doves *(Streptopelia risoria)*. *J. Comp. Physiol. Psychol.* **88**, 256–263.

Clark, J. H., and Peck, E. J., Jr. (1977). Steroid hormone receptors: basic principles and measurement. *In* "Receptor and Hormone Action" (B. W. O'Malley, and L. Birnbaumer, eds.), Vol. 1, pp. 383–410. Academic Press, New York.

Clifton, P. G., and Andrew, R. J. (1981). A comparison of the effects of testosterone on aggressive responses by the domestic chick to the human hand and to a large sphere. *Anim. Behav.* **29**, 610–620.

Clifton, P. G., Andrew, R. J., and Gibbs, M. E. (1979). Testosterone and a passive avoidance task. *Abstr. Int. Ethol. Conf. 16th.*

Cohen, J. (1980). "Midbrain and Motor Control of Courtship Behavior in Male and Female Ring Doves *(Streptopelia risoria)*. "Ph.D Thesis, Rutgers State Univ., Newark, New Jersey.

Cohen, J., and Cheng, M. F. (1979). Role of vocalizations in the reproductive cycle of ring doves *(Streptopelia risoria)*: effects of hypoglossal nerve section on the reproductive behavior and physiology of the female. *Horm. Behav.* **13**, 113–127.

Cohen, J., and Cheng, M. F. (1981). The role of the midbrain in courtship behavior of the female ring dove *(Streptopelia risoria)*: evidence from radiofrequency lesion and hormone implant studies. *Brain Res.* **207**, 279–301.

Collias, N. E. (1944). Aggressive behavior among vertebrate animals. *Physiol. Zool.* **17**, 83–123.

Collias, N. E. (1946). Some experiments on broody behavior in fowl and pigeons. *Anat. Rec.* **96**, 572.

Collias, N. E. (1950). Hormones and behaviour with special reference to birds and the mechanism of hormone action. *In* "A Symposium on Steroid Hormones" (E. S. Gordon, ed.), pp. 277–329. Univ. of Wisconsin Press, Madison.

Collias, N. E., Frumkes, P. J., Brooks, D. S., and Barfield, R. J. (1961). Nest building and breeding behavior of castrated village weaverbirds *(Textor cucullatus)*. *Am. Zool.* **1**, 349.

Colombo, L., and Colombo-Belvedere, P. (1978). Biosynthesis of 5β-reduced and 11-oxygenated androgens by the ovary of the Japanese quail at different stages of the follicular cycle. *Gen. Comp. Endocrinol.* **34**, 99.

Colombo-Belvedere, P., Colombo, L., and Deviche, P. (1981). Attivita steroido 11β-e 21-idrossilasiche nel testicolo di *Anas platyrhynchos*. *Boll. Zool.* **48** (Suppl. 1981), 38.

Connell, G. M., Connell, C. J., and Eik-Nes, K. B. (1966). Testosterone synthesis by the two-day-old chick testis *in vitro*. *Gen. Comp. Endocrinol.* **7**, 158–165.

Crawford, W. C., and Glick, B. (1975). The function of the preoptic, mammilaris lateralis and ruber nuclei in normal and sexually inactive male chickens. *Physiol. Behav.* **15**, 171–175.

Croix, D., Hendrick, J. C., Balthazart, J., and Franchimont, P. (1974). Dosage radioimmunologique de l'hormone folliculo-stimulante (FSH) hypophysaire de canard à l'aide d'un système de mammifères. *C. R. Seances Soc. Biol. Ses Fil.* **168**, 136–140.

Crook, J. H., and Butterfield, P. A. (1968). Effects of testosterone propionate and luteinizing hormone on agonistic and nest building behavior of *Quelea quelea*. *Anim. Behav.* **16**, 370–384.

Crook, J. H., and Butterfield, P. A. (1969). Effects of testosterone propionate and luteinizing hormone on agonistic and nest building behaviour of *Quelea quelea*. *Anim. Behav.* **16**, 370–384.

Cummins, R. A., Budtz-Olsen, O. E., Walsh, R. N., and Worsley, A. (1974). Testosterone, early experience and behavioural arousal in a novel environment. *Horm. Behav.* **5**, 283–288.

Cunningham, D. L., Siegel, P. B., and Van Krey, H. P. (1977). Androgen influences on mating behavior in selected lines of Japanese quail. *Horm. Behav.* **8**, 166–174.

Curio, E. (1960). Ontogenese und Phylogenese einiger Triebaüsserungen von Fliegenschnäppern. *J. Ornithol.* **101**, 291–309.

Cusick, E. K., and Wilson, F. E. (1972). On the control of spontaneous testicular regression in tree sparrows *(Spizella arborea)*. *Gen. Comp. Endocrinol.* **19**, 441–456.

Damassa, D. A., Smith, E. R., Tennent, B., and Davidson, J. M. (1977). The relationship between circulating testosterone levels and male sexual behavior in rats. *Horm. Behav.* **8**, 275–286.

Davidson, J. M. (1969). Effects of estrogen on the sexual behavior of male rats. *Endocrinology (Baltimore)* **84**, 1365–1372.

Davies, D. T. (1980). The neuroendocrine control of gonadotrophin release in the Japanese quail. III. The role of the tuberal and anterior hypothalamus in the control of ovarian development and ovulation. *Proc. R. Soc. London Ser. B* **206**, 421–437.

Davies, D. T., and Follett, B. K. (1975a). The neuroendocrine control of gonadotrophin release in the Japanese quail. I. The role of the tuberal hypothalamus. *Proc. R. Soc. London Ser. B* **191**, 285–301.

Davies, D. T., and Follett, B. K. (1975b). The neuroendocrine control of gonadotrophin release in the Japanese quail. II. The role of the anterior hypothalamus. *Proc. R. Soc. London Ser. B* **191**, 303–315.

Davies, D. T., and Follett, B. K. (1975c). Electrical stimulation of the hypothalamus and luteinizing hormone secretion in Japanese quail. *J. Endocrinol.* **67**, 431–438.

Davies, D. T., and Follett, B. K. (1980). Neuroendocrine regulation of gonadotrophin-releasing hormone secretion in the Japanese quail. *Gen. Comp. Endocrinol.* **40**, 220–225.

Davies, D. T., Massa, R., and James, R. (1980). Role of testosterone and of its metabolites in regulating gonadotrophin secretion in the Japanese quail. *J. Endocrinol.* **84**, 211–222.

Davis, D. E. (1957). Aggressive behavior in castrated starlings. *Science (Washington, D.C.)* **126**, 253.

Davis, D. E., and Domm, L. V. (1941). The sexual behavior of hormonally treated domestic fowl. *Proc. Soc. Exp. Biol. Med.* **48**, 667–669.

Davis, D. E., and Domm, L. V. (1943). The influence of hormones on the sexual behavior of the fowl. *In* "Essays in Biology," pp. 171–181. Univ. of California Press, Berkeley.

Davis, P. G., McEwen, B. S., and Pfaff, D. W. (1979). Localized behavioral effects of tritiated estradiol implants in the ventromedial hypothalamus of female rats. *Endocrinology (Baltimore)* **104**, 898–903.

Delius, J. D., Craig, B., and Chaudoir, C. (1976). Andrenocorticotropic hormone, glucose and displacement activities in pigeons. *Z. Tierpsychol.* **40**, 187–193.

De Reviers, M. (1971). Le développement testiculaire chez le coq. I. Croissance pondérale des testicules et développement des tubes séminifères. *Ann. Biol. Anim. Biochim. Biophys.* **11**, 519–530.

Desjardins, C., and Turek, F. W. (1977). Effects of testosterone on spermatogenesis and luteinizing hormone release in Japanese quail. *Gen. Comp. Endocrinol.* **33**, 293–303.

Dessi-Fulgheri, F., Lucarini, N., and Lupo de Prisco, C. (1976). Relationships between testosterone metabolism in the brain, other endocrine variables and intermale aggression in mice. *Aggressive Behav.* **2**, 223–231.

Deviche, P. (1976). Behavioural effects of ACTH or corticosterone administration to adult male domestic ducks, *Anas platyrhynchos* L. *J. Comp. Physiol.* **110**, 357–366.

Deviche, P. (1979a). Behavioral effects of castration and testosterone propionate replacement combined with ACTH in the male domestic duck *(Anas platyrhynchos* L.*)*. *J. Exp. Zool.* **207**, 471–480.

Deviche, P. (1979b). Effects of testosterone propionate and pituitary–adrenal hormones on the social behaviour of male ducklings *(Anas platyrhynchos)* in two test situations. *Z. Tierpsychol.* **49**, 77–86.

Deviche, P. (1982). Are gonadotrophins directly involved in the control of avian activities? In "Aspects of the Endocrinology of Birds: Practical and Theoretical Implications" (C. G. Scanes, M. A. Ottinger, A. D. Kenny, J. Balthazart, J. Cronshaw, and I. Chester Jones, eds.), pp. 105–116. Texas Tech. Univ. Press, Lubbock.

Deviche, P., and Balthazart, J. (1976). Behavioural and morphological effects of testosterone and gonadotrophins in the young male domestic duck *(Anas platyrhynchos* L.*)*. *Behav. Processes* **1**, 217–232.

Deviche, P., and Delius, J. D. (1981). Short-term modulation of domestic pigeon *(Columba livia* L.*)* behaviour induced by intraventricular administration of ACTH. *Z. Tierpsychol.* **55**, 335–342.

Deviche, P., and Schumacher, M. (1981). Behavioural and morphological dose-responses to testosterone and to 5α-dihydrotestosterone in the castrated male Japanese quail. *Behav. Processes*.

Deviche, P., Heyns, W., Balthazart, J., and Hendrick, J. C. (1979). Inhibition of LH plasma levels by corticosterone administration in the male ducklings *(Anas platyrhynchos)*. *IRCS Med. Sci. Libr. Compend.* **7**, 622.

Deviche, P., Bottoni, L., and Balthazart, J. (1982a). 5β-Dihydrotestosterone is weakly androgenic in the adult Japanese quail *(Coturnix coturnix japonica)*. *Gen. Comp. Endocrinol.* **48**, 421–424.

Deviche, P., Massa, R., Bottoni, L., and Hendrick, J. C. (1981b). Effects of corticosterone on the hypothalamo–pituitary gonadal system of male Japanese quails exposed to short or to long photoperiod. *J. Endocrinol.* **95**, 165–173.

De Vlaming, V. L. (1979). Actions of prolactin among the vertebrates. In "Hormones and Evolution" (E. J. W. Barrington, ed.), pp. 561–642. Academic Press, New York.

Döhler, K. D., von zur Mühlen, A., Gärtner, K., and Döhler, U. (1977). Effect of various blood sampling techniques on serum levels of pituitary and thyroid hormones in the rat. *J. Endocrinol.* **74**, 341–342.

Doi, O., Takai, T., Nakamura, T., and Tanabe, Y. (1980). Changes in the pituitary and plasma LH, plasma and follicular progesterone and estradiol, and plasma testosterone and estrone concentrations during the ovulatory cycle of the quail *(Coturnix coturnix japonica)*. *Gen. Comp. Endocrinol.* **41**, 156–163.

Domm, L. V., Davis, D. E., and Blivaiss, B. B. (1942). Observations on the sexual behavior of hormonally treated brown leghorn fowl. *Anat. Rec.* **84**, 481–482.

Donham, R. S. (1979). Annual cycle of plasma luteinizing hormone and sex hormones in male and female mallards *(Anas platyrhynchos)*. *Biol. Reprod.* **21**, 1273–1285.

Donham, R. S., Dane, C. W., and Farner, D. S. (1976). Plasma luteinizing hormone and the development of ovarian follicles after loss of clutch in female mallards *(Anas platyrhynchos)*. *Gen. Comp. Endocrinol.* **29**, 152–155.

Donovan, B. T. (1980). Role of hormones in perinatal brain differentiation. In "The Endocrine Function of the Brain" (M. Motta, ed.), pp. 117–141. Raven, New York.

Drent, R. (1975). Incubation. In "Avian Biology" (D. S. Farner, J. R, King, and K. C. Parkes, eds.), Vol. 5, pp. 333–420. Academic Press, New York.

Dubé, J. Y., and Tremblay, R. R. (1974). Androgen binding proteins in cock's tissues: properties of ear lobe protein and determination of binding sites in head appendages and other tissues. *Endocrinology (Baltimore)* **95**, 1105–1112.

Dudley, S. D., Blaustein, J. D., Gray, J. M., Roy, E. J., and Wade, G. N. (1979). Long-term retention of estradiol by brain cell nuclei and female rat sexual behavior. *Abstr. Conf. Reprod. Biol. 11th*, p. 10.

Dupuy, G. M., Roberts, K. D., Bleau, G., and Chapdelaine, A. (1977). Inhibition of prostatic 5α-reductase and 3α-hydroxysteroid dehydrogenase by two antiandrogens. *J. Steroid Biochem.* **8**, 1145–1151.

Dupuy, G. M., Roberts, K. D., Bleau, G., and Chapdelaine, A. (1978). Steroidal inhibitors of prostatic 5α-reductase: structure–activity relationships. *J. Steroid Biochem.* **9**, 1043–1047.

Egge, A. S., and Chiasson, R. B. (1963). Endocrine effects of diencephalic lesions in the white leghorn hen. *Gen. Comp. Endocrinol.* **3**, 346–361.

Eisner, E. (1960). The relationship of hormones to the reproductive behaviour of birds, referring especially to parental behavior: a review. *Anim. Behav.* **8**, 155–179.

El Halawani, M. E., Burke, W. H., and Dennison, P. T. (1980). Effect of nest-deprivation on serum prolactin level in nesting female turkeys. *Biol. Reprod.* **23**, 118–123.

Erickson, C. J. (1970). Induction of ovarian activity in female ring doves by androgen treatment of castrated males. *J. Comp. Physiol. Psychol.* **71**, 210–215.

Erickson, C. J., and Hutchison, J. B. (1977). Induction of nest-material collecting in male Barbary doves by intracerebral androgen. *J. Reprod. Fertil.* **50**, 9–16.

Erickson, C. J., and Martinez-Vargas, M. C. (1975). The hormonal basis of cooperative nest-building. *In* "Neural and Endocrine Aspects of Behaviour in Birds" (P. Wright, P. G. Caryl, and D. M. Vowles, eds.), pp. 92–109. Elsevier, Amsterdam.

Erickson, C. J., Bruder, R. H., Komisaruk, B. R., and Lehrman, D. S. (1967). Selective inhibition by progesterone of androgen-induced behavior in male ring doves *(Streptopelia risoria)*. *Endocrinology (Baltimore)* **81**, 39–44.

Erpino, M. J. (1969). Hormonal control of courtship behaviour in the pigeon *(Columba livia)*. *Anim. Behav.* **17**, 401–405.

Etches, R. J., Garbutt, A., and Middleton, A. L. (1979). Plasma concentrations of prolactin during egg laying and incubation in the ruffed grouse *(Bonasa umbellus)*. *Can. J. Zool.* **57**, 1624–1627.

Etienne, A. (1964). Der Einfluss von Testosteron auf das Verhalten junger Stockerpel *(Anas platyrhynchos* L.*)*. *Z. Tierpsychol.* **21**, 348–358.

Etienne, A., and Fischer, H. (1964). Untersuchung über das Verhalten kastrierter Stockenten *(Anas platyrhynchos* L.*)* und dessen Beeinflussung durch Testosteron. *Z. Tierpsychol.* **21**, 348–358.

Farner, D. S. (1975). Photoperiodic controls in the secretion of gonadotrophins in birds. *Am. Zool.* **15** (Suppl. 1), 117–135.

Farner, D. S., and Wingfield, J. C. (1980). Reproductive endocrinology of birds. *Annu. Rev. Physiol.* **42**, 457–472.

Feder, H. H. (1978). Specificity of steroid hormone activation of sexual behaviour in rodents. *In* "Biological Determinants of Sexual Behaviour" (J. B. Hutchison, ed.), pp. 395–424. Wiley, Chichester.

Feder, H. H., Storey, A., Goodwin, D., Reboulleau, C., and Silver, R. (1977). Testosterone and "5α-dihydrotestosterone" levels in peripheral plasma of male and female ring doves *(Streptopelia risoria)* during the reproductive cycle. *Biol. Reprod.* **16**, 666–677.

Feder, H. H., Blaustein, J. D., and Nock, B. L. (1979a). Oestrogen-progestin regulation of female sexual behavior in guinea pigs. *J. Steroid Biochem.* **11**, 873–877.

Feder, H. H., Landau, I. T., and Walker, W. A. (1979b). Anatomical and biochemical substrates of the actions of estrogens and antiestrogens on brain tissues that regulate female sex behavior of rodents. *In* "Endocrine Control of Sexual Behavior" (C. Beyer, ed.), pp. 317–340. Raven, New York.

Fevold, H. R., and Eik-Nes, K. B. (1962). Progesterone metabolism by testicular tissue of the

English sparrow *(Passer domesticus)* during the annual reproductive cycle. *Gen. Comp. Endocrinol.* **2**, 506–515.
Fevold, H. R., and Eik-Nes, K. B. (1963). Progesterone metabolism by testicular tissue of the English sparrow *(Passer domesticus)*. *Gen. Comp. Endocrinol.* **3**, 335–345.
Fevold, H. R., and Pfeiffer, E. W. (1968). Androgen production *in vitro* by phalarope gonadal tissue homogenates. *Gen. Comp. Endocrinol.* **10**, 26–33.
Follett, B. K. (1973). Circadian rhythms and photoperiodic time measurement in birds. *J. Reprod. Fertil.* (Suppl. 19), 5–18.
Follett, B. K. (1976). Plasma follicle-stimulating hormone during photoperiodically induced sexual maturation in male Japanese quail. *J. Endocrinol.* **69**, 117–126.
Follett, B. K. (1978). Photoperiodism and seasonal breeding in birds and mammals. *In* "Control of Ovulation" (G. E. Lamming and D. B. Creighton, eds.), pp. 267–293. Butterworth, London.
Follett, B. K., and Davies, D. T. (1975). Photoperiodicity and the neuroendocrine control of reproduction in birds. *Symp. Zool. Soc. London* **35**, 199–224.
Follett, B. K., and Maung, S. L. (1978). Rate of testicular maturation, in relation to gonadotrophin and testosterone levels, in quail exposed to various artificial photoperiods and to natural day lengths. *J. Endocrinol.* **78**, 267–280.
Follett, B. K., and Robinson, J. E. (1980). Photoperiod and gonadotrophin secretion in birds. *Prog. Reprod. Biol.* **5**, 39–61.
Follett, B. K., Scanes, C. G., and Cunningham, F. J. (1972). A radioimmunoassay for avian luteinizing hormones. *J. Endocrinol.* **52**, 359–398.
Follett, B. K., Davies, D. T., and Gledhill, B. (1977). Photoperiodic control of reproduction in Japanese quail: changes in gonadotrophin secretion on the first day of induction and their pharmacological blockade. *J. Endocrinol.* **74**, 449–460.
Foreman, M. M., and Moss, R. L. (1977). Effects of subcutaneous injection and intrahypothalamic infusion of releasing hormones upon lordotic response to repetitive coital stimulation. *Horm. Behav.* **8**, 219–234.
Fox, C. A., Ismail, A. A. A., Love, D. N., Kirkham, K. E., and Loraine, J. A. (1972). Studies on the relationship between plasma testosterone levels and human sexual activity. *J. Endocrinol.* **52**, 51–58.
Friedman, M. B. (1977). Interactions between visual and vocal courtship stimuli in the neuroendocrine response of female doves. *J. Comp. Physiol. Psychol.* **91**, 1408–1416.
Furr, B., and Thomas, B. (1970). Estimation of testosterone in plasma of the domestic fowl. *J. Endocrinol.* **48**, XLII.
Furuyama, S., Nayes, D. M., and Nugent, C. A. (1970). A radioimmunoassay for plasma testosterone. *Steroids* **16**, 415–428.
Gallagher, T. F., and Koch, F. C. (1935). The quantitative assay for the testicular hormone by the comb-growth reaction; second communication. *J. Pharmacol. Expl. Ther.* **55**, 97–117.
Galli, F. E., Irusta, O., and Wasserman, G. F. (1973). Androgen production by testes of *Gallus domesticus* during postembryonic development. *Gen. Comp. Endocrinol.* **21**, 262–266.
Gardner, J. A., and Fisher, A. E. (1968). Induction of mating in male chicks following preoptic implantation of androgen. *Physiol. Behav.* **3**, 709–712.
Garnier, D. (1971). Variations de la testostérone du plasma périphérique chez le canard Pékin au cours du cycle annuel. *C. R. Hebd. Seances Acad. Sci. Ser. D* **272**, 1665–1668.
Gentle, M. J., Wood-Gush, D. G. M., and Goron, J. (1978). Behavioural effects of hyperstriatal ablation in Gallus domesticus. *Behav. Processes* **3**, 137–148.
Ghraf, R., Lax, E. R., and Schriefers, H. (1979). Action of oestrogens and antioestrogens on oestrogen-inducible cytoplasmic and androgen-inducible microsomial activity of 3α-hydroxysteroid dehydrogenase in male rat kidney. *J. Endocrinol.* **83**, 157–163.

Gibson, M. J., and Cheng, M. F. (1979). Neural mediation of estrogen-dependent courtship behavior in female ring doves. *J. Comp. Physiol. Psychol.* **93**, 855–867.

Giorgi, E. P. (1980). The transport of steroid hormones into animal cells. *Int. Rev. Cytol.* **65**, 49–115.

Gispen, W. H., and Isaacson, R. L. (1981). ACTH-induced excessive grooming in the rat. *Pharmacol. Ther.* **12**, 209–246.

Gledhill, B., and Follett, B. K. (1976). Diurnal variation and the episodic release of plasma gonadotrophins in Japanese quail during a photoperiodically-induced gonadal cycle. *J. Endocrinol.* **71**, 245–257.

Gogan, F. (1968). Sensibilité hypothalamique à la testostérone chez le canard. *Gen. Comp. Endocrinol.* **11**, 316–327.

Goldsmith, A. R., and Williams, D. M. (1980). Incubation in mallards *(Anas platyrhynchos)*: changes in plasma levels of prolactin and luteininzing hormone. *J. Endocrinol.* **86**, 371–379.

Goldsmith, A. R., Edwards, C., Koxrucu, M., and Silver, R. (1981). Concentrations of prolactin and luteinizing hormone in relation to incubation and development of the crop gland. *J. Endocrinol.* **90**, 437–443.

Goodale, H. D. (1913). Castration in relation to the secondary sexual characters in brown leghorns. *Am. Nat.* **47**, 159–169.

Gorman, M. L. (1974). The endocrine basis of pair-formation behaviour in the male eider, *Somateria mollissima*. *Ibis* **116**, 451–465.

Gorski, R. A. (1976). The possible neural sites of hormonal facilitation of sexual behavior or the female rat. *Psychoneuroendocrinology* **1**, 371–387.

Gosney, D., and Hinde, R. A. (1975). An oestrogen-mediated effect of photoperiod in the reproductive behaviour of the budgerigar. *J. Reprod. Fertil.* **45**, 547–548.

Gray, G. D., Davis, H. N., Kenny, A. McM., and Dewsbury, D. A. (1976). Effect of mating on plasma levels of LH and progesterone in montane voles *(Microtus montanus)*. *J. Reprod. Fertil.* **47**, 89–91.

Grunt, J. A., and Young, W. C. (1953). Consistency of sexual behaviour patterns in individual male guinea pigs following castration and androgen therapy. *J. Comp. Physiol. Psychol.* **46**, 138–144.

Gschwendt, M. (1980). The general validity of the subunit model of the progesterone receptor from chick oviduct appears questionable. Comparison of progesterone and estrogen receptor. *Mol. Cell. Endocrinol.* **19**, 57–67.

Guhl, A. M. (1949). Heterosexual dominance and mating behaviour in chickens. *Behaviour* **2**, 102–120.

Guichard, A., Cedard, L., Mignot, Th. M., Scheib, D., and Haffen, K. (1977). Radioimmunoassays of steroids produced by cultured chick embryo gonads: differences according to age, sex and side. *Gen. Comp. Endocrinol.* **32**, 255–265.

Guichard, A., Scheib, D., Haffen, K., Mignot, Th. M., and Cedard, L. (1980). Comparative study in steroidogenesis by quail and chick embryonic gonads in organ culture. *J. Steroid Biochem.* **12**, 83–87.

Gurney, M. E., and Konishi, M. (1980). Hormone-induced sexual differentiation of brain and behavior in zebra finches. *Science (Washington, D.C.)* **208**, 1380–1383.

Güttinger, H. R. (1979). The integration of learnt and genetically programmed behaviour: a study of hierarchical organization in songs of canaries, green finches and their hybrids. *Z. Tierpsychol.* **49**, 285–303.

Guyomarc'h, J. C. (1974). "Les Vocalisations des gallinacés. Structure des sons et des répertoires. Ontogenése motrice et acquisition de leur sémantique. Thèse, Univ. of Rennes, France.

Guyomarc'h, J. C., and Thibout, E. (1969). Rythmes et cycles dans l'émission du chant chez la caille japonaise *(Coturnix coturnix japonica)*. *Rev. Comp. Anim.* **3**, 37–49.

Gwinner, E. (1973). Circannual rhythms in birds: their interaction with circadian rhythms and envrionmental photoperiod. *J. Reprod. Fertil.* (Suppl. 19), 51–65.

Gwinner, E. (1974). Testosterone induces "splitting" of circadian locomotor activity rhythms in birds. *Science (Washington, D.C.)* **185**, 72–74.

Gwinner, E. (1975a). Circadian and circannual rhythms in birds. *In* "Avian Biology" (D. S. Farner, J. R. King, and K. C. Parkes, eds.), Vol. 5, pp. 221–285. Academic Press, New York.

Gwinner, E. (1975b). Effects of season and external testosterone on the freerunning circadian activity rhythm of European starlings *(Sturnus vulgaris)*. *J. Comp. Physiol.* **103**, 315–328.

Gyermek, L., Genther, G., and Fleming, L. (1967). Some effects of progesterone and related steroids on the central nervous system. *Int. J. Neurochem.* **6**, 191–198.

Hamilton, J. B. (1938). Precocious masculine behaviour following administration of synthetic male hormone substance. *Endocrinology (Baltimore)* **23**, 53–57.

Hamilton, J. B., and Golden, W. R. C. (1939). Responses of the female to male hormone substances. *Endocrinology (Baltimore)* **25**, 737–748.

Hammer, N. M. (1968). The photorefractory period of the house finch. *Ecology* **49**, 211–227.

Hansen, E. W. (1966). Squab-induced crop growth in ring dove foster parents. *J. Comp. Physiol. Psychol.* **62**, 120–122.

Hansen, E. W. (1971a). Squab-induced crop growth in experienced and inexperienced ring dove *(Streptopelia risoria)* foster parents. *J. Comp. Physiol. Psychol.* **77**, 375–381.

Hansen, E. W. (1971b). Responsiveness of ring dove foster parents to squabs. *J. Comp. Physol. Psychol.* **77**, 382–387.

Harding, C. F., and Follett, B. K. (1979). Hormone changes triggered by aggression in a natural population of blackbirds. *Science (Washington, D. C.)* **203**, 918–920.

Harlan, R. E., Gordon, J. H., and Gorski, R. A. (1979). Sexual differentiation of the brain: implications for neuroscience. *Rev. Neurosci.* **4**, 31–71.

Harrison, R. W., and Toft, D. O. (1975). Estrogen receptors in the chick oviduct. *Endocrinology (Baltimore)* **96**, 199–205.

Harvey, S., Merry, B. J., and Phillips, J. G. (1980). Influence of stress on the secretion of corticosterone in the duck *(Anas platyrhynchos)*. *J. Endocrinol.* **87**, 161–171.

Haynes, R. L., and Glick, B. (1974). Hypothalamic control of sexual behavior in the chicken. *Poult. Sci.* **53**, 27–38.

Hazelwood, R. L. (1972). The intermediary metabolism of birds. *In* "Avian Biology" (D. S. Farner, J. R. King, and K. C. Parkes, eds.), Vol. 2, pp. 472–526. Academic Press, New York.

Hinde, R. A. (1958). The nest building behavior of domesticated canaries. *Proc. Zool. Soc. London* **131**, 1–48.

Hinde, R. A. (1965). Interaction of internal and external factors in integration of canary reproduction. *In* "Sex and Behavior" (F. A. Beach, ed.), pp. 381–415. Wiley, New York.

Hinde, R. A. (1966). "Animal Behaviour: A Synthesis of Ethology and Comparative Psychology." McGraw-Hill, London.

Hinde, R. A. (1970). "Animal Behaviour: A Synthesis of Ethology and Comparative Psychology," 2nd ed. McGraw-Hill Kogakusha, Tokyo.

Hinde, R. A. (1973). Behavior *In* "Avian Biology" (D. S. Farner, J. R. King, and K. C. Parkes, eds.), Vol. 3, pp. 480–535. Academic Press, New York.

Hinde, R. A., and Steel, E. A. (1962). Selection of nest material by female canaries. *Anim. Behav.* **10**, 67–75.

Hinde, R. A., and Steel, E. A. (1966). Integration of reproductive behaviour in female canaries. *Symp. Soc. Exp. Biol.* **20**, 401–426.

4. HORMONAL CORRELATES OF BEHAVIOR

Hinde, R. A., and Steel, E. A. (1975). The dual role of day length in controlling canary reproduction. *Symp. Zool. Soc. London* **35**, 245–259.

Hinde, R. A., and Steel, E. A. (1976). The effect of male song on an estrogen-dependent behavior pattern in the female canary. *(Serinus canarius). Horm. Behav.* **7**, 293–304.

Hinde, R. A., and Steel, E. A. (1978). The influence of day length and male vocalizations on the estrogen-dependent behavior of female canaries and budgerigars, with discussion of data from other species. *Adv. Study Behav.* **8**, 39–73.

Hinde, R. A., Bell, R. Q., and Steel, E. A. (1963). Changes in sensitivity of the canary brood patch during the natural breeding season. *Anim. Behav.* **11**, 553–560.

Hinde, R. A., Steel, E. A., and Follett, B. K. (1974). Effect of photoperiod on oestrogen-induced nest-building in ovariectomized or refractory canaries *(Serinus canarius). J. Reprod. Fertil.* **40**, 383–399.

Hirano, H., Nakamura, T., and Tanabe, Y. (1978). Changes in plasma LH and testosterone concentrations of Japanese quail from hatching to sexual maturity. *Nippon Kakin Gakkaishi (Jpn. Poult. Sci.)* **15**, 242–247.

Hiroshige, T. (1974). Circadian rhythm of corticotrophin-releasing activity in the rat hypothalamus: an attempt at physiological validation. *In* "Biological Rhythms in Neuroendocrine Activity" (M. Kawakami, ed.), pp. 267–280. Igaku Shoin, Tokyo.

Höhn, E. O. (1947). Sexual behaviour and seasonal changes in the gonads and adrenals of the mallard. *Proc. Zool. Soc. London* **117**, 281–304.

Höhn, E. O. (1960). Seasonal changes in the mallard's penis and their hormonal control. *Proc. Zool. Soc. London* **134**, 547–555.

Höhn, E. O., and Cheng, S. C. (1967). Gonadal hormones in Wilson's phalarope *(Steganopus tricolor)* and other birds in relation to plumage and sex behavior. *Gen. Comp. Endocrinol.* **8**, 1–11.

Holzbauer, M. (1971). *In vivo* production of steroids with central depressant action by the ovary of the rat. *Br. J. Pharmacol.* **43**, 560–569.

Horn, G. (1979). Imprinting—in search of neural mechanisms. *Trends Neurosci.* 1979, 219–222.

Horn, G., McCabe, B. J., and Bateson, P. P. G. (1979). An autoradiographic study of the chick brain after imprinting. *Brain Res.* **168**, 361–373.

Horst, H. J., and Paulke, E. (1977). Comparative study of androgen uptake and metabolism in domestic and wild mallard drakes. *Gen. Comp. Endocrinol.* **32**, 138–145.

Hutchison, J. B. (1967). Initiation of courtship by hypothalamic implants of testosterone propionate in castrated doves *(Streptopelia risoria). Nature (London)* **216**, 591–592.

Hutchison, J. B. (1969). Changes in hypothalamic responsiveness to testosterone in male barbary doves *(Streptopelia risoria). Nature (London)* **222**, 176–177.

Hutchison, J. B. (1970a). Differential effects of testosterone and oestradiol on male courtship in Barbary doves *(Streptopelia risoria). Anim. Behav.* **18**, 41–51.

Hutchison, J. B. (1970b). Influence of gonadal hormones on the hypothalamic integration of courtship behaviour in the Barbary dove. *J. Reprod. Fertil.* (Suppl. 11), 15–41.

Hutchison, J. B. (1971). Effects of hypothalamic implants of gonadal steroids on courtship behaviour in Barbary doves *(Streptopelia risoria). J. Endocrinol.* **50**, 97–113.

Hutchison, J. B. (1974a). Post-castration decline in behavioural responsiveness to intra-hypothalamic androgen in doves. *Brain Res.* **81**, 169–181.

Hutchison, J. B. (1974b). Effect of photoperiod on the decline in behavioural responsiveness to intra-hypothalamic androgen in doves *(Streptopelia risoria). J. Endocrinol.* **63**, 583–584.

Hutchison, J. B. (1975). Target cells for gonadal steroids in the brain: studies on steroid-sensitive mechanisms of behaviour. *In* "Neural and Endocrine Aspects of Behaviour in Birds" (P. Wright, P. G. Caryl, and D. M. Vowles, eds.), pp. 123–137. Elsevier, Amsterdam.

Hutchison, J. B. (1976). Hypothalamic mechanisms of sexual behavior, with special reference to birds. *Adv. Study Anim. Behav.* **6**, 159–200.

Hutchison, J. B. (1978). Hypothalamic regulation of male sexual responsiveness to androgen. *In* "Biological Determinants of Sexual Behaviour" (J. B. Hutchison, ed.), pp. 277–317. Wiley, Chichester.

Hutchison, J. B., and Katongole, C. (1975). Plasma testosterone and aggressive courtship in the male Barbary dove *(Streptopelia risoria). J. Endocrinol.* **65**, 275–276.

Hutchison, R. E. (1978). Hormonal differentiation of sexual behavior in Japanese quail. *Horm. Behav.* **11**, 363–387.

Hutchison, R. E., Hinde, R. A., and Bendon, B. (1968). Oviduct development and its relation to other aspects of reproduction in domesticated canaries *J. Zool.* **155**, 87–102.

Idler, D. R. (ed) (1972). "Steroids in Nonmammalian Vertebrates." Academic Press, New York.

Illius, A. W., Haynes, N. B., and Lamming, G. E. (1976). Effects of ewe proximity on peripheral plasma testosterone levels and behaviour in the ram. *J. Reprod. Fertil.* **48**, 25–32.

Immelmann, K. (1971). Ecological aspects of periodic reproduction. *In* "Avian Biology" (D. S. Farner, J. R. King, and K. C. Parkes, eds.), Vol. 1, pp. 341–389. Academic Press, New York.

Irving, R. A., Mainwaring, W. I. P., and Spooner, P. M. (1976). The regulation of haemoglobin synthesis in cultured chick blastoderms by steroids related to 5β-androstane. *Biochem. J.* **154**, 81–93.

Ishii, S., and Tsutsui, K. (1981). Hormonal control of aggressive behavior in male Japanese quail. *In* "Aspects of the Endocrinology of Birds: Practical and Theoretical Implications" (C. G. Scanes, M. A. Ottinger, A. D. Kenny, J. Balthazart, J. Cronshaw, and I. Chester Jones, eds.), pp. 125–131. Texas Tech. Univ. Press, Lubbock.

Ismail, A. A. A., Niswender, G. D., and Midgley, A. R. (1972). Radioimmunoassay of testosterone without chromatography. *J. Clin. Endocrinol. Metab.* **34**, 177–184.

Jacob, J., Balthazart, J., and Schoffeniels, E. (1979). Sex differences in the chemical composition of uropygial gland waxes in domestic ducks. *Biochem. Syst. Ecol.* **7**, 149–153.

Jaffe, B. M., and Behrman, H. R. (eds) (1974). "Methods of Hormone Radioimmunoassay." Academic Press, New York.

Jaffe, R. B. (1969). Testosterone metabolism in target tissues: hypothalamic and pituitary tissues of the adult rat and human fetus and in the immature rat epiphysis. *Steroids* **14**, 483–498.

Jallageas, M., and Assenmacher, I. (1970). Testostéronémie du canard photo-stimulé ou soumis à des injections répétes de testostérone. *C. R. Seances Soc. Biol. Ses Fil.* **164**, 2338–2341.

Jallageas, M., and Assenmacher, I. (1973). Effets du jeûne et de la castration sur la cinétique du métabolisme de la testostérone chez le canard. *Gen. Comp. Endocrinol.* **20**, 405–406.

Jallageas, M., and Attal, J. (1968). Dosage par chromatographie en phase gazeuse de la testostérone plasmatique chez le canard, la caille, le pigeon. *C. R. Hebd. Seances Acad. Sci. Ser. D* **267**, 341–347.

Jallageas, M., Assenmacher, I., and Follett, B. K. (1974). Testosterone secretion and plasma luteinizing hormone concentration during a seasonal cycle in the Pekin duck, and after thyroxine treatment. *Gen. Comp. Endocrinol.* **23**, 472–475.

James, H. (1962). Imprinting with visual flicker: effects of testosterone cyclopentylpropionate. *Anim. Behav.* **10**, 341–346.

Jellinck, P. H., Krey, L., Davis, P. G., Kamel, F., Luine, V., Parsons, B., Roy, E. J., and McEwen, B. S. (1981). Central and peripheral action of estradiol and catecholestrogens administered at low concentration by constant infusion. *Endocrinology (Baltimore)* **108**, 1848–1854.

Jensen, E. V., and Desombre, E. R. (1973). Estrogen-receptor interaction. *Science (Washington, D.C.)* **182**, 126–134.

Johns, J. E. (1964). Testosterone induced nuptial feathers in phalaropes. *Condor* **66**, 449–454.
Johnson, A. L., and van Tienhoven, A. (1980). Plasma concentrations of six steroids and LH during the ovulatory cycle of the hen, *Gallus domesticus*. *Biol. Reprod.* **23**, 386–393.
Jolles, J., Rompa-Barendregt, J., and Gispen, W. H. (1979). ACTH-induced excessive grooming in the rat: the influence of environmental and motivational factors. *Horm. Behav.* **12**, 60–72.
Jones, R. E. (1969). Effect of prolactin and progesterone on gonads of breeding California quail. *Proc. Soc. Exp. Biol. Med.* **134**, 172–174.
Jouan, P., and Samperez, S. (1980). Metabolism of steroid hormones in the brain. *In* "The Endocrine Functions of the Brain" (M. Motta, ed.), pp. 95–115. Raven, New York.
Juhasz, L. P., and van Tienhoven, A. (1964). Effect of electrical stimulation of the telencephalon on ovulation and oviposition in the hen. *Am. J. Physiol.* **207**, 286–290.
Kaltenhäuser, D. (1971). Über Evolutionsvorgänge in der Schwimmentenbalz. *Z. Tierpsychol.* **29**, 481–540.
Kao, L. W. L., and Weisz, J. (1979). Inhibition of steroid 5α-reductase *in vivo*: effect on suppression of luteinizing hormone and stimulation of accessory sex organs by testosterone in the orchidectomized rat. *J. Endocrinol.* **81**, 209–220.
Karten, H. J. (1967). The organization of the ascending auditory pathway in the pigeon *(Columba livia)*. I. Diencephalic projections of the inferior colliculus *(Nucleus mesencephali lateralis, pars dorsalis)*. *Brain Res.* **6**, 409–427.
Karten, H. J., and Hodos, W. (1967). "A Stereotaxic Atlas of the Brain of the Pigeon *(Columba livia).*" John Hopkins Press, Baltimore, Maryland.
Katongole, C. B., Naftolin, F., and Short, R. V. (1971). Relationship between blood levels of luteinizing hormone and testosterone in bulls and the effects of social stimulation. *J. Endocrinol.* **50**, 457–466.
Katzenellenbogen, B. S. (1980). Dynamics of steroid hormone receptor action. *Annu. Rev. Physiol.* **47**, 17–35.
Kawashima, M., Kamiyoshi, M., and Tanaka, K. (1978). A cytoplasmic progesterone receptor in the hen pituitary and hypothalamic tissues. *Endocrinology (Baltimore)* **102**, 1207–1213.
Kawashima, M., Kamiyoshi, M., and Tanaka, K. (1979a). Cytoplasmic progesterone receptor concentrations in the hen hypothalamus and pituitary: difference between laying and nonlaying hens and changes during the ovulatory cycle. *Biol. Reprod.* **20**, 581–585.
Kawashima, M., Kamiyoshi, M., and Tanaka, K. (1979b). Effects of progesterone, estradiol and testosterone on cytoplasmic progesterone receptor concentrations in the hen hypothalamus and pituitary. *Biol. Reprod.* **21**, 639–646.
Kawashima, M., Kamiyoshi, M., and Tanaka, K. (1980). Relationship between the changes in cytoplasmic progesterone receptor concentration and in nuclear progesterone binding sites in the hen hypothalamus and pituitary during the ovulatory cycle. *Endocrinol. Jpn.* **27**, 667–670.
Kerlan, J. T., and Jaffe, R. B. (1974). Plasma testosterone levels during the testicular cycle of the redwinged blackbird *(Agelaius phoeniceus)*. *Gen. Comp. Endocrinol.* **22**, 428–432.
Kim, Y. S., Stumpf, W. E., Sar, M., and Martinez-Vargas, M. C. (1978). Estrogen and androgen target cells in the brain of fishes, reptiles and birds: phylogeny and ontogeny. *Am. Zool.* **18**, 425–433.
Kime, D. E. (1980). Comparative aspects of testicular androgen biosynthesis in non-mammalian vertebrates. *In* "Steroids and Their Mechanism of Action in Nonmammalian Vertebrates" (G. Delrio, and J. Brachet, eds.), pp. 17–31. Raven, New York.
King, J. R., Follett, B. K., Farner, D. S., and Morton, M. L. (1966). Annual gonadal cycles and pituitary gonadotrophins in *Zonotrichia leucophrys gambelii*. *Condor* **68**, 476–487.
King, R. J. B. (1976). Intracellular reception of steroid hormones. *Essays Biochem.* **12**, 41–76.

Klinghammer, E., and Hess, E. H. (1964). Parental feeding in ring doves *(Streptopelia roseogrisea)*: innate or learned? *Z. Tierpsychol.* **21**, 338–347.
Klint, T. (1980). Influence of male nuptial plumage on mate selection in the female mallard *(Anas platyrhynchos)*. *Anim. Behav.* **28**, 1230–1238.
Kluijver, H. N. (1933). Bijdrage tot de biologie en de ecologie van den spreeuw. *Versl. Plziekt. Dienst. Wageningen* **69**, 1–146.
Kniewald, Z., Massa, R., and Martini, L. (1971). Conversion of testosterone into 5α-androstan-17β-ol-3-one at the anterior pituitary and hypothalamic level. *In* "Hormonal Steroids" (V. H. T. James, and L. Martini, eds.), pp. 784–791. Excerpta Medica, Amsterdam.
Komisaruk, B. (1967). Effects of local brain implants of progesterone on reproductive behavior in ring doves. *J. Comp. Physiol. Psychol.* **64**, 219–224.
Komisaruk, B. R. (1978). The nature of the neural substrate of female sexual behaviour in mammals and its hormonal sensitivity: review and speculations. *In* "Biological Determinants of Sexual Behavior" (J. B. Hutchison, ed.), pp. 349–393. Wiley, Chichester.
Kon, O. L., Webster, R. A., and Spelsberg, T. C. (1980). Isolation and characterization of the estrogen receptor in hen oviduct: evidence for two molecular species. *Endocrinology (Baltimore)* **107**, 1182–1191.
Korenbrot, C. C., Schomberg, D. W., and Erickson, K. J. (1974). Radioimmunoassay of plasma estradiol during the breeding cycle of ring doves. *Endocrinology (Baltimore)* **94**, 1126–1132.
Kubli-Garfias, C., Medrano-Conde, L., Beyer, C., and Bondani, A. (1979). *In vitro* inhibition of rat uterine contractility induced by 5α and 5β progestins. *Steroids* **34**, 609–617.
Kuusisto, P. (1941). Studien über die Okologie und Tagesrhythmik von *Phylloscopus trochilus acredula* (L.) mit besonderer Berücksichtigung der Brutbiologie. *Acta Zool. Fenn.* **31**, 1–120.
Lahr, E., and Riddle, O. (1938). Proliferation of crop-sac epithelium in incubating and in prolactin-injected pigeons studied with the colchicine method. *Am. J. Physiol.* **123**, 614–619.
Landau, T. I. (1980). Facilitation of male sexual behavior in adult male rats by the aromatization inhibitor, 1,4,6-androstatriene-3,17–dione (ATD). *Physiol. Behav.* **25**, 173–177.
Landsberg, J. W., and Weiss, J. (1977). Stress and increase in the corticosterone level prevent imprinting in ducklings. *Behaviour* **57**, 173–189.
Lane, S. E., Gidari, A. S., and Levere, R. D. (1975). Cytoplasmic receptor protein for etiocholanolone in chick embryo liver. *J. Biol. Chem.* **250**, 8209–8213.
Larsson, K., Södersten, P., and Beyer, C. (1973). Sexual behavior in male rats treated with estrogen in combination with dihydrotestosterone. *Horm. Behav.* **4**, 289–299.
Lawrence, L. K. (1953). Nesting life and behaviour of the Red-eyed Vireo. *Can. Field Nat.* **67**, 67–77.
Laws, D. F., and Farner, D. S. (1960). Prolactin and the photoperiodic testicular response in white-crowned sparrows. *Endocrinology (Baltimore)* **67**, 279–281.
Lazarus, J., and Crook, J. H. (1973). The effects of luteinizing hormone, oestrogen and ovariectomy on the agonistic behaviour of female *Quelea quelea*. *Anim. Behav.* **21**, 49–60.
Legait, N. (1959). "Contribution à l'étude morphologique et expérimentale du système hypothalamo-neurohypophysaire de la poule Rhode-Islande." Thèse, Univ. of Louvain, Belgium.
Lehrman, D. S. (1955). The physiological basis of parental feeding in the ring dove *(Streptopelia risoria)*. *Behaviour* **7**, 241–286.
Lehrman, D. S. (1958a). Induction of broodiness by participation in courtship and nest-building in the ring dove *(Streptopelia risoria)*. *J. Comp. Physiol. Psychol.* **51**, 32–36.

Lehrman, D. S. (1958b). Effect of female sex hormones on incubation behavior in the ring dove *(Streptopelia risoria)*. *J. Comp. Physiol. Psychol.* **51**, 142–145.
Lehrman, D. S. (1961). Hormonal regulation of parental behavior in birds and infrahuman mammals. *In* "Sex and Internal Secretions" (W. C. Young, ed.), pp. 1268–1382. Williams & Wilkins, Baltimore, Maryland.
Lehrman, D. S. (1963). On the initiation of incubation behaviour in doves. *Anim. Behav.* **11**, 433–438.
Lehrman, D. S. (1964a). The reproductive behavior of ring doves. *Sci. Am.* **211**, 48–54.
Lehrman, D. S. (1964b). Control of behavior cycles in reproduction. *In* "Social Behavior and Organization among Vertebrates" (W. Etkin, ed.), pp. 143–166. Univ. of Chicago Press, Chicago, Illinois.
Lehrman, D. S. (1965). Interaction between internal and external environments in the regulation of the reproductive cycle of the ring dove. *In* "Sex and Behavior" (F. A. Beach, ed.), pp. 355–380. Wiley, New York.
Lehrman, D. S., and Brody, P. N. (1961). Does prolactin induce incubation behaviour in the ring dove. *J. Endocrinol.* **22**, 269–275.
Lehrman, D. S., and Brody, P. N. (1964). Effect of prolactin on established incubation behavior in the ring dove. *J. Comp. Physiol. Psychol.* **57**, 161–165.
Lehrman, D. S., Brody, P. N., and Wortis, R. P. (1961). The presence of the mate and of nesting material as stimuli for the development of incubation behavior and for gonadotropin secretion in the ring dove *(Streptopelia risoria)*. *Endocrinology (Baltimore)* **68**, 507–516.
Lieberburg, I., and Nottebohm, F. (1979). High-affinity androgen binding proteins in syringeal tissues of songbirds. *Gen. Comp. Endocrinol.* **37**, 286–293.
Lincoln, G. A., Peet, M. J., and Cunningham, R. A. (1977). Seasonal and circadian changes in the episodic release of follicle-stimulating hormone. luteinizing hormone and testosterone in rams exposed to artificial photoperiods. *J. Endocrinol.* **72**, 337–349.
Lincoln, G. A., Racey, P. A., Sharp, P. J., and Klandorf, H. (1980). Endocrine changes associated with spring and autumn sexuality of the Rook, *Corvus frugileus*. *J. Zool.* **190**, 137–153.
Lisk, R. D. (1962). Diencephalic placement of estradiol and sexual receptivity in the female rat. *Am. J. Physiol.* **203**, 493–496.
Lisk, R. D. (1978). The regulation of sexual "heat." *In* "Biological Determinants of Sexual Behaviour" (J. B. Hutchison, ed.), pp. 425–466. Wiley, Chichester.
Lofts, B. (1975). Environmental control of reproduction. *Symp. Zool. Soc. London* **35**, 177–197.
Lofts, B., and Marshall, A. J. (1956). The effect of prolactin administration on the internal rhythm of reproduction in male birds. *J. Endocrinol.* **13**, 101–106.
Lofts, B., and Murton, R. K. (1973). Reproduction in birds. *In* "Avian Biology" (D. S. Farner, J. R. King, and K. C. Parkes, eds.), Vol. 3, pp. 1–107. Academic Press, New York.
Lofts, B., Murton, R. K., and Westwood, N. J. (1966). Gonadal cycles and the evolution of breeding seasons in British Columbidae. *J. Zool.* **150**, 249–272.
Lofts, B., Follett, B. K., and Murton, R. K. (1970). Temporal changes in the pituitary-gonadal axis. *Mem. Soc. Endocrinol.* **18**, 545–575.
Lott, D. G., and Comerford, S. (1968). Hormonal initiation of parental behavior in inexperienced ring doves. *Z. Tierpsychol.* **25**, 71–75.
Lücke, J., and Haase, E. (1980). Autoradiographische Untersuchungen am Gehirn von Bergfinken *(Fringilla montifringilla* L.*)* nach Injektion von ^{3}H-testosteron. *J. Hirnforsch.* **21**, 369–380.
Luine, V., Nottebohm, F., Harding, C., and McEwen, B. S. (1980). Androgen affects cholinergic enzymes in syringeal motor neurons and muscles. *Brain Res.* **192**, 89–107.

Lumia, A. R. (1972). The relationships among testosterone, conditioned aggression, and dominance in male pigeons. *Horm. Behav.* **3**, 277–286.

Luttge, W. G., Hall, N. R., Wallis, C. J., and Campbell, J. C. (1975). Stimulation of male and female sexual behaviour in gonadectomized rats with estrogen and androgen therapy and its inhibition with concurrent antihormone therapy. *Physiol. Behav.* **14**, 65–73.

MacLusky, N. J., and McEwen, B. S. (1980). Progestin receptors in rat brain: distribution and properties of cytoplasmic progestin-binding sites. *Endocrinology (Baltimore)* **106**, 192–202.

MacLusky, N. J., and Naftolin, F. (1981). Sexual differentiation of the central nervous system. *Science (Washington, D.C.)* **211**, 1294–1303.

Maley, M. J. (1969). Electrical stimulation of agonistic behavior in the mallard. *Behaviour* **34**, 138–160.

March, G. L., and McKeown, B. A. (1973). Serum and pituitary prolactin changes in the band-tailed pigeon *(Columba fasciata)* in relation to the reproductive cycle. *Can. J. Physiol. Pharmacol.* **51**, 583–589.

Marler, P., Kreith, M., and Willis, E. (1962). An analysis of testosterone-induced crowing in young domestic cockerels. *Anim. Behav.* **10**, 48–54.

Marshall, A. J., and Serventy, D. L. (1956). The breeding cycle of the short-tailed shearwater, *Puffinus tenuirostris* (Temmeck) in relation to transequatorial migration and its environment. *Proc. Zool. Soc. London* **127**, 489–510.

Martin, B. (1980). Steroid-protein interactions in nonmammalian vertebrates: distribution, origin, regulation and physiological significance of plasma steroid binding proteins. In "Steroids and Their Mechanism of Action in Nonmammalian Vertebrates" (G. Delrio, and J. Brachet, eds.), pp. 63–73. Raven, New York.

Martin, J. T. (1975). Hormonal influences in the evolution and ontogeny of imprinting behavior in the duck. *Prog. Brain Res.* **42**, 357–366.

Martin, J. T. (1978a). Embryonic pituitary adrenal axis, behavior development and domestication in birds. *Am. Zool.* **18**, 489–499.

Martin, J. T. (1978b). Imprinting behavior: pituitary–adrenocortical modulation of the approach response. *Science (Washington, D.C.)* **200**, 565–567.

Martinez-Vargas, M. C. (1974). Nest building in the ring dove *(Streptopelia risoria)*: hormonal and social factors. *Behaviour* **50**, 123–151.

Martinez-Vargas, M. C., and Erickson, C. J. (1973). Social and hormonal determinants of nest building in the ring dove *(Streptopelia risoria)*. *Behaviour* **45**, 12–37.

Martinez-Vargas, M. C., Sar, M., and Stumpf, W. E. (1974). Brain targets for androgens in the dove *(Streptopelia risoria)*. *Am. Zool.* **14**, 1285.

Martinez-Vargas, M. C., Gibson, D. B., Sar, M., and Stumpf, W. E. (1975). Estrogen target sites in the brain of the chick embryo. *Science (Washington, D.C.)* **190**, 1307–1308.

Martinez-Vargas, M. C., Stumpf, W. E., and Sar, M. (1975b). Estrogen localization in the dove brain. Phylogenetic considerations and implications for nomenclature. *Anat. Neuroendocrinol. Int. Conf.*, pp. 166–175.

Martinez-Vargas, M. C., Stumpf, W. F., and Sar, M. (1976). Anatomical distribution of estrogen target cells in the avian CNS: a comparison with the mammalian CNS. *J. Comp. Neurol.* **167**, 83–104.

Massa, R., and Sharp, P. J. Conversion of testosterone to 5β-reduced metabolites in the neuroendocrine tissues of the maturing cockerel. *J. Endocrinol.* **88**, 263–269.

Massa, R., Cresti, L., and Martini, L. (1977). Metabolism of testosterone in the anterior pituitary gland and the central nervous system of the European starling *(Sturnus vulgaris)*. *J. Endocrinol.* **75**, 347–354.

Massa, R., Davies, D. T., and Bottoni, L. (1980). Cloacal gland of the Japanese quail: androgen dependence and metabolism of testosterone. *J. Endocrinol.* **84**, 223–230.

Mathewson, S. F. (1961). Gonadotropic hormones affect aggressive behavior in starlings. *Science (Washington, D.C.)* **134,** 1522–1523.
McBride, G., Parer, I. P., and Foenander, F. (1969). The social organization and behaviour of the feral domestic fowl. *Anim. Behav. Monogr.* **2,** 127–181.
McCabe, B. J., Horn, G., and Bateson, P. P. G. (1979). Effects of rhythmic hyperstriatal stimulation on chicks' preferences for visual flicker. *Physiol. Behav.* **23,** 137–140.
McCollom, R. E., Siegel, P. B., and Van Krey, H. P. (1971). Responses to androgen in lines of chickens selected for mating behavior. *Horm. Behav.* **2,** 31–42.
McDonald, P. A. (1979). Luteinizing hormone-releasing factor (LRF) and reproductive behavior in male doves exposed to long and short photoperiods. *Neuroendocrinology* **28,** 151–154.
McDonald, P. A., and Liley, N. R. (1978). The effects of photoperiod on androgen-induced reproductive behavior in male ring doves, *Streptopelia risoria*. *Horm. Behav.* **10,** 85–96.
McEwen, B. S. (1978). Sexual maturation and differentiation: the role of the gonadal steroids. *Prog. Brain Res.* **48,** 291–307.
McEwen, B. S. (1980). Binding and metabolism of sex steroids by the hypothalamic–pituitary unit: physiological implications. *Annu. Rev. Physiol.* **42,** 97–110.
McEwen, B. S. (1981). Neural gonadal steroid actions. *Science (Washington, D.C.)* **211,** 1303–1311.
McEwen, B. S., Lieberburg, I., Chaptal, C., Davis, P. G., Krey, L. C., MacLusky, N. J., and Roy, E. J. (1979). Attenuating the defeminization of the neonatal rat brain: mechanisms of action of cyproterone acetate, 1,4,6-androstratriene-3,17-dione and a synthetic progestin, R 5020. *Horm. Behav.* **13,** 269–281.
McEwen, B. S., Davis, P. G., Jellinck, P. H., Krey, L. C., Lieberburg, I., Luine, V. N., MacLusky, N. J., Parsons, B., and Roy, E. J. (1980). Steroid hormone receptors, brain cell function, and the neuroendocrine system. *In* "Receptors for Neurotransmitters and Peptide Hormones" (G. Pepeu, M. J. Kuhar, and S. J. Enna, eds.), pp. 383–390. Raven, New York.
McKinney, F. (1969). The behaviour of ducks. *In* "The Behaviour of Domestic Animals" (E. S. E. Hafez, ed.), pp. 593–626. Baillière, Tindall & Cassell, London.
McNeilly, A. S., Etches, R. J., and Friessen, H. G. (1978). A heterologous radioimmunoassay for avian prolactin: application to the measurement of prolactin in the turkey. *Acta Endocrinol. (Copenhagen)* **89,** 60–69.
Meier, A. H., and Davis, K. B. (1967). Diurnal variations in the fattening response to prolactin in the White-throated Sparrow, *Zonotrichia albicollis*. *Gen. Comp. Endocrinol.* **8,** 110–114.
Meier, A. H., and Dusseau, J. W. (1968). Prolactin and the photoperiodic gonadal response in several avian species. *Physiol. Zool.* **41,** 95–103.
Meier, A. H., Burns, J. T., Davis, K. B., and John, T. M. (1971). Circadian variations in sensitivity of the pigeon crop-sac to prolactin. *J. Interdiscip. Cycle Res.* **2,** 161–172.
Meyer, C. C. (1972). Inhibition of precocial copulation in the domestic chick by progesterone brain implants. *J. Comp. Physiol. Psychol.* **79,** 8–12.
Meyer, C. C. (1973). Testosterone concentration in the male chick brain: an autoradiographic survey. *Science (Washington, D.C.)* **180,** 1381–1383.
Meyer, C. C., and Angelo, R. L. (1975). Suppression of precocial copulation by progesterone implants in the male chick forebrain. *J. Comp. Physiol. Psychol.* **88,** 687–692.
Meyer, C. C., Parker, D. M., and Salzen, E. A. (1976). Androgen sensitive midbrain sites and visual attention in chicks. *Nature (London)* **259,** 689–690.
Meyerson, B. J., Palis, A., and Sietnieks, A. (1979). Hormone–monoamine interactions and sexual behavior. *In* "Endocrine Control of Sexual Behavior" (C. Beyer, ed.), pp. 389–404. Raven, New York.

Michael, R. F., and Saayman, G. S. (1968). Differential effects on behaviour of the subcutaneous and intravaginal administration of oestrogen in the rhesus monkey *(Macaca mulata). J. Endocrinol.* **41,** 231–246.
Millikan, G. C. (1972). The development of filial behavior in ducklings. *Behaviour* **43,** 13–17.
Mills, J. (1966). Human circadian rhythms. *Physiol. Rev.* **46,** 128–171.
Moguilewsky, M., and Raynaud, J. P. (1977). Progestin binding sites in the rat hypothalamus, pituitary and uterus. *Steroids* **30,** 99–109.
Moguilewsky, M., and Raynaud, J. P. (1979). The relevance of hypothalamic and hypophyseal progestin receptor regulation in the induction and inhibition of sexual behavior in the female rat. *Endocrinology (Baltimore)* **105,** 516–521.
Morel, G., and Bourlière, F. (1956). Recherches écologiques sur les *Quelea quelea* (L.), de la basse vallée du Sénégal. II. La reproduction. *Alauda* **24,** 97–122.
Mori, M., Suzuki, K., and Tamaoki, B. (1974). Testosterone metabolism in rooster comb. *Biochim. Biophys. Acta* **337,** 118–128.
Moss, R. L., and Dudley, C. A. (1981). Hormonal, electrophysiological and behavioral actions exerted by luteinizing hormone-releasing hormone. *Abstr. Conf. Reprod. Biol. 13th,* p. 24.
Moss, R. L., and McCann, S. M. (1973). Induction of mating behavior in rats by luteinizing hormone releasing factor. *Science (Washington, D.C.)* **181,** 177–179.
Moss, R. L., and McCann, S. M. (1975). Action of luteinizing hormone-releasing factor (LRF) in the initiation of lordosis behavior in the estrone-primed ovariectomized female rat. *Neuroendocrinology* **17,** 309–318.
Moss, R. L., Dudley, C. A., and Schwartz, N. B. (1977). Coitus-induced release of luteinizing hormone in the proestrus rat: fantasy or fact? *Endocrinology (Baltimore)* **100,** 394–397.
Murton, R. K., and Westwood, N. J. (1975). Integration of gonadotrophin and steroid secretion, spermatogenesis and behaviour in the reproductive cycle of male pigeon species. *In* "Neural and Endocrine Aspects of Behaviour in Birds" (P. Wright, P. G. Caryl, and D. M. Vowles, eds.), pp. 51–89. Elsevier, Amsterdam.
Murton, R. K., and Westwood, N. J. (1977). "Avian Breeding Cycles." Oxford Univ. Press (Clarendon), Oxford.
Murton, R. K., Thearle, R. J. P., and Lofts, B. (1969). The endocrine basis of breeding behaviour in the feral pigeon *(Columba livia).* I. Effects of exogenous hormones on the preincubation behaviour of intact males. *Anim. Behav.* **17,** 286–306.
Naftolin, R., Ryan, K. J., Davies, I. J., Reddy, V. V., Flores, F., Petro, Z., and Kuhn, M. (1975). The formation of estrogens by central neuroendocrine tissues. *Rec. Prog. Horm. Res.* **31,** 295–319.
Nakajo, S., and Tanaka, K. (1956). Prolactin potency of the cephalic and caudal lobe of the anterior pituitary in relation to broodiness in the domestic fowl. *Poult. Sci.* **35,** 990–994.
Nakamura, T., and Tanabe, Y. (1972). *In vitro* steroidogenesis by testes of the chicken *(Gallus domesticus). Gen. Comp. Endocrinol.* **19,** 432–440.
Nakamura, T., and Tanabe, Y. (1974). *In vitro* metabolism of steroid hormones by chicken brain. *Acta Endocrinol. (Copenhagen)* **75,** 410–416.
Nalbandov, A. V., Hochhauser, M., and Dugas, M. (1945). A study of the effect of prolactin on broodiness and on cock testes. *Endocrinology (Baltimore)* **36,** 251–258.
Neumann, F., and Steinbeck, H. (1974). Antiandrogens. *Hand. Exp. Pharmacol.* **35,** 235–484.
Niswender, G. D., and Midgley, A. R., Jr. (1970). Hapten-radioimmunoassay for steroid hormones. *In* "Immunologic Methods in Steroid Determination" (F. G. Peron, and B. V. Caldwell, eds.), pp. 149–173. Appleton-Century-Crofts, New York.
Noble, G. K., and Wurm, M. (1940a). The effect of testosterone propionate on the blackcrowned night heron. *Endocrinology (Baltimore)* **26,** 837–850.

Noble, G. K., and Wurm, M. (1940b). The effect of hormones on the breeding behavior of the laughing gulls. *Anat. Rec.* **78**, (Suppl.), 50–51.

Noble, G. K., and Zitrin, A. (1942). Induction of mating behaviour in male and female chicks following injection of sex hormones. *Endocrinology (Baltimore)* **30**, 327–334.

Noble, R. (1972). The effects of estrogen and progesterone on copulation in female quail *(Coturnix coturnix japonica)* housed in continuous dark. *Horm. Behav.* **3**, 199–204.

Nottebohm, F. (1969). The "critical period" for song learning. *Ibis* **111**, 386–387.

Nottebohm, F. (1975). Vocal behavior in birds. *In* "Avian Biology" (D. S. Farner, J. R. King, and K. C. Parkes, eds.), Vol. 5, pp. 287–332. Academic Press, New York.

Nottebohm, F. (1980). Testosterone triggers growth of brain vocal control nuclei in adult female canaries. *Brain Res.* **189**, 429–436.

Nottebohm, F., and Arnold, A. P. (1976). Sexual dimorphism in vocal control areas of the songbird brain. *Science (Washington, D.C.)* **194**, 211–213.

Nottebohm, F., Stokes, T. M., and Leonard, C. M. (1976). Central control of song in the canary. *Serinus canaria. J. Comp. Neurol.* **165**, 457–486.

O'Connell, M. E., Reboulleau, C., Feder, H. H., and Silver, R. (1981). Social interactions and androgen levels in birds. *Gen. Comp. Endocrinol.* **44**, 454–463.

Ottinger, M. A., and Bakst, M. (1981). Peripheral androgen concentrations and testicular morphology in embryonic and young male Japanese quail. *Gen. Comp. Endocrinol.* **43**, 170–177.

Ottinger, M. A., and Brinkley, H. J. (1978). Testosterone and sex-related behavior and morphology: relationship during maturation and in the adult Japanese quail. *Horm. Behav.* **11**, 175–182.

Ottinger, M. A., and Brinkley, H. J. (1979a). Testosterone and sex-related physical characteristics during maturation of the male Japanese quail *(Coturnix coturnix japonica). Biol. Reprod.* **20**, 905–909.

Ottinger, M. A., and Brinkley, H. J. (1979b). The ontogeny of crowing and copulatory behavior in Japanese quail *(Coturnix coturnix japonica). Behav. Processes* **4**, 43–51.

Ozon, R. (1965). Mise en évidence d'hormones stéroides oestrogènes dans le sang de la poule adulte et chez l'embryon de poulet. *C. R. Hebd. Seances Acad. Sci. Ser. D* **261**, 5664–5666.

Ozon, R. (1972). Androgens in fishes, amphibians, reptiles and birds. *In* "Steroids in Nonmammalian Vertebrates" (D. R. Idler, ed.), pp. 328–389. Academic Press, New York.

Palka, Y. S., Ramirez, V. D., and Sawyer, C. H. (1966). Distribution and biological effects of triatiated estradiol implanted in the hypothalamus–hypophyseal region of female rats. *Endocrinology (Baltimore)* **78**, 487–489.

Parsons, B., MacLusky, N. J., Krey, L., Pfaff, D. W., and McEwen, B. S. (1980). The temporal relationship between estrogen-inducible progestin receptors in the female rat brain and the time course of estrogen activation of mating behavior. *Endocrinology (Baltimore)* **107**, 774–779.

Parvizi, N., and Ellendorff, F. (1980). Recent views on endocrine effects of catecholestogens. *J. Steroid Biochem.* **12**, 331–335.

Patel, M. D. (1936). The physiology of the formation of "pigeon's milk." *Physiol. Zool.* **9**, 129–152.

Payne, R. B. (1972). Mechanisms and control of molt. *In* "Avian Biology" (D. S. Farner, J. R. King, and K. C. Parkes, eds.), Vol. 2, pp. 103–155. Academic Press, New York.

Peczely, P., and Pethes, G. Y. (1979). Alterations in plasma sexual steroid concentrations in the collared dove *(Streptopelia decaocto)* during the sexual maturation and reproduction cycle. *Acta Physiol. Acad. Sci. Hung.* **54**, 161–170.

Pfaff, D. W. (1970). Nature of sex hormone effects on rat sex behavior: specificity of effects and individual patterns of response. *J. Comp. Physiol. Psychol.* **73**, 349–358.

Pfaff, D. W. (1973). Luteinizing hormone-releasing factor potentiates lordosis behavior in hypophysectomized ovariectomized female rats. *Science (Washington, D.C.)* **182**, 1148–1149.

Phillips, R. E. (1964). "Wildness" in the mallard duck. Effects of brain lesions and stimulation on escape behaviour and reproduction. *J. Comp. Neurol.* **127**, 89–100.

Phillips, R. E., and Barfield, R. J. (1977). Effects of testosterone implants in midbrain vocal areas of capons. *Brain Res.* **122**, 378–381.

Phillips, R. E., and McKinney, F. (1962). The role of testosterone in the displays of some ducks. *Anim. Behav.* **10**, 244–246.

Phillips, R. E., and Van Tienhoven, A. (1962). Some physiological correlates of pintail reproductive behavior. *Condor* **64**, 291–299.

Phillips, R. E., and Youngren, O. M. (1971). Brain stimulation and species-typical behaviour: activities evoked by electrical stimulation of the brains of chickens *(Gallus gallus)*. *Anim. Behav.* **19**, 757–779.

Pietras, R. J., and Wenzel, B. M. (1974). Effects of androgens on body weight, feeding and courtship behavior in the pigeon. *Horm. Behav.* **5**, 289–302.

Popa, G. T., and Popa, F. G. (1933). Certain functions of the midbrain in pigeons. *Proc. R. Soc. London Ser. B* **113**, 191–195.

Potash, L. M. (1970). Neuroanatomical regions relevant to production and analysis of vocalization within the avian *torus semicircularis*. *Experientia* **26**, 1104–1105.

Pröve, E. (1974). Der Einfluss von Kastration und Testosteronsubstitution auf das Sexualverhalten männlicher Zebra finken *(Taeniopygia guttata castanotis* Gould). *J. Ornithol.* **115**, 338–347.

Pröve, E. (1978). Quantitative Untersuchungen zu Wechselbeziehungen zwischen Baltzaktivität und Testosterontitern bei männlichen Zebrafinken *(Taeniopygia guttata castanotis* Gould). *Z. Tierpsychol.* **48**, 47–67.

Purvis, K., and Haynes, N. B. (1974). Short term effects of copulation, human chorionic gonadotrophin injection and non-tactile association with a female on testosterone levels in the male rat. *J. Endocrinol.* **60**, 429–439.

Puts, C. (1978). "Action des androgènes sur le comportement d'empreinte du poussin domestique." Mémoire de licence, Univ. of Liège, Belgium.

Raitasuo, K. (1964). Social behaviour of the mallard, *Anas platyrhynchos*, in the course of the annual cycle. *Riistatieteellisia Julkaisuga* **24**, 1–72.

Ralph, C. L. (1959). Some effects of hypothalamic lesions on gonadotrophin release in the hen. *Anat. Rec.* **134**, 411–432.

Ralph, C. L., and Fraps, R. M. (1959). Effect of hypothalamic lesions on progesterone induced ovulation in the hen. *Endocrinology (Baltimore)* **65**, 819–824.

Ravona, H., Snapir, N., and Perek, M. (1973). The effect on the gonadal axis in cockerels of electrolytic lesions in various regions of the basal hypothalamus. *Gen. Comp. Endocrinol.* **20**, 112–124.

Reddy, V. V. R., Rajan, R., and Daly, M. J. (1981). Oestrogen metabolism in adult rat's brain. *Acta Endocrinol. (Copenhagen)* **96**, 7–14.

Rhees, R. W., Abel, J. H., Jr., and Haack, D. W. (1972). Uptake of triatiated steroids in the brain of the duck *Anas platyrhynchos)*. An autoradiographic study. *Gen. Comp. Endocrinol.* **18**, 292–300.

Richdale, L. E. (1952). "Post-Egg Period in Albatrosses." Privately Published, Dunedin, New Zealand.

Riddle, O. (1935). Aspects and implications of the hormonal control of the maternal instinct. *Proc. Am. Phil. Soc.* **75**, 521–525.

Riddle, O. (1937). Physiological responses to prolactin. *Cold Spring Harbor Symp. Quant. Biol.* **5**, 218–228.
Riddle, O. (1963). Prolactin or progesterone as key to parental behaviour: a review. *Anim. Behav.* **11**, 419–432.
Riddle, O., and Bates, R. W. (1933). Concerning anterior pituitary hormones. *Endocrinology (Baltimore)* **17**, 689–698.
Riddle, O., and Dikshorn, S. W. (1932). Secretion of crop-milk in the castrate pigeon. *Proc. Soc. Exp. Biol. Med.* **29**, 1213–1215.
Riddle, O., and Lahr, E. L. (1944). On broodiness of ring doves following implants of certain steroid hormones. *Endocrinology (Baltimore)* **35**, 255–260.
Riddle, O., Bates, R. W., and Dykshorn, S. W. (1932). A new hormone of the anterior pituitary. *Proc. Soc. Exp. Biol. Med.* **29**, 1211–1212.
Riddle, O., Bates, R. W., and Dykshorn, S. W. (1933). The preparation, identification and assay of prolactin. A hormone of the anterior pituitary. *Am. J. Physiol.* **105**, 191–216.
Riddle, O., Bates, R. W., and Lahr, E. L. (1935). Prolactin induces broodiness in fowl. *Am. J. Physiol.* **111**, 352–360.
Rogers, L. J. (1974). Persistence and search influenced by natural levels of androgens in young and adult chickens. *Physiol. Behav.* **12**, 197–204.
Rohwer, S. (1975). The social significance of avian winter plumage variability. *Evolution* **29**, 593–610.
Rohwer, S., and Rohwer, F. C. (1978). Status signaling in Harris' sparrows: experimental deceptions achieved. *Anim. Behav.* **26**, 1012–1022.
Rosenblatt, J. S. (1978). Behavioral regulation of reproductive physiology: a selected review. *Comp. Endocrinol. Proc. Int. Symp. 8th*, pp. 177–188.
Rothstein, R. (1967). Some observations on the nesting behavior of Japanese quail *(Coturnix coturnix japonica)* in pseudo-natural conditions. *Poult. Sci.* **46**, 260–262.
Rowe, P. H., Racey, P. A., Lincoln, G. A., Ellwood, M., Lehane, J., and Shenton, J. C. (1975). The temporal relationship between the secretion of luteinizing hormone and testosterone in man. *J. Endocrinol.* **64**, 17–26.
Rubin, B. S., and Barfield, R. J. (1980). Priming of estrous responsiveness by implants of 17β-estradiol in the ventromedial hypothalamic nucleus of female rats. *Endocrinology (Baltimore)* **106**, 504–509.
Saad, R., Goldsmith, A., Follett, B. K., and Silver, R. (1981). Effects of diverse androgens on sexual behavior, suppression of LH secretion, and tissue morphology in male ring doves.
Sachs, B. D. (1969). Photoperiodic control of reproductive behavior and physiology of the male Japanese quail *(Coturnix coturnix japonica)*. *Horm. Behav.* **1**, 7–24.
Saeki, Y., and Tanabe, Y. (1955). Changes in prolactin content of fowl pituitary during broody periods and some experiments on the induction of broodiness. *Poult. Sci.* **34**, 909–919.
Sandor, T., and Idler, D. R. (1972). Steroid methodology. In "Steroids in Nonmammalian Vertebrates" (D. R. Idler, ed.), pp. 6–36. Academic Press, New York.
Scanes, C. G., Chadwick, A., and Bolton, N. J. (1976). Radioimmunoassay of prolactin in the plasma of the domestic fowl. *Gen. Comp. Endocrinol.* **30**, 12–20.
Scanes, C. J., Sharp, P. J., and Chadwick, A. (1977). Changes in plasma prolactin concentration during ovulatory cycle of the chicken. *J. Endocrinol.* **72**, 401–402.
Scanes, C. G., Chadwick, A., Sharp, P. J., and Bolton, N. J. (1978). Diurnal variation in plasma luteinizing hormone levels in the domestic fowl *(Gallus domesticus)*. *Gen. Comp. Endocrinol.* **34**, 45–49.
Scanes, C. G., Merrill, G. F., Ford, R., Mauser, P., and Horowitz, C. (1980). Effects of stress (hypoglycemia, endotoxin and ether) on the peripheral circulating concentration of corticosterone in the domestic fowl *Gallus domesticus*). *Comp. Biochem. Physiol. C.* **66**, 183–186.

Schanbacher, B. D., Gomes, W. R., and Vandermark, N. L. (1974). Diurnal rhythm in serum testosterone levels and thymidine uptake by testes in the domestic fowl. *J. Anim. Sci.* **38**, 1245–1248.
Schein, M. W., and Hale, E. B. (1959). The effect of early experience on male sexual behaviour of androgen injected turkeys. *Anim. Behav.* **7**, 189–200.
Schleidt, W. M. (1970). Precocial sexual behaviour in turkeys *(Meleagris gallopavo*L.*).* *Anim. Behav.* **18**, 760–761.
Schommer, M. (1977). "On the Social Behaviour of Gadwall *(Anas strepera)*: Displays, Pair Bonds and Effects of Testosterone Injections." Ph.D. Thesis, Univ. of Leicester, England.
Schommer, M. (1978). Development of vocalizations in gadwall. *Biophon.* **6**, 4–5.
Schrader, W. T., and O'Malley, B. W. (1972). Progesterone-binding components of chick oviduct. IV. Characterization of purified subunits. *J. Biol. Chem.* **247**, 51–59.
Schrader, W. T., and O'Malley, B. W. (1980). Mechanism of action of steroid hormones in chicken oviduct. *In* "Steroids and Their Mechanism of Action in Nonmammalian Vertebrates" (G. Delrio, and J. Brachet, eds.), pp. 179–187. Raven, New York.
Schumacher, M., and Balthazart, J. (1983). The post-natal differentiation of sexual behavior in the Japanese quail *(Coturnix coturnix japonica). Behav. Processes* **8**, 189–195.
Schutz, F. (1975). Der Einfluss von Testosteron auf die Partnerwahl bei geprägt aufgezogenen Stockentenweibchen: Nachweis lätenter Sexualprägung. *Verh. Dtsch. Zool. Ges.* **67**, 339–344.
Sefton, A. E., and Siegel, P. B. (1975). Selection for mating ability in Japanese quail. *Poult. Sci.* **54**, 788–794.
Selinger, H. E., and Bermant, G. (1967). Hormonal control of aggressive behavior in Japanese quail *(Coturnix coturnix japonica). Behaviour* **28**, 255–268.
Selmanoff, M. K., Brodkin, L. D., Weiner, R. I., and Siiteri, P. K. (1977). Aromatization and 5α-reduction of androgens in discrete hypothalamic and limbic regions of the male and female rat. *Endocrinology (Baltimore)* **101**, 841–848.
Sharp, P. J., and Follett, B. K. (1969). The effect of hypothalamic lesions on gonadotrophin release in Japanese quail *(Coturnix coturnix japonica). Neuroendocrinology* **5**, 205–218.
Sharp, P. J., and Massa, R. (1980). Coversion of progesterone to 5α- and 5β-reduced metabolites in the brain of the hen and its potential role in the induction of the preovulatory release of luteinizing hormone. *J. Endocrinol.* **86**, 459–464.
Sharp, P. J., Culbert, J., and Wells, J. W. (1977). Variations in stored and plasma concentrations of androgens and luteinizing hormone during sexual development in the cockerel. *J. Endocrinol.* **74**, 467–476.
Sharp, P. J., Scanes, C. J., Williams, J. B., Harvey, S., and Chadwick, A. (1979). Variations in concentrations of prolactin, luteinizing hormone, growth hormone, and progesterone in the plasma of broody bantams *(Gallus domesticus). J. Endocrinol.* **80**, 51–57.
Sheridan, P. J. (1981). Unaromatized androgen is taken up by the neonatal rat brain: two receptor systems for androgen. *Dev. Neurosci. (Basel)* **4**, 46–54.
Sherman, M. R., Corvol, P. L., and O'Malley, B. W. (1970). Progesterone-binding components of chick oviduct. Preliminary characterization of cytoplasmic components. *J. Biol. Chem.* **245**, 6085–6096.
Shoemaker, H. H. (1939). Social hierarchy in flocks of the canary. *Auk* **56**, 381–406.
Siegel, P. B. (1965). Genetics of behavior: selection for mating ability in chicken. *Genetics* **52**, 1269–1277.
Siegel, P. B. (1972). Genetic analysis of male mating behavior in chickens *(Gallus domesticus).* I. Artificial selection. *Anim. Behav.* **20**, 564–570.
Silver, R. (1977). Effects of the antiandrogen cyproterone acetate on reproduction in male and female ring doves. *Horm. Behav.* **9**, 371–379.

Silver, R., and Barbière, C. (1977). Display of courtship and incubation behaviour during the reproductive cycle of the male ring dove *(Streptopelia risoria)*. *Horm. Behav.* **8**, 8–21.
Silver, R., and Buntin, J. (1973). Role of adrenal hormones in incubation behaviour of male ring doves Streptopelia risoria). *J. Comp. Physiol. Psychol.* **84**, 453–463.
Silver, R., and Feder, H. H. (1973). Role of gonadal hormones in incubation behavior of male ring doves *(Streptopelia risoria)*. *J. Comp. Physiol. Psychol.* **84**, 464–471.
Silver, R., Feder, H. H., and Lehrman, D. S. (1973). Situational and hormonal determinants of courtship, aggressive and incubation behavior in male ring doves *(Streptopelia risoria)*. *Horm. Behav.* **4**, 163–172.
Silver, R., Reboulleau, C., Lehrman, D. S., and Feder, H. H. (1974). Radioimmunoassay of plasma progesterone during the reproductive cycle of male and female ring doves *(Streptopelia risoria)*. *Endocrinology (Baltimore)* **94**, 1547–1554.
Silver, R., O'Connell, M., and Saad, R. (1979). Effect of androgen on the behavior of birds. *In* "Endocrine Control of Sexual Behavior" (C. Beyer, ed.), pp. 223–278. Raven, New York.
Silver, R., Goldsmith, A. R., and Follett, B. K. (1980). Plasma luteinizing hormone in male ring doves during the breeding cycle. *Gen. Comp. Endocrinol.* **42**, 19–24.
Silverin, B. (1975). Reproductive organs and breeding behaviour of the male pied flycatcher *Ficedula hypoleuca* (Pallas). *Ornis Scand.* **6**, 15–26.
Silverin, B. (1978a). Circannual rhythms in gonads and endocrine organs of the great tit *Parus major* in south-west Sweden. *Ornis Scand.* **9**, 207–213.
Silverin, B. (1978b). Enzyme histochemical studies in testes and seminal vesicles of the pied flycatcher *(Ficedula hypoleuca)* during the breeding season. *J. Zool.* **186**, 335–345.
Silverin, B. (1980a). Effects of prolactin on the gonad and body weight of the male pied flycatcher during the breeding period. *Endokrinologie* **76**, 45–50.
Silverin, B. (1980b). Effects of long-acting testosterone treatment on free–living pied flycatcher, *Ficedula hypoleuca* during the breeding period. *Anim. Behav.* **28**, 906–912.
Smith, E. R., Damassa, D. A., and Davidson, J. M. (1977a). Plasma testosterone and sexual behavior following intracerebral implantation of testosterone propionate in castrated male rat. *Horm. Behav.* **8**, 77–87.
Smith, E. R., Damassa, D. A., and Davidson, J. M. (1977b). Hormone administration: peripheral and intracranial implants. *In* "Methods in Psychobiology" (R. D. Myers, ed.), Vol. 3, pp. 259–279. Academic Press, New York.
Södersten, P. (1972). Mounting behavior in female rats during the estrous cycle, after ovariectomy, and after estrogen or testosterone administration. *Horm. Behav.* **3**, 307–320.
Södersten, P. (1973). Estrogen-activated sexual behavior in male rats. *Horm. Behav.* **4**, 247–256.
Södersten, P. (1974). Mounting behavior in pregnant and pseudo-pregnant female rats. *Horm. Behav.* **5**, 345–354.
Södersten, P. (1979). Role of estrogen in the display and development of sexual behavior in male rats. *In* "Endocrine Control of Sexual Behavior" (C. Beyer, ed.), pp. 305–315. Raven, New York.
Södersten, P., and Gustafsson, J. A. (1980). A way in which estradiol might play a role in the sexual behavior of male rats. *Horm. Behav.* **14**, 271–274.
Södersten, P., and Hansen, S. (1978). Effect of castration and testosterone, dihydrotestosterone or oestradiol replacement treatment in neonatal male rats on mounting behaviour in the adult. *J. Endocrinol.* **76**, 251–260.
Södersten, P., Damassa, D. A., and Smith, E. R. (1977). Sexual behavior in developing male rats. *Horm. Behav.* **8**, 320–341.
Södersten, P., Eneroth, P., and Ekberg, P. H. (1980). Episodic fluctuations in concentration of

androgen in serum of male rats: possible relationship to sexual behaviour. *J. Endocrinol.* **87**, 463–471.

Sossinka, R. (1975). Quantitative Untersuchungen zur sexuellen Reifung des Zebrafinken, *Taeniopygia castanotis* Gould. *Verh. Dtsch. Zool. Ges.* **1974**, 344–347.

Sossinka, R., Pröve, E., and Kalberlah, H. H. (1975). Der Einfluss von Testosteron auf den Gesangsbeginn beim Zebrafinken *(Taeniopygia guttata castanotis). Z. Tierpsychol.* **39**, 259–264.

Sossinka, R., Pröve, E., and Immelman, K. (1980). Hormonal mechanisms in avian behavior. In "Avian Endocrinology" (A. Epple, ed.), pp. 533–547. Academic Press, New York.

Steel, E. A., Hinde, R. A. (1963). Hormonal control of brood patch and oviduct development in domesticated canaries. *J. Endocrinol.* **26**, 11–24.

Steel, E. A., and Hinde, R. A. (1966). Effect of exogenous serum gonadotrophin (PMS) on aspects of reproductive development in female domesticated canaries. *J. Zool.* **149**, 12–30.

Steel, E. A., and Hinde, R. A. (1972a). Influence of photoperiod on PMSG-induced nest-building in canaries. *J. Reprod. Fertil.* **31**, 425–431.

Steel, E. A., and Hinde, R. A. (1972b). Influence of photoperiod on oestrogenic induction of nest building in canaries. *J. Endocrinol.* **55**, 265–278.

Steel, E. A., and Hinde, R. A. (1976). Effect of a skeleton photoperiod on the day length dependent response to oestrogen in canaries *(Serinus canarius). J. Endocrinol.* **70**, 247–254.

Steel, E. A., Gosney, S., and Hinde, R. A. (1977). Effect of male vocalizations on the nest-occupation response of female budgerigars to oestrogen and prolactin. *J. Reprod. Fertil.* **49**, 123–125.

Steimer, T., and Hutchison, J. B. (1980). Aromatization of testosterone within a discrete hypothalamic area associated with the behavioural action of androgen in the male dove. *Brain Res.* **192**, 586–591.

Steimer, T., and Hutchison, J. B. (1981). Metabolic control of the behavioural action of androgens in the dove brain: testosterone inactivation by 5β-reduction. *Brain Res.* **209**, 189–204.

Steinbeck, H., Elger, W., and Neumann, F. (1967). Sexual activity of male rats under the influence of oestradiol and antiandrogens and recurrence of libido after cessation of treatment. *Acta Endocrinol. (Copenhagen)* **119** (Suppl.), 63 (Abs.).

Stern, J. M. (1972). Androgen accumulation in hypothalamus and anterior pituitary of male ring doves; influence of steroid hormones. *Gen. Comp. Endocrinol.* **18**, 439–449.

Stern, J. M. (1974). Estrogen facilitation of progesterone-induced incubation behavior in castrated male ring doves. *J. Comp. Physiol. Psychol.* **87**, 332–337.

Stern, J. M., and Lehrman, D. S. (1969). Role of testosterone in progesterone-induced incubation behaviour in male ring doves *Streptopelia risoria). J. Endocrinol.* **44**, 13–22.

Stetson, M. H. (1972). Feedback regulation by estradiol of ovarian function in Japanese quail. *J. Reprod. Fertil.* **31**, 205–213.

Stettenheim, P. (1972). The integument of birds. In "Avian Biology" (D. S. Farner, J. R. King, and K. C. Parkes, eds.), Vol. 2, pp. 1–63. Academic Press, New York.

Stevens, V. C. (1961). Experimental study of nesting by *Coturnix* quail. *J. Wildl. Manage.* **25**, 99–101.

Stokes, T. M., Leonard, C. M., and Nottebohm, F. (1974). The telencephalon, diencephalon and mesencephalon of the canary, *Serinus canaria* in stereotaxic coordinates. *J. Comp. Neurol.* **156**, 337–374.

Stratton, L. G., Ewing, L. L., and Desjardins, C. (1973). Efficacy of testosterone-filled polydimethylsiloxane implants in maintaining plasma testosterone in rabbits. *J. Reprod. Fertil.* **35**, 235–244.

Stumpf, W. E. (1971). Autoradiographic techniques and the localization of estrogen, androgen and glucocorticoid in the pituitary and brain. *Am. Zool.* **11,** 725–739.

Stumpf, W. E., and Sar, M. (1975). Autoradiographic techniques for localizing steroid hormones. *In* "Methods in Enzymology" (B. W. O'Malley, and J. G. Hardman, eds.), Vol. 36, pp. 135–156. Academic Press, New York.

Tanabe, Y., Nakamura, T., Fujioka, K., and Doi, O. (1979). Production and secretion of sex steroid hormones by the testes, the ovary and the adrenal glands of embryonic and young chickens *(Gallus domesticus) Gen. Comp. Endocrinol.* **39,** 26–33.

Tanabe, Y., Nakamura, T., Omiya, Y., and Yano, T. (1980). Changes in the plasma LH, progesterone, and estradiol during the ovulatory cycle of the duck *(Anas platyrhynchos domestica)* exposed to different photoperiods. *Gen. Comp. Endocrinol.* **41,** 378–383.

Temple, S. A. (1974). Plasma testosterone titers during the annual reproductive cycle of starlings *(Sturnus vulgaris). Gen. Comp. Endocrinol.* **22,** 470–479.

Tenniswood, M., Bird, C. E., and Clark, A. F. (1981). Kinetics of *in vivo* 5α-androstane-3β,17β diol metabolism in adult normal and castrated male rats. *J. Steroid Biochem.* **14,** 199–204.

Teyler, T. J., Vardaris, R. M., Lewis, D., and Rawitch, A. B. (1980). Gonadal steroids: effects on excitability of hippocampal pyramidal cells. *Science (Washington, D.C.)* **209,** 1017–1019.

Thapliyal, J. P., and Saxena, R. (1964). The effect of prolactin on the hypophysis and testes of an Indian weaverbird *(Ploceus philippinus). Gen. Comp. Endocrinol.* **4,** 119–123.

Tixier-Vidal, A., and Follett, B. K. (1973). The adenohypophysis. *In* "Avian Biology" (D. S. Farner, J. R. King, and K. C. Parkes, eds.), Vol. 3, pp. 109–182. Academic Press, New York.

Toran-Allerand, C. D. (1976). Sex steroids and the development of the newborn mouse hypothalamus and preoptic area *in vitro*: implications for sexual differentiation. *Brain Res.* **106,** 407–412.

Toran-Allerand, C. D. (1978). Gonadal hormones and brain development: cellular aspects of sexual differentiation. *Am. Zool.* **18,** 553–565.

Toran-Allerand, C. D., Gerlach, J. L., and McEwen, B. S. (1980). Autoradiographic localization of (^{3}H)-estradiol related to steroid responsiveness in cultures of the newborn mouse hypothalamus and preoptic area. *Brain Res.* **184,** 517–522.

Tsutsui, K., and Ishii, S. (1981). Effects of sex steroids on aggressive behaviors of adult male Japanese quail. *Gen. Comp. Endocrinol.* **44,** 480–486.

Turek, F. W., Desjardins, C., and Menaker, M. (1976). Antigonadal and progonadal effects of testosterone in male house sparrows. *Gen. Comp. Endocrinol.* **28,** 395–402.

Van Doorn, E. J., and Clark, A. F. (1973). *In vitro* studies on the inhibition of pig liver steroid Δ^4-5β-reductase activity by naturally occurring and synthetic estrogens. *Biochim. Biophys. Acta* **309,** 254–262.

Van Tienhoven, A., and Juhasz, L. P. (1962). The chicken telencephalon, diencephalon and mesencephalon in stereotaxic coordinates. *J. Comp. Neurol.* **118,** 185–197.

Verhoeven, G. (1980). Effects of neurotransmitters and follicle stimulating hormone on the aromatization of androgens and the production of adenosine 3',5'-monophosphate by cultured testicular cells. *J. Steroid Biochem.* **12,** 315–322.

Vidal, J. M. (1971). Precocial sexual behaviour. Ontogeny of sexual behaviour in the domestic cock *(Gallus domesticus). Behaviour* **39,** 20–38.

Voigt, W., Fernandez, E. P., and Hsia, S. (1970). Transformation of testosterone into 17β-hydroxy-5α-androstan-3-one by microsomal preparations of human skin. *J. Biol. Chem.* **245,** 5594–5599.

Vowles, D. M., and Harwood, D. (1966). The effect of exogenous hormones on aggressive and defensive behaviour in the ring dove *(Streptopelia risoria). J. Endocrinol.* **36,** 35–51.

Wada, M. (1979). Photoperiodic control of LH secretion in Japanese Quail with special reference to the photoinducible phase. *Gen. Comp. Endocrinol.* **39,** 141–149.

Wade, G. N. (1976). Sex hormones, regulatory behaviours and body weight. *Adv. Study Behav.* **6,** 201–279.

Warren, R. P., and Hinde, R. A. (1959). The effect of oestrogen and progesterone on the nest-building of domesticated canaries. *Anim. Behav.* **7,** 209–213.

Warren, R. P., and Hinde, R. A. (1961a). Does the male stimulate estrogen secretion in female canaries? *Science (Washington, D.C.)* **133,** 1354–1355.

Warren, R. P., and Hinde, R. A. (1961b). Roles of the male and the nest-cup in controlling reproduction in female canaries. *Anim. Behav.* **9,** 64–67.

Watson, A., and Moss, R. (1971). Spacing as affected by territorial behavior, habitat and nutrition in red grouse *(Lagopus l. scoticus).* In "Behavior and Environment: The Use of Space by Animals and Men" (A. H. Esser, ed.), pp. 92–111. Plenum, New York.

Weniger, J.-P., and Zeis, A. (1982). Conversion of 5α-dihydrotestosterone to oestrone and oestradiol by the chick embryo ovary in organ culture. *J. Steriod Biochem.* **17,** 573–576.

Wenzel, B. M. (1971). Olfaction in birds. *Hand. Sensory Physiol.* **4,** 432–448.

Wenzel, B. M. (1973). Chemoreception. *In* "Avian Biology" (D. S. Farner, J. R. King, and K. C. Parkes, eds.), Vol. 3, pp. 389–415. Academic Press, New York.

Weston. H. G. (1947). Breeding behavior of the Black-headed Grosbeak. *Condor* **49,** 54–73.

Whalen, R. E., and Edwards, D. A. (1969). Effects of the antiandrogen cyproterone acetate on mating behavior and seminal vesicle tissue in male rats. *Endocrinology (Baltimore)* **84,** 155–156.

White, S. J. (1975a). Effects of stimuli emanating from the nest on the reproductive cycle in the ring dove. I. Pre-laying behaviour. *Anim. Behav.* **23,** 854–868.

White, S. J. (1975b). Effects of stimuli emanating from the nest on the reproductive cycle in the ring dove. II. Building during the pre-laying period. *Anim. Behav.* **23,** 869–882.

White, S. J. (1975c). Effects of stimuli emanating from the nest on the reproductive cycle in the ring dove. III. Building in the post-laying period and effects on the success of the cycle. *Anim. Behav.* **23,** 883–888.

Whitsett, J. M., Irvin, E. W., Edens, F. W., and Thaxton, J. P. (1977). Demasculinization of male Japanese quail by prenatal estrogen treatment. *Horm. Behav.* **8,** 254–263.

Wilke, J. T. (1977). Ultradian biological periodicities in the integration of behavior. *Int. J. Neurosci.* **7,** 125–143.

Willems, J. (1978). "Le Contrôle de la sécrétion des hormones gonadotropes chez le canard domestique *(Anas platyrhyhchos* L.). Mémoire de licence, Univ. of Liège, Belgium.

Wilson, E. K., Rogler, J. C., and Erb, R. E. (1979). Effect of sexual experience, location, malnutrition, and repeated sampling on concentrations of testosterone in blood plasma of *Gallus domesticus* roosters. *Poult. Sci.* **58,** 178–186.

Wilson, J. A., and Glick, B. (1970). Ontogeny of mating behavior in the chicken. *Am. J. Physiol.* **218,** 951–955.

Wilson, S. C., and Sharp, P. J. (1975). Episodic release of luteinizing hormone in the domestic fowl. *J. Endocrinol.* **64,** 77–86.

Wimersma Greidanus, Tj. B. Van, Bohus, B., and DeWied, D. (1974). Effects of peptide hormones on behavior. Differential localization of the influence of lysine vasopressine and of ACTH 1–10 on avoidance behavior. A study with rats bearing lesions in the parafascicular nuclei. *Neuroendocrinology* **14,** 280–288.

Wingfield, J. C., and Farner, D. S. (1975). The determination of five steroids in avian plasma by radioimmunoassay and competitive protein binding. *Steroids* **26,** 315–327.

Wingfield, J. C., and Farner, D. S. (1978). The annual cycle of plasma irLH and steroid hormones in feral populations of the White-crowned Sparrow, *Zonotrichia leucophrys gambelii. Biol. Reprod.* **19,** 1046–1056.

Wingfield, J. C., and Farner, D. S. (1979). Some endocrine correlates of renesting after loss of clutch or brood in the White-crowned Sparrow, *Zonotrichia leucophrys gambelii*. *Gen. Comp. Endocrinol.* **38**, 322–331.

Wingfield, J. C., and Farner, D. S. (1980). Control of seasonal reproduction in temperate-zone birds. *Prog. Reprod. Biol.* **5**, 62–101.

Wingfield, J. C., Newman, A., Hunt, G. L., and Farner, D. S. (1980). Androgen in high concentrations in the blood of female western gulls, *Larus occidentalis*. *Naturwissenschaften* **67**, 514–515.

Witschi, E. (1954). Vertebrate gonadotropins. *Comp. Physiol. Reprod.* **4**, 149–163.

Witschi, E. (1961). Sex and secondary sexual characters. *In* "Biology and Comparative Physiology of Birds" (A. J. Marshall, ed.), pp. 115–168. Academic Press, New York.

Wood-Gush, D. G. M. (1963). The relationship between hormonally-induced sexual behaviour in male chicks and their adult sexual behaviour. *Anim. Behav.* **11**, 400–402.

Wood-Gush, D. G. M. (1975). Nest construction by the domestic hen: some comparative and physiological considerations. *In* "Neural and Endocrine Aspects of Behaviour in Birds" (P. Wright, P. G. Caryl, and D. M. Vowles, eds.), pp. 35–49. Elsevier, Amsterdam.

Wood-Gush, D. G. M., and Gilbert, A. B. (1964). The control of the nesting behaviour of the domestic hen. II. The role of the ovary. *Anim. Behav.* **12**, 451–453.

Wood-Gush, D. G. M., and Gilbert, A. B. (1969). Oestrogen and the pre-laying behaviour of the domestic hen. *Anim. Behav.* **17**, 586–589.

Wood-Gush, D. G. M., and Gilbert, A. B. (1970). The nesting behaviour of hens with ovarian transplants. *Anim. Behav.* **18**, 52–54.

Wood-Gush, D. G. M., and Gilbert, A. B. (1973). Some hormones involved in the nesting behavior of hens. *Anim. Behav.* **21**, 98–103.

Wood-Gush, D. G. M., Langley, G. A. S., Leitch, A. F., Gentle, M. J., and Gilbert, A. B. (1977). An autoradiographic study of sex steroids in the chicken telencephalon. *Gen. Comp. Endocrinol.* **31**, 161–168.

Woods, J. E., and Brazzill, D. M. (1981). Plasma 17β-estradiol levels in the chick embryo. *Gen. Comp. Endocrinol.* **44**, 37–43.

Woods, J. E., Simpson, R. M., and Moore, P. L. (1975). Plasma testosterone levels in the chick embryo. *Gen. Comp. Endocrinol.* **27**, 543–547.

Yahr, P. (1979). Data and hypothesis in the tales of dihydrotestosterone. *Horm. Behav.* **13**, 92–96.

Yahr, P., and Gerling, S. A. (1978). Aromatization and androgen stimulation of sexual behavior in male and female rats. *Horm. Behav.* **10**, 128–142.

Young, C. E., and Rogers, L. J. (1978). Effects of steroidal hormones on sexual, attack and search behavior in the isolated male chick. *Horm. Behav.* **10**, 107–117.

Young, W. C. (1961). The hormones and mating behavior. *In* "Sex and Internal Secretions" (W. C. Young, ed.), Vol. 2, pp. 1173–1239. Williams & Wilkins, Baltimore, Maryland.

Zeier, H., and Karten, H. J. (1971). The archistriatum of the pigeon: organization of afferent and efferent connections. *Brain Res.* **31**, 313–326.

Zigmond, R. E. (1975). Target cells for gonadal steroids in the brain: studies on hormone binding and metabolism. *In* "Neural and Endocrine Aspects of Behaviour in Birds" (P. Wright, P. G. Caryl, and D. M. Vowles, eds.), pp. 111–121. Elsevier, Amsterdam.

Zigmond, R. E., Stern, J. M., and McEwen, B. S. (1972). Retention of radioactivity in cell nuclei in the hypothalamus of the ring dove after injection of ^{3}H-testosterone. *Gen. Comp. Endocrinol.* **18**, 450–453.

Zigmond, R. E., Nottebohm, F., and Pfaff, D. W. (1973). Androgen-concentrating cells in the midbrain of a song bird. *Science (Washington, D.C.)* **179**, 1005–1007.

Zucker, I. (1966). Effects of an antiandrogen on the mating behavior of male guinea pigs and rats. *J. Endocrinol.* **35**, 209–210.

Chapter 5

THE BIOLOGY OF AVIAN PARASITES: HELMINTHS

Robert L. Rausch

I.	Introduction	367
	Historical Considerations	368
II.	Characteristics of Helminths	371
	A. Phylum Platyhelminthes	371
	B. Phylum Aschelminthes	373
	C. Phylum Acanthocephala	373
III.	Occurrence of Helminths in Birds	374
	Prevalence	375
IV.	Specificity of Helminths	406
	Specific and Nonspecific Helminths in Birds	410
V.	Acquisition of Helminths by Birds	422
	A. Diet	422
	B. Age and Acquisition of Helminths	426
	C. Migration of Birds and Acquisition of Helminths	428
VI.	Localization of Helminths in the Avian Host	429
VII.	Pathogenicity of Helminths in Birds	430
	References	432

I. Introduction

Organisms, usually designated parasites, living on or in the body of birds include representatives of diverse taxonomic groups. The most important of these are protozoa (Sporozoa and Mastigophora), helminths (Cestoda, Trematoda, Nematoda, and Acanthocephala), and arthropods. The last include the mites and ticks (Acarina) and insects (Mallophaga, Siphonaptera, and Diptera; less commonly, members of other orders). Rather than provide minimal information concerning all of these, this chapter will consider the biology of a single group of symbiotes, the helminths, in somewhat greater detail. Helminth–host interactions are discussed in a general way, with no attempt to provide an exhaustive review of the relevant literature.

The literature concerning helminths in wild birds is so extensive that it obscures the limitations that exist in our knowledge of this group of organisms. Taxonomic studies, initiated in central Europe about two centuries ago, were long restricted geographically, and even now helminth faunas are known best only in some regions in Europe and the Soviet Union. A few economically important birds, mainly of the orders Anseriformes and Galliformes, have been studied relatively intensively, but on the whole little or no information concerning helminths is available for a great number of the approximately 9000 Recent avian species. Developmental cycles have been defined for comparatively few helminths in birds, and the investigation of fundamental helminth–host interactions has only recently begun. Possibilities for such studies will be increasingly diminished in the future, as habitat destruction and other factors continue to have an adverse effect on avian populations.

Historical Considerations

Compared with the larger, more obvious forms in man and synanthropic mammals, helminths in birds are rather obscure, and it is not remarkable that none were mentioned in the earliest biomedical writings. The first report of such an organism from an avian host was perhaps that of Demetrios Pepagomenos (thirteenth century), who observed worms (spirurids?) under the nictitating membrane of falcons used in hunting (Huber, 1906).

The study of helminths in birds began in Europe during the late eighteenth century. One of the first significant works was that of Goeze (1782), who described several species from avian hosts in his *Versuch einer Naturgeschichte der Eingeweidewürmer thierischer Körper*. Also in 1782, additional species from birds were described by Bloch. Batsch (1786) compiled earlier published records of helminths from animals and contributed a classification of the cestodes. In a work supplemental to the volume by Goeze, Zeder (1800) devised a classification of the helminths, including those from birds. Numerous species were described in the publications of Rudolphi, of which the most important was his *Entozoorum Synopsis* (1819). The volumes by Dujardin (1845) and Diesing (1850) demonstrate the accomplishments in helminthology during the first half of the nineteenth century. After about 1850, with rapidly improving technology and increasing emphasis on experimental investigation, much effort was devoted to the study of the anatomy and developmental cycles of helminths in man and domestic animals. Names prominently associated with such work are those of F. Küchenmeister, C. Th. von Siebold, P. J. van Beneden, and Rudolf Leuckart. Krabbe's important study of cestodes in birds appeared in 1869, and during this period also, von Linstow (1878) published his *Compendium der Helminthologie*, followed

5. BIOLOGY OF AVIAN PARASITES

by a supplement in 1889. In these, all known helminths from animals were listed by host, usually with an indication of the site of localization.

Numerous publications concerning helminths in birds appeared during the early decades of the present century. Among the most important was Fuhrmann's (1908) *Die Cestoden der Vögel*, which was based in part on the study of early collections preserved in European museums. This work appeared in revised form as *Les Ténias des Oiseaux* (Fuhrmann, 1932). The volume by Joyeux and Baer (1936), in the series *Faune de France*, was an important source of information on cestodes, including those in birds. Dubois's *Monographie des Strigeida (Trematoda)* appeared in 1938, followed by supplements in 1953, 1968, and 1970. The publication in 1914 of Skriabin's *Vogelcestoden aus Russisch Turkestan* marked the beginning of intensive investigation of the helminth fauna of the Soviet Union. The Union Helminthological Expeditions, of which more than 350 were conducted under the direction of K. I. Skriabin (the first in 1919) have been of particular importance. The first important work concerning cestodes in southeast Asia was that of Southwell (1930), in *The Fauna of British India*.

In North America, the name of Joseph Leidy is associated with the earliest studies concerning helminths. His contributions, beginning in 1846, have been reviewed by J. Leidy, Jr. (1904). Important studies on helminths in birds were made by H. B. Ward and his students during the early decades of the twentieth century. These include publications by Ransom (1909; cestodes), Harrah (1922; trematodes), Mayhew (1925; cestodes), and others. Linton published important work on trematodes and cestodes of birds in 1927 and 1928. Cycles of trematodes occurring in birds were defined by H. W. Stunkard and G. A. LaRue and their students. During this period also, H. J. Van Cleave described numerous species of acanthocephalans, and E. B. Cram (1927) published the first comprehensive work on nematodes in birds.

In the southern hemisphere, as compared with the Holarctic, investigations have been less comprehensive both geographically and faunistically. In Africa, conditions initially favored the study of helminths in mammals because of the availability of material from specimens that died in European zoos. An important early work concerned in part with trematodes in birds was Looss's (1896) *Recherches sur la faune parasitaire de l'Égypte*. Such trematodes were included also by Odhner (1910) in his *Nordostafrikanische Trematoden*, based on materials collected by the Swedish Zoological Expedition to Egypt and the White Nile in 1901. During the early twentieth century, helminths from African birds were studied by J. G. Baer, O. Fuhrmann, C. E. Joyeux, H. A. Baylis, H. O. Mönnig, R. J. Ortlepp, and others. Information concerning helminths in South America first appeared in the works of Rudolphi and Diesing, based on materials collected by early explorers. Of particular significance were the collections made in Brasil by Johann Natterer during the period 1817–1836. During the early decades of the twentieth century, important contributions were made by H. von Ihering, A. Lutz, K. Wolffhügel, L. Travassos, E. Vogelsang, and others. After about 1930, numerous papers concerning helminths in birds were published by E. Caballero y C. and co-workers in Mexico.

In Australia, the names of T. H. Johnston and P. M. Mawson are associated with early studies. The first opportunities to make collections in regions of high southern latitude were provided by the various national expeditions to the Antarctic. Helminths from birds were described by von Linstow (1888, 1906), Rennie and Reid (1912), Railliet and Henry (1912), Leiper and Atkinson (1914, 1915), and others.

Current Trends

Since about the middle of the present century, the volume of literature concerning helminths in animals other than man has expanded at an increasing rate, attributable in part to the effort of growing numbers of helminthologists worldwide and in part to a diversification of research made possible by the application of technological advances in such relevant fields as biochemistry and immunology. During this period also, increasing effort has been directed toward more precise definition of the relationships of helminths and their hosts, such as coevolution and *Wirtskreiserweiterung (sensu* Osche, 1957), host specificity, localization of helminths, intra- or interspecific competition among helminths, and dynamics of populations of the organisms making up specific assemblages. Such investigations are perhaps hampered by the current emphasis on parasitism *sensu lato* as a phenomenon, without adequate consideration of the diversity of organisms that may be designated *parasites* under the broadest application of the term (see Price, 1980).

Taxonomy, systematics, and knowledge of cycles remain basic to investigations involving helminths. Publications in these areas have also increased in number since about 1950, although they make up a decreasing proportion of the helminthological literature. Concerning only helminths in birds, some of the more comprehensive works published since 1960 include (see References for original titles): Bykhovskaia-Pavlovskaia (1962; trematodes of birds of the Soviet Union), Sultanov (1963; helminths of birds in Uzbekistan), Spasskaia (1966; hymenolepidid cestodes of birds of the Soviet Union), Dubinina (1966; cestodes of the family Ligulidae), Spasskaia and Spasskii (1971; cestodes of birds of Tuva), Bona (1975; dilepidid cestodes of ciconiiform birds); Illescas Gomez (1977; helminths of birds in the Province of Granada), Spasskaia and Spasskii (1977; dilepidid cestodes of birds of the Soviet Union), and Baruš *et al.* (1978; nematodes of piscivorous birds of the Pal-

5. BIOLOGY OF AVIAN PARASITES

aearctic). In addition, numerous monographs on the taxonomy and systematics of helminths include those occurring in birds. The literature concerning developmental cycles and other biological characteristics of such organisms is with few exceptions widely scattered.

II. Characteristics of Helminths

A few comments concerning the characteristics of helminths found in birds are included here, but details of classification, some of which are controversial, are not considered.

A. Phylum Platyhelminthes

1. Class Trematoda

Trematodes of the subclass Digenea occur in the adult (sexually reproductive) stage in the avian final host in which, depending on species, they localize in the alimentary canal or in other organs. With the exception of members of the families Schistosomatidae and Ornithobilharziidae, digenetic trematodes are monoecious. Trematodes in birds are usually flattened dorsoventrally, and rounded to elongate in outline. However, they range in shape from nearly filiform to subspherical, and some have more or less bipartite bodies. Most trematodes in birds are relatively small, ranging from less than 0.5 mm to only several millimeters in length. When alive, most are whitish or slightly tinged with color. The opaque internal organs or the egg-filled uterus may be discernible, and in many species of trematodes, these structures impart a characteristic appearance. Asexual reproduction typically involves a sequence of developmental stages in an aquatic mollusk, or less commonly in a terrestrial mollusk. The asexual generations develop following the establishment of the ciliated miracidium, which emerges from the egg expelled by the final host. The result of this reproductive process, which may continue for the life of the first intermediate host, is the production of large numbers of cercariae. The cercaria usually emerges from the mollusk and has a short aquatic existence. It may infect the final host directly, or further development to the metacercarial stage may take place. The infective metacercaria is encysted in or on the second intermediate host (invertebrate or vertebrate), or on emergent vegetation, shells of mollusks, or other objects in water. Further development (to the sexually reproducing stage) depends on the ingestion of the metacercaria by the final host. In this host, the trematode attains maturity after a specific period, and egg production begins.

2. Class Cestoidea

In the strobilar (sexually reproductive) stage, cestodes inhabit the alimentary canal of the final host. From 5 to 12 or more orders of cestodes may be distinguished, depending on the system of classification accepted. Cestodes in birds fall into only 3 orders: Pseudophyllidea, Cyclophyllidea, and Aporidea. In the strobilar stage, cestodes are polyzoic animals consisting of an apparently anterior organ of attachment (scolex), followed in sequence by a short proliferative zone and by the strobila derived from it. The strobila is composed of few to many segments, which increase in age posteriad from the proliferative zone. Thus, immature, mature (with fully developed reproductive organs), and gravid segments are distinguished. Cestodes are typically monoecious. Certain genera of the family Dioecocestidae contain dioecious forms, all of which occur in birds. Strobilae of cestodes are usually dorsoventrally flattened, and segmentation may or may not be strongly defined. Alive or preserved, they are usually white and often somewhat translucent; some have a slight gray or brownish tinge. Compared with cestodes in mammals, those in birds do not attain large size. In general, their length ranges from a few millimeters to several centimeters, but there are exceptions; one that attains a meter or more in length is *Houttuynia struthionis* (Houttuyn, 1773), in the Ostrich (*Struthio camelus*). The cycle of cyclophyllidean cestodes involves a single intermediate host, typically an invertebrate. In the case of certain species of cestodes in falconiform and strigiform birds, the larval stage occurs in mammals. The infective embryo, enclosed by egg membranes, is expelled in the feces of the avian final host, and must be ingested by the intermediate host. Thereafter, development of the infective larva (a cysticercoid or modification thereof) takes place, and the cycle is completed with its ingestion by the final host. The cycle of pseudophyllidean cestodes is more complex, involving two aquatic intermediate hosts. The cestode egg is unembryonated when expelled by the final host, and after a period of development in water, the motile, ciliated embryo emerges and is eaten by a small crustacean, in which the first larval stage (procercoid) is produced. When the latter is ingested by a fish, further development of the larva to the stage infective for the final host takes place. The cycle is completed when the plerocercoid in the fish is eaten by the final host. If, instead, another fish consumes the second intermediate host, the plerocercoid penetrates the wall of the stomach or intestine and relocalizes. With respect to the cycle, the second fish (paratenic host) functions as the second intermediate host.

B. Phylum Aschelminthes

Nematodes are arranged in two classes, Adenophorea and Secernentia, the latter of which includes most of the species that are found in birds. In their adult stage, nematodes occur in various organs, but in the avian host, those in the alimentary canal are perhaps the most frequently noticed. Nematodes are usually more or less cylindrical with tapered ends, but they differ widely in proportions of the body. The sexes are separate, and males are typically smaller than females. Species in birds range from a few millimeters to several centimeters in length. Many are white and relatively opaque when alive; others exhibit specific patterns imparted by the arrangement of their internal organs or by the nature of their intestinal content. Most nematodes produce eggs which are expelled by the final host. The cycle typically includes four larval stages, each of which is followed by a molt. Third-stage larvae are usually infective for their respective hosts, but exceptions exist. Cycles of nematodes differ in pattern, but may be generally categorized as direct or indirect, the latter involving one or two intermediate hosts. In some cases, infective larvae are transmitted to the final host by hematophagous intermediate hosts.

C. Phylum Acanthocephala

Acanthocephalans are the least common of the helminths found in birds. Four classes are distinguished in the phylum. Three of these, Archiacanthocephala, Palaeacanthocephala, and Apororhynchida (sometimes included in the Archiacanthocephala), include species for which birds serve as final host. Localization is in the intestine of the host. The relatively few species known from birds have been listed by Petrochenko (1958) and Golvan (1960, 1962). Acanthocephalans, when distended, are more or less cylindrical, with proportions of the body differing with species. Superficially, the body consists of an anterior proboscis armed with spines, an adjacent neck, and a relatively large trunk. Small spines may be present also on the neck and trunk. Most specimens in birds range from 1 to 2 cm in length, but some are considerably longer and more slender in proportion. The proboscis may be inverted into the anterior part of the body, but acanthocephalans are usually found with this structure firmly embedded in the intestinal wall of the host. The proboscis in some species [e.g., *Filicollis anatis* (Schrank, 1788)] characteristically penetrates the full thickness of the host's intestinal wall, with the distal end, which is often inflated, situated externally and enclosed by a thin adventitia covered by serosa. Living specimens observed *in situ* are usually flattened and may be quite flaccid; in this state they may resemble cestodes. Acanthocephalans are often white, but they may be stained by the intestinal content of the host. For example, specimens found in Glaucous-winged Gulls (*Larus glaucescens*) collected in March 1955 near the mouth of the Anchor River, Kenai Peninsula, were stained bright orange, as were some host tissues, by pigments from tunicates on which the gulls were feeding. Acanthocephalans have rather simple life cycles, as indicated by the relatively few that have been defined. The egg, containing the fully developed acanthor, is ingested by an arthropod, which serves as intermediate host. The next stage of development (within the intermediate host) is the acanthella, which is followed by the infective stage, the cysticanth. The cysticanth develops to maturity in the final host. In animals other than those serving as final host, it may relocalize and persist in the infective stage. Such *paratenic* hosts function as second intermediate hosts, and for some species they appear to be essential for infection of the final host.

III. Occurrence of Helminths in Birds

That establishment of the taxonomic status of helminths in free-living animals necessarily precedes the scientific investigation of their biological characteristcs is self-evident. Surveys of helminths of birds of a single species or of closely related species (e.g., members of a single genus or family) in a given geographic area produce useful information in the form of local lists or inventories, but they do not provide any perspective concerning the host range of the respective species or any indication of ecological or other factors influencing their occurrence. It is much more informative to survey helminths in entire avian faunas, so far as it is practicable to do so. The investigations on which the present summary is partly based were undertaken to obtain a better understanding of the natural occurrence of helminths in wild birds and mammals and to obtain evidence for or against the hypothesis that pathogenicity could rarely be demonstrated in such animals. The surveys were conducted by me mainly in the north-central United States during the period 1943–1948, and thereafter in Alaska with the participation of colleagues. With the inclusion of comparatively few specimens collected in Wyoming and southern Canada (Manitoba), birds examined in the central states numbered 2388, with 232 species representing 140 genera, 44 families, and 15 orders. The series in Alaska consisted of 3089 birds of 178 species, representing 117 genera, 36 families, and 13 orders. Most of the birds were migratory species, collected during migration or on the breeding grounds, but resident species were also included.

These investigations were also undertaken at a time when the adverse

5. BIOLOGY OF AVIAN PARASITES

impact of human activities (destruction of habitat and environmental pollution) on avian populations had not become generally apparent. Even so, a considerable proportion of the specimens, particularly of the more uncommon species, were obtained by means other than collecting. Federal and state agencies made available numerous dead or injured birds, including for example the three Trumpeter Swans (*Cygnus buccinator*) examined in Alaska. Most of the raptors obtained in the north-central states were those that had been routinely trapped on one of the state game farms. In Alaska, Golden Eagles (*Aquila chrysaetos*) caught incidentally in traps set for wolves were made available over several years by the Nunamiut Eskimos. Similarly, a series of 123 Snowy Owls (*Nyctea scandiaca*) consisted almost entirely of birds that had been trapped for food by the Eskimos at Wainwright and Point Barrow in autumn of the years when dense breeding populations had been present in conjunction with high numerical densities of lemmings (*Lemmus sibiricus*). With diminishing possibilities for investigative surveys in many geographic regions in the future, but with the recognition of the lack of knowledge of aspects of avian biology, an even greater effort should be made to derive maximum information from each specimen.

Prevalence

Each species of bird may be assumed to have a characteristic, but not specific, helminth fauna which can be defined by sampling the population. For some avian species, a large sample is required for this purpose, because the prevalence of helminths may differ widely with species of host. The adequacy of surveys also depends on the thoroughness with which birds are examined, for helminths in organs other than the alimentary canal may be overlooked unless dissections are complete, and even then larval stages in tissues often are not found.

In general, birds acquire helminths through the ingestion of the infective larvae in organisms that characteristically compose a part of their diet. The availability of such larvae, however, depends on the interaction of component members of helminth–host assemblages at a level sufficient to sustain cycles of the respective species of helminths. This requirement evidently is not met in biotopes of some avian species. Rates of infection in some canopy-dwelling birds, such as the Vireonidae, appear to be extremely low. In contrast, high rates of infection with helminths of diverse species are characteristic of birds associated with aquatic habitat, such as the Rallidae and Anatidae, in which various factors contribute to a high intensity of trophic interactions, at least seasonally.

Data concerning the prevalence of helminths in birds collected in the

TABLE I

Prevalence of Helminths in Birds in the North-Central United States (Southern Series)[a]

Family	Number of species	Total number of birds	Number infected	Trematoda	Cestoda	Nematoda	Acanthocephala
Gaviidae	1	2	2	2	2	—	—
Podicipedidae	4	21	21	7	14	—	—
Phalacrocoracidae	1	4	4	4	2	4	1
Ardeidae	7	47	42 (89)	25 (53)	31 (66)	25 (53)	16 (34)
Anatidae	18	124	91 (73)	55 (44)	77 (62)	22 (18)	26 (21)
Cathartidae	1	2	—	—	—	—	—
Accipitridae	9	138	114 (83)	96 (70)	25 (18)	57 (41)	4 (3)
Pandionidae	1	3	3	1	1	1	—
Falconidae	3	11	3	2	—	1	—
Tetraoninae	3	6	4	3	2	—	—
Phasianinae	2	23	10	1	2	10	—
Rallidae	4	87	63 (72)	36 (41)	34 (40)	5 (6)	24 (28)
Charadriidae	3	32	17	6	13	4	—
Scolopacidae	16	132	82 (62)	19 (14)	73 (55)	7 (5)	1 (<1)
Phalaropodinae	2	7	6	6	3	—	—
Laridae	6	83	30 (36)	22 (27)	14 (17)	3 (4)	2 (2)
Columbidae	1	9	2	—	2	—	—
Cuculidae	2	11	2	—	2	—	—
Tytonidae	1	6	2	—	—	2	—

Family							
Strigidae	8	116	83 (72)	48 (41)	16 (14)	50 (43)	4 (3)
Caprimulgidae	2	9	4	2	4	—	—
Apodidae	1	7	3	—	3	—	—
Trochilidae	2	3	—	—	—	—	—
Alcedinidae	1	16	—	8	—	3	—
Picidae	8	102	10	—	30 (29)	3 (3)	—
Tyrannidae	12	56	30 (29)	2 (4)	6 (11)	1 (2)	1 (2)
Alaudidae	1	30	10 (18)	—	3 (10)	—	—
Hirundinidae	7	85	3 (10)	5 (6)	45 (53)	(>2)	—
Corvidae	6	77	48 (56)	5 (6)	27 (35)	12 (16)	1 (>1)
Paridae	5	68	34 (44)	—	9 (13)	2 (3)	—
Sittidae	2	32	11 (16)	—	3 (9)	—	—
Certhiidae	1	9	3 (9)	—	—	—	—
Troglodytidae	4	12	—	—	3	—	—
Mimidae	2	13	3	1	1	1	4
Muscicapidae	10	128	4	6 (5)	56 (44)	17 (13)	8 (6)
Motacillidae	1	2	68 (53)	—	—	—	—
Bombycillidae	1	8	—	—	—	—	—
Laniidae	1	6	—	—	—	—	—
Sturnidae	1	26	18	—	17	2	5
Vireonidae	5	27	—	—	—	—	—
Parulidae	20	103	19 (18)	—	15 (15)	4 (4)	—
Icteridae	10	172	39 (23)	13 (8)	24 (14)	10 (6)	17 (10)
Fringillidae	11	149	4 (3)	—	3 (2)	—	1 (<1)
Emberizidae	25	384	89 (23)	6 (<2)	67 (17)	17 (4)	9 (2)

[a]Percentages in parentheses.

TABLE II

PREVALENCE OF HELMINTHS IN BIRDS IN ALASKA (NORTHERN SERIES)[a]

Family	Number of species	Total number of birds	Number infected	Trematoda	Cestoda	Nematoda	Acanthocephala
Gaviidae	4	65	54 (83)	9 (14)	54 (83)	15 (23)	4 (6)
Podicipedidae	2	62	61 (98)	20 (32)	49 (79)	20 (32)	4 (6)
Phalacrocoracidae	3	46	26 (57)	7 (15)	—	10 (22)	22 (48)
Ardeidae	1	1	—	—	—	—	—
Anatidae	27	584	386 (66)	75 (13)	388 (66)	22 (4)	62 (11)
Accipitridae	10	64	41 (64)	12 (19)	21 (33)	18 (28)	4 (6)
Tetraoninae	6	258	152 (59)	80 (31)	33 (13)	61 (24)	—
Gruidae	1	13	8	4	4	—	—
Haematopodidae	1	7	4	—	4	—	—
Charadriidae	6	62	34 (55)	5 (8)	30 (48)	1 (2)	1 (2)
Scolopacidae	21	260	183 (70)	42 (16)	179 (69)	4 (2)	48 (18)
Phalaropodinae	2	52	21 (40)	—	21 (40)	—	2 (4)
Stercorariidae	3	89	31 (35)	5 (6)	26 (29)	4 (4)	3 (3)
Laridae	9	352	249 (71)	61 (17)	220 (63)	13 (4)	7 (2)
Alcidae	11	159	44 (28)	4 (3)	30 (19)	12 (8)	6 (4)
Strigidae	7	201	72 (36)	2 (1)	51 (25)	44 (22)	—

Family							
Alcedinidae	1	9	—	1	—	—	—
Picidae	4	22	5	—	5	1	—
Tyrannidae	5	11	2	2	1	1	—
Alaudidae	1	4	2	—	2	—	—
Hirundinidae	2	5	1	—	1	—	—
Corvidae	5	183	32 (17)	9 (5)	16 (9)	7 (4)	—
Paridae	2	20	3	—	3	—	—
Sittidae	1	1	1	—	1	—	—
Certhiidae	1	1	—	—	—	—	—
Cinclidae	1	7	—	—	—	—	—
Troglodytidae	1	3	1	—	1	—	—
Muscicapidae	10	174	113 (65)	61 (35)	92 (53)	16 (9)	16 (9)
Motacillidae	4	78	30 (38)	7 (9)	25 (32)	—	1 (1)
Bombycillidae	1	4	2	1	1	—	—
Laniidae	1	15	1	—	1	—	—
Vireonidae	1	1	—	—	—	—	—
Parulidae	6	35	12	5	4	3	—
Icteridae	1	9	3	3	—	1	—
Fringillidae	5	43	3 (7)	1 (2)	2 (5)	—	—
Emberizidae	11	189	97 (51)	29 (15)	94 (50)	6 (3)	1 (<1)

[a]Percentages in parentheses.

north-central states and in Alaska (hereafter designated southern and northern series, respectively) are summarized by family in Tables I and II. As has been typical of most surveys of this kind, species of birds were often represented by too few specimens to establish natural levels of infection and to permit definition of helminth faunas. For the two series of birds considered, it may be assumed that all helminths in the alimentary canal were recorded, but some proportion of trematodes and possibly of nematodes occurring in other organs, such as the kidneys, was no doubt overlooked when examinations were made under field conditions. The recorded rates of infection therefore may have been somewhat lower than the actual rates. Few larval stages were found. Their prevalence could have been precisely determined only through artificial digestion of tissues, including skeletal musculature.

The overall rates of infection by helminths were 41 and 54% for the southern and northern series, respectively. With reference to the four major groups of organisms represented, rates were similar for trematodes (16 and 14%), nematodes (11 and 8%), and acanthocephalans (5 and 6%). Different rates (26 and 44%) were obtained for cestodes, which were the most commonly occurring helminths in both series. For comparison, similar data from surveys conducted in various regions of Eurasia are summarized in Table III. Detailed information concerning findings in individual species of birds is available from the original publications. Comparable surveys have not been made in the southern hemisphere.

In general, overall rates of infection by helminths as well as the relative representation of the major groups thereof depend on the species composition of the series of birds examined. Findings in birds of a single family from two or more regions may not be comparable, because of intergeneric differences in rates of infection and in composition of helminth faunas. Such differences are particularly apparent among passerine birds, and appear to be attributable mainly to ethological and ecological factors that influence the range of infective larval stages to which birds of the respective genera are exposed, rather than to specificity of the helminths involved. Various biotic factors may influence the composition of helminth faunas of birds under conditions of different latitudinal zones (see Bykhovskaia-Pavlovskaia, 1962).

High rates of infection are typical of birds that feed on freshwater organisms during the warmer months of the year. Such birds are also characteristically infected by helminths of numerous species. Lists of helminths recorded from more than 4000 birds representing 107 species from aquatic habitats and marshes in the Ukraine have been published by Smogorzhevskaia (1976). Comparable rates of infection may be observed in marine birds of some families, but, usually, in these cases only a few species of helminths are represented.

Some examples will provide an indication of the prevalence of helminths in birds of the more prominent families in the Holarctic.

TABLE III
Prevalence of Helminths in Birds in Various Regions of Eurasia[a]

Region	No. of birds	No. of species	Total infected	Trematoda	Cestoda	Nematoda	Acanthocephala	Reference
Prov. of Granada, Spain, 1970s	1714	89	551 (32)	189 (11)	306 (18)	190 (11)	32 (2)	Illescas Gomez (1977)
Baden, Switzerland (some exotic species included), 1890s	630	73	450 (71)	124 (20)	231 (37)	252 (40)	41 (7)	Wolffhügel (1900)
Southern Georgia (Gruziia), USSR, 1948	558	66	257 (46)	61 (11)	171 (31)	118 (21)	14 (2)	Ryzhikov (1951)
Chany Lake, Novosibirsk Oblast', USSR, 1946	799	103	479 (60)	215 (27)	345 (43)	193 (24)	38 (5)	Mozgovoi et al. (1951)
Komi ASSR, 1947	961	81	547 (56)	314 (32)	352 (36)	163 (17)	4 (0.4)	Pod'iapol'skaia et al. (1951)
Kirgizia, USSR, 1945	549	60	270 (49)	100 (18)	180 (32)	135 (24)	30 (5)	Matevosian (1951)
Kazakh SSR, 1949	170	32	154 (91)	115 (68)	117 (69)	104 (61)	9 (5)	Gagarin (1951)
Lake Baikal, 1949	1036	136	643 (62)	253 (24)	434 (42)	219 (21)	17 (2)	Ryzhikov and Sudarikov (1951)
Amur River Basin, 1958	2097	182	1126 (54)	514 (24)	713 (34)	413 (20)	61 (3)	Ryzhikov et al. (1961b)
Primorsk Krai (Soviet Far East), 1948	558	87	353 (63)	138 (25)	225 (40)	136 (24)	1 (0.2)	Sadokov (1951)
Iakut ASSR, 1953	436	71	277 (63)	127 (29)	197 (45)	141 (32)	19 (4)	Mozgovoi et al. (1956)
Chukotka, 1961	982	71	828 (84)	549 (56)	655 (67)	300 (31)	153 (16)	Spasskii et al. (1963)
Kamchatka Peninsula, 1960	739	110	560 (76)	304 (41)	320 (43)	180 (24)	71 (10)	Spasskii et al. (1962)
Kamchatka Peninsula, 1961	806	79	620 (77)	417 (52)	363 (45)	188 (23)	78 (10)	Spasskii et al. (1963)

[a] Percentages indicated in parentheses.

Gaviidae. Of 65 loons of four species examined in Alaska, 54 (83%) were infected by helminths, with rates for the major groups as follows: trematodes, 14%; cestodes, 83%; nematodes, 23%; and acanthocephalans, 6%. All of 15 Red-throated Loons (*Gavia stellata*) examined in Kamchatka by Spasskii et al. (1962) were infected, as were 16 of this species and 8 Arctic Loons (*G. arctica*) collected in Chukotka (Spasskii et al., 1963).

Podicipedidae. Ten of 15 Pied-billed Grebes (*Podilymbus podiceps*) collected in the north-central states were infected by trematodes and/or cestodes. Of 52 grebes of the genus *Podiceps* examined in Alaska, 51 harbored helminths, with rates as follows: trematodes, 33%; cestodes, 94%; nematodes, 38%; and acanthocephalans, 8%. Rybicka (1961) found all of 35 Great Crested Grebes (*Podiceps cristatus*) from Drużno Lake, in Poland, to be infected by cestodes. Species of helminths were enumerated by group by Smogorzhevskaia (1976) for 170 grebes representing five species. The most detailed information concerning helminths in grebes has been published by Vaidova (1978), who examined 244 specimens from three regions of Azerbaidzhan. Vaidova's data are summarized as follows: Great Crested Grebe—90 of 95 infected, with 32 species of helminths recorded; Little Grebe (*Tachybaptus ruficollis*)—94 of 104 specimens infected, with 27 species of helminths; Eared Grebe (*P. nigricollis*)—all of 22 infected, with 8 species of helminths; and Red-necked Grebe (*P. grisegena*)—16 of 23 infected, with 11 species of helminths.

Phalacrocoracidae. Records in the literature indicate that rates of infection in cormorants are high, as is the diversity of helminth faunas in these birds. Of 44 Pelagic Cormorants (*Phalacrocorax pelagicus*) collected in Alaska, 52% harbored helminths, as follows: trematodes, 16%; nematodes, 20%; and acanthocephalans, 50%. Cestodes were not found. Tsimbaliuk and Belogurov (1964) examined 38 Pelagic Cormorants from islands in the Bering Sea, and reported that 36 (94%) were infected with nematodes. Belopol'skaia (1952) discussed the occurrence of helminths in European Shags (*P. aristotelis*). Cormorants of two species were investigated by Vaidova (1978) in Azerbaidzhan. There, of 115 specimens of Great Cormorant (*P. carbo*), 110 were infected with helminths representing 29 species, and of 137 Pygmy Cormorants (*P. pygmaeus*), 127 were infected; helminths of 17 species were recorded.

Ardeidae. High rates of infection with helminths of numerous species are characteristic of herons. In the north-central states, 41 of 46 birds representing seven species were infected. The major groups of helminths were represented as follows: trematodes, 54%; cestodes, 65%; nematodes, 57%; and acanthocephalans, 35%. In the basin of the Amur River, 12 of 17 Gray

Herons (*Ardea cinerea*) harbored helminths (Ryzhikov *et al.*, 1961b). In Azerbaidzhan, Vaidova (1978) examined 102 Gray Herons and found 85 to be infected; 44 species of helminths were involved. There also, helminths of 35 species were found in 71 of 92 Purple Herons (*A. purpurea*). Vaidova provided comparable data for *Ardeola ralloides*, *Bubulcus ibis*, *Casmerodius albus*, *Egretta garzetta*, *Botaurus stellaris*, *Ixobrychus minutus*, and *Nycticorax nycticorax*.

Ciconiidae. Comparatively little information is available concerning the prevalence of helminths in storks. Species recorded from both the White Stork (*Ciconia ciconia*) and the Black Stork (*C. nigra*) were listed by Szidat (1940). Of 29 White Storks examined by Vaidova (1978) in Azerbaidzhan, 15 were infected with helminths representing 10 species.

Threskiornithidae. High rates of infection involving diverse species of helminths appear to be characteristic of ibises. Bush and Forrester (1976) found all of 140 White Ibises (*Eudocimus albus*) in Florida to be infected with helminths representing 42 species. In Azerbaidzhan, 36 (80%) of 45 Glossy Ibises (*Plegadis falcinellus*) harbored helminths representing 14 species (Vaidova, 1978).

Phoenicopteridae. Comparatively few species of helminths are known from flamingos. In Azerbaidzhan, Vaidova (1978) recorded five species from 13 of 18 Greater Flamingos (*Phoenicopterus ruber*).

Anatidae. The helminths of ducks and geese are comparatively well known in the Holarctic. The list compiled by McDonald (1969) included just over 1000 species, of which the major groups were represented as follows: trematodes, 47%; cestodes, 28%; nematodes, 16%; and acanthocephalans, 9%. High rates of infection involving diverse species of helminths are characteristic of anatids. In Alaska, 419 (73%) of 584 anatids of 27 species were found to be infected. In the Ukraine, Smogorzhevskaia (1976) recorded 161 species from 818 birds of 22 species. Trematodes were present in 303 (72%) of 417 anatids of 14 species examined by Bykhovskaia-Pavlovskaia (1953) in the Barabinsk forest–steppe (western Siberia). Rates of infection based on relatively large series of birds have been reported by Ryzhikov *et al.* (1951; basin of the Amur River), Spasskii and Sonin (1961; Kamchatka), Mozgovoi *et al.* (1951; Chany Lake, western Siberia), Spasskii *et al.* (1963; Kamchatka), and others. Detailed information concerning helminths of the four groups found in anatids of 26 species in Azerbaidzhan was provided by Vaidova (1978).

Accipitridae. Intergeneric, and perhaps infrageneric, differences exist in rates of infection and in composition of helminth faunas of birds of this family. For 138 hawks of the genera *Accipiter, Buteo,* and *Circus* examined in the north-central states, the overall rate was 83%. Of 76 Red-tailed Hawks (*B. jamaicensis*) in this series, 66 (87%) harbored helminths, as follows: trematodes, 82%; cestodes, 20%; nematodes, 46%; and acanthocephalans, ~ 1%. Helminths occurred in 22 (81%) of 27 Cooper's Hawks (*A. cooperi*) but neither cestodes nor acanthocephalans were present. Of 48 birds of the genera *Accipiter, Buteo, Aquila,* and *Haliaeetus* examined in Alaska, 32 (67%) were infected: trematodes, 15%; cestodes, 33%; nematodes, 27%; and acanthocephalans, 8%. Twenty-four (89%) of 27 accipitrids of the genera *Accipiter, Buteo, Aquila,* and *Circus* examined by Mozgovoi *et al.* (1951) in the vicinity of Chany Lake were infected. Ryzhikov and Sudarikov (1951) reported similar findings around Lake Baikal. Smogorzhevskaia (1976) recorded 18 species of helminths from 43 birds of the genera *Circus, Milvus,* and *Haliaeetus* in the Ukraine. Numbers of birds and rates of infection reported by Vaidova in Azerbaidzhan were as follows: Hen Harrier (*C. cyaneus*), 8 of 13; Eurasian Marsh-Harrier (*C. aeruginosus*), 54 (63%) of 85; Pallid Harrier (*C. macrourus*), 1 of 1; Black Kite (*M. migrans*), 89 (70%) of 125; Red Kite (*M. milvus*), 2 of 2; European Sparrowhawk (*Accipiter nisus*), 17 (57%) of 30; White-tailed Eagle (*Haliaeetus albicilla*), 20 of 21; Tawny Eagle (*Aquila rapax*), 13 of 14; Booted Eagle (*Hieraeetus pennatus*), 2 of 2; Lesser Spotted Eagle (*Aquila pomarina*), 6 of 10; Common Buzzard (*Buteo buteo*), 30 (83%) of 36; Rough-legged Hawk (*B. lagopus*), 1 of 1; Long-legged Buzzard (*B. rufinus*), 1 of 3; Short-toed Eagle (*Circaetus gallicus*), 12 of 13; and Eurasian Honey-Buzzard (*Pernis apivorus*), 8 of 10. Findings in a large series of raptors collected along the Caspian Sea were reported by Zablotskii (1962).

Pandionidae. The available records indicate that Ospreys (*Pandion haliaetus*) commonly harbor helminths. All of three specimens examined by our group in Ohio and Wyoming, and five examined by Smogorzhevskaia (1976) in the Ukraine, were infected. Vaidova (1978) examined 25 Ospreys in Azerbaidzhan, of which 22 (88%) were infected. From these, helminths of 25 species were obtained.

Falconidae. Among the Falconiformes, somewhat lower rates of infection appear to be typical of birds of the genus *Falco*. Only 3 of 11 birds obtained in the north-central states were infected, and only trematodes and nematodes were present. Helminths were recorded from 7 of 13 falcons examined in Alaska, among which all of 5 Gyrfalcons (*F. rusticolus*) were infected. Rates were comparatively low in small series of birds examined by Mozgovoi *et al.* (1951) near Chany Lake, and by Pod'iapol'skaia *et al.* (1951) on the Pechora River (Komi ASSR). A large series of falcons was examined by Vaidova over a period of 20 years (1955–1975). Vaidova found 68 (94%) of 72

Peregrine Falcons (*F. peregrinus*) to be infected, but only 12 species of helminths were recorded. Rates for falcons of other species examined were as follows: *F. tinnunculus*, 69 (83%) of 83; *F. naumanni*, 84 (59%) of 143; *F. subbuteo*, 3 of 5; and *F. vespertinus*, 1 of 1.

Tetraoninae. Helminths of the more common gallinaceous birds are well known, as a result of numerous investigations conducted since the beginning of the present century (see Leslie and Shipley, 1912). Morphological descriptions and other information for all species recorded from such birds were provided by Kasimov (1956). Since that time, numerous studies have contributed to the knowledge of helminths in the Galliformes.

The prevalence of helminths in birds of the subfamily Tetraoninae is considered here to be representative of the order. Rates were relatively low in grouse of the genera *Canachites, Bonasa, Lagopus,* and *"Pedioecetes"* [now *Tympanuchus*] examined in Alaska. Helminths were found in only 152 (37%) of 410 birds, with representation of the major groups as follows: trematodes, 20%; cestodes, 8%; and nematodes, 15%. The rates of infection were similar (27 to 33%) for four species: Spruce Grouse (*C. canadensis*), Ruffed Grouse (*B. umbellus*), Willow Ptarmigan (*L. lagopus*), and White-tailed Ptarmigan (*L. leucurus*). Higher rates (65 and 56%, respectively) were obtained for Rock Ptarmigan (*L. mutus*) and Sharp-tailed Grouse (*T. phasianellus*). The occurrence of helminths of four species in 338 Willow Ptarmigan collected at two localities in Norway has been described by Wissler and Halvorsen (1977). Informative studies have also been published by Gushanskaia (1952) and Madsen (1952). A survey of helminths in Ruffed Grouse in eastern North America was conducted over a 10-year period by Davidson *et al.* (1977), who recorded 23 species from 327 birds examined. Data from earlier studies were published by Boughton (1937), Dorney and Kabat (1960), and others. Ryzhikov and Sudarikov (1951) found 19 (51%) of 37 Common Hazel Grouse (*Tetrastes bonasia*) collected near Lake Baikal to be infected. Helminths were found in 56 of 101 Common Hazel Grouse in the Amur River basin (Ryzhikov *et al.*, 1961b). Rates of infection were relatively low: trematodes, 6%; cestodes, 37%; and nematodes, 2%. Only 41 (38%) were infected of 108 Common Hazel Grouse examined by Spasskii *et al.* (1958) in the Tuvinsk Autonomous Oblast'. The helminths included trematodes (6%), cestodes (19%), and nematodes (31%). In contrast, 39 were infected of 40 Black Grouse (*Tetrao tetrix*) collected in the same region. In these, trematodes were present in 2.5%, cestodes in 30%, and nematodes in 97%. There also, 19 of 20 Capercaillies (*T. urogallus*) harbored helminths, with trematodes in 5%, cestodes in 45%, and nematodes in 90%.

Gruidae. The most thorough data concerning helminths of cranes were reported by Forrester *et al.* (1974), who found 73 of 74 (Greater) Sandhill

Cranes (*Grus canadensis tabida*), obtained on the wintering grounds in Florida, to be infected. Helminths of 14 species (trematodes and nematodes) were identified. Of 13 specimens of *G. canadensis* examined in Alaska, 8 were infected (4 with trematodes and 4 with cestodes). Three cranes of the same species were examined in Chukotka by Spasskii et al. (1963); all harbored helminths, among which trematodes, nematodes, and acanthocephalans were represented. All of 3 European Cranes (*G. grus*) were found to be infected in the Tuvinsk Autonomous Oblast' (Spasskii et al., 1958). There also, 6 of 8 White-naped Cranes (*G. vipio*) harbored helminths, which included cestodes, nematodes, and acanthocephalans. Helminths of 3 species were recorded from 2 European Cranes by Vaidova (1978) in Azerbaidzhan.

Rallidae. Among the rallids, birds of the genus *Fulica* have been rather intensively investigated. An early survey of helminths in American Coots (*F. americana*) in Iowa was made by Roudabush (1942), who examined 17 specimens. Seventeen species of helminths were recorded from 60 coots (59 infected) examined in Florida by Kinsella (1973). These included trematodes of 8 species, cestodes of 2 species, nematodes of 6 species, and acanthocephalans of one species. Of 68 coots examined by our group in the north-central states, 56 (82%) were infected, with trematodes in 49%, cestodes in 31%, nematodes in 4%, and acanthocephalans in 34%. In Poland (Dru\.zno Lake), Rybicka (1961) examined 140 specimens of European Coot (*F. atra*) over a 3-year period. All were infected, and all harbored cestodes; other helminths were not reported. From a series of 150 coots examined in the Ukraine, Smogorzhevskaia (1976) identified trematodes of 28 species, cestodes of 3 species, nematodes of 6 species, and acanthocephalans of 2 species. In western Siberia, Bykhovskaia-Pavlovskaia (1953) found 55 of 75 coots to be infected by trematodes, of which 20 species were represented. Vaidova (1978) reported that helminths were present in 67% of 191 specimens of *F. atra* examined in Azerbaidzhan. Twenty-one species were identified. In Florida, Kinsella and Hon (1973) compared helminth faunas of 56 Common Gallinules (*Gallinula chloropus*) and 52 Purple Gallinules (*Porphyrula martinica*). Of the 23 species of helminths recorded, 12 were shared. Eight species occurring in gallinules were also reported from coots in Florida (Kinsella, 1973). Of 13 specimens of *G. chloropus* examined by Vaidova (1978) in Azerbaidzhan, 6 harbored helminths, consisting only of trematodes of 2 species. In eastern North American, trematodes of 26 species and cestodes of 2 species were recorded by Heard (1970) from 126 Clapper Rails (*Rallus longirostris*). No information concerning nematodes or acanthocephalans was provided. In Iowa, Redington and Ulmer (1966) examined 15 Soras (*Porzana carolina*) and 31 Virginia Rails (*R. limicola*), from which 8 and 9 species of helminths were recorded, respectively. Little information is available concerning helminths in rails of other species in Eurasia.

Charadriidae. Surveys of helminths in plovers have usually involved small series of birds. In Alaska, we found 34 (55%) of 62 birds representing the genera *Charadrius*, *Pluvialis*, *Aphriza*, and *Arenaria* infected. In the north-central states, helminths were present in 16 (59%) of 27 Killdeers (*C. vociferus*). Rates were 19% for trematodes, 44% for cestodes, and 15% for nematodes. In the Tuvinsk Autonomous Oblast', Spasskii *et al.* (1958) found 21 of 25 birds representing the genera *Pluvialis*, *Charadrius*, and *Vanellus* to be infected. Of 10 Little Ringed Plovers (*C. dubius*) collected on the Pechora River by Pod'iapol'skaia *et al.* (1951), 8 were infected by cestodes and/or nematodes. Twenty-two birds of the same species were examined in Azerbaidzhan by Vaidova (1978), who found 12 to be infected by trematodes representing a single species. Smogorzhevskaia (1976) recorded helminths of 24 species from 132 European Lapwings (*V. vanellus*) examined in the Ukraine, and provided lists of species identified from other charadriids, of which smaller series were collected.

Scolopacidae. Helminths of scolopacids have been rather intensively studied in Eurasia. A few examples of recorded rates are given here. In Alaska, 70% of 260 scolopacids representing 10 genera were infected. Trematodes occurred in 16%, cestodes in 69%, nematodes in 4%, and acanthocephalans in 3%. However, acanthocephalans were found in only 2 birds other than Purple Sandpipers (*Calidris maritima*), of which 14 of 27 were infected. Of 221 birds of six genera collected in the basin of the Amur River, 167 (76%) harbored helminths (Ryzhikov *et al.*, 1961a). Near Lake Baikal, 222 (79%) of 281 birds were infected (Ryzhikov and Sudarikov, 1951). In Kamchatka, the rate was 85% for 175 birds (Spasskii *et al.*, 1963). Smogorzhevskaia (1976) recorded helminths of 32 species from 106 Common Redshanks (*Tringa totanus*), and of 19 species from 92 Wood Sandpipers (*T. glareola*) in the Ukraine. Findings in numerous additional species of scolopacids, of which smaller series were examined, were also listed.

Phalaropodinae. Little information is available concerning helminths in phalaropes. Of 35 Red Phalaropes (*Phalaropus fulicaria*) obtained in Alaska, 19 harbored helminths consisting of cestodes (50%) and acanthocephalans (6%); of 17 Red-necked Phalaropes (*P. lobatus*), 2 had cestodes. Helminths of 10 species (4 of cestodes, 3 of trematodes, 3 of nematodes) were reported by Schmidt and Frantz (1972) from 90 Wilson's Phalaropes (*P. tricolor*) collected in Montana and Colorado. In Chukotka, Spasskii *et al.* (1963) found helminths in 18 of 20 Red Phalaropes, and in 38 (79%) of 48 Red-necked

Phalaropes. Rates for members of all of the major groups were somewhat higher in the latter species. Smogorzhevskaia (1976) recorded helminths of 12 species from 26 Red-necked Phalaropes examined in the Ukraine.

Stercorariidae. The information available suggests that rather low rates of infection and low diversity of helminth faunas are characteristic of jaegers. The overall rate for 89 birds examined in Alaska was only 35%. Fifteen (31%) of 48 Pomarine Jaegers (*Stercorarius pomarinus*) were infected. Helminths were found in only 3 of 17 Parasitic Jaegers (*S. parasiticus*), but 13 of 24 Long-tailed Jaegers (*S. longicaudus*) were infected. Acanthocephalans were not recorded from Pomarine Jaegers, nor were trematodes found in Long-tailed Jaegers.

Laridae. Birds of this family exhibit well-defined intergeneric differences in rates of infection and in composition of helminth faunas. The helminths of gulls have been much studied, perhaps reflecting the abundance and availability of these birds.

In Alaska, we examined 295 gulls representing eight species in four genera, and 56 terns of a single species. Helminths of all four groups usually were present in gulls of the genus *Larus*. Of 54 Glaucous Gulls (*L. hyperboreus*), 47 (87%) were infected, with trematodes in 11%, cestodes in 85%, nematodes in 4%, and acanthocephalans in 9%. Fifty-five (73%) of 75 Glaucous-winged Gulls (*L. glaucescens*) were infected, with trematodes in 37%, cestodes in 52%, nematodes in 12%, and acanthocephalans in ~1%. Of 74 Mew Gulls (*L. canus*), 67 (91%) harbored helminths. The four groups were represented as follows: trematodes, 19%; cestodes, 86%; nematodes, none; acanthocephalans, ~1%. Only cestodes were recorded from two species, the Black-legged Kittiwake (*Rissa tridactyla*), with 32 (65%) of 49 infected, and Sabine's Gull (*Xema sabini*), with 16 (73%) of 22 infected. Series of birds of other species were small. Cestodes alone were present in 8 of 9 Bonaparte's Gulls (*L. philadelphia*). No helminths were found in 5 Ross' Gulls (*Rhodostethia rosea*).

In California, Young (1950) found 56 (85%) of 66 gulls to be infected. His series included 25 Western Gulls (*Larus occidentalis*) and 24 California Gulls (*L. californicus*). In eastern North America, helminths of Herring Gulls (*L. argentatus*) were investigated in New Brunswick by Pomeroy and Burt (1964), who reported 6 species from the 17 birds examined. In Newfoundland, Threlfall (1968a) recorded 35 species (11 of trematodes; 10 of cestodes; 11 of nematodes; and 3 of acanthocephalans) from a series of 410 Herring Gulls. Also in Newfoundland, Threlfall (1968b) reported findings in 32 Great Black-backed Gulls (*L. marinus*) and 72 Black-legged Kittiwakes. In Alberta, Hair and Holmes (1970) examined 33 Bonaparte's Gulls, from which they recorded helminths of 7 species. Szidat (1964) compared the helminth faunas of Great Black-backed Gulls and Common Black-headed Gulls (*L. ridibundus*) with those of 2 South American species, the Kelp Gull (*L. dominicanus*) and the Brown-hooded Gull (*L. maculipennis*). Szidat concluded that the latter shared a few species with gulls in the Nearctic, but none with those in the Palaearctic.

Numerous surveys of helminths in gulls have been made in Eurasia. In Great Britain, Pemberton (1963) found 97% of 146 Black-headed Gulls to be infected. Helminths of 27 species were represented (11 of trematodes, 8 of cestodes, 6 of nematodes, and 1 of acanthocephalans). Pemberton also reported findings in smaller series of Herring Gulls and Lesser Black-backed Gulls (*Larus fuscus*). Data concerning the occurrence of helminths in 305 gulls of 3 species (Herring Gull, Great Black-backed Gull, and Lesser Black-backed Gull) in Great Britain were published by Williams and Harris (1965). A series of papers on helminths in Mew Gulls in Norway has been published by Bakke and co-workers (1972).

Farther to the east, in Azerbaidzhan, rates of infection for gulls were reported by Vaidova (1978) as follows: Herring Gull, 90% of 31; Mew Gull, 83% of 18; Great Black-headed Gull (*Larus ichthyaetus*), 100% of 23; Common Black-headed Gull, 85% of 88; and Little Gull (*L. minutus*), 80% of 20. Shigin (1961) reported findings in gulls and terns collected around the Rybinsk Reservoir north of Moscow. Smogorzhevskaia (1976) provided data concerning helminths in 1475 birds of the suborder Lari in the Ukraine. That series included gulls of nine species: Black-legged Kittiwake, Great Black-backed Gull, Lesser Black-backed Gull, Common Black-headed Gull, Mediterranean Black-headed Gull (*L. melanocephalus*), Little Gull, Herring Gull, Mew Gull, and Slender-billed Gull (*L. genei*). Both Vaidova and Smogorzhevskaia provided lists of helminths recorded from each species of bird. Surveys made in the Soviet Far East, most of which have been cited earlier, usually involved relatively small series of gulls in which findings were comparable to those mentioned above.

Both intergeneric and geographic differences appear to exist in rates of infection in terns. Helminths were found in only 8 of 26 Common Terns (*Sterna hirundo*) collected in the north-central states (trematodes in 8; cestodes in 2). Only 4 (12%) of 33 Black Terns (*Chlidonias niger*) in the same region harbored helminths (trematodes in 3; one each with cestodes and nematodes). In Alaska, 17 (30%) of 56 Arctic Terns (*S. paradisaea*) were infected. The rate for trematodes was 14%, and for cestodes, 18%. In Chukotka, Spasskii *et al.* (1963) recorded helminths in only 4 (16%) of 25 Common Terns, and in 13 of 21 Arctic Terns. Spasskii and Sonin (1961) found 9 of 28 Common Terns and 8 of 12 Aleutian Terns (*S. aleutica*) to be infected in Kamchatka.

Higher rates have usually been reported for terns collected farther south in Eurasia. In Azerbaidzhan, Vaidova (1978) examined terns representing 8 species. Sixteen (50%) of 32 Whiskered Terns (*Chlidonias hybridus*) were infected, with helminths of 14 species represented. Of 90 White-winged Terns (*C. leucopterus*), 51 (56%) were infected. Helminths of 13 species were recorded. Thirty-two (43%) of 75 Common Terns had helminths representing 14 species. High rates were reported in birds collected by Leonov (1958) on the Black Sea; 95% of 56 Sandwich Terns (*Sterna sandvicensis*), 90% of 61 Common Terns, 67% of Little Terns (*S. albifrons*), and 95% of 37 Gull-billed Terns (*S. nilotica*). In contrast, Ryzhikov (1951) found helminths in only 10 (23%) of 44 Black Terns examined in southern Gruziia. Of terns collected in the Ukraine by Smogorzhevskaia (1976), 5 species were represented by comparatively large series. For these, helminths identified numbered as follows: 315 Black Terns, 35; 45 Gull-billed Terns, 21; 88 Sandwich Terns, 30; 278 Common Terns, 50; and 66 Little Terns, 23. In western Siberia (Barabinsk Steppe), Bykhovskaia-Pavlovskaia (1953) found 37 (77%) of 48 Black Terns, and 72 (56%) of 129 Common Terns to be infected.

Alcidae. Little information exists concerning helminths in most birds of the family Alcidae. In terms of numbers, the most comprehensive survey was that of Threlfall (1971), who examined 1001 alcids representing 6 species from Newfoundland and other localities in the Atlantic Ocean. Rates of infection were as follows: 49 Razor-billed Auks (*Alca torda*), 25%; 674 Common Murres (*Uria aalge*), 37%; 62 Thick-billed Murres (*U. lomvia*), 60%; 48 Dovekies (*Alle alle*), 12%; and 71 Atlantic Puffins (*Fratercula arctica*), 8%. In Alaska, the 159 alcids examined were of 11 species, for which the overall rate of infection was 28%. The major groups of helminths were represented as follows: trematodes, 3%; cestodes, 19%; nematodes, 8%; and acanthocephalans, 4%. The highest rate was obtained in Thick-billed Murres, with 11 of 21 birds infected. Helminths were present in about 25–30% of Common Murres (12 specimens); Pigeon Guillemot (*Cepphus columba*, 28 specimens); Crested Auklet (*Aethia cristatella*, 21 specimens); Horned Puffin (*F. corniculata*, 20 specimens); and Tufted Puffin (*Lunda cirrhata*, 16 specimens). None of 29 Least Auklets (*Aethia pusilla*) was infected.

In Eurasia, 52 Thick-billed Murres were collected by Markov (1941) on Novaia Zemlia, where very large colonies occur, and there, the rate of infection was 58%, with helminths of only four species represented. In Kamchatka, Spasskii *et al.* (1963) examined a small series of alcids, of which 5 of 12 Thick-billed Murres and 4 of 8 Common Murres harbored helminths. Belogurov *et al.* (1968) examined 56 alcids representing four species, on the shores of the Sea of Okhotsk. The major groups of helminths were represented as follows: trematodes, four species; nematodes, two species; and acanthocephalans, one species. The greatest diversity (five species) was observed in Tufted Puffins. Tsimbaliuk and Belogurov (1964) examined 80 Tufted Puffins and 17 Horned Puffins from islands in the Bering Sea. Thirty (37%) of the former were infected by nematodes representing four species; 5 of the latter were infected by nematodes of two species. In a synopsis of trematodes of birds in the Soviet Union, Bykhovskaia-Pavlovskaia (1962) listed only two species each for Thick-billed Murre, Common Murre, Black Guillemot (*Cepphus grylle*), Spectacled Guillemot (*C. carbo*), and Atlantic Puffin. One species each was listed for Ancient Murrelet (*Synthliboramphus antiquus*) and Tufted Puffin. In all, only eight species of trematodes were recorded.

Columbidae. In general, low rates of infection and low diversity of helminths appear to be characteristic of doves. In Colorado, helminths from 609 Band-tailed Pigeons (*Columba fasciata*) were studied by Olsen and Braun (1980). Of these, 535 were adults, of which only 76 (14%) were infected. Cestodes were found in 52 birds (10%), nematodes in 24 (4.5%). The two groups of helminths were represented by two and four species, respectively. Helminths were not present in young birds. In the north-central states, 2 of the 9 Mourning Doves (*Zenaida macroura*) examined were infected by cestodes.

A higher prevalence of helminths has been recorded in doves collected in southern Eurasia. In Azerbaidzhan, cestodes were the only helminths found by Vaidova (1978) in 10 infected Wood Pigeons (*Columba palumbus*), of which 23 were examined. Vaidova also reported that 59 (46%) of 128 European Turtle-Doves (*Streptopelia turtur*) were infected exclusively with cestodes, with eight species represented. For 129 doves of five species examined by Sultanov (1963) in Uzbekistan, the overall rate was 35%. Only cestodes and nematodes were recorded, with rates of 33 and 2%, respectively. Oshmarin and Demshin (1972) examined 64 doves of three species in the Democratic Republic of Vietnam. Of these, 22 (42%) of 53 Spotted Doves (*S. chinensis*) were infected. Trematodes occurred in 4 (8%), cestodes in 16 (30%), and nematodes in 4. Trematodes recorded from doves of four species in the Soviet Union were listed by Bykhovskaia-Pavlovskaia (1962).

Cuculidae. Again, comparatively little information is available concerning the prevalence of helminths in cuckoos and related birds. In the north-central states, we examined 8 Yellow-billed Cuckoos (*Coccyzus americanus*) and 3 Black-billed Cuckoos (*C. erythropthalmus*). Two of the former harbored cestodes. Data concerning helminths in cuculids in Eurasia have usually been based on small series of birds. Of 15 cuckoos of two species, *Cuculus canorus* and *C. saturatus*, collected in Iakutia by Ryzhikov *et al.* (1974), 1 was infected with cestodes and 4 with trematodes. Twenty specimens of *C. canorus* were examined by Spasskii *et al.* (1958) in the Tuvinsk Autonomous Oblast'. Trematodes were found in a single bird.

Tytonidae. Apparently few records of helminths from owls of this family have been published. In the north-central states, two of six Barn Owls (*Tyto alba*) were found to be infected with nematodes.

Strigidae. Helminths occurring in some of the more common owls are comparatively well known, but these make up only a small proportion of the approximately 128 species in the family Strigidae. An overall rate of 72% was obtained for 116 owls, representing eight species, in the southern series. Major groups of helminths were represented as follows: trematodes, 41%; cestodes, 14%; nematodes, 43%; and acanthocephalans, ~1%. Of 79 Great Horned Owls (*Bubo virginianus*), 57 (72%) harbored helminths. Rates for the major groups were: trematodes, 51%; cestodes, 14%; and nematodes, 36%. Twelve specimens each of Common Screech-Owls (*Otus asio*) and Barred Owls (*Strix varia*) were examined; 8 of each harbored helminths. Cestodes were present in 1 of 5 Burrowing Owls (*Athene cunicularia*).

Owls in the northern series (Alaska) exhibited different rates of infection and differences in the composition of helminth faunas. Helminths were found in only 72 (36%) of 201 birds representing seven species. Trematodes occurred in ~1%; cestodes, in 21%; and nematodes, in 17%. Eleven of 12 Great Horned Owls were infected, with nematodes in all, and with trematodes in 1 as well. The helminths in 39 (32%) of 123 Snowy Owls consisted of only cestodes and nematodes. Rates were 27 and 7%, respectively. Only 3 of 33 Short-eared Owls (*Asio flammeus*) had any helminths. These included cestodes in 1 bird and nematodes in 2. Higher rates were recorded for Northern Hawk-Owls (*Surnia ulula*), of which 15 of 24 were infected. Trematodes occurred in 1, cestodes in 5, and nematodes in 11.

Ryzhikov *et al.* (1974) determined rates of 17 and 32%, respectively, for trematodes and cestodes in 60 owls of seven species collected in Iakutia. Data for other helminths were not included. In Uzbekistan, Sultanov (1963) examined 18 owls of four species and found 11 to be infected by nematodes representing nine species. Five species were recorded from the Little Owl (*Athene noctua*). Helminths of other groups did not occur in this series of owls. Of 26 owls of seven species examined by Spasskii *et al.* (1958) in the Tuvinsk Autonomous Oblast', 17 were infected. Trematodes occurred in 1, cestodes in 7, nematodes in 14, and acanthocephalans in 1.

Caprimulgidae. Little information is available concerning the occurrence of helminths in Caprimulgiformes. Four of seven Common Nighthawks (*Chordeiles minor*) collected in the north-central states were infected. Of these, all had cestodes, and two had trematodes as well. No helminths were found in two Whip-poor-wills (*Caprimulgus vociferus*). Small numbers of birds have been examined at various localities in Eurasia. With respect to trematodes obtained from caprimulgids in the Soviet Union, Bykhovskaia-Pavlovskaia (1962) listed four species from *Caprimulgus indicus* from the Primorsk region, and three species from *C. europaeus*.

Apodidae. Scattered records from small numbers of birds indicate that swifts often harbor helminths. Of 7 Chimney Swifts (*Chaetura pelagica*) collected in Ohio, 3 had cestodes. In the region of Lake Baikal, Ryzhikov and Sudarikov (1951) examined 20 European Swifts (*Apus apus*) and 16 White-throated Needletails (*Hirundapus caudacutus*), all of which harbored helminths. Of the former species, trematodes occurred in 12 and cestodes in 17; of the latter, trematodes were found in 7 and cestodes in 15.

Trochilidae. Almost no information exists concerning helminths in hummingbirds. Cestodes have been described from a few species.

Alcedinidae. As might be predicted on the basis of their diet, kingfishers have distinctive helminth faunas. Ten of 16 Belted Kingfishers (*Ceryle alcyon*) collected in the north-central states were infected. Trematodes occurred in 8 birds, and nematodes in 2. One of 9 specimens collected in Alaska harbored trematodes. In Azerbaidzhan, Vaidova (1978) examined 60 European Kingfishers (*Alcedo atthis*), of which 29 (48%) were infected. Two species of trematodes (one a larval stage) and one species of nematode were reported. Smogorzhevskaia (1976) reported two species of trematodes (one a larval stage) from 15 European Kingfishers examined in the Ukraine, but rates of infection were not given. Only one of 5 European Kingfishers examined in the basin of the Amur River harbored helminths consisting only of trematodes. High rates of infection were recorded in southeast Asia, where kingfishers of two species were investigated in the Democratic Republic of Vietnam by Oshmarin and Demshin (1972). Forty-three (90%) of 48 European Kingfishers were infected, and all of the major groups of helminths were represented: trematodes in 24 (50%), cestodes in 38 (79%), nematodes in 34 (71%), and acanthocephalans in 2 (4%). Helminths found in 7 of 25 White-breasted Kingfishers (*Halcyon smyrnensis*) consisted only of cestodes. All of the latter as well as those in 10 of the European Kingfishers were the second larval stage (sparganum) of a diphyllobothriid.

Meropidae. As indicated by the following examples, rates of infection in bee-eaters appear not to be high. Of 24 specimens of *Merops apiaster* examined by Sultanov (1963) in Uzbekistan, only 2 (8%) harbored helminths. Cestodes were present in 2 birds and nematodes in 1, with each group represented by single species. In Azerbaidzhan, 5 (10%) of 51 bee-eaters examined by Vaidova (1978) were infected. Five species of helminths were recorded.

Coraciidae. The following examples suggest that prevalence and diversity of helminths may be high in rollers. Seventy-one of 100 specimens of European Rollers (*Coracias garrulus*) examined by Vaidova (1978) in Azerbaidzhan were infected, in which helminths of 11 species were recorded: 1 species of trematodes, 3 of cestodes, 6 of nematodes, and 1 of acanthocephalans. Twenty birds were investigated by Sultanov (1963) in Uzbekistan. Eighteen harbored helminths: trematodes in 1, cestodes in 11, and nematodes in 18. In all, 10 species were identified.

Picidae. The prevalence of helminths is often low in woodpeckers, but intergeneric differences may be expected, depending on diets. The southern series of birds included 102 woodpeckers representing eight species, in which the overall rate was 29%, and helminths of only two groups were recorded. All of the 30 infected birds had cestodes, but nematodes were recorded only from Northern Flickers (*Colaptes auratus*). Of 23 birds of this species, 3 harbored nematodes and 14, cestodes. Cestodes were found also in 4 of 43 Downy Woodpeckers (*Picoides pubescens*), in 2 of 12 Hairy Woodpeckers (*P. villosus*), and in 7 of 13 Yellow-bellied Sapsuckers (*Sphyrapicus varius*). The northern series included 22 woodpeckers of four species. Of these, 5 were infected with cestodes, and 1 (a flicker) with nematodes. Comparatively high rates have sometimes been recorded in Eurasia. In the Primorsk region (Soviet Far East), Sadokov (1951) collected 63 woodpeckers representing five species. Sixteen of 29 Gray-headed Woodpeckers (*Picus canus*) were infected; trematodes were present in 3, cestodes in 9, and nematodes in 6. Of 27 White-backed Woodpeckers (*Picoides leucotos*), 25 were stated to have been infected, but an error was made in figures given for the respective groups of helminths. In the Tuvinsk Autonomous Oblast', Spasskii *et al.* (1958) examined 86 birds, of which three species were represented by adequate series. Of 26 Three-toed Woodpeckers (*P. tridactylus*), only 2 were infected (cestodes in 1, nematodes in 1). In contrast, 16 of 26 Great Spotted Woodpeckers (*Picoides major*) were infected: trematodes in 4; and cestodes in 15. Of 28 Lesser Spotted Woodpeckers (*Picoides minor*), 2 harbored termatodes and 2, cestodes. Farther south, in Azerbaidzhan, a higher prevalence of helminths was recorded by Vaidova (1978), who found that 9 of 10 Green Woodpeckers (*Picus viridis*) were infected with cestodes representing two species. Seventy-four (82%) of 90 Middle Spotted Woodpeckers (*Picoides medius*) had helminths consisting of one species of trematodes, two of cestodes, and one of nematodes. One species each of trematodes, cestodes, and nematodes were recorded from 17 of 29 examples of *P. syriacus*.

Tyrannidae. Little is known concerning the prevalence of helminths in birds of this family. The southern series included 56 tyrannids, representing 12 species. The overall rate of infection was very low (18%), with the major groups represented as follows: trematodes in 2 (4%), cestodes in 5 (9%), nematodes in 1 (2%), and acanthocephalans in 1. Of 14 specimens of Eastern Kingbird (*Tyrannus tyrannus*), only 4 were infected (trematodes in 2; cestodes and nematodes in 1 each). In 9 Eastern Phoebes (*Sayornis phoebe*), 1 acanthocephalan only was found. Cestodes occurred in 1 of 9 Eastern Wood Pewees (*Contopus virens*); no helminths were recorded from 7 Western Wood Pewees (*C. sordidulus*). The series included only 12 birds of the genus *Empidonax*, of which 1 harbored cestodes. Of 7 birds of this genus examined in Alaska, 2 were infected with trematodes and 1 with cestodes. MacKenzie *et al.* (1979) compared helminths occurring in the Eastern Kingbird and the Western Kingbird (*T. verticalis*). Rates were 83% for 38 specimens of the former species and 94% for 33 of the latter. From the 2 species combined, 9 species of helminths were identified.

Alaudidae. Of 30 Horned Larks (*Eremophila alpestris*), collected in the north-central states during winter and early spring, only 3 harbored helminths (cestodes). Higher rates have been obtained in other geographic regions. In Spain, Illescas Gomez (1977) found 35% of Skylarks (*Alauda arvensis*) to be infected, as follows: trematodes in 4 (7%), cestodes in 15 (21%), and nematodes in 6 (10%). Of 20 Greater Short-toed Larks (*Calandrella cinerea*), trematodes occurred in 2, cestodes in 3, and nematodes in 9. Of 32 Crested Larks (*Galerida cristata*) examined in Bulgaria by Paspalev and Zheliazkova-Paspaleva (1965), only 2 were infected (1 each with cestodes and nematodes). Spasskii *et al.* (1958) collected 118 birds representing three species in the Tuvinsk Autonomous Oblast'. Thirty-one of 60 Greater Short-toed Larks were infected, with cestodes in 5 (8%), nematodes in 26 (43%), and acanthocephalans in 1. Of 35 Horned Larks, 22 were infected. Cestodes occurred in 4, and nematodes in 20. The highest rate was recorded in Skylarks, in which 21 of 23 were infected. The major groups of helminths were represented as follows: trematodes in 1, cestodes in 6, nematodes in 18, and acanthocephalans in 3. Vaidova (1978) examined 117 Skylarks in Azerbaidzhan. Sixty-one (52%) were infected, but only two species of helminths were present (a trematode and an acanthocephalan).

Hirundinidae. A high prevalence of helminths is usually observed in swallows. Eighty-five birds representing seven species were examined in the north-central states. Of these, 48 (56%) were infected, with trematodes in 5 (6%), cestodes in 45 (53%), and nematodes in 2. The series were too small to discern intergeneric differences in prevalence. In Bulgaria, Paspalev and Zheliazkova-Paspaleva (1965) found helminths in 20 of 47 Barn Swallows

(*Hirundo rustica*); trematodes occurred in 8, cestodes in 17, and nematodes and acanthocephalans in 1 each. Only 9 of 22 House Martins (*Delichon urbica*) had helminths, all of which were cestodes. All of 16 House Martins examined by Illescas Gomez (1977) in Spain harbored helminths. Trematodes occurred in 13, cestodes in 12, and nematodes in 1. All of 6 House Martins examined (R. L. Rausch, unpublished observations) at Lake Shikaribetsu in central Hokkaido had both trematodes and cestodes. In the Tuvinsk Autonomous Oblast', Spasskii et al. (1958) found that 12 of 28 Bank Swallows (*Riparia riparia*) and 11 of 12 House Martins were infected.

Corvidae. Helminths of the more common species of corvids are comparatively well known. A high prevalence of helminths is characteristic of birds of some species, but well-defined intergeneric and geographic differences are evident. Only a few examples from the numerous surveys of corvids are included here.

Morgan and Waller (1941) appear to have been the first to investigate the parasites of the Common Crow (*Corvus brachyrhynchos*). Rates of infection were indicated for each of the 7 species of helminths recorded (1 of trematodes, 2 of cestodes, and 4 of nematodes). In the north-central states, 14 of 16 crows were found to be infected, with helminths of all of the major groups represented. Andrews and Threlfall (1975) examined 99 Common Crows in Newfoundland, of which 94 were infected with helminths representing 12 species: 3 of trematodes, 3 of cestodes, 5 of nematodes, and 1 of acanthocephalans. In Texas, Naderman and Pence (1980) found 65 (96%) of 68 crows to be infected. They identified 8 species of helminths. Cestodes only were present in 4 of 6 Common Ravens (*C. corax*) collected by us in Wyoming. Of 76 ravens examined in Alaska, only 6 (8%) harbored helminths (trematodes in 4; cestodes in 2). We found helminths in 10 of 11 Northwestern Crows (*C. caurinus*) on Kodiak Island. Trematodes were present in 2, cestodes in 8.

Numerous surveys of corvids have been conducted in Eurasia. The helminth fauna of the Rook (*Corvus frugilegus*) has been investigated in detail by Baruš et al. (1972), in Czechoslovakia. They discussed the occurrence of the 21 species of helminths recorded from 327 birds. In Azerbaidzhan, Vaidova (1978) found helminths in 29 (62%) of 47 Rooks; seven species were recorded. For 38 Common Jackdaws (*C. monedula*) examined in Spain by Illescas Gomez (1977), major groups of helminths were represented as follows: trematodes in 9 (24%), cestodes in 10 (26%), nematodes in 12 (31%), and acanthocephalans in 1 (3%). Helminths representing all of the major groups were recorded by Spasskii et al. (1958) from 56 (86%) of 65 Jackdaws in the Tuvinsk Autonomous Oblast'. In Uzbekistan, Sultanov (1963) found that 32 (80%) of 40 Carrion Crows (*C. corone corone*) were infected. Helminths of all groups but acanthocephalans were represented. Helminths were present in 12 of 13 Hooded Crows (*C. corone cornix*) collected on the Pechora River by Pod'iapol'skaia *et al.* (1951). In contrast, only 13 (17%) of 76 specimens examined by Vaidova (1978), in Azerbaidzhan, were infected. Vaidova reported eight species of helminths from this series.

Three Black-billed Magpies (*Pica pica*) collected by us in Wyoming were infected (trematodes in 1; cestodes in 1), but cestodes alone were found in 2 of 8 birds collected in Alaska. Helminths of all groups but acanthocephalans were represented in 8 infected magpies of 13 examined by Sultanov (1963) in Uzbekistan. Thirty-three (82%) of 40 of these birds examined in Bulgaria by Paspalev and Zheliazkova-Paspaleva (1965) were infected. In Azerbaidzhan, Vaidova (1978) recorded helminths from 106 (66%) of 161 birds; 15 species were represented. Twenty-three (77%) of 30 Azure-winged Magpies (*Cyanopicus cyanus*) collected by Sadokov (1951) in the Primorsk region of the Soviet Far East were infected. Trematodes occurred in 5, cestodes in 20, and nematodes in 6.

The first survey of helminths in the Blue Jay (*Cyanocitta cristata*) was made by Boyd *et al.* (1956), who found 80 of 100 birds collected in the northeastern states to be infected. Helminths of 16 species were recorded (4 of trematodes, 2 of cestodes, and 10 of nematodes). All but 1 of 50 Blue Jays examined by Cooper and Crites (1974a) on one of the islands in Lake Erie were infected. Helminths of 25 species were identified from this series of birds. Helminths were present in 16 (35%) of 46 Eurasian Jays (*Garrulus glandarius*) collected by Vaidova (1978) in Azerbaidzhan. Five species of helminths were reported. Spasskii *et al.* (1958) obtained helminths from 20 of 23 jays examined in the Tuvinsk Autonomous Oblast'. Twenty-three of 30 birds collected by Sadokov (1951) in the Primorsk region of eastern Siberia were infected. Sixteen each had trematodes and nematodes; 1 had cestodes.

The prevalence of helminths appears to be low in jays of the genus *Perisoreus*. In Alaska, only 12 (14%) of 85 Gray Jays (*P. canadensis*) were found to be infected. Trematodes occurred in 3 birds, cestodes in 3, and nematodes in 7. In Iakutia, helminths (cestodes) were present in only 1 of 9 specimens of *P. infaustus* examined by Mozgovoi *et al.* (1956). None was found in 7 specimens collected by Ryzhikov and Sudarikov (1951) at Lake Baikal, nor in 2 examined by Ryzhikov *et al.* (1961b) in the basin of the Amur River. In the Tuvinsk Autonomous Oblast', Spasskii *et al.* (1958) recorded helminths from 2 of 12 birds (1 each with cestodes and nematodes).

Relatively little information is available concerning helminths in nutcrackers. Of 10 specimens of Clark's Nutcracker (*Nucifraga columbiana*) examined in Wyoming, only 1 was infected (with a nematode). Twenty of 29 Eurasian Nutcrackers (*N. caryocatactes*) examined near Lake Baikal by Ryzhikov and Sudarikov (1951) were infected. The helminth fauna consisted of trematodes (in 2), cestodes (in 12), nematodes (in 7), and acanthocephalans (in 1). Pod'iapol'skaia et al. (1951) found helminths in 11 of 18 birds collected on the Pechora River (trematodes in 4, cestodes in 1, nematodes in 8). Nothing was found in 3 nutcrackers examined (R. L. Rausch, unpublished observations) on the upper Kolyma River, in the Soviet Far East.

Aegithalidae. Little is known of the helminths occurring in Long-tailed Tits. Only 1 (with cestodes) of 25 specimens of *Aegithalos caudatus* collected by Ryzhikov et al. (1961b) in the basin of the Amur River was infected. None of six specimens examined by Spasskii et al. (1958) in the Tuvinsk Autonomous Oblast' was infected.

Paridae. A low prevalence of helminths appears to be characteristic of the various species of tits. Sixty-eight birds representing five species of the genus *Parus* were examined in the north-central states. Helminths occurred in 11 (16%), as follows: cestodes in 9 (13%) and nematodes in 2 (3%). Of a series of 37 Black-capped Chickadees (*P. atricapillus*), 6 (16%) were infected (cestodes in 4; nematodes in 2). Two of 19 Tufted Titmice (*P. bicolor*) harbored helminths (cestodes). In Alaska, cestodes were found in 3 of 16 Black-capped Chickadees. In the Tuvinsk Autonomous Oblast', an overall rate of 13% was obtained for 121 birds collected by Spasskii et al. (1958). For 77 Willow Tits (*P. montanus*) the rate was 13%; only cestodes were present in the 10 infected birds. For 19 Great Tits (*P. major*) and 24 Azure Tits (*P. cyanus*), rates of 21 and 8%, respectively, were obtained. For 79 birds of five species collected in the Amur River basin by Ryzhikov et al. (1961b), the overall rate was 20%. This series included 35 Willow Tits, of which 8 (23%) were infected. Forty-six birds of the same species were examined in Kamchatka by Spasskii and Sonin (1961), who found 7 (15%) to harbor helminths. Trematodes were present in 5 (11%) and cestodes in 3 (6%). Published data concerning findings in other species of *Parus* are comparable.

Sittidae. Helminths (cestodes) were found in 1 of 22 White-breasted Nuthatches (*Sitta carolinensis*) and in 2 of 10 Red-breasted Nuthatches (*S. canadensis*) in the north-central states. The overall rate was 9%. Of a series of 43 Eurasian Nuthatches (*S. europaea*) examined by Spasskii et al. (1958) in the Tuvinsk Autonomous Oblast', 7 (16%) were infected (trematodes in 1; cestodes in 7). Nineteen birds of the same species were collected on the Amur River by Ryzhikov et al. (1961b), who found 8 to be infected. Trematodes occurred in 6 and cestodes and nematodes in 1 each.

Certhiidae. Apparently nothing is known concerning the prevalence of helminths in creepers. None was present in nine specimens of Brown Creeper (*Certhia americana*) collected in the north-central states. Trematodes were found in one of three birds of this species examined by Mozgovoi and Popova (1951) in the Belovezha Forest.

5. BIOLOGY OF AVIAN PARASITES   399

Cinclidae. Helminths of several species have been described from Dippers, but data concerning their prevalence are not extensive. The occurrence of trematodes of a single species in *Cinclus mexicanus* in the Pacific Northwest was reported by Macy and Strong (1967), who evidently examined at least 20 birds. No helminths were found in 7 Dippers examined in arctic Alaska. Ten of 15 Eurasian Dippers (*C. cinclus*) were examined by Vaidova (1978) in Azerbaidzhan. Helminths of two species were found in the 10 infected birds. Seventeen Brown Dippers (*C. pallasii*) were examined by Ryzhikov *et al.* (1962) in the basin of the Amur River. Helminths occurred in 12 (cestodes in 7; nematodes in 9).

Troglodytidae. The literature did not provide extensive data concerning the prevalence of helminths in wrens. The southern series (i.e., north-central states) included only 12 birds, representing four species, of which 3 were infected with cestodes [1 House Wren (*Troglodytes aedon*) collected in Ohio and 2 Rock Wrens (*Salpinctes obsoletus*) collected in Wyoming]. Cestodes were also found in 1 of 3 Winter Wrens (*T. troglodytes*) collected in Alaska.

Mimidae. Little information is available concerning helminths in even the common mimids. Peet and Ulmer (1970) discussed the prevalence of trematodes obtained from 19 specimens of Brown Thrasher (*Toxostoma rufum*) collected in Iowa, but helminths of other groups were not considered. Two Brown Thrashers of 7 collected in the north-central states were infected (trematodes in 1, nematodes in 1, and acanthocephalans in 2). Of 6 Gray Catbirds (*Dumetella carolinensis*) from the same region, 2 harbored helminths (cestodes in 1; acanthocephalans in 2).

Muscicapidae. Birds of a few commonly occurring species in this family have been rather intensively studied by helminthologists, whose findings indicate that the prevalence of helminths is often high, and that considerable diversity exists in the composition of helminth faunas. Only a few examples are given here.

In North America, we examined 257 birds representing 11 species of the subfamily Turdinae, for which the overall rate of infection by helminths was 51%. The highest rate was recorded in Varied Thrushes (*Zoothera naevia*) in Alaska, in which 49 (96%) of 51 birds were infected. Major groups were represented as follows: trematodes in 65%, cestodes in 71%, nematodes in 16%, and acanthocephalans in 29%. Of 102 American Robins (*Turdus migratorius*; specimens from both series combined), 85 were infected (trematodes in 29, cestodes in 69, nematodes in 22, and acanthocephalans in 7). In Colorado, Slater (1967) found 58 (94%) of 62 robins to be infected, with 11 species of helminths recorded. Thrushes examined by Vaidova (1978) in

Azerbaidzhan included 102 European Blackbirds (*T. merula*), of which 78 were infected. Helminths of 26 species were reported from these: 9 of trematodes, 8 of cestodes, 3 of nematodes, and 6 of acanthocephalans. Thrushes collected by Spasskii *et al.* (1958) in the Tuvinsk Autonomous Oblast' included 52 Fieldfares (*T. pilaris*) and 49 Black-throated Thrushes (*T. ruficollis atrogularis*), for which rates of 84 and 86%, respectively, were recorded. Data for these and other species examined in other regions are comparable. The occurrence of trematodes in thrushes representing five species of *Turdus* in Poland has been discussed by Machalska (1980).

Little information is available for some of the other common thrushes in North America. My series included only small numbers of birds of the genera *Catharus*, *Hylocichla*, and *Sialia* (seven species in all). Of a total of 55 specimens, 29 (53%) harbored helminths.

No helminths were found in 7 Bluethroats (*Erithacus* [*Luscinia*] *svecicus*) collected in Alaska near the eastern limits of the range of this species. Five of 13 birds examined near Chany Lake were infected (Mozgovoi *et al.*, 1951), as were 10 of 14 in the Tuvinsk Autonomous Oblast' (Spasskii *et al.*, 1958). In Kamchatka, helminths occurred in 17 of 32 Rubythroats (*E.* [*L.*] *calliope*) collected by Spasskii *et al.* (1958), but only 1 of 11 was found to be infected by Spasskii and Sonin (1961).

Sixteen (38%) of 42 Northern Wheatears (*Oenanthe oenanthe*) collected in arctic Alaska were infected. Five birds had cestodes; 1 each had trematodes, nematodes, and acanthocephalans. On the Pechora River, 5 of 16 birds were infected (trematodes in 4; cestodes in 1) (Pod'iapol'skaia *et al.*, 1951).

In the subfamily Sylviinae, helminths (cestodes) were found in 2 of 16 Arctic Warblers (*Phylloscopus borealis*) in northern Alaska. Trematodes were the only helminths present in 5 of 12 of these birds examined by Pod'iapol'skaia *et al.* (1951) on the Pechora River. There also, they found 13 of 37 Willow Warblers (*P. trochilus*) to be infected (trematodes in 11; cestodes and nematodes in 1 each). Findings in 51 birds representing six species of *Phylloscopus* along the Amur River have been reported by Ryzhikov *et al.* (1961b).

Little information exists concerning the prevalence of helminths in kinglets (*Regulus* spp.). Nothing was found in 15 Golden-crowned Kinglets (*R. satrapa*) and 8 Ruby-crowned Kinglets (*R. calendula*) collected in the north-central states.

Motacillidae. The helminths occurring in wagtails and pipits are comparatively well known. In Alaska, 19 (32%) of 59 Yellow Wagtails (*Motacilla flava*) were infected, as follows: trematodes in 6 (10%), cestodes in 14 (24%), and acanthocephalans in 1 (~2%). Two White Wagtails (*M. alba*) were free of helminths. On the Amur River, Ryzhikov *et al.* (1962) found 9 of 21 Yellow

Wagtails and 12 of 19 White Wagtails to be infected. For the same two species in Kamchatka, rates were 15 of 24 and 7 of 13, respectively (Spasskii et al., 1962). Wagtails of three species, one represented by few specimens, were examined by Pod'iapol'skaia et al. (1951) on the Pechora River. Fifty-three (64%) of 83 White Wagtails harbored helminths: trematodes in 34 (41%), cestodes in 35 (42%), and nematodes in 3 (4%). Of 31 Citrine Wagtails (M. citreola), 20 were infected (trematodes in 16; cestodes in 7; nematodes in 4). Spasskii et al. (1958) found helminths in 22 (31%) of 71 White Wagtails examined in the Tuvinsk Autonomous Oblast'. Trematodes occurred in 5 (7%), cestodes in 5, nematodes in 15 (21%), and acanthocephalans in 3 (4%). Ryzhikov et al. (1961b) reported findings in 96 wagtails, representing four species, from the region of the Amur River. Helminths occurred in 8 of 21 White Wagtails and in 22 (31%) of 72 Gray Wagtails (M. cinerea). Other published data concerning rates in these and other species of Motacilla are similar.

In Alaska 5 of 15 Water Pipits (Anthus spinoletta) harbored helminths (cestodes only). Cestodes and trematodes were present in 1 of 2 Red-throated Pipits (A. cervinus) examined. In the Tuvinsk Autonomous Oblast', 40 of 82 Brown Tree-Pipits (A. trivialis) collected by Spasskii et al. (1958) were infected. Trematodes occurred in 11 (13%), cestodes in 19 (23%), and nematodes in 25 (30%). Sadokov (1951) found 8 of 13 Richard's Pipits (A. novaeseelandiae) to be infected. In Kamchatka, helminths were found in 53 of 57 Olive Tree-Pipits (A. hodgsoni) collected by Spasskii et al. (1963). Trematodes were recorded in 48 (84%), cestodes in 22 (39%), and nematodes and acanthocephalans in 4 each (7%). Seventeen (31%) of 55 Olive Tree-Pipits examined on the Amur River by Ryzhikov et al. (1961b) were infected.

Bombycillidae. The prevalence of helminths in waxwings appears to be rather low. None was found in 8 Cedar Waxwings (*Bombycilla cedrorum*) collected in the north-central states. Of 4 Bohemian Waxwings (*B. garrulus*) in Alaska, 2 were infected (trematodes and cestodes in 1 each). On the Pechora River, Pod'iapol'skaia et al. (1951) found helminths in 7 of 11 Bohemian Waxwings (trematodes in 7; cestodes in 1). Of 3 Japanese Waxwings (*B. japonica*) collected by Ryzhikov et al. (1961a) on the Amur River, 1 harbored nematodes. There appear to be no data for large series of waxwings.

Laniidae. The prevalence of helminths is comparatively high in shrikes of some species. Only 1 of 15 Northern Shrikes (*Lanius excubitor*) in Alaska was infected (cestodes). Nematodes were found in 1 of 2 Northern Shrikes examined by Mozgovoi et al. (1951). No helminths were found in 6 Loggerhead Shrikes (*L. ludovicianus*) examined in Ohio. Brown Shrikes (*L. cristatus*) have been investigated at various localities in Eurasia. Ryzhikov et al. (1961b) recorded helminths from 44 of 79 birds collected along the Amur

River, as follows: trematodes in 19 (24%), cestodes in 23 (29%), and nematodes in 28 (35%). Other examples of rates in Brown Shrikes are 23% (31 birds) in southern Gruziia (Ryzhikov, 1951) and 67% (42 birds) in the Tuvinsk Autonomous Oblast' (Spasskii et al., 1958). Helminths were present in all of 40 Long-tailed Shrikes (*L. schach*) collected by Oshmarin and Demshin (1972) in the Democratic Republic of Vietnam. Trematodes occurred in 4 (10%), cestodes in 29 (73%), nematodes in 10 (25%), and acanthocephalans in 1 (3%).

Sturnidae. The European Starling (*Sturnus vulgaris*) is helminthologically one of the best known of the passerine birds. According to the compilation of records by Hair and Forrester (1970), the helminth fauna of the starling includes 81 species representing 51 genera. The first survey of parasites of this bird in North America was understaken by Boyd (1951), approximately 60 years after its introduction; 90% of 300 birds harbored helminths, with 15 species represented. In England, Owen and Pemberton (1962) examined 358 starlings. They also recorded helminths of 15 species: 4 trematodes, 5 cestodes, 5 nematodes, and 1 acanthocephalan. Vaidova (1978) in Azerbaidzhan found 105 (56%) of 186 starlings to be infected by helminths representing 10 species (1 of trematodes, 4 of cestodes, 2 of nematodes, and 3 of acanthocephalans). Farther east, in the Tuvinsk Autonomous Oblast', Spasskii et al. (1958) found 17 of 19 starlings to be infected; at Chany Lake, Mozgovoi et al. (1951) recorded helminths from 15 of 31 birds. Little information is available concerning sturnids of other species. On the Amur River, Ryzhikov et al. (1961a) found helminths in 4 of 8 Daurian Starlings (*S. sturninus*) and in 8 of 9 White-cheeked Starlings (*S. cineraceus*). The few records suggest that rates in the Rose-colored Starling (*S. roseus*) are low.

Vireonidae. Helminths are uncommon in Vireos. None were recorded from 27 birds, representing five species of the genus *Vireo* in the north-central states.

Parulidae. A low prevalence of helminths usually appears to be characteristic of Wood Warblers. In the north-central states, 103 birds were examined, representing 20 species in seven genera (*Mniotilta; Vermivora*, 2 spp.; *Dendroica*, 10 spp.; *Seiurus*, 2 spp.; *Geothlypis*, 2 spp.; *Wilsonia*, 2 spp.; and *Setophaga*). Only 19 (18%) of these birds harbored helminths (cestodes in 15; nematodes in four). In Alaska, 35 warblers representing 6 species in four genera (*Vermivora; Dendroica*, 3 spp.; *Seiurus; Wilsonia*) were examined. Eleven birds (31%) were infected with helminths: trematodes in 5 (14%), cestodes in 4 (11%), and nematodes in 3 (9%). Nineteen of 24 Northern Waterthrushes (*Seiurus noveboracensis*) examined by Jewer and Threlfall (1978) in Newfoundland were infected. Helminths of 5 species were recorded.

Ploceidae. Information concerning prevalence of helminths is available for only a few species of Weaver Finches. Helminths of the House Sparrow (*Passer domesticus*) have been investigated at various localities; two examples of findings in this synanthropic bird are included here. In Spain, Illescas Gomez (1977) recorded a rate of 27% for 226 House Sparrows examined. Trematodes occurred in 11 (5%), cestodes in 54 (24%), and nematodes in 2 (~1%). Of 149 birds collected by Vaidova (1978) in Azerbaidzhan, 96 (64%) harbored helminths, of which only three species were recorded (one of trematodes; two of cestodes).

Two species of Weaver Finches in addition to House Sparrows were investigated by Paspalev and Zheliazkova-Paspaleva (1965) in Bulgaria. Helminths occurred in 5 (6%) of 85 European Tree Sparrows (*Passer montanus*): trematodes in 4 birds; cestodes in 2. Fifteen (17%) of 87 Spanish Sparrows (*Passer hispaniolensis*) were infected: trematodes in 12 (14%), cestodes in 3 (~3%), and nematodes in 1 (~1%). Vaidova (1978) reported findings in Weaver Finches of two species in Azerbaidzhan. Fifty-four of 109 European Tree Sparrows were infected with helminths representing four species (two each of trematodes and cestodes). Of 27 Rock Sparrows (*Petronia petronia*), 6 (22%) harbored helminths, of which two species were recorded (one each of cestodes and acanthocephalans). To the east, in the Tuvinsk Autonomous Oblast', Spasskii *et al.* (1958) found helminths in 35 of 96 European Tree Sparrows (cestodes in 34; nematodes in 5).

Icteridae. Numerous species of helminths have been reported from icterids, but data concerning their prevalence are not extensive. The southern series included 164 icterids representing 10 species in eight genera (*Euphagus*, 2 spp.; *Icterus; Dolichonyx; Sturnella*, 2 spp.; *Agelaius; Xanthocephalus; Quiscalus;* and *Molothrus*). Helminths were recorded from 102 (62%) of these birds. So far as could be determined from the small series of the respective species, rates of infection were highest in the Red-winged Blackbird (*A. phoeniceus;* 73% infected of 64) and in the Eastern Meadowlark (*S. magna;* 10 infected of 14). Only 1 of 9 Rusty Blackbirds (*E. carolinus*) harbored helminths (nematodes); 5 of 19 Brewer's Blackbirds (*E. cyanocephalus*) were infected (cestodes in 5; acanthocephalans in 2). Trematodes alone were present in 2 of 9 Northern Orioles (*I. galbula*). The Yellow-headed Blackbird (*X. xanthocephalus*) appeared to have the most diversified helminth fauna; 4 of 9 birds were infected, with trematodes in 3, cestodes in 1, nematodes in 2, and acanthocephalans in 2. Rates obtained for other species of icterids in the southern series of birds were moderately low. Three of 9 Rusty Blackbirds examined in Alaska were infected (trematodes in 3; nematodes in 1).

Several investigators have reported information concerning helminths in the Red-winged Blackbird. Helminths of 3 species (1 each of trematodes, cestodes, and acanthocephalans) were recorded from 61 birds examined by Spory (1965) in Ohio. Trematodes occurred in 27 birds (44%), cestodes in 33 (54%), and acanthocephalans in 2 (3%). Cooper and Crites (1974b) found only 9 of 40 birds in Maryland to be infected. Helminths of 5 species were recorded. Findings in both Red-winged Blackbirds and Common Grackles (*Quiscalus quiscula*) were reported by Stanley and Rabalais (1971). Forty-eight of 50 Common Grackles collected by Cooper and Crites (1974c) on one of the islands in Lake Erie were infected; from this series of birds, helminths representing 15 species were recorded.

Emberizidae. The prevalence of helminths was relatively low in the 403 birds of this family included in my southern series. Twenty-seven species were represented, allocated among 14 genera (*Calcarius; Plectrophenax; Zonotrichia*, 4 spp.; *Melospiza*, 2 spp.; *Junco*, 2 spp.; *Ammodramus*, 3 spp.; *Spizella*, 5 spp.; *Pooecetes; Pipilo; Spiza; Pheucticus*, 2 spp.; *Cardinalis; Passerina;* and *Piranga*, 2 spp.). For these emberizids the overall rate was only 23%, with major groups of helminths represented as follows: trematodes in 6 (~1%), cestodes in 67 (17%), nematodes in 17 (4%), and acanthocephalans in 9 (2%). Prevalence in birds of species represented by series of 20 or more individuals are examplified by the following: Vesper Sparrow (*Pooecetes gramineus*), 6 (26%) of 23; Tree Sparrow (*Spizella arborea*), 6 (19%) of 32; White-throated Sparrow (*Z. albicollis*), 21 (57%) of 37; Song Sparrow (*M. melodia*), 5 (11%) of 46; Swamp Sparrow (*M. georgiana*), 4 of 20.

Generally higher rates of infection were recorded for emberizids collected in Alaska, where 189 specimens represented 11 species in eight genera (*Calcarius*, 2 spp.; *Plectrophenax; Zonotrichia*, 3 spp.; *Melospiza; Passerella; Junco; Ammodramus;* and *Spizella*). The overall rate for the 189 birds was 51% (97 infected), with the major groups represented as follows: trematodes in 29 (15%), cestodes in 94 (50%), nematodes in 6 (3%), and acanthocephalans in 1 (<1%). Prevalences recorded in birds of representative species were: Lapland Longspur (*C. lapponicus*) 20 (48%) of 42; Snow Bunting (*Plectrophenax nivalis*), 12 (26%) of 46 [helminths (cestodes) were found in all of 5 Snow Buntings of the endemic race (*P. nivalis hyperboreus*) on St. Matthew Island, in the Bering Sea]; Fox Sparrow (*Passerella iliaca*), 42 (86%) of 49. Rates of helminths in the Fox Sparrows were: trematodes in 25 (51%), cestodes in 41 (84%), and nematodes in 6 (12%).

In North America, few reports have concerned the prevalence of helminths in emberizids. Hunter and Quay (1953) examined 100 Seaside Sparrows (*Ammodramus maritimus*) in North Carolina, where high rates of infection were recorded: trematodes in 43%, cestodes in 52%, nematodes in 62%, and acanthocephalans in 87%. The helminths fauna consisted of 21 species. Jewer and Threlfall (1978) reported findings in 20 Fox Sparrows collected in

Newfoundland. Eighteen birds harbored helminths, of which 11 species were represented. Large series of emberizids have been examined for helminths in Eurasia, and some examples of findings are included here. On the Pechora River, Pod'iapol'skaia et al. (1951) collected birds of several species, of which 4 were represented by adequate series. Rates were 19 (27%) of 70 Little Buntings (*Emberiza pusilla*), 13 (38%) of 34 Yellow-breasted Buntings (*E. aureola*), 10 of 19 Yellowhammers (*E. citrinella*), and 19 (25%) of 76 Reed Buntings (*E. schoeniclus*). Comparatively low rates were recorded by Ryzhikov et al. (1961b), who examined 248 birds representing 9 species of *Emberiza* on the Amur River. Fifty-eight birds (23%) harbored helminths (trematodes in 7%, cestodes in 15%, nematodes in 4%, and acanthocephalans in 4%). Spasskii et al. (1963) collected emberizids of several species on Kamchatka. Of 64 Yellow-breasted Buntings, 39 (61%) were infected.

Little is known about helminths in tanagers (subfamily Thraupinae). Of 10 Scarlet Tanagers (*Piranga olivacea*) collected in the north-central states, only 3 were infected (trematodes and cestodes in 1 each; nematodes in 2). A single acanthocephalan was the only helminth recorded from a series of 9 Western Tanagers (*P. ludoviciana*) in Wyoming.

Fringillidae. The prevalence of helminths was very low in fringillids examined in North America; the southern series included 149 birds representing 11 species in seven genera (*Carduelis*, 3 spp.; *Leucosticte*; *Carpodacus*, 3 spp.; *Pinicola*; *Loxia*, 2 spp.; and *Coccothraustes*). Only 4 birds (3%) harbored helminths (cestodes in 3; acanthocephalans in 1). Findings were similar in Alaska, where only 43 fringillids were examined. These represented 5 species in four genera (*Carduelis*, 2 spp.; *Leucosticte*; *Pinicola*; and *Loxia*). Three birds harbored helminths (trematodes in 1; cestodes in 2). Prevalences usually have been low also in fringillids investigated in Eurasia. In Spain, Illescas Gomez (1977) found helminths in only 4 of 100 Serins (*Serinus serinus*; cestodes in 1; nematodes in 3). However, 19% of 35 Chaffinches (*Fringilla coelebs*) were infected (cestodes and acanthocephalans in 2 each; nematodes in 3). Fifty-seven fringillids representing 9 species were examined by Ryzhikov et al. (1961b) on the Amur River, of which 12 (21%) were infected. For 87 birds of 19 species collected in the Tuvinsk Autonomous Oblast' by Spasskii et al. (1958), the overall rate was 8%. The highest rate as well as the greatest diversity of helminths were recorded in Bramblings (*F. montifringilla*). Four of 17 birds were infected, with trematodes in 2, and cestodes, nematodes, and acanthocephalans in 1 each.

The foregoing data indicate that some proportion of any avian population can be expected to harbor helminths. Any apparent exception would probably be eliminated by adequate sampling, because helminths in birds characteristically exhibit specificity only at the level of family or order, if at all. Many helminths are holarctic in distribution, occurring in the same or closely related species of avian hosts in North America and Eurasia. Within the poorly defined limits imposed by host specificity, the composition of helminth faunas and the prevalence of helminths of the respective species are influenced by age of the host, season, latitudinal zone, and other factors, in addition to fundamental differences in diet. Some of these factors are briefly discussed.

IV. Specificity of Helminths

That animals of various species have characteristic helminths was apparently first perceived by Pallas and Bremser nearly two centuries ago. This concept was concisely expressed by Bremser (1819), "*Manche Thiere haben, ihnen ganz eigene, Eingeweidewürmer, die in anderen Thieren nicht gefunden werden* [p. 10]." Krabbe (1869) recognized that different groups of birds have characteristic species of cestodes. The concept was refined by Fuhrmann (1908), who stated, "*Betrachtet man die Verteilung der zahlreichen Taenien-Arten in den verschiedenen Vogelgruppen, so beobachtet man die sehr charakteristische Erscheinung, dass eine bestimmte Art immer nur in einer bestimmten Vogelgruppe vorkommt und so für dieselbe typisch ist* [p. 3]." That conclusion forms the basis for Fuhrmann's Rule, which, having been extended by K. I. Skriabin to include helminths of the other groups, is often designated the Fuhrmann–Skriabin Rule. Fuhrmann's Rule has been discussed by Bykhovsky (Bykhovskii) (1961), Mačko (1968), and others.

As applied here, the term *host specificity* refers to the phylogenetically based limitation of the range of hosts of a given species of helminth. The etymology of the term requires that helminths be designated specific or nonspecific (Mačko, 1968). Nonetheless, lack of specificity in a helminth of a given species does not imply that its range of potential hosts is unrestricted. As currently used, terms such as "ecologic specificity" and "ethologic specificity" are often confusing, and it seems illogical to use them to denote the range of hosts of a nonspecific helminth as determined by shared opportunity for exposure to the infective stage (i.e., animals that have similar diets). For example, the trematode *Maritrema afanassjewi* Belopol'skaia, 1952, which has been recorded from birds and mammals of several species, is apparently capable of development to sexual maturity in most homeothermic animals that feed on its intermediate host, the marine amphipod *Orchestia ochotensis* Brandt (see Rausch *et al.*, 1979). Often, host specificity has been considered only from the helminthological aspect, without regard for the equally important involvement of the hosts in any given assemblage in which specificity is perceived [see review by Dogiel (Dogel') (1963, p.

434)]. The importance of the host with respect to specificity was appropriately emphasized by Ernst Mayr, who, in reference to coevolution of parasites and their hosts, remarked (1957) that "the association between parasite and host goes far back into geologic history, the evolution of the parasite being as closely correlated with that of the host as if it were an organ of the host [p. 7]." Host specificity is readily demonstrable among helminths, but its fundamental nature is not understood.

One concept that seeks to explain these associations is that of a *continuum*, formed with respect to compatibility with the helminth by the potential hosts of a given species of helminth in the sexually reproductive stage. That is, hosts range from those required to perpetuate the helminth, through those in which the helminth becomes established but may not mature, to those in which the helminth never becomes established (or, in other terms: required hosts; suitable hosts, but inadequate to perpetuate the helminth; and unsuitable hosts) (Holmes, 1979). The examples given by Holmes and earlier by Kennedy (1975) to illustrate this pattern of helminth–host relationships involved acanthocephalans of two species, *Metechinorhynchus salmonis* (Müller, 1780) and *Pomphorhynchus laevis* (Müller, 1776), respectively, in populations of fishes. The concept has been demonstrated to have validity in application to host relationships of acanthocephalans in fishes, but it is not wholly applicable to acanthocephalans, which simply are not usually host specific, occurring in birds and mammals. The acanthocephalan *Corynosoma strumosum* (Rudolphi, 1802), which is found commonly in pinnipeds, provides an example of the lack of specificity in such organisms. In Alaska, *C. strumosum* was recorded not only from seals and sea lions, but also from mammals of phylogenetically distinct groups, and from birds. These included Arctic Foxes (*Alopex lagopus*; often infected); Red Foxes (*Vulpes vulpes*; several records); domestic dog (several records); Sea Otter (*Enhydra lutris*); man; Bald Eagle (*Haliaeetus leucocephalus;* 2 records); and Glaucous Gull (*Larus hyperboreus*). It appears simply that *C. strumosum* becomes successfully established in piscivorous or carrion-eating birds or mammals that ingest the infective larvae in fishes of certain species. The range of hosts of species of acanthocephalans occurring in birds and mammals is evident from the list given by Petrochenko (1956, p. 375).

That host specificity is not characteristic of acanthocephalans occurring in birds and mammals has long been recognized. On this basis, Fuhrmann (1908) criticized the hypothesis of von Ihering (1902) that acanthocephalans are useful as zoogeographic indicators in animals of these groups. With reference to von Ihering's conclusions, Fuhrmann remarked (1908), "*v. Ihering stützt sich zur Aufstellung dieser Sätze auf die Acanthocephalen der Säugetiere und Vögel. Ein genaues Studium dieser Parasiten bei Vögeln hat aber gezeigt, dass unter allen Helminthen keine Gruppe so wenig geeignet ist wie diese, obige Sätze zu stützen. Da dieselben also auf die Acanthocephalen basiert, stehen sie auf sehr schwachen Füssen* [p. 21]." Fuhrmann's comments are relevant also to the use of acanthocephalans as examples in formulating concepts concerning host specificity in helminths of other groups. Holmes's (1979) assertion to the contrary, it appears that a *positive–negative dichotomy* comes closer to an adequate representation of host-specificity than does the concept of a continuum.

If host-specific helminths (cestodes and nematodes in homeothermic vertebrates) commonly became established in other than natural hosts, or those in which they become established with or without the attainment of sexual maturity, one would expect to record them frequently in such animals in the course of faunistic surveys. But such occurrences are so rare as to be noteworthy events. My findings in some thousands of wild birds and mammals (the majority consisting of mammals) examined mainly in North America and Eurasia suggest that cestodes and nematodes are almost without variance restricted to their natural hosts, which usually represent a phylogenetically determined range of taxa. In some cases involving mammals of monotypic genera, a species of cestode may be limited to hosts of a single species. More frequently, helminths of a single species or a single genus are restricted in occurrence to mammals of a single genus. Concerning cestodes in mammals, it appears that a species is nonspecific if its host range extends beyond a group of closely related genera. In birds, species of helminths are rarely limited to congeneric hosts, but more commonly occur in members of one or more families or of an order.

Nonetheless, ostensibly host-specific helminths of homeothermic vertebrates are sometimes found in animals of species that are phylogenetically disparate with respect to the natural host(s). The few examples given here include the finding of a cestode later described as *Schistotaenia tenuicirrus* (Chandler, 1948) in a Common Crow (*Corvus brachyrhynchos*) collected in October 1944 in Ohio, at a distance of several kilometers from habitat suitable for grebes, for which cestodes of the genus *Schistotaenia* are specific (Rausch, 1970). The specimen, associated with other cestodes [*Variolepis variabilis* (Mayhew, 1925)] was fully developed and was found with its rostellum deeply embedded in the wall of the crow's intestine. In Alaska, where Hoary Marmots (*Marmota caligata*) commonly harbor cestodes of a single species, two poorly developed cestodes of another species were found in the intestine of an animal collected in June 1963. These were determined to be *Monoecocestus americanus* (Stiles, 1895), a common, host-specific cestode of the Porcupine (*Erethizon dorsatum*). Most of the Hoary Marmots examined were collected in areas where Porcupines were also numerous, but *M. americanus* occurred in only 1 of 75 animals examined. Also in Alaska, the nematode *Soboliphyme baturini* (Petrov, 1930), a nematode of mustelids, particularly of the genus *Martes*, was recorded in a lynx (*Felis lynx*). In the north-central states, *Dictyocaulus viviparous* (Bloch, 1782), a nematode of ruminants, was reported from 1 of 648 voles of the genus *Microtus* (Rausch and Tiner, 1949). Similarly, another nematode of ruminants, *Trichostrongylus axei* (Cobbold, 1879), was recorded by McGee (1980) from 2 of 209 Richardson's Ground Squirrels (*Citellus richardsoni*) collected in Saskatchewan. McGee also cited an earlier record of this nematode from ground squirrels of the same species. Perhaps more remarkable was the finding of the trematode *Clonorchis sinensis* (Cobbold, 1875) in a domestic duck in China by Komiya and Kondo (1951). *Clonorchis sinensis* occurs in mammals, including man, domestic pig, synanthropic carnivores, and others. The specimens from the duck were small and incompletely developed. However, Komiya and Kondo cited an earlier finding of *C. sinensis* in a Black-crowned Night-Heron (*Nycticorax nycticorax*), in which the eggs contained fully developed miracidia. Numerous other records of this kind no doubt exist in the literature, but such occurrences nonetheless are rare.

Unusual findings of this kind perhaps can be attributed in part to some abnormality or defect in immune or other physiological systems that influence the individual vertebrate's susceptibility to infection by helminths. It would be difficult on other grounds to account, as an example, for the occasional occurrence of larval stages of taeniid cestodes in carnivores of species that constitute their natural final hosts (R. L. Rausch, unpublished data). Alternatively, comparable alterations might occur in helminths.

An understanding of the range of final hosts of helminths ultimately depends on sound taxonomy. Erroneous determinations of both helminths and their vertebrate hosts have no doubt led to misconceptions concerning host specificity. Some particular problems are associated, in this connection, with trematodes which, like acanthocephalans, are often at a more advanced state of development (i.e., reproductive organs are already well developed) when they enter the final host. With such an attribute, they may be able to mature and begin to produce eggs within a short time in a range of hosts, as was demonstrated for *Microphallus opacus* (Ward, 1894) (Rausch, 1947). Trematodes also may exhibit host-associated morphological modification to a degree not observed in helminths of other groups. Such variation has been investigated by several helminthologists; a single example suffices here. A wide range of host-related variation was demonstrated by Blankespoor (1974) in trematodes identified as *Plagiorchis noblei* Park, 1936 [= *P. elegans* (Rud., 1802)], obtained experimentally in birds and mammals of 17 species (of the 51 species exposed) that had received infective metacercariae derived from a single trematode from a naturally infected Red-winged Blackbird. Blankespoor did not attempt to correlate the respective morphological types with the numerous nominal species (>100) allocated to the genus *Pla-*

giorchis, but he demonstrated that most of the taxonomic characters previously accepted as valid for distinguishing species of the genus were subject to host-associated variation. The implications of these findings in defining host specificity of trematodes of some groups are obvious.

Reviews of more recent work concerning indices of host specificity, dynamics of parasite–host populations, or modeling of parasite–host systems, information concerning these subjects, and additional references, are to be found in publications by Hair and Holmes (1975), Kennedy (1975), Esch (1977), Rohde (1980), and Price (1980).

A complex terminology has developed with reference to hosts of helminths (and of other commensal or parasitic organisms). In addition to *intermediate* host and *final* or *definitive* host, the term *paratenic* host (= reservoir host; transport host; secondary intermediate host; *hôte d'attente*; *Stapelwirt*; *Transportwirt*; *reservuarnyi khoziain*) refers to animals that fundamentally take the place of an intermediate host but that are not usually required for completion of the cycle (see later). A more elaborate terminology to denote specific roles of paratenic hosts in cycles of helminths has been proposed by Odening (1969). Two terms proposed by Chabaud (1965) refer to the adaptation of nematodes to hosts. *Parasite transfuge* designates a nematode which develops without morphologic modification in a host other than the natural host. *Parasite de capture* is used in reference to a nematode that becomes isolated in a new host and undergoes morphological and/or biological change (speciation). This phenomenon appears to be identical to that designated under the term *Wirtskreiserweiterung* by Osche (1957). Odening (1976) has discussed in detail the origins and applications of the vocabulary relating to hosts.

Specific and Nonspecific Helminths in Birds

Synthesis of even a small part of the voluminous data available concerning the composition of helminth faunas of birds of the respective families will not be attempted here; instead, a consideration of each of the major groups of helminths will provide some general indications of the complexity and diversity of helminth–host relationships. Emphasis has been placed on specificity or lack thereof in the sexually reproducing stage of helminths. The range of hosts of the larval stages as well can be broad or narrow.

1. Digenetic Trematodes

Among the helminths, digenetic trematodes have the most complex cycles. The range of first intermediate hosts, in which asexual reproduction takes place (usually mollusks of a single species or genus), is very narrow.

The host-finding ability of the motile miracidium released from the egg depends on the interaction of various behavioral and physiological factors (see the review by Christensen, 1980) that ensures entry into the required mollusk. Invasion of the mollusk by the miracidium, by whatever means, is followed by the development of asexual reproducing stages (sporocyst, redia). The sequence of asexual stages is subject to considerable variation, depending on the group of trematodes involved.

In all cases, however, asexual reproduction leads to the formation of cercariae in the first intermediate host. Beginning with the cercarial stage, the cycle may exhibit various patterns. The cercaria may remain within the first intermediate host, forming there the metacercaria (which is a somewhat modified cercaria) that is infective for the final host (e.g., *Leucochloridium* in passeriform birds). In aquatic cycles, the cercaria may leave the mollusk and be ingested directly by the final host (fishes); it may directly invade the final host (members of the family Ornithobilharziidae, in birds); it may encyst on vegetation, shells, snails, etc. [e.g., *Zygocotyle lunata* (Diesing, 1836) in anseriform birds]; or, more commonly, it may invade a second intermediate host, an invertebrate or vertebrate, in which the metacercaria encysts. Trematodes that develop from progenetic metacercariae are typically nonspecific for the final host (e.g., Microphallidae; see Rausch, 1947).

In strigeoid trematodes, the cercariae undergo a process of metamorphosis, leading to more advanced larval stages in second or third intermediate hosts. Paratenic hosts are often involved in the cycles of these trematodes. In species of the family Strigeidae, most of which occur in the intestine of birds, the cercaria may transform to a metacercarial stage (tetracotyle) in the second intermediate host or, in obligate four-host cycles, it forms a mesocercaria in the second intermediate host (e.g., tadpole), after which development to the infective tetracotyle takes place in vertebrates that consume infected tadpoles or frogs. These in turn are the source of infection for the final host (e.g., raptors). In the family Diplostomatidae (e.g., *Neodiplostomum* in birds), the cycle is similar. The cercaria may transform to the infective metacercarial stage (diplostomulum) in the second intermediate host. In the case of some species, the metacercaria is capable of infecting the final host, but when ingested by other vertebrates that are unsuitable the diplostomula reencyst and persist in the infective state until they are ingested by a suitable host.

Metacercariae of strigeids are the most commonly recorded larval trematodes in birds. According to the list published by McDonald (1969), the tetracotyle of *Strigea falconis* Szidat, 1928 alone has been reported from birds of more than 100 species, representing several orders. The range of final hosts of these trematodes is correspondingly wide. Cycles of various trematodes have been described in detail by Ginetsinskaia (1968).

Comprehensive studies such as that by Bykhovskaia-Pavlovskaia (1962), involving trematodes of 521 species, provide information in great detail concerning the occurrence of these helminths in birds. Bykhovskaia-Pavlovskaia pointed out that the preponderant numbers of species were recorded in birds of the lower orders, of which most are associated with wet or aquatic habitats, as compared with birds occurring in terrestrial habitats (including granivorous and insectivorous species). For birds that are associated with wet habitats, the species of trematodes recorded by order numbered as follows: Podicipediformes, 42; Pelecaniformes, 37; Ciconiiformes, 70; Anseriformes, 145; Falconiformes, 53; Galliformes, 64; Gruiformes, 56; Charadriiformes, 225. For the second avian group, numbers were: Columbiformes, 12; Cuculiformes, 9; Strigiformes, 16; Caprimulgiformes, 7; Coraciiformes, 17; Piciformes, 16; Apodiformes, 12; and Passeriformes, 99. The total number of species recorded from birds of the first group was 401, representing 26 families, as compared with 126 species representing 15 families from the second group.

Of the 521 species of trematodes considered, 68.8% occurred in birds of single orders, and more than half of this component were recorded individually from birds of single species. Of the remainder, 17.8% were found in birds representing two orders; 6.6%, three orders; 2.4%, four orders; and 1.6%, five orders. From these data, it would appear that not more than about 180 species (each recorded from birds of single species) are host specific. Approximately 31% (~160 species), in birds representing two or more orders, can be considered nonspecific. The status of the remainder, approximately 180 species, is uncertain. Bykhovskaia-Pavlovskaia (1962, Table 5) summarized numbers of species of trematodes by family with respect to the orders of their avian hosts. All species were listed by host as well. Comparable data concerning the occurrence of trematodes in birds are available from the publications of Smogorzhevskaia (1976), Vaidova (1978), and others.

2. *Cestodes*

Cestodes in birds represent three orders, Pseudophyllidea, Cyclophyllidea, and Aporidea. The last, which includes cestodes of three genera occurring in anseriform birds, is not considered here.

a. Pseudophyllidean Cestodes. Cestodes of the order Pseudophyllidea are represented in birds by species of the genus *Diphyllobothrium* in the family Diphyllobothriidae, and of the genera *Ligula*, *Digramma*, and *Schistocephalus* in the family Ligulidae. The cycles of all involve two intermediate hosts, of which the first is an aquatic crustacean and the second, a fish.

5. BIOLOGY OF AVIAN PARASITES

The identity of some of the nominal species of *Diphyllobothrium* reported from avian hosts has been uncertain, but Freze (1977) recognized only three species in birds in Europe. The best-known species, and the one considered here, is *D. dendriticum* (Nitzsch, 1824), which occurs widely in piscivorous birds and mammals in the Holarctic. The motile coracidium, after emergence from the egg, is eaten by calanoid copepods, in which the procercoid develops. The second larval stage, infective for the final host, occurs in plankton-feeding fishes or in the predatory fishes that serve as paratenic hosts (Fig. 1). In lakes on the Kenai Peninsula, *Diaptomus pribilofensis* Juday and Muttkowski is the common copepod, the probable first intermediate host of *D. dendriticum*. In that region, sticklebacks (*Gasterosteus aculeatus* and *Pungitius pungitius*) serve as second intermediate hosts, as was confirmed experimentally in gulls (Rausch and Hilliard, 1970). Rainbow Trout (*Salmo gairdneri*) are abundant in the same lakes and serve as paratenic hosts, but they do not harbor plerocercoids until they become large enough to feed on sticklebacks (i.e., when they attain a length of about 260 mm and more). Therefore, plerocercoids accumulate in trout, with numbers increasing with age, so that massive infections are typical of large specimens. Consequently, either sticklebacks or trout may serve as the source of infection for birds and mammals in that region. In other regions of Alaska, plerocercoids of *D. dendriticum* (identity confirmed from cestodes obtained experimentally in gulls and dogs) were found in grayling (*Thymallus arcticus*) and in burbots (*Lota lota*). *Diphyllobothrium dendriticum* is not host specific, and it occurs in a considerable range of hosts among piscivorous birds and mammals, including humans (Rausch and Hilliard, 1970).

Dubinina (1966, Table 3) determined experimentally that various calanoid and cyclopoid copepods serve as first intermediate hosts of cestodes of the family Ligulidae. Among these, *Eudiaptomus gracilis* (Sars) and *Cyclops strenuus* Fisch appeared to be equally suitable for all of the cestodes investigated [*Ligula intestinalis* Linnaeus, 1758, *L. colymbi* Zeder, 1803, *Digramma interrupta* (Rud., 1810), *Schistocephalus solidus* (Müller, 1776), and *S. pungitii* Dubinina, 1959]. Second intermediate hosts of the better known ligulids have been identified as well. The plerocercoid of *L. intestinalis* occurs in cyprinids, whereas that of *L. colymbi* is usually found in cobitids. Fishes of the genera *Carassius* and *Abramis* serve as second intermediate host of *D. interrupta*. The plerocercoids of *S. solidus* and *S. pungitii* appear to be restricted to development in *Gasterosteus aculeatus* and *Pungitius pungitius*, respectively (Dubinina, 1966).

These cestodes in the strobilar stage are nonspecific. *Ligula intestinalis* occurs in various piscivorous birds, but most commonly in gulls. *Ligula colymbi* is found most commonly in grebes (*Podiceps cristatus* and *P. grisegena*), but it has been recorded also from herons, ibises, mergansers,

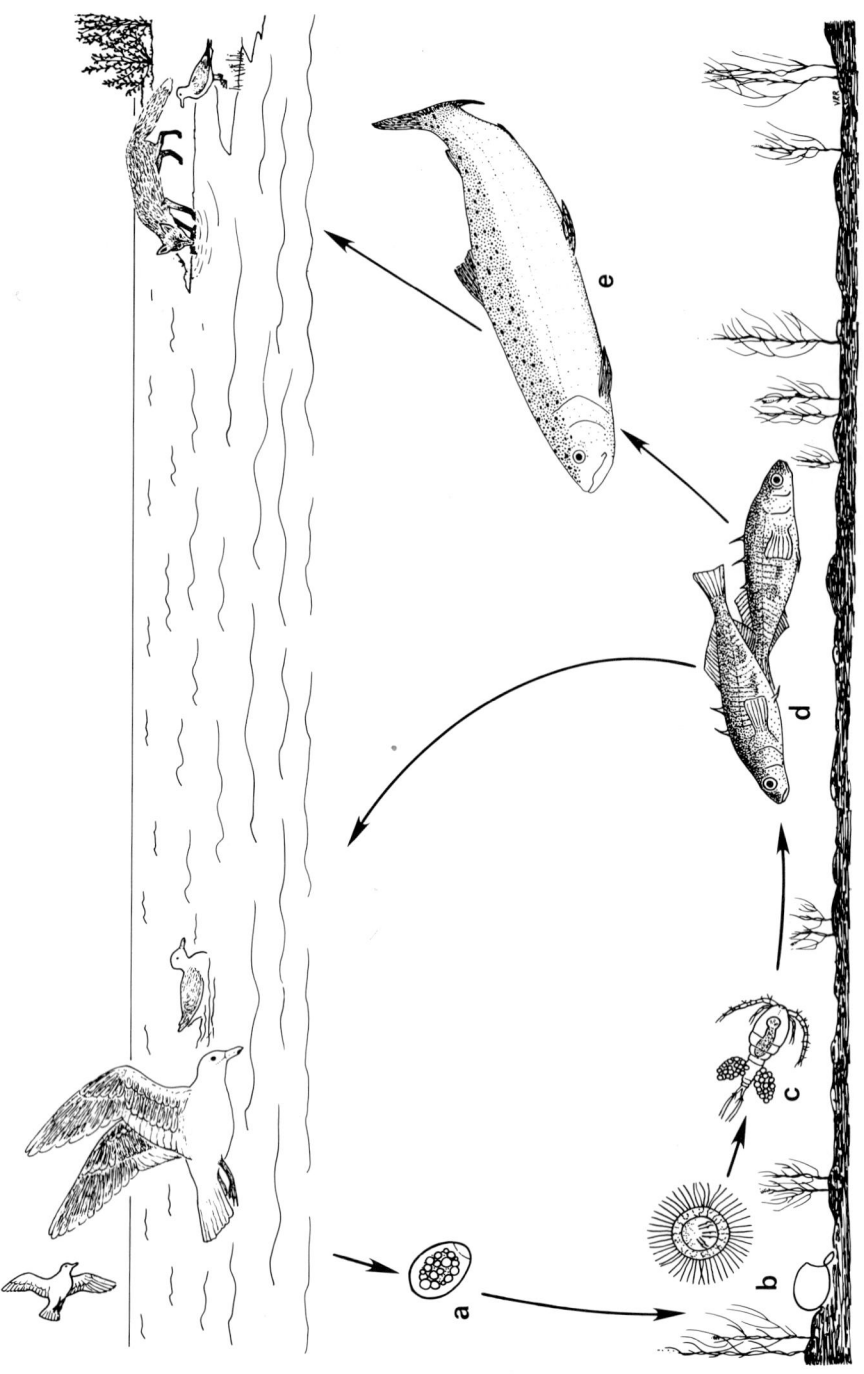

FIG. 1. Cycle of *Diphyllobothrium dendriticum*. a, Egg from final host; b, coracidium; c, procercoid in first intermediate host, a copepod; d, plerocercoid in second intermediate host; e, paratenic host.

and harriers (Dubinina, 1966). It was obtained on one occasion from a Pied-billed Grebe (*Podilymbus podiceps*) collected by me in Ohio. *Digramma interrupta* is commonly found in gulls, but it occurs also in piscivorous ducks and in grebes.

Schistocephalus spp. occur in birds of various species and, occasionally, in mammals. In Alaska, records of these cestodes were from Common Loon (*Gavia immer*), Red-throated Loon (*G. stellata*), Arctic Loon (*G. arctica*), Red-necked Grebe (*Podiceps grisegena*), Red-breasted Merganser (*Mergus serrator*), Sandhill Crane (*Grus canadensis*), Wandering Tattler (*Heteroscelus incanus*), Long-tailed Jaeger (*Stercorarius longicaudus*), Pomarine Jaeger (*S. pomarinus*), Herring Gull (*Larus argentatus*), Mew Gull (*L. canus*), and Arctic Tern (*Sterna paradisaea*). *Schistocephalus* spp. were found most commonly in Red-necked Grebes, among the birds listed.

The record from the Wandering Tattler is of interest, because Baer (1962) questioned the origin of cestodes labeled as having come from a Purple Sandpiper (*Calidris maritima*) in Iceland, on the grounds that *Schistocephalus solidus* was present in the vial, and Baer thought it unlikely that a sandpiper would eat an infected stickleback. Two of four migrating tattlers collected in May 1966 at Chandler Lake, in the Brooks Range, were infected, and, as well, remains of nine-spined sticklebacks were present in the intestine. Because the ground was still snow covered, the narrow zone of open water at the margin of the lake provided the only possibility for feeding.

Also in Alaska, *Schistocephalus* spp. were recorded from humans, domestic dogs (frequently), Mink (*Mustela vison*), and River Otter (*Lutra canadensis*).

b. *Cyclophyllidean Cestodes.* As is evident from Section III,A, cyclophyllidean cestodes usually make up the largest component of helminth faunas of birds in most families. Cestodes characteristically occurring in birds of the respective orders have been reviewed by Baer (1957).

Cyclophyllidean cestodes in birds represent numerous families arranged in several suborders. Thirteen families were listed by Yamaguti (1959) (one of which, Taeniidae, is not represented in avian hosts). A larger number of families is recognized by helminthologists in the Soviet Union, where most of the major taxonomic revisions of the Cestoda have been undertaken. According to Movsesian (1977), the suborder Davaineata, with three families, includes 340 species, of which most occur in avian hosts (recorded from 359 species of birds in 23 orders). In the Soviet Union alone, the family Hymenolepididae in the suborder Hymenolepidata is represented in birds by 240 species in 51 genera (Spasskaia, 1966). In the same suborder, in the superfamily Dilepidoidea, the family Dilepididae includes about 400 species, of which more than 160 have been recorded from birds in the Soviet Union (Spasskaia and Spasskii, 1977). Other suborders have comparatively few species. In Tetrabothriata, only 52 species, restricted to marine birds and mammals, are recognized (Temirova and Skriabin, 1978).

The cycle of cyclophyllidean cestodes characteristically involves a single intermediate host (Fig. 2). However, the cycle of the Tetrabothriata is unknown (there are indications that the infective larva occurs in fishes), and that of the Mesocestoidata may require two intermediate hosts. In the Mesocestoidata, the stage infective for the final host (two species occur in birds) is a distinctive morphological type designated a tetrathyridium. Otherwise, the larval stage is a cysticercoid or a modification thereof. In the Hymenolepididae, for example, seven types of cysticercoids were distinguished by Spasskaia (1966), based on the system established earlier by Skriabin, and Matevosian. Additional terms have been proposed by Bondarenko and Kontrimavichus (1976) to designate larvae of three types in the hymenolepidid genus *Aploparaksis*. Whereas distinct morphologic groups of larvae are discernible in this genus, they are not correlated with the morphological groups distinguishable in the strobilar stage.

Invertebrates representing a wide range of taxonomic groups characteristically serve as intermediate hosts of cyclophyllidean cestodes. Again,

FIG. 2. Cycle of a cyclophyllidean cestode. a, Gravid segment from final host; b, egg; c, cysticercoid in the intermediate host, an oligochaete.

there are some exceptions; in *Mesocestoides* (Mesocestoidata), the infective larva occurs most commonly in mammals, but it also occurs in birds, reptiles, and amphibians. Mammals serve as intermediate host of cestodes of two genera, *Cladotaenia* (Dilepidoidea) and *Paruterina* (Paruterinoidea; certain species only), which occur respectively in falconiform and strigiform birds. The larval stages of these cestodes are typical cysticercoids, and the cycles seem to exemplify a process of *Wirtskreiserweiterung* whereby mammals, in place of insects or other invertebrates, have acquired the function of intermediate host. In the genus *Paruterina*, two species, *P. candelabraria* (Goeze, 1782) and *P. rauschi* Freeman, 1957, utilize rodents as intermediate hosts (see Rausch, 1949). Two additional species in owls may also do so, but their cycles have not been elucidated. The remaining 16 species in the genus *Paruterina* occur in birds of the orders Apodiformes, Coraciiformes, Cuculiformes, and Passeriformes, and, presumably, their larval stages develop in invertebrates (insects?).

The cycles have been defined for numerous cestodes in the suborders Hymenolepidata and Davaineata, of which most species occur in avian final hosts. The respective larval stages appear to be specific for invertebrates representing various taxonomic groups, and organisms of a single species may serve as intermediate host for cestodes of different species of one or more genera. However, the individual invertebrates rarely seem to have concurrent infections involving larvae of two or more species of cestodes. Crustaceans and annelids are important as intermediate hosts of cestodes that occur in charadriiform and anseriform birds. Some examples are given here to demonstrate the range of organisms involved in cyclophyllidean cycles and to provide a better understanding of the complexity of helminth–host assemblages in aquatic biocenoses.

Information concerning intermediate hosts of cestodes in the family Hymenolepididae has been published by Spasskaia (1966), among others. Spassikaia's data are summarized here, with genera of cestodes correlated with taxonomic group of intermediate host: in copepods—*Drepanidotaenia, Fimbriaria, Fimbriarioides, Microsomacanthus, Diploposthe, Echinocotyle, Myxolepis, Retinometra, Tscherkovilepis, Sobolevicanthus, Dicranotaenia,* and *Diorchis;* in ostracods—*Dicranotaenia, Sobolevicanthus, Retinometra, Echinocotyle, Diorchis, Diploposthe, Fimbriaria, Fimbriarioides, Cloacotaenia,* and *Drepanidolepis;* in amphipods—*Fimbriaria, Microsomacanthus, Drepanidolepis, Tschertkovilepis, Myxolepis,* and *Hymenolepis* s.l.; in oligochaetes—*Limnolepis, Monorcholepis,* and *Aploparaksis;* in leeches—*Microsomacanthus* [but leeches were considered by Zajíček and Valenta (1969) to be a paratenic host of *Microsomacanthus*]; in gastropods—*Nadejdolepis;* in myriopods—*Hispaniolepis;* in Coleoptera—*Echinolepis* and *Staphylepis;* in Diptera—*Hispaniolepis* and *Echinolepis;* in Ephemeroptera—

Colymbilepis; in Orthoptera—*Passerilepis.* Gastropods of various species were listed by Spasskaia as paratenic hosts of larval cestodes of several of the aforementioned genera. At Chaunsk Gulf, in Chukotka, Bondarenko and Kontrimavichus (1976) found oligochaetes to be infected naturally by cysticercoids representing 11 species of *Aploparaksis,* of which the strobilar stages occurred in charadriiform and anseriform birds. Gvozdev and Maksimova (1979) determined that a branchiopod (*Artemia*) serves as intermediate host for *Gynandrotaenia stammeri* (Fuhrmann, 1936), which occurs in flamingos, and their findings indicated that the affinities of this cestode are with the Hymenolepididae. Additional information concerning intermediate hosts of hymenolepidid cestodes is provided in the publications by Golikova (1959), Karmanova (1968), Podesta and Holmes (1970), Tolkacheva (1975), and Dobrokhotova (1975).

Examples of intermediate hosts of several dilepidid cestodes (Dilepidoidea) have been established by Tomilkovskaia (1975), in Chukotka: oligochaetes—*Fuhrmannolepis, Anomotaenia,* and *Sacciuterina;* tipulids (Diptera)—*Dichoanotaenia;* chironomids (Diptera)—*Paricterotaenia* and *Trichocephaloides.* Bondarenko and Tomilovskaia (1979) also determined that oligochaetes serve as intermediate hosts for the single species in the genus *Rauschitaenia.* The strobilar stages of these cestodes also occur in birds associated with wet habitats.

Some information concerning the intermediate hosts of cestodes of the suborder Davaineata has been provided by Artiukh (1966), as follows: in *Nereis*—*Ophryocotyle;* in mollusks—*Davainea* and *Raillietina;* in insects (various orders)—*Raillietina.*

3. *Nematodes*

The cycles of nematodes may be direct (sometimes with the involvement of paratenic hosts), or indirect, involving intermediate or paratenic hosts, of which some actively transmit infective larvae to the final host. Birds therefore may acquire nematodes in several ways: by ingesting embryonated eggs (or invertebrates which themselves have ingested such eggs); by feeding on animals, invertebrate or vertebrate, which contain infective larvae; or through the inoculation of infective larvae by hematophagous arthropods.

Nematodes of the subclass Secernentia make up the great majority of species occuring in avian hosts. There appears to be no discernible relationship between pattern of cycle and specificity, or lack thereof, for the final host. For reasons defined by Chabaud (1957), it is difficult to assess specificity in nematodes, and Chabaud considered that it is not a phylogenetic specificity, such as that in cestodes, but rather it is *"une spécificité par affinité de métabolisme* [p. 240]." Nematodes may be specific for birds of a restricted taxonomic group, or they may be nonspecific. Few species are shared both by birds and mammals. One exception is *Trichinella pseudospiralis* Garkavi, 1972 (Adenophorea), a nematode of carnivorous mammals that has also been recorded from the Rook (*Corvus frugilegus*) by Shaikenov (1980). Heteroxenous nematodes exhibit considerable variation in range of intermediate hosts, but larval stages are often limited to development in a comparatively narrow range of organisms. Specificity in nematodes has been discussed in detail by Osche (1957), Chabaud (1957, 1965), and Inglis (1971).

The range of avian hosts for nematodes of the more commonly occurring species can best be appreciated by reviewing data obtained by means of surveys such as those conducted by Smogorzhevskaia (1976) and Vaidova (1978). The work of Baruš et al. (1978) summarizes host records for nematodes occurring in piscivorous birds. Cycles of some representative nematodes are outlined here.

The genus *Capillaria* (Adenophorea: Capillariidae) is represented by numerous species occurring in both terrestrial and aquatic birds. Infection usually takes place through the ingestion of embryonated eggs. For some species, paratenic hosts are involved in transmission.

Nematodes of the genus *Eustrongylides* (Dioctophymidae) occur in aquatic birds (loons, cormorants, herons, etc.). In the case of *E. excisus* Jägerskiöld, 1909, the eggs embryonate and hatch in water, after which development of the larvae takes place in oligochaetes. Fishes serve as second intermediate hosts, but paratenic hosts (other fishes, amphibians, and reptiles) may be involved (Karmanova, 1968).

A greater diversity in patterns of cycles exists among nematodes of the subclass Secernentia. Two examples might be taken from the order Strongylida. *Amidostomum anseris* (Zeder, 1800) (Amidostomidae), a nematode of anseriform birds, has been extensively studied. The eggs of the nematode are expelled in the feces of the host, after which the larvae develop and hatch. The birds become infected through ingestion of the larvae incidentally as they feed. However, Enigk and Dey-Hazra (1968) demonstrated experimentally that the infective larvae also are capable of percutaneous invasion of the host, after which migration takes place via the lungs. The adult worms inhabit the ventriculus.

The cycle of *Syngamus trachea* (Montagu, 1811) (Syngamidae), which inhabits the trachea of birds of various orders, is also well known. The larvae develop in eggs expelled in the feces of the host, after which they may or may not hatch. Infection of the avian host may be direct, or more often may involve feeding on paratenic hosts (earthworms, mollusks, insects, etc.).

Three examples might be taken from the family Anisakidae (Ascaridida), of which numerous representatives occur in the alimentary canal of birds. *Contracaecum microcephalum* (Rud., 1809) is found in piscivorous birds of various orders. Larvae develop after the eggs are expelled by the host, and hatch in water. Cyclopoid crustaceans of the genera *Cyclops* and *Macrocyclops* serve as first intermediate host, after which development to the infective stage takes place in larvae of dragonflies, *Coenagrion* and *Agrion*. Fishes apparently serve as paratenic hosts (Mozgovoi *et al.*, 1965).

The genus *Porrocaecum* is represented by numerous species in birds. The cycle of *P. ensicaudatum* (Zeder, 1800), which occurs commonly in thrushes, involves earthworms as intermediate host. Two species, *P. angusticolle* (Molin, 1860) and *P. depressum* (Zeder, 1800), are common nematodes in falconiform and strigiform birds, respectively (cf. Morgan and Schiller, 1950a,b). Infective larvae are found encysted subcutaneously or in the body cavity of shrews, *Sorex* spp., and other insectivores, on which such birds prey. The first intermediate host (invertebrate) of these nematodes apparently has been eliminated through a process of *Wirtskreiserweiterung*, involving the intercalation of insectivorous mammals in the cycle (for a detailed discussion see Osche, 1957).

The order Spirurida includes many species that occur in avian hosts. *Avioserpens mosgovoyi* Supriaga, 1965 (Dracunculidae) usually occurs in grebes and coots, in which it forms large, entangled aggregations at specific anatomical sites. Calanoid and cyclopoid crustaceans of various genera serve as intermediate hosts. Experimental studies by Supriaga (1971) have shown that larval dragonflies, fishes, and amphibians serve as paratenic hosts.

Synhimantus (Dispharynx) nasutus (Rud., 1819) (Acuariidae), which inhabits the proventriculus, is an excellent example of a nonspecific nematode, because it apparently will develop in any bird that ingests infective larvae. *Synhimantus nasutus* occurs most commonly in birds that feed on the ground (e.g., Galliformes and Passeriformes), but it has been reported from birds of various other orders. Terrestrial isopods of the genera *Armadillidium* and *Porcellio* serve as intermediate host. The host specificity and evolution of acuariid nematodes, with particular reference to those occurring in waterbirds, have been discussed by Sergeeva (1978).

In birds, nematodes of the superfamily Filarioidea occur in subcutaneous tissues, in organs such as the heart, and in other loci. Larvae released by the females circulate in the blood (microfilariae), and are ingested by hematophagous insects, in which development to the stage infective for the final host takes place. In the case of *Splendidofilaria fallisensis* (Anderson, 1954) (Onchocercidae), a nematode of anseriform birds, blackflies, *Simulium* spp., were determined by Anderson (1956) to be the intermediate host. Infective larvae were transmitted through the biting of the insects. *Sarconema eurycerca* Wehr, 1939 is usually found in vessels of the myocardium in swans. Seegar *et al.* (1976) established that a biting louse (Mallophaga), *Trinoton anserinum* (Fabricius), is the intermediate host and vector.

Nematodes of the family Diplotriaenidae (Diplotriaenoidea) inhabit the air sacs of birds, particularly of passeriforms. In the case of *Diplotriaena* spp., eggs released by the females in the air sacs reach the lungs, are passed up the trachea, swallowed, and expelled in the feces of the host. Experimental studies by Anderson (1957, 1962) have shown that insects of the order Orthoptera (Acrididae) serve as intermediate hosts of *Diplotriaena bargusinica* Skriabin, 1917 and of *Diplotriaenoides translucidus* Anderson, 1956.

4. Acanthocephalans

As noted earlier, acanthocephalans in birds represent three of the four classes recognized in the phylum. Of these, the class Palaeacanthocephala includes a much larger number of species as compared with the class Archiacanthocephala, which in birds is represented by members of only three genera, and the class Apororhynchida, which includes a single genus with five species. According to Petrochenko (1958), about 190 species of acanthocephalans have been reported from birds, of which about 450 species representing 25 orders have been recorded as final hosts of these organisms. As shown by Petrochenko (1958, p. 409, Table 16), the most numerous species have been found in birds of the following orders: Passeriformes, 63 species; Charadriiformes, 44; Anseriformes, 29; Falconiformes, 28; Ciconiiformes, 22; Strigiformes, 15. For other orders, numbers of species ranged from 12 down to 1. As has also been noted, acanthocephalans appear to be essentially nonspecific in homeothermic final hosts, although numerous species occurring in birds have not been reported from mammals, and vice versa. Ecological factors may be important in segregating species between vertebrates of the two classes.

Cycles of acanthocephalans typically involve a single intermediate host. On the basis of the compilation by Golvan (1960, 1962), it appears that crustaceans and insects serve as intermediate hosts for acanthocephalans of the classes Palaeacanthocephala and Archiacanthocephala, respectively. A comparable segregation is evident in known paratenic hosts. Fishes of many species have been reported to be paratenic hosts for members of the Palaeacanthocephala, where amphibians and reptiles as such are poorly represented. In contrast, only amphibians (batrachians) and reptiles were listed by Golvan (1962) as paratenic hosts for the Archiacanthocephala.

Petrochenko (1958) also listed birds and mammals of several species as paratenic hosts of small numbers of species of acanthocephalans (only two were from birds). In the Pacific Northwest, we have frequently recorded immature acanthocephalans in mammals, particularly insectivores, in which they were found in the body cavity, attached to the mesentery or omentum. Specimens from a Spotted Skunk (*Spilogale putorius*) and from Pacific Shrews (*Sorex pacificus*) were identified as *Centrorhynchus conspectus* Van

Cleave and Pratt, 1940. Specimens from a Coast Mole (*Scapanus orarius*) were identified as *Centrorhynchus* sp. *Centrorhynchus conspectus* is a characteristic, if rather uncommon, acanthocephalan in hawks and owls in North America [it has been recorded from Eurasia by Khokhlova (1978), from a Mountain Hawk-Eagle (*Spizaetus nipalensis*)]. It was found, for example, in two of four Broad-winged Hawks (*Buteo platypterus*), in one of 74 Red-tailed Hawks (*B. jamaicensis*), and in two of 12 Barred Owls (*Strix varia*), all collected in the north-central states. With respect to the infection of hawks and owls by this acanthocephalan, the conditions appear to be analogous to those involving the two species of *Porrocaecum* in raptors, mentioned above. The Spotted Skunk is well within the range of size of suitable prey for large owls, such as the Great Horned Owl (*Bubo virginianus*).

V. Acquisition of Helminths by Birds

Helminths are acquired by birds mainly as the result of ingesting infective stages. Other patterns of transmission, which involve comparatively few species of helminths, include (1) direct invasion of the final host by cercariae of trematodes of the family Ornithobilharziidae; (2) incidental ingestion of embryonated eggs or free larvae (rarely is there direct invasion by such larvae), in the case of nematodes having direct cycles; (3) incidental ingestion of metacercariae of trematodes on vegetation or other objects; and (4) transmission by hematophagous arthropods. Emphasis is placed here on infection resulting from the ingestion of infective larval stages.

A. Diet

Within the broader constraints imposed by host specificity (relative to helminths of some groups), the composition and diversity of helminth faunas are related to the kinds and diversity of animals that compose the natural diets of birds of the respective species. Relevant information concerning this subject is available from numerous publications, including those of Bykhovskaia-Pavlovskaia (1962; trematodes), Ryšavý (1966; cestodes), Baruš *et al.* (1972; helminths in the Rook), and Smogorzhevskaia (1976; all helminths). A review of the findings of Smogorzhevskaia will demonstrate the relationship between diet and the composition and diversity of helminth faunas in birds.

In the Ukraine, Smogorzhevskaia distinguished three trophoecological groups of birds, of which most were associated to some extent with wet

5. BIOLOGY OF AVIAN PARASITES

habitats. Group 1 consisted of birds whose diets included both plants and animals [i.e., all rallids except the Corncrake (*Crex crex*) and most anseriforms, excluding some species of *Aythya*, Oldsquaw (*Clangula hyemalis*), and a few others]. Group 2 included birds whose diets consisted of animals, predominantly invertebrates [i.e., Corncrake; all shorebirds; gulls and terns of certain species; all grebes except the Great Crested Grebe (*Podiceps cristatus*); a small number of anseriforms; and certain waders, including the Glossy Ibis (*Plegadis falcinellus*), White Stork (*Ciconia ciconia*), and the Squacco Heron (*Ardeola ralloides*)]. Group 3 was composed of birds whose diets consisted also of animals, predominantly vertebrates [i.e., piscivorous gulls and terns; all loons; Great Crested Grebe; mergansers; certain waders, including egrets and bitterns; all falconiform birds; and the European Kingfisher (*Alcedo atthis*)].

Helminths recorded from birds of the first group included 182 species (86 of trematodes, 65 of cestodes, 28 of nematodes, and 3 of acanthocepahalans). Although the cycles of numerous species had not been elucidated, it could be demonstrated that helminths of 158 species were acquired through the consumption of animals that served as intermediate or paratenic hosts. The specific sources of the helminths were determined to be of the following proportions for birds of group 1: crustaceans, 43.9%; mollusks, 23.1%; and vertebrates (fishes and amphibians), 8.2%. [Data for other groups of less importance were summarized by Smogorzhevskaia (1976, p. 289)]. Trematodes of 3 species had invasive cercariae, and metacercariae of 9 species had been ingested with vegetation or other materials submerged in water. Nematodes of 12 species had direct cycles.

Helminths of 328 species were recorded from birds of the second group (124 species of trematodes, 141 of cestodes, 53 of nematodes, and 10 of acanthocephalans). For birds of this group, the majority of species of helminths (214) had unknown cycles but, by analogy with known cycles of closely related species, it was concluded that 318 species had been acquired through the ingestion of intermediate or paratenic hosts. Percentages of total helminths transmitted by animals of respective groups were as follows: crustaceans, 29.6%; insects, 17.7%; and vertebrates (fishes and amphibians), 19.8%. Trematodes of 2 species had invasive cercariae, and metacercariae of 3 species occurred on plants or other objects. Nematodes of 5 species had direct cycles.

From birds of the third group, 266 species of helminths were recorded (131 of trematodes, 40 of cestodes, 49 of nematodes, and 6 of acanthocephalans). Cycles of 113 species had not been elucidated, but it could be established by Smogorzhevskaia that helminths representing about 217 species were transmitted by intermediate or paratenic hosts. Percentages of helminths acquired from animals of the more important groups were as follows: vertebrates (fishes, amphibians, and mammals), 65%; crustaceans, 15.5%. Trematodes of 2 species had invasive cercariae, and metacercariae of 2 species were associated with aquatic vegetation. Nematodes of 5 species had direct cycles.

As a source of infection for birds other than hawks and owls, mammals have little importance. The opportunistic feeding on rodents by birds such as Glaucous Gulls and Pomarine Jaegers does not usually result in the acquisition of helminths. In areas of northern Alaska, large breeding concentrations of Pomarine Jaegers are present in years when Brown Lemmings (*Lemmus sibiricus*) attain high numerical densities. Under such conditions, the jaegers are obligate predators on lemmings (Maher, 1974), but all helminths found

FIG. 3. Four-host cycle of a strigeid trematode. a, Egg from final host; b, miracidium; c, first intermediate host; d, cercaria; e, mesocercaria in second intermediate host; f, metacercaria in third intermediate host.

in them are of species that are characteristic of birds of the suborder Lari. Under the same conditions, the lemmings serve as intermediate host for a cestode found commonly in the Snowy Owl (*Nyctea scandiaca*) and for at least four species in the Arctic Fox (*Alopex lagopus*).

Birds may serve as second intermediate hosts or paratenic hosts for strigeid trematodes, such as *Strigea falconis* Szidat, 1929, for which owls and hawks are final hosts (Fig. 3).

1. Enhanced Transmission Mediated by Infective Larval Stages

Larval stages of helminths of numerous species are to some degree pathogenic for their hosts. Asexual stages of digenetic trematodes may cause mechanical injury or otherwise affect adversely the mollusks in which they become established (cf. Bertman, 1980; Sluiters, 1981). Fishes may become blind or immobilized as a consequence of the localization of metacercariae in the eyes or brain, respectively (Szidat, 1964; for a general review of the effects of such larval trematodes on fishes see Dogiel *et al.*, 1961). We have observed that rodents infected naturally by proliferative larvae of the taeniid cestodes *Echinococcus multilocularis* Leuckart, 1863 and *Taenia polyacantha* Leuckart, 1856 may exhibit pathological changes in organs and hampered locomotion. The literature contains numerous descriptions of such phenomena, which in many cases have the effect of increasing the vulnerability of affected animals to predation. Under such conditions, the transmission of the infective larvae to the final host is enhanced.

Of those helminths occurring in birds, pathogenicity of larval stages in intermediate hosts has been observed among digenetic trematodes, ligulid cestodes, and acanthocephalans. Pathological effects may be much more common than has been recognized, because their manifestations often may not be readily discerned in invertebrates. The progenetic plerocercoids of *Schistocephalus* spp. (Ligulidae) attain such large size in the body cavity of the second intermediate host (fishes of the genera *Gasterosteus* and *Pungitius*) that organ function is disturbed and ability to swim is impaired. Arme and Owen (1967) concluded that, in addition to the extreme distension of the body caused by the larvae, organ size is decreased and reproduction may be inhibited. Lester (1971) found that infected fishes have an increased oxygen demand, which may be a contributing factor to their seeking sheltered places (e.g., shallow lake margins) and remaining near the surface of the water. Such behavior obviously increases the exposure of infected individuals to predation.

Holmes and Bethel (1972) established experimentally that gammarids infected by cysticanths of an acanthocephalan, *Polymorphus paradoxus* Connell and Corner, 1957, typically occurring in aquatic rodents, exhibit altered behavior with respect to phototaxis and evasive action and that their susceptibility to predation by ducks was thereby increased. Moore (1981) determined that isopods (*Armadillidium*) infected by cysticanths of another acanthocephalan, *Plagiorhynchus formosus* (Van Cleave, 1918), commonly occurring in ground-feeding birds, were overrepresented in the diet of nestling European Starlings. The specific mechanism responsible for the disproportionate predation on infected isopods was not discerned.

That pathogenesis by such larval helminths in the intermediate host is adaptive in origin is uncertain. In many cases, helminth-induced vulnerability to predators is associated only with massive infections in the intermediate host, whereas animals with light infections do not appear to be affected. The obligatory predator–prey relationship existing between final and intermediate hosts of many species of helminths ensures completion of the cycle, even when only a small proportion of the prey population harbors the larval stage. High prevalences of helminths are often typical of the final hosts, when rates of infection by the corresponding larval stages in the intermediate hosts are very low. This relationship is particularly evident in assemblages involving cestodes and their mammalian hosts.

In contrast, the prevalence of the larval stage of a helminth in the population of the intermediate host may be very high. Berra and Au (1978), for example, found that 89% of 4175 fishes representing 29 species, from a stream in Ohio, were infected by metacercariae of *Uvulifer ambloplitis* (Hughes, 1927), a host-specific trematode of the Belted Kingfisher (*Ceryle alcyon*). The high observed prevalence in fishes, if it is general, does not, however, necessarily lead to massive infections in the final host. In a small series of kingfishers (16 specimens) collected in the same region of North America, I found *Uvulifer* in only 8. In the 7 birds for which counts of trematodes were recorded, numbers ranged from 3 to 65 (mean = 26).

Under some conditions, high rates of infection in the intermediate host may ultimately have an adverse (pathogenic) effect on the final host. Szidat (1956) observed that terns driven into the Kurisch Haff (East Prussia) by a prolonged, severe storm consumed large numbers of smelt (*Osmerus eperlanus*), which had died as a consequence of invasion by cercariae of a trematode, *Cotylurus variegatus* [=*C. erraticus* (Rud., 1809)]. The terns became massively infected by trematodes, and the resulting mortality was of such magnitude that some years elapsed before their population could recover. In this case, however, the terns were considered already to have been in a severely weakened state.

2. Age and Acquisition of Helminths

The examination of nestlings or of recently hatched precocious young will disclose that acquisition of helminths by birds usually begins immediately after hatching. For example, one finds that nestling Northern Flickers (*Colaptes auratus*) typically have massive infections by cestodes of the genus *Raillietina* when they are still so young that no feathers have appeared (R. L. Rausch, unpublished). The composition of helminth faunas of young birds before the first migration is essentially the same as that of the conspecific adults. As well, helminths of some species occur only in young birds. Trematodes [*Episthmium bursicola* (Creplin, 1830); *Prosthogonimus* spp.] that inhabit the bursa Fabricii of young birds are lost by the time that this organ disappears (Dogiel, 1963).

In birds of some groups, specialized processes involved in the feeding of the young may prevent or inhibit acquisition of helminths by them. For example, Olsen and Braun (1980), in Colorado, did not observe any to be infected of 74 Band-tailed Pigeons (*Columba fasciata*) less than 9 months old, whereas 12.5% of 535 older birds harbored helminths. It seems probable that wild doves begin to acquire helminths only after their diets are like those of the adults, but this remains to be confirmed.

Several investigations have been concerned with the relationship between composition of helminth faunas and age of the avian hosts. Creutz and Gottschalk (1969) reported findings in 457 Common Black-headed Gulls (*Larus ridibundus*), of which 434 were of known age, in the Oberlausitz region of the German Democratic Republic. The first helminths (a single species of trematode) appeared in birds 4 to 6 days old. By an age of 15 days, 6 species were recorded. After an age of 4 months, 20–21 species were present. Between the second and sixth years of life, the gulls harbored the largest numbers of helminths, representing 4–6 species. Creutz and Gottschalk concluded that both numbers of species present and intensity of infection decreased with age. In the young birds, the sequence with which helminths of the respective groups were acquired was as follows: trematodes, first; cestodes, second; and nematodes, third. Bakke (1972) did not find statistically significant differences in prevalence of helminths among juvenile, immature, and adult Mew Gulls (*L. canus*) in Norway. However, mean numbers increased with age. The earliest infection recorded was in a bird less than 4 days old (trematode).

Cornwell and Cowan (1963) investigated the acquisition of helminths in 85 young Canvasbacks (*Aythya valisineria*). Helminths were found to be few during the first week of life, with trematodes predominating. Numbers increased rapidly thereafter, and one bird harbored 3590 cestodes and 1869 trematodes at an age of 1–2 weeks. The largest numbers of helminths occurred in birds 8–9 weeks old, in which a maximum of 40,464 individuals was recorded from a single bird. Cornwell and Cowan considered that qualities of initial infections were determined primarily by the location of the nest, around which the young birds first fed. Findings in Black Ducks (*Anas rubripes*) have been discussed with reference to age by McLaughlin and Burt (1973). Some early observations on helminths in Canada Geese (*Branta canadensis*) were made by Wehr and Herman (1954), who concluded that the young birds acquired most infections during their first week of life.

Studies involving galliform and charadriiform birds were undertaken by Bondarenko (1975), who concluded that young grouse harbored a group of nonspecific helminths not found in adults, since birds such as Willow Ptarmigan (*Lagopus lagopus*) subsist mainly on invertebrates during the first 2 weeks of life, after which they gradually shift to a diet of green vegetation (see Mikheev, 1948). Species occurring in adult grouse consisted of those that have direct cycles, those that are transmitted by hematophagous arthropods, or those that are acquired when small invertebrates are ingested incidentally with vegetation. The helminth fauna of adult grouse was considered to be more diversified, and composed mainly of species characteristic of galliform birds. In contrast, young shorebirds had less diversified faunas than did adults, and trematodes predominated. Diversity of species increased with age, as a result of the gradual broadening of the range of invertebrate organisms consumed.

The relationships of helminth faunas to age of host has been discussed in detail by Bykhovskaia-Pavlovskaia (1962, p. 195; trematodes), Dogiel (1963, p. 254), and Smogorzhevskaia (1976, p. 305).

3. Migration of Birds and Acquisition of Helminths

Von Ihering (1902) was perhaps the first to perceive the possibility of some relationship between the distribution of helminths and the dispersal of their avian and mammalian hosts. One of the earliest studies to consider the geographic origins of helminths in migratory birds was that of Szidat (1940), concerning storks, *Ciconia ciconia* and *C. nigra*, which nested in East Prussia and migrated to and from wintering grounds in Africa. From an analysis of the helminth faunas of these birds, Szidat determined that characteristic species were acquired in northern Europe. However, because such helminths, particularly trematodes, had their closest affinities with species occurring in storks in Africa, Szidat concluded that this group of birds was probably of African origin.

Dogiel (1963; original version published in 1941) separated helminths in migratory birds into four groups: (1) species that occur in the host throughout the year; (2) species of southern origin, acquired on the wintering grounds; (3) species of northern origin, acquired on the breeding grounds; and (4) species acquired along routes of migration. As remarked by Dogiel, the dynamics of helminth infections in birds are complicated by migration, in combination with factors such as seasonal influences, composition of diets, cycles of intermediate hosts, and ecological characteristics of the final hosts.

In general, the largest component of helminth faunas of migratory birds is acquired on the breeding grounds, where the diet of the avian host is most diversified. When birds migrate to the wintering grounds for the first time, their helminth faunas are composed mainly of species acquired early in life. Numbers of species representing all groups of helminths are gradually reduced during migration and over winter, partly as a consequence of the short life spans of helminths of many species. When biotic conditions on the wintering grounds are similar to those on the breeding grounds (i.e., in the case of birds that make only short migrations), there is often little change in the composition of helminth faunas throughout the year. Some helminths acquired on the wintering grounds, particularly trematodes occurring in organs other than the intestine, may persist over the summer in the host and still be present at the time of the next migration.

Relatively few studies have been undertaken to define characteristics of the helminth faunas of migratory birds during winter. Such investigations, involving species whose helminths on the breeding grounds are well known, would provide useful information concerning rates at which helminths are lost or acquired, as well as a better understanding of changes that occur during migration. For a detailed discussion of this subject, with examples of helminths involved, see Dogiel (1963) and Bykhovskaia-Pavlovskaia (1962).

VI. Localization of Helminths in the Avian Host

In birds, the habitat of cestodes and acanthocephalans is the intestine. In mammals, a few species occur in the common bile duct or, as in the case of *Pyramicocephalus phocarum* (Fabricius, 1780) (Diphyllobothriidae) in the Bearded Seal (*Erignathus barbatus*), attach within the stomach. Depending on the species involved, cestodes may attach at any level in the intestine, from the pylorus posteriad, but most are found in the lower duodenum and jejunum. *Cloacotaenia megalops* (Nitzsch, 1829) (Hymenolepididae) occurs in the cloaca of anseriform birds. Acanthocephalans are usually found in the lower intestine. Representatives of the genus *Aporhynchus*, of which five species are known (Okulewicz and Maruszewski, 1980) are restricted to the cloaca of passeriform birds.

When repeated observations are made, it is usually found that each species of cestode occupies a specific biotope in the intestine of the avian host. However, the extent of possible density-related modifications is not understood. We have observed repeatedly that cestodes of the genus *Echinococcus* (Taeniidae) localize in the jejunum of the canine final host when their numbers are comparatively low. With greater numbers present, the cestodes are distributed anterior and posterior to the apparently optimal biotope in the jejunum, and with massive infections they occur uniformly from the pylorus to the cecum. Hair and Holmes (1975) have studied complex aggregations of helminths, consisting mainly of cestodes, in ducks. They observed that the respective species occupied characteristic biotopes, but that patterns of distribution exhibited some variation. They suggested that variation in sites of localization might be attributable to interactions with helminths of other species. With reference to localization of helminths, Rohde (1979) stated,

> Circumstantial evidence shows that an important factor responsible for niche restriction in parasites is selection to increase intraspecific contact and thus mating. Niche diversification is self-augmenting, and in a continuously expanding niche space populations would be diluted to such a degree that mating would become impossible without the counteracting selection for niche restriction. The probability that two species show complete niche coincidence is infinitesimally small even without competition, and it is therefore not permissible to use niche differences as proof for competition [p. 667].

It appears that controlled studies involving concurrent infections with cestodes of not more than two or three species would provide useful information concerning density-related and other interactions. The definition of the spatial extent of biotopes of given species in the intestine is not sufficient for such investigations, because cestodes and other helminths also exhibit a characteristic three-dimensional distribution which may be at least as important with respect to interspecific interactions.

Species-specific biotopes are also characteristic of trematodes and nematodes, with the majority of species of both groups occurring in the alimentary canal. In addition, these helminths occur in diverse cavities, in ducts, and in the vascular system. Nematodes frequently inhabit subcutaneous tissues. In both groups, members of respective genera or families usually exhibit uniformity in site of localization in the final host.

VII. Pathogenicity of Helminths in Birds

Pathogenicity may be erroneously attributed to helminths in birds in the absence of any other perceivable agents that might be implicated as a cause of morbidity or mortality. During the earlier decades of the present century, much effort was devoted to the investigation of the presumed role of helminths in periodic fluctuations in numbers of birds of economically important species (e.g., tetraonids), but with little success. Most birds when found dead are not in a suitable condition to permit the detection of any but the most obvious organisms, and it is perhaps not remarkable that such deaths have frequently been attributed to "parasitism." Nonetheless, helminths of certain species, at least under some conditions, are indeed pathogenic. Examples of mortality in birds, sometimes on a large scale, as a result of infection by helminths, have been described frequently in the literature. A few cases are considered here to illustrate the diversity of pathogenic processes that may be involved.

There is little evidence that cestodes have an adverse effect on the avian final host, except perhaps in domesticated waterfowl (see Shevtsov and Zaskind, 1960). Acanthocephalans usually evoke some degree of local tissue response around the embedded proboscis, with intensification of the reaction following the death of the helminth (Pflugfelder, 1956).

Trematodes may produce ulceration of epithelial surfaces or other, usually minor lesions. For example, *Pegosomum* spp. (Echinostomatidae), which inhabit the gall bladder of bitterns, cause marked hyperplasia of the epithelium and fibrosis of the underlying tissues as the result of chronic irritation by the tegumental spines of the trematodes. Similar lesions in the gall bladder or bile ducts of mammals are often produced by spinose trematodes of other groups (e.g., Campulidae, in marine mammals). Trematodes have occasionally been identified as a cause of morbidity or mortality in birds. Gibson et al. (1972) concluded that inflammation leading to necrosis of the ceca was the cause of death in ducks of various species infected by *Cyathocotyle bushiensis* Khan, 1962 (Cyathocotylidae), in Quebec. The deaths of more than 700 American Coots were evidently caused by *Sphaeridiotrema* cf. *globulus* (Rud., 1814) (Psilostomidae) in Wisconsin (Trainer and Fischer, 1963). An unusual occurrence was reported by Claugher (1976), who estimated that more than 200,000 White-faced Storm-Petrels (*Pelagodroma marina*) died in the vicinity of the Chatham group of islands (New Zealand), as a result of the entanglement of their feet and immobilization by filamentous metacercariae of the trematode *Distomum filiferum* [=*Syncoelium filiferum* (Sars, 1885) (Syncoeliidae)].

Local lesions, particularly in the alimentary canal, may be caused in birds by nematodes. In many cases, the severity of the effects of such infections is directly correlated with numbers of nematodes present. Under adverse conditions, most often involving captive or domestic birds, enhanced transmission of nematodes may lead to massive infections followed by severe losses. However, pathogenicity of nematodes has frequently been demonstrated as well in free-living birds (see Wehr, 1971).

Synhimantus (*Dispharynx*) *nasutus* (Acuariidae) is a nonspecific nematode that may produce minimal to severe lesions in birds of numerous species. The infective larval stage, as noted above, occurs in terrestrial isopods. The nematodes attach to the mucosa of the proventriculus, where they cause ulceration and often evoke a severe inflammatory process. In the case of a Song Sparrow (*Melospiza melodia*) that I collected in central Ohio in December 1944, 46 of these nematodes were present in the proventriculus, which showed an approximately 10-fold increase in size. The walls of the organ were much thickened, and the lumen was partially occluded. The bird probably would not have survived the winter. In contrast, single nematodes of this species found in each of 4 Killdeers (*Charadrius vociferus*) of 27 examined in the north-central states appeared to be of negligible significance.

Findings in wild birds are often equivocal with respect to pathogenicity of nematodes. For example, an Eastern Meadowlark (*Sturnella magna*) examined in southern Wisconsin in April 1948 appeared to have been retarded in development, as indicated by a body weight of only 72.2 g, as compared to a mean weight of 114.7 g for six normal birds collected in the same area during the months of March and April. The helminth fauna of this bird consisted of four specimens of *Diplotriaena* sp. (Nematoda: Diplotriaenidae), situated just ventral to the heart; a single specimen of another nematode, *Capillaria* sp. (Capillariidae), in the small intestine; and three cestodes, *Anonchotaenia globata* (von Linstow, 1879) (Anonchotaeniidae). The bird, a female, appeared to be normal in development, except for size, and in reproductive condition. In a case of this kind, it is tempting to attribute some degree of pathogenicity to the nematodes, with respect to cardiac function.

Depending on the species involved, nematodes of the genus *Capillaria* may be without apparent effect on the avian host, or they may cause severe morbidity and death. Clausen and Gudmundsson (1981) reported that infections by *C. contorta* (Creplin, 1839) were the cause of death of 13 of 36 Gyrfalcons (*Falco rusticolus*) found dead in Iceland. This nematode causes inflammatory changes in the mucosa of the crop and esophagus in birds of various species (see Wehr, 1971).

Nematodes of the genus *Eustrongylides* (Dioctophymidae) have frequently been reported as a cause of fatal peritonitis in piscivorous birds. Wiese et al. (1977), for example, reported mortalities of as much as 84% in nestling ardeids of five species on an island in Delaware. In this case, the infective larvae had been acquired through consumption of killifish (*Fundulus heteroclitus*).

In view of the ubiquity of helminths in birds, the proportion of species to which overt pathogenicity can be ascribed is very small. To this group might be added some unknown number which perhaps causes disorders of a subclinical nature. On the basis of present knowledge, most helminths must be regarded as symbiotes that have no discernible adverse effect on their avian hosts.

REFERENCES

Anderson, R. C. (1956). The life cycle and seasonal transmission of *Ornithofilaria fallisensis* Anderson, a parasite of domestic and wild ducks. *Can. J. Zool.* **34**, 485–525.

Anderson, R. C. (1957). Observations of the life cycles of *Diplotriaenoides translucidus* Anderson and members of the genus *Diplotriaena*. *Can. J. Zool.* **35**, 15–24.

Anderson, R. C. (1962). On the development, morphology, and experimental transmission of *Diplotriaena bargusinica* (Filarioidea: Diplotriaenidae). *Can. J. Zool.* **40**, 1175–1186.

Andrews, S. E., and Threlfall, W. (1975). Parasites of the common crow (*Corvus brachyrhynchos* Brehm, 1822) in insular Newfoundland. *Proc. Helminthol. Soc. Wash.* **42**, 24–28.

Arme, C., and Owen, R. W. (1967). Infections of the three-spined stickleback, *Gasterosteus aculeatus* L., with the plerocercoid larvae of *Schistocephalus solidus* (Müller, 1776), with special reference to pathological effects. *Parasitology* **57**, 301–314.

Artiukh, E. S. (1966). Daveneaty—lentochnye gel'minty dikikh i domashnikh zhivotnykh. *Osn. Tsestodol.* **6**, 5–511.

Baer, J. G. (1957). Répartition et endémicité des cestodes chez les Reptiles, Oiseaux et Mammifères. *Symp. Spécificité Parasit. Parasites Vertébr. C. R. 1st*, pp. 270–291.

Baer, J. G. (1962). Cestoda. In "The Zoology of Iceland" (E. Bertelsen *et al.*, eds.), Vol. 2, Part 12, pp. 1–63. Munksgaard, Copenhagen and Reykjavik.

Bakke, T. A. (1972). Studies on the helminth fauna of Norway XXIII. The common gull, *Larus canus* L., as final host for Digenea (Platyhelminthes). II. The relationship between infection and sex, age and weight of the common gull. *Norw. J. Zool.* **20**, 189–204.

Baruš, V., Ryšavý, B., Groschaft, J., and Folk, Č. (1972). The helminth fauna of *Corvus frugilegus* L. (Aves, Passeriformes) in Czechoslovakia and its ecological analysis. *Prirodoved. Pr. Ustavu Cesk. Akad. Ved Brne* **6**, 1–53.

Baruš, V., Sergeeva, T. P., Sonin, M. D., and Ryzhikov, K. M. (1978). "Helminths of Fish-Eating Birds of the Palaearctic Region," Vol. 1. Junk, The Hague.

Batsch, A. J. G. C. (1786). "Naturgeschichte der Bandwurmgattung überhaupt und ihrer Arten inbesondere, nach den neueren Beobachtungen in einem systematischen Auszuge." Johann Jacob Gebauer, Halle.

Belogurov, O. I., Leonov, V. A., and Zueva, L. S. (1968). Gel'mintofauna ryboiadnykh ptits (chaek i chistikov) poberezh'ia Okhotskogo moria. In "Gel'minty zhivotnykh Tikhogo okeana" (K. I. Skriabin and Iu. L. Mamaev, eds.), pp. 105–124. Akad. Nauk SSSR, Nauka, Moscow.

Belopol'skaia, M. M. (1952). Parazitofauna morskikh vodoplavaiushchikh ptits. *Uch. Zap. Leningr. gos. Univ. im. A.A. Zhdanova Ser. Biol. Nauk*, No. 28, pp. 127–180.

Berra, T. M., and Au, R.-J. (1978). Incidence of black spot disease in fishes in Cedar Fork Creek, Ohio. *Ohio J. Sci.* **78**, 318–322.

Bertman, M. (1980). Histopathology of hepatopancreas of *Lymnaea stagnalis* L. infected with strigeid sporocysts (Trematoda). *Acta Parasitol. Pol.* **27**, 437–442.

Blankespoor, H. D. (1974). Host-induced variation in *Plagiorchis noblei* Park, 1936 (Plagiorchiidae: Trematoda). *Am. Midl. Nat.* **92**, 415–433.

Bloch, M. E. (1782). "Abhandlung von der Erzeugung der Eingeweidewürmer und den Mitteln wider dieselben. Eine von der Königlich Dänischen Societät der Wissenschaften zu Copenhagen gekrönte Preisschrift." Siegismund Friedrich Heffe, Berlin.

Bona, F. V. (1975). Étude critique et taxonomique des Dilepididae Fuhrm., 1907 (Cestoda) Parasites des Cinconiiformes. Considérations sur la spécificité et la spéciation. *Monit. Zool. Ital., N.S., Monogr.* No. 1.

Bondarenko, S. K. (1975). Vliianie vozrasta na gel'mintofaunu kurinykh i kulikov. In "Paraziticheskie organizmy severo-vostoka Azii" (V. L. Kontrimavichus, ed.), pp. 128–133. Akad. Nauk SSSR, Vladivostok.

Bondarenko, S. K., and Kontrimavichus, V. L. (1976). Polymorphism of larvae of the genus *Aploparaksis* Clerc, 1903 (Hymenolepididae). *Folia Parasitol. (Prague)* **23**, 39–44.

Bondarenko, S. K., and Tomilovskaia, N. S. (1979). Novyi rod dilepidid—*Rauschitaenia* gen. nov. i zhiznennyi tsikl *R. ancora* (Mamaev, 1959) comb. nov.—parazita bekasov. In "Ekologiia i morfologiia gel'mintov pozvonochnykh Chukotki" (M. D. Sonin, ed.), pp. 29–37. Nauka, Moscow.

Boughton, R. V. (1937). Endoparasitic infestations in grouse, their pathogenicity and correlation with meteoro-topographical conditions. *Minn. Agric. Exp. Stn. Tech. Bull.*, No. 121.
Boyd, E. M. (1951). A survey of parasitism of the starling *Sturnus vulgaris* L. in North America. *J. Parasitol.* **37**, 56–84.
Boyd, E. M., Diminno, R. L., and Nesslinger, C. (1956). Metazoan parasites of the blue jay, *Cyanocitta cristata* L. *J. Parasitol.* **42**, 332–345.
Bremser, J. G. (1819). "Über lebende Würmer im lebenden Menschen. Ein Buch für ausübende Ärzte." Schaumburg, Vienna.
Bush, A. O., and Forrester, D. J. (1976). Helminths of the white ibis in Florida. *Proc. Helminthol. Soc. Wash.* **43**, 17–23.
Bykhovskaia-Pavlovskaia, I. E. (1953). Fauna sosal'shchikov ptits zapadnoi Sibiri i ee dinamika. *Parazitol. Sb.* **15**, 5–116.
Bykhovskaia-Pavlovskaia, I. E. (1962). "Trematody ptits fauny SSSR. Ekologi-geograficheskii obzor." Akad. Nauk SSSR, Moscow and Leningrad.
Bykhovsky, B. (1961). "Monogenetic Trematodes. Their Systematics and Phylogeny," English transl. Amer. Inst. Biol. Sci., Washington, D.C.
Chabaud, A. (1957). Spécificité parasitaire chez les nématodes parasites de vertébrés. *Symp. Spécificité Parasit. Parasites Vertébr. C. R. 1st*, pp. 230–242.
Chabaud, A. (1965). Spécificité parasitaire. I. Chez les Nématodes parasites de vertébrés. *In* "Traité de Zoologie. Anatomie, Systématique, Biologie" (P.-P. Grassé, ed.), Vol. 4, pp. 548–557. Masson, Paris.
Christensen, N. Ø. (1980). A review of the influence of host- and parasite-related factors and environmental conditions on the host-finding capacity of the trematode miracidium. *Acta Trop.* **37**, 303–318.
Claugher, D. (1976). A trematode associated with the death of the white-faced storm petrel (*Pelagodroma marina*) on the Chatham Islands. *J. Nat. Hist.* **10**, 633–641.
Clausen, B., and Gudmundsson, F. (1981). Causes of mortality among free-ranging gyrfalcons in Iceland. *J. Wildl. Dis.* **17**, 105–109.
Cooper, C. L., and Crites, J. L. (1974a). The helminth parasites of the blue jay, *Cyanocitta cristata bromia*, from South Bass Island, Ohio. *Can. J. Zool.* **52**, 1421–1423.
Cooper, C. L., and Crites, J. L. (1974b). A study of the helminth parasites of overwintering Red-winged Blackbirds (*Agelaius phoeniceus*) from Laurel, Maryland. *J. Parasitol.* **60**, 962.
Cooper, C. L., and Crites, J. L. (1974c). Helminth parasites of the Common Grackle, *Quiscalus quiscula versicolor*, from South Bass Island, Ohio. *Proc. Helminthol. Soc. Wash.* **41**, 233–237.
Cornwell, G. W., and Cowan, A. B. (1963). Helminth populations of the canvasback (*Aythya valisineria*) and host–parasite–environmental relationships. *Trans. North Am. Wildl. Nat. Resour. Conf. 28th*, pp. 173–199.
Cram, E. B. (1927). "Bird Parasites of the Nematode Suborders Strongylata, Ascaridata, and Spirurata." *Bull. U.S. Natl. Mus.*, No. 140.
Creutz, G., and Gottschalk, C. (1969). Endoparasitenbefall bei Lachmöwen in Abhängigkeit vom Alter. *Angew. Parasitol.* **10**, 80–91.
Davidson, W. R., Doster, G. L., Pursglove, S. R., Jr., and Prestwood, A. K. (1977). Helminth parasites of ruffed grouse (*Bonasa umbellus*) from the eastern United States. *Proc. Helminthol. Soc. Wash.* **44**, 156–161.
Diesing, C. M. (1850). "Systema Helminthum," Vol. 1. Braumüller, Vindobonae.
Dobrokhotova, O. V. (1975). Rasprostranenie Cyclopoida v vodoemakh Kazakhstana i ikh rol' v tsirkuliatsii vozbuditelei tsestodozov ryb i ptits. *In* "Ekologiia parazitov vodnykh Zhivotnykh" (E. V. Gvozdev, ed.), pp. 108–141. Akad. Nauk Kazakh SSR, Nauka, Alma-Ata.
Dogiel, V. A.(1963). "Allgemeine Parasitologie," revised by G. I. Polianskii, and E. M. Cheissin. Fischer, Jena.

Dogiel, V. A., Petrushevski, G. K., and Polyanski, Yu. I. (1961). "Parasitology of Fishes." Oliver & Boyd, Edinburgh and London.

Dorney, R. S., and Kabat, C. (1960). Relation of weather, parasitic disease and hunting to Wisconsin ruffed grouse populations. *Wisc. Conserv. Dep. Tech. Bull.*, No. 20.

Dubinina, M. N. (1966). "Remnetsy Cestoda: Ligulidae Fauny SSSR. Monograficheskoe issledovanie." Akad. Nauk SSSR, Nauka, Moscow & Leningrad.

Dubois, G. (1938). Monographie des Strigeida (Trematoda). *Mem. Soc. Sci. Nat. Neuchâtel* **6**, 5–535.

Dubois, G. (1953). Systématique des Strigeida. Complément de la Monographie. *Mem. Soc. Sci. Nat. Neuchâtel* **8**, Fas. 1, 5–141.

Dubois, G. (1968). Synopsis des Strigeidae et des Diplostomatidae (Trematoda). *Mem. Soc. Sci. Nat. Neuchâtel* **10**, 5–259.

Dubois, G. (1970). Synopsis des Strigeidae et des Diplostomatidae (Trematoda). *Mem. Soc. Sci. Nat. Neuchâtel* **10**, Fasc. 2, 259–727.

Dujardin, F. (1845). "Histoire naturelle des Helminthes ou vers intestinaux." Crapelet, Paris.

Enigk, K., and Dey-Hazra, A. (1968). Die perkutane Infektion bei *Amidostomum anseris* (Strongyloidea, Nematoda). *Z. Parasitenkd.* **31**, 155–165.

Esch, G. W. (1977). "Regulation of Parasite Populations." Academic Press, New York.

Forrester, D. J., Bush, A. O., Williams, L. E., Jr., and Weiner, D. J. (1974). Parasites of greater sandhill cranes (*Grus canadensis tabida*) on their wintering grounds in Florida. *Proc. Helminthol. Soc. Wash.* **41**, 55–59.

Freze, V. I. (1977). Lentetsy Evropy (Eksperimental'noe izuchenie polimorfizma). *Tr. Gel'mintol. Lab. Akad. Nauk SSSR* **27**, 174–204.

Fuhrmann, O. (1908). Die Cestoden der Vögel. *Zool. Jahrb. (Suppl.)* **10**, 1–232.

Fuhrmann, O. (1932). Les Ténias des Oiseaux. *Mem. Univ. Neuchâtel*, Vol. 8.

Gagarin, V. G. (1951). Rabota 273-i soiuznoi gel'mintologicheskoi ekspeditsii 1949 g. v Kustanaiskoi oblasti Kazakhskoi SSR. *Tr. Gel'mintol. Lab. Akad. Nauk SSSR* **5**, 293–298.

Gibson, G. G., Broughton, E., and Choquette, L. P. E. (1972). Waterfowl mortality caused by *Cyathocotyle bushiensis* Khan, 1962 (Trematoda: Cyathocotylidae), St. Lawrence River, Quebec. *Can. J. Zool.* **50**, 1351–1356.

Ginetsinskaia, T. A. (1968). "Trematody—ikh zhiznennye tsikly, biologiia i evoliutsiia." Akad. Nauk SSSR, Nauka, Leningrad.

Goeze, J. A. E. (1782). "Versuch einer Naturgeschichte der Eingeweidewürmer thierischer Körper." Philipp Adampape, Blankenburg.

Golikova, M. N. (1959). Ekologo-parazitologicheskoe izuchenie biotsenoza nekotorykh ozer Kaliningradskoi oblasti. *In* "Ekologicheskaia parazitologiia" (Iu. I. Polianskii, ed.), pp. 150–194. Univ. of Leningrad, Leningrad.

Golvan, Y.-J. (1960). Le phylum des Acanthocephala. Troisième note. La classe des Palaeacanthocephala (Meyer, 1931). *Ann. Parasitol. Hum. Comp.* **35**, 138–165, 350–386, 573–593, 713–723.

Golvan, Y.-J. (1962). Le phylum des Acanthocephala. (Quatrième note). La classe des Archiacanthocephala (A. Meyer 1931). *Ann. Parasitol. Hum. Comp.* **37**, 1–72.

Gushanskaia, L. Kh. (1952). K gel'mintofaune dikikh kurinykh ptits SSSR. *Tr. Gel'mintol. Lab. Akad. Nauk SSSR* **6**, 175–222.

Gvozdev, E. V., and Maksimova, A. P. (1979). Morfologiia i tsikl razvitiia tsestody *Gynandrotaenia stammeri* (Cestoidea: Cyclophyllidea), parazitiruiushchei u flamingo. *Parasitologiia* **13**, 56–60.

Hair, J. D., and Forrester, D. J. (1970). The helminth parasites of the starling (*Sturnus vulgaris* L.): a checklist and analysis. *Am. Midl. Nat.* **83**, 555–564.

Hair, J. D., and Holmes, J. C. (1970). Helminths of Bonaparte's gulls, *Larus philadelphia*, from Cooking Lake, Alberta. *Can. J. Zool.* **48**, 1129–1131.

Hair, J. D., and Holmes, J. C. (1975). The usefulness of measures of diversity, niche width and niche overlap in the analysis of helminth communities in waterfowl. *Acta Parasitol. Pol.* **23**, 253–269.

Harrah, E. C. (1922). North American monostomes primarily from fresh water hosts. *Ill. Biol. Monogr.* **7**, 225–324.

Heard, R. W., III. (1970). Parasites of the clapper rail, *Rallus longirostris* Boddaert. II. Some trematodes and cestodes from *Spartina* marshes of the eastern United States. *Proc. Helminthol. Soc. Wash.* **37**, 147–153.

Holmes, J. C. (1979). Parasite populations and host community structure. *In* "Host–Parasite Interfaces" (B. B. Nickol, ed.), pp. 27–46. Academic Press, New York.

Holmes, J. C., and Bethel, W. M. (1972). Modification of intermediate host behavior by parasites. *Zool. J. Linn. Soc.* **51** (Suppl. 1), 123–149.

Huber, J. Ch. (1906). Demetrios Pepagomenos über die Würmer in den Augen der Jagdfalken. *Zool. Annal.* **2**, 71–73.

Hunter, W. S., and Quay, T. L. (1953). An ecological study of the helminth fauna of Macgillivray's seaside sparrow, *Ammospiza maritima macgillivraii* (Audubon). *Am. Midl. Nat.* **50**, 407–413.

Illescas Gomez, M. P. (1977). "Helmintos parásitos de las aves de la Provincia de Granada." Copartgraf, Maracena–Grenada.

Inglis, W. G. (1971). Speciation in parasitic nematodes. *Adv. Parasitol.* **9**, 185–223.

Jewer, D. D., and Threlfall, W. (1978). Parasites of the fox sparrow (*Passerella iliaca*) and northern waterthrush (*Seiurus noveboracensis*) in Newfoundland, Canada. *Proc. Helminthol. Soc. Wash.* **45**, 270–272.

Joyeux, Ch., and Baer, J. G. (1936). "Faune de France. Cestodes." Lechevalier, Paris.

Karmanova, E. M. (1968). Vodnye oligokhety, kak khoziaeva gel'mintov. *Tr. Astrakh. Zapoved.*, No. 11, pp. 141–176.

Kasimov, G. B. (1956). "Gel'mintofauna okhotnych'e-promyslovykh ptits otriada kurinykh." Akad. Nauk SSSR, Moscow.

Kennedy, C. R. (1975). "Ecological Animal Parasitology." Wiley, New York.

Khokhlova, I. G. (1978). Taksonomicheskii obzor akantotsefalov ptits SSSR. *Tr. Gel'mintol. Lab. Akad. Nauk SSSR* **28**, 121–166.

Kinsella, J. M. (1973). Helminth parasites of the American coot, *Fulica americana americana*, on its winter range in Florida. *Proc. Helminthol. Soc. Wash.* **40**, 240–242.

Kinsella, J. M., and Hon, L. T. (1973). A comparison of the helminth parasites of the common gallinule (*Gallinula chloropus cachinnans*) and the purple gallinule (*Porphyrula martinica*) in Florida. *Am. Midl. Nat.* **89**, 467–473.

Komiya, Y., and Kondo, S. (1951). *Anas domestica* as a definitive host and *Ophiocephalus argus* as a second intermediate host of *Clonorchis sinensis*. *Jpn. Med. J.* **4**, 157–161.

Krabbe, H. (1869). "Bidrag til Kundskab Fuglenes Baendelorme." Lunos, Copenhagen.

Leidy, J. (1904). Researches in helminthology and parasitology, with a bibliography of his contributions to science, arranged by Joseph Leidy, Jr. *Smithson. Misc. Collect.* **46**, 4–281.

Leiper, R. T. and Atkinson, E. L. (1914). Helminths of the British Antarctic Expedition, 1910–1913. *Proc. Zool. Soc. London* **1**, 222–226.

Leiper, R. T., and Atkinson, E. L. (1915). Parasitic worms with a note on a free-living nematode. *Br. Antarct. (Terra Nova) Expedition 1910 Nat. Hist. Rep. Zool.* **2**, 19–60.

Leonov, V. A. (1958). Gel'mintofauna chaikovykh ptits Chernomorskogo zapovednika i sopredel'noi territorii Khersonskoi oblasti. *Uch. Zap. Gor'k. gos. Pedagog. Inst. im. M. Gor'kogo* **20**, 266–295.

Leslie, A. S., and Shipley, A. E. (1912). "The Grouse in Health and Disease, Being the Popular Edition of the Report of the Committee of Inquiry on Grouse Disease." Smith & Elder, London.

Lester, J. G. (1971). The influence of *Schistocephalus* plerocercoids on the respiration of *Gasterosteus* and a possible resulting effect on the behavior of the fish. *Can. J. Zool.* **49,** 361–366.

Linton, E. (1927). Notes on cestode parasites of Birds. *Proc. U.S. Natl. Mus.* **70,** 1–73.

Linton, E. (1928). Notes on trematode parasites of birds. *Proc. U.S. Natl. Mus.* **73,** 1–36.

Looss, A. (1896). Recherches sur la faune parasitaire de l'Égypte. Première partie. *Mém. Inst. Égypt.* **3,** 1–252.

Machalska, J. (1980). Helminth fauna of birds of the genus *Turdus* L., examined during their spring and autumn migration. I. Digenea. *Acta Parasitol. Pol.* **27,** 153–172.

MacKenzie, D. I., McKenzie, C. E., and Brownlie, L. W. (1979). Comparison of the helminth fauna of eastern and western kingbirds at Delta Marsh, Manitoba. *Can. J. Zool.* **57,** 1143–1149.

Mačko, J. K. (1968). Zur Frage der Spezifität in der Parasitologie. *Helminthologia* **9,** 315–373.

Macy, R. W., and Strong, G. L. (1967). *Laterotrema cascadensis* sp. n. (Trematoda: Stomylotrematidae) from the dipper, *Cinclus mexicanus*. *J. Parasitol.* **53,** 584–586.

Madsen, H. (1952). A study on the nematodes of Danish gallinaceous game-birds. *Dan. Rev. Game Biol.* **2,** 1–126.

Maher, W. J. (1974). "Ecology of the Pomarine, Parasitic, and Long-Tailed Jaegers in Northern Alaska." Pacific Coast Avifauna, No. 37. Cooper Ornithol. Soc., Los Angeles.

Markov, G. S. (1941). Parasitic worms of birds of Bezymannaya Bay (Novaya Zemlya). *C. R. (Dok.) Acad. Sci. URSS* **30,** 579–582.

Matevosian, E. M. (1951). Rabota 250-i soiuznoi gel'mintologicheskoi ekspeditsii 1945 g. v Kirgizskoi SSR. *Tr. Gel'mintol. Lab. Akad. Nauk SSSR* **5,** 186–194.

Mayhew, R. L. (1925). Studies on the avian species of the cestode family Hymenolepididae. *Ill. Biol. Monogr.* **10,** 1–125.

Mayr, E. (1957). Evolutionary aspects of host specificity among parasites of vertebrates. *Symp. Spécificité Parasit. Parasites Vertébr. C. R. 1st*, pp. 7–14.

McDonald, M. E (1969). Catalogue of Helminths of Waterfowl (Anatidae). *Spec. Sci. Rep. U.S. Bur. Sport Fish. Wildl*, No. 126.

McGee, S. G. (1980). Helminth parasites of squirrels (Sciuridae) in Saskatchewan. *Can. J. Zool.* **58,** 2040–2050.

McLaughlin, J. D., and Burt, M. D. B. (1973). Changes in the cestode fauna of the black duck, *Anas rubripes* (Brewster). *Can. J. Zool.* **51,** 1001–1006.

Mikheev, A. V. (1948). "Belaia kuropatka." Sov. Ministry RSFSR, Glav. uprav. zapoved., Moscow.

Moore, J. (1981). Preferential predation by starlings on isopods harboring cysticanths of *Plagiorhynchus* (*Prosthorhynchus*) *formosus*. *Program Abstr. Annu. Meeting Am. Soc. Parasitol. 56th*, Abstr. No. 126.

Morgan, B. B., and Schiller, E. L. (1950a). A note on *Porrocaecum depressum* (Zeder, 1800) (Nematoda: Anisakinae). *Trans. Am. Microsc. Soc.* **69,** 210–213.

Morgan, B. B., and Schiller, E. L. (1950b). *Porrocaecum angusticolle* (Nematoda) in North American hawks. *Trans. Am. Microsc. Soc.* **69,** 371–372.

Morgan, B. B., and Waller, F. F. (1941). Some parasites of the eastern crow (*Corvus brachyrhynchos brachyrhynchos* Brehm). *Bird-Banding* **12,** 17–22.

Movsesian, S. O. (1977). "Tsestody fauny SSSR i sopredel'nykh territorii. (Daveneaty)." Nauka, Moscow.

Mozgovoi, A. A., and Popova, T. I. (1951). Rabota 264-i soiuznoi gel'mintologicheskoi ekspeditsii 1947 g. v gosudarstvennom zapovednike "Belovezhskaia pushcha." *Tr. Gel'mintol. Lab. Akad. Nauk SSSR* **5**, 220–231.

Mozgovoi, A. A., Spasskii, A. A., and Popova, T. I. (1951). Rabota 257-i soiuznoi gel'mintologicheskoi ekspeditsii 1946 g. na ozere Chany Novosibirskoi oblasti. *Tr. Gel'mintol. Lab. Akad. Nauk SSSR* **5**, 195–206.

Mozgovoi, A. A., Ryzhikov, K. M., Sudarikov, V. E., and Leikina, E. S. (1956). Rabota 290-i soiuznoi gel'mintologicheskoi eksepditsii 1953 g. v Iakutskoi ASSR. *Tr. Gel'mintol. Lab. Akad. Nauk SSSR* **8**, 51–76.

Mozgovoi, A. A., Semenova, M. K., and Shakhmatova, V. I. (1965). Tsikl razvitiia *Contracaecum microcephalum* (Ascaridata: Anisakidae)—nematody vodoplavaiushchikh ptits. *Mater. Nauchn. Konf. Vses. Oba. Gel'mintol.*, Part I, pp. 154–159.

Naderman, J., and Pence, D. B. (1980). Helminths of the common crow, *Corvus brachyrhynchos* Brehm, from West Texas. *Proc. Helminthol. Soc. Wash.* **47**, 100–105.

Odening, K. (1969). Obligate und additionale Wirte der Helminthen. *Angew. Parasitol.* **10**, 21–36.

Odening, K. (1976). Conception and terminology of hosts in parasitology. *Adv. Parasitol.* **14**, 1–93.

Odhner, T. (1910). Nordostafrikanische Trematoden grösstenteils vom Weissen Nil (von der schwedischen zoologischen Expedition gesammelt). I. Fascioliden. *Results Sw. Zool. Exped. Egypt White Nile 1901 Jägerskiöld* **23A**, 1–170.

Okulewicz, J., and Maruszewski, W. (1980). *Aporhynchus silesiacus* sp. n. (Apororhynchidae, Acanthocephala)—a parasite of passerine birds (Passeriformes). *Acta Parasitol. Pol.* **27**, 459–469.

Olsen, O. W., and Braun, C. E. (1980). Helminth parasites of band-tailed pigeons in Colorado. *J. Wildl. Dis.* **16**, 65–66.

Osche, G. (1957). Die "Wirtskreiserweiterung" bei parasitischen Nematoden und die sie bedingenden biologisch–ökologischen Faktoren. *Z. Parasitenkd.* **17**, 437–489.

Oshmarin, P. G., and Demshin, N. I. (1972). Gel'minty domashnikh i nekotorykh dikikh zhivotnykh Demokraticheskoi respubliki V'etnam. *Tr. Biol. Pochv. Inst. Nov. Ser.* **11**, 5–115.

Owen, R. W., and Pemberton, R. T. (1962). Helminth infection of the starling (*Sturnus vulgaris* L.) in northern England. *Proc. Zool. Soc. London* **139**, 557–587.

Paspalev, G. V., and Zheliazkova-Paspaleva, A. (1965). Izsledvaniia v'rkhu khelmintofaunata na divi ptitsi ot Trakiia. I. Trematodi (Trematoda). *In* "Fauna na Trakiia" (G. Paspalev, ed.), Vol. 2, pp. 5–46. Bulgarian Akad. Nauk, Sofia.

Peet, S., and Ulmer, M. J. (1970). Trematode parasites of the brown thrasher, *Toxostoma rufum*, from Dickinson County, Iowa. *Proc. Iowa Acad. Sci.* **77**, 196–199.

Pemberton, R. T. (1963). Helminth parasites of three species of British gulls, *Larus argentatus* Pont., *L. fuscus* L. and *L. ridibundus* L. *J. Helminthol.* **37**, 57–88.

Petrochenko, V. I. (1956). "Akantotsefaly (skrebni) domashnikh i dikikh zhivotnykh" (K. I. Skriabin, ed.), Vol. 1. Akad. Nauk SSSR, Moscow.

Petrochenko, V. I. (1958). "Akantotsefaly (skrebni) domashnikh i dikikh zhivotnykh" (K. I. Skriabin, ed.), Vol. 2. Akad. Nauk SSSR, Moscow.

Pflugfelder, O. (1956). Abwehrreaktion der Wirtstiere von *Polymorphus boschadis* Schr. (Acanthocephala). *Z. Parasitenkd.* **17**, 371–382.

Podesta, R. B., and Holmes, J. C. (1970). Hymenolepidid cysticercoids in *Hyalella azteca* of Cooking Lake, Alberta: life cycles and descriptions of four new species. *J. Parasitol.* **56**, 1124–1134.

Pod'iapol'skaia, V. P., Spasskii, A. A., and Ryzhikov, K. M. (1951). Rabota 265-i soiuznoi

gel'mintologicheskoi expeditsii 1947 g. na reke Pechore (Komi ASSR). *Tr. Gel'mintol. Lab. Akad. Nauk SSSR* **5**, 232–251.
Pomeroy, M. K., and Burt, M. D. B. (1964). Cestodes of the herring gull, *Larus argentatus* Pontoppidan, 1763, from New Brunswick, Canada. *Can. J. Zool.* **42**, 959–973.
Price, P. W. (1980). "Evolutionary Biology of Parasites." Princeton Univ. Press, Princeton, New Jersey.
Railliet, A., and Henry, A. (1912). Helminthes recueillis par l'Expédition Antarctique Française du Pourquoi-Pas. I. Cestodes d'oiseaux. *Bull. Mus. Natl. Hist. Nat.* **18**, 35–39.
Ransom, B. H. (1909). The taenioid cestodes of North American birds. *Bull. U.S. Natl. Mus.* **69**, 1–141.
Rausch, R. L. (1947). Some observations on the host relationships of *Microphallus* opacus (Ward, 1894) (Trematoda). *Trans. Am. Microsc. Soc.* **66**, 59–63.
Rausch, R. L. (1949). Observations on the life cycle and larval development of *Paruterina candelabraria* (Goeze, 1782) (Cestoda: Dilepididae). *Am. Midl. Nat.* **42**, 713–721.
Rausch, R. L. (1970). Studies on the helminth fauna of Alaska. XLV. *Schistotaenia srivastavai* n. sp. (Cestoda: Amabiliidae) from the red-necked grebe, *Podiceps grisegena* (Boddaert). *In* "H. D. Srivastava Commemoration Volume" (S. Singh and B. K. Tandan, eds.), pp. 109–115. Prem, Lucknow.
Rausch, R. L., and Hilliard, D. K. (1970). Studies on the helminth fauna of Alaska. XLIX. The occurrence of *Diphyllobothrium latum* (Linnaeus, 1758) (Cestoda: Diphyllobothriidae) in Alaska, with notes on other species. *Can. J. Zool.* **48**, 1201–1219.
Rausch, R. L., and Tiner, J. D. (1949). Studies on the parasitic helminths of the north central states. II. Helminths of voles (*Microtus* spp.). Preliminary report. *Am. Midl. Nat.* **41**, 665–694.
Rausch, R. L., Krechmar, A. V., and Rausch, V. R. (1979). New records of helminths from the brown bear, *Ursus arctos* L., in the Soviet Far East. *Can. J. Zool.* **57**, 1238–1243.
Redington, B. C., and Ulmer, M. J. (1966). Helminth parasites of rails and host–parasite relationships of the trematode *Protoechinostoma mucronisertulatum*. *Proc. Iowa Acad. Sci.* **73**, 391–405.
Rennie, J., and Reid, A. (1912). The cestoda of the Scottish National Antarctic Expedition. *Trans. R. Soc. Edinburgh* **48**, 441–453.
Rohde, K. (1979). A critical evaluation of intrinsic and extrinsic factors responsible for niche restriction in parasites. *Am. Nat.* **114**, 648–671.
Rohde, K. (1980). Host specificity indices of parasites and their application. *Experientia* **36**, 1370.
Roudabush, R. L. (1942). Parasites of the American coot (*Fulica americana*) in central Iowa. *Iowa State Coll. J. Sci.* **16**, 437–441.
Rudolphi, C. A. (1819). "Entozoorum synopsis cui accedunt mantissa duplex et indices locupletissimi." Augustus Rücker, Berolini.
Rybicka, K. (1961). The tapeworms of Drużno Lake birds (the Anseriformes excepted). *Acta Parasitol. Pol.* **6**, 143–178 (Engl. transl. Natl. Sci. Found., Washington, D.C.).
Ryšavý, B. (1966). The occurrence of cestodes in the individual orders of birds and the influence of food on the composition of the fauna of bird cestodes. *Folia Parasitol. (Prague)* **13**, 158–169.
Ryzhikov, K. M. (1951). Rabota 268-i soiuznoi gel'mintologicheskoi ekspeditsii 1948 g. v zapadnoi Gruzii. *Tr. Gel'mintol. Lab. Akad. Nauk SSSR* **5**, 252–260.
Ryzhikov, K. M., and Sudarikov, V. E. (1951). Rabota 272-i soiuznoi gel'mintologicheskoi ekspeditsii 1949 g. v raione ozera Baikal. *Tr. Gel'mintol. Lab. Akad. Nauk SSSR* **5**, 270–292.

Ryzhikov, K. M., Kadenatsii, A. N., Akhmerov, A. Kh., and Kontrimavichus, V. L. (1961a). Rabota Amurskoi gel'mintologicheskoi ekspeditsii (314 SGE) v 1959 g. *Tr. Gel'mintol. Lab. Akad. Nauk SSSR* 11, 393–413.

Ryzhikov, K. M., Pavlov, A. V., Akhmerov, A. Kh., and Kontrimavichus, V. L. (1961b). Rabota Amurskoi gel'mintologicheskoi ekspeditsii (314 SGE) v 1958 g. *Tr. Gel'mintol. Lab. Akad. Nauk SSSR* 11, 373–392.

Ryzhikov, K. M., Kadenatsii, A. N., Akhmerov, A. Kh., and Kontrimavichus, V. L. (1962). Rabota Amurskoi gel'mintologicheskoi ekspeditsii (314 SGE) v 1960 g. *Tr. Gel'mintol. Lab. Akad. Nauk SSSR* 12, 120–138.

Ryzhikov, K. M., Gubanov, N. M., Tokacheva, L. M., Khokhlova, I. G., Zinov'eva, E. N., and Sergeeva, T. P. (1974). "Gel'minty Ptits Iakutii i Sopredel'nykh Territorii." Nauka, Moscow.

Sadokov, S. B. (1951). Rabota 270-i soiuznoi gel'mintologicheskoi ekspeditsii 1948 g. v Primorskom krae. *Tr. Gel'mintol. Lab. Akad. Nauk SSSR* 5, 261–269.

Schmidt, G. D., and Frantz, D. W. (1972). Helminth parasites of Wilson's phalarope, *Steganopus tricolor* Vieillot, 1819, in Montana and Colorado. *Proc. Helminthol. Soc. Wash.* 39, 269–270.

Seegar, W. S., Schiller, E. L., Sladen, W. J. L., and Trips, M. (1976). A Mallophaga, *Trinoton anserinum*, as a cyclodevelopmental vector for a heartworm parasite of waterfowl. *Science (Washington, D.C.)* 194, 739–741.

Sergeeva, T. P. (1978). Stanovlenie gostal'noi spetsifichnosti gel'mintov chaikovykh ptits (na primere nematod semeistv Acuariidae i Streptocaridae). *Tr. Gel'mintol. Lab. Akad. Nauk SSSR* 28, 38–46.

Shaikenov, B. (1980). Spontaneous infection of birds with *Trichinella pseudospiralis* Garkavi, 1972. *Folia Parasitol. (Prague)* 27, 227–230.

Shevtsov, A. A., and Zaskind, L. N. (1960). "Gel'minty i gel'mintozy domashnikh vodoplavaiushchikh ptits." Khar'kov. Gosudar. Univ., Khar'kov.

Shigin, A. A. (1961). Gel'mintofauna chaikovykh ptits Rybinskogo vodokhranilishcha. *Tr. Darvinskogo Gos. Zapov.* 7, 309–362.

Skriabin, K. I. (1914). Vogelcestoden aus Russisch Turkestan. *Zool. Jahrb. Abt. Syst.*, 37, 411–492.

Slater, R. L. (1967). Helminths of the robin, *Turdus migratorius* Ridgway, from northern Colorado. *Am. Midl. Nat.* 77, 190–199.

Sluiters, J. F. (1981). Development of *Trichobilharzia ocellata* in *Lymnaea stagnalis* and the effects of infection on the reproductive system of the host. *Z. Parasitenkd.* 64, 303–319.

Smogorzhevskaia, L. A. (1976). "Gel'minty vodoplavaiushchikh i bolotnykh ptits fauny Ukrainy." Naukova Dumka, Kiev.

Southwell, T. (1930). "The Fauna of British India, including Ceylon and Burma," Vol. 2. Taylor & Francis, London.

Spasskaia, L. P. (1966). "Tsestody ptits SSSR—Gimenolepididy." Akad. Nauk SSSR, Nauka, Moscow.

Spasskaia, L. P., and Spasskii, A. A. (1971). "Tsestody ptits Tuvy." Akad. Nauk Moldav. SSR, Shtiintsa, Kishinev.

Spasskaia, L. P., and Spasskii, A. A. (1977). "Tsestody ptits SSSR. Dilepididy sukhoputnykh ptits." Akad. Nauk, SSSR, Nauka, Moscow.

Spasskii, A. A., and Sonin, M. D. (1961). Rabota Kamchatskoi gel'mintologicheskoi ekspeditsii (317 SGE) v 1959 g. *Tr. Gel'mintol. Lab. Akad. Nauk SSSR* 11, 414–431.

Spasskii, A. A., Ivashkin, V. M., Bogoiavlenskii, Iu. K., and Sonin, M. D. (1958). Rabota 306-i soiuznoi gel'mintologicheskoi ekspeditsii 1956–1957 gg. v Tuvinskoi avtonomoi oblasti. *Rab. Exp. Gel'mintol. Lab*, pp. 73–103.

Spasskii, A. A., Freze, V. I., Bogoiavlenskii, Iu. K., and Roitman, V. A. (1962). Rabota Kamchatskoi gel'mintologicheskoi ekspeditsii (317-ia SGE) v 1960 g. *Tr. Gel'mintol. Lab. Akad. Nauk SSSR* **12**, 201–221.

Spasskii, A. A., Bogoiavlenskii, Iu. K., Kontrimavichus, V. L., and Paramonov, B. B. (1963). Rabota Kamchatskoi gel'mintologicheskoi ekspeditsii (317-ia SGE) v 1961 g. *Tr. Gel'mintol. Lab. Akad. Nauk SSSR* **13**, 369–381.

Spory, G. R. (1965). Some internal and external parasites of the redwinged blackbird, *Agelaius phoeniceus phoeniceus* L., from central Ohio; including descriptions of three new feather mites. *Ohio J. Sci.* **65**, 49–59.

Stanley, J. G., and Rabalais, F. C. (1971). Helminth parasites of the red-winged blackbird, *Agelaius phoeniceus*, and common grackle, *Quiscalus quiscula*, in northwestern Ohio. *Ohio J. Sci.* **71**, 302–303.

Sultanov, M. A. (1963). "Gel'minty domashnikh i okhotnich'e-promyslovykh ptits Uzbekistana." Akad. Nauk Uzbek. SSR, Tashkent.

Supriaga, A. M. (1971). Biologicheskii tsikl *Avioserpens mosgovoyi* (Camallanata; Dracunculidae)—nematody vodoplavaiushchikh ptits. *In* "Sbornik rabot po gel'mintologii posviashchen 90-letiiu so dnia rozhdeniia Akademika K. I. Skriabina" (A. M. Iarnykh, ed.), pp. 374–383. Kolos, Moscow.

Szidat, L. (1940). Die Parasitenfauna des weissen Storches und ihre Beziehungen zu Fragen der Ökologie, Phylogenie und der Urheimat der Störche. *Z. Parasitenkd.* **11**, 563–592.

Szidat, L. (1956). Geschichte, Anwendung und einige Folgerungen aus den parasitogenetischen Regeln. *Z. Parasitenkd.* **17**, 237–268.

Szidat, L. (1964). Vergleichende helminthologische Untersuchungen an den argentinischen Grossmöwen *Larus marinus dominicanus* Lichtenstein und *Larus ridibundus maculipennis* Lichtenstein nebst neuen Beobachtungen über die Artbildung bei Parasiten. *Z. Parasitenkd.* **24**, 351–414.

Temirova, S. I., and Skriabin, A. S. (1978). Podotriad Tetrabothriata (Ariola, 1899) Skriabin, 1940. *Osn Tsestodol.* **9**, 8–117.

Threlfall, W. (1968a). Studies on the helminth parasites of the American herring gull (*Larus argentatus* Pont.). *Can. J. Zool.* **46**, 1119–1126.

Threlfall, W. (1968b). The helminth parasites of three species of gulls in Newfoundland. *Can. J. Zool.* **46**, 827–830.

Threlfall, W. (1971). Helminth parasites of alcids in the northwestern North Atlantic. *Can. J. Zool.* **49**, 461–466.

Tolkacheva, L. M. (1975). Rakoobraznye—promezhutochnye khoziaeva tsestod vodnykh i bolotnykh ptits Karasukskikh ozer. *In* "Parazity v prirodnykh kompleksakh severnoi Kulundy" (K. M. Ryzhikov and S. S. Folitarek, eds.), pp. 114–143. Akad. Nauk SSSR (Sibir. Otdel.), Novosibirsk.

Tomilovskaia, N. S. (1975). Lichinki tsestod dilepidid ptits Chaunskoi nizmennosti. *In* "Paraziticheskie organizmy severo-vostoka Azii" (V. L. Kontrimavichus, ed.), pp. 224–232. Akad. Nauk SSSR, Vladivostok.

Trainer, D. O., and Fischer, G. W. (1963). Fatal trematodiasis of coots. *J. Wildl. Manage.* **27**, 483–486.

Tsimbaliuk, A. K., and Belogurov, O. I. (1964). O faune nematod ryboiadnykh ptits ostrovov Beringova moria. *Nauchn. Dokl. Vyssh. Shk. Biol. Nauki*, No. 4, pp. 7–11.

Vaidova, S. M. (1978). "Gel'minty ptits Azerbaidzhan." Akad. Nauk Azerbaidzh., SSR. Elm, Baku.

von Ihering, H. (1902). Die Helminthen als Hilfsmittel der zoogeographischen Forschung. *Zool. Anz.* **26**, 42–51.

von Linstow, O. (1878). "Compendium der Helminthologie. Ein Verzeichnis der bekannten Helminthen." Hahn'sche, Hannover.

von Linstow, O. (1888). Report on the Entozoa. Zoology of the Challenger Expedition, Challenger Reports (Zool.) **23**, 1–18.

von Linstow, O. (1889). "Compendium der Helminthologie. Nachtrag. Die Litteratur der Jahre 1878–1889." Hahn'sche, Hannover.

von Linstow, O. (1906). Nematodes of the Scottish National Antarctic Expedition 1902–1904. *Proc. R. Soc. Edinburgh* **26**, 464–472.

Wehr, E. E. (1971). Nematodes. *In* "Infectious and Parasitic Diseases of Wild Birds" (J. W. Davis, R. C. Anderson, L. Karstad, and D. O. Trainer, eds.), pp. 185–233. Iowa State Univ. Press, Ames.

Wehr, E. E., and Herman, C. M. (1954). Age as a factor in acquisition of parasites by Canada geese. *J. Wildl. Manage.* **18**, 239–247.

Wiese, J. H., Davidson, W. R., and Nettles, V. F. (1977). Large scale mortality of nestling ardeids caused by nematode infection. *J. Wildl. Dis.* **13**, 376–382.

Williams, I. C., and Harris, M. P. (1965). The infection of the gulls *Larus argentatus* Pont., *L. fuscus* L. and *L. marinus* L. with cestoda on the coast of Wales. *Parasitology* **55**, 237–256.

Wissler, K., and Halvorsen, O. (1977). Helminths from willow grouse (*Lagopus lagopus*) in two localities in North Norway. *J. Wildl. Dis.* **13**, 409–413.

Wolffhügel, K. (1900). "Beitrag zur Kenntnis der Vogelhelminthen." Inaug. Dissertation, Univ. Buchdruckerei, Univ. of Basel, Switzerland.

Yamaguti, S. (1959). "Systema Helminthum. The Cestodes of Vertebrates," Vol. 2. Wiley (Interscience), New York.

Young, R. T. (1950). Cestodes of California gulls. *J. Parasitol.* **36**, 9–12.

Zablotskii, V. I. (1962). Materialy k gel'mintofaune khishchnykh ptits poberezhii Kaspiiskogo moria. *Tr. Astrakh. Zapov.*, No. 6, pp. 91–114.

Zajíček, D., and Valenta, A. (1969). *Erpobdella octoculata* L. (Hirudinea), the reservoir host of *Microsomacanthus parvula* (Kowalewski, 1904) in Czechoslovakia. *Vestn. Cesk. Spol. Zool.* **33**, 272–277.

Zeder, J. G. H. (1800). "Erster Nachtrag zur Naturgeschichte der Eingeweidewürmer, mit Zufässen und Anmerkungen herausgegeben." Lebrecht Crusius, Leipzig.

Chapter 6

BURSA OF FABRICIUS

Bruce Glick

I.	Introduction	443
II.	Morphology of the Bursa	444
III.	Origin of Bursal Lymphocytes	455
IV.	Bursal Microenvironment	458
V.	Bursal Kinetics	460
	A. Testosterone Control of Growth	460
	B. Steroid Receptors	461
	C. Adrenal Control of Growth	462
	D. Disease, Nutrient, and Pesticide Influences on Bursal Growth	463
	E. Lymphocyte Life Span	464
VI.	The Bursa's Influence on Other Organs	465
	A. Adrenal Gland	465
	B. Thymus	466
VII.	Regulation of Immunoglobulin Synthesis	467
	A. The Early Period	467
	B. Surgical Bursectomy and Antibody Synthesis	468
	C. Hormonal Bursectomy and Antibody Synthesis	468
	D. Ontogeny of Antibody Synthesis	472
	E. Bursectomy and Immunoglobulin Synthesis	473
	F. Immunosuppression	480
VIII.	Summary	481
	References	484

I. Introduction

The bursa of Fabricius is a dorsal diverticulum of the proctodeal region of the cloaca. From the first description of the bursa by Hieronymus Fabricius in the seventeenth century (Adelmann, 1967) until the middle of the twentieth century, the function of this gland was an enigma. Fabricius believed it to be present only in females where it functioned as a semen reservoir (Adelmann, 1967). The numerous descriptive names applied to the bursa attest to its having been considered a functional part of the reproductive system (egg reservoir, genital apparatus, seminal vesicle, prostate, Cowper's gland, and female pouch receiving sperm), digestive system (third caecum), or excretory system (anal gland, anal pouch, urinary vessel, and bladder) (Retterer and Lelievre, 1913). My approach to understanding the function of the bursa was to remove the bursa during its most rapid growth period. These experiments set the stage for a fellow graduate student (T. S. Chang) accidentally to select bursectomized birds to be immunized for a student laboratory. The failure of these birds to produce antibody prompted us to conduct a series of experiments which clearly demonstrated the dependency of antibody production on the bursa of Fabricius. These initial studies led to the identification of two distinct lymphocyte populations that held the key to understanding immunity.

The observation of a supposed humoral immune response following the removal of the bursa in the perinatal chick was a bit of serendipity that unlocked a Pandora's box from which numerous scientific disciplines have sampled. The samplings helped to establish the discipline of immunobiology, extended our understanding of the origin of certain hemopoietic cells, sharpened our perception of the fine structure of lymphoid tissue, and emphasized the diverse roles that a biological structure (i.e., the bursa) may play. Several of these roles will be discussed in this chapter.

II. Morphology of the Bursa

Advancements into embryology were recorded into the latter part of the sixteenth and early part of the seventeenth centuries by Girolamo Fabrici or Fabrizio (Hieronymus Fabricius), born in 1533 in Aquapendente, Italy. Although Fabricius was an excellent teacher when he desired to be, he more often directed his energies toward research (Adelmann, 1967). The manuscript *De Formatione Ovi et Pulli* (1621), published after Fabricius's death in 1619, included his description of the bursa.

> The third thing which should be noted in the podex is the double sac (bursa) which in its lower portion projects toward the pubic bone and appears visible to the observer as soon as the uterus, already mentioned, presents itself to view [Adelmann, 1967, p. 147].

The bursa of Fabricius, named for Fabricius, is descriptively a cul-de-sac or diverticulum that is easily removed because of its superficial location in the body cavity (Fig. 1). The bursa is round or oval in the chicken (Fig. 2), but it is elongated in the Pekin duck (Glick, 1963) and European Starling (*Sturnus vulgaris*; B. Glick, unpublished observations). The smooth, whitish outer surface of the chicken's bursa is in direct contrast with its plicated luminal surface (Fig. 2). The number and size of the plicae varies with the age and physiological condition of the bird. There are usually between 11 and 14 primary plicae and 6–7 secondary plicae (Jolly, 1915; Ackerman and Knouff,

6. BURSA OF FABRICIUS 445

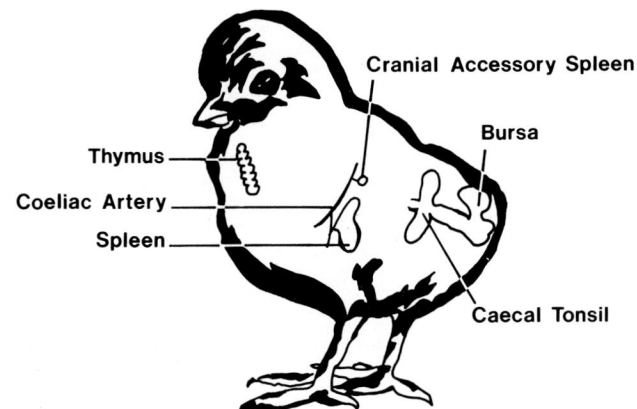

FIG. 1. Central lymphomyeloid tissues, thymus, bursa of Fabricius, and several peripheral lymphoid tissues of the chicken. (From Glick, 1970a.)

1959). Plicae are not evident in all birds, for example, the European Starling (Fig. 3). A fine-structure study of the starling's bursa is underway in our laboratory. Embryologically, the plicae of the chicken's bursa experience a reorganization of surface epithelial cells, which extend into the tunica propria as epithelial buds (Fig. 4). The epithelial buds that develop into bursal follicles appear in most strains of chickens between 12 and 15 days of embryonic development. The bursal follicles, more apparent after hatching than before, are divisible into a cortex and medulla (Fig. 5). The medulla, epithelial in origin, contains reticular cells and fibers, small numbers of macrophages and plasma cells, and numerous lymphoblasts and lymphocytes

FIG. 2. Bursa of Fabricius from a 2-week-old chicken. The bursa on the right has been everted to show the plicae. (From Glick, 1964.)

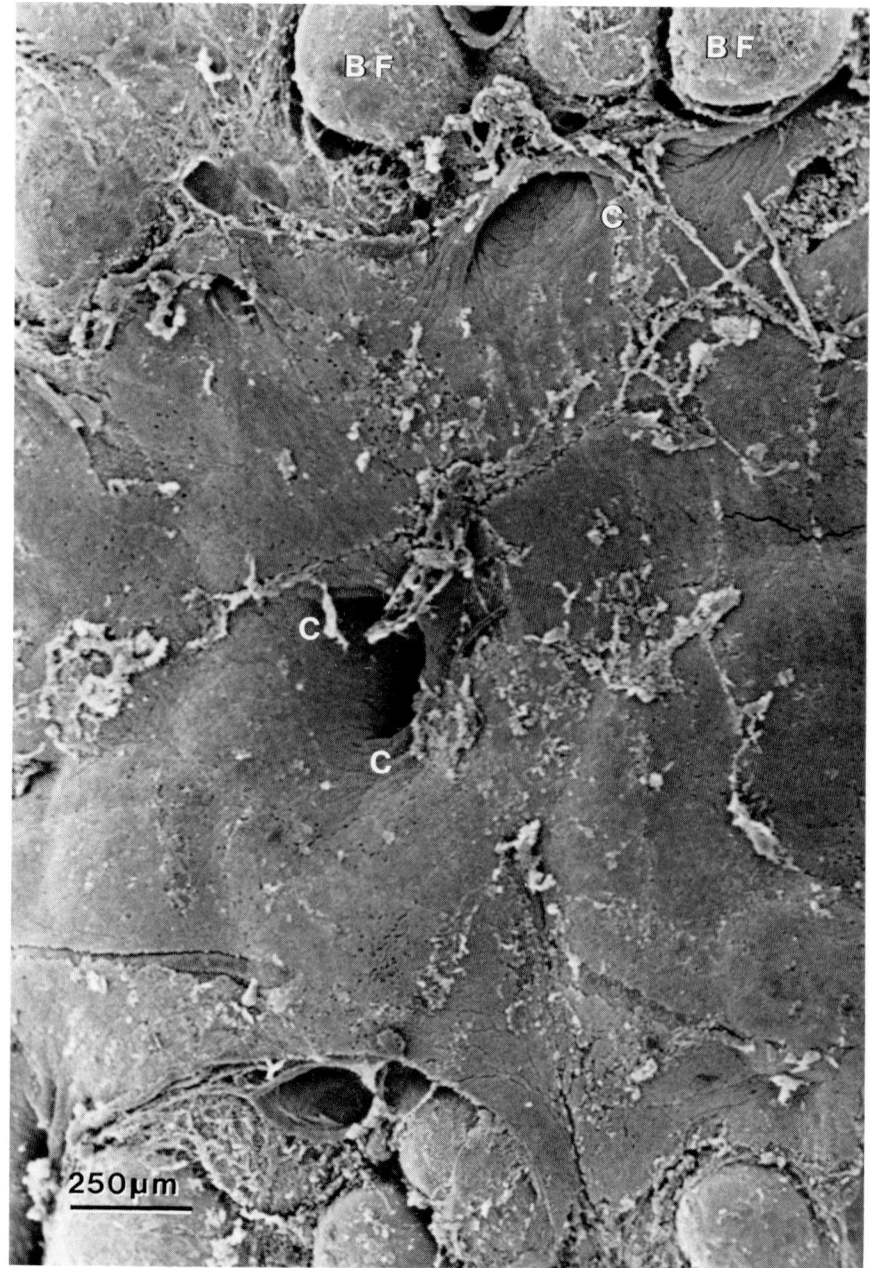

FIG. 3. Scanning electron micrograph of a bursa from a male European Starling (*Sturnus vulgaris*) before fledging. The luminal surface of the bursa lacks plicae but reveals channels (C) that lead to the bursal follicles (BF). ×40. (From Glick and Olah, 1982.)

6. BURSA OF FABRICIUS   447

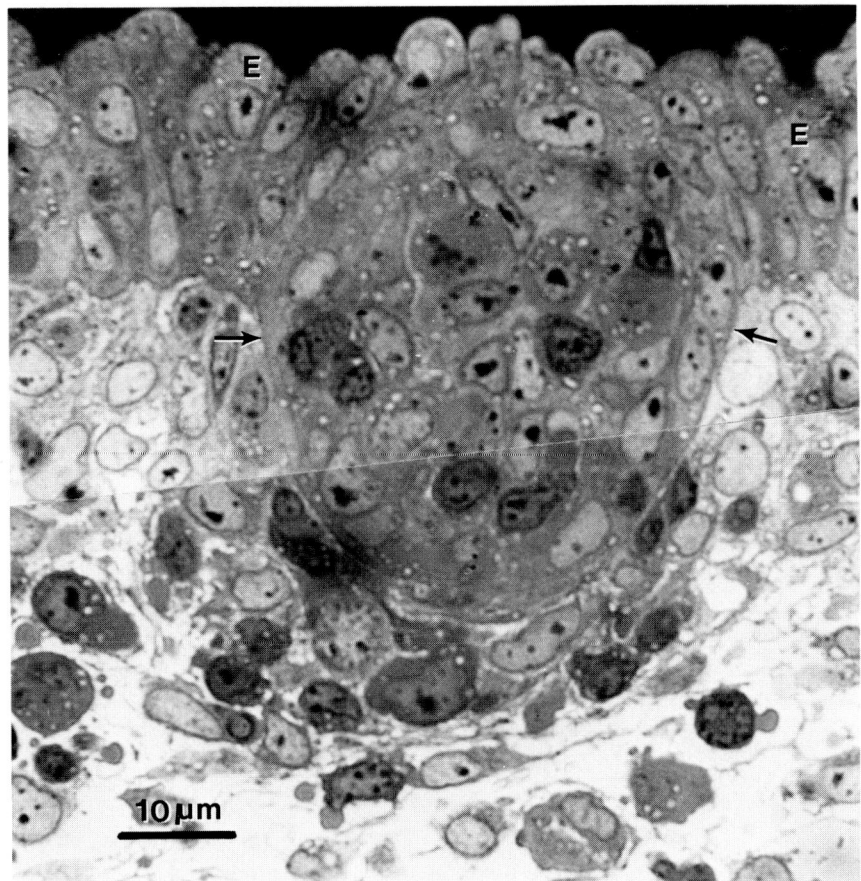

FIG. 4. Bursal epithelium of the embryo reorganize and extend into the tunica propria as epithelial buds. E, Epithelium; arrow, bud. ×1400.

(Olah et al., 1975). The cortex lacks reticular cells, but it possesses lymphoblasts, lymphocytes, macrophages, and plasma cells.

The pudendal arteries and veins and mesenteric vein are the major vessels of the bursa (Pintea et al., 1967). Lymphatics have been identified draining the bursa (Dransfield, 1945; Ekino et al., 1980). Sympathetic fibers, the pelvic nerve, and the intestinal nerves that enter the first bursa cloacal ganglion at the anterior pole of the bursa innervate the bursa (Pintea et al., 1967; Cordier, 1969).

Separating the cortex from the medulla are epithelial cells whose basal lamina is continuous with that of the epithelial cells of the plicae (Jolly, 1915; Boyden, 1922; Ackerman and Knouff, 1959; Olah et al., 1975). The medullary component of the bursal follicle is in direct contact with the luminal

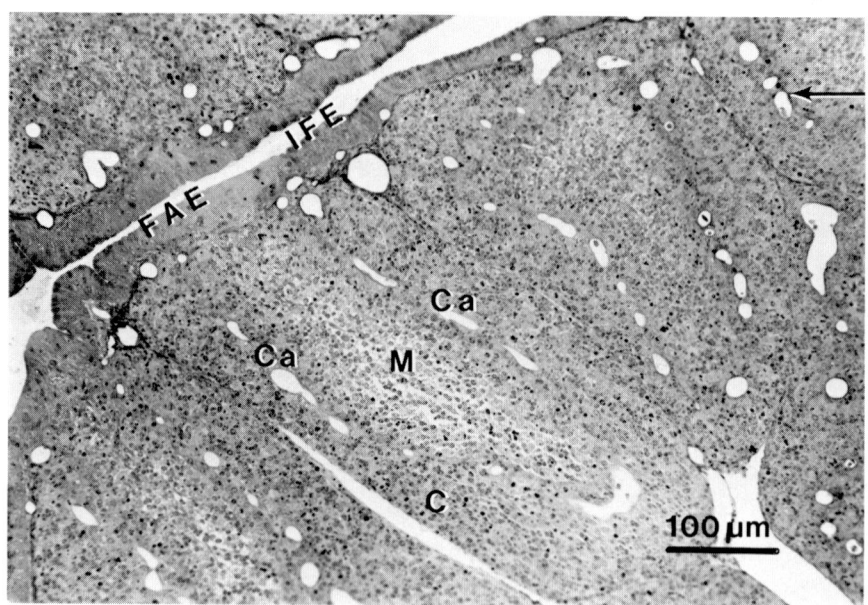

Fig. 5. Subsequent to perfusing the posthatch chick one can clearly see the capillaries (Ca) within the cortex (C) of the bursal follicle (arrow). FAE, Follicle associated epithelium; IFE, interfollicular epithelium; M, medulla. ×140.

epithelial cells (Fig. 5). Scanning electron microscopy (SEM) has revealed that the epithelium covering the plicae is divisible into follicle-associated epithelium (FAE), covering the follicle, and that located between the follicles, the interfollicular epithelium (IFE) (Bockman and Cooper, 1973; Holbrook et al., 1974). The polygonal FAE cells exhibit a well-delineated border and few evenly spaced microvilli that are indistinct at hatching (Fig. 6), are more numerous and slender within the first week (Fig. 7), and become shorter and thicker with age (Fig. 8). On the other hand, the IFE cells are distinct from the time of hatching (Fig. 9). The bursal follicles are jug-shaped and extend deep into the substance of the plica (Fig. 10). During the period of bursal regression, a distinct separation occurs between the FAE and IFE, with an increase in the convolution of the IFE (Fig. 11). These observations complement our original identification of projecting follicles (PF) and button-like follicles (BLF) (Holbrook et al., 1974). The BLF were similar to those in Fig. 11. Subsequently, we showed that the PF (Fig. 12) appeared before the BLF in late embryonic development (Glick et al., 1977a). This suggests that BLF originated by aging from PF.

Viewing the luminal surface of the bursa with SEM, we were impressed with the large number of FAE areas. Because the medullary component of the bursa is located beneath the FAE, one should be able to estimate the number of bursal follicles by counting the FAE areas. We estimated this number by following the procedure of Schaffner *et al.* (1974), who applied carbon to the chick's vent and observed it internalized to the bursa with the subsequent delineation of the FAE. It had been demonstrated that pinocytosis was a characteristic of FAE and not IFE (Bockman and Cooper, 1973). Within 2 hours after applying 0.1 ml of a suspension of carbon particles (64 mg/ml) to the vent of 4-week-old chickens, we recorded an average of 800 FAE areas/fold (Olah and Glick, 1978a). This value multiplied by the number of plicae (10–15) revealed the presence of 8,000–12,000 FAE areas/bursa. Thus the bursa contains more follicles than previously reported (Wenckebach, 1896; Boyden, 1922). Applying geometry to the question of the number of bursal follicles, we outlined a trapezoid on a given plica and found its area to be 6.6 mm^2 (Olah and Glick, 1978a). The mean of each FAE

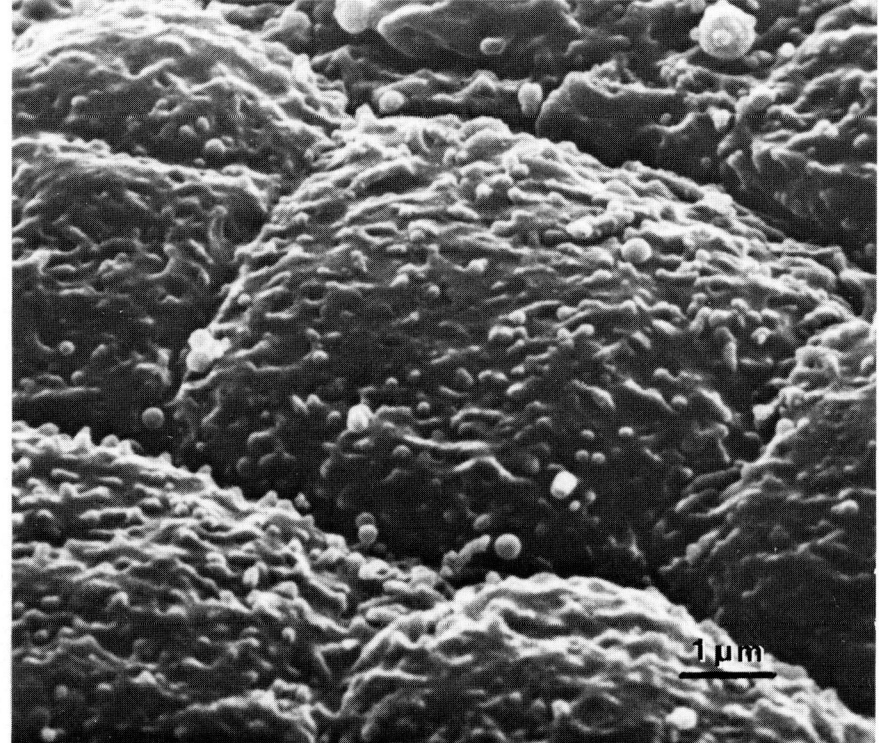

FIG. 6. The polygonal follicle-associated epithelial cells exhibit a well-delineated border and few evenly spaced microvilli which are indistinct at hatching. ×8500.

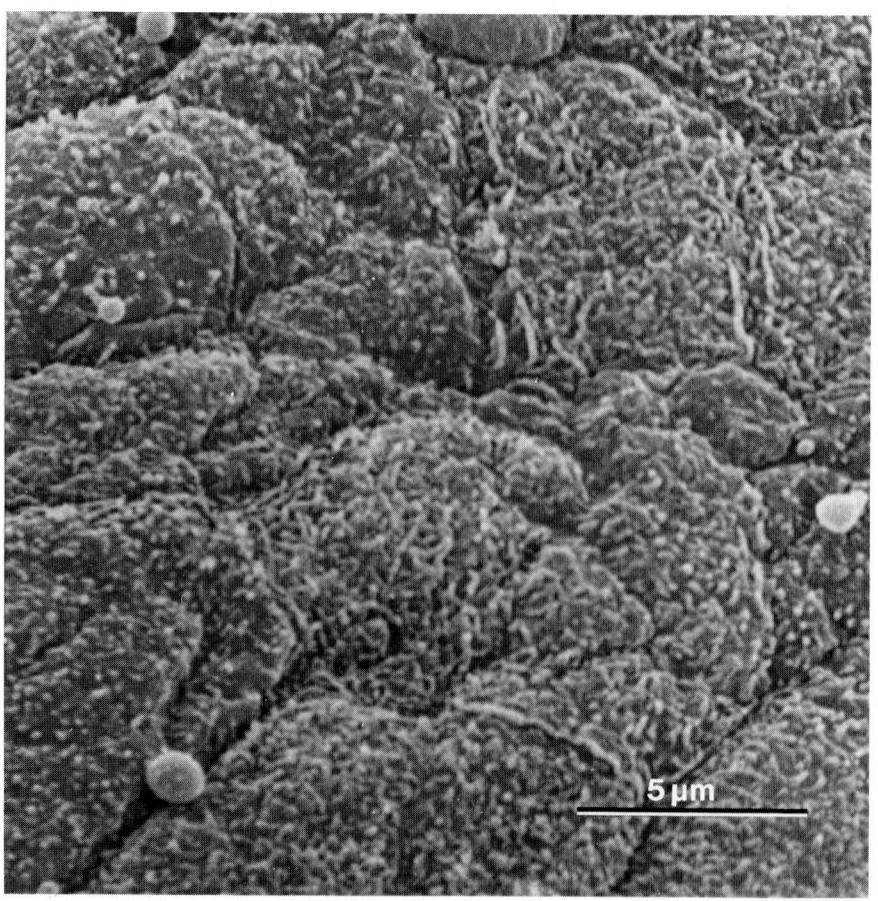

Fig. 7. The microvilli of the polygonal follicle-associated epithelium are more numerous and slender within the first week after hatching. ×3500.

within this geometric area was 0.6 mm^2; therefore, the relative amounts of FAE and IFE were 10 and 90, respectively. The bursal surface area was calculated to be 10 cm^2. Because the FAE is capable of sampling the luminal environment and is in direct contact with medullary cells, one might conclude that approximately 1 cm^2 of the bursal surface area (the FAE) is immunologically oriented.

The rate of passage of colloidal carbon and tritiated thymidine (^{3}HTdR) from the anal lip to the FAE and bursal medulla has been determined (Schaffner *et al.*, 1974; Sorvari and Sorvari, 1977; Sorvari *et al.*, 1977; Glick,

1977b, 1981). Colloidal carbon applied to the anal lip appeared in the bursal lumen in 2 minutes, FAE in 15 minutes, medulla in 6 hours, and cortex in 72 hours (Sorvari and Sorvari, 1977). The application of ^{3}HTdR (100 μci, 0.1 ml, specific activity 6.7 ci/mmol) to the vent of day-old chicks labeled the medulla within 20 minutes (Glick, 1981). The number and location of labeled cells increased over a 48-hour period (Glick, 1977b, 1981).

On the basis of our published (Holbrook *et al.*, 1974) and unpublished (K. A. Holbrook, W. Perkins, and B. Glick, 1981) data, we suggest that the bursal follicle is divided into two structural units: a medullary lymphocyte–FAE unit and an extramedullary unit. The latter is in contact with the vascular system and connective tissue. The basal lamina of the IFE is continuous with that of the medullary epithelium and separates the medulla from

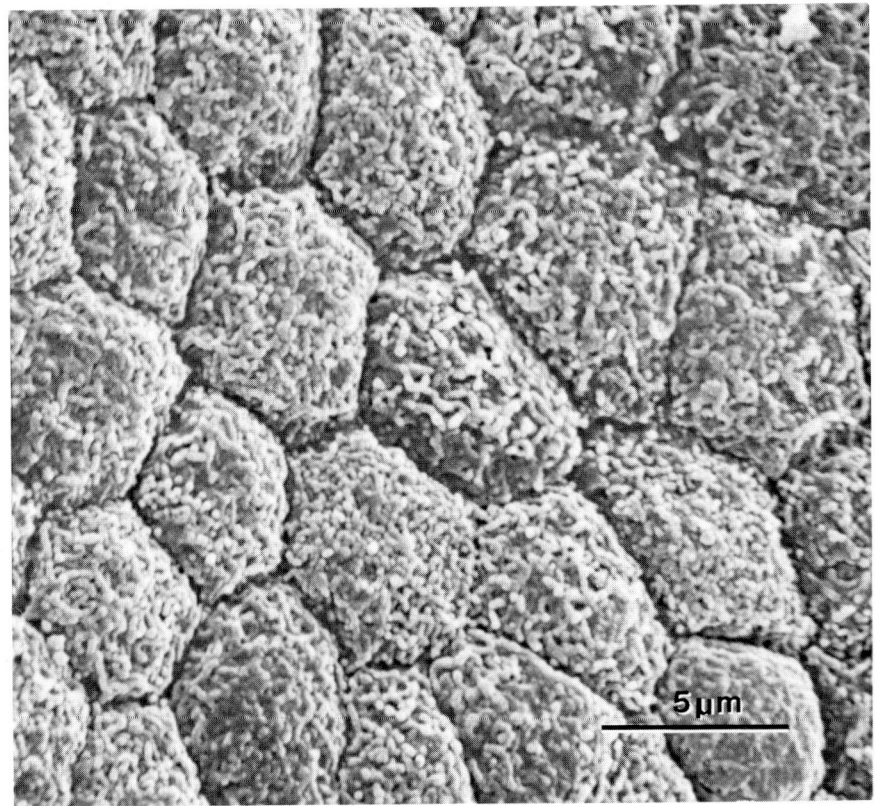

FIG. 8. The microvilli of the follicle-associated epithelium become shorter and thicker with age (7 weeks). ×3400.

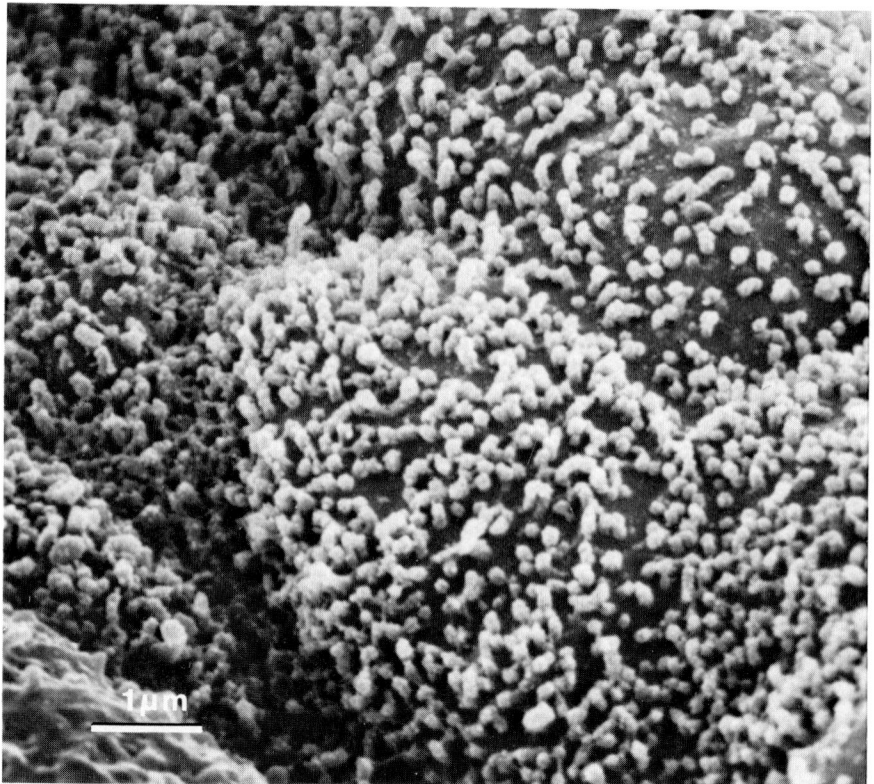

Fig. 9. In contrast to the follicle-associated epithelium, the microvilli of the interfollicular epithelium are distinct from the time of hatching. ×8500.

the cortex. The medulla is also separated from the bursal lumen by FAE. However, whereas the medullary lymphocyte–FAE unit is surrounded by epithelial cells, lymphocytes of this unit are not sequestered because bursal luminal contents may enter the unit through the FAE. Such a migration into the medullary lymphocyte–FAE unit could modify the microenivronment of the unit and be translated into cell activation or inhibition. The observation of lymphocytes on the surface of bursal plicae (Holbrook et al., 1974) might have been in response to the luminal transport of unknown substances by the FAE. The observation that the medullary lymphocyte–FAE unit can be stimulated to produce antibody by anal or intraluminal application of antigen (Van Alten and Meuwissen, 1972; Sorvari et al., 1975; Sorvari and Sorvari, 1977) and that the intraluminal application of horseradish peroxidase will be concentrated in the FAE and released by the FAE into the lumen (Bockman

6. BURSA OF FABRICIUS   453

and Stevens, 1976) demonstrate the influence of the bursal lumen on the medullary lymphocyte–FAE unit. Cloacal applications of plutonium-239 (Schaffner *et al.*, 1976) or intramuscular injections of cyclophosphamide (Cy) to newly hatched chicks (Sachs *et al.*, 1979) prevented the uptake of carbon by the bursal plicae, and these treatments would be expected to prevent the previously observed antibody synthesis. The identification of antibody to

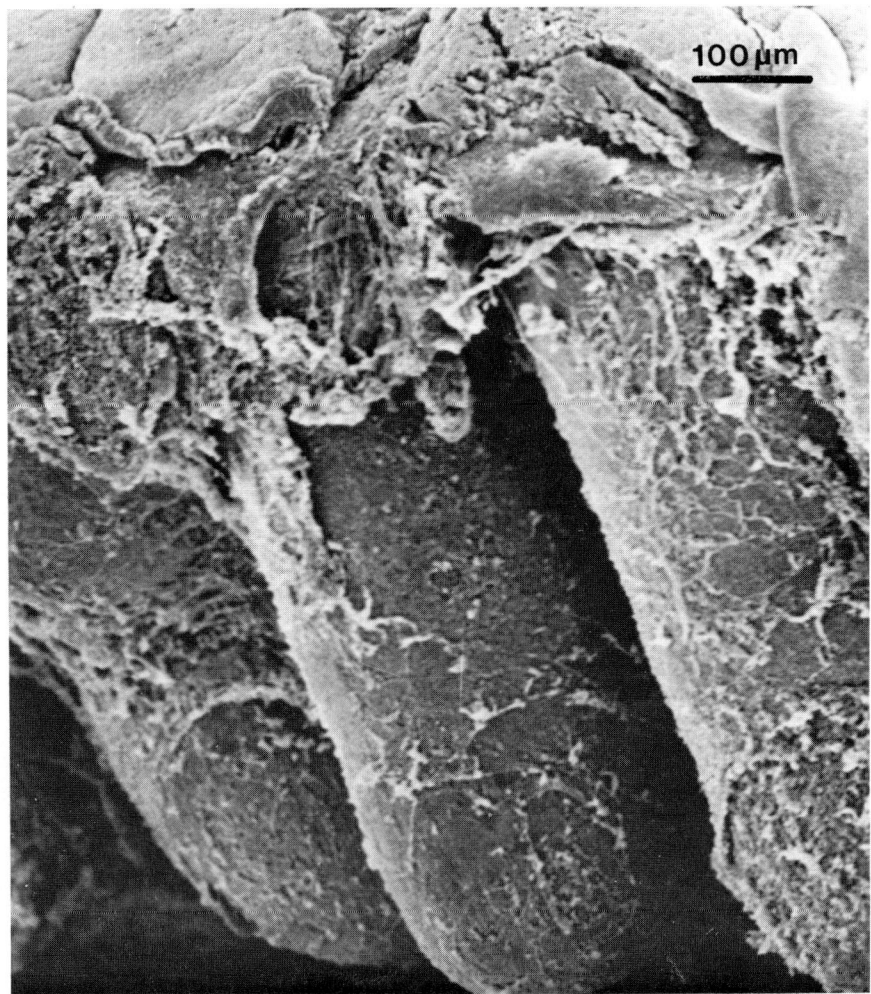

FIG. 10. The elongated, jug-shaped bursal follicles are revealed in this sagittal section from a bursal plica of an 8-week-old bird. ×85.

Brucella abortus in the lumen but not in the medulla subsequent to the application of *B. abortus* to the anal lip (Sorvari *et al.*, 1977) suggests a role for the lymphocytes identified on the plical surface. Our SEM studies revealed these extruded lymphocytes to be smooth (Holbrook *et al.*, 1974). SEM studies of single suspensions of bursal (B) and thymic (T) lymphocytes revealed that more than 90% of the lymphocytes from both tissues were smooth (Stinson and Glick, 1978). These data and others (Nazerian *et al.*, 1976) emphasize the surface similarity between T and B lymphocytes, but they do not rule out surface differences between T and B cells from peripheral lymphoid tissue (Stinson and Glick, 1978). The significance of tubular structures (anastomosing tubules in endoplasmic reticulum) observed in B lymphocytes from 12-day-old chickens (Matos and Sousa, 1978) is unclear.

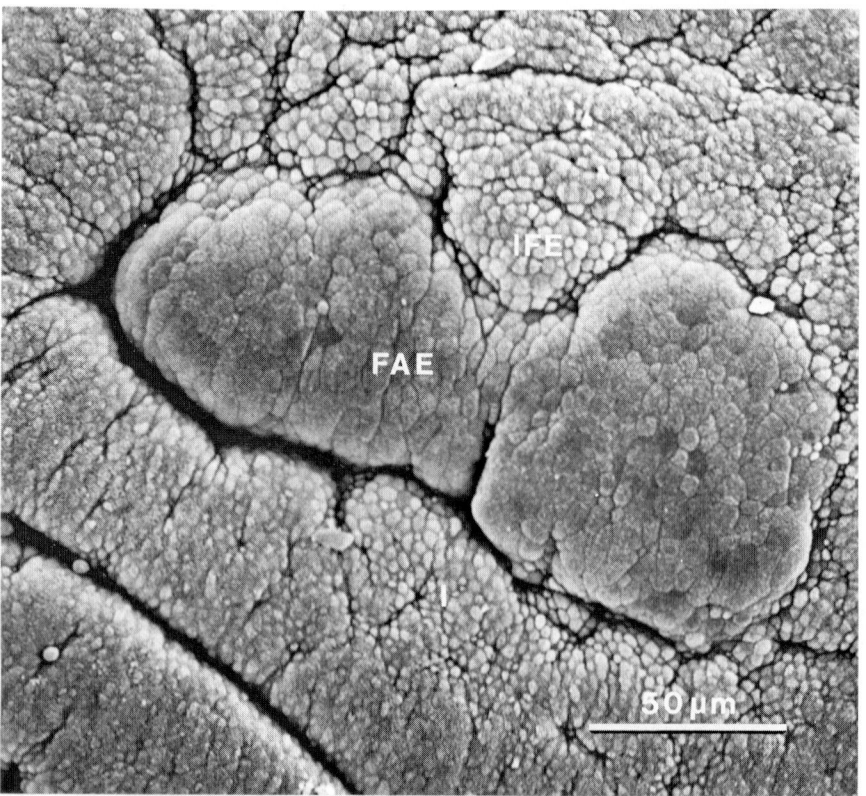

FIG. 11. Button-like follicles appear shortly after hatching but are the predominant follicles in regression as revealed in this preparation from a 12-week-old bird. As the medullary lymphocytes disappear, the follicle-associated epithelium (FAE) sinks into the follicle creating the crater between the FAE and interfollicular epithelium (IFE). ×350.

6. BURSA OF FABRICIUS

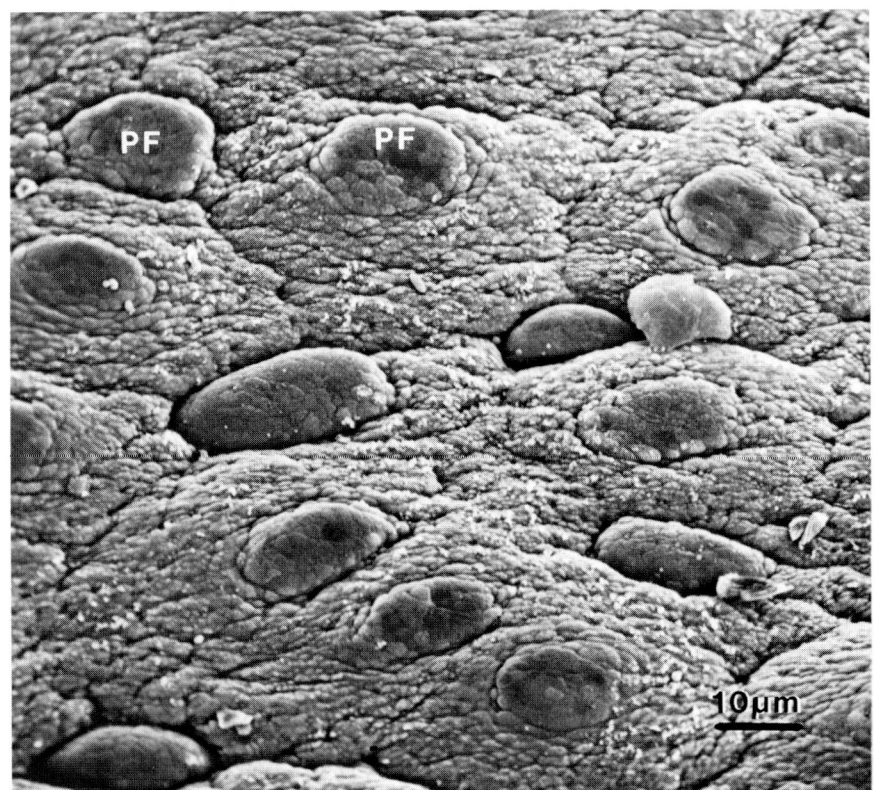

Fig. 12. The projecting follicle (PF), characterized by a raised follicle-associated epithelium predominates during the late embryonic and posthatch periods. ×200.

Are these structures restricted to B cells, or can they be found in T cells that occupy diffuse lymphatic tissue? Because diffuse lymphatic tissue has been identified dorsal to the bursal duct, do cells of this region possess tubular structures (Odend'hal and Breazile, 1979)?

III. Origin of Bursal Lymphocytes

Early explanations for the origin of B lymphocytes were differentiation, within the endodermal environments of the bursa, of primary hemocytoblasts to lymphocytes (Jolly, 1915) and a direct transformation of epithelial cells to medullary lymphocytes. Also, cortical lymphocytes originated, in part, from stellate mesenchymal cells and epithelial-cell migration (Ackerman and Knouff, 1959; Ackerman, 1962). Evidence for a blood-borne-progenitor-cell origin for the B lymphocyte (Jaffe and Fechheimer, 1966; Moore and Owen, 1965, 1966, 1967a) has been reviewed (Glick, 1977a). The elegant experiments of LeDouarin (LeDouarin and Jotereau, 1973a,b, 1975; LeDouarin *et al.*, 1975; Houssain *et al.*, 1976) took advantage of the difference between the interphase nuclei of the Japanese Quail (*Coturnix* sp.) and chicken to confirm and extend the blood-borne thesis. The quail nucleus is distinguished by the presence of a large clump of heterochromatin that is absent from the nucleus of the chicken (LeDouarin, 1973a,b; LeDouarin and Jotereau, 1975). The transplantation of endomesodermal bursal tissue from 5- to 6-day-old quail embryos to the somatopleure of 3-day-old chick embryos for a period of 13–14 days revealed quail chromatin pattern in the epithelial and reticular cells of the bursa and chick or host chromatin pattern in the lymphocytes. When bursal primordia from 7- to 10-day-old quail embryos were transplanted to 3-day-old chick somatopleure, the developing lymphocytes exhibited chimerism in that both quail (donor) and chick (host) chromatin patterns were present. Bursal implants from 11-day-old quail embryos resulted in a lymphocyte population with only quail chromatin, suggesting that the flow of cells to the quail's bursa begins about day 7 and may cease by day 11 of embryonic development. Reciprocal experiments (chick bursal primordium to quail somatopleure) demonstrated basophilic stem-cell migration to the chick's bursa between 8 and 15 days of embryonic development (DE). Evidence is available suggesting a limited number of stem cells colonizing an individual bursal follicle (LeDouarin *et al.*, 1975; Lydyard *et al.*, 1976).

Is the stem cell circulating prior to its appearance in the bursa and thus awaiting acceptance by the bursa, or does the bursa await the genesis and release of the stem cell? LeDouarin *et al.* (1976) demonstrated the former to be true. The origin of the stem cell was reported to be the yolk sac (Moore and Owen, 1967a,b; Metcalf and Moore, 1971). Additional evidence implicates an intraembryonic origin for stem cells (Dieterlen-Lievre *et al.*, 1976; Lassila *et al.*, 1978, 1979, 1980; Martin *et al.*, 1979; Toivanen *et al.*, 1979; Beaupain *et al.*, 1979). Prior to the fourteenth somite stage or before the establishment of circulation, a quail-area pelucida (embryo) replaced that of a chick and thus resulted in a yolk sac (chick) embryo chimera (Martin *et al.*, 1978). The lymphoid cells of the thymus and bursa possessed the nuclear characteristics of the quail transplant and, therefore, were of intraembryonic origin. Interestingly, by 13 DE, the yolk sac of some embryos contained all quail erythropoietic cells, suggesting a migration of intra-embryonic cells to the yolk sac. An embryo of 2 days homozygous for immunoglobulin G (IgG) allotype G1^a replaced a similarly aged embryo (area pelucida) of IgG allotype

G1^b (Martin et al., 1979). The serum IgG allotype at 3, 7, and 20 weeks of age matched that of the grafted embryo, G1^a, not that of the host, therefore reconfirming the intraembryonic origin of stem cells. The prebursal stem cell has been identified within the mesenchymal-cell population of the embryo (Lassila et al., 1979). These investigators arrested lymphoid development of the bursa by treating the embryo with Cy at 15, 16, and 17 DE and then intravenously injected on day 18 with 10^7 intraembryonic mesenchymal cells obtained from 7-day-old embryos. The intraembryonic mesenchymal-cell transplant was successful in restoring bursal and splenic histogenesis and antibody response to human gamma globulin and *Brucella abortus*.

B-Cell development in the bursa subsequent to the establishment of prebursal cells occurs in three maturation stages (A. Toivanen et al., 1972; Toivanen and Toivanen, 1973; P. Toivanen et al., 1972a,b,c, 1974a,b,c). Bursal stem cells appear late in embryonic development and the first few days after hatching. When B cells in this initial stage are injected into normal or Cy-treated birds, they home to the bursa and are capable of restoring its lymphoid compartment. They require the cytoarchitecture of the bursal reticulum for maturation. Early postbursal cells, representing a second maturational stage, may be harvested from the bursa of 3- to 6-week-old birds. They require no further differentiation in the bursa and are fully competent in restoring antibody potential to immunosuppressed chickens. These cells may home to the bursa. A third stage included postbursal cells that appear in the bursa after 7 weeks. They are unable to restore bursal structure or germinal centers in immunosuppressed birds, but they restore antibody competence in these chickens. It is interesting that the largest number of contact sites between lymphocytes and lymphocytes and marchophages in the cortex of bursal follicles occurs at 5 weeks, and the fewest prior to 3 weeks and between 5 and 8 weeks (Holbrook et al., 1977). The functional activity of the B lymphocytes may relate to these contact sites. The immunocompetence of B cells from immunologically mature birds is further seen by their ability to enhance the anti-mouse-red-blood-cell response of newly hatched chicks (Seto, 1976). The differentiation from bursal stem cell to postbursal stem cell has been experimentally identified (Vainio and Toivanen, 1979). Allogeneic and syngeneic bursal stem cells (from 4-day-old chicks) were transplanted to 4-day-old Cy-treated chicks. The B lymphocytes are absent in Cy-treated birds (See Section VI,F). Then, after 6 weeks cells were harvested and called syngeneically educated or allogeneically educated bursal cells. They were then transplanted to a second group of 4-day-old Cy-treated birds. Because the transplanted syngeneically and allogeneically educated bursal cells failed to repopulate the Cy-treated birds's bursa but did reconstitute their antibody potential, one may conclude that they were no longer bursal stem cells but had matured in the first host to postbursal cells. These data also reveal that maturation of B cells will take place in an allogeneic environment.

In vitro experiments with 10-, 11-, or 12-day-old embryonic bursal explants emphasized that epithelial thickening precedes the appearance of basophilic follicles and lymphoid follicles (Ritter and Lebacq, 1977).

A large body of data supports the conclusion that multipotential stem cells populate the thymus and bursa where they develop into T and B cells, respectively (Moore and Owen, 1966; LeDouarin *et al.*, 1977). In a series of well-designed experiments, Weber and Foglia (1979) showed the existence of pre-T and pre-B cells, which appear destined for the thymus and bursa, respectively (Weber and Mausner, 1977; Weber and Alexander, 1978). Chromosomally marked bone-marrow cells transferred from normal or testosterone propionate-injected (TPI) individuals (14 DE) to irradiated recipients of the same age proliferated in the hosts' thymus and bursa and contributed to the T- and B-cell pool. On the other hand, chromosomally marked bone-marrow cells transferred from 2-day-old normal or TPI birds proliferated in the host thymus and contributed to the T-cell pool but did not proliferate in the bursa or contribute to the B-cell pool. These data suggest a separate lineage for PT and PB cells, restriction of PB cells to embryos, and the presence of PT cells embryonically and posthatch.

IV. Bursal Microenvironment

Because the microenvironment of the bursa should reflect its cellular composition, a study of B-cell types and their products would be instructive in understanding the bursal microenvironment. Also, an endocrine role might be considered for the bursa if bursal products could be implicated in the regulation of other biological units (Glick, 1980). A hormonal substance, bursapoietin, has been tentatively identified in the bursa (Brand *et al.*, 1976; Goldstein *et al.*, 1977). Bursapoietin will induce specific bone-marrow cells to differentiate into bursacytes. The endocrine status of the bursa and thymus has been reviewed by Glick (1980).

We have identified a secretory cell in the bursal medulla that, like the thymic macrophage of Beller and Unanue (1977, 1978), is concentrated at the corticomedullary border (Olah and Glick, 1978b; Olah *et al.*, 1979). The secretory cell became evident following Cy injection, which eliminated the small lymphocyte (Fig. 13). The application of colchicine to the anal lip produced necrosis of B lymphocytes and FAE, but cells associated with the corticomedullary border (secretory cells) were less affected (Romppanen and Sorvari, 1980). Identifying a role for the secretory cell in the bursa's microenvironment has been a major goal of our laboratory.

6. BURSA OF FABRICIUS

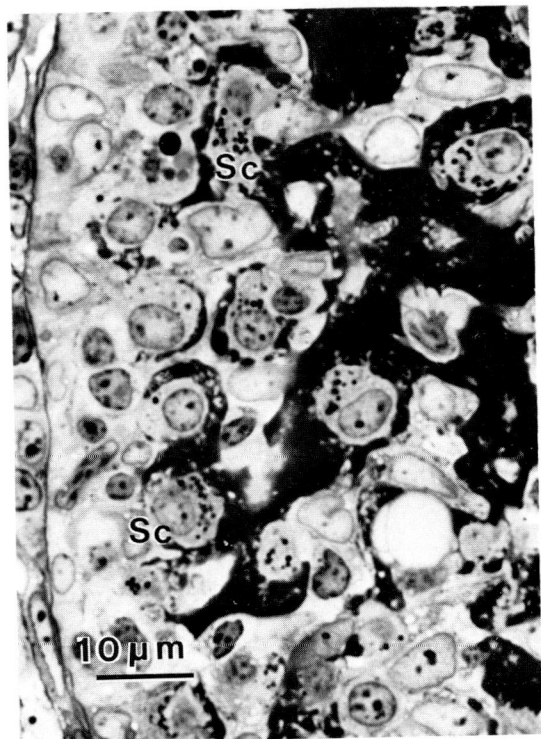

Fig. 13. Secretory cells (Sc), located at the corticomedullary border of follicles from untreated birds, are amplified following the injection of cyclophosphamide. [From Olah and Glick (1979). *Dev. Comp. Immunol.* **3**. Copyright 1979, Pergamon Press, Ltd.]

Possibly the first lymphokine identified in birds was a mitogenic factor that appeared to be controlled by both the bursa and thymus (Oates *et al.*, 1972). A substance inhibiting the mitogenic response of spleen cells to phytohemagglutinin was extracted from the bursa (Danielson and Van Alten, 1974; Van Alten *et al.*, 1976). Our laboratory has identified lymphokine elaboration by B and T cells. T and B cells will migrate from capillary tubes within 4 and 18 hours, respectively (Subba Rao and Glick, 1977). B cells from chickens sensitized to purified protein derivative (PPD) will not migrate in the presence of PPD (Subba Rao and Glick, 1977). Supernatants obtained by incubating sensitized B cells in the presence of PPD will inhibit the migration of naive B cells. This substance has been named lymphocyte inhibitory factor (LyIF). The elaboration of LyIF by B cells (and T cells) is age dependent. Birds were sensitized weekly to PPD, and 3 weeks later the ability of B cells to migrate in the presence of PPD was assessed. B cells elaborated peak levels of LyIF between 5 and 10 weeks of age; thereafter,

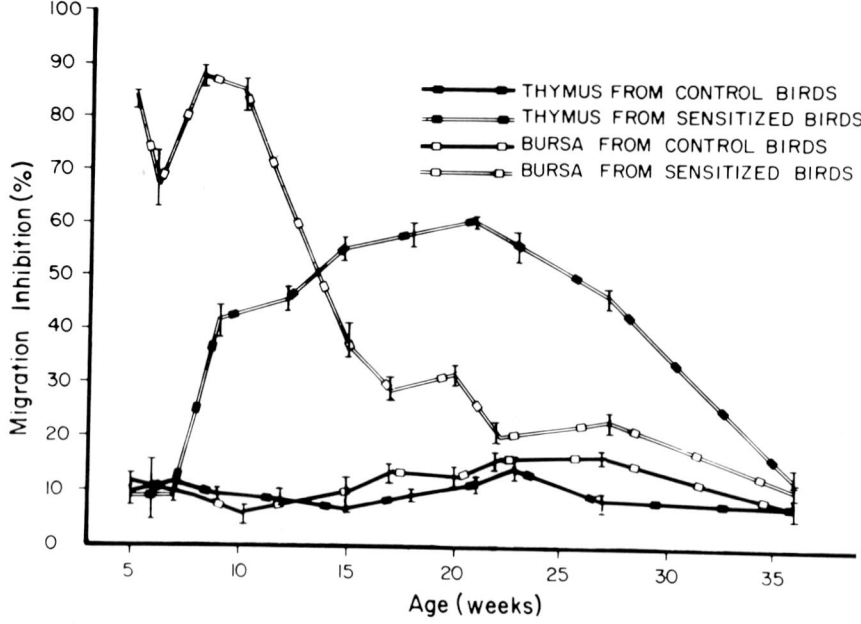

FIG. 14. The release of lymphocyte inhibitory factor by bursal and thymus cells. Percentage migration inhibition *in vitro* of bursal and thymic cells from normal and sensitized birds in the presence of sensitizing agent, purified protein derivative (PPD). Inhibition (%) = area of migration with PPD/area of migration without PPD × 100. Each mean (point) represents eight observations and is accompanied by SEM. (From Subba Rao and Glick, 1977.)

the synthesis or release of LyIF declined and approached control values by 22 weeks of age (Fig. 14). The avian thrombocyte, like the T cell, migrates rapidly out of a capillary tube (Stinson *et al.*, 1979). Thrombocyte migration was inhibited by sensitized T and B cells in the presence of the sensitizing agent as well as by activated supernatants from these cells. The treatment of sensitized B lymphocytes with anti-T serum plus complement did not reduce the B cells's ability to produce a thrombocyte inhibitory factor (ThrIF). The production of LyIF, ThrIF, and bursapoietin by the bursa suggests an advanced level of cell maturity within the microenvironment of the bursa.

V. Bursal Kinetics

A. TESTOSTERONE CONTROL OF GROWTH

Functional investigation of the bursa was aided by studies that characterized its growth rate (Glick, 1955a) and indeed made possible the early observation of the bursa's control of antibody potential (Glick *et al.*, 1956). Our observation that the bursa grows most rapidly between hatching and 3 weeks of age (Glick, 1956, 1960a; Landreth and Glick, 1973) has been confirmed (Wolfe *et al.*, 1962; Freeman *et al.*, 1966; Dieter and Breitenbach, 1968; Vriend *et al.*, 1975). Following the period in which the bursa grows faster than body weight, the bursa plateaus in its growth. The plateau period is followed by regression that precedes sexual maturity and occurs as early as 2 months of age (Glick, 1956, 1960a; Wolfe *et al.*, 1962; Dieter and Breitenbach, 1968; Hoffman-Fezer and Lade, 1972; Landreth and Glick, 1973; Vriend *et al.*, 1975).

The endocrine system has been shown to play an important role in the kinetics of bursal growth (Glick, 1980). Following the observation that caponization would enhance bursal size (Jolly and Pezard, 1928; Glick, 1957a; Wolfe *et al.*, 1962), the administration of androgens was shown to initiate bursal regression in the chick (Kirkpatrick and Andrews, 1944; Glick 1957a, 1970; Zarrow *et al.*, 1961; Schomberg *et al.*, 1964; Dieter and Breitenbach, 1970, 1971) and embryo (Meyer *et al.*, 1959; Warner and Burnet, 1961; Aspinall *et al.*, 1961; Glick and Sadler, 1961; Glick, 1961; Rao *et al.*, 1962; Warner *et al.*, 1962; Glick *et al.*, 1977b). Bursal regeneration occurred approximately 5 weeks after the last testosterone propionate (TP) injection of neonates (Glick, 1970b). However, the bursas of embryos exposed to testosterone do not regenerate (Glick *et al.*, 1977a). Although high levels of testosterone Propionate inhibit development of the embryonic bursa, there is some evidence that low doses (20–200 μg) may stimulate bursal growth (Norton and Wira, 1977). The size of bursal lymphocytes declines with increasing embryonic age (Peterson and Good, 1965; Sherman and Auerbach, 1966; Kulkarni *et al.*, 1971), suggesting that the decrease in bursal size is, in part, a consequence of rapid cell division.

B. STEROID RECEPTORS

The exquisite sensitivity of the bursa to testosterone is apparent when one recognizes the resistance of the embryonic thymus to testosterone treatment (Glick, 1961; Warner *et al.*, 1962). These observations suggested either a receptor difference between B and T lymphocytes or a failure of testosterone to enter the thymic milieu. We utilized scintillation and autoradiography to show that B lymphocytes bind more tritiated testosterone than do T lymphocytes (Glick and Schwarz, 1975). Tritiated testosterone has been localized in the nuclei of mesenchymal B cells (Gasc *et al.*, 1979) and, along with other steroids, in the cytoplasm of B cells (Sullivan and Wira, 1978). Cytosol androgen receptors were identified in endodermal and mesenchymal components of the embryonic bursa, with the greatest concentration of receptors identified in the endodermal tissue (LeDouarin et al., 1980). Bursae were trypsinized and dissociated into endodermal (epithelium) and mesenchymal components. Chimeric combinations in embryos of endodermal and mesenchymal components lead to normal bursal development. Testosterone treatment of endoderm but not mesenchymal tissue prevented normal bursal histogenesis, whereas the reverse led to normal ontogeny. Therefore, the absence of *in ovo* bursal growth in the presence of testosterone may be explained by testosterone's disruption of endodermal components.

C. ADRENAL CONTROL OF GROWTH

An enlarged adrenal gland and an involuted bursa have been observed to follow the stress of muscular fatigue (Garren and Shaffner, 1954a,b), typhoid infection (Garren and Barber, 1955), cold (Garren and Shaffner, 1956), and restraint of chicks (Newcomer and Connally, 1960). Because an earlier paper had demonstrated that adrenal extracts affect bursal development (Selye, 1943), it was logical to conclude that the adrenal gland exerted a direct control of bursal growth. Further research did identify a negative correlation between the adrenal gland and the bursa (Glick, 1960a) and demonstrated bursal regression in the presence of cortisone and corticosterone (Glick, 1957a,b, 1959, 1960b, 1967, 1972; Zarrow et al., 1961; Bellamy and Leonard, 1964; Dieter and Breitenbach, 1970, 1971; Sato and Glick, 1970) and ACTH (Siegel, 1961; Breitenbach, 1962; Sato and Glick, 1964a). Using a method of specific binding of [^{3}H]corticosterone more receptors were located in B lymphocytes (1200/cell) than in T cells (600/cell; Schaumburg and Crone, 1971). Clarification of the physiological control of bursal growth by the adrenal gland awaits successful bilateral adrenalectomy of perinatal chicks.

Selection experiments leading to changes in bursal size have been successful (Jaap, 1958, 1960; Temple and Jaap, 1961; Glick and Dreesen, 1976; Muir and Jaap, 1969). Heritability estimates for bursal weight during the rapid period of growth have been calculated (Glick, 1955a,b, Muir and Jaap, 1969), with full sib analyses ranging between 0.61 and 0.78 for heavy and leghorn types, respectively (Muir and Jaap, 1969). The subtle influence of adrenal control of bursal growth was observed in selection experiments leading to the production of large-bursa (LB) and small-bursa (SB) lines (Glick and Dreesen, 1967). The birds of the SB line possessed adrenal glands that were significantly larger than the LB-line adrenal glands. Growing SB-line B cells in tissue culture in the absence of serum or in the presence of LB-line serum enhanced their replication, whereas the replication of LB-line B cells was suppressed in the presence of SB-line serum (Kulkarni et al., 1971). These data suggested that the selection had affected an extrinsic change, that is, increased adrenal activity in the SB line. On the other hand, an intrinsic bursal change was realized, because the replication of LB-line bursae exceeded that of the SB-line bursae in the absence of serum (Kulkarni et al., 1971).

D. DISEASE, NUTRIENT, AND PESTICIDE INFLUENCES ON BURSAL GROWTH

Various diseases, nutrients, and pesticides have been reported to affect bursal size. *Eimeria tenella* infection may either decrease (Challey, 1962; Rouse, 1967; Visco, 1973) or increase (Pierce and Long, 1965) the size of the bursa. Hypertrophy of the bursa was observed in birds recovering from *E. acervalina* and *E. necatrix* (Panda and Combs, 1964). Age and breed differences may account for these variable results, because bursal growth is dependent on breed (Glick, 1956). Infectious bursal disease (IBD) is a viral illness that produces acute pathognomic changes in the bursa (Cosgrove, 1962; Hitchner, 1970). Within 36 hours after IBD infection, the bursal medulla exhibits lymphocyte degeneration and necrosis. By 48 hours, almost complete loss of lymphocytes occurs in the medulla, accompanied by blast-cell mitosis, macrophagocytosis, and plasmacytosis (Cheville, 1967). The increased weight of the bursa at this time is due to edema, hyperemia, and the accumulation of heterophils. Perinatal chicks infected with IBD virus exhibit immunosuppression (Faragher et al., 1972; Allan et al., 1972; Giambrone et al., 1977), which appears to result from a direct action of the IBD virus on B cells (Hirai et al., 1979; Sivanandan and Maheswaran, 1980).

Larger bursae occurred in chicks fed penicillin (Glick, 1957b, 1959), but the growth rates of the bursa and thymus of chicks fed zinc bacitracin or oxytetracycline were not consistently affected over an 8-week period (Franti et al., 1971). Pyridoxine-deficient diets during the first 2 weeks of life significantly reduce bursal and thymic size (Asmar et al., 1968). These results might be translated into a depressed immune response, because pyridoxine deficiencies in mammals impair cell-mediated immunity (Willis and Schwartz, 1975). Amino acid deficiencies modify bursal development and humoral immunity in the chicken (Glick et al., 1981). Feeding polychlorinated biphenyls to hens reduced bursal weight in the offspring (Harris et al., 1976). On the other hand, mirex or DDT fed at high levels (200–800 ppm) failed to reduce bursal or thymic weights (Glick, 1974). However, birds receiving 400 ppm but not 200 ppm of DDT and starved for 7 days experienced a significant reduction in precipitin response.

E. Lymphocyte Life Span

The weight of the bursa is a reflection of its cell population, the interaction of cells and humoral substances, and the exogenous influence of other biological units, such as adrenal glands. The latter influences have already been discussed. The cell population contributing to bursal weight has an *in situ* and blood-borne origin. The blood-borne origins of lymphocytes have been described. The dynamics of the B lymphocyte were neglected until 1976–1977, when we identified short-lived and long-lived (slowly turning over) lymphocytes within the bursa (Glick, 1976, 1977b). Application of ^{3}HTdR to the air cell of incubating eggs is an effective way to label the majority of small lymphocytes (SL) in the lymphomyeloid compartments of the chicken (Glick and Schwarz, 1975; Glick, 1976). Tritiated thymidine (0.1 ml, 100 μci, specific activity 6.7 ci/mmol) applied at 18 DE labeled 100% of bursal SL within 2–3 days (Fig. 15). All labeled SL disappeared between 9 and 15 days after ^{3}HTdR application. These data emphasized the rapid rate of lymphocyte turnover within the bursa (Kulkarni *et al.*, 1971, 1972). Although multiple application of ^{3}HTdR at 14, 16, and 18 days of incubation reemphasized the presence of short-lived lymphocytes in the bursa, it also identified a small

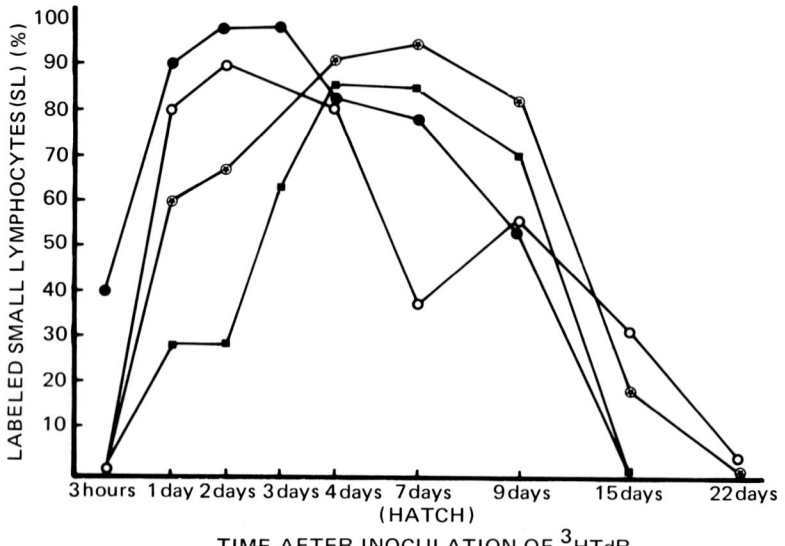

Fig. 15. Tritiated thymidine applied to the air cell at 18 days of embryonic development labeled 100% of bursal small lymphocytes (SL) within 2–3 days. ●, Bursa; ○, bone marrow; ◉, spleen; ■, thymus. [From Glick (1976), by permission of the publisher from Lymphocyte lifespan in chickens. *In* "Phylogeny of Thymus and Bone Marrow—Bursa Cells" (R. K. Wright and E. L. Cooper, eds.). Copyright 1976 by Elsevier/North-Holland, Inc.]

percentage of long-lived lymphocytes. The spleen and bone marrow contained the longest-lived population of SL. The origin of the bone marrow's short- and long-lived lymphocytes is not known. Elimination of the embryonic bursa delayed the appearance of normal numbers of bone marrow SL in posthatch chickens (Glick and Rosse, 1981). Bursectomized–splenectomized birds possessed an increased number of bone-marrow SL which included Ig^+ cells (B. Glick, unpublished data). The migration of long-lived lymphocytes from the spleen or bone marrow to the bursa and thymus seems doubtful because the bursa and thymus lacked labeled cells, although they were present in the spleen, bone marrow, and peripheral blood. These data agree with previous evidence that precursor cells from bone marrow do not migrate to the bursa (Weber, 1975).

VI. The Bursa's Influence on Other Organs

A. ADRENAL GLAND

Observations of the inverse relationship between the bursa and testes (Glick, 1955a) led to the identification of the testes' suppressive role of the bursa, but they failed to identify a bursal influence on the testes. Similarly, the inverse relationship between the bursa and adrenal growth (Glick, 1960a) has been experimentally explained by the antagonistic action of corticosteroids on bursal growth. Bursal control of adrenal function was first implicated by a series of experiments originating in the laboratories of Professor Perek (Perek *et al.*, 1959; Perek and Eckstein, 1959; Perek and Eilat, 1960). These investigators observed that administration of ACTH did not deplete adrenal ascorbic (AA) acid of neonatal chickens, but it did so in adults. Absence of a bursa in the adult suggested a bursal-directed influence of AA acid and led to the demonstration that AA acid of neonatally bursectomized chicks but not sham-operated chicks would be depleted by ACTH injections (Fig. 16). The sensitization of neonatal adrenals to ACTH in the absence of the bursa suggests that the bursa produces a humoral factor that functions as an antireceptor interfering with (1) the binding of ACTH to adrenal cells and thus (2) the depletion of AA acid. Accompanying receptor binding by a *bursal humoral factor* would be the modulation of biosynthetic pathways, which might be translated into lower levels of corticosteroid synthesis in neonatal birds relative to adults. The increased levels of corticoids in neonatal chickens treated with testosterone suggest an inhibitory action of the bursa on hormonal production by the adrenal gland (Breitenbach, 1962). Preliminary data from our laboratory (Taylor and Glick, 1980) and a personal

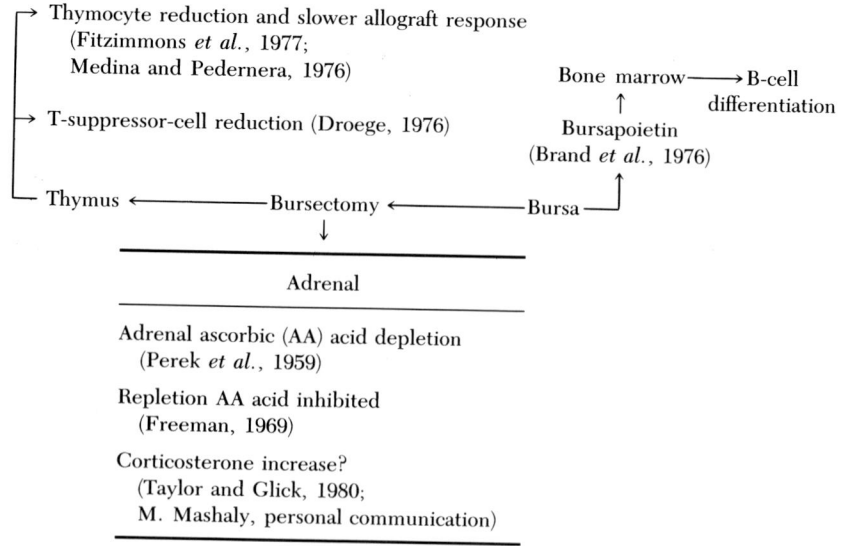

Fig. 16. A scheme depicting the known or putative influence of the bursa and bursectomy on the adrenal thymus and bone marrow. (Modified from Glick, 1980.)

communication from Dr. Magdi Mashaly revealed a higher level of corticosterone in 10-week-old bursectomized birds, (Fig. 16). However, *in ovo* bursectomy at 68 hours reduced corticosterone secretion, but it increased adrenal gland and testes weights and the percentage of Leydig cells in 17-day-old embryos (Pedernera *et al.*, 1980). Bursal tissue grafted to the chorioallantoic membrane of 9.5-day-old embryos restored these variables to normal. Experiments by Freeman (1967, 1969, 1970) stressed the positive control of AA acid by the bursa. Freeman's data demonstrated that AA-acid depletion initially exceeds repletion in the presence of a small stimulus, such as handling, whereas with a stronger stimulus, handling plus ACTH, repletion is equal to or greater than depletion. Repletion fails to occur in bursectomized birds. Therefore, Freeman's thesis is that AA acid repletion is under bursal control (Fig. 16).

B. THYMUS

The bursa is not restricted in its interaction to the endocrines. Additional experiments have implicated a putative bursal influence on thymic function. Posthatching bursectomy interfered with the T-suppressor-cell population of the thymus (Droege, 1976). B cells or humoral factors may be involved in the latter reaction. Surgical bursectomy of the 70-hour-old embryo reduces the allograft response (Medina and Pedernera, 1976; Fitzsimmons et al., 1977), which is dependent on the normal differentiation of the thymus (Aspinall et al., 1963; Cooper et al., 1966). In ovo bursectomy before 72 hours or hormonal bursectomy between 2 and 5 days of incubation significantly reduces the thymocyte count (Fitzsimmons et al., 1977). Because hormonal bursectomy after 5 days of incubation did not influence the thymocyte count, one may conclude that a humoral factor, rather than B lymphocytes, was involved in the bursal control of thymic function. A putative bursal humoral factor, bursapoietin, has been identified (Brand et al., 1976; Goldstein et al., 1977). Mature B cells were generated from bone-marrow cells in the presence of bursapoietin. Bursal, endocrine, and thymic relationships are diagrammatically represented in Fig. 16.

VII. Regulation of Immunoglobulin Synthesis

A. THE EARLY PERIOD

The "enigmatic" bursa would have been an appropriate way of referring to the bursa prior to 1950. My studies with the bursa of Fabricius began in 1952. The first step was to characterize its growth pattern, which led to our original observations on bursal function (Glick, 1955a; Glick et al., 1956). Having found that the bursa's rapid growth period occurs during the first 3 weeks after hatching, I bursectomized birds only during this period rather than later, as in previous investigations (Woodward, 1931; Taibel, 1938). Several of these bursectomized birds and shams were randomly obtained in 1954 by T. S. Chang for the purpose of demonstrating to a class the production of antibody to Salmonella antigen. Uncharacteristically, the majority of the birds injected failed to produce antibody, and these birds proved to be ones that had been bursectomized (Glick, 1955a). This observation was followed by two experiments that adequately documented the original chance observations. At 7 weeks of age, 7 of 10 birds bursectomized at 12 days of age and 2 of 10 controls were unable to produce antibody to S. typhimurium O antigen (Glick, 1955a). In the second experiment the suppressive influence of bursectomy on antibody synthesis was shown to be independent of breed or sex (Chang et al., 1955; Glick et al., 1956). Although the significance of these experiments was not immediately apparent to us, they were responsible for the end of the bursa's "enigmatic" period.

B. Surgical Bursectomy and Antibody Synthesis

The early bursectomy experiments showed that the bursa was involved in antibody potential and suggested that its control of immunity was related to its growth pattern. The latter became apparent when bursectomy in 2-week-old chicks led to a greater suppression of antibody synthesis in response to sheep red blood cells (SRBCs) than bursectomy at ages during the plateau period (Chang et al., 1957). These observations were confirmed with the soluble antigen bovine serum albumin (BSA; Mueller et al., 1960; Graetzer et al., 1963). Chicks bursectomized at hatching failed to make antibody in response to a primary injection of human O red cells, but they did respond to secondary and tertiary injections (Isakovic et al., 1963; Jankovic and Isakovic, 1966a,b). The appearance of natural agglutinins to rabbit, rat, and human erythrocytes was delayed in birds bursectomized at hatching (Graetzer et al., 1963; Jankovic and Isakovic, 1967). Numerous investigators reported that bursectomy at hatching did not eliminate a primary response to a variety of antigens (Kemenes and Pethes, 1963; Eychmans et al., 1968; Rose and Orlans, 1968; Maranova and Hasek, 1969; Cain et al., 1969; Alm, 1970; Sato and Glick, 1972). When the class of antibody was measured by these investigators, it was found to be immunoglobulin M (IgM). Failure of the posthatching bursectomies to eliminate antibody synthesis in all birds was responsible for (1) our original suggestion that other sites might be involved in immunoglobulin synthesis (Glick, 1968a) and (2) the investigations of Meyer et al., (1959), who helped to define a functional embryonic period for the bursa.

C. Hormonal Bursectomy and Antibody Synthesis

Utilizing the known regressive influence of androgens on the bursa of growing birds (Kirkpatrick and Andrews, 1944; Glick, 1957a), Meyer et al. (1959) demonstrated that the injection of 19-nortestosterone (0.63 mg) into an egg on the fifth day of incubation would arrest bursal development. Treatment with 19-nortestosterone on day 5 eliminated the chicken's ability to produce precipitin in response to BSA, whereas similar hormonal treatment on day 13 or 14 allowed a minimal response to BSA (Mueller et al., 1960, 1962). The percentage of antibody responders to human gamma globulin (HGG) in birds hormonally bursectomized by an injection of TP at 12 DE was 16, 48, and 60, following the primary, secondary, and tertiary injections, respectively (Warner et al. 1969). Failure to produce antibody does not interfere with the birds' ability to become tolerant, and appears to enhance such tolerance (Grebenau and Thorbecke, 1978).

A novel method for introducing steroids into an incubating egg was patented by Seltzer (1956). The egg was dipped into a solution containing a steroid and either ethanol or a surface-active agent. This method was purported to cause a sex reversal. In an attempt to verify this claim, we immersed fertile eggs to a depth of 2.5 cm (pointed end) for 5 second into an ethanol solution containing various concentrations of TP or diethylstilbestrol. Although we were unable to confirm the sex-reversal claims, we did observe the elimination of the bursa and a reduced antibody potential in chicks hatched from eggs dipped in 100 ml of ethanol containing 2 g of TP (TPD) (Glick and Sadler, 1961; Glick, 1961; Wilson and Glick, 1966). Later experiments revealed that the antibody response of chicks hatched from eggs dipped in various concentrations of TP (80, 160, 2500 mg/100 ml) reflected the presence of bursal follicles (May and Glick, 1964; Sato and Glick, 1964b). The precipitin response of 8-week-old chicks to BSA was 13.1, 95.4, and 207.5 µg of antibody nitrogen (AbN) per ml serum for TPD birds lacking bursal follicles, TPD birds with some atrophic follicles, and controls with follicles, respectively (Sato and Glick, 1964b). Also, weight reduction of the bursa by steroids influenced the antibody response (May and Glick, 1964; Claflin et al., 1966). Claflin et al. (1966) demonstrated that chicks hormonally bursectomized with 0.04 mg of 19-nortestosterone on 5 DE and possessing a bursa weighing less than 200 mg at the time of analysis were significantly compromised in their ability to produce antibody to *Brucella abortus*. The antibody was primarily IgM with IgG antibody increasing slightly after the secondary injection. Testosterone propionate administered on day 8 of incubation eliminated the bursa in 19-day-old embryos (Warner and Burnet, 1961). When TP was administered into the allantoic cavity (2.5 mg) of 12-day-old embryos, it reduced the thymic cortex and medulla in 10% of the chicks (Szenberg and Warner, 1962) and prevented the synthesis of antibody to BSA, HGG, *Brucella abortus*, and an endotoxin from *Salmonella adelaide* (Warner et al., 1962). These investigators formulated the concept of the dissociation of immunological responsiveness (Glick, 1977a) after observing a delayed homograft response in only those chickens whose thymic cortex had been thinned by the TP treatment (Szenberg and Warner, 1962).

1. Thymic Cell Contribution to Humoral Immunity

The dissociation of immunological responsiveness attributed the control of antibody-mediated immunity to the bursa and cell-mediated immunity (transplantation immunity) to the thymus. Although this was a useful concept it soon became apparent that this was too parochial a view. Thymectomized or thymectomized–irradiated birds experienced a reduced antibody potential (Graetzer et al., 1963; Cooper et al., 1966; Ivanyi and Salerno,

1971). The dosage of human serum albumin (HSA) influenced the response of thymectomized–irradiated chickens in that 100 mg of HSA improved titers significantly above those of the group injected with 0.5 mg of HSA (Ivanyi and Salerno, 1971). The increased concentration of antigen improved the likelihood of activating the reduced population of T cells. In retrospect, initial evidence for T-helper cells in the chicken was contained in a hapten-conjugate study (Rouse and Warner, 1972). Chickens exhibiting a delayed allograft response were selected from a group thymectomized at hatching and subsequently injected with antithymocyte serum. Antibody titer to the hapten-conjugate DNP–BSA was depressed in these birds, suggesting the absence of the T-helper cells that are known to be necessary for the carrier effect in eliciting the antibody to hapten (Miller and Mitchell, 1968). Further evidence for the specific activation of T cells by antigen was the observation that spleen cells from bursectomized–irradiated agammaglobulinemic chickens immunized with keyhole limpet hemocyanin (KLH) proliferated *in vitro* in the presence of KLH (Jacobs et al., 1972). The presence of T but not B cells in these birds indicated an involvement of T cells in a KLH response. The decisive experiments for the T-helper role in chickens came from the laboratories of Dr. Jeanette Thorbecke. The transfer to irradiated chicks of spleens from either normal birds sensitized to trinitrophenyl (TNP) KLH or from agammaglobulinemic birds (B negative) sensitized to BSA failed to confer a plaque-forming (PF) response against TNP-coated SRBCs when the spleen cells transferred were accompanied by TNP–BSA. However, the combination of cells from spleens of normal chickens previously injected with TNP–KLH (B-cell sensitized) and of agammaglobulinemic birds previously immunized with BSA (T-cell sensitized) promoted the PF-cell response in TNP–BSA-sensitized irradiated chicks (Weinbaum et al., 1973). Synergism was shown in the PF-cell response to SRBCs with thymus and bone marrow and T- and B-cell transfers (McArthur et al., 1972). Failure of the thymus and bone-marrow combinations to enhance PF cells before 14 days of age suggested the dependency of the bone marrow on B-cell export. A third cell or macrophage improved the repeatibility of the results.

2. *Bursal Cell Contributions to Cell-Mediated Immunity*

The graft versus host response (Warner and Szenberg, 1962, 1963; Jaffe, 1965; Cain et al., 1968; Potworowski et al., 1971; Weber, 1973) and allograft rejection (Aspinall et al., 1963; Isakovic et al., 1963; Meyer et al., 1964; Perey et al., 1967) are T-cell mediated. Also, delayed hypersensitivity (DH) is dependent on T cells (Szenberg and Warner, 1963; Jankovic and Isvaneski, 1963; Okuyama 1965a,b; Warner et al., 1971; Morita and Soekawa,

1972). Delayed hypersensitivity was normal in agammaglobulinemic birds (Theis and Thorbecke, 1972). Because B cells but not T cells are absent in agammaglobulinemic birds, DH was attributed to the T cell. Reciprocal experiments, that is, the bursa's influence on cell-mediated immunity, have been published. Oxazalone (33 mg/kg) was applied to the neck of 9-week-old chickens previously bursectomized at hatching (Caruso et al., 1977). The wattles of these chickens on challenge to oxazalone swelled less than those of controls. These results are in contrast to those of Weidanz et al. (1976), who failed to find agammaglobulinemic birds defective in their response to the contact dermatitis elicited by oxazalone. These differences might be attributed to concentration and timing of chemical application; or the surgical bursectomy may have allowed a small population of suppressor B cells to remain, which the chemical bursectomy performed by Weidanz et al. may have eliminated.

Interaction of T and B cells in an allograft response have been suggested. Hatched K-line ducklings were made tolerant to P line by receiving an intraperitoneal injection of P-line serum and a skin graft from the P line (Karakoz et al., 1976). These tolerant ducklings reject the allograft upon receiving immune sera to the P line if both bursa and thymus are intact. Removal of either markedly reduces the ducklings' ability to mount an allograft response. Thymic implants but not bursal implants enhance their return to normal. Allograft survival differed significantly between birds bursectomized *in ovo* at 62 hours and their sham controls (Medina and Pedernera, 1976). However, it was only 40 days postallograft that more *in ovo* bursectomized birds retained their allograft (4 of 14) than comparable controls (1 of 21). These data implicated a cooperation between T and B cells in a cell-mediated response. However, other interpretations may be entertained. Bursectomy of SC (B^2B^2)-line birds interfered with their ability to mount a DH response to HGG and BSA (Palladino et al., 1978). The addition of immunoglobulin or of spleen cells improved the survival and the DH response of the agammaglobulinemic SC-line birds suggesting that stress might interfere with T-cell–macrophage interactions. Strain differences may also explain differing results, because bursectomy of FC ($B^{15}B^{21}$)-line birds did not interfere with the response to HGG or BSA.

3. Phagocytosis

The bursa and thymus of neonatal chickens failed to clear an intravenous injection of India ink, whereas the ink concentrated in the liver, spleen, and bone marrow (Glick et al., 1964). These experiments identified the thrombocyte as a major phagocytic cell in avian blood. The phagocytic ability of heterophils, monocytes, and thrombocytes was not compromised by bursectomy at the time of hatching. We measured the postengulfment phase of phagocytosis by collecting relatively pure populations of thrombocytes from normal chickens and chickens bursectomized at hatching and measured their O_2 uptake in the presence of carbon (Yarbrough et al., 1971). Oxygen uptake of thrombocytes increased in the presence of carbon but was unaffected by the bursectomy. A reduced clearance rate of lipid emulsion has been reported in bursectomized birds (Waltenbaugh et al., 1976). The deficiency in the bursectomized birds was attributed primarily to a missing serum factor and not directly to the phagocytic cell.

D. Ontogeny of Antibody Synthesis

The ontogeny of immune responses to a variety of antigens has been assessed by bursectomy at various ages. The earlier the bursectomy required to suppress antibody synthesis to a particular antigen, the earlier would be the development of immunity to that antigen. Suppression of SRBCs, rabbit red blood cells, Japanese encephalomyelitis virus, influenza virus, *Pasteurella multocida*, and *Escherichia coli* B occurred only following hormonal bursectomy or bursectomy at the time of hatching (Hirota and Bito, 1975). Antigen-binding cells to SRBCs were detected in the bursa on 18 DE, whereas the bursa on 16 DE exhibited binding of KLH and TGAL (Lydyard et al., 1976). Therefore, immunocompetence to these antigens in chickens develops during the incubation period. Suppression of the BSA response occurred after bursal removal at the time of hatching and at 4 but not 7 days of age. Competence to BSA develops by the first week after hatching. In a third group of antigens, *Brucella abortus*, *Staphylococcus*, and ovalbumin, the development of competence is delayed until as late as 2 weeks after hatching, as shown by antibody suppression in birds bursectomized between 1 and 7 days but not in those bursectomized 12 days after hatching. Immunocompetence to Newcastle disease virus (NDV) develops before hatching, because NDV antibody production was not influenced by bursectomy at 12 or 14 days (Cho, 1963; Glick, 1964), but was suppressed by bursectomy at 2 days (Sato, 1966) and eliminated by bursectomy *in ovo* or at the time of hatching (Feace and Poli, 1967; Matsuda and Bito, 1973). The antigen to which the chicken develops the latest competence appears to be *Salmonella pullorum*. Bursectomy as late as at 14 days suppressed the primary antibody response but not the secondary response to *S. pullorum* (Matsuda and Bito, 1973). The IgG antibody response to SRBCs and IgG and IgM antibody to *Bordetella pertussis* was eliminated by bursectomy at the time of hatching (Ivanyi, 1975).

E. BURSECTOMY AND IMMUNOGLOBULIN SYNTHESIS

1. The Bursa as a Site of Antibody Synthesis

The bursa as a site of antibody synthesis was first investigated by Kerstetter et al. (1962). Using immunofluorescence (IF), they showed that the bursa synthesized antibody to bovine immunoglobulin. These positive results were followed by a series of negative data in which the bursa of chickens failed to produce PF cells to SRBCs (Dent and Good, 1965), antibody to soluble antigens (Glick and Watley, 1967; Jankovic and Mitrovic, 1967; Choi and Good, 1973), or agglutinins to SRBCs by either intraperitoneal or intravenous routes (Abramoff and Brien, 1968; Keily and Abramoff, 1969; Lagerlof and Sundelin, 1970). Intrathymic or intravenous injections led to PF cells in the bursa (Jankovic et al., 1972). Van Alten and Meuwissen (1972) demonstrated the production of PF cells in the bursa following an inoculation of SRBCs into the cloaca but not by intramuscular or intrabursal routes. Also, bursal lymphocytes bound *Escherichia coli*, a normal cloacal isolate, but not γ-streptococci, which are not cloacal isolates (Waltenbaugh and Van Alten, 1974). A plaque assay identified a few bursal cells (0.025%) secreting antibody against the flagella of *E. coli*. The application of BSA to the vent in turkeys stimulated anti-BSA synthesis in the bursa and caecal tonsil (Burns, 1980). These data suggest a sentinel role for the bursa (Glick and Olah, 1981). There appears to be some selectivity in the bursal response to antigens presented via intrabursal and cloacal routes since the bursa of 8- to 12-week-old birds did not synthesize antibody to bovine gamma globulin presented by these routes (Choi and Good, 1973). Alternatively, one might suggest that these birds were challenged too late, that is, during bursal regression (Glick 1956) and past the expected peak response (Waltenbaugh and Van Alten, 1974). It is tempting to theorize that the lymphocytes that we observed on the plical surface of the bursa by SEM (Holbrook et al., 1974) might be effector cells responding to luminal antigens. The presence of carbon in the bursal medulla following its internalization by the cloaca suggests that circulating antibody to *Brucella abortus* following application of this bacterium to the vent was a direct result of the bacterium entering the bursal microenvironment (Sorvari et al., 1975).

2. Biosynthesis of Immunoglobulin

Although we failed to identify antibody to BSA in the bursa, we did observe that bursal cells stained intensely with a fluorescein-conjugated rabbit anti-chicken immunoglobulin (Glick and Whatley, 1967). Acrylamide-gel electrophoresis revealed the presence of IgM in the bursa of 1-week-old chickens. IgM had been tentatively identified after culturing the bursa from 16-day-old embryos (Marinkovich and Baluda, 1966). In a more extensive study, Thorbecke et al. (1968) reported (1) IgM in the bursa at 18 DE, (2) IgG in the bursa at the time of hatching, (3) a more active synthesis of immunoglobulin in the bursa than in the spleen up to the first week of life, and (4) that IgM synthesis in the bursa was not antigen dependent. The latter was reemphasised by Choi and Good (1972a), who reported that antigen modifies the secretion of immunoglobulin in the spleen but does not influence its synthesis by bursal cells. B cells from adult spleens and 2-day-old bursal cells synthesized and released immunoglobulin at similar rates suggesting that age may not alter the B cells' capacity for immunoglobulin biosynthesis (Lifter and Choi, 1977). A population of bursal cells was identified that did not secrete IgM (Choi and Good, 1972a). This IgM had faster migrating heavy chains and differed in its carbohydrate moiety from the secretory immunoglobulin of bursal cells. The heavy chains of secretory IgM contained more glucosamine and galactose than the nonsecretory immunoglobulin that appeared at 14 DE (Choi and Good, 1972b).

The dependency on the bursa for the switch from IgM to IgG was first promulgated by a series of experiments in the laboratories of R. A. Good and Max Cooper. *In ovo* bursectomies performed early (17 DE) or late (19–21 DE) impaired IgG synthesis (Van Alten et al., 1968; Cooper et al., 1969; Moticka and Van Alten, 1971, 1972a,b). Kincade and Cooper (1971) utilized a fluorescein-conjugated anti-μ or anti-γ sera that identified IgM (anti-μ) cells in the bursal medulla at 14 DE and IgG (anti-γ) cells in this area at 21 DE. Both of these immunoglobulins appeared later in other lymphomyeloid tissues, with IgM observed at 17 DE in the caecal tonsils and 19 DE in the spleen. IgG appeared in these tissues after hatching. The medulla of a given follicle that was positive for IgG always contained cells positive for IgM. These data and the observations that occasional bursal cells reacted to both anti-μ and anti-γ conjugates emphasized that the switch from IgM to IgG occurred in the bursa.

IgA, a third immunoglobulin synthesized by the chicken (Lebacq-Verheyden et al., 1972a, 1974; Orlans and Rose, 1972; Bienenstock et al., 1973; Leslie and Martin, 1973a,b) is the predominant immunoglobulin synthesized by the chicken's intestinal mucosa (Lebacq-Verheyden et al., 1972b; Bienenstock et al., 1973; Leslie and Martin, 1973b). Treatment *in ovo* with anti-μ serum followed by surgical bursectomy at the time of hatching eliminated the synthesis of IgM, IgG, and IgA (Kincade and Cooper, 1971; Leslie and Martin 1973a, 1974). IgA was delayed in birds bursectomized at hatching (Martin and Leslie, 1973). IgM and IgG but not IgA were present in birds bursectomized *in ovo* at 17 DE (Kincade and Cooper, 1973). These data suggested that the sequence of immunoglobulin conversion in the bursa is IgM → IgG → IgA (Bienenstock et al., 1973; Kincade and Cooper, 1973). Bursectomy or thymectomy lead to a deficiency of IgA and IgG, demonstrating the necessity of both T and B lymphocytes for the ontogeny of these immunoglobulins (Perey and Bienenstock, 1973). The absence of IgA but not IgM or IgG in the bursectomized–thymectomized birds suggested that synthesis of IgG involved more than the bursal environment. The suppression of IgM and IgA but not IgY (IgG) in bursectomized birds treated with anti-α and anti-μ sera suggested that IgY cells may not be precursors for IgA cells (Martin and Leslie, 1974). The adult levels of IgY (50 days: >3.4 mg/ml) and IgM (40 days: 1.1 mg/ml) occurred earlier than the adult levels of IgA (94 days: 0.67 mg/ml) (Martin and Leslie, 1973). The IgY and IgM of 56-day-old bursectomized birds was slightly lower and markedly higher, respectively, than normal levels, whereas the IgA levels of bursectomized birds were not different at this age. Compensatory increase of IgY or IgM did not occur in the presence of IgA deficiency following bursectomy at the time of hatching

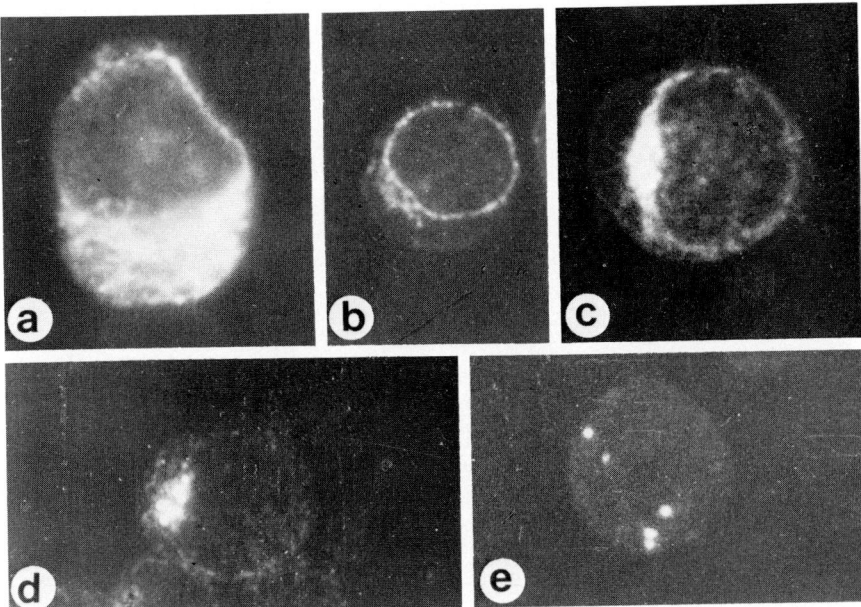

FIG. 17. Developmental patterns of cytoplasmic immunoglobulin M (cIgM) in bursal lymphocytes. Type 1 cells (a) diffuse cIgM predominates from the twelfth to sixteenth day of incubation and in the cortex of mature follicles. Type 2 includes perinuclear (b) and Golgi (c and d) distributions of cIgM, prominent between the sixteenth day of incubation and 2 weeks of age. Type 3 cells (e) show cytoplasmic droplets possessing cIgM, appearing in the medulla of mature follicles after hatching. ×2100. [From Grossi et al. (1977), copyright 1977 The Williams & Wilkins Co., Baltimore.]

(Leslie and Martin, 1974). The IgG of chickens differs physicochemically from mammalian IgG in that it has a large molecular weight, more carbohydrate, and displays an unusual heavy–light (H–L) chain interaction (Leslie and Clem, 1969). These authors refer to avian IgG as IgY.

Bursal cells may be classified by their IgM patterns. At 12 DE, bursal cells exhibited cytoplasmic IgM (cIgM) and surface IgM (sIgM) (Grossi et al., 1977). Cytoplasmic IgM at this time was diffuse (type I, Fig. 17). Low levels of anti-μ were effective in removing sIgM on 15 DE, but higher levels were necessary with increased age. This correlated with the type of IgM pattern. Type 3 (single or multiple droplets of immunoglobulin) was the most difficult to modulate. Cortical bursal cells were type 1 and, therefore, easier to modulate than the medullary cells, which were type 3. The latter type developed in the medulla independently of antigen. Two weeks after hatching none of the spleen, bone marrow, caecal tonsil, thymic, or the small lymphocytes of peripheral blood had detectable cIgM, whereas bursal cells at this age showed a type 3 cIgM. Cytoplasmic immunoglobulin from splenic B cells had a 10-fold greater turnover rate than sIg, suggesting that in the chicken the biosynthesis of sIg may be different from that of cIg (Choi, 1976). Also, circulating immunoglobulin differs in its turnover rate. The half-life of IgM in dysgammaglobulinemic and normal chickens was similar (1.5 days), whereas the half-life of IgG from normal birds (2.1 days) was less than dysgammaglobulinemic or hypogammaglobulinemic birds (3.5 days) (Frommel et al., 1970).

3. Bursal Control of Immunoglobulin Synthesis

The bursa was initially implicated in the control of immunoglobulin synthesis when Long and Pierce (1963) reported a reduced level of circulating gammaglobulin in birds hatched from TPI eggs. The hormonal bursectomy resulting from TP injection may lead to reduced levels of IgG (Carey and Warner, 1964; Warner et al., 1969) and occasional agammaglobulinemia (Warner et al., 1969). Surgical bursectomy, like TP injection, leads to reduced levels of IgG but may elicit higher than normal levels of IgM (Ortega and Der, 1964; Cooper et al., 1965; Morgan and Glick, 1972; Martin and Leslie, 1973, 1974). Irradiation of the surgically bursectomized birds may eliminate IgG and IgM synthesis (Cooper et al., 1966), reduce IgG synthesis (Alm and Peterson, 1969), or alter immunoelectrophoretic patterns of either immunoglobulin (Ivanyi et al., 1969). A restricted electorphoretic mobility of residual IgG was observed in bursectomized birds (Pierce et al., 1966; Ivanyi et al., 1969), whereas the L- or H-chain patterns of IgG were not different in bursectomized birds (Gold and Benedict, 1967). In ovo bursectomy between 17 and 21 DE reduces immunoglobulin synthesis, with IgG being more sensitive than IgM (Van Alten et al., 1968; Cooper et al., 1969; Blythman and White, 1977). Hormonal bursectomy resulting from dipping eggs (TPD) prior to the fifth day of embryonic development suppresses IgG and enhances the synthesis of IgM (Merkenschlager et al., 1966; Glick, 1968b; Bruggeman et al., 1969; Lerner et al., 1971). These selected examples of the regulation of immunoglobulin synthesis by several methods of bursectomy have led to the concept of a bursa-dependent role for IgG synthesis and a bursa-independent role for IgM synthesis.

Bursa-Dependent and Bursa-Independent Immunoglobulin Synthesis. Sixty-four percent of hormonally bursectomized (TPI) birds produced antibody to a secondary injection of human gammaglobulin, whereas only 40% responded to a tertiary injection of *Brucella abortus* (Warner et al., 1969). The sera of 8 of these 16 TPI birds failed to develop IgG or IgM precipitin arcs in immunoelectrophoretic analysis. Suppression of IgM and IgG occurred in chickens treated as 13-day-old embryos with anti-μ serum and bursectomized at the time of hatching (Kincade et al., 1970, 1973; Cooper et al., 1972). Delaying the administration of anti-μ serum until after bursectomy at hatching suppressed IgM but not IgG synthesis (Kincade et al., 1970). These results complemented the previously discussed data in which (1) immunoglobulin synthesis occurred initially in the bursa, (2) IgG synthesis in the bursa was a later event than IgM synthesis, and (3) dual synthesis of IgG and IgM was occasionally seen in an individual bursal cell (Kincade and Cooper 1971), and they were responsible for the conclusion that the bursa has an obligatory role in immunoglobulin synthesis. The concept of sequential immunoglobulin synthesis and the dependency of antibody synthesis on the bursa gained credence after multiple antigenic injections into bursectomized birds yielded predominantly IgM antibody (Jankovic and Isakovic, 1966a, 1967; Rose and Orlans, 1968; Eychmans et al., 1968; Maranova and Hasek, 1969; Alm, 1970). However, the very production of antibody in hormonally bursectomized (TPI and TPD) birds in which embryonic bursal development was abolished (Glick and Sadler, 1961; Claflin et al., 1966; Glick, 1968a) led us to formulate the thesis that "other sites in the chicken were capable of conditioning or supplying immunocompetent cells [Glick, 1968a, p. 1057]." The concept of a bursa-independent IgM route was reinforced after observing elevated IgM levels in TPD birds (Merkenschlager et al., 1966; Glick, 1968b; Morgan and Glick, 1972; Hoffman-Fezer and Losch, 1973) and the failure by some to find agammaglobulinemia by irradiation of bursectomized birds (Ivanyi et al., 1969; Van Meter et al., 1969). We then designed experiments to test the possibility of IgM synthesis in the absence of a bursa. Our experiments demonstrated that TPD birds possessed (1) either no IgG or reduced levels, (2) higher than normal levels of IgM, (3) significant levels of IgM antibody after the fourth immunization of SRBCs, and (4) reduced numbers of germinal centers and plasma cells in the caecal tonsil and spleen (Lerner et al., 1971). The failure of hormonal (TPD) bursectomy to eliminate the embryonic bursa has been raised to explain the presence of immunoglobulin in these birds. However, a comparison of the TPD and TPI methods revealed the presence of a bursa with plicae in the TPI embryos, whereas 75% of the TPD embryos lacked a visible bursa (Glick and McDuffie, 1974). Between 15 and 21 DE, 6 of 21 TPI bursae and 3 of 20 TPD bursae or *bursal areas* contained buds or bursal follicles (Glick and McDuffie, 1974, 1975). Therefore, the TPD method is more effective in suppressing bursal development than is the TPI method. Chickens deprived of their bursa as embryos should not be capable of immunoglobulin synthesis (Kincade et al., 1970; Kincade and Cooper, 1971). However, higher than normal immunoglobulin levels have been reported for TPD birds (Glick, 1968b; Lerner et al., 1971; Subba Rao et al., 1978). We suggest that precursor cells destined for the bursa in TPD birds take up residence at other sites where they synthesize IgM (Lerner et al., 1971), or precursor cell association with the bursal area may enhance their ability to synthesize IgM (Glick and McDuffie, 1975). Elevated IgM levels in TPD birds may represent antibodies to viral agents (Warner et al., 1969). We addressed this possibility by raising TPD and control birds in pathogen-free environments (Subba Rao et al., 1978). Our results do indeed offer some evidence for the role of viruses in stimulating IgM antibody and IgM in TPD birds (Fig. 18). However, the TPD birds raised in the pathogen-free environment still maintained a threefold higher IgM level than did comparable controls (330 mg/dl versus 100 mg/dl). Judging from the ability of infused IgG to lower the IgM levels of controls, we theorized that the elevated level of IgM in TPD birds may reflect, in part, the absence or near absence of circulating IgG (Subba Rao et al., 1978).

Some TPD birds have been reported to be agammaglobulinemic. We proposed that the bursa is steroid-sensitive during the initial week of embryonic development but both the bursa and precursor cell are steroid sensitive during week 2 of embryonic development (Glick and McDuffie, 1974). Therefore, bursal exposure during the first week of development would influence the bursal microenvironment and not the precursor cell, which would retain its innate IgM potential. Sex steroid administration during the second week of embryonic development would disrupt both the bursal microenvironment and precursor cell leading to potential agammaglobulinemia in these TPI birds. Support for our bursa-independent IgM pathway comes from numerous laboratories (Bruggeman et al., 1969; Losch, 1971; Fitzsimmons et al., 1973; Bryant et al., 1973; Sato and Abe, 1975; Albini and Wick, 1975; Jankovic et al., 1977; Hirota et al., 1980). Particularly convincing are

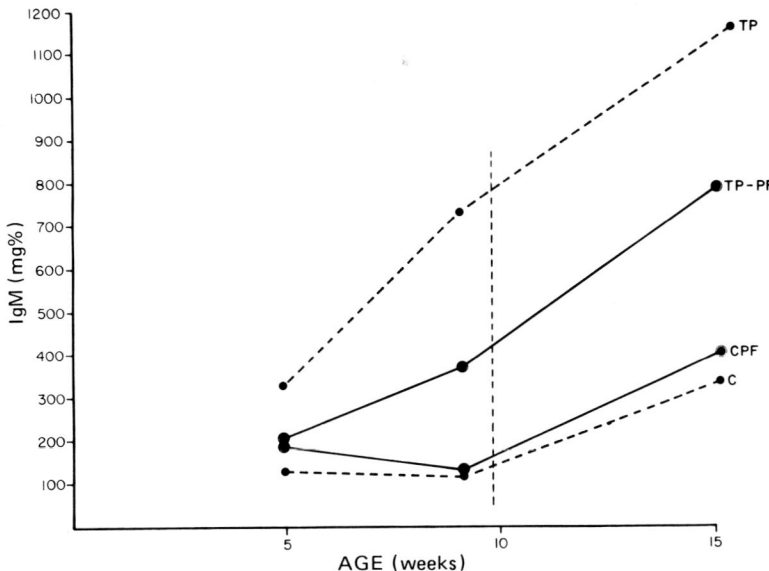

FIG. 18. A comparison of the immunoglobulin M (IgM) levels of chicks hatched from eggs dipped in testosterone propionate (TP) and control (C) chicks raised in conventional or pathogen-free (PF) environments and injected with SRBCs at 1, 4, 9, and 14 weeks of age. All birds were removed from the PF environment and returned to the conventional environment at 10 weeks of age. [From Subba Rao et al. (1978), copyright 1978 The Williams and Wilkins Co., Baltimore.]

the *in ovo* bursectomies performed prior to 72 hours of incubation (Fitzsimmons et al., 1973; Jankovic et al., 1975, 1977). *In ovo* bursectomy at 72 hours (17- to 18-somite stage) eliminated the bursa and one-third of the large intestine in the hatched chick (Fitzsimmons et al., 1973). These birds produced agglutinins to SRBCs, and six of eight made immunoglobulin (possibly IgM). Surgical removal of the caudal portion of the tail bud before 64 hours of embryonic development (29- to 32-somite stage) did not eliminate lymphopoiesis of cells possessing bursal membrane determinants (Jankovic et al., 1975). These bursal cells were evident in spleen, bone marrow, and thymus of 21-day-old bursaless embryos. These authors concluded "that stem cells can differentiate into BU (bursa) cells in the absence of the bursa and that there are non-bursal sites of BU cell formation [p. 658]." A modified B cell may be present in the spleen of TPI-treated birds. The treatment of normal spleens with anti-B serum inhibited their ability to adoptively transfer IgM antibody potential to SRBCs, whereas the same treatment of spleens from TPI-treated birds did not abolish their IgM potential (Hirota et al., 1980). This argues for the presence in TPI birds of B cells that lack the normal B cell surface antigens. The *in ovo* bursectomized embryos produced IgG and IgM antibodies to guinea pig erythrocytes during their third week of posthatching life (Jankovic et al., 1976). IgM was present in a few birds bursectomized *in ovo* (52–64 hours; Jankovic et al., 1977). The nonbursal environment may function only in the absence of the bursal environment. A neuroendocrine control may exist for the bursa (Jankovic et al., 1978). The presence of cells positive for anti-L chains but not anti-γ in the yolk sac at 8 DE and cIgM$^+$ cells in the yolk sac at 3 DE is additional evidence that a nonbursal site may supply Ig$^+$ cells (Lebacq and Ritter, 1979).

F. IMMUNOSUPPRESSION

Inhibition of immunoglobulin synthesis and bursal ontogeny by *in ovo* exposure to sex steroids has been discussed in Section VIIB,C, and E. A more effective method of immunosuppression was introduced by Lerman and Weidanz (1970). Their treatment of chicks with 4 or 6 mg of Cy for each of the first 3 days of life suppressed the primary and secondary responses to BSA, SRBCs, and *Salmonella typhimurium* as well as reduced the levels of IgG and IgM. Large, pale, reticular cells rather than the basophilic lymphocytes occupied the bursal follicles in 2-day-old chicks treated in the prior 2 days with 6 mg Cy/day (Glick, 1971). The bursae of these chicks did not regenerate, but those chicks receiving a single injection of Cy revealed an occasional normal bursal follicle by 5 weeks of age. The thymic cortex was disrupted 12 hours after the last Cy injection, but it returned to normal by 2 weeks of age (Glick, 1971; Linna et al., 1972). Cy produced agammaglobulinemia (Glick, 1971) and a deficient graft versus host capability during the first 3 weeks of age (P. Toivanen et al., 1972c), but not later (Lerman and Weidanz, 1970; Linna et al., 1972). The toxicity of Cy varies with the breed and strain of chickens, and some birds may regain their immunocompetence (Rouse and Szenberg, 1974). Reconstitution of immunoglobulin synthesis in Cy-treated and in bursectomized birds has been reviewed elsewhere (Glick, 1977a).

Immunosuppression by cellular transfer in the chicken has been reported. Irradiation of recipient chickens for 3 consecutive days and infusion with bone marrow from agammaglobulinemic birds resulted in a suppression of antibody to KLH and immunoglobulin, whereas antibody and immunoglobulin of irradiated birds injected with bone marrow from normal birds were not suppressed (Blaese et al., 1974). It was theorized that the suppression was lymphocyte mediated. Surgically bursectomized irradiated chickens (2–3 weeks of age) failed to recover their ability to produce antibody or immunoglobulin following the adoptive transfer of splenic lymphocytes from agammaglobulinemic chickens (Palladino *et al.*, 1976). The transferred T cells from agammaglobulinemic chickens may have developed into suppressor cells after exposure to B cell antigens in the recipient. That the T cell of agammaglobulinemic birds is naive to B cells and will respond to the B cell by developing suppressor potential (Palladino *et al.*, 1976) was supported by Grebenau *et al.* (1976), who presented evidence that spleen cells from agammaglobulinemic birds transferred with *Brucella abortis* into recipients will not suppress antibody to *B. abortis*. However, the adoptive *B. abortis* response was suppressed by spleen cells from agammaglobulinemic birds when these birds were previously injected with histocompatible B cells. Treatment of spleens from agammaglobulinemic birds with anti-T plus complement abolished the suppression. The ability of spleen cells from agammaglobulinemic birds to suppress antibody response is age dependent and may occur spontaneously in birds over 15 weeks of age (Grebenau *et al.*, 1979). Bursal cells transferred to agammaglobulinemic birds less than 8 weeks old do not induce suppressor cells, but they do so in birds older than 8 weeks. This is an apparent *autoimmune response* of agammaglobulinemic birds to B cells that are the target cells in the suppressed recipient (Lerman *et al.*, 1980). Spleen cells primed to SRBCs and transferred to culture generate suppressor cells if immediately exposed to SRBCs, whereas helper-cell induction is evident if the addition of antigen is delayed for at least 2 days (Chi *et al.*, 1980). Similar *in vitro* responses are evident for spleens from agammaglobulinemic birds. These responses are specific, because immunocompetence to *Brucella abortus in vitro* is not affected. The absence of T-suppressor cells in the peripheral blood lymphocytes suggests a sequestering of this cell in the agammaglobulinemic bird (Kermani-Arab and Leslie, 1977). Suppressor T cells have been identified in an inherited 7-S immunoglobulin deficiency (Gershwin *et al.*, 1979; Benedict *et al.*, 1980; Chann *et al.*, 1980).

VIII. Summary

The bursa of Fabricius is a dorsal diverticulum of the cloaca. In the chicken the bursa is round or oval, but it is elongated in the White Pekin duck and European Starling. The luminal surface of the bursa has numerous folds in the chicken and duck, but these structures are absent in the starling. A reorganization of the surface epithelium of the folds into epithelial buds occurs between 12 and 15 days in the chick embryo. These buds develop into follicles that contain bursal (B) lymphocytes. The bursal follicle is divided into cortex and medulla by epithelial cells whose basal lamina is continuous with that of the surface epithelial cells of the folds. The medulla is in direct contact with the follicle-associated epithelium (FAE). Each cell of the FAE is polygonal and possesses few microvilli. Follicle-associated epithelial cells are pinocytotic and account for 1 cm^2 or 10% of the bursal surface area. The pinocytotic activity of the FAE helps to explain the ability of the medullary lymphocytes to produce antibody to an anal or intraluminal application of antigen. Also, transfer of materials from lumen to medulla via FAE could modify the bursal microenvironment and explain the occurrence of lymphocytes on the luminal surface. Scanning electron microscopy has not revealed a surface difference between B and thymic (T) lymphocytes. The remaining 90% of the bursal surface area is covered by interfollicular epithelium (IFE), which has a dense covering of microvilli and releases mucin into the bursal lumen.

The folds of the late embryo and perinate reveal projecting follicles. With the loss of lymphocytes from the medulla, the FAE sinks into the follicle and separates from the IFE, producing a button-like follicle.

Reciprocal transplants of endomesodermal bursae between Japanese Quail and chickens took advantage of distinct heterochromatin characteristics of each species to demonstrate that stem-cell migration to the chick's bursa occurs between 8 and 15 days of embryonic development (DE), whereas such a migration occurs between 7 and 11 days in the quail. The stem cell has been shown to be circulating prior to its appearance in the bursa. Thus its entry into the bursa may be dependent on a change in the bursal microenvironment. More recent evidence indicates an intraembryonic origin for the stem cell.

The prebursal stem cell, of mesenchymal origin, is followed by 3 maturational stages of B cells. The initial stage occurs late in embryonic development and the first few days after hatching. These bursal stem cells may restore the lymphoid compartment of a regressed bursa, but they require further maturation before reconstituting an antibody-deficient animal. A second maturational stage includes early postbursal cells (3- to 6-week-old birds), which home to the bursa and may restore antibody potential to immunosuppressed birds. The third maturational stage of postbursal cells appears after 7 weeks and is not able to restore bursal structure, but it does restore antibody competence.

Although a multipotential stem cell may seed the thymus and bursa, there is evidence for pre-B and pre-T cells that are destined for the bursa and thymus, respectively. The bursal microenvironment may regulate the entry of pre-B cells, their differentiation, and their kinetics.

The bursa has been shown to produce several humoral substances. Bursapoietin induces the maturation of specific bone-marrow cells to bursacytes. Lymphocyte inhibitory factor and thrombocyte inhibitory factor are released by bursacytes and inhibit the correspondingly named cells. A special cell, the secretory cell, has been identified in the bursa and may contribute to its microenvironment.

The bursa grows most rapidly between hatching and 3 weeks of age, plateaus between 3 and 8 weeks, and regresses thereafter.

The weight of the bursa is a reflection of its cell population, which may be influenced by humoral substances and exogenous factors. Caponization enhances bursal growth, whereas pharmacological levels of testosterone produce regression. There is some evidence that low doses of testosterone (<200 µg) may stimulate bursal growth. Bursal lymphocytes are more efficient in binding testosterone than the T lymphocytes. Cytosol androgen receptors were identified in endodermal and mesenchymal components of the embryonic bursa, with the greatest concentration of receptors identified in the endodermal tissue.

Bursal regression in the presence of stressor agents is mediated by the hypothalamic–pituitary–adrenal axis. The selection for large-bursa and small-bursa (SB) lines led to a significantly larger adrenal in the SB-line birds. This reinforces the importance of the adrenal glands in regulating bursal growth. The bursa appears to influence adrenal function by controlling adrenal ascorbic acid repletion. The majority of bursal small lymphocytes have been found to turn over rapidly (i.e., they are short-lived) when labeled with radiotracers. The spleen and bone marrow possess the longest-lived population of small lymphocytes. The number of bone-marrow small lymphocytes is influenced by the bursa, but the reverse is not true.

Removing the bursa during its rapid-growth period leads to a marked reduction in antibody synthesis. Failure of *in ovo* hormonal bursectomy and posthatching surgical bursectomy to eliminate antibody synthesis in all birds was responsible for our original suggestion that other sites may be involved in regulating antibody production. More recent experiments have revealed that bursectomy interfered with the T-suppressor-cell population of the thymus.

Embryonically, the bursa appears to be the first lymphoid tissue to produce immunoglobulin M (IgM; 13 DE) with the switch from IgM to immunoglobulin G (IgG) occurring initially in the bursa at 21 DE. IgM and IgG but not immunoglobulin A (IgA) are present in birds that have been bursectomized *in ovo* at 17 DE. These data suggested that the sequence of immunoglobulin conversion in the embryonic bursa may be IgM → IgG → IgA. Dipping eggs into testosterone propionate (TP) solutions prior to 3 DE enhances IgM synthesis, whereas in injection of TP into the allantoic cavity at 12 DE suppresses IgG and IgM synthesis. Removal of a caudal segment from an embryo before 72 hours of incubation allows the chick to hatch without a bursa, but with the ability to produce antibody, Ig, and B cells. All of these observations reinforce the concept of a bursa-dependent role for IgG synthesis and a bursa-independent role for IgM synthesis.

The dissociation of immunological responsiveness was suggested after observing a delayed transplantation response in only those chickens whose thymic cortex had been reduced by TP injection at 12 DE. This concept attributed the control of antibody-mediated immunity to the bursa and cell-mediated immunity to the thymus. Although this was a useful concept, it became apparent that this was too parochial a view, because T cells (T-helper cells) were identified that cooperated with B cells in the production of antibody. Also, it has been suggested that B cells cooperate with T cells in a cell-mediated response.

The ontogeny of the immune response to a variety of antigens has been assessed by bursectomy at various ages. Immunocompetence develops (1) embryonically to sheep red-blood cells, rabbit red-blood cells, several viruses, and *Escherichia coli*, (2) during the first week after hatching to bovine serum albumin, and (3) during the second week after hatching to *Salmonella pullorum*. Also B cells exhibit an ontogeny of IgM patterns. Bursal cells from 12-day-old embryos possess both cytoplasmic IgM (cIgM) and surface IgM (sIgM). Cytoplasmic IgM may be diffuse (12–16 DE), perinuclear in location (16 DE to 2 weeks after hatching), or present as prominent droplets in the medullary cells of mature follicles. Cytoplasmic IgM turns over at a faster rate than sIgM.

Immunosuppression by cellular transfer in the chicken has been reported. Surgically bursectomized and irradiated chickens failed to recover their ability to produce antibody or immunoglobulin following the adoptive transfer of splenic or bone-marrow lymphocytes from agammaglobulinemic birds. The transferred T cells from agammaglobulinemic chickens may develop into suppressor cells after exposure to B-cell antigens.

The bursa's central role in immunity, its response to and possible influence on different endocrines, the presence of special cell types, and the synthesis of numerous humoral substances illustrate the important role the bursa of Fabricius plays in physiological homeostasis.

ACKNOWLEDGMENT

Sincere appreciation is extended to Doris Thompson (laboratory assistance), Beverly Beck (reprint file), Patricia McCluskey (typing), the NIH for past and present support (ROI CA 20169), and MAFES (journal article number 4896).

REFERENCES

Abramoff, P., and Brien, N. B. (1968). Studies of the chicken immune response. I. Correlation of the cellular and humoral Immune response. *J. Immunol.* **100**, 1204–1209.

Ackerman, G. A. (1962). Electron microscopy of the bursa of Fabricius of the embryonic chick with particular reference to the lymphoepithelial nodules. *J. Cell Biol.* **13**, 127–147.

Ackerman, G. A., and Knouff, R. A. (1959). Lymphocytopoiesis in the bursa of Fabricius. *Am. J. Anat.* **140**, 163–205.

Adelmann, H. B. (1967). "The Embryological Treatises of Hieronymus Fabricius of Aquapendente." Cornell Univ. Press, Ithaca, New York.

Albini, B., and Wick, G. (1975). Ontogeny of lymphoid cell surface determinants in the chicken. *Int. Arch. Allergy Appl. Immunol.* **48**, 513–529.

Allan, W. H., Faragher, J. T., and Cullen, G. A. (1972). Immunosuppression by the infectious bursal agent in chickens immunized against new castle disease. *Vet. Rec.* **90**, 511–516.

Alm, G. V. (1970). The *in vivo* spleen response to sheep erythrocytes in bursectomized–irradiated chickens. *Acta Pathol. Microbiol. Scand. Sect.* **78**, 641–646.

Alm, G. V., and Peterson, R. D. A. (1969). Antibody and immunoglobulin production at the cellular level in bursectomized–irradiated chickens. *J. Exp. Med.* **129**, 1247–1259.

Asmar, J. A., Oaghir, H. J., and Azar, H. A. (1968). Effects of pyridoxine deficiency on the lymphatic organs and certain blood components of the neonatal chicken. *J. Nutr.* **95**, 153–159.

Aspinall, R. L., Meyer, R. K., Graetzer, M. A., and Wolfe, H. R. (1961). Effect of various steroids on the development of the bursa Fabricii in chick embryos. *Endocrinology (Baltimore)* **68**, 944–949.

Aspinall, R. L., Meyer, R. K., Graetzer, M. A., and Wolfe, H. R. (1963). Effect of thymectomy and bursectomy on the survival of skin homografts in chickens. *J. Immunol.* **90**, 872–879.

Beaupain, D., Martin, C., and Dieterlen-Lievre, F. (1979). Are developmental hemoglobin changes related to the origin of stem cells and site of erythropoiesis? *Blood* **53**, 212–225.

Bellamy, D., and Leonard, R. A. (1964). Effect of cortisol on the growth of chicks. *Gen. Comp. Endocrinol.* **5**, 402–410.

Beller, D. I., and Unanue, E. R. (1977). Thymic maturation *in vitro* by a secretory product from macrophages. *J. Immunol.* **118**, 1780–1787.

Beller, D. I., and Unanue, E. R. (1978). Thymic macrophages modulate one stage of T cell differentiation *in vitro*. *J. Immunol.* **121**, 1861–1864.

Benedict, A. A., Tam, L. E., Chann, T. C., Pollard, L. W., Abplanalp, H., and Gershwin, M. E. (1980). Inherited 7 S immunoglobulin deficiency in chickens: late onset dysgammaglobulinemia and evaluation of the lymphocyte proliferative responses and other immunologic functions. *Clin. Immunol. Immunopathol.* **17**, 1–14.

Bienenstock, J., Gauldie, J., and Perey, D. Y. E. (1973). Synthesis of IgG, IgA, IgM by chicken tissues: immunoflourescent and 14 C amino acid incorporation studies. *J. Immunol.* **111**, 1112–1118.

Blaese, R. M., Weiden, P. L., Koski, I., and Dooley, N. (1974). Infectious agammaglobulinemia: transmission of immunodeficiency with grafts of agammaglobulinemic cells. *J. Exp. Med.* **140**, 1097–1101.

Blythman, H., and White, R. (1977). Effect of early bursectomy on germinal centre and immunoglobulin production in chickens. *Immunology* **33**, 671–677.

Bockman, D. E., and Cooper, M. D.(1973). Pinocytosis by epithelium associated with lymphoid follicles in the bursa of Fabricius, appendix, and Peyer's patches. An electron microscopic study. *Am. J. Anat.* **136**, 455–478.

Bockman, D. E., and Stevens, W. (1976). Gut-associated lymphoepithelial tissue bidirectional transport of tracer by specialized epithelial cells associated with lymphoid follicles. *J. Reticuloendothel. Soc.* **21**, 245–254.

Boyden, E. A. (1922). The development of the cloaca in birds with special reference to the origin of the bursa of Fabricius. The formation of a urodaeal sinus, and the regular occurrence of a cloacal fenestra. *Am. J. Anat.* **30**, 163–202.

Brand, A., Gilmour, D. G., and Goldstein, G. (1976). Lymphocyte-differentiating hormone of bursa of Fabricius. *Science (Washington, D.C.)* **193**, 319–321.

Breitenbach, R. P. (1962). The effect of ACTH on adrenocortical secretion and ascorbic acid depletion in normal and testosterone treated cockerels. *Poult. Sci.* **41,** 1318–1324.

Bruggeman, J. B., Friedrich, B., and Losch, U. (1969). A comparative study of the immunoglobulin patterns of naturally dysgammaglobulinemic and hormonally bursectomized chickens. *Z. Tierphysiol. Tiernaehr. Futtermittelkd.* **24,** 317–330.

Bryant, B. J., Adler, H. E., Cordy, D. R., Shifrine, M., and Damassa, A. J. (1973). The avian bursa-independent humoral immune system. Serologic and morphologic studies. *Eur. J. Immunol.* **3,** 9–15.

Burns, R. B. (1980). Cloacal immunization of the turkey using a soluble antigen. *Br. Poult. Sci.* **21,** 249–252.

Cain, W. A., Cooper, M. D., and Good, R. A. (1968). Cellular immune competence of spleen bursa and thymus cells. *Nature (London)* **217,** 87–89.

Cain, W. A., Cooper, M. D., Van Alten, P. J., and Good, R. A. (1969). Development and function of the immunoglobulin producing system. II. Role of the bursa in the development of humoral immunological competence. *J. Immunol.* **102,** 671–678.

Carey, J., and Warner, N. L. (1964). Gamma-globulin synthesis in hormonally bursectomized chickens. *Nature (London)* **203,** 198–199.

Caruso, C., Bellavia, A., Bellina, L., and Salerno, A. (1977). Delayed-type skin reactions in bursectomized or thymectomized chickens. *Experientia* **33,** 1105–1126.

Challey, J. R. (1962). The role of the bursa of Fabricius in adrenal response and mortality due to *Eimeria tennella* infections in the chicken. *J. Parasitol.* **48,** 352–358.

Chang, T. S., Glick, B., and Winter, A. R. (1955). The significance of the bursa of Fabricius of chickens in antibody production. *Poult. Sci.* **34,** 1187 (Abs.).

Chang, T. S., Rheins, M. S., and Winters, A. R. (1957). The significance of the bursa of Fabricius on antibody production in chickens. *Poult. Sci.* **36,** 735–739.

Chann, T. C., Benedict, A. A., Tam, L. Q., Abplanalp, H., and Gershwin, M. E. (1980). Inherited 7S immunoglobulin deficiency in chickens: presence of suppressor T cells that suppress synthesis of 7S immunoglobulin but not IgM. *J. Immunol.* **125,** 108–114.

Cheville, N. F. (1967). Studies on the pathogenesis of Gumboro disease in the bursa of Fabricius, spleen and thymus of the chicken. *Am. J. Pathol.* **51,** 527–553.

Chi, D. S., Grebenau, M. D., and Thorbecke, G. J. (1980). Antigen-induced helper and suppressor T cells in normal and agammaglobulinemic chickens. *Eur. J. Immunol.* **10,** 203–209.

Cho, B. R. (1963). The effect of bursectomy of chickens in antibody response to Newcastle disease virus. *Am. J. Vet. Res.* **24,** 832–834.

Choi, Y. S. (1976). Biosynthesis of membrane bound Ig and secretion of Ig by chicken lymphoid cells. *Biochemistry* **15,** 1037–1042.

Choi, Y. S., and Good, R. A. (1972a). Development of chicken lymphoid system. I. Synthesis and secretion of immunoglobulins by chicken lymphoid cells. *J. Exp. Med.* **135,** 1133–1150.

Choi, Y. S., and Good, R. A. (1972b). Development of chicken lymphoid system. II. Synthesis of primordial immunoglobulin M by the bursa cells of chick embryos. *J. Exp. Med.* **136,** 8–20.

Choi, Y. S., and Good, R. A. (1973). Biosynthesis and secretion of antibody by chicken lymphoid cells. *J. Immunol.* **110,** 1485–1491.

Claflin, A. J., Smithies, O., and Meyer, R. K. (1966). Antibody responses in bursa-deficient chickens. *J. Immunol.* **97,** 693–700.

Cooper, M. D., Peterson, R. D. A., and Good, R. A. (1965). Delineation of the thymic and bursal lymphoid systems in the chickens. *Nature (London)* **205,** 143–146.

Cooper, M. D., Peterson, R. D. A., South, M. A., and Good, R. A. (1966). The functions of the thymus system and bursa system in the chicken. *J. Exp. Med.* **123,** 75–103.

Cooper, M. D., Cain, W. A., Van Alten, P. J., and Good, R. A. (1969). Development and function of the immunoglobulin producing system. I. Effect of bursectomy at different stages of development on germinal centers, plasma cells, immunoglobulins, and antibody production. *Int. Arch. Allergy* **35**, 242–252.

Cooper, M. D., Lawton, A. R., and Kincade, P. W. (1972). A two-stage model for development of antibody-producing cells. *Clin. Exp. Immunol.* **11**, 143–149.

Cordier, A. (1969). Study of innervation of the bursa of Fabricius in the embryonic and adult chick. *Acta Anat.* **73**, 38–47.

Cosgrove, A. A. (1962). An apparently new disease of chickens—avian nephrosis. *Avian Dis.* **6**, 385–389.

Danielson, J. R., and Van Alten, P. J. (1974). Lymphocyte proliferation inhibited by cells and by effector substances obtained from bursal lymphocytes. *Immunol. Cancer Prog. Exp. Tumor Res.* **19**, 194–202.

Dent, P. B., and Good, R. A. (1965). Absence of antibody production in the bursa of Fabricius. *Nature (London)* **207**, 491–493.

Dieter, M. P., and Breitenbach, R. P. (1968). The growth of chicken lymphoid organs, testes, and adrenals in relation to the oxidation state and concentration of adrenal and lymphoid organ vitamin C. *Poult. Sci.* **47**, 1463–1469.

Dieter, M. P., and Breitenbach, R. P. (1970). A comparison of the lympholytic effects of corticosterone and testosterone propionate in immature cockerels. *Proc. Soc. Exp. Biol. Med.* **133**, 357–364.

Dieter, M. P., and Breitenbach, R. P. (1971). Vitamin C in lymphoid organs of rats and cockerels treated with corticosterone or testosterone. *Proc. Soc. Exp. Biol. Med.* **137**, 341–346.

Dieterlen-Lievre, R. D., Beaupain, D., and Martin, C. (1976). Origin of erythropoietic stem cells in avian development: shift from the yolk sac to an intraembryonic site. *Ann. Immunol. (Paris)* **127**, 857–863.

Dransfield, J. W. (1945). The lymphatic system of the domestic fowl. *Br. Vet. J.* **101**, 171–179.

Droege, W. (1976). The antigen-inexperienced thymic suppressor cells: a class of lymphocytes in the young chicken thymus that inhibits antibody production and cell-mediated immune responses. *Eur. J. Immunol.* **6**, 279–287.

Ekino, S., Nawa, Y., Tanaka, K., Matsuno, K., Fujii, H., and Kotani, M. (1980). Suppression of immune response by isolation of the bursa of Fabricius from environmental stimuli. *Aust. J. Exp. Biol. Med. Sci.* **58**, 289–296.

Eychmans, L., Schonne, E., and Eyssen, H. (1968). Antibody production in bursectomized chickens after injection Bacteriophage ϕX124. *Life Sci.* **7**, 161–169.

Faragher, J. T., Allan, W. H., and Cullen, G. A. (1972). Immunosuppressive effect of the infectious bursal agent in the chicken. *Nature New Biol.* **237**, 118–119.

Feace, A., and Poli, G. (1967). Effetto della vaccumazione contro la pseudopeste in polli privati ormonicamente della borsa di Fabrizio. *Arch. Vet. Ital.* **18**, 241–265.

Fitzsimmons, R. C., Garrod, E. M., and Garnett, I. (1973). Immunological responses following early embryonic surgical bursectomy. *Cell. Immunol.* **9**, 377–383.

Fitzsimmons, R. C., Dixon, D., and Kocal, E. (1977). The bursal–thymic interrelationship and ontogeny of the immune response in the chick embryo. *Dev. Immunobiol. Proc. Symp.*, pp. 387–394.

Franti, C. E., Adler, H. E., and Julian, L. M. (1971). Antibiotic growth promotion: effects of bacitracin and oxytetracycine on intestines and selected lymphoid tissues of New Hampshire cockerels. *Poult. Sci.* **50**, 97–99.

Freeman, B. M. (1967). Effect of stress on the ascorbic acid content of the adrenal gland of *Gallus domesticus*. *Comp. Biochem. Physiol.* **23**, 303–305.

Freeman, B. M. (1969). The bursa of Fabricius and adrenal cortical activity in *Gallus domesticus. Comp. Biochem. Physiol.* **29,** 639–646.

Freeman, B. M. (1970). The effects of adrenocorticotrophic hormone on adrenal weight and adrenal ascorbic acid in the normal and bursectomized fowl. *Comp. Biochem. Physiol.* **32,** 755–761.

Freeman, B. M., Chubb, L. G., and Pearson, A. W. (1966). Some effects of adrenocorticotrophic hormone on bursectomized and intact chickens. *In* "Physiology of the Domestic Fowl" (C. Horton-Smith, and E. C. Amoroso, eds.), pp. 103–112. Oliver & Boyd, Edinburgh.

Frommel, D., Perey, D. Y. E., and Good, R. A. (1970). Metabolism of γG and γM immunoglobulins in normal and hypogammaglobulinemic chickens. *J. Immunol.* **105,** 1–6.

Garren, H. W., and Barber, C. W. (1955). Endocrine and lymphatic gland changes occurring in young chickens with fowl typhoid. *Poult. Sci.* **34,** 1250–1259.

Garren, H. W., and Shaffner, C. S. (1954a). Factors concerned in the response of young New Hampshires to muscular fatigue. *Poult. Sci.* **33,** 1095–1104.

Garren, H. W., and Shaffner, C. S. (1954b). Young turkeys are more resistant to muscular fatigue than young chickens. *Poult. Sci.* **33,** 866–867.

Garren, H. W., and Shaffner, C. S. (1956). How the period of exposure to different stress stimuli affects the endocrine and lymphatic gland weights of young chickens. *Poult. Sci.* **35,** 266–273.

Gasc, J. M., Sar, M., and Stumpf, W. E. (1979). Androgen target cells in the bursa of Fabricius of the chicken embryo: autoradiographic localization. *Proc. Soc. Exp. Biol. Med.* **160,** 55–58.

Gershwin, M. E., Erickson, K., Montero, J., Abplanalp, H., Eklund, J., Benedict, A. A., and Ikeda, R. M. (1979). The immunopathology of spontaneously acquired dysgammaglobulinemia in chickens. *Clin. Immunol. Immunopathol.* **17,** 15–30.

Giambrone, J. J., Anderson, W. I., Reid, M. W., and Eidson, C. S. (1977). Effect of infectious bursal disease on the severity of *Eimeria tenella* infection in broiler chicks. *Poult. Sci.* **56,** 243–246.

Glick, B. (1955a). "Growth and Function of the Bursa of Fabricius in the Domestic Fowl." Ph.D. Dissertation, Ohio State Univ., Columbus.

Glick, B. (1955b). Growth and function of the bursa of Fabricius. *Poult. Sci.* **34,** 1196.

Glick, B. (1956). Normal growth of the bursa of Fabricius in chickens. *Poult. Sci.* **35,** 843–851.

Glick, B. (1957a). Experimental modification of the growth of the bursa of Fabricius. *Poult. Sci.* **36,** 18–23.

Glick, B. (1957b). The effect of penicillin and cortisone on the bursa of Fabricius. *Poult. Sci.* **36,** 1038–1041.

Glick, B. (1959). Experimental production of the stress picture with cortisone and the effect of penicillin in the young chicken. *Ohio J. Sci.* **59,** 81–86.

Glick, B. (1960a). Growth of the bursa of Fabricius and its relationship to the adrenal gland in the White Pekin duck, White Leghorn, outbred and inbred New Hampshire. *Poult. Sci.* **39,** 130–139.

Glick, B. (1960b). The effect of bovine growth hormone, desoxycorticosterone, and cortisone on the weight of the bursa of Fabricius, adrenal glands, heart, and body weight of young chickens. *Poult. Sci.* **39,** 1527–1533.

Glick, B. (1961). The influence of dipping eggs in hormonal solutions on the lymphatic tissue and antibody response of chickens. *Endocrinology (Baltimore)* **69,** 984–985.

Glick, B. (1963). The effect of surgical and chemical bursectomy in the White Pekin Duck. *Poult. Sci.* **42,** 1106–1113.

Glick, B. (1964). The bursa of Fabricius and the development of immunologic competence. *In*

"The Thymus in Immunobiology" (R. A. Good, and A. Gabrielsen, eds.), pp. 343–358, Harper (Hoeber), New York.

Glick, B. (1967). Antibody and gland studies in cortisone and ACTH injected birds. *J. Immunol.* **98**, 1076–1084.

Glick, B. (1968a). The immune response of bursaless birds as influenced by antibiotics and age. *Proc. Soc. Exp. Biol. Med.* **127**, 1054–1057.

Glick, B. (1968b). Serum protein electrophoresis patterns in acrylamide gel: patterns from normal and bursaless birds. *Poult. Sci.* **47**, 807–814.

Glick, B. (1970a). The bursa of Fabricius: the central issue. *BioScience* **20**, 602–604.

Glick, B. (1970b). The immune response of testosterone propionate treated chicks. *Int. Arch. Allergy Appl. Immunol.* **38**, 93–103.

Glick, B. (1971). Morphological changes and humoral immunity in cyclophosphamide-treated chicks. *Transplantation* **11**, 433–439.

Glick, B. (1972). Cortisone, age and antibody-mediated immunity. *Int. Arch. Allergy Appl. Immunol.* **43**, 766–771.

Glick, B. (1974). Antibody-mediated immunity in the presence of Mirex and DDT. *Poult. Sci.* **53**, 1476–1485.

Glick, B. (1976). Lymphocyte lifespan in chickens. *In* "Phylogeny of Thymus and Bone Marrow—Bursa Cells" (R. K. Wright, and E. L. Cooper, eds.), pp. 237–246. Elsevier/North-Holland, New York.

Glick, B. (1977a). The bursa of Fabricius and immunoglobulin synthesis. *Int. Rev. Cytol.* **48**, 345–402.

Glick, B. (1977b). Lymphocyte life span and migration of bursal lymphocytes. *In* "Developmental Immunology" (J. B. Solomon and J. D. Horton, eds.), pp. 371–378. Elsevier/North-Holland, New York.

Glick, B. (1978). The immune response in the chicken: lymphoid development of the bursa of Fabricius and thymus and avian immune response role for the gland of Harder. *Poult. Sci.* **57**, 1441–1444.

Glick, B. (1980). The thymus and bursa of Fabricius: endocrine organs? *In* "Avian Endocrinology" (A. Epple, and M. H. Stetson, eds.), pp. 209–230. Academic Press, New York.

Glick, B. (1981). B-cell kinetics within the bursa of Fabricius. *Fed. Proc. Fed. Am. Soc. Exp. Biol.* **40**, 1102.

Glick, B., and Olah, I. (1982). The morphology of the starling (*Sturnus vulgaris*) bursa of Fabricius. *Anat. Rec.* **204**, 341–348.

Glick, B., and Dreesen, L. J. (1967). The influence of selecting for large and small bursa size on adrenal, spleen, and thymus weights. *Poult. Sci.* **46**, 396–402.

Glick, B., and McDuffie, F. C. (1974). Embryonic bursal development in the presence of testosterone. *Proc. World's Poult. Congr. 15th*, pp. 519–521.

Glick, B., and McDuffie, F. C. (1975). Immunoglobulin and the bursa of Fabricius. *J. Reticuloendothel. Soc.* **17**, 119–125.

Glick, B., and Olah, I. (1981). Gut-associated-lymphoid tissue of the chicken. *In* "Scanning Electron Microscopy III," pp. 99–108. SEM, Chicago.

Glick, B., and Rosse, C. (1981). Cellular composition of the bone marrow in the chicken. II. The effect of age and the influence of the bursa of Fabricius on the size of cellular compartments. *Anat. Rec.* **200**, 471–479.

Glick, B., and Sadler, C. R. (1961). The elimination of the bursa of Fabricius and reduction of antibody production in birds from eggs dipped in hormone solutions. *Poult. Sci.* **40**, 185–189.

Glick, B., and Schwarz, M. R. (1975). Thymidine and testosterone binding by bursal and thymic lymphocytes. *Immunol. Commun.* **4**, 123–137.

Glick, B., and Whatley, S. (1967). The presence of immunoglobulin in the bursa of Fabricius. *Poult. Sci.* **46**, 1587–1589.

Glick, B., Chang, T. S., and Jaap, R. G. (1956). The bursa of Fabricius and antibody production in the domestic fowl. *Poult. Sci.* **35**, 224–225.

Glick, B., Sato, K., and Cohenour, F. (1964). Comparison of the phagocytic ability of normal and bursectomized birds. *J. Reticuloendothel. Soc.* **1**, 442–449.

Glick, B., Perkins, W. D., and Holbrook, K. A. (1977a). Scanning and electron microscopy of the bursa of Fabricius from normal and testosterone-treated embryos. *J. Dev. Comp. Immunol.* **1**, 41–46.

Glick, B., Subba Rao, D. S. V., and McDuffie, F. C. (1977b). Identifying lympholytic and androgenic effects of androgenic steroids. *Gen. Comp. Endocrinol.* **31**, 133–137.

Glick, B., Thompson, D., and Day, E. J. (1981). Calorie–protein deficiencies and the immune response of the chicken. I. Humoral immunity. *Poult. Sci.* **60**, 2494–2500.

Gold, E. F., and Benedict, A. A. (1967). Comparison of polypeptide chains of γG-globulin from bursectomized and normal chickens. *Proc. Soc. Exp. Biol. Med.* **125**, 535–538.

Goldstein, G., Scheid, M., Boyse, E. A., Brand, A., and Gilmour, D. G. (1977). Thymopoietin and bursopoietin: induction signals regulating early lymphocyte differentation. *Cold Spring Harbor Symp. Quant. Biol.* **41**, 5–8.

Graetzer, M. A., Wolfe, H. R., Aspinall, R L., and Meyer, R. K. (1963). Effect of thymectomy and bursectomy on precipitin and natural hemagglutinin production in the chicken. *J. Immunol.* **90**, 878–887.

Grebenau, M., and Thorbecke, G. (1978). T cell tolerance in the chicken. I. Parameters affecting tolerance induction to human γ-globulin in agammaglobulinemic and normal chickens. *J. Immunol.* **120**, 1046–1053.

Grebenau, M. D., Lerman, S. P., Palladino, M. A., and Thorbecke, G. J. (1976). Suppression of adoptive antibody responses by addition of spleen cells from agammaglobulinemic chickens "immunized" with histocompatible bursa cells. *Nature (London)* **260**, 46–48.

Grebenau, M. D., Lerman, S. P., Chi, D. S., and Thorbecke, G. J. (1979). Transfer of agammaglobulinemia in the chicken. I. Generation of suppressor activity by injection of bursa cells. *Cell. Immunol.* **51**, 92–108.

Grossi, C. E., Lydyard, P. M., and Cooper, M. D. (1977). Ontogeny of B cells in the chicken. II. Changing patterns of cytoplasmic IgM expression and of modulation requirements for surface IgM by anti-μ antibodies. *J. Immunol.* **119**, 749–756.

Harris, S., Cecil, H. C., Bitman, J., and Lillie, R. J. (1976). Antibody response and reduction in bursa of Fabricius and spleen weights of progeny of chickens fed PCBs. *Poult. Sci.* **55**, 1933–1940.

Hirai, K., Kunchiro, K., and Shimakura, S. (1979). Characterization of immunosuppression in chickens by infectious bursal disease virus. *Avian Dis.* **24**, 950–965.

Hirota, Y., and Bito, Y. (1975). Diverse effects of bursectomy on humoral immune responses in the chicken. *Poult. Sci.* **54**, 1524–1538.

Hirota, Y., Suzuki, T., and Bito, Y. (1980). The development of unusual B-cell functions in the testosterone-propionate-treated chicken. *Immunology* **39**, 29–36.

Hitchner, S. B. (1970). The differentiation of infectious bursal disease (Gumboro) and avian nephrosis. *Proc. World's Poult. Congr. 14th*, p. 447.

Hoffman-Fezer, G., and Lade, R. (1972). Postembryonic development and involution of the bursa Fabricii in the chicken. *Z. Zellforsch. Mikrosk. Anat.* **124**, 406–418.

Hoffman-Fezer, G., and Losch, U. (1973). Comparative histological and immunological studies on immunologically defective hens. I. Hormonally bursectomized hens. *Zentralbl. Veterinaermed. Reiche A* **20**, 586–595.

Holbrook, K. A., Perkins, W., and Glick, B. (1974). The fine structure of the bursa of Fabricius.

"B" cell surface configuration and lymphoepithelial organization revealed by scanning and transmission electron microscopy. *J. Reticuloendothel. Soc.* **16**, 300–311.

Holbrook, K. A., Perkins, W. D., and Glick, B. (1977). Contact sites between lymphoid cells of the bursa of Fabricius, *in vivo* and *in vitro. Anat. Rec.* **189**, 567–576.

Houssain, E., Belo, M., and Le Douarin, N. M. (1976). Investigations on cell lineage and tissue interactions in the developing bursa of Fabricius through interspecific chimeras. *Dev. Biol.* **53**, 250–264.

Isakovic, K., Jankovic, B. D., Popeskovic, L., and Milosevic, D. (1963). Effect of neonatal thymectomy, bursectomy, and thymo–bursectomy on haemagglutinin production in chickens. *Nature (London)* **200**, 273–274.

Ivanyi, J. (1975). Immunodeficiency in the chicken I. Disparity in suppression of antibody responses to various antigens following surgical bursectomy. *Immunology* **28**, 1007–1013.

Ivanyi, J., and Salerno, A. (1971). Impairment of humoral antibody response in neonatally thymectomized and irradiated chickens. *Eur. J. Immunol.* **1**, 227–230.

Ivanyi, J., Marvanova, H., and Skamene, E. (1969). Immunoglobulin synthesis and lymphocyte transformation by anti-immunoglobulin sera in bursectomized chickens. *Immunology* **17**, 325–331.

Jaap, R. G. (1958). Large bursae Fabricii in Leghorn type baby chickens. *Poult. Sci.* **37**, 1462–1464.

Jaap, R. G. (1960). Heritabilities, gene interaction and correlations for growth of glands associated with antibody formation in the young chicken. *Poult. Sci.* **34**, 557–560.

Jacobs, R. P., Blaese, R. M., and Oppenheim, J. J. (1972). Inhibition of antigen-stimulated *in vitro* proliferation of thymic dependent chicken spleen cells by specific antibody. *J. Immunol.* **109**, 324–333.

Jaffe, W. P. (1965). The bursa of Fabricius and homograft immunity. *Poult. Sci.* **44**, 1615–1616.

Jaffe, W. P., and Fechheimer, N. S. (1966). Cell transport and the bursa of Fabricius, *Nature (London)* **212**, 92.

Jankovic, B. D., and Isakovic, K. (1966a). Effect of hyperimmunization on haemagllutinin production in neonatally bursectomized chickens. *Iugosl. Physiol. Pharmacol. Acta* **2**, 159–161.

Jankovic, B. D., and Isakovic, K. (1966b). Antibody production in bursectomized chickens given repeated injections of antigen. *Nature (London)* **211**, 202–203.

Jankovic, B. D., and Isakovic, K. (1967). Natural haemagglutinins in surgically bursectomized and thymectomized chickens. *Folia Biol. (Prague)* **13**, 401–405.

Jankovic, B. D., and Isvaneski, M. (1963). Experimental allergic encephalomyelitis in thymectomized bursectomized and normal chickens. *Int. Arch. Allergy Appl. Immunol.* **23**, 188–206.

Jankovic, B. D., and Mitrovic, K. (1967). Antibody producing cells in the chicken, as observed by fluorescent antibody technique. *Folia Biol. (Prague)* **13**, 406–410.

Jankovic, B. D., Isakovic, K., and Petrovic, S. (1972). Direct stimulation of lymphoid tissue of the chicken. I. Antibody producing, antibody-transferring and plaque forming capacity of the thymus, spleen and bursa of Fabricius following intrathymic and intravenous injection of antigen. *Eur. J. Immunol.* **2**, 18–25.

Jankovic, B. D., Knezevic, Z., Isakovic, K., Mitrovic, K., Markovic, M. B., and Rascevic, M. (1975). Bursa lymphocytes and IgM containing cells in chicken embryos bursectomized at 52–64 hours of incubation. *Eur. J. Immunol.* **5**, 656–659.

Jankovic, B. D., Isakovic, K., Markovic, B. M., Rascevic, M., and Knezevic, Z. (1976). Non bursal origin of humoral immunity: immune capacity and cytomorphological changes in chickens bursectomized as 52–64 hour-old embryos. *Exp. Hematol. (Copenhagen)* **4**, 246–255.

Jankovic, B. D., Isakovic, K., Markovic, B. M., and Rascevic, M. (1977). Immunological capacity of the chicken embryo. II. Humoral immune responses in embryos and young chickens bursectomized and sham-bursectomized at 52–64 h of incubation. *Immunology* **32**, 689–699.

Jankovic, B. D., Isakovic, K., and Knezevic, Z. (1978). Ontogeny of the immuno– neuro–endocrine relationship. Changes in lymphoid tissues of chick embryos surgically decapitated at 33–38 hours of incubation. *Dev. Comp. Immunol.* **2**, 479–492.

Jolly, J. (1915). La bourse de Fabricius et les organes lympho–epitheliaux. *Arch. Anat. Microsc. Morphol. Exp.* **16**, 363–546.

Jolly, J., and Pezard, A. (1928). La castration retarde l'involution de la bourse de Fabricius. *C. R. Seances Soc. Biol. Ses Fil.* **98**, 379.

Karakoz, I., Hasek, M., and Kohoutova, L. (1976). The role of bursa of Fabricius and thymus in antibody-mediated allograft rejection in ducks. *Folia Biol. (Prague)* **22**, 304–411.

Keily, D., and Abramoff, P. (1969). Studies of the chicken immune response. III. Cellular and humoral antibody production in the splenectomized chickens. *J. Immunol.* **102**, 1058–1063.

Kemenes, F., and Pethes, G. (1963). Further evidence for the role of the bursa of Fabricius in antibody production in chickens. *Z. Immun. Allergieforsch.* **125**, 446–458.

Kermani-Arab, V., and Leslie, G. A. (1977). Suppression of immunoglobulin synthesis by transplantation of T cells from anti-μ bursectomized chickens into normal recipients. *J. Immunol.* **119**, 530–536.

Kerstetter, T. H., Buss, I. O., and Went, H. A. (1962). Antibody producing function of the bursa of Fabricius of the Ring-necked Pheasant. *J. Exp. Zool.* **149**, 233–237.

Kincade, P. W., and Cooper, M. D. (1971). Development and distribution of immunoglobulin containing cells in the chicken. An immunofluorescent analysis using purified antibodies to μ, γ and light chains. *J. Immunol.* **106**, 371–382.

Kincade, P. W., and Cooper, M. D. (1973). Immunoglobulin A: site and sequence expression in developing chicks. *Science (Washington, D.C.)* **179**, 398–400.

Kincade, P. W., Lawton, A. R., Bockman, D. E., and Cooper, M. D. (1970). Suppression of immunoglobulin G synthesis as a result of antibody mediated suppression of immunoglobulin M synthesis in chickens. *Proc. Natl. Acad. Sci. USA* **67**, 1918–1925.

Kincade, P. W., Self, K. S., and Cooper, M. D. (1973). Survival and function of bursa-derived cells in bursectomized chickens. *Cell. Immunol.* **8**, 93–102.

Kirkpatrick, C. M., and Andrews, F. N. (1944). The influence of sex hormones on the bursa of Fabricius in the ring-necked pheasant. *Endocrinology (Baltimore)* **34**, 340–345.

Kulkarni, P. G., Dreesen, L. J., Love, J., and Glick, B. (1971). *In vitro* activity of bursal lymphocytes from two lines of chickens. *Exp. Cell Res.* **66**, 124–128.

Kulkarni, P., Dreesen, L. J., and Glick, B. (1972). The metabolic activity of bursal and thymic cells. *Immunol. Commun.* **1**, 375–384.

Lagerlof, B., and Sundelin, P. (1970). Hemolysin production in spleen and peripheral blood cells from normal and leukemic chickens. *Int. J. Cancer* **5**, 364–369.

Landreth, K. S., and Glick, B. (1973). Differential effect of bursectomy on antibody production in a large and small bursa line of New Hampshire chickens. *Proc. Soc. Exp. Biol. Med.* **144**, 501–505.

Lassila, O., Eskola, J., Toivanen, P., Martin, C., and Dieterlen-Lievre, F. (1978). The origin of lymphoid stem cells studied in chick yolk sac–embryo chimaeras. *Nature (London)* **272**, 353–354.

Lassila, O., Eskola, J., and Toivanen, P. (1979). Prebursal stem cells in the intraembryonic mesenchyme of the chick embryo at 7 days of incubation. *J. Immunol.* **123**, 2091–2094.

Lassila, O., Eskola, J., Toivanen, P., and Dieterlen-Lievre, F. (1980). Lymphoid stem cells in the intraembryonic mesenchyme of the chicken. *Scand. J. Immunol.* **11**, 445–448.

Lebacq, A.-M., and Ritter, M. (1979). B-cell precursors in early chicken embryos. *Immunology* **37**, 123–134.

Lebacq-Verheyden, A. M., Vaerman, A. M., and Heremans, H. F. (1972a). A possible homologue of mammalian IgA in chicken serum and secretions. *Immunology* **22**, 165–172.

Lebacq-Verheyden, A. M., Vaerman, J. P., and Heremans, J. F. (1972b). Immunohistologic distribution of the chicken immunoglobulins. *J. Immunol.* **109**, 652–654.

Lebacq-Verheyden, A. M., Vaerman, J. P., and Heremans, J. F. (1974). Quantification and distribution of chicken immunoglobulins IgA, IgM & IgG in serum and secretions. *Immunology* **27**, 683–692.

LeDouarin, N. M., and Jotereau, F. V. (1973a). Recherches sur l'origine embryologique des lymphocytes du thymus chez l'embryon d'oiseau. *C. R. Hebd. Seances Acad. Sci. Ser. D* **276**, 629–632.

LeDouarin, N. M., and Jotereau, F. V. (1973b). Origin and renewal of lymphocytes in avian embryo thymuses. *Nature New Biol.* **246**, 25–27.

LeDouarin, N. M., and Jotereau, F. V. (1975). Tracing of cells of the avian thymus through embryonic life in interspecific chimeras. *J. Exp. Med.* **142**, 17–40.

LeDouarin, N. M., Houssaint, E., Jotereau, F. V., and Belo, M. (1975). Origin of hemopoietic stem cells in embryonic bursa of Fabricius and bone marrow studied through interspecific chimeras. *Proc. Natl. Acad. Sci. USA* **72**, 2701–2705.

LeDouarin, N. M., Jotereau, F. V., and Houssaint, E. (1976). The lymphoid stem cells in the Avian Embryo. *In* "Phylogeny of Thymus and Bone Marrow—Bursa Cells" (R. K. Wright, and E. L. Cooper, eds.), pp. 117–226. Elsevier/North-Holland, Amsterdam.

LeDouarin, N. M., Houssaint, E., and Jotereau, F. V. (1977). Differentiation of the primary lymphoid organs in avian embryos: origin and homing of the lymphoid stem cells. *In* "Avian Immunology" (A. A. Benedict, ed.), pp. 29–37. Plenum, New York.

LeDouarin, N. M., Michel, G., and Baulieu, E. E. (1980). Studies of testosterone-induced involution of the bursa of Fabricius. *Dev. Biol.* **75**, 288–302.

Lerman, S. P., and Weidanz, W. D. (1970). Selective suppression of humoral immunity. The effect of cyclophosphamide on the ontogeny of the humoral immune response in chickens. *J. Immunol.* **105**, 614–619.

Lerman, S. P., Grebenau, M. D., Chi, D. S., Palladino, M. A., Galton, J., and Thorbecke, G. J. (1980). Transfer of agammaglobulinemia in the chicken. II. Characterization of the target of suppression. *Cell. Immunol.* **51**, 109–128.

Lerner, K. G., Glick, B., and McDuffie, F. C. (1971). Role of the bursa of Fabricius in IgG and IgM production in the chicken: evidence for the role of a non-bursal site in the development of humoral immunity. *J. Immunol.* **107**, 493–511.

Leslie, G. A., and Clem, L. W. (1969). Phylogeny of immunoglobulin structure and function. III. Immunoglobulins of the chicken. *J. Exp. Med.* **130**, 1337–1352.

Leslie, G. A., and Martin, L. N. (1973a). Suppression of chicken immunoglobulin ontogeny by F(ab')$_2$ fragments of anti-μ chain and by anti-L chain. *Int. Arch. Allergy* **45**, 429–438.

Leslie, G. A., and Martin, L. N. (1973b). Modulation of immunoglobulin ontogeny in the chicken: effect of purified antibody specific for μ chain on IgM, IgY, and IgA production. *J. Immunol.* **110**, 959–967.

Leslie, G. A., and Martin, L. N. (1974). The secretory immunologic system of fowl. IV. Serum and salivary immunoglobulins in normal, agammaglobulinemic and dysgammaglobulinemic chickens. *Int. Arch. Allergy* **46**, 834–841.

Lifter, J., and Choi, Y. (1977). Detergent solubilization of B-lymphocyte immunoglobulin. *In* "Avian Immunology" (A. A. Benedict, ed.), pp. 99–107. Plenum, New York.

Linna, T. J., Frommel, D., and Good, R. A. (1972). Effect of early cyclophosphamide treatment on the development of lymphoid organs and immunological function in the chicken. *Int. Arch. Allergy* **42**, 20–39.

Long, P. L., and Pierce, A. E. (1963). Role of cellular factors in the mediation of immunity to avian coccidiosis (*Eimeria tenella*). *Nature (London)* **200**, 426–427.

Losch, V. (1971). Investigation on dysgammaglobulinemia hens. *Arch. Gefluegelkd.* **1**, 25–28.

Lydyard, P. M., Grossi, C. E., and Cooper, M. D. (1976). Ontogeny of B cells in the chicken. I. Sequential development of clonal diversity in the bursa. *J. Exp. Med.* **144**, 79–97.

Maranova, H., and Hasek, P. (1969). The influence of bursectomy and thymectomy on the primary antibody formation to various antigens. *Folia Microbiol. (Prague)* **14**, 171–178.

Marinkovich, V. A., and Baluda, M. A. (1966). In vitro synthesis of M-like globulin by various chick embryonic cells. *Immunology* **10**, 383–397.

Martin, C., Beaupain, D., and Dieterlen-Lievre, F. (1978). Developmental relationships between vitelline and intra-embryonic haemopoiesis studied in avian 'yolk sac chimaeras'. *Cell Differ.* **7**, 115–130.

Martin, C., Lassila, O., Nurmi, T., Eskola, J., Dieterlen-Lievre, F., and Toivanen, P. (1979). Intraembryonic origin of lymphoid stem cells in the chicken: studies with sex chromosome and IgG allotype markers in histocompatible yolk sac–embryo chimaeras. *Scand. J. Immunol.* **10**, 333–338.

Martin, L. N., and Leslie, G. A. (1973). Ontogeny of IgA in normal and neonatally bursectomized chickens, with corroborative data on IgY and IgM. *Proc. Soc. Exp. Biol. Med.* **143**, 241–243.

Martin, L. N., and Leslie, G. A.(1974). IgM-forming cells as the immediate precursor of IgA-producing cells during ontogeny of the immunoglobulin-producing system of the chicken. *J. Immunol.* **113**, 120–126.

Matos, A., and Sousa, R. (1978). Tubuloreticular structures in chicken bursa Fabricii lymphocytes. *Experientia* **34**, 1212–1219.

Matsuda, H., and Bito, Y. (1973). Different effects of bursectomy of chickens on immune response to newcastle disease virus and *Salmonella pullorum* antigen. *Poult. Sci.* **52**, 1042–1052.

May, D., and Glick, B. (1964). Weight of the bursa of Fabricius and antibody response of chicks hatched from eggs dipped in varying concentrations of testosterone propionate. *Poult. Sci.* **43**, 450–454.

McArthur, W. P., Gilmour, D. G., and Thorbecke, G. J. (1972). Immunocompetent cells in the chicken. *Cell. Immunol.* **8**, 103–111.

Medina, E., and Pedernera, E. (1976). Effect of early bursectomy on allografts survival in chickens. *Experientia* **33**, 274–275.

Merkenschlager, M., Riedel, G., Kirchner, B., and Losch, U. (1966). Die Immunoglobuline bursektomierter Küken in der Immunoelectrophorese (IE). *Naturwissenschaften* **16**, 408–409.

Metcalf, D., and Moore, M. A. S. (1971). Embryonic aspects of haemopoiesis. *Front. Biol.* **24**, 172–271.

Meyer, R. K., Rao, M. A., and Aspinall, R. L. (1959). Inhibition of the development of the bursa of Fabricius in the embryo of the common fowl by 19-nortestosterone. *Endocrinology (Baltimore)* **64**, 890–898.

Meyer, R. K., Aspinall, R. L., Graetzer, M. A., and Wolfe, H. R. (1964). Effect of corticosterone on the skin homograft reaction and on precipitin and hemagglutinin production in thymectomized and bursectomized chickens. *J. Immunol.* **92**, 446–452.

Miller, J. F. A. P., and Mitchell, G. F. (1968). Cell to cell interaction in the immune response. I. Hemolysin-forming cells in neonatally thymectomized mice reconstituted with thymus or thoracic duct lymphocytes. *J. Exp. Med.* **128**, 801–820.

Moore, M. A. S., and Owen, J. J. T. (1965). Chomosome marker studies on the development of the haemopoietic system in the chick embryo. *Nature (London)* **208**, 956, 989–990.

Moore, M. A. S., and Owen, J. J. T. (1966). Experimental studies in the development of the bursa of Fabricius. *Dev. Biol.* **14,** 40–51.
Moore, M. A. S., and Owen, J. J. T. (1967a). Stem cell migration in developing myeloid and lymphoid systems. *Lancet* **2,** 658–659.
Moore, M. A. S., and Owen, J. J. T. (1967b). Chromosome marker studies in the irradiated chick embryo. *Nature (London)* **215,** 1081–1082.
Morgan, G. W., Jr., and Glick, B. (1972). A quantitative study of serum proteins in bursectomized and irradiated chickens. *Poult. Sci.* **51,** 771–778.
Morita, C., and Soekawa, M. (1972). Effect of thymectomy and bursectomy on migration inhibition test of splenic cells in chickens. *Poult. Sci.* **51,** 1133–1136.
Moticka, E. J., and Van Alten, P. J. (1971). Alterations in the kinetics of hemagglutinin formation. *J. Immunol.* **107,** 512–517.
Moticka, E. J., and Van Alten, P. J. (1972a). Cellular haemolysin response of chickens bursectomized as embryos. *Folia Biol. (Prague)* **18,** 331–335.
Moticka, E. J., and Van Alten, P. J. (1972b). Natural antibody to sheep erythrocytes in bursectomized chickens. *Proc. Soc. Exp. Biol. Med.* **141,** 295–297.
Mueller, A. P., Wolfe, H. R., and Meyer, R. K. (1960). Precipitin production in chickens. 21. Antibody production in bursectomized chickens and in chickens injected with 19-nor-testosterone on the fifth day of incubation. *J. Immunol.* **85,** 172–180.
Mueller, A. P., Wolfe, H. R., Meyer, R. K., and Aspinall, R. L. (1962). Further studies on the role of the bursa of Fabricius in antibody production. *J. Immunol.* **88,** 354–361.
Muir, F. V., and Jaap, R. G. (1969). Response of selection for bursa of Fabricius weight at hatching in the domestic fowl. *Poult. Sci.* **48,** 185–191.
Nazerian, K., Ackerson, A., and Hopper, G. (1976). Scanning electron microscopy in the study of chicken T and B cells from marek's disease tumours. *Avian Pathol.* **5,** 135–145.
Newcomer, W. S., and Connally, J. D. (1960). The bursa of Fabricius as an indicator of chronic stress in immature chickens. *Endocrinology (Baltimore)* **67,** 264–266.
Norton, J., and Wira, C. (1977). Dose-related effects of the sex hormones and cortisol on the growth of the bursa of Fabricius in chick embryos. *J. Steroid Biochem.* **8,** 985–987.
Oates, C. M., Bissenden, J. F., Maini, R. N., Payne, L. N., and Dumonde, D. C. (1972). Thymus and bursa dependence of lymphocyte mitogenic factor in the chicken. *Nature New Biol.* **239,** 137–139.
Odend'hal, S., and Breazile, J. E. (1979). Diffusely infiltrated area of lymphoid cells in the cloacal bursa. *J. Reticuloendothel. Soc.* **25,** 315–324.
Okuyama, S. I. (1965a). Immunological status of chickens with neonatal thymectomy and bursectomy. I. General aspects of neonatal thymectomy and bursectomy in chickens. *Sci. Rep. Res. Inst. Tohoka Univ. Ser. C* **12,** 275–284.
Okuyama, S. I. (1965b). Immunological status of chickens with neonatal thymectomy and bursectomy. III. Delayed hypersensitivity in chickens with neonatal thymectomy and bursectomy. *Sci. Rep. Res. Inst. Tohoka Univ. Ser. C* **12,** 297–302.
Olah, I., and Glick, B. (1978a). The number and size of the follicular epithelium (FE) and follicles in the bursa of Fabricius. *Poult. Sci.* **57,** 1445–1450.
Olah, I., and Glick, B. (1978b). Secretory cells in the medulla of the bursa of Fabricius. *Experienta* **34,** 1642–1643.
Olah, I., and Glick, B. (1979). Light and electron microscope structure of secretory cells in the medulla of bursal follicles of normal and cyclophosphamide-treated chickens. *Dev. Comp. Immunol.* **3,** 101–115.
Olah, I., Rohlich, P., and Toro, I. (1975). "Ultrastructure of Lymphoid Organs." Lippincott, Philadelphia, Pennsylvania.

Olah, I., Glick, B., McCorkle, F., and Stinson, R. (1979). Light and electron microscope structure of secretory cells in the medulla of bursal follicles of normal and cyclophosphamide-treated chickens. *J. Dev. Comp. Immunol.* **3**, 101–115.

Orlans, E., and Rose, M. E. (1972). An IgA-like immunoglobulin in the fowl. *Immunochemistry* **9**, 833–838.

Ortega, L. G., and Der, B. K. (1964). Studies of agammaglobulinemia induced by ablation of the bursa of Fabricius. *Fed. Proc. Fed. Am. Soc. Exp. Biol.* **23**, 546.

Palladino, M. A., Lerman, S. P., and Thorbecke, G. J. (1976). Transfer of hypogammaglobulinemia in two inbred chicken strains by spleen cells from bursectomized donors. *J. Immunol.* **116**, 1673–1676.

Palladino, M. A., Grebenau, M. D., and Thorbecke, G. J. (1978). Requirements for induction of delayed hypersensitivity in the chicken. *Dev. Comp. Immunol.* **2**, 121–132.

Panda, B., and Combs, G. F. (1964). The effect of coccidiosis in different glands of the growing chick. *Avian Dis.* **8**, 7–13.

Pedernera, E. A., Romano, M., Besedovsky, H. O., and Aguilar, M. D. C. (1980). The bursa of Fabricius is required for normal endocrine development in chicken. *Gen. Comp. Endocrinol.* **42**, 413–419.

Perek, M., and Eckstein, B. (1959). The adrenal ascorbic acid content of molting hens and the effect of ACTH on the thermal ascorbic acid content of laying hens. *Poult. Sci.* **38**, 996–999.

Perek, M., and Eilat, A. (1960). The bursa of Fabricius and adrenal ascorbic acid depletion following ACTH injections in chicks. *J. Endocrinol.* **20**, 284–285.

Perek, M., Eckstein, B., and Eshkol, Z. (1959). The effect of ACTH on adrenal ascorbic acid in laying hens. *Endocrinology (Baltimore)* **64**, 831–832.

Perey, D. Y. E., and Bienenstock, J. (1973). Effects of bursectomy and thymectomy on ontogeny of fowl IgA, IgG and IgM. *J. Immunol.* **111**, 633–637.

Perey, D. Y., Cooper, M. D., and Good, R. A. (1967). Normal second set wattle homograft reaction in agammaglobulinemic chickens. *Transplantation* **5**, 615–623.

Peterson, R. D. A., and Good, R. A. (1965). Morphological and developmental differences between the cells of the chicken's thymus and bursa of Fabricius. *Blood* **26**, 269–280.

Pierce, A. F., and Long, P. L. (1965). Studies on acquired immunity to coccidiosis in bursaless and thymectomized fowls. *Immunology* **9**, 427–440.

Pierce, A. E., Chubb, R. C., and Long, P. L. (1966). The significance of the bursa of Fabricius in relation to the synthesis of 7S and 19S immune globulins and specific antibody activity in the fowl. *Immunology* **10**, 321–338.

Pintea, V., Constantinescu, Gh. M., and Radu, C. (1967). Vascular and nervous supply of bursa of Fabricius in the hen. *Acta Vet. Acad. Sci. Hung.* **17**, 263–268.

Potworowski, E. F., Zavallone, J. D., Gilker, J. C., and Lamoureux, G. (1971). Inhibition of graft-versus-host reaction by thymus-specific antibodies. *Rev. Eur. Etud. Clin. Biol.* **16**, 155–157.

Rao, M. A., Aspinall, R. L., and Meyer, R. K. (1962). Effect of dose and time of administration of 19-nortestosterone on the differentiation of lymphoid tissue in the bursa of Fabricii of chick embryos. *Endocrinology (Baltimore)* **70**, 159–167.

Retterer, E., and Lelievre, E. (1913). Homologies de la bourse de Fabricius. *C. R. Soc. Biol.* **74**, 382.

Ritter, M. A., and Lebacq, A.-M. (1977). Embryonic bursa development in vitro. *Eur. J. Immunol.* **7**, 468–475.

Romppanen, T., and Sorvari, T. E. (1980). Chemical bursectomy of chickens with colchicine applied to the anal lips. *Am. J. Pathol.* **100**, 193–208.

Rose, M. E., and Orlans, E. (1968). Normal immune responses of bursaless chickens to a secondary antigenic stimulus. *Nature (London)* **217**, 231–235.

Rouse, B. T., and Szenberg, A. (1974). Functional and morphological observations on the effect of cyclophosphamide on the immune response of the chicken. *Aust. J. Exp. Biol. Med. Sci.* **52**, 873–885.

Rouse, B. T., and Warner, N. L. (1972). Depression of humoral antibody formation in the chicken by thymectomy and antilymphocyte serum. *Nature New Biol.* **236**, 79–80.

Rouse, T. C. (1967). "The Effect of Neonatal Bursectomy, Thymectomy and Splenectomy on Resistance to Caecal Coccidiosis (*Eimeria tenella*) in White Leghorn Cockerels." Ph.D. Dissertation, Univ. of Wisconsin, Madison (Diss. Abs. No. 28,1286).

Sachs, H. G., Beezhold, D. H., and Van Alten, P. J. (1979). The effect of cyclophosphamide on the structure and function of the bursal epithelium. *J. Reticuloendothel. Soc.* **26**, 1–9.

Sato, K. (1966). The relation of the bursa of Fabricius to the immune response against new castle disease virus. *Igaku no Ayumi* **57**, 183.

Sato, K., and Abe, S. (1975). The possible presence of a bursa-independent, IgM-producing system in chicks. *Immunology* **28**, 293–299.

Sato, K., and Glick, B. (1964a). Effect of ACTH on total adrenal cholesterol in bursectomized chicks. *Am. J. Physiol.* **207**, 47–49.

Sato, K., and Glick, B. (1964b). Bursa histology and antibody response. *J. Miss. Acad. Sci.* **10**, 213–218.

Sato, K., and Glick, B. (1970). Antibody and cell mediated immunity in corticosteroid-treated chicks. *Poult. Sci.* **49**, 982–986.

Sato, K., and Glick, B. (1972). Production of mercaptoethanol sensitive antibody in bursectomized chicks. *Poult. Sci.* **51**, 1358–1360.

Schaffner, T., Mueller, J., Hess, M. W., Cottier, H., Sordat, B., and Ropke, C. (1974). The Bursa of Fabricius: a central organ providing for contact between the lymphoid system and intestinal content. *Cell. Immunol.* **13**, 304–312.

Schaffner, T., Herring, J., Gerber, K., and Cottier, H. (1976). Bursa of Fabricius: uptake of radio-active particles and radiotoxic "sealing" of bursal follicles. *Adv. Exp. Med. Biol.* **66**, 33–39.

Schaumburg, B. P., and Crone, M. (1971). Binding of corticosterone by thymus cells, bursa cells and blood lymphocytes from the chicken. *Biochim. Biophys. Acta* **237**, 494–501.

Schomberg, D. W., Stob, M., and Andrews, F. N. (1964). Effects of 6α-methyl-17-acetoxyprogesterone, 17α-ethynyl-19-nortestosterone, progesterone, and testosterone propionate on the adrenals, gonads, and bursa of Fabricius of chicken. *Gen. Comp. Endocrinol.* **4**, 54–60.

Seltzer, W. (1956). The method of controlling the sex of the avian embryo, improving embryo hatchability and improving viability of the hatched chick. U.S. Patent No. 2,734,482.

Selye, H. (1943). Morphological changes in the fowl following chronic over-dosage with various steroids. *J. Morphol.* **73**, 401–421.

Seto, F. (1976). Enhancement of the hemagglutinin response in chickens by transferred thymocytes, bone marrow cells and bursacytes. *In* "Phylogeny of Thymus and Bone Marrow—Bursa Cells" (R. K. Wright, and E. L. Cooper, eds.), pp. 227–236, Elsevier/North-Holland, New York.

Sherman, J., and Auerbach, R. (1966). Quantitative characterization of chick thymus and bursa development. *Blood* **27**, 371–379.

Siegel, H. S. (1961). Age and sex modification of response to adrenocorticotropin in young chickens. I. Changes in adrenal and lymphatic gland weights. *Poult. Sci.* **40**, 1263–1274.

Sivanandan, V., and Maheswaran, S. K. (1980). Immune profile of infectious bursal disease. I. Effect of infectious bursal disease virus on peripheral blood T and B lymphocytes of chickens. *Avian Dis.* **24**, 715–725.

Sorvari, R., and Sorvari, T. (1977). Bursa Fabricii as a peripheral lymphoid organ, transport of various materials from the anal lips to the bursal lymphoid follicles with reference to its immunological importance. *Immunology* **32**, 499–505.

Sorvari, R., Naukkarinen, A., and Sorvari, T. E. (1977). Anal sucking-like movements in the chicken and chick embryo followed by the transportation of environmental material to the bursa of Fabricius, caeca and caecal tonsils. *Poult. Sci.* **56**, 1426–1429.

Sorvari, T., Sorvari, R., Ruotsalainen, P., Toivanen, A., and Toivanen, P. (1975). Uptake of environmental antigens by the bursa of Fabricius. *Nature (London)* **253**, 217–219.

Stinson, R. S., and Glick, B. (1978). Scanning electron microscopy of chicken lymphocytes: a comparative study of thymic, bursal and splenic lymphocytes. *Dev. Comp. Immunol.* **2**, 311–318.

Stinson, R. S., Mashaly, M., and Glick, B. (1979). Thrombocyte migration and the release of thrombocyte inhibitory factor (ThrIF) by T and B cells in the chicken. *Immunology* **36**, 769–774.

Subba Rao, D. S. V., and Glick, B. (1977). The production of the lymphocyte inhibitory factor (LyIF) by bursal and thymic lymphocytes. *In* "Avian Immunology" (A. A. Benedict, ed.), pp. 87–89. Plenum, New York.

Subba Rao, D. S. V., McDuffie, F. C., and Glick, B. (1978). The regulation of IgM production in the chicken: roles of the bursa of Fabricius, environmental antigens, and plasma IgG. *J. Immunol.* **120**, 783–787.

Sullivan, D. A., and Wira, C. R. (1978). Sex hormone and glucocorticoid receptors in the bursa of Fabricius of immature chicks. *J. Immunol.* **122**, 2617–2623.

Szenberg, A., and Warner, N. L. (1962). Dissociation of immunological responsiveness in fowls with a hormonally arrested development of lymphoid tissue. *Nature (London)* **194**, 146–147.

Szenberg, A., and Warner, N. L. (1963). Breakdown of polyvalent tolerance in the chicken of thymic grafts. *Nature (London)* **198**, 1012–1013.

Taibel, A. M. (1938). Effecto della bursectomia sul timo in *Gallus domesticus. Riv. Biol.* **24**, 364–372.

Taylor, R., and Glick, B. (1980). Corticosterone and serum protein levels in male bursectomized New Hampshire chickens. *Poult. Sci.* **59**, 1666.

Temple, R. W., and Jaap, R. G. (1961). Age of dam and response to selection for increased weight of the bursa of Fabricius in day-old chicks. *Poult. Sci.* **40**, 1305–1319.

Theis, G. A., and Thorbecke, G. J. (1972). Suppression of delayed hypersensitivity reaction in bursectomized chickens by passively administered anti-immunoglobulin antisera. *J. Immunol.* **110**, 91–97.

Thorbecke, G. J., Warner, N. L., Hochwald, G. M., and Ohanian, S. H. (1968). Immune globulin production by the bursa of Fabricius of young chickens. *Immunology* **15**, 123–134.

Toivanen, A., Toivanen, P., and Good, R. A. (1972). Transplantation of cells from bursa of Fabricius into surgically bursectomized chicks. *Int. Arch. Allergy* **43**, 588–599.

Toivanen, P., and Toivanen, A. (1973). Bursal and postbursal stem cells in chickens. Functional characteristics. *Eur. J. Immunol.* **3**, 585–595.

Toivanen, P., Toivanen, A., and Good, R. A. (1972a). Ontogeny of bursal function in chicken. I. Embryonic stem cell for humoral immunity. *J. Immunol.* **109**, 1058–1070.

Toivanen, P., Toivanen, A., Linna, T. J., and Good, R. A. (1972b). Ontogeny of bursal function in chicken. II. Postembryonic stem cell for humoral immunity. *J. Immunol.* **109**, 1071–1080.

Toivanen, P., Toivanen, A., and Good, R. A. (1972c). Ontogeny of bursal function in chicken. III. Immunocompetent cell for humoral immunity. *J. Exp. Med.* **136**, 816–831.

Toivanen, P., Toivanen, A., and Vainio, D. (1974a). Complete restoration of bursa-dependent immune system after transplantation of semiallogeneic stem cells into immunodeficient chicks. *J. Exp. Med.* **139**, 1344–1349.

Toivanen, P., Toivanen, A., and Sorvari, T. (1974b). Incomplete restoration of the bursa-dependent immune system after transplantation of allogeneic stem cells into immunodeficient chicks. *Proc. Natl. Acad. Sci. USA* **71**, 957–961.

Toivanen, P., Toivanen, A., Molnar, G., and Sorvari, T. (1974c). Bursal and postbursal cells in the chicken. Age dependence of germinal center formation in spleen. *Int. Arch. Allergy* **47**, 749–761.

Toivanen, P., Lassila, O., Martin, C., Dieterlen-Lievre, F., Nurmi, T., and Eskola, J. (1979). Intraembryonic mesenchymal as a source of lymphoid stem cells. *Folia Biol. (Prague)* **25**, 299–300.

Vainio, O., and Toivanen, A. (1979). Maturation of bursal stem cells within allogeneic or syngeneic bursal microenvironment: acquisition of post-bursal maturity. *J. Immunol.* **123**, 1960–1964.

Van Alten, P. J., and Meuwissen, H. J. (1972). Production of specific antibody by lymphocytes of the bursa of Fabricius. *Science (Washington, D.C.)* **176**, 45–48.

Van Alten, P. J., Cain, W. A., Good, R. A., and Cooper, M. D. (1968). Gamma globulin production and antibody synthesis in chickens bursectomized as embryos. *Nature (London)* **217**, 358–360.

Van Alten, P. J., Waltenbaugh, C., Jaesson, L., and Danielson, J. (1976). Suppression of lymphocyte mitogenesis by a factor derived from bursal lymphocytes. In "Neoplasm Immunity: Mechanisms" (R. G. Trispen, ed.), pp. 1–14.

Van Meter, R., Good, R. A., and Cooper, M. D. (1969). Ontogeny of circulating immunoglobulins in normal, bursectomized and irradiated chicks. *J. Immunol.* **102**, 370–374.

Visco, R. J. (1973). The effect of *Eimeria tenella* infection and testosterone treatment on the weight of the bursa of Fabricius in young chickens. *Poult. Sci.* **52**, 1034–1042.

Vriend, J., Oishi, T., and Domey, R. G. (1975). Effects of light environments on the weight of bursa of Fabricius in white rock chicks. *Growth* **39**, 53–66.

Waltenbaugh, C. R., and Van Alten, P. J. (1974). The production of antibody by bursal lymphocytes. *J. Immunol.* **113**, 1079–1084.

Waltenbaugh, C. R., Allen, C., Molnar, J., Sabet, T. Y., and Van Alten, P. S. (1976). Impairment of clearance by the reticuloendothelial system of bursectomized chickens. *J. Reticuloendothel. Soc.* **19**, 3–9.

Warner, N. L., and Burnet, F. M. (1961). The influence of testosterone treatment on the development of the bursa of Fabricius in the chick embryo. *Aust. J. Biol. Sci.* **14**, 380–387.

Warner, N. L., and Szenberg, A. (1962). Effect of neonatal thymectomy on the immune response in the chicken. *Nature (London)* **96**, 784.

Warner, N. L., and Szenberg, A. (1963). Immunological reactivity of bursaless chickens in graft versus host reactions. *Nature (London)* **199**, 43–44.

Warner, N. L., Szenberg, A., and Burnet, F. M. (1962). The immunological role of different lymphoid organs in the chicken. I. Dissociation of immunological responsiveness. *Aust. J. Exp. Biol. Med. Sci.* **40**, 373–388.

Warner, N. L., Uhr, J. W., Thorbecke, G. J., and Ovary, Z. (1969). Immunoglobulins, antibodies and the bursa of Fabricius: induction of agammaglobulinemia and the loss of all antibody-forming capacity of hormonal bursectomy. *J. Immunol.* **103**, 1317–1330.

Warner, N. L., Ovary, Z., and Kantor, F. S. (1971). Delayed hypersensitivity reactions in normal and bursectomized chickens. *Int. Arch. Allergy* **40**, 719–728.

Weber, W. T. (1973). Selective proliferation of T cells in the mixed lymphocyte interaction. *Cell. Immunol.* **12**, 487–492.

Weber, W. T. (1975). Avian B lymphocyte subpopulations: origins and functional capacities. *Transplant. Rev.* **24,** 113–158.

Weber, W. T., and Alexander, J. E. (1978). The potential of bursa-immigrated hematopoietic precursor cells to differentiate to functional B and T cells. *J. Immunol.* **121,** 653–657.

Weber, W. T., and Foglia, L. M. (1979). Evidence for the presence of precursor B cells in normal and in hormonally bursectomized chick embryos. *Cell. Immunol.* **52,** 84–94.

Weber, W. T., and Mausner, R. (1977). Migration patterns of avian embryonic bone marrow cells and their differentiation to functional T and B cells. *In* "Avian Immunology" (A. A. Benedict, ed.), pp. 47–59. Plenum, New York.

Weidanz, W. P., Weber, W. T., and Maguire, H. C., Jr. (1976). Allergic contact dermatitis in the B-cell deficient chicken. *Int. Arch. Allergy Appl. Immunol.* **50,** 775–758.

Weinbaum, F. I., Gilmour, D. G., and Thorbecke, G. J. (1973). Immunocompetent cells of the chicken. III. Cooperation of carrier sensitized T cells from agammaglobulinemic donors with hapten immune B cells. *J. Immunol.* **110,** 1434–1436.

Wenckebach, K. (1896). Die Follikel der Bursa Fabricii. *Anat. Anz.* **2,** 150–160.

Willis, J. I., and Schwartz, M. R. (1975). The circulating life span, immunocompetence and ^{3}H-uridine uptake of small lymphocytes from thymus-grafted, neonatally thymectomized rats. *J. Immunol.* **115,** 734–738.

Wilson, J. A., and Glick, B. (1966). Temperature control of testosterone propionate solutions used for reducing embryonic development of the bursa of Fabricius. *Poult. Sci.* **45,** 892–896.

Wolfe, H. R, Sheridan, S. A., Bilstad, N. M., and Johnson, M. A. (1962). The growth of lymphoidal organs and the testes of chickens. *Anat. Rec.* **142,** 485–493.

Woodward, M. (1931). "Studies in Bursectomized and Thymectomized Chickens." M.S. Dissertation, Kansas State Coll.,

Yarbrough, J., Wells, M., and Glick, B. (1971). Phagocytosis: the metabolism of thrombocytes from intact and bursaless birds. *J. Reticuloendothel. Soc.* **9,** 248–253.

Zarrow, M. X., Greenman, D. L., and Peters, L. E. (1961). Inhibition of the bursa of Fabricius and the stilboestrol stimulated oviduct of the domestic chick. *Poult. Sci.* **40,** 87–94.

AUTHOR INDEX

Numbers in italics refer to pages on which the complete references are listed.

A

Abe, S., 478, *497*
Abel, J. H., Jr., 276, 285, 286, 287, *358*
Abplanalp, H., 481, *485*, *486*, *488*
Abraham, G. E., 290, *335*
Abramoff, P., 473, *484*, *492*
Abs, M., 223, *335*
Abul-Hajj, Y. J., 326, *335*
Ackerman, G. A., 444, 447, 456, *484*, *485*
Ackerson, A., 454, *495*
Adelmann, H. B., 443, 444, *485*
Adigweme, A. I., 326, *335*
Adkins, E. K., 228, 230, 231, 234, 259, 260, 261, 262, 268, 269, 271, 280, 289, 301, 302, 310, 311, 312, 313, 322, 323, 324, 325, 327, *335*, *336*
Adkins-Regan, E., 230, 231, 311, 325, *336*
Adler, H. E., 463, 478, *486*, *487*
Adler, N. T., 228, 234, 259, 260, 261, 268, 269, 271, 310, 311, 313, 323, 324, *336*
Afton, A. D., 176, *212*
Aguilar, M. D. C., 466, *496*
Ainley, D.G., 170, 171, *218*
Åkerman, B., 285, 286, *336*
Akhmerov, A. Kh., 383, 385, 387, 397, 398, 399, 400, 401, 402, 405, *440*
Albini, B., 478, *485*
Alcock, J., 133, 202, 204, *145*, *212*
Alconero, B. B., 90, 94, *145*
Alexander, J. E., 458, *500*
Allan, W. H., 463, *485*, *487*
Allee, W. C., 228, 229, 230, 234, *336*
Allen, C. R., 472, *499*
Allen, J. M., 21, *74*
Alm, G. V., 268, 476, 477, *485*
Alway, J. H., 143, *145*
Amadon, D., 170, 172, *212*
Anderson, R.C., 400, 421, *432*, *433*
Anderson, W. I., 463, *488*
Andreev, A., 192, *212*
Andrew, R. J., 115, *145*, 223, 224, 254, 255, 257, 261, 265, 284, 293, 308, 311, *336*, *337*, *342*

Andrews, F. N., 461, 468, *492*, *497*
Andrews, S. E., 396, *433*
Angelo, R. L., 287, 330, *355*
Ankney, C. D., 61, 72, 176, 177, 178, 191, *212*
Annison, E. F., 27, *72*
Anthony, R., 167, *212*
Ar, A., 5, 6, 79, 169, *218*
Archer, J., 255, *337*
Aren-Engelbrektsson, B., 268, *337*
Arme, C., 425, *433*
Arnold, A. P., 118, *145*, 224, 228, 230, 232, 260, 276, 282, 283, 284, 285, 286, 305, *337*
Arscott, G. H., 168, *212*
Artiukh, E. S., 418, *433*
Aschoff, J., 169, 174, 193, *212*
Ashkenazie, S., 176, 192, 194, 196, 198, *212*
Ashmole, N. P., 141, *145*
Asmar, J. A., 463, *485*
Aspinall, R. L., 461, 467, 468, 469, *485*, *494*, *495*, *496*
Assenmacher, I., 22, *72*, 262, 264, *350*
Atkinson, E. L., 370, *436*
Attal, J., 262, *350*
Attardi, B., 231, *337*
Au, R.-J., 426, *433*
Auckland, J. N., 54, *72*
Auerbach, R., 461, *497*
Aulie, A., 20, 21, 56, *72*
Austin, G. T., 7, 8, 16, 26, *72*, 194, *212*
Azar, H. A., 463, *485*

B

Baer, J. G., 369, 415, *433*, *436*
Bailey, R. E., 250, *337*
Baker, D. H., 27, *73*
Bakhuis, W. L., 101, *145*
Bakke, T. A., 389, 427, *433*
Bakst, M., 293, *357*
Balda, R. P., 18, 19, *73*, 164, *219*
Baldwin, S. P., 3, 17, 31, 55, *72*, 76, 172, *212*

501

Balmer, R. T., 18, 72
Balthazart, J., 224, 228, 230, 231, 232, 233, 236, 237, 240, 250, 252, 255, 257, 258, 259, 260, 261, 262, 263, 267, 273, 274, 275, 285, 290, 291, 293, 295, 296, 297, 298, 299, 300, 301, 302, 303, 305, 306, 307, 320, 321, 322, 323, 324, 325, 326, 328, 331, 332, 334, 337, 338, 339, 341, 342, 344, 350, 360
Balthazart-Raze, C., 323, 325, 326, 338
Baltin, S., 102, 145
Baltz, D. M., 171, 212
Baluda, M. A., 474, 494
Bammann, A. R., 163, 192, 194, 216
Baptista, L. F., 121, 122, 124, 145, 146
Barash, D. P., 204, 213
Barber, C. W., 462, 488
Barbière, C., 245, 249, 250, 252, 361
Barfield, R. J., 258, 276, 277, 279, 280, 281, 282, 284, 286, 287, 303, 339, 342, 358, 359
Barraud, E. M., 128, 146
Barry T. W., 177, 213
Bartelmez, G. W., 171, 213
Bartholomew, G. A., 20, 72, 86, 146, 172, 218
Bartke, A., 300, 339
Barton, A. D., 35, 76
Baruš, V., 370, 396, 419, 422, 433
Baskett, T. S., 20, 73
Bates, R. W., 250, 252, 300, 339, 359
Bateson, P. P. G., 85, 86, 103, 104, 105, 106, 107, 108, 109, 112, 146, 159, 285, 339, 340, 349, 355
Batsch, A. J. G. C., 368, 433
Baulieu, E. E., 462, 493
Beach, F. A., 142, 143, 146, 230, 231, 232, 273, 285, 292, 339
Beardmore, J. A., 111, 147
Beaupain, D., 456, 485, 487, 494
Beaver, P. W., 107, 146
Beck, B., 141, 146
Bedard, J., 171, 213
Beecher, I. M., 115, 146
Beecher, M. D., 115, 146
Beer, C. G., 115, 146, 147
Beezhold, D. H., 453, 497
Behrman, H. R., 290, 350
Bekoff, M., 88, 147
Bell, D. J., 27, 72
Bell, R. Q., 240, 339, 349

Bellairs, R., 2, 72
Bellamy, D., 462, 485
Bellavia, A., 471, 486
Beller, D. I., 458, 485
Bellina, L., 471, 486
Belo, M., 456, 491
Belogurov, O. I., 382, 390, 391, 433, 441
Belopol'skaia, M. M., 382, 406, 433
Bendon, B., 226, 350
Benedict, A. A., 476, 481, 485, 486, 488, 490
Bennett, A. F., 86, 148, 201, 213
Bennett, M. A., 229, 339
Benoff, F. H., 224, 308, 331, 339, 340
Bergtold, W. H., 2, 72
Bermant, G., 228, 360
Bernstein, M. H., 19, 72
Berra, T., M., 426, 433
Berry, H. H., 296, 340
Berthold, A. D., 222, 230, 340
Berthold, P., 140, 147
Bertman, M., 425, 433
Bertolini, A., 257, 340
Bertram, B., 114, 123, 147
Besedovsky, H. O., 466, 496
Bethel, W. M., 425, 436
Beyer, C., 273, 274, 326, 352
Biebach, H., 179, 180, 183, 213
Bienenstock, J., 474, 475, 485, 496
Bilby, L. W., 9, 72
Bilstad, N. M., 461, 500
Birchard, G. F., 86, 147
Bird, C. E., 327, 363
Bird, M. W., 106, 157
Birkhead, T. R., 134, 147, 204, 213
Bischof, H.-J., 108, 147
Bissenden, J. F., 459, 495
Bissonnette, T. H., 167, 213
Bitman, J., 463, 490
Bito, Y., 472, 478, 479, 490, 494
Bixby, J. L., 101, 160
Black, C. P., 172, 176, 192, 194, 196, 197, 215
Blackmore, F. H., 189, 213
Blaese, R. M., 470, 480, 485, 491
Blankespoor, H. D., 409, 433
Blaustein, J. D., 235, 265, 267, 331, 332, 334, 338, 340, 344, 345
Bleau, G., 324, 345
Blem, C. R., 185, 213
Blivaiss, B. B., 22, 72

AUTHOR INDEX

Blohm, T. R., 324, *340*
Bloch, M. F., 409, *433*
Blow, W. L., 45, 47, 48, 50, 77, *82*
Blyth, J. S. S., 61, *72*
Blythman, H., 477, *485*
Boag, D. A., 143, 145, *147*
Boag, P. T., 47, *72*
Bockman, D. E., 448, 449, 452, 453, 477, 478, *485*, 492
Boersma, P. D., 61, *72*
Bogoiavlenskii, Iu. K., 381, 382, 383, 385, 386, 387, 389, 390, 392, 394, 395, 396, 397, 398, 400, 401, 402, 403, 405, *440*, *441*
Bohus, B., 257, *340*
Bolton, N. J., 250, *359*
Bona, F. V., 370, *433*
Bondani, A., 326, *352*
Bondarenko, S. K., 416, 418, 428, *433*
Böni, A., 18, *73*
Boomgardt, J., 27, *73*
Boop, J. J., 230, 231, 262, 301, 302, 324, *336*
Boorman, K. N., 27, *73*, 190, *213*
Boss, E. R., 224, 227, 230, *340*
Bossert, W. H., 69, *74*
Bottoni, L., 275, 319, 320, 321, 322, 323, 328, *338*, *340*, *344*, *355*
Boughton, R. V., 385, *433*
Bourlière, F., 236, *356*
Bowen, D. E., 69, *75*
Boyd, E. M., 397, 402, *434*
Boyd, R., 194, *213*
Boyden, E. A., 447, 449, *485*
Boyden, T. C., 194, *213*
Boyse, E. A., 458, 467, *490*
Brackenbury, J., 275, *340*
Bradley, P., 285, *340*
Brand, A., 458, 466, 467, *485*, *490*
Braun, C. E., 391, 427, *438*
Brazzill, D. M., 311, *365*
Breazile, J. E., 455, *495*
Breitenbach, R. P., 20, 73, 254, *340*, 461, 462, *486*, *487*
Bremser, J. G., 406, *434*
Breneman, W. R., 250, *340*
Brenner, F. J., 172, *213*
Brien, N. B., 473, *484*
Brinkley, H. J., 223, 291, 292, 298, *357*
Brockway, B. F., 237, *340*
Brodkin, L. D., 268, 316, *360*

Brody, P. N., 238, 246, 251, 303, *353*
Brody, S., 3, 36, 37, *73*
Brooks, D. S., 258, *342*
Brooks, W. S., 100, *147*
Brosset, A., 112, *147*
Broughton, E., 431, *434*
Brown, J. L., 116, 118, 120, *147*, 282, *340*
Brown, R. T., 107, *147*
Brown, W. O., 28, *79*
Brownlie, L. W., 395, *437*
Bruder, R. H., 287, 300, 330, *345*
Bruggeman, J. B., 477, 478, *486*
Bryant, B. J., 478, *486*
Bryant, D. M., 16, 24, 25, 59, 60, 61, *73*, 185, 187, 192, *213*, 215
Budtz-Olsen, O. E., 255, *343*
Buntin, J. D., 247, 248, 250, 252, *341*, *361*
Burger, A. E., 192, 194, *218*, 296, *341*
Burger, J., 12, *73*
Burghardt, G., 88, *147*
Burke, W. H., 250, 251, 254, *341*, *345*
Burkitt, J. P., 249, *341*
Burley, N., 204, *213*
Burnet, F. M., 461, 469, *499*
Burns, J. T., 250, *355*
Burns, R. B., 473, *486*
Bursian, A. V., 94, *147*
Burt, M. D. B., 388, 428, *437*, *439*
Busbee, E. L., 129, *147*
Bush, A. O., 383, 385, *434*, *435*
Buss, I. O., 473, 492
Butler, E. J., 27, *73*
Butterfield, P. A., 229, 253, 258, 259, *341*, *342*
Bykhovskaia-Pavlovskaia, I. E., 370, 380, 383, 386, 390, 391, 393, 412, 422, 428, 429, *434*
Bykhovsky, B., 406, *434*

C

Cade, T. J., 141, *147*
Cain, B. W., 170, *213*
Cain, W. A., 468, 470, 474, 477, *486*, *487*, *499*
Calder, W. A., 20, 73, 162, 163, 164, 170, 173, 192, 193, 201, *213*
Caldwell, B. W., 300, *339*
Callard, G. V., 280, 311, 316, 318, 327, 329, *341*
Campbell, J. C., 231, *354*
Canters, K., 192, *214*

Capranica, R. R., 101, *160*
Card, L. E., 22, 78
Carey, C., 6, 73, 170, 171, 172, *214*
Carey, J., 476, *486*
Caridoit, F., 274, *341*
Carpenter, C. R., 233, *341*
Carpenter, F. L., 165, 166, 192, 194, 196, 210, *214*, *215*
Caruso, C., 471, *486*
Case, T. J., 11, 63, 65, 66, 67, 68, 73
Casey, D. W., 50, 78
Cecil, H. C., 463, *490*
Cedard, L., 292, 311, *347*
Cekan, S. Z., 290, *341*
Chabaud, A., 418, 419, *434*
Chadwick, A., 250, 254, 298, *359*, *360*
Challey, J. R., 463, *486*
Champy, C., 244, *341*
Chang, T. S., 461, 467, 468, *486*, *490*
Chann, T. C., 481, 485, *486*
Chapdelaine, A., 324, *345*
Chapnick, M. H., 167, *213*
Chaptal, C., 231, 327, *355*
Chatonnet, J., 192, *217*
Chaudoir, C., *343*
Cheng, M. F., 226, 228, 230, 231, 232, 234, 235, 236, 237, 238, 240, 247, 248, 252, 258, 259, 260, 267, 269, 270, 271, 281, 283, 284, 293, 295, 300, 301, 303, 304, 323, 324, 325, 326, 331, 332, 334, *338*, *341*, *342*, *346*
Cheng, S. C., 227, 229, 274, *349*
Cherfas, J. J., 108, *147*
Cheville, N. F., 463, *486*
Chi, D. S., 481, *486*, *490*, *493*
Chiasson, R. B., 281, *345*
Chilgren, J. D., 187, 188, 189, 190, 191, *214*
Cho, B. R., 472, *486*
Choi, Y. S., 473, 474, 476, *486*, *493*
Choquette, L. P. E., 431, *435*
Christensen, N. O., 411, *434*
Chubb, L. G., 461, *488*
Chubb, R. C., 476, *496*
Claflin, A. J., 469, 477, *486*
Clark, A. B., 61, 73
Clark, A. F., 327, 328, *363*
Clark, G. A., Jr., 102, *147*
Clark, J. H., 331, *342*
Clark, L., 18, 19, 73
Claugher, D., 431, *434*
Clausen, B., 432, *434*

Clem, L. W., 476, *493*
Clifton, P. G., 257, 265, *342*
Coates, M. E., 27, 73
Cock, A. G., 3, 31, 41, 51, 73
Cockeram, C. C., 45, 47, 48, 82
Cody, M. L., 65, 73
Cohen, J., 283, 284, 304, *342*
Cohenour, F., 471, *490*
Coles, R. B., 95, *157*
Colle, P., 244, *341*
Collias, N. E., 106, *147*, 228, 229, 230, 234, 249, 254, 258, 261, 262, *336*, *342*
Collopy, M. W., 163, 192, 194, *216*
Colombo, L., 291, *342*
Colombo-Belvedere, P., 291, *342*
Combs, G. F., 463, *496*
Connally, J. D., 462, *495*
Connell, C. E., 170, *214*
Connell, C. J., 291, *342*
Connell, G. M., 291, *342*
Constantinescu, Gh. M., 447, *496*
Cooch, F. G., 111, *147*
Cooke, F., 111, *147*
Cooper, C. L., 397, 404, *434*
Cooper, J., 6, 82
Cooper, M. D., 448, 449, 456, 467, 468, 469, 470, 472, 474, 475, 476, 477, 478, 485, *486*, 487, *490*, *492*, *494*, *496*
Coppinger, R. P., 134, *147*
Cordier, A., 447, *487*
Cordy, D. R., 478, *486*
Cornwell, G. W., 427, *434*
Corvol, P. L., 331, *360*
Cosgrove, A. A., 463, *487*
Cottier, H., 449, 450, 453, *497*
Cowan, A. B., 427, *434*
Cowan, P. J., 127, *147*
Cox, D. C., 274, *340*
Craig, B., *343*
Cram, E. B., 369, *434*
Crawford, R. D., 12, 58, 73
Crawford, W. C., 280, *342*
Cresti, L., 318, *354*
Creutz, G., 427, *434*
Crites, J. L., 397, 404, *434*
Croix, D., 290, *342*
Crone, M., 462, *497*
Crook, J. H., 229, 253, 258, 259, *341*, *342*, *352*
Cruze, W. W., 130, *147*
Csapo, A., 21, 73
Culbert, J., 223, 292, *360*

Cullen, G. A., 463, *485*, *487*
Cullen, J. M., 6, 7, 8, 11, 15, 16, 24, 25, 26, 68, 70, 80, 126, *148*, 170, 184, 185, *218*
Cummins, R. A., 255, *343*
Cunningham, D. L., 308, *343*
Cunningham, F. J., 290, *346*
Cunningham, R. A., 300, *353*
Curio, E., 88, 89, *148*, *159*, 292, *343*
Cusick, E. K., 277, *343*
Custer, T. W., 192, 194, 197, *214*
Cuthbert, N. L., 172, *214*

D

Daan, S., 23, 24, 25, 63, 66, 68, *74*
Daly, M. J., 329, *358*
Damassa, A. J., 478, *486*
Damassa, D. A., 261, 262, 287, 301, 306, *343*, *361*
Dane, C. W., 302, *344*
Danielson, J. R., 459, *487*, *499*
Davidson, J. M., 261, 262, 274, 287, 301, 306, *343*, *361*
Davidson, W. R., 385, 432, *434*, *442*
Davies, D. T., 225, 277, 281, 298, 320, 321, 322, 323, 324, *343*, *346*, *355*
Davies, I. J., 316, 322, *356*
Davies, N. B., 129, 130, 131, 134, 144, *148*
Davies, O. L., 39, *74*
Davis, D. E., 228, 229, 230, 232, 233, 234, 318, *343*
Davis, E. A., Jr., 189, *214*
Davis, G. K., 22, *83*
Davis, H. N., 303, *347*
Davis, K. B., 250, *355*
Davis, P. G., 231, 281, 288, 327, 329, 330, *343*, *350*, *355*
Dawson, W. R., 21, 67, *74*, 76, 86, *146*, *148*
Day, E. J., 463, *490*
Decker, J. D., 93, *148*
DeGraw, W. A., 191, *214*
Delditte, P., 192, *216*
Delius, J. D., 141, *148*, 257, *343*, *344*
Demshin, N. I., 391, 393, 402, *438*
DeMuth, R. E., 33, *79*
Dennison, P. T., 250, 251, 254, *341*, *345*
Dent, P. B., 473, *487*
Der, B. K., 476, *496*
De Reviers, M., 223, *343*
De Rycker, C., 255, 285, *337*
Desjardins, C., 262, 301, *343*, *362*, *363*

Desombre, E. R., 315, *350*
Dessi-Fulgheri, F., 328, *343*
DeSteven, D., 58, *74*
Deviche, P., 228, 230, 232, 233, 257, 258, 259, 261, 273, 275, 291, 296, 298, 300, 301, 302, 305, 306, 320, 321, 322, 323, 327, 337, *342*, *343*
De Vlaming, V. L., 250, 254, *344*
Dewsbury, D. A., 303, *347*
Dey-Hazro, 419, *435*
Diamond, A. W., 113, *148*
Diamond, J. M., 122, *148*
Dickerson, J. W. T., 21, 54, *74*
Diehl, B., 185, *214*
Diesing, C. M., 368, 411, *434*
Dieter, M. P., 461, 462, *487*
Dieterlen-Lievre, F., 156, 157, *485*, *492*, *494*, *499*
Dieterlen-Lievre, R. D., 456, *487*
Dietrich, K., 121, *148*
Dilger, W. C., 90, *148*
Diminno, R. L., 397, *434*
Dittus, W. P. J., 119, 120, *148*
Dixon, D., 466, 467, *487*
Dmi'el, R., 200, *219*
Dobrokhotova, O. V., 418, *434*
Döhler, K. D., 302, *344*
Döhler, U., 302, *344*
Dogiel, V. A., 406, 425, 427, 428, *429*, *434*, *435*
Doi, O., 248, 292, 298, 311, 326, *344*, *363*
Dolnik, V. R., 23, 28, 76, 188, 189, 190, 200, *214*, *215*, *216*
Domey, R. G., 461, *499*
Domm, L. V., 228, 230, 232, 233, 234, *343*, *344*
Donham, R. S., 296, 302, *344*
Donovan, B. T., 312, *344*
Dooley, N., 480, *485*
Dooling, R., 123, *148*
Dorney, R. S., 385, *435*
Doster, G. L., 385, *434*
Dougherty, J. E., 56, 57, *76*
Dransfield, J. W., 447, *487*
Dreesen, L. J., 461, 462, 463, 464, *489*, *492*
Drent, R. H., 23, 24, 25, 59, 63, 66, 68, *74*, *82*, 171, 192, *214*, 235, 236, 251, *344*
Drobney, R. D., 170, 174, 175, 176, 177, *214*
Droege, W., 466, 467, *487*
Dror, Y., 52, 53, *78*
Drost, R., 113, *148*

Dubé, J. Y., 331, *344*
Dubinina, M. N., 370, 413, 415, *435*
Dubois, G., 369, *435*
Dudley, C. A., 258, 303, *356*
Dudley, S. D., 265, 340, *344*
Dugas, M., *356*
Dujardin, F., 368, *435*
Dumonde, D. C., 459, *495*
Dunn, E. H., 18, 19, 23, 24, 25, 28, 63, 69, 74, 184, 185, *214*
Dunn, E. K., 16, 18, 59, *74*
Dunnet, G. M., 12, 58, *78*
Dupuy, G. M., 324, *345*
Dusseau, J. W., 250, *355*
Dwyer, T. J., 192, 194, *214*
Dykshorn, S. W., 247, 250, 252, *359*

E

Eastzer, D. H., 88, 90, 116, *152*, *160*
Ebbinge, B., 192, *214*
Eckstein, B., 465, 466, *496*
Edens, F. W., 280, 311, *364*
Edwards, C., 252, *347*
Edwards, D. A., 231, *364*
Egge, A. S., 281, *345*
Eidson, C. S., 463, *488*
Eik-Nes, K. B., 291, *342*, *345*, *346*
Eilat, A., 465, *496*
Eisner, E., 249, 251, 254, *345*
Ekberg, P. H., 261, *361*
Ekino, S., 447, *487*
Eklund, J., 481, *488*
Elger, W., 231, *362*
El Halawani, M. E., 250, 251, *345*
Ellendorff, F., 329, *357*
Ellwood, M., 300, *359*
El-Wailly, A. J., 172, 181, 182, *214*
Ely, C. A., 113, *148*
Emlen, S. T., 110, 114, 136, 137, 145, *148*
Eneroth, P., 261, *361*
Enigk, K., 419, *435*
Erb, R. E., 300, 302, *364*
Erickson, C. J., 237, 238, 286, 287, 300, 330, *345*, *354*
Erickson, K. J., 291, 293, 295, 301, 352, *481*, *488*
Ernst, U., 88, 89, *148*, *159*
Erpino, M. J., 230, 232, 233, *345*
Esch, G. W., 410, *435*
Eshkol, Z., 465, 466, *496*

Eskola, J., 456, 457, *492*, *494*, *499*
Etches, R. J., 250, 254, *345*, *355*
Etienne, A., 224, 228, 230, 232, *345*
Ettinger, A. O., 163, 176, 192, 194, 195, 196, 198, 208, *214*
Evans, R. M., 115, 127, *147*, *148*
Ewald, P. W., 165, 166, *215*
Ewing, L. L., 301, *362*
Eychmans, L., 468, *477*
Eyssen, H., 468, 477, *487*
Eyster, M. B., 189, *215*

F

Fabens, A. J., 40, *74*
Fabricius, E., 105, *149*
Fagan, 140, 143, 145, *149*
Falconer, A. J., 40, *74*
Falconer, D. S., 43, 50, *74*
Falconer, I. R., 22, 45, *74*
Faragher, J. T., 463, *;85*, *487*
Farner, D. S., 18, 76, 167, 169, 191, *216*, 225, 226, 228, 243, 245, 250, 262, 275, 290, 291, 296, 302, 303, 335, *344*, *345*, *351*, *352*, *364*, *365*
Feace, A., 472, *487*
Fechheimer, N. S., 456, *491*
Fedak, M., 200, *219*
Feder, H. H., 235, 247, 248, 249, 252, 262, 267, 291, 293, 295, 301, 303, 331, 332, 333, 334, *338*, *340*, *345*, *357*, *361*
Fernandez, E. P., 330, *363*
Ferrari, W., 257, *340*
Ferrell, B. R., 22, *77*
Fevold, H. R., 291, *345*, *346*
Ficken, M. S., 140, 141, 142, 143, 145, *149*
Findell, P. R., 179, 180, *215*
Fischer, G. J., 108, 127, *149*
Fischer, G. W., 431, *441*
Fischer, H., 27, 74, 228, 230, 232, *345*
Fisher, A. E., 261, 262, 280, *346*
Fitzsimmons, R. C., 466, 467, 478, 479, *487*
Fjeldså, J., 16, *74*
Fleming, L., 330, *348*
Foelix, R. F., 91, 93, *149*
Foenander, F., 236, *355*
Foglia, L. M., 458, *500*
Folk, C., 211, *215*, 370, 396, 422, *433*
Follett, B. K., 22, *82*, 167, 169, *216*, 225, 228, 232, 243, 250, 252, 262, 277, 290, 291, 293, 295, 298, 300, 302, 303, 310,

321, 323, 325, 342, 343, 347, 348, 349, 350, 351, 353, 359, 360, 361, 363
Ford, R., 302, 359
Foreman, M. M., 258, 346
Forrester, D. J., 383, 385, 402, 434, 435
Forsyth, I. A., 250, 252, 341
Foster, M. S., 28, 29, 74
Fox, C. A., 303, 346
Franti, C. E., 463, 487
Frantz, D. W., 387, 440
Fraps, R. M., 281, 358
Freeman, B. M., 18, 20, 27, 72, 74, 461, 466, 487, 488
Fretwell, S. D., 69, 74, 208, 215
Freze, V. I., 381, 382, 401, 413, 435, 441
Friedman, M. B., 303, 346
Friedmann, H., 114, 149
Friedrich, B., 477, 478, 486
Friessen, H. G., 250, 355
Frommel, D., 476, 480, 488, 493
Frost, P. G. H., 166, 192, 194, 215, 218
Frost, S. K., 166, 192, 194, 215
Frumkes, P. J., 258, 342
Fry, C. H., 134, 149
Fuelling, D. E., 53, 83
Fuhrmann, O., 369, 406, 407, 435
Fujii, H., 447, 487
Fujioka, K., 248, 292, 311, 326, 363
Furr, B., 262, 346
Furuyama, S., 290, 346

G

Gadgil, M., 69, 74
Gärtner, K., 302, 344
Gagarin, V. G., 381, 435
Gallagher, J. E., 104, 107, 108, 109, 149
Gallagher, T. F., 274, 346
Galli, F. E., 291, 346
Galton, J., 481, 493
Garbutt, A., 254, 345
Garcia-Austt, E., 95, 159
Gardiner, A., 16, 24, 25, 73, 185, 213
Gardner, J. A., 261, 262, 280, 346
Garlick, P. J., 24, 77
Garnett, I., 478, 479, 487
Garnett, M. C., 47, 75
Garnier, D., 262, 346
Garren, H. W., 462, 488
Garrod, E. M., 478, 479, 487
Gasc, J. M., 461, 488

Gates, G. R., 95, 157
Gauldie, J., 474, 475, 485
Gavrilov, V. M., 23, 28, 76, 188, 189, 190, 200, 214, 215
Genther, G., 330, 348
Gentle, M. J., 276, 285, 346, 365
Gerlach, J. L., 272, 363
Gerling, S. A., 327, 365
Gershwin, M. E., 481, 485, 486, 488
Gessa, G. L., 257, 340
Gessaman, J. A., 179, 180, 215
Ghraf, R., 319, 346
Giambrone, J. J., 463, 488
Gibb, J., 192, 215
Gibbs, M. E., 257, 265, 342
Gibson, G. C., 431, 435
Gibson, M. J., 281, 346
Gibson-Hill, C. A., 143, 149
Gidari, A. S., 326, 352
Gilbert, A. B., 276, 285, 365
Gilker, J. C., 470, 496
Gill, F. B., 165, 192, 194, 215
Gilman, A. P., 170, 171, 218
Gilmour, D. G., 458, 466, 467, 470, 485, 490, 494, 500
Ginetsinskaia, T. A., 411, 435
Giorgi, E. P., 267, 315, 316, 318, 347
Gispen, W. H., 257, 347, 351
Gittleman, J. L., 133, 149
Gladstone, D. E., 202, 215
Glazener, E. W., 45, 47, 48, 50, 77, 82
Gledhill, B., 225, 298, 300, 346, 347
Glick, B., 22, 74, 280, 311, 342, 348, 364, 444, 445, 446, 448, 449, 450, 451, 452, 454, 456, 457, 458, 459, 460, 461, 462, 463, 464, 465, 466, 467, 468, 469, 471, 472, 473, 476, 477, 478, 480, 486, 488, 489, 492, 493, 494, 495, 496, 497, 498, 500
Gobert, A., 137, 152
Gochfeld, M., 128, 149
Goeze, J. A. E., 368, 417, 435
Gogan, F., 277, 347
Gold, E. F., 476, 490
Golden, W. R. C., 229, 348
Goldsmith, A. R., 228, 232, 250, 252, 254, 293, 295, 303, 323, 325, 347, 359
Goldstein, G., 458, 466, 467, 485, 490
Golikova, M. M., 418, 435
Golvan, Y.-J., 373, 421, 435
Gomes, W. R., 298, 360

Gompertz, 35, 75
Good, R. A., 457, 461, 468, 469, 470, 473, 474, 476, 477, 480, *486*, *487*, *488*, *493*, *496*
Goodale, H. D., 228, 233, *347*
Goodman, D., 69, 75
Goodwin, D., 252, 262, 291, 293, 295, 303, 333, *345*
Gordon, J. H., 312, *348*
Gordon, M. H., 2, 17, 79
Gorman, M. L., 225, 233, *347*
Goron, J., 285, *346*
Gorski, R. A., 281, 312, *347*, *348*
Gosney, D., 310, *347*
Gosney, S., 310, *362*
Goto, J., 96, 97, 127, *159*
Gottlieb, G., 90, 91, 93, 94, 95, 96, 98, 99, 106, 107, 108, 109, 115, *149*, *150*, *157*
Gottschalk, C., 427, *434*
Gowe, R. O., 61, 77
Graber, R. R., 192, *215*
Graetzer, M. A., 461, 467, 468, 469, *485*, *490*, *494*
Grant, P. R., 47, 72
Grau, C. R., 170, 171, *218*
Grav, H. J., 56, 72
Graves, H. B., 108, *150*
Gray, G. D., 303, *347*
Gray, J. M., 265, *340*, *344*
Grebenau, M. D., 468, 481, *486*, *490*, *493*
Greeley, F., 167, *215*
Green, L., 133, *150*
Green, R. E., 129, 130, 131, 134, 144, *148*
Greenlaw, J. S., 194, *215*
Greenman, D. L., 461, 462, *500*
Greenwood, P. J., 22, 75
Groschaft, J., 370, 396, 422, *433*
Grossi, C. E., 456, 472, 475, 476, *490*, *494*
Grunt, J. A., 305, *347*
Gschwendt, M., 331, *347*
Gubanov, N. M., 391, 392, *440*
Gudmundsson, F., 432, *434*
Güttinger, H. R., 257, *347*
Guhl, A. M., 268, *347*
Guichard, A., 292, 311, *347*
Gurney, M. E., 284, 301, 313, *347*
Gushanskaia, L. Kh., 373, 385, *435*
Gustafsson, J. A., 273, 327, *361*
Guyomarc'h, J. C., 224, 298, *347*
Gvozdev, E. V., 418, *435*
Gwinner, E., 136, 138, 139, 142, 143, *150*, *160*, 225, 256, *348*

Gyermek, L., 330, *348*

H

Haack, D. W., 276, 285, 286, 287, *358*
Haase, E., 276, 282, 285, 286, *353*
Hadley, N. F., 164, *219*
Haffen, K., 292, 311, *347*
Haftorn, S., 185, *215*
Hahn, D. C., 7, 33, 61, *75*, *80*
Hahn, S., 115, *146*
Hailman, J. P., 89, 105, 106, 125, 126, 127, 142, *150*, *152*
Hails, C. J., 187, 192, *215*
Hainsworth, F. R., 37, *80*, 192, 194, *215*
Hair, J. D., 388, 402, 410, 430, *435*, *436*
Halbersleben, 61, 75
Hale, C., 133, *150*
Hale, E. B., 224, 233, *360*
Hall, N. R., 231, *354*
Halvorsen, O., 385, *442*
Hamburger, V., 90, 91, 92, 93, 98, 99, 100, *148*, *150*
Hamilton, J. B., 224, 229, *348*
Hammer, N. M., 250, *348*
Hansen, E. W., 252, 253, *341*, *348*
Hansen, S., 314, *361*
Hanson, H. C., 177, *215*
Harding, C. F., 275, 291, 303, 326, *348*
Harlan, R. E., 312, *348*
Harrah, E. C., 369, *436*
Harris, J. H., Jr., 176, *215*
Harris, M. P., 52, 75, 112, *150*, 389, *442*
Harris, S., 463, *490*
Harrison, R. W., 331, *348*
Hart, J. S., 193, 199, *213*, *215*
Hartmann, F. A., 170, 171, 172, *215*
Hartree, A. S., 22, 75
Harvey, P. H., 58, 75
Harvey, S., 250, 254, 302, *348*, *360*
Harwood, D., 228, 229, 265, 267, *363*
Hasek, M., 471, *492*
Hayashida, T., 22, 75
Haynes, N. B., 303, *350*, *358*
Haynes, R. L., 280, *348*
Hazelwood, B. S., 22, 75
Hazelwood, R. L., 22, 75, 257, *348*
Heard, R. W., III, 386, *436*
Heaton, M. B., 94, 96, 97, *150*
Henrick, J. C., 224, 230, 231, 232, 233, 257, 258, 259, 262, 274, 290, 291, 296, 297,

AUTHOR INDEX

298, 300, 301, 305, 306, 307, 320, 321, 325, 337, 338, 342, 344
Henry, A., 370, 439
Heremans, J. F., 474, 493
Herman, C. M., 428, 442
Herrmann, H., 21, 74, 75
Hespenheide, H. A., 69, 75
Hess, E. H., 87, 97, 103, 104, 105, 107, 108, 109, 146, 151, 157, 253, 273, 352
Hess, M. W., 449, 450, 497
Heyns, W., 230, 232, 301, 344
Hilliard, D. K., 413, 439
Hinde, R. A., 87, 104, 105, 131, 151, 226, 227, 230, 236, 240, 241, 242, 243, 244, 272, 273, 280, 292, 308, 310, 339, 347, 348, 349, 350, 362, 364
Hindman, J. L., 107, 151
Hirai, K., 463, 490
Hirano, H., 291, 349
Hiroshige, T., 222, 300, 349
Hirota, Y., 472, 478, 479, 490
Hirsch, K. V., 170, 171, 218
Hirschberg, D., 224, 262, 263, 301, 302, 322, 323
Hitchner, S. B., 463, 490
Hochhauser, M., 356
Hochwald, G. M., 474, 498
Hodos, W., 284, 351
Höhn, E. O., 225, 226, 227, 229, 274, 349
Hörlyk, N.-O., 126, 151
Hoffman-Fezer, G., 461, 477, 490
Hoffmeister, D. F., 141, 151
Hogan, J. A., 128, 151, 211, 218
Hogan-Warburg, A. J., 128, 151
Holbrook, K. A., 448, 451, 452, 454, 457, 473, 490, 491
Holmes, J. C., 388, 407, 408, 410, 418, 425, 430, 436, 438
Holmes, R. T., 171, 172, 176, 192, 194, 196, 197, 215, 217
Holzbauer, M., 330, 349
Hon, L. T., 386, 436
Hopper, G., 454, 495
Horn, G., 108, 146, 285, 339, 340, 349, 355
Horowitz, C., 302, 359
Horst, H. J., 316, 349
Horton, E. W., 27, 75
Houssaint, E., 456, 458, 491, 493
Houston, D. C., 29, 75
Howarth, B., Jr., 22, 75
Howe, H. F., 60, 61, 62, 75, 202, 215
Hsia, S., 330, 363

Hubbard, J. D., 176, 181, 185, 192, 194, 197, 198, 206, 215
Huber, J. Ch., 368, 436
Hudec, K., 211, 215
Hunt, G. L., 335, 365
Hunter, W. S., 404, 436
Hurley, J. F., 274, 340
Hutchison, J. B., 227, 228, 229, 231, 232, 237, 258, 259, 260, 261, 262, 264, 265, 266, 274, 277, 286, 305, 308, 309, 310, 312, 316, 318, 319, 321, 325, 327, 328, 329, 330, 331, 345, 349
Hutchison, R. E., 223, 259, 329, 350
Huxley, J. S., 2, 41, 75, 79

I

Idler, D. R., 291, 359
Illius, A. W., 303, 350
Immelmann, K., 103, 105, 109, 110, 112, 114, 117, 121, 151, 225, 302, 303, 350
Illescas Gomez, M. P., 370, 381, 395, 396, 403, 405, 436
Immelman, K., 260, 265, 267, 302, 304, 362
Impekoven, M., 95, 100, 101, 102, 151
Inglis, W. G., 419, 436
Ingold, P., 115, 151
Inman, N. G., 230, 232, 339
Irashkin, V. M., 385, 386, 387, 392, 394, 395, 396, 397, 398, 400, 401, 402, 403, 405, 440
Irusta, O., 291, 346
Irvin, E. W., 280, 311, 364
Irving, R. A., 326, 350
Isaacson, R. L., 257, 347
Isakovic, K., 468, 473, 477, 478, 479, 480, 491, 492
Ishii, S., 228, 306, 350, 363
Ismail, A. A. A., 290, 303, 346, 350
Isvaneski, M., 470, 491
Ivanyi, J., 469, 470, 472, 476, 477, 491

J

Jaap, R. G., 461, 462, 467, 490, 491, 495, 498
Jackson, J. A., 64, 67, 82
Jacob, J., 275, 350
Jacobs, R. P., 470, 491
Jaeckel, J. B., 85, 86, 146
Jaesson, L., 459, 499
Jaffe, B. M., 290, 350

Jaffe, R. B., 296, 318, 350, 351
Jaffee, W. P., 456, 470, 491
Jallageas, M., 262, 264, 350
James, H., 255, 350
James, R., 320, 321, 322, 323, 324, 343
Jankovic, B. D., 468, 470, 473, 478, 479, 480, 491, 492
Jehl, J. R., Jr., 171, 216
Jellinck, P. H., 281, 329, 330, 350, 355
Jenkins, P. F., 122, 151
Jensen, E. V., 315, 350
Jewer, D. D., 402, 404, 436
John, T. M., 250, 355
Johns, J. E., 227, 351
Johnson, A. L., 298, 351
Johnson, M. A., 461, 500
Johnson, O. W., 168, 170, 216
Johnson, S. R., 172, 216
Johnston, D. W., 171, 216
Jolles, J., 257, 351
Jolly, J., 444, 447, 455, 461, 492
Jones, F. M., 133, 151
Jones, P. J., 177, 216
Jones, R. E., 250, 351
Jotereau, F. V., 456, 458, 493
Jouan, P., 268, 318, 351
Joyeux, Ch., 369, 436
Juhasz, L. P., 285, 351
Julian, L. M., 463, 487

K

Kabat, C., 385, 435
Kadenatsii, A. N., 387, 399, 401, 402, 440
Kalberlah, H. H., 224, 362
Kale, H. W., 170, 172, 214, 216
Kaltenbäuser, D., 292, 351
Kamel, F., 329, 350
Kamiyoshi, M., 267, 331, 332, 334, 351
Kantor, F. S., 470, 499
Kao, L. W. L., 324, 351
Karakoz, I., 471, 492
Karl, D. P., 170, 214
Karmanova, E. M., 418, 419, 436
Karten, H. J., 282, 284, 286, 351, 365
Kasimov, G. B., 385, 436
Katongole, C. B., 262, 303, 350, 351
Katzenellenbogen, B. S., 315, 351
Kaufman, L., 2, 15, 75
Kavanaugh, A. J., 41, 75
Kawashima, M., 267, 331, 332, 334, 351

Kear, J., 97, 126, 129, 130, 151, 152
Keeton, W. T., 136, 137, 138, 152
Keily, D., 473, 492
Kemenes, F., 468, 492
Kendeigh, S. C., 3, 17, 18, 20, 23, 28, 55, 72, 75, 76, 172, 181, 189, 190, 212, 216
Kennedy, C. R., 410, 436
Kenny, A. McM., 303, 347
Kerlan, J. T., 296, 351
Kermani-Arab, V., 481, 492
Kern, M. D., 172, 216
Kerstetter, T. H., 473, 492
Khokhlova, I. G., 391, 392, 422, 436
Kilgore, D. L., Jr., 86, 147
Kim, Y. S., 276, 280, 351
Kime, D. E., 291, 351
Kincade, P. W., 474, 477, 478, 487, 492
King, A. P., 88, 90, 116, 152, 160
King, A. S., 20, 76
King, C. R., 21, 22, 23, 76
King, D. B., 21, 22, 23, 76
King, J. R., 18, 20, 73, 76, 122, 146, 162, 163, 164, 167, 169, 172, 174, 175, 176, 181, 188, 189, 190, 191, 192, 194, 195, 196, 198, 201, 208, 210, 211, 213, 214, 216, 217, 225, 351
King, R. J. B., 315, 331, 352
Kinney, T. B., 46, 51, 76
Kinsella, J. M., 386, 436
Kirchner, B., 477, 494
Kirkham, K. E., 303, 346
Kirkpatrick, C. M., 15, 76, 461, 468, 492
Klandorf, H., 225, 296, 303, 353
Kleiber, M., 56, 57, 76, 163, 168, 216
Klinghammer, E., 253, 273, 352
Klint, T., 111, 152, 274, 352
Klomp, H., 60, 76
Klopfer, P. H., 105, 106, 113, 152
Kluijver, H. N., 172, 216, 249, 352
Knezevic, Z., 479, 480, 491, 492
Kniewald, Z., 318, 352
Knouff, R. A., 444, 447, 456, 485
Kocal, E., 466, 467, 487
Koch, F. C., 274, 346
Köhler, W., 108, 160
Koelink, A. F., 185, 216
Kohoutova, L., 471, 492
Komisaruk, B. R., 233, 287, 300, 330, 333, 345, 352
Komiya, Y., 409, 436
Kon, O. L., 331, 352

AUTHOR INDEX

Kondo, S., 409, *436*
Konishi, M., 95, 119, 120, 121, *152, 154*, 284, 301, 313, *347*
Kontrimavichus, V. L., 381, 382, 383, 386, 387, 389, 390, 397, 398, 399, 400, 401, 402, 405, 416, 418, *433, 440*
Koplin, J. R., 163, 192, 194, *216*
Korenbrot, C. C., 291, 293, 295, 301, *352*
Korschgen, C. E., 176, *216*
Koski, I., 480, *485*
Koskimies, J., 17, 56, *76*
Kotani, M., 447, *487*
Koutnik, D. L., 230, 231, 262, 301, 302, 311, 324, *336*
Kovach, J. K., 90, 95, 97, 108, 127, *152*
Krabbe, H., 368, 406, *436*
Krapu, G. L., 176, 177, *214, 216*
Krechmar, A. V., 406, *439*
Kreith, M., 224, *354*
Krey, L. C., 231, 267, 281, 316, 327, 329, 330, *350, 355, 357*
Kroodsma, D. E., 116, 117, 123, 144, *152*
Kruijt, J., 143, *153*
Ku, J. Y., 39, *74*
Kubli-Garfias, C., 326, *352*
Kuenen, D. J., 124, *158*
Kuhlmann, F., 13, *76*
Kuhn, M., 316, 322, *356*
Kulkarni, D. G., 461, 463, 464, *492*
Kunchiro, K., 463, *490*
Kuo, Z.-Y., 88, 90, 91, 93, 95, 98, *150, 153*
Kushlan, J. A., 170, 185, 192, *216*
Kuusisto, P., 249, *352*

L

Lack, D., 3, 26, 43, 52, 56, 59, 60, 63, 65, 66, 68, *76*, 114, *153*, 171, *216*
Lack, E., 59, *76*
Lade, B. I., 115, *153*
Lade, R., 461, *490*
Lagerlof, B., 473, *492*
Lahr, E. L., 245, 250, *339, 352, 359*
Lahr, R. L., 250, *339*
Lahti, L., 17, *76*
Lamming, G. E., 303, *350*
Lamoureux, G., 470, *496*
Landau, I. T., 267, 327, *345, 352*
Landreth, K. S., 461, *492*
Landsberg, J.-W., 108, 109, *153, 160*, 256, *352*

Lanerolle, de, N., 255, 284, 337
Langley, G. A. S., 276, 285, *365*
Laird, A. K., 35, *76*
Lane, S. E., 326, *352*
Lanyon, W. E., 115, 118, 121, *153*
Larsson, K., 268, 273, 274, *337, 352*
Lasiewski, R. C., 67, *76*
Lassila, O., 456, 457, *492, 494, 499*
Latimer, H. B., 2, 8, 15, *76*
Laughlin, M. E., 324, *340*
Lawrence, J. M., 170, 171, *216, 218*
Lawrence, L. K., 249, *352*
Laws, D. F., 250, *352*
Lawton, A. R., 477, 478, *487, 492*
Lax, E. R., 319, *346*
Lazarus, J., 259, *352*
Lebacq, A.-M., 458, 480, *493, 496*
Lebacq-Verheyden, A. M., 474, *493*
LeDouarin, N. M., 456, 458, 462, *491, 493*
LeFebvre, E. A., 163, 192, 194, *219*
Legait, N., 250, *352*
Lehane, J., 300, *359*
Lehrman, D. S., 98, *153*, 226, 228, 230, 231, 232, 233, 236, 237, 238, 245, 246, 247, 248, 249, 250, 252, 252, 253, 254, 260, 269, 270, 271, 273, 287, 291, 293, 295, 300, 301, 303, 324, 330, 332, 333, *342, 345, 352, 353, 362*
Leidy, J., 369, *436*
Leiper, R. T., 370, *436*
Leitch, A. F., 276, 285, *365*
Lelievre, E., 444, *496*
LeMaho, Y., 192, *217*
Lemmon, W. B., 109, 110, *159*
Lemon, R. E., 119, 120, 122, *148, 153*
Leonard, C. M., 282, 283, 286, *357, 362*
Leonard, R. A., 462, *485*
Leonov, V. A., 390, *433, 436*
Lerman, S. P., 480, 481, *490, 493, 496*
Lerner, I. M., 3, *76*
Lerner, K. G., 477, 478, *493*
Leslie, A. S., 385, *437*
Leslie, G. A., 474, 475, 476, 481, *492, 493, 494*
Lester, J. G., 426, *437*
Levenson, H., 163, 192, 194, *216*
Levere, R. D., 326, *352*
Levin, H. L., 97, *156*
Levinson, G., 273, *339*
Lewin, V., 168, 169, *217*
Lewis, D., 27, 73, 190, *213*, 267, 316, *363*

Li, C. H., 22, 77
Libby, D. A., 22, 76
Lieberburg, I., 231, 276, 281, 323, 326, 327, 330, 331, 353, 355
Lifter, J., 474, 493
Ligon, J. D., 129, 153
Liley, N. R., 310, 355
Lill, A., 202, 217
Lillie, R. J., 463, 490
Lin, C. Y., 50, 78
Lincoln, G. A., 225, 296, 300, 303, 353, 359
Lind, H., 100, 126, 151, 153
Linna, T. J., 480, 493
Linton, E., 369, 437
Lisk, R. D., 233, 276, 353
Liversidge, R., 114, 153
Ljunggren, L., 167, 217
Lloyd, C. S., 58, 77
Löhrl, 113, 143, 153
Lofts, B., 225, 232, 234, 249, 250, 259, 275, 353, 356
Long, P. L., 463, 476, 494, 496
Looss, A., 369, 437
Loraine, J. A., 303, 346
Lorenz, K., 98, 103, 109, 131, 153
Losch, U., 477, 478, 486, 490, 494
Lott, D. G., 253, 273, 353
Louw, G. N., 296, 340
Love, D. N., 303, 346
Love, J., 461, 463, 464, 492
Lucarini, N., 328, 343
Lucas, A. M., 17, 77
Lücke, J., 276, 282, 285, 286, 353
Luine, V. N., 275, 281, 326, 329, 330, 350, 354, 355
Lumia, A. R., 229, 354
Lundgren, H. P., 189, 220
Lupo de Prisco, C., 328, 343
Lustick, S., 172, 187, 188, 200, 216
Lutherman, C. Z., 229, 336
Lutjen, A., 119, 154
Luttge, W. G., 231, 354
Lydyard, P. M., 456, 472, 475, 476, 490, 494

M

McArthur, W. P., 470, 494
McBride, G., 236, 355
McCabe, B. J., 285, 339, 349, 355
McCance, R. A., 21, 54, 77

McCann, S. M., 258, 356
McCollom, R. E., 232, 308, 355
McCorkle, F., 458, 496
MacDonald, G. E., 126, 128, 158
McDonald, M. E., 383, 411, 437
McDonald, P. A., 258, 310, 355
McDuffie, F. C., 461, 477, 478, 489, 490, 493, 498
McEwen, B. S., 231, 258, 265, 267, 272, 287, 288, 312, 316, 327, 329, 330, 331, 334, 343, 350, 354, 355, 357, 363, 365
McGee, S. G., 409, 437
Machalska, J., 400, 437
MacInnes, D. D., 176, 177, 212
McKenzie, C. E., 395, 437
MacKenzie, D. I., 395, 437
Mackie, R. J., 167, 217
McKinney, F., 224, 230, 232, 233, 236, 355, 358
Mačko, J. K., 406, 437
McLaughlin, J. D., 428, 437
MacLean, S. F., Jr., 29, 81, 171, 217
Maclusky, N. J., 231, 267, 281, 312, 314, 316, 327, 330, 354, 355, 357
MacMillen, R. E., 165, 166, 192, 214
McNabb, F. M. A., 19, 21, 23, 77, 81
McNabb, R. A., 19, 21, 77, 81
McNeilly, A. S., 250, 355
McRoberts, M. R., 54, 77
Macy, R. W., 399, 437
Madsen, H., 385, 437
Maher, W. J., 424, 437
Maheswaran, S. K., 463, 497
Maguire, H. C., Jr., 471, 500
Maini, R. N., 459, 495
Mainwaring, W. I. P., 303, 350
Maksimova, A. P., 418, 435
Malacarne, G., 302, 338
Maley, M. J., 286, 354
Manning, E., 155
Maranova, H., 468, 477, 494
March, G. L., 250, 354
Markov, G. S., 390, 437
Markovic, M. B., 478, 479, 480, 491, 492
Marks, H. L., 22, 49, 50, 75, 77
Marler, P., 89, 115, 117, 119, 121, 122, 123, 154, 156, 224, 354
Marshall, A. J., 2, 77, 244, 250, 353, 354
Martin, A. C., 170, 217
Martin, A. R., 58, 75

AUTHOR INDEX

Martin, B., 314, 354
Martin, C., 456, 457, 485, 487, 494, 499
Martin, D. J., 129, 153
Martin, G. A., 50, 77
Martin, J. T., 256, 257, 354
Martin, L. N., 474, 475, 476, 493, 494
Martinez-Vargas, M. C., 237, 238, 276, 280, 281, 282, 284, 286, 345, 351, 354
Martini, L., 318, 352, 354
Maruszewski, W., 429, 438
Marvanova, H., 476, 477, 491
Mashaly, M., 460, 498
Mason, P., 230, 336
Massa, R., 228, 299, 301, 306, 318, 319, 320, 321, 322, 323, 324, 328, 329, 338, 340, 343, 344, 352, 354, 355, 360
Matevosian, E. M., 381, 437
Mathewson, S. F., 258, 355
Matos, A., 454, 494
Matsuda, H., 472, 494
Matsuno, K., 447, 487
Maung, S. L., 228, 321, 346
Mauser, P., 302, 359
Mausner, R., 458, 500
Maxson, S. J., 165, 170, 176, 192, 194, 196, 217
May, D., 469, 494
Mayhew, R. L., 369, 408, 437
Mayr, E., 407, 437
Medina, E., 466, 467, 471, 494
Medrano-Conde, L., 326, 352
Meier, A. H., 22, 77, 250, 355
Meites, J., 22, 76
Menaker, M., 262, 301, 363
Merkenschlager, M., 477, 494
Merrill, G. F., 302, 359
Merritt, E. S., 61, 77
Merry, B. J., 302, 348
Mertens, J. A. L., 18, 58, 77, 181, 182, 183, 217
Messmer, E., 115, 154
Messmer, I., 115, 154
Metcalf, B. W., 324, 340
Metcalf, D., 456, 494
Metfessel, M., 119, 154
Mewaldt, L. R., 113, 156
Meyer, C. C., 157, 265, 276, 280, 282, 284, 287, 330, 355
Meyer, R. K., 167, 215, 254, 340, 461, 467, 468, 469, 477, 485, 486, 490, 494, 495
Meyerson, B. J., 335, 356

Michael, R. F., 275, 356
Michel, G., 462, 493
Middleton, A. L., 254, 345
Midgley, A. R., Jr., 290, 350, 356
Mignot, Th., 292, 311, 347
Mikheev, A. V., 428, 437
Millar, R. P., 296, 340, 341
Miller, J. F. A. P., 470, 494
Millikan, G. C., 255, 356
Mills, J., 300, 356
Millward, D. J., 24, 77
Milosevic, D., 468, 491
Mitchell, G. F., 470, 494
Mitrovic, K., 473, 479, 491
Moguilewsky, M., 267, 316, 332, 356
Molnar, G., 457, 499
Molnar, J., 472, 499
Moltz, H., 106, 109, 154
Montero, J., 481, 488
Montevecchi, W. A., 7, 33, 61, 77, 80
Moore, F. R., 140, 154
Moore, J., 426, 437
Moore, M. A. S., 456, 458, 494, 495
Moore, P. L., 311, 365
Moreau, R. E., 143, 154
Morejohn, G. V., 171, 212
Morel, G., 236, 356
Morgan, B. B., 396, 420, 437
Morgan, G. W., Jr., 476, 477, 495
Morgan, K. K., 171, 217
Mori, M., 318, 319, 356
Morita, C., 470, 495
Morrell, G. M., 133, 154
Morris, J. B., 230, 231, 262, 301, 302, 324, 336
Morse, D. H., 205, 217
Morse, K., 54, 77
Morton, E. S., 28, 77
Morton, M. L., 121, 124, 146, 167, 169, 216, 225, 351
Moss, F. P., 21, 77
Moss, R., 192, 217, 228, 258, 346
Moss, R. L., 258, 303, 356
Mosteller, F., 39, 77
Moticka, E. J., 474, 495
Moudgal, N. R., 22, 77
Movsesian, S. O., 415, 437
Moynihan, M., 115, 154
Mozgovoi, A. A., 383, 384, 398, 400, 420, 438
Mrosovsky, N., 211, 218

Mueller, A. P., 468, *495*
Mueller, H. C., 129, 130, 133, *154*
Mueller, J., 449, 450, *497*
Mugaas, J. N., 172, 176, 192, 194, 195, 196, 198, 208, 210, 211, *217*
Muir, F. V., 462, *495*
Mulligan, J. A., 116, 117, *154*
Mundinger, P. C., 115, 117, 122, *154*
Murdock, L. C., 194, *217*
Murton, R. K., 225, 226, 232, 234, 243, 249, 259, 275, 296, *353*, *356*
Mussehl, F. E., 61, *75*
Musto, N., 300, *339*
Myrcha, A., 185, *214*

N

Naderman, J., 396, *438*
Nagy, K. A., 163, *220*
Naftolin, F., 312, 314, *354*
Naftolin, R., 316, 322, *356*
Nair, K. R., 38, *78*
Nakajo, S., 254, *356*
Nakamura T., 248, 291, 292, 298, 311, 318, 326, 329, *344*, *349*, *356*, *363*
Nalbandov, A. V., 22, *78*, 250, 254, *356*
Naukkarinen, A., 450, 454, *498*
Nawa, Y., 447, *487*
Nayes, D. M., 290, *346*
Nazerian, K., 454, *495*
Neff, M., 15, *78*
Negri-Cesi, P., 228, 299, 301, 306, 320, 321, 322, 324, 328, *338*
Nelson, A. L., 170, *217*
Nelson, J. B., 26, 60, *78*, 141, *155*
Nesheim, M. C., 28, *81*
Nesslinger, C., 397, *434*
Nettles, V. F., 432, *442*
Neumann, F., 231, *356*, *362*
Newman, A., 335, *365*
Newcomer, W. S., 462, *495*
Newton, I., 187, 189, 200, *217*
Nice, M. M., 4, 5, 6, 10, 13, 70, *78*, 86, 96, 118, 143, *154*
Nicki, R. M., 106, *155*
Nicolai, J., 121, *155*
Nir, I., 52, 53, *78*
Nisbet, I. C. T., 61, *78*
Nishiyama, H., 96, 97, 127, *159*
Niswender, G. D., 290, *350*, *356*
Nitsan, Z., 52, 53, *78*

Noble, G. K., 224, 228, 230, 234, 236, 249, *357*
Noble, R., 234, *357*
Nobukuni, K., 96, 97, 127, *159*
Nock, B. L., 230, 231, 234, 235, 260, 268, 324, *336*, *345*
Nolan, V., Jr., 61, *78*
Nordskog, A. W., 49, 50, *78*, *82*
North, M. E. W., 122, *158*
Norton, D. W., 171, *217*
Norton, J., 461, *495*
Nottebohm, F., 114, 115, 117, 118, 119, 120, 123, 124, 143, *152*, *155*, 257, 265, 275, 276, 282, 283, 284, 285, 286, 288, 301, 323, 326, 331, 337, *353*, *354*, *362*, *365*
Nottebohm, M. E., 119, *155*
Nowotna, A., 15, *75*
Nugent, C. A., 290, *346*
Nurmi, T., 456, 457, *494*, *499*

O

Oaghir, H. J., 463, *485*
Oates, C. M., 459, *495*
Oberholser, H. C., 31, *72*
Ockleford, E. M., 101, *155*
O'Connell, M. E., 223, 229, 231, 248, 249, 251, 253, 254, 295, 296, 303, *357*, *361*
O'Connor, R. J., 9, 16, 18, 20, 23, 26, 33, 42, 44, 55, 58, 60, 61, 62, 63, *78*
Odend'hal, S., 455, *495*
Odening, K., 410, *438*
Odhner, T., 369, *438*
Odum, E. P., 20, *78*, 170, *214*
Ohanian, S. H., 474, *498*
Ohnu, S., 231, *337*
Oishi, T., 461, *499*
Okulewicz, J., 429, *438*
Okuyama, S. I., 470, *495*
Olah, I., 447, 449, 458, 459, 473, *489*, *495*, *496*
Ollason, J. C., 15, 58, *78*
Olsen, O. W., 391, 427, *438*
O'Malley, B. W., 315, 331, *360*
Omiya, Y., 298, *363*
Oniki, Y., 12, 65, *79*
Oppenheim, J. J., 470, *491*
Oppenheim, R. W., 90, 91, 92, 93, 94, 95, 97, 99, 100, 101, 102, 127, *149*, *155*, *156*
Orians, G. H., 110, 135, *156*

AUTHOR INDEX

Oring, L. W., 110, *148*, 165, 170, 176, 192, 194, 196, *217*
Orlans, E., 468, 474, 477, *496*, *497*
Ortega, L. G., 476, *496*
Osbaldiston, G. W., 56, 57, 79
Osche, G., 370, 410, 419, 420, *438*
Oshmarin, P. G., 391, 393, 402, *438*
Ottinger, M. A., 223, 291, 292, 298, 301, 302, 321, 322, 339, 357
Ovary, Z., 468, 470, 476, 477, 478, *499*
Owen, R. B., 192, *220*
Owen, J. J. T., 456, 458, *494*, *495*
Owen, R. W., 402, 425, *433*, *438*
Ozon, R., 289, 357

P

Paganelli, C. V., 5, 6, 79, 169, 171, *217*, *218*
Paladino, F., V., 199, 200, *217*
Palis, A., 335, *356*
Palka, Y. S., 287, *357*
Palladino, M. A., 471, 481, *490*, *493*, *496*
Panda, B., 463, *496*
Papkoff, H., *341*
Paramonov, B. B., 381, 382, 383, 386, 387, 389, 390, 410, 405, *441*
Parer, I. P., 236, *355*
Parisi, P., 6, 73, 170, 171, 172, *214*
Parker, A., 141, 143, *156*
Parker, D. M., 276, 280, 282, 284, *355*
Parker, J. E., 168, *212*
Parr, C. P., 170, *214*
Parsons, B., 267, 281, 316, 329, 330, *350*, *355*, *357*
Parsons, J., 61, 62, 79
Parvizi, N., 329, *357*
Paspalev, G. V., 395, 397, 403, *438*
Patel, M.D., 245, *357*
Patten, S. M., 170, 171, *218*
Paulke, E., 316, *349*
Paulson, G. W., 108, *156*
Pavlov, A. V., 383, 385, 397, 398, 400, 401, 405, *440*
Payne, L. N., 459, *495*
Payne, R. B., 110, 114, 123, *156*, 172, *217*, *274*, *357*
Pearce, J., 28, 79
Pearson, A. W., 461, *488*
Pearson, O. P., 165, 192, 194, *217*
Peck, E. J., Jr., 331, *342*
Peczely, P., 252, 296, *357*

Pedernera, E. A., 466, 467, 471, *494*, *496*
Peet, M. J., 300, *353*
Peet, S., 399, *438*
Pemberton, R. T., 389, 402, *438*
Pembrey, M. S., 2, 17, 79
Pence, D. B., 396, *438*
Perdeck, A. C., 125, *158*
Perek, M., 188, 277, 465, 466, *217*, *358*, *496*
Perey, D. Y. E., 470, 474, 475, 476, *485*, *488*, *496*
Perkins, W. D., 448, 451, 452, 454, 457, 473, *490*, *491*
Perrins, C. M., 58, 75
Peters, L. E., 461, 462, *500*
Peters, S. S., 44, 48, 51, 58, 63, *80*, *81*, 123, *154*, *156*
Petersen, A., 168, 169, *218*
Petersen, M. R., 170, 171, *218*
Peterson, R. D. A., 467, 469, 476, *485*, *486*, *496*
Pethes, G. Y., 252, 296, *357*, 468, *492*
Petro, Z., 280, 311, 316, 318, 322, 327, 329, *341*, *356*
Petrochenko, V. I., 373, 407, 421, *438*
Petrovic, S., 473, *491*
Petrushevski, G. K., 425, *435*
Pezard, A., 461, *492*
Pfaff, D. W., 258, 265, 267, 268, 276, 279, 280, 281, 282, 284, 285, 286, 288, 316, 337, 339, *343*, *357*, *358*, *365*
Pfeiffer, E. W., 291, *346*
Pflugfelder, O., 431, *438*
Phillips, J. G., 302, *348*
Phillips, R. E., 224, 226, 230, 232, 233, 277, 284, 285, 286, *358*
Pickett, P., 311, *336*
Pickert, R., 117, 144, *152*
Picozzi, N., 143, *160*
Pierce, A. E., 463, 476, *494*, *496*
Pietras, R. J., 257, *358*
Pinkowski, B. C., 172, 194, *218*
Pintea, V., 447, *496*
Pniewski, E. E., 230, 231, 262, 301, 302, 323, 324, 327, *336*
Podesta, R. B., 418, *438*
Pod'iapol'skaia, V. P., 381, 384, 387, 397, 398, 400, 401, 405, *438*
Pohl, H., 169, 174, 193, 199, 200, *212*, *218*
Poli, G., 472, *487*
Pollard, L. W., 481, *485*
Polyanski, Yu. I., 425, *435*

Pomeroy, M. K., 388, *439*
Poole, E. L., 170, *218*
Popa, F. G., 282, *358*
Popa, G. T., 282, *358*
Popeskovic, L., 468, *491*
Popovo, T. I., 381, 383, 384, 398, 400, 401, 402, *438*
Portmann, A., 2, 6, 7, 9, *79*
Potash, L. M., 282, *358*
Potworowski, E. F., 470, *496*
Poulsen, H., 119, *156*
Poulson, T. L., 172, *218*
Price, P. W., 117, *156*, 370, 410, *439*
Pröve, E., 224, 230, 232, 260, 265, 267, 291, 302, 303, 304, 306, *358, 362*
Provine, R. R., 98, *156*
Pruitt, K. M., 33, *79*
Pütter, A., 36, *79*
Pun, C. F., 61, *72*
Pursglove, S. R., Jr., 385, *434*
Purvis, K., 303, *358*
Puts, C., 255, 285, 338, *358*

Q

Quay, T. L., 404, *436*
Querner, U., 140, *147*
Quinney, T. E., 135, *156*

R

Rabalais, F. C., 404, *441*
Rabinowitch, V. E., 129, *156*
Racey, P. A., 225, 296, 300, 303, *353, 359*
Radu, C., 447, *496*
Raheja, K. L., 23, *79*
Rahn, H., 5, 6, 73, *79*, 169, 170, 171, 172, *214, 217, 218*
Railliet, A., 370, *439*
Raitasuo, K., *358*
Raitt, R. J., 168, 169, *218*
Rajan, R., 329, *358*
Ralph, C. J., 113, 281, *156, 358*
Ramirez, V. D., 287, *357*
Ransom, B. H., 369, *439*
Rao, M. A., 461, 468, *494, 496*
Rascevic, M., 478, 479, 480, *491, 492*
Ratti, J. T., 37, *82*
Rausch, R. L., 406, 408, 409, 411, 413, 417, *439*
Rausch, V. R., 406, *439*
Raveling, D. G., 170, 176, 177, *218*
Ravona, H., 277, *358*
Rawitch, A. B., 267, 316, *363*
Ray, T. S., 109, 110, *159*
Raynaud, J. P., 267, 316, 332, *356*
Reader, M., 101, *159*
Reboulleau, C., 236, 237, 247, 252, 262, 291, 293, 295, 301, 302, 303, 332, 333, *338, 345, 357, 361*
Recher, H. F., 135, *156*
Recher, J. A., 135, *156*
Redington, B. C., 386, *439*
Reddy, V. V. R., 316, 322, 329, *356, 358*
Reese, E. P., 105, *146*
Reeve, E. C. R., 41, *79*
Reid, A., 370, *439*
Reid, M. W., 463, *488*
Reineke, E. P., 23, *81*
Rennie, J., 370, *439*
Retterer, E., 444, *496*
Rhees, R. W., 276, 285, 286, 287, *358*
Rheins, M. S., 468, *486*
Rice, J. O., 122, *156*
Richards, F. J., 37, *79*
Richards, O. W., 41, *75*
Richdale, L. E., 52, *79*, 223, *358*
Richter, W., 61, *79*
Ricker, W. E., 33, *79*
Ricklefs, R. E., 6, 7, 8, 9, 11, 12, 13, 14, 15, 16, 17, 18, 20, 21, 23, 24, 25, 26, 28, 32, 33, 34, 35, 37, 38, 40, 41, 44, 48, 51, 55, 58, 60, 61, 63, 64, 65, 66, 67, 68, 69, 70, 71, *79, 80,* 86, *156,* 170, 184, 185, *218*
Riddle, O., 245, 247, 250, 252, 253, 339, *352, 358, 359*
Riedel, G., 477, *494*
Riesen, A. H., 108, *157*
Ringer, R. K., 23, *81*
Ringrose, A. T., 61, 62, *81*
Ripley, K. L., 94, 157
Ritter, M. A., 458, 480, *493, 496*
Robel, R. J., 170, *214*
Roberts, B., 142, *157*
Roberts, K. D., 324, *345*
Robertson, T. B., 34, 41, *80*
Robinson, J. E., 228, 321, *346*
Rogers, J. A., 106, *155*
Rogers, L. J., 224, 254, 255, 323, 326, 337, *359, 365*

Rogler, J. C., 300, 302, *364*
Rohde, K., 410, 430, *439*
Rohlf, J., 45, *81*
Rohlich, P., 447, *495*
Rohwer, F. C., 229, 274, *359*
Rohwer, S., 229, 274, *359*
Roitman, V. A., 381, 382, 401, *441*
Romano, M., 466, *496*
Romanoff, A. J., 168, 170, *218*
Romanoff, A. L., 168, 170, *218*
Rompa-Barendregt, J., 257, *351*
Romppanen, T., 458, *496*
Ronay, G., 276, 279, 280, 282, 284, 286, *339*
Ropke, C., 449, 450, *497*
Rose, M. E., 468, 474, 477, *496*, *497*
Rose, S. P. R., 108, *146*, 285, *339*
Rosenblatt, J. S., 304, *359*
Rosenblum, L. A., 109, *154*
Ross, H. A., 40, 58, 59, *80*
Rosse, C., 465, *489*
Rothstein, R., 236, *359*
Roudabush, R. L., 386, *439*
Roudybush, T. E., 170, 171, *218*
Rouse, B. T., 470, 480, *497*
Rouse, T. C., 463, *497*
Rowe, B., 170, *214*
Rowe, P. H., 300, *359*
Roy, E. J., 231, 265, 281, 327, 329, 330, *340*, *344*, *350*, *355*
Royama, T., 58, *80*
Rubin, B. S., 281, 287, *359*
Rudolphi, C. A., 360, *439*
Ruotsalainen, P., 452, 473, *498*
Ryan, K. J., 280, 311, 316, 318, 322, 327, 329, *341*, *356*
Rybicka, K., 382, 386, *439*
Ryšavý, B., 370, 396, 422, *433*, *439*
Ryzhikov, K. M., 381, 383, 384, 385, 387, 390, 391, 392, 393, 397, 398, 399, 400, 401, 402, 405, *433*, *438*, *439*

S

Saad, R., 223, 228, 229, 231, 232, 248, 249, 251, 253, 254, 295, 296, 323, 325, *359*, *361*
Saayman, G. S., 275, *356*
Sabet, T. Y., 472, *499*
Sachs, B. D., 226, 228, 260, 265, 266, *359*
Sachs, H. G., 453, *497*

Sadler, C. R., 461, 469, 477, *489*
Sadokov, S. B., 381, 394, 397, 401, *440*
Saeki, Y., 254, *359*
Safriel, U. N., 176, 192, 194, 196, 198, *212*
Sagher, R. M., 56, *81*
Salerno, A., 469, 470, 471, *486*, *491*
Salt, G. W., 20, *81*
Salzen, E. A., 157, 276, 280, 282, 284, *355*
Samperez, S., 268, 318, *351*
Sandor, T., 291, *359*
Sang, J. H., 61, *72*
Sar, M., 276, 280, 281, 282, 284, 286, *351*, *354*, *363*, 461, *488*
Sato, K., 462, 468, 469, 471, 472, *490*, *497*
Sauer, E. G. F., 136, *157*
Sauer, E. M., 136, *157*
Sauer, F., 136, 142, *157*
Saunders, J. C., 95, *157*
Sawyer, C. H., 287, *357*
Saxena, R., 250, *363*
Scanes, C. G., 250, 290, 298, 302, *346*, *359*
Scanes, C. J., 250, 254, *360*
Schaefer, H. H., 105, *157*
Schaffner, T., 449, 450, 453, *497*
Schaible, J., 22, *76*
Schanbacher, B. D., 298, *360*
Scharloo, W., 47, *82*
Schartz, R. L., 192, 194, *218*
Schatzman, G. L., 324, *340*
Schaumburg, B. P., 462, *497*
Scheib, D., 292, 311, *347*
Scheid, M., 458, 467, *490*
Schein, M. W., 224, 233, *360*
Scheiss, L. R., 15, *81*
Schifferli, 42, 61, 72, *81*
Schiller, E. L., 420, *437*, *440*
Schleidt, W. M., 119, *157*, 224, 261, 291, 293, *360*
Schmalhausen, I., 2, 70, *81*
Schmekel, L., 25, *81*
Schmidt, G. D., 387, *440*
Schmidt-Nielson, K., 200, *219*
Schneirla, T. C., 87, 98, *157*
Schoener, A., 39, 40, *81*
Schoener, T. W., 39, 40, *81*
Schönwetter, M., 170, 171, *218*
Schoffeniels, E., 275, 338, *350*
Schomberg, D. W., 291, 293, 295, 301, *352*, 461, *497*
Schommer, M., 224, 233, *360*
Schonne, E., 468, 477, *487*

Schrader, W. T., 315, 331, *360*
Schreiber, R. W., 170, 171, *216*, *218*
Schriefers, 319, *346*
Schumacher, M., 261, 275, 301, 302, 321, 322, 327, *339*, *344*, *360*
Schutz, F., 107, 108, 111, 112, *154*, *157*, 256, *360*
Schwartz, N. B., 303, *356*
Schwarz, M. R., 461, 463, 464, *489*, *500*
Scott, D. M., 177, 178, *212*
Scott, M. L., 27, 28, 29, *81*
Scoville, R., 115, *157*
Seaburne-May, G., 108, *146*
Searcy, M., 123, *148*
Searcy, W. A., 123, *156*
Seastedt, T. R., 29, *81*
Sedláček, J., 97, *157*
Seegar, K. C., 61, 62, *81*
Seegar, W. S., 420, *440*
Sefton, A. E., 308, *360*
Self, K. S., 477, *492*
Selinger, H. E., 228, *360*
Selmanoff, M. K., 268, 316, *360*
Seltzer, W., 469, *497*
Selye, H., 462, *497*
Semenova, M. K., 420, *438*
Sergeeva, T. P., 419, 420, *433*, *440*
Serventy, D. L., 113, *157*, 244, *354*
Seto, F., 457, *497*
Setzer, H. W., 141, *151*
Shaffner, C. S., 462, *488*
Shaikenov, B., 419, *440*
Shakhmatova, V. I., 420, *438*
Shapira, N., 52, 53, *78*
Shapiro, A. J., 106, 107, *158*
Sharp, P. J., 223, 225, 250, 254, 277, 296, 298, 300, 302, 303, 321, 328, 329, *353*, *359*, *360*, *364*
Shenton, J. C., 300, *359*
Sheridan, P. J., *360*
Sheridan, S. A., 461, *500*
Sherman, J., 461, *497*
Sherman, M. R., 331, *360*
Sherry, D. F., 86, *157*, 211, *218*
Sherry, T. W., 172, 176, 192, 194, 196, 197, *215*
Shettleworth, S. J., 134, *157*
Shevtsov, A. A., 431, *440*
Shifrine, M., 478, *486*
Shigin, A. A., 389, *440*
Shilov, I. A., 18, 19, *81*

Shimakura, S., 463, *490*
Shipley, A. E., 385, *437*
Shoemaker, H. H., 308, *360*
Shrout, P. E., 107, *146*
Siegel, H. S., 462, *497*
Siegel, P. B., 45, 49, 50, *81*, 108, *150*, 232, 308, 331, *339*, *340*, *343*, *355*, *360*
Siegfried, W. R., 6, *82*, 192, 194, *218*
Sietnieks, A., 335, *356*
Siiteri, P. K., 268, 316, *360*
Silver, R., 223, 228, 229, 230, 231, 232, 235, 238, 239, 245, 247, 248, 249, 250, 251, 252, 253, 254, 262, 291, 293, 295, 296, 300, 301, 303, 323, 325, 332, 333, *342*, *345*, *347*, *359*, *360*, *361*
Silverin, B., 225, 226, 228, 249, 250, 254, *361*
Simpson, R. M., 311, *365*
Singh, A., 23, *81*
Sivanandan, V., 463, *497*
Sjölander, S., 110, 111, *158*
Sjoerdsma, A., 324, *340*
Skamene, E., 476, 477, *491*
Skinner, R. W., 171, *219*
Skoglund, W. C., 61, 62, *81*
Skriabin, A. S., 416, *441*
Skriabin, K. I., 369, 421, *440*
Skutch, A. F., 4, *81*
Sladen, W. J. L., 420, *440*
Slater, R. L., 399, *440*
Sluckin, W., 104, 106, *157*
Sluiters, J. F., 425, *440*
Smith, E. R., 261, 262, 287, 301, 306, *343*, *361*
Smith, F. V., 105, 106, *157*
Smith, J. N. M., 47, *81*
Smith, R. W., 113, *148*
Smith, S. M., 89, 112, 113, 115, 128, 129, 131, 132, 134, 141, *157*, *158*
Smithies, O., 469, 477, *486*
Smogorzhevskaia, L. A., 380, 382, 383, 384, 386, 387, 388, 389, 390, 393, 412, 419, 423, 428, *440*
Snapir, N., 277, *358*
Snapp, B. D., 127, *158*
Snedecor, J. G., 23, *79*
Snow, B. K., 4, *81*
Snow, D. W., 27, *81*
Södersten, P., 261, *268*, 269, 271, 273, 274, 314, 327, *337*, *352*, *361*
Soekawa, M., 470, *495*

Sokal, R. R., 45, *81*
Solomon, J., 22, *81*
Sonin, M. D., *383*, *385*, 386, 387, 389, 392, 394, 395, 396, 397, 398, 400, 401, 402, 403, 405, 419, *433*, *440*
Sonneman, P., 110, 111, *158*
Sordat, B., 449, 450, *497*
Sorvari, R., 450, 451, 452, 454, 473, *498*
Sorvari, T. E., 450, 451, 452, 454, 457, *458*, 473, *496*, *498*
Sossinka, R., 223, 224, 260, 265, 267, 302, 304, *362*
Sousa, R., 454, *494*
South, M. A., 467, 469, 476, *486*
Southwell, T., 369, *440*
Southwick, E. E., 164, *218*
Southern, W. E., 138, 140, *158*
Sparling, D. W., 118, *158*
Spasskaia, L. P., 370, 415, 416, 417, *440*
Spasskii, A. A., 370, 381, 382, 383, 384, 386, 387, 389, 390, 392, 394, 395, 396, 397, 398, 400, 401, 402, 403, 405, 416, *438*, *440*, *441*
Spelsburg, T. C., 331, *352*
Spiers, E. D., 19, 23, *81*
Spooner, P. M., 303, *350*
Spory, G. R., 404, *441*
Spring, L., 171, *218*
Stanley, J. G., 404, *441*
Stanwood, C. J., 2, *81*
Steel, E. A., 240, 241, 242, 243, 244, 273, 308, 310, *348*, *349*, *362*
Steele, R. E., 300, *339*
Steen, J. B., 21, 56, 72
Steimer, T., 264, 316, 318, 319, 321, 325, 328, 329, *362*
Steinbeck, H., 231, *356*, *362*
Stern, J. M., 247, 265, 300, 301, 330, 331, 332, 333, *362*, *365*
Stetson, M. H., 281, *362*
Stettenheim, P., 17, 77 274, *362*
Stevens, M., 224, 233, 259, 267, 293, 300, 301, *338*
Stevens, V. C., 236, *362*
Stevens, W., 452, 453, *485*
Stevenson, R. E., 210, *218*
Stewart, P. A., 171, 172, *219*
Stieve, H., 172, *219*
Stiles, F. G., 113, *158*, 165, 192, 194, *219*
Stinson, R. S., 458, 460, *496*, *498*
Stob, M., 461, *497*

Stokes, A. W., 127, *158*
Stokes, T. M., 282, 283, 286, *357*, *362*
Stonehouse, B., 141, *158*
Stonehouse, S., 141, *158*
Stoner, E. A., 142, *158*
Storey, A. E., 106, 107, *158*, 252, 262, 291, 293, 295, 303, 333, *345*
Stratton, L. G., 301, *362*
Street, M., 29, 54, *82*
Streich, G., 15, *82*
Strobel, M. G., 126, 128, *158*
Strong, G. L., 399, *437*
Stumpf, W. E., 276, 280, 281, 282, 284, 286, *351*, *354*, *362*, *362*, 461, *488*
Sturkie, P. D., 22, *82*
Subba Rao, D. S. V., 459, 460, 461, 478, *490*, *498*
Sudarikov, V. E., 381, 383, 384, 385, 387, 393, 397
Sullivan, D. A., 461, *498*
Sulman, F., 188, *217*
Sultanov, M. A., 370, 391, 392, 393, 394, 396, 397, *441*
Summers, K. R., 59, *82*
Sumner, E. L., Jr., 2, *82*, 141, *158*
Sundelin, P., 473, *492*
Supriaga, A. M., 420, *441*
Sutter, E., 7, 15, *82*
Suzuki, K., 318, 319, *356*
Suzuki, T., 478, 479, *490*
Swetosarow, E., 15, *82*
Szenberg, A., 461, 469, 470, 480, *497*, *498*, *499*
Szidat, L., 383, 388, 389, 425, 426, 428, *441*

T

Taibel, A. M., 467, *498*
Takai, T., 298, *344*
Tam, L. E., 481, *485*
Tam, L. Q., 481, *486*
Tamaoki, B., 318, 319, *356*
Tamura, M., 121, 122, *154*
Tanabe, Y., 23, *82*, 248, 254, 291, 292, 298, 311, 318, 326, 329, *344*, *349*, *359*, *363*
Tanaka, K., 254, 267, 331, 332, 334, *351*, 447, 487, *356*
Tangl, F., 172, *219*
Tarboton, W. R., 192, 194, *219*
Taylor, C. R., 200, *219*
Taylor, R., 465, 466, *498*

Temirova, S. I., 416, *441*
Temple, R. W., 462, *498*
Temple, S. A., 225, 228, 296, *363*
Tennent, B., 306, *343*
Tenniswood, M., 327, *363*
Teyler, T. J., 267, 316, *363*
Thapliyal, J. P., 250, *363*
Thaxton, J. P., 280, 311, *364*
Thearle, R. J., P., 232, 259, *356*
Theis, G. A., 471, *498*
Thomas, B., 262, *346*
Thomas, C. H., 45, 47, 48, *82*
Thompson, A. L., 145, *158*
Thompson, C. F., 61, *78*
Thompson, D., 463, *490*
Thompson, W. L., 122, *156*
Thorbecke, G. J., 468, 470, 471, 474, 476, 477, 478, 481, 490, *486*, *490*, *493*, *494*, *496*, *498*, *500*
Thorpe, W. H., 89, 105, 115, 117, 121, 122, 142, 143, 145, *151*, *153*, *158*
Threlfall, W., 388, 390, 396, 402, 404, *433*, *436*, *441*
Tinbergen, N., 124, 125, 131, *151*, *158*
Tiner, J. D., 409, *439*
Tinkle, D. W., 164, *219*
Tixier-Vidal, A., 22, *82*, 250, *363*
Toft, D. O., 331, *348*
Toivanen, A., 452, 457, 473, *498*
Toivanen, P., 452, 456, 457, 473, 480, *492*, *494*
Tolhurst, B., 101, *159*
Tolkacheva, L. M., 391, 392, 418, *440*, *441*
Tolman, C. W., 128, *159*
Tolman, H. S., 50, *78*
Tomilovskaia, N. S., 418, *433*, *441*
Toran-Allerand, C. D., 272, 312, 314, *363*
Toro, I., 447, *495*
Torre-Bueno, J. R., 164, 200, *219*
Toufar, J., 211, *215*
Tovar, H., 141, *145*
Trainer, D. O., 431, *441*
Trembley, R. R., 331, *344*
Trips, M., 420, *440*
Trivers, R. L., 202, 204, *219*
Trollope, J., 143, *159*
Tsimbaliuk, A. K., 382, 391, *441*
Tsutsui, K., 228, 306, *350*, *363*
Tucker, V. A., 199, *219*
Tukey, J. W., 39, *77*
Turek, F. W., 262, 301, *343*, *363*

Turner, J. R. G., 133, 134, *154*, *159*
Turner, M. E., Jr., 33, *79*
Tyler, S. A., 35, *76*

U

Uhr, J. W., 468, 476, 477, 478, *499*
Ulmer, M. J., 386, 399, *438*, *439*
Unanue, E. R., 458, *485*
Upp, C. W., 61, *82*
Utter, J. M., 163, 192, 194, *219*

V

Vaerman, A. M., 474, *493*
Vaerman, J. P., 474, *493*
Vaidova, S. M., 382, 383, 384, 386, 387, 389, 390, 391, 393, 394, 395, 396, 397, 399, 402, 419, *441*
Vainio, O., 457, *499*
Valenta, A., 417, *442*
Van Alten, P. J., 452, 453, 459, 468, 472, 473, 474, 477, *486*, *487*, *495*, *497*, *499*
Van Balen, J. H., 47, 59, *82*
Vandeputte-Poma, J., 29, *82*
Vandermark, N. L., 298, *360*
Vander Wall, S. B., 164, *219*
Van Doorn, E. J., 328, *363*
Van Krey, H. P., 232, 308, 331, 339, *340*, *343*, *355*
Van Meter, R., 477, *499*
Van Noordwijk, 47, *82*
Van Tienhoven, A., 226, 284, 298, *351*, *358*, *363*
Vanzulli, A., 95, *159*
Vardaris, R. M., 267, 316, *363*
Vauclair, J., 108, *159*
Veghte, J. H., 164, *219*
Verbeek, 194, *219*
Verghese, M. W., 49, 50, *82*
Vergoni, W., 257, *340*
Verhoeven, G., 322, *363*
Vidal, J. M., 108, *159*, 223, *363*
Vieth, W., 88, 89, 148, *159*
Vince, M. A., 94, 95, 98, 99, 100, 101, 102, 105, 108, 125, 128, 131, *151*, *155*, *159*
Visco, R. J., 463, *499*
Visintini, F., 95, *159*
Vleck, C. M., 179, 180, 183, *219*
Vohra, P., 54, *77*
Voight, W., 330, *363*

AUTHOR INDEX
521

Voitkevich, A. A., 22, 23, 82
von Bertalanffy, L., 35, 82
von Ihering, H., 407, 428, 441
von Linstow, O., 368, 369, 370, 442
von St. Paul, U., 131, 153
von zur Mühlen, A., 302, 344
Vowles, D. M., 228, 229, 265, 267, 363
Vriend, J., 461, 499

W

Wada, M., 96, 97, 127, 159, 298, 363
Wade, G. N., 257, 265, 340, 344, 364
Wainwright, A. A. P., 108, 146
Wakely, J. S., 192, 219
Walker, W. A., 267, 345
Waller, F. F., 396, 437
Wallis, C. J., 231, 354
Walsberg, G. E., 165, 166, 179, 180, 181, 183, 186, 192, 194, 197, 198, 202, 209, 210, 219
Walsh, R. N., 255, 343
Waltenbaugh, C. R., 459, 472, 499
Walter, M. J., 110, 111, 159
Ward, P., 177, 216
Ward, W. H., 189, 220
Warner, N. L., 461, 468, 469, 470, 474, 476, 477, 478, 497, 486, 498, 499
Warren, R. P., 2, 17, 79, 240, 241, 364
Warriner, C. C., 109, 110, 159
Waser, M. S., 119, 154
Wasserman, G. F., 291, 346
Watley, S., 473, 490
Watson, A., 228, 364
Weathers, W. W., 163, 220
Weber, W. T., 458, 465, 470, 471, 499, 500
Webster, R. A., 331, 352
Wehr, E. E., 428, 431, 432, 442
Weidanz, W. D., 471, 480, 493, 500
Weiden, P. L., 480, 485
Weidmann, R., 125, 160
Weidmann, U., 106, 125, 126, 127, 160
Weinbaum, F. I., 470, 500
Weiner, R. I., 268, 316, 360
Weir, D., 143, 160
Weiss, J., 108, 109, 153, 160, 256, 352
Weisz, J., 324, 351
Wekstein, D. R., 20, 82
Weller, M. W., 176, 220
Wells, H., 121, 146
Wells, J. W., 223, 292, 360

Wells, M., 472, 500
Wemmer, C., 113, 131, 160
Wenckebach, K., 449, 500
Weniger, J.-P., 327, 364
Went, H. A., 473, 492
Wenzel, B. M., 257, 275, 358, 364
Werschkul, D. B., 64, 67, 82
West, G. C., 20, 82, 189, 199, 200, 218, 220
West, M. J., 88, 90, 116, 152, 160
Westbrook, W. H., 231, 339
Westerterp, K., 185, 220
Westwood, N. J., 225, 226, 232, 234, 243, 259, 296, 353, 356
Weston, H. G., 249, 364
Whalen, R. E., 231, 364
White, S. C., 6, 8, 11, 15, 16, 24, 25, 26, 27, 28, 29, 32, 37, 40, 41, 68, 70, 80, 82, 86, 156, 170, 184, 185, 218
White, S. J., 236, 238, 364
White, R., 477, 485
Whitsett, J. M., 280, 311, 364
Wick, G., 478, 485
Widdowson, E. M., 9, 72
Wiens, J. A., 194, 220
Wiese, J. H., 432, 442
Wiley, W. H., 61, 82
Wilhelmsson, M., 268, 337
Wilke, J. T., 364
Wilkinson, R., 115, 160
Willems, J., 300, 318, 338, 364
Williams, A. J., 6, 82
Williams, D. M., 250, 254, 347
Williams, G. C., 69, 82
Williams, I. C., 389, 442
Williams, J. B., 250, 254, 360
Willis, E., 224, 354
Willis, J. I., 463, 500
Wilson, D. S., 61, 73
Wilson, E. K., 300, 302, 364
Wilson, F. E., 277, 343
Wilson, G. F., 128, 160
Wilson, J. A., 280, 311, 364, 469, 500
Wilson, S. C., 300, 302, 364
Wiltschko, W., 136, 138, 160
Wimersma Greidanus, Tj. B. Van, 257, 364
Winchester, C. F., 22, 83
Wingfield, J. C., 225, 226, 228, 245, 250, 262, 275, 302, 303, 335, 345, 364, 365
Winters, A. R., 467, 468, 486
Wira, C. R., 461, 495, 498
Wissler, K., 385, 442

Withers, P. C., 192, 196, *220*
Witschi, E., 258, 274, 275, *365*
Wolf, L. L., 165, 192, 194, *215*, *220*
Wolfe, H. R., 461, 467, 468, 469, *485*, *490*, *494*, *495*, *500*
Wolffhügel, K., 381, *442*
Wood, H. B., 187, *220*
Wood-Gush, D. G. M., 224, 236, 276, 285, *346*, *365*
Woods, J. E., 311, *365*
Woodward, M., 467, *500*
Wooley, J. B., 192, *220*
Woolf, N. K., 101, *160*
Worley, L. G., 31, *72*
Worsley, A., 255, *343*
Wortis, R. P., 238, 246, 303, *353*
Wright, M. H., 167, *220*
Wright, P. L., 167, *220*
Wurm, M., 228, 230, 234, 236, 249, *357*

Y

Yahr, P., 273, 325, 327, *365*
Yamaguti, S., 415, *442*
Yano, T., 298, *363*
Yarbough, J., 472, *500*

Yarbrough, C. G., 18, 19, *83*
Young, C. E., 224, 255, 323, 326, *365*
Young, R. T., 28, *81*, 388, *442*
Young, W. C., 268, 305, *347*, *365*
Youngren, O. M., 286, *357*
Yule, W. J., 53, *83*

Z

Zablotskii, V. I., 384, *442*
Zach, R., 47, *81*
Zajiček, D., 417, *442*
Zarrow, M. X., 461, 462, *500*
Zaskind, L. N., 431, *440*
Zavallone, J. D., 470, *496*
Zeder, J. G. H., 368, *442*
Zeier, H., 286, *365*
Zeis, A., 327, *364*
Zheliazkova-Paspaleva, A., 395, 397, 403, *438*
Zigmond, R. E., 265, 276, 282, 286, 288, 329, *365*
Zimmerman, J. L., 192, 194, *218*
Zinov'eva, E. N., 391, 392, *440*
Zitrin, A., 224, *357*
Zolman, J. F., 20, *82*
Zuker, I., 231, *365*
Zuevo, L. S., 390, *433*

INDEX TO BIRD NAMES

A

Accipiter
  cooperii, 384
  nisus, 384
Accipitridae, 376, 378, 383
Acrocephalus scirpaceus, 129, 130, 134, 144
Actitis macularia, 165, 170, 176, 192, 196, 205, 207
Aegithalidae, 398
Aegithalos caudatus, 398
Aegolius acadicus, 192
Aethia
  cristatella, 390
  pusilla, 390, 391
Agapornis
  fischeri, 90
  roseicollis, 90
Agelaius
  phoeniceus, 26, 133–135, 167, 172, 181, 202, 227, 296, 303, 403, 404, 409
  tricolor, 172
Aimophila carpalis, 7, 8, 26
Aix sponsa, 95, 170, 174–177
Alauda arvensis, 395
Alaudidae, 377, 379, 395
Albatross, Royal, *see Diomedea epomophora*
Alca torda, 59, 115, 390
Alcedinidae, 377, 379, 393
Alcedo atthis, 393, 423
Alcidae, 6, 11, 71, 378, 390, 391
Alectoris chukar, 167
Alectura lathami, 102
Alle alle, 390
Amandava amandava, 121, 122
Amazilia fimbriata, 199
Ammodramus maritimus, 404
Anas, 16
  acuta, 224, 226
  clypeata, 176
  platyrhynchos, 5, 15, 28, 90, 91, 93, 95, 96, 98–100, 103, 106, 107, 111, 168, 170, 174–177, 224, 228, 231–233, 236, 255–259, 267, 272, 274, 276, 285, 287, 293, 296–301, 305–307, 310, 318, 319, 431, 444, 471, 481
  rubripes, 192, 427
  strepera, 192, 223, 224
Anatidae, 375, 376, 378, 383
Anis, *see Crotophaga*
Anous minutus, 4
Anser
  anser, 170
  caerulescens, 111, 177, 428
Anseriformes, 118, 174, 228, 251, 368, 412, 421, 423
Anthus
  cervinus, 401
  hodgsoni, 401
  novaeseelandiae, 401
  spinoletta, 192, 401
  trivialis, 401
Aphriza, 387
Apodidae, 377, 393
Apodiformes, 412, 417
Aptenodytes forsteri, 192
Apteryx australis, 169, 170, 173, 174
Apus, 16
  apus, 52, 56, 393
Aquila, 384
  chrysaetos, 375
  pomarina, 384
  rapax, 384
Ardea
  cinerea, 383
  herodias, 135
  purpurea, 383
Ardeidae, 376, 378, 382
Ardeola ralloides, 383, 423
Arenaria interpres, 170, 387
Asio
  flammeus, 190, 392
  otus, 192
Athene
  cunicularia, 392
  noctua, 392
Auk, Razor-billed, *see Alca torda*
Auklet
  Crested, *see Aethia cristatella*
  Least, *see Aethia pusilla*
  Rhinoceros, *see Cerorhinca monocerata*
Avadavat, Red, *see Amandava amandava*

Aythya, 423
  americana, 224
  valisineria, 427

B

Bee-eater, Red-throated, *see Merops bulocki*
Bellbird, 16
  Bearded, *see Procnias averano*
Bittern, 423, 431
Blackbird, 89, 125
  Brewer's, *see Euphagus cyanocephalus*
  European, *see Turdus merula*
  Red-winged, *see Agelaius phoeniceus*
  Rusty, *see Euphagus carolinus*
  Yellow-headed, *see Xanthocephalus xanthocephalus*
Blackcap, *see Sylvia atricapilla*
Bombycilla
  cedrorum, 401
  garrulus, 401
  japonica, 401
Bombycillidae, 377, 379, 401
Bonasa umbellus, 254, 385
Botaurus stellaris, 383
Brambling, *see Fringilla montifringilla*
Branta
  canadensis, 170, 174, 177, 428
  leucopsis, 192
Brushturkey, Australian, *see Alectura lathami*
Bubo virginianus, 392, 422
Bubulcus ibis, 170, 383
Budgerigar, *see Melopsittacus undulatus*
Bullfinch, European, *see Pyrrhula pyrrhula*
Bunting
  Indigo, *see Passerina cyanea*
  Little, *see Emberiza pusilla*
  Reed, *see Emberiza schoeniclus*
  Snow, *see Plectrophenax nivalis*
  Yellow-breasted, *see Emberiza aureola*
Buteo
  buteo, 384
  jamaicensis, 384, 422
  lagopus, 384
  platypterus, 422
  regalis, 192, 384
  rufinus, 384
Buzzard
  Common, *see Buteo buteo*
  European, *see Buteo buteo*
  Long-legged, *see Buteo rufinus*

C

Cairina moschata, 256
Calandrella cinerea, 395
Calcarius, 404
  lapponicus, 29, 192, 197, 404
Calidris
  alpina, 24, 171
  bairdii, 171
  maritima, 387, 415
  mauri, 171
  pusilla, 176, 192, 196, 198
Calliope, 400
Calypte anna, 142, 165, 166, 192
Campylorhynchus brunneicapillus, 26
Canachites, 385
  canadensis, 143, 385
Canary, *see Serinus canarius*
Canvasback, *see Aythya valisineria*
Capercaillie, *see Tetrao urogallus*
Caprimulgidae, 10, 377, 392, 393
Caprimulgiformes, 412
Caprimulgus
  europaeus, 393
  indicus, 393
  vociferus, 392
Cardinal, Common, *see Cardinalis cardinalis*
Cardinalis, 119, 404
  cardinalis, 119
Carduelis, 405
  chloris, 131
  flammea, 199
Cariamidae, 6
Carpodacus, 405
  mexicanus, 172, 174
Casmerodius albus, 170, 183
Casuarius casuarius, 170
Catbird, Gray, *see Dumetella carolinensis*
Cathartidae, 376
Catharus, 400
  guttatus, 172
Cepphus
  carbo, 391
  columba, 24, 171, 185, 390
  grylle, 171, 391
Cerorhinca monocerata, 59
Certhia americana, 398
Certhiidae, 377, 379, 398
Ceryle alcyon, 393, 426
Chaetura pelagica, 393

INDEX TO BIRD NAMES 525

Chaffinch, see *Fringilla coelebs*
Charadridae, 64, 376, 378, 387
Charadriiformes, 5, 15, 174, 412, 421
Charadrius
 dubius, 387
 vociferus, 387, 432
Chickadee, Black-capped, see *Parus atricapillus*
Chicken, see *Gallus gallus*
Chionis minor, 296
Chlidonias
 hybridus, 390
 leucopterus, 390
 niger, 389, 390
Chordeiles minor, 392
Ciconia
 ciconia, 383, 423, 428
 nigra, 383, 428
Ciconiidae, 383
Ciconiiformes, 412, 421
Cinclidae, 379, 399
Cinclus
 cinclus, 399
 mexicanus, 399
 pallasii, 399
Circaetus gallicus, 384
Circus
 aeruginosus, 384
 cyaneus, 384
 macrourus, 384
Cistothorus
 palustrias, 117, 123, 172
 platensis, 123
Clangula hyemalis, 423
Coccothraustes, 405
Coccyzus americanus, 391
Colaptes auratus, 172, 394, 427
Colibri coruscans, 192
Colinus virginianus, 96, 97, 170
Columba, 16
 fasciata, 391, 427
 livia, 94–97, 109, 137, 138, 171, 174, 223, 225, 226, 230, 233, 244, 259
 palumbus, 167, 225, 391
Columbidae, 252, 376, 391
Columbiformes, 118, 174, 228, 250, 251, 284, 412
Contopus virens, 395
Coot, European, see *Fulica atra*
Coracias garrulus, 394
Coraciidae, 394

Coraciiformes, 16, 412, 417
Cormorant, 419
 Cape, see *Phalacrocorax capensis*
 Double-crested, see *Phalacrocorax auritus*
 Great, see *Phalacrocorax carbo*
 Pelagic, see *Phalacrocorax pelagicus*
 Pygmy, see *Phalacrocorax pygmaeus*
Corncrake, see *Crex crex*
Corvidae, 131, 377, 379, 396–398
Corvus, 142
 brachyrhynchos, 396, 408
 caurinus, 396
 corax, 142, 396
 corone cornix, 397
 frugilegus, 225, 296, 396, 419
 monedula, 396
Coturnix
 chinensis, 19
 coturnix, 15, 170, 174
Cowbird, Brown-headed, see *Molothrus ater*
Crabplovers, see Dromadidae
Crane, 16
 European, see *Grus grus*
 Greater Sandhill, see *Grus canadensis tabida*
 Sandhill, see *Grus canadensis*
 White-naped, see *Grus vipio*
Creeper, Brown, see *Certhia americana*
Crex crex, 423
Crotophaga, 12
Crow, 142
 Common, see *Corvus brachyrhynchos*
 Hooded, see *Corvus corone cornix*
 Northwestern, see *Corvus caurinus*
Cuckoo
 European, see *Cuculus canorus*
 Old World, 236
 Yellow-billed, see *Coccyzus americanus*
Cuculidae, 376, 391, 392
Cuculiformes, 412, 417
Cuculus
 canorus, 114
 saturatus, 114, 391
Cyanocitta cristata, 132, 134, 135, 397
Cyanopicus cyanus, 397
Cygnus buccinator, 375

D

Delichon urbica, 24, 26, 41, 61, 62, 185, 187, 192, 396
Dendrocygna autumnalis, 170

Dendroica, 402
  caerulescens, 192, 197, 204
  petechia, 172
Diomedea epomophora, 223
Dipper
  Brown, see *Cinclus pallasii*
  Eurasian, see *Cinclus cinclus*
Dolichonyx, 403
Dove, 29, 112, 231, 234, 267, 273, 286, 287, 295, 326, 328, 329, 332, 333
  Collared, see *Streptopelia decaocto*
  Mourning, see *Zenaida macroura*
  Ring, 172, 174, 226–230, 232–236, 238, 243, 245–248, 251–253, 258, 259, 261, 262, 264, 266, 267, 269–271, 273, 276, 277, 280–282, 284, 286, 293–296, 298, 300, 301, 303–305, 308–310, 316, 318, 323–325, 328, 330, 331, 334
  Spotted, see *Streptopelia chinensis*
Dovekie, see *Alle alle*
Dromadidae, 5
Duck, see *Anas platyrhynchos*
  Black, see *Anas rubripes*
  Muscovy, see *Cairina moschata*
  Pekin, 97
  Wood, see *Aix sponsa*
Dumetella carolinensis, 132, 399
Dunlin, see *Calidris alpina*

E

Eagle
  Booted, see *Hieraeetus pennatus*
  Golden, see *Aquila chrysaetos*
  Lesser Spotted, see *Aquila pomarina*
  Short-toed, see *Circaetus gallicus*
  Tawny, see *Aquila rapax*
  White-tailed, see *Haliaeetus albicilla*
Egret, 423
Egretta
  caerulea, 135
  garzetta, 383
  thula, 172
  tricolor, 170
Eider, Common, see *Somateria mollissima*
Elaenia flavogaster, 28
Elaenia, Yellow-bellied, see *Elaenia flavogaster*
Elanoides forficatus, 143

Elanus
  caeruleus, 192
  leucurus, 192
Emberiza
  aureola, 405
  citrinella, 405
  pusilla, 405
Emberizidae, 377, 379, 404, 405
  Thraupinae, 405
Emerald, Glittering-throated, see *Amazilia fimbriata*
Empidonax
  minimus, 176, 192, 196, 204
  traillii, 176, 192, 196, 198, 204, 205, 207
Eremophila alpestris, 395
Erithacus svecicus calliope, 400
Estrildidae, Viduinae, 114
Eudocimus albus, 170, 185, 192, 383
Eumomota superciliosa, 89, 132, 134, 135, 141, 145
Euphagus
  carolinus, 172, 403
  cyanocephalus, 172, 403

F

Falco, 384
  naumanni, 385
  peregrinus, 385
  rusticolus, 384, 432
  sparverius, 129, 130, 133, 180, 192
  subbuteo, 385
  tinnunculus, 385
  vespertinus, 385
Falcon, Peregrine, see *Falco peregrinus*
Falconidae, 376, 384
Falconiformes, 6, 17, 284, 372, 412, 421, 423
Ficedula
  albicollis, 113
  hypoleuca, 19, 225, 226, 228, 244
Finch, 122
  Bengalese, see *Lonchura striata*
  Medium Ground, see *Geospiza fortis*
  Strawberry, see *Amadava amadava*
  Zebra, see *Poephila guttata*
Flamingo, Greater, see *Phoenicopterus ruber*
Flicker, Northern, see *Colaptes auratus*

INDEX TO BIRD NAMES 527

Flycatcher
  Collared, *see Ficedula albicollis*
  Pied, *see Ficedula hypoleuca*
  Willow, *see Empidonax traillii*
Fratercula
  arctica, 390
  corniculata, 390
Fregata
  aquila, 141
  minor, 143
Frigatebird
  Ascension, *see Fregata aquila*
  Great, *see Fregata minor*
Fringilla
  coelebs, 115, 117, 119, 121, 123, 124, 188–190, 265, 276, 282, 283, 405
  montifringilla, 276, 282, 285, 286, 405, 406
Fringillidae, 377, 379, 405, 406
Fulica atra, 105, 127, 386

G

Gadwall, *see Anas strepera*
Galerida cristata, 395
Galliformes, 6, 12, 29, 64, 70, 118, 126, 174, 228, 250, 251, 310, 368, 385, 412, 420
Gallinula chloropus, 105, 127, 386
Gallinule
  Common, *see Gallinula chloropus*
  Purple, *see Porphyrula martinica*
Gallus, 16
  gallus, 8, 20–22, 27, 46–54, 56, 57, 61, 62, 85, 87, 90–92, 94–98, 100, 105–109, 115, 127, 130, 168, 170, 174, 188, 211, 224, 229, 230, 236, 248, 253–255, 257, 261–264, 268, 274, 276–282, 285, 286, 292, 293, 302, 308, 311–313, 316, 317, 322, 323, 325–327, 330, 331, 444–484
Gannet, Northern, *see Sula bassana*
Garrulus glandarius, 397
Gavia
  arctica, 382, 415
  immer, 415
  stellata, 382, 415
Gaviidae, 376, 378, 382
Geospiza fortis, 47
Geothlypis, 402
Godwit, Black-tailed, *see Limosa limosa*, 100

Goose, 103
  Blue, *see Anser caerulescens*
  Canada, *see Branta canadensis*
  Lesser Snow, *see Anser caerulescens*
Grackle, Common, *see Quiscalus quiscula*
Gracula religiosa, 114
Grebe
  Crested, *see Podiceps cristatus*
  Eared, *see Podiceps nigricollis*
  Great Crested, *see Podiceps cristatus*
  Pied-billed, *see Podilymbus podiceps*
  Red-necked, *see Podiceps grisegena*
Greenfinch, *see Carduelis chloris*
Grosbeak, Black-headed, *see Pheucticus melanocephalus*
Grouse, 5
  Black, *see Tetrao tetrix*
  Common Hazel, *see Tetrastes bonasia*
  Red, *see Lagopus lagopus scoticus*
  Ruffed, *see Bonasa umbellus*
  Sharp-tailed, *see Tympanuchus phasianellus*
  Spruce, *see Canachites canadensis*
Gruidae, 6, 378, 385
Gruiformes, 6, 412
Grus
  canadensis, 381
    tabida, 386
  grus, 385
  vipio, 385
Guillemot
  Black, *see Cepphus grylle*, 391
  Pigeon, *see Cepphus columba*, 24, 390
  Spectacled, *see Cepphus carbo*, 391
Gull, 5, 11, 62, 64, 70, 71, 112, 115, 230, 423
  Bonaparte's, *see Larus philadelphia*
  Brown-hooded, *see Larus maculipennis*
  California, *see Larus californicus*
  Common Black-headed, *see Larus ridibundus*
  Glaucous, *see Larus hyperboreus*
  Glaucous-winged, *see Larus glaucescens*
  Great Black-headed, *see Larus ichthyaetus*
  Herring, *see Larus argentatus*
  Kelp, *see Larus dominicanus*
  Laughing, *see Larus atricilla*
  Lesser Black-backed, *see Larus fuscus*
  Little, *see Larus minutus*

Mediterranean Black-headed, *see Larus melanocephalus*
Mew, *see Larus canus*
Ring-billed, *see Larus delawarensis*
Sabine's, *see Xema sabini*
Slender-billed, *see Larus genei*
Western, *see Larus occidentalis*
Gygis alba, 4
Gymnorhinus cyanocephalus, 129
Gyps, 29
Gyrfalcon, *see Falco rusticolus*

H

Haematopodidae, 378
Haematopus ostralegus, 126
Halcyon smyrnensis, 393
Haliaeetus albicilla, 384
Harrier
 Hen, *see Circus cyaneus*
 Pallid, *see Circus macrourus*
Hawk, 4, 5, 424
 Broad-winged, *see Buteo platypterus*
 Cooper's, *see Accipiter cooperii*
 Red-tailed, *see Buteo jamaicensis*
 Rough-legged, *see Buteo lagopus*
Hawk-Eagle, Mountain, *see Spizaetus nipalensis*
Hawk-Owl, Northern, *see Surnia ulula*
Heron, 4, 5, 419
 Gray, *see Ardea cinerea*
 Great Blue, *see Ardea herodias*
 Little Blue, *see Egretta caerulea*
 Purple, *see Ardea purpurea*
 Squacco, *see Ardeola ralloides*
Heteroscelus incanus, 415
Hieraeetus pennatus, 384
Hirundapus caudacutus, 393
Hirundinidae, 377, 379, 395, 396
Hirundo rustica, 26, 396
Hoatzin, *see Opisthocomus hoazin*
Honey-Buzzard, Eurasian, *see Pernis apivorus*
Hummingbird
 Andean Hillstar, *see Oreotrochilus estella*
 Anna's, *see Calypte anna*
Hylocichla, 400

I

Ibis
 Glossy, *see Plegadis falcinellus*
 White, *see Eudocimus albus*

Icteridae, 377, 379, 403
Icterus galbula, 403
Iiwi, *see Vestiaria coccinea*
Indigobird, Village, *see Vidua chalybeata*
Ixobrychus minutus, 383

J

Jackdaw, Common, *see Corvus monedula*
Jaeger
 Long-tailed, *see Stercorarius longicaudus*
 Parasitic, *see Stercorarius parasiticus*
 Pomarine, *see Stercorarius pomarinus*
Jay, 142
 Blue, *see Cyanocitta cristata*
 Eurasian, *see Garrulus glandarius*
 Gray, *see Perisoreus canadensis*
 Piñon, *see Gymnorhinus cyanocephalus*
Junco, 404

K

Kestrel, American, *see Falco sparverius*
Killdeer, *see Charadrius vociferus*
Kingbird
 Eastern, *see Tyrannus tyrannus*
 Western, *see Tyrannus verticalis*
Kingfisher
 Belted, *see Ceryle alcyon*
 European, *see Alcedo atthis*
 White-breasted, *see Halcyon smyrnensis*
Kinglet, Golden-crowned, *see Regulus satrapa*
Kiskadee, Great, *see Pitangus sulphuratus*
Kite
 Black, *see migrans*
 Red, *see Milvus milvus*
 Swallow-tailed, *see Elanoides forficatus*
Kittiwake, Black-legged, *see Rissa Tridactyla*
Kiwi, 175
 Brown, *see Apteryx australis*

L

Lagopus
 lagopus, 20, 189, 192, 385, 428
  scoticus, 143, 228
 leucurus, 192, 385
 mutus, 192, 385

INDEX TO BIRD NAMES

Laniidae, 377, 379, 401, 402
Lanius
  collurio, 131, 185
  excubitor, 131, 401
  ludovicianus, 112, 129, 130, 132, 142, 401
  schach, 402
  senator, 131
Lapwing, European, *see Vanellus vanellus*
Laridae, 126, 174, 376, 378, 388
Lark
  Crested, *see Galerida cristata*
  Greater Short-toed, *see Calandrella cinerea*
  Horned, *see Eremophila alpestris*
Larus, 125, 388
  argentatus, 24, 61, 113, 125, 126, 129, 140, 141, 171, 174, 185, 224, 228, 389, 415
  atricilla, 61, 171, 174, 125, 234
  californicus, 388
  canus, 171, 388, 389, 415, 427
  delawarensis, 129, 140, 171
  dominicanus, 389
  fuscus, 389
  glaucescens, 171, 374, 388
  hyperboreus, 171, 388, 407
  ichthyaetus, 141, 388, 389
  maculipennis, 389
  marinus, 171
  melanocephalus, 389
  minutus, 389
  occidentalis, 171, 174, 388
  philadelphia, 388
  ridibundus, 125, 389, 427
Leucosticte, 405
Limosa lapponica, 171
Lonchura
  malabarica, 110
  striata, 110, 111, 121
Longspur, Lapland, *see Calcarius lapponicus*
Loon, 11, 16, 419, 423
  Arctic, *see Gavia arctica*
  Common, *see Gavia immer*
  Red-throated, *see Gavia stellata*
Lophortyx, 169
  californicus, 167, 168
  gambelii, 168
Loxia, 405
Lunda cirrhata, 171, 390
Lyrebird, 123

M

Magpie
  Azure-winged, *see Cyanopicus cyanus*
  Black-billed, *see Pica pica*
Mallard, *see Anas platyrhynchos*
Marsh-Harrier, Eurasian, *see Circus aeruginosus*
Marsh-Wren, 144
  Long-billed, *see Cistothorus palustris*
Martin, House, *see Delichon urbica*
Meadowlark
  Eastern, *see Sturnella magna*
Megapodiidae, 5, 11, 16, 102
Meleagris gallopavo, 28, 54, 119, 170, 174, 224, 292
Melopsittacus undulatus, 16, 199, 236, 243, 244, 254, 310
Melospiza
  georgiana, 123, 404
  melodia, 13, 47, 86, 116, 117, 121, 123, 172, 174, 404, 431
Menuridae, 123
Merganser, Red-breasted, *see Mergus serrator*
Mergus serrator, 415
Meropidae, 393, 394
Merops
  apiaster, 393
  bulocki, 134
Milvus
  migrans, 384
  milvus, 384
Mimidae, 377, 399
Mimus, 123
  polyglottos, 118, 192
Mniotilta, 402
Mockingbird, Common, *see Mimus polyglottos*
Molothrus, 403
  ater, 61, 88, 90, 116, 119, 172, 177, 188
Motacilla
  alba, 400, 401
  cinerea, 401
  citreola, 401
  flava, 400, 401
Motacillidae, 377, 379, 400, 401
Motmot, Turquoise-browed, *see Eumomota superciliosa*
Murre
  Common, *see Uria aalge*
  Thick-billed, *see Uria lomvia*

Murrelet, Ancient, *see Synthliboramphus antiquus*, 391
Muscicapidae, 377, 379, 399, 400
Myna, Hill, *see Gracula religiosa*

N

Nectarinia
  famosa, 192
  olivacea, 192
  reichenowi, 165, 192
Needletail, White-throated, *see Hirundapus caudacutus*
Nighthawk, Common, *see Chordeiles minor*
Night-Heron, Black-crowned, *see Nycticorax nycticorax*
Noddy, Black, *see Anous minutus*
Nucifraga
  caryocatactes, 192, 397
  columbiana, 164, 397
Nutcracker
  Clark's *see Nucifraga columbiana*
  Eurasian, *see Nucifraga caryocatactes*
Nuthatch
  Eurasian, *see Sitta europaea*
  Red-breasted, *see Sitta canadensis*
  White-breasted, *see Sitta carolinensis*
Nyctea scandiaca, 375, 392, 425
Nycticorax nycticorax, 228, 236, 249, 383, 408

O

Oceanodroma leucorhoa, 15, 170, 185
Oenanthe oenanthe, 400
Oilbird, *see Steatornis caripensis*
Oldsquaw, *see Clangula hyemalis*
Opisthocomus hoazin, 4, 5, 11, 16
Oreotrochilus estella, 165, 166, 192, 196, 210
Oriole, Northern, *see Icterus galbula*
Osprey, *see Pandion haliaetus*
Ostrich, *see Struthio camelus*
Otus asio, 392
Owl, 5, 88, 424
  Barn, *see Tyto alba*
  Barred, *see Strix varia*
  Burrowing, *see Athene cunicularia*
  Great Horned, *see Bubo virginianus*
  Little, *see Athene noctua*
  Short-eared, *see Asio flammeus*
  Snowy, *see Nyctea scandiaca*
Oxyura maccoa, 192
Oystercatcher, European, *see Haematopus ostralegus*

P

Pandion haliaetus, 384
Pandionidae, 376, 384
Paridae, 377, 379, 398
Parulidae, 377, 379, 402
Parus, 398
  atricapillus, 118, 132, 398
  bicolor, 398
  caeruleus, 26
  cyanus, 398
  major, 19, 47, 61, 62, 128, 172, 182, 183, 225, 226, 398
  montanus, 398
Passer, 16
  domesticus, 26, 44, 63, 134, 135, 172, 185, 188–190, 403
  hispaniolensis, 403
  montanus, 172, 403
Passerculus sandwichensis princeps, 59
Passerella iliaca, 404, 405
Passeriformes, 5, 11, 12, 16, 18, 26, 29, 37, 44, 62, 69, 115, 119, 131, 142–144, 174, 177, 228, 250, 251, 284, 310, 412, 417, 420, 421
Passerina, 404
  cyanea, 122, 136, 137
Pelagodroma marina, 431
Pelecaniformes, 16, 412
Pelecanus occidentalis, 135, 170, 174
Pelican, Brown, *see Pelecanus occidentalis*
Perisoreus
  canadensis, 397
  infaustus, 397
Pernis apivorus, 384
Petrel, 11
Petrochelidon pyrrhonota, 192, 196
Petronia petronia, 403
Phainopepla, 186, 209, 210
  nitens, 165, 166, 192, 197, 198, 204, 206
Phalacrocoracidae, 376, 382
Phalacrocorax, 186
  aristotelis, 382
  auritus, 16, 24, 185
  capensis, 296
  carbo, 382

pelagicus, 382
pygmaeus, 382
Phalarope
  Red, *see Phalaropus fulicaria*
  Red-necked, *see Phalaropus lobatus*
  Wilson's, *see Phalaropus tricolor*
Phalaropodinae, 376, 378, 387, 388
Phalaropus, 227
  fulicaria, 171, 387
  lobatus, 171, 387
  tricolor, 227, 387
Phasianidae
  Phasianinae, 376, 378
  Tetraoninae, 376, 378, 385
Phasianus colchicus, 15, 28, 127, 167, 170
Pheasant, Ring-necked, *see Phasianus colchicus*
Pheucticus, 404
  melanocephalus, 249
Phoebe, Eastern, *see Sayornis phoebe*
Phoenicopteridae, 383
Phoenicopterus ruber, 383
Phylloscopus
  borealis, 400
  trochilus, 139, 249, 400
Pica pica, 172, 176, 192, 195, 196, 198, 204, 205, 208, 210, 397
Picidae, 377, 379, 394
Piciformes, 16, 412
Picoides
  major, 394
  medius, 394
  minor, 394
  pubescens, 394
  tridactylus, 394
  villosus, 394
Picus
  canus, 394
  syriacus, 394
  viridis, 394
Pigeon, *see Columba livia*
  Band-tailed, *see Columba fasciata*
  Wood, *see Columba palumbus*
Pinicola, 405
Pintail, Northern, *see Anas acuta*
Pipilo, 404
Pipit
  Red-throated, *see Anthus cervinus*
  Richard's, *see Anthus novaeseelandiae*
  Water, *see Anthus spinoletta*

Piranga, 404
  ludoviciana, 405
  olivacea, 405
Pitangus sulphuratus, 89, 132, 134, 135
Plectrophenax nivalis hyperboreus, 404
Plegadis falcinellus, 383, 423
Ploceidae, 403
Ploceus cucullatus, 258
Plover, Little Ringed, *see Charadrius dubius*
Pluvialis, 387
Podiceps
  cristatus, 5, 10, 16, 382, 413, 423
  grisegena, 382, 415
  nigricollis, 382
Podicipedidae, 376, 378, 382
Podicipediformes, 412
Podilymbus podiceps, 170, 415
Poephila guttata, 111, 117, 121, 129, 172, 179, 180, 223, 224, 228, 230, 232, 265, 267, 275, 276, 282, 283, 285, 286, 302, 303, 305, 306, 313, 326
Pooecetes gramineus, 404
Porphyrula martinica, 386
Porzana carolina, 170, 386
Procellariiformes, 6, 17, 24, 26, 169
Procnias averano, 5, 28, 69
Progne subis, 192, 204
Ptarmigan, 28
  Rock, *see Lagopus mutus*
  White-tailed, *see Lagopus leucurus*
  Willow, *see Lapopus lagopus*
Ptychorampus aleuticus, 171
Puffin
  Atlantic, *see Fratercula arctica*
  Horned, *see Fratercula corniculata*
  Tufted, *see Lunda cirrhata*
Puffinus tenuirostris, 113, 244
Pyrrhula pyrrhula, 121, 187

Q

Quail, 5, 28, 71, 103, 232, 292, 311–313, 318, 321, 323–328
  Bobwhite, *see Colinus virginianus*
  European, *see Coturnix coturnix*
  Japanese, 7, 8, 13–16, 19, 49, 50, 54, 55, 61, 90, 96, 101, 110, 127, 143, 223, 224, 226, 228, 230, 234, 236, 260–262, 268, 269, 271, 274, 275,

281, 291, 298, 300, 306, 310–313, 319, 320, 322, 323, 328, 456, 482
Painted, *see Coturnix chinensis*
Quelea quelea, 177, 229, 236, 258, 259
Quiscalus quiscula, 60–62, 96, 97, 172, 403, 404

R

Rail, 5, 64
  Clapper, *See Rallus longirostris*
  Virginia, *see Rallus limicola*
Rallidae, 375, 376, 386
Rallus
  limicola, 170, 386
  longirostris, 386
Ratite, 16
Raven, Common, *see Corvus corax*
Redhead, *see Aythya americana*
Redpoll, Common, *see Carduelis flammea*
Redshank, Common, *see Tringa totanus*
Regulus satrapa, 400
Riparia riparia, 115, 396
Rissa tridactyla, 125, 171, 388, 389
Robin
  American, *see Turdus migratorius*
  Clay-colored, *see Turdus grayi*
Roller, European, *see Coracias garrulus*
Rook, *see Corvus frugilegus*
Rubythroat, *see Erithacus svecicus calliope*
Rynchops nigra, 171

S

Salpinctes obsoletus, 399
Sandpiper
  Purple, *see Calidris maritima*
  Spotted, *see Actitis macularia*
  Wood, *see Tringa glareola*
Sapsucker, Yellow-bellied, *see Sphyrapicus varius*
Sayornis phoebe, 172, 395
Scolopacidae, 64, 376, 378, 387
  Phalaropodinae, 376, 378, 387, 388
Screech-Owl, Common, *see Otus asio*
Seiurus noveboracensis, 402
Serinus
  canarius, 119, 226, 236, 240–244, 273, 275, 282, 283, 286, 310
  serinus, 405

Setophaga, 402
Shag, European, *see Phalacrocorax aristotelis*
Shearwater, 52
  Short-tailed, *see Puffinus tenuirostris*
Sheathbill, Black-faced (Lesser), *see Chionis minor*
Shrike
  Brown, *see Lanius cristatus*
  Great Grey, *see Lanius excubitor*
  Loggerhead, *see Lanius ludovicianus*
  Long-tailed, *see Lanius schach*
  Northern, *see Lanius excubitor*
  Red-backed, *see Lanius collurio*
  Woodchat, *see Lanius senator*
Sialia, 400
Silverbill, Common, *see Lonchura malabarica*
Sitta
  canadensis, 398
  carolinensis, 398
  europaea, 398
Sittidae, 377, 379, 398
Skylark, *see Alauda arvensis*
Somateria mollissima, 142, 225
Sora, *see Porzana carolina*
Sparrow
  Chipping, *see Spizella passerina*
  European Tree, *see Passer montanus*
  Fox, *see Passerella iliaca*
  Harris's, *see Zonotrichia querula*
  House, *see Passer domesticus*
  Ipswich, *see Passerculus sandwichensis princeps*
  Rock, *see Petronia petronia*
  Rufous-winged, *see Aimophila carpalis*
  Seaside, *see Ammodramus maritimus*
  Song, *see Melospiza melodia*
  Spanish, *see Passer hispaniolensis*
  Swamp, *see Melospiza georgiana*
  Tree, *see Spizella arborea*
  Vesper, *see Pooecetes gramineus*
  White-crowned, *see Zonotrichia leucophrys*
  White-throated, *see Zonotrichia albicollis*
Sparrowhawk, European, *see Accipiter nisus*
Sphenisciformes, 6
Sphyrapicus varius, 394
Spiza, 404
  americana, 192
Spizaetus nipalensis, 422

INDEX TO BIRD NAMES

Spizella, 404
  arborea, 277, 404
  passerina, 113
Starling
  Daurian, see *Sturnus sturninus*
  European, see *Sturnus vulgaris*
  Starling, Rose-colored, see *Sturnus roseus*
  White-cheeked, see *Sturnus cineraceus*
Steatornis caripensis, 26, 28, 69
Stellula calliope, 192
Stercorariidae, 378, 388
Stercorarius
  longicaudus, 388, 415
  parasiticus, 388
  pomarinus, 388, 415
Sterna, 125
  albifrons, 171, 390
  aleutica, 389
  fuscata, 16, 24–26, 68
  hirundo, 7, 16, 24–26, 71, 128, 389, 390
  maxima, 171
  nilotica, 390
  paradisaea, 125, 171, 389, 415
  sandvicensis, 126, 171, 390
Stork
  Black, see *Ciconia nigra*
  White, see *Ciconia ciconia*
Storm-Petrel
  Leach's, see *Oceanodroma leucorhoa*
  White-faced, see *Pelagodroma marina*
Streptopelia
  chinensis, 391
  decaocto, 296
  turtur, 391
Strigidae, 377, 378, 392
Strigiformes, 372, 412, 421
Strix varia, 392, 422
Struthio camelus, 372
Sturnella, 121
  magna, 403, 432
Sturnidae, 377, 402
Sturnus, 16
  cineraceus, 402
  roseus, 402
  sturninus, 402
  vulgaris, 7, 8, 13–15, 49, 58, 61, 63, 123, 167, 168, 172, 174, 180, 185, 200, 225, 229, 249, 256, 258, 296, 318, 319, 402, 426, 444, 446, 481
Sula, 26
  bassana, 7, 8, 141, 170

Surnia ulula, 392
Swallow
  Bank, see *Riparia riparia*
  Barn, see *Hirundo rustica*
Swan, Trumpeter, see *Cygnus buccinator*
Swift
  Chimney, see *Chaetura pelagica*
  European, see *Apus apus*
Sylvia
  atricapilla, 140
  borin, 138–140, 142

T

Tanager
  Scarlet, see *Piranga olivacea*
  Western, see *Piranga ludoviciana*
Tattler, Wandering, see *Heteroscelus incanus*
Tauraco fischeri, 143
Tern, 5, 11, 64, 70, 71, 141, 423
  Aleutian, see *Sterna aleutica*
  Arctic, see *Sterna paradisaea*
  Black, see *Chlidonias niger*
  Common, see *Sterna hirundo*
  Gull-billed, see *Sterna nilotica*
  Little, see *Sterna albifrons*
  Sandwich, see *Sterna sandvicensis*
  Sooty, see *Sterna fuscata*
  Whiskered, see *Chlidonias hybridus*
  White, see *Gygis alba*
  White-winged, see *Chlidonias leucopterus*
Tetrao
  tetrix, 385
  urogallus, 385
Tetraoninae, 376, 378, 385
Tetrastes bonasia, 385
Thrasher, Brown, see *Toxostoma rufum*
Thraupinae, 405
Threskiornithidae, 383
Thrush
  Black-throated, see *Turdus ruficollis atrogularis*
  Song, see *Turdus philomelos*
  Varied, see *Zoothera naevia*
Tinamidae, 16
Tit
  Blue, see *Parus caeruleus*
  Great, see *Parus major*
Titmouse, 131
Toxostoma rufum, 399

Tree-Pipit
  Brown, see *Anthus trivialis*
  Olive, see *Anthus hodgsoni*
Tringa
  glareola, 387
  totanus, 387
Trochilidae, 377, 393
Troglodytes
  aedon, 20, 55, 181, 399
  troglodytes, 399
Troglodytidae, 377, 379, 399
Turaco, Fischer's, see *Tauraco fischeri*
Turdus, 16
  grayi, 28
  migratorius, 172, 399
  merula, 88, 125, 400
  philomelos, 125
  ruficollis atrogularis, 400
Turkey, see *Meleagris gallopavo*
Turtle-Dove, European, see *Streptopelia turtur*
Tympanuchus phasianellus, 385
Tyrannidae, 377, 379, 395
Tyrannus
  tyrannus, 395
  verticalis, 395
Tyto alba, 143, 391
Tytonidae, 376, 391

U

Uria
  aalge, 171, 390, 391
  lomvia, 390, 391

V

Vanellus vanellus, 387
Vermivora, 402
Vestiaria coccinea, 165, 166, 192
Vidua chalybeata, 100
Viduinae, 114
Vireo, 402
  olivaceus, 176, 192, 197, 204, 249
Vireonidae, 375, 377, 379, 402
Vulture, Griffon, see *Gyps*

W

Wagtail
  Citrine, see *Motacilla citreola*
  Gray, see *Motacilla cinerea*
  White, see *Motacilla alba*
  Yellow, see *Motacilla flava*
Warbler
  Arctic, see *Phylloscopus borealis*
  Garden, see *Sylvia borin*
  Reed, see *Acrocephalus scirpaceus*
  Willow, see *Phylloscopus trochilus*
Waterthrush, Northern, see *Seiurus noveboracensis*
Waxwing
  Bohemian, see *Bombycilla garrulus*
  Cedar, see *Bombycilla cedrorum*
  Japanese, see *Bombycilla japonica*
Weaverbird, Village, see *Ploceus cucullatus*
Weaver-Finch, see Ploceidae
Wheatear, Northern, see *Oenanthe oenanthe*
Whip-poor-will, see *Caprimulgus vociferus*
Wilsonia, 402
Woodpecker
  Downy, see *Picoides pubescens*
  Gray-headed, see *Picus canus*
  Great Spotted, see *Picoides major*
  Green, see *Picus viridis*
  Hairy, see *Picoides villosus*
  Lesser Spotted, see *Picoides minor*
  Middle Spotted, see *Picoides medius*
  Three-toed, see *Picoides tridactylus*
Wood-Pewee, Eastern, see *Contopus virens*
Wood-Stork, 24
Wren
  Cactus, see *Campylorhynchus brunneicapillus*
  House, see *Troglodytes aedon*
  Rock, see *Salpinctes obsoletus*
  Sedge, see *Cistothorus platensis*
  Winter, see *Troglodytes troglodytes*

X

Xanthocephalus xanthocephalus, 172, 403
Xema sabini, 171, 388

INDEX TO BIRD NAMES

Y

Yellowhammer, *see Emberiza citrinella*, 405

Z

Zenaida macroura, 7, 8, 20, 172, 174, 391
Zonotrichia, 113, 186, 404
  albicollis, 404
  leucophrys, 117, 119–123, 167, 168, 172, 181, 185, 188, 189, 191, 192, 195, 197, 199, 200, 204, 206, 207, 225, 226, 262, 293, 297, 302, 303
  pugetensis, 245
  querula, 123, 274, 275
Zoothera naevia, 399

SUBJECT INDEX

A

Acanthocephala
  characteristics, 373, 374
  specificity, 421, 422
ACTH, effect on behavior, 257
Adrenal gland, effect
  of bursa, 465, 466
  on growth of bursa, 462, 463
Agammaglobulinemia, 470–481
Aggression, effect of hormones, 227–230
Air sac, 19
Amino acids
  effect on growth, 27–30
  molting requirements, 189–191
Androgen, 221, see also specific substances
  effect on aggression, 228
    on incubation, 249
    on nest building, 236
    sensory, 273
  metabolism, 316–329
  sensitivity, 308–310
  specificity, 268–272
Antibiotics, effect on growth of bursa, 463
Antibody synthesis, 468–473
Aschelminthes, characteristics, 373

B

Behavior, 12, 13, see also specific types
  effect of hormones, 221–335, see also
    specific hormones
  of embryos, see Embryo
  hatching, see Hatching
  nature–nurture debate, 87–90
  ontogeny, 85–145
  parental, effect of hormones on care of
    young, 251–254
  play, see Play behavior
  posthatching, 103–144
Bioenergetics, see Energetics
Blood sugar, 18
Body mass, energy expenditure, 191–195
Body temperature, see Thermoregulation

Brain
  effect of hormones, 276–288
    anterior hypothalamus, 276–281
    mechanisms, 304–334
      individual differences, 304–308
      sex differences and sexual
        differentiation, 310–314
      steroid sensitivity, 308–310
      strain differences, 308
    midbrain, 281–285
    posterior hypothalamus, 276–281
    preoptic area, 276–281
    telencephalic structures, 285, 286
  metabolism
    steroid receptors, 330–334
    of steroids, 314–330
Breeding, 49
  seasonal effect of hormones, 225, 226
Brood
  size
    effect on growth, 60, 61
    energetics, 185–187
  variations, effect on growth, 62, 62
Bursa of Fabricius, 443–484
  cell-mediated immunity, 470, 471
  effect on adrenal gland, 465, 466
    on thymus, 466, 467
  follicles, 445, 449–455
  growth
    effect of adrenal, 462, 463
      of disease, 463
      of nutrients, 463
      of pesticides, 463
      of testosterone, 460, 461
  immunosuppression, 480, 481
  kinetics, 460–465
  lymphocytes, see Lymphocyte, bursal
  microenvironment, 458–460
  morphology, 444–455
  phagocytosis, 471, 472
  regulation of immunoglobulin synthesis,
    467–481
  site of antibody synthesis, 468–473
  steroid receptors, 461, 462
Bursapoietin, 458

538 SUBJECT INDEX

Bursectomy
  antibody synthesis, 468–472
    ontogeny, 472
  hormonal, and antibody synthesis,
    468–481

C

Calcium, bone development, 29
Carbohydrates, effect on growth, 28
Castration, 223, 232
Cestodes, see also specific types
  characteristics, 372
  cyclophyllidean, specificity, 415–418
  pseudophyllidean, specificity, 412–415
Circannual rhythm, 138, 139
  effect of hormones, 256, 257
Cold exposure, 56, 57, see also
    Homeothermy
Corticosterone
  effect on brain, 285
    on imprinting, 256
Courtship
  effect of hormones, 230–233
    of testosterone, 266, 267
Crop milk, 252, 253
Cyproterone acetate, effect on sexual
    behavior, 230, 231

D

Defeminization, 311–314
Defense
  resource, energetics, 165–167
  territorial, energetics, 165–167
Demasculinization, 311–314
Development
  allometric growth, 41, 42
  altricial, 3–17
  anatomical, 13–17, see also specific organ
    precocial versus altricial, 13–15
  characteristics, 5
  determinant, 163, 164
  energetics, see Energetics
  environmental factors, 10, 11, 51–57
  evolution of types, 9–12
  feeding behavior, 124–135
  growth equation, 30–43, see also Growth
    data
  heritability, 45–48
  homeothermy, see Homeothermy

hormones, 22, 23
interspecific variation, 63–71
intraspecific variation, 58–63
lipid reserves, 25–27
migration, see Migration
nidicolous, 3, 4
nidifugous, 3, 4
nutrient requirements, 27–30, see also
    Nutrition
orientation, see Orientation
phases, 13
phenotypic variance, 43–48
  of play behavior, 140–144
precocial, 3–17
postnatal, 1–71
semialtricial, 4, 5
semiprecocial, 4, 5
sensory, 94–99
somatomotor, 91–94
songs and calls, 114–124
Dexamethasone, effect on incubation, 247
Diethylstilbestrol, effect on nest building,
    238

E

Ecology, environment, effect on
    development, 10, 11
Ectothermy, 163
Egg
  altricial species, 5–7
  characteristics, 5–7
  composition, 169–173
  mass, 169
  precocial species, 5–7
  size, effect on growth, 61, 62
  synthesis
    efficiency, 174, 175
    energetics, 169–175
    energy sources, 175–178
  yolk, 6
Embryo
  behavior, 90–99
    sensory capacity, 94–99
    somatomotor activity, 91–94
  nervous system, 93
Endothermy, 163
Energetics
  of activities, see specific activity
  annual cycles, 208–210
  ecological, 161–212
    characteristics, 163, 164

SUBJECT INDEX

growth efficiency, 24, 25
metabolism, 28, 179, 180, 188, 189, 193
  maintenance, 195–199
  of postnatal development, 23–27
  species differences, 192–210
  total expenditure, of free-living birds, 191–210
Energy acquisition rates, 205–208
Erythrocyte, 18
Estradiol
  dose–response relationships, 260, 261
  effect on brain, 281, 285
    on courtship, 232
    on nest building, 236, 237, 240, 241
    on vocalization, 284
  measurement, 291
  metabolism, 329, 330
  in reproductive cycle, 293–298
  sensory effects, 273, 274
  sexual maturation, 224
  specificity, 268–271
Estrogen, 221
  effect on brain, 280, 281
    on nest building, 241–243
    on senses, 273
    on sexual behavior, 233–235
  sex differences and sexual differentiation, 311
  specificity, 268–272
Evolution, development type, 9–12

F

Fasting, 55, 56
Fat deposition, *see* Lipid reserves
Feathers, *see* Plumage
Fecundity, interspecies variations, 66, 67
Feeding, 124–135, *see also* Food, Nutrition, Force feeding, Self feeding
  color choices, 126, 127
  intermittent, 53, 54
  parent's role, 124–128
Flight, 12, 13
  energy expenditure, 193, 194
Follicle-stimulating hormone
  effect on behavior, 258–260
    on courtship, 232
  in reproductive cycle, 293–298
Food, *see also* Feeding, Nutrition
  deprivation, chronic, 55, 56
  in development, 124–135

foraging, 166, 205–208
handling techniques, 130, 131
noxious prey, 133–135
recognition, 128–130
role in evolution, 11
supply, effect on growth, 59, 60, 68, 69
warning coloration, 133–135
Foraging, energetics, 205–208
Force feeding, 52, 53
Fruit, as dietary source, 28

G

Genetics, of growth, 43–51
  heritability, 45–48
    estimation, 46–48
  phenotypic variance, 43–48
    estimation, 45
  response to selection, 49, 50, 51
Glycogen level in liver, 18
Gonadal growth, 223
  energetics, 167, 168
Gonadotropin, 221
  effect on behavior, 258–260
    on nest building, 241
    on sexual behavior, 231, 235
  plasma, 323
Grain, as dietary source, 28, 29
Growth, *see also* Development
  heterogonic, 8
  isomorphic, 7–10
Growth data, 30–43
  analyses, 31–33
    allometric, 41, 42
    increment, 40, 41
    multivariate, 42, 43
  curve fitting, 33–41
  longitudinal correlations, 42
  models, 33–37
    fitting, 38, 39
  parameters, statistical inference, 39, 40
  transformed time scales, 37
  types, 30, 31
Growth hormone, 22, 23

H

Habitat, effect on growth, 58, 59
Hatching, 12
  acceleration, 101
  behavior, 99–102
    differences, 102

post- , behavior, 103–144
  stages, 99, 100
Hearing, 95
Helminths, 367–432, see also specific type
  acquisition, 422–429
    age, 426–428
    diet, 422–429
    infective larval stages, 425
    migration, 428, 429
  characteristics, 371–374
  current trends, 370, 371
  historical considerations, 368–370
  localization in host, 429, 430
  occurrence, 374–406
  pathogenicity, 430–432
  prevalence
    Alaska, 378, 379
    Eurasia, 381
    north-central United States, 376, 377
  specificity, 406–422
Hemoglobin, 18
Hierarchy establishment, effect of hormones, 227–230
Homeothermy, 11, 12, 56, 57, 163
  air sacs, development, 19, 20
  blood chemistry, 18
  development, 17–21
  nervous and hormonal control, 20, 21
  plumage as insulation, 19
  surface-to-volume ratio, 18, 19
  thermogenesis, 20
Hormone, see also specific type
  changes in effectors, 275, 276
  dose–response relationships, 260–264
  effect on behavior, 221–335
    on brain, 276–288
  fate, 288–304
  latency, 264–267
  measurement, 288–304
  mode of action, 272–288
  nonsteroidal, 257–260
  peripheral effects, 272–276
  production of physiological levels of steroids, 300–302
  role in development, 22, 23
  social effects, 274, 275
  specificity, 268–272
  synthesis, 260–288
Host specificity, 406, 407

I

Immunoglobulin
  biosynthesis, 473–476
  immunosuppression, 480, 481
  effect of bursa, 467–481
Immunosuppression, 480, 481
Imprinting, 103–114
  criteria, 104
  effect of hormones, 255, 256
  filial, 104
  home area, 113
  irreversibility, 105, 109
  sensitive period, 105, 107–109
  sexual, 104
  species recognition, 109–112
  surrogates, 105–107
Incubation
  effect of hormones, 244–251
  energetics, 178–183
    biophysical models, 181–183
    estimates, 179–181
  metabolism, 179, 180
Infantile sexuality, 292

L

Lipid reserves, 25–27
Locomotion, pedal, 12, 13
Luteinizing hormone, in reproductive cycle, 293–298
Luteinizing hormone releasing hormone, effect on behavior, 258–260
Lymphocyte, bursal, 454, 455
  life span, 464, 465
  origin, 455–458
Lymphokine, 459

M

Metabolism, effect of assimilation on growth rate, 69, 70
Migration
  acquisition of helminths, 428, 429
  celestial clues, 136, 137
  learning, 135–140
  magnetic clues, 138
Minerals, effect on growth, 29
Molting
  amino acid requirements, 189–191

SUBJECT INDEX

energetics, 187–191
metabolism, 188, 189
Mortality, interspecies variations, 66, 67

N

Nematodes, *see also* specific type
specificity, 418–421
Neonate
altricial, 7–9
characteristics, 7–9
body composition, 7–9
organ size, 8
taxonomic distribution, 5
development, *see* Development
precocial, 7–9
Nervous system, development, effect of hormones, 272
Nest building, effect of hormones, 235–244
Nestling
care of, energetics, 184–187
parental energetics, 203, 204
Nutrition
effect on growth rate, 69
postnatal development, 27–30
restricted intake, 54, 55

O

Orientation, learning, 135–140
Ovarian growth, energetics, 168, 169
Oviducal growth, energetics, 168, 169

P

Parasites, 367–432, *see also* specific types
Parental care, 12, 13
Pecking, 98
Peripheral organ, effect of hormones, 272–278
Persistence, effect of hormones, 254, 255
Phagocytosis, 471, 472
Photic sensitivity, 94
Pipping, 100–102
Plasma steroid
covariations of, and behavior, 291–300
ontogeny, 291–293
daily and ultradian variations, 298–300
measurement, 289, 290
reproductive cycle and annual variations, 293–298

Platyhelminthes, characteristics, 371, 372
Play behavior, 140–144
characteristics, 140, 141
locomotory, 142
objects, 141, 142
social, 143
Plumage
development, 16, 17
downy, 16
feathers, 16, 17
effect of hormones, 227, 274, 275
as insulation, 19
synthesis, 189
Predator, 131–133
Progesterone
effect on aggression, 229
on brain, 285
on incubation, 247–249
on nest building, 238, 241
on parental behavior, 253
on sexual behavior, 235
measurement, 291
metabolism, 329, 330
in reproductive cycle, 293–298
Progestin, 221
Prolactin, 221
effect on aggression, 229
on parental behavior, 252–254
in incubation, 245, 246, 250, 251
sensory effects, 273
Proprioception, 95, 98
Protein, effect on growth, 27, 28
Pupillary reflex, 96, 97

R

Radioimmunoassay, 290
Reproduction
adult effect, on growth rate, 69
energetics, 201–204
hormones, 221
plasma steroid levels, 293–298

S

Season, effect on growth, 58, 59
Self-feeding, 12, 13, 126–135
Senses, *see also* specific type
chemical, 97, 98
development, 94–99

Sexual behavior
  effect of hormones
    female, 233–235
    male, 230, 231
Sexual maturation, effect of hormones, 223–225
Sibling competition, interspecies variations, 67, 68
Skeletal muscles, development, 21, 22
Song, *see also* Vocalization
  development, 114–124
  learning, 120–124
  between mates, 122
Species variation, growth patterns, 63–71
Steroid, *see also* specific substance
  brain metabolism, 314–330
  cellular mechanism, 315
  radioimmunoassay, 290, 291
  receptors, brain metabolism, 330–334
  sensitivity, 308–310
Subsong, 143, 144

T

Tactile sensitivity, 94, 95
Taxonomy, neonatal classification, 5, 6
Territorial defense, effect of hormones, 227–230
Testosterone
  daily variations, 298–300
  dose–response relationships, 260–264
  effect on aggression, 228, 229
    on brain, 276–280, 285
    on circadian rhythms, 256, 257
    on courtship, 232, 233
    on growth of bursa, 460, 461
    on imprinting, 255, 256
    on nest building, 236
    on persistence, 255
    on sexual behavior, 230, 231, 235
    on vocalization, 117, 120, 223, 224, 281–285
    on weight, 275
  gonadal growth, 223
  latency, 265–267
  measurement, 291
  metabolism, 316–329
    biological properties of metabolites, 322–326

  enzymatic activity, 319–322
  reproductive behavior, 323–326
  secondary sexual characteristics, 322, 323
ontogeny, 291–293
in reproductive cycle, 293–298
sex differences and sexual differentiation, 311
specificity, 268–272
strain differences, 308
Thermogenesis, 20
Thermoregulation, energetics, 199–201
Thymic cell, humoral immunity, 469, 470
Thymus, effect of bursa, 466, 467
Thyroid hormone, 22, 23
Tissue allocation, effect on growth rate, 70, 71
Trematodes, *see also* specific type
  characteristics, 371
  digenetic, specificity, 410, 411

V

Vestibular sensitivity, 95
Vision, 96, 97
Vocalization, 114–124, *see also* Song
  development, 114–124
  effect of deafening, 118–120
    of estradiol, 284
    of testosterone, 117, 120, 223, 224, 281–285
  hypoglossal dominance, 124
  learning sequence, 118
  recognition, 115, 116
  role of brain, 282–285
  sensitive period, 116–118
  subsong, 118

W

Weather, effect on growth, 59, 60

Z

Zugunruhe, 139, 140

THE LIBRARY
ST. MARY'S COLLEGE OF MARYLAND
ST. MARY'S CITY, MARYLAND 20686